Origami Design Secrets

Origami Design Secrets

Mathematical Methods for an Ancient Art

Robert J. Lang

A K Peters Ltd.

Editorial, Sales, and Customer Service Office

A K Peters, Ltd.
63 South Avenue
Natick, MA 01760
www.akpeters.com

Library of Congress Cataloging-in-Publication Data

Lang, Robert J.
Origami design secrets : mathematical methods for an ancient art / Robert J. Lang.
p. cm.
Includes index.
ISBN 1-56881-194-2
1. Origami–Design. 2. Origami–Mathematics. I. Title.

TT870.L2614 2003
736'.982–dc21 2003043317

Printed in India at Replika Press Pvt. Ltd.
07 06 05 10 9 8 7 6 5 4 3

Table of Contents

Acknowledgements

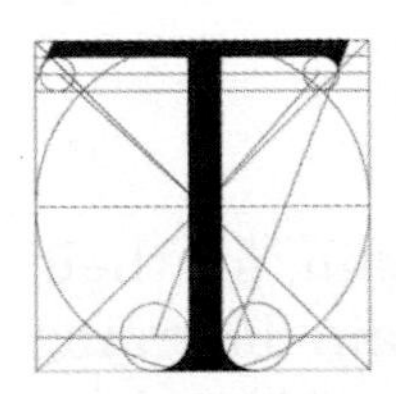

This book was a labor of many years. It is both my earliest book and my latest book; I began writing up my ideas on how to design when I began my first book in the early 1980s, but not until recently have I developed the framework for those ideas. Over the years, I have been influenced by many scientists and artists, both inside and outside of origami, all of whom contributed, one way or another, to the present tome. It is impossible for me to identify everyone who has contributed to my work, but some of the larger pieces come from the following, whom I thank:

Neal Elias, for his encouragement and for introducing me to the magic of box pleating and the realization that anything was possible in origami.

Lillian Oppenheimer and Alice Gray, for introducing me to the wide, wild world of origami fanatics.

Akira Yoshizawa, who started it all, then showed that there was more to origami art than just clever designs.

Dave Brill, who showed that you could have both clever design and high art in the same model.

John Montroll, who took origami design to an unequaled level and who has been a constant source of inspiration and friendship.

Michael LaFosse, who took origami art to an unequaled level and Richard Alexander; both have been equally great friends.

John Smith, James Sakoda, and especially David Lister for sharing a wealth of information about the history of origami, both privately and on the origami-L mailing list; David Lister, as well for numerous private comments and corrections with respect to origami history.

Toshiyuki Meguro, Jun Maekawa, and Fumiaki Kawahata, who developed circle and tree methods in Japan and who all provided crucial insights to my own work along the way.

Marshall Bern, who encouraged me to write my first origami computer science technical paper.

Barry Hayes, who, with Marshall, proved mathematically that origami is really, really hard (lest there be any doubt).

Erik and Martin Demaine, who have been friends and collaborators in computational origami; in particular, the mathematical theory in Chapter 12 is as much theirs as mine.

Thomas Hull, who, as the focal point of origami math, has done more to bring origamists and mathematicians together than anyone else.

Koshiro Hatori, who provided translations of several of the references.

Dave Mitchell, for his One-Crease Elephant.

Dr. Emmanuel Mooser, for his Train.

Raymond W. McLain and Raymond K. McLain, for their generous permission to reproduce the latter's Train diagrams and Raymond K.'s recollections of the early days of American origami.

In addition to the above, numerous other insights, encouragement, ideas, concepts, and criticisms came from Peter Engel, Robert Geretschläger, Chris Palmer, Paulo Barreto, Helena Verrill, Alex Bateman, Brian Ewins, Jeremy Shafer, Issei Yoshino, Satoshi Kamiya, Hideo Komatsu, Masao Okamura, and Makoto Yamaguchi. A particular thank you goes to Toshi Aoyagi who for many years acted as matchmaker and translator between me and many of my Japanese colleagues.

I am particularly indebted to Peter Engel, Marc Kirschenbaum, and Diane Lang for proofreading the text and the diagrams and making numerous suggestions for corrections and improvements. Needless to say, any errors that remain are entirely my own.

I would like to thank my editor, Heather Holcombe, for helpful suggestions, corrections, and especially patience.

Last, but most important, I must thank my wife, Diane, for her constant support and encouragement.

1 Introduction

In 1988, a French artist named Alain Georgeot prepared an exhibition of 88 elephants. They were made of folded paper, each different, and each one an example of *origami*, the Japanese art of paper folding. An art exhibition devoted entirely to origami is rare; one devoted to elephants is extremely unusual; and one devoted entirely to origami elephants was entirely unprecedented.

A display of 88 paper elephants illustrates both the remarkable attraction origami has for some people—after all, how many people would take the time to fold 88 versions of the same thing?—and the remarkable versatility of the art. Georgeot's collection of elephants represented only the tiniest fraction of the modern origami repertoire. Tens of thousands of paper designs exist for animals, plants, and objects, a regular abecedarium of subject matter. There are antelopes, birds, cars, dogs, elephants (of course), flowers and gorillas; horses, ibexes, jays, and kangaroos; lions, monkeys, nautiluses, octopi, parrots, quetzalcoatls, roses, sharks, trains, ukuleles, violinists, whelks, xylophones, yaks, and zebras, the last complete with stripes.

Innumerable innovations have been wrought upon the basic theme of folded paper. There are action figures: birds that flap their wings, violinists who bow their violins, inflatable boxes, clapping monkeys, snapping jaws. There are paper airplanes that fly—one won an international contest—and airplanes that don't fly, but are replicas of famous aircraft: the space shuttle, the SR-71 Blackbird, and the venerable Sopwith Camel. In some models, a single piece of paper is folded into several figures (a bull, bullfighter, and cape, for example) and in others, many identical pieces of

Figure 1.1.
A herd of origami elephants.

paper are assembled into enormous multifaceted polyhedra. If you can think of an object either natural or man-made, someone, somewhere, has probably folded an origami version.

The art of origami was originally Japanese, but the 88 elephants and the tens of thousands of other designs come from all over the world. Many figures originated in Japan, of course, but the U.S.A., England, France, Germany, Belgium, Argentina, Singapore, Australia, and Italy are major centers of origami activity. The designs range from simple figures consisting of only two or three folds to incredibly complex "test pieces" requiring hours to fold. Most of these thousands of designs have one thing in common, however: Nearly all were invented in the last 50 years.

Thus, origami is both an old art and a young art. Its youth is somewhat surprising. After all, folded paper has been an art form for some 15 centuries. It is ancient; one would not expect 98 percent of the innovation to come in the last 2 percent of the art's existence! Yet it has. Fifty years ago, all of the different origami designs in the world could have been catalogued on a single typed sheet of paper, had anyone had the inclination to do so. No model would have run over about 20 or 30 steps. Most could be folded in a few minutes, even by a novice. This is no longer the case. Today, in books, journals, and personal archives, the number of recorded origami designs runs well into the thousands; the most sophisticated designs have hundreds of steps and take several hours for an experienced folder to produce. The past 60 years in Japan, and 40 years worldwide, have seen a renaissance in the world of origami and an acceleration of its evolution.

And this has happened in the face of stringent barriers. The traditional rules of origami—one sheet of paper, no cuts—are daunting. It would appear that only the simplest abstract shapes are feasible with such rules. Yet over hundreds of years, by trial and error, two to three hundred designs were developed. These early designs were for the most part simple and stylized. Complexity and realism—insects with legs, wings, and antennae—were not possible until the development of specialized design methods in the latter part of the 20th century.

Although there are now many thousands of origami designs, there are not thousands of origami designers. In fact, there is only a handful of designers who have gone beyond basics, only a handful who can and do design sophisticated models. Although there is far more exchange of completed designs now than there used to be, there is not a similar exchange of design techniques.

This imbalance arises because it is much easier to describe how to replicate an origami figure than how to design one. Origami designs spread through publication of their folding sequence—a set of step-by-step instructions. The folding sequence, based on a simple code of dashed and dotted lines and arrows devised by the great Japanese master Akira Yoshizawa, transcends language boundaries and has led to the worldwide spread of origami.

While thousands of folding sequences have been published in books, magazines, and conference proceedings, a step-by-step folding sequence does not necessarily communicate how the model was designed. The folding sequence is usually optimized for ease of folding, not to show off design techniques or the structure of the model. In fact, some of the most enjoyable folding sequences are ones that obscure the underlying design of the model so that the appearance of the final structure comes as a surprise. "How to fold" is rarely "how to design." Folding sequences are widespread but relatively few of the design techniques of origami have ever been set down on paper.

Over the last 35 years I have designed some 400+ original figures. The most common question I am asked is, "How do you come up with your designs?" Throughout the history of origami, most designers have designed by "feel," by an intuition of which steps to take to achieve a particular end. My own approach to design has followed what I suspect is a not uncommon pattern; it evolved over the years from simply playing around with the paper, through somewhat more directed playing, to systematic folding. Nowadays, when I set out to fold a new subject, I have a pretty good idea about how I'm going to go about folding it and can usually produce a fair approximation of my subject on the first try.

Hence the perennial question: How do you do that? The question is asked as if there were a recipe for origami design somewhere, a cookbook whose steps you could follow to reliably produce any shape you wanted from the square of paper. I don't think of origami design as a cookbook process so much as a bag of tricks from which I select one or more in the design of a new model. Here is a base (a fundamental folding pattern) with six legs: I'll use it to make a beetle. Here is a technique for adding a pair of points to an existing base: I'll combine these to make wings. Some designers have deeper bags of tricks than others; some, like John Montroll, have a seemingly bottomless bag of tricks. I can't really teach *the* way to design origami, for there is no single way to design; but what I can and will try to do in this book, is to pass on some of the tricks from my bag. Origami design can indeed be pursued in a systematic fashion. There are now simple, codified mathematical and geometric techniques for developing a desired structure.

This book is a collection of those techniques. It is not a step-by-step recipe book for design. Origami is, first and foremost, an art form, an expression of creativity, and it is the nature of creativity that it cannot be taught directly. However, it can be developed through example and practice. As in other art forms, you can learn techniques that serve as a springboard for creativity.

The techniques of origami design that are described in this book are analogous to a rainbow of colors on an artist's palette. You don't need a broad spectrum, but while one can paint beautiful pictures using only black and white, the introduction of other colors immeasurably broadens the scope of what is possible. And yet, the introduction of color itself does not make a painting more artistic; indeed, quite the opposite can happen. So it is with origami design. The use of sophisticated design techniques—sometimes called "technical folding," or *origami sekkei*—makes the resulting model neither artistic nor unartistic. But having a richer palette of techniques from which to choose can allow the origami artist to more fully express his or her artistic vision. That vision could include elements of the folding sequence: Does it flow naturally? Is the revelation of the finished form predictable or surprising? It could include elements of the finished form: Are the lines harmonious or jarring? Does the use of folded edges contribute to or detract from the appearance? Does the figure use paper efficiently or waste it? The aesthetic criteria to be addressed are chosen by the artist. Any given technique may contribute to some criteria (and perhaps degrade others). By learning a variety of design techniques, the origami artist can pick and choose to apply those techniques that best contribute to the desired effect.

These techniques are not always strict; they are sometimes more than suggestions, but less than commandments. In some cases, they are vague rules of thumb: "Beyond eight flaps, it is more efficient to use a middle flap." But they can also be as precise as a mathematical equation. In recent years, origami has attracted the attention of scientists and mathematicians, who have begun mapping the "laws of nature" that underlie origami, and converting words, concepts, and images into mathematical expression. The scientific fields of computer science, number theory, and computational geometry support and illuminate the art of origami; even more, they provide still more powerful techniques for origami design that have resulted in further advances of the art in recent years. Many design rules that on the surface apply to rather mundane aspects of folding, for example, the most efficient arrangements of points in a base, are actually linked to deep mathematical questions. Just a few of the subjects that bear on the process of origami design include the obvious ones of geometry and trigonometry, but also number theory, coding theory, the study of binary numbers, and linear algebra as well. Surprisingly, much of the theory is accessible and requires no more than high school mathematics to understand. I will, on occasion, bring out deeper connections to mathematics where they are relevant and interesting and I will provide some mathematical derivations of important concepts, but in most cases I will refrain from formal mathematical proofs. My emphasis throughout this work will be upon usable rules rather than mathematical formality.

As with any art, ability comes with practice, whether the art is origami folding or origami design. The budding origami designer develops his or her ability by designing and seeing the result. Design can start simply by modifying an existing fold. Make a change; see the result. The repeated practice builds circuits in the brain linking cause and effect, independent of formal rules. Many of today's origami designers develop their folds by a process they often describe as intuitive. They can't describe how they design: "The idea just comes to me." But one can create pathways for intuition to take hold by starting with small steps of design. The great leap between following a path and making one's own path arises from the development of an understanding of why: Why did the designer do it that way? Why does the first step start with a diagonal fold rather than a square fold? Why do the first creases hit the corners? Why, in another model, do the first creases miss the corners only by a little bit? Why does a group of creases emanate from a spot in the interior of the paper? If you are a beginning designer, you should realize that *no design is sacred*. To learn to design, you must disregard reverence for another's model, and be willing to pull it apart, fold it differently, change it and see the effects of your changes.

Small ideas lead to big ideas; the concepts of design build upon one another. So do the chapters of this book. In each chapter, I introduce a few design principles and their associated terms. Subsequent chapters build on the ideas of earlier chapters. Along the way you will see some of my own designs, each chosen to illustrate the principles introduced in the chapter in which it appears.

Chapter 2 introduces the fundamental building blocks of origami: the basic folds. If you have folded origami before, you may already be familiar with the symbols, terms, and basic steps, but if not, it is essential that you read through this section. Chapter 2 also introduces a key concept: the relationship between the crease pattern and the folded form, a relationship that we will use and cultivate throughout the book.

Chapter 3 initiates our foray into design by examining a few designs. The first stage of origami design is modification of an existing design; in this chapter, you will have an opportunity to explore this approach by devising simple modifications to a few figures.

Chapter 4 introduces the concept of a base, a fundamental form from which many different designs may be folded. You will learn the traditional bases of origami, a number of variations on these bases, and several methods of modifying the traditional bases to alter their proportions.

Chapter 5 expands upon the idea of modifying a base by focusing upon modifications that turn a single point into two, three, or more simply by folding. This technique, called point-splitting, has obvious tactical value in designing, but it also serves as an introduction to the concept of modifying portions of a base while leaving others unchanged.

Chapter 6 introduces the concept of grafting: modifying a crease pattern as if you had spliced into it additional paper for the purpose of adding structural elements to an existing form. Grafting is the simplest incarnation of a broader idea, that the crease patterns for origami bases are composed of separable parts.

Chapter 7 then expands upon the idea of grafting and shows how multiple intersecting grafts can be used to create patterns and textures within a figure—scales, plates, and other textures. This set of techniques stands somewhat independently, as almost any figure can be "texturized."

Chapter 8 generalizes the concept of grafting to a set of techniques called tiling: cutting up and reassembling different pieces of crease patterns to make new bases. This chapter defines both tiles and matching rules that apply to the edges of tiles to insure that the assemblies of tiles can be folded into a flat shape. Chapter 8 also introduces the powerful concept of a uniaxial base—a family of structures that encompasses both the traditional origami bases and many of the most complex modern bases.

Chapter 9 shows how the tile decorations that enforce matching can be expanded into a design technique in their own right: the circle/river method, in which the solution of an origami base can be derived from packing circles into a square box. Circle/river packing is one of the most powerful design techniques around, capable of constructing figures with arbitrary configurations of flaps, and yet it can be employed using nothing more than a pencil and paper.

Chapter 10 explores more deeply the crease patterns within tiles; those that fit within circle/river designs are called molecules. The chapter presents the most common molecules, which are sufficient to construct full crease patterns for any uniaxial origami base.

Chapter 11 presents a different formulation of the circle/river packing solution for origami design, called tree theory, in which the design of the base is related to an underlying stick figure and the packing problem is related to a set of conditions applying to paths along the stick figure. Although equivalent to circle/river packing, the approach shown here is readily amenable to computer solution. It is the most mathematical chapter, but is in many ways the culmination of the ideas presented in the earlier chapters for designing uniaxial bases.

Chapter 12 then introduces a particular style of origami called box pleating, which has been used for some of the most complex designs ever constructed. Box pleating in some ways goes beyond uniaxial bases; in particular, it can be used to construct fully three-dimensional figures. But it can also be implemented mathematically, and I will show how tree theory can be generalized to encompass box-pleated designs.

Chapter 13 continues to move beyond uniaxial bases, introducing the idea of hybrid bases, which combine elements from uniaxial bases with other non-uniaxial structures. The world of origami designs is enormously larger than the uniaxial bases that are the focus of this book, but as this chapter shows, elements from uniaxial bases can be combined with other structures, expanded, and extended, to yield ever-greater variety in origami figures.

For specialists or readers who wish to go farther, Chapter 14 presents a formal mathematical treatment of tree theory which will be best appreciated by those with a background in linear algebra and optimization theory. The References section provides references and commentary organized by chapter with citations for material from both the mathematical and origami literature related to the concepts in each chapter.

Each chapter includes step-by-step folding instructions for one or more of my origami designs chosen to illustrate the design concepts presented in the chapter. I encourage you to fold them as you work your way through the book. Most have not been previously published. I have also in several chapters presented crease patterns and bases of models whose instructions have been published elsewhere; you will find sources for their full folding sequences in the References section.

The concepts presented here are by and large my own discoveries, developed over some 35 years of folding. They were not developed in isolation, however. Throughout the book I have pointed out sources of influence and/or ideas I have adopted. In several cases others have come up with similar ideas independently (an event not without precedent in both origami and the sciences). Where I am aware of independent invention by others, I have attempted to identify it as such. However, the formal theory of origami design is very much in its infancy. Sources of design techniques are often unpublished and/or widely scattered in sometimes obscure sources. This work is not intended to be a comprehensive survey of origami design, and if it seems that I have left out something or someone, no slight was intended.

Technical folding, *origami sekkei*, is an edifice of concepts, with foundations, substructure, and structure. Because the organization of this book mirrors this structure, I encourage you to read the book sequentially. Each chapter provides the foundation to build concepts in the next. Let's start building.

2 The Building Blocks of Origami

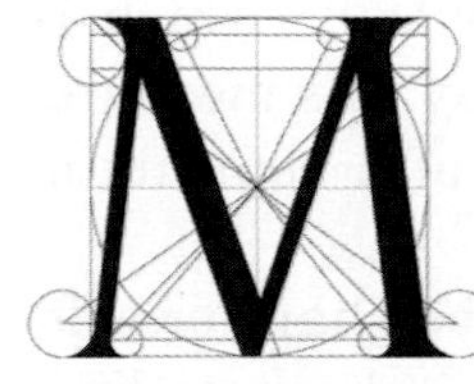

Much of the charm of origami lies in its simplicity: There is the square, there are the folds. There are only two types of folds: mountain folds (which form a ridge) and valley folds (which form a trough). So, square + mountain folds + valley folds is the recipe for nearly all of origami. How simple can you get?

But is it true that there are even two types of fold? Indeed, the mountain fold can be turned into a valley fold merely by turning the paper over. So, instead of there being two types of fold, perhaps there is only one.

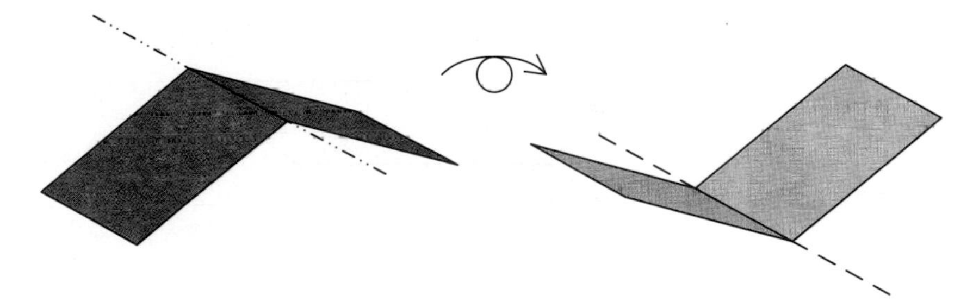

Figure 2.1.
A mountain fold is the same as a valley fold turned over.

On the other hand, perhaps there are *three* types of fold: valley folds, mountain folds, and unfolds. If we fold the paper in half and unfold it, we will be left with a line on the paper—a crease—which is also a type of fold. Creases are sometimes merely artifacts, leftover marks from the early stages of folding, but they can also be useful tools. Creases can provide reference points ("fold this point to that crease") and in the purest style of folding (no measuring devices, such as rulers, allowed) creases, folded edges, and their intersections are the only things that can serve as reference points. Creases are also commonly made in preparation for a complex maneuver. Origami diagrammers attempt to break folding instructions into a sequence of simple steps, but some maneuvers are inherently complex and require you to bring 5 or 6 (or 10 or 20) folds together at once. For such pleasant challenges, it's a big help to have all the creases already in place. Precreasing helps tame the dragon.

Valley, mountain, and crease are the three types of folds from which all origami springs. But even a valley fold is not necessarily the same as another valley fold if the layers of paper do not lie flat. When models move into three dimensions, both valley and mountain folds can vary in another way: the fold angle, which can take on many values. If you draw two lines perpendicular to the fold line, then the angle between the two lines in the third dimension can vary continuously, from 0° (for a valley fold) to 180° (which is no fold at all) to 360° (for a mountain fold). By this measure, valley, mountain, and crease, are all part of a continuum of fold angle.

There is yet more variation: A fold can be sharp or soft. The mathematical model of a "fold" is an infinitely sharp line, but with real paper, the sharpness of the fold is something the folding artist can choose. Sharp creases are not always desir-

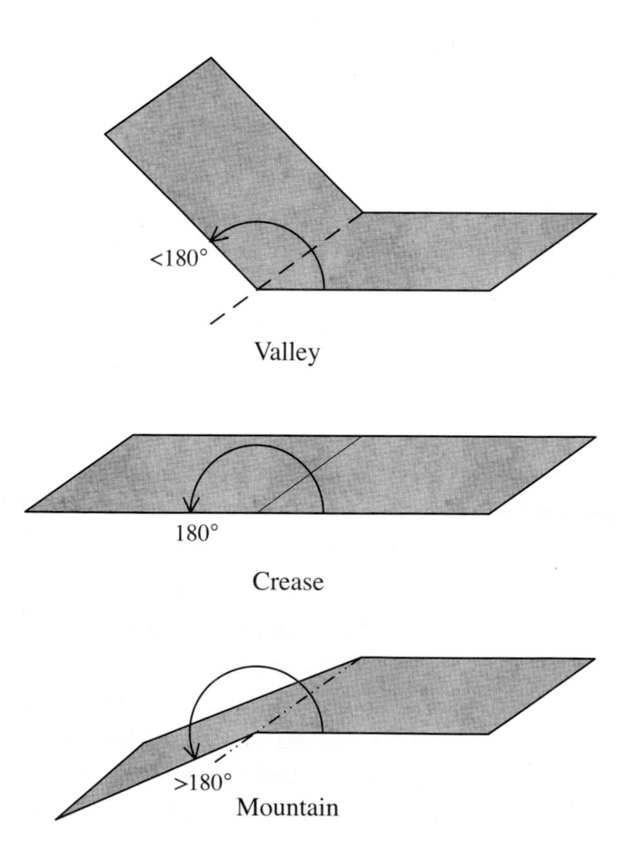

Figure 2.2.
Valley folds, creases, and mountain folds are all part of a continuum.

able. In a complex model with many folds, sharp creases can weaken the paper to the point that the paper rips. In a model of a natural subject, sharp lines can be harsh and unlifelike, whereas soft, rounded folds can convey an organic quality, a sense of life. On the other hand, when precision is called for, sharp folding may be required to avoid a crumpled mess down the road. Consequently, most models call for a mix of sharp and soft folding, and while the distinction can sometimes be given in the folding diagrams, in most cases, the artist must simply develop through experience a feel for how sharp a given crease must be.

2.1. Symbols and Terms

Origami instruction is conveyed through diagrams—a system of lines, arrows, and terms that has become the *lingua franca* (or perhaps *lingua japonica*) of the worldwide arena. The modern system of origami diagrams was first devised by the great Japanese master Akira Yoshizawa in his books of the 1940s and 1950s, and was subsequently adopted (with minor variations) by the two early Western origami authors Samuel L. Randlett (United States) and Robert Harbin (U. K.). Despite occasional attempts by others at establishing a rival notation (e.g., Isao Honda, who used dashed lines everywhere, but distinguished mountain folds by a "P" next to the line), the Yoshizawa/Randlett/Harbin system caught on and has become the sole international system in the origami world.

No system is perfect, and over the years, various diagrammers have made their own additions to the system. Some, like open and closed arrows (to denote open and closed sink folds), died a quiet death; others, like Montroll's "unfold" arrow, have become firmly established in the origami diagrammatic lexicon (symbolicon?). Every author has his or her particular quirks of diagramming, but the core symbols and terms are nearly universal.

Odds are that you already have some familiarity with origami and have encountered the Yoshizawa diagramming system. It will, however, serve us to run through the basic symbols and terms, both to establish a fixed starting point and to start the wheels turning for origami design, which is as much a way of looking at origami as it is a set of codified tools.

The first thing to run through are origami terms, which include names, directions, and positions. Origami diagrams are ideally drawn so that the diagrams themselves are sufficient to enable the reader to fold the model (which allows people the world over to fold from them; a Japanese or Russian folder can fold from English diagrams and vice-versa). Nevertheless, many people find folding instructions more readily comprehensible with a verbal instruction attached, and so in the instructions in this book, you will find both words and pictures.

Origami verbal instructions are given as if the paper were flat on the page before you. Thus, words that say "fold the flap upward" mean that if you orient the working model the same way as the diagram on the page, you will fold the flap toward the top of the page. "Up," "down," and "to the side" all refer to directions with respect to the printed page. While directions are always given as if the paper were flat on the page, you may find it easier to pick the model up, fold in midair, or even turn it over to make the fold (mountain folds are commonly made by turning the paper over and forming a valley fold). If you do this, be sure that you always return it to the orientation shown in the next diagram.

As the folded model begins to accumulate multiple layers of paper, it becomes necessary to distinguish among the layers. By convention, the term "near" refers to the layers closest to you (i.e.; those on top) and "far" layers are those on the bottom (thus, reserving the words "top" and "bottom" for directions with respect to the page).

Origami paper typically has a white side and a colored side. The two colors are featured in some models—there are origami skunks, pandas, and even zebras and chessboards whose coloration derives from skillful usage of the two sides of the paper. Even if only one side is visible at the end, it is helpful in keeping track of what's going on to show the two sides as distinct colors, and that is what I have done here.

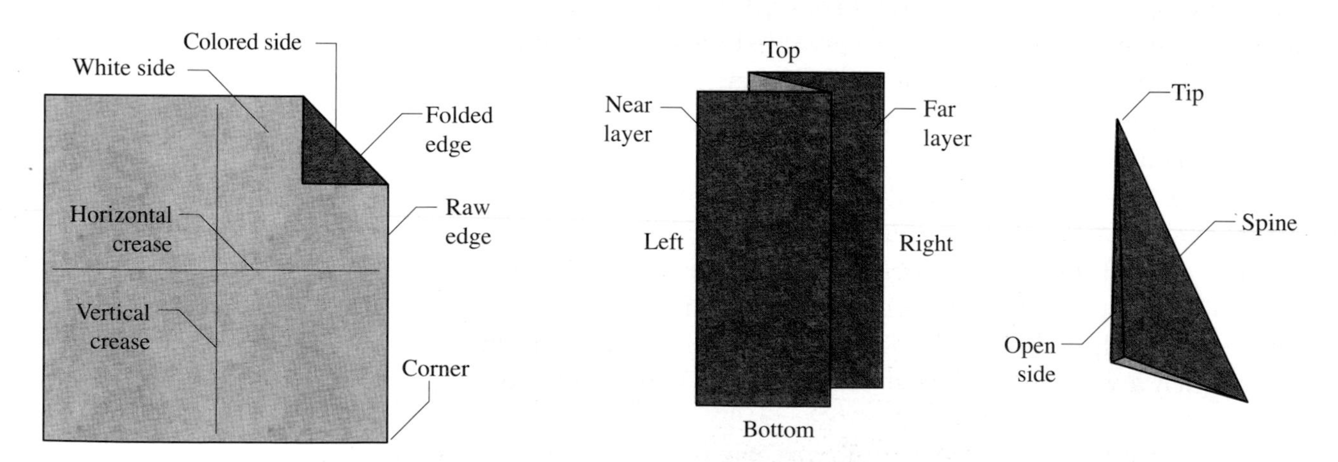

Figure 2.3.
Verbal terms that apply to origami diagrams.

Brightly colored origami paper often comes precut to squares. One of the small ironies of the art is that when precut square origami paper was introduced in Japan near the turn of the 20th century, it was made from inexpensive European machine-made paper, since handmade Japanese *washi* was far too expensive for most purposes. Thus, the origami paper that is considered the most authentically Japanese wasn't even originally from Japan!

For your own folding, there is no special requirement on paper other than it hold a crease and not easily rip. Traditional origami paper—available from most art and craft stores, via the Internet, and at many stores in the Japanese quarter of large cities—is relatively inexpensive and conveniently precut to squares. (However, it may not be precisely square. Like most machine-made papers, prepackaged origami paper has a definite grain and will change proportion slightly with humidity; a square in Florida will probably be a rectangle in Nevada.) Other papers that are useful are thin artist's foil (also available from art stores), foil wrapping papers, and various thin art papers you may run across with names like *unryu*, *kozo*, and *lokta*.

Origami diagrams are usually line drawings. Even in this day of three-dimensional computer rendering, line drawings convey the information of folding as well as anything (and they don't require a $10,000 workstation to create!). There are five types of lines that are used for different features of the folded shape. Paper edges, either raw (an original edge of the paper) or folded, are indicated by a solid line. Creases are indicated by a thinner line, and will often stop before they reach the edge of the paper. Valley folds are indicated by a dashed line; mountain folds by a chain (dot-dot-dash) line. The "X-ray line," a dotted line, is used to indicate anything hidden behind other layers, and could be used to represent a hidden edge (most often), fold, or arrow. It will

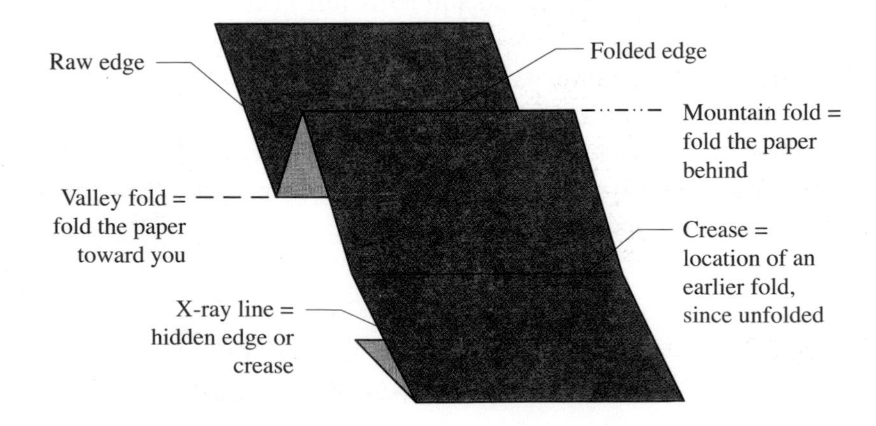

Figure 2.4.
The five types of lines used in origami diagrams.

usually be clear from context what the X-ray line was meant to represent.

Actions are indicated by arrows, both showing the motion of the paper as a fold is made, and sometimes manipulations of the entire model. An open hollow arrow is used to show the application of pressure (usually in connection with a reverse or sink fold). See Figures 2.21–2.23 and 2.40–2.47 for examples.

Figure 2.5.
A hollow arrow indicates to "push here."

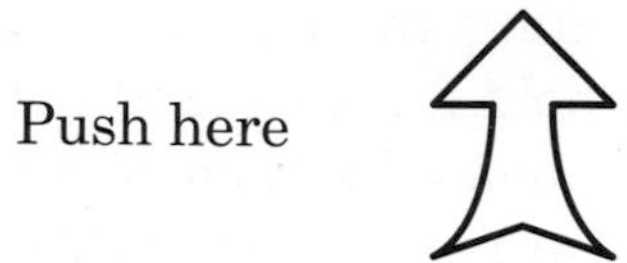

An arrow with a loop through it indicates to turn the paper over—either from side to side (like turning the pages of a book) or from top to bottom (like flipping forward or backward in a wall calendar), with the direction specified by the orientation of the arrow.

Turn the paper over
from side to side

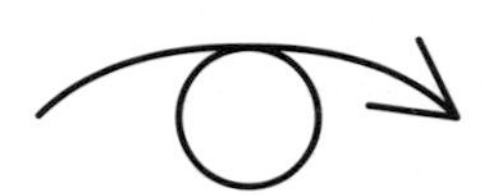

Turn the paper over
from top to bottom

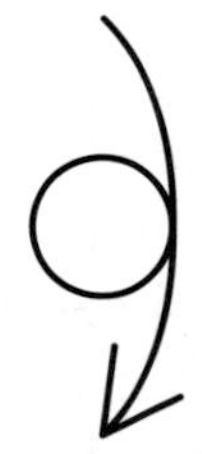

Figure 2.6.
A looped arrow indicates to turn the paper over.

If the model is to be rotated in the plane of the page, that is indicated by a fraction enclosed in two arrows showing the direction of rotation. The number inside the arrows is the fraction of a circle through which the rotation takes place. "1/2" is a half turn, i.e., the top becomes the bottom and vice-versa; "1/4" indicates a quarter-turn. Sometimes the amount of rotation is not a simple fraction; rather than putting something unwieldy like "21/34" in the arrows, I'll usually round it to the nearest quarter-turn and you can use the subsequent diagram to pin down the orientation precisely.

Rotate the paper

Figure 2.7.
A fraction inside a circle formed from two arrows indicates to rotate the paper.

Most origami is folded flat at every step. However, when a model becomes three-dimensional, either because the final model is 3-D or one or more intermediate steps are 3-D, it frequently becomes necessary to show multiple views of the model to fully convey what is going on. In such cases, a small stylized eye indicates the vantage point from which a subsequent view is taken.

View from this vantage point

Figure 2.8.
An eye with a dotted line indicates the sightline used to specify a new point of view.

The next symbol indicates one of the most dreaded instructions in all of origami: repetition. You have worked through a long, tortuous sequence of folds, you think you're coming to the end, and there it is: "repeat steps 120–846 on the other 7 flaps." The bad news is usually conveyed in words, but for those who fold from the diagrams alone, repetition is conveyed by a symbol as well. Harbin, the great Western popularizer of origami, devised an arrow with hash marks to indicate repetition; however, this symbol is unnecessarily ambiguous, and I have preferred to use a boxed leader enclosing the range of steps to be repeated, as shown in Figure 2.9.

Repeat a range of steps

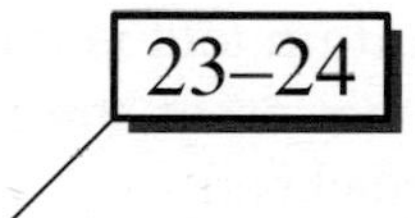

Figure 2.9.
A range of steps to be repeated is indicated by a boxed sequence of the numbered steps to be repeated.

Lastly, it frequently arises that a fold is to be made at 90° to another fold or to a folded edge. When this takes place and it is not obvious that the fold is at 90°, I will indicate it by a small right-angle symbol next to (and aligned with) the relevant intersection.

Right angle

Figure 2.10.
A right angle is indicated by the geometer's symbol of a right angle located next to the relevant crease.

2.2. Basic Folding Steps

Now, we turn to the basic folds of origami—single folds, or combinations of a few folds that occur over and over in origami figures. Most of these combinations date back hundreds of years in Spain and Japan as concepts, if not as recognized steps. These are, however, the building blocks from which nearly all origami models arise. The names are of much more recent vintage and vary from country to country, but in

English-speaking countries, the names given here are widely accepted.

The first basic fold is the generic valley fold—a fold made with a single straight line, with the fold made concave toward the folder. The fold itself is indicated by a dashed line, which divides the paper into two parts, one stationary (usually), one moving. A symmetric double-headed arrow is used to indicate which part moves and the direction of motion. The moving part almost always must rotate up and out of the plane of the page; this motion is conveyed by curving the arrow.

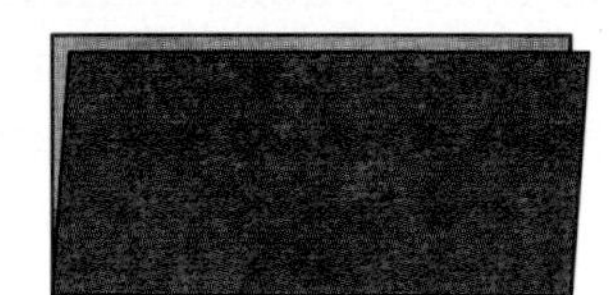

Figure 2.11.
A valley fold, as diagrammed, and the result.

The opposite of a valley fold is a mountain fold, which is called for when a portion of the paper is to be folded behind. The mountain fold is indicated by a chain line (dot-dot-dash), and the motion of the paper is indicated by a hollow single-sided arrowhead.

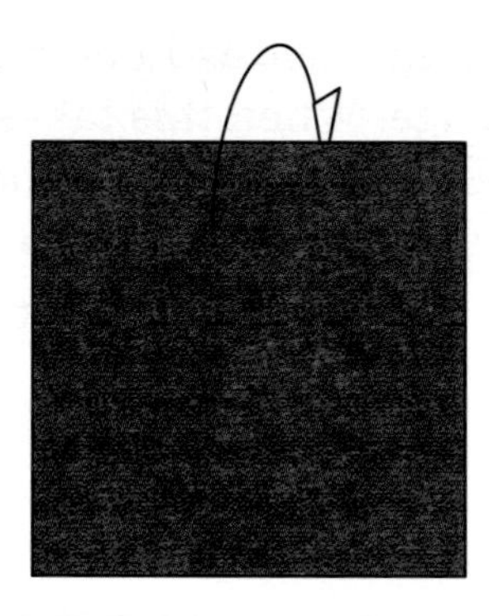

Figure 2.12.
A mountain fold, as diagrammed, and the result.

Quite often, a mountain fold is shown as a bit of shorthand for "turn the paper over, make a valley fold, and then turn it back to the original orientation," as in the example in Figure 2.12. However, mountain folds are frequently used to tuck paper into a pocket or between layers, situations where turning the paper over will not necessarily make a valley fold possible.

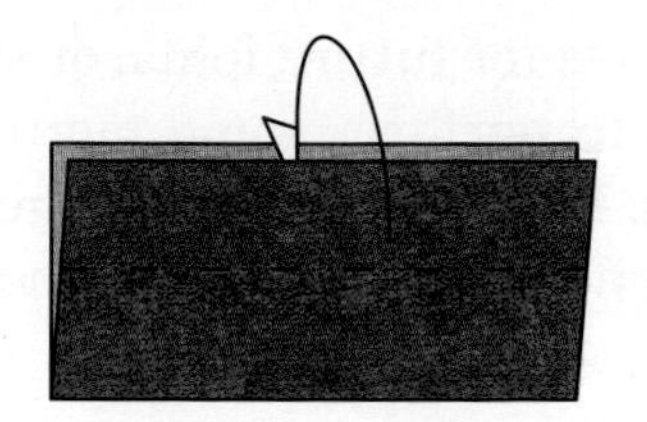
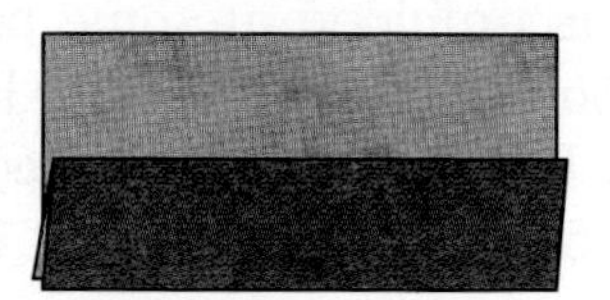

Figure 2.13.
A mountain fold is not always amenable to "turn the paper over and make a valley fold."

When a mountain fold (or, less often, a valley fold) is used to tuck one layer between two others, the layers will be separated as in Figure 2.13 and the arrow will be drawn between the two layers. If, when folding, you find that a flap can be folded into more than one location, examine the drawing closely, as the arrow will likely show where the layer should go.

Quite often, both a mountain fold and a valley fold will be called for on parallel layers, a maneuver that is commonly used for thinning legs and other appendages. This step is shown with two arrows and, if possible, both the mountain and valley fold. You may perform both a mountain and a valley fold if you wish, but many folders actually form both folds as mountain folds, making one from each side of the paper.

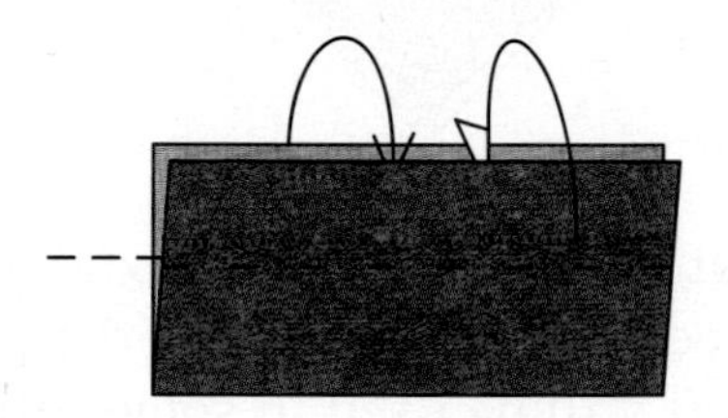

Figure 2.14.
Mountain and valley folds used to thin a flap.

Figure 2.14 illustrates several common subtleties of origami diagrams. The valley fold on the far layer is made clear by extending the fold line (the dashed line) beyond the edge of the paper. The valley fold is understood to run completely along the far layer of paper, even though it is not shown. (I could use an X-ray line to indicate the extension of the valley fold, but I don't in this figure because it would get mixed up with the overlaid mountain fold line). Both the mountain and valley fold layers get tucked into the middle of the model, which you can tell by observing that both arrowheads travel between the two layers. The resultant figure—the drawing to the right—shows the disposition of the layers along its edge, which makes this example unambiguous. It is often not possible to show such layers, however; you must rely upon the arrows between the layers as in the figure on the left.

Folds, once made, do not always persist to the end of the model. It is a fairly frequent occurrence that folds are made to establish reference points or lines for future folds, or that a model is unfolded at some point to perform some manipulation upon hidden or interior lines. In either case, folds get unfolded. Unfolding is indicated by a symmetric hollow-headed arrow, as shown in Figure 2.15.

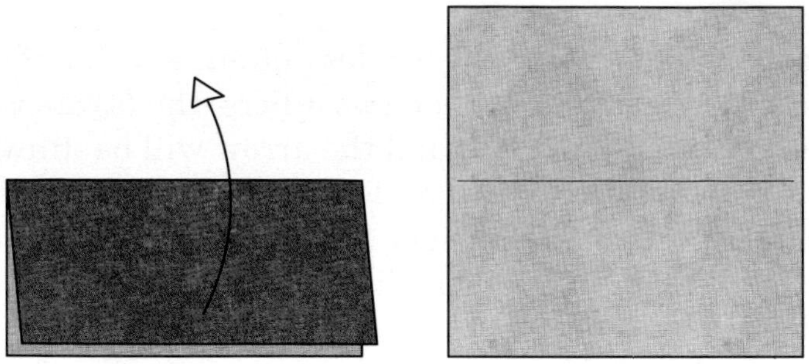

Figure 2.15.
The unfold arrow.

The same symbol is used to indicate when paper is to be pulled out from an interior pocket, as shown in Figure 2.16.

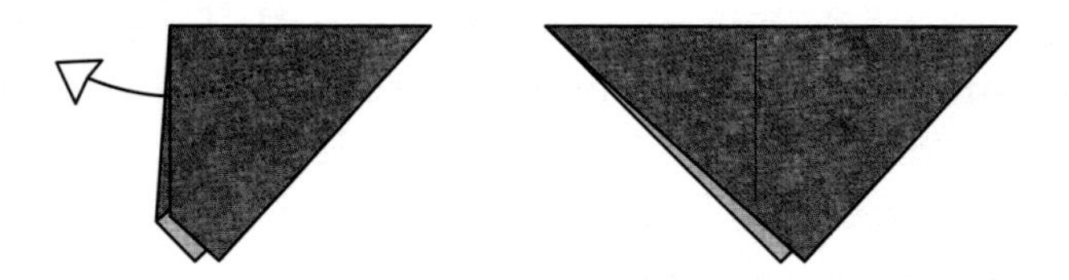

Figure 2.16.
The unfold arrow used to show pulling paper out from inside the model.

Particularly in the early stages of folding a model, one will make a fold and then immediately unfold it, for the purpose of establishing a crease that will be used in some future (usually more complicated) step. To keep the diagrams fairly compact, the fold-and-unfold action is commonly expressed in a single figure, and is indicated by a single double-headed arrow that combines the fold arrow (valley fold) and unfold arrow in a single arrow.

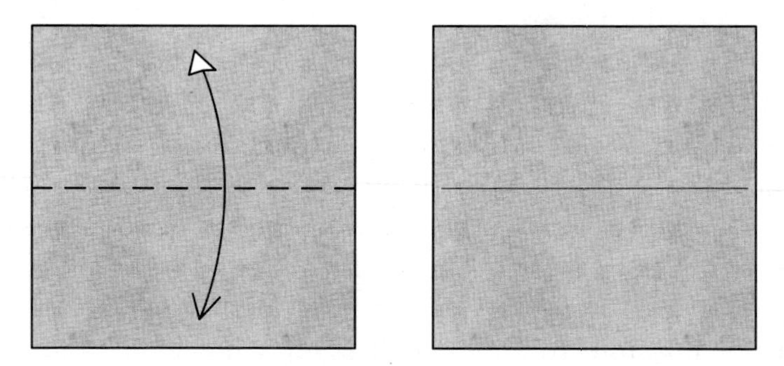

Figure 2.17.
Fold-and-unfold is indicated by a double-headed arrow that combines the "fold" and "unfold" arrowheads.

Most of the time, the fold in a fold-and-unfold step will be a valley fold, but on occasion, the desired crease is a mountain fold. Rather than diagramming this in three steps (turn the paper over, valley-fold-and-unfold, turn the paper back over), I will use the mountain fold arrow in combination with the unfold arrow, as shown in Figure 2.18. It should be understood that what is intended is to fold the moving flap behind, make the crease, and then unfold.

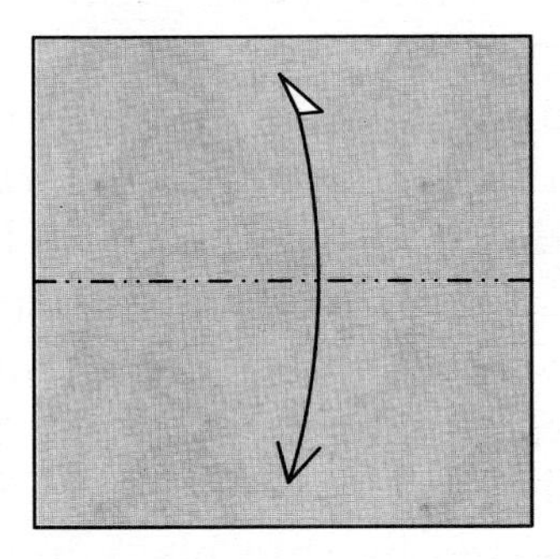

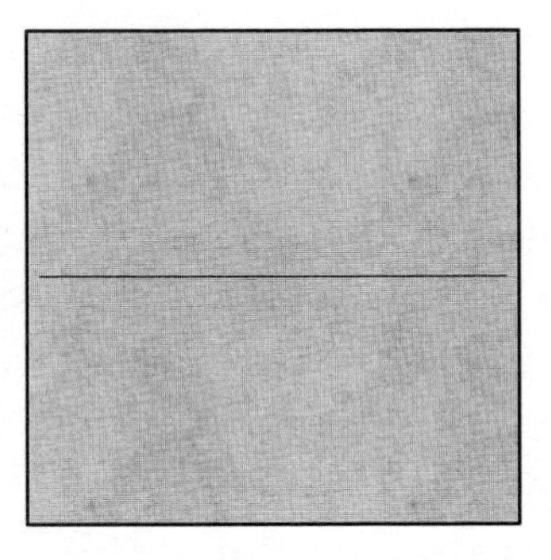

Figure 2.18.
Mountain-fold and unfold.

In the study of origami design, the crease pattern of the finished figure or a subset of same provides a great deal of information about the structure of the model—often more information than the sequence of folding instructions, because it shows the entire model at once. The simplest form of the crease pattern simply shows all creases as crease lines, as in Figure 2.19, which shows the crease pattern for the traditional Japanese flapping bird.

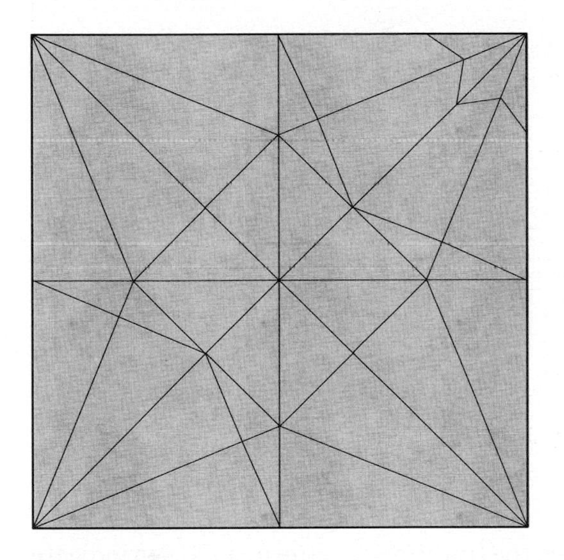

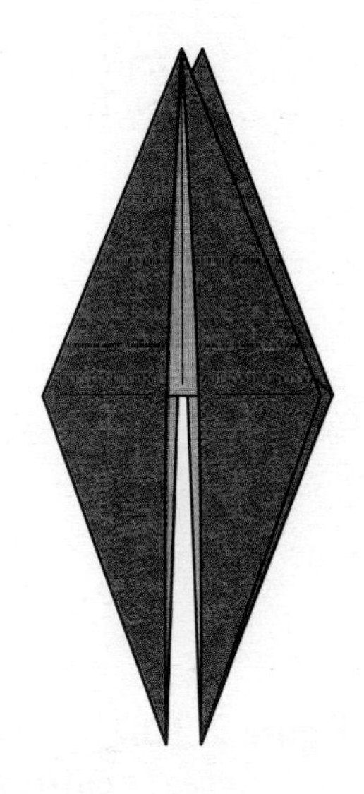

Figure 2.19.
Crease pattern, base, and folded model of the traditional Japanese flapping bird.

However, knowing just the location of the creases is not as useful as it could be; it is far more useful to know the directions of the creases, i.e., whether they are valley or mountain folds.

("More useful" is a bit of an understatement. In 1996, Marshall Bern and Barry Hayes proved that figuring out crease directions from a generic crease pattern is computationally part of a class of problems known as "NP-complete." As such problems grow in size, they quickly outstrip the abilities of any computer to solve.)

Thus, it is more helpful to give the direction—or crease assignment—of the creases: mountain, valley, or crease (that is, not folded at all). The traditional mountain and valley lines—chain and dashed—tend to lose their distinction in large crease patterns, dissolving into a morass of confusing clutter. Thus, in crease patterns, I will adopt a different convention. Creases that are valley fold lines will be indicated by solid colored lines, while mountain folds will be solid black lines. Creases that lie flat will be indicated by thin black lines as usual. (Flat creases that don't play an important role are not shown at all, but it is sometimes helpful to show creases that were important to the construction of the base.) A point where two or more creases come together is called a *vertex* of the crease pattern. To see the difference between the two line styles, compare the two examples in Figure 2.20.

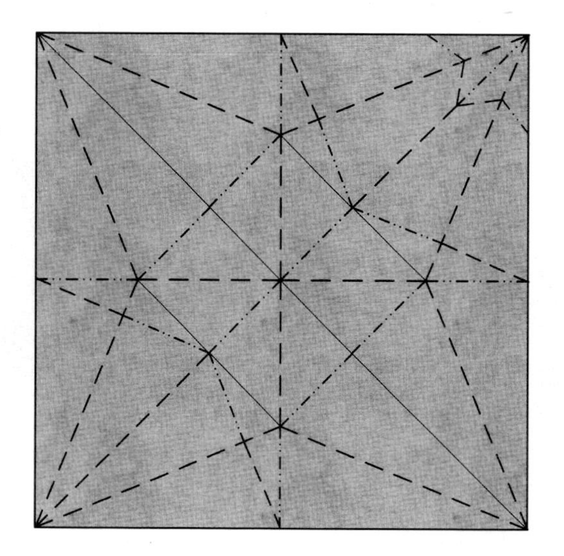

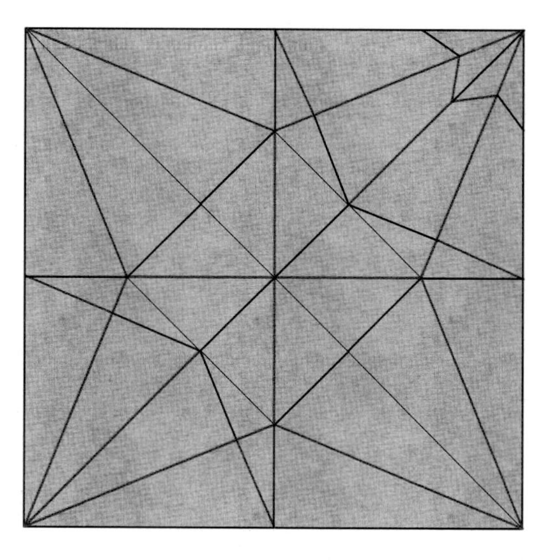

Figure 2.20.
Left: a crease pattern using the traditional patterned lines to indicate mountain and valley folds.
Right: the same pattern using colored lines.

While all origami models are created entirely from mountain and valley folds, they often occur in distinct combinations, combinations that occur often enough that they have been given names of their own.

The first and simplest combination fold is the inside reverse fold, which is a fold used to change the direction of a flap. While either a mountain or valley fold could usually be used in the same place, a reverse fold combines both mountain and valley and is usually more permanent, since the tension of the paper tends to keep the reverse fold together. A reverse fold always takes place on a flap consisting of at least two layers of paper. In an inside reverse fold, the mountain fold line occurs on the near layer, a valley fold occurs on the far layer, and the "spine" above the fold lines is turned inside-out. It is indicated by a push arrow, since to form the reverse fold, the spine must be pushed and turned inside-out. If the far edges are visible, then the valley fold may be shown extending from the visible edge, as in Figure 2.21.

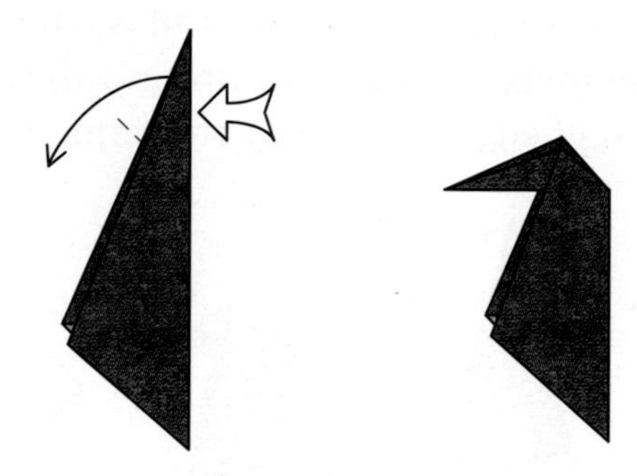

Figure 2.21.
The inside reverse-fold.

In the inside reverse fold, the tip of the flap ends up pointing away from the spine; in Figure 2.21, the spine is the right side of the flap, so the tip must point to the left. If you wanted it to point to the right, then you would use the other type of reverse fold, the outside reverse fold, which is illustrated in Figure 2.22. Again, there is a mountain fold and a valley fold, but in the outside reverse fold, the valley fold occurs on the near layers and the mountain fold on the far layers, opposite from what happens in the inside reverse fold. The outside reverse fold is also indicated by a push arrow, because it is typically made by pushing at the spine with one's thumb while wrapping the edges of the paper around to the right. Like the inside reverse fold, it is much more permanent than a simple mountain or valley fold would be.

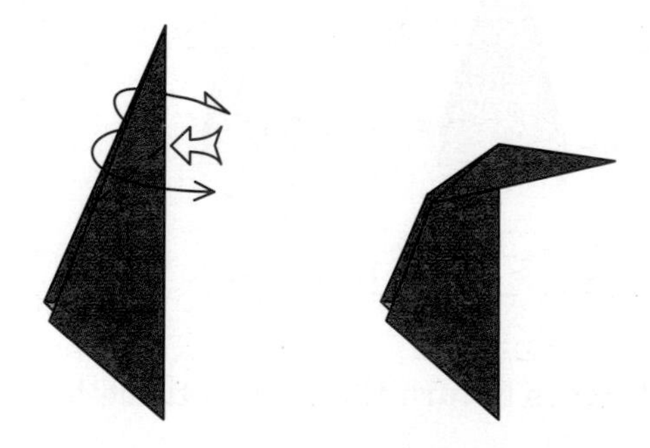

Figure 2.22.
The outside reverse fold.

In the verbal instructions, the term "reverse fold" (without an "inside" or "outside" qualifier) will generally mean "inside reverse fold."

A simple flap with two layers has only two possible types of reverse fold: inside or outside. More complicated flaps with multiple layers can have multiple possibilities or even combinations of the two; for example, the triangular shape shown in Figure 2.23 (made by folding a square in thirds at one corner) can be either inside- or outside-reverse-folded to either the left or right; in addition, it is possible to make a sort of hybrid reverse fold that combines aspects of both. The silhouettes of all three shapes (and for that matter, the mountain- or valley-folded equivalents) are the same; they differ only in their crease patterns. In diagrams throughout the book, they will be distinguished by the presence or absence of push arrows (distinguishing reverse folds from mountain or valley folds) and/or the configuration of edges shown in subsequent diagrams.

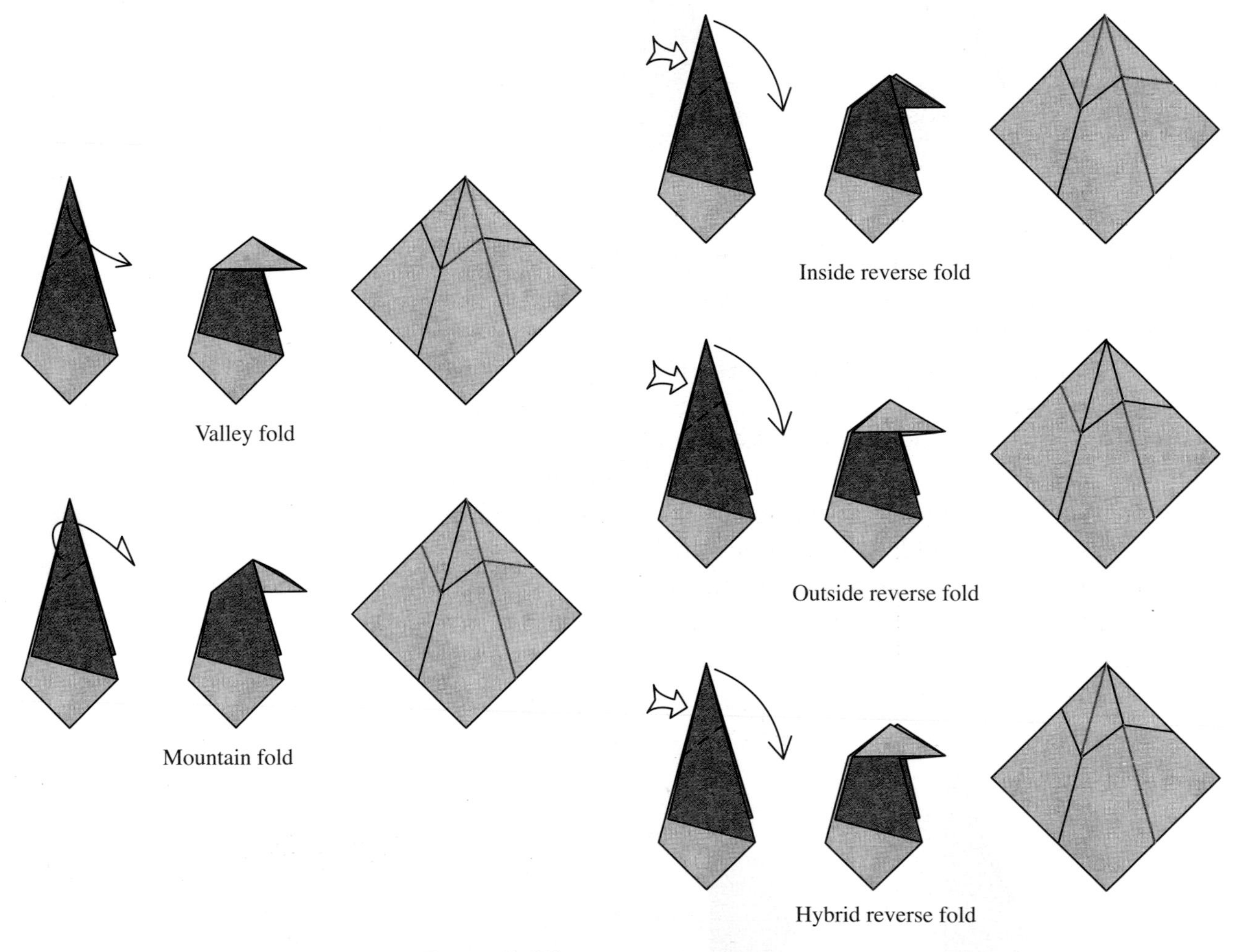

Figure 2.23.
The five possible ways to turn the tip of a three-layer flap.

Another combination fold that occurs with some regularity is the rabbit-ear fold (which acquired its name from some rabbit design long since lost in the mists of antiquity). The rabbit-ear fold is almost always performed on a triangular flap, and is characterized by three valley folds along the angle bisectors of the triangle with a fourth fold, a mountain fold, extending from the point of intersection perpendicularly to one side.

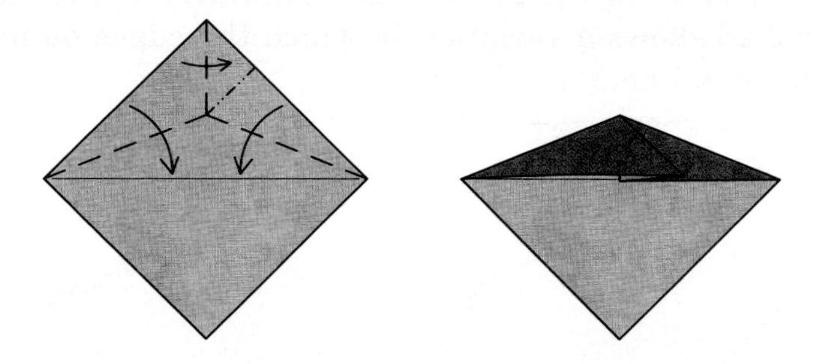

Figure 2.24.
The rabbit-ear fold.

When a rabbit-ear fold is formed, all of the edges lie on a common line. Remarkably, this procedure works for a triangle of any shape—or perhaps it is not so remarkable, since the rabbit ear is merely a demonstration of Euclid's theorem that the angle bisectors of any triangle meet at a common point.

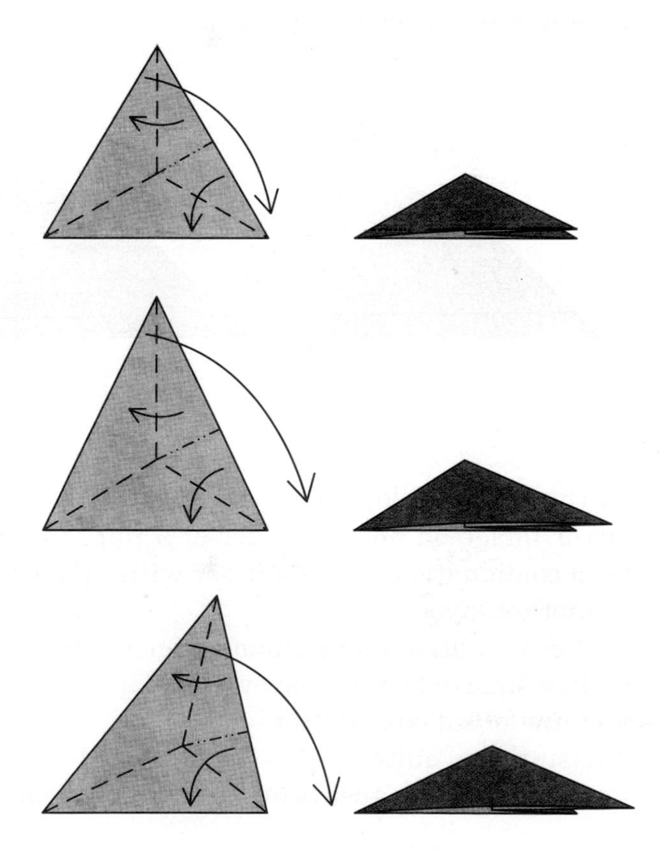

Figure 2.25.
The rabbit ear can be folded from any triangle.
Top: equilateral.
Middle: isosceles.
Bottom: scalene.

Rabbit-ear folds occur not only on isolated flaps. Bringing all the edges to lie on a common line is a special property; the rabbit ear is the simplest example of a molecule, which is the name for any crease pattern with this property. We will encounter rabbit-ear crease patterns and molecules in much detail and many guises as we delve more deeply into systematic design.

In addition to the simple, straightforward rabbit ear, there are two variations that are regularly encountered. Figure 2.26 shows a variation in which the edges do not lie on a common line.

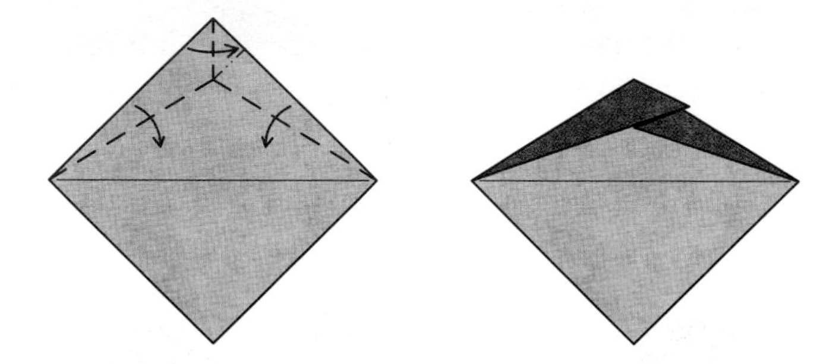

Figure 2.26.
A variation of a rabbit ear.

Figure 2.27 shows a combination of two rabbit ears made from the near and far layers of a two-layered flap. Known, appropriately, as a double rabbit ear, it is typically formed by pinching the near and far layers of the flap into rabbit ears and then swinging the tip over to the side.

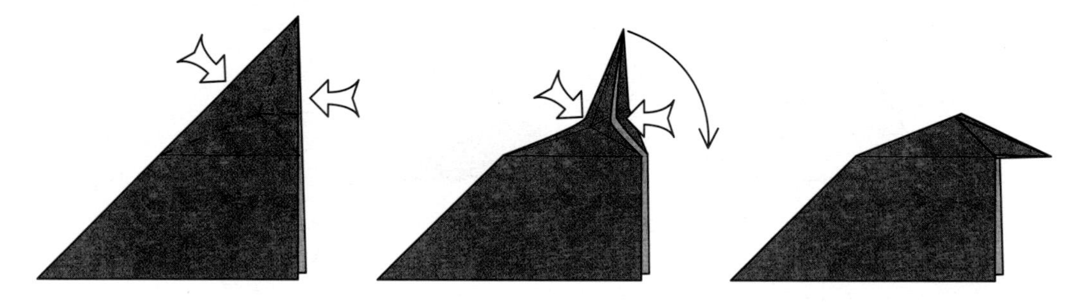

Figure 2.27.
A double rabbit ear.

Just as the reverse fold is a combination of a valley fold with its mirror image on another layer of a flap, the double rabbit ear is a combination of a rabbit ear with its mirror image also on another layer.

The next combination fold commonly encountered is the squash fold. In a squash fold, the layers of a flap are spread to the sides and the folded edge flattened.

The squash fold is quite easy to perform (and sometimes very satisfying). It is nearly always formed symmetrically, that

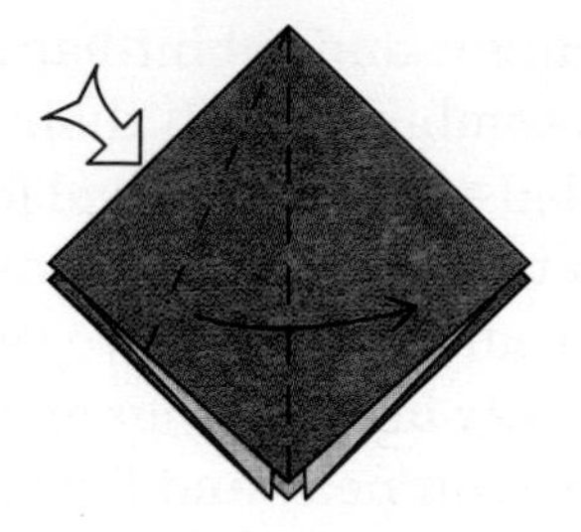
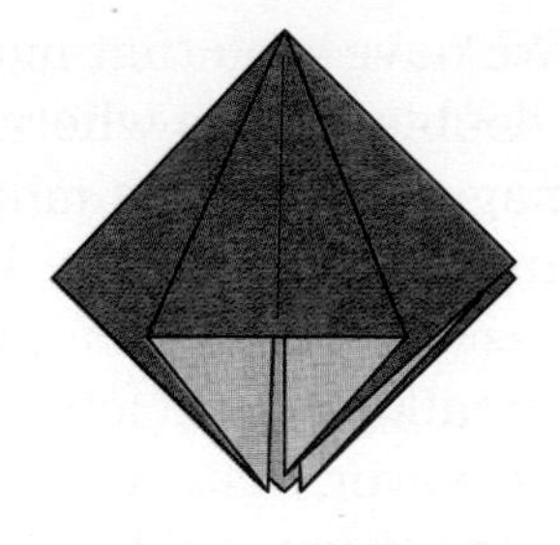

Figure 2.28.
The squash fold.

is, making equal angles on both the left and right. In the symmetric form, the crease that used to be the folded edge will be lined up with one or more raw edges underneath, as in Figure 2.28. It is also possible to squash-fold a point, as shown in Figure 2.29. Squash-folded points are harder to keep symmetric, because the point covers up the layers underneath, but you can make them symmetric by turning the paper over and checking the alignment on the other side before you make the creases sharp.

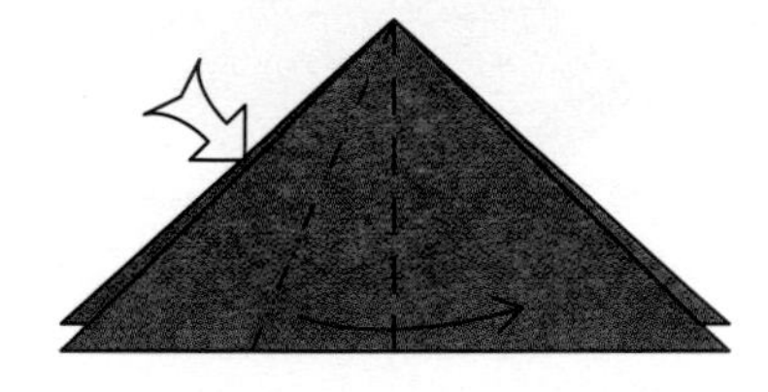
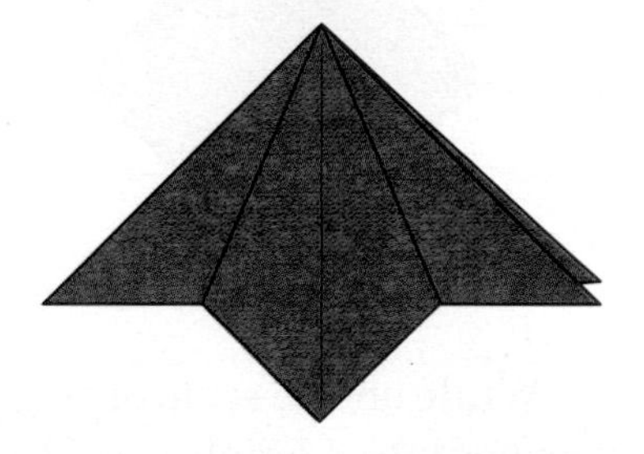

Figure 2.29.
Another version of a squash fold.

There are four creases involved in a squash fold: two valleys on each side of two mountains (usually, only one of each is visible on the near side of the flap). All four creases come together at a point. Most of the time, the two valley folds are side-by-side and the squash fold is symmetric about the valley fold. However, a squash fold can be made asymmetrically and it sometimes happens that the two valley folds are not side-by-side. When that happens, a portion of the visible flap can be seen to rotate (about the intersection of all the creases). This asymmetric version of a squash fold occurs often enough that it is given its own name: a swivel fold.

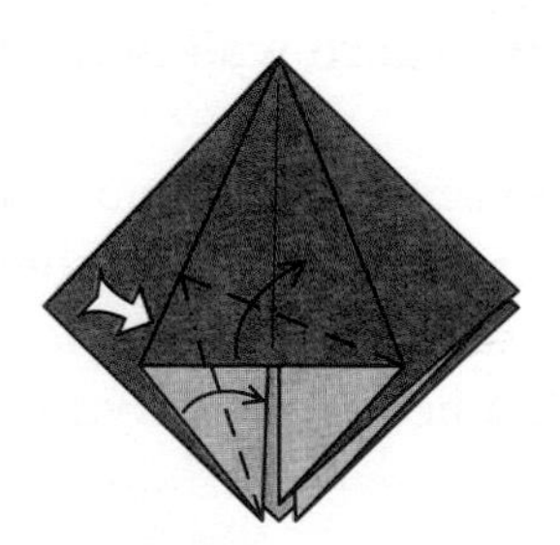
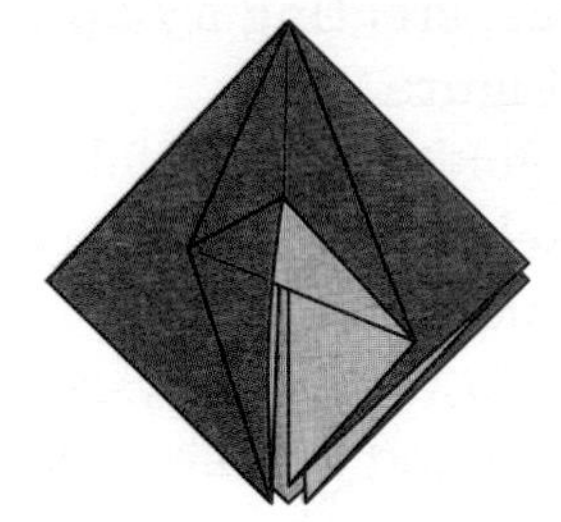

Figure 2.30.
A swivel fold.

We have seen that mountain, valley, and rabbit-ear folds have doubled forms where they are combined with their mirror images. Are there similarly doubled squash or swivel folds? The answer, surprisingly, is yes, and the combination is as difficult as the squash fold is easy. The combination of two swivel folds is called a petal fold (it is commonly used in origami flowers). However, instead of being formed on near and far layers (as in the reverse folds and double rabbit-ear fold), the two mirror-image swivel or squash folds are formed side by side. The petal fold is a very famous fold; it is the key step in the traditional Japanese flapping bird. It is diagrammed as two side-by-side squash folds that share a common valley fold.

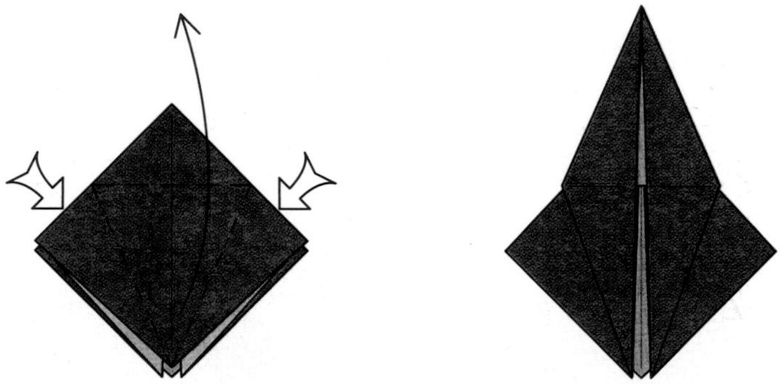

Figure 2.31.
The petal fold.

While on the scale of origami difficulty (which runs simple, intermediate, complex, really complex!), the petal fold is only considered an intermediate maneuver, it is usually quite challenging for an origami novice to perform, and so is commonly broken down into several steps with some precreasing, as shown in Figure 2.32.

When you are a beginning folder, it is helpful to make the precreases as in steps 2 and 3 in Figure 2.32. However, as you become comfortable with folding, it's better to not precrease the sides as in step 2, because it is difficult to make the creases through both layers run precisely through the corners. It is neater to simply form the bisectors in each layer individually.

Petal-folding is usually performed on a flap to make it simultaneously narrow and longer. It is also possible to petal-fold an edge, creating a flap where there was none before, as shown in Figure 2.33.

Petal folds, squash folds, reverse folds, and rabbit ears are all closely related to each other. It is often possible to reach the same end by more than one means. For example, the petal fold shown in Figure 2.33 can also be realized by making two reverse folds and a valley fold.

1. The most common petal fold starts with this shape, called the Preliminary Fold.

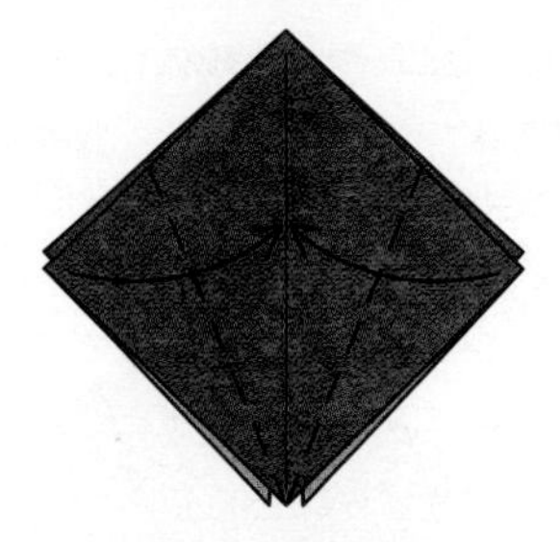

2. Fold the sides in so that the raw edges lie along the center line.

3. Fold the top point down over the other two flaps.

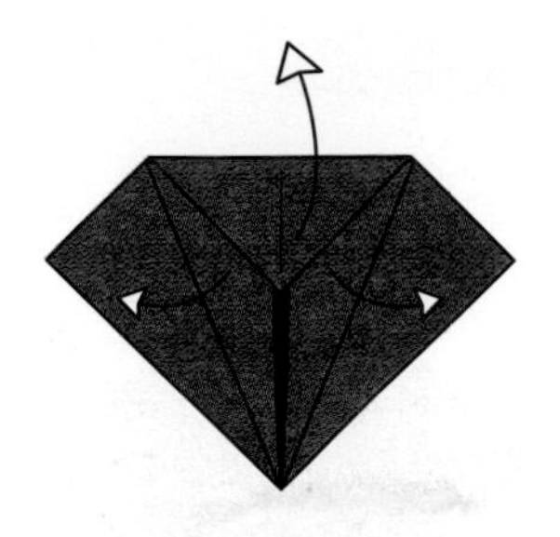

4. Unfold all three flaps.

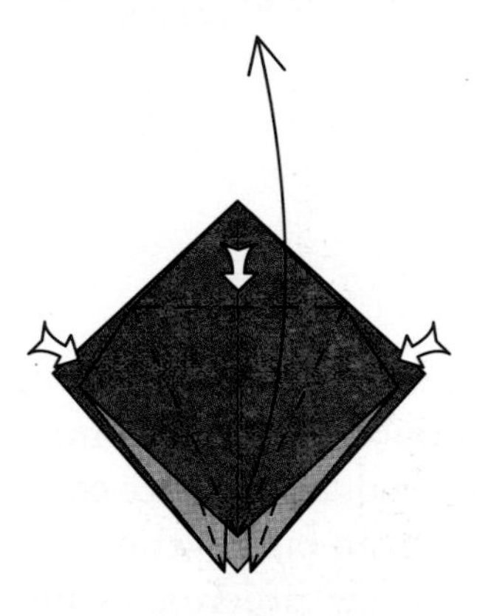

5. To make the petal fold, lift up the first layer of the bottom corner while holding down the top of the model just above the horizontal crease. Allow the sides to swing in.

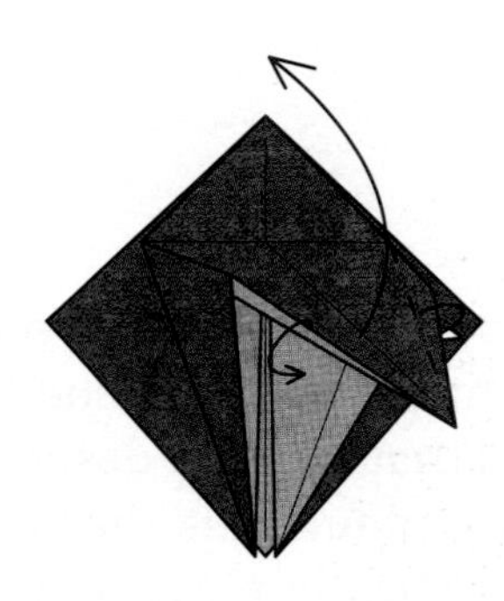

6. Continue lifting up the point; reverse the direction of the two creases running to its tip, changing valley folds to mountain folds.

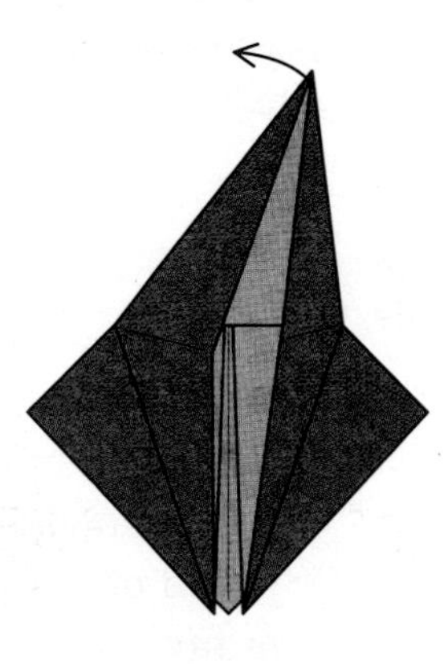

7. Continue lifting the point all the way; then flatten.

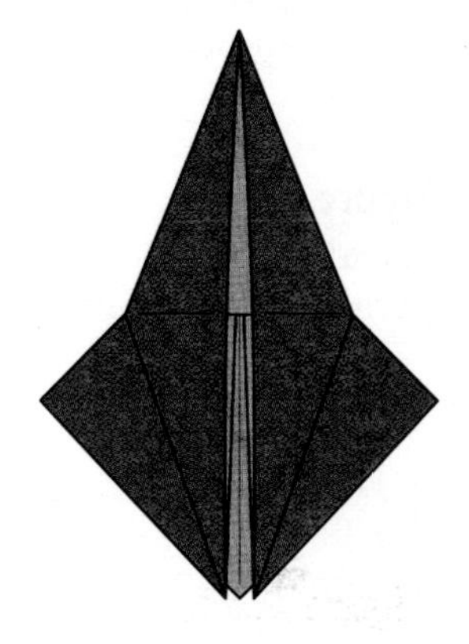

8. Finished petal fold.

Figure 2.32.
The sequence to make a petal fold.

And if you were to cut apart the finished petal fold along the center line (cutting both slightly left and right of the center line to be sure to sever all layers that touch the center line), the petal-folded flap would turn out to be two rabbit ears!

Thus, the various combination folds are not distinct entities so much as convenient ways of getting two or four creases to come together at once. What is important in origami

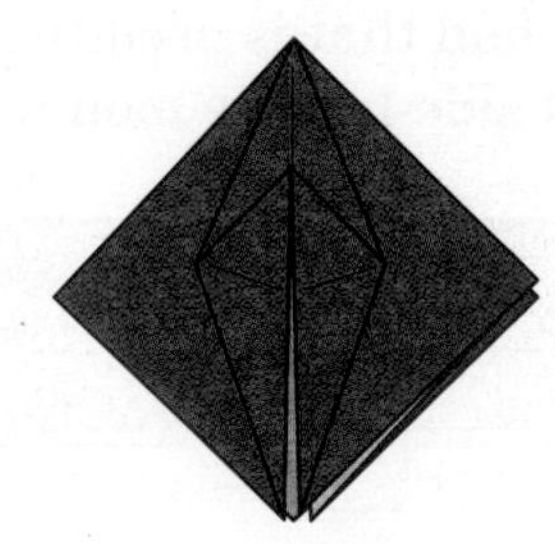

Figure 2.33.
Petal-folding an edge.

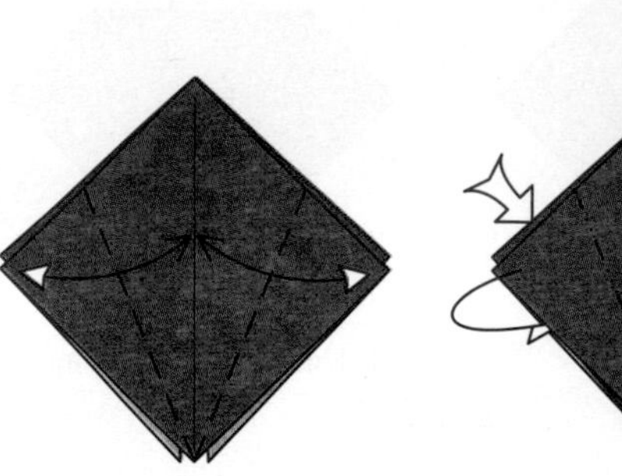

1. Fold the sides in to lie along the center line and unfold.

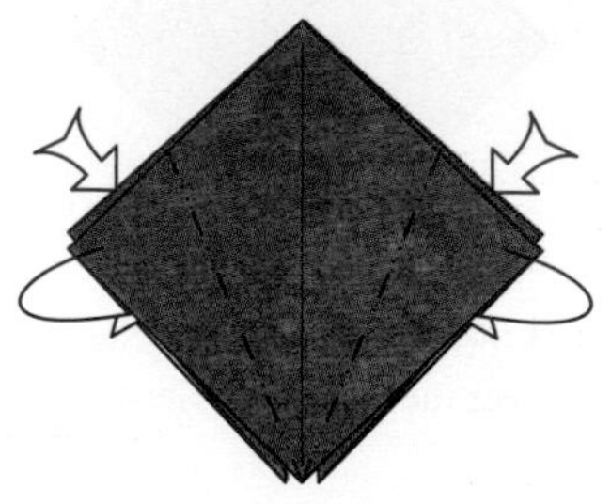

2. Reverse-fold the edges inside using the creases you just made.

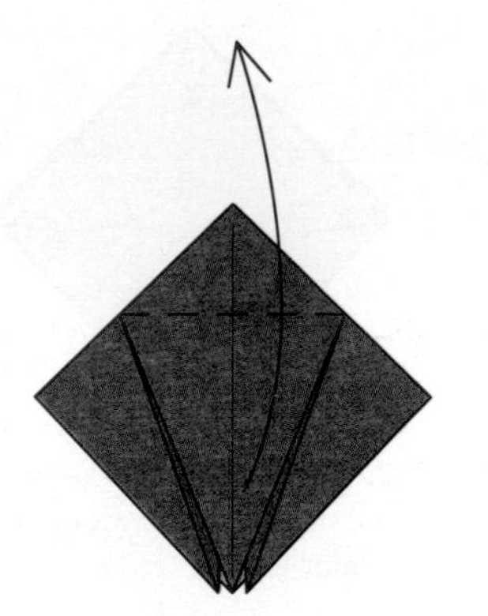

3. Lift up the frontmost flap.

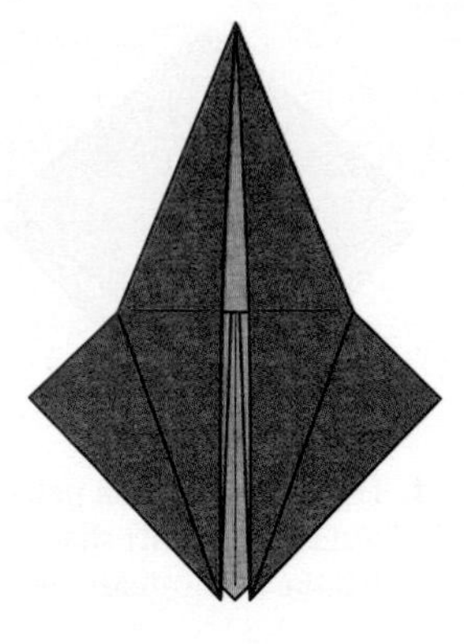

4. Finished petal fold.

Figure 2.34.
An alternative way to make a petal fold using reverse folds.

design is the underlying structure, not the specific sequence of steps one takes to get to the finished model (although it must be acknowledged that once the design is fixed, a sequence composed of simple combinations that flows from one to the next is far more aesthetically pleasing than a few precreases followed by, "Make these 150 creases come together at once").

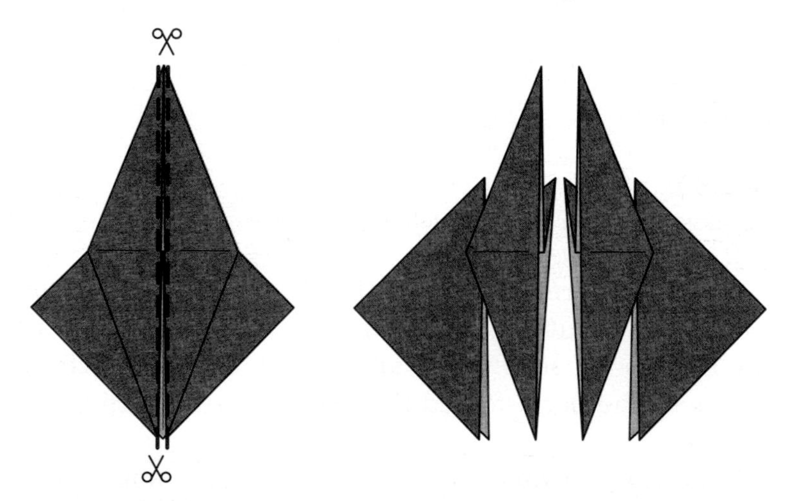

Figure 2.35.
A bisected petal fold reveals that it is composed of two rabbit-ear folds.

Reverse folds are commonly used to change the direction of a flap, for example, to do the final shaping. Another combination fold that is used to shape flaps is the pleat, which consists of side-by-side mountain and valley folds.

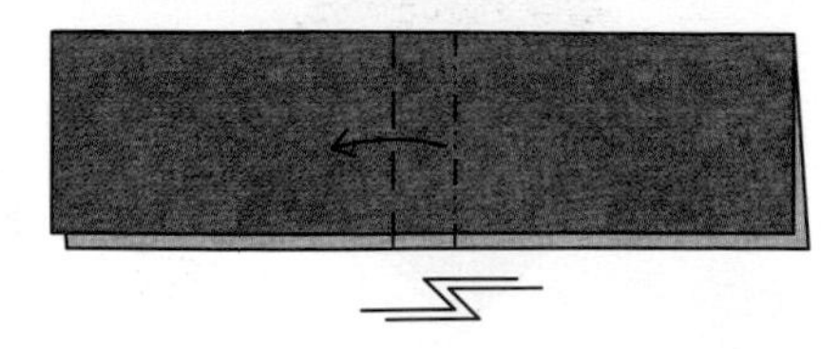

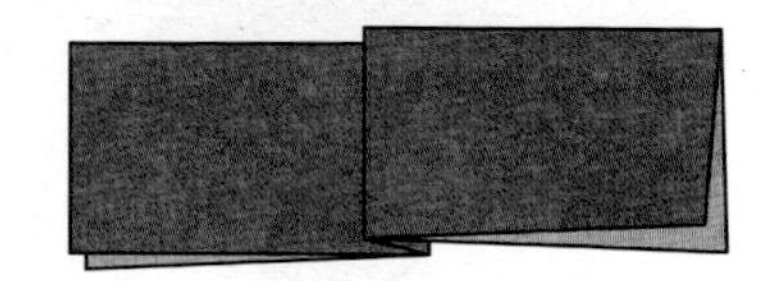

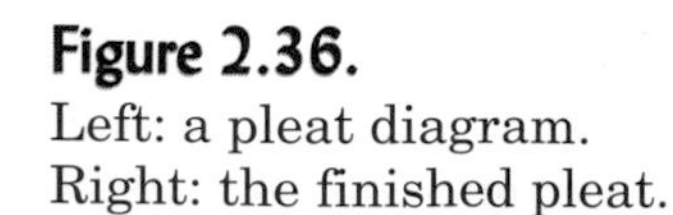

Figure 2.36.
Left: a pleat diagram.
Right: the finished pleat.

A pleat formed through a single layer of paper is unambiguous. However, when there are multiple layers present, there is a closely related fold, illustrated in Figure 2.37, which is called a crimp.

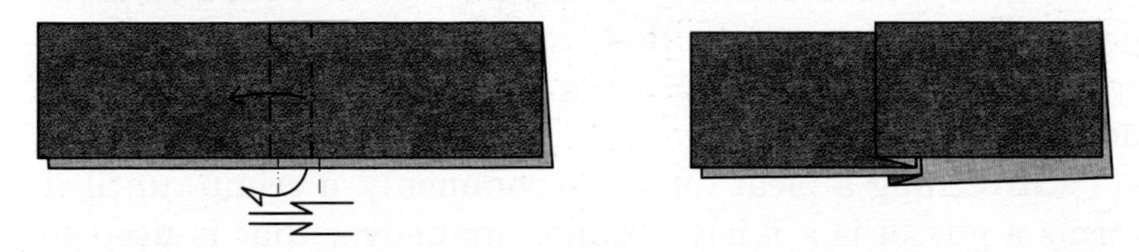

Figure 2.37.
Left: a crimp diagram.
Right: the finished crimp.

The crimp is a combination of a pleat with its mirror image on the far layer of paper. Thus, a crimp bears the same relationship to a pleat that an inside reverse fold bears to a mountain fold (or an outside reverse fold to a valley fold). Just as reverse folds do not come undone as easily as mountain or valley folds, crimps are more permanent than pleats. Both crimps and pleats are diagrammed by showing the fold lines on the near layers of paper; they can be distinguished by examining the edges of the flap. Sometimes it is not practical to show the edges, so a small set of zigzag lines is drawn next to the edge (as in Figures 2.36 and 2.37), which represents an edge-on view of the finished crimp or pleat.

The two folds of a pleat or crimp are often parallel, but they need not be. If they are not parallel, then the flap will change direction, with the net change of direction equal to twice the difference between the angles of the two creases.

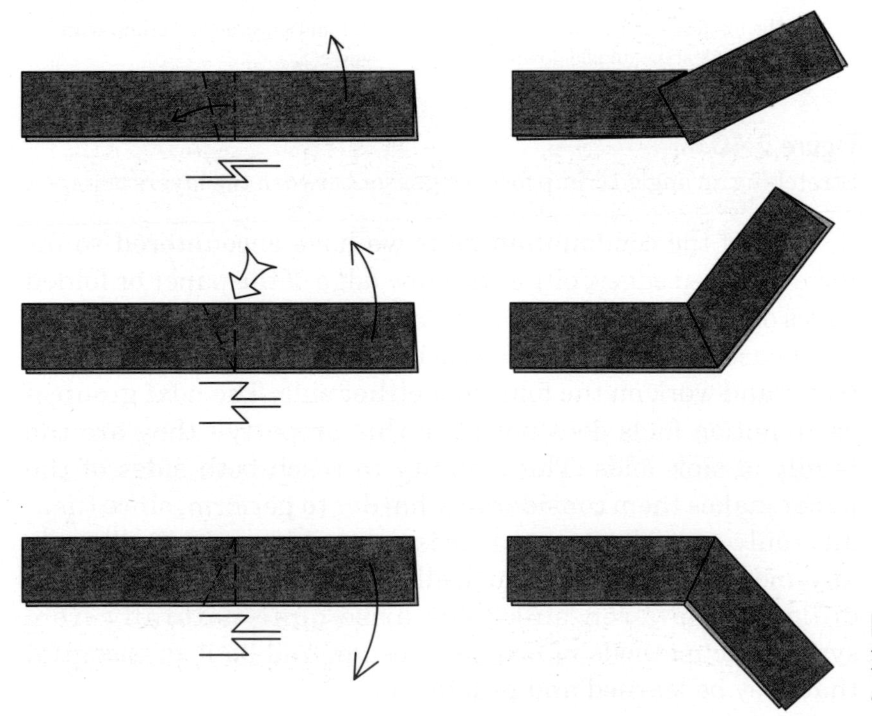

Figure 2.38.
Examples of angled pleats (top) and crimps (center, bottom).

The valley and mountain folds that make up a pleat or crimp can meet each other at one edge of the flap or the other but cannot meet in the interior of the paper without adding additional creases. If you try to make them meet in the interior, which you can do by stretching the ends of an angled pleat or crimp away from each other, you will find that a small gusset must form that extends from the intersection point to the adjacent edges.

Stretching a pleat (or more commonly, a crimp) until it forms a gusset is a fairly common maneuver that is used to soften the change of angle to realize a more natural, rounded form; but stretching gussets is also the basis of some of the most powerful design techniques that we will see.

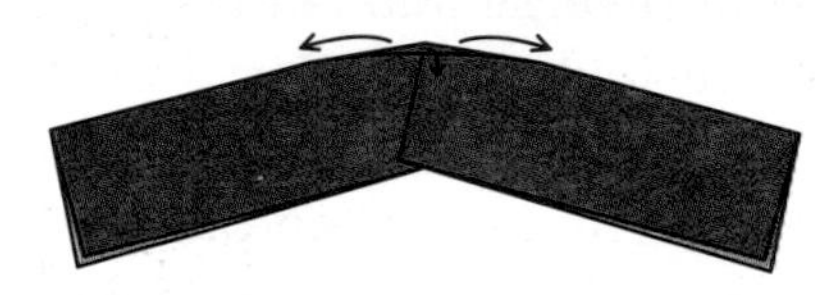

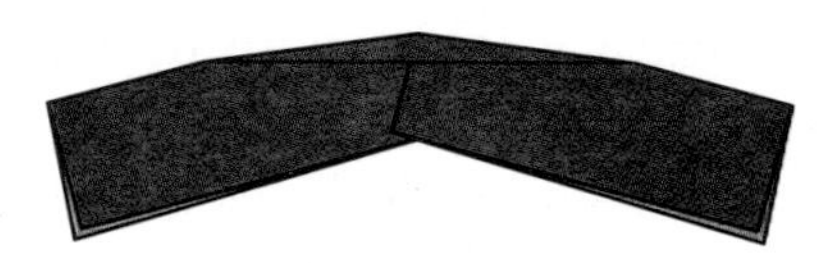

1. Example of a stretched pleat. Pull the two sides apart, keeping the angle fixed.

2. The top will form a little hood on either the front or back side.

3. Finished stretched pleat with gusset.

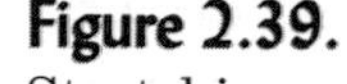

Figure 2.39.
Stretching an angled pleat forms a gusset on either the near or far layers.

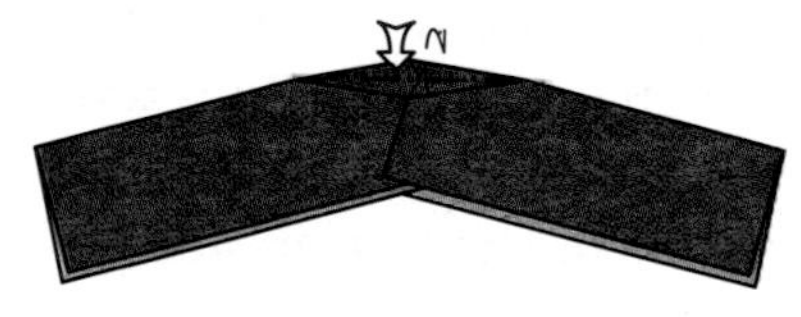

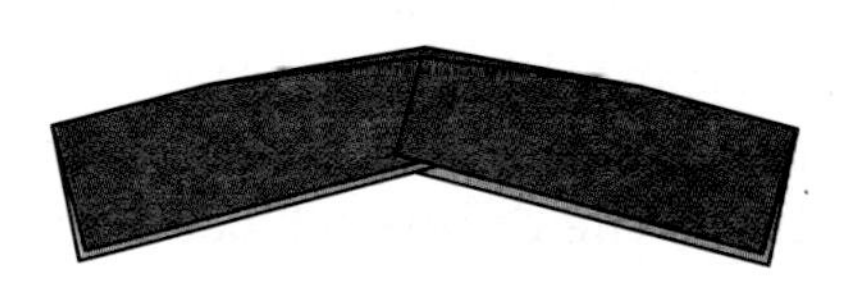

1. Example of a stretched crimp. Pull the two sides apart, keeping the angle fixed.

2. The top forms a narrow diamond; dent the middle down between the layers.

3. Finished stretched crimp with gusset.

Figure 2.40.
Stretching an angled crimp forms a gusset between the layers of paper.

All of the combination folds we have encountered so far have involved edges, either the raw edge of the paper or folded edges on which the creases terminate. Their formation is somewhat eased by the ability to reach around behind each layer of paper and work on the fold from either side. The next group of combination folds does not have this property—they are the family of sink folds. The inability to reach both sides of the paper makes them considerably harder to perform, since (usually) only one side of the paper is accessible, and usually puts any model including them well into the complex rating of difficulty. However, sink folds arise quite naturally from systematic methods of origami design, and so it is essential that they be learned and practiced.

The simplest of the various sink folds is the spread sink, which is only marginally more difficult than a squash fold. It works the same way; a flap is lifted up, its edges are spread symmetrically, and the result is flattened. What distinguishes a spread sink from a squash fold is that in the spread sink, at least two layers—an outer one and an inner one—are simultaneously squashed while remaining joined. Spread sinks are very satisfying to make; you start by flattening the very tip of the flap, then as the edges are stretched to the sides, the flattened region grows and reaches its maximum size when the paper is completely flat.

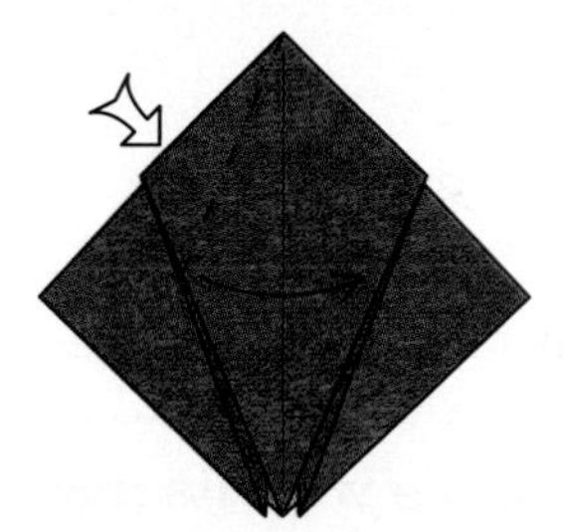

Figure 2.41.
A spread sink.

Spread sinks are most often formed from triangular corners, but there are analogous structures that form convex polygons of any size and shape.

The next member of the sink family is the conventional, or open, sink. The open sink is a simple inversion of a corner formed from a region in the interior of the paper. Conceptually, it is quite simple: The line of the sink is a mountain fold, which runs all the way around the point being sunk like a road girdling a mountain peak. All of the creases above the sink line get converted to the opposite parity, mountain to valley, valley to mountain.

What makes an open sink "open" is that the part of the paper being sunk can (usually) be opened out entirely flat, which allows a relatively straightforward strategy for its formation: stretch the edges apart so that the tip of the point to be sunk flattens out, pinch a mountain fold all the way around, then push the middle down into the model and flatten the model. The creases in the sunk region will (again, usually) fall into the right place.

Figure 2.42 shows this process, including the intermediate stage, and the crease pattern of the result.

It is sometimes possible to make an open sink by performing a spread sink first, as Figure 2.43 shows.

The example in Figure 2.43 is for a four-sided sink—one in which the point has four ridges coming down from it (and

Figure 2.42.
The open sink, formation and crease pattern.

the polygon outlined by the mountain folds “going around the mountain” is a quadrilateral), but you can form three-, five-, and higher-sided sinks in a similar way.

As we have seen, a valley fold can combine with its mirror image to make a reverse fold, a squash fold can combine with its mirror image to make a petal fold, and a rabbit ear can combine with its mirror image to make a double rabbit ear. Can a sink fold be combined with its mirror image? Yes, in

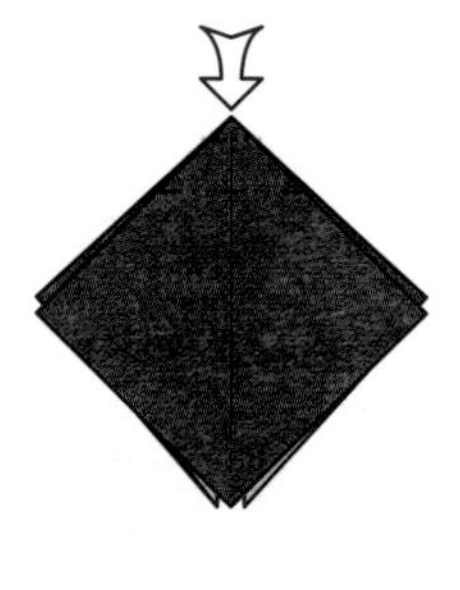

1. A sink fold can sometimes easily be made as a spread-sink, as this sequence shows.

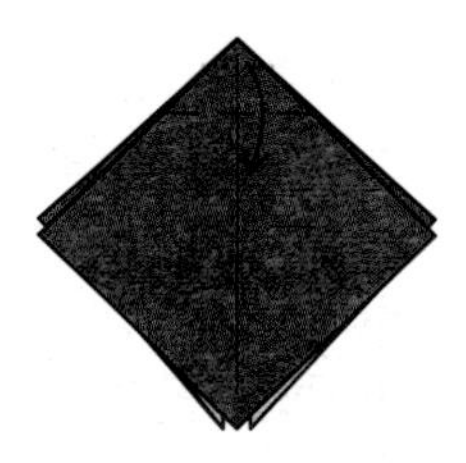

2. Fold the point down along the sink line.

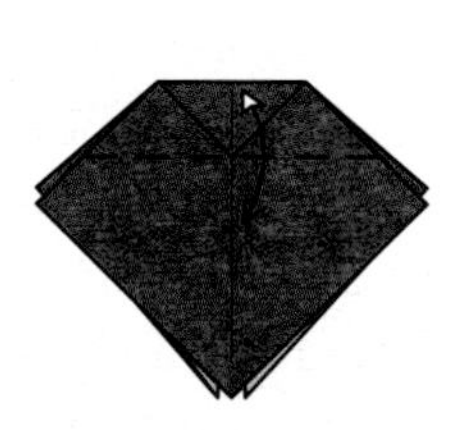

3. Fold and unfold along a crease that just touches the tip of the point.

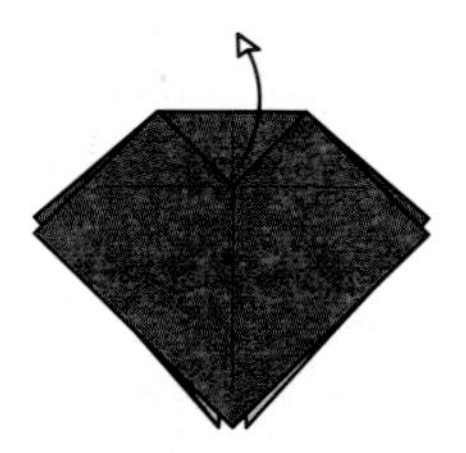

4. Unfold the point.

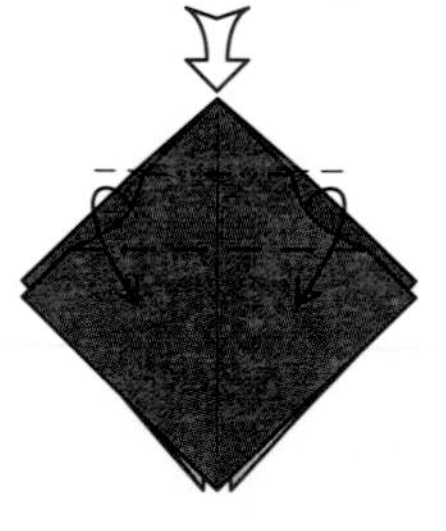

5. Grasp the sides and fold them down while simultaneously stretching and pushing down on the top flap.

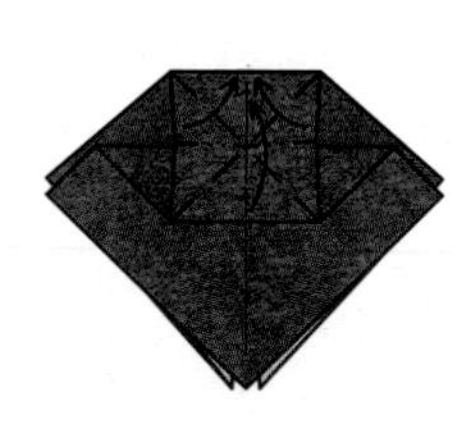

6. Bring the middle of the sides of the square region together at the top.

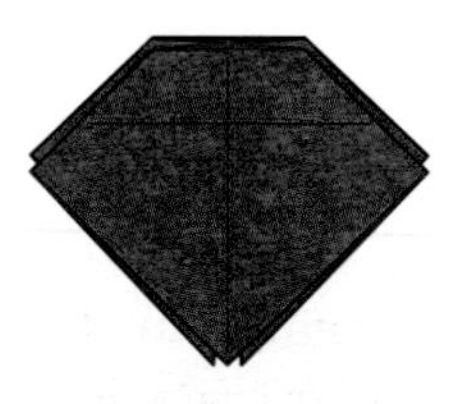

7. Completed sink fold.

Figure 2.43.
Sequence for making a sink fold using a spread sink.

multiple ways, but the most common way happens when a point is sequentially sunk down and back up. The maneuver is called a double sink (or triple or quadruple sink, for more complicated generalizations).

Although a multiple sink can be made sequentially—make the lowest sink, then reach inside and sink the point back upward—it's usually easier to make them all together, first pinching the mountain folds around the point, then pinching the valley folds around before attempting to close up the model.

Sinks were recognized as distinct origami steps in the late 1950s and early 1960s. However, it took until the 1980s for a new variant to become common, the closed sink (whose recognition forced the division of sinks into "open" and "closed" varieties). A closed sink is also an inversion of a point, but in such a way that it is not possible to open the point flat while performing the maneuver. This makes closed sinks extremely hard to perform. In fact, from a strictly mathematical viewpoint, it is impossible to perform a closed sink using a finite number of folds (and what is impossible in mathematics is usually pretty hard in reality). That we can make closed sinks

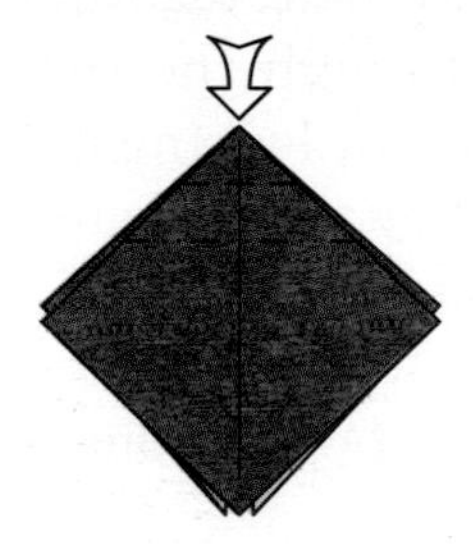

1. A double sink is indicated by parallel mountain and valley creases with a push arrow.

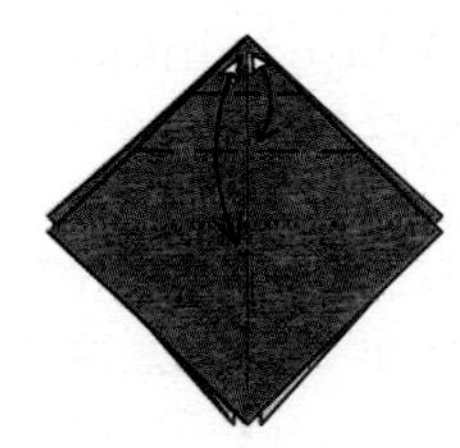

2. Always pre-crease all lines of a multiple sink before opening.

3. Form both the mountain and valley folds running all the way around the flattened polygon.

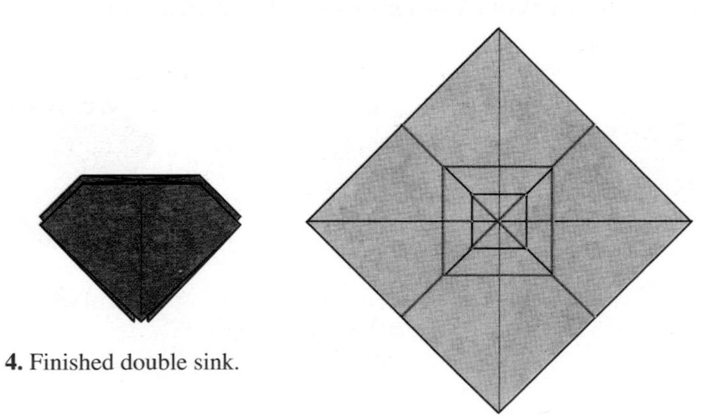

4. Finished double sink.

Figure 2.44.
A double sink, how to make it, and its crease pattern.

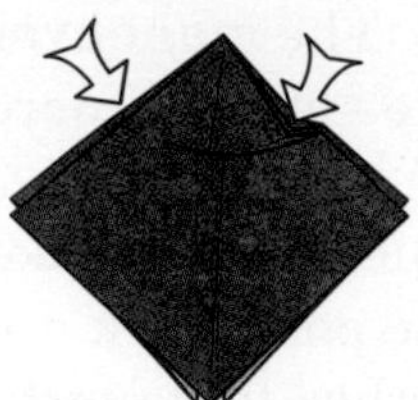

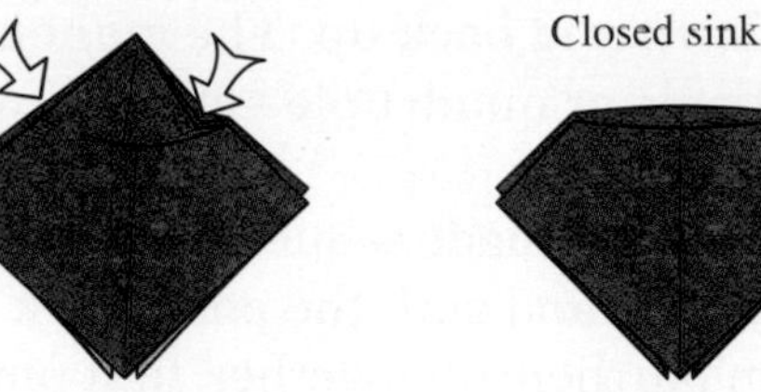

1. A closed sink is also indicated by a push arrow, but is formed differently.

2. Open the point into a cone; starting at one side, start inverting the cone.

3. Flatten when fully inverted. In a closed sink, some edges are trapped at the top of the sink.

4. In an open sink, all edges are visible at the top of the sink.

Figure 2.45.
Formation of a closed sink. Right: the edges of an open sink for comparison.

at all is due to the ability to "roll" a crease through one or more folded layers of paper.

Superficially, a closed sink is diagrammed the same way as an open sink: a push arrow and a mountain fold. However, in the closed sink, instead of forming the mountain fold all the way around every layer, some of the layers are held together, forming a cone, and the point is inverted through the cone without opening it out. Closed sinks are useful for locking layers together, as the edges of the pocket formed by a closed sink, unlike that of an open sink, cannot usually be opened up. The finished result can be distinguished by the presence of pleated layers inside the pocket of an open sink versus few or none in a closed sink.

In general, the more acute the point of a closed sink, the harder it is to carry out; anything narrower than a right angle is usually so difficult that it's more efficient to do it in two steps, as shown in Figure 2.46. First, fold the point into a rabbit ear, closed-sink the top of the rabbit ear, then once the sink is started, fully invert the rabbit ear back into the shape of the original point.

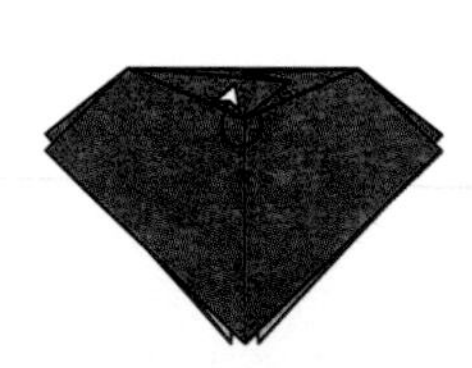

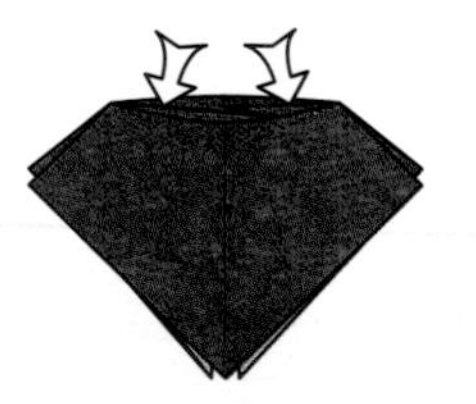

1. Another way to make a closed sink is to fold down the point and fold a rabbit ear from it.

2. Bring two layers of paper in front of the rabbit ear.

3. Push down inside the pocket, opening the point back up.

4. Finished closed sink.

Figure 2.46.
How to make a closed sink from a sharp point.

For any given corner, there is only one way of making an open sink, but there are multiple ways of forming closed sinks; in fact, a sink can be open at one end and closed at the other, an arrangement called a mixed sink. The different varieties are not always distinguishable from the outside, as different arrangements of interior (hidden) layers can have the same outward appearance. For a quadrilateral sink—one with four ridges running down from the top point—there are nine distinct configurations. They and their crease patterns are shown in Figure 2.47.

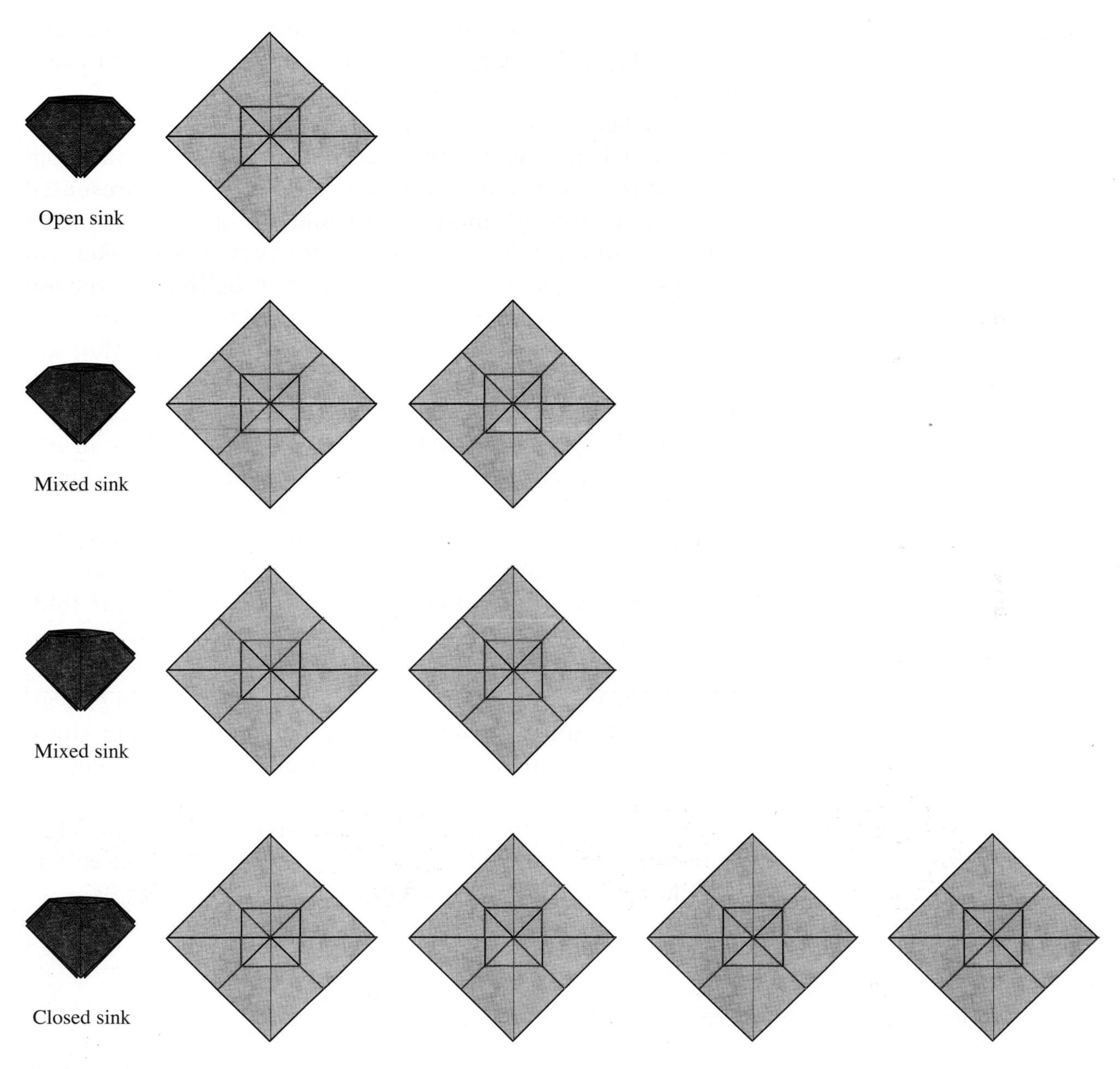

Figure 2.47.
The nine distinct types of sink for a four-ridged point.

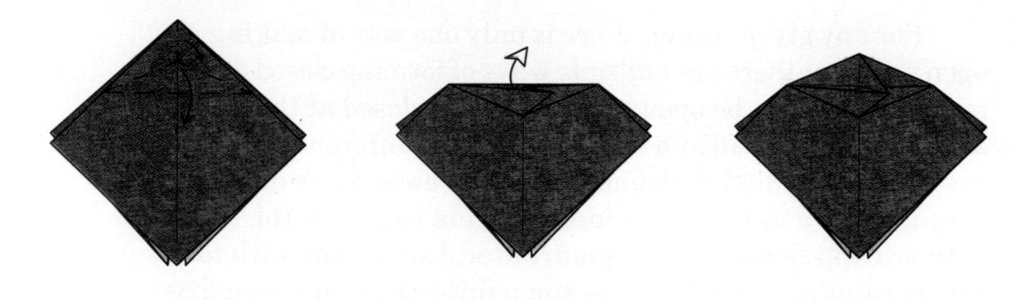

Figure 2.48.
An unsink fold.

In diagrams, which version of sink is desired is usually conveyed by the arrangement of edges in the subsequent views and/or by cut-away views of the interior layers.

The last—and by many accounts, the most challenging—of the sink folds goes by the name of unsink. As the name suggests, it is a reversal of a sink fold. That is, you are presented with an apparently sunken point and the object is to invert the point upward. The challenge here is that while you can always push a point downward to sink it, pulling a layer upward is problematic when there is nothing to grab onto.

Unsink folds come in open and closed varieties that are analogous to their similarly named sink brethren. The unsink is the youngest of the sink combination folds: It only began to be used in the late 1980s, and since then, only sporadically. It is not hard to imagine why. Most of the other combination folds arise naturally from the process of "playing with" the paper. If you want to change the direction of a point, the reverse fold naturally follows. Stretch a point to make it longer, and you are likely to (re)discover the petal fold. Shorten a flap—crimps and pleats fill the bill. And removal or rounding of a corner will lead you to reverse folds and sinks, both open and closed. But the unsink is something of an anomaly. It's unlikely to arise from simple doodling or shaping. But it does arise very directly from systematic origami design. In this chapter, we are—fortunately—still far away from being forced to learn to unsink, but we now, having enumerated the basic folds of origami, are ready to make our first forays into origami design.

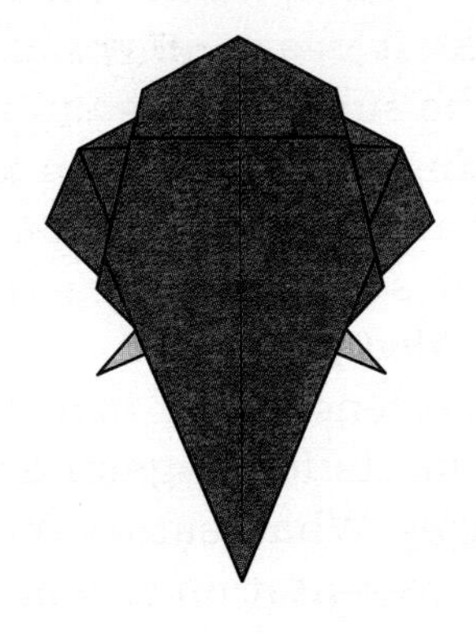

3 Elephant Design

In the beginning—at least, according to some mythologies—there was the Elephant. And so it is with the elephant that we begin our foray into origami design. The elephant—the subject of Georgeot's exhibition—is one of the most common subjects for origami. Presumably, this is because it is so readily suggested. Almost any large shape with a trunk is recognizable as an elephant. If the shape has four legs and large, floppy ears, so much the better. But all these features aren't needed; in fact, it is possible to fold an elephant using a single fold, as Figure 3.1 shows (designed by Dave Mitchell).

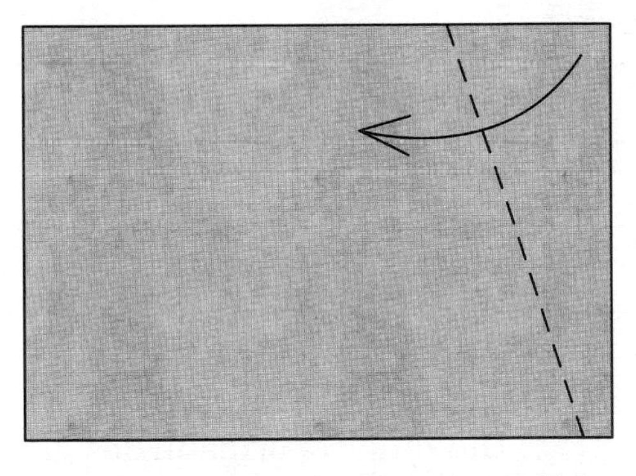

1. Begin with a sheet of writing paper. Fold the upper right corner down along an edge.

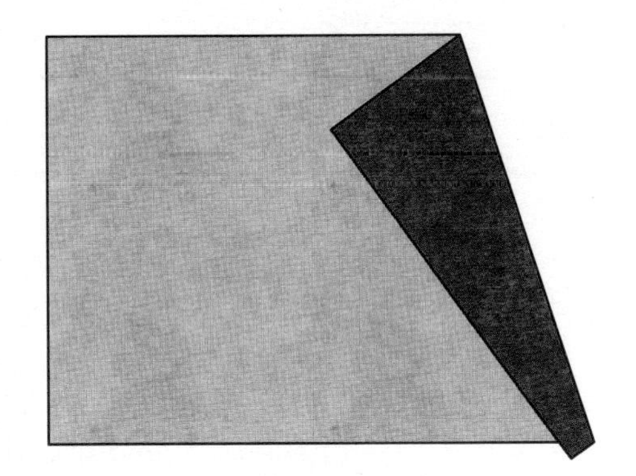

2. Finished One-Crease Elephant.

Figure 3.1.
Dave Mitchell's One-Crease Elephant.

Do you see it? The elephant is facing to the right.

Yes? Perhaps? This simple model—about as simple as you can get—illustrates one of the most important characteristics of origami models: They simplify the subject. Nearly all origami design is representational, but unlike, say, painting, the constraints of folding with no cuts make it nearly impossible to produce a truly accurate image of the subject. Origami is, as origami artist and architect Peter Engel has noted, an art of suggestion. Or put another way, it is an art of abstraction. The challenge to the origami designer is to select an abstraction of the subject that can be realized in folded paper.

You can also select a subject that lends itself to abstraction. Elephants are also popular subjects for origami design because they offer a range of challenges. What features do you include in the design? Is it a spare representation relying on a few lines to suggest a form, or is it necessary to capture all of the features of the subject? Getting the head and trunk may be sufficient for some folders, while others will be satisfied with nothing less than tusks, tail, and toenails. A somewhat more detailed elephant is shown in Figure 3.2.

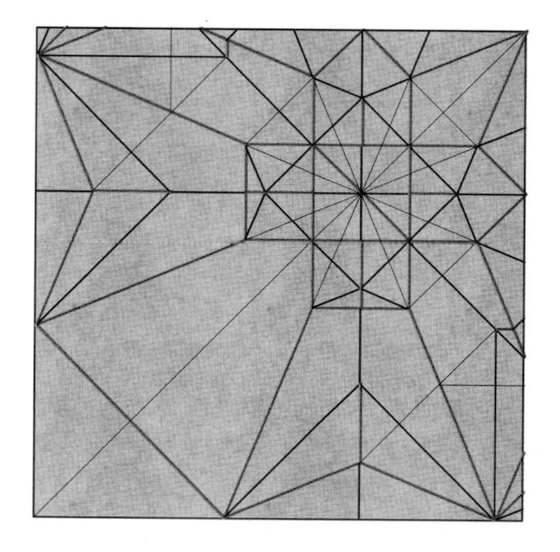

Figure 3.2.
Base crease pattern and folded model of my African Elephant.

These two designs illustrate the range of origami design: Every origami design falls somewhere along a continuum of complexity. Arguably, the one-crease elephant is the simplest possible origami elephant. But the complex elephant is almost assuredly *not* the most complex elephant possible. Complexity in origami is an open-ended scale; the title of "most complex" origami design (for any subject) is always transitory.

Furthermore, complexity carries with it a special burden. We do not denigrate the one-crease model for its abstraction; indeed, its abstract nature is part of its elegance and charm.

But a complex model creates a certain level of expectation in the viewer: an expectation that the model will convey a richer vision. The more folds we have in the model, the more we can reasonably expect from it. And thus, we must make every fold in the design count for something in the end result if elegance is to be attained.

Georgeot's exhibition consisted of 88 elephants ranging from simple to very complex indeed. But elephants, like rabbits, have a way of multiplying. Once Alain became known as "the origami elephant guy," origami elephants continued to come his way. He writes that he has accumulated 155 different designs to date. Many folders have sent more than one, up to eight different designs from a single folder.

If you were to pick any two of Georgeot's elephants, you would find that they differ in many ways: One could be flat, the other three-dimensional; one in profile, one in front view. They might differ in the orientation of the paper relative to the model, in the number of appendages, or in what part of the paper those appendages come from. They may differ in the level of abstraction versus verisimilitude, in cartoonism versus realism, even in the use of curved versus straight lines (and which lines are chosen). All of these features are decisions that the designer makes along the way, whether consciously or unconsciously.

Of all the artistic criteria that may be applied to origami, one of the most important, yet elusive, is elegance. Elegance as it applies to origami is a concept not easily described. It implies a sense of fitness, of economy of effort. In origami, an elegant fold is one whose creases seem to go together, in which there is no wasted paper, whose lines are visually pleasing. Elegance cannot be easily quantified, but there is a property closely related to elegance that can be: efficiency.

While elegance is a subjective measure of the quality of a design, efficiency is an objective measure. An efficient model is one in which all of the paper gets used for something; nothing is tucked out of the way. Inefficient models are those with unnecessary layers of paper. Such models are thick and bulky, difficult to fold, and usually less aesthetically pleasing than a model without unnecessary layers of paper.

The most efficient models are the largest possible for a given size sheet of paper. If you have folded two figures from ten-inch squares of paper and one figure is three inches across and the other is two inches across, then the smaller figure must by necessity have more layers of paper on average in any given flap. The smaller model will generally be thicker; it will hold together less well; and it will show more edges, which will break up the lines of the model. In short, the less efficient a model is, the poorer its visual appeal. Thus, efficiency is an aesthetic goal as well as a mathematical goal. For a base with a fixed number of flaps, the most efficient base is that base in which the flaps are as large as possible.

The tools of origami design cannot (yet) directly address elegance, but they can address its close relative, efficiency, by quantifying what is possible and impossible and providing direction for maximally efficient structures. To wield the tools of origami design, one must have some tools to start with. The way to build a set of tools is to examine some examples of design and deconstruct the model, identifying and isolating specific techniques. To get started and to illustrate some basic principles of origami design, let's add three more elephants to the roster.

3.1. Elephant Design 1

The first design shown in Figure 3.3 is for an Elephant's Head. It is very simple and takes only five creases.

This is very simple—it's perhaps one step up from the one-crease elephant, although, you might note, it took five steps. Can you devise an elephant using exactly two creases? Exactly three?

3.2. Elephant Design 2

On the scale of origami complexity, both the One-Crease Elephant and the Elephant's Head fall into the "simple" category. But as we add more features to a model, it almost invariably increases in complexity. As an illustration, let's take the same basic design as the Elephant's Head and add a pair of tusks to it.

The folding required increased substantially, just to create two tiny points for tusks. But I also added a few steps to give definition to the ears (step 9). Why? Why not just leave the face a flat surface as we did in the previous model? Two reasons. In the first Elephant's Head, the ears came almost for free—there were two flaps (the corners of the square) available to work with. But in this design, we needed to create side flaps (in steps 8 and 9) to define the ears, which required more folding.

There's a second reason, however, which is a bit more subtle. There is an aesthetic balance that needs to be maintained across an origami design. The tusks introduce some small, fine features into the model. The contrast between those

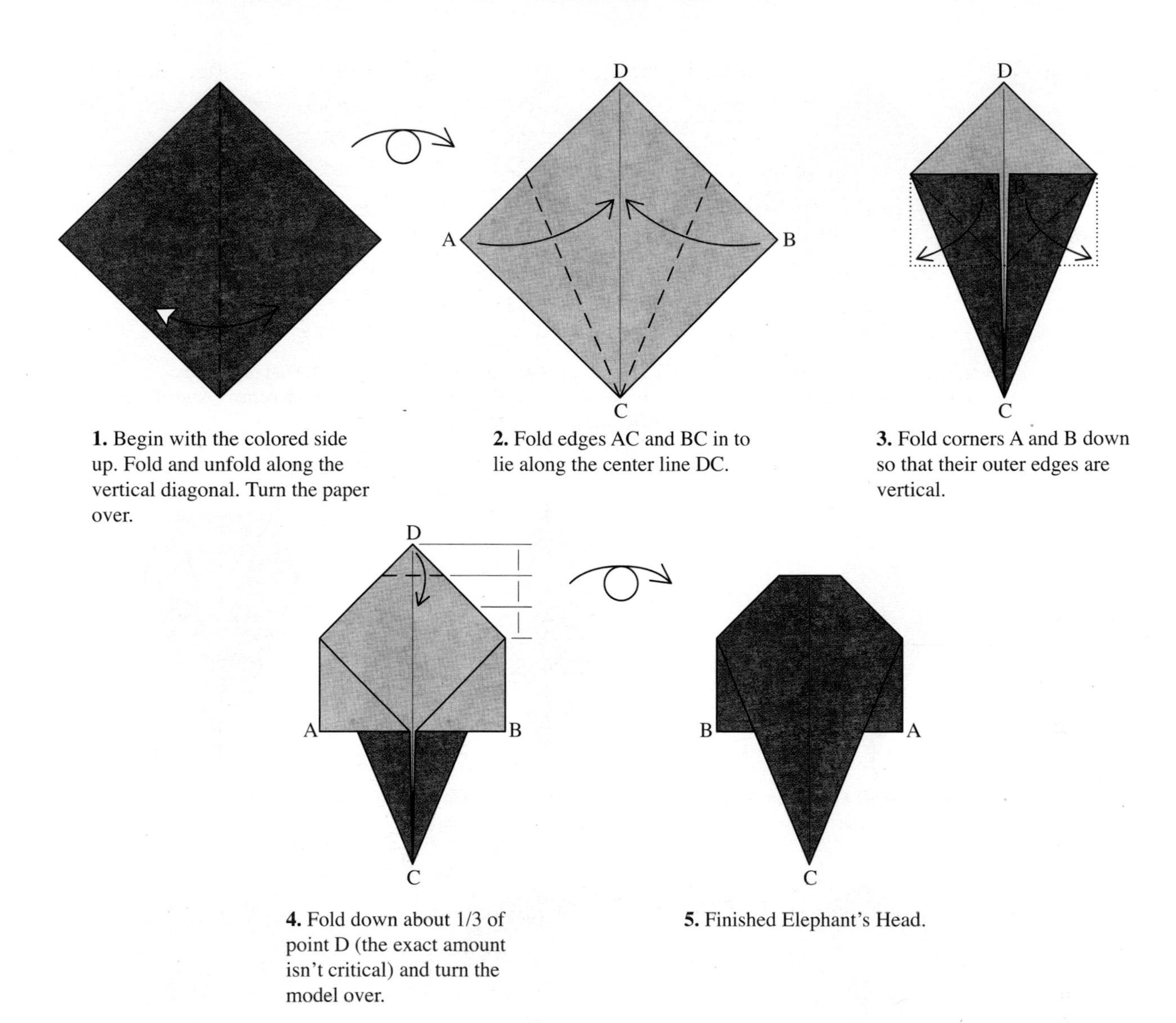

D
A
B
C
D
C
1. Begin with the colored side up. Fold and unfold along the vertical diagonal. Turn the paper over.
2. Fold edges AC and BC in to lie along the center line DC.
3. Fold corners A and B down so that their outer edges are vertical.
D
A
B
C
B
A
C
4. Fold down about 1/3 of point D (the exact amount isn't critical) and turn the model over.
5. Finished Elephant's Head.

Figure 3.3.
Folding sequence for an Elephant's Head.

fine features and the broad, flat, featureless expanse of the face is jarring, so we introduced two folds to break up the surface of the face a bit and bring some balance to the lines of the model.

3.3. Elephant Design 3

We can take another step up the ladder of complexity. Now we'll make the tusks a bit longer.

These three models depict the same subject, but with progressively greater anatomical accuracy (although they still leave a lot to be desired—like a body). They are simple, but

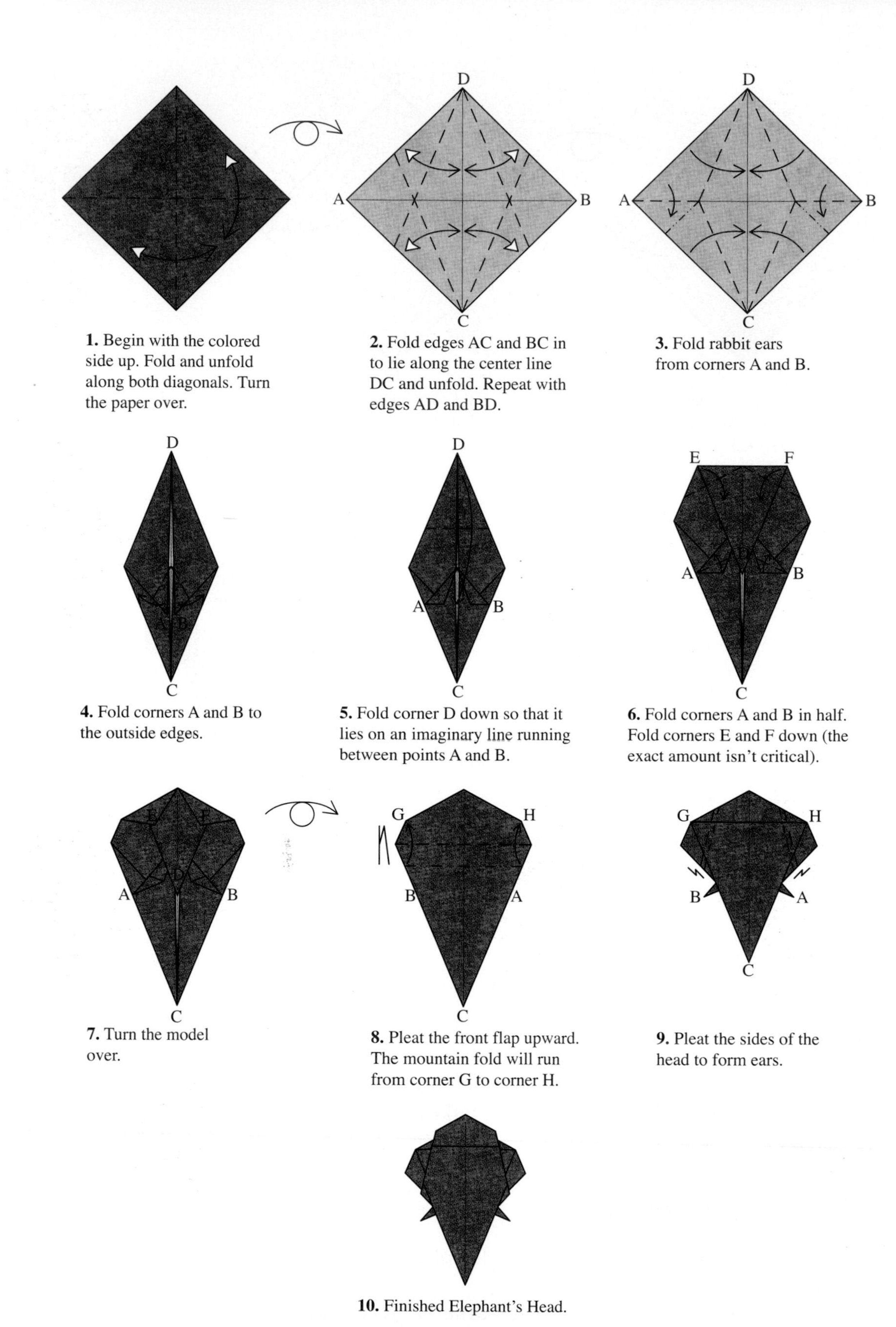

1. Begin with the colored side up. Fold and unfold along both diagonals. Turn the paper over.
2. Fold edges AC and BC in to lie along the center line DC and unfold. Repeat with edges AD and BD.
3. Fold rabbit ears from corners A and B.
4. Fold corners A and B to the outside edges.
5. Fold corner D down so that it lies on an imaginary line running between points A and B.
6. Fold corners A and B in half. Fold corners E and F down (the exact amount isn't critical).
7. Turn the model over.
8. Pleat the front flap upward. The mountain fold will run from corner G to corner H.
9. Pleat the sides of the head to form ears.
10. Finished Elephant's Head.

Figure 3.4.
Folding sequence for the more complex Elephant's Head.

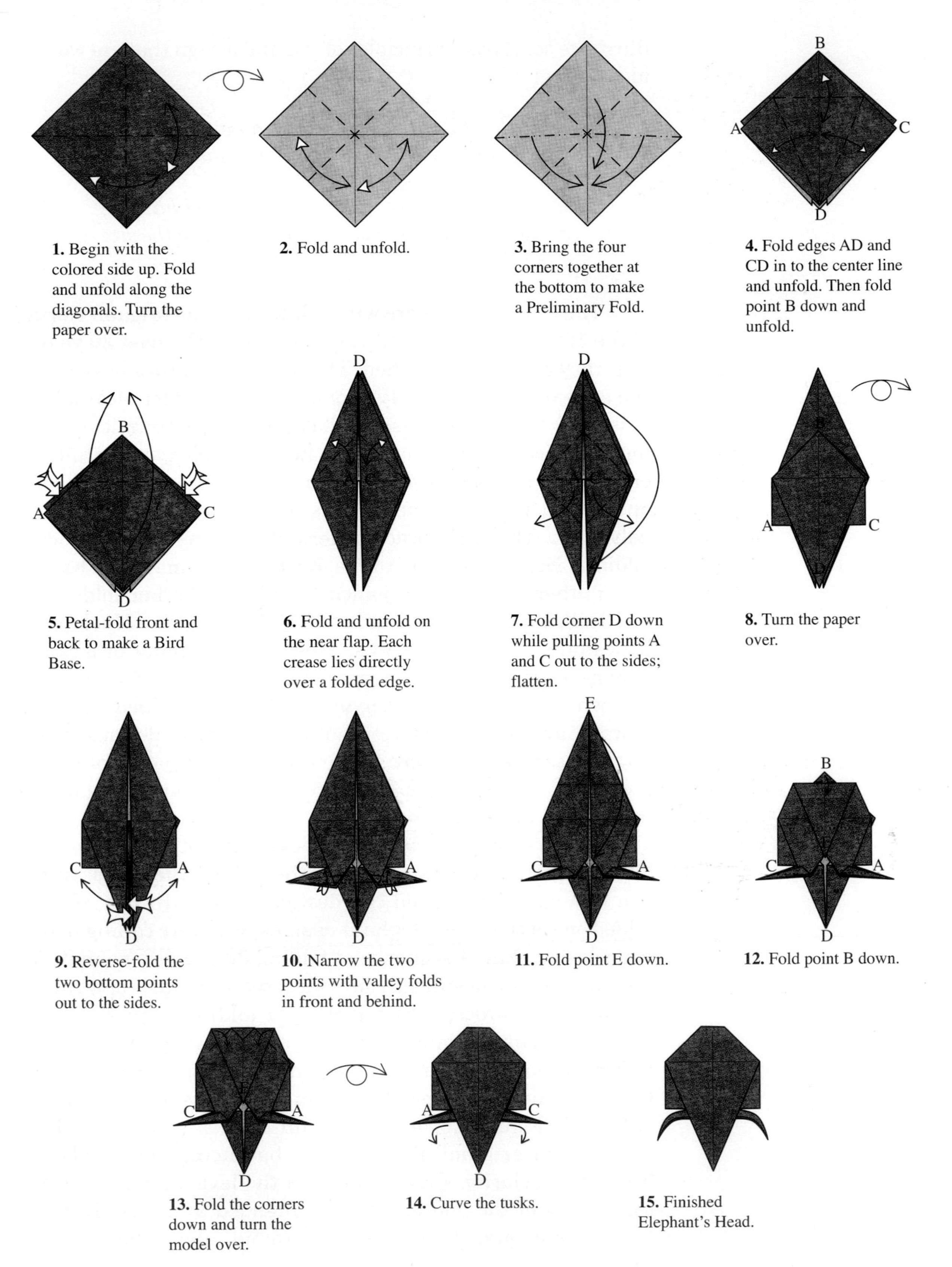

1. Begin with the colored side up. Fold and unfold along the diagonals. Turn the paper over.

2. Fold and unfold.

3. Bring the four corners together at the bottom to make a Preliminary Fold.

4. Fold edges AD and CD in to the center line and unfold. Then fold point B down and unfold.

5. Petal-fold front and back to make a Bird Base.

6. Fold and unfold on the near flap. Each crease lies directly over a folded edge.

7. Fold corner D down while pulling points A and C out to the sides; flatten.

8. Turn the paper over.

9. Reverse-fold the two bottom points out to the sides.

10. Narrow the two points with valley folds in front and behind.

11. Fold point E down.

12. Fold point B down.

13. Fold the corners down and turn the model over.

14. Curve the tusks.

15. Finished Elephant's Head.

Figure 3.5.
Folding sequence for yet another Elephant's Head.

illustrate some basic principles of origami design that are worth identifying:

> *Generally, the more long points a model has, the more complex its folding sequence must be.*
>
> *Generally, the more long points a model has, the smaller the final model will be relative to the size of the square.*

These principles were widely known in the origami world of the 1960s and 1970s, but it was not until the past 20 years that they could be quantified. The past two decades have seen the appearance of a new type of origami, the "technical fold." It is hard to define precisely what constitutes technical folding; technical folding tends to be fairly complex and detailed, encompassing insects, crustaceans, and other point-ridden animals. It is often geometric, as in box-pleated models and polyhedra. The early practitioners of what we call technical folding—Neal Elias, Max Hulme, Kosho Uchiyama, and a handful of others—have been joined by a host of other folders—Montroll, Engel, and myself in the U.S., Fujimoto, Maekawa, Kawahata, Yoshino, Kamiya, Meguro, and many others in Japan—over the past 20 years. In fact, technical folding has its own name in Japan: *origami sekkei*. It is difficult to pin down a unique characteristic of a model that defines it as *origami sekkei*, but I have a candidate criterion: A fold is a technical fold when its structure shows clear evidence of intentional design.

The first steps of design, however, do not require use of any specialized techniques or mathematical theorems. Anyone who can fold origami can design origami. In fact, if you folded one of the three elephant designs, you were calling upon your design skill. A sequence of folding diagrams—no matter how detailed—can still only provide a set of samples of what is a continuous process. In following a folding sequence, the reader must interpolate; he must connect the steps in his mind to form a continuous process. Depending on the amount of detail into which the steps are broken down, this process can be easy, as in Figure 3.6, or it can be difficult, as in Figure 3.7.

A good origami diagrammer, balancing the needs of brevity and clarity, strives to match the level of detail to the complexity of the fold and to the intended audience. In this book, I have aimed for a middle ground, along the lines of Figure 3.8.

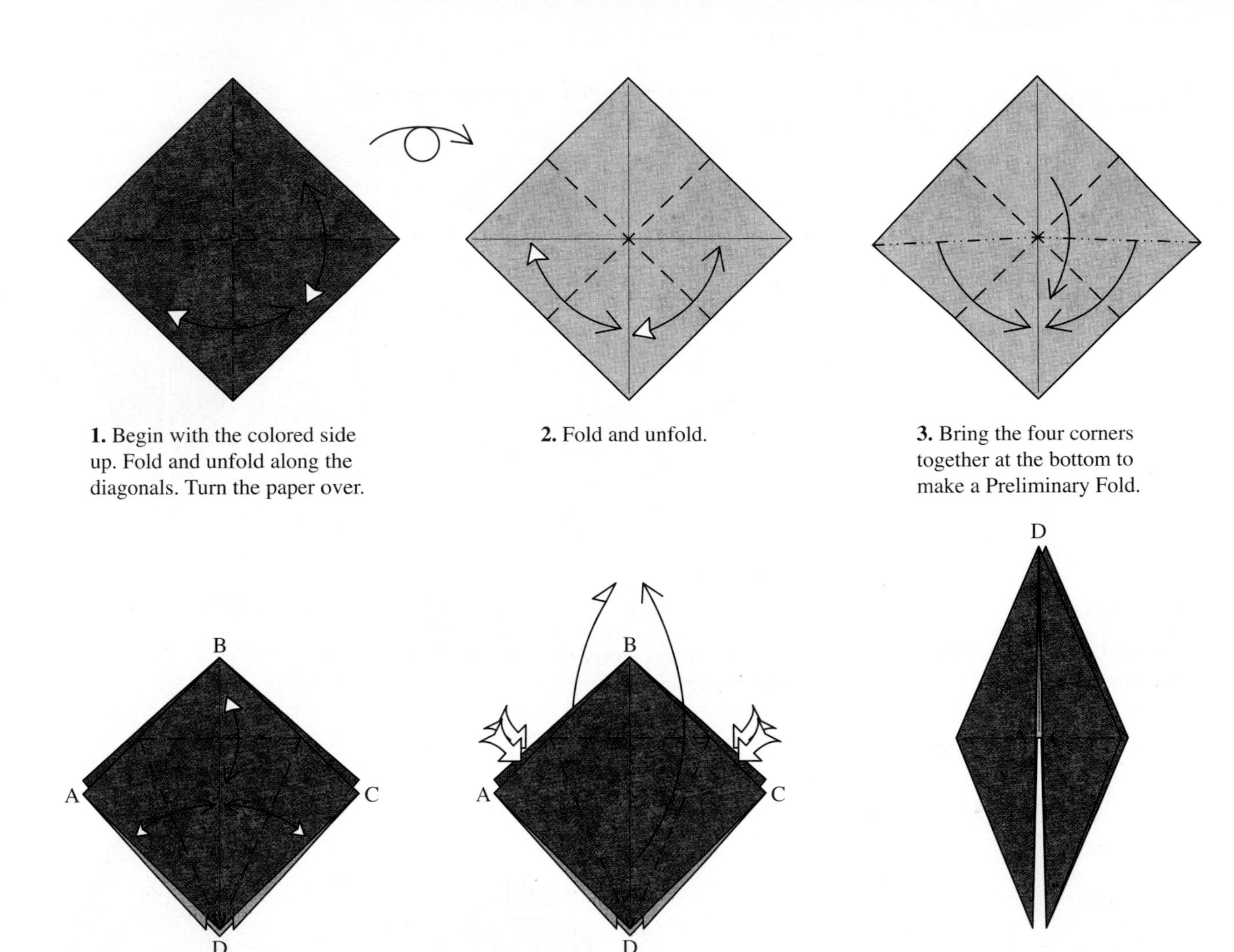

1. Begin with the colored side up. Fold and unfold along the diagonals. Turn the paper over.
2. Fold and unfold.
3. Bring the four corners together at the bottom to make a Preliminary Fold.
B
A
C
D
4. Fold edges AD and CD in to the center line and unfold. Then fold point B down and unfold.
B
A
C
D
5. Petal-fold front and back to make a Bird Base.
D
6. The Bird Base.

Figure 3.6.
Detailed sequence for folding a Bird Base.

When you begin following diagrams, you require each instruction to be broken down into the smallest possible steps. As you gain experience in following diagrams, the jumps between steps become larger. Instead of seeing every individual crease, the creases start to come in groups of two or three. As we have seen, the most common groups of creases have been given names: reverse folds, rabbit-ear folds, petal folds. More advanced folds may have groups of 10 or 20 creases that must be all brought together at once, or several different folds must occur simultaneously, or not all creases may be visible in the diagram. Following such a sequence is even more a process of design. Following a folding sequence is, in effect, resolving

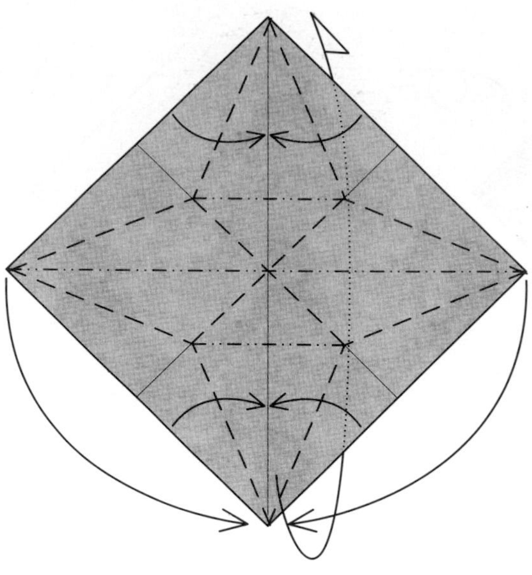

1. Divide the square in half vertically and horizontally with creases. Crease all angle bisectors at the corners, then assemble using the creases shown.

2. The Bird Base.

Figure 3.7.
Compact sequence for folding a Bird Base.

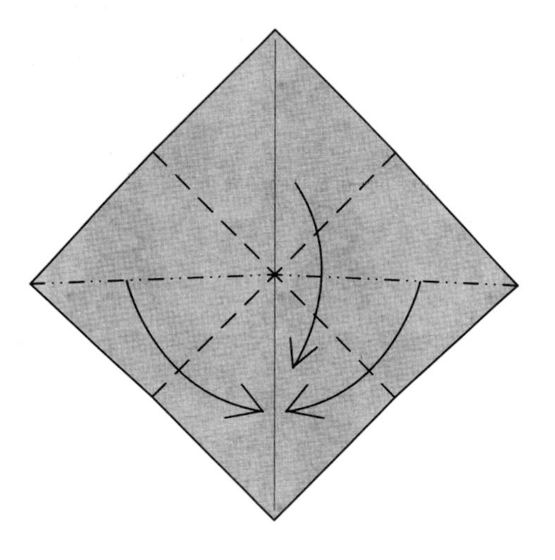

1. Begin with a square with creased diagonals. Bring the four corners together at the bottom and flatten.

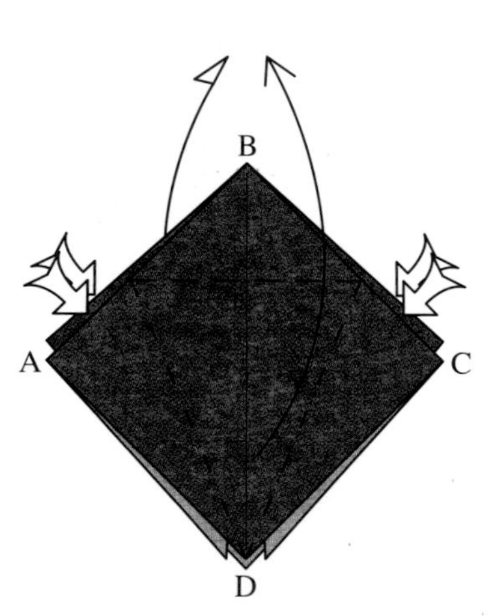

2. Petal-fold front and back.

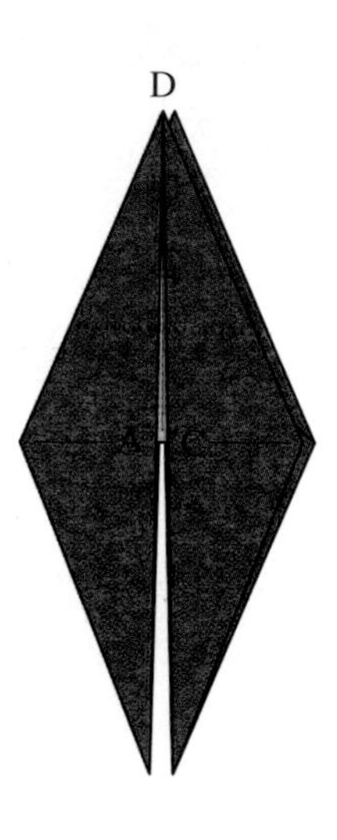

3. The Bird Base.

Figure 3.8.
Intermediate sequence for folding a Bird Base.

a series of small design problems going from one configuration of the paper to the next. Designing an entirely new model is the same task, merely scaled up.

Origami design runs along a continuous scale ranging from minor modification of an existing design to the "ground-up" creation of an entirely new model. Just as a beginning

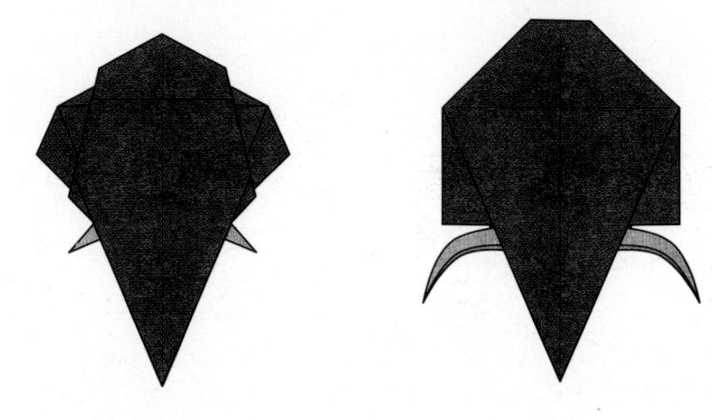

Figure 3.9.
Two variations on the Elephant's Heads.

folder should begin to fold simple models from diagrams, the beginning designer should choose simple shapes to design.

And now is as good a time as any to start. The elephants in Figures 3.4 and 3.5 have colored tusks. Can you find a way to alter each model so that the tusks become white as shown in Figure 3.9? (Hint: Turn a flap inside-out.)

The first stage of origami design is to modify someone else's work, as you can with the elephants. Origami design is, in large part, built on the past. The origami designers of the present have created new techniques, but in doing so, they used techniques of those anonymous Japanese folders of history (as well as those of their contemporaries, of course). It behooves us to spend some time studying how prior generations of folders designed their models.

4 Traditional Bases

The design of an origami model may be broken down into two parts, folding the base, and folding the details. A *base* is a regular geometric shape that has a structure similar to that of the subject, although it may appear to bear very little resemblance to the subject. The *detail folds*, on the other hand, are those folds that transform the appearance of the base into the final model. The design of a base must take into account the entire sheet of paper. All the parts of a base are linked together and cannot be altered without affecting the rest of the paper. Detail folds, on the other hand, usually affect only a small part of the paper. These are the folds that turn a flap into a leg, a wing, or a head. Converting a base into an animal using detail folds requires tactical thinking. Developing the base to begin with requires strategy.

The traditional Japanese designs were, by and large, derived from a small number of bases that could be used to make different types of birds, flowers, and various other figures. For much of the 20th century, most new origami designs were also derived from these same basic shapes.

Bases have been both a blessing and a curse to inventive folding: a blessing because the different bases can each serve as a ready-made starting point for design, a curse because by luring the budding designer onto the safe, well-trodden path of using an existing base, he or she starts to feel that there's nothing new to do and never explores the wilds of base-free origami design.

We will, by the end of this book, do both. However, we will start with the traditional bases—first, to understand what our origami designer forebears had to work with, and second, because the traditional bases, despite being picked over by scores of origami designers for decades, still have some surprising life in them. While they may seem like unique constructions, the traditional origami bases are actually specific embodiments of quite broad and general design principles. By thoroughly understanding the traditional bases, we are prepared to understand the deeper principles of origami design.

4.1. The Classic Bases

So what, exactly, are the standard bases of the origami repertoire? Now, it must be admitted that any labeling scheme that dubs certain structures "the standard bases" is going to be somewhat arbitrary. But there are four shapes, all hundreds of years old in Japan, which are the basis of several traditional models, and have a particularly elegant relationship with one another that takes on a special significance in origami design. They are often called the four *Classic Bases* of origami and are named for the most famous models that can be folded from them: the Kite, Fish, Bird, and Frog Base.

Perhaps not surprisingly, in many cases, more of the structure of an origami model is evident in the crease pattern than in the folded base. For one thing, in the crease pattern, all parts of the paper are visible, while in the folded model only the outermost layers are visible—perhaps 90% or more of them are hidden. Furthermore, certain structures appear over and over in a crease pattern, which you can recognize as features of the finished model. (Do a lot of creases come together at a single point? That point probably becomes the tip of a flap of the model.) With practice, you can learn to read the structure of a model in the crease pattern as if it were the entire folding sequence. The crease patterns, bases, and a representative model from each of the four Classic Bases are shown in Figure 4.1.

We have already encountered three of these in the Elephant's Head series—the Kite, Fish, and Bird Bases. (Challenge: Can you design an elephant that makes full use of the flaps of a Frog Base?) There is no precise definition of a base; perhaps a good working definition is "a geometric form with the same general shape and/or number of flaps as the desired subject."

In origami, a *flap* is a region of paper that can be manipulated relatively independently of other parts of the model. In origami design, bases supply flaps; major flaps on a base then

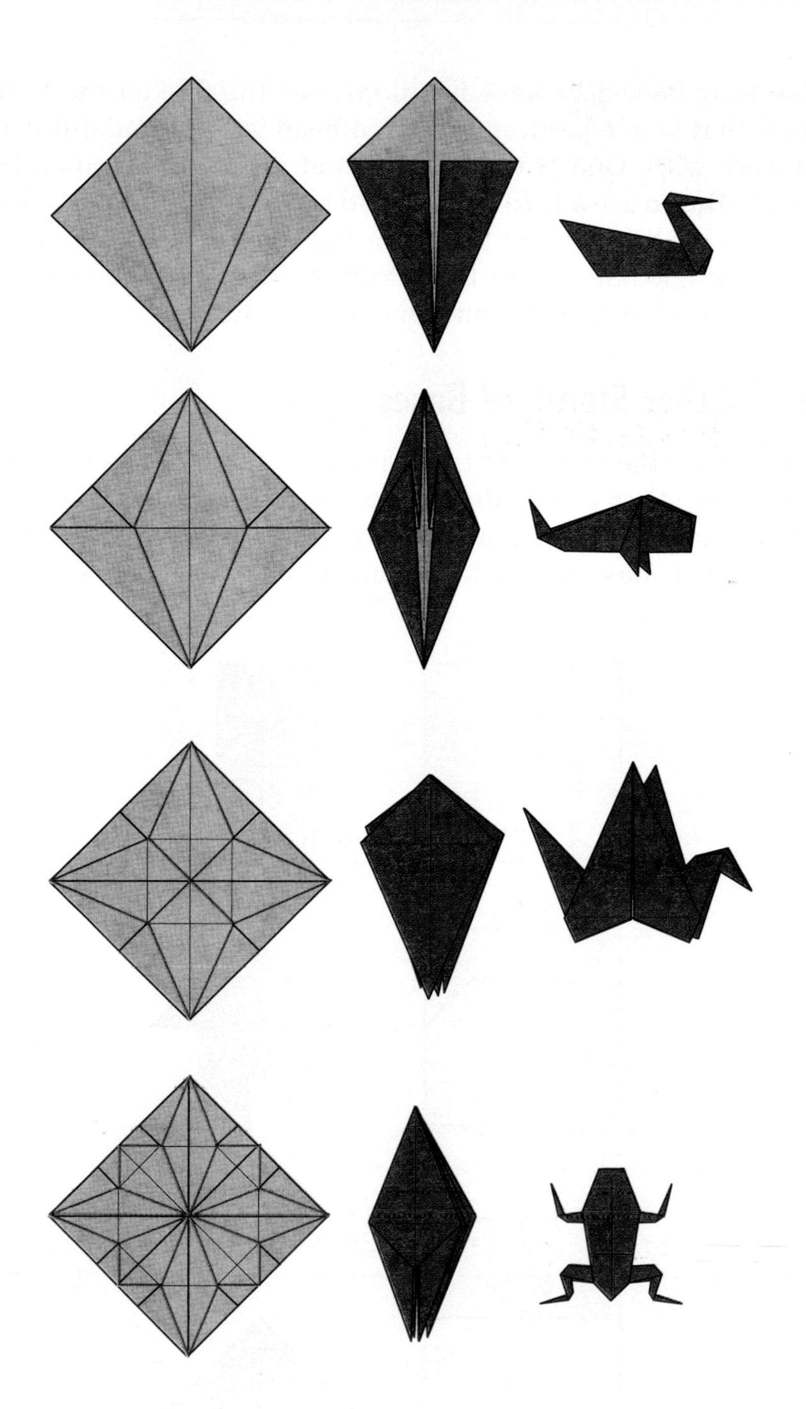

Figure 4.1.
Crease pattern, base, and a representative model for (top to bottom): Kite Base; Fish Base; Bird Base; Frog Base.

get turned into major appendages of a final model. The Kite, Fish, Bird, and Frog bases have, respectively, one, two, four, and five large flaps and one, two, one, and four smaller flaps. To fold an animal, you usually need to start with a base that has the same number of flaps as the animal has appendages. A simple fish has two large flaps (head and tail) and two small ones (pectoral fins), which is why the Fish Base is so appropriate and so named. The average land-dwelling vertebrate has five major appendages (four legs and a head), which suggests the use of the Frog Base and rules out a long tail.

The Frog Base does have five flaps, but the flap on the Frog Base that is in a position to form a head is thick and difficult to work with. One of the four flaps of a Bird Base would be easier. But to use a Bird Base to fold a four-legged animal, you would have to represent two of the legs (usually the rear legs) with a single flap. In the 1950s and 1960s, there were a lot of three-legged origami animals hobbling around.

4.2. Other Standard Bases

The Classic Bases are not the only bases in regular use. There are a few other candidates for standard bases: the so-called Preliminary Fold (a precursor to the Bird and Frog Bases), the Waterbomb Base (obtainable from the Preliminary Fold by

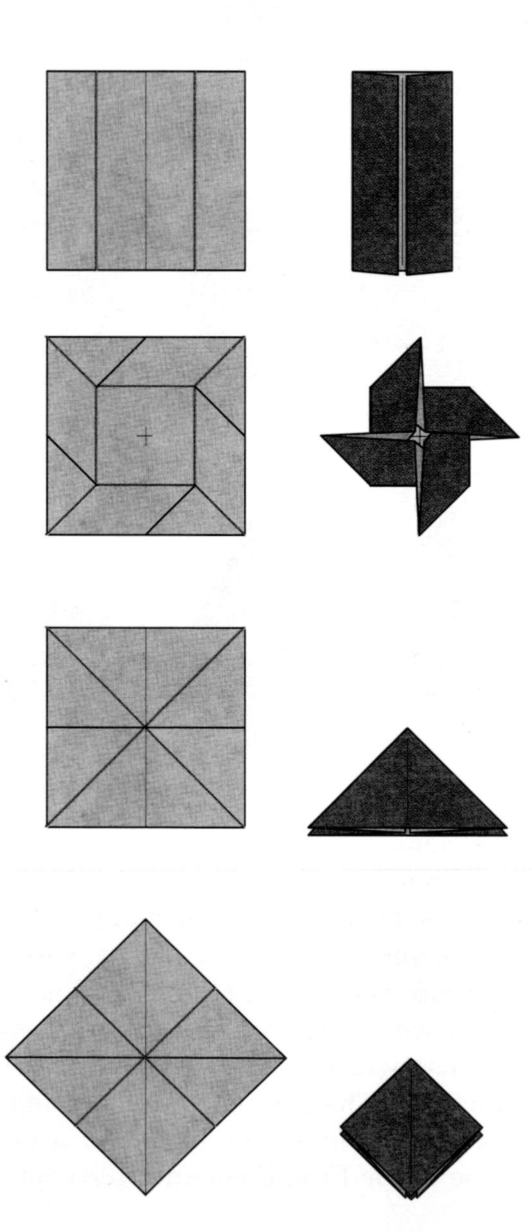

Figure 4.2.
Top to bottom: the Cupboard Base, Windmill Base, Waterbomb Base, and Preliminary Fold.

turning it inside-out), the Cupboard Base (consisting of only two folds), and the Windmill Base (also known as the Double-Boat Base in Japan).

The Preliminary Fold was named Fold rather than Base by Harbin since it was a precursor to other bases, a somewhat artificial distinction that has stubbornly persisted in the English-speaking origami world.

Up through the 1970s, origami designers combined these bases with other procedures, known variously as blintzing, stretching, offsetting, and so forth—and we will learn some of these as well—resulting in a proliferation of named bases. It was not unheard-of to find, for example, a "double stretched Bird Base (type II)" as the starting form for a model. (Rhoads's Bat, *Secrets of Origami*). Of all the possible variants, two are sufficiently noteworthy as to deserve attention: the stretched Bird Base and the blintzed Bird Base are fairly versatile treatments of the classic Bird Base that have seen heavy use in modern times. Both are shown in Figure 4.3.

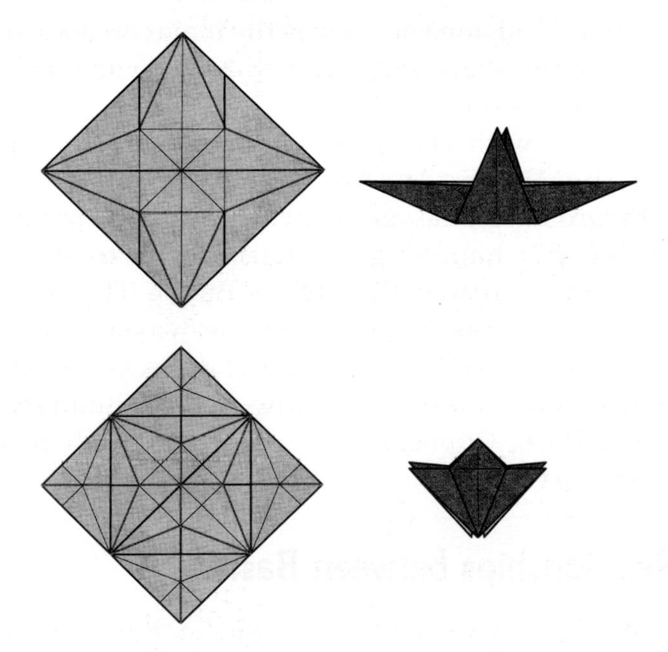

Figure 4.3.
Top: stretched Bird Base.
Bottom: blintzed Bird Base.

The stretched Bird Base is derived from the traditional Bird Base. It is obtained by pulling two opposite corners of the Bird Base as far apart as possible and flattening the result. Harbin recognized several variants of the stretched Bird Base, but the version shown in Figure 4.3 is the most common.

The blintzed Bird Base is also derived from the traditional Bird Base. It is obtained by folding the four corners to the center of a square, folding a Bird Base from the reduced square, and then unwrapping the extra paper to form new flaps. There are several ways of unwrapping the corners to make use of the extra flaps. The procedure of folding four corners to the center is called *blintzing*, named after the *blintz* pastry in which the four corners of a square piece of dough are folded to the center. For many years blintzing a base has been recognized as a straightforward way of increasing the number of flaps in a base. Take a square, fold the four corners to the center, fold a base from the result, then unwrap the corners, creating a new set of extra flaps from the loose paper.

Yet another named base system had been developed in Japan by Michio Uchiyama in the 1930s and thereafter. His system, carried on by his son Kosho, recognized two broad families of bases, one characterized by diagonal or radiating folds (type A) and the other by predominantly rectilinear folds (type B). Figures 4.4 and 4.5 show both families of bases. I have labeled them with Uchiyama's original numbering but rearranged them to better illustrate the relationships between bases.

Note that Uchiyama only gives the major creases for each base; to make the shape flat, you will have to add additional creases on some of the patterns.

Beginning with the development of subject-specific bases in the 1970s (Animal Base, Flying Bird Base, Human Figure Base), the variety of bases quickly proliferated to the point that naming every base began to seem a bit silly (the Great Crested Flycatcher Base). The net result was that most names were left by the wayside. Different authorities recognize different groups of bases as the standard set, but the four Classic Bases plus the Preliminary Fold, Waterbomb Base, Cupboard Base and Windmill Base are common to most.

4.3. Relationships between Bases

The standard bases are not wholly independent; some can be derived from others, as was suggested by Uchiyama's classification system and is illustrated more explicitly in Figure 4.6. Arrows indicate derivation. The square can be folded into a Cupboard Base, which can be further transformed into a Windmill Base. Similarly, the Kite Base is but a way station on the path to a Fish Base. The Preliminary Fold and Waterbomb Base are the same thing—one is just the inverse of the other—but while the Preliminary Fold alone can be turned directly into a Bird Base, either the Waterbomb Base or the Preliminary Fold can be used to make a Frog Base.

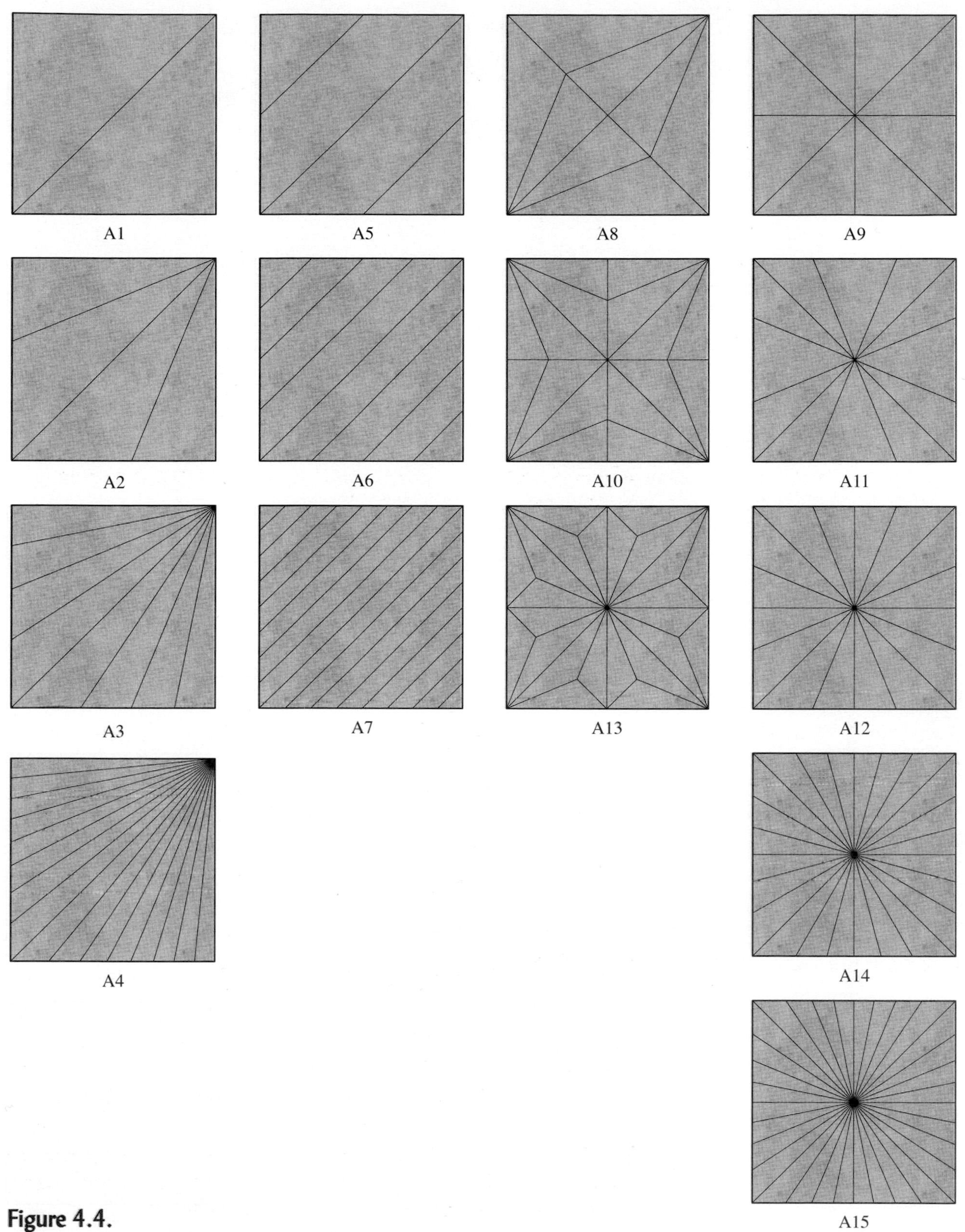

Figure 4.4.
The Uchiyama system of A bases, which are based primarily upon diagonal and/or radial folds. Note that the Kite Base, Fish Base, Bird Base, and Frog Base are among them.

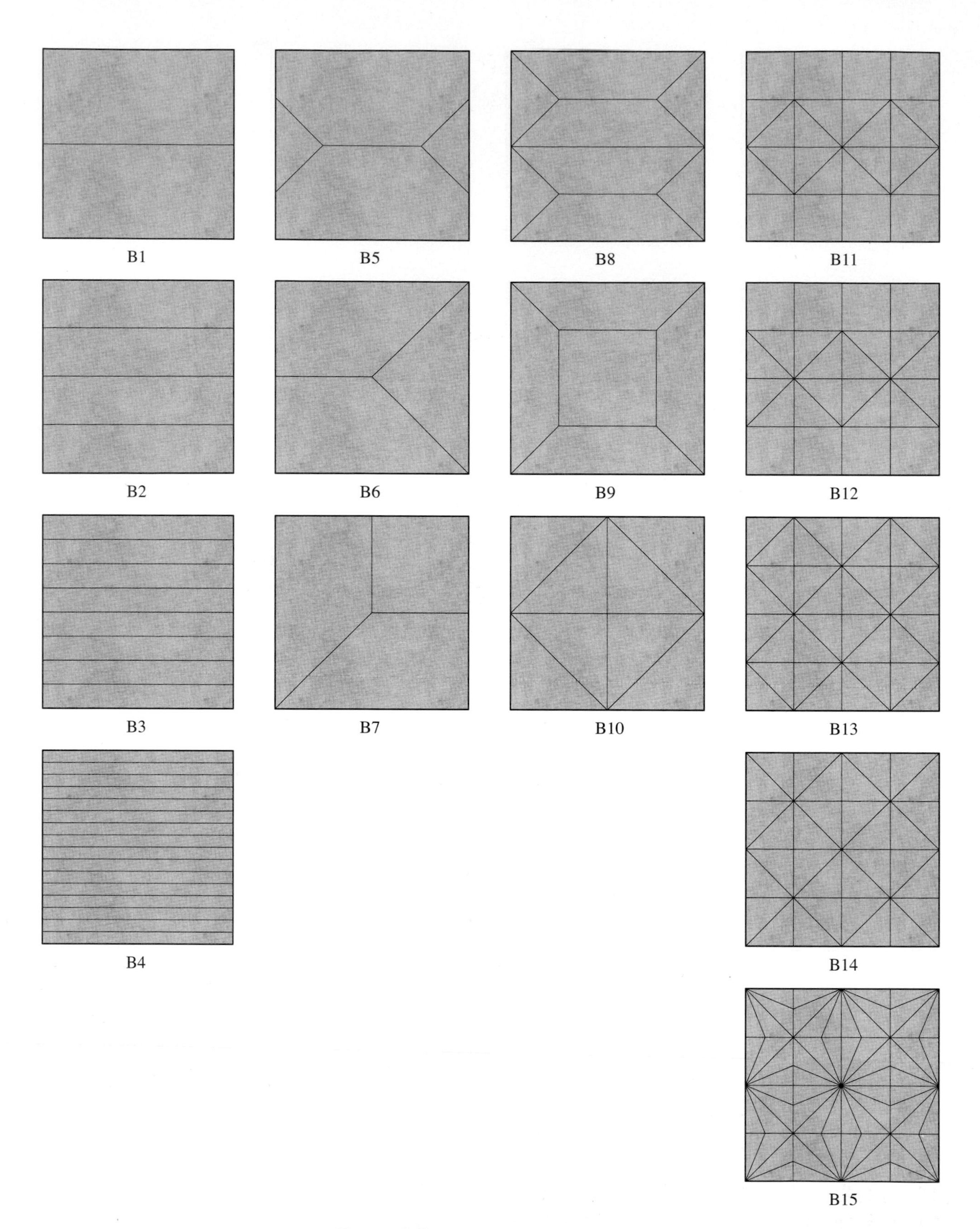

Figure 4.5.
Uchiyama's B bases.

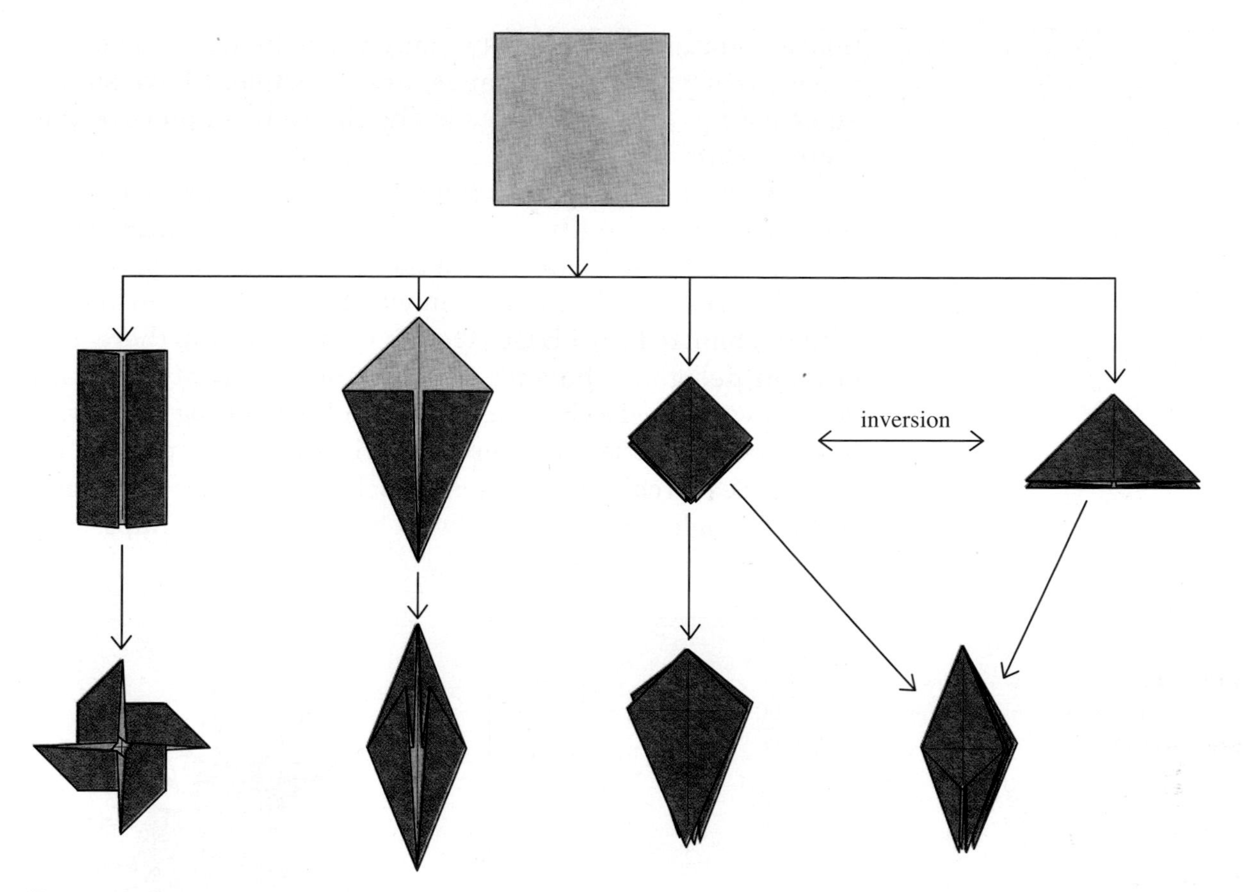

Figure 4.6.
Family tree of the Standard Bases.

But the four Classic Bases—Kite, Fish, Bird, and Frog—share a deeper similarity that is only evident when one examines their crease patterns. In these four bases, the same fundamental pattern appears in multiples of two, four, eight, and sixteen.

This reappearing shape is an isosceles right triangle with two creases in it; Figure 4.7 shows how it appears in each base in successively smaller sizes. Although the crease direc-

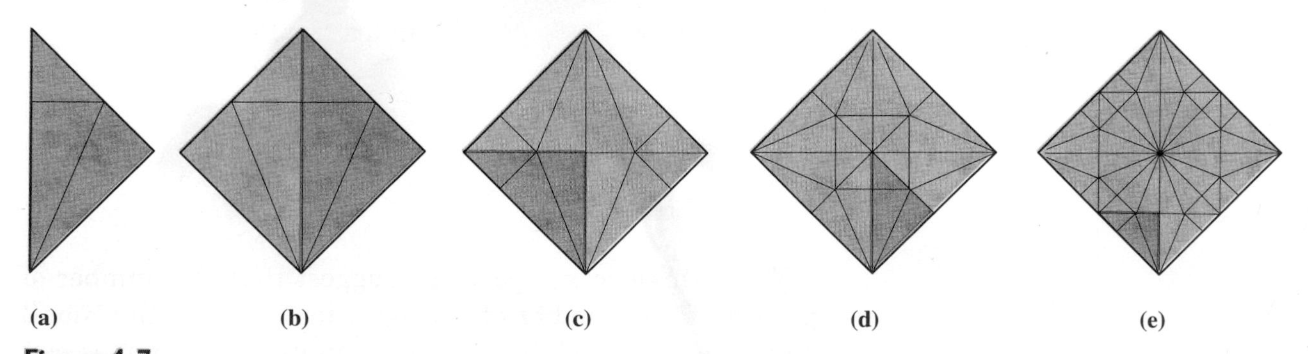

Figure 4.7.
(a) The basic triangle. (b) Kite Base. (c) Fish Base. (d) Bird Base. (e) Frog Base.

tions (mountain versus valley) may vary, the locations of the two creases within each triangle are the same. I have shown all creases as generic creases in the figure to emphasize this commonality.

Two of these isosceles triangles can be assembled into a square, yielding the Kite Base. Four give the Fish Base. Eight give the Bird Base. Sixteen give the Frog Base. The pattern is clear. We could easily go to 32, in which case we would end up with the blintzed Bird Base. There's no need to stop there, and origami designers haven't. In the mid-20th century Akira Yoshizawa devised a Crab based on the blintzed Frog Base, with 64 copies of the triangle; more recently, the crease pattern for my own Sea Urchin (Figure 4.8), which incorporates 128 copies of this triangle, creates a base with 25 equal-length flaps.

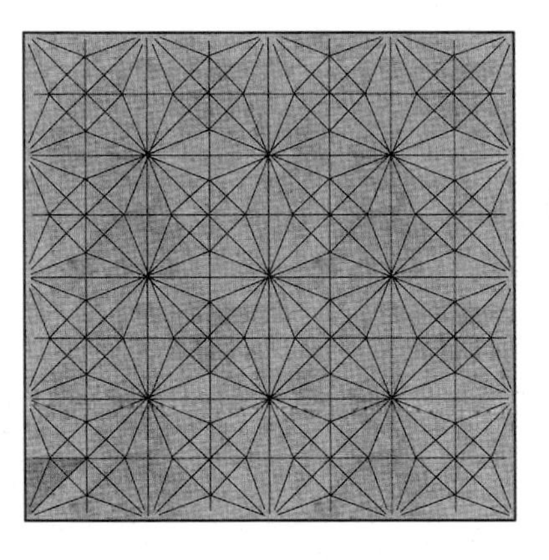

Figure 4.8.
Crease pattern and folded form of Sea Urchin.

The repeating pattern of triangles—first observed by Kenneway in his column in *British Origami* magazine, "The ABCs of Origami"—is more than a geometrical curiosity. As we increase the number of triangles, we also increase the number of long flaps in the resulting base. The first three crease patterns suggest a simple relationship between triangles and flaps:

Base	Triangles	Flaps
Kite	2	1
Fish	4	2
Bird	8	4

These three crease patterns suggest that the number of flaps is half the number of triangles in the base. But small numbers can be deceiving. A small number of examples can masquerade as many possible sequences—for the very next base breaks the pattern:

Base	Triangles	Flaps
Frog	16	5

So the Frog has five, rather than eight flaps, as the simple pattern would suggest. And the Sea Urchin really messes things up:

Base	Triangles	Flaps
Urchin	128	25

You might find it an interesting experiment to fold the crease patterns that lie between the Frog and Sea Urchin into bases and count the number of flaps in each (Hint: start with a blintzed Bird Base and blintzed Frog Base, but you will have to perform some additional manipulations to free the flaps).

So, there isn't a simple relationship between the number of triangles and the number of flaps. But there is a relationship nonetheless. Let us draw an arc of a circle in the triangular unit; then draw each arc in the unit as it appears in the crease pattern of the base.

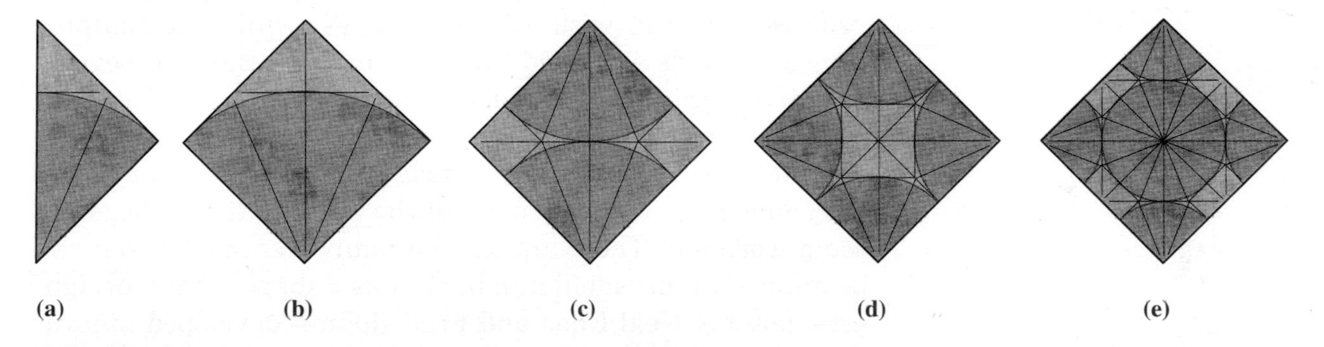

Figure 4.9.
(a) The triangle unit, with inscribed circle. (b) Kite Base. (c) Fish Base. (d) Bird Base. (e) Frog Base.

The basic triangle unit contains 1/8 of a circle. When the units are combined, however, the circular arcs combine with each other to form quarter-, half-, and whole circles. If we count the number of distinct circular pieces, we get in the Kite Base, one quarter-circle; in the Fish Base, two quarter-circles; in the Bird Base, four quarter-circles; and in the Frog Base, four quarter-circles plus one whole circle, making five sections in all. One, two, four, and five—these are the same numbers as the number of long flaps in each of the Classic Bases. (If we do the same to the Urchin pattern, we will find

25 circles or circular segments—and of course there were 25 flaps as well.) Clearly, there is some relationship here between circles and flaps. But why circles? And what about the paper that is not part of any circle? Circles seem rather innocuous, but by drawing them onto a crease pattern, we have touched on a connection to the underlying structure of origami, which we will soon explore.

4.4. Designing with Bases

The Japanese designers of the past—and most of their modern successors—did not worry about units and circles, of course. For most of the history of folding, the Classic Bases were nothing more than starting points for origami design; you picked the base that had the right number of flaps. For a bird with folded wings, use the Bird Base. A human figure, with two arms, two legs, and a head, use the Frog Base. But what about more complex figures? Insects and arthropods, with six, eight, ten, or more legs, wings, horns, pincers, and other appendages, became an enormous challenge. As early as the 1950s, far-sighted origami designers made forays into these more complex bases. Yoshizawa, using a multiply blintzed base, produced his remarkable crab with 12 appendages, while the sculptor George Rhoads exploited the blintzed Bird Base for several distinctive animals, including his famous Elephant. But these were the exceptions.

And so, the early days of origami design saw the use of the same bases over and over, to the point that they began to seem worn out. There are only so many treatments that can be applied to this small number of basic shapes. A few designers—notably Neal Elias and Fred Rohm—developed innovative manipulations of the Classic Bases that opened up rich new veins of origami source material. For the most part, however, the Classic Bases are pretty well picked over.

Still, one occasionally finds a shiny nugget of originality among the tailings of the Classic Bases. Sometimes, a model's structure simply calls for the flaps and proportions of a Classic Base, as in the designs shown in Figures 4.10–4.15, which are folded from the Windmill, Kite, Bird, and Frog Bases. Take, for example, the Stealth Fighter shown in Figure 4.10 as crease pattern and folded model. It is folded from the Windmill Base, which can be seen in its crease pattern.

Or can it? The crease pattern, which typically shows the major creases of the model, contains more creases than just those of the base. But if you focus on the longer creases, you

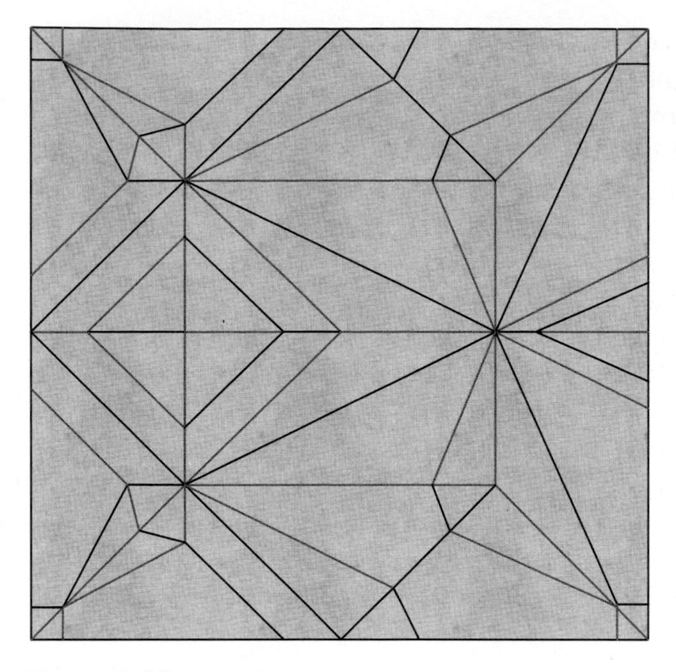
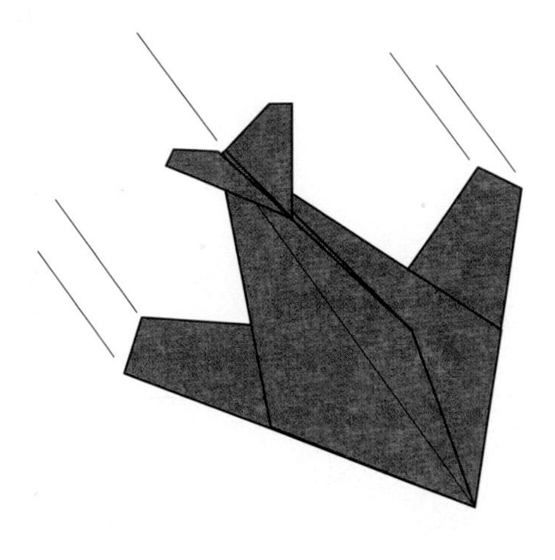

Figure 4.10.
Crease pattern and model of Stealth Fighter, from a Windmill Base.

can probably pick out the creases of a Windmill Base with some of the flaps flopped around; perhaps it will help to show just the creases of the modified base.

When a model is folded from a base (whether one of the Classic Bases, or a new special-purpose base designed just

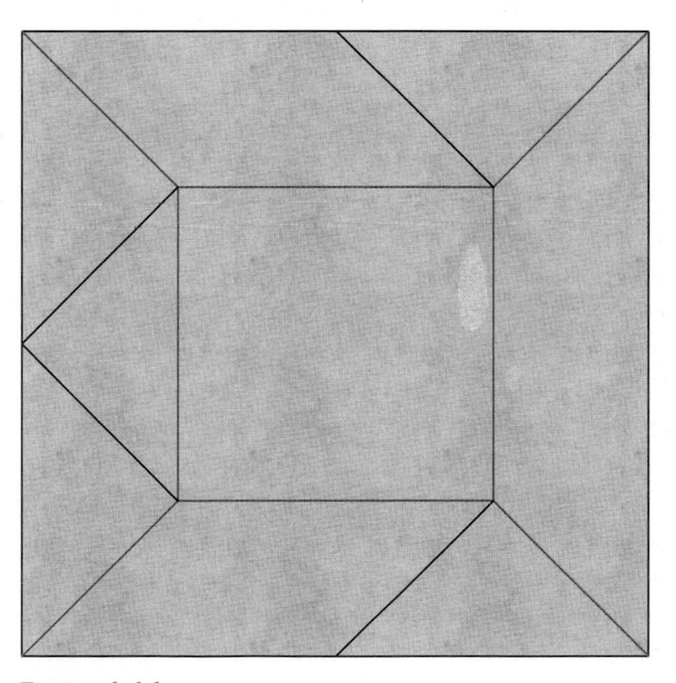
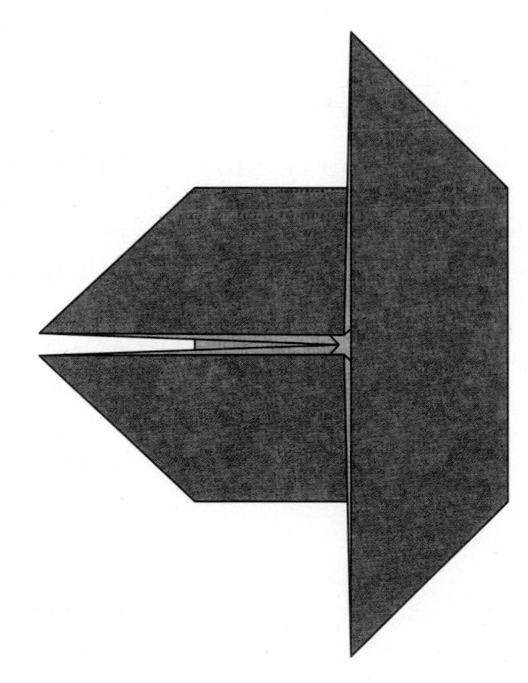

Figure 4.11.
Crease pattern and folded form of the modified Windmill Base from which the fighter is constructed.

for that model), if it has a linear, step-by-step folding sequence (as most designs do), the base creases are made first and usually persist throughout the folding of the model. By examining the crease pattern for a model and identifying the base creases, you can gain information about the folding sequence for the model because the earlier steps will be devoted to construction of the base. The base creases tend to be longer than later folds in the sequence. Thus, for example, in the Snail shown in Figure 4.12, we can pick out several long creases in the pattern that identify the probable base, which is shown in Figure 4.13.

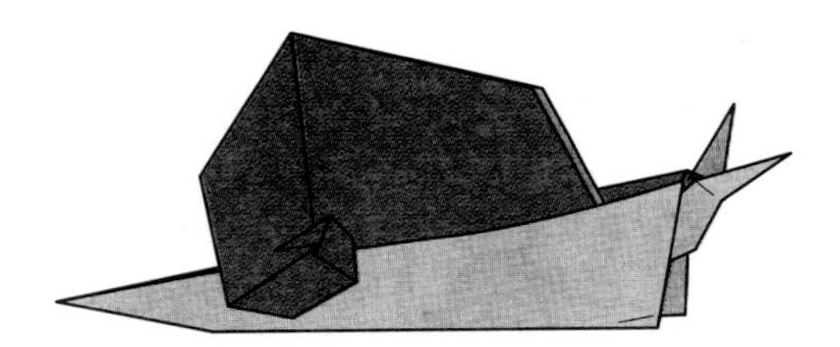

Figure 4.12.
Crease pattern and model of Snail, from a Kite Base.

In this pattern, composed of only six creases, you can already see the basic structure of the snail: the tail, the two corners that become the antennae, and the long colored point that becomes the shell.

The full pattern obtained by unfolding the folded model is often too cluttered to clearly discern the structure. It is more useful to show just the major creases, typically those when the base is complete but before the final shaping has begun. Throughout this book I will show crease patterns at this intermediate stage of folding, along with a drawing of the shape that corresponds to the crease pattern and the folded model.

Interestingly, the more complicated the base is, the easier it often is to recognize in the crease pattern because its creases form a distinctive pattern. In the Valentine in Figure 4.14, it

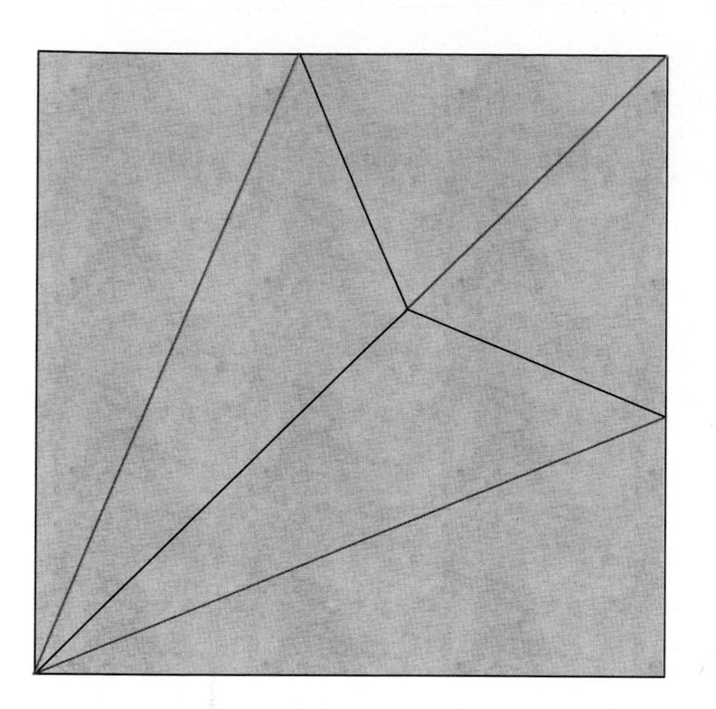

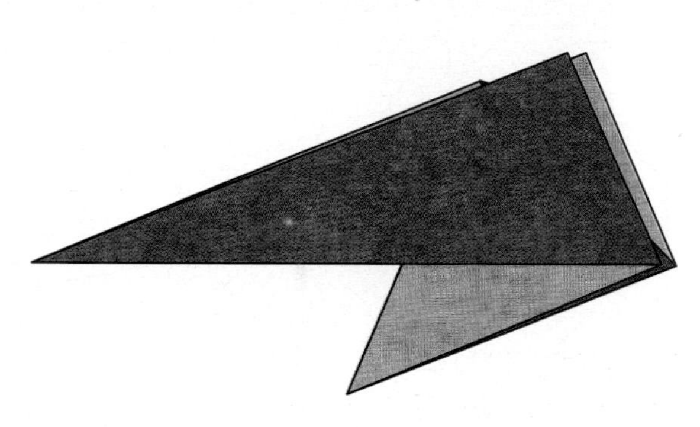

Figure 4.13.
The base creases (a Kite Base plus a single reverse fold) for the Snail.

is not particularly clear from the folded model where it came from, but in its crease pattern, its Bird Base heritage is unmistakably present.

Similarly, the outlines of a Frog Base are clearly visible in the lines of the Hummingbird in Figure 4.15.

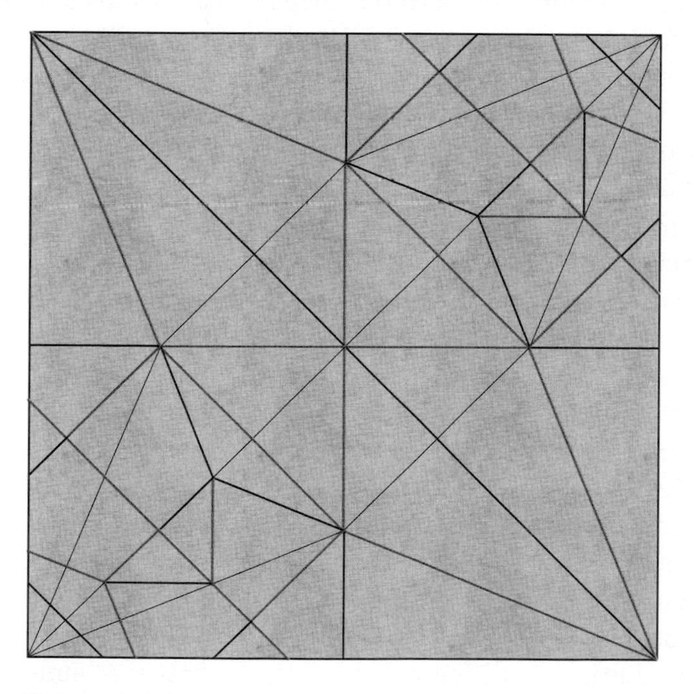

Figure 4.14.
Crease pattern, base, and folded model of Valentine, from a Bird Base.

Figure 4.15.
Crease pattern, base, and folded model of Hummingbird, from a Frog Base.

4.5. Simple Variations on Bases

To move out of the confines of the standard bases, in what I will call the early exploratory period of origami, the 1950s and 1960s, many folders experimented with alterations and combinations of the bases. A not-uncommon tactic was to fold a square half as one base and half as another: thus, a half-Bird, half-Frog Base, for example. However, with only a handful of bases around to start with, one quickly exhausts the possibilities of the technique.

The 1960s and 1970s saw another variation that can also lead to new structures: Use the crease pattern of the original base, but distort it in some controllable way. The most common application of this second approach is to offset the base, that is, shift the nexus of creases that typically arises at the center of the paper away from the true center, either toward an edge or toward a corner. Since the amount of shift was something that could be continuously varied, this technique provides a greater range of possibility than discrete combinations of fractional bases.

As an example, the crease pattern of a Bird Base can be shifted in two distinctly different ways that preserve some symmetry, as shown in Figure 4.16. The crease pattern can be shifted toward a corner or an edge.

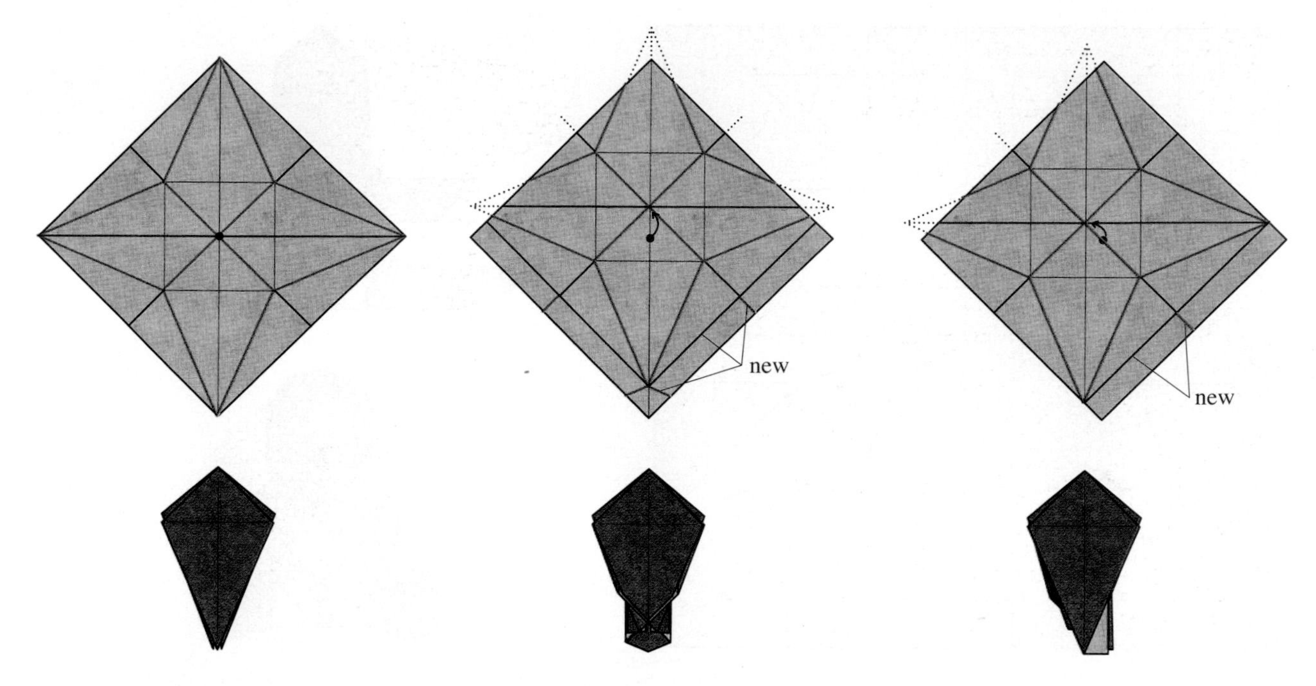

Figure 4.16.
Left: a Bird Base. Middle: the crease pattern shifted toward a corner. Right: the crease pattern shifted toward an edge.

Neither of the two variations is aesthetically pleasing; the creases terminate in rather arbitrary locations, resulting in several points with ragged edges and others with a parallel-edged strip running along one side. These excess bits will usually detract from the model unless they can be incorporated into the design—that is, used to create one or more additional features of the model. That this incorporation can be done successfully is illustrated by the Baby in Figure 4.17, which is based on an offset Waterbomb Base, and which uses the extra strip to realize the color-changed diaper. Can you identify the creases of the Waterbomb Base in the crease pattern? It will be a bit harder, because portions of the base creases have changed direction or been smoothed out in the course of folding the model, so they are not as evident. Nevertheless, you should be able to pick out the creases of a Waterbomb Base shifted toward an edge.

It is also possible to offset the central crease cluster while preserving the points where the creases all come together at the corners of the square. This eliminates the ragged points of the previous offsetting technique, but now the edges of the four points are no longer aligned. That may or may not be a drawback; two of the points are now longer than the other two, making the base perhaps better suited to other subjects. Such a base is said to be a *distorted base*.

Figure 4.17.
Crease pattern, base, and folded model of Baby, from an offset Waterbomb Base.

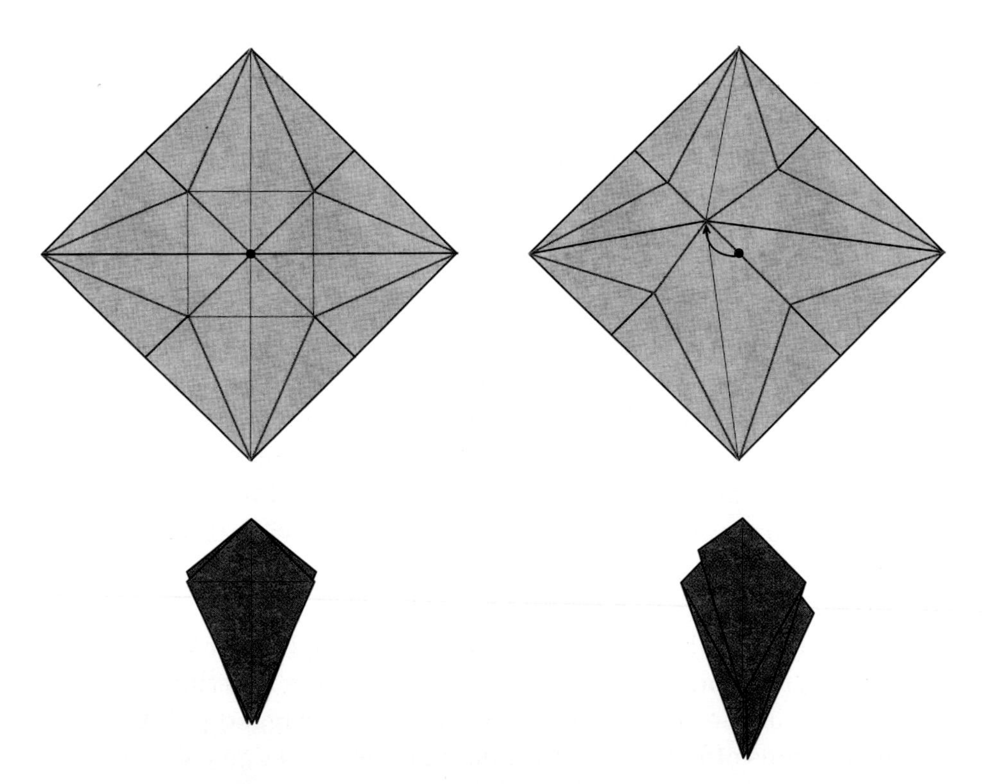

Figure 4.18.
Left: an ordinary Bird Base.
Right: a distorted Bird Base with the corners preserved at the corners of the square.

There are many more distortions that can be performed on the Classic Bases and their combinations, but they don't change the basic structure. A Bird Base has four long flaps; a distorted Bird Base still has four longish flaps, even if one or two are now a bit longer than the others. Offsetting and other distortions can vary the distribution of edges around a flap, but they don't create entirely new flaps. Even if they did not constitute such well-trodden turf, the Classic Bases don't have enough variety among them to serve as a starting point for all origami subjects. Quite often, however, the origami designer will find that a Classic Base suffices—almost. The situation will arise when you need just a bit more—an extra flap, a single longer flap, a cluster of points where one exists. In such cases, we'll need to deviate from the standard bases, which we can do in several ways. We can convert single points to multiples, we can add extra paper to an existing base, and finally, we can design an entirely new custom-purpose base starting from the structural form of the model. Each of these three approaches is a stage of origami design, each moving farther into new design territory.

Folding Instructions

Stealth Fighter

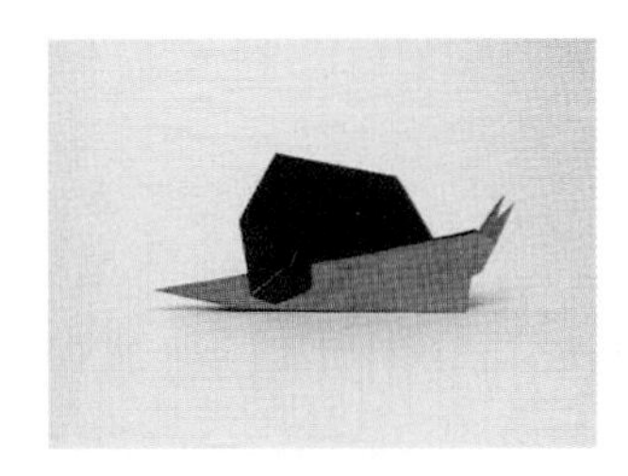

Snail

Valentine

Hummingbird

Baby

Stealth Fighter

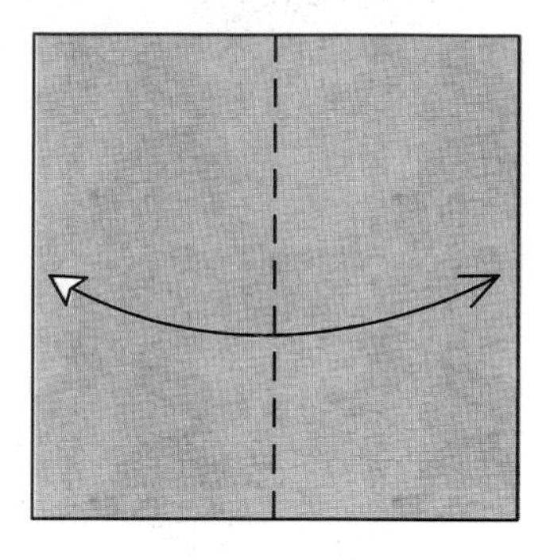

1. Begin with the white side up. Fold the paper in half from side to side and unfold.

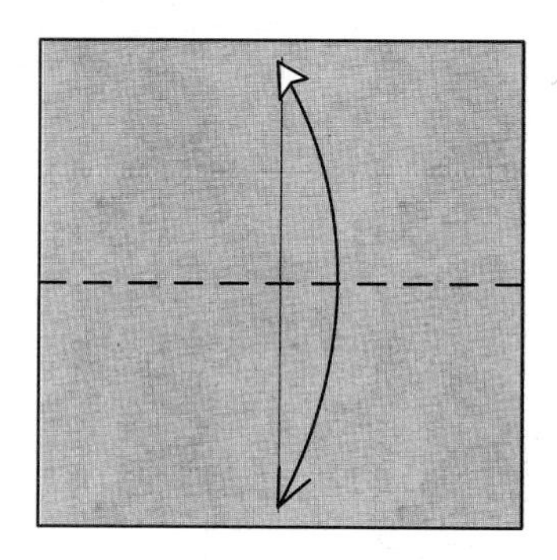

2. Fold the paper in half from top to bottom and unfold.

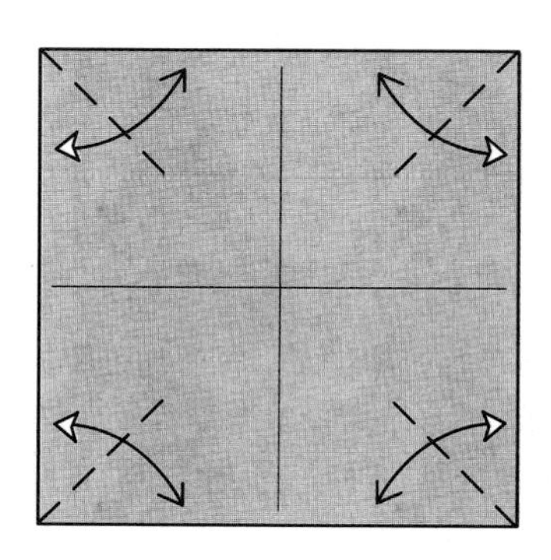

3. Fold each corner in half along an angle bisector and unfold.

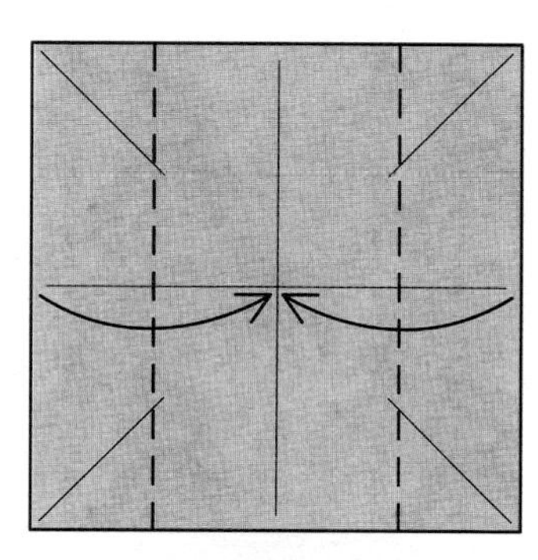

4. Fold the side edges in to lie along the vertical center line.

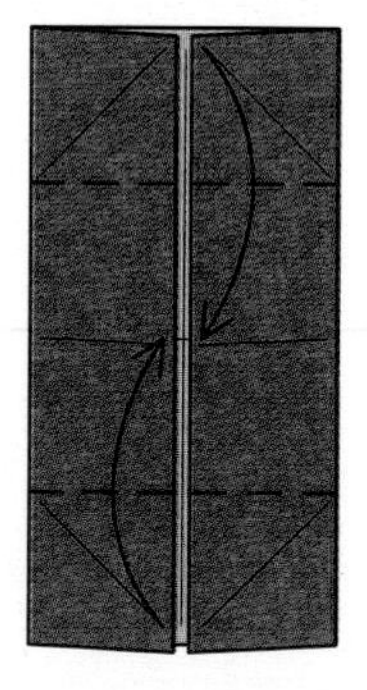

5. Fold the top and bottom edges to the center of the paper.

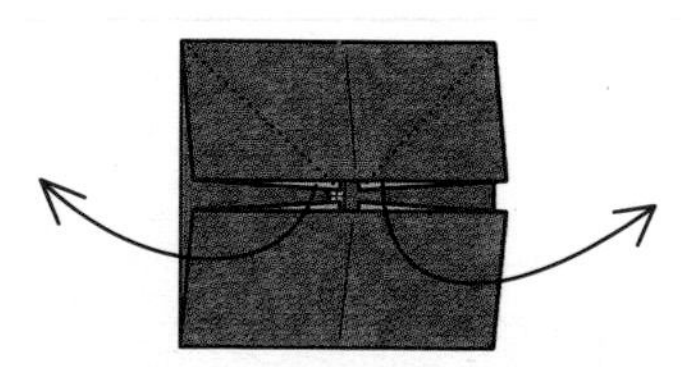

6. Pull the corners out to the sides and pinch them in half.

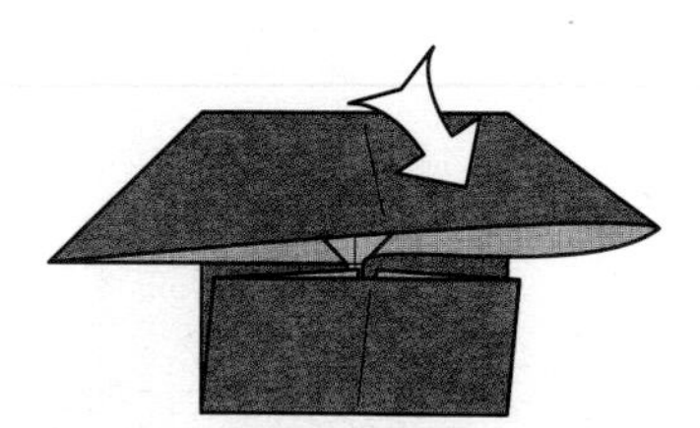

7. Flatten the corners, making a sharp point (the left side is shown completed here).

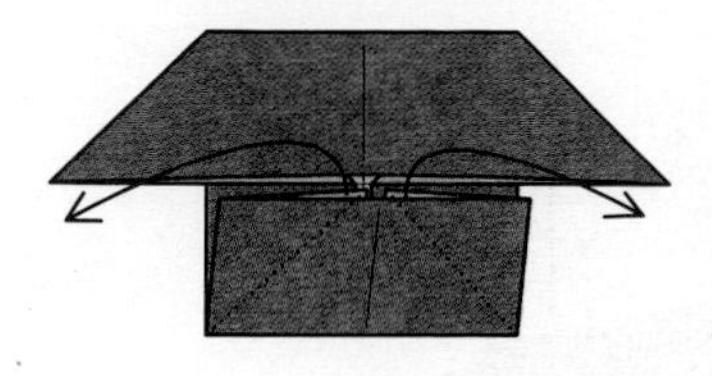

8. Repeat on the bottom two corners.

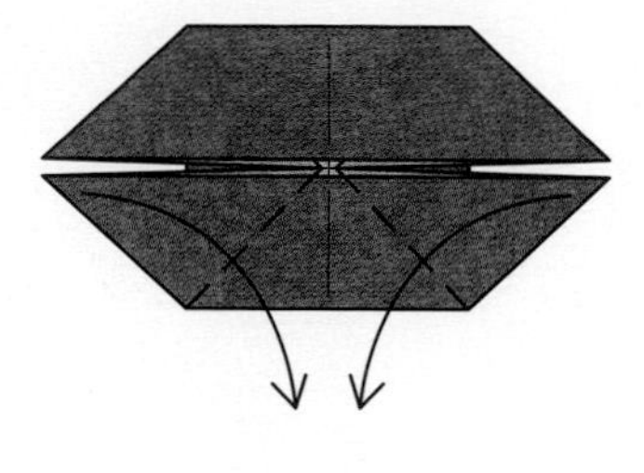

9. Fold the two lower points down as far as they will go.

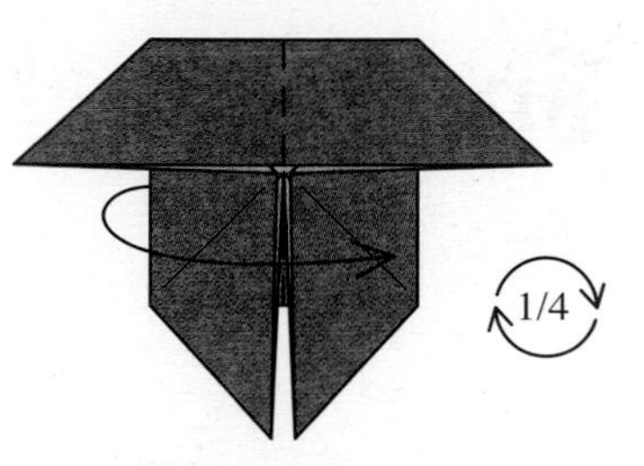

10. Fold the model in half and rotate 1/4 turn clockwise.

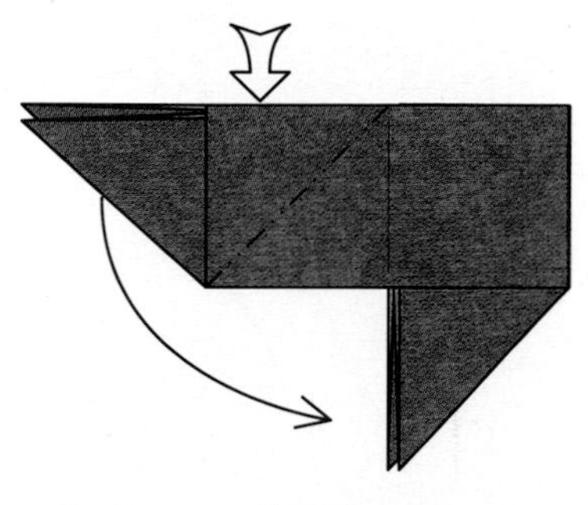

11. Reverse-fold the left side downward symmetrically.

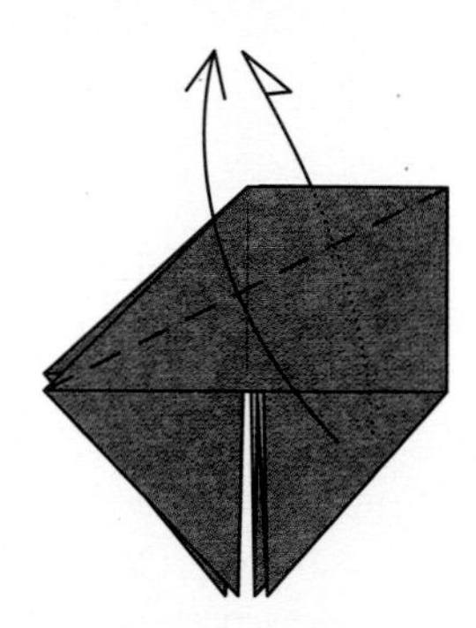

12. Fold one layer up in front and one to the rear behind.

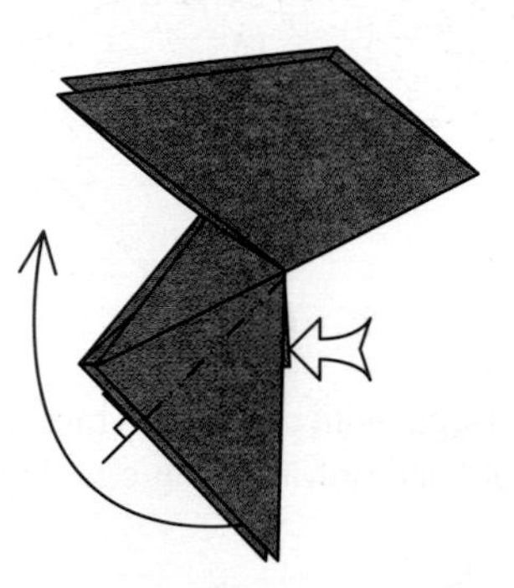

13. Reverse-fold the pair of flaps as if they were a single flap (they are joined about halfway down).

14. Reverse-fold the tips of each corner so that the raw edges line up.

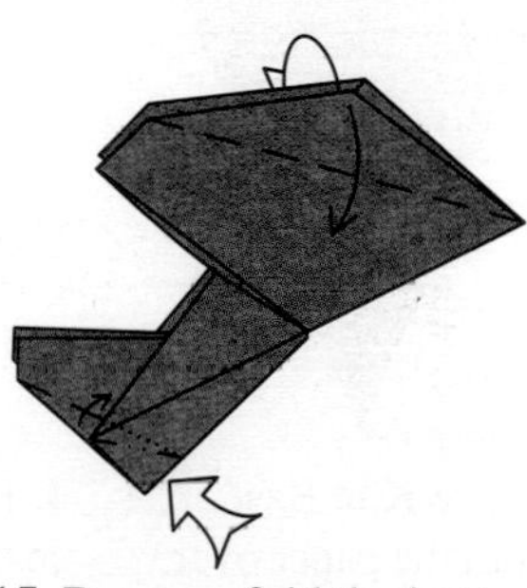

15. Reverse-fold the bottom corners. Valley-fold the leading edge of the near wing; repeat behind.

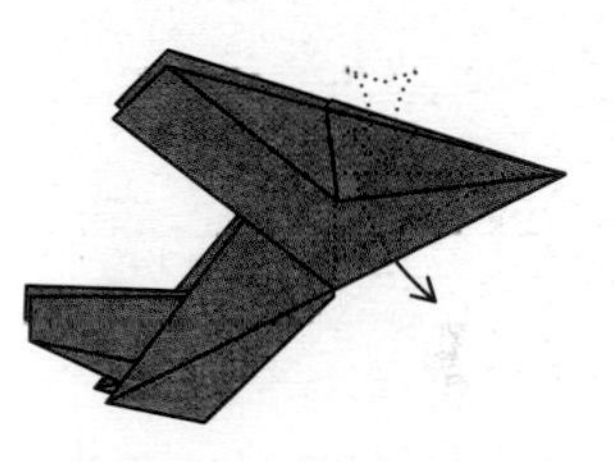

16. Reverse-fold a hidden layer down from the inside of the bottom of the model.

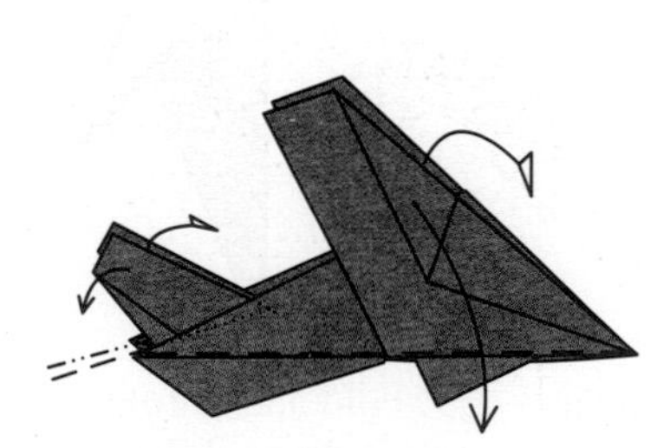

17. Spread the wings and tail.

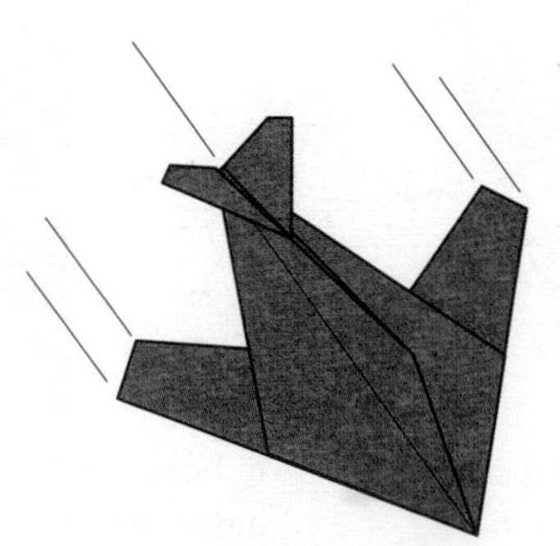

18. Finished Stealth Fighter.

Snail

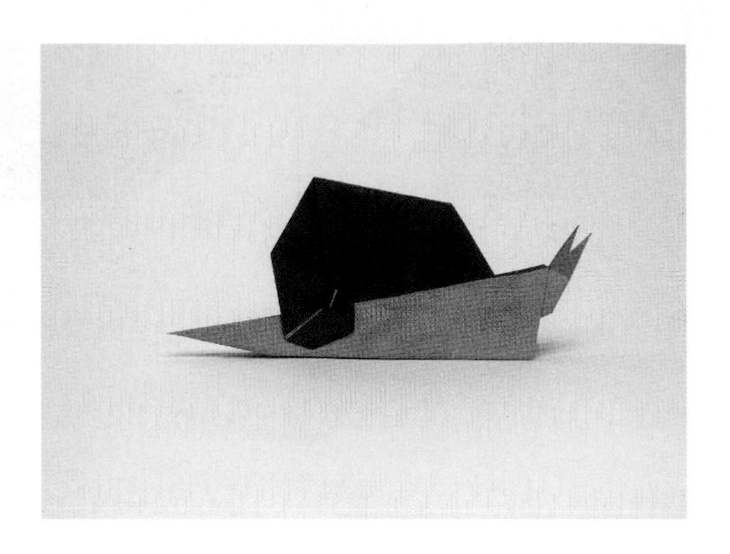

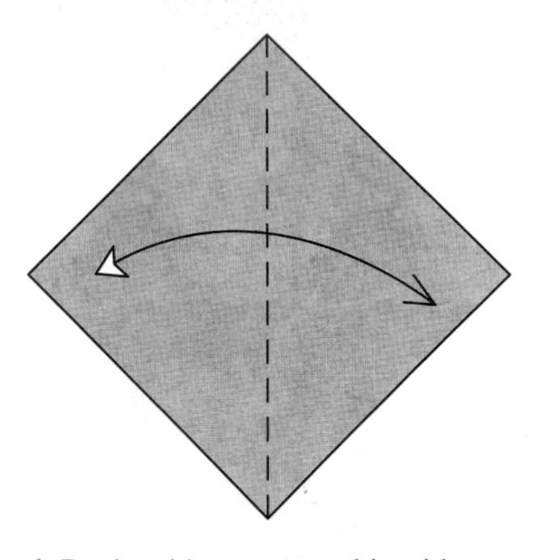

1. Begin with a square, white side up. Fold and unfold along one diagonal.

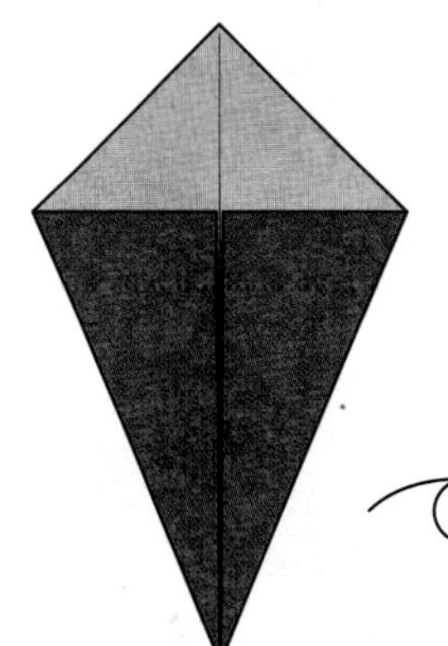

2. Fold the edges to the diagonal crease.

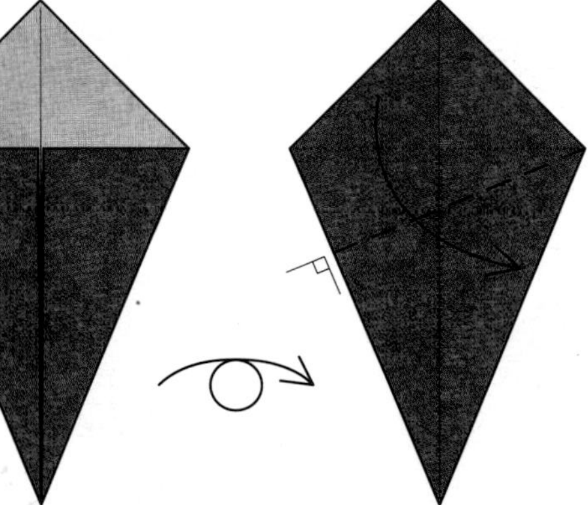

3. The Kite Base. Turn the paper over.

4. Fold the top flap down so that the left edges are aligned and the crease goes through the right corner.

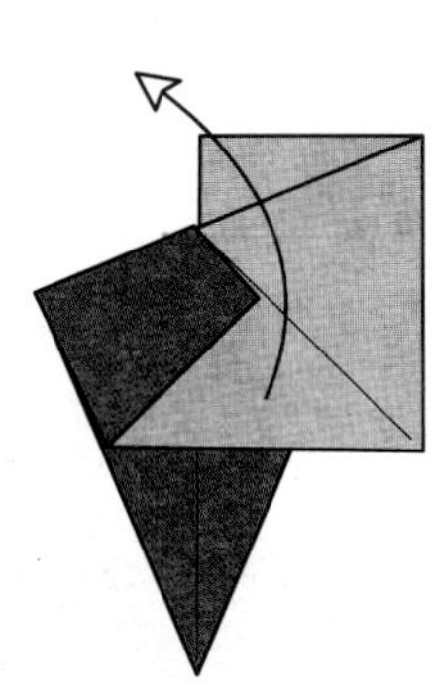

5. Unfold.

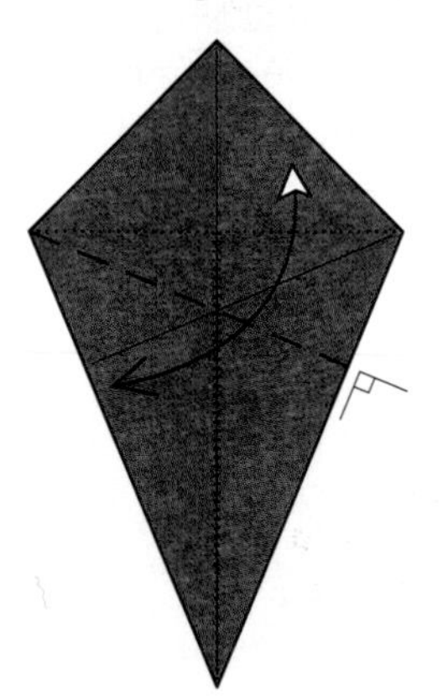

6. Repeat steps 4–5 in the other direction.

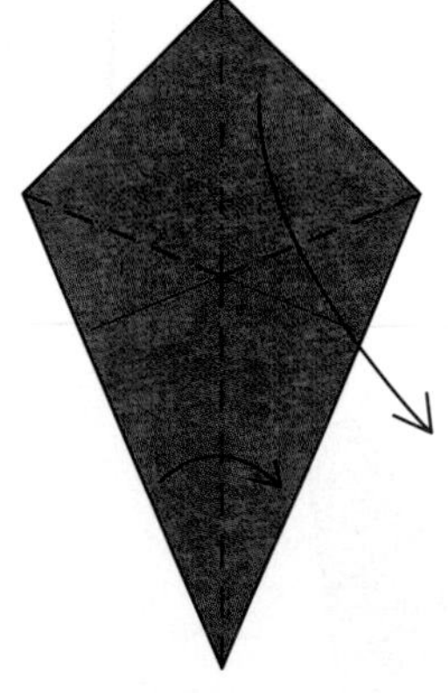

7. Fold the paper in half while reverse-folding the top corner downward.

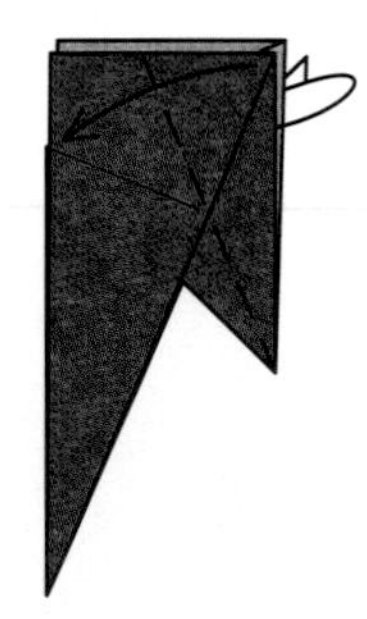

8. Fold the near edge along the angle bisector (the corner touches the edge at an existing crease). Repeat behind.

9. Fold and unfold along the angle bisector, creasing lightly. Repeat behind.

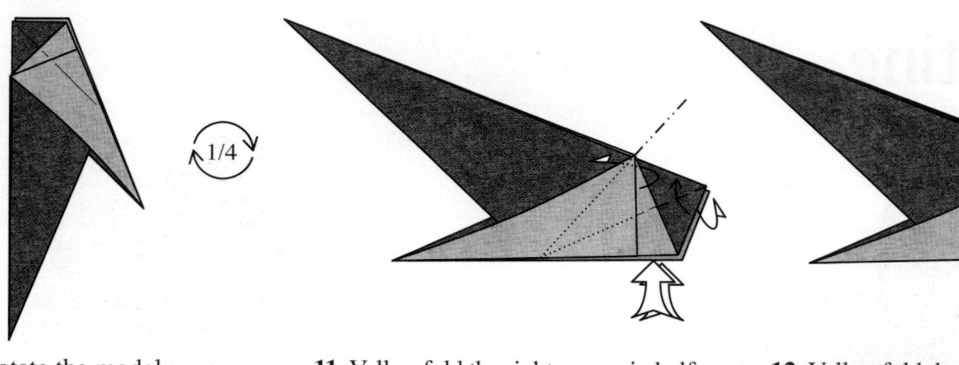

10. Rotate the model clockwise slightly more than a quarter turn.

11. Valley-fold the right corner in half while reverse-folding all of the excess paper up as high as possible. Repeat behind.

12. Valley-fold the near flap upward. Repeat behind.

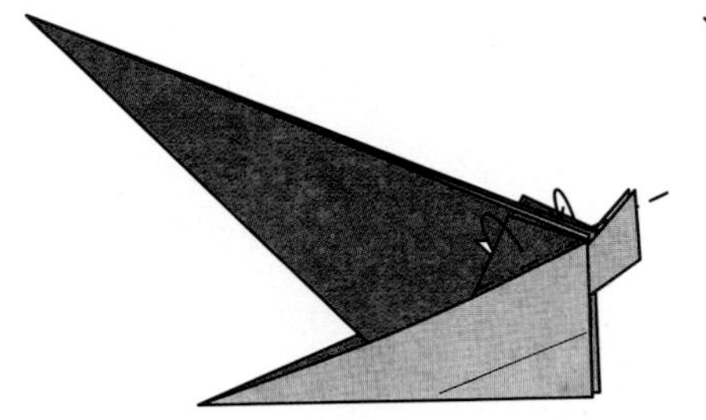

13. Mountain-fold the colored corner behind. Repeat behind.

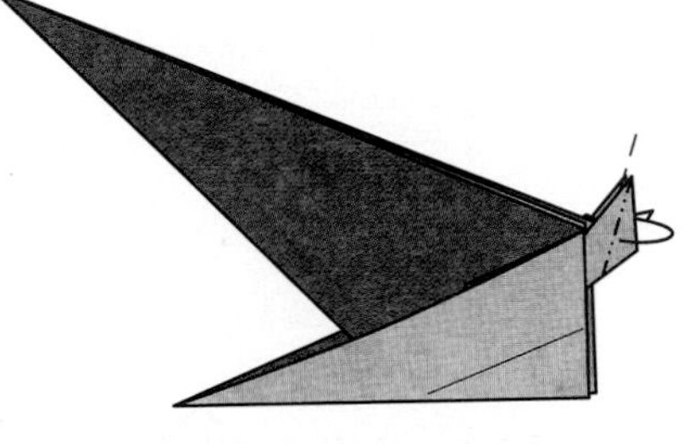

14. Mountain-fold the edge of the near flap behind. Repeat behind.

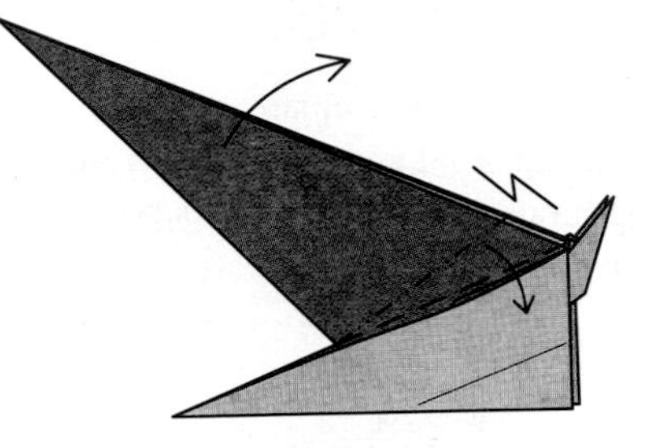

15. Pleat the colored flap.

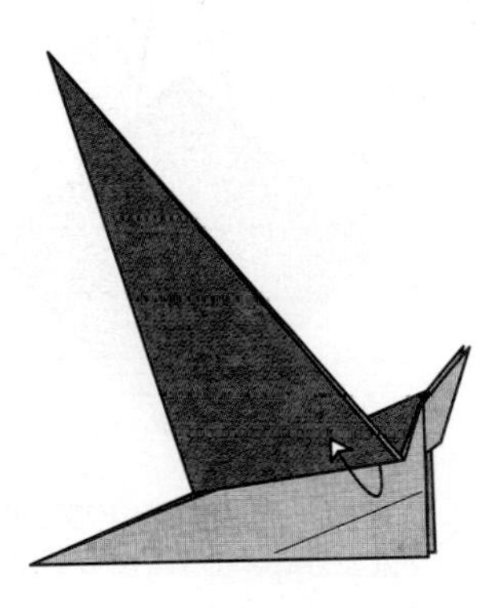

16. Tuck the pleat under the white layer.

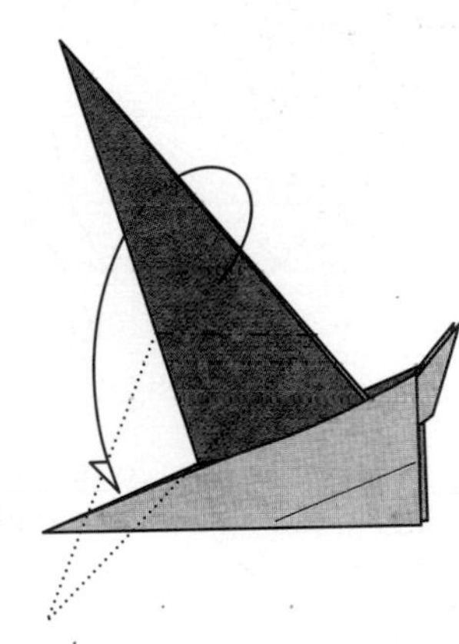

17. Mountain-fold the colored flap behind and then bring it in front of the white point at the lower left.

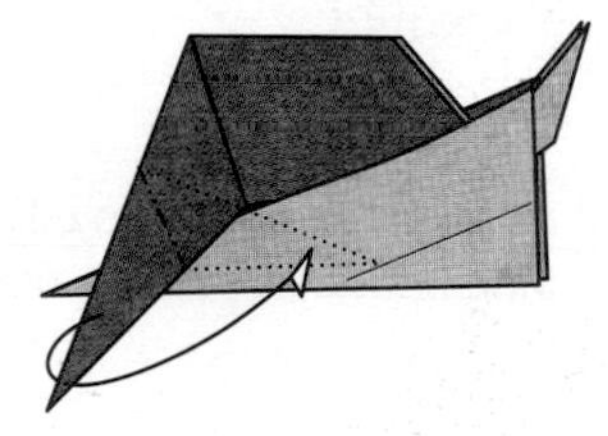

18. Mountain-fold the colored flap around again.

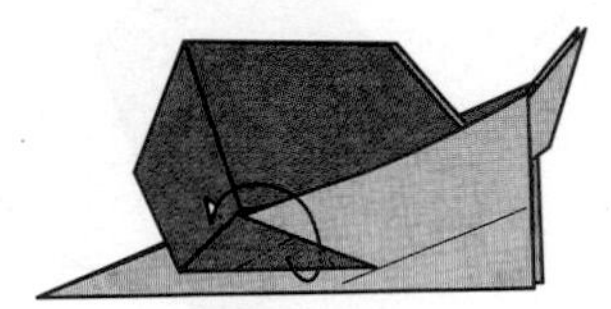

19. Mountain-fold two more times in the same way.

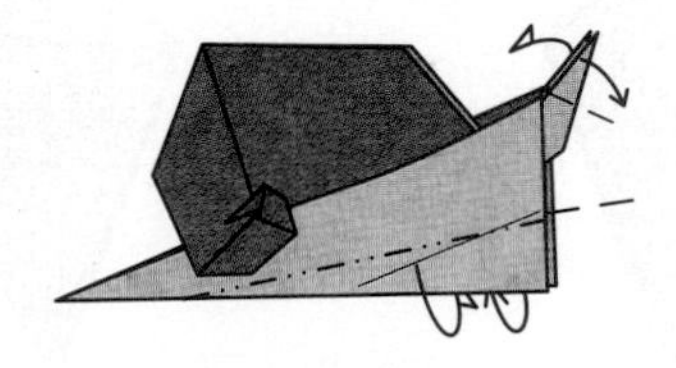

20. Mountain-fold the bottom edges. Valley-fold the antennae out to the sides.

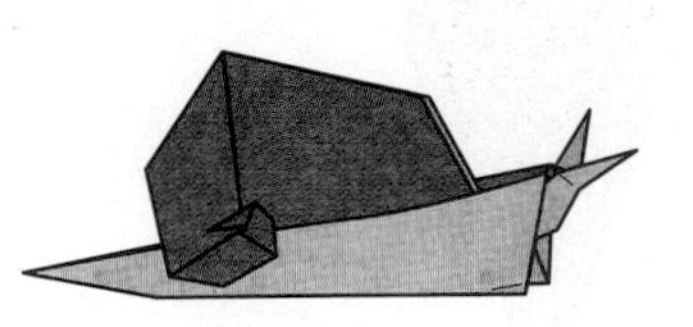

21. Finished Snail.

Valentine

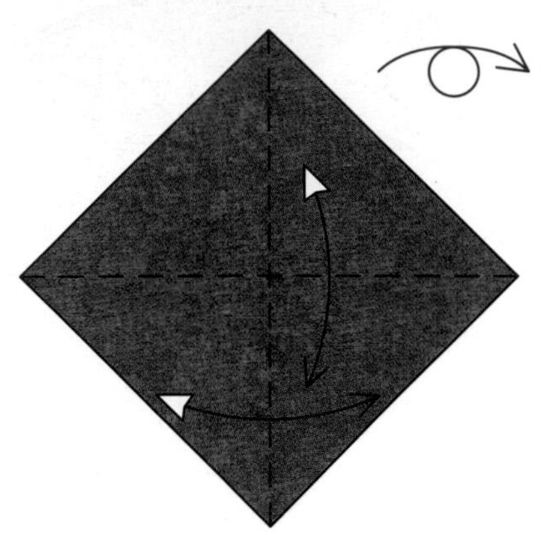

1. Begin with a square, colored side up. Fold and unfold along the diagonals. Then turn the paper over.

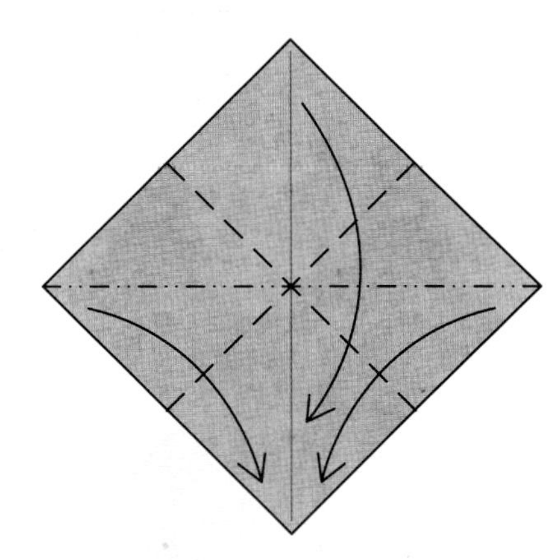

2. Fold the paper in half from edge to edge and unfold. Repeat the other direction.

3. Bring all four corners together at the bottom, forming a Preliminary Fold.

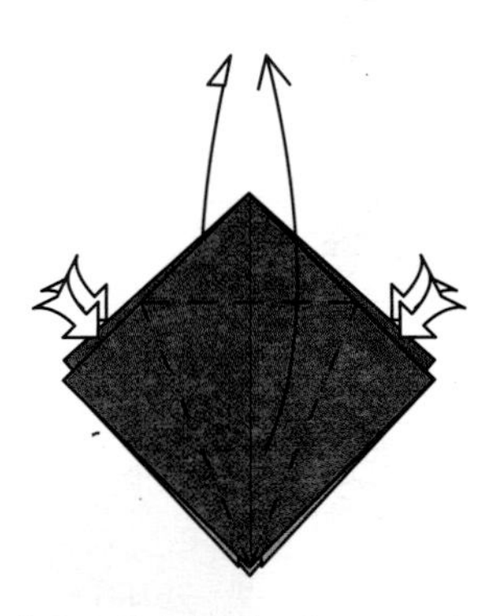

4. Petal-fold the front and back flaps to form a Bird Base.

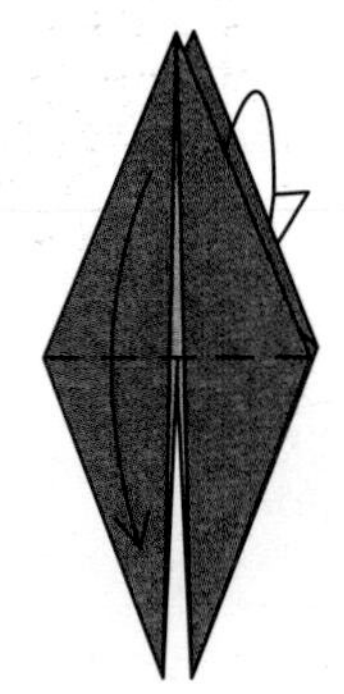

5. Fold one flap down in front and one behind.

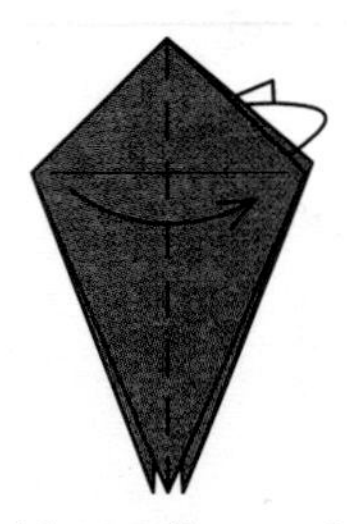

6. Fold one flap to the right in front and one to the left behind.

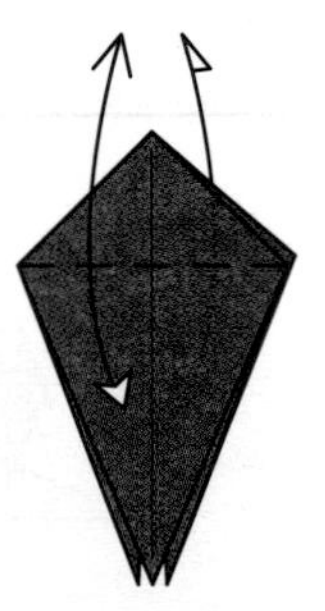

7. Fold one flap up in front and back down. Repeat behind.

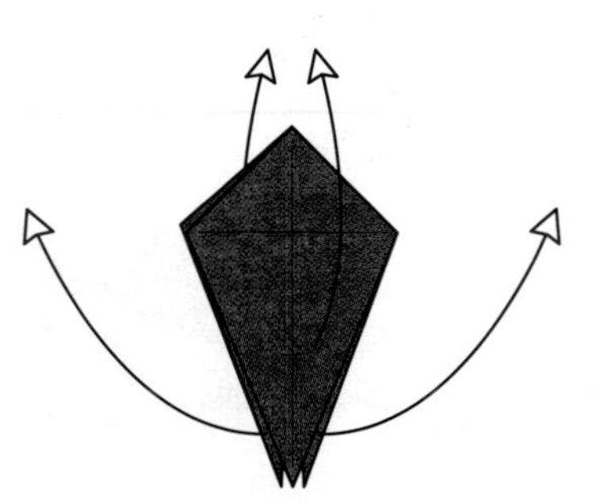

8. Unfold the model completely.

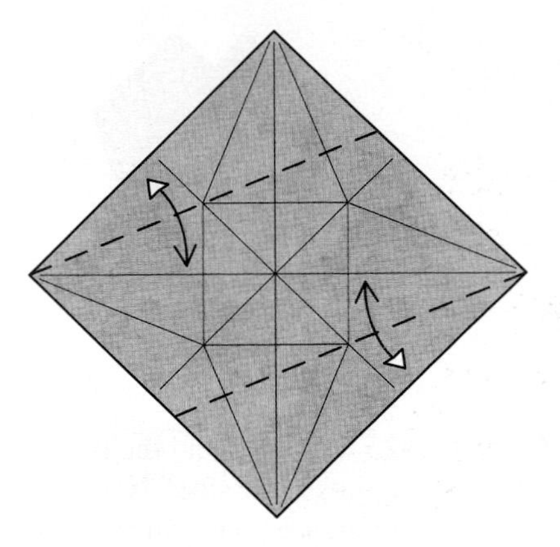

9. Extend an existing crease to the lower left edge. Repeat above.

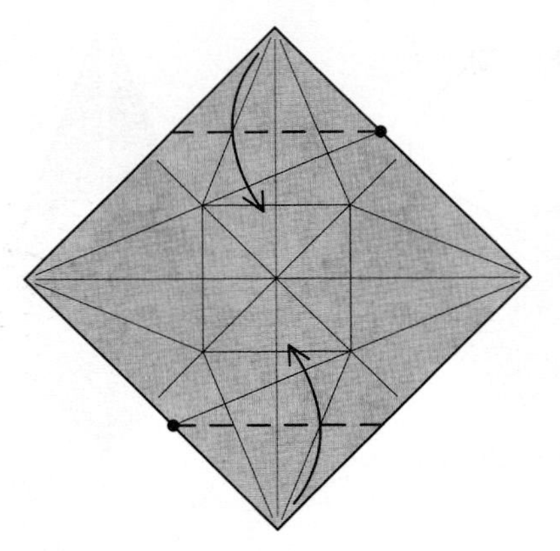

10. Fold the bottom corner up; the crease hits the edge at the same place as the last crease. Repeat with the top corner.

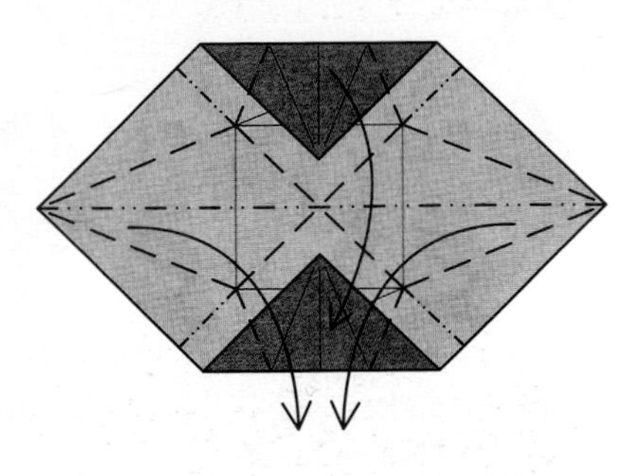

11. Refold the Bird Base, using the existing creases (although you will have to make new creases through the colored corners).

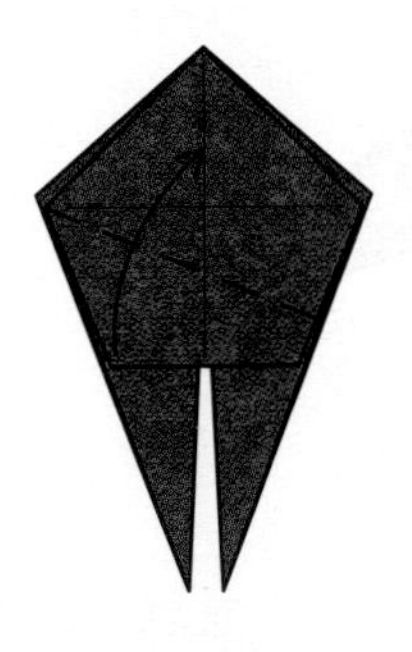

12. Fold the bottom left corner up to the center line; the fold runs along an existing crease. You don't need to make this fold sharp.

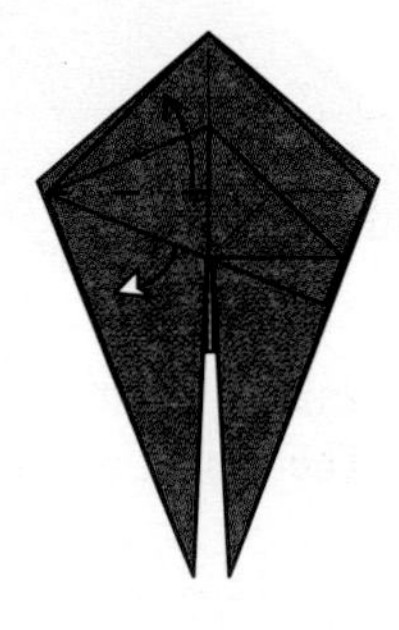

13. Shift some paper upward as far as possible, releasing the trapped paper underneath the flap.

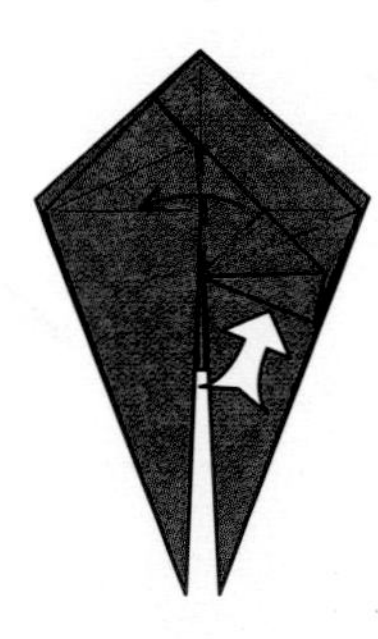

14. Squash-fold the corner, swinging the excess paper over to the left.

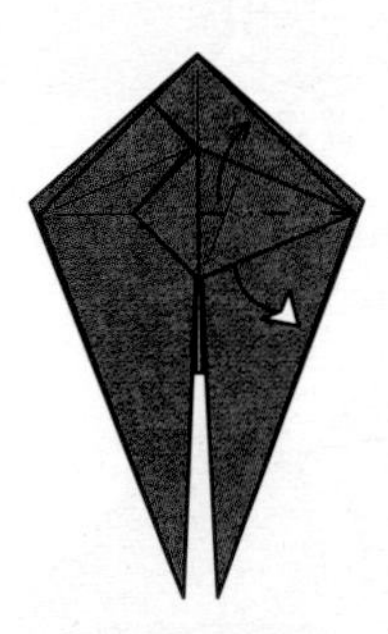

15. As you did in step 13, shift some paper upward, releasing the trapped paper underneath.

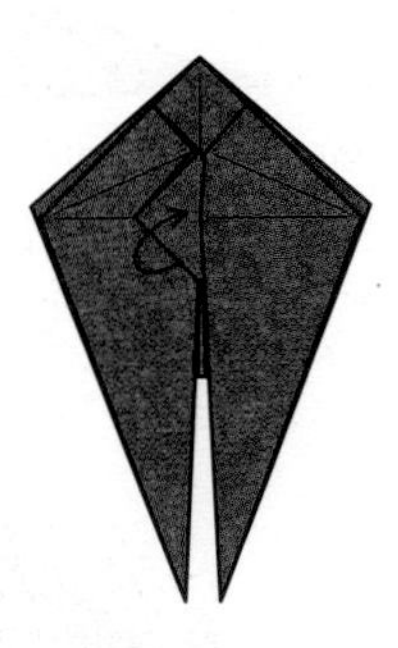

16. Bring a raw edge in front of the flap.

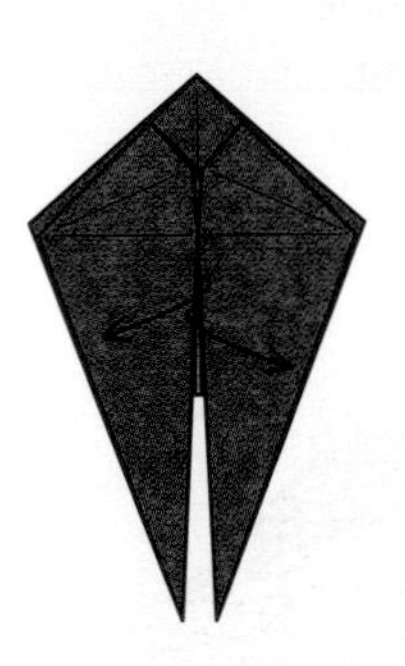

17. Open out the raw edges slightly.

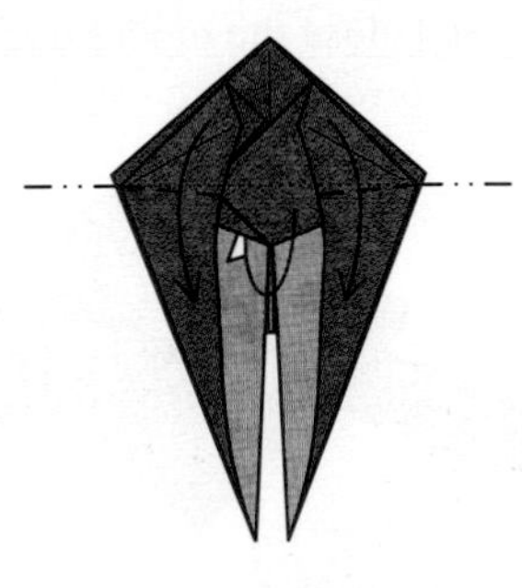

18. Fold the two top near flaps downward while folding the blunt interior flap underneath. Close the model and flatten firmly.

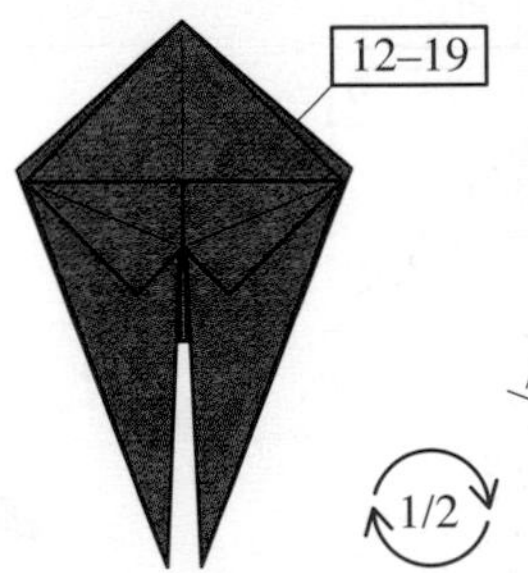

19. Repeat steps 12–19 behind. Then rotate the model 1/2 turn.

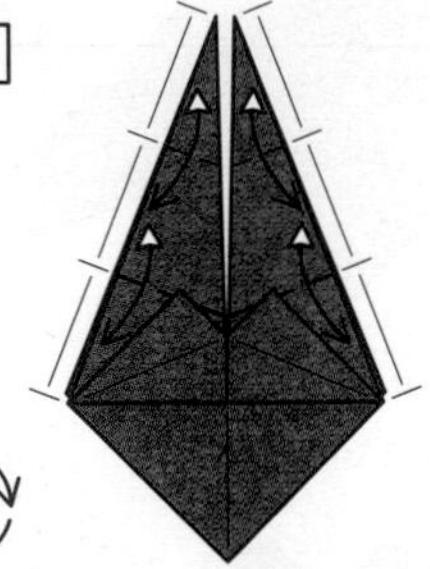

20. Divide each vertical flap into thirds with valley folds.

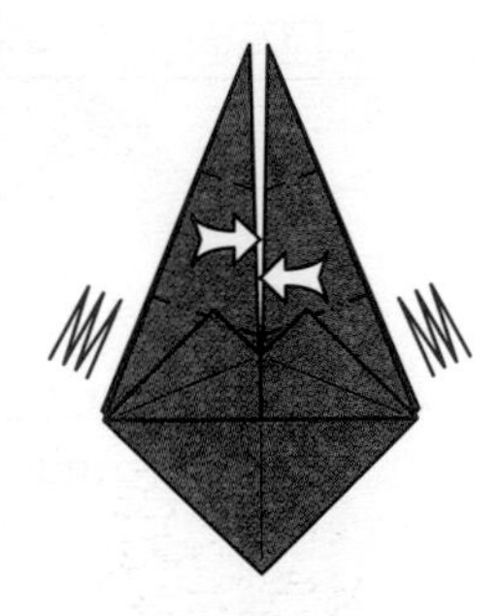

21. Reverse-fold each flap up and down on the existing creases.

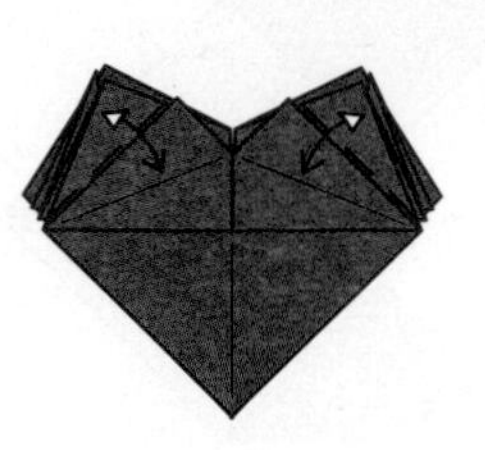

22. Fold and unfold through the near layers only. Repeat behind.

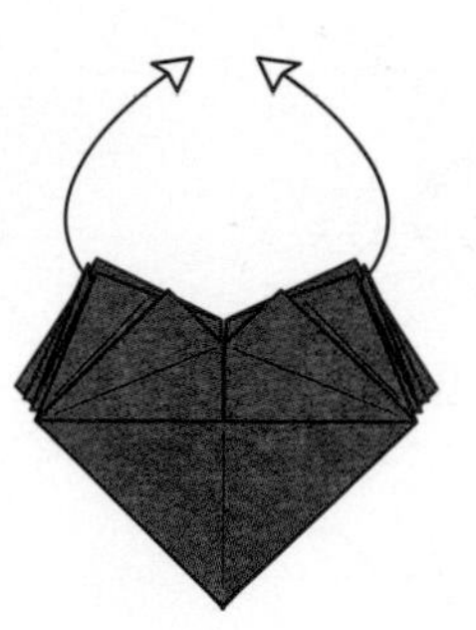

23. Unfold to step 20.

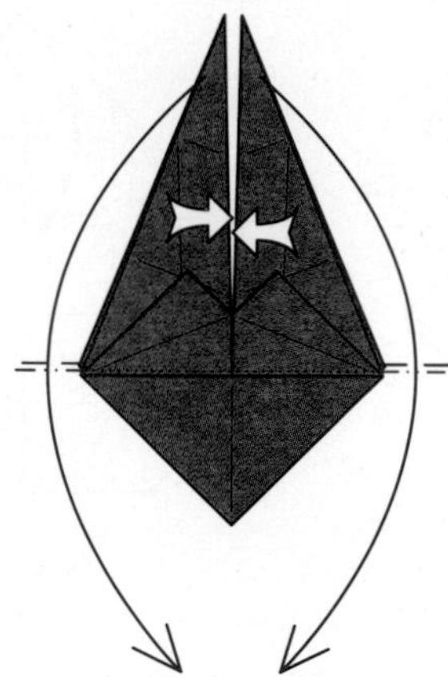

24. Reverse-fold both points as far downward as possible.

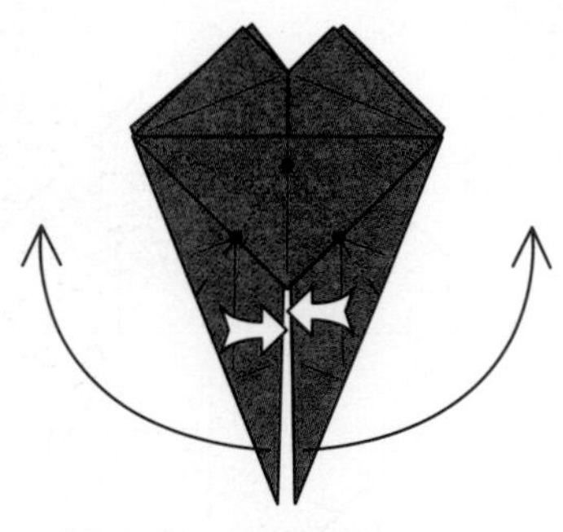

25. Reverse-fold the two points to the sides. Note the reference points for the two reverse folds.

26. Reverse-fold two edges upward on each side.

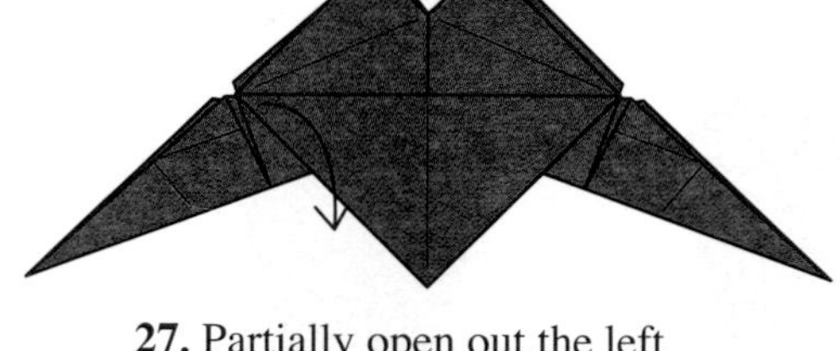

27. Partially open out the left side. The model will not lie flat.

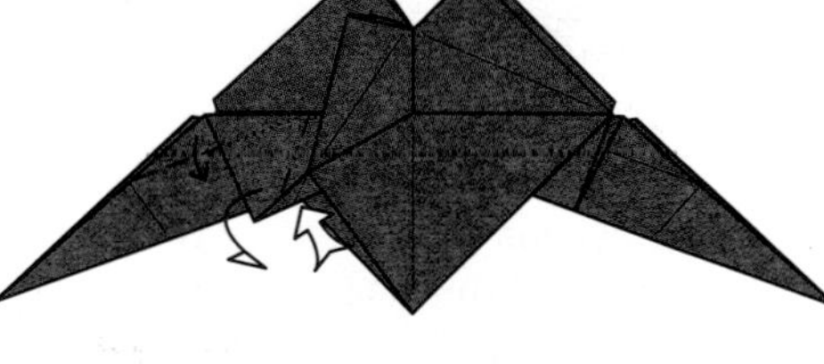
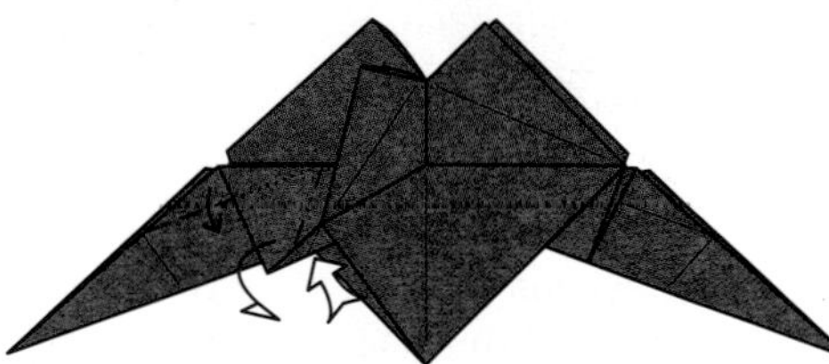

28. Valley-fold the flap on the existing crease; simultaneously flatten the interior pleat and mountain-fold its edge underneath. The model will not lie flat.

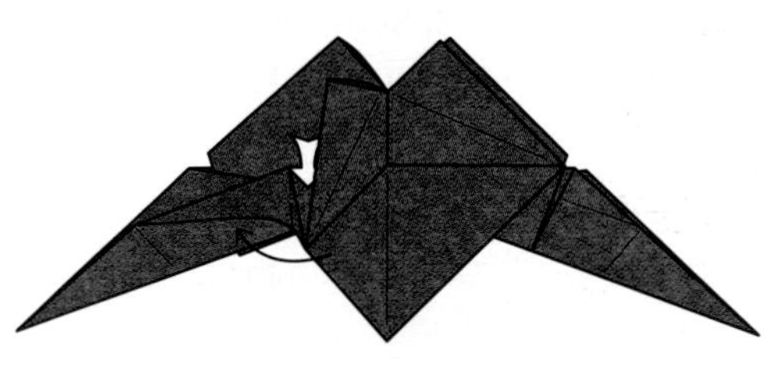

29. Push·down on the triangle and close the model back up.

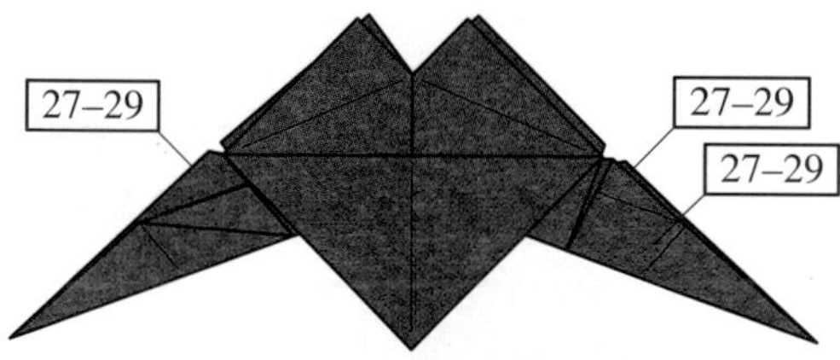

30. Repeat steps 27–29 behind and on the right.

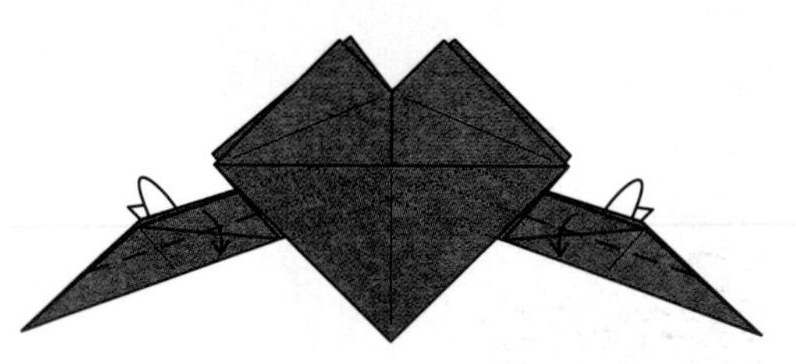

31. Valley-fold the near edge of each flap. Repeat behind.

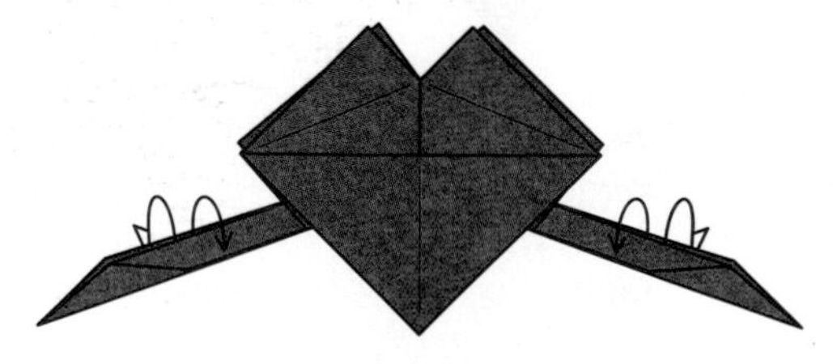

32. Wrap a raw edge from the inside of each flap to the front; repeat behind.

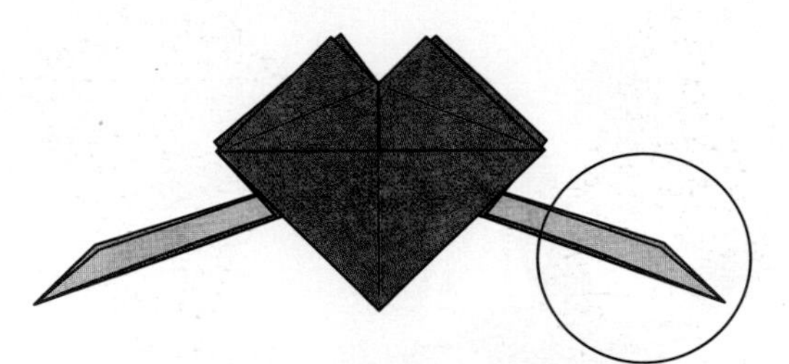

33. Steps 34–41 focus on the tips of the white flaps.

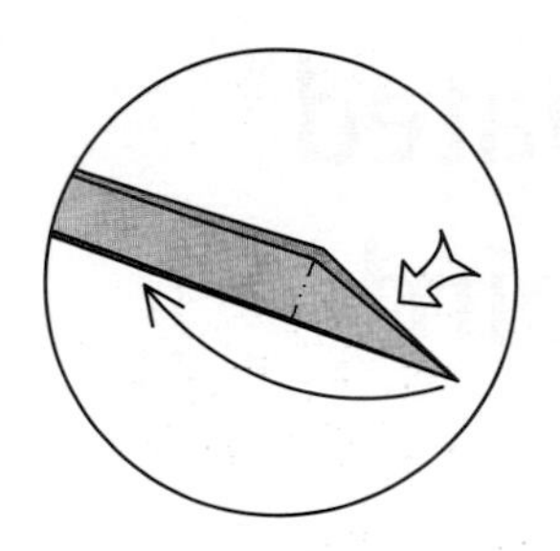

34. Double-reverse-fold the tip inside.

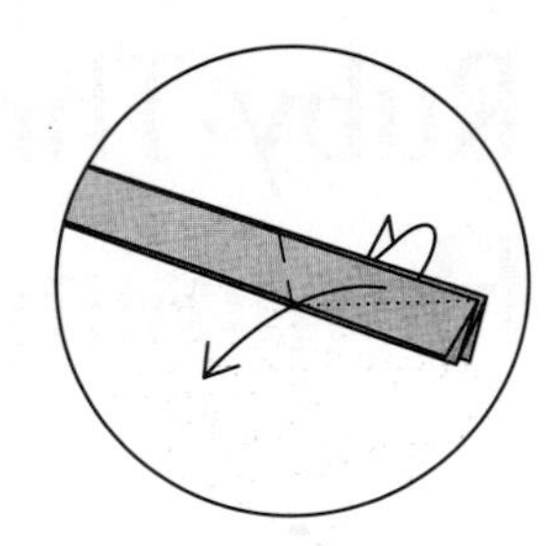

35. Outside reverse-fold the flap.

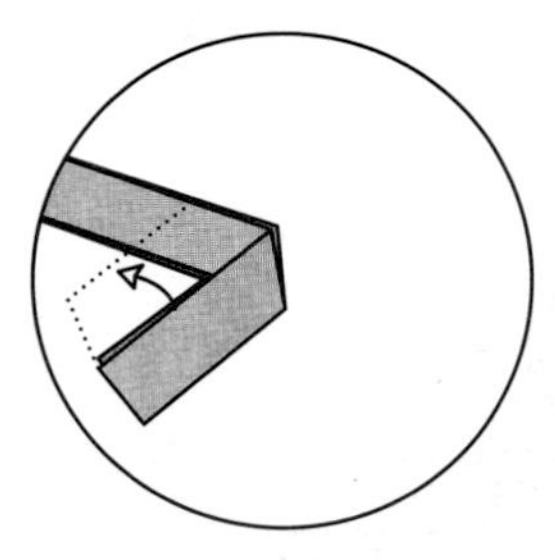

36. Pull out some loose paper.

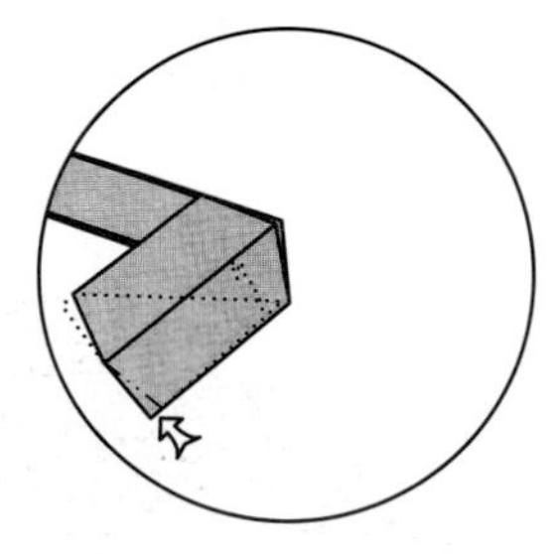

37. Shift the interior point so that its tip lies along the center line.

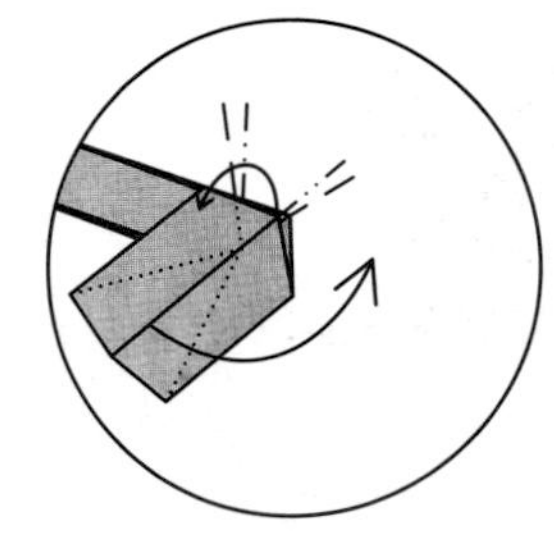

38. Crimp and rotate the end of the flap so that it aligns with the shaft emanating to the left.

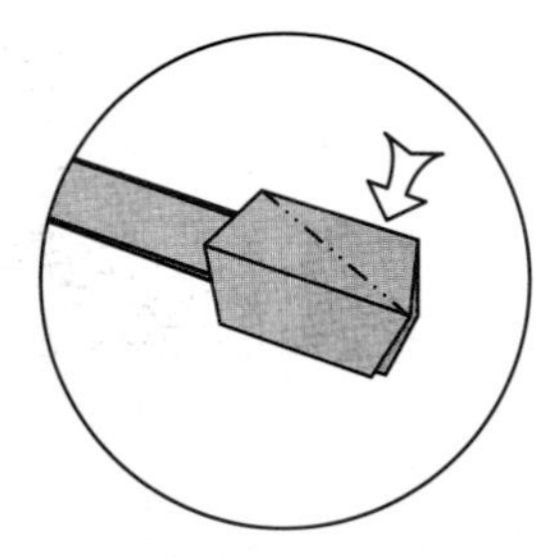

39. Closed-sink the corner.

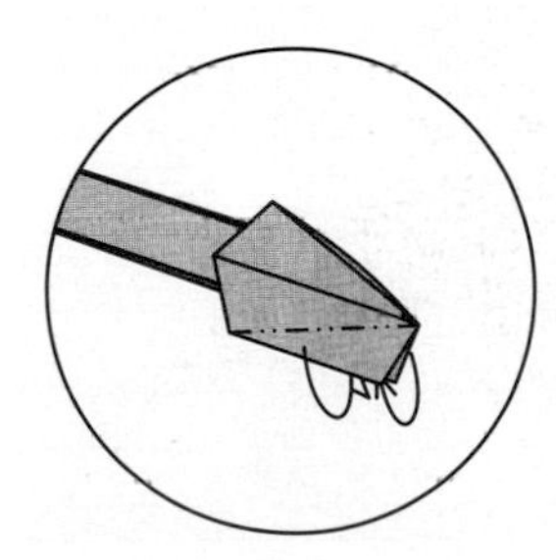

40. Mountain-fold the bottom edge underneath. Repeat behind.

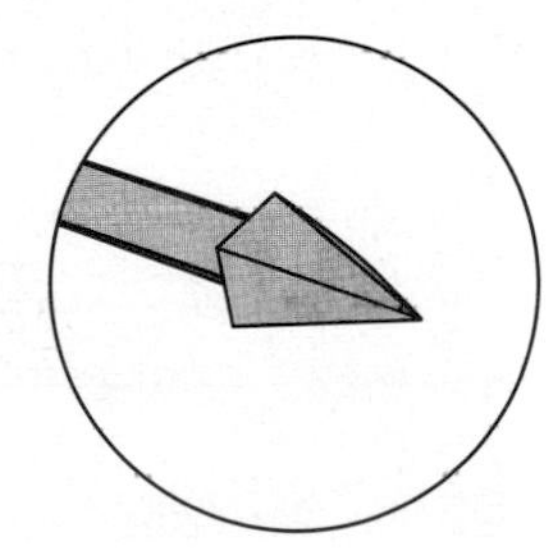

41. Like this.

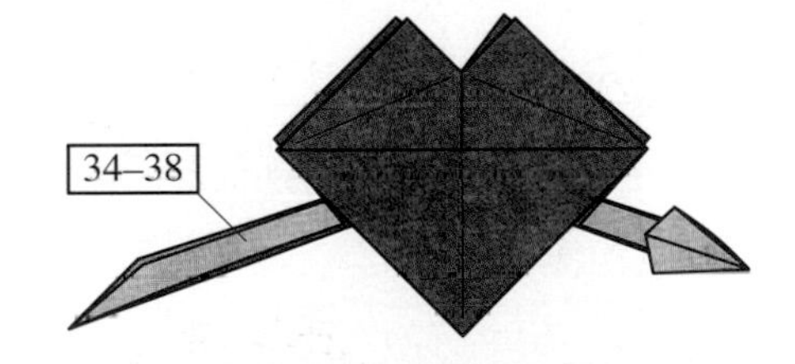

42. Repeat steps 34–38 (stop at 38!) on the right.

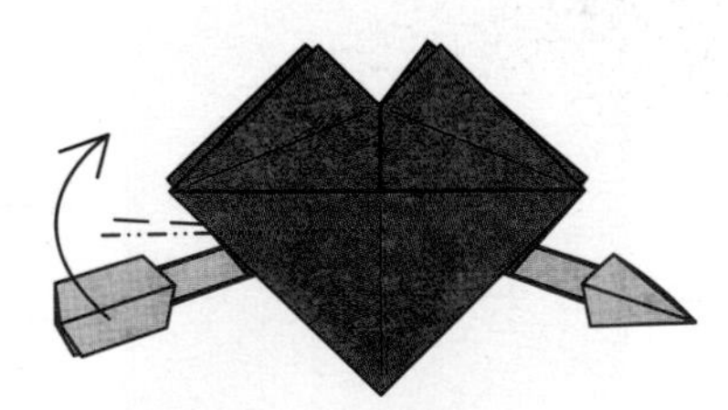

43. Crimp the left side upward so that the two halves of the white shaft line up.

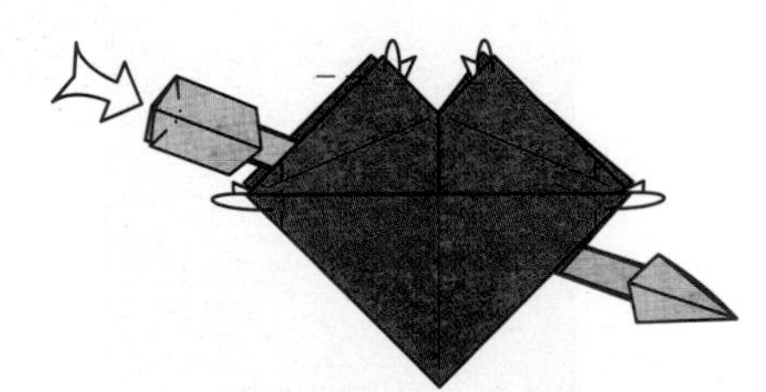

44. Round the heart with mountain folds. Dent the tail of the arrow to make it slightly three-dimensional.

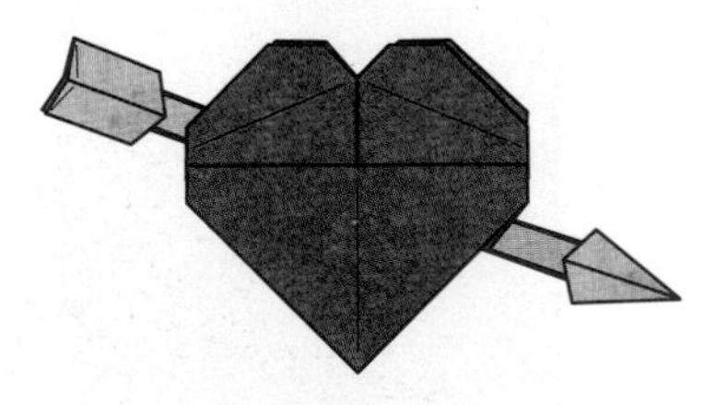

45. Finished Valentine.

Ruby-Throated Hummingbird

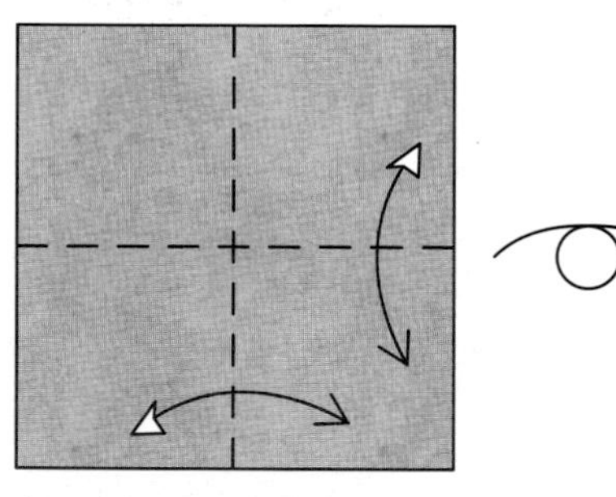

1. Begin with a square, white side up. Fold the paper in half in both directions and unfold. Then turn the paper over.

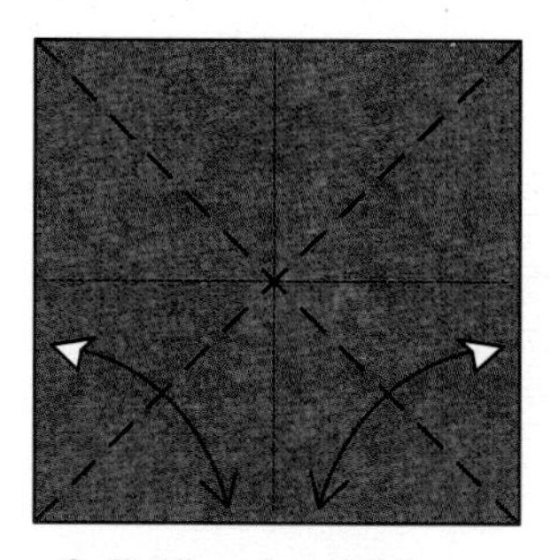

2. Fold and unfold along the diagonals.

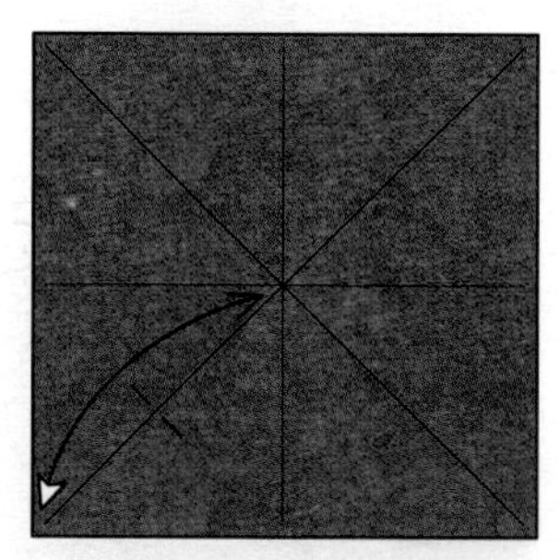

3. Fold the bottom left corner to the center and unfold, making a small pinch along the diagonal.

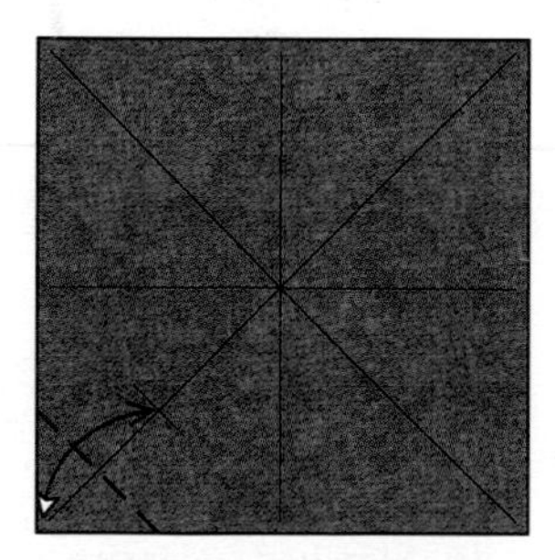

4. Fold the bottom left corner to the point you just made and unfold.

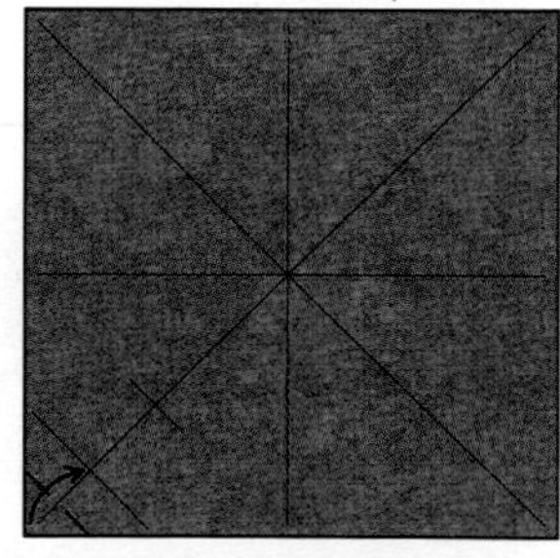

5. Fold the corner up along the diagonal; the crease hits the mark you just made.

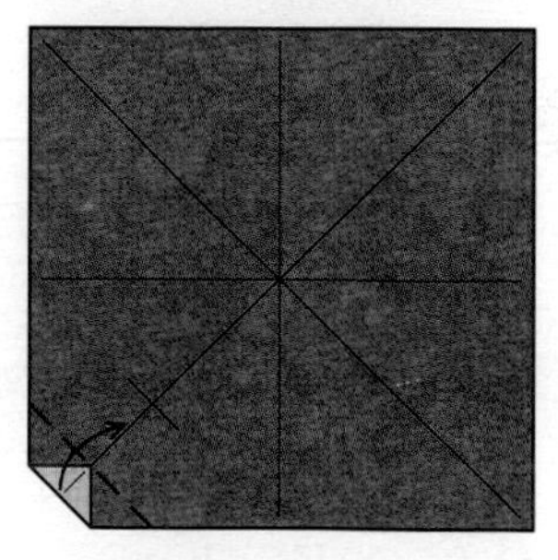

6. Fold the corner up again on the crease you made in step 4.

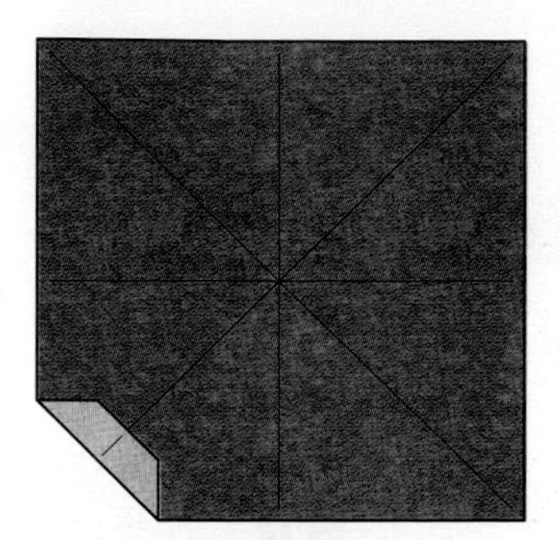

7. Turn the paper over from top to bottom.

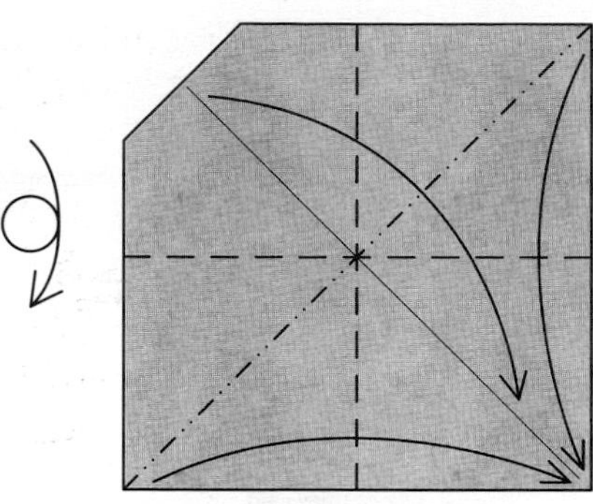

8. Fold a Preliminary Fold. Rotate the paper 1/8 turn clockwise.

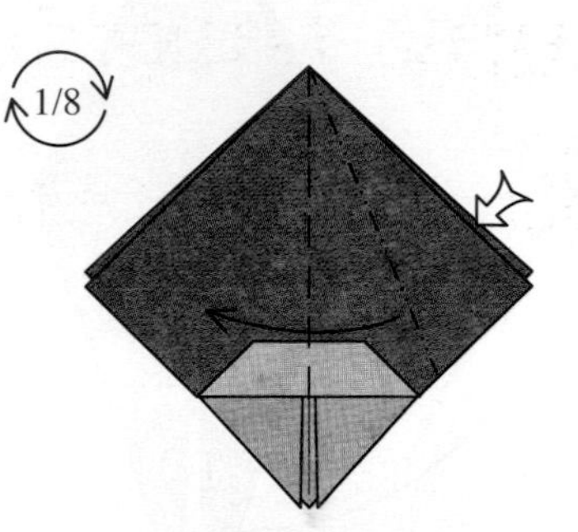

9. Squash-fold the near right flap.

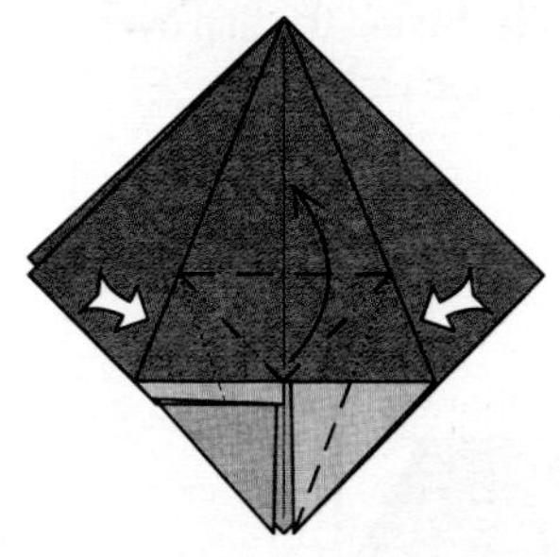

10. Petal-fold the edge.

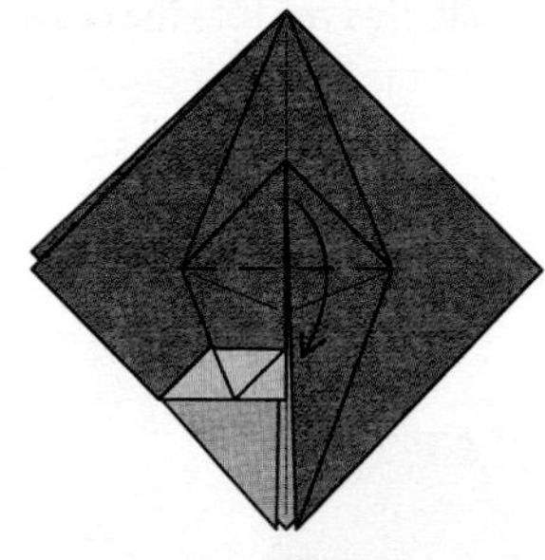

11. Fold the flap down.

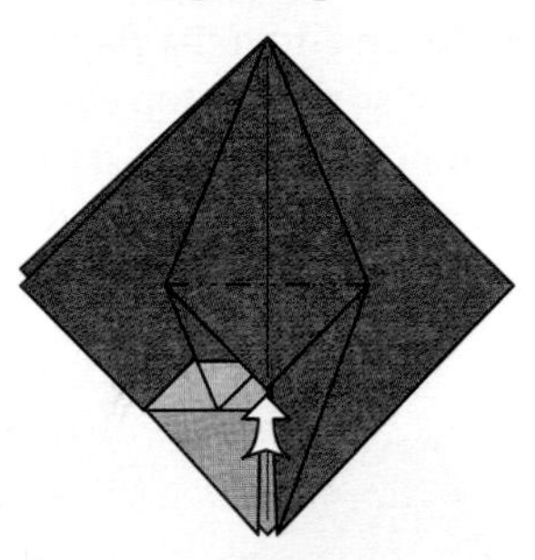

12. Sink the point up inside the model.

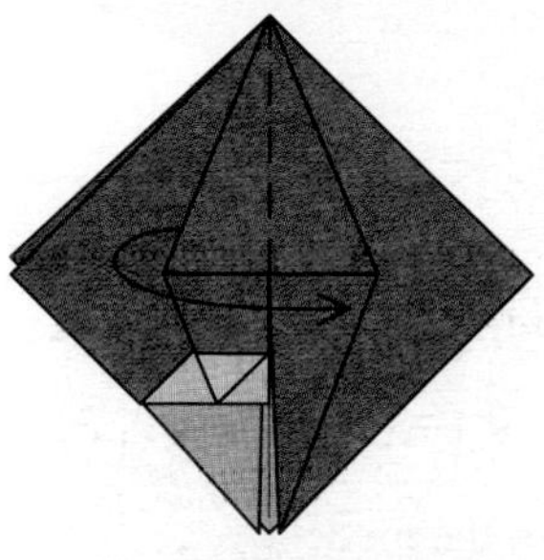

13. Fold one layer to the right.

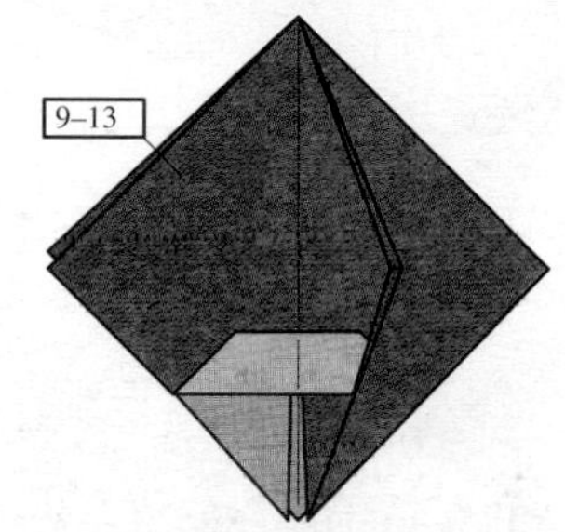

14. Repeat steps 9–13 on the left.

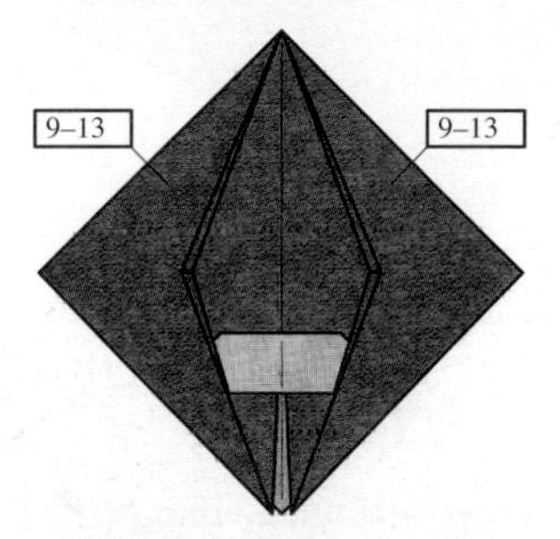

15. Repeat steps 9–13 on both sides behind.

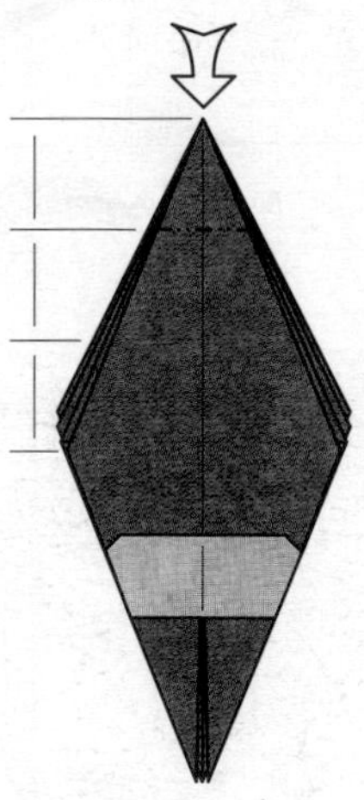

16. Sink 1/3 of the top point.

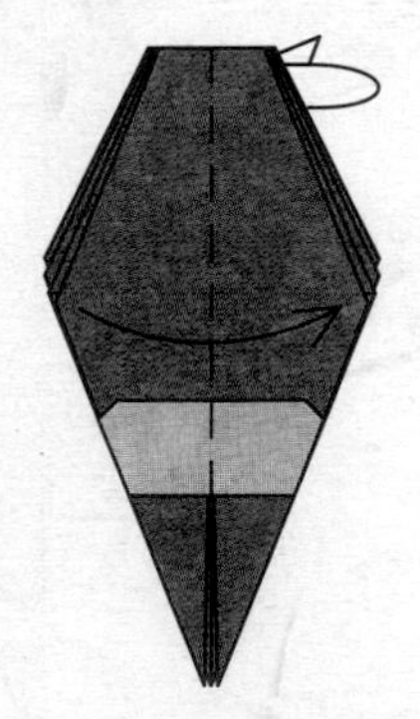

17. Fold one layer to the right in front and one to the left in back.

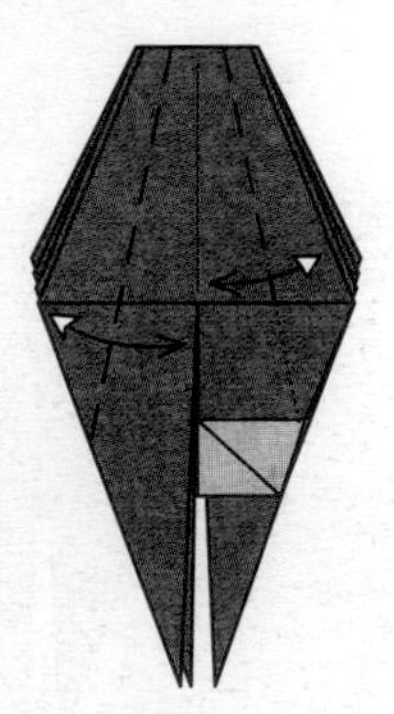

18. Fold the near layer of the left edge to the center line and unfold. Do the same on the right, but don't make the crease sharp below the horizontal edge.

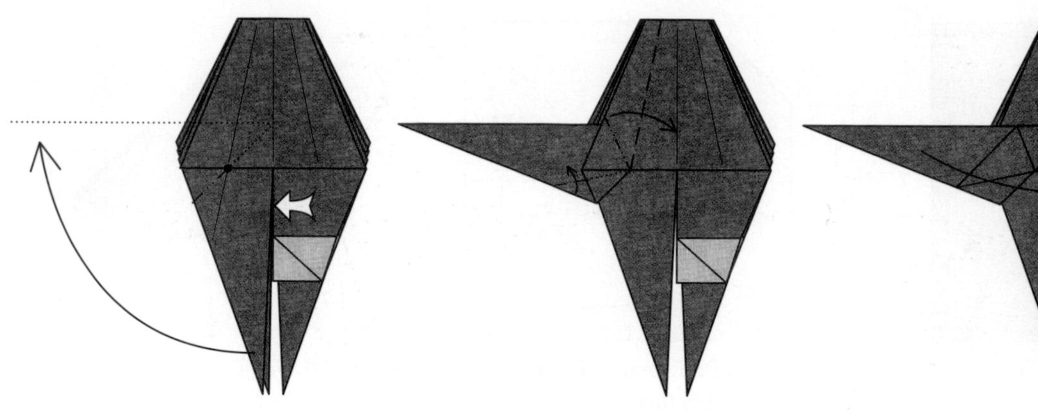

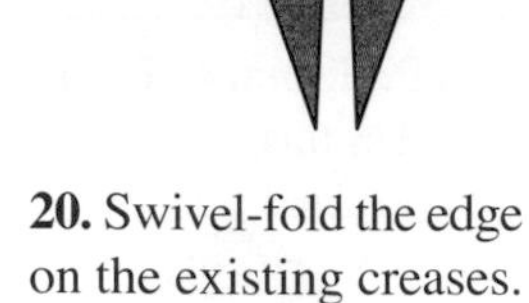

19. Reverse-fold the flap out to the side so that its top edge is horizontal. Note where the reverse fold goes under the folded edge.

20. Swivel-fold the edge on the existing creases.

21. Swing the flap over to the right.

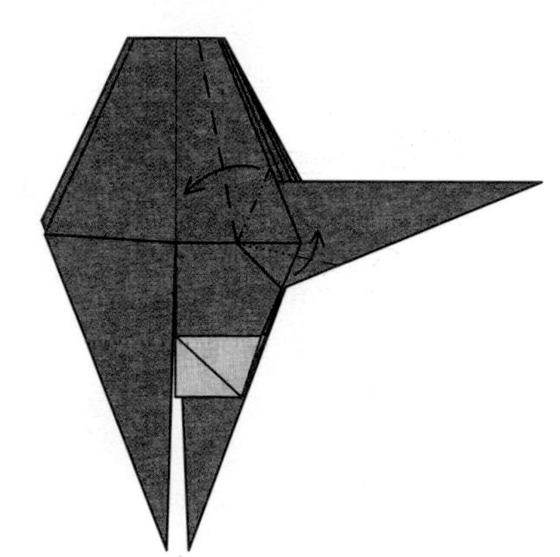

22. Swivel-fold the flap to match its other side.

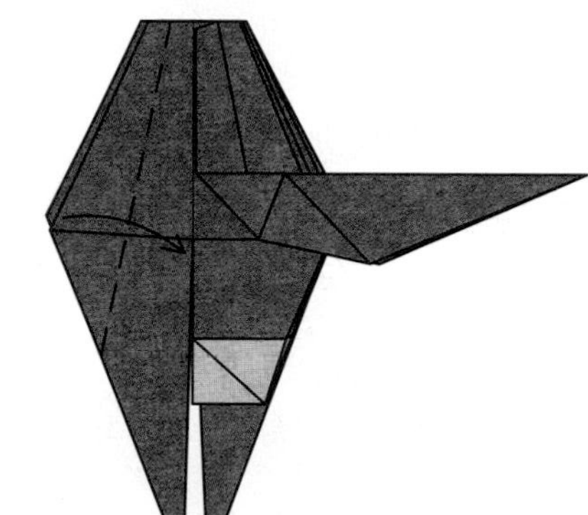

23. Valley-fold the edge to the center line.

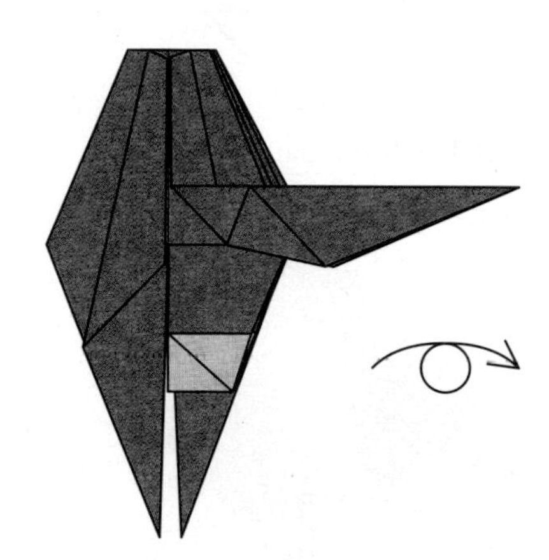

24. Turn the model over.

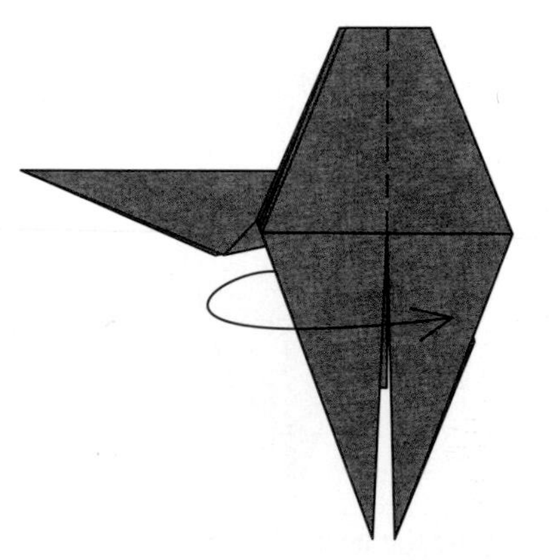

25. Fold a pair of layers over to the right.

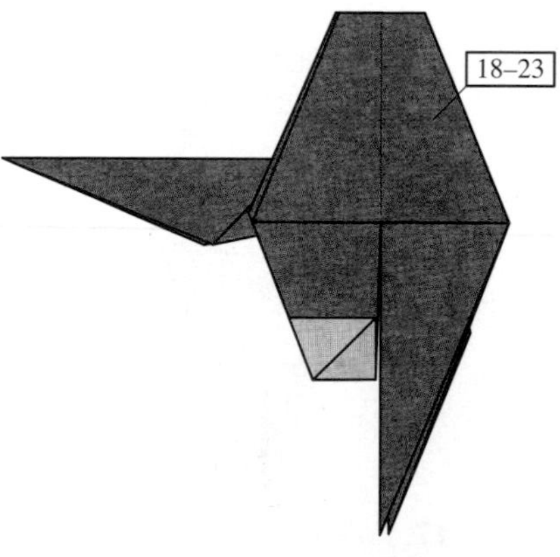

26. Repeat steps 18–23 on this side.

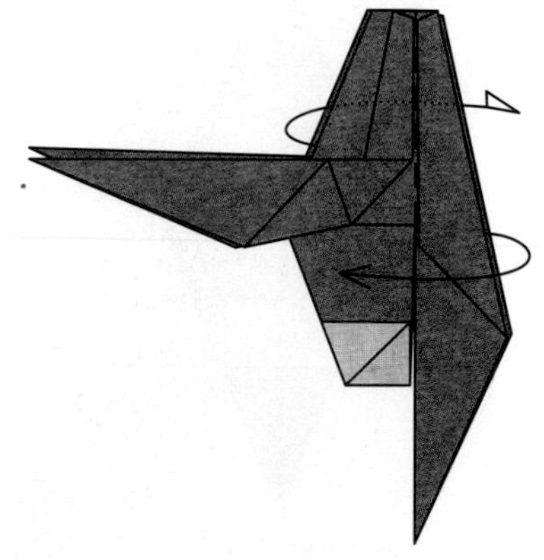

27. Fold one layer to the left in front and two layers in back to make the model symmetric.

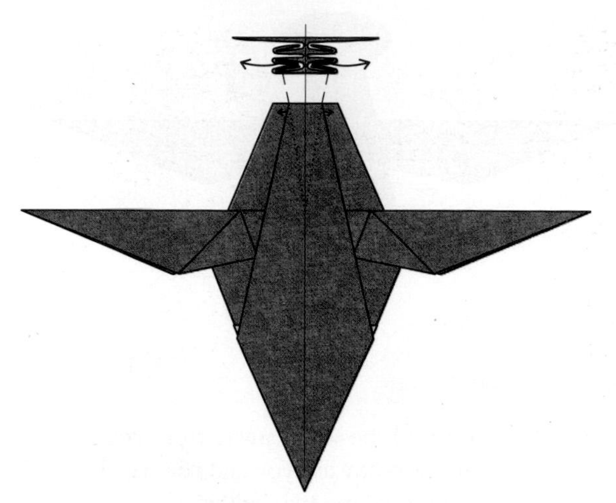

28. Fold one layer out to the sides. The edge view at the top shows which layers come out.

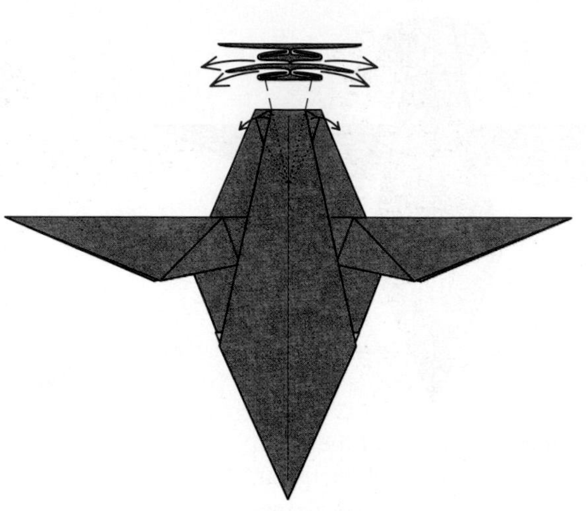

29. Pull the layers out farther to the sides.

30. Turn the model over from side to side.

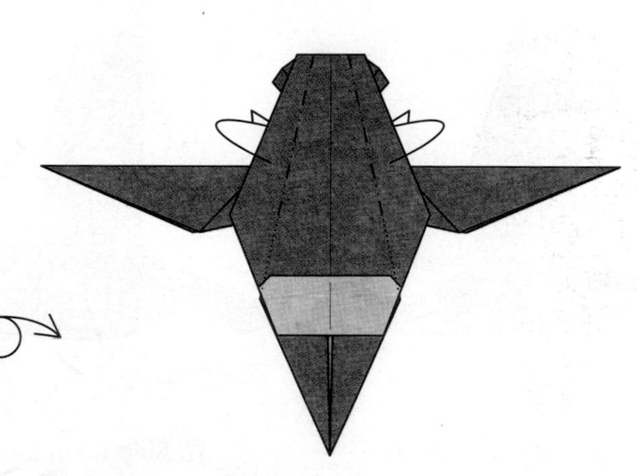

31. Mountain-fold the edges underneath, but only make the creases sharp on the upper part of the flap; the lower part should be rounded and three-dimensional.

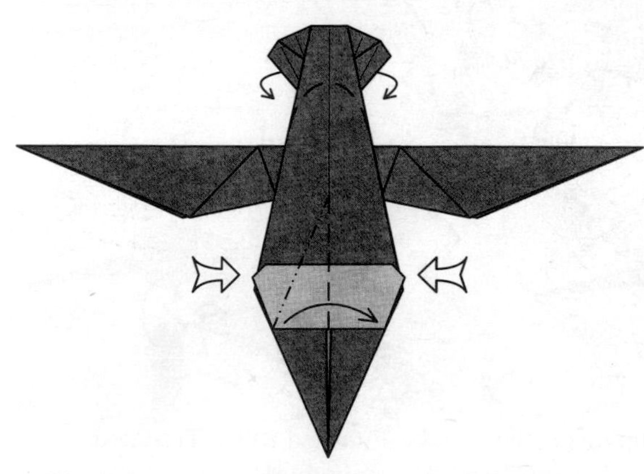

32. Form a pleat in the flap so that the upper part bulges upward; make a smooth valley crease at the base of the tail, rounding it slightly.

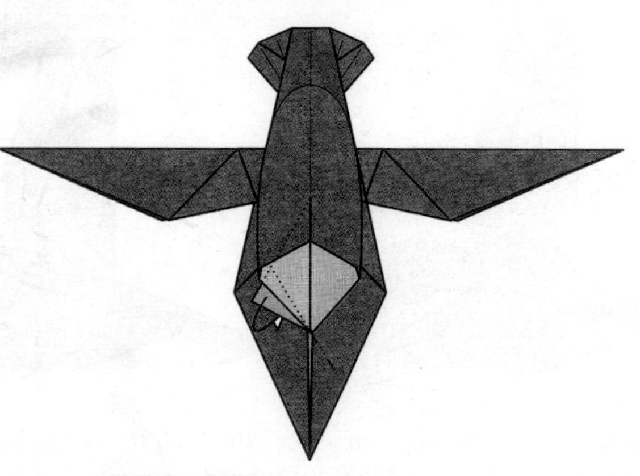

33. Mountain-fold the flap underneath to lock the pleat.

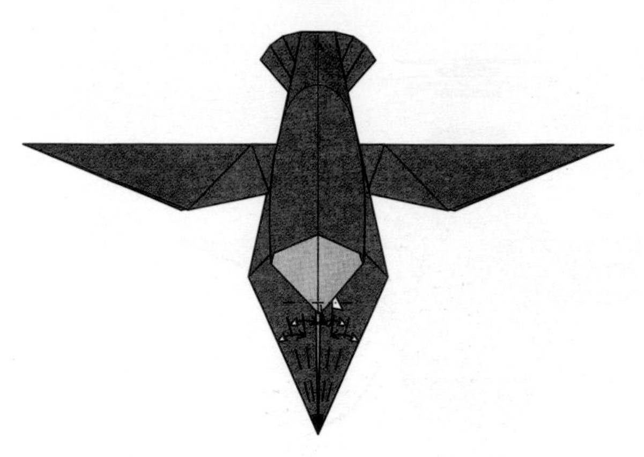

34. Mountain-fold a bit of the tip underneath to further lock the pleat. Crease the bottom point into sixths.

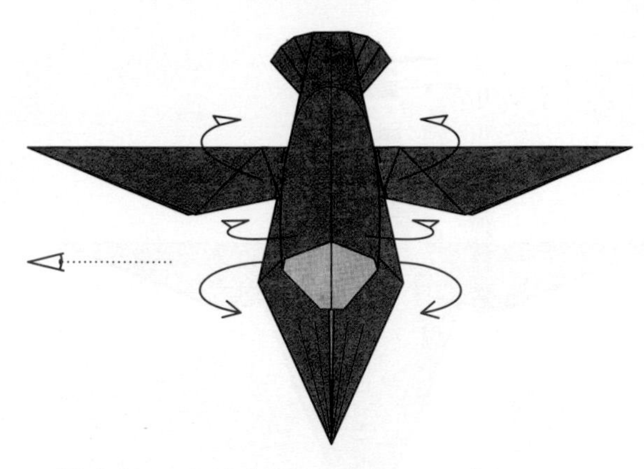

35. Mountain-fold the wings away from you. Curve the far layers toward you and near layers away from you, forming two halves of a cylinder. The next view is from the side.

36. Sink the belly to round it. Crimp the head downward and crimp the point in and out to form a beak.

37. Sink 1/3 of the beak, using the existing creases.

38. Round the leading edge of the wing. Reverse-fold the wing tips. Reverse-fold the corner of the head near the beak. Repeat behind.

39. Mountain-fold the edges of the beak underneath. Round the wing tips further with tiny reverse folds.

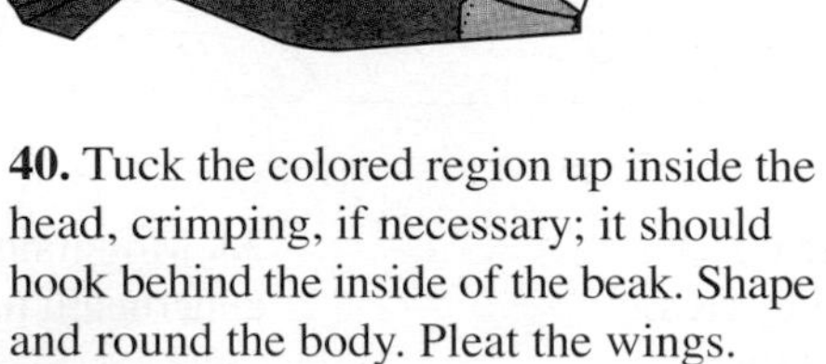

40. Tuck the colored region up inside the head, crimping, if necessary; it should hook behind the inside of the beak. Shape and round the body. Pleat the wings.

41. Finished Ruby-Throated Hummingbird.

Baby

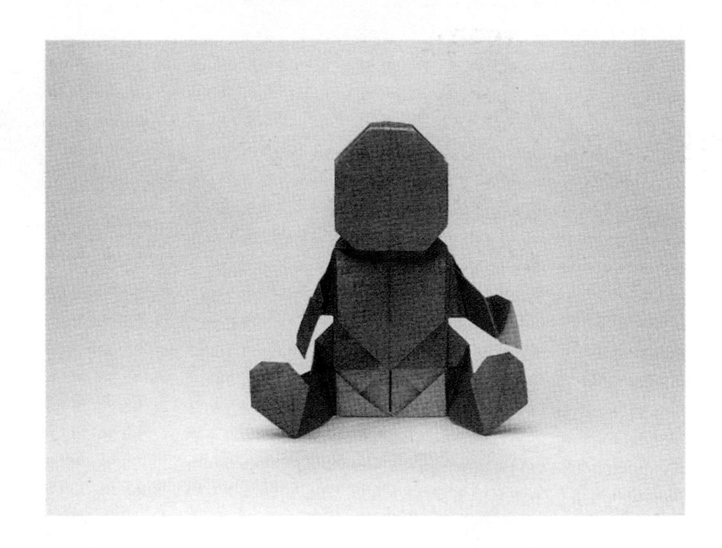

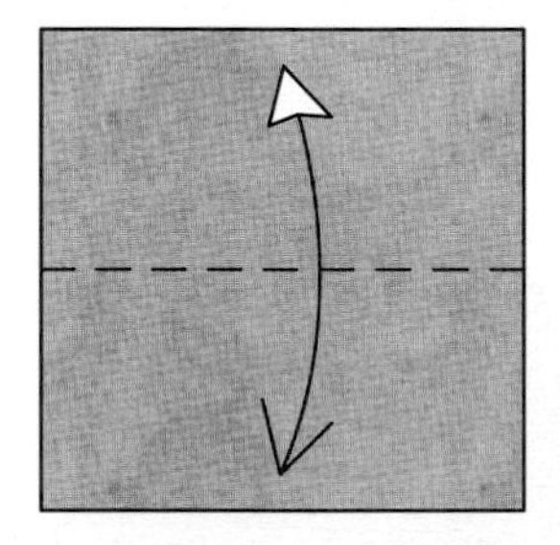

1. Fold the paper in half and unfold.

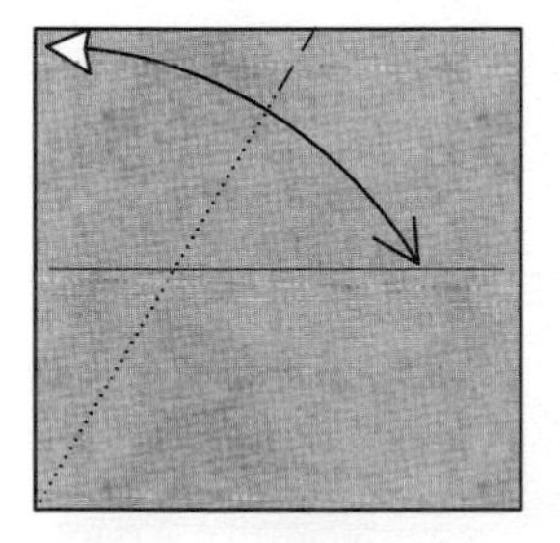

2. Fold the top left corner down to touch the horizontal crease; make a pinch along the top edge and unfold.

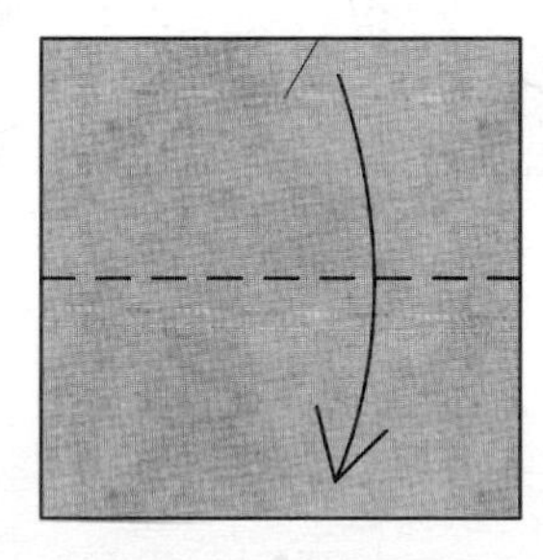

3. Fold the paper in half on the existing crease.

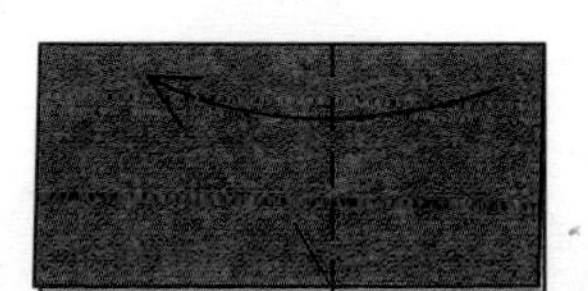

4. Fold the right side over to the left, making the crease hit the edge at the same place as the mark.

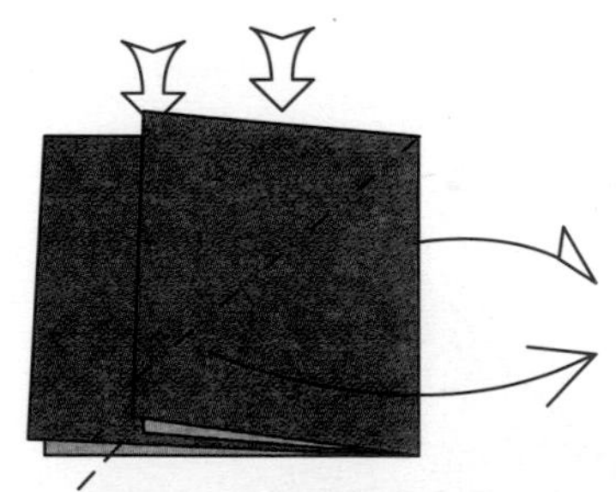

5. Squash-fold both flaps symmetrically.

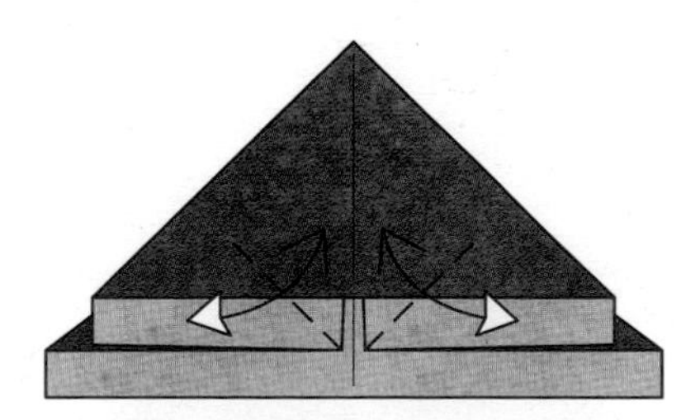

6. Crease the angle bisectors.

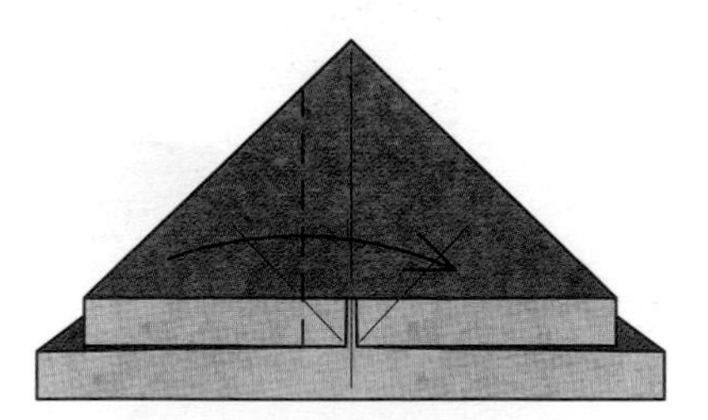

7. Fold the left flap over along a vertical crease that runs through the intersection of the horizontal edge with the crease you just made.

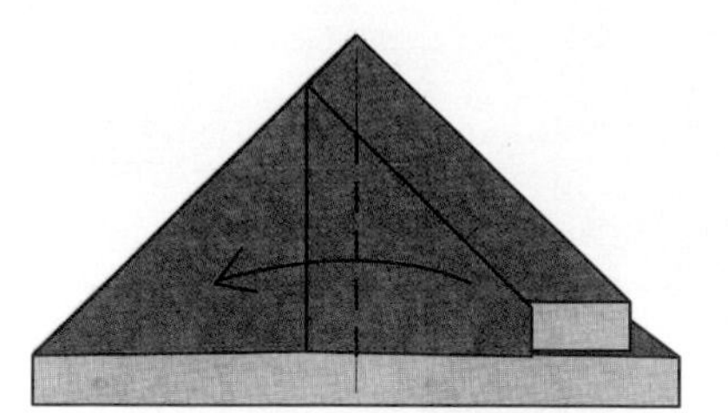

8. Fold it back to the left along a vertical crease that lines up with the center line of the model.

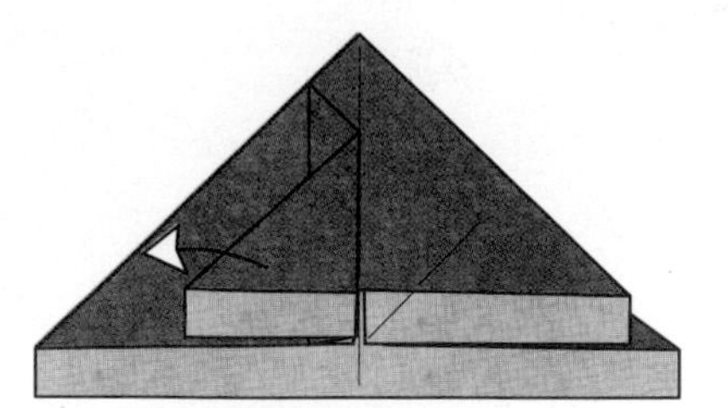

9. Unfold to step 6.

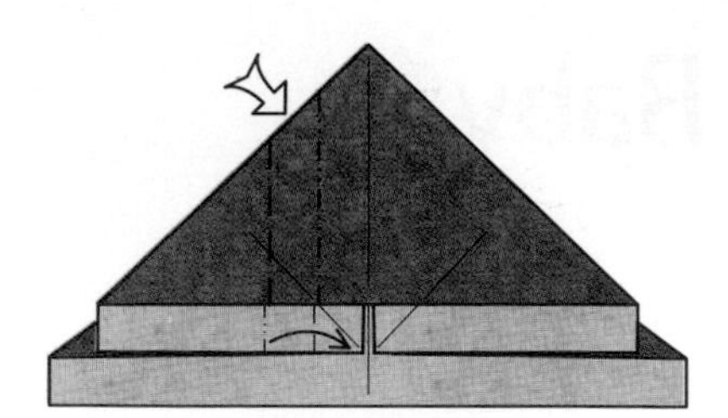

10. Reverse-fold the flap in and back out, using the creases you made in steps 6–8.

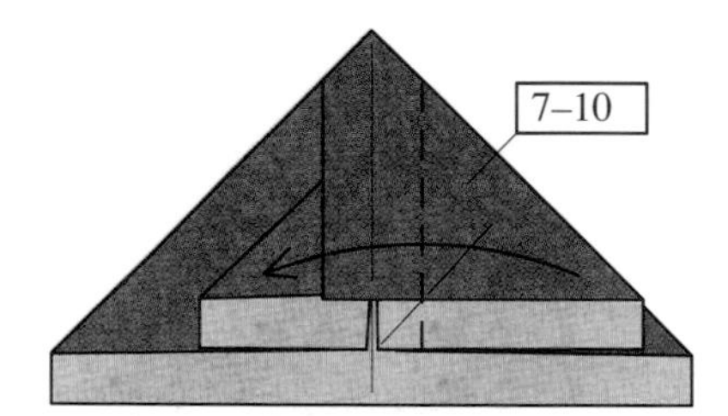

11. Repeat steps 7–10 on the right.

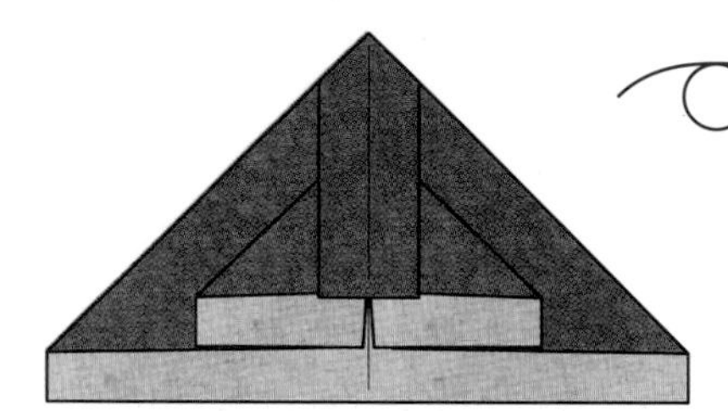

12. Turn the paper over.

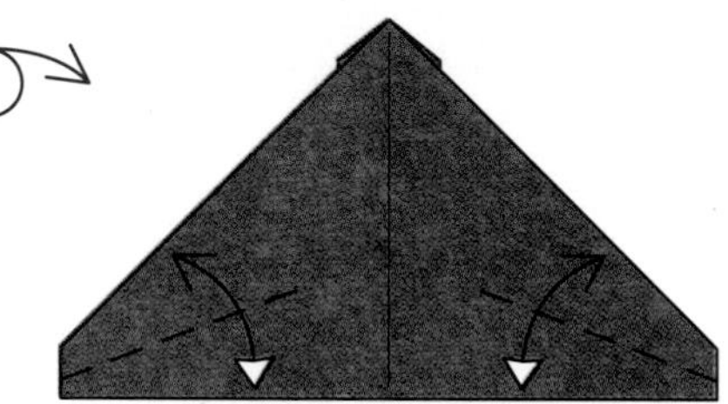

13. Fold the bottom edge up to the left diagonal, crease only as far as shown, and unfold. Repeat on the right.

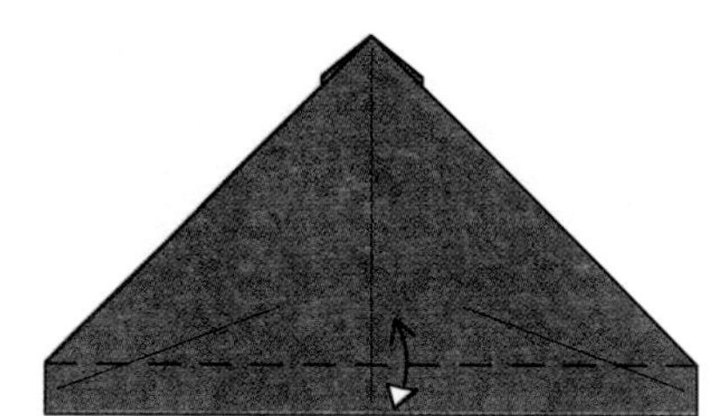

14. Fold the bottom edge up and unfold.

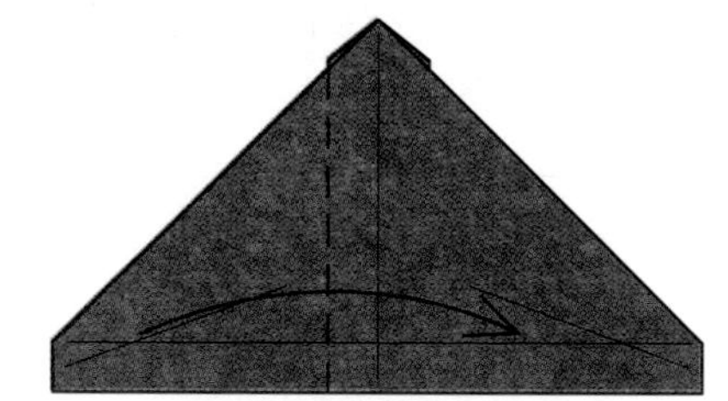

15. Fold the left flap over along a vertical crease that lines up with the edge behind it.

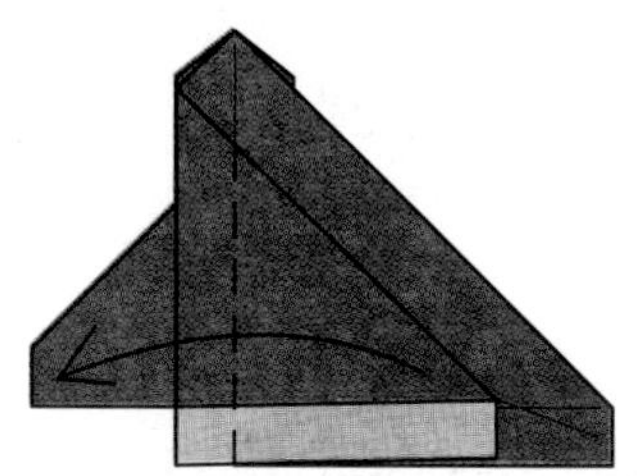

16. Fold it back to the left along a vertical crease that lines up with the center line. The flap also lines up with the flap behind it.

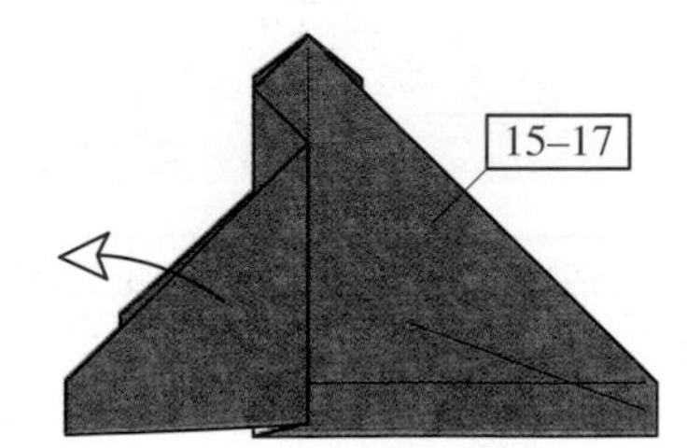

17. Unfold to step 15, and repeat steps 15–17 on the right.

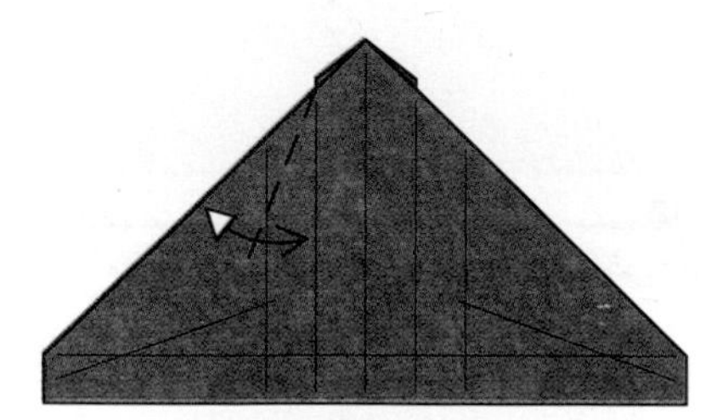

18. Fold and unfold along an angle bisector.

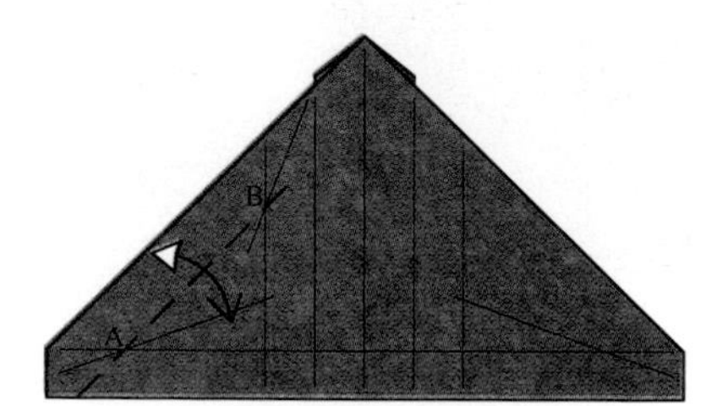

19. Make a crease that connects points A and B.

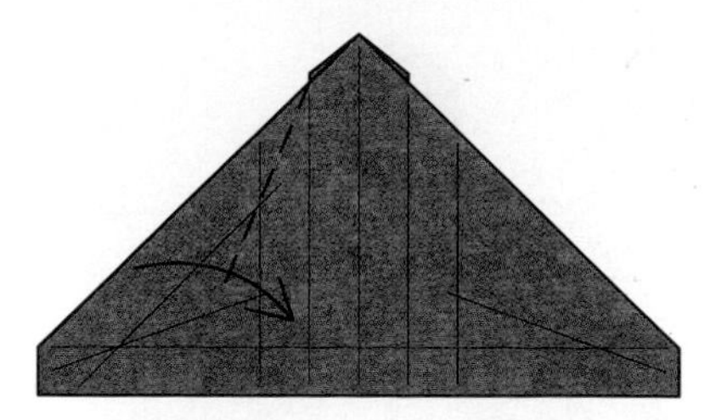

20. Begin to fold the left edge over along the existing crease, but don't make the crease sharp at the bottom.

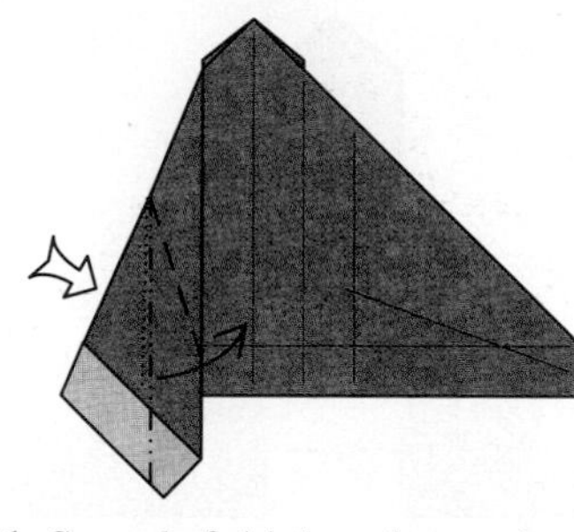

21. Squash-fold the edge to the right. The mountain fold and the hidden valley fold (indicated by an X-ray line) are on existing creases; the visible valley fold forms when you flatten the model.

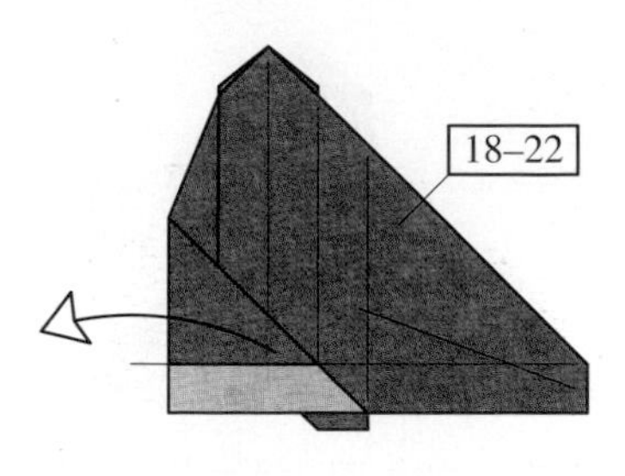

22. Unfold to step 18 and repeat steps 18–22 on the right.

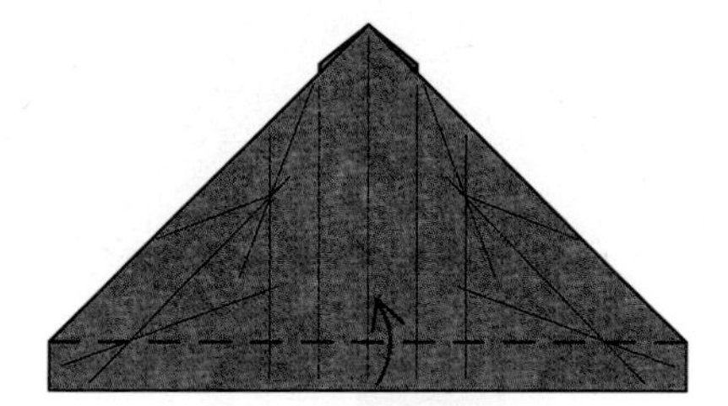

23. Fold the bottom edge up.

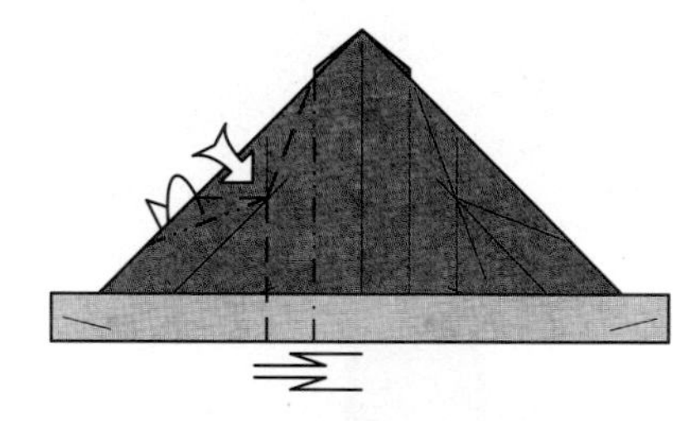

24. Crimp the flap using the existing creases and incorporating the small pleat along the diagonal.

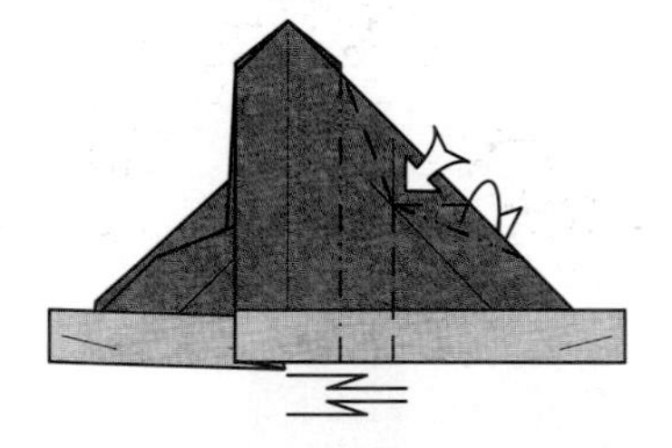

25. Repeat step 24 on the right.

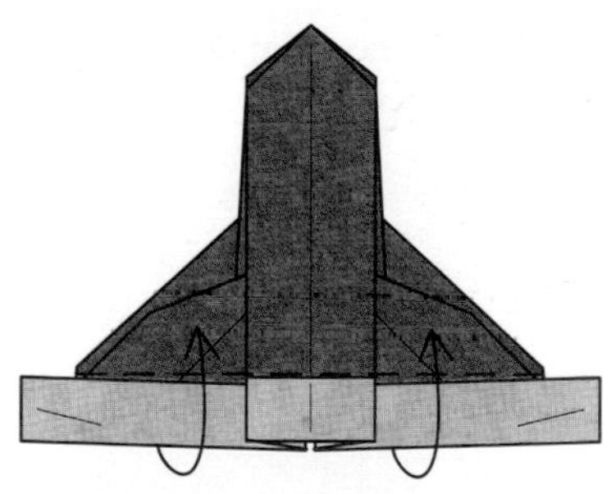

26. Fold a single layer of the bottom edge up, but don't flatten the paper.

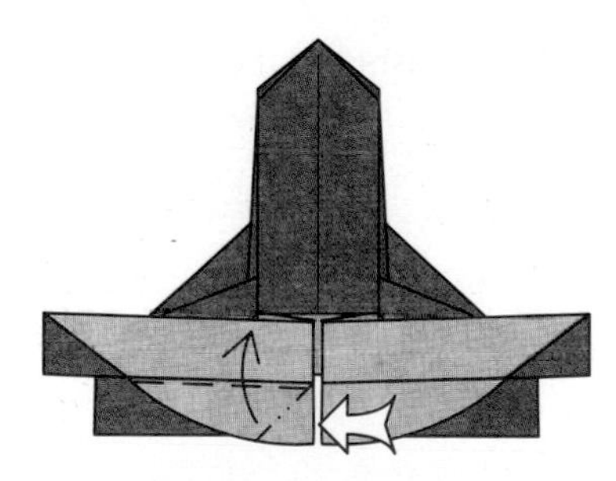

27. Squash-fold the exposed corner.

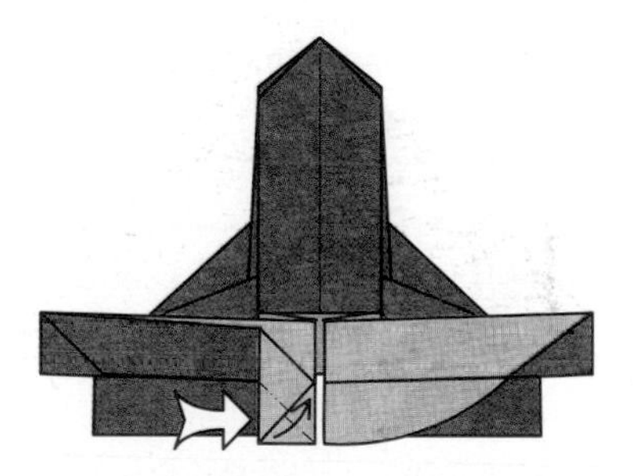

28. Reverse-fold the corner.

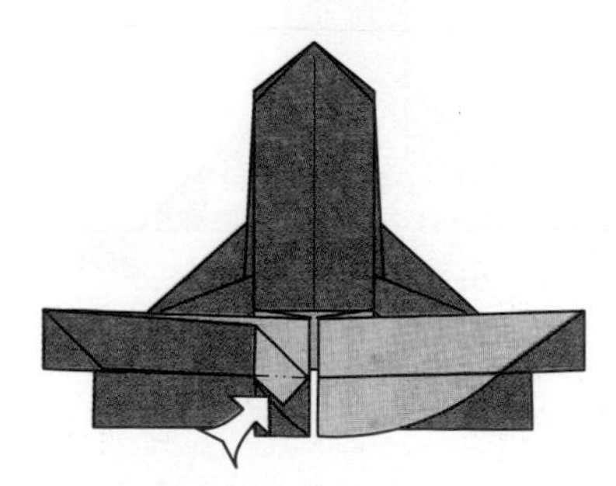

29. Reverse-fold the corner up inside the model.

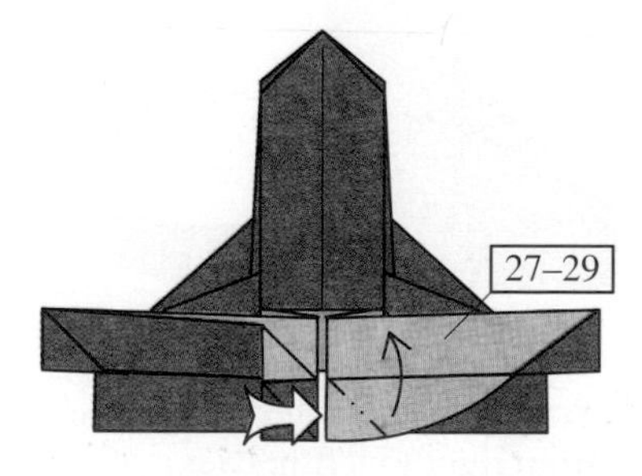

30. Repeat steps 27–29 on the right.

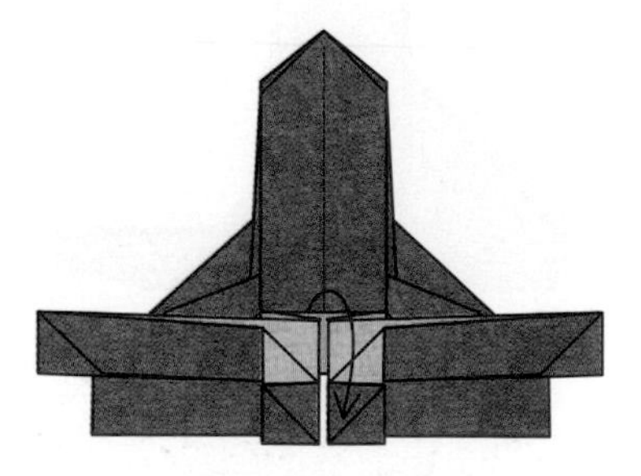

31. Fold the bottom edge back down.

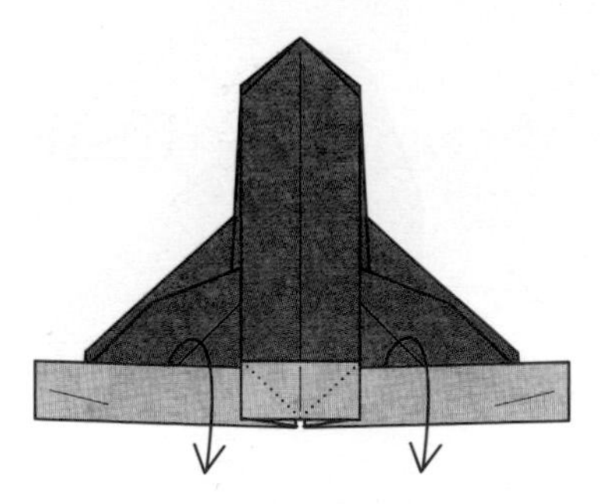

32. Pull the loose edge as far down as possible on both the left and right flaps.

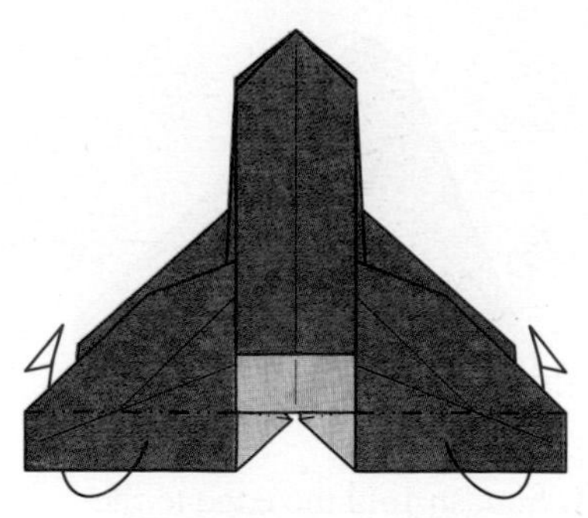

33. Mountain-fold the edges underneath.

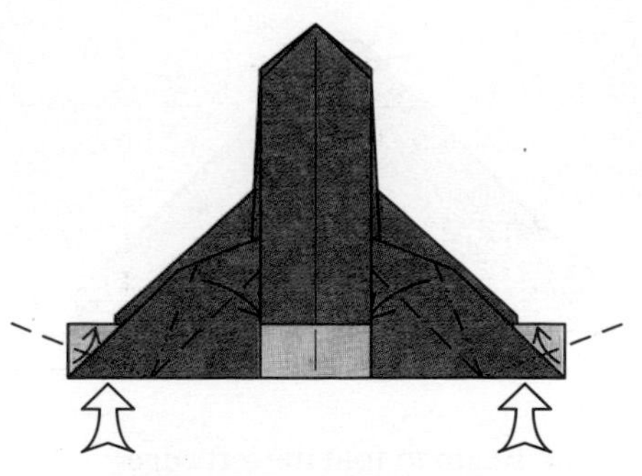

34. Squash-fold each corner, making the valley folds on existing creases.

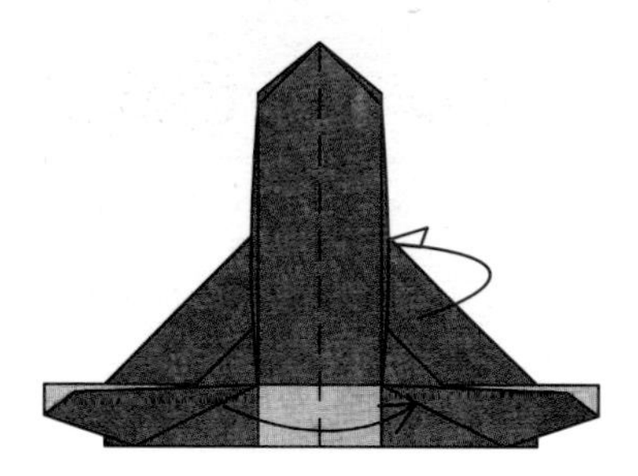

35. Fold one flap from left to right in front, and one from right to left behind.

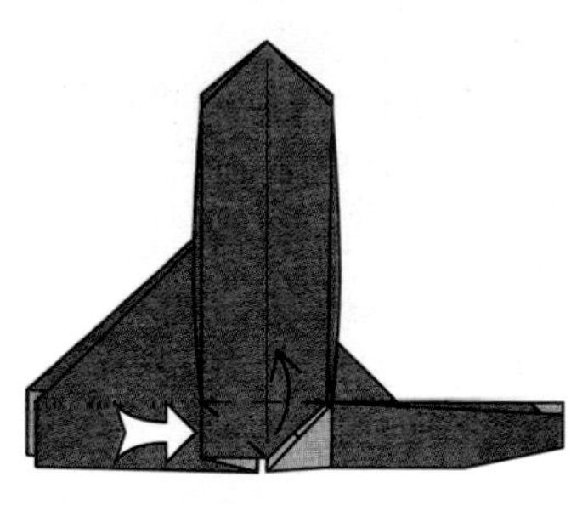

36. Squash-fold the corner. Repeat behind.

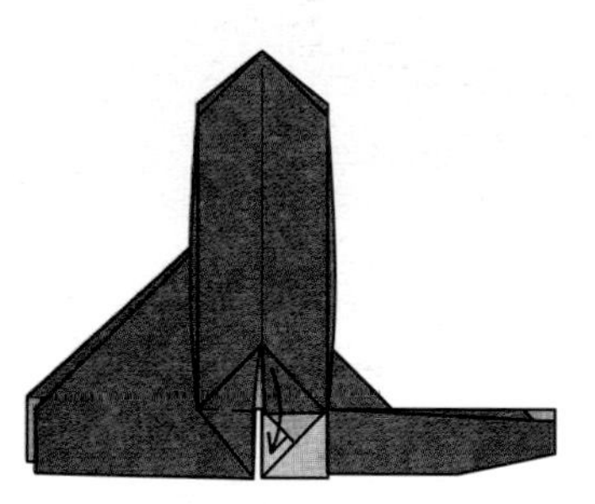

37. Fold the point down. Repeat behind.

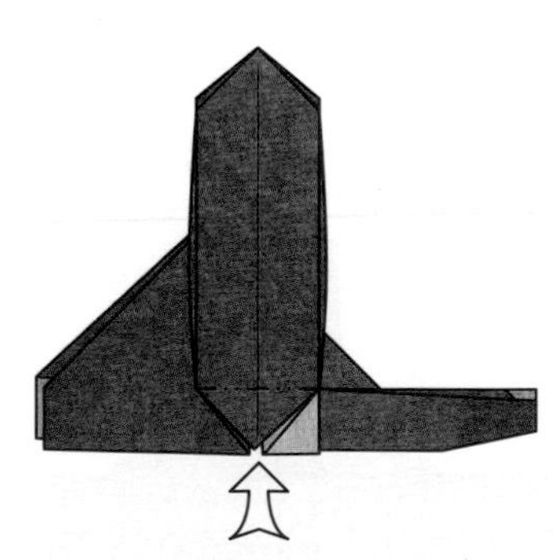

38. Sink the point up into the model. Repeat behind.

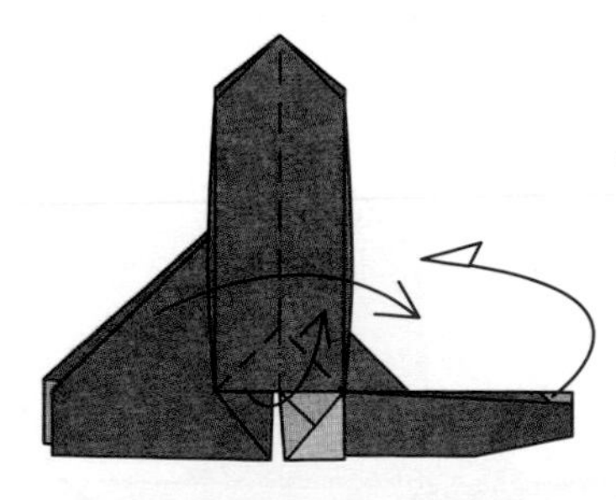

39. Swing one flap from left to right in front and one from right to left in back, incorporating the reverse fold shown on each side.

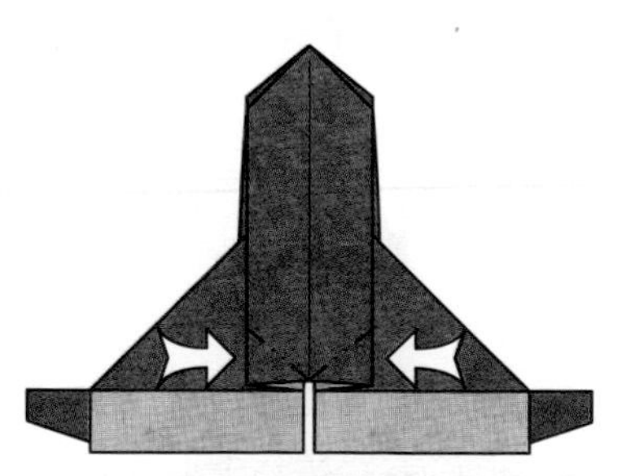

40. Reverse-fold the two corners.

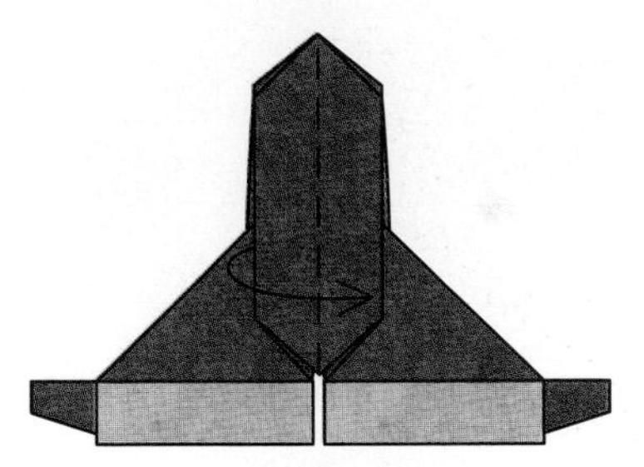

41. Fold one narrow layer over to the right; the top of the model will not lie flat.

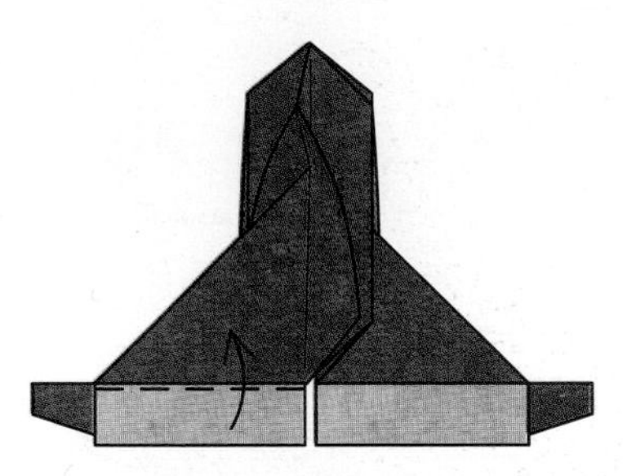

42. Fold the white layer up.

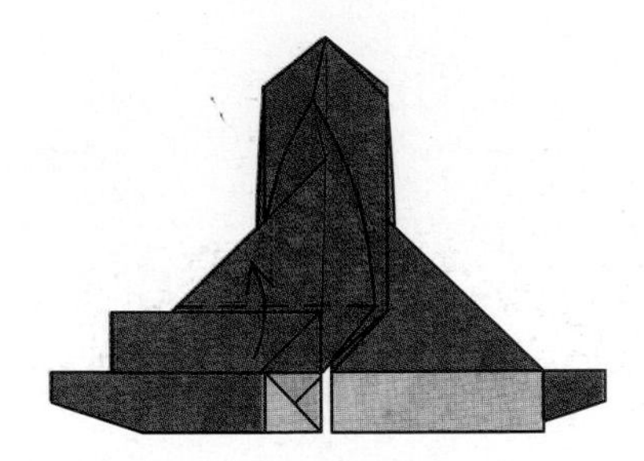

43. Fold up one more layer.

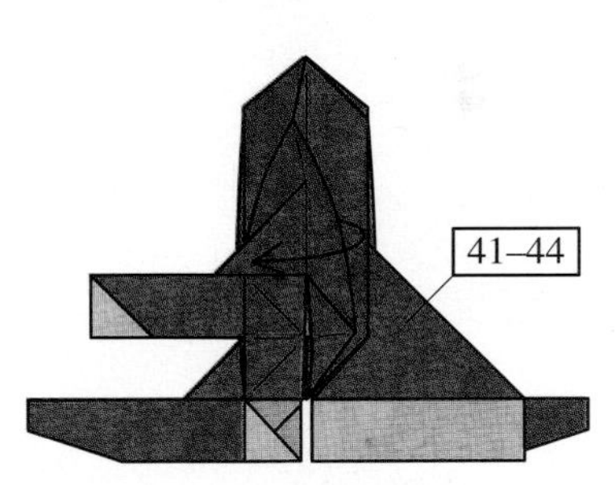

44. Close the left side up and repeat steps 41–44 on the right.

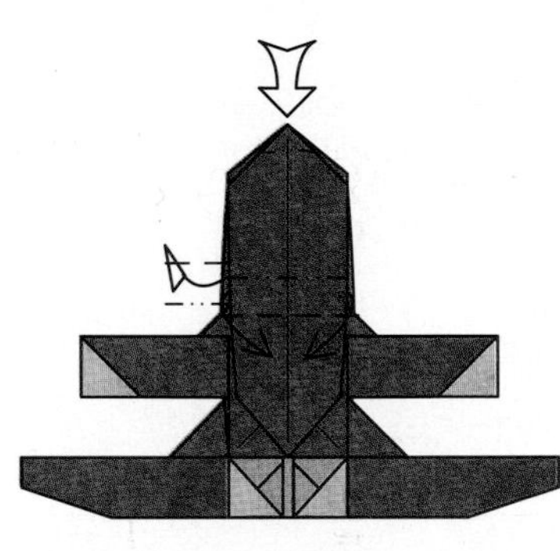

45. Crimp the head downward in front and, symmetrically, in back; simultaneously push in the top of the head to make it three-dimensional.

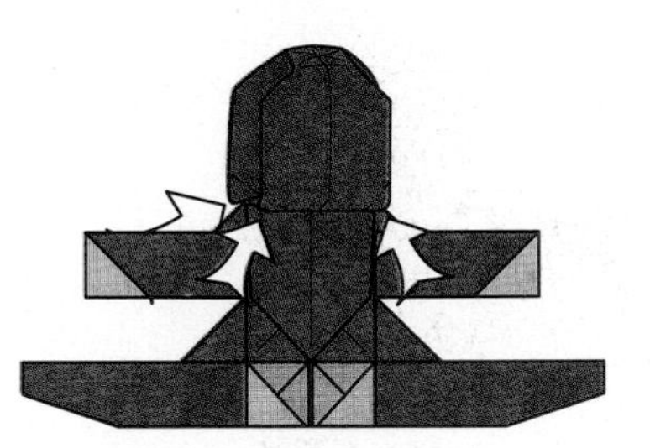

46. Reverse-fold the corners of the head to round it; repeat behind.

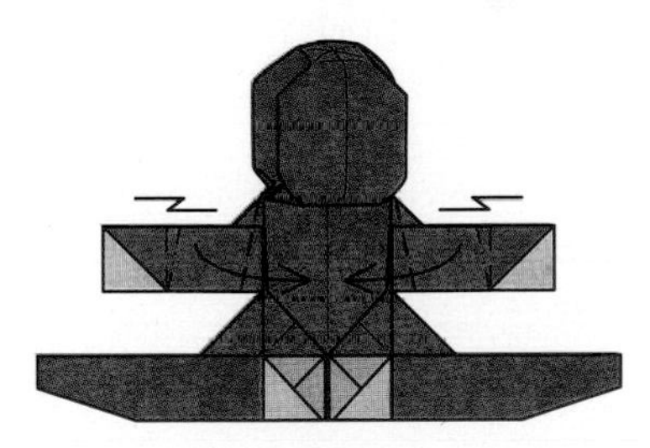

47. Pleat the hands and fold them toward the front.

48. Fold the legs forward and squash-fold the feet.

49. Shape the feet with mountain folds. Put your finger up inside the model and round the body slightly. Position the arms.

50. The finished Baby.

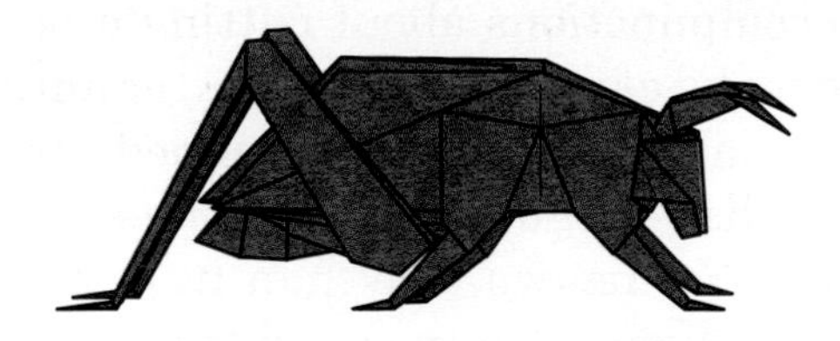

5 Splitting Points

The currency with which the origami designer most often deals is the point (or flap). Subjects can be classified by the number of significant points that they have. A snake has two (a head and a tail); a standing bird has four (six, if the wings are outstretched). A mammal has six (four legs, a head, and a tail), while a spider has eight. A lobster may have twelve; a centipede, one hundred. The number of points in the subject dictates the number of flaps needed in the base from which it is folded.

The number of points assigned to the subject depends not only on the subject, but also upon what the designer defines as a "significant" point. Quite often, minor features such as ears may be derived from small amounts of excess paper in the model and may be safely ignored in the initial stages of design. However, the larger the point is, the more important it is to include it in the ground stages of design.

Most people's first designs are modifications of another model. One's first structural design might very well be a modification of an existing base, either a Classic Base or perhaps a more recent base from the origami literature. A prime reason to modify a base is to obtain one or more extra flaps. Short of redesigning the base entirely, it is often possible to convert one flap into two, three, or more flaps by folding alone, a process I call *point-splitting*. The ability to split a point—without cutting—is a useful tactic to have in the designer's arsenal, and it also provides tangible evidence of the mutability of origami bases.

5.1. The Yoshizawa Split

Of course, there is always one way to split a point—cut it in two. Traditional Japanese designs quite often did split points in this way (the custom of one-piece, no-cut folding is a relatively modern restriction) and many of the designers of the 1950s and 1960s had no compunctions about cutting a point into two or more pieces to make ears, antlers, wings, or antennae. Isao Honda—who for many in the West defined 1960s origami design via his English-language publications—used cuts as a matter of routine. Yoshizawa, the man from whom Honda derived many of his designs, also on occasion used cuts, but even in the 1950s had developed what is to my knowledge the first technique of splitting a point into two by folding alone. This procedure is illustrated in Figure 5.1, on a Kite Base flap. (I encourage you to fold a Kite Base and try it out.)

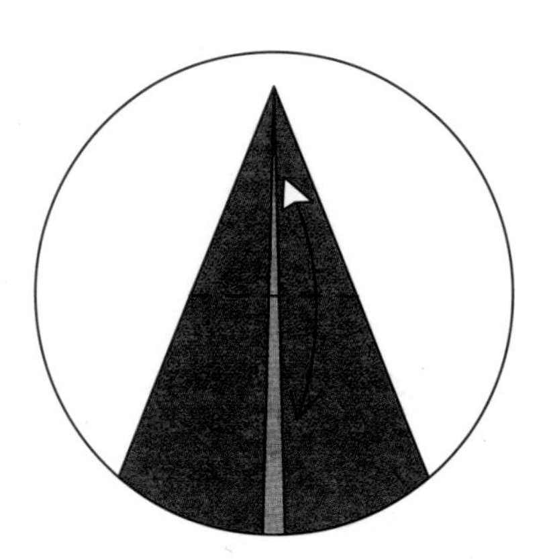

1. Fold the top of the flap down and unfold.

2. Sink the tip on the existing creases.

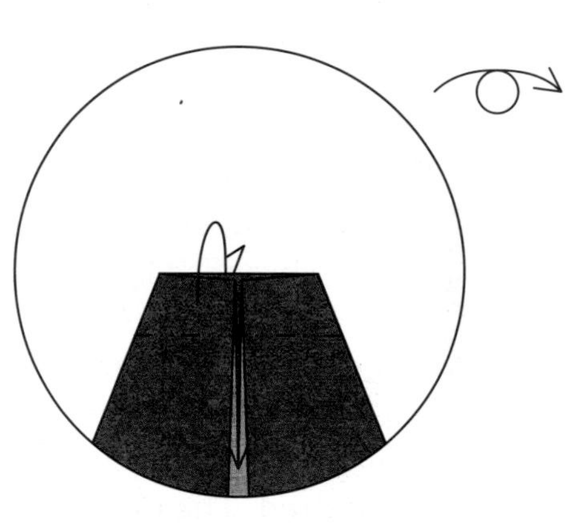

3. Mountain-fold a portion of the flap behind. The exact amount isn't critical. Then turn the paper over.

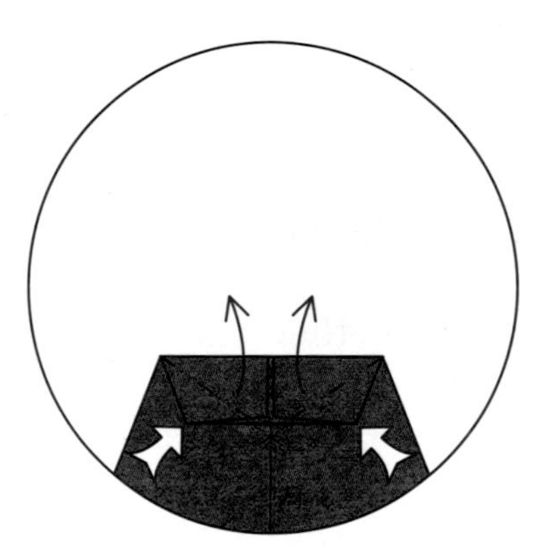

4. Fold the flaps up and spread-sink the corners.

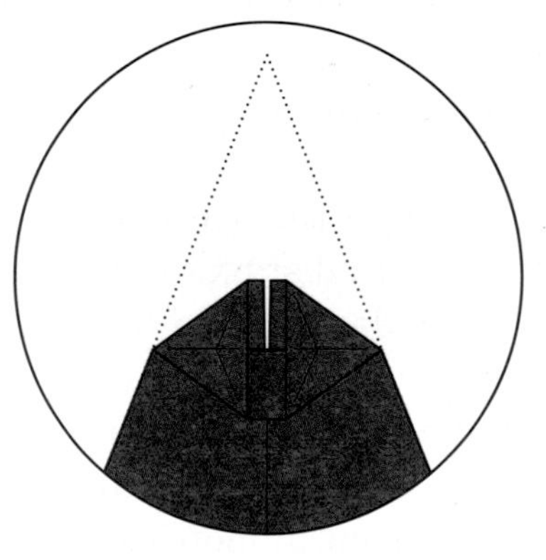

5. Finished split flap. The dashed line shows the outline of the original flap.

Figure 5.1.
Sequence for splitting a Kite Base flap into two smaller points.

Where once there was one flap, now there are two flaps, albeit considerably shorter than the one we started with (which is indicated by the dashed line). This maneuver is particularly nice for turning one long flap into a short pair of ears, which is precisely what Yoshizawa used it for.

Of course, the resulting points are smaller than the one we started with; nothing comes for free. Our folding sequence was not particularly directed at making them long, which leads to the question: How long could they get?

That question actually raises a more fundamental one: How does one *quantitatively* define the length of a flap? Intuition suggests that we define the length of a flap as the distance from the baseline to the farthest point of the flap, but this only pushes the question down a level: What would we mean by the baseline of the flap? Let us define the baseline of a flap as an imaginary line drawn across the flap, above which the flap can freely move. While this still permits some wiggle room in the definition, it allows us to define a baseline for the two flaps in the split point above as the obvious horizontal crease; in this case, we can clearly define the length of the two flaps as shown in Figure 5.2.

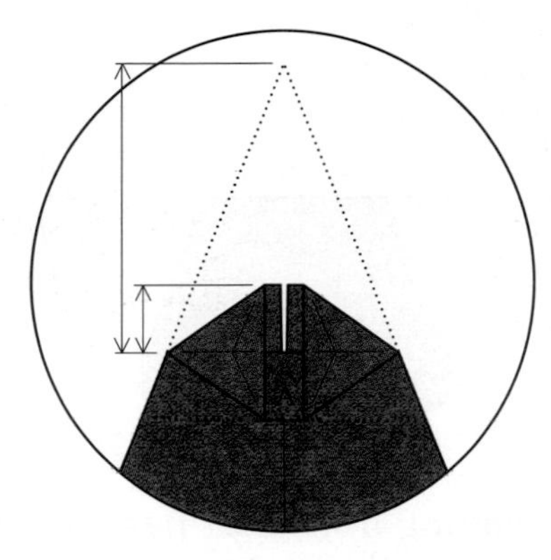

6. The new flaps are shorter than the original we started with.

Figure 5.2.
Comparison of the original and newly created flaps.

The two new flaps are less than 1/4 of the length of the original—not a very good tradeoff, it would seem. That gets us back to our question: How long can we make the two flaps? If you have folded an example to play with, you can answer this experimentally, by taking flap A and gently pulling it lower down while massaging the spread-sunk triangles and allowing them to expand toward each other, as shown in Figure 5.3. (It is actually easier to do this before having ever pressed the two triangles flat.)

As Yoshizawa pointed out in his opus, *Origami Dokuhon I*, the optimum length is attained when the distance from the top edge down to the valley fold is equal to half the width of the top edge.

Clearly, these are the longest flaps we can make, at least, by this technique. Note that the baseline of the flaps moved downward in the process. Quite often, we have the baseline of the flap already defined and we'd like to make the longest

Figure 5.3.
Construction of the maximum-length pair.

7. Grasp flap A, and pull it lower while expanding the spread-sunk corners. Flatten when the two spread-sunk triangles meet in the middle of the paper.

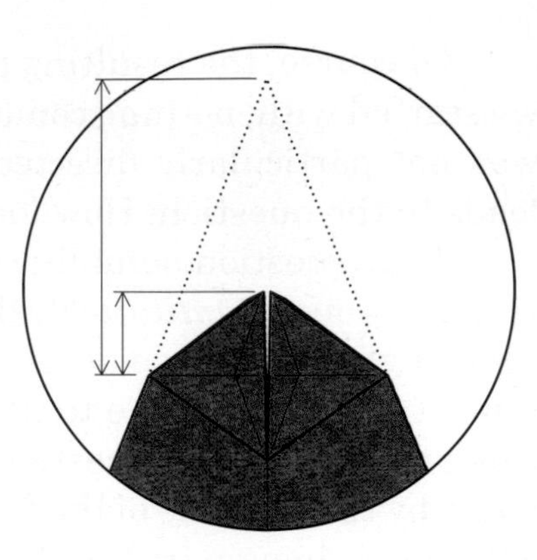

8. The maximum length pair of points.

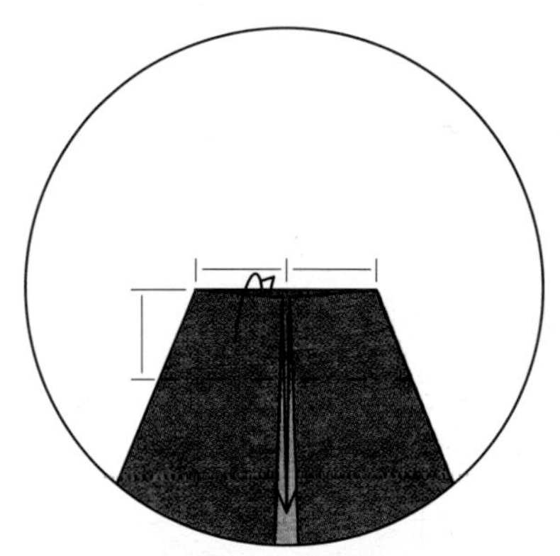

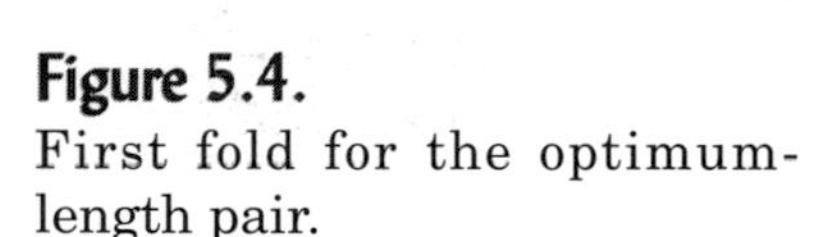

Figure 5.4.
First fold for the optimum-length pair.

points possible extending from that baseline. Examination of the geometry of the point pair shows that, with a few precreases, we can go straight to the optimum-length fold, as shown in Figure 5.5.

Even in the optimum-length case, the two flaps you end up with are much shorter than the original flap you started with. The ratio of their lengths can be worked out using a bit of trigonometry.

$$\frac{\text{short flap}}{\text{long flap}} = \frac{\tan(33.75°)}{\tan(67.5°)} \approx 0.277.$$

Each of the short flaps is 28% of the length of the long flap; in other words, we've given up almost a factor of four in length. This seems unnecessarily wasteful. One might think that the length of a long flap could somehow be divided up when we split the flap; one might think we should be able to divide a long flap of length 1 into two flaps of length 1/2, or three of length 1/3, and so forth.

And in fact, we can do better than the Yoshizawa split. This procedure is quick and (relatively) simple, and it's

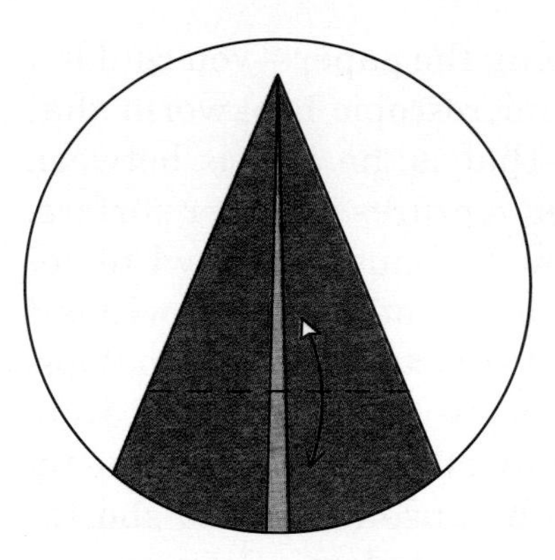

1. Fold and unfold to define the desired location of the base of the two flaps.

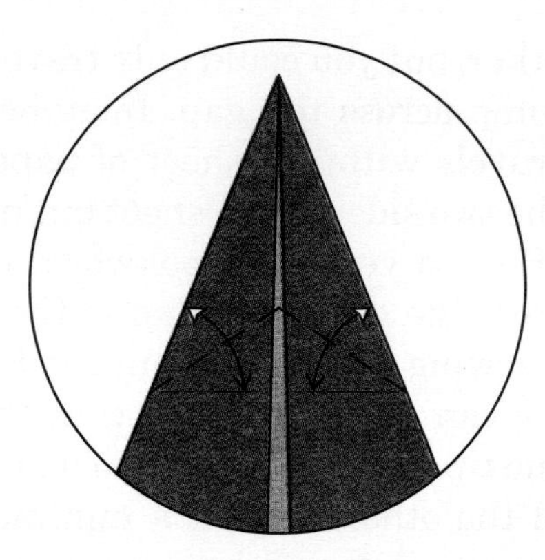

2. Fold and unfold along angle bisectors.

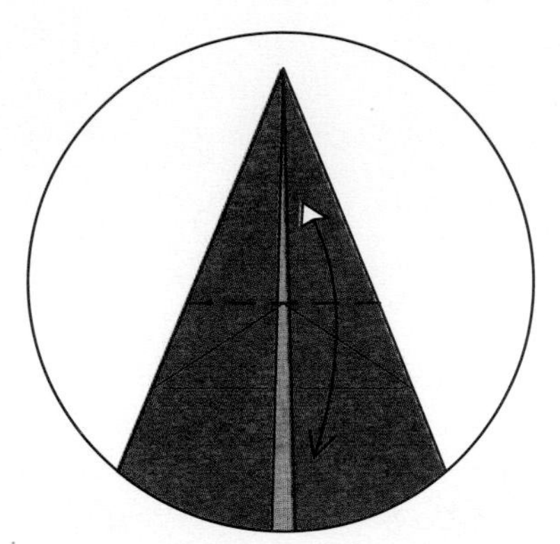

3. Fold and unfold along a horizontal crease that passes through the intersection of the two bisectors.

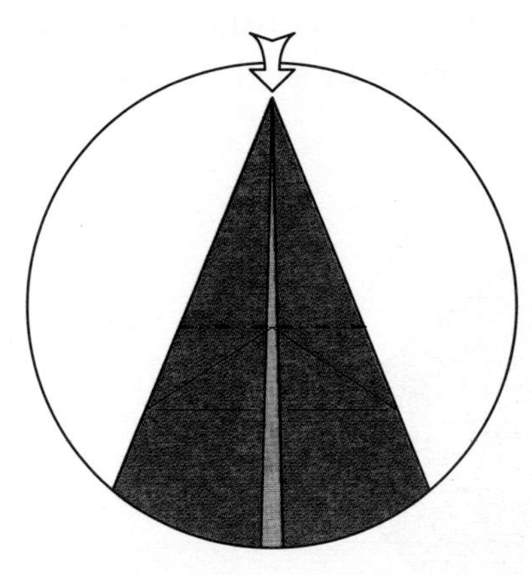

4. Sink the point on the crease you just made.

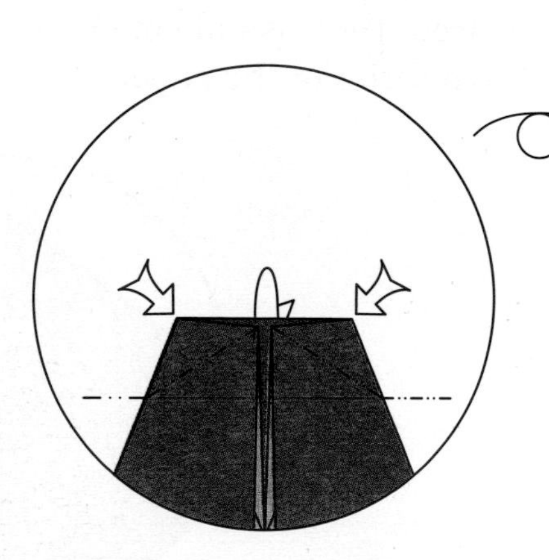

5. Mountain-fold the far edge behind while spread-sinking the corners. Turn the model over.

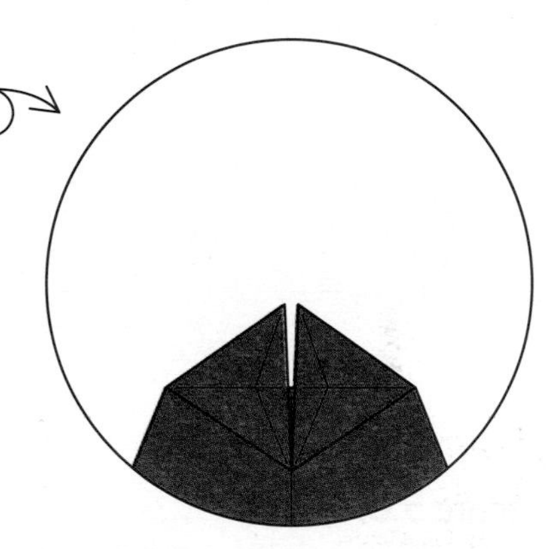

6. Finished pair of flaps.

Figure 5.5.
Alternate folding sequence for the optimum-length pair of points.

certainly good enough to generate a short pair of ears, but it would be nice to do better—for example, to take a three-legged Bird Base animal and give it that elusive fourth leg.

5.2. The Ideal Split

The key to making two longer flaps is to ignore the foreground and examine the background; that is, turn our attention away from the flaps themselves and instead, look at the space around the flaps. The thing that makes two flaps two instead of one is not the paper making up the flaps; it's the space we've created between the flaps that defines the pair. What limits the length of the flap is the length of the gap. And that is significant because the gaps as well as the flaps consume paper—and so we must allocate paper for both.

A small thought experiment will bring this out. Suppose you wished to travel from the tip of one flap to the tip of the other, but you could only travel along the paper—you couldn't jump across the gap. Imagine a microscopic bookworm that travels within a sheet of paper—that is, he crawls between the two sides of the sheet but never ventures to either surface. (He is a very shy bookworm.) How far must he crawl to get from the tip of one flap to the tip of the other? Even without knowing anything about the folded structure of the two flaps, we can say for certain that the bookworm must travel from the tip of one flap down to its baseline, then back out to the tip of the other flap (at a minimum) because there's no shorter path that doesn't require the bookworm to leave the paper, as shown in Figure 5.6. So the bookworm must travel the sum of the lengths of both flaps. (And since it may not be possible to go from the baseline of one flap directly to the other flap via the interior of the paper, the journey could be even longer.)

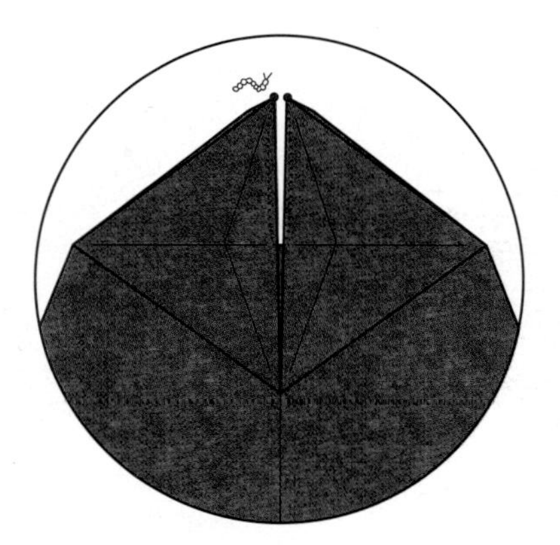

Figure 5.6.
Path followed by an origami bookworm.

Suppose, for the moment, that our bookworm were further restricted to traveling only along folded or raw edges. Then there is only a small number of paths he could travel along. The two shortest paths, labeled A and B, are shown in Figure 5.7 by dashed lines (in some cases, he is traveling along hidden layers of paper). A third path, labeled C, is shown that does not follow existing folds.

It helps to distinguish the different paths by simultaneously examining the crease pattern and the model with the paths drawn on each. These are shown together in Figure 5.7.

Of course, paths A and B are only two of the possible paths the bookworm could take, but these are the two shortest paths that travel along folded or raw edges of the paper. Neither, however, is the shortest possible path from the bookworm's point of view, which is the same whether the paper is folded or unfolded. That shortest path is easy to draw on the crease pattern; it's a straight line. It's a bit harder to work out what it is on the folded model, lying as it does in hidden layers of paper, but it is shown as path C in Figure 5.7.

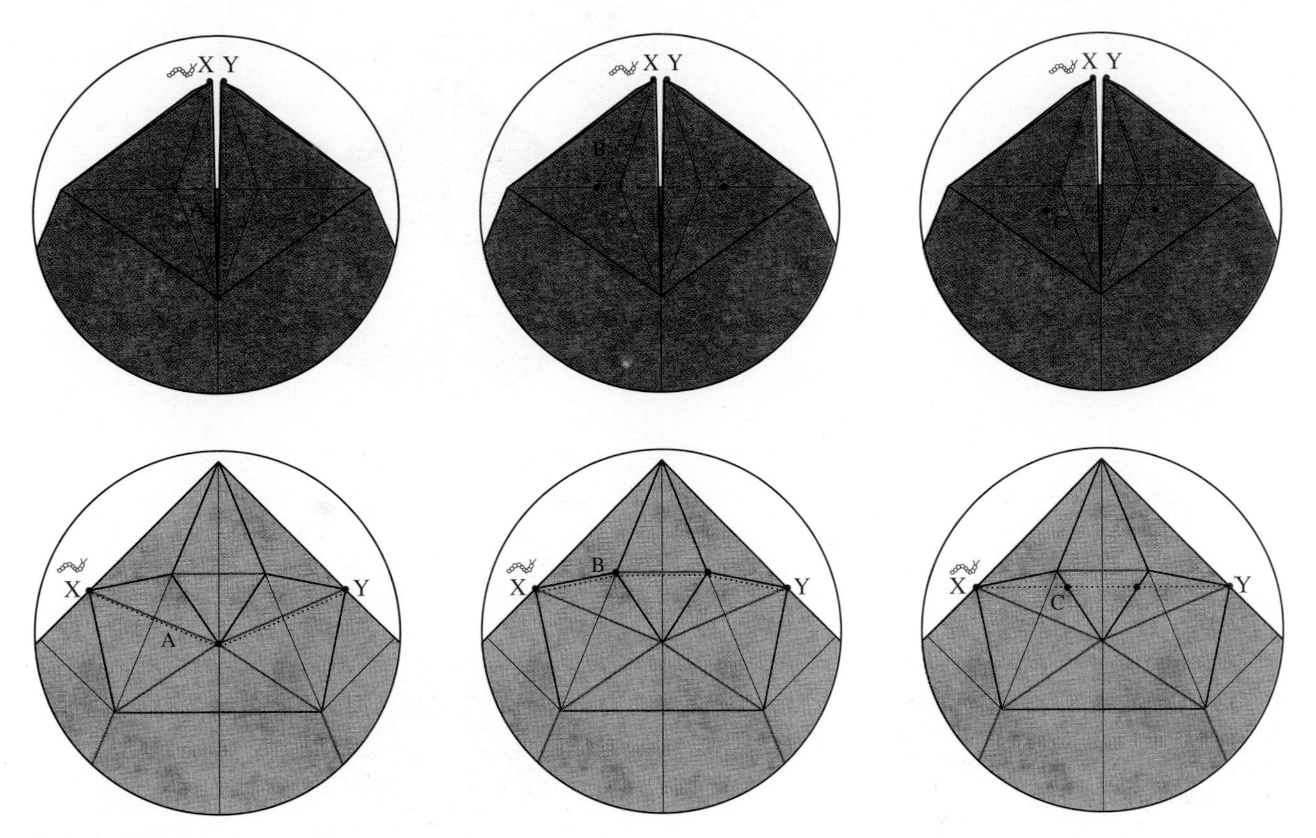

Figure 5.7.
Upper row: path in the folded form.
Bottom: path traveling along the surface of the paper.

Now, what's interesting is that although path C is the shortest path from tip to tip in the unfolded paper, it's clearly not the shortest possible path in the folded model. As can be seen in the figure, the bookworm backtracks a bit and actually travels somewhat below the baseline of the two points. This means that we've devoted more paper to the gap than we really needed to—paper that could have been used to make longer points.

In the most efficient possible structure, the amount of paper that is used to create a gap between two points would be as close as possible to the minimum required. In other words, if we compared the tip-to-tip path in the folded model and the crease pattern, they would look something like Figure 5.8.

Of course, we don't know what the rest of the crease pattern looks like or even what the folded model looks like. But we've identified several salient features of both. We know where the tips of the two points are (indicated by the black dots), and we know how deep the gap is in the folded model (half the distance between the point tips on the crease pattern). Knowing the depth of the gap, we also know where the baseline of the two flaps must be, and we can make corresponding creases on the paper.

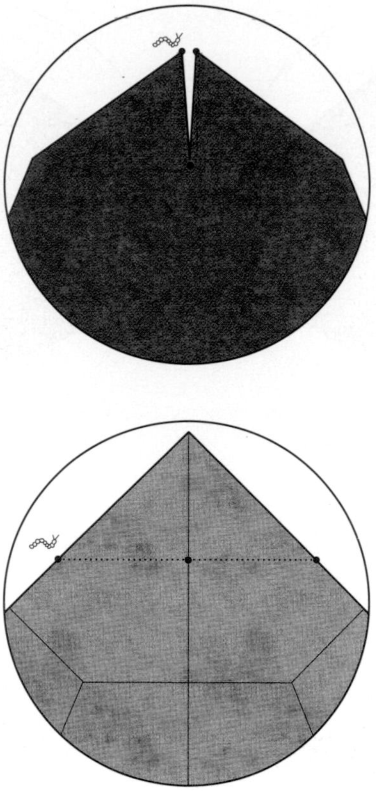

Figure 5.8.
The optimum tip-to-tip path. Top: path in the folded model. Bottom: path through the paper. The creases (and the exact shape of the folded model) are not yet specified.

Now, we have two points with a gap between them, and the shortest bookworm path on the crease pattern is also the shortest bookworm path on the folded model. This strikes precisely the right balance between paper devoted to the flaps and paper devoted to the gap. The paper saved from the gap can go into making longer flaps. And indeed, a comparison of the folded and original flap in Figure 5.10 shows that the two flaps are indeed longer than in the simpler split.

The ratio of the lengths of the new and original flaps is:

$$\frac{\text{short flap}}{\text{long flap}} = \frac{\tan(45°)}{\tan(67.5°)} \approx 0.414,$$

which is almost 50% longer than that obtained by simply sinking and spread-sinking the corners. This is, in fact, the longest possible pair of flaps that can be made from a standard Bird Base corner flap, and so I call it an *ideal split*.

One quibble you may have: In this form, the two flaps overlap each other while the Yoshizawa-split flaps have daylight between them. So the structures are not perfectly comparable. It is possible to further sink and squash the ideal split to put a gap between the two flaps (at the cost of a slight reduction of gap depth). But if you fold the two pairs in half (as one might, for example, in making a pair of ears), then the two arrangements can be compared directly (see Figure 5.11).

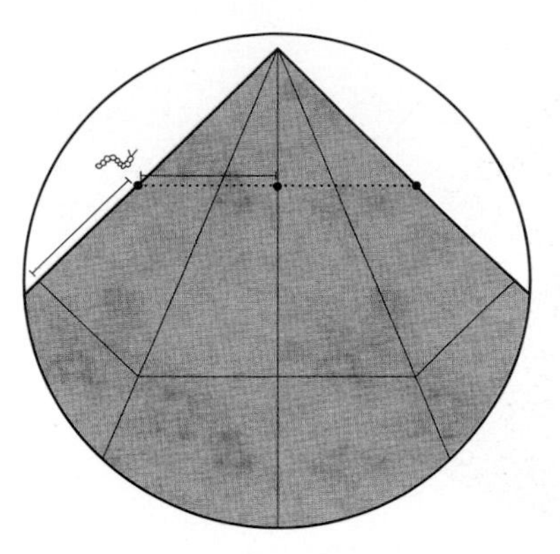

1. For an optimally efficient point split, the point where the base crease hits the edge is separated from the desired point tip by a distance equal to half the separation between the point tips.

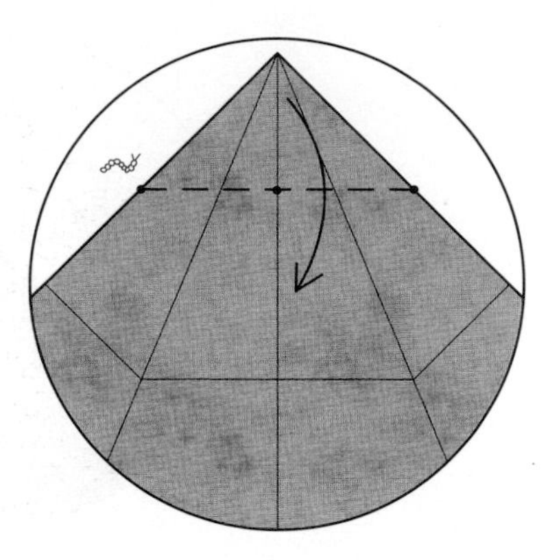

2. Now we'll refold the model. The top point isn't used; fold it down.

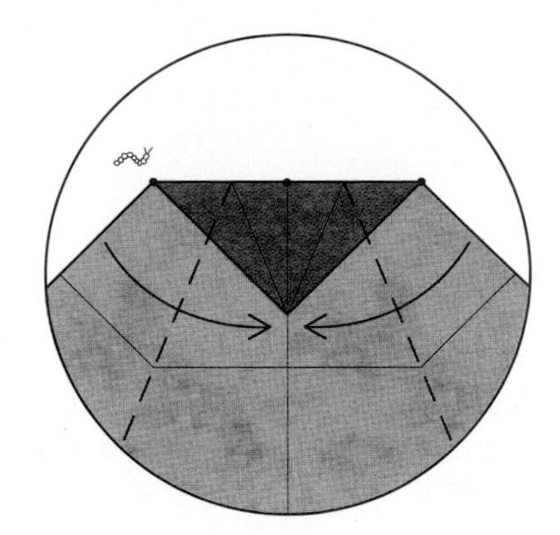

3. Fold the sides in.

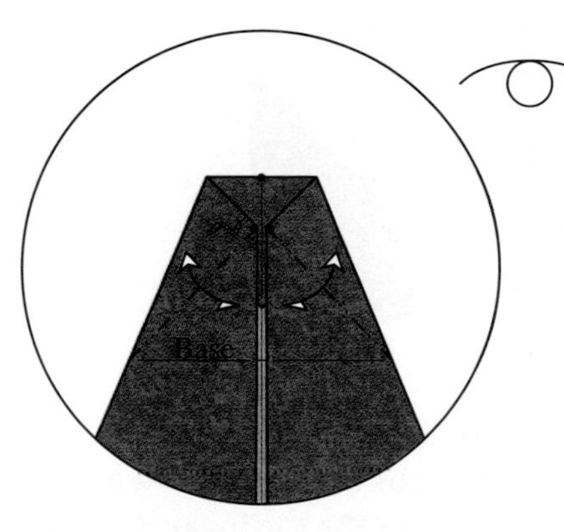

4. Make creases that connect the edges of the base with the points that will become the tips of the two flaps. Turn the model over.

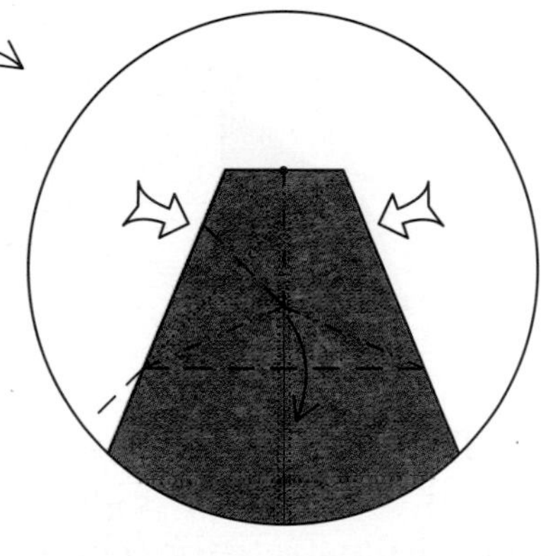

5. Squeeze the sides in so that the extra paper swings downward, using the creases on the far layers that you made in the previous step.

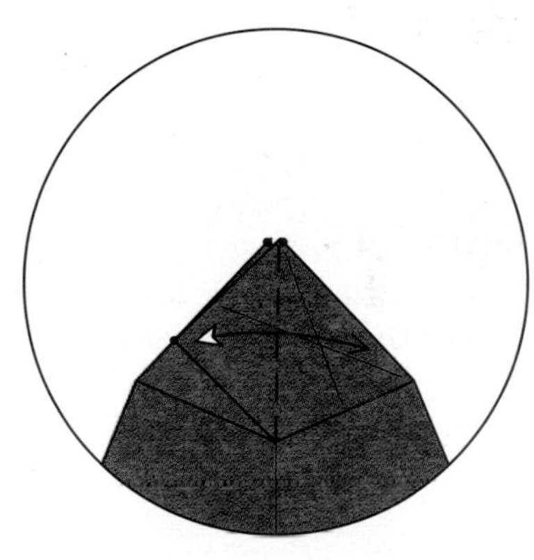

6. Fold the excess paper from side to side. Flatten firmly.

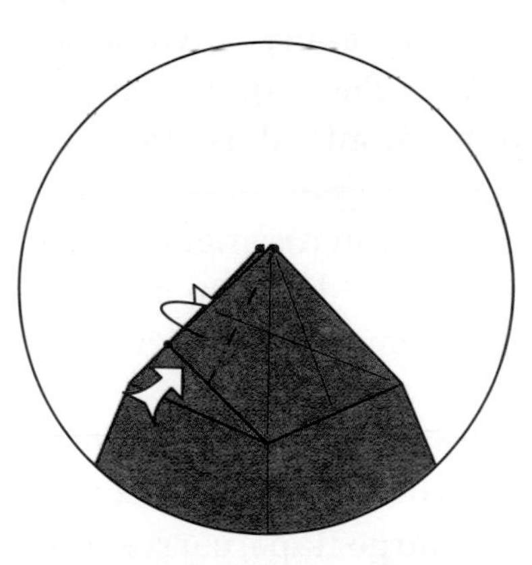

7. Reverse-fold the edge inside. Observe that the edge of the reverse fold (the black dot) lines up with the previously defined base.

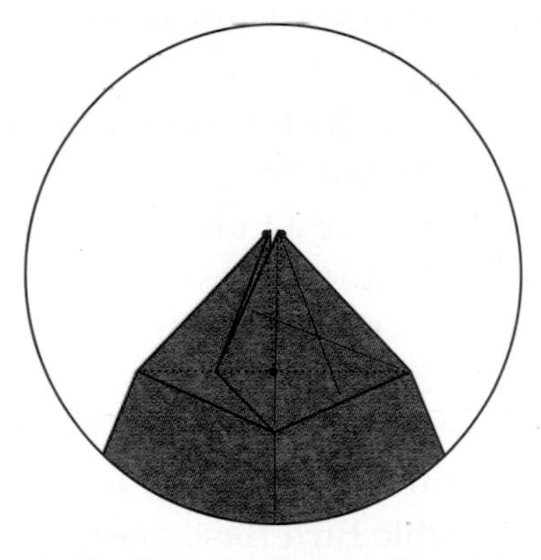

8. The finished split point. Now the shortest path on the crease pattern is also the shortest path on the folded model.

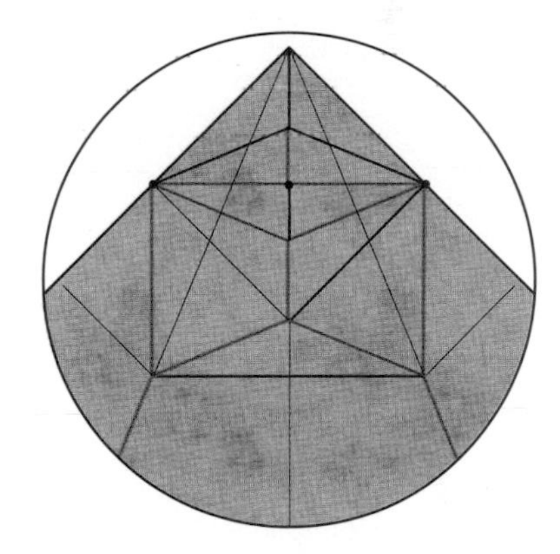

9. Crease pattern.

Figure 5.9.
Folding sequence for the ideal split.

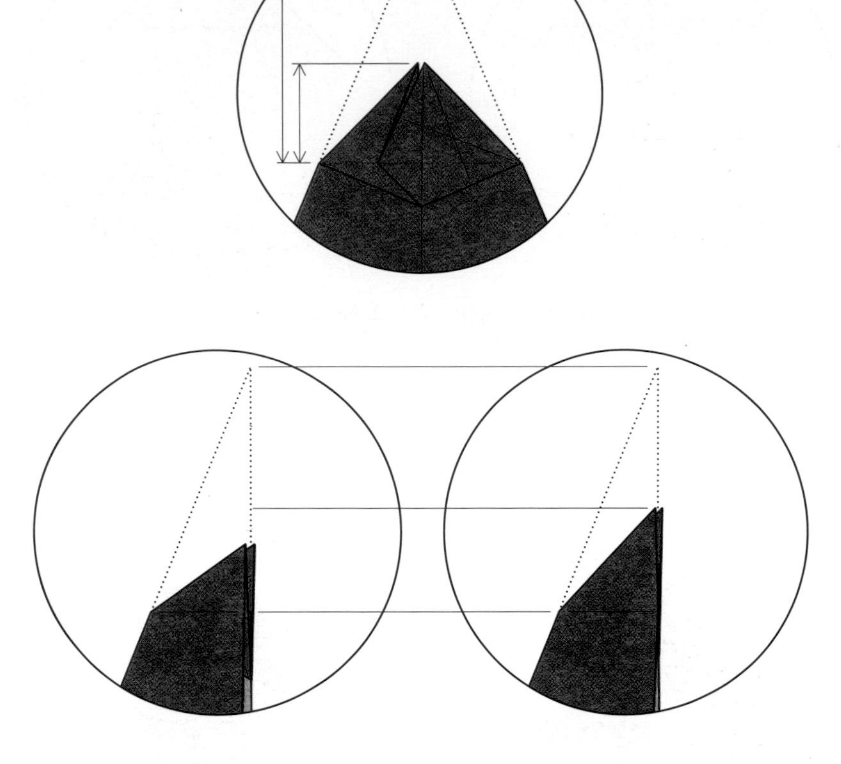

Figure 5.10.
The finished point pair compared to the original flap length.

Figure 5.11.
The ideal split is about 50% longer than the Yoshizawa split.

The ideal split takes more folds to perform if you fold it the way I showed in Figure 5.9. But that sequence was designed to illustrate the connection between the paths and crease pattern. That's not necessarily the most straightforward way to fold. Once you've worked out the crease pattern for a model or technique, it's worthwhile going back and experimenting with different ways of folding. There are many ways of performing an ideal split on a standard flap. The sequence in Figure 5.12, which was developed by John Montroll, is one of the most elegant.

There are numerous variations, both in arrangements of layers (note that this sequence has a slightly different arrangement of the layers) and in the folding sequence that gets you to the finish.

Point-splitting can be used to breathe new life into old structures. For example, few shapes are as picked-over as the venerable Bird Base, possessed of four large flaps, corresponding to head, tail, and two wings. But by splitting the tail point, we can create two legs instead of a tail; by splitting the head point, we can create a head with an open beak, a head with a crest, or quite another flying beast altogether: a Pteranodon.

You will find folding instructions for this figure at the end of this chapter. It includes both ideal and Yoshizawa splits.

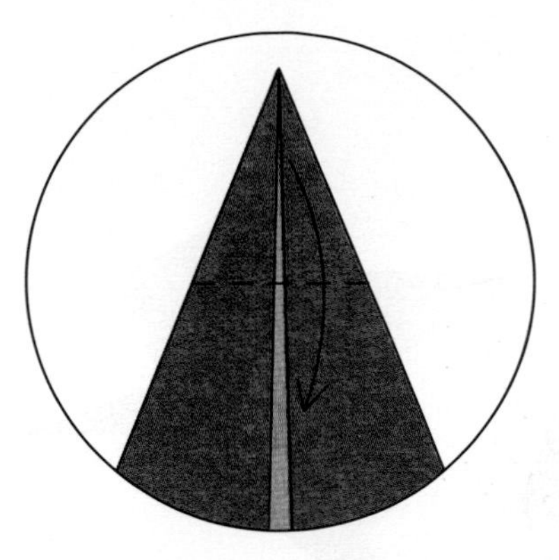

1. Fold the top of the flap down.

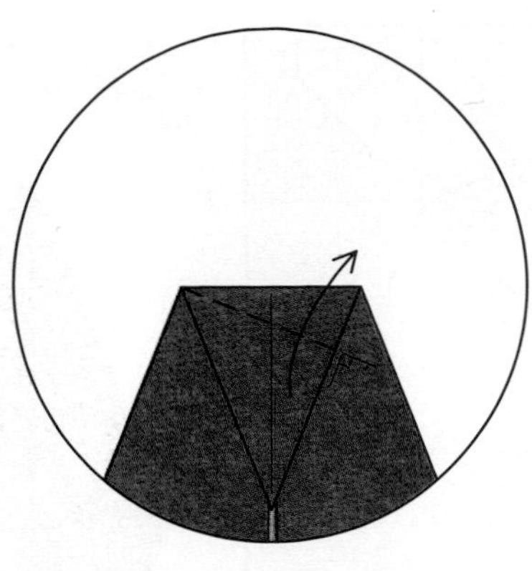

2. Fold the flap up so that its right edge is aligned with the layer underneath.

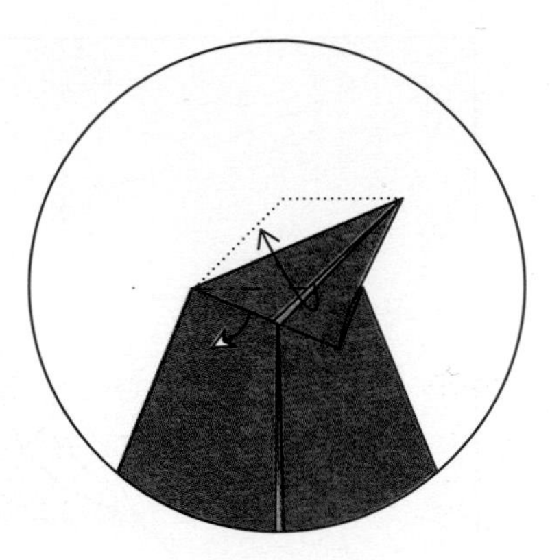

3. Pull out the loose edge as far as possible.

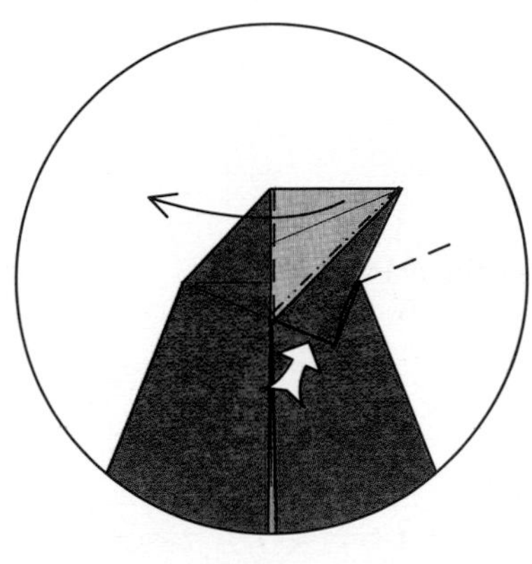

4. Squash-fold the edge and swing the flap over to the left on a vertical fold.

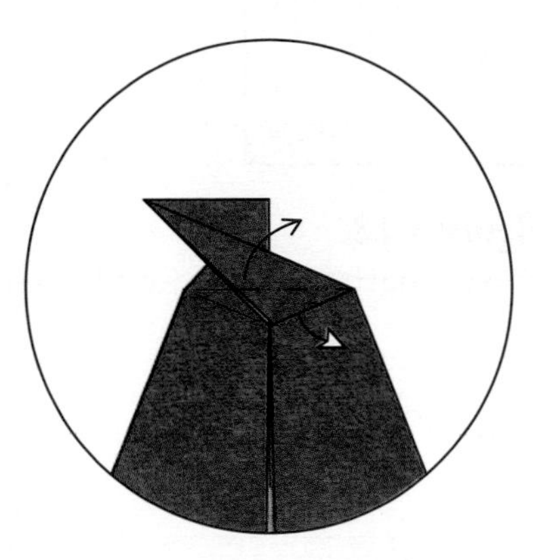

5. Pull up the loose edge as far as possible, releasing the trapped paper under the flap.

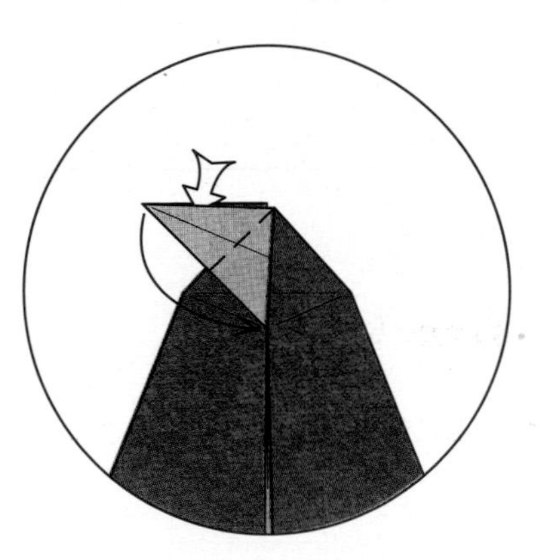

6. Outside-reverse-fold the white point.

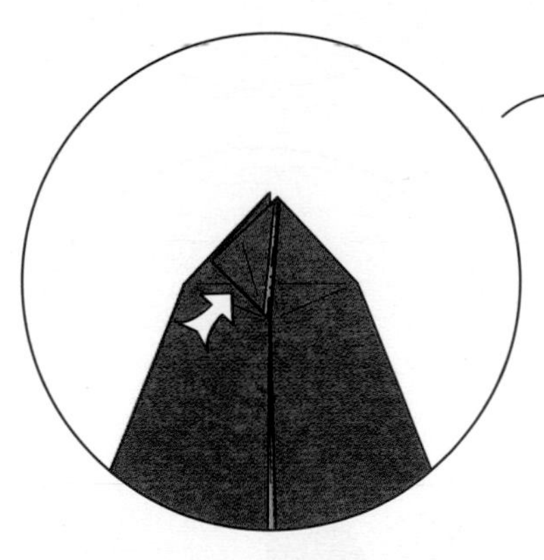

7. Reverse-fold the flap through the model. Then turn the model over.

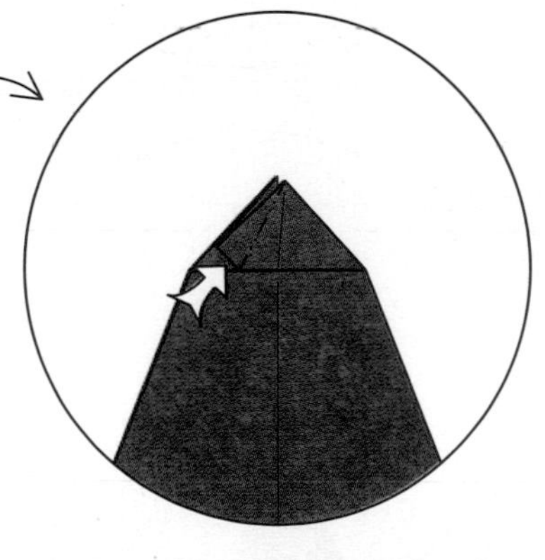

8. Reverse-fold the flap back to the center line.

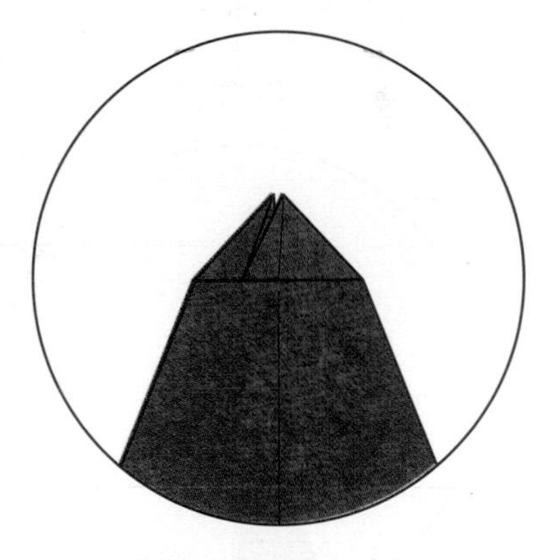

9. Finished ideal split.

Figure 5.12.
Folding sequence for the ideal split, after Montroll.

Splits can be used for more than point multiplication; as an auxiliary benefit, by splitting the large central point with a Yoshizawa split, we can reduce its height while preserving

Figure 5.13.
Crease pattern, base, and folded model for a Pteranodon.

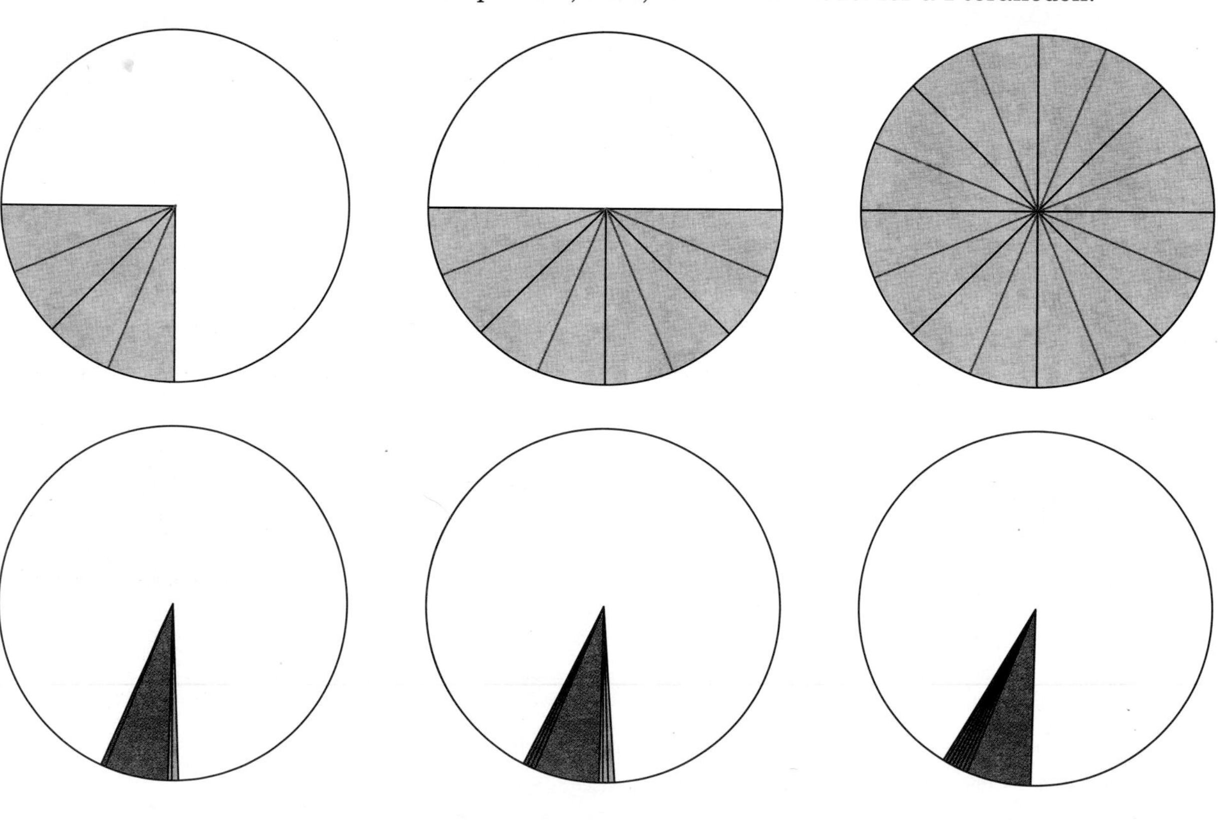

1. Top: crease pattern for a corner flap. Bottom: the folded flap.

2. Top: crease pattern for an edge flap. Bottom: the folded flap.

3. Top: crease pattern for a middle flap. Bottom: the folded flap.

Figure 5.14.
Crease pattern (upper row) and folded flap (lower row) for three types of points: (left) corner flap, (middle) edge flap, (right) middle flap.

the flapping action, so this Pteranodon flaps its wings when you pull its head and legs like the traditional flapping bird.

5.3. Splitting Edge and Middle Flaps

The folding sequence I showed in Figure 5.12 works for a *corner flap*, a flap formed from a corner of the square, which describes the main flaps of the four Classic Bases. In addition to four corner flaps, the Frog Base possesses a different type of flap: Its central point comes from the middle of the paper, a so-called middle flap. Furthermore, the smaller points on a Frog Base come from the edge of the paper, an edge flap. You might ask, is it possible to split edge and middle flaps?

Indeed it is, although the layers get thicker and less manageable as you move from corner to edge to middle flap. The process is relatively straightforward for an edge flap. An edge flap still has two raw edges, and so it can be treated like a corner flap and split into two points in the same way as a corner point. In fact, the folding sequence illustrated in Figure 5.12 may be used as well on an edge flap.

However, a middle flap has no raw edges, which played a prominent role in the folding sequence of Figure 5.12. But

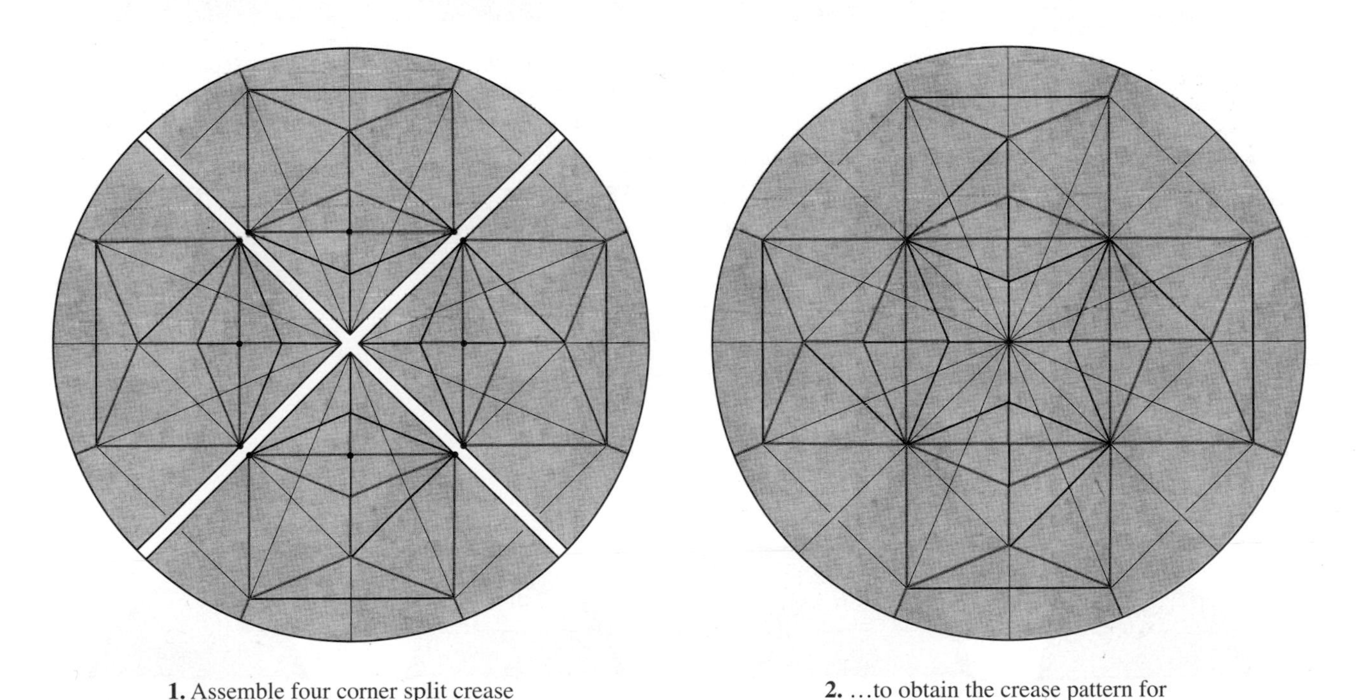

1. Assemble four corner split crease patterns ...

2. ...to obtain the crease pattern for a middle split.

Figure 5.15.
A middle flap can be constructed by stitching together four corner flaps (and their crease patterns) along their raw edges.

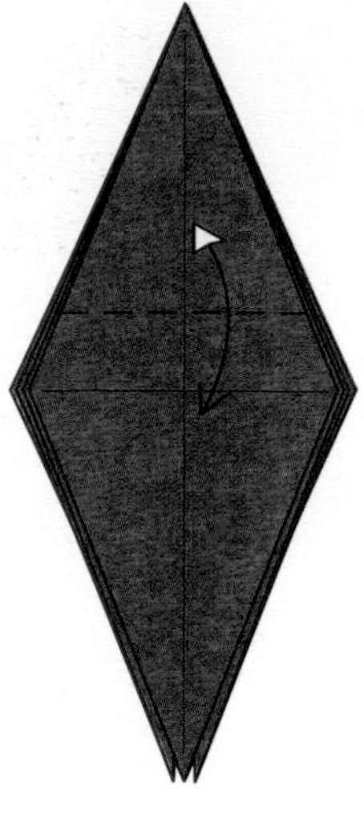

1. Begin with a Frog Base. Fold and unfold to define the baseline.

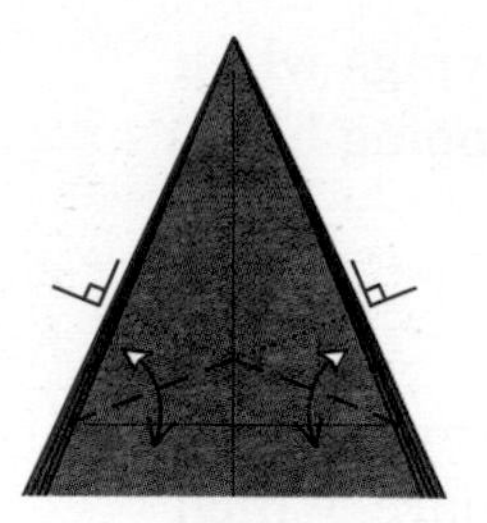

2. Top of the model. Fold and unfold.

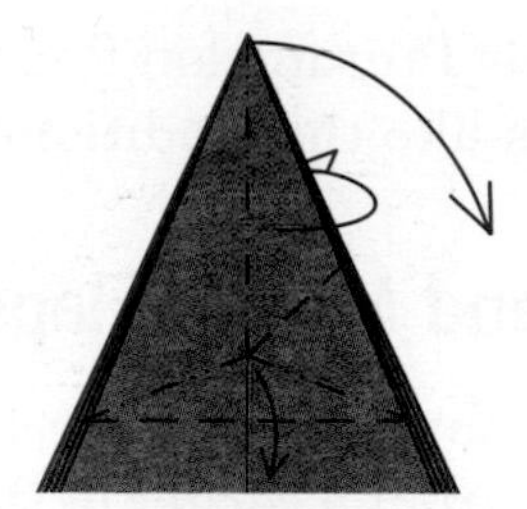

3. Pleat and fold a rabbit ear from the thick point using the existing creases.

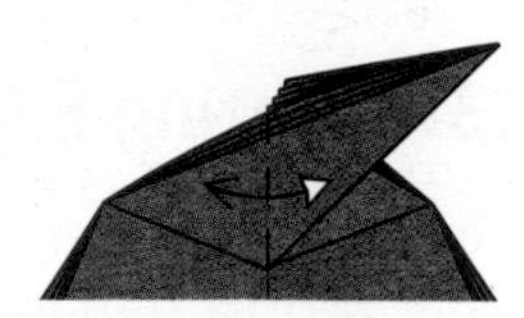

4. Fold and unfold through all layers.

5. Wrap one layer of paper to the front.

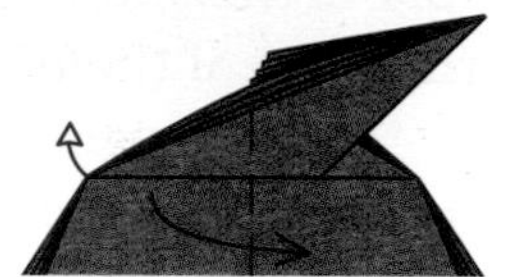

6. Fold one layer to the right, releasing the trapped paper at the left that links it to the next layer.

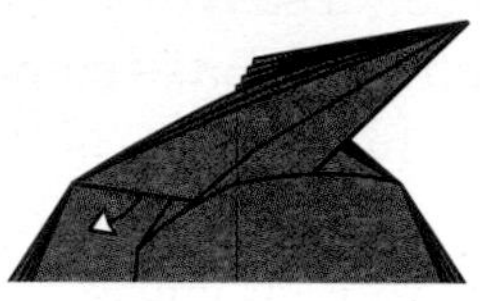

7. In progress. Pull paper out from here.

8. Close the model back up.

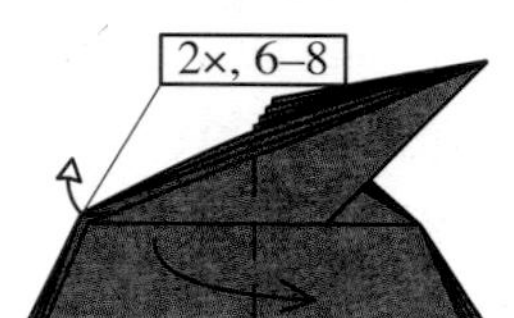

9. Repeat steps 6–8 on the next two layers.

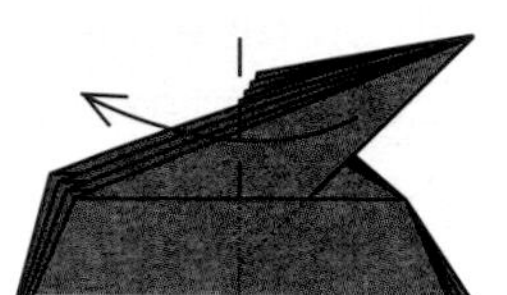

10. Swing the point over to the left.

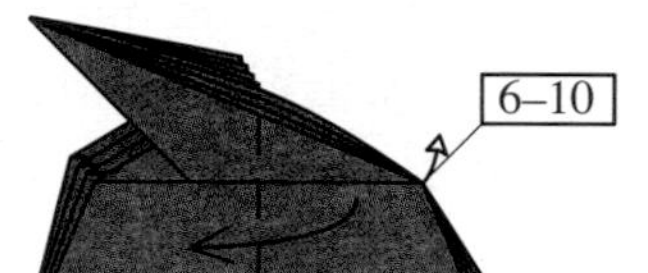

11. Repeat steps 6–10 on the right.

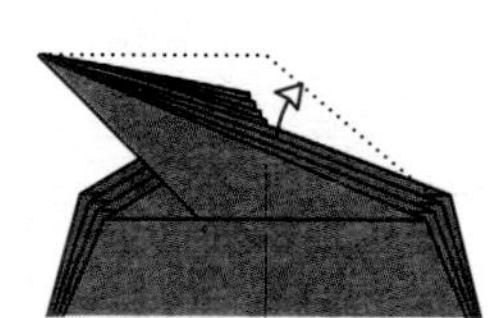

12. There are four edges on the top right; pull out as much of the loose paper between the third and fourth edges as possible. A hidden pleat disappears in the process.

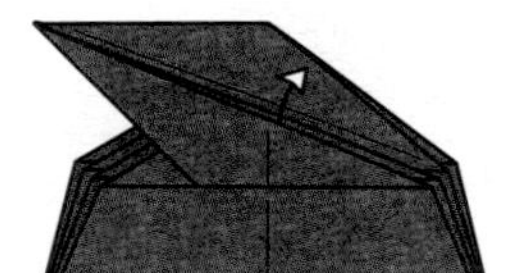

13. Pull out the loose paper between the first and second layer.

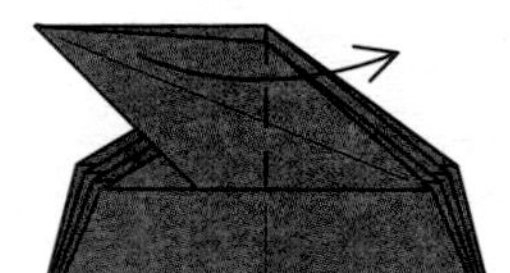

14. Swing the point back to the right.

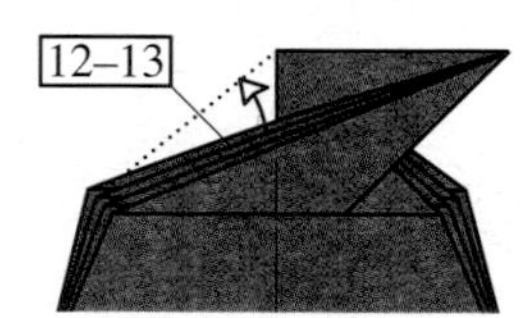

15. Repeat steps 12–13 on the left.

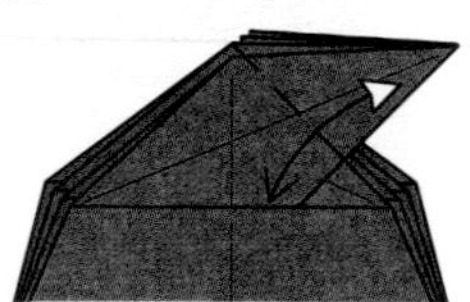

16. Fold and unfold.

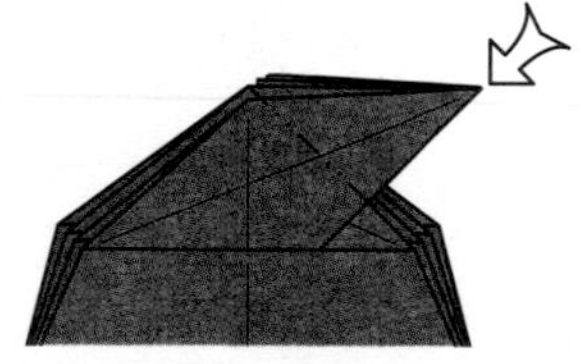

17. Open-sink the point. You will have to open out the top of the model somewhat to accomplish this.

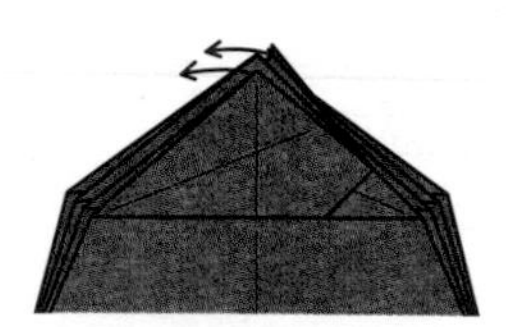

18. Rearrange the layers at the top so that the central square forms a Preliminary Fold.

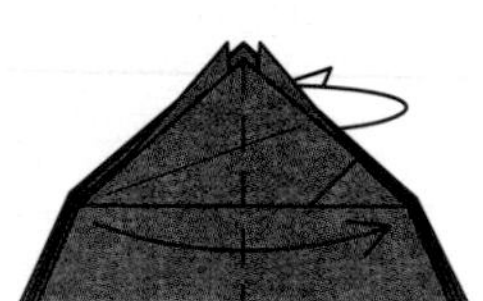

19. Fold one layer to the right in front, and one to the left behind. The model should be symmetric as in step 1.

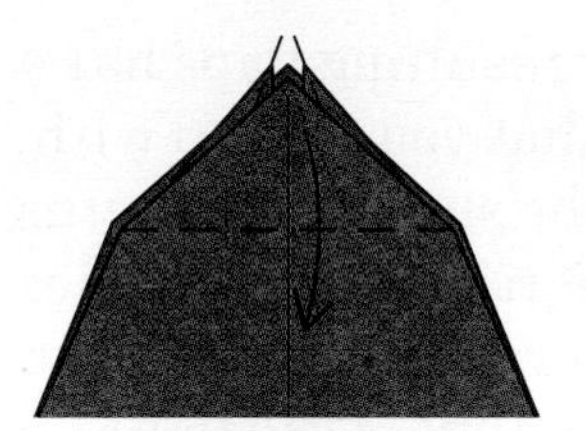

20. Petal-fold.

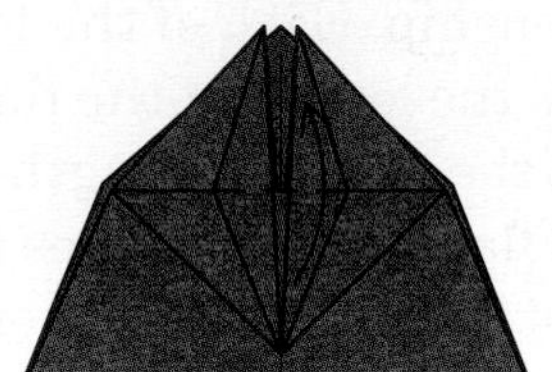

21. Fold the flap back upward.

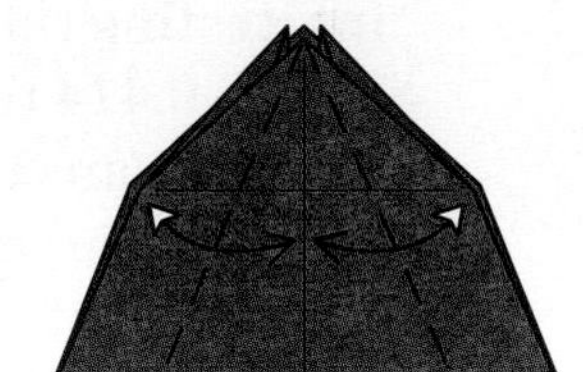

22. Fold and unfold.

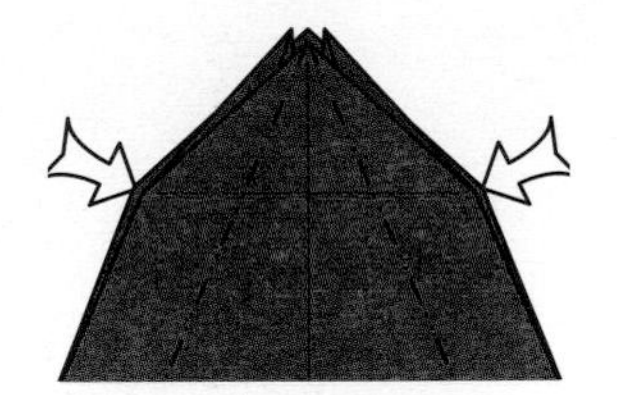

23. Closed-sink the edges on the creases you just made.

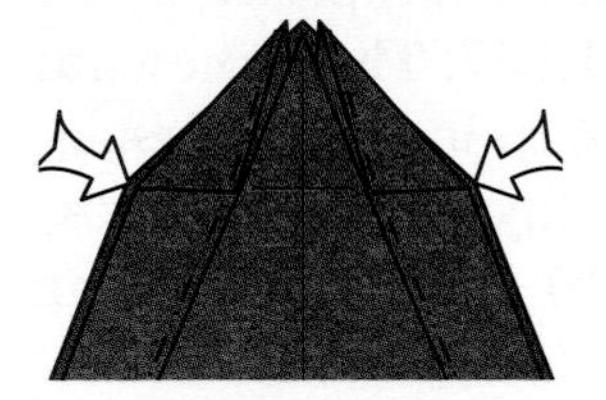

24. Closed-sink the next pair of edge.

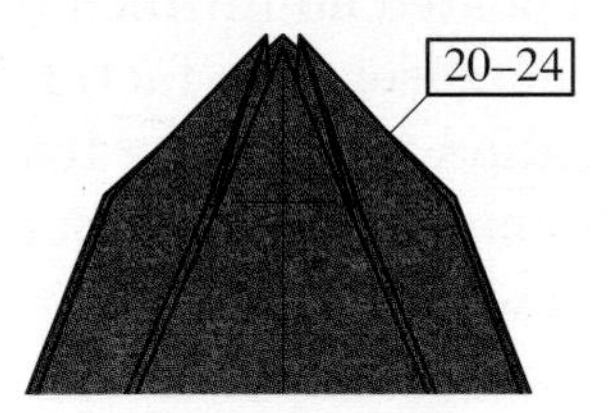

25. Repeat steps 20–24 behind.

26. Finished four points.

Figure 5.16.
Folding sequence to split a middle flap into four smaller flaps.

we can get an idea of what is possible by examining the crease patterns of corner, middle, and edge flaps.

As Figure 5.14 shows, the crease pattern for an edge flap is simply that of a pair of corner flaps, and that of a middle flap is four corner flaps; edge and middle flaps themselves can be made by sewing together corner flaps along their raw edges.

It thus makes sense to see if the analogy continues. For example, is it possible to put together the crease patterns for four split corner flaps to get the crease pattern for a split middle flap?

It is, and the result is shown in Figure 5.15. This can be folded up (although some of the crease directions will have to be reversed from what is shown). The sequence to fold it directly from a middle flap is quite challenging, however, which perhaps accounts for its rarity in published origami designs. One possible sequence is shown in Figure 5.16. You can try this sequence on the top of a Frog Base.

When you split a middle flap, you obtain four smaller flaps. (Similarly, an edge point can be split into either two or three flaps, depending on how you orient the creases with respect to the edge.) The split middle flap gives rise to an interesting relationship. Each of the four resulting flaps has a length 0.414 times the length of the flap that you started with, which means that the *total* flap length—the sum of the lengths of the four created flaps—is four times that ratio, or 1.657 times the length of the starting flap. In other words, the total length of the created flaps is actually longer than that of the flap you started with.

Since the result of a split middle flap is four more middle flaps, these may be thinned and split again by the same process, and the process repeated ad infinitum. At each repetition, the total flap length increases by a factor of 1.657. The somewhat surprising implication is that the total flap length increases without limit. Of course, the number of layers increases extremely rapidly as well, setting a practical upper limit to flap multiplication.

5.4. More Complex Splits

One can split multiple flaps to different depths to achieve different effects and further variation. We saw in the flapping Pteranodon that splitting two of its flaps could be used to extend the life of the Bird Base. A Bird Base, of course, has a total of four flaps. All four can be split, yielding structures that let us move away from birds and into other kingdoms. Splitting all four points of a Bird Base gives enough small flaps

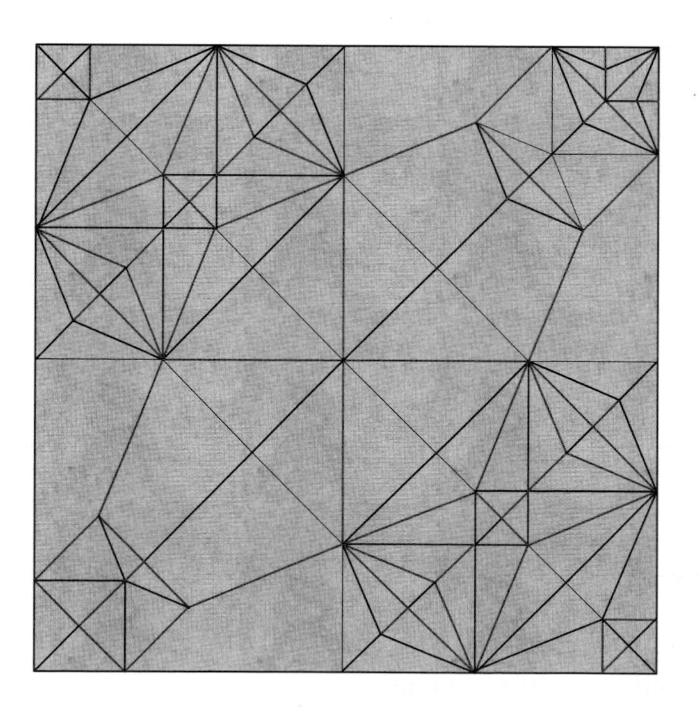

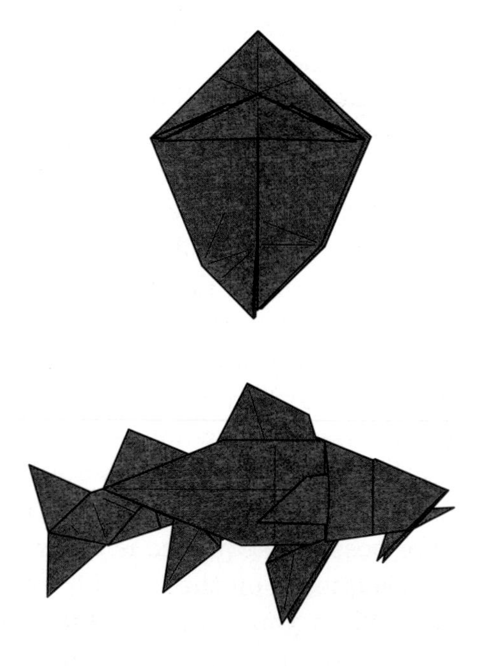

Figure 5.17.
Crease pattern, base, and folded model of the Goatfish.

to make a variety of fish with dorsal, ventral, anal, caudal, and pectoral fins, as the example shown in Figure 5.17 illustrates.

By using this structure and varying the depth of the splits and the shaping and orientation of the resulting flaps, you can create quite a wide variety of fish. The Goatfish also illustrates another type of split—splitting one flap into three (which are used for the lower jaw and barbels). You will find a folding sequence for this new split in the instructions for the Goatfish at the end of this chapter.

In fact, there are many ways to split a point, and a point can be split into two, three, four, five, or more points. We've already seen two ways to split a point into two; here is another way in Figure 5.18, which also readily generalizes to three or four smaller points.

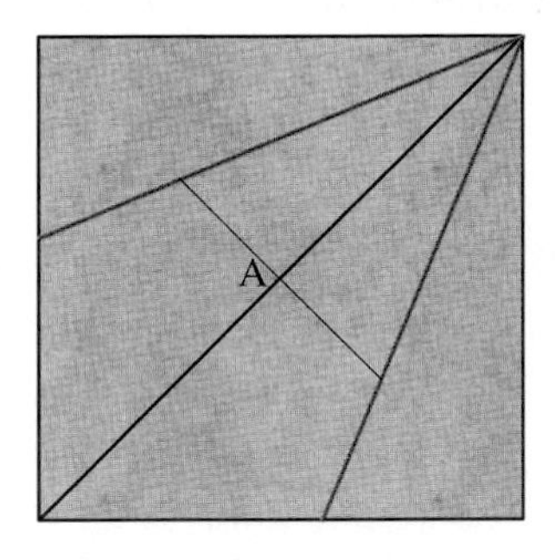

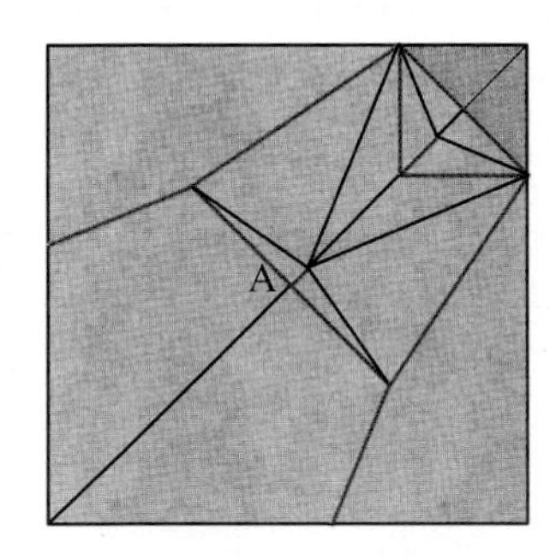

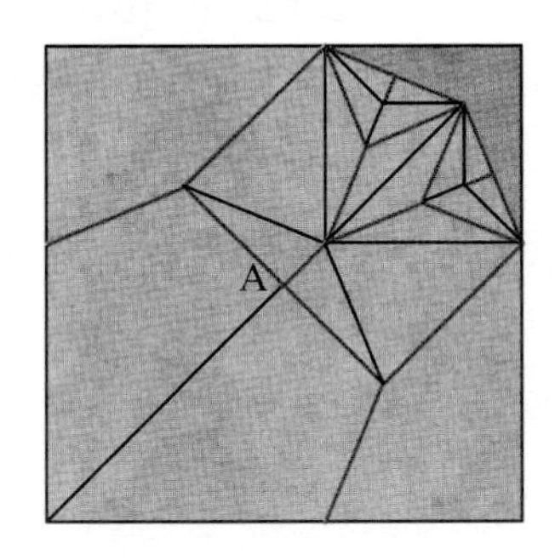

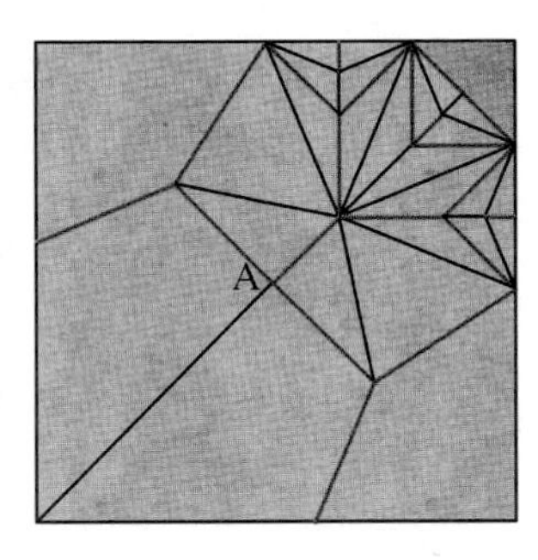

Figure 5.18.
Crease patterns for a kite flap and the flap split into two, three, and four points. To compare scales, the downward diagonal crease A is in the same place in all four patterns.

The shaded regions in these patterns go unused and would typically be folded underneath. Rather than drawing in the creases they would incur, I've simply left them blank to emphasize the common structure of the three splits.

These three patterns are part of a family that can easily be extended to larger numbers of points. You can, of course, use these three patterns as recipes to be called upon whenever two, three, or four points are needed, but it is much more useful to examine their structure, to break them down into components and understand the contribution of each component.

The first part of this examination is to identify the regions of the crease patterns that become the tips of the various points. Since the folded flaps come to sharp points, their tips correspond to single points on the crease pattern. These points are identified in Figure 5.19 by dots.

Now, as we saw with the Yoshizawa split, what defines two or more points is not so much the points themselves,

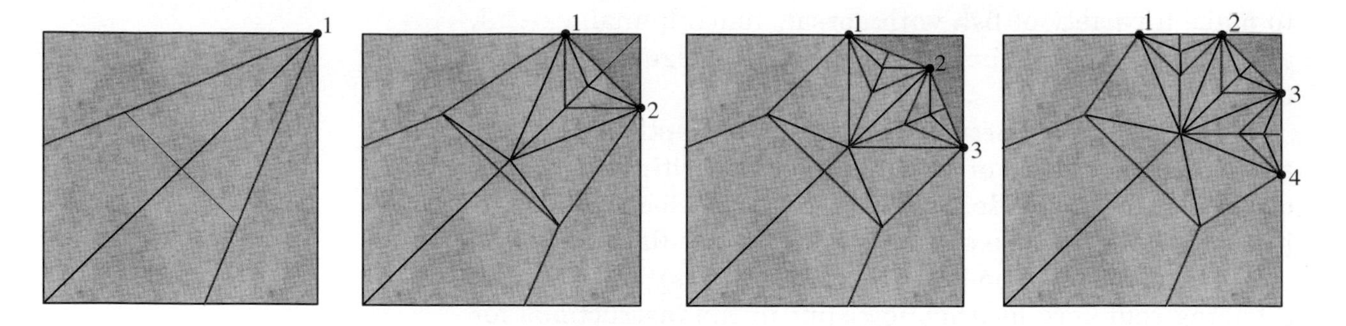

Figure 5.19.
Split points with the point tips identified.

but the presence of a gap between them. If we examine the crease patterns for the two-, three-, and four-point patterns, we see a common wedge of creases that appears in every pattern. Figure 5.20 shows one instance shaded in each of the three patterns.

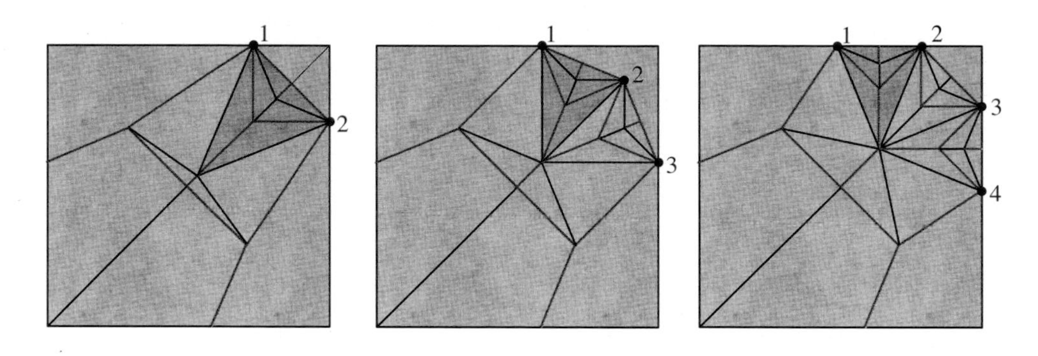

Figure 5.20.
The three patterns with the common wedge of creases.

If we cut out just this wedge from any one of the patterns and fold it up, we get the structure shown in Figure 5.21.

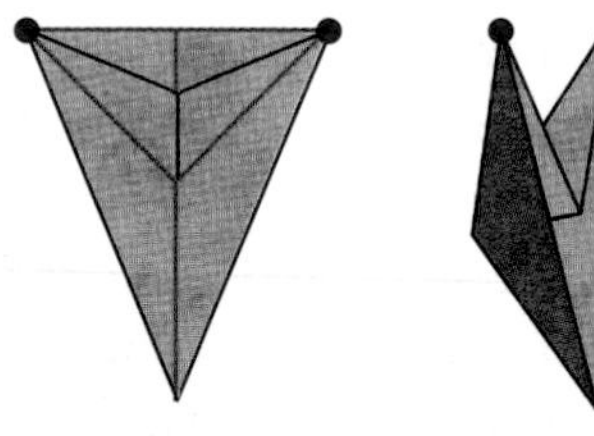

Figure 5.21.
Left: the crease pattern on the wedge.
Right: the folded structure.

Two points have one copy of this wedge, three points have two, four points have three. Crease patterns with progressively larger numbers of points include progressively larger numbers of copies of a basic element. This is not particularly surprising. What is interesting, and perhaps just a bit unexpected, is that the basic unit of replication is not a single flap;

instead, the unit that is replicated is two half-flaps with a gap between them. As with the Yoshizawa split, the gap contains the secret to the structure.

So two, three, or four points may be constructed by using one, two, or three wedges. It should also be clear that the diagonal creases inside the wedge don't have any particular significance. They're there to make the points narrow, and they divide the angle into equal divisions so that in the folded result, the edges line up. But we could easily have used fewer or more diagonal creases and gotten fatter or skinnier points, as shown in Figure 5.22.

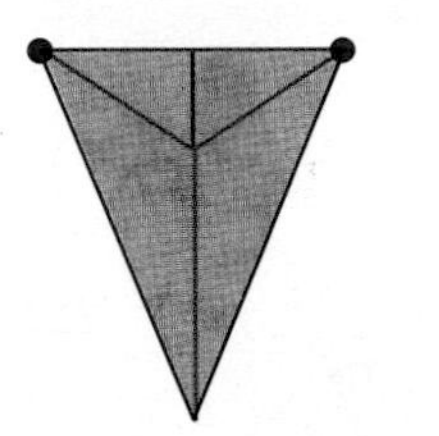
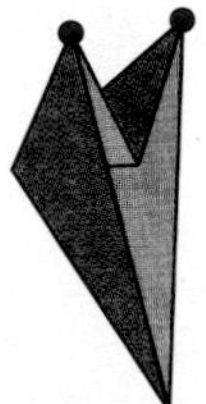
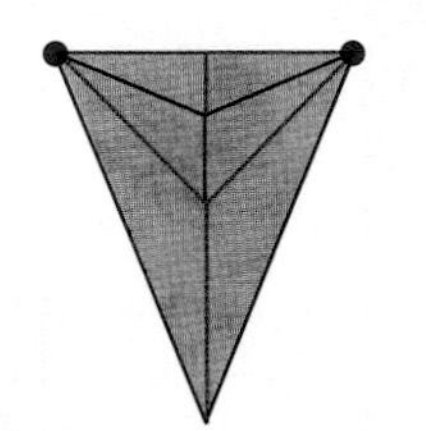
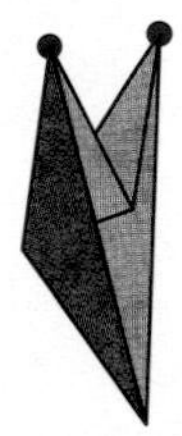
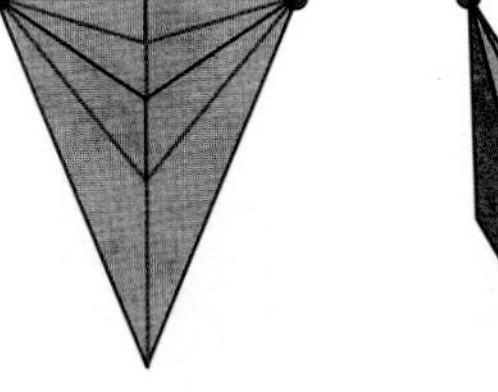

Figure 5.22.
The wedge unit with the two points divided into successively greater numbers of divisions.

We can build the number of points we need by assembling the number of wedges we need: For N points, we use $N-1$ wedges.

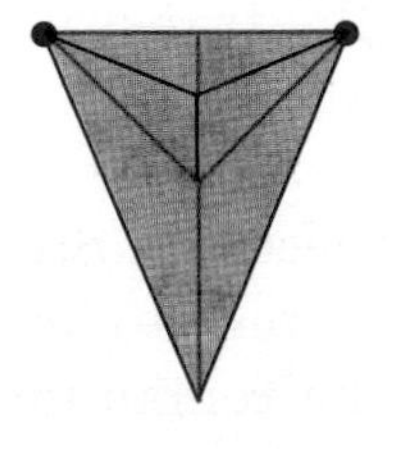
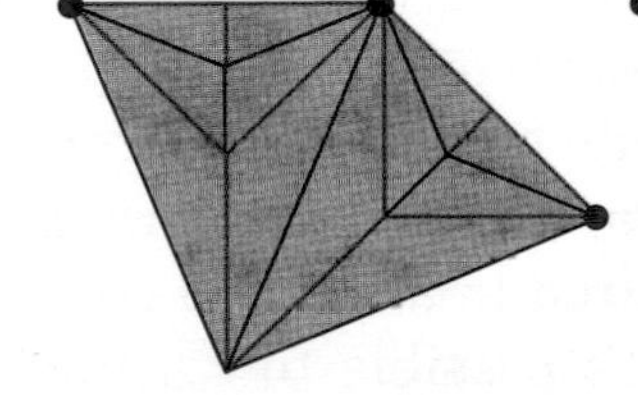
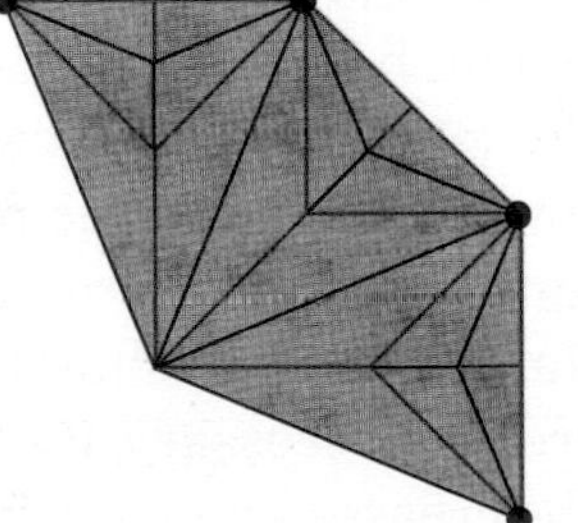

Figure 5.23.
One, two, and three wedges define gaps between two, three, and four points.

But now, we need to overlay this structure on the original flap and somehow connect the creases so that the whole thing folds flat. Note that the wedge assembly itself folds flat no matter how many we use, so the only thing we need to do is find the creases that enable the multipoint wedge group to connect to the rest of the flap.

Let's use the three-wedge (four-point) module as an example. If I simply overlay this pattern onto the original Kite Base, it is clear that with no other changes, the pattern will not fold flat (if it is not clear from the crease pattern, draw the pattern and try folding it.)

However, since we'll want the mountain folds at the edges of the wedge group to line up with the raw edges of the square, it's clear that we'll need a crease to bring those two lines together. That forces the two valley folds shown in Figure 5.24, each of which necessarily bisects the angle between the raw edge of the paper and the boundary of the wedge group.

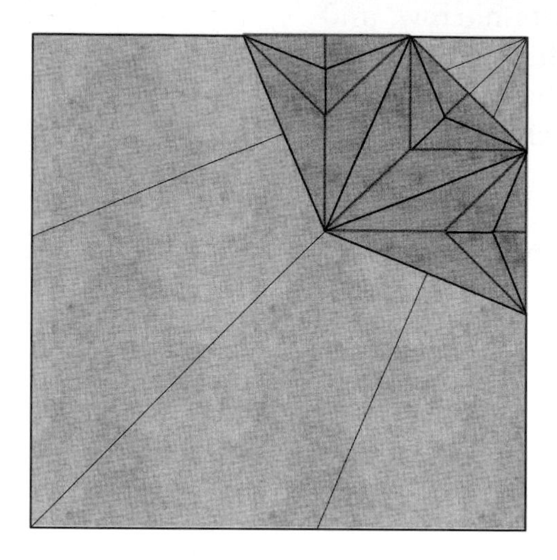

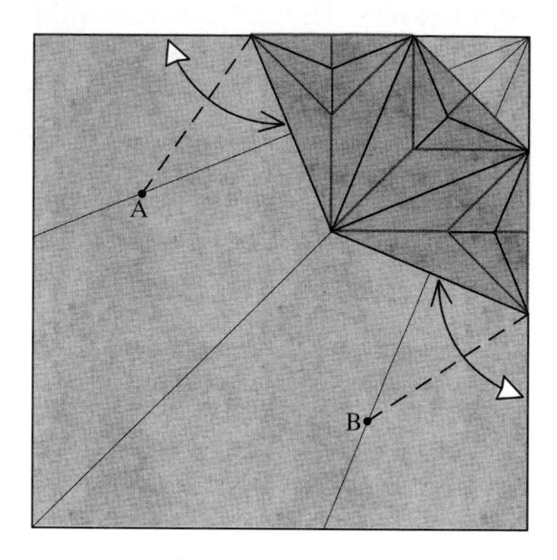

Figure 5.24.
Left: the wedge group, superimposed on a kite flap.
Right: Add two valley folds to make the raw edges line up with the edges of the wedge group.

The points where the valley folds hit the original kite folds—marked A and B in Figure 5.24—mark the transition region between the creases of the kite flap and the creases of the point group. The kite creases don't propagate any farther toward the tip of the paper than these two intersections. Still more creases are required to allow the crease pattern to fold flat, however. While it's possible to calculate the remaining creases necessary, it's far easier to simply fold the model with the creases known so far and force it flat; the necessary creases will fall into just the right place, giving the final crease pattern shown in Figure 5.25.

This can be generalized to larger numbers of points, but you will find that with five or more points, you will have to use narrower wedges (with an apex angle of 30° rather than 45°) in order to put the outermost points on the raw edge of the paper and not cut off some of the inner points, as shown in Figure 5.26.

Extending these creases to connect to kite flap creases is left as an exercise for the reader. Here's another exercise: Can you use this technique on a Frog Base to create a frog with four toes on the forelegs and five on the back legs?

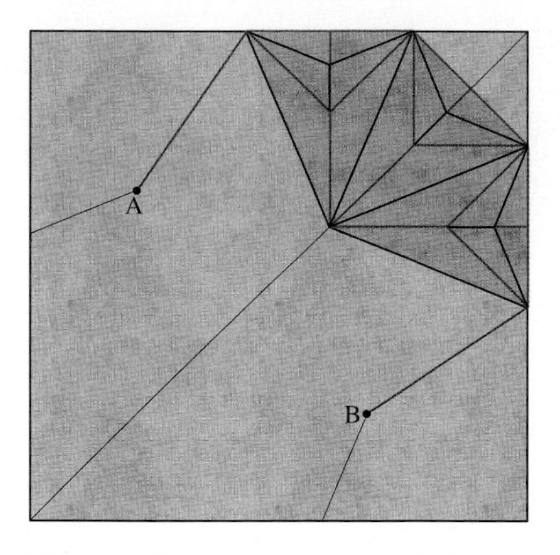

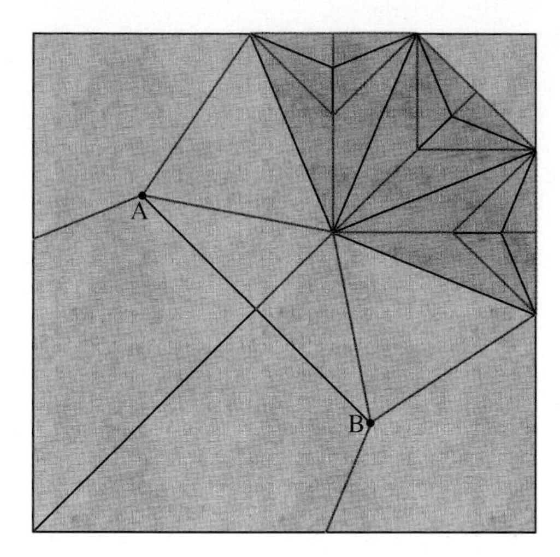

Figure 5.25.
Left: The kite creases terminate at points A and B.
Right: The completed crease pattern. (The corner goes unused and should be folded down before making the creases.)

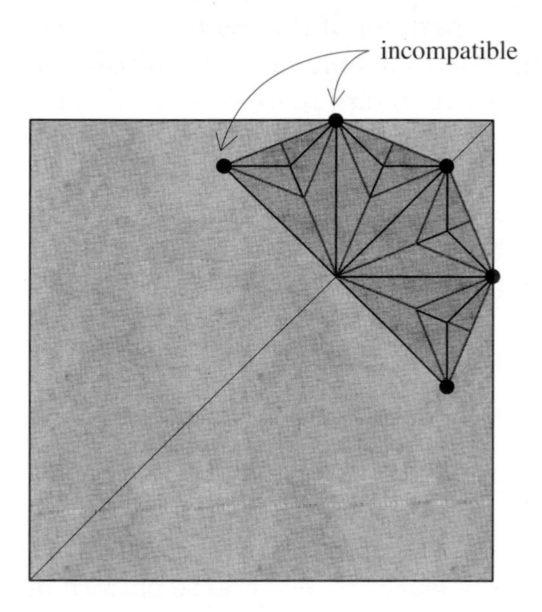

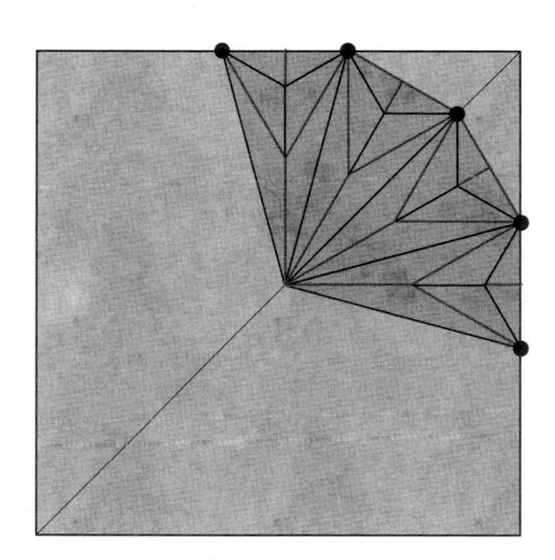

Figure 5.26.
Left: Four wedges don't let you put the outermost points on the raw edge of the flap.
Right: Using narrower wedges permits this construction to be extended to five or more points.

Another family of splits works particularly well for odd numbers of points. It's shown in Figure 5.27 for three, five, and seven points (what would you ever use seven small points for?), but this method, like the other, generalizes in an obvious way.

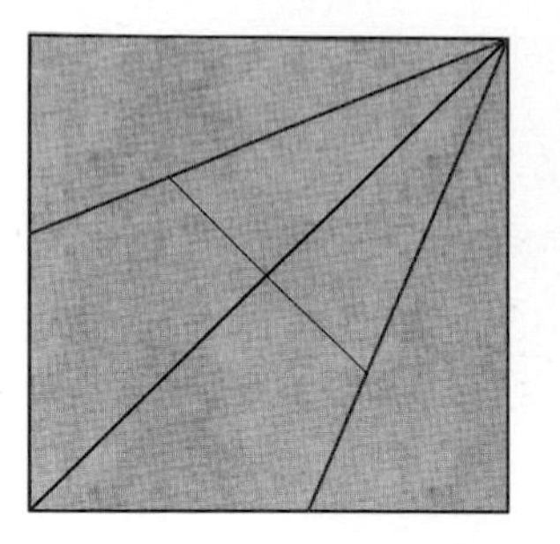
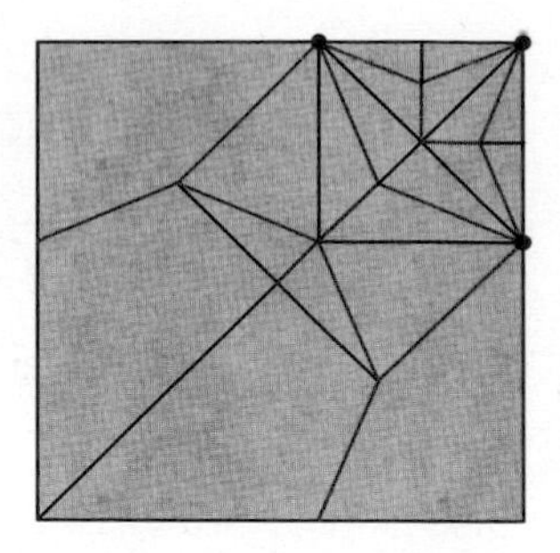
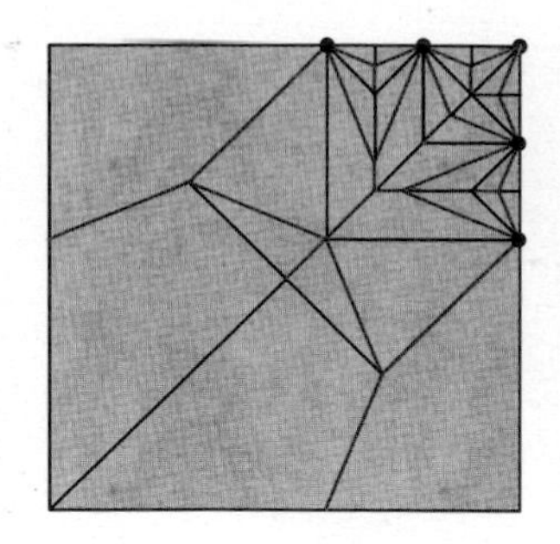
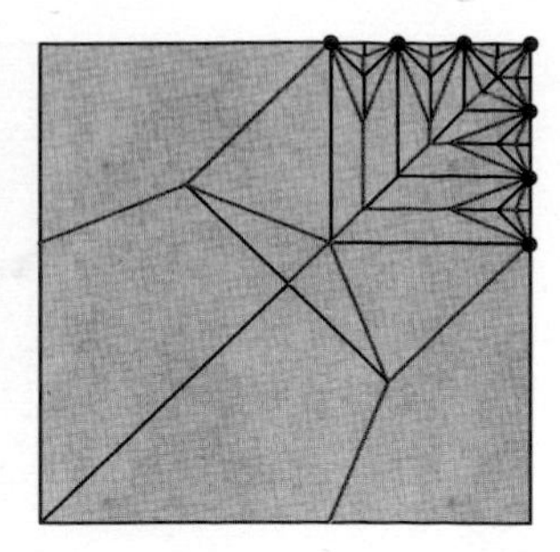

Figure 5.27.
Crease patterns for splitting a point into three, five, and seven smaller points.

There is also a variation of this pattern that works for even numbers of points, which you might enjoy trying to discover. Which of the two families is better? It depends on the model. The two families may be distinguished by the major mountain folds: In the previous example, they radiate from a point; in this family, they are parallel. If the group of points is to be spread out (a technique that enhances the illusion of length), the radial family seems to fan more neatly; it's ultimately a personal choice dictated by the aesthetics of the model.

I would encourage you to fold up a few bases and try out the different splitting techniques; then unfold them and examine the crease patterns. Most point-splitting sequences have a distinct pattern of creases in which the converging creases that form a flap suddenly stop at an obtuse triangular pleat that then radiates outward with creases that form two, three, or more points.

5.5. More Applications of Splitting

Once you become familiar with point-splitting, you can use it in many ways to form pairs of features. The crease pattern in Figure 5.28 is the base for a Walrus. Can you elucidate its structure from the crease pattern? It is recognizably a version of a Bird Base (to be precise, a stretched Bird Base), but two opposite flaps have been split—to make the tusks at one end and the tail flippers at the other. Figure 5.28 shows the crease pattern, the base, and the folded model. From these clues, you should be able to reconstruct the model entirely. (If you can't fold the model from the crease pattern, base, and folded model, full instructions may be found in books cited in the references.)

A more sophisticated form of point-splitting is employed in the model whose crease pattern is shown in Figure 5.29. The Grasshopper is also clearly from a Bird Base, but with

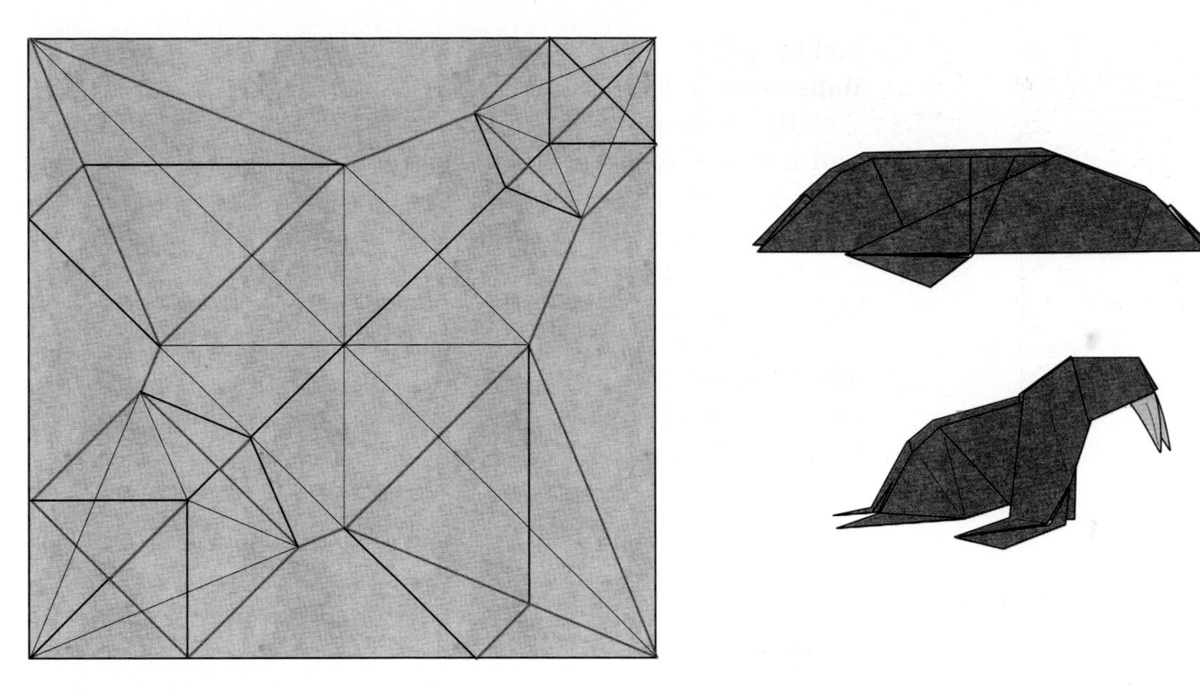

Figure 5.28.
Crease pattern, base, and folded model of the Walrus.

Figure 5.29.
Crease pattern, base, and folded model of the Grasshopper.

three splits. The central point has been sunk and Yoshizawa-split; then two of the four long flaps have also been split into three points each. This, too, is a challenge: Can you figure out how to perform the splitting functions on a Bird Base to yield the Grasshopper's base? (And then, shape the resulting flaps into the folded model.)

By using point-splitting, you can add extra appendages and features to models made from existing bases. However, there is a cost in layers, and a limit on size. As we have seen, even the theoretically optimum two-point split results in flaps less than half the size of the flap you started with. Point-splitting cannot double a flap at the same size, or increase the length of a flap, or add more points to the end of the flap without shortening it. For that, one must find a way to add paper to the model while preserving its basic structure. We'll find out how in the next chapter.

Folding Instructions

Pteranodon

Goatfish

Pteranodon

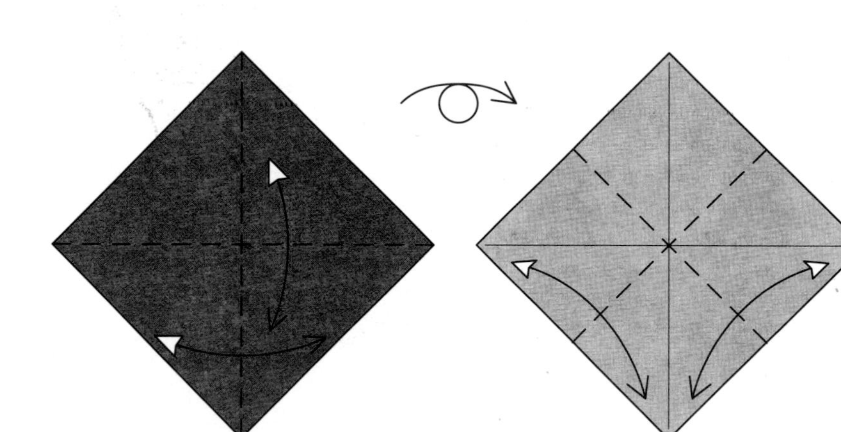

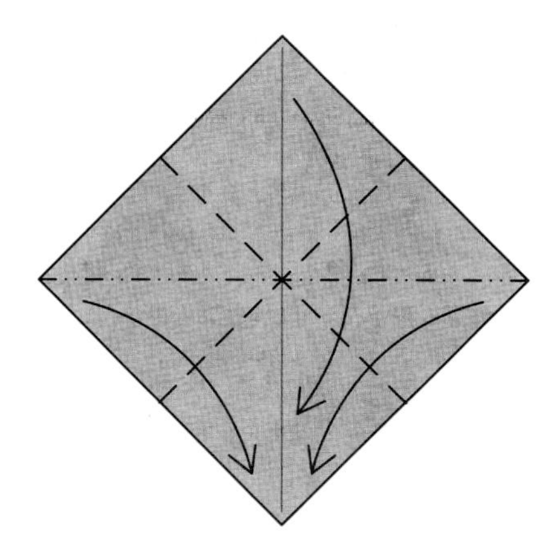

1. Begin with a square, colored side up. Fold and unfold along the diagonals. Turn the paper over.

2. Fold the paper in half from edge to edge and unfold. Repeat the other direction.

3. Bring all four corners together at the bottom, forming a Preliminary Fold.

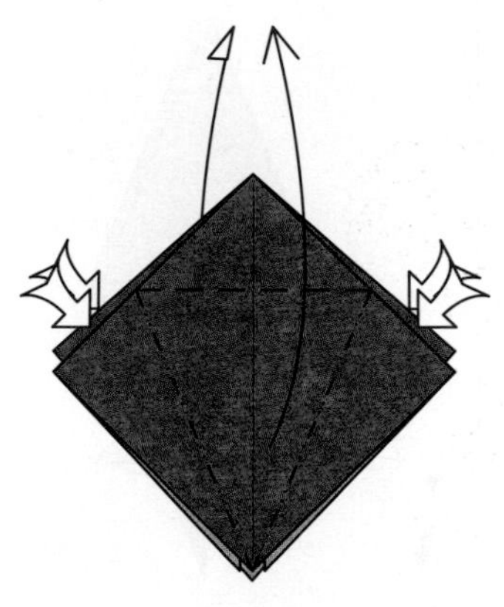

4. Petal-fold the front and back flaps to form a Bird Base.

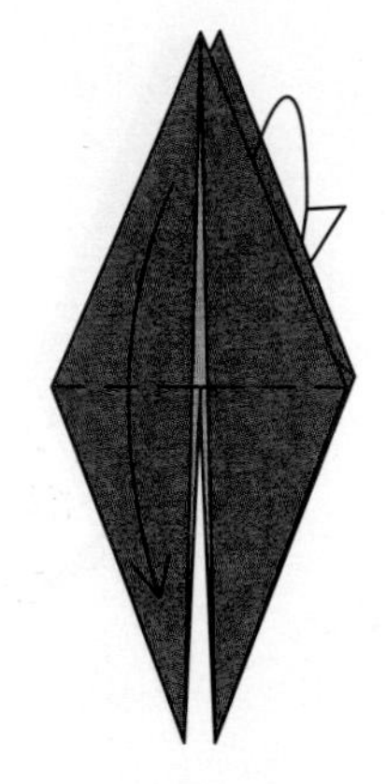

5. Fold one flap down in front and one behind.

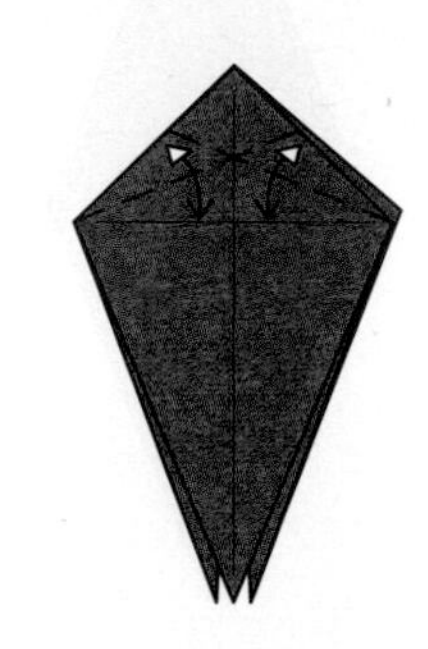

6. Fold the upper left edge down to the horizontal crease and unfold. Repeat with the upper right edge.

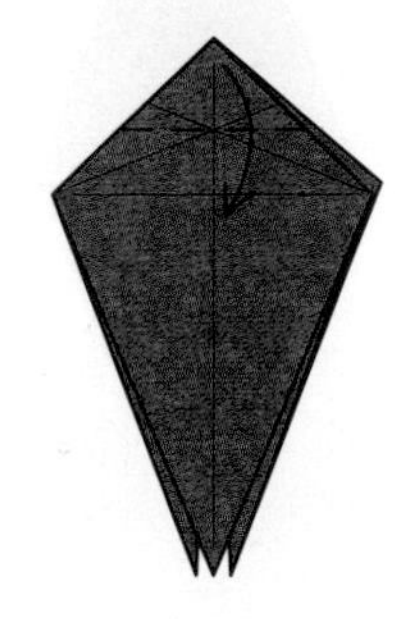

7. Fold the top point down along a crease that passed through the crease intersection along the center line.

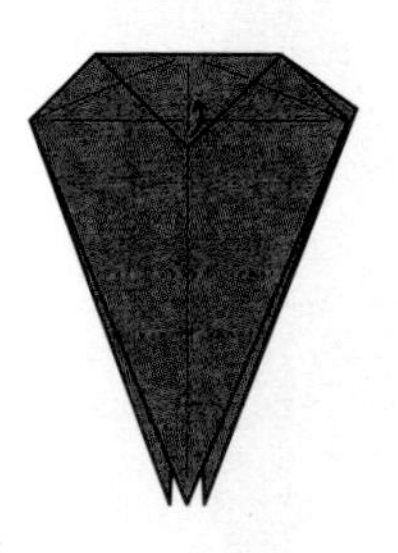

8. Fold up the tip of the corner along a crease aligned with the existing horizontal crease.

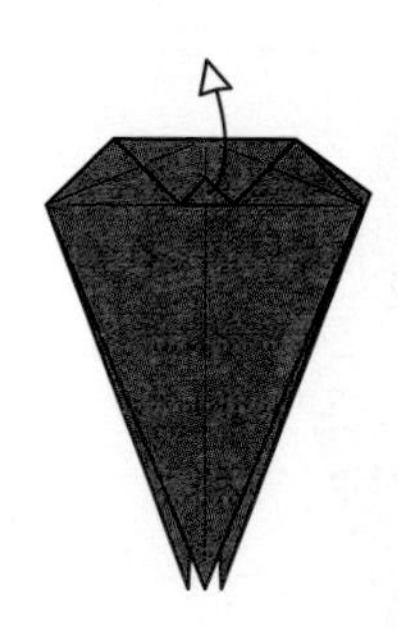

9. Unfold to step 6.

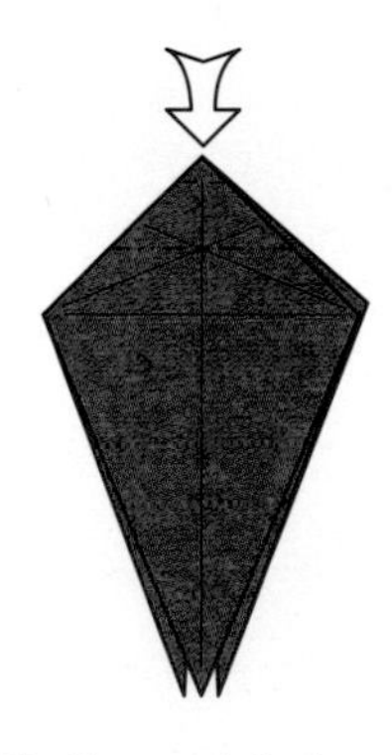

10. Open-sink the point in and out on the existing creases.

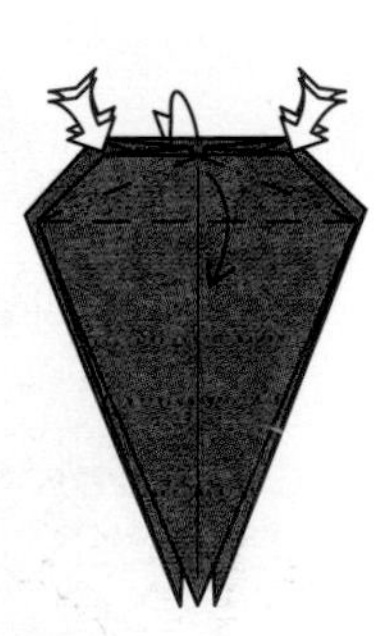

11. Spread-sink the two top near corners. Repeat behind.

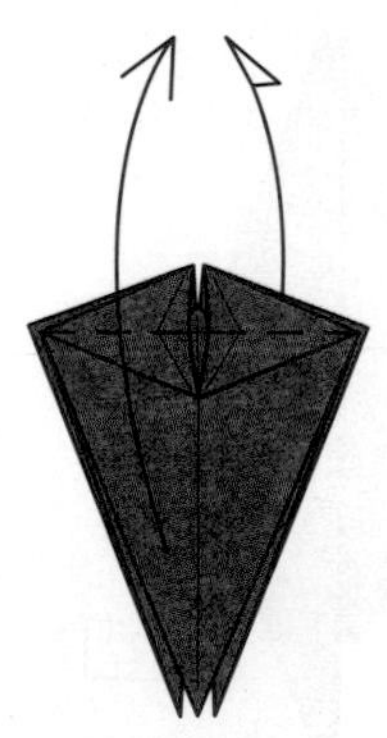

12. Fold one long flap up in front and one up behind.

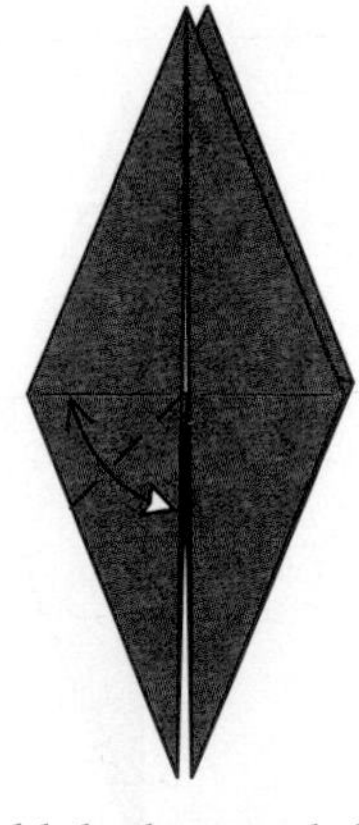

13. Fold the bottom left flap up to lie along the horizontal crease; crease lightly and unfold.

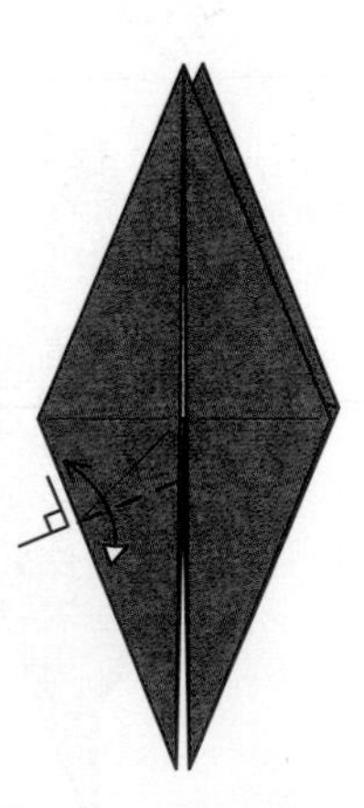

14. Fold the flap up again so that the crease hits the edge at the same point, but now the left edges are aligned. Crease firmly and unfold.

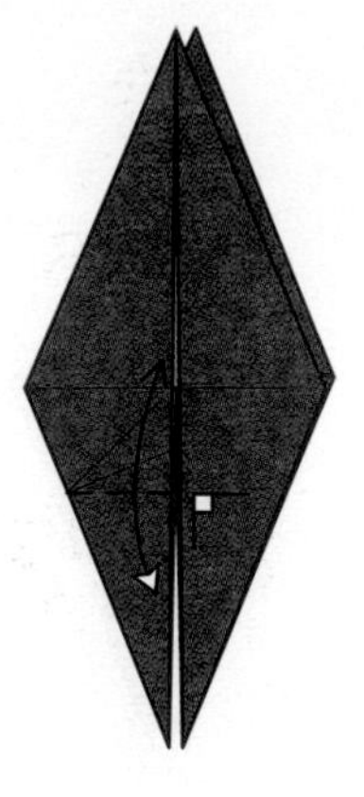

15. Again, fold the flap up with the crease hitting the left edge in the same place, but now the right edges are aligned. Crease firmly and unfold.

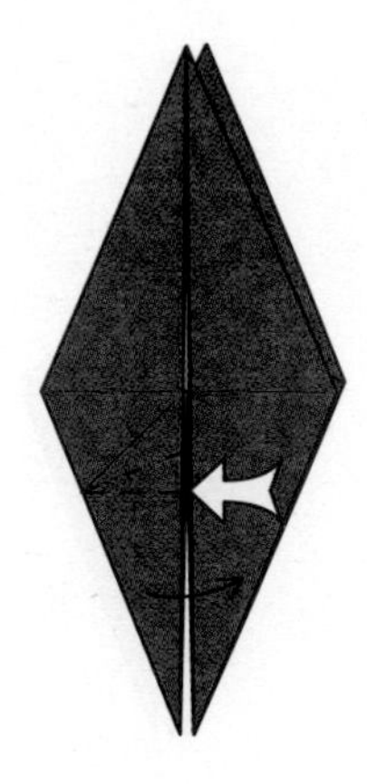

16. Crimp using the existing creases.

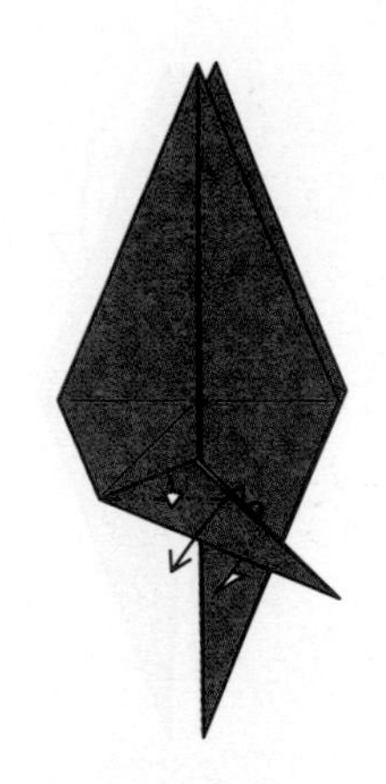

17. Pull out the loose paper completely on both the near and far sides of the flap.

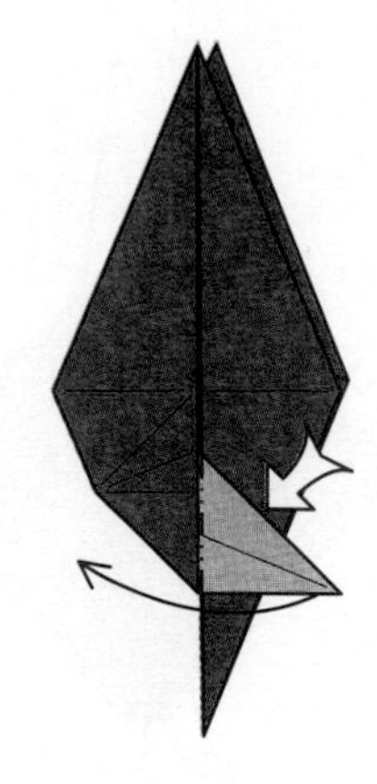

18. The raw edges should be perfectly horizontal; if they aren't, adjust the crimp and flatten. Then reverse-fold the white flap to the left.

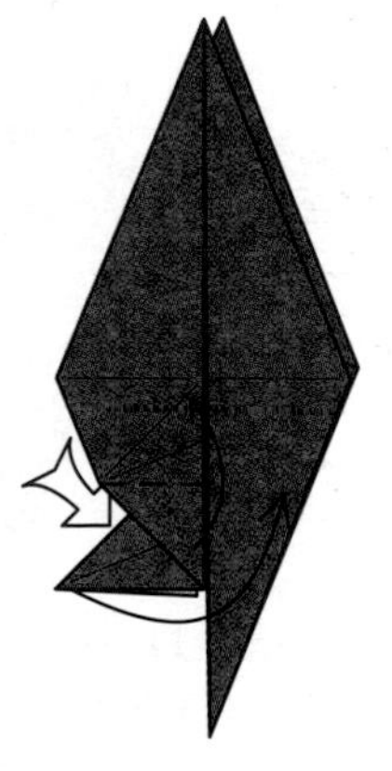

19. Squash-fold the flap over to the right.

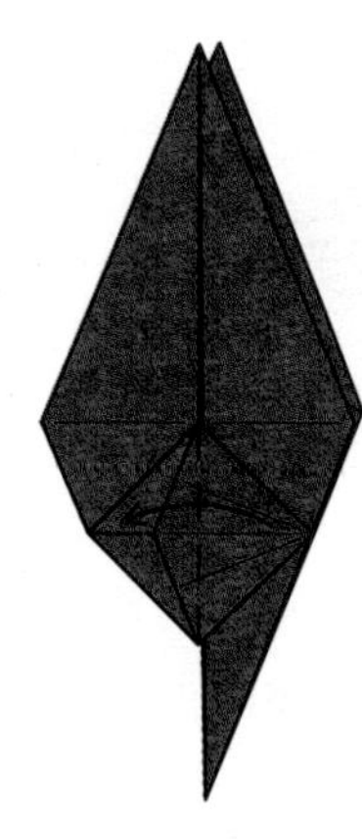

20. Fold the corner over to the left.

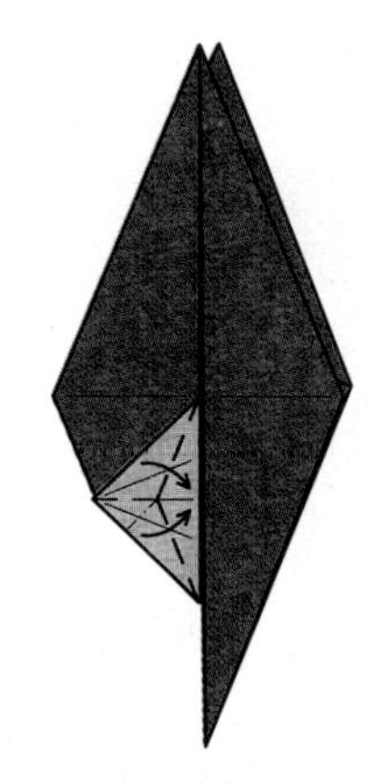

21. Fold a rabbit ear.

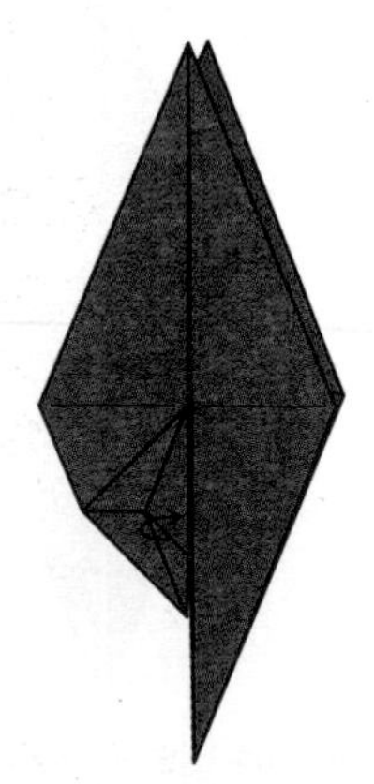

22. Bring one layer in front, thus hiding the tip of the rabbit ear.

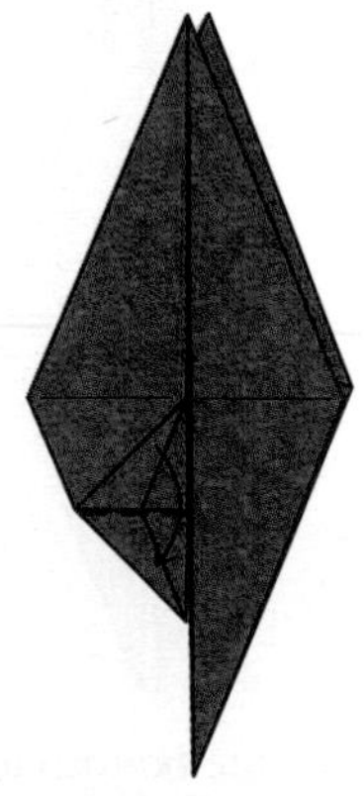

23. Fold one flap down.

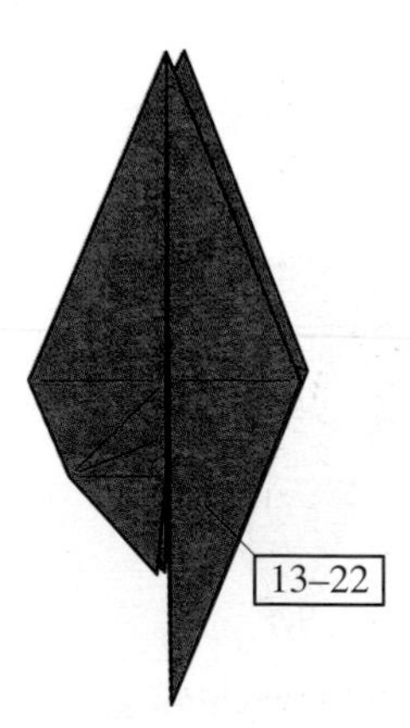

24. Repeat steps 13–22 on the right.

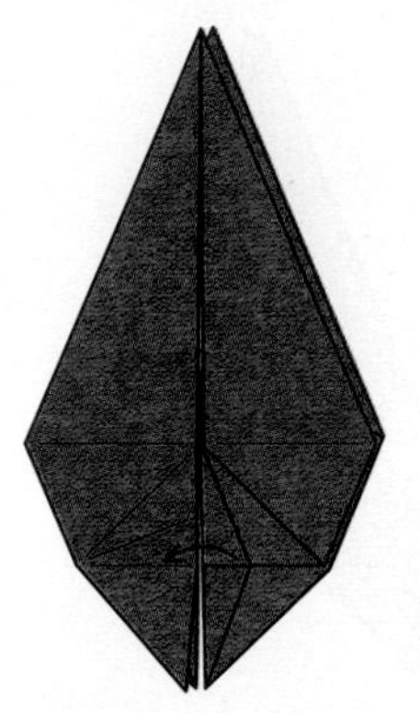

25. Fold one layer to the left.

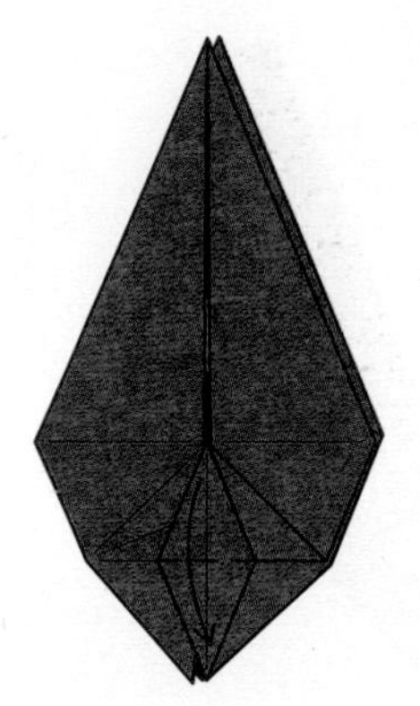

26. Fold one flap down.

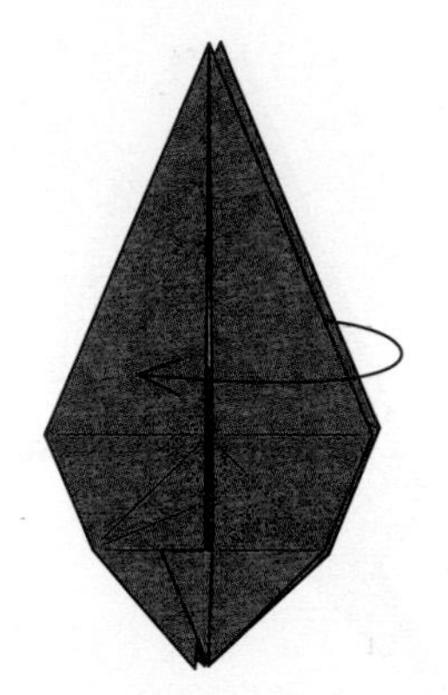

27. Open the right side of the model, which causes the hidden edge to squash downward.

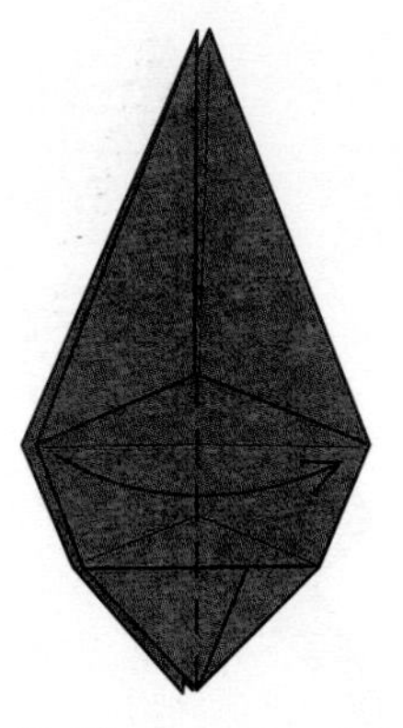

28. Close the model back up.

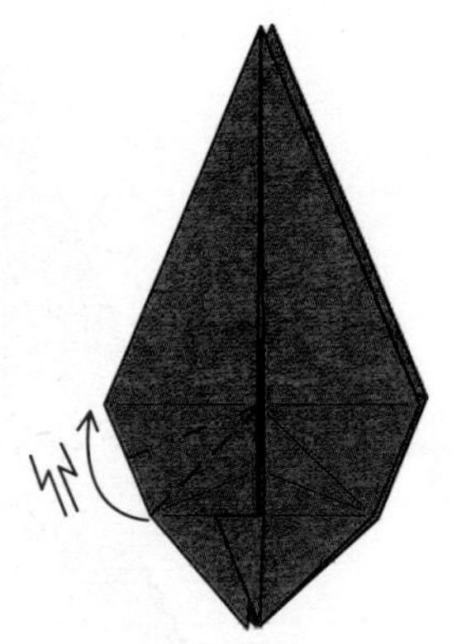

29. Crimp the left side upward.

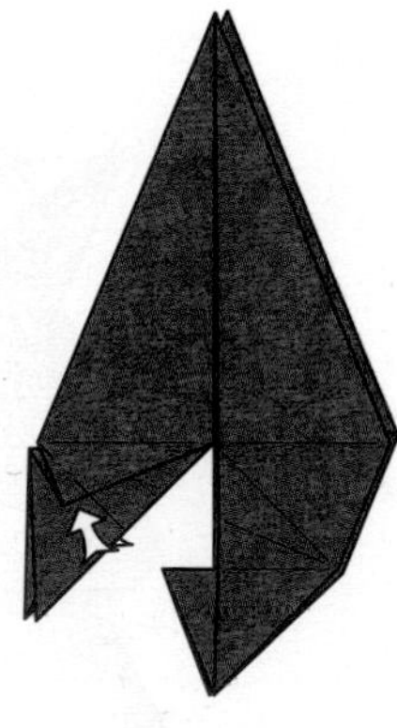

30. Reverse-fold the corner. Note that the mountain fold aligns with a valley crease on the layer beneath. Repeat behind.

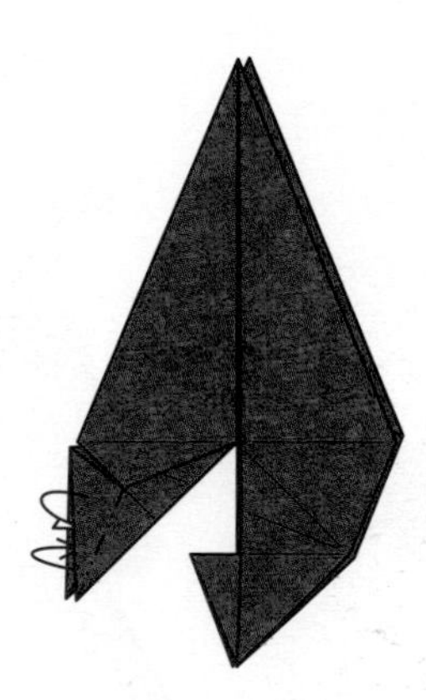

31. Mountain-fold an edge behind. Repeat behind.

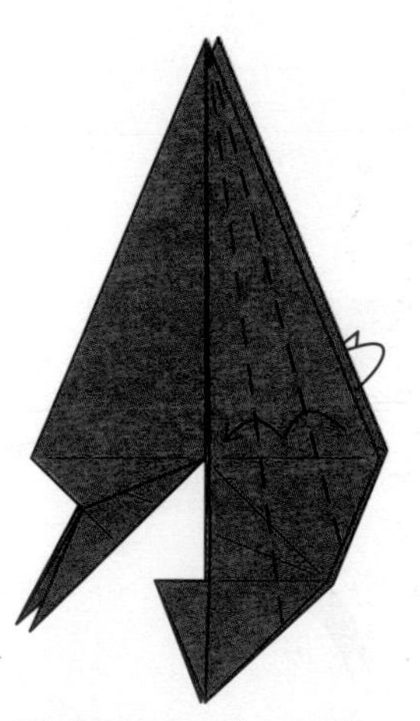

32. Fold the right edge over and over, dividing the angle in thirds. Repeat behind.

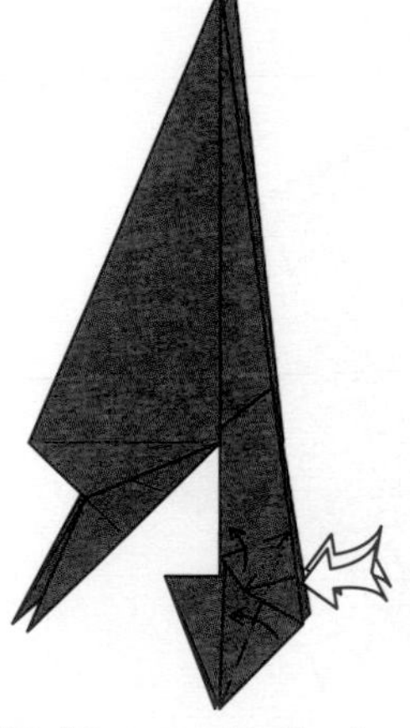

33. Narrow the flap by forming a rabbit ear. Repeat behind.

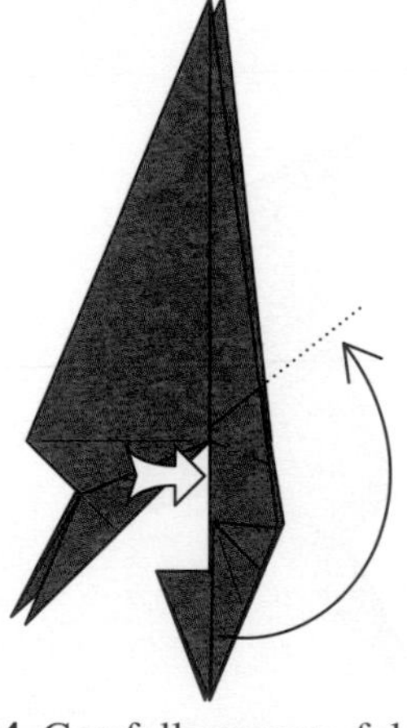

34. Carefully reverse-fold through the thick layers. The top edge of the reverse fold hits the marked spot.

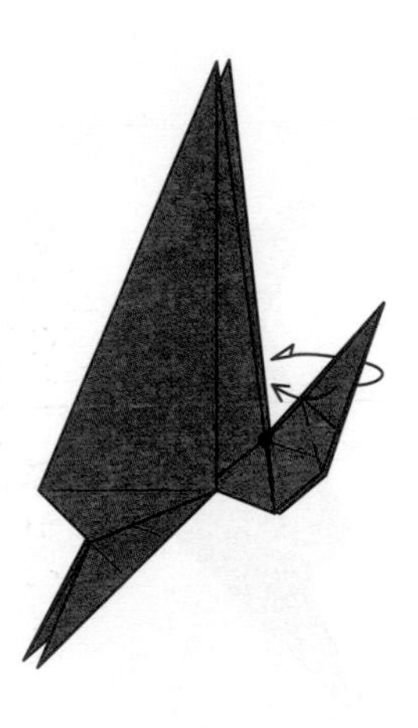

35. Outside-reverse-fold the layers.

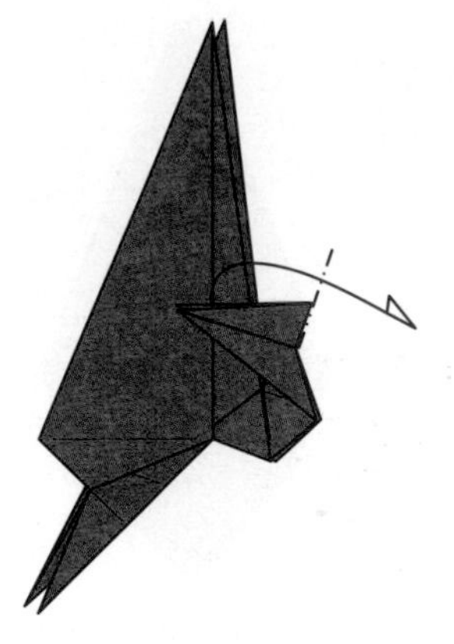

36. Mountain-fold the rear flap to the front as far as possible.

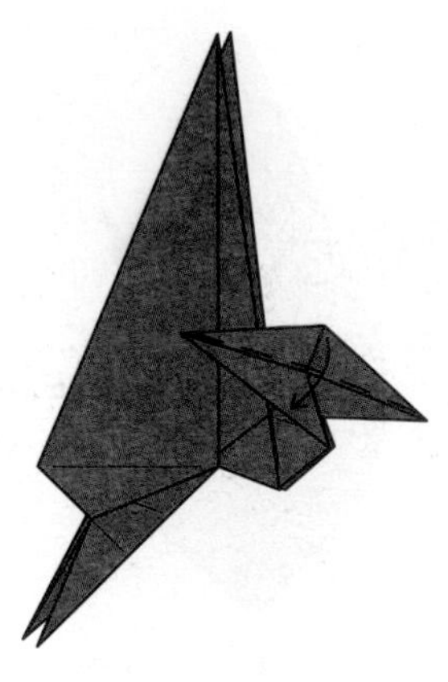

37. Fold the top edge down.

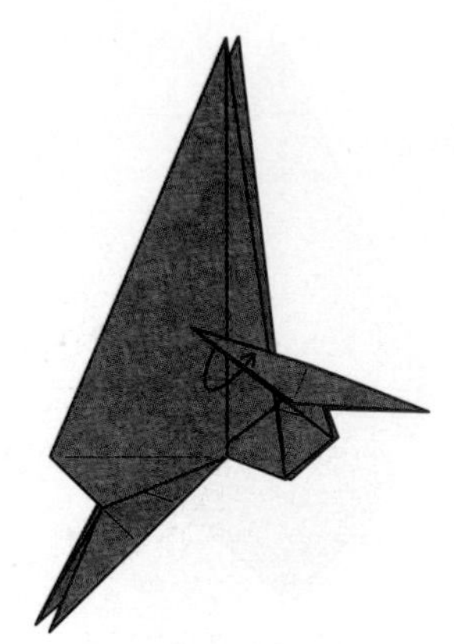

38. Bring the near wing in front of the head.

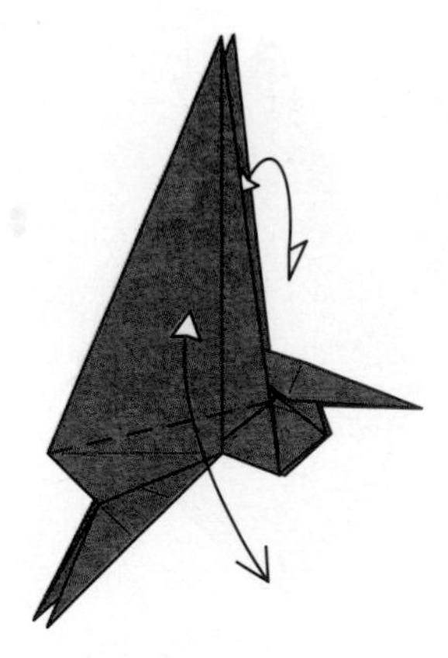

39. Fold the wing down and unfold. Repeat behind.

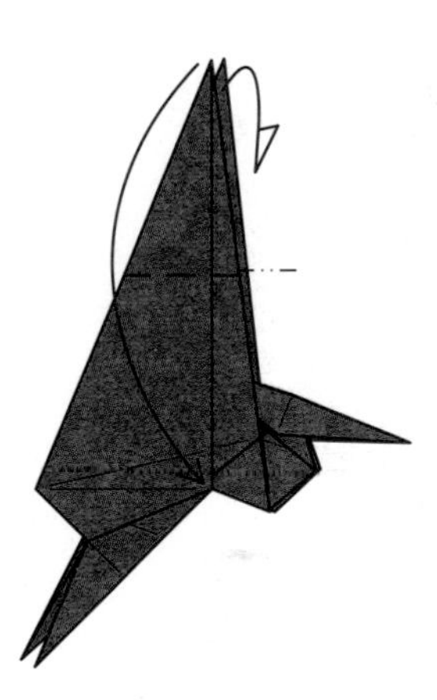

40. Fold the wing tip down. Repeat behind.

41. Fold the wing tip back up so that the left edge is aligned with the layer beneath. Repeat behind.

42. Unfold to step 40.

43. Crimp the wing on the existing creases. Repeat behind.

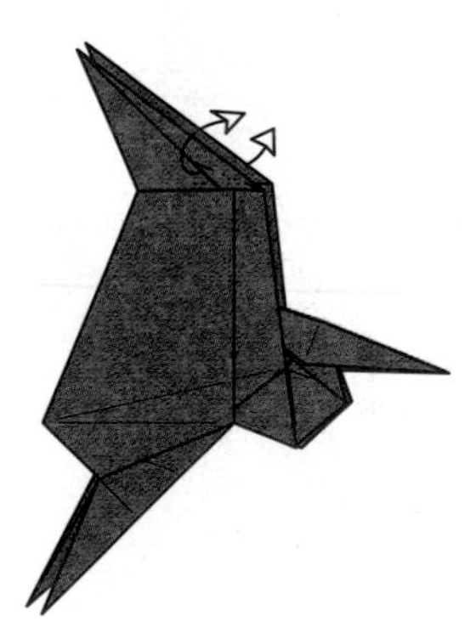

44. Pull out some loose paper from the leading edge of the wing. Repeat behind.

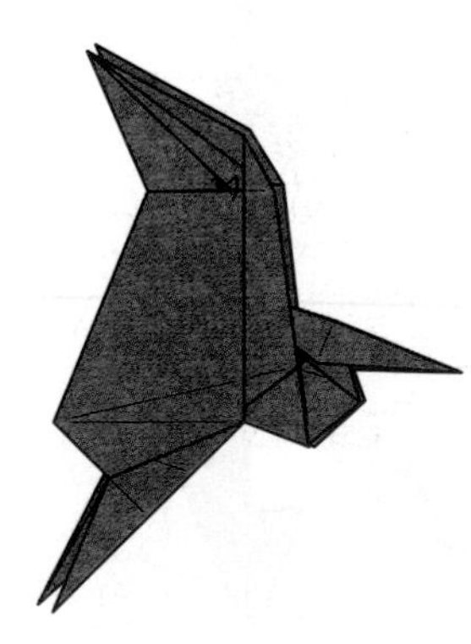

45. Tuck the small corner under a raw edge. Repeat behind.

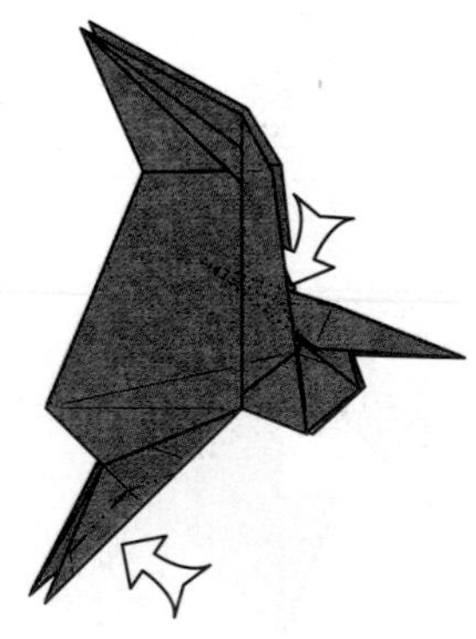

46. Shape the head and legs.

47. Finished Pteranodon. Hold the neck and feet and pull to make him flap his wings.

Goatfish

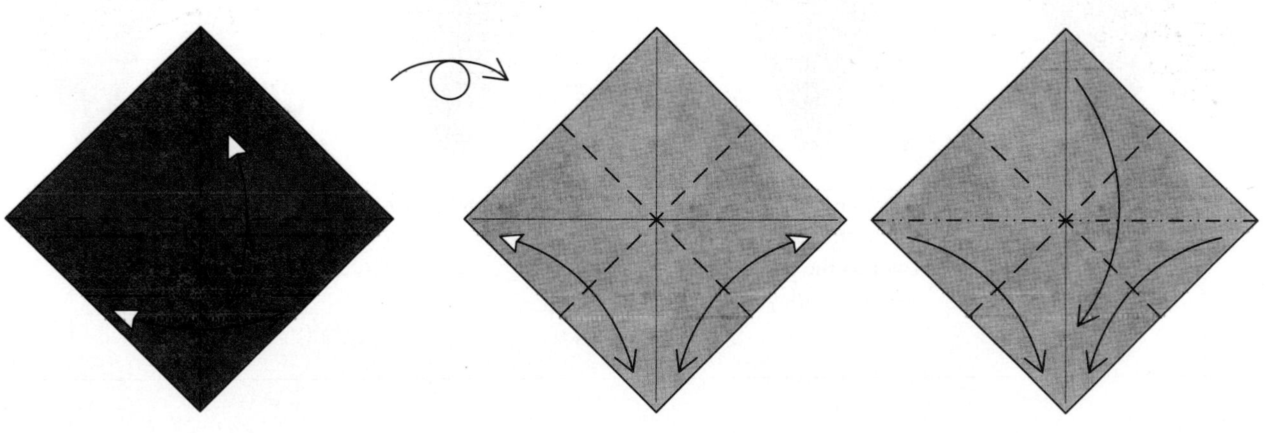

1. Begin with a square, colored side up. Fold and unfold along the diagonals. Turn the paper over.

2. Fold the paper in half from edge to edge and unfold. Repeat the other direction.

3. Bring all four corners together at the bottom, forming a Preliminary Fold.

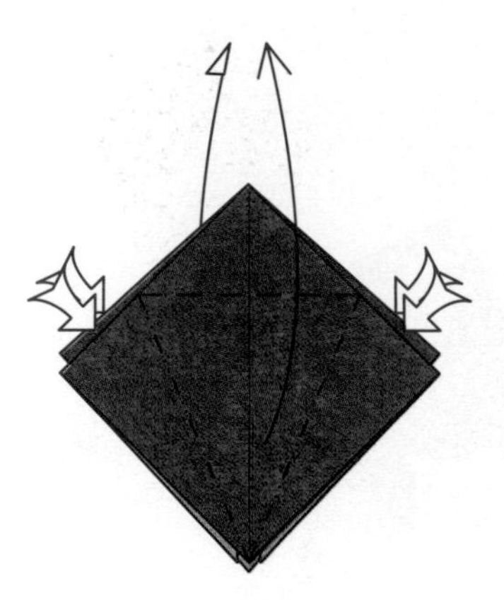

4. Petal-fold the front and back flaps to form a Bird Base.

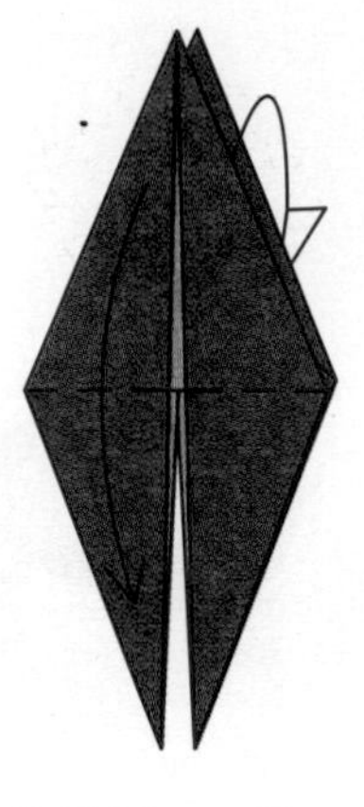

5. Fold one flap down in front and one behind.

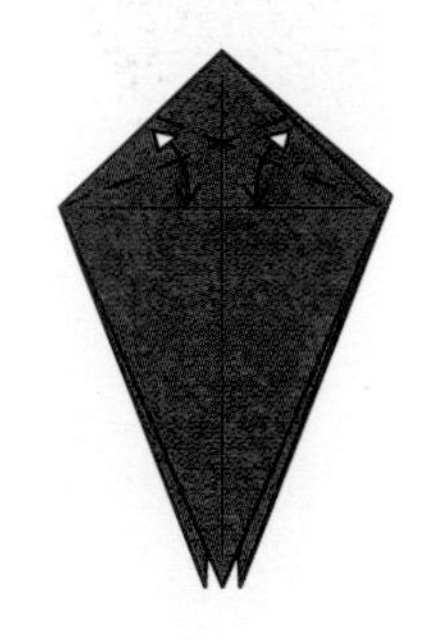

6. Fold the upper left edge down to the horizontal crease and unfold. Repeat with the upper right edge.

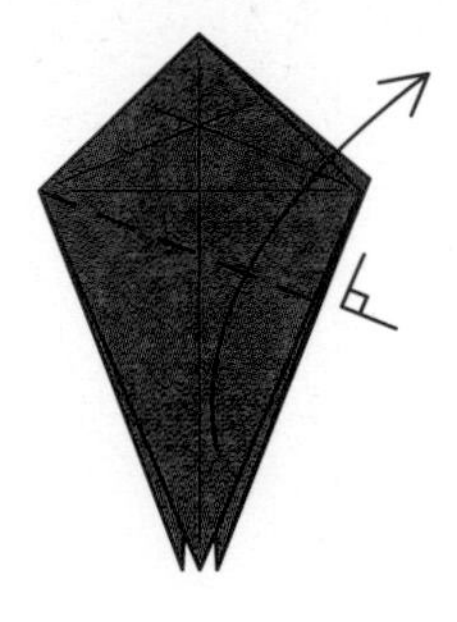

7. Fold one flap up to the right so that the right edge is aligned with the layer underneath.

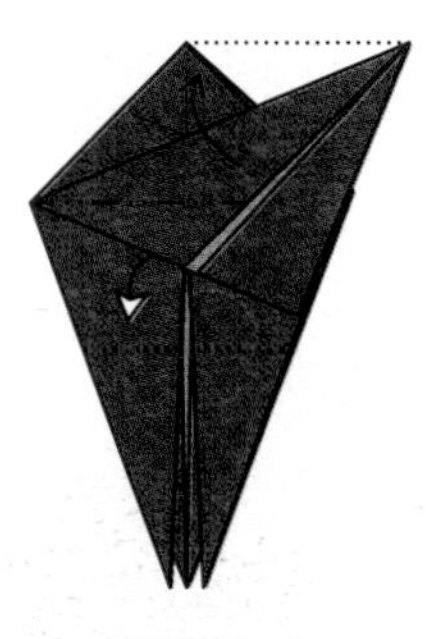

8. Pull out some loose paper as far as possible.

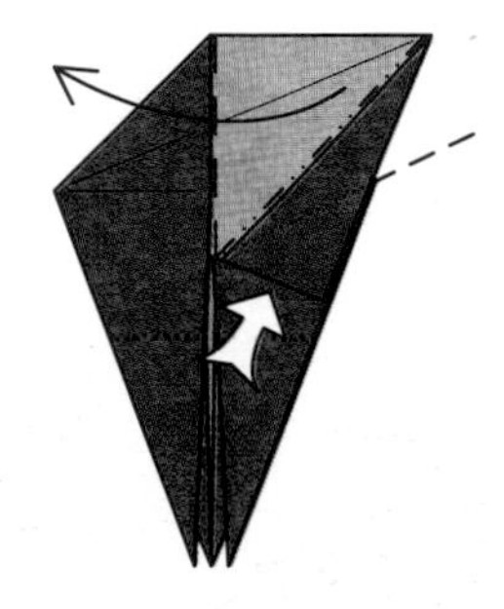

9. Squash-fold the corner, swinging the white flap over to the left along a vertical valley fold.

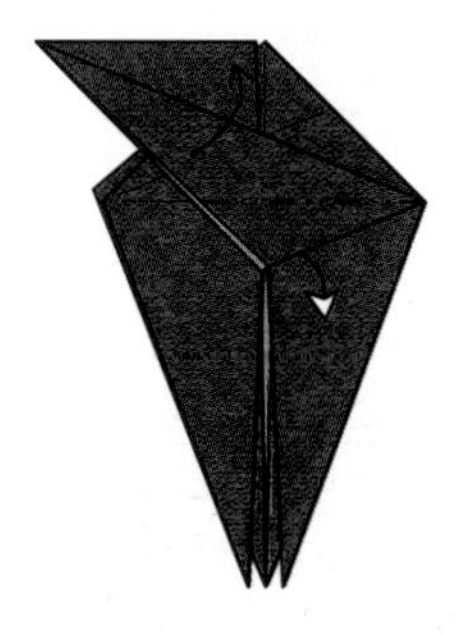

10. Pull out some loose paper on the right as in step 8.

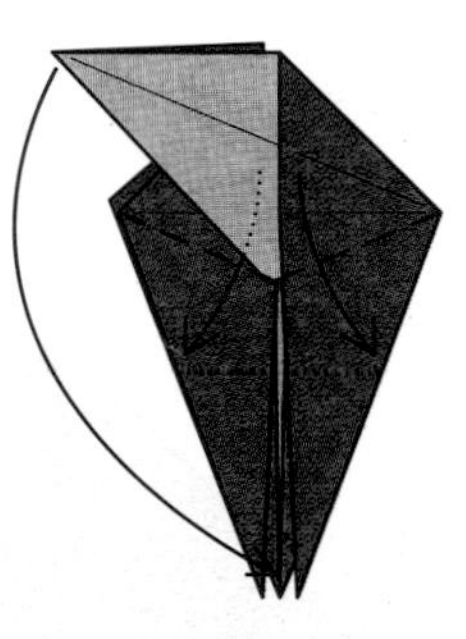

11. Open out the flap and flatten.

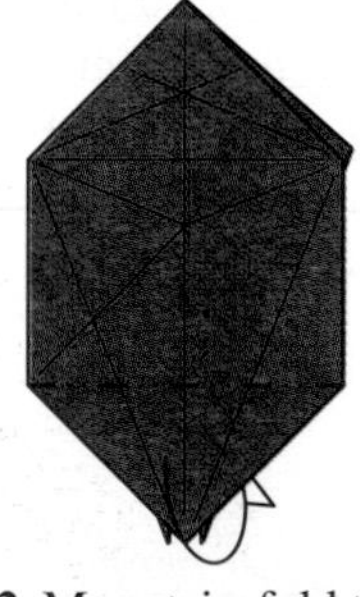

12. Mountain-fold the edge underneath.

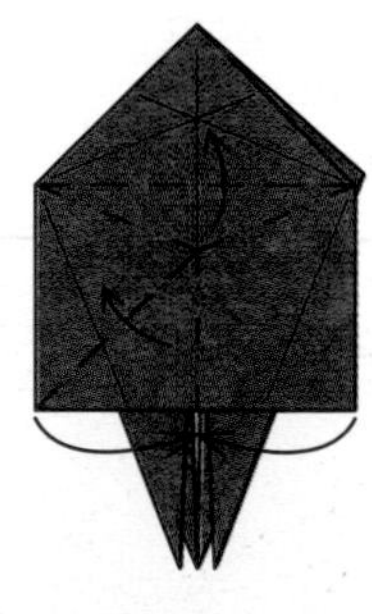

13. Using the existing creases, bring the two bottom corners together while buckling the middle upward.

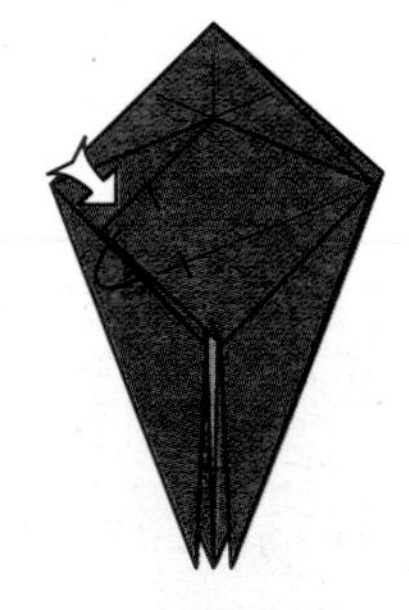

14. Reverse-fold the edge inside.

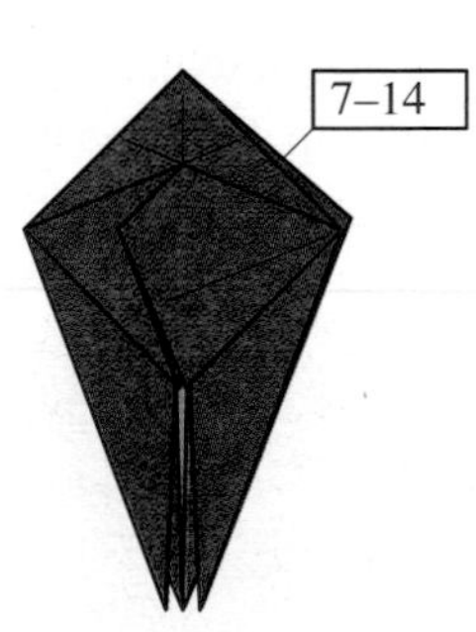

15. Repeat steps 7–14 behind.

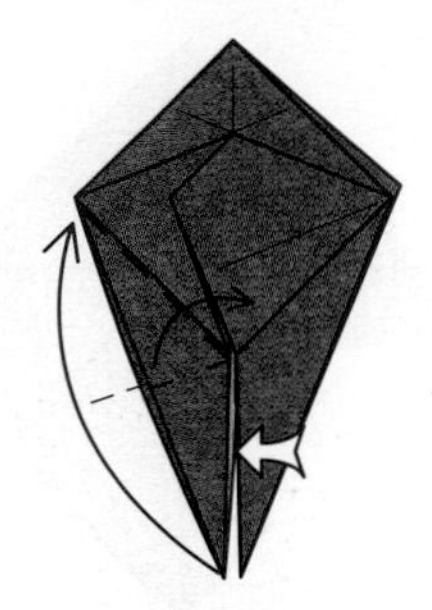

16. Squash-fold one flap.

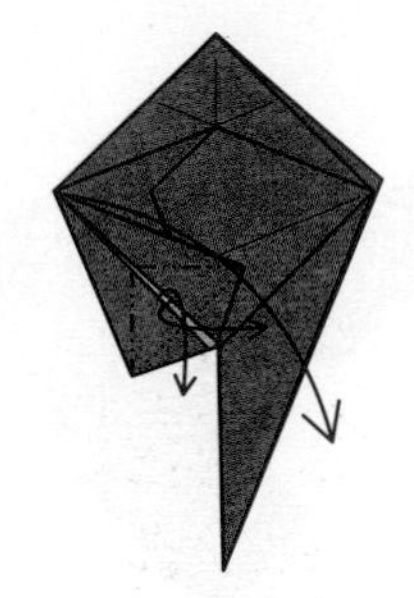

17. Open out the raw edges and fold the flap down and to the right.

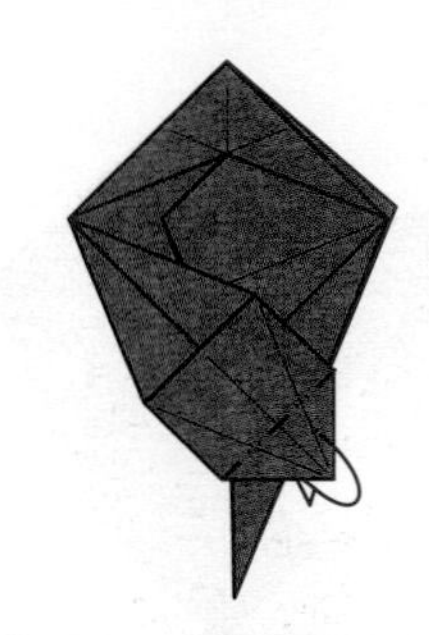

18. Mountain-fold the corner underneath.

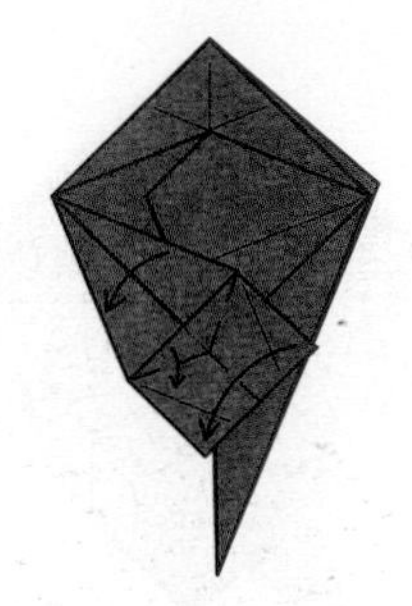

19. Fold the flap in half, forming a small reverse fold as you close it.

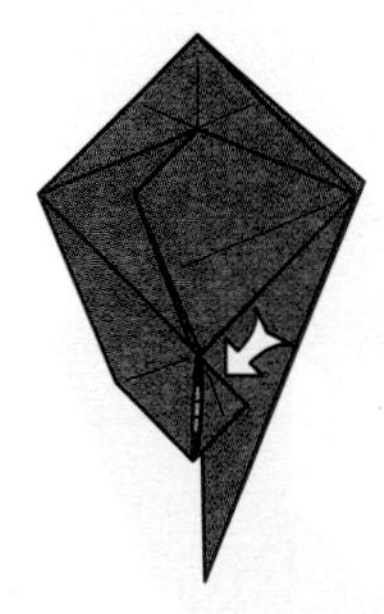

20. Reverse-fold the corner.

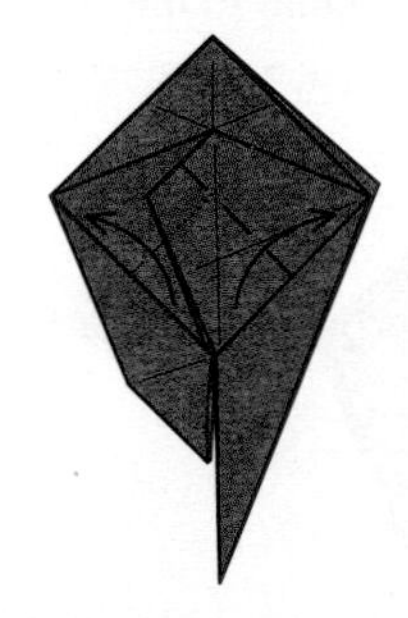

21. Fold the two points up and out to the sides, one to the left, one to the right. Repeat behind.

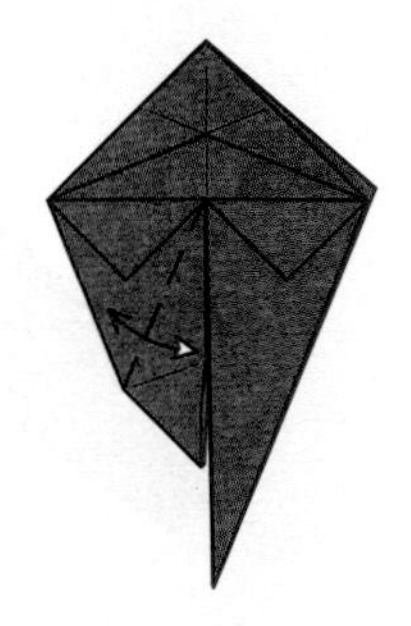

22. Fold and unfold the split point.

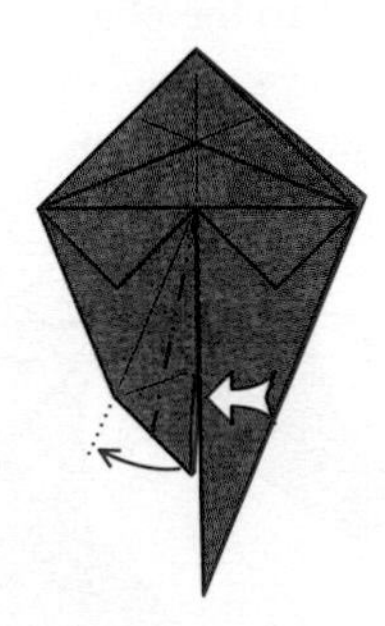

23. Reverse-fold the pair of points to the left, so that the folded edge lines up with the crease you made in step 21.

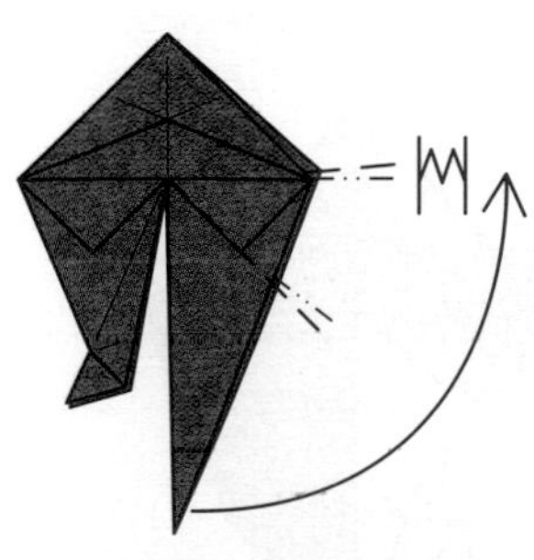

24. Crimp the bottom right point upward so that its bottom edge is horizontal.

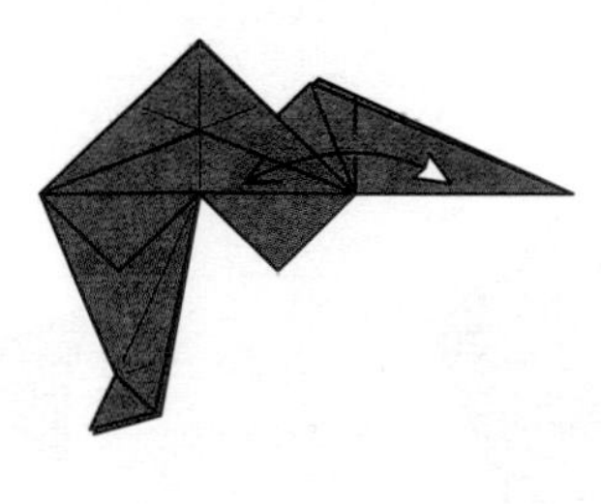

25. Fold and unfold.

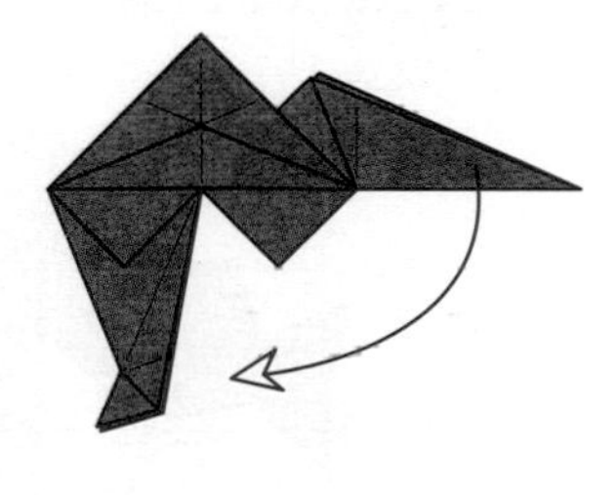

26. Unfold to step 24.

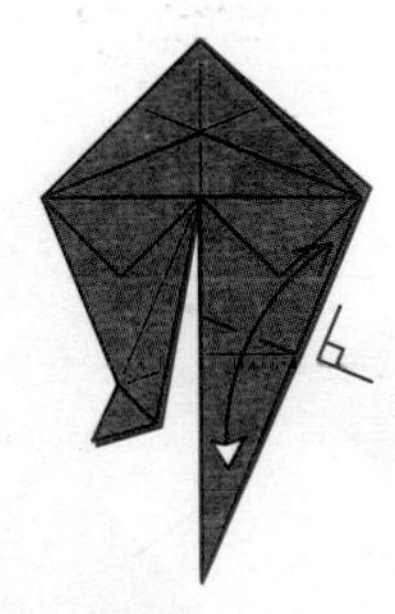

27. Fold and unfold along a crease perpendicular to the right edge.

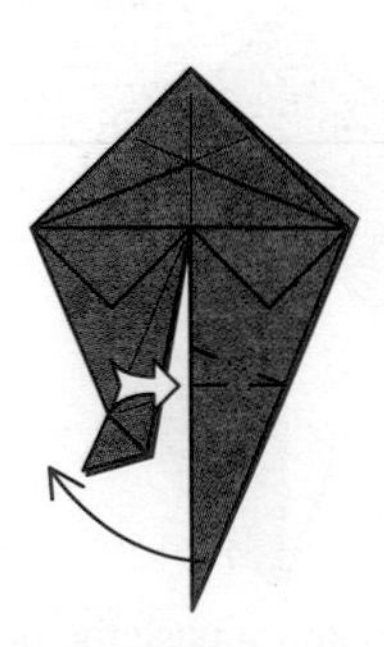

28. Crimp the point using the existing creases.

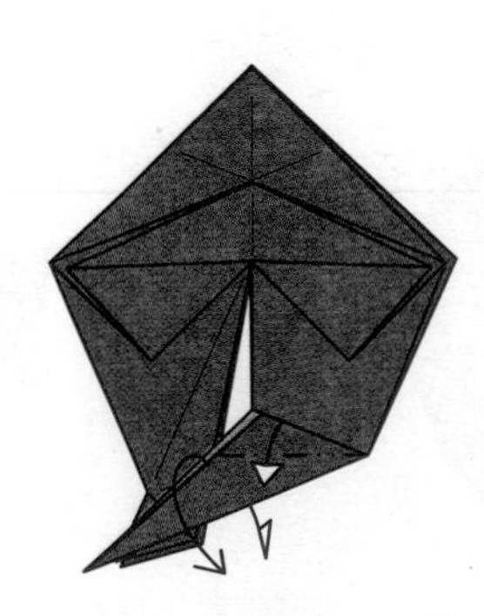

29. Pull out the loose paper completely.

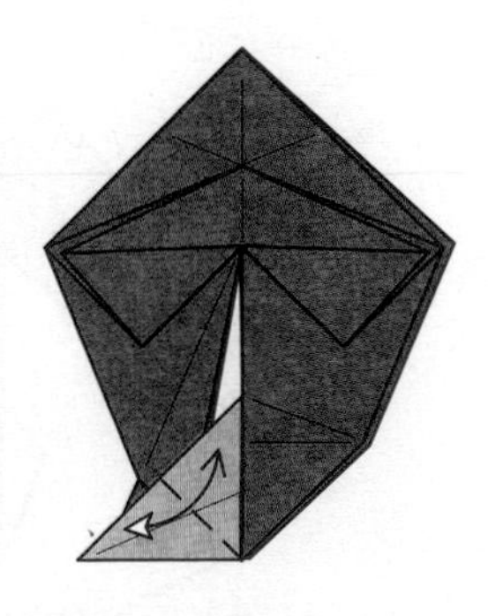

30. Fold and unfold.

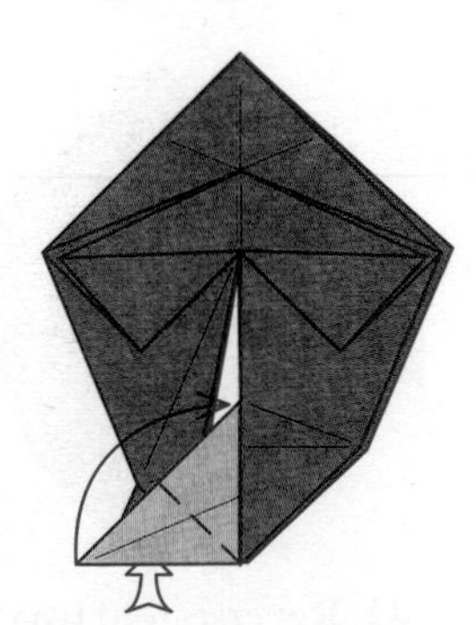

31. Outside-reverse-fold the flap, then put it behind the left flap.

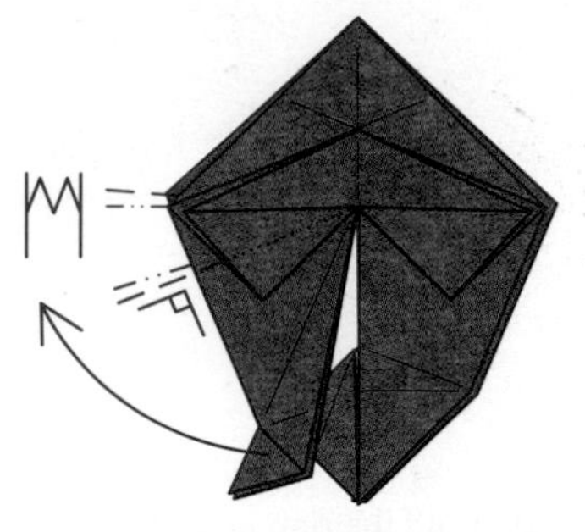

32. Crimp the left point upward; note that the lower creases are perpendicular to the edge.

33. Fold the flap up along the folded edge.

34. Fold the flap back to the left along a crease perpendicular to the edge.

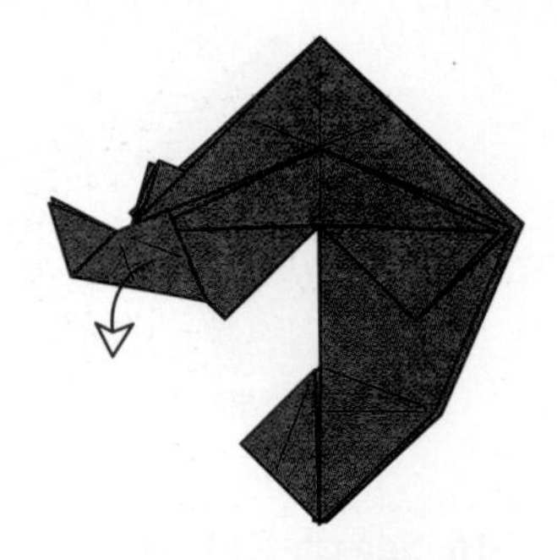

35. Unfold to step 33.

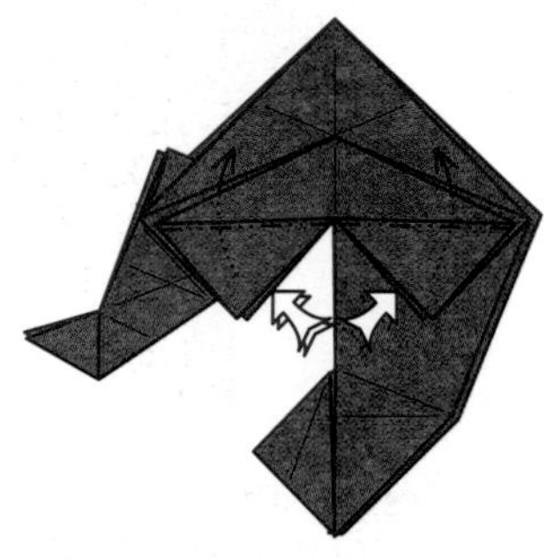

36. Reverse-fold four corners upward.

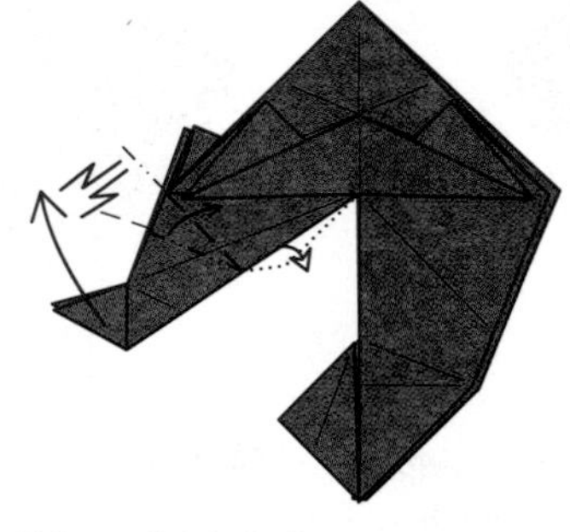

37. Crimp the left flap upward on the existing creases; at the same time, push down some paper from the underside of the flap to the right of the crimp.

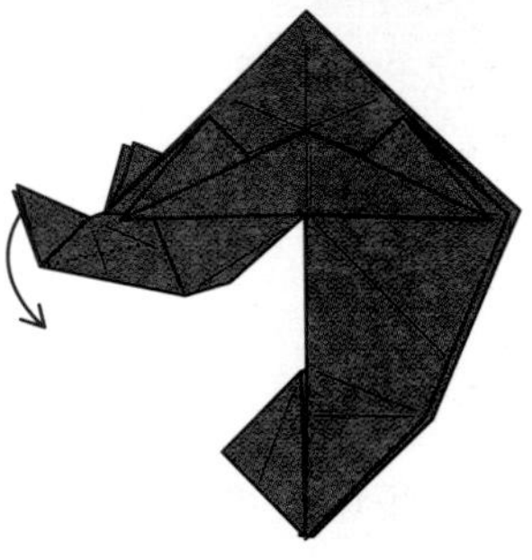

38. Mountain-fold the inside flap as far down as possible.

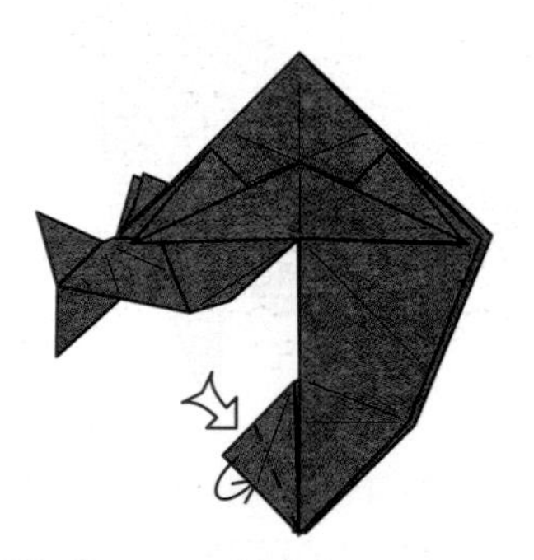

39. Reverse-fold the corner.

40. Unwrap a single layer.

41. Reverse-fold the corner.

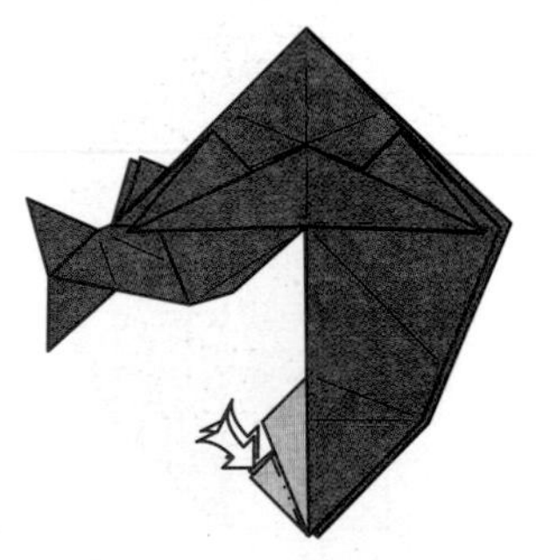

42. Reverse-fold two corners to align with the folded edges.

43. Reverse-fold two corners to align with the folded edges.

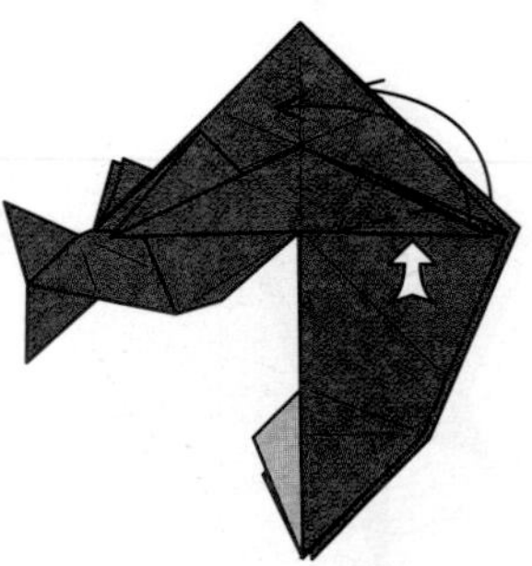

44. Reverse-fold and squash the near flap up and over to the left, forming a long gusset in the layer behind the flap. Repeat behind.

45. Reverse-fold one edge. Do not repeat behind.

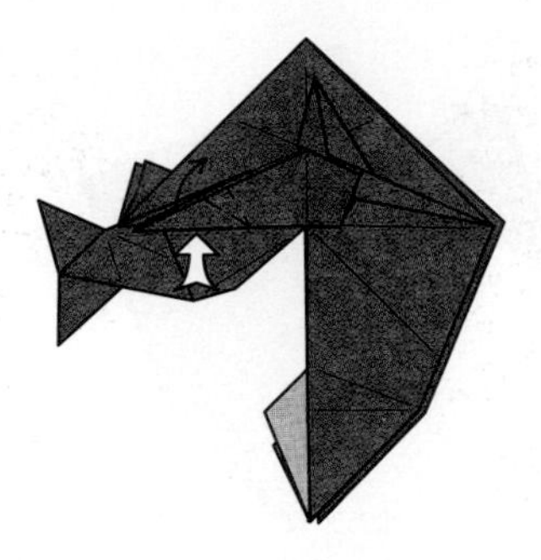

46. Squash-fold the flap upward.

47. Turn the paper over.

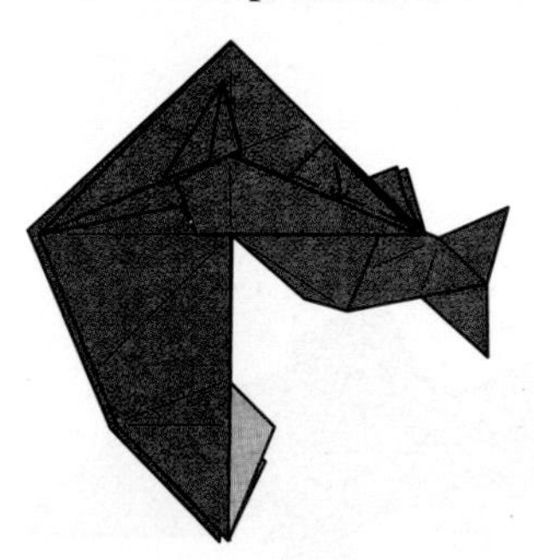

48. Tuck the edge into the pocket.

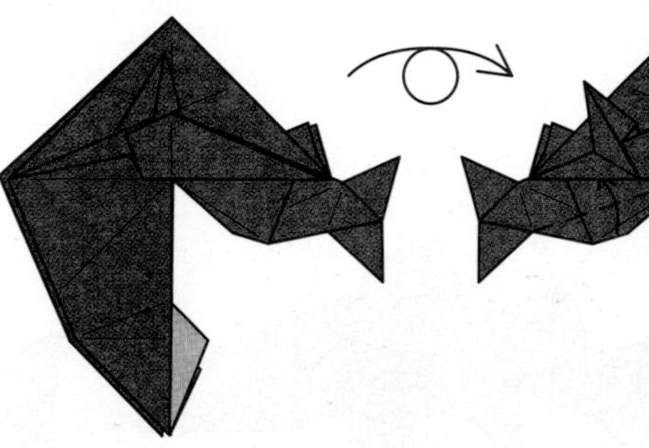

49. Turn the paper back over.

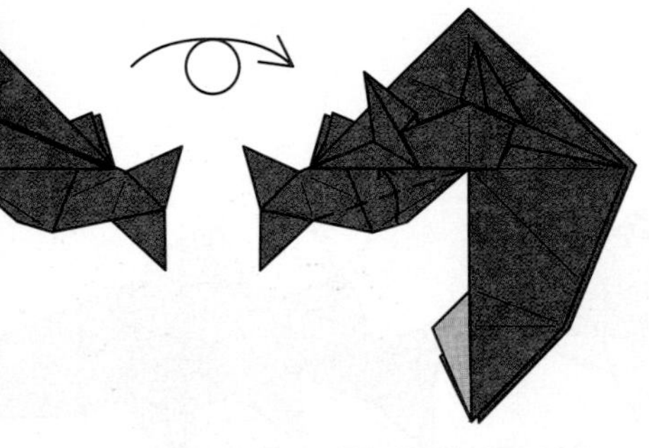

50. Fold the thick edge upward.

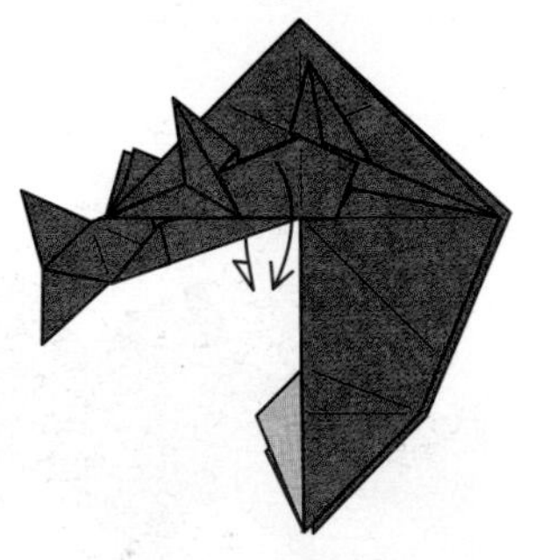

51. Fold one flap down in front and behind.

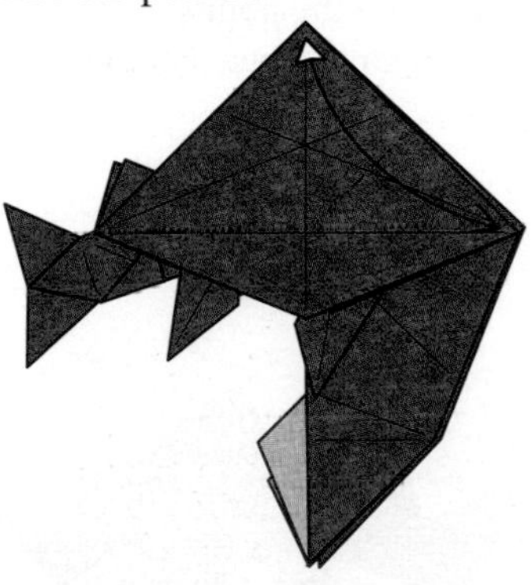

52. Fold and unfold, making a light pinch through the diagonal crease.

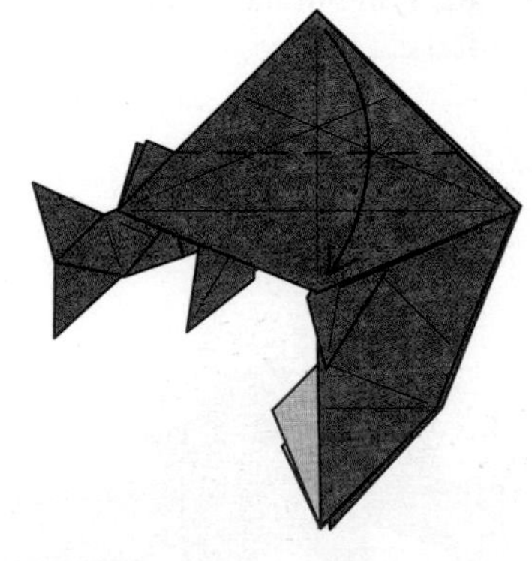

53. Fold the top point down along a horizontal crease that runs through the pinch you just made.

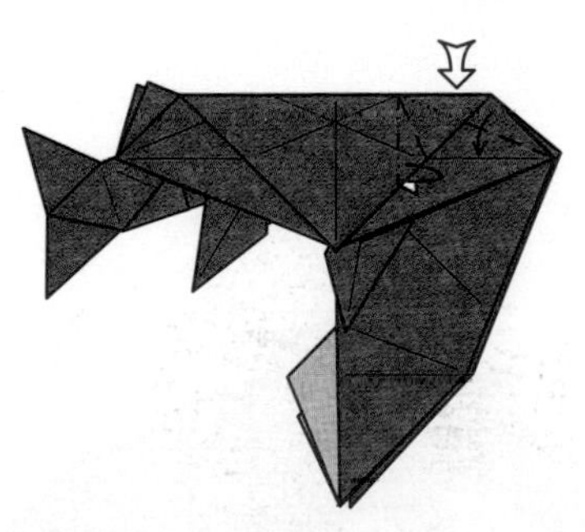

54. Reverse-fold the edge, tucking the excess paper underneath using a vertical mountain fold.

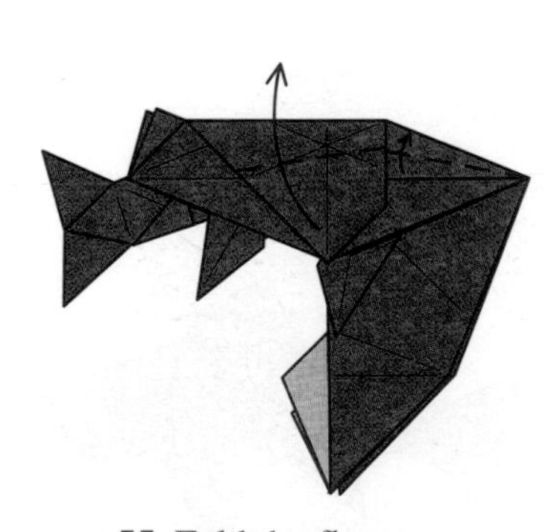

55. Fold the flap up.

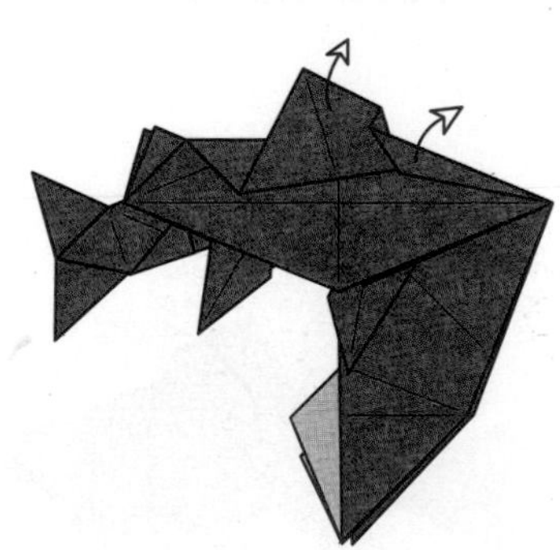

56. Unfold to step 53.

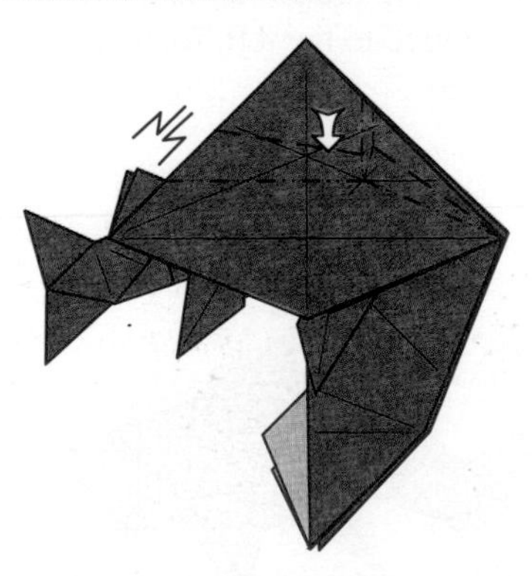

57. Refold, making the creases on the front side match the creases on the far side.

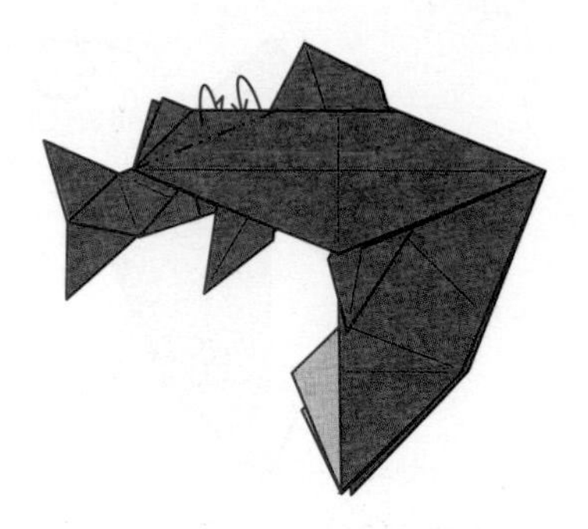

58. Mountain-fold the edges inside in front and behind.

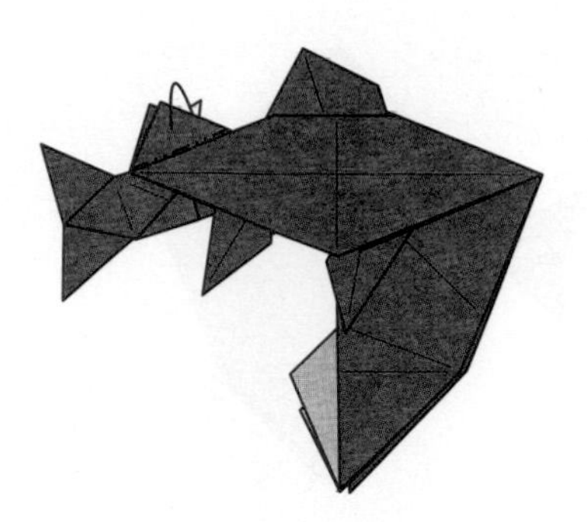

59. Mountain-fold one of the two paired flaps inside.

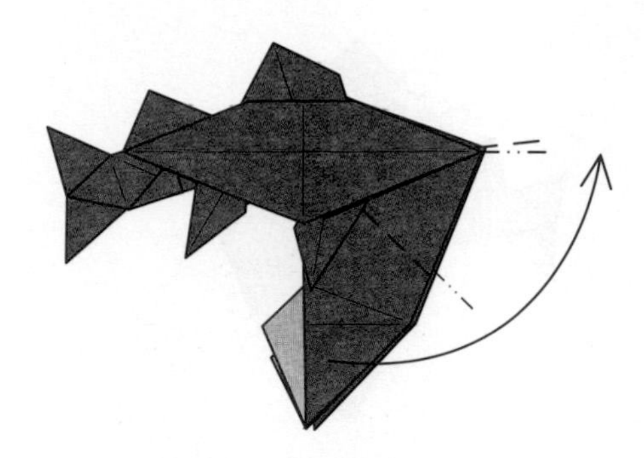

60. Refold the crimp you undid in step 26.

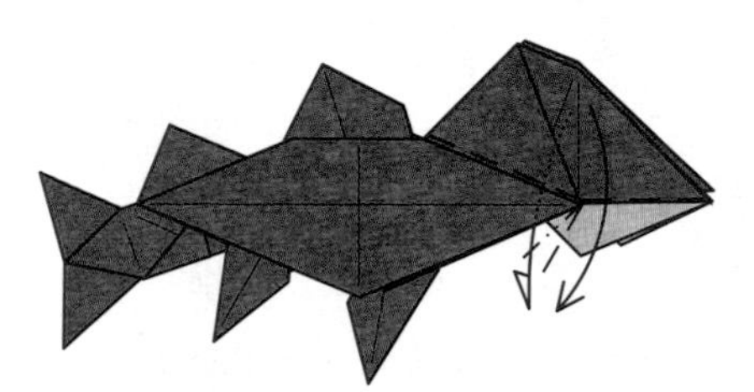

61. Outside reverse-fold the right side.

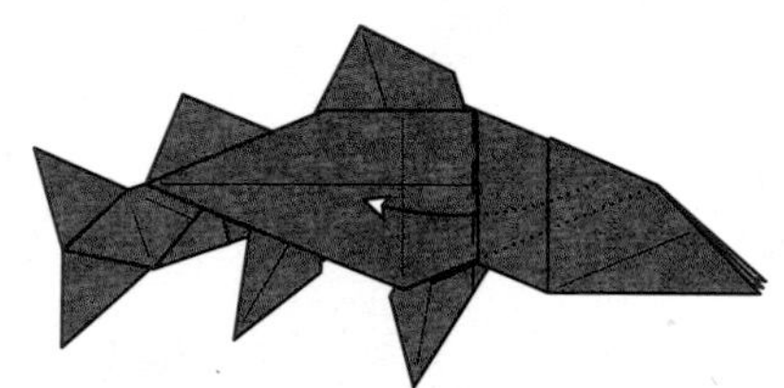

62. Pull out the corner. Repeat behind.

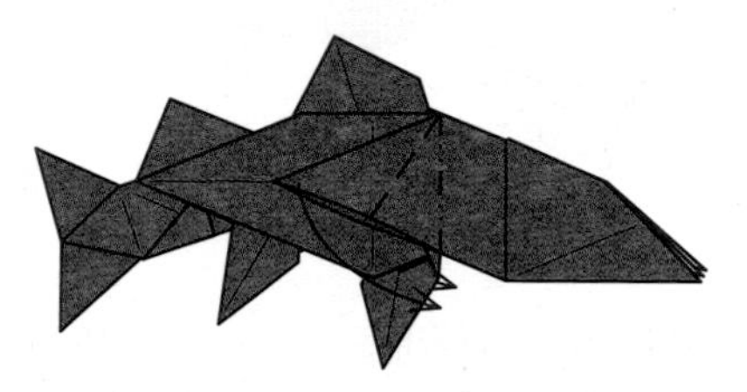

63. Crimp the point downward. Repeat behind.

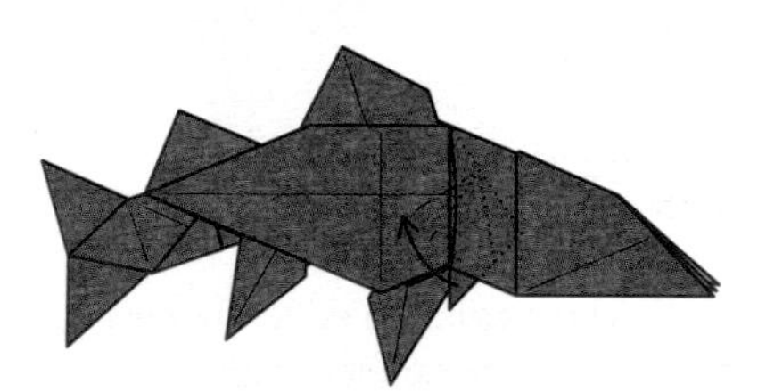

64. Crimp the corner and twist it upward to the left. Repeat behind.

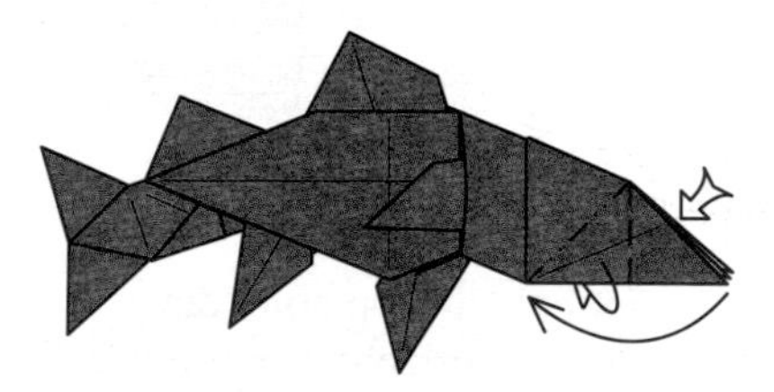

65. Crimp the right side underneath.

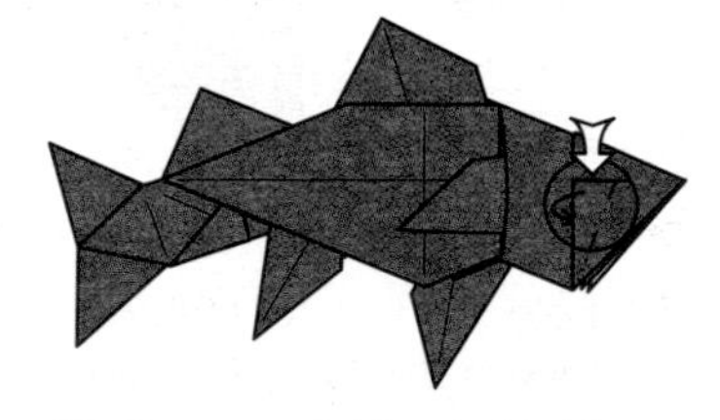

66. Reverse-fold two inside corners created by the crimp.

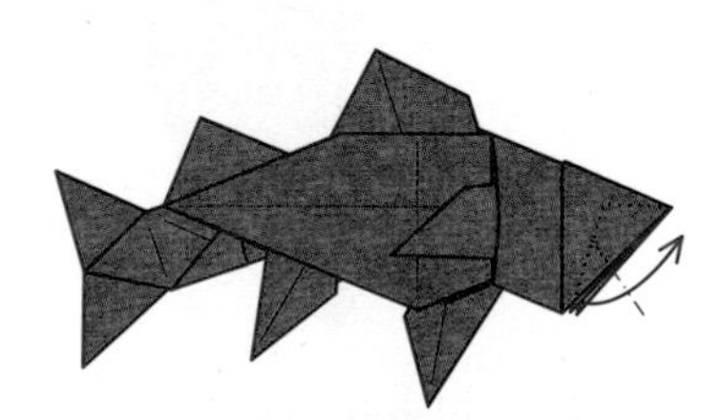

67. Reverse-fold the middle point (of three) upward.

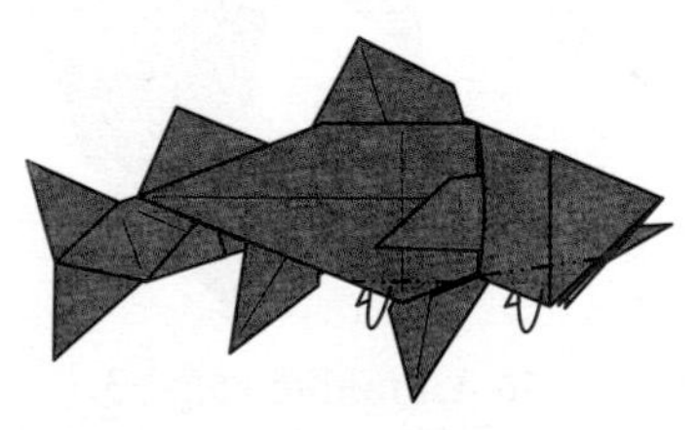

68. Mountain-fold the bottom edges of the jaw and belly.

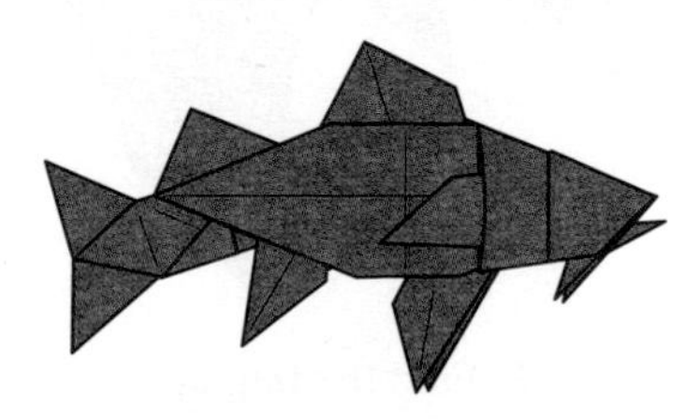

69. Finished Goatfish.

6 Grafting

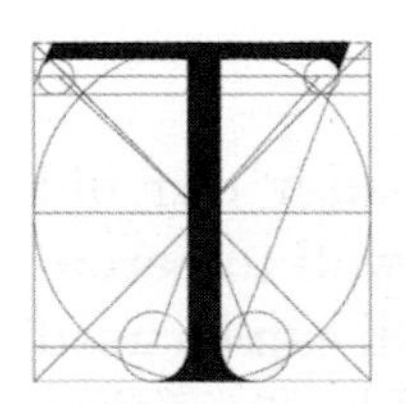

The initial stages of origami design are usually modifications of existing designs. This modification can take two forms. The simplest is that which every folder does consciously or unconsciously: simply altering proportions of the folding sequence while still following the designer's instructions. You could change the proportions of particular creases, change the crease firmness from sharp to smooth (or vice versa), add or remove creases, straighten what is curved and curl what is straight. It is very easy to change a model in the final shaping folds. In fact, it is very difficult to make a precise replica of someone else's fold, particularly if the design is fairly complex. And precise duplication is rarely desirable; an artist must develop his or her own vision of the folded model even when following someone else's design, and therefore must not be afraid to deviate from the original folding sequence.

However, changing proportions of an existing model is very limiting to the origami designer. You can only work with the existing structure; you have the same number of flaps, the same lengths, the same relative positions. Techniques such as point-splitting can turn one flap into two, or three, or more, but only, as we have seen, with a substantial penalty in flap length.

Quite often, what is needed—or at least desired—in a derivative origami design is not just a rearrangement of the existing paper, but actually a bit more paper somewhere: a longer leg, an extra set of appendages, another petal on the flower, another horn on the beetle. At such times, you might have a nearly complete design (either your own or someone else's) to which you would like to add a bit more structure, but there's no more paper from which to make the new bits. At such times—particularly if you've already put a great deal of work into the model—the prospect of starting over from scratch can be downright depressing.

Now, if it were allowed, one might be tempted to simply glue on a bit of extra paper, just as in the previous chapter the obvious solution to splitting a point was to cut it in two. But just as we found ways to achieve the same result as cutting while preserving one-piece no-cut folding, it turns out that it is quite often possible to achieve the same result as gluing—to add a bit of paper to a particular location without redesigning the entire model—while preserving the square that we started from. This process is remarkably versatile: It can transform a run-of-the-mill design into something special or even extraordinary and has been utilized by many of the world's top origami designers. I call it *grafting*.

6.1. Border Grafts

For a concrete example of grafting, let's consider our old standby, the Bird Base, which lends itself very well to perching birds. The Bird Base has four long flaps, which can be used for a head, tail, and two legs. The simplest bird I know of that can be folded from the Bird Base is a traditional design, the Crow: Narrow two flaps to make legs, then reverse-fold one flap and crimp the two narrowed flaps to create a head, legs, and feet, as shown in Figure 6.1.

Now, as origami designers have done for decades, you can use the Bird Base to realize a wide variety of perching birds, so long as you don't need open wings, by using this basic design. By adding more folds—extra reverse folds, crimps, rounding folds—it's possible to make many distinct and recognizable species with suggestions of wings, feathers, and even eyes and other features.

But one thing that most of these origami birds have in common is that the foot is represented by a single toe, and real birds, of course, have four toes. This simplification is not intrinsically bad; all origami is somewhat abstract, and in the overall design of a model, there should be a balance in the level of abstraction. A simple, clean-lined model can succeed perfectly well with simplified feet.

However, it must be said that a distinctive feature of many birds is their splayed feet, whether standing on the ground or grasping a branch or twig. There are occasions when it would be quite desirable to have full, four-toed feet on a perching bird design.

One approach, of course, would simply be to turn the foot flap into four flaps using the point-splitting techniques of the

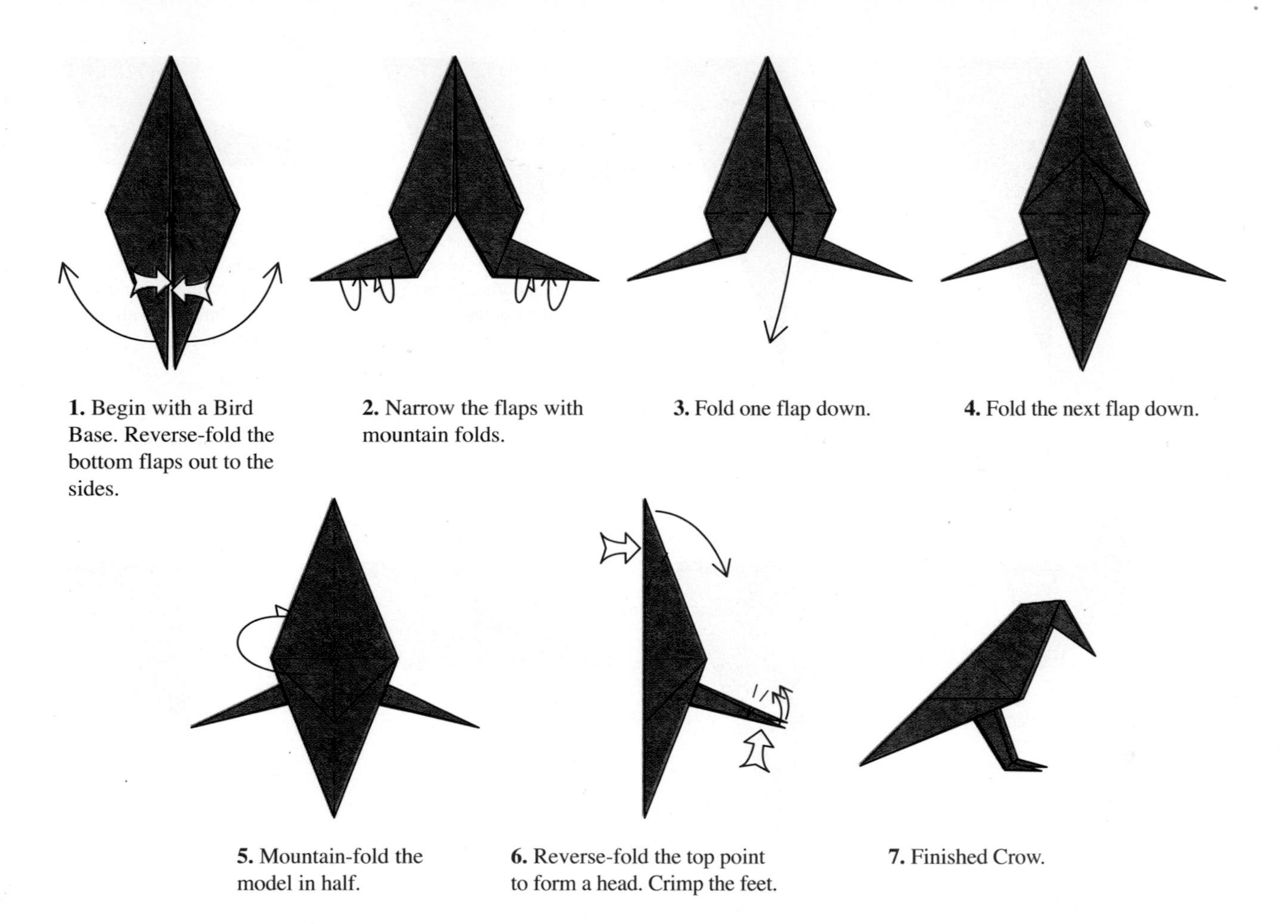

1. Begin with a Bird Base. Reverse-fold the bottom flaps out to the sides.
2. Narrow the flaps with mountain folds.
3. Fold one flap down.
4. Fold the next flap down.
5. Mountain-fold the model in half.
6. Reverse-fold the top point to form a head. Crimp the feet.
7. Finished Crow.

Figure 6.1.
Folding instructions for a simple Crow.

previous chapter. This change comes at a cost, of course: In order to obtain four flaps at the tip, the main flap gets substantially consumed. It is possible to obtain four toes using the sequence shown in Figure 6.2 (a radial four-point split), but the legs that are left are short, fairly wide, and sufficiently thick so that narrowing them to approximate a bird's stick-like legs is rather difficult.

For a perching bird, it would be desirable to keep the legs long but to replace the single point at the tip of each leg with four points without reducing the length of the leg. That requires a net addition of paper.

If it weren't for those one-piece rules, we could simply glue on an extra bit of paper at the feet as in Figure 6.3. We could make a pair of four-toed feet from two tiny bird bases, glue them onto the larger bird, and presto! We're done.

But from a single sheet, what can we do? Well, we could see if it's possible to somehow obtain the functionality of three

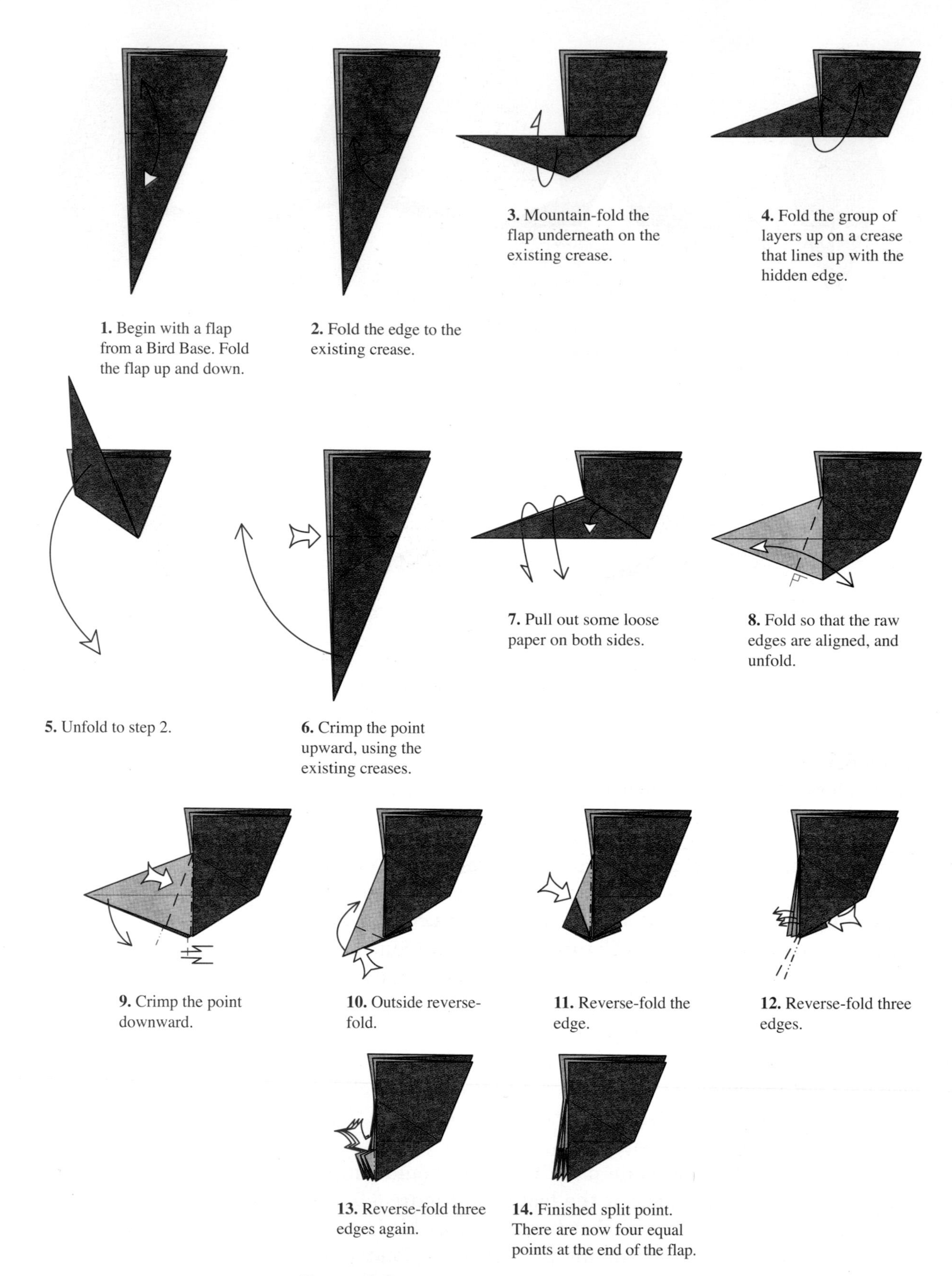

1. Begin with a flap from a Bird Base. Fold the flap up and down.
2. Fold the edge to the existing crease.
3. Mountain-fold the flap underneath on the existing crease.
4. Fold the group of layers up on a crease that lines up with the hidden edge.
5. Unfold to step 2.
6. Crimp the point upward, using the existing creases.
7. Pull out some loose paper on both sides.
8. Fold so that the raw edges are aligned, and unfold.
9. Crimp the point downward.
10. Outside reverse-fold.
11. Reverse-fold the edge.
12. Reverse-fold three edges.
13. Reverse-fold three edges again.
14. Finished split point. There are now four equal points at the end of the flap.

Figure 6.2.
Folding sequence for splitting one point into four smaller points.

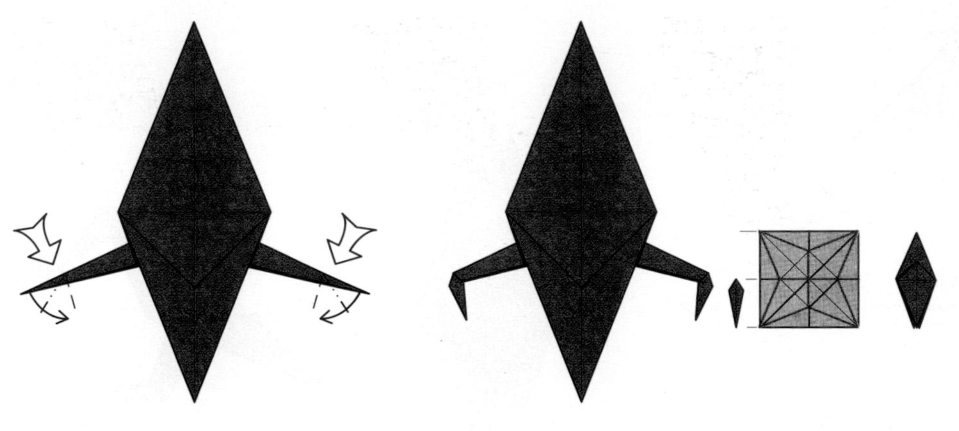

5. Squash-fold the feet downward.

6. Fold two tiny Bird Bases from squares whose side is twice the length of the squashed point.

7. Attach the Bird Bases to the two flaps.

8. Now the Bird Base has four toes on each foot.

Figure 6.3.
Adding toes to a Bird Base by gluing.

squares—two small and one big—from a single sheet. And since the feet folded from the small squares are attached to the tips of the leg flaps in the big square, it makes sense to try attaching the small squares to the corners of the big square that correspond to the leg flaps.

To do this, we'll need to identify the relationship between the square (and its crease pattern) and the folded model. You can do this in practice by coloring flaps of the folded model and then unfolding it to a crease pattern and noting where the colored bits fall. With practice, however, you'll be able to keep track of such points as you unfold the model without coloring.

Figure 6.4 shows the unfolded Crow—which we will take to represent a generic bird—and identifies which parts of the square make up which parts of the bird.

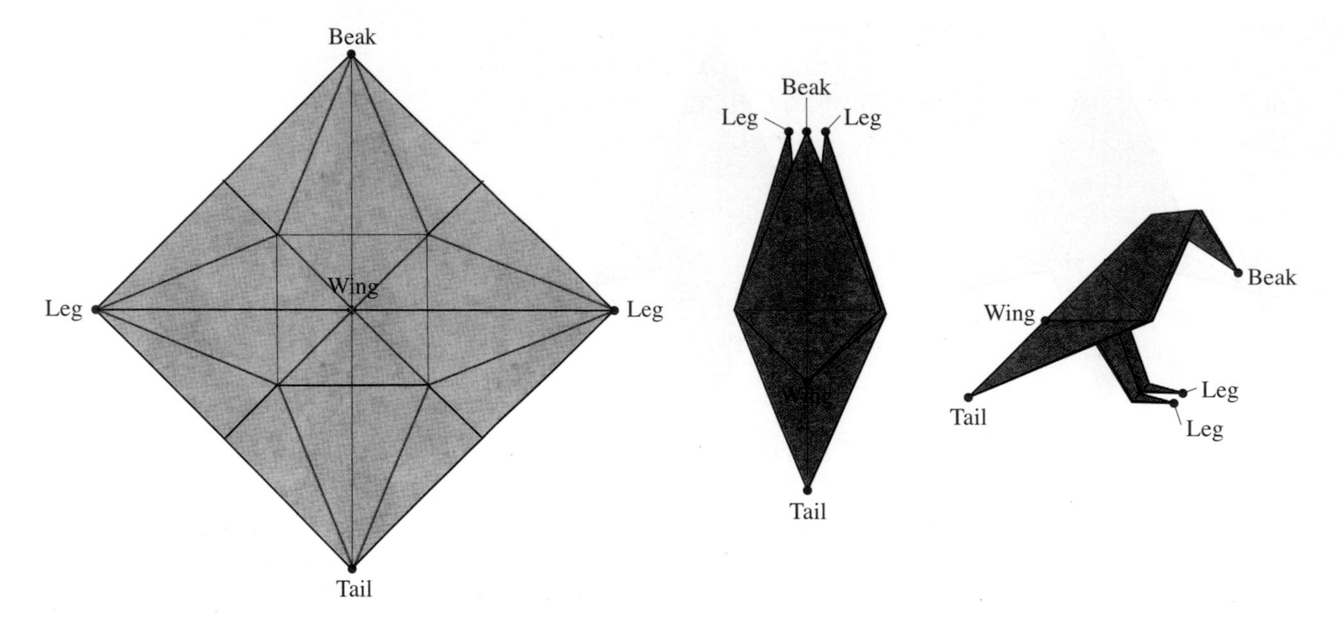

Figure 6.4.
Left: the crease pattern for the Crow.
Middle: the base.
Right: the completed model.

The two side corners become the legs. We would thus add the small squares that form Bird Bases to the corners of the larger square as in Figure 6.5.

Now, if we had three squares actually joined at their corners, we could certainly fold a four-footed bird from this unusual shape. The practice of folding from corner-joined squares is not unknown in origami (a 1797 publication by

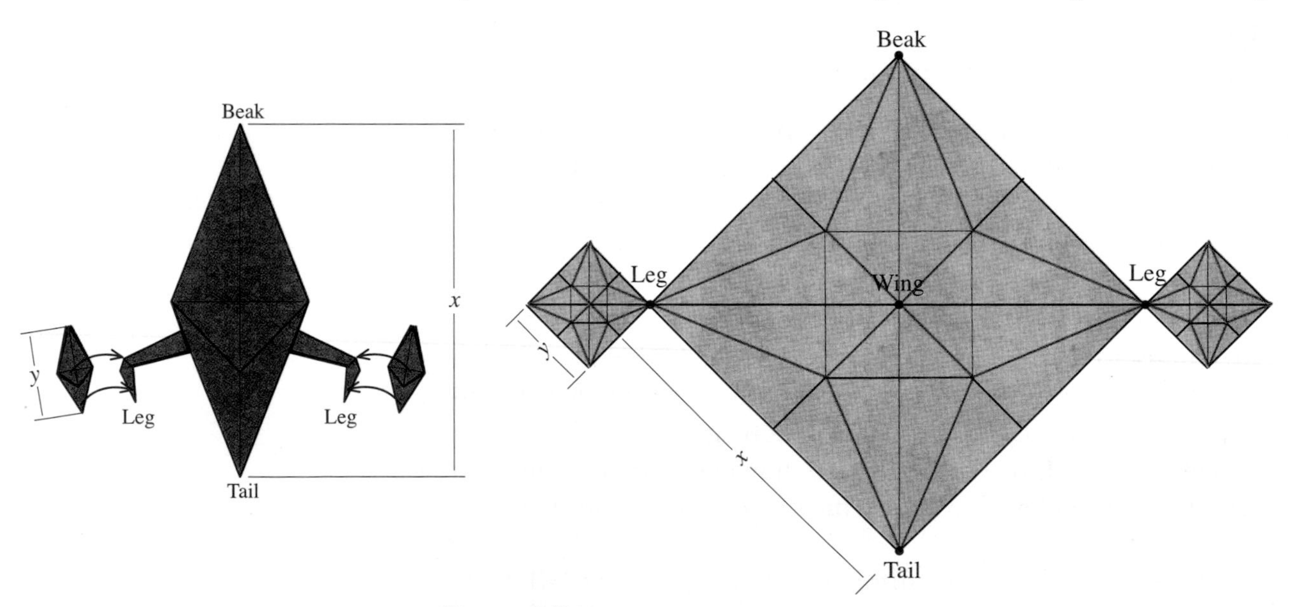

Figure 6.5.
Two squares attached to another at their corners.

Roko-an displayed numerous examples of joined cranes folded from paper cut in such a fashion) but we will attempt to fold our design from a single square. Thus, we need to obtain all three shapes as portions of a single square. The easiest way to turn the trio of squares back into a single square is to extend the sides of the smaller squares until they join, forming a larger square as shown in Figure 6.6.

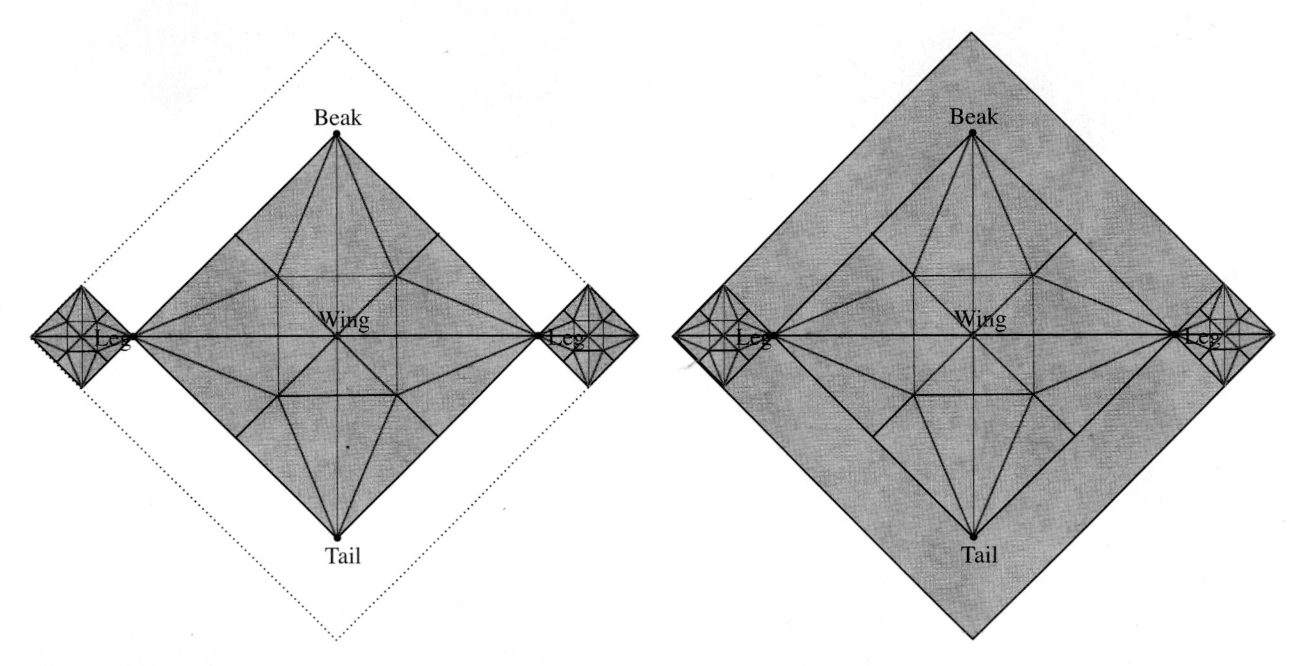

Figure 6.6.
The three squares embedded within a larger square.

Now, we have embedded all three squares in a larger square, which can be used—we hope—to fold a bird with four-toed feet. Does it work? Let's try it out.

The folding sequence shown in Figure 6.7 gives a Bird Base with a small square at each of the corners. However, this square is attached to the larger Bird Base along its full diagonal. Is it possible to fold the small square into another Bird Base? Yes, but with a slightly modified sequence to accommodate the fact that the small square is joined to the larger flap, which limits how it can be manipulated.

The resulting Bird Base can be used to make conventional bird feet, although it is desirable to narrow some of the flaps and redistribute their layers, as shown in the Songbird model at the end of the chapter.

What we have done here is to add some more paper to the square while keeping it square, by grafting on more paper, in this case, a border running all the way around the outside of the square. We call this a *border graft*. Grafting

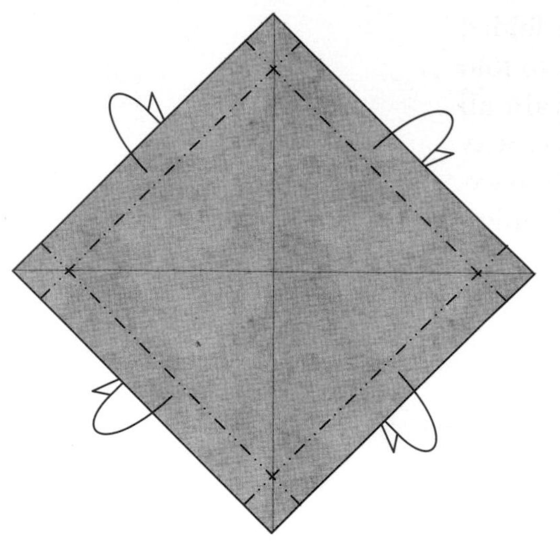

1. Begin with a square with the diagonals creased. Mountain-fold a bit of the edge behind all the way around.

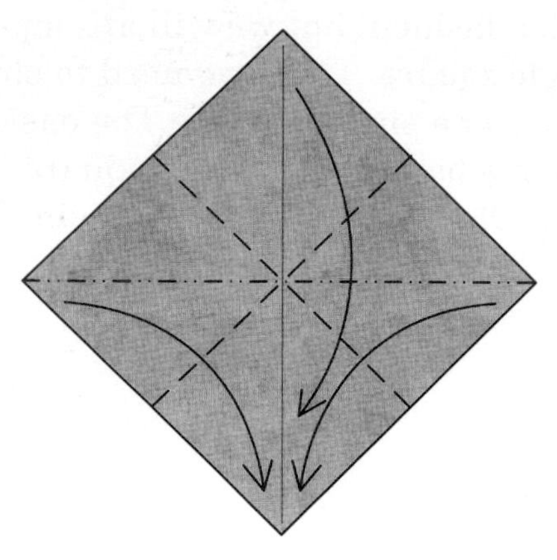

2. Fold a Preliminary Fold.

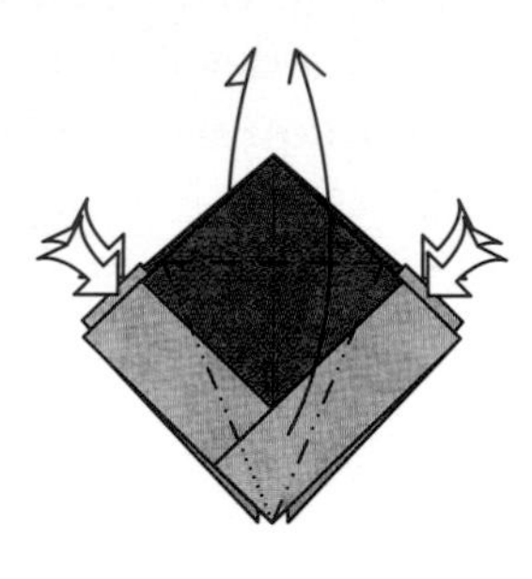

3. Petal-fold to form a Bird Base.

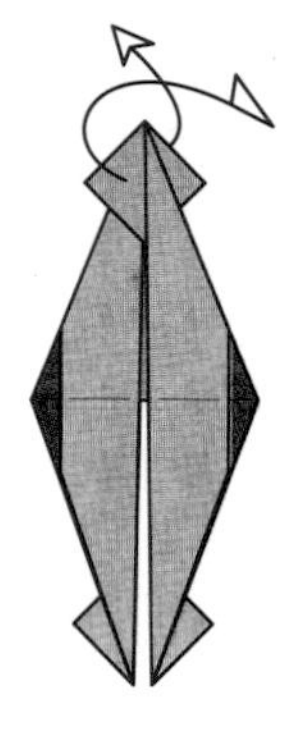

4. Unwrap the loose paper from the top near flap.

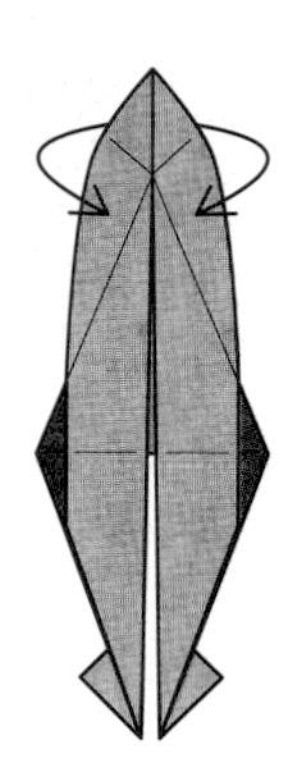

5. Pop the corner inside-out so that it becomes concave upward.

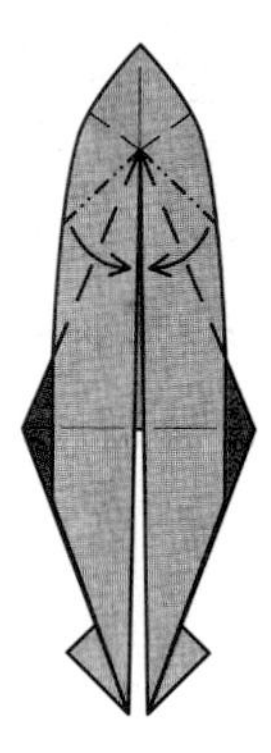

6. Pleat the sides downward and squash-fold the top.

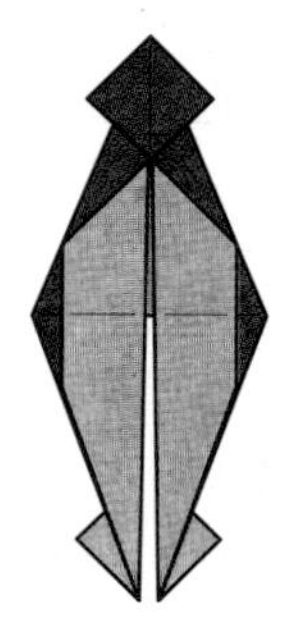

7. The small square at the top is one of the squares we added at the corners.

Figure 6.7.
Folding sequence for a Bird Base with two small squares at opposite corners.

can be a powerful technique for adding both large and small features to an origami model.

However, grafting carries with it a risk: inefficiency. All we really needed to add to this model were the two small squares in the corners. But now, we've added a wide border all the way around the square; most of this extra paper will go unused, adding to the thickness of the model without adding anything to the design.

Thus, there is a bit of an art to using grafting in design; while the graft may have been inspired by the desire to add a small bit of paper in one or two places, the challenge is to mini-

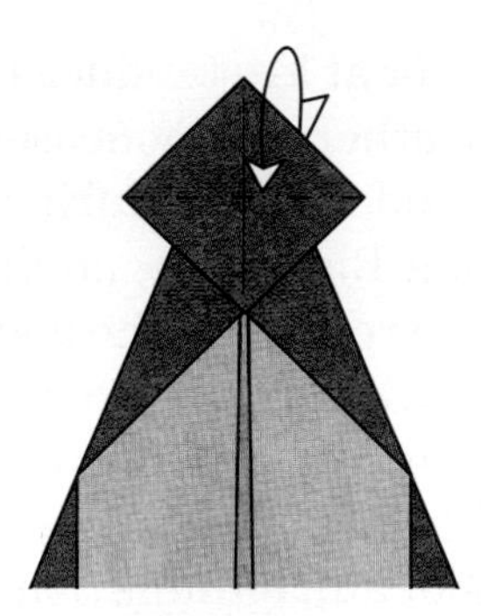

8. Continuing on (from Figure 6.7). Mountain-fold the flap behind with the crease running from corner to corner and unfold.

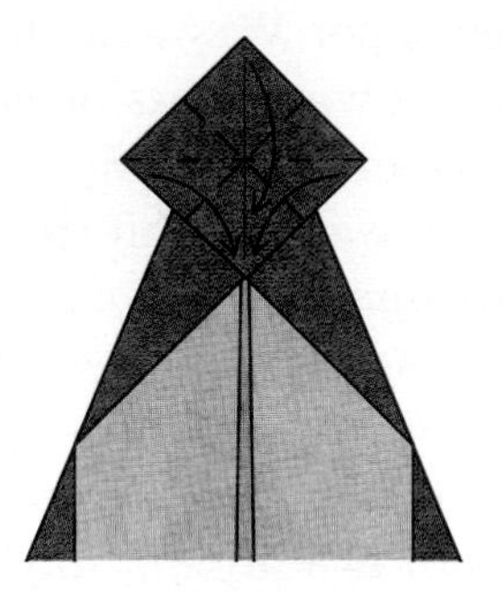

9. Form a Preliminary Fold with the ncar layers.

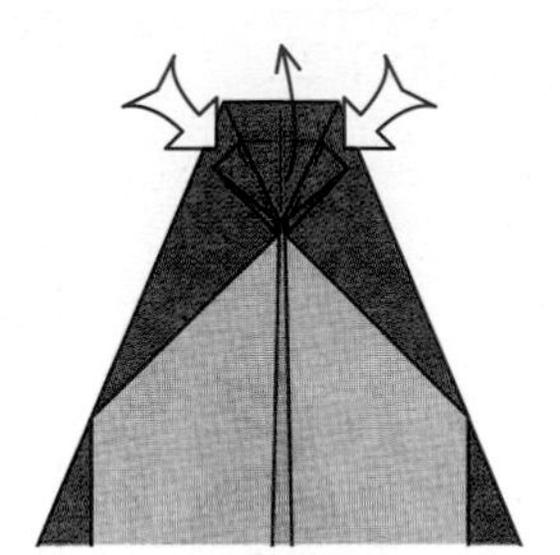

10. Petal-fold the top corner.

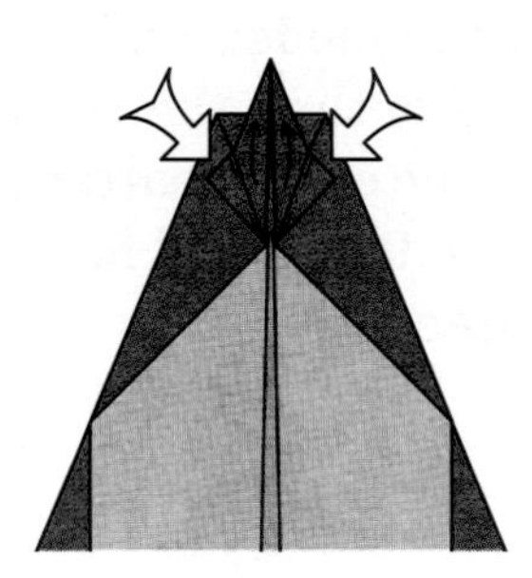

11. Petal-fold the two points together with the trapped corner inside.

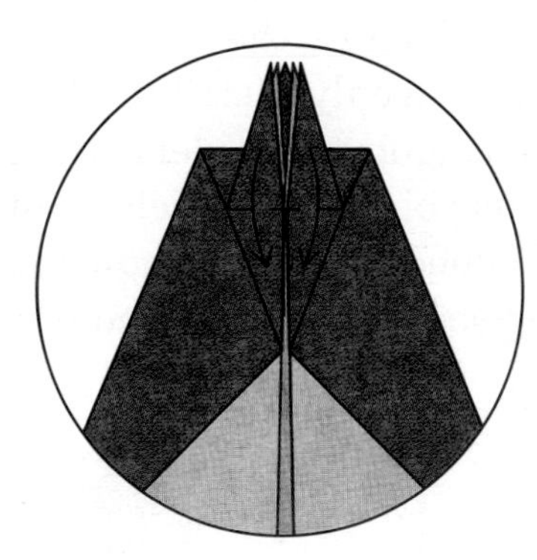

12. Fold the two points back down.

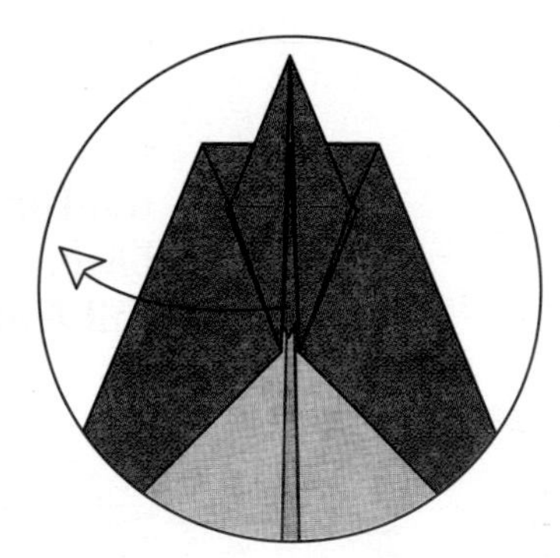

13. Pull the corner out completely.

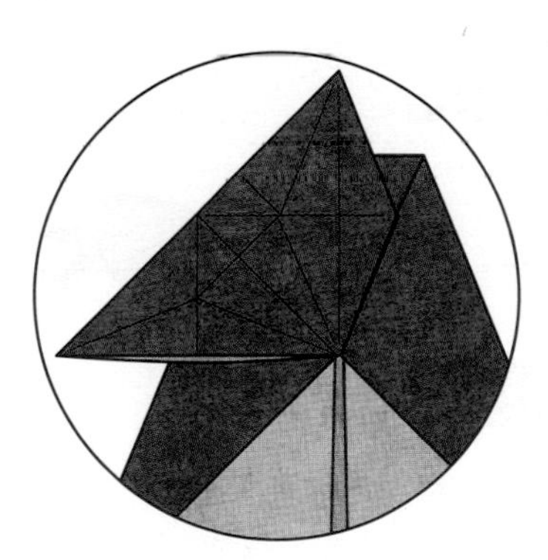

14. Like this. The flap is now a Bird-Base-creased square, folded in half along the diagonal.

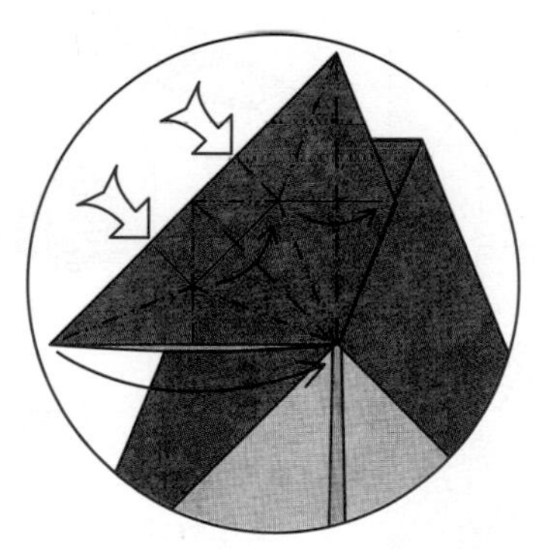

15. Use the existing creases to collapse the corner into a Bird Base.

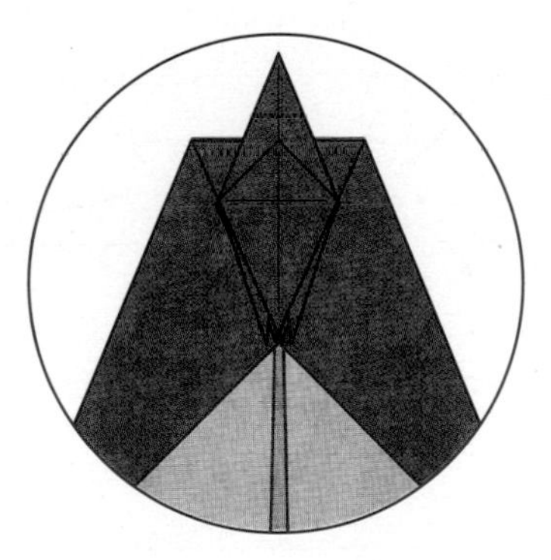

16. The finished four-pointed Bird Base at the tip of the larger Bird Base.

Figure 6.8.
Folding sequence for making the smaller square into a Bird Base.

mize wasted paper. Waste can be avoided—or at least reduced—by figuring out ways of using some of the otherwise nonessential added paper. For example, when we add feet by grafting a border all the way around the outside of a Bird Base, adding paper at two opposite corners of the square results in the addition of paper at the other two corners, which become the head and tail. We don't need to add four toes to either the head or tail, obviously, but if we can put that additional paper to good use, the result is a further improved model and elimination of the waste. As it turns out, the paper at the head end can be used to make a double (i.e., open) beak, while the paper at the tail end can be used to make the tail longer, wider, or a bit of both. Furthermore, it's also possible to use some of the border that runs between adjacent corners to make a more fully rounded body. Thus, the excess paper goes to good use: The layers can be evenly distributed through the model, and the result is a songbird considerably more lifelike than the original Bird Base bird from which we started. The crease pattern, base, and folded model are shown in Figure 6.9; you will find a full folding sequence at the end of the chapter.

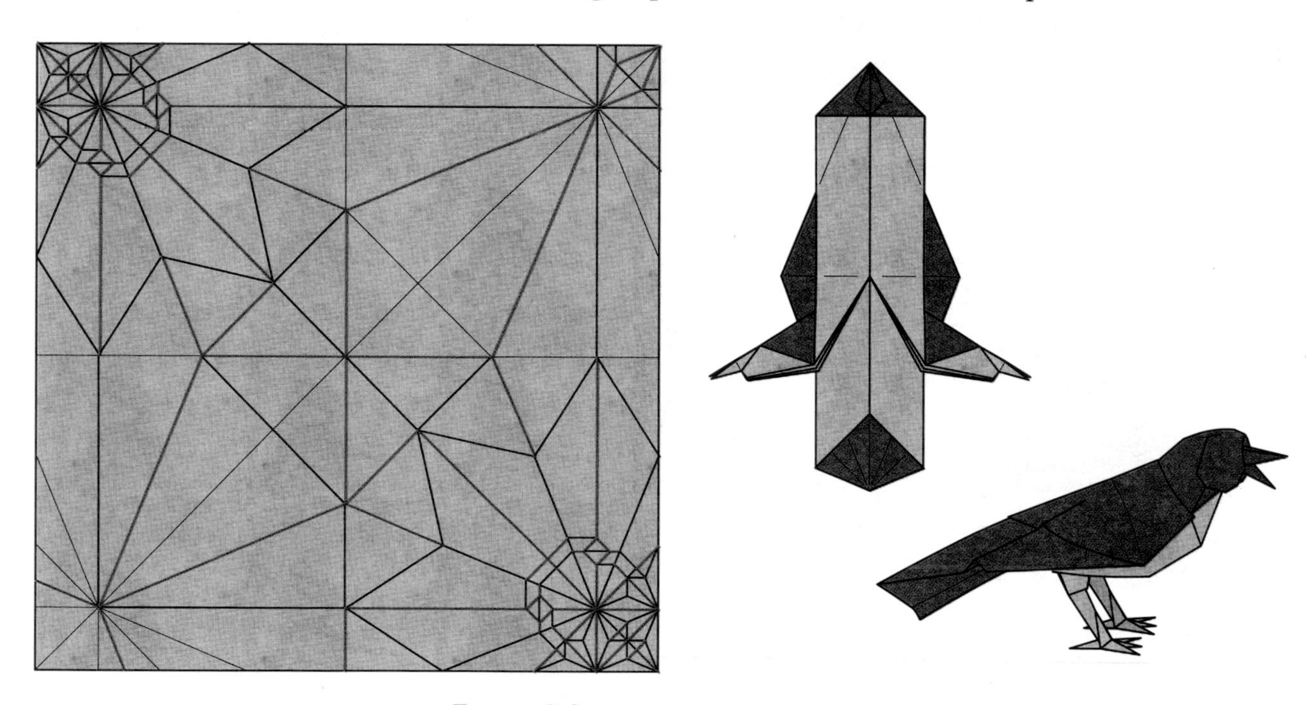

Figure 6.9.
Crease pattern, base, and folded model of the Songbird.

As an extra bonus, this configuration allows us to make the legs and breast with the opposite side of the paper showing, creating a nice two-toned effect.

A border graft need not run all the way around the square; if you only need to add paper to one end, you can simply add paper to two sides, creating a new square at one corner. The risk, as with all grafted bases, is that the paper you've added to complete the square is essentially wasted unless you can find some other use for it.

As an example, some years ago a composite (multisheet) origami model had become quite popular by combining the head from a dragon by Kunihiko Kasahara (itself a three-piece composite model) with the body, wings, legs, and tail of a simple one-sheet dragon by Robert Neale as shown in Figure 6.10. The combination became known as the Kasahara-Neale Dragon. Kasahara's head was folded from a Bird Base, while Neale's Dragon was folded from another, larger Bird Base, and the two would be joined with glue. It was a picture-perfect scenario to make the combination from a single sheet using grafting.

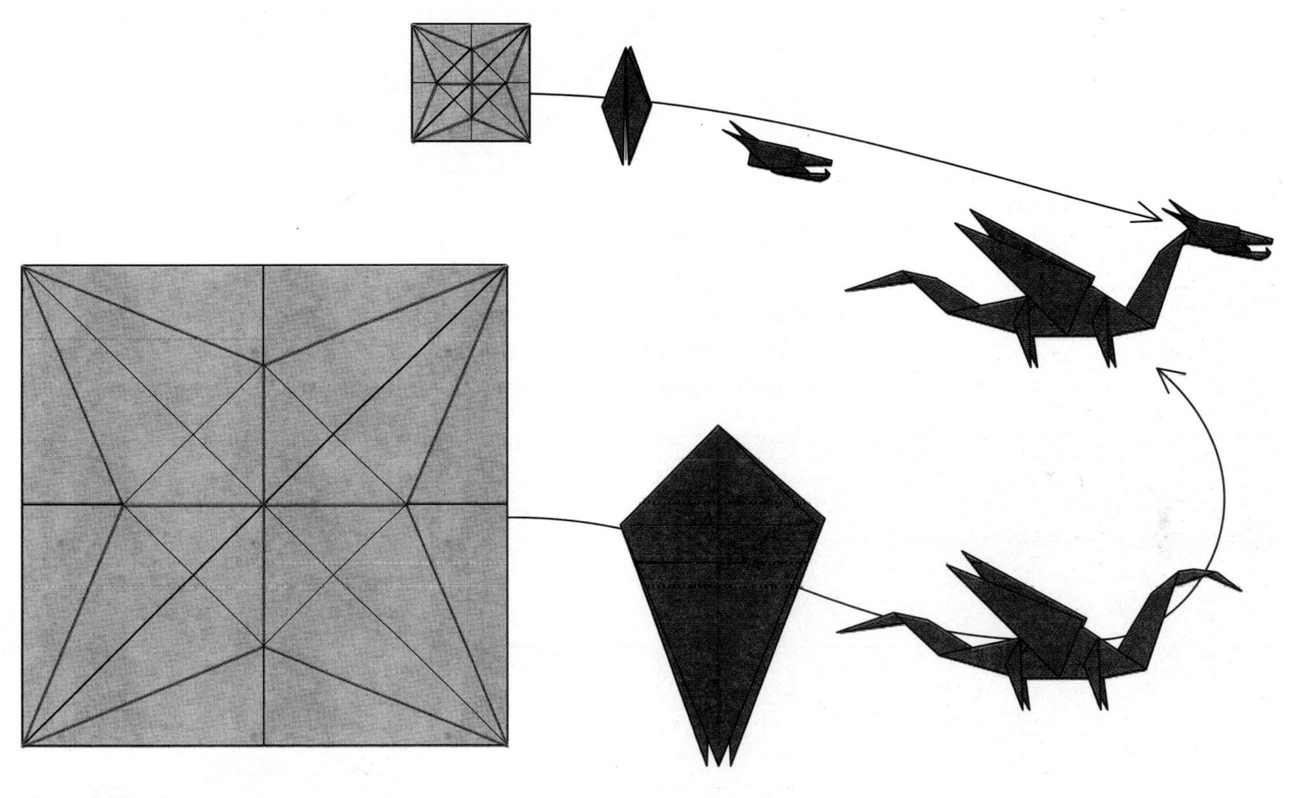

Figure 6.10.
Assembly of two different-sized Bird Bases into the Kasahara-Neale Dragon.

Since the small square would be joined to a corner of the larger square, we can use the border grafting technique. However, as Figure 6.11 shows, we are adding a fair amount of paper just to get that one little square in the upper corner.

Fortunately, the extra paper becomes part of the two wing flaps, and so it can be used to give the Dragon somewhat

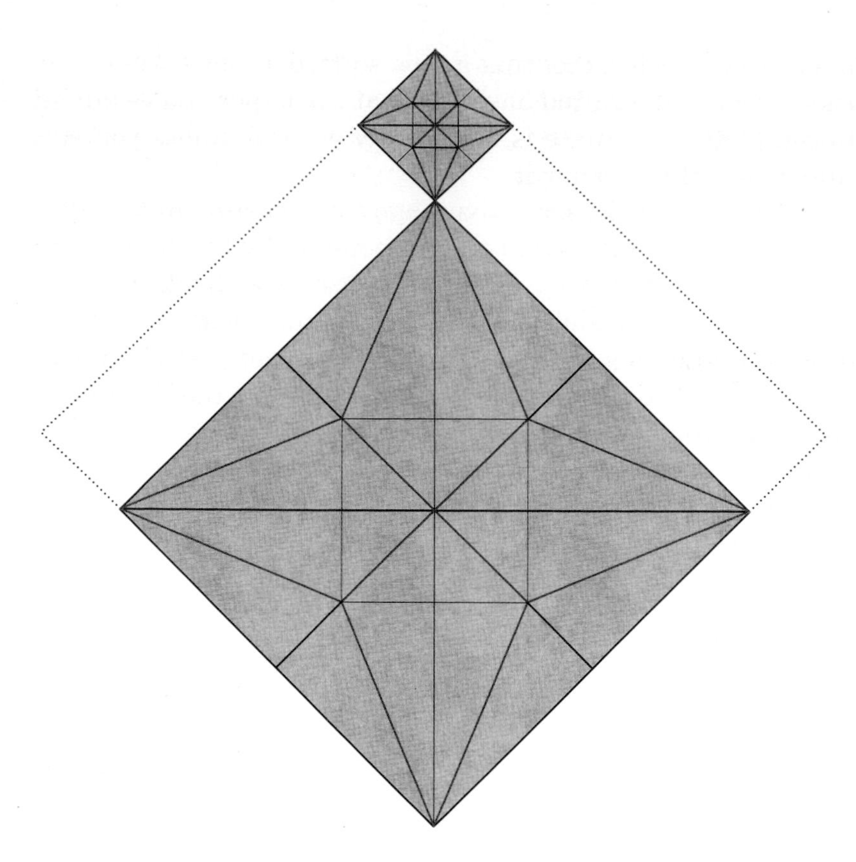

Figure 6.11.
Position of the two squares within a larger square.

larger wings than the original Neale Dragon from which it is derived. In this way, efficiency is preserved, and the model benefits from a bit of serendipity. As with the Songbird, there is some further work to find a workable folding sequence. Put

Figure 6.12.
Crease pattern, base, and folded model of the KNL Dragon.

it all together, and the result is a charming little dragon that stands on its own. I call the new model the Kasahara-Neale-Lang Dragon—KNL Dragon, for short. You will find its folding sequence at the end of this chapter.

Figure 6.12 shows the crease pattern, base, and folded model of the KNL Dragon. You should be able to pick out the two Bird Bases as well as the boundary of the border graft.

6.2. Strip Grafting

Grafting does not always put the added paper around the outside of the model; if that's all that there were to grafting, we would quickly exhaust the possibilities of the technique. But we can add grafts in the interior of the paper as well, by cutting patterns apart and reassembling them with our new additions—a far more powerful and versatile technique.

If, for example, we wished to add feet without adding excess paper at the head or tail, we could add the additional paper in a strip running across the middle of the square. Imagine, for example, cutting the Bird Base in half horizontally and pulling the two ends apart. Then the two "foot" squares could be joined by a strip that cuts across the middle of the paper, and the result inserted into the gap, as shown in Figure 6.13.

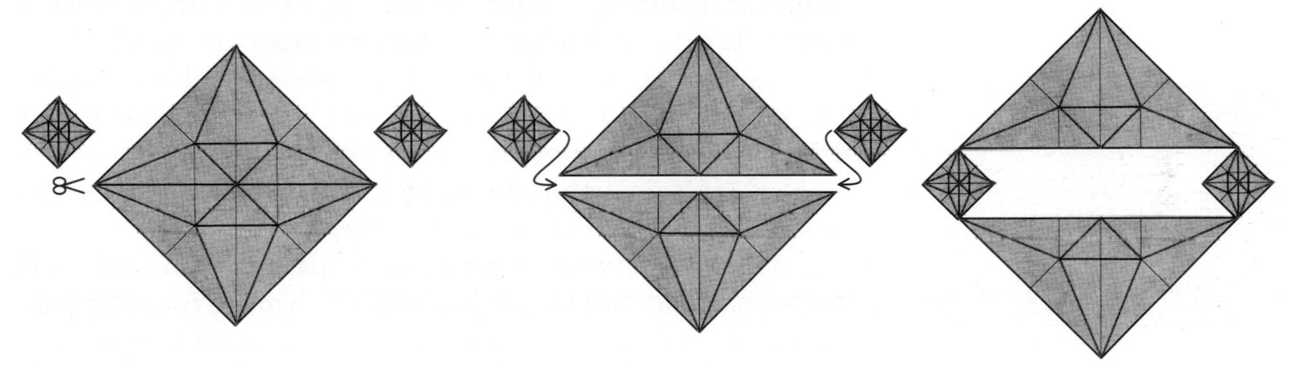

Figure 6.13.
Construction of a strip-grafted model; two squares are joined by a strip and inserted along a cut across the square.

But a problem arises; when the creases are connected across the strip, one of the four Bird Base points is no longer freely accessible. Fortunately, the fifth point in a Bird Base, which comes from the center of the square, can be pressed into service as the desired fourth point.

The result can be folded into many different types of birds, but because the extra layers in the legs are evenly distributed, I find this structure particularly suitable for a long-

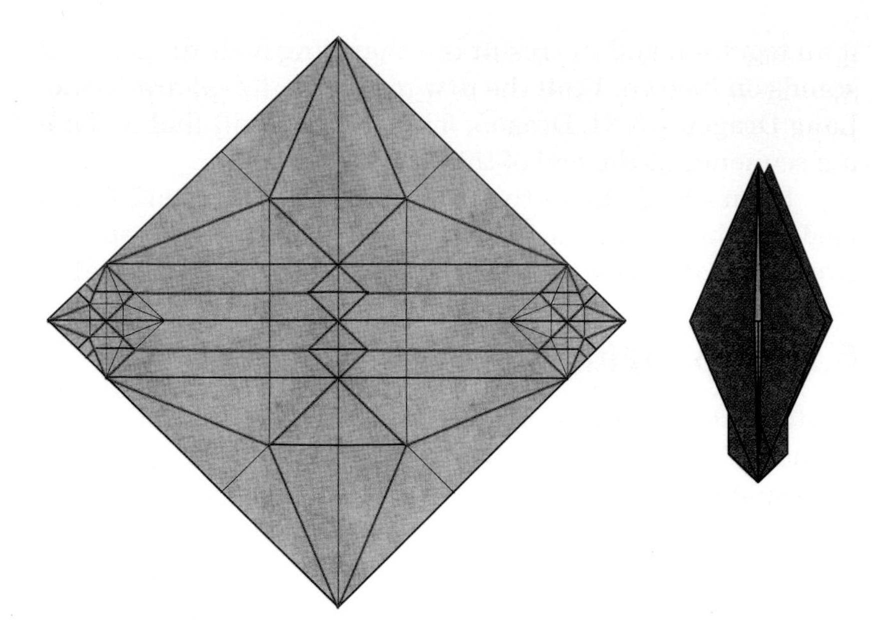

Figure 6.14.
Crease pattern for a Bird Base with a strip graft. Note that the creases around the fourth point of the tiny Bird Base are no longer used and the point is not free.

legged wading bird, realization of which I shall leave as a challenge for the reader.

So, as these two models illustrate, one can create paper to add features by augmenting a square in one of two ways; you can add paper around the outside, or you can add a strip cutting across the model. Of course, in the second case, there's no need to actually cut the square and paste in a strip. You simply design in the strip from the very beginning. How wide a strip? It depends on how much extra paper you need.

It's also possible to add multiple strips. If, for example, you wanted a bird with four-toed feet and a split beak and more paper in the tail, then you could add two strips: one running side to side, one running up and down. Can you design and fold such a bird using two crossing strips?

A straightforward application of strip grafting arises if you wish to add toes to four limbs that are made from the four corners of the square. An example that is difficult to resist is a multitoed frog, and the logical model to start from is the traditional Japanese Frog, which is, of course, folded from the Frog Base. Now, as we saw in the previous chapter, you can make a multitoed frog by splitting the four leg flaps, but that approach unavoidably shortens the flaps. We can also use grafting to add toes to all four limbs of the traditional Frog to realize a model in which the toes are more prominent and the legs remain relatively long.

"Relatively" is the key concept here. Grafting, like point-splitting, shortens the limb flaps. If the final size of the square is fixed, we need to shrink the pregraft base to allow room for the grafted paper. As the saying goes, there's no such thing as a free flap. But there is a difference between the size reduction arising from point-splitting and that from grafting; in point-splitting, the split flaps get reduced in size, but the rest of the model (the body, in the case of the Frog) remains the same size. In grafting, on the other hand, the entire model is shrunk in proportion to accommodate the graft, so the basic proportions of the model are unaltered from their pregrafted values.

There are two ways we've seen to augment a square at its corners: We could add a border graft—a strip running all the way around the outside—or we could add two strips crossing in the middle. Both could be used (and I encourage you to try both yourself), but the crossing-strips configuration offers an extra bonus of creating some extra paper in the middle of the paper, as shown in Figure 6.15. Why is that a bonus? In the traditional Frog, the middle of the paper winds up at the head. It's always nice to have some extra paper at the head of an animal where it can be used for facial features—mouth, tongue, teeth—or, in the case of a tree frog, prominent eyes. We may not have started designing a frog with eyes, but if the opportunity presents itself, we'll take it.

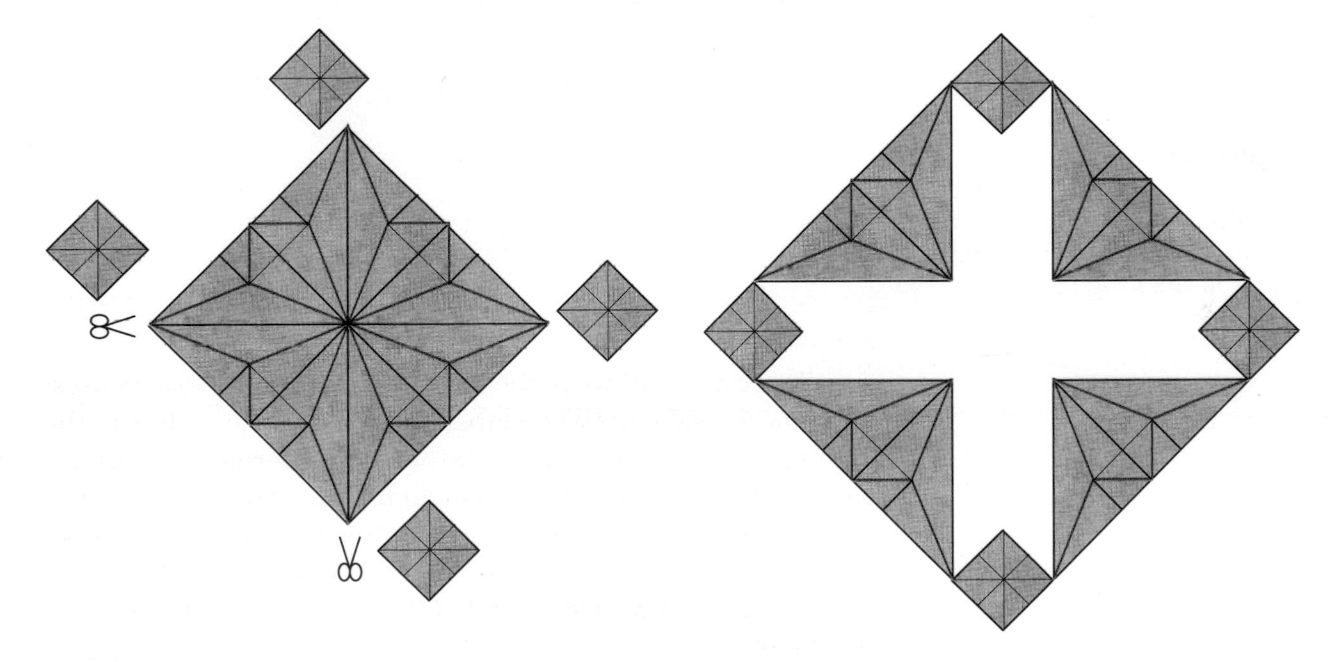

Figure 6.15.
Adding four squares to a Frog Base by cutting along the diagonals.

Now, in designing a strip-grafted model, there is a decision to be made: How large should the small squares be, or equivalently, how wide should the strip be? You can, of course, simply use trial and error: Try wider and narrower strips and see if the feet come out too big or too small. But there is another factor that should be taken into account. To keep the lines of the model clean, it is desirable to make the edges line up as much as possible, which means that features we add by grafting would ideally line up with features that are already there in the pregrafted base. In the case of the Frog, if we make the added strip the same width as the flaps of the Frog Base, the new layers will be exactly half the width of the Frog Base flaps. Narrowing the flaps by folding them in half will result in all the edges lining up, giving a neater appearance. Thus, the resulting crease pattern from which we start would be something like Figure 6.16:

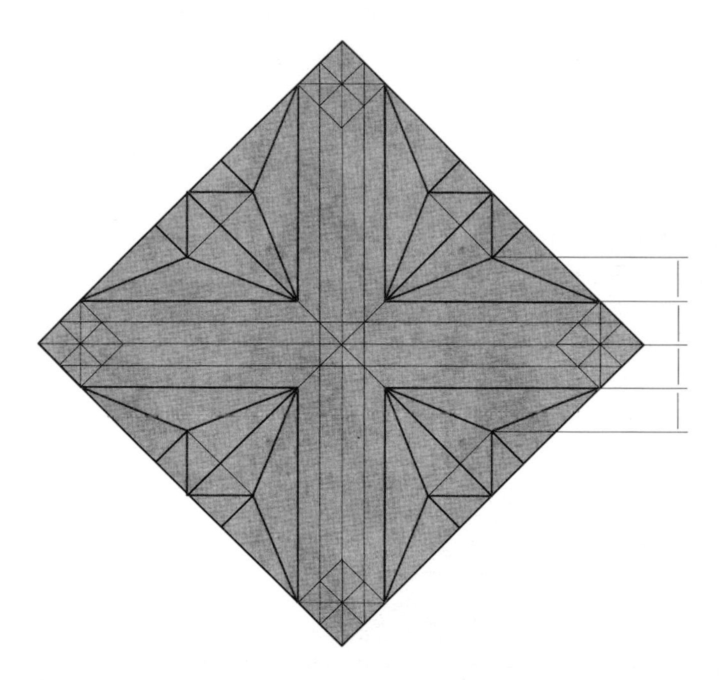

Figure 6.16.
An elegant proportion arises if the dimensions of the smaller square are matched to dimensions of the Frog Base.

There are still two problems: How do we find these creases, and how do we actually collapse the pattern (i.e., what is the folding sequence)? The coordinates of the reference points can be numerically calculated from their geometric relationship; you can measure and plot their location. (It is also possible to devise folding sequences for any reference point, but that problem, which is quite rich in and of itself, is beyond the scope of this book.)

For finding the folding sequence, a good way to start is to make the paper resemble a base that you already know how to fold—in this case, the traditional Frog Base. It often works when a base has been augmented by strips simply to fold the strips so that the crease pattern looks like the ungrafted base, then proceed with the square fold as if it were all one sheet of paper as shown in Figure 6.17.

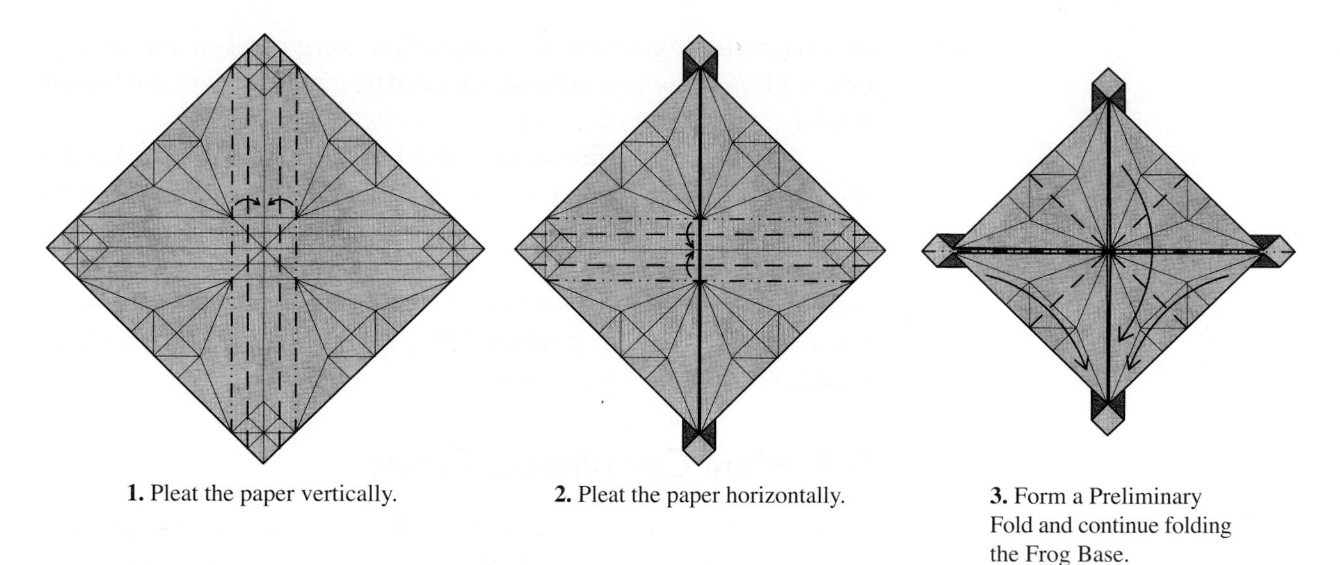

1. Pleat the paper vertically.

2. Pleat the paper horizontally.

3. Form a Preliminary Fold and continue folding the Frog Base.

Figure 6.17.
Initial folding sequence to construct the augmented Frog Base.

This puts all the creases in the right place, but you will often find that some of the layers get trapped; this happens with the thick layers formed by the crossing pleats in the center of the paper. In such cases, you will probably have to partially unfold the model to disentangle the layers (a process dubbed *decreeping* by origami artist Jeremy Shafer). Decreeping accomplishes two things: It makes all of the layers accessible so that they can be turned into other features,

Figure 6.18.
Crease pattern, base, and folded model of the Tree Frog.

and by reducing or eliminating folds composed of many layers, it allows the layers to stack neatly, giving a cleaner folded model.

When you are designing, it's both reasonable and common to fold many layers together in order to put creases into the right place. Once you know where the creases are, you can search for alternate folding sequences that permit a more sequential assembly; such a sequence for the Tree Frog of Figure 6.18 is shown in the folding sequence at the end of the chapter.

6.3. More Complicated Grafts

Thus far we've used grafting to add paper to one or more corners of a square. We can do this in two ways: by adding a border graft (a strip of paper running all or partway around the square), or by adding a strip graft (a strip of paper cutting across the crease pattern). The strip graft necessitates that we cut the crease pattern into two or more pieces to insert the strip. It may seem vaguely disquieting to cut up the origami square, but you should get used to the idea: more complicated cuts, instigated by more complicated grafts, are shortly to come.

In any event, all we've looked at so far is adding features to the corners of a square, but since there are only four corners on a square, it's pretty easy to enumerate all possible ways of using border and strip grafts to augment corners. However, it's also possible to use grafting to add paper in the middle of an edge.

Why might we want to do this? Well, for one thing, not all models derive their flaps from the corners of the paper. One of the most straightforward applications of grafting is to add extra paper to the end of a flap, as we've done for toes, for example. If the tip of the flap in question comes from the edge of the paper, rather than the middle, then we should add paper in the middle of the edge.

And there is much more variety in adding paper to a spot along the edge of the square, since there are an infinite number of locations along the square where we might perform our surgery. And it will be surgery of the strip-grafting sort; as we will see, border grafts are far more limiting than strip grafts when it comes to adding paper along edges of preexisting crease patterns.

As a concrete example and to have something to work with, let's take the simple lizard shown in Figure 6.19 (and whose folding sequence is given at the end of the chapter). This model fits together quite neatly; it's questionable whether one should even try to add feet. What you give up in aesthetics may very well not be compensated by what one gains in adding paper to the appendages. But for sake of illustration, let's assume that we wanted to add some paper to the four legs to obtain feet.

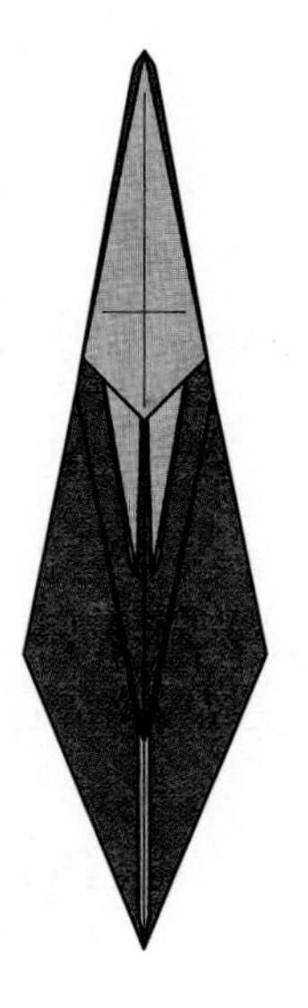

Figure 6.19.
Crease pattern, base, and folded model for the lizard.

Now, before we dive into slicing and dicing this or any crease pattern, let me point out that the simpler the crease pattern is, the easier it is to visualize the structure of the resulting base. It is therefore worthwhile to eliminate as many unnecessary creases as possible from the pattern you start with. The pattern in Figure 6.19 (as is the case with all crease patterns I show) doesn't show every single crease in the model, which would be far too cluttered, but only the creases used to fold the base (which, in the case of the lizard, is step 36 of the folding sequence).

The base is obviously not an entire lizard, but it has all of the essential features: the head, tail, body, and four legs. Even so, the crease pattern is still quite busy with creases, which is because by step 36, we have made all the points fairly narrow and introduced a lot of creases in the process.

If we look even earlier in the folding sequence, we see another version of the base that still captures all of the essential elements (it has the same number and length of flaps as the skinny version) but has a much simpler crease pattern, shown in Figure 6.20.

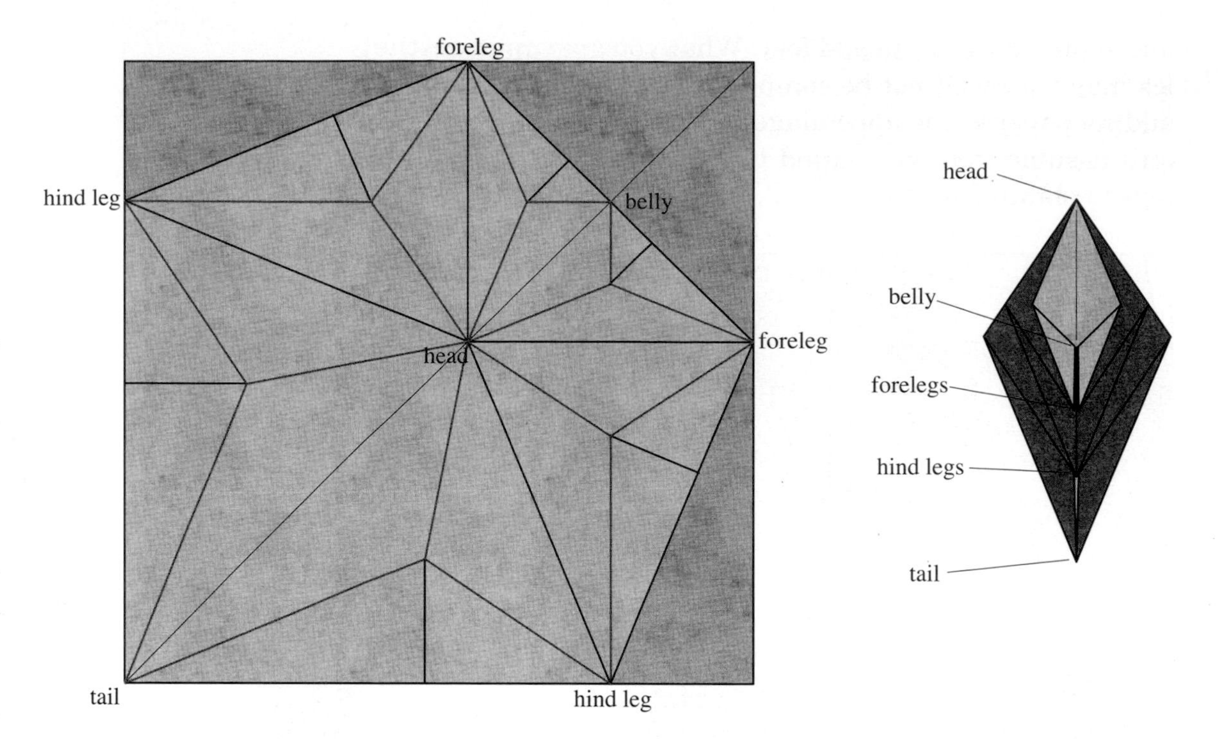

Figure 6.20.
Crease pattern and folded example of the simplified lizard base.

One other thing I've done in simplifying the crease pattern is to ignore those parts of the square that aren't essential to the base, by coloring them in and not drawing any creases in them. In this case, three of the four corners go unused.

What do we mean by "not essential to the base"? A simple definition of nonessential would be if you can cut the paper away and don't lose any flaps. In the lizard, the pair of scalene triangles at upper left and lower right in Figure 6.20 clearly fits this definition (which you can verify by direct experimentation: Fold the base, cut off the corners, and refold it). The upper right corner is questionable; it creates the white underbelly of the lizard, and the raw corner can be used to make a lower jaw and/or tongue for the lizard (try it), but if we take the major features of the base to be head, body, legs, and tail, we can assuredly cut away the corner and still obtain these features in the same sizes and locations.

Of course, such nonessential corners, even if they create no new flaps, still add to the thickness of the existing flaps. This can be either a feature or a bug in the design, depending on whether the extra paper adds necessary stiffness to the flap (feature) or causes the flap to split, splay apart, or unfold (bug).

In the crease pattern, I've also added labels that show which parts of the crease form which features of the base. We use these to define where we wish to add extra paper to create the toes.

Now, as we've seen, there are several different ways to add toes to this lizard. We could split the leg points, albeit at a cost in length. We could do it with a border graft. And we could do it with a strip graft, in more than one way, as it turns out. But let's first look at trying a border graft, not because it works best, but because it doesn't work very well at all. In origami design, understanding why a design technique doesn't work can sometimes be more valuable than folding one that works.

Proceeding as with the bird's feet of the previous section, let's try adding a small Bird Base to the tip of each of the four legs as shown in Figure 6.21, which, reasoning by analogy, should give us four toes on each foot.

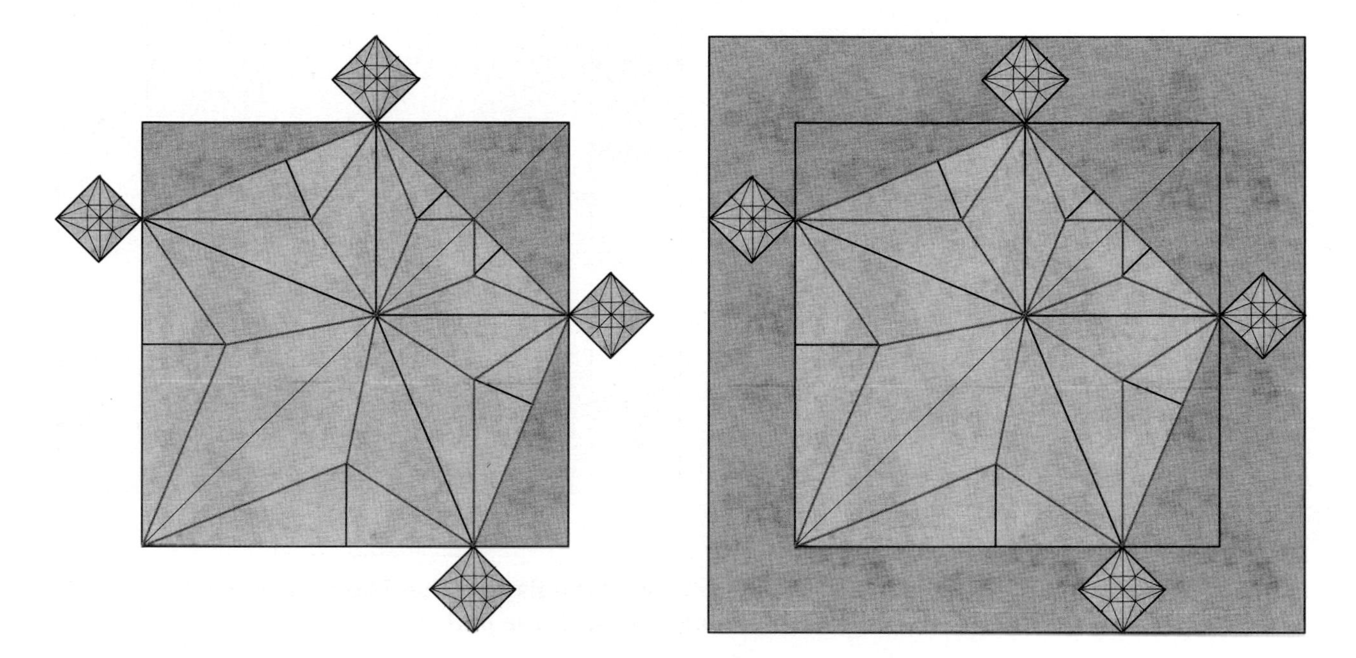

Figure 6.21.
Left: adding four small Bird Bases to the feet of the lizard base. Right: the shape embedded in a larger square.

The first thing that stands out from this construction is that we've added a whopping great quantity of nonessential paper to the pattern (in addition to the nonessential paper that was already there at three of the four corners). Basically, all of the colored region on the right in Figure 6.21 is nonessential. The second thing of importance—which doesn't stand out, but can be ascertained from examination of the pattern—is that it turns out to be impossible to add creases to the colored region in any way that allows the four "toes" (the tips of the bird base) to come together.

How, you may ask, can one be so sure of impossibility? By a small thought experiment: an imaginary manipulation of the base, as if such a base actually existed. If we had such a base (the same as the lizard base, but with all four toes together at the tip of the feet) then we would be able to manipulate the flaps the same way as the flaps of the original lizard base. In particular, we should be able to manipulate the flaps into the same arrangement as step 19 of the folding sequence for the Lizard. This arrangement is shown in Figure 6.22.

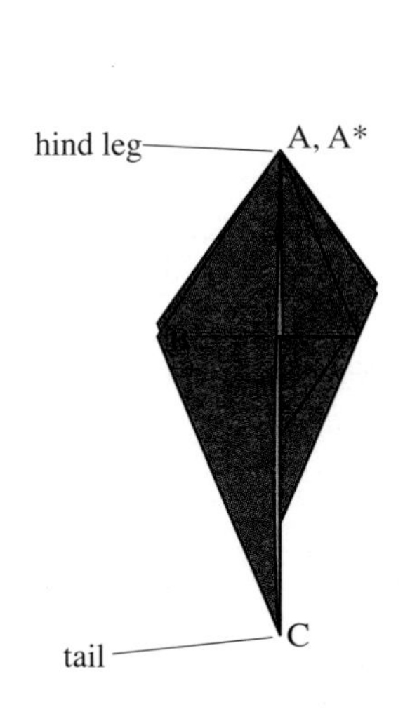

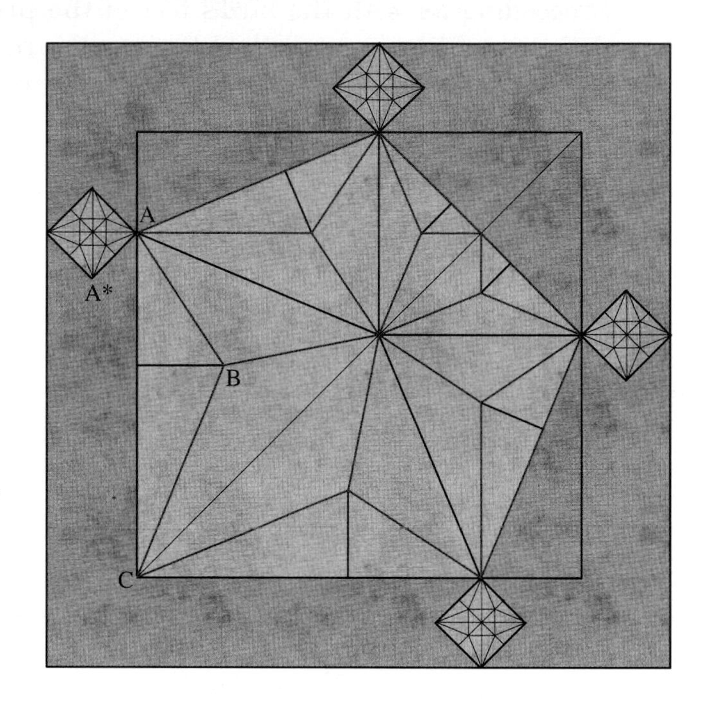

Figure 6.22.
Left: a configuration of the flaps of the lizard base.
Right: a possible partial crease pattern.

Now, the image on the left in Figure 6.22 shows one possible arrangement of the base, with one of the triangles in the crease pattern indicated by its corners A, B, and C. The base might not look exactly like this, of course; the extra paper we've added might create more layers that cover up or conceal the original lizard base flaps, but somewhere inside the extended base, we'd have the original lizard base. And in this folded configuration, triangle ABC, which is part of the original base, is flat.

Let's ask a question: How far is it from point C (the tail) to point A (tip of the leg/one of the toes)? The answer is evident from the figure; since triangle ABC is flat in both the crease pattern and the folded base, the distance is equal to the length of line segment AC. Call this distance x.

If we've assumed that all the hind-leg toes are together at point A in the folded base, then the toe marked A* in the crease pattern must be one of them. So it, too, must be separated from the tail point C by the same distance x—in the folded form.

Now let's look back at the crease pattern. It's clear that point A* is somewhat closer to point C in the crease pattern than point A. So in the crease pattern, the distance from A* to C is less than x. In the folded form, it's equal to x. So whatever the crease pattern in the colored region is, it has to increase the separation between points A* and C.

But this is impossible. Short of stretching or cutting, there is no way to fold a sheet of paper that *increases* the distance between two points. Folding can only reduce this distance. The goal is impossible to attain; there is no set of creases added to the border graft that allows all four bird base points to come together at the tip of the leg flaps.

So, while border grafts allow you to add a Bird Base to the corners of a square, they don't allow one to be added to the edge in the same way.

More importantly, we have touched on a very deep concept in origami design: the relationship between distances in the folded and unfolded form. As we saw in point-splitting, where it was key to the design of the ideal split, examining distances in the folded base and on the unfolded pattern can show what is possible and impossible and provides guidance for the location of important creases. And as we will see in later chapters, this relationship forms the underpinning of a full mathematical theory of efficient origami design.

In the example described above, the point marked A* was the one that caused the problem. What if we only wanted two toes? Then could we use a border graft that incorporated the two-flap Fish Base? The embedded crease pattern is shown in Figure 6.23.

This pattern does not incorporate any contradictory assumptions, unlike the previous example, and if you draw it out, you will find that you can add creases to realize a lizard base with two flaps at the tip of each foot.

But as noted earlier, the border graft is fairly inefficient anyhow, and we have one other possible way of performing a graft: the strip graft. We can graft a strip into a model by cutting it apart and inserting the strip(s) between the cut edges.

In the strip graft that we used in the Tree Frog, we cut each flap down the middle—more specifically, along an obvious line of symmetry. Where should we cut this pattern? The

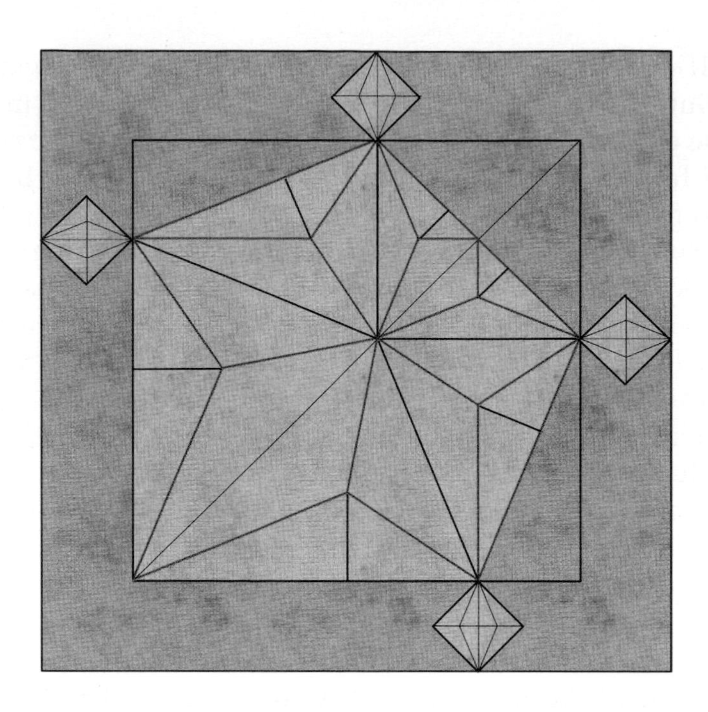

Figure 6.23.
Embedded crease pattern for a border graft using four Fish Base points.

lizard flaps don't have quite the same left-right symmetry, but in the essential crease pattern (the central light pentagon), each flap has the same general structure as the flap of the Tree Frog, as shown in Figure 6.24.

Each flap is composed of four facets separated by alternating mountain and valley folds. In the Tree Frog, we split

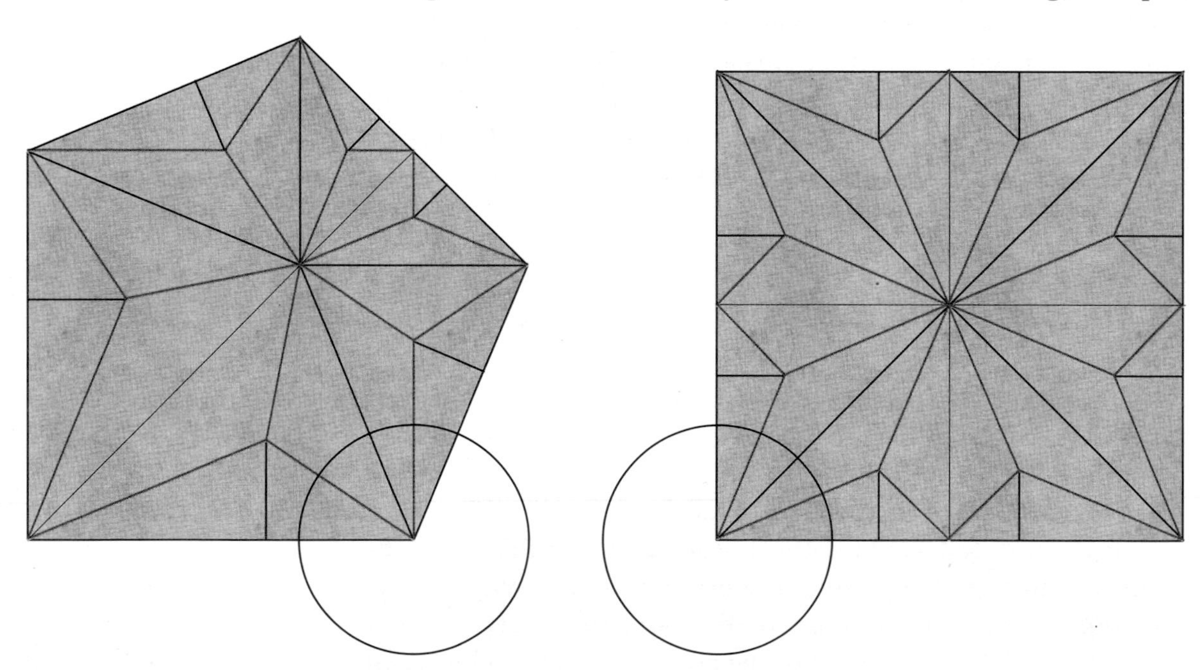

Figure 6.24.
Left: crease pattern for the lizard base.
Right: crease pattern for the Frog Base. Observe that the flap in each case is composed of four wedges.

each flap along its central mountain fold. In our lizard example, we can do the same. If we split a flap along a crease, then there is a very simple construction of creases to impose on the inserted strip that is easily generalizable to any number of flaps, which is shown in Figure 6.25. Each strip is divided into parallel pleats, one for each gap between toes. The ends of the pleats are then reverse-folded to separate the individual toes.

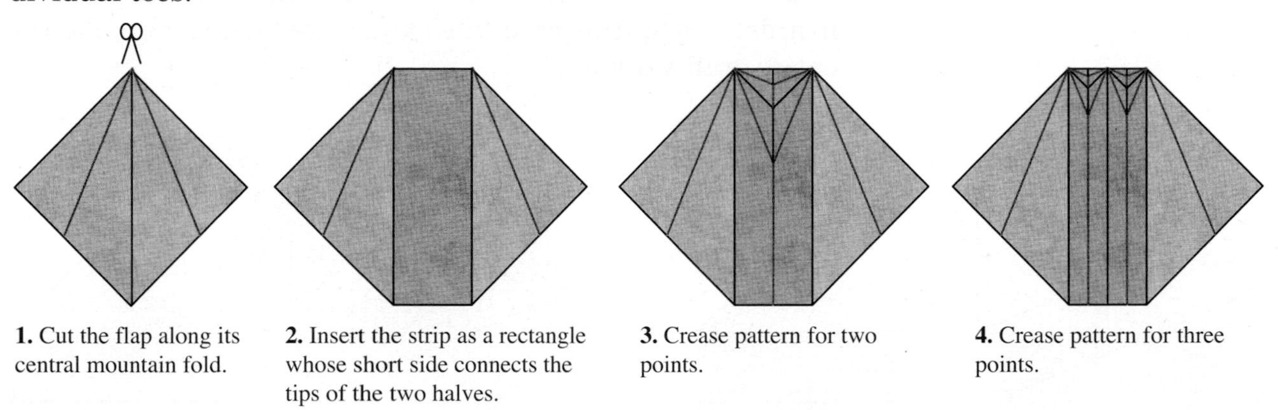

1. Cut the flap along its central mountain fold.

2. Insert the strip as a rectangle whose short side connects the tips of the two halves.

3. Crease pattern for two points.

4. Crease pattern for three points.

Figure 6.25.
Cut and insert a strip of paper to split the flap into two or three smaller points.

So, all we need to do is cut along the mountain folds in each flap and insert rectangular strips, which will be subdivided into as many points as we want toes.

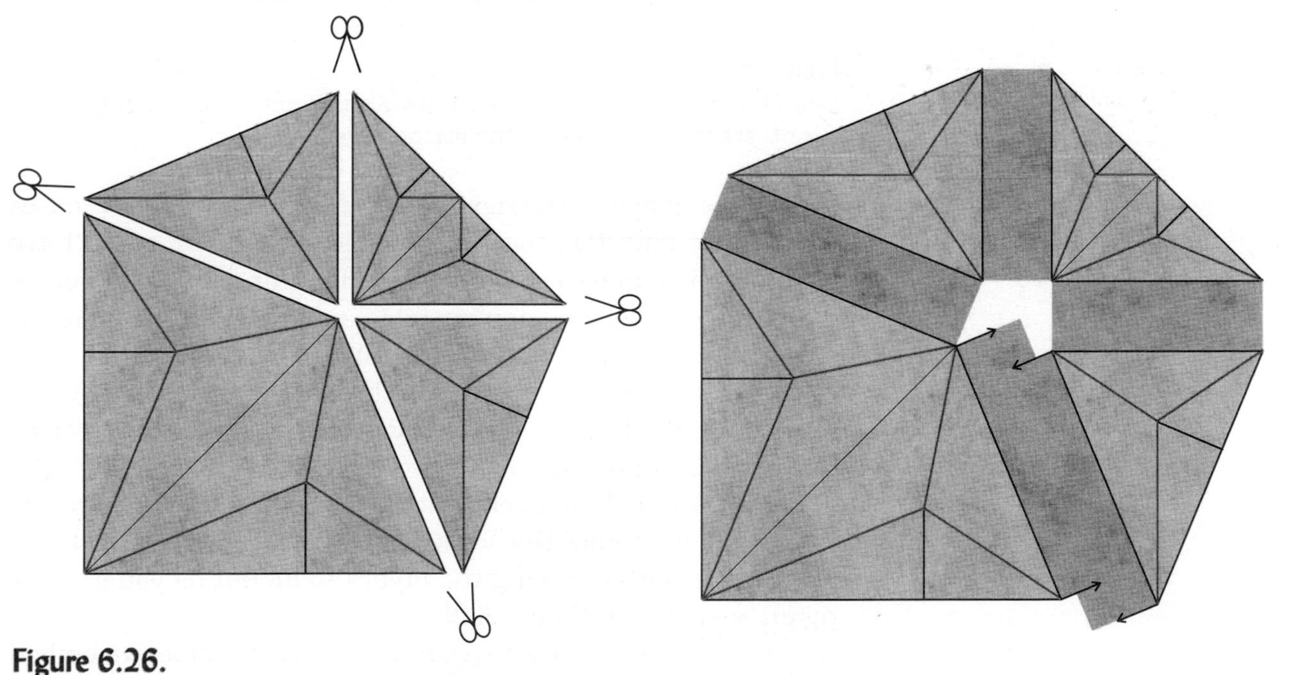

Figure 6.26.
Left: the lizard base, cut along mountain folds.
Right: with strips inserted.

The images in Figure 6.26 illustrate the cut-and-insert process. However, one problem arises: It's not possible to get all of the strips to line up. As you see in the right image, only three out of the four rectangular strips can be aligned to the pieces of the crease pattern.

This problem is fixable, however, by making one more cut and adding a strip down the middle of the tail. We don't need to divide the tail (forked tails being relatively rare among lizards), but in this case, the strip is necessary to make the entire graft work out.

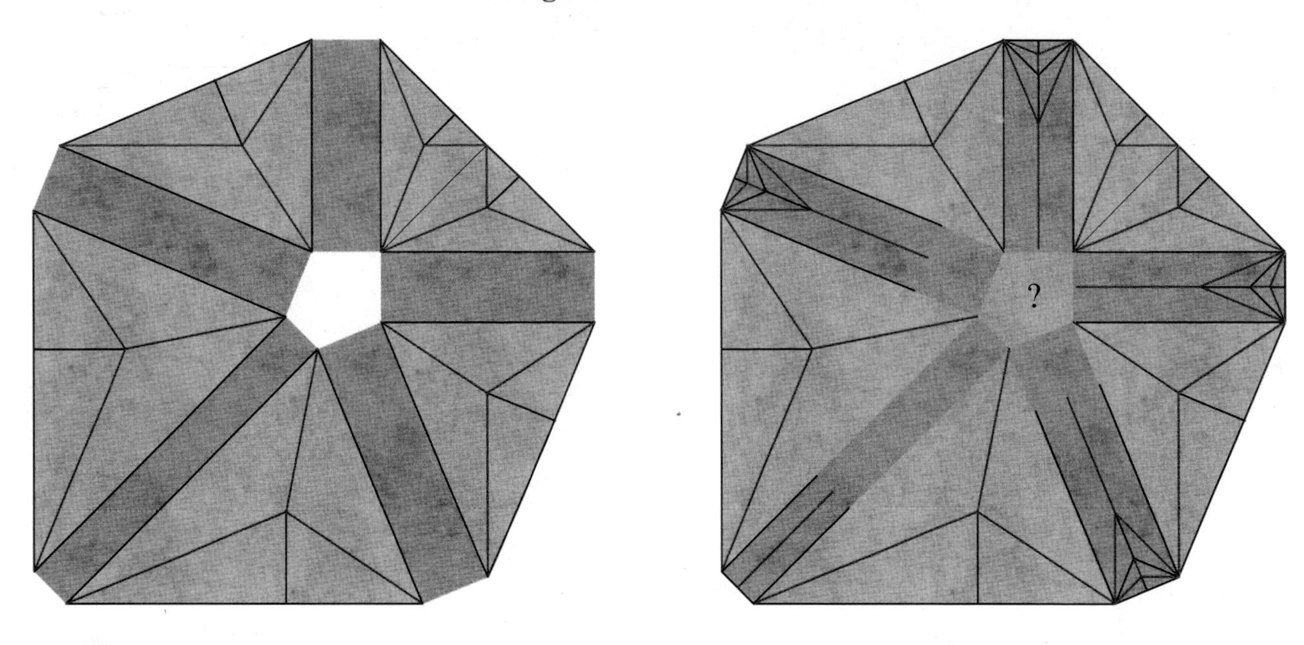

Figure 6.27.
Left: the dissected pattern with an additional strip down the tail. Right: starting to draw in the strip creases.

In the image on the right in Figure 6.27, I've added creases that create only two toes for simplicity, but you could have easily used strip patterns for three, four, or five toes if you so desired. (And since we don't need to split the tail, I have left it as a pleat.) An open question is: What happens in the very center of the pattern when all the creases come together? Obviously, all of the pleats that we're making from the strips have to terminate at each other somehow. In this case, the easiest thing to do is to cut out the crease pattern, make the folds that we know the location of, and then extend them toward the center, forcing the layers to lie flat as you go. The result is shown in Figure 6.28.

Finally, to get back to a square starting shape, we embed this unusual polygon into a square, which results in the pattern on the right in Figure 6.28. I will leave the folding of

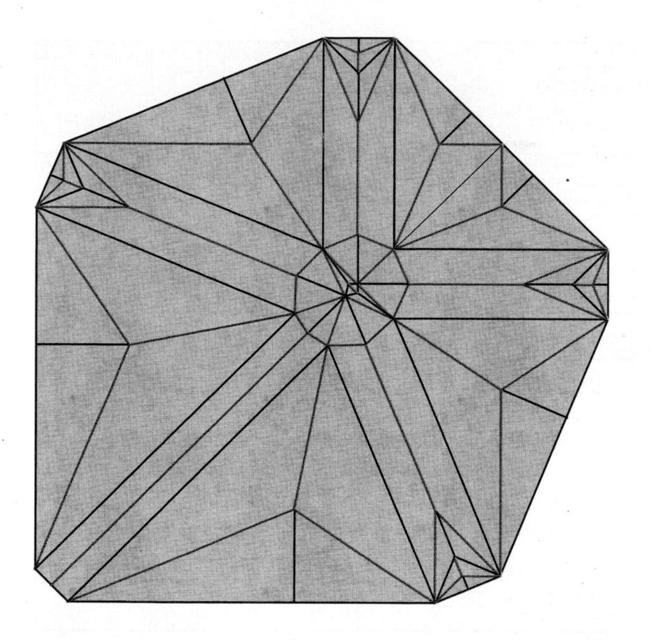

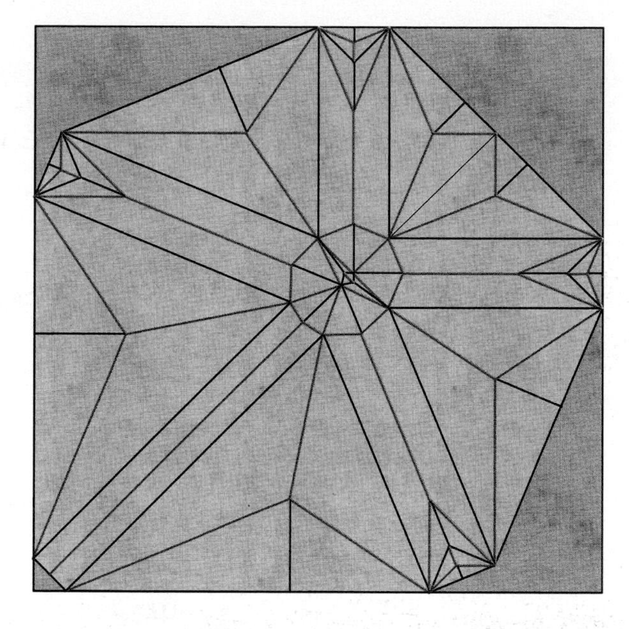

Figure 6.28.
Left: crease pattern with the strip creases extended to the center.
Right: resulting pattern, embedded within a square.

this pattern into a toed base as an exercise for you. One more exercise you might enjoy is working out how the strip crease patterns for larger numbers of points meet where they come together.

As we found with the Tree Frog, the place where the pleats come together creates several small middle flaps; these could easily be turned into eyes or other facial features. In fact, the middle flaps don't have to be just a byproduct; you can create middle flaps intentionally by adding strip grafts to a crease pattern.

And this isn't the only way to add strip grafts. A weakness in this crease pattern is that while all of the points at the end of the forelegs lie on the raw edge of the square, some of the points on the hind legs are middle flaps—they come from the interior of the paper, which means they are twice as thick as the others. If we use a strip graft along an edge to get a collection of points, the strip must be perpendicular to the edge (as it is in the forelegs) to keep all the points on the edge.

Well, there's nothing that says we have to cut along existing creases to insert a strip graft. It's perfectly acceptable to cut across creases, form the pleats of the strip creases, and then fold the original model, as illustrated in the crease pattern in Figure 6.29.

But this is more wasteful than it needs to be. A pleated strip, once started, has to keep propagating in the same direction until it hits something else; you can't change the

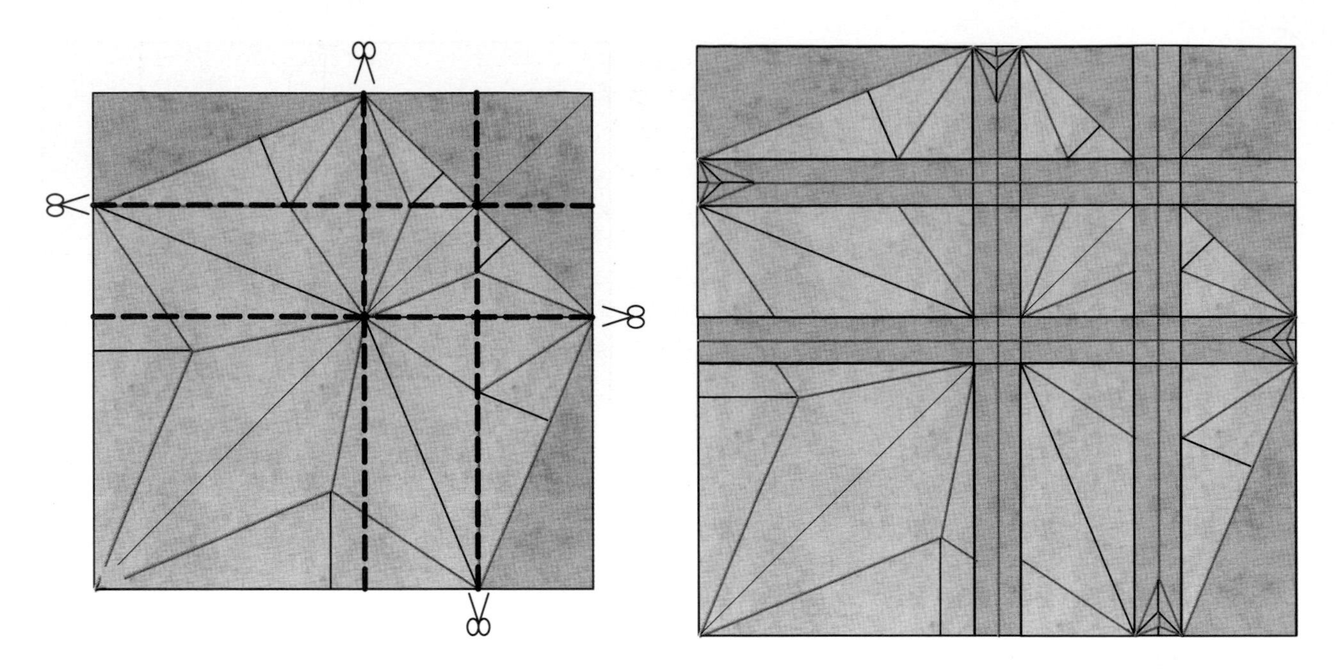

Figure 6.29.
Left: cut lines for strip grafts perpendicular to the edge at each foot.
Right: the embedded pattern, with partial strip creases.

direction of an isolated pleat. Pleats can certainly cross without changing direction, as is shown in Figure 6.29. But when two pleats collide, that's an opportunity for them to coalesce into a single pleat running in a different direction, which reduces the total amount of added paper. Thus, we can create a

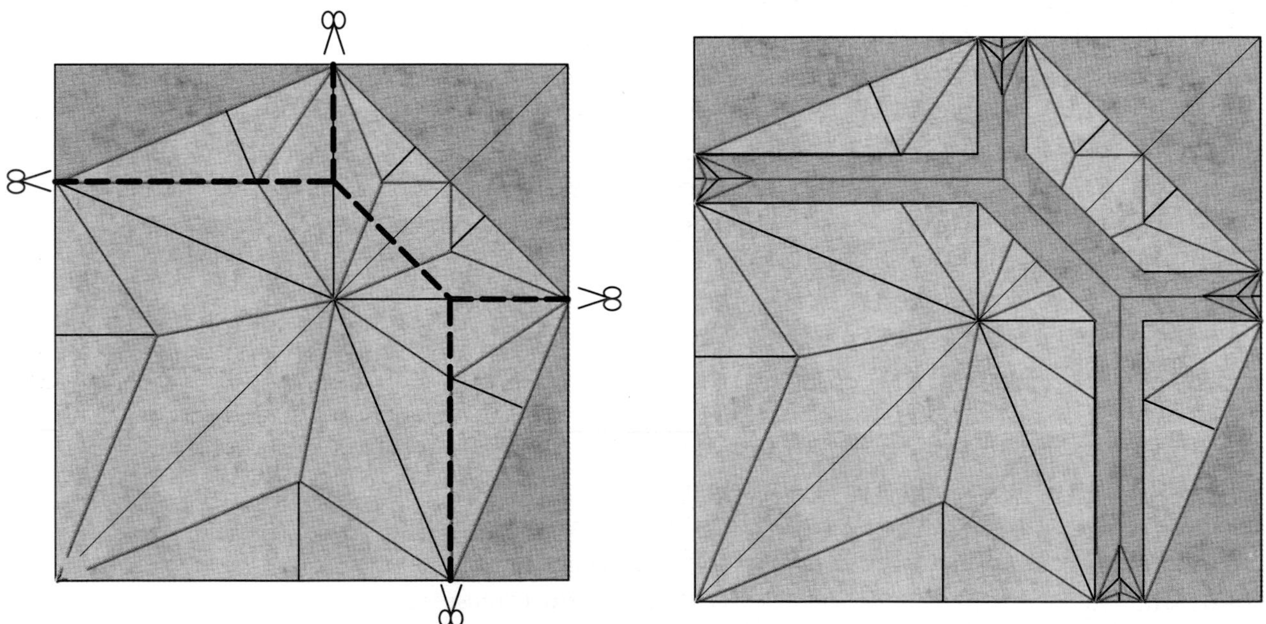

Figure 6.30.
Left: cut lines for a simplified strip graft.
Right: grafts inserted.

much simpler set of strip grafts by propagating pleats inward from the edges of the square and noting that when they meet, we can send off a single new pleat that connects both pleat intersections. The resulting pattern, which is much more efficient with paper, is illustrated in Figure 6.30.

I should point out that the crease lines in the original base actually propagate into the pleats; I have left those out of Figure 6.30 in the interest of clarity. Again, I would encourage you to draw up this pattern and fold it into a lizard and/or to extend the pattern to larger numbers of toes.

6.4. More Applications of Grafts

One of the more enjoyable uses of border and strip grafts is to breathe new life into an old model. There is a shrimp design in the traditional Japanese repertoire, folded from a Bird Base, that is elegant but spare. Simply adding a border graft on two sides allows one to add the larger tail and split claws that make a respectable crawfish, as shown in Figure 6.31. I encourage you to fold it and try it out yourself. The structure is simple enough that you should be able to make it from the crease pattern alone. The folded model shown in Figure 6.31 is still quite minimal; by narrowing the claws and adding further shaping folds, you should be able to produce quite a realistic model.

Figure 6.31.
Crease pattern, base, and folded model of the Crawfish.

Strip grafts can get fairly complicated and can actually comprise most of the paper in the model. The crease pattern in Figure 6.32 shows the base for a treehopper, a type of insect; this strip graft is used to create three points from one flap at each end (note the resemblance of the crease pattern to that of an ideal split). I have highlighted the strip graft in the pattern. If you cut out the strip and butt the two halves of the remainder together, you will observe the underlying base: a simple modification of the stretched Bird Base.

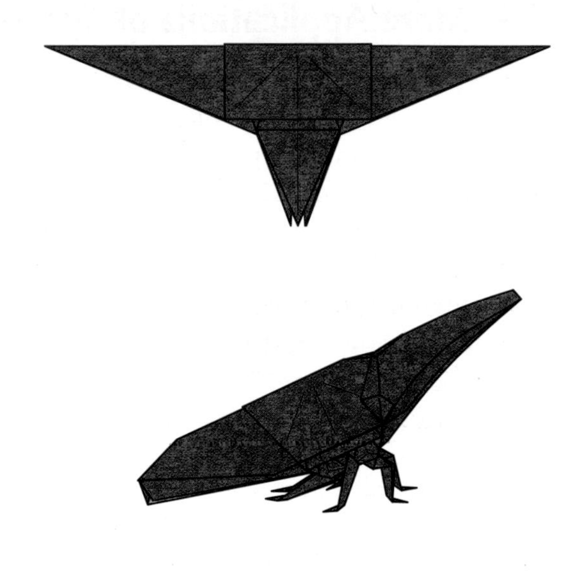

Figure 6.32.
Crease pattern, base, and folded model of the Treehopper.

Figure 6.33 incorporates two strip grafts into a shape otherwise composed of half of a blintzed Waterbomb Base and half of a blintzed Frog Base. The transformation from base to folded model is more complex than most, but you should have no trouble in going from the crease pattern to the base. The extra paper in the split gets used to form the split in the wings, the mesocentrum (the triangle in the middle of the back), and both the pair of horns on the thorax and four horns on the head (yes, they really look like that. These beetles are popular as pets in Japan).

Now if you have worked your way through this chapter so far, you may quite likely feel that the concept of toes has been thoroughly pummeled into submission. And it is true, there is only a finite number of designs that can benefit from the addition of clusters of small points to preexisting flaps.

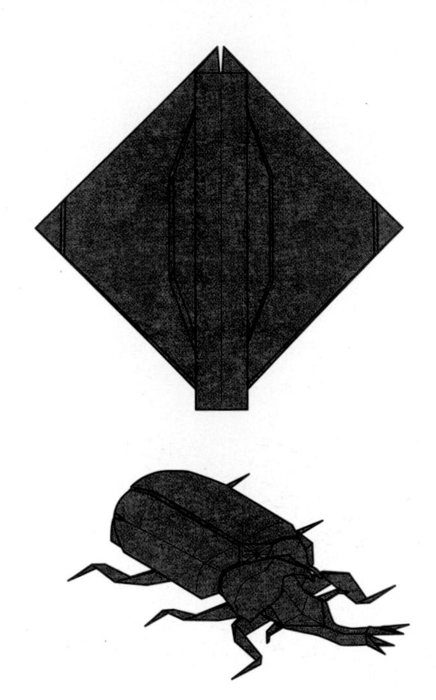

Figure 6.33.
Crease pattern, base, and folded model of the Japanese Horned Beetle.

Toes, claws, hands, feathers, and horns all have a place, but leave the great majority of origami subject material untouched. But by working through these specific examples, you've now seen the basic concepts behind the much larger world of grafted bases.

Grafting effectively allows both cutting and gluing in an origami-acceptable way. But it does more: To make use of grafting, one needs to start looking at crease patterns and pieces of crease patterns as distinct entities that bring a particular function to the origami model: a flap, a set of flaps, an open space. By cutting and assembling pieces of existing bases into new bases, you can break out of the rigid hierarchy of the traditional bases and realize entirely new custom bases; in addition, you can selectively add patterns and textures to all or part of a model. We will learn techniques for both of these in the next section.

Folding Instructions

Songbird 1

KNL Dragon

Lizard

Tree Frog

Songbird I

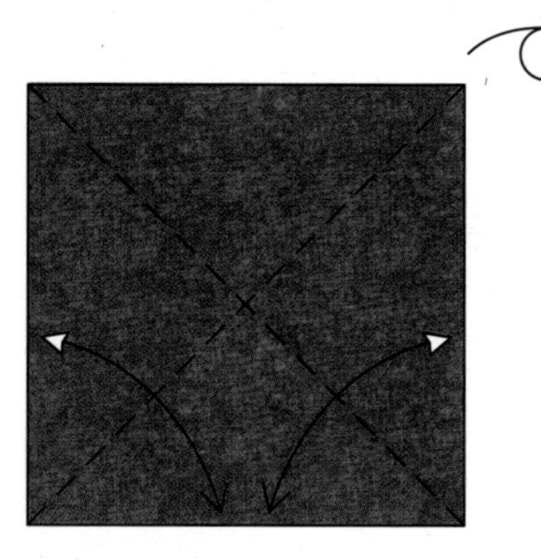

1. Begin with a square, colored side up. Fold and unfold along the diagonals, then turn the paper over.

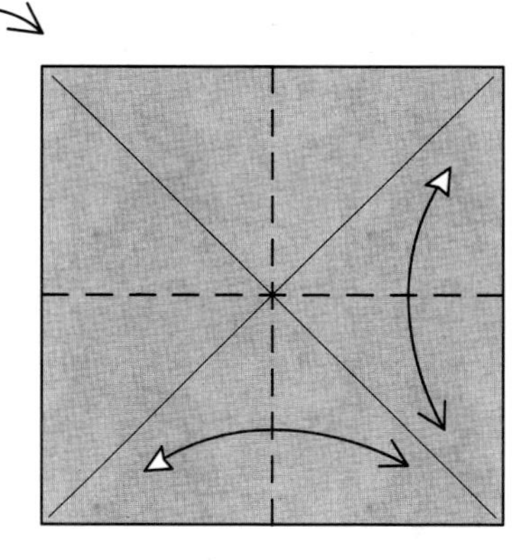

2. Fold and unfold in half vertically and horizontally.

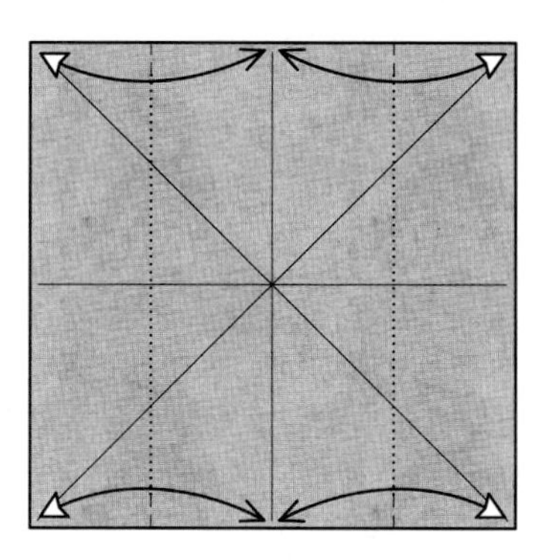

3. Fold the sides in to the middle and make pinches along the top and bottom edges; unfold.

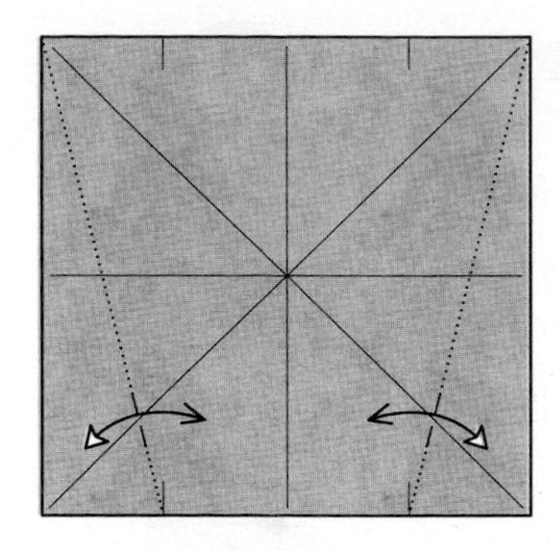

4. Fold the left edge along a fold connecting the top corner with the bottom pinch mark; make a pinch where the fold crosses the diagonal and unfold. Repeat on the right.

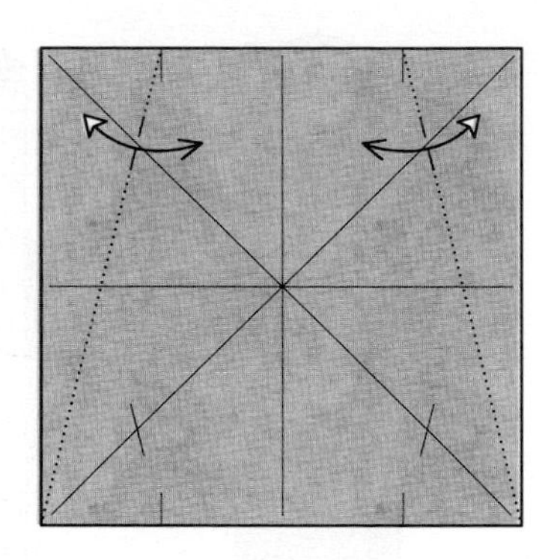

5. Repeat step 4, using the bottom corners and top pinch marks.

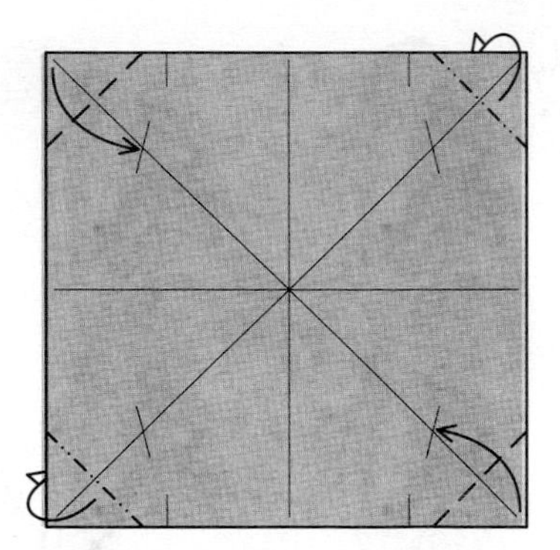

6. Valley-fold two diagonally opposite corners to the crease intersections in front; mountain-fold the other two corners to the crease intersections behind.

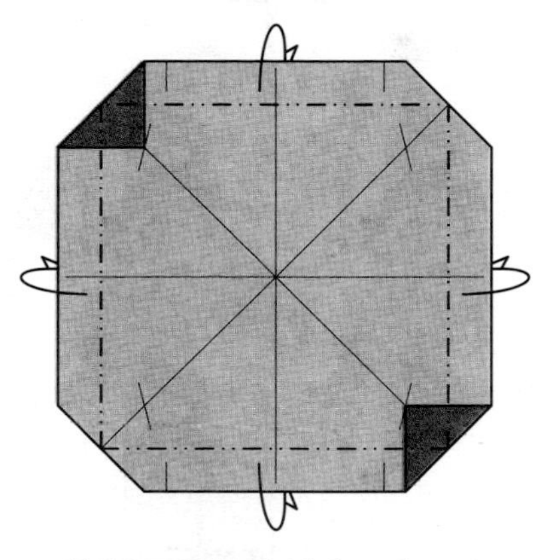

7. Mountain-fold the edges behind all the way around.

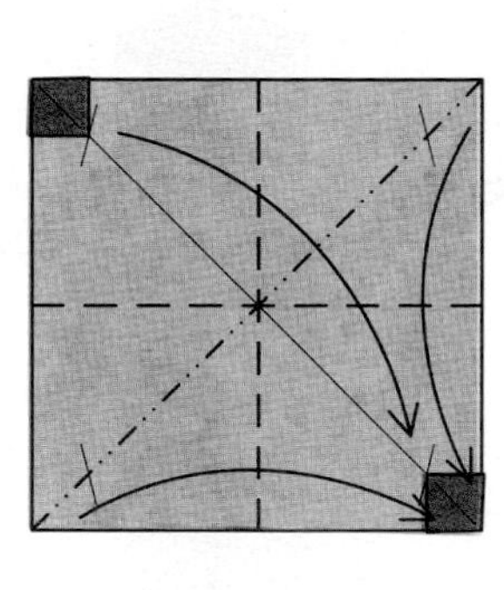

8. Form a Preliminary Fold. Then rotate 1/8 turn clockwise to put the four corners at the bottom.

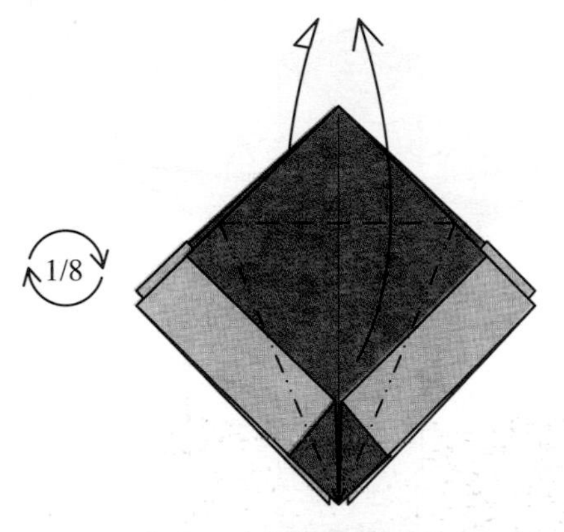

9. Petal-fold both front and rear to form a Bird Base.

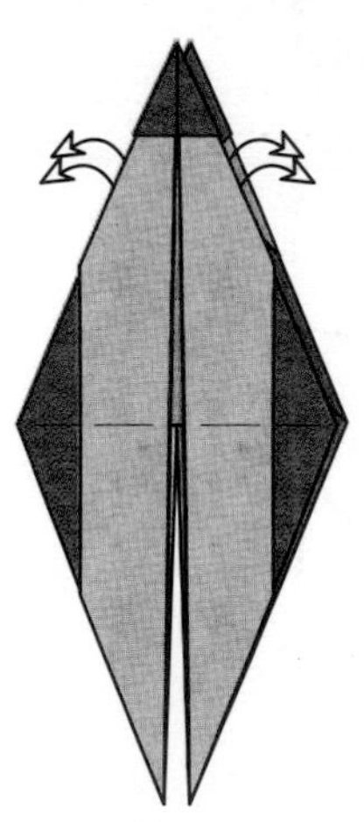

10. Release the loose layers of paper on both the front and rear flaps.

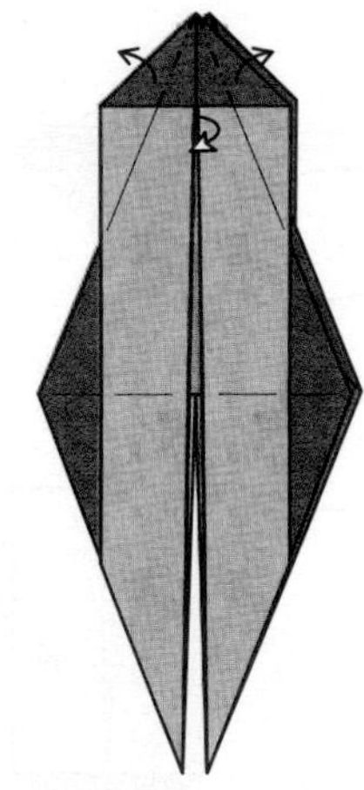

11. Release some more paper from the interior and spread the edges as far apart as possible. Do not repeat behind.

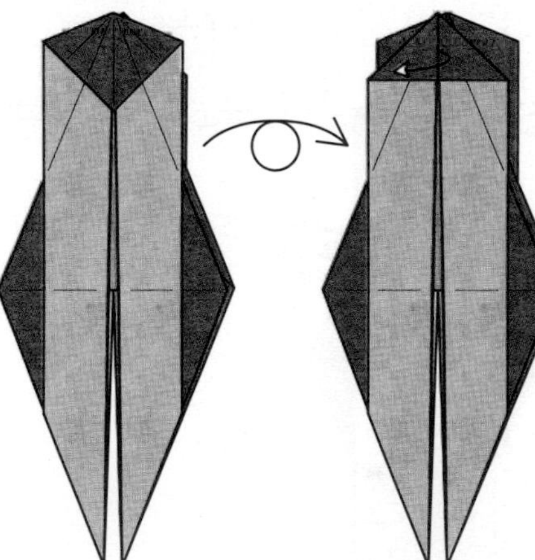

12. Turn the model over.

13. Release some paper from the interior and swing all the excess over to the left.

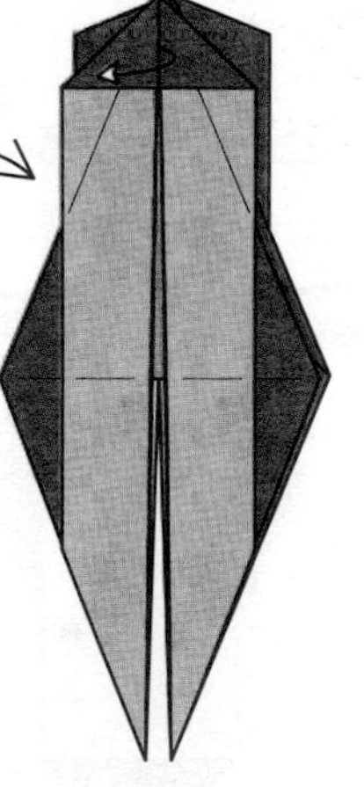

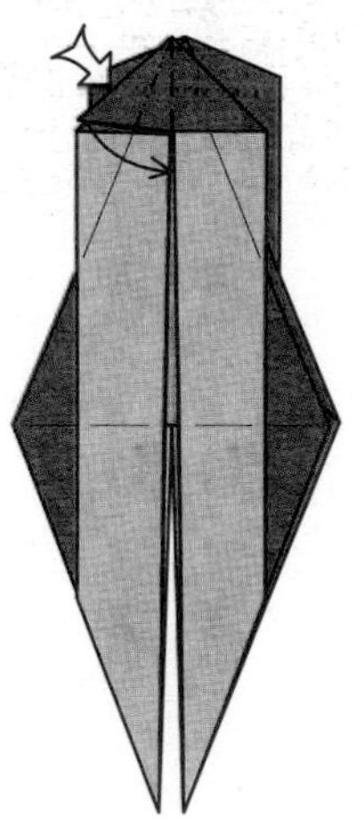

14. Squash-fold the flap symmetrically.

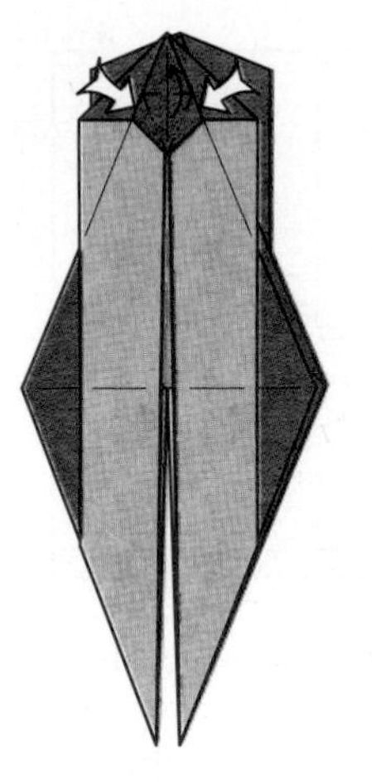

15. Petal-fold the tip of the flap.

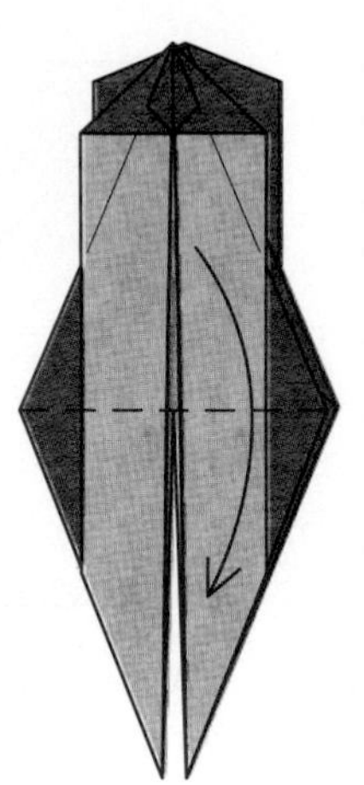

16. Fold the flap down.

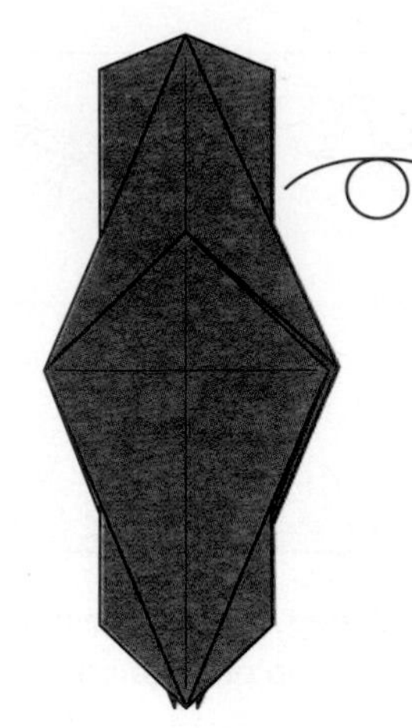

17. Turn the model over from side to side.

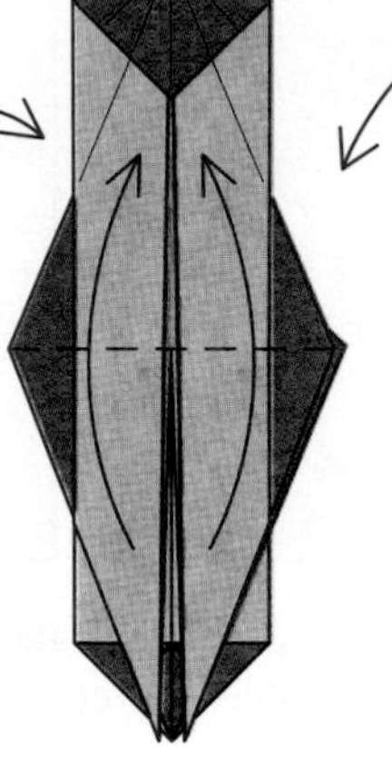

18. Fold the two points upward and flatten firmly.

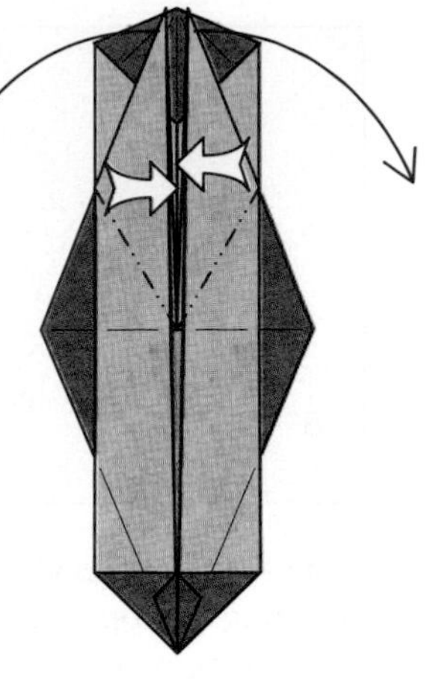

19. Reverse-fold two points out to the sides.

20. Wrap two layers from inside to outside on each leg.

21. Wrap two layers from inside to outside again on each leg.

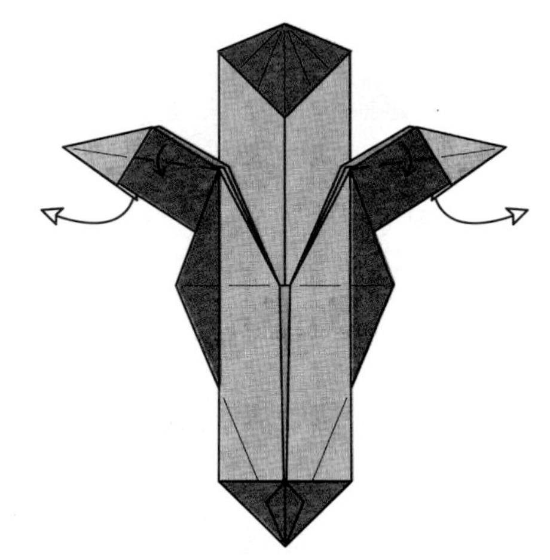

22. Pull the white corners as far out to the sides as possible.

23. Open each leg, spreading the layers symmetrically.

24. Steps 25–46 will focus on the left leg.

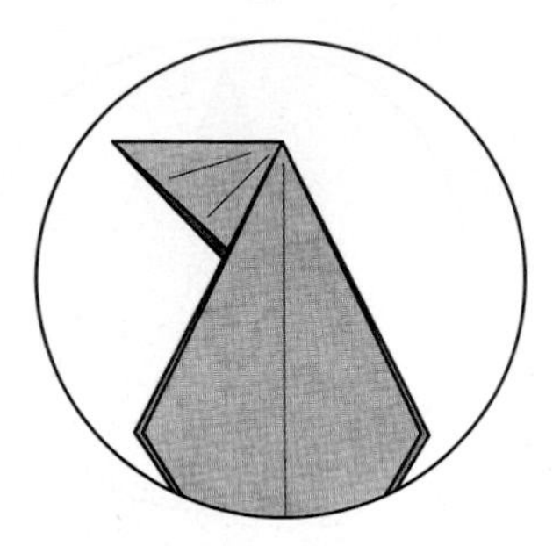

25. Turn the paper over and rotate so that axis of the leg is up-and-down.

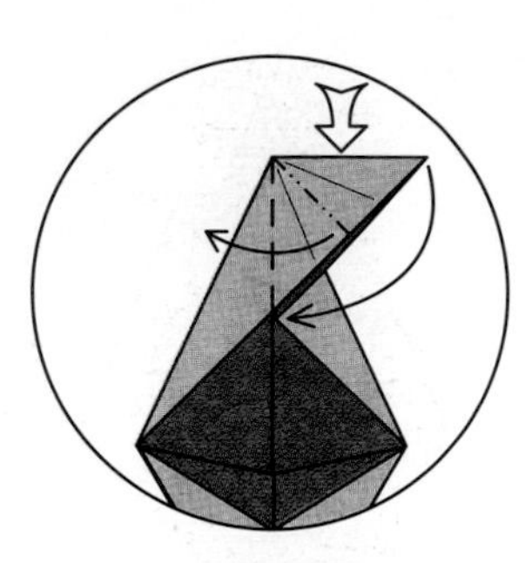

26. Squash-fold the flap symmetrically.

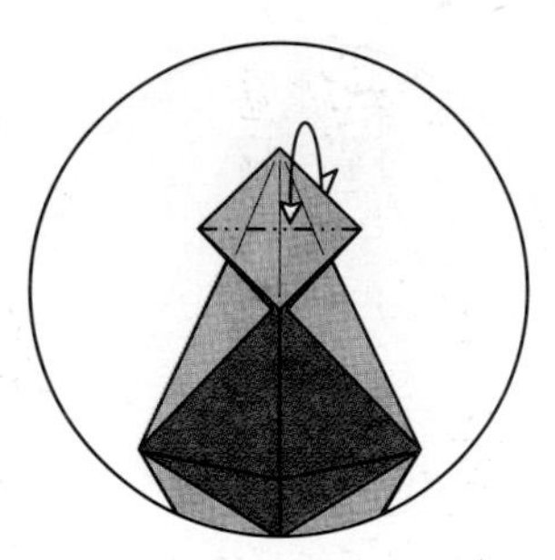

27. Mountain-fold the top half of the small square behind and unfold.

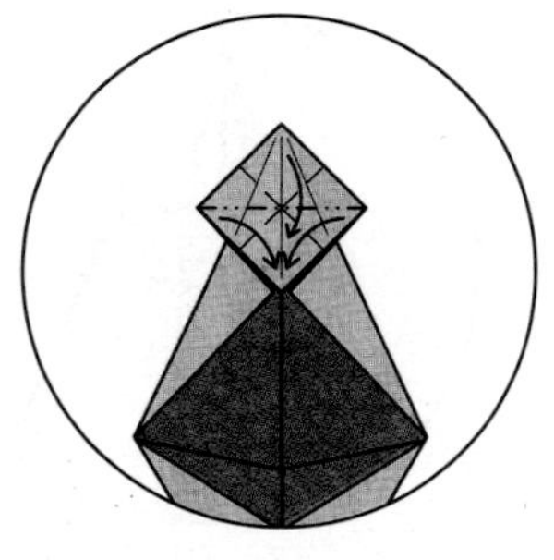

28. Fold a tiny Preliminary Fold from the square.

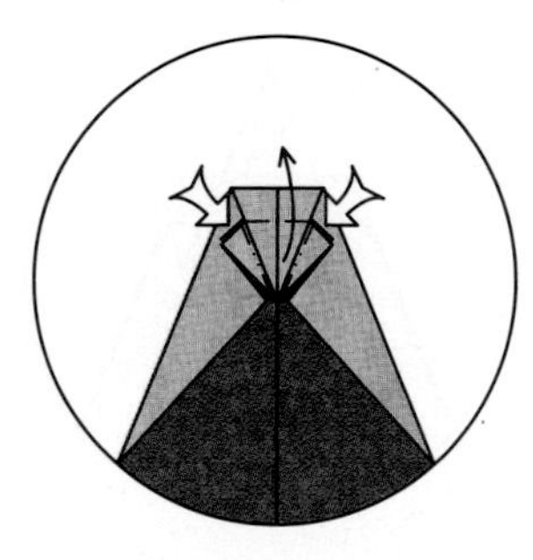

29. Petal-fold the near point.

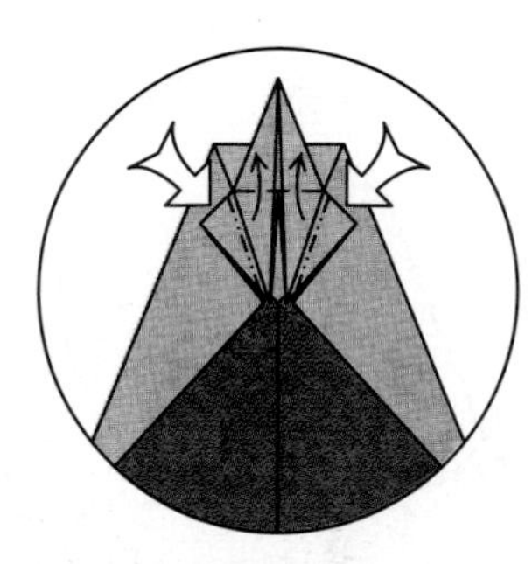

30. Petal-fold the remaining pair of flaps, including the trapped point along with them.

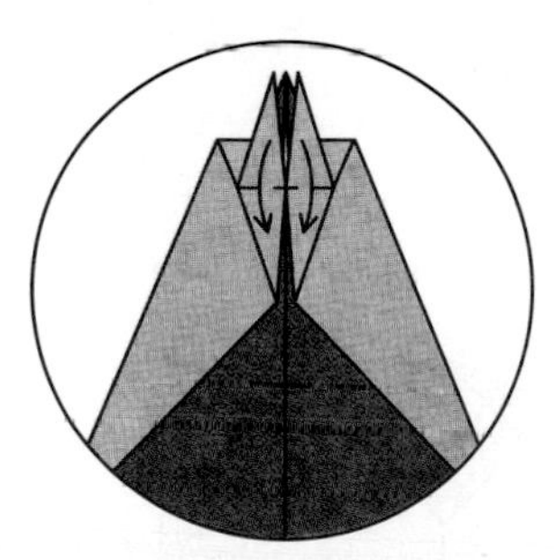

31. Fold the three points back down.

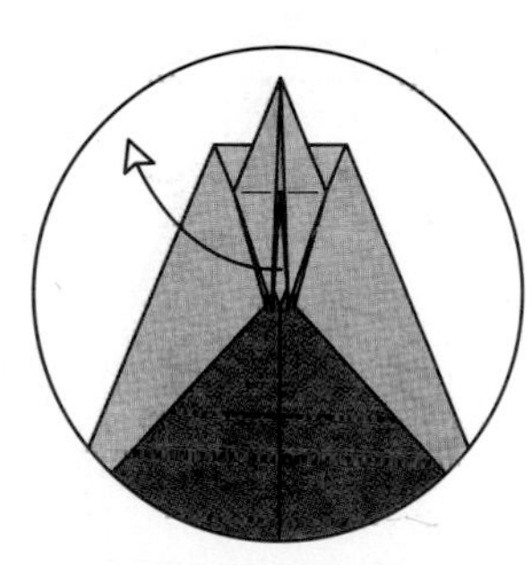

32. Carefully release and pull out the trapped corner.

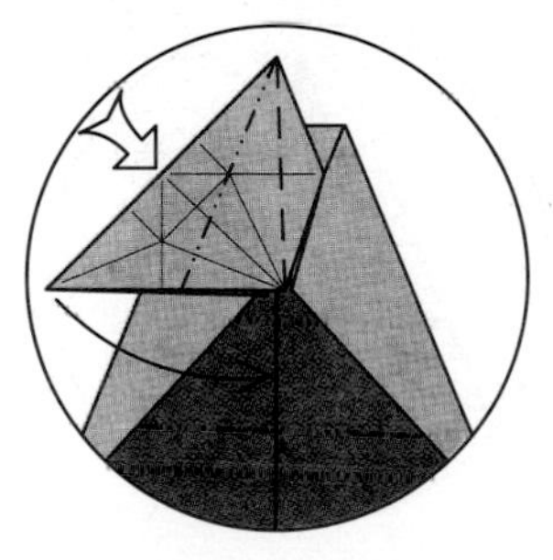

33. Squash-fold the flap.

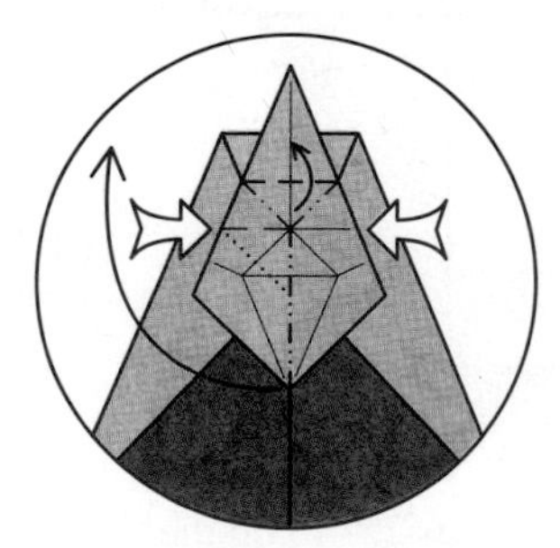

34. Pinch the flap in half and swing it up to the left, using the existing creases.

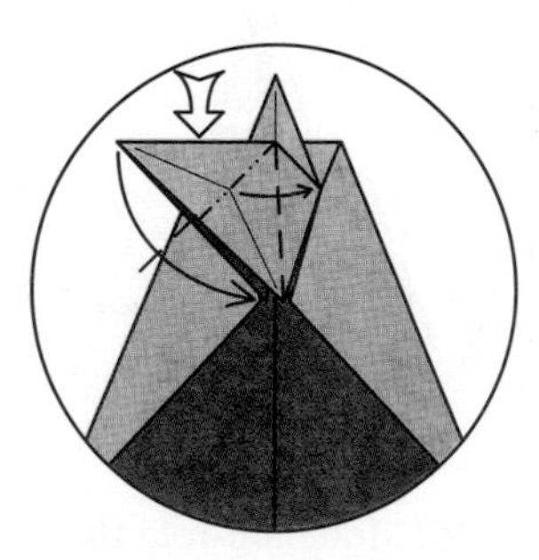

35. Squash-fold the flap.

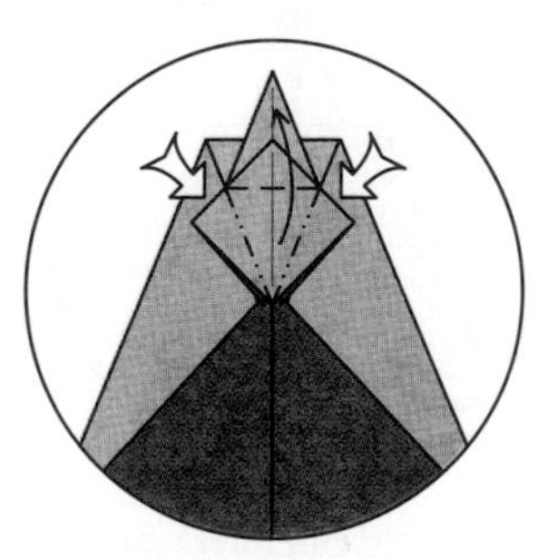

36. Petal-fold the flap.

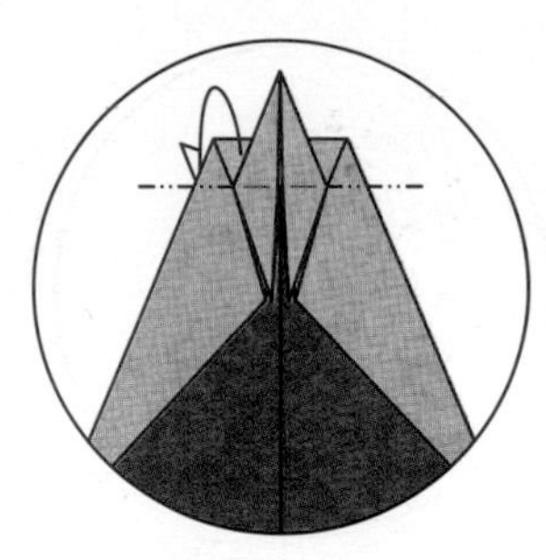

37. Mountain-fold the thick flap behind. Turn the paper over.

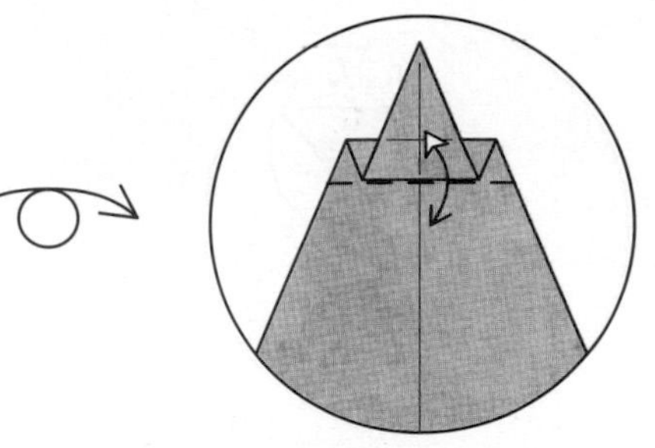

38. Fold and unfold through all layers.

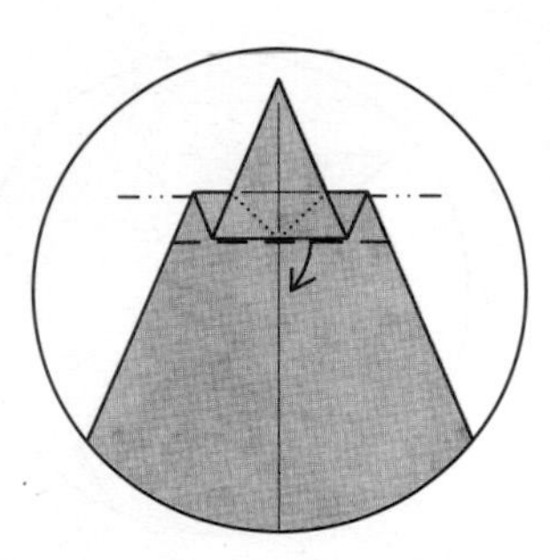

39. Pull out a double layer of paper, forming gussets in the interior.

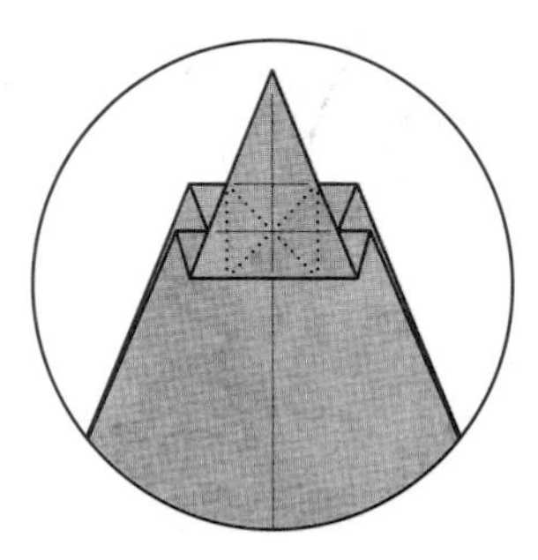

40. Like this. The gussets are indicated by the hidden lines.

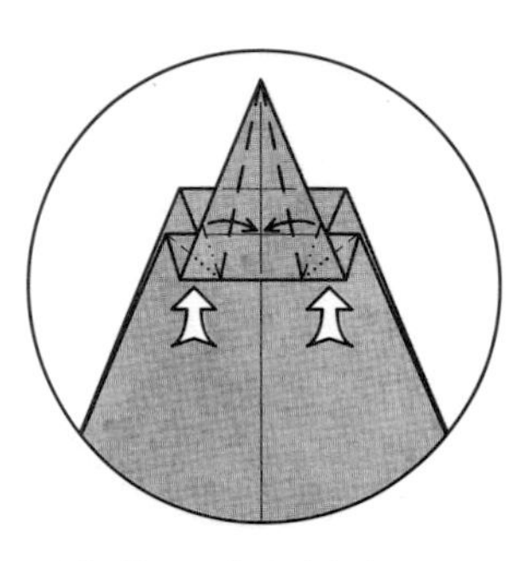

41. Squash-fold the corners.

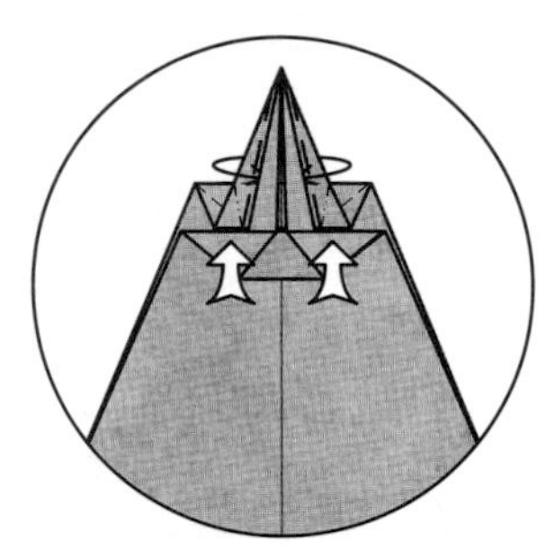

42. Repeat on the next layer up, tucking the edges into pockets.

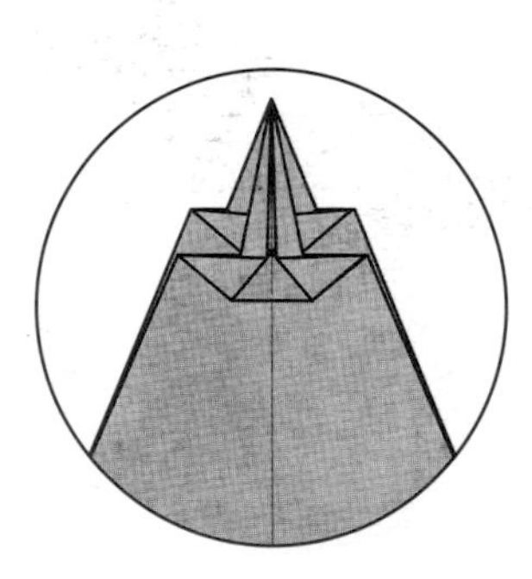

43. Turn the paper back over.

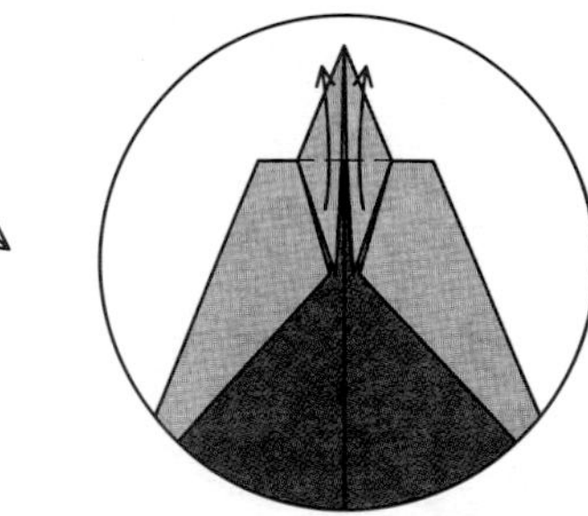

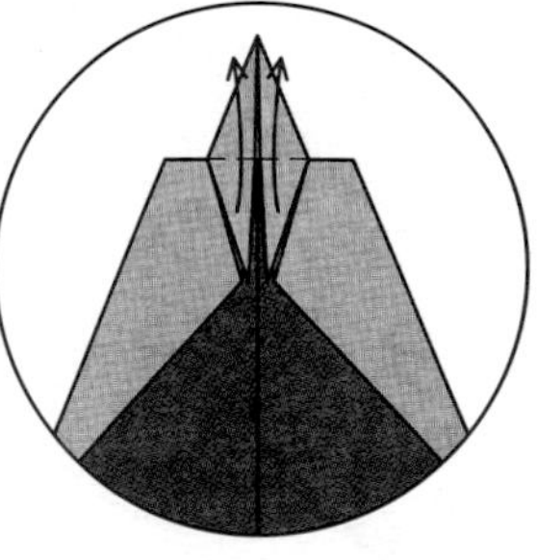

44. Fold two points up.

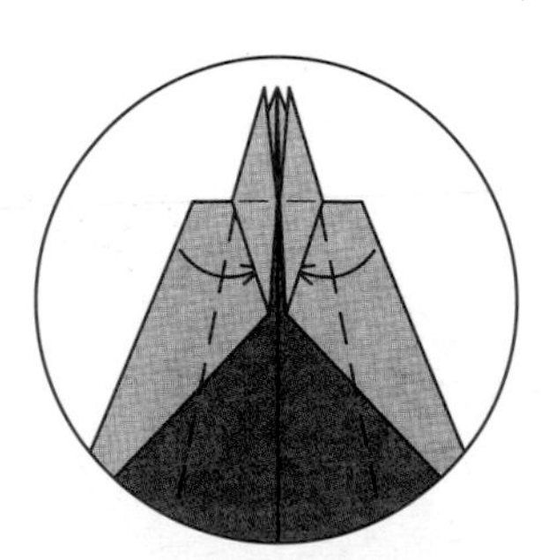

45. Fold one layer on each side to the center, tucking the top into the pockets.

46. Mountain-fold the leg in half (edges away from the head; see the next step).

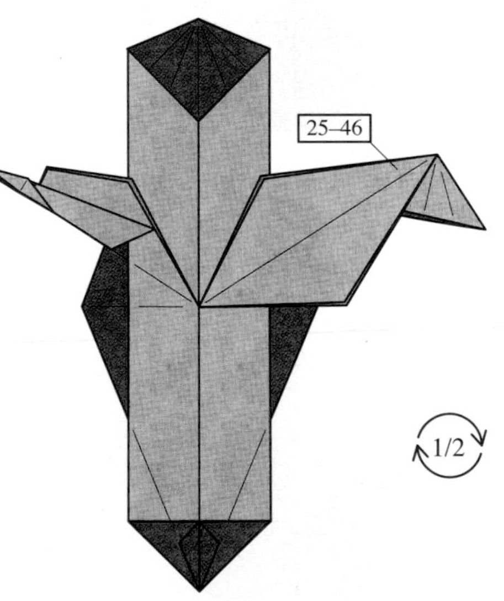

47. Repeat steps 25–46 on the right. Then rotate the model 1/2 turn.

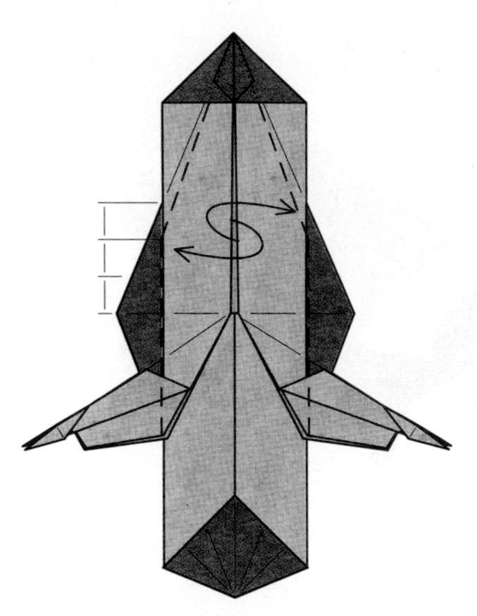

48. Open out the white edges. Note the location of the angle in the valley fold; above the fold, the model will not lie flat.

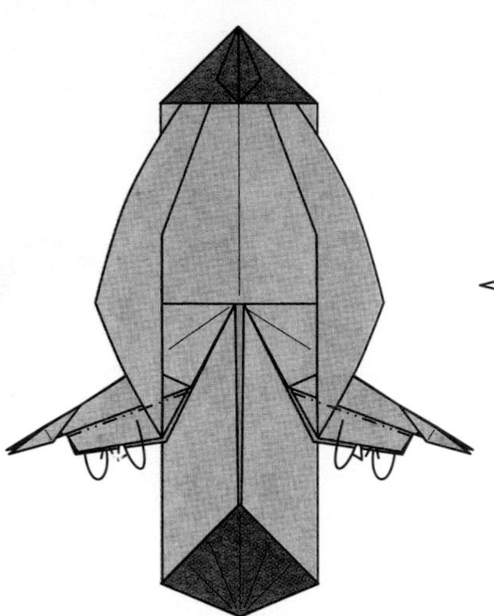

49. Mountain-fold the near layer of each leg inside and valley-fold the far layer.

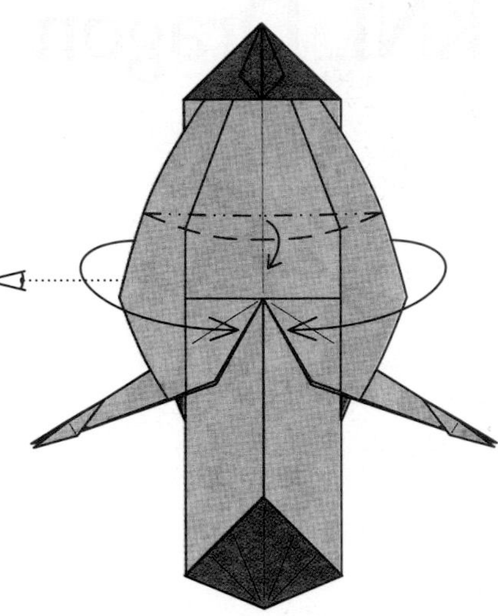

50. Crimp the body through the angle change and curve the sides around, so that the cross-section is an inverted "U."

51. From here onward, you may vary the folds as you like to alter the shape of the bird. Crimp the head downward, keeping it rounded.

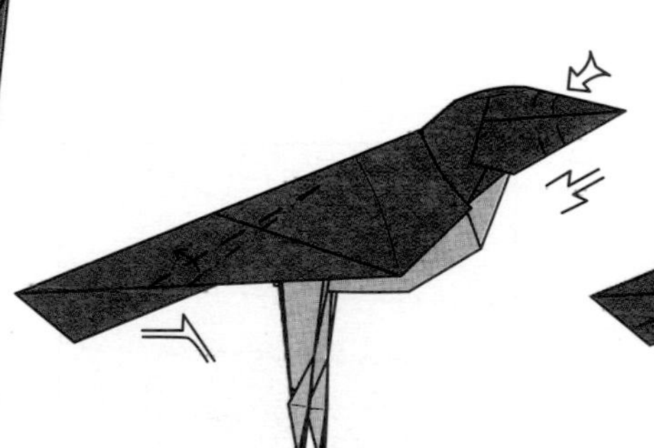

52. Crimp the beak. Pleat the tail; repeat behind.

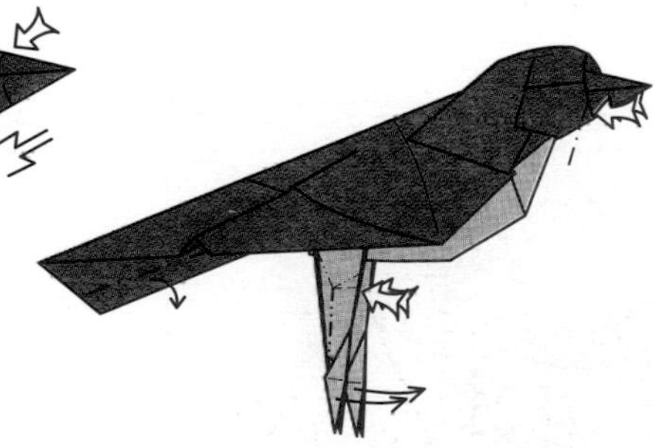

53. Round the tail. Double-rabbit-ear the legs. Reverse-fold the cheek.

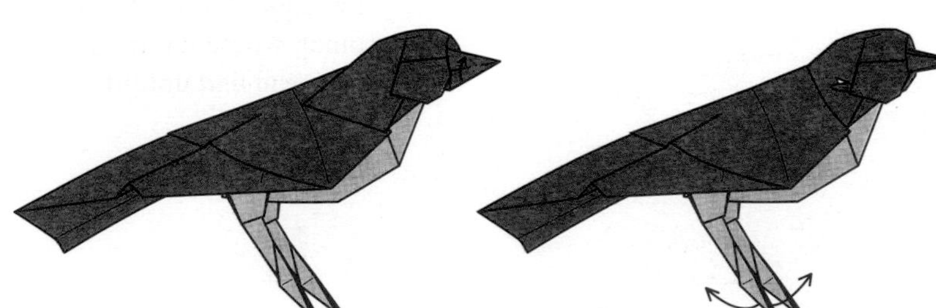

54. Valley-fold the near edge of the beak upward. Repeat behind.

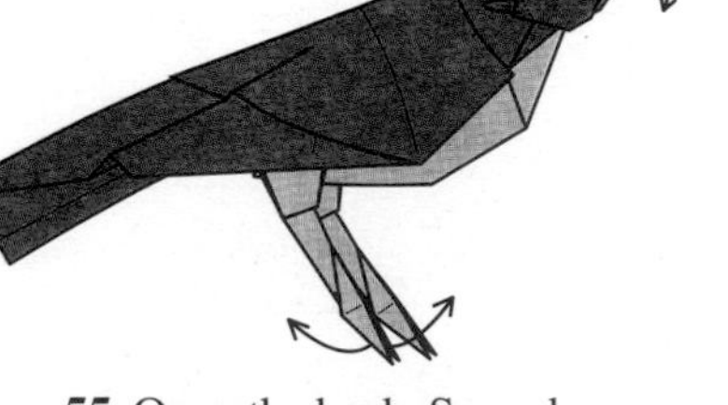

55. Open the beak. Spread the toes. Shape the rest of the body.

56. Finished Songbird.

KNL Dragon

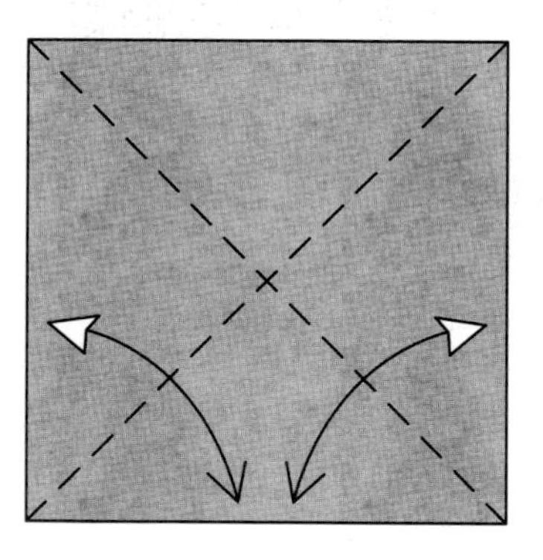

1. Begin with a square, white side up. Fold and unfold along both diagonals.

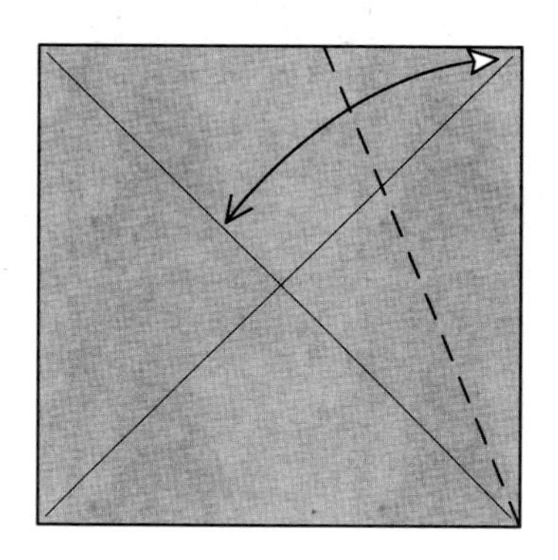

2. Fold the right edge to the downward diagonal and unfold.

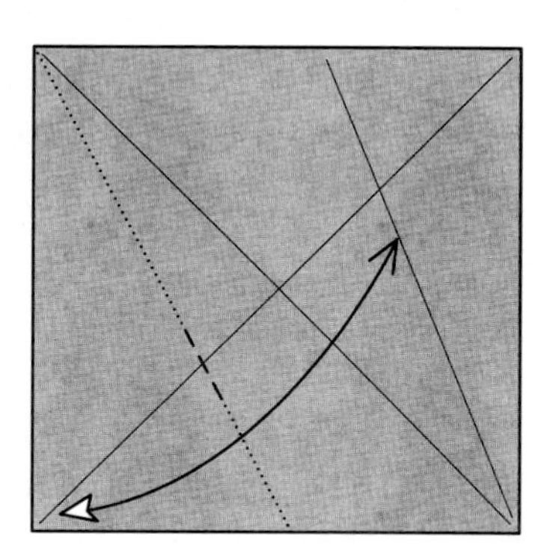

3. Fold the bottom left corner to the crease you just made so that the crease passes through the upper left corner; make a pinch where it crosses the diagonal and unfold.

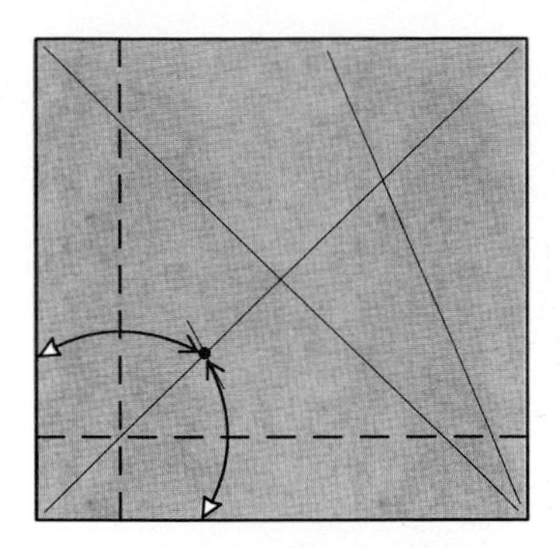

4. Fold the left edge over to the crease intersection and unfold. Repeat with the bottom edge.

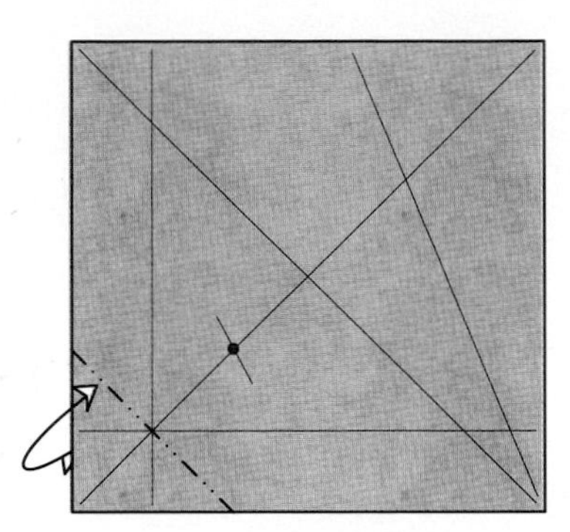

5. Mountain-fold through the intersection of the last two creases (the corner touches the marked crease intersection) and unfold.

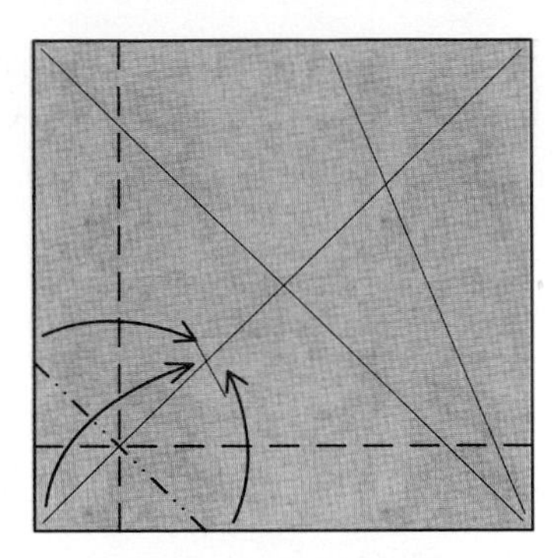

6. Make all three folds at once, forming a small Preliminary Fold in the lower left corner.

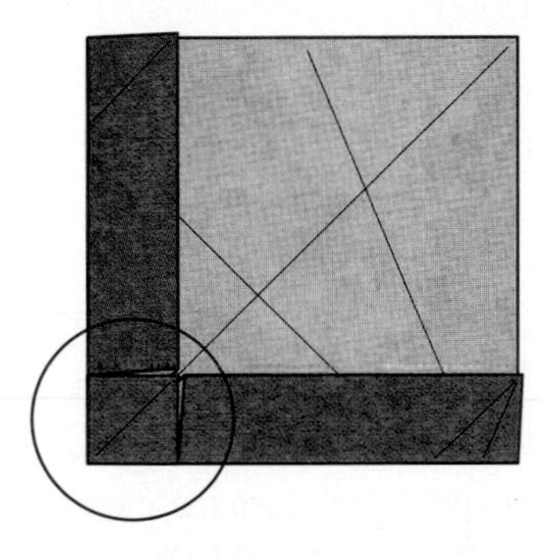

7. Steps 8–16 will focus on the lower left corner.

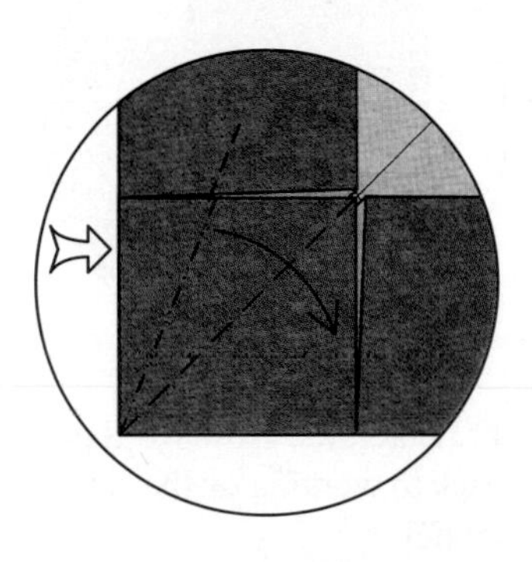

8. Squash-fold the corner.

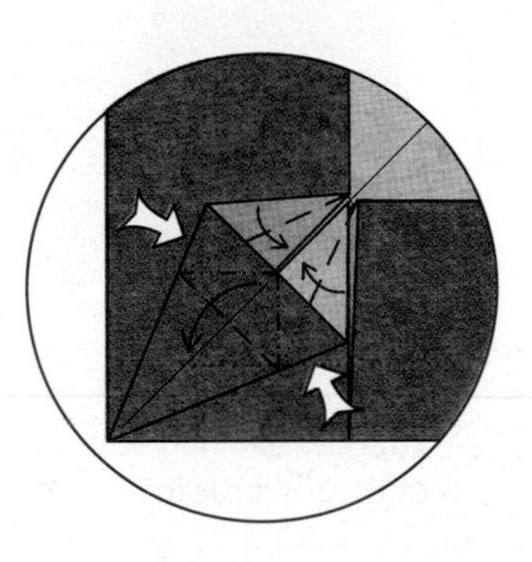

9. Petal-fold the edge.

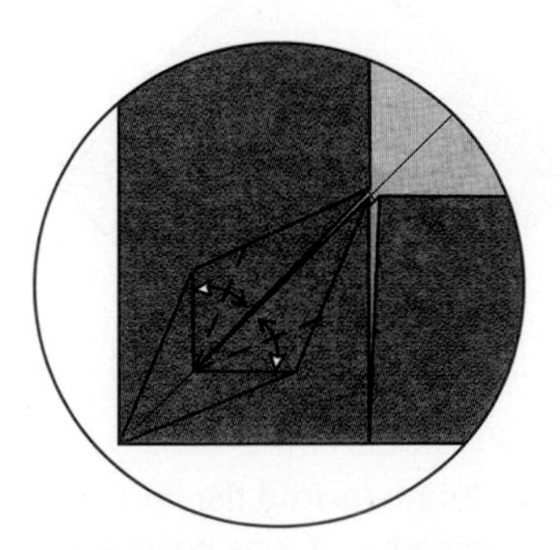

10. Fold and unfold along angle bisectors.

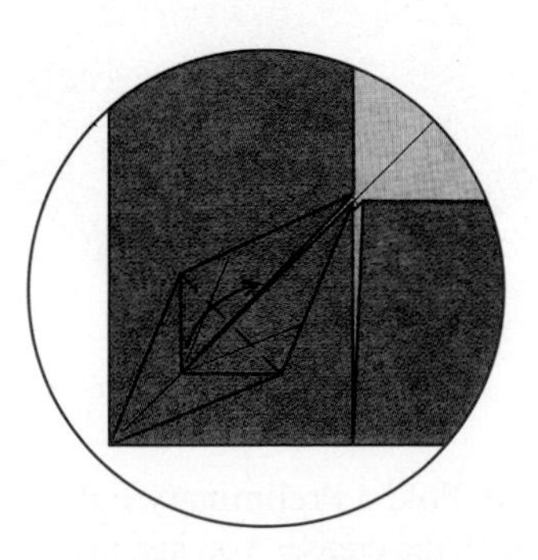

11. Fold the corner up and to the right.

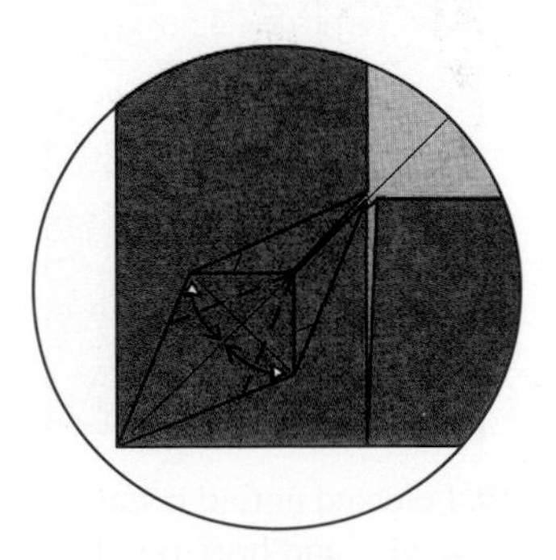

12. Fold and unfold.

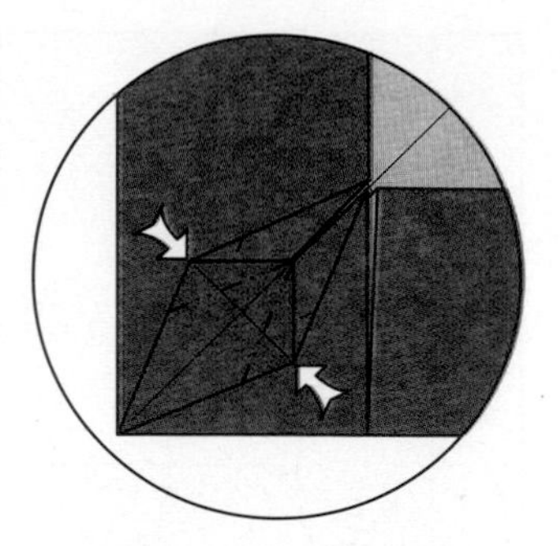

13. Open-sink the corners using the existing creases.

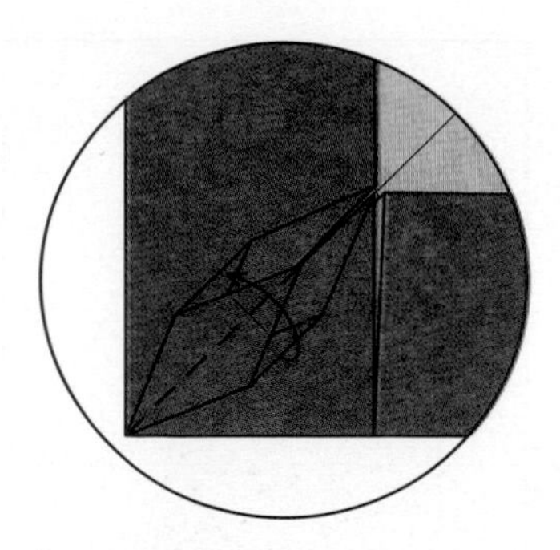

14. Fold one layer upward.

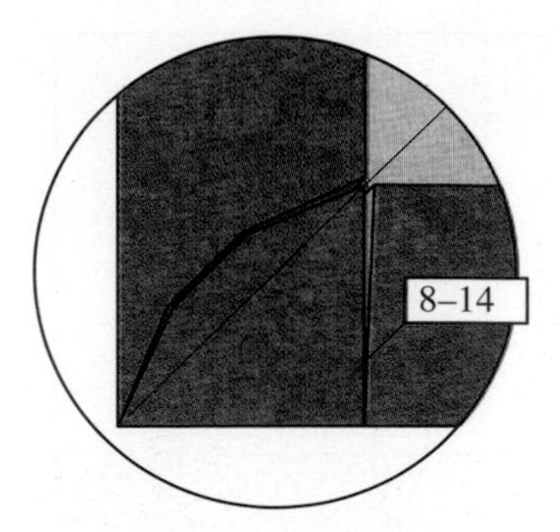

15. Repeat steps 8–14 on the other flap.

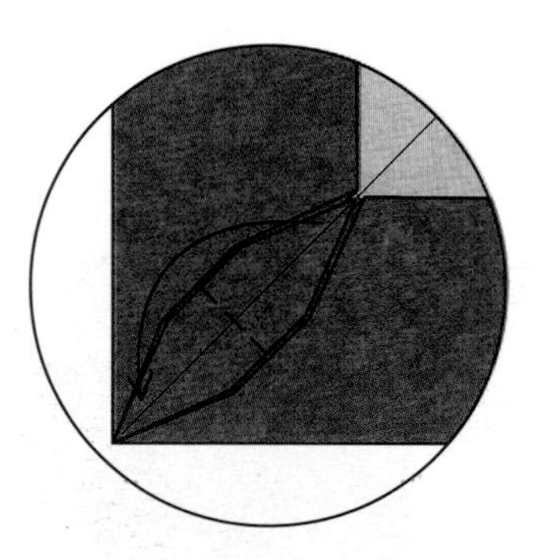

16. Fold one flap down to the lower left.

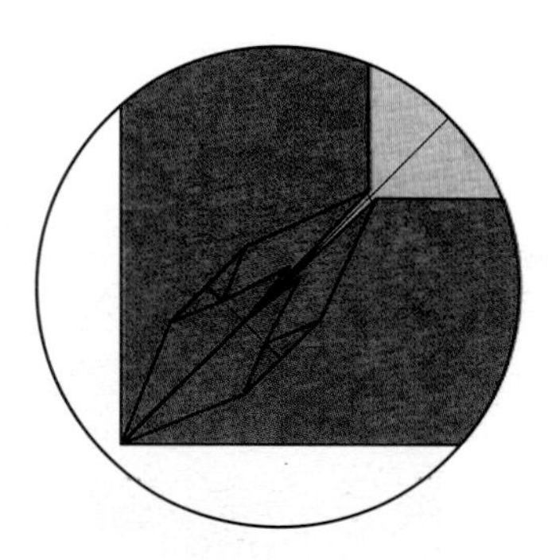

17. Like this. Now we'll go back to working on the entire model.

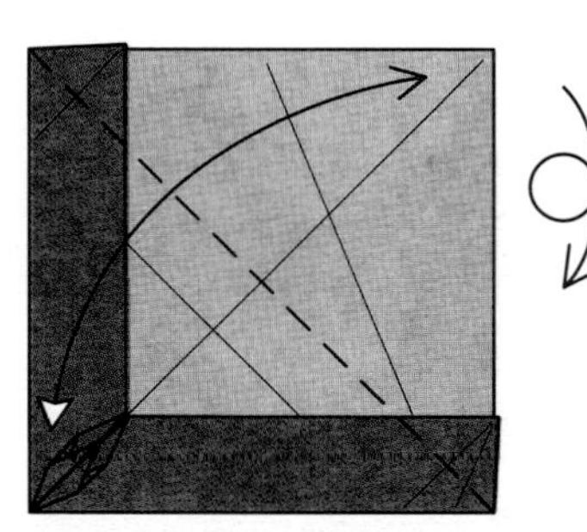

18. Fold and unfold along the diagonal. Then turn the paper over from top to bottom.

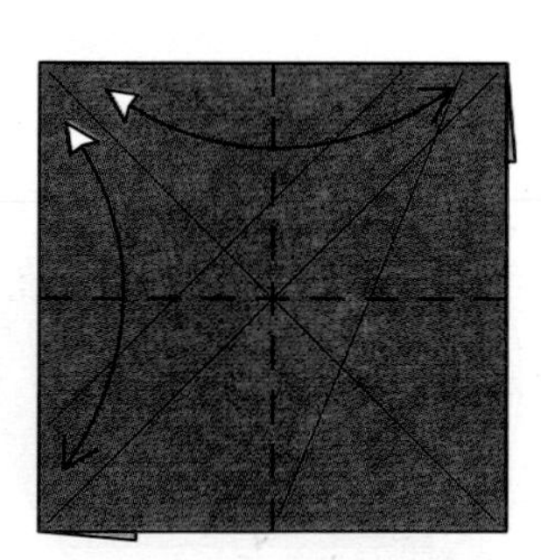

19. Fold and unfold in half both vertically and horizontally. Then rotate the paper 1/8 turn clockwise.

20. Fold a Preliminary Fold with the creases you just made.

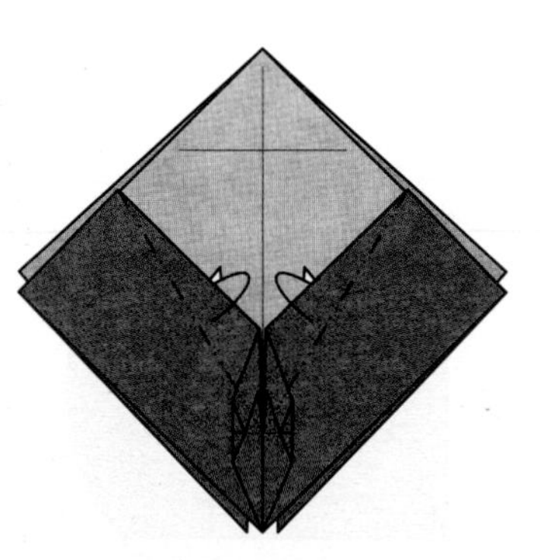

21. Mountain-fold the corners underneath as far as possible.

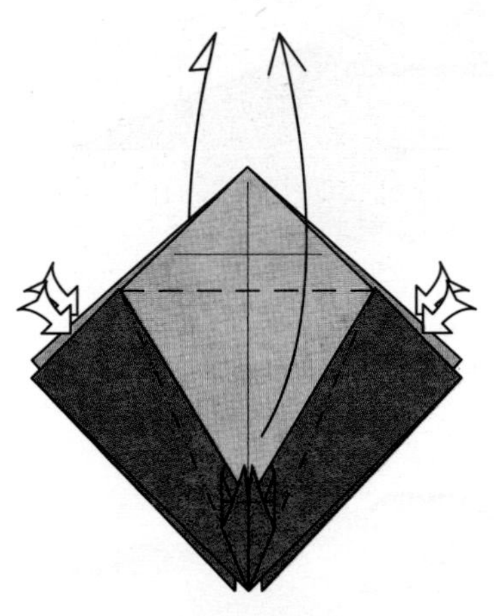

22. Petal-fold front and rear.

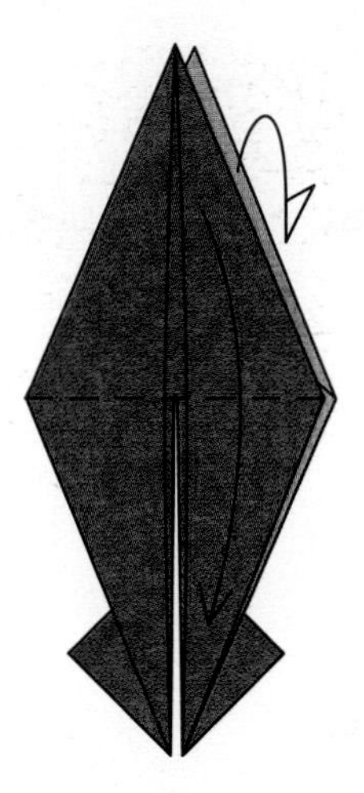

23. Fold the two flaps back down.

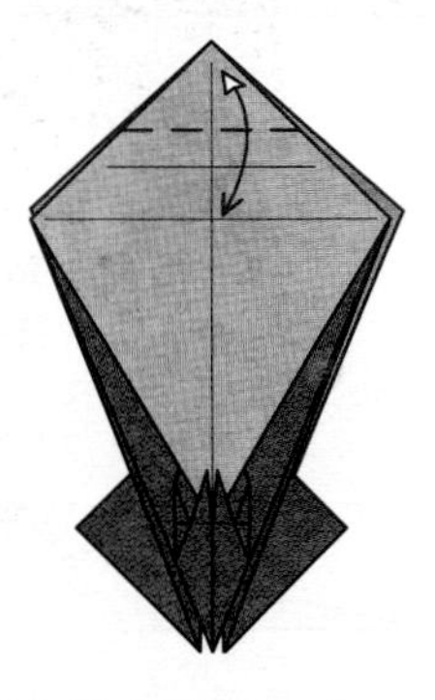

24. Fold the tip down to the horizontal crease and unfold.

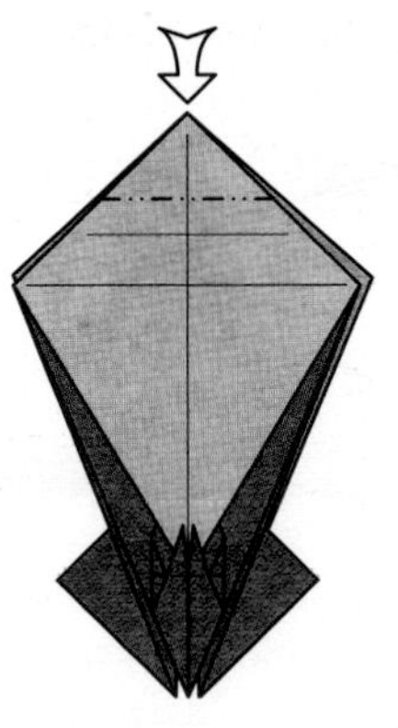

25. Open-sink the top corner.

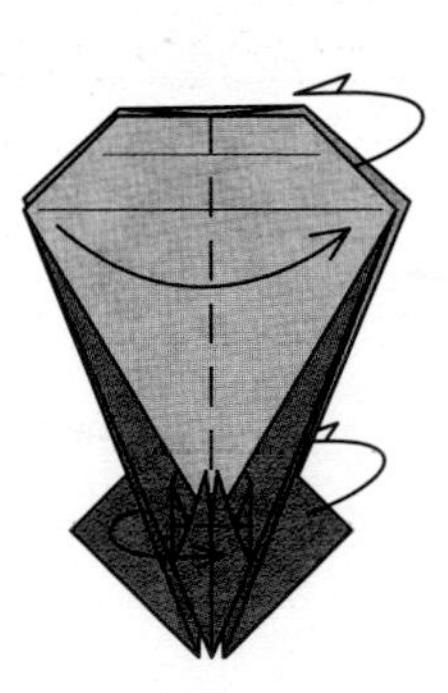

26. Fold one layer to the right in front and one to the left behind.

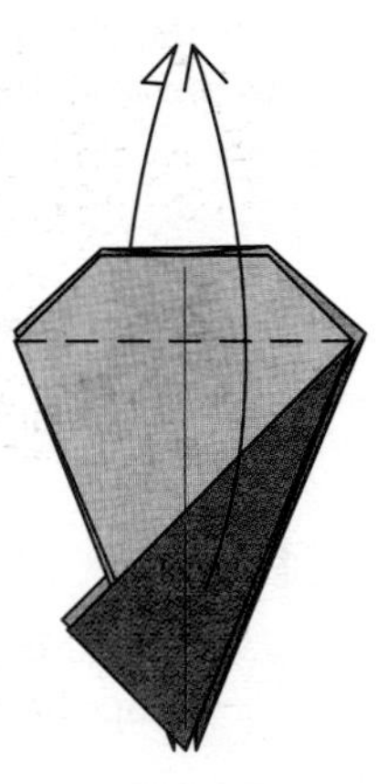

27. Fold one flap up in front and behind.

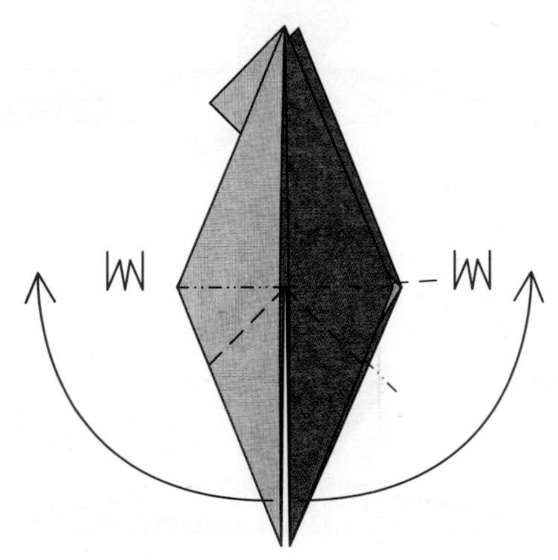

28. Crimp the two points out to the sides.

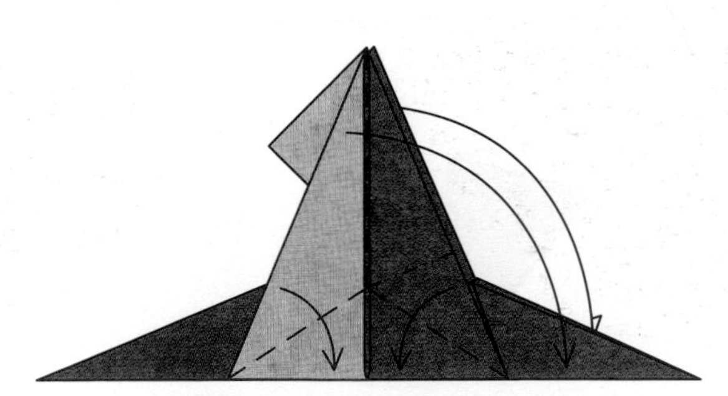

29. Fold a rabbit ear from the upright flap and swing it over to the right. Repeat behind.

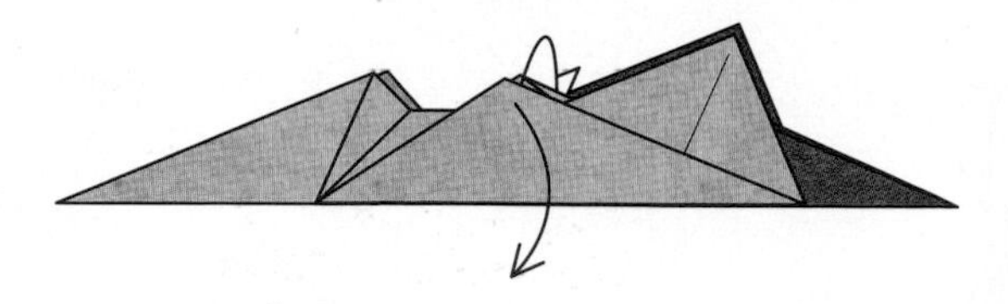

30. Fold a layer down in front. Repeat behind.

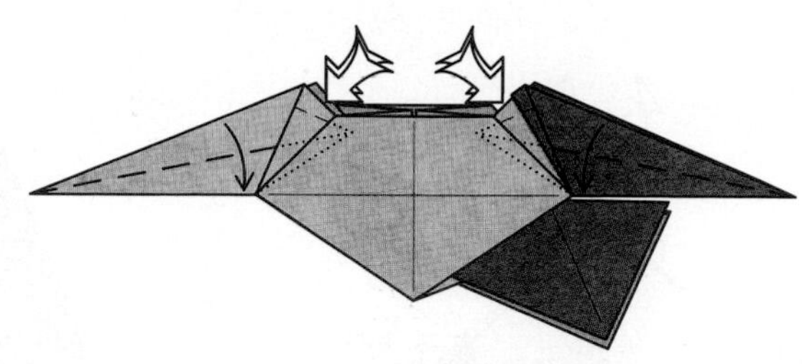

31. Reverse-fold each of four edges down to lie along the horizontal edges.

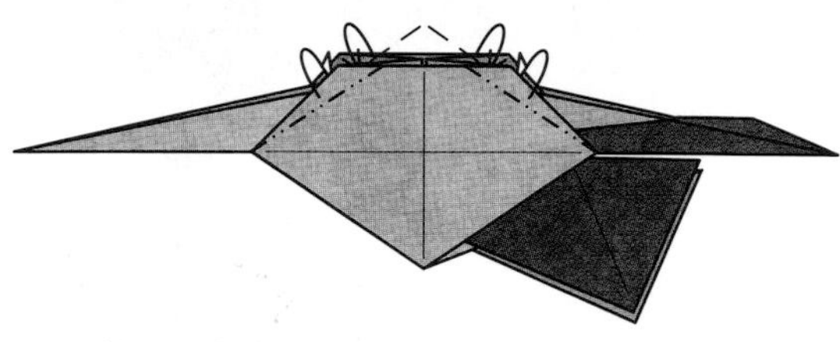

32. Mountain-fold the corners to line up with the edges of the reverse folds from the previous steps.

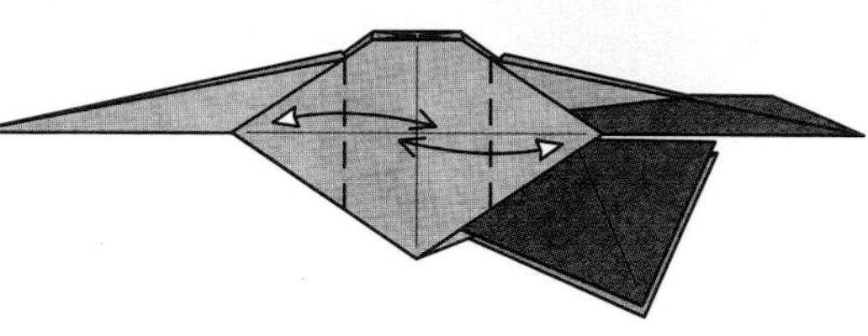

33. Fold the left white point as far to the right as possible and unfold. Repeat with the left point and with both points behind.

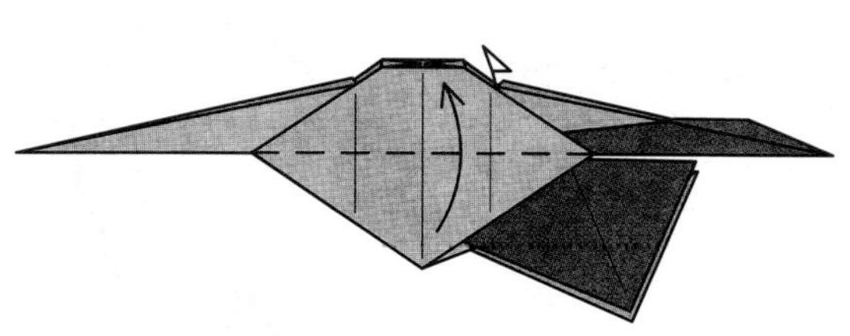

34. Fold the front and back flaps back upward.

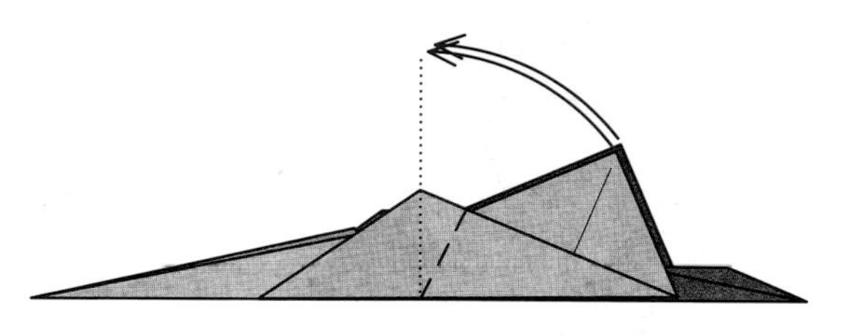

35. Fold the front flap up so that the raw corner is directly over the middle of the model. Repeat behind.

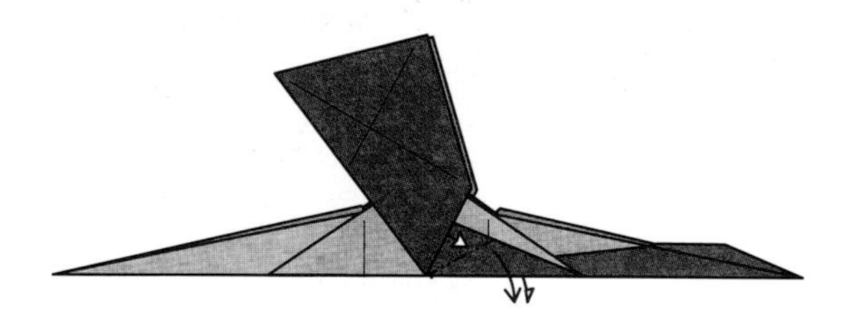

36. Pull out some loose paper. Repeat behind.

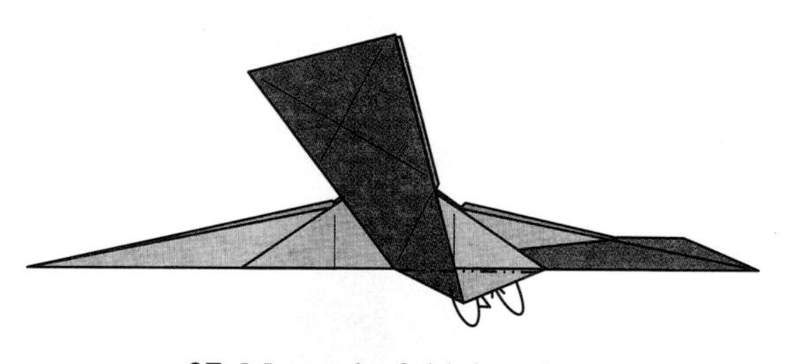

37. Mountain-fold the edge inside the near pocket.

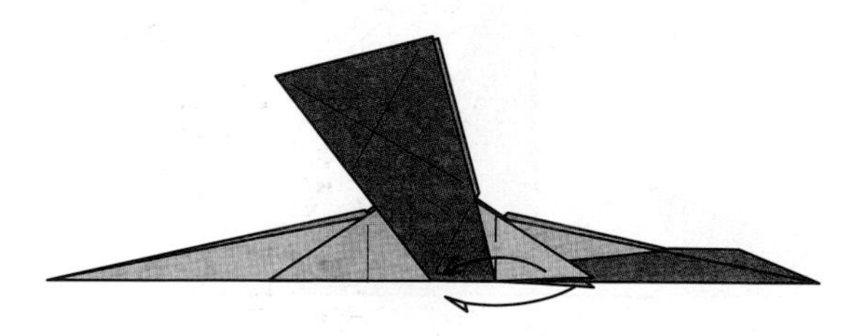

38. Fold the two points front and rear to the left.

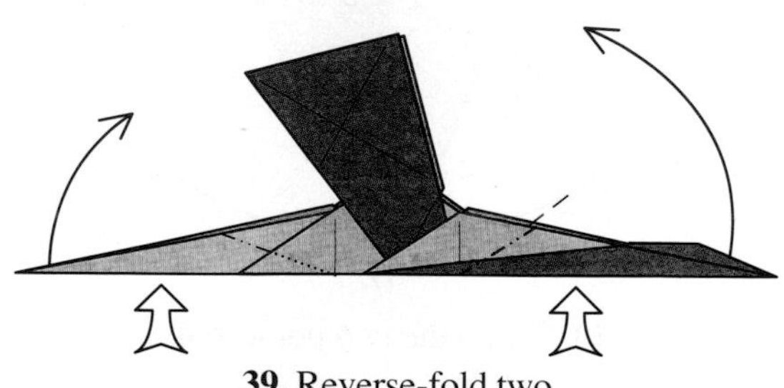

39. Reverse-fold two points up.

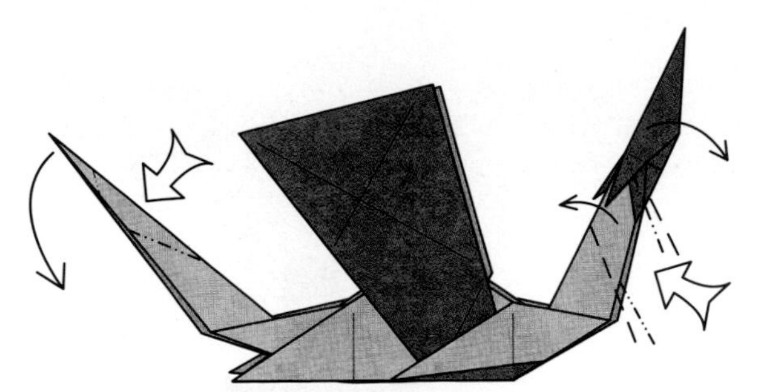

40. Reverse-fold the head and tail twice each.

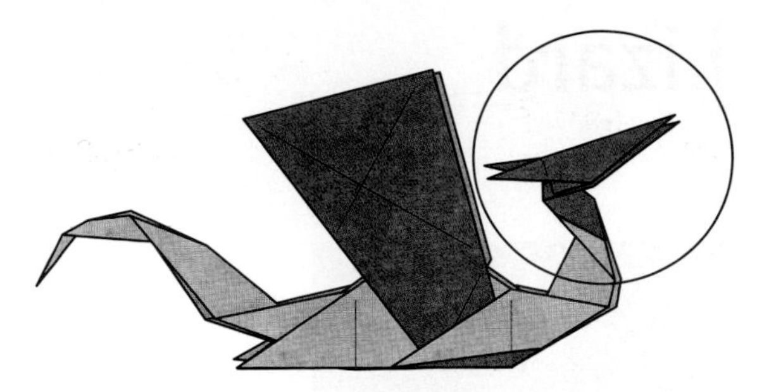

41. Steps 42–46 will focus on the head.

42. Crimp the top layer.

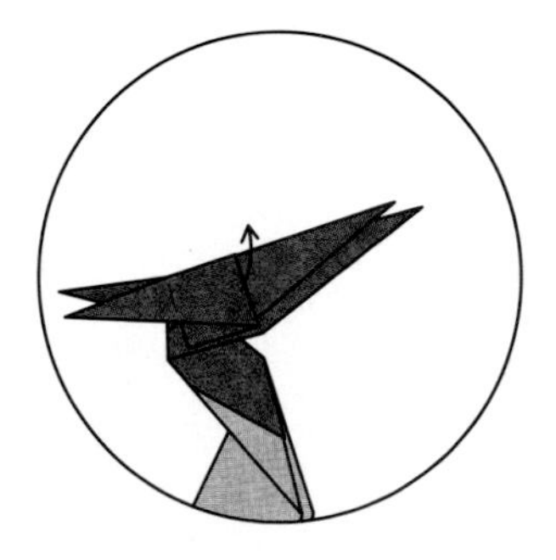

43. Swivel the top flap upward slightly.

44. Outside-reverse-fold the bottom jaw. Outside-reverse-fold the nose.

45. Outside-reverse-fold the bottom jaw in and out to make a tongue.

46. Finished head.

47. Fold four rabbit ears to make legs.

48. Pleat the wings.

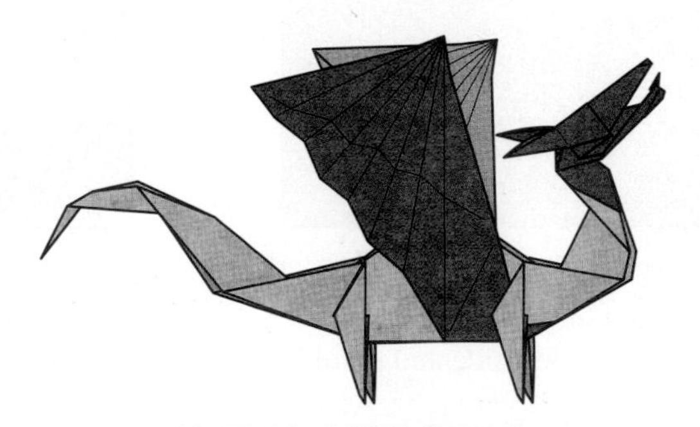

49. Finished KNL Dragon.

Lizard

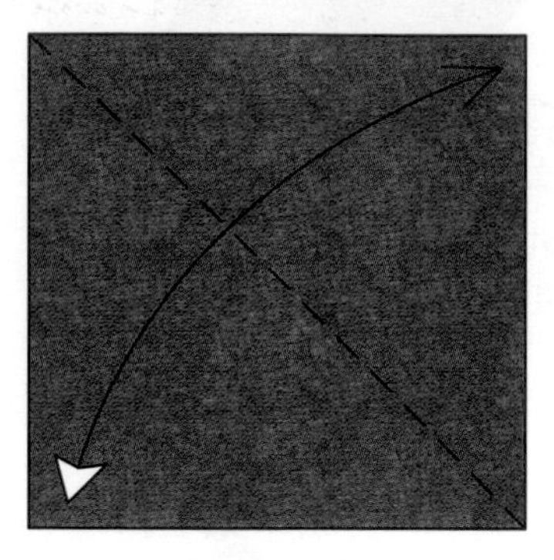

1. Begin with a square, colored side up. Fold and unfold along one diagonal.

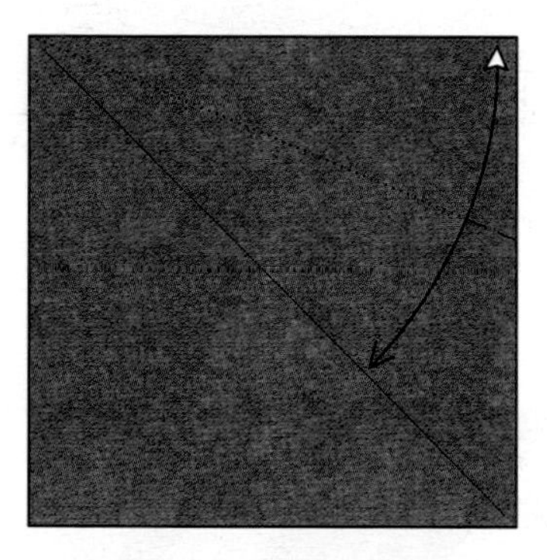

2. Fold the top edge down to the diagonal and make a pinch along the right edge.

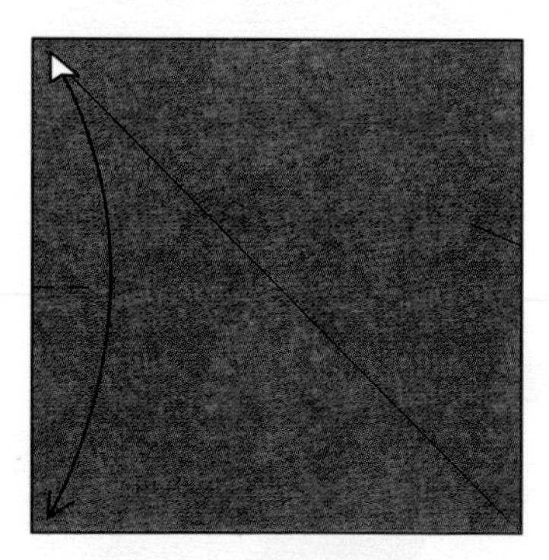

3. Fold the top left corner down to the bottom; make a pinch along the left side and unfold.

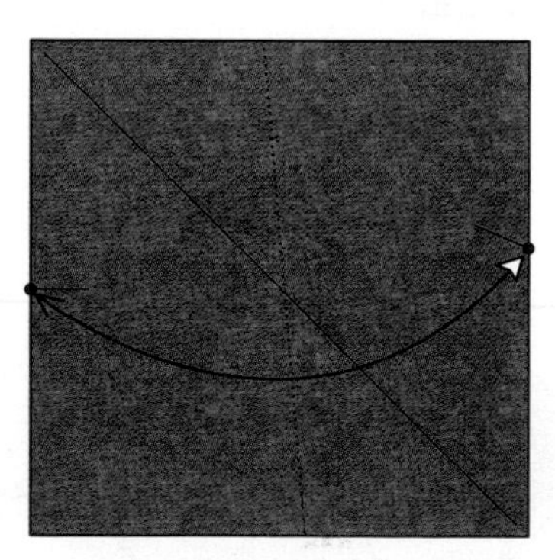

4. Bring the two indicated points together and make a pinch along the bottom edge.

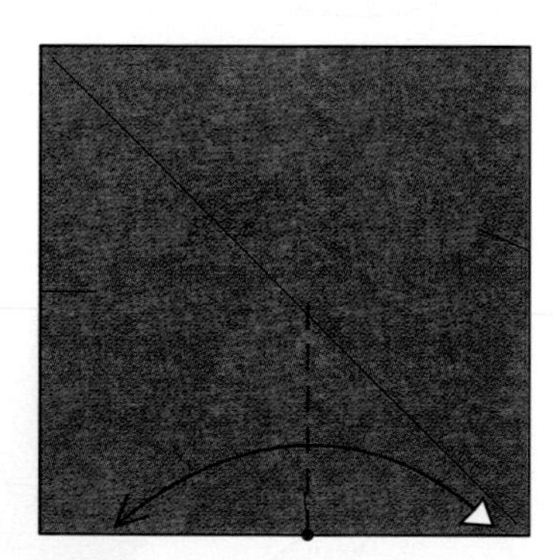

5. Fold the bottom right corner over to the left so that the edges line up and the crease goes through the mark you just made; make the crease sharp up to the diagonal and unfold.

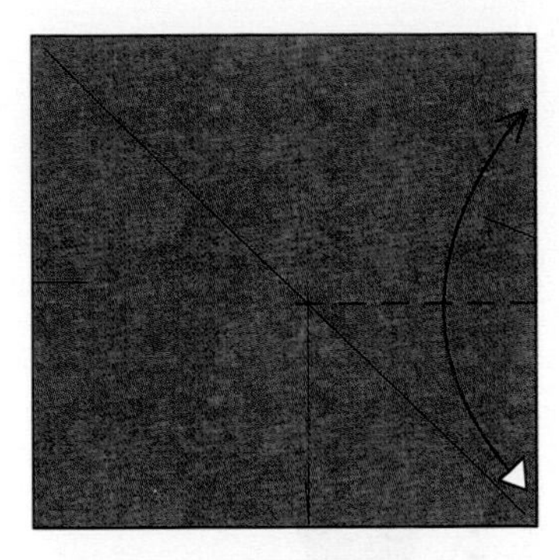

6. Fold the corner up so that the edges line up and the crease hits the diagonal at the same place; make the crease sharp from the diagonal to the right edge and unfold.

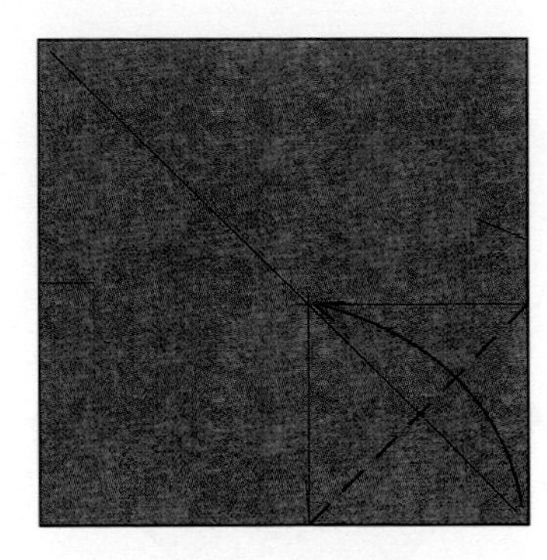

7. Fold the bottom right corner up along the diagonal so that the crease hits the edge at the same place as the creases you just made.

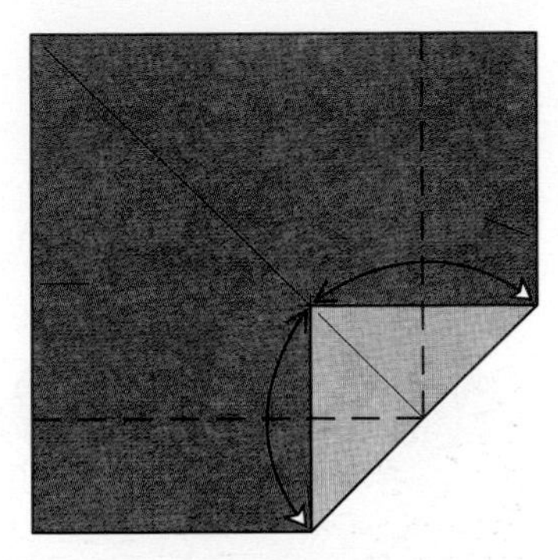

8. Fold the bottom edge up so that the corner touches the white corner at the diagonal; crease all the way across and unfold. Repeat with the right edge.

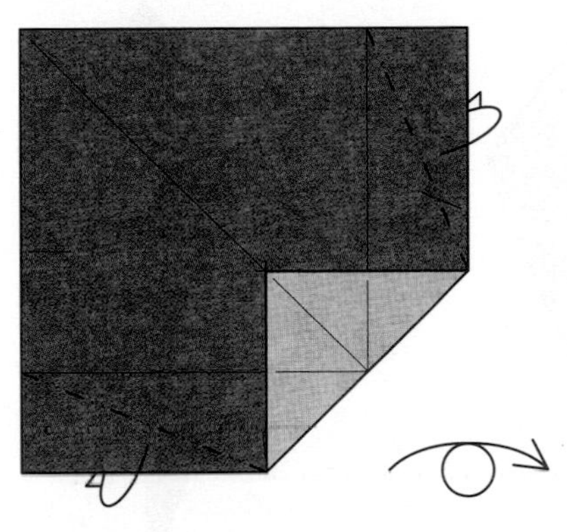

9. Mountain-fold two corners behind. Then turn the paper over and rotate it so that the colored corners are on the sides.

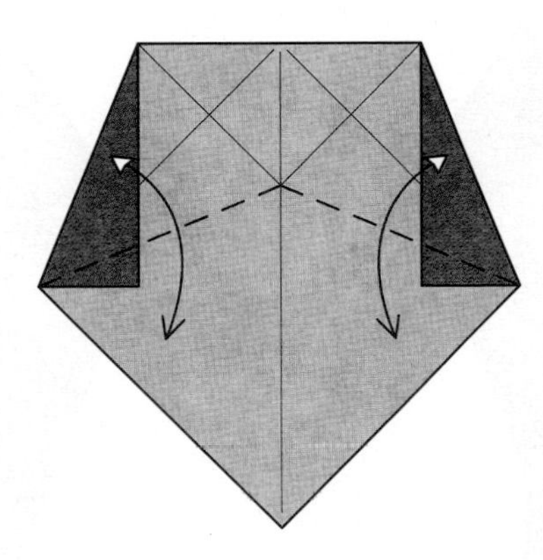

10. Fold and unfold along creases connecting the corners to the crease intersection in the center.

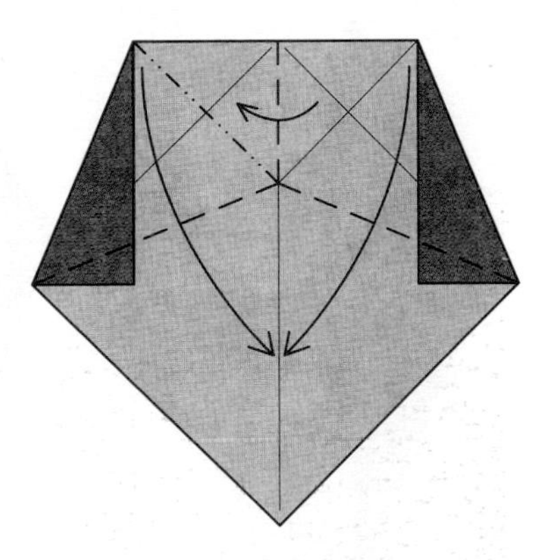

11. Fold a rabbit ear, bringing the two top corners together along the center line.

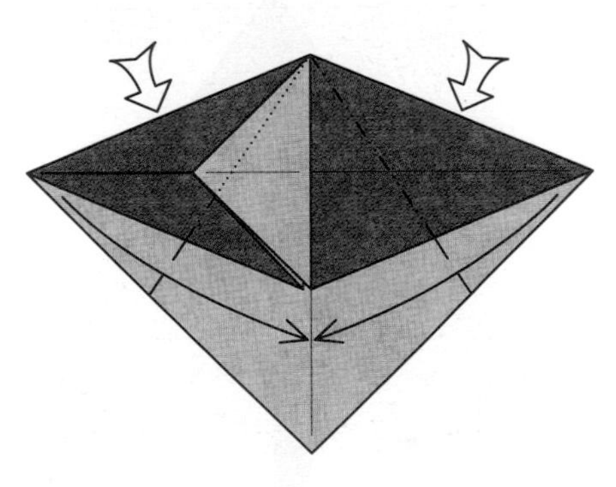

12. Reverse-fold the side points down to lie along the center line.

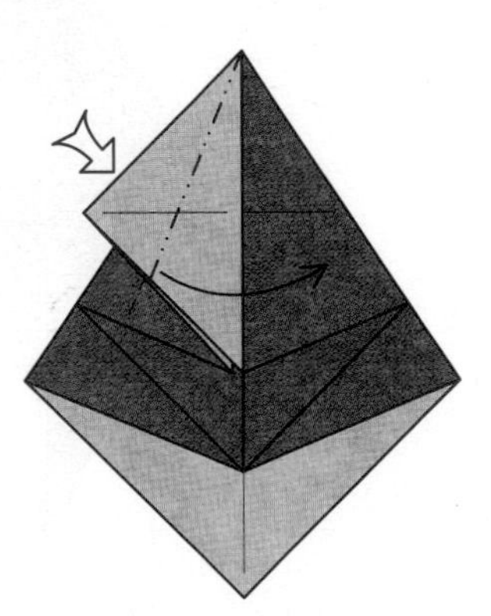

13. Squash-fold the white flap symmetrically.

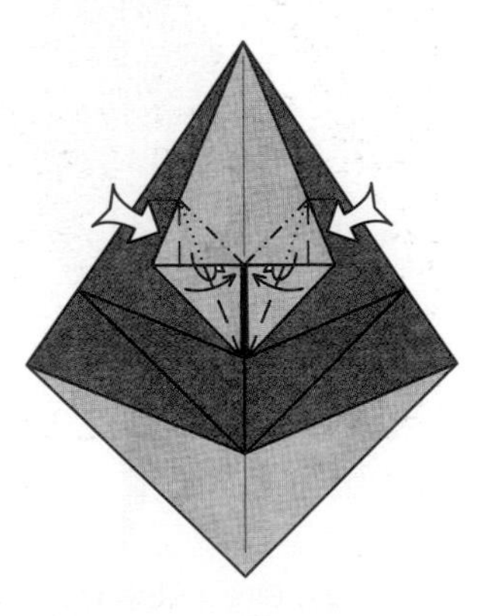

14. Reverse-fold the edges.

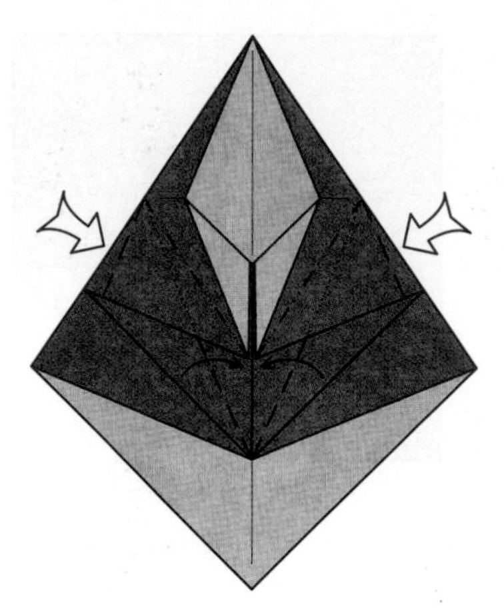

15. Reverse-fold the next pair of edges.

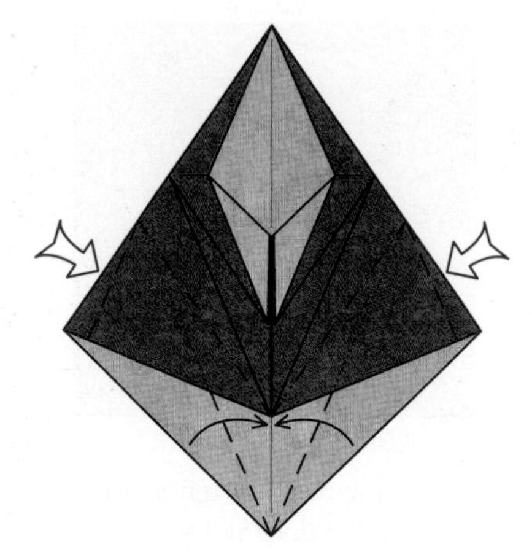

16. Reverse-fold the remaining pair of edges.

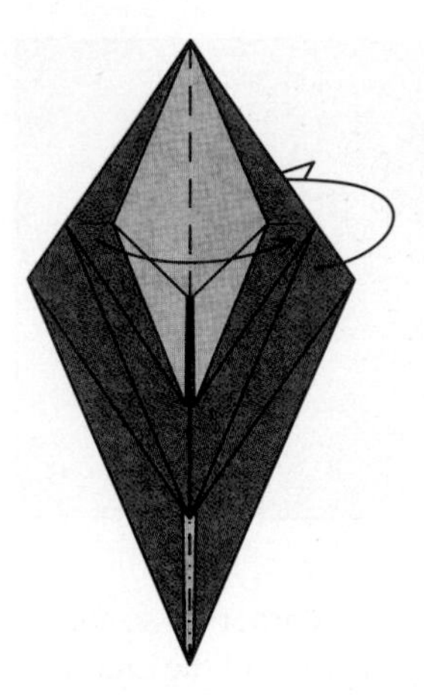

17. Fold two layers to the right in front and one to the left behind.

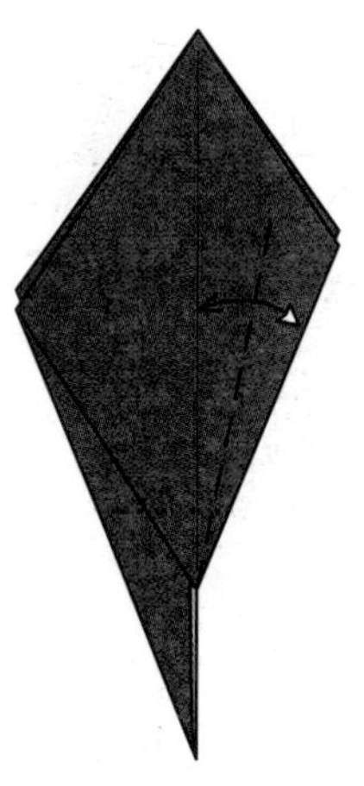

18. Fold the edge over to the vertical crease.

19. Fold the point up to the top of the model.

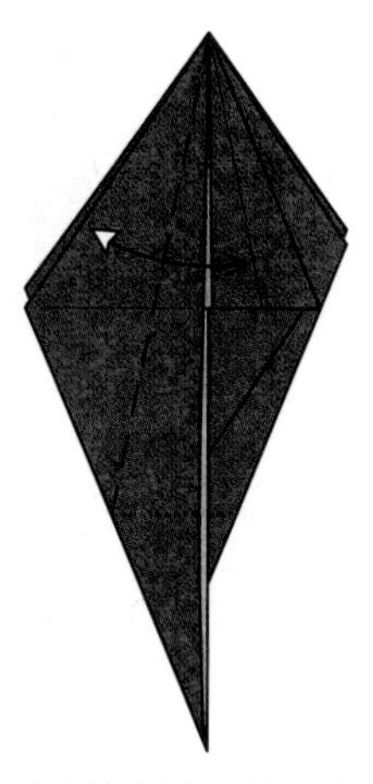

20. Fold the left edge over to the crease and unfold.

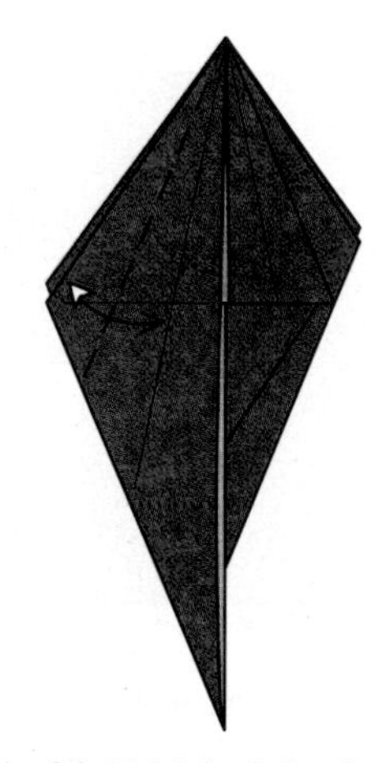

21. Fold the left edge over to the crease you just made and unfold.

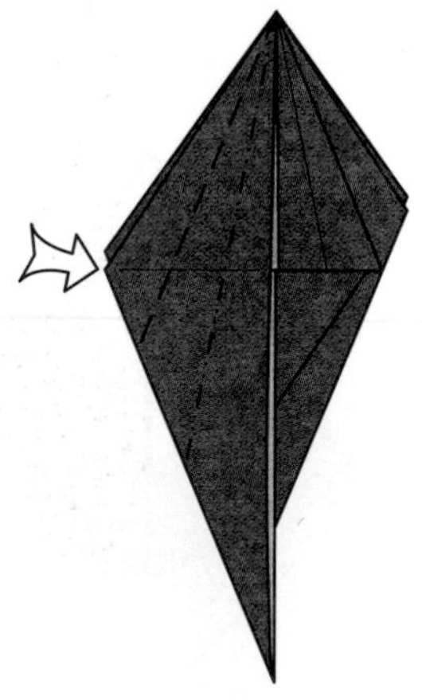

22. Open-sink the corner in and out on the existing creases.

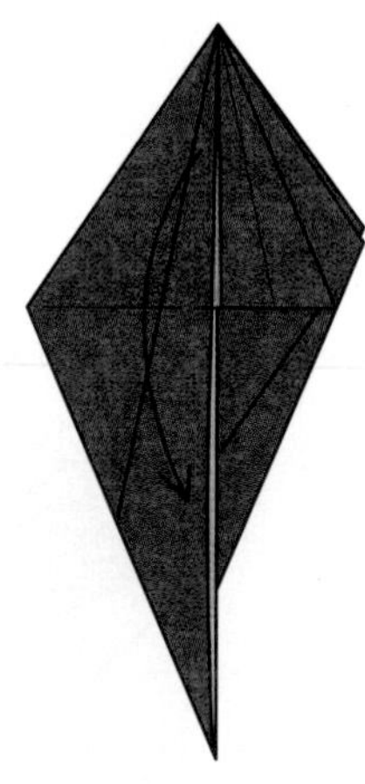

23. Fold the point down.

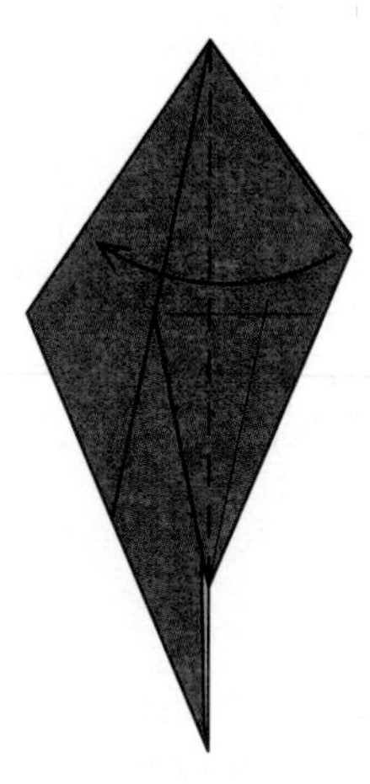

24. Fold one flap to the left.

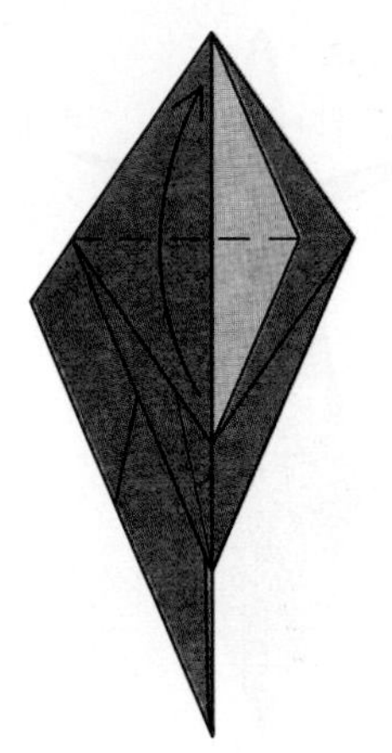

25. Fold one flap up as far as possible.

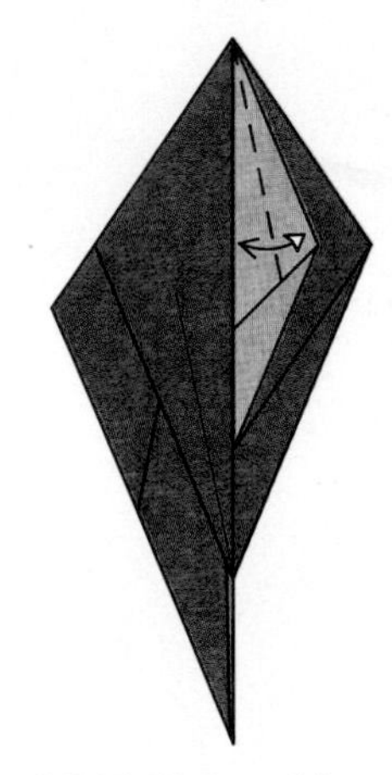

26. Fold the white edge to the center line and unfold.

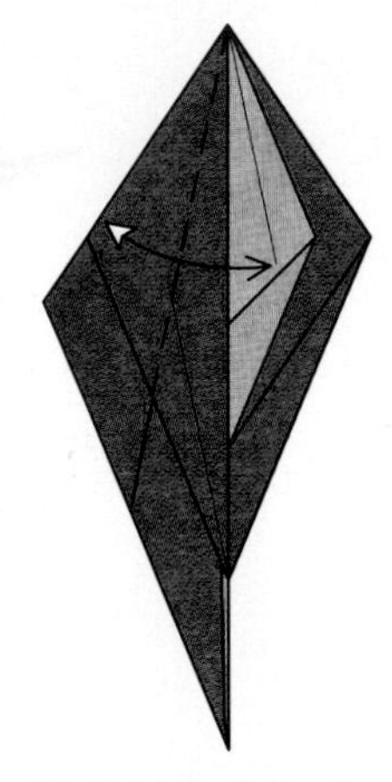

27. Fold the left edge to the crease you just made and unfold.

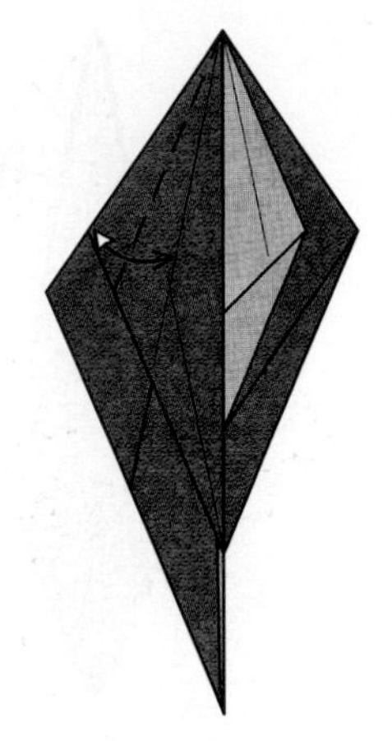

28. Fold the left edge to the crease you just made and unfold.

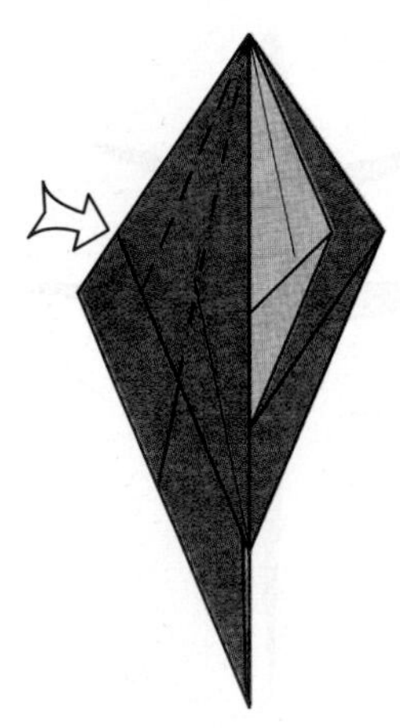

29. Open-sink the corner in and out on the existing creases.

30. Spread-sink the corner.

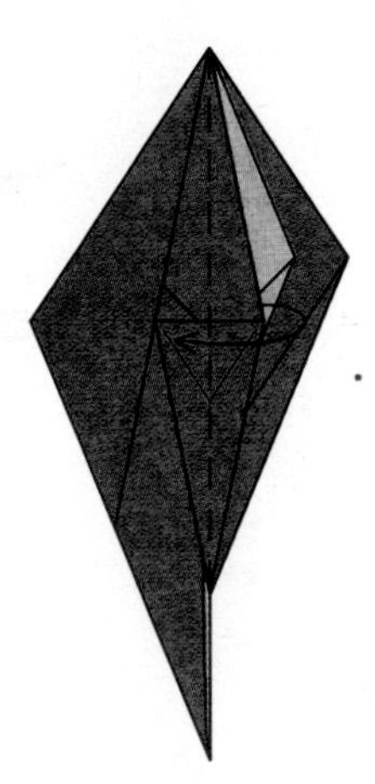

31. Close up the flap.

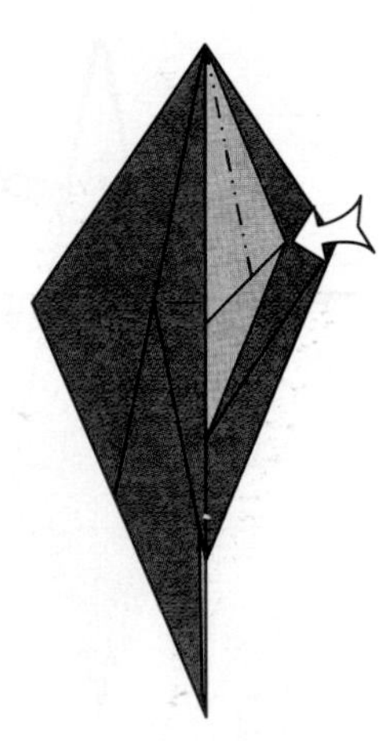

32. Open-sink the corner on the existing creases.

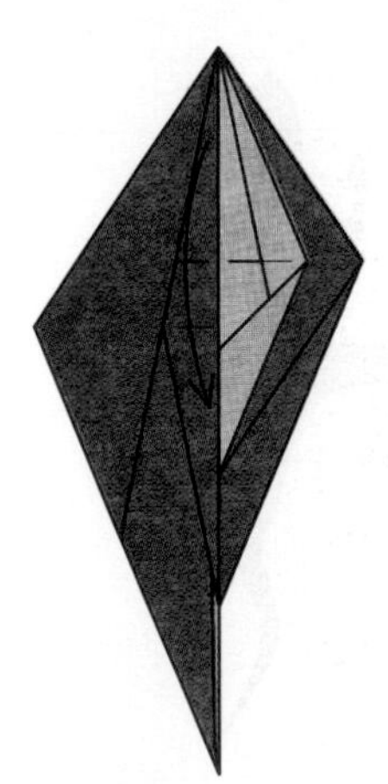

33. Fold the flap down.

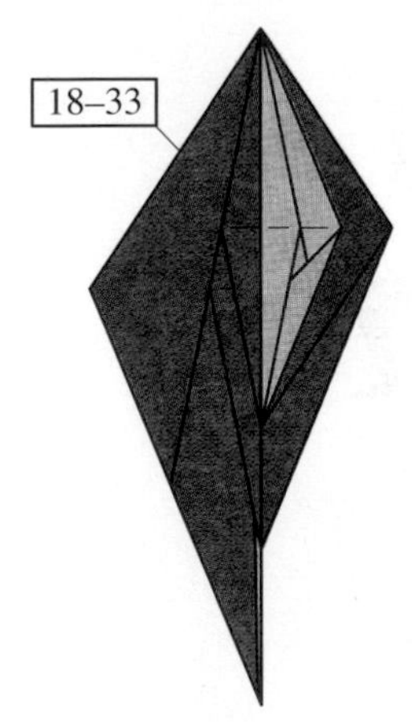

34. Repeat steps 18–33 behind.

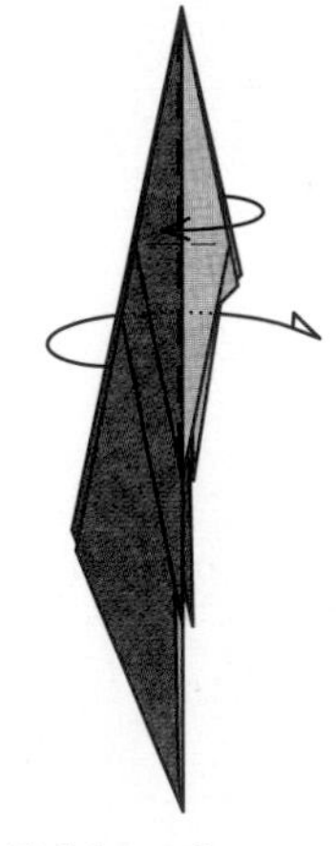

35. Fold one layer to the left in front and two to the right behind, leaving the model symmetric.

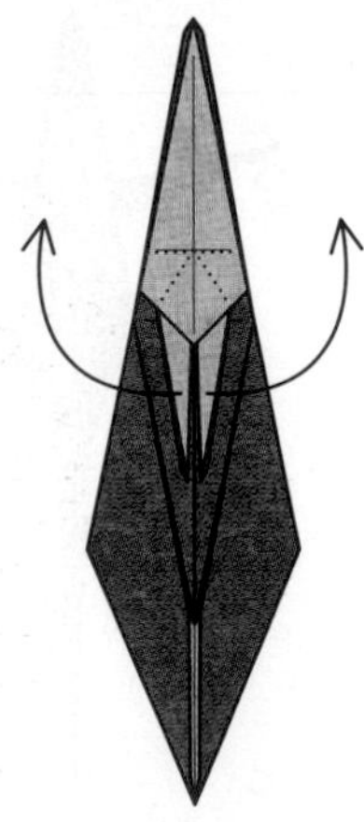

36. Crimp the two points out to the sides and slightly upward.

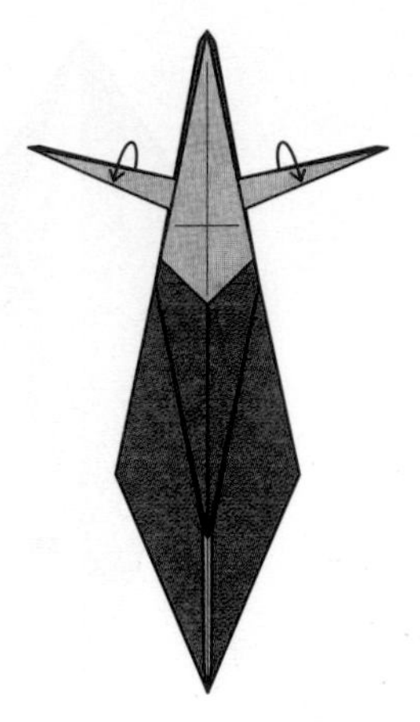

37. Wrap one colored layer from inside to outside.

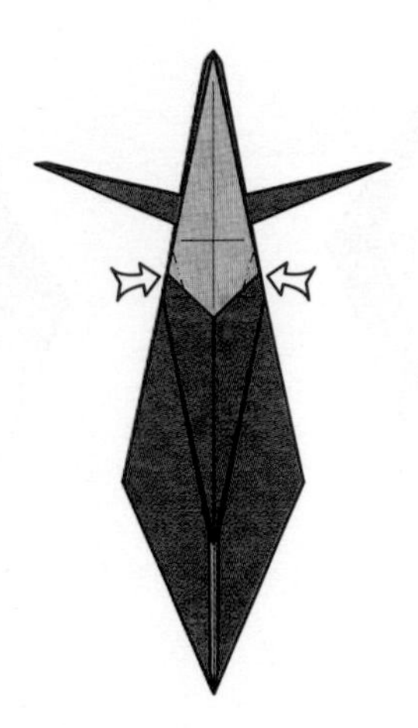

38. Sink the white corners.

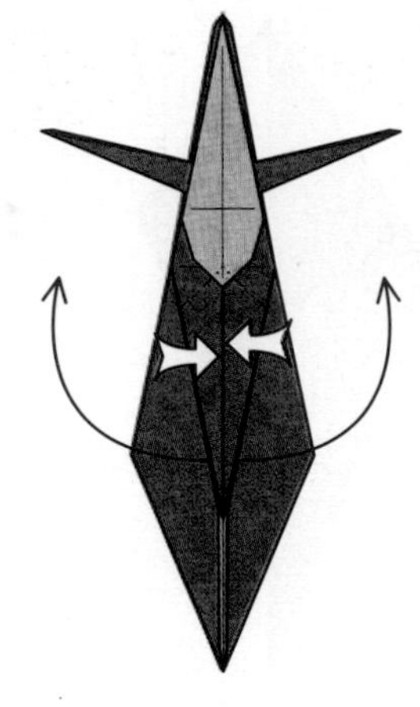

39. Reverse-fold the hind legs out to the sides.

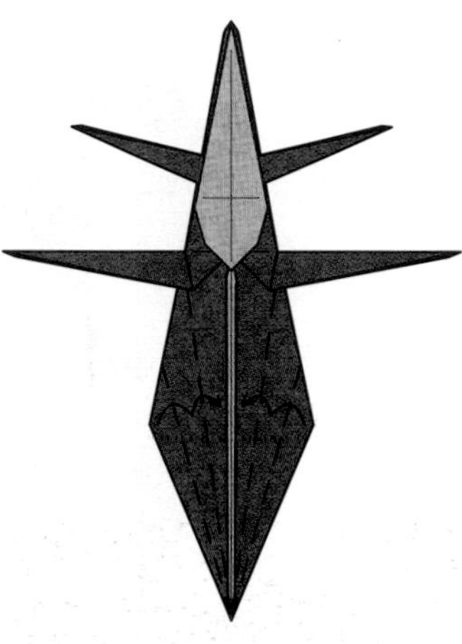

40. Narrow the tail by folding each side over and over in thirds.

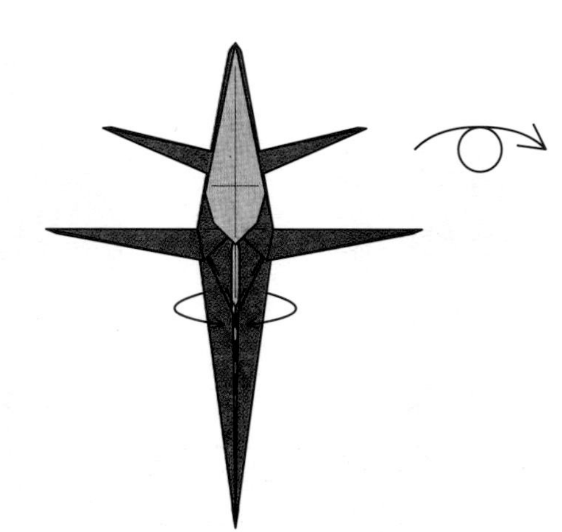

41. Partially rabbit-ear the tail and turn the model over.

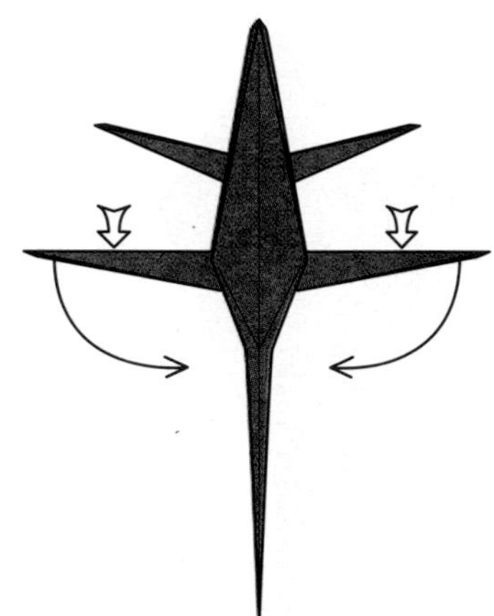

42. Reverse-fold the hind legs.

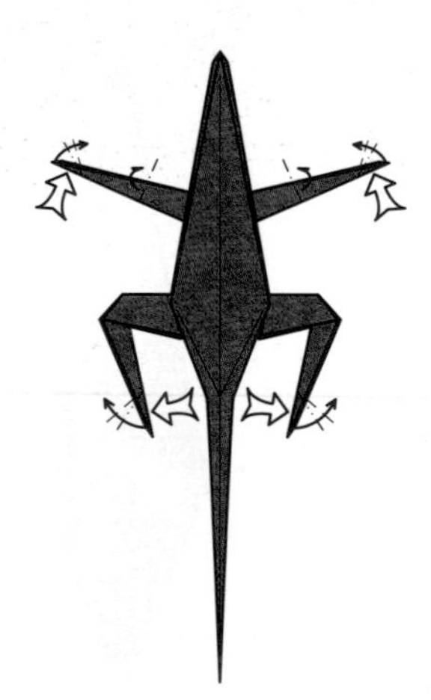

43. Reverse-fold the hind feet. Squash-fold the front feet and mountain-fold the forelegs away from you.

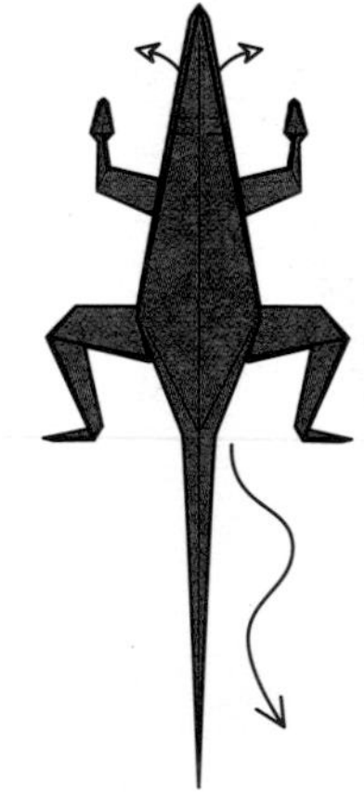

44. Crimp the head slightly and pull out the layers on the sides of the head. Curve the tail.

45. Finished Lizard.

Tree Frog

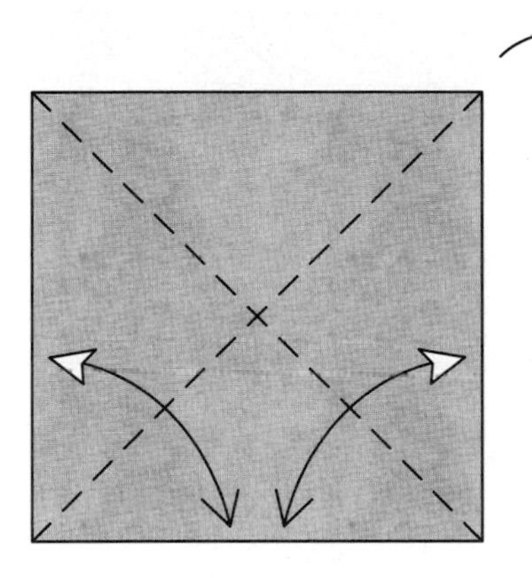

1. Begin with a square, white side up. Fold and unfold along the diagonals. Turn the paper over.

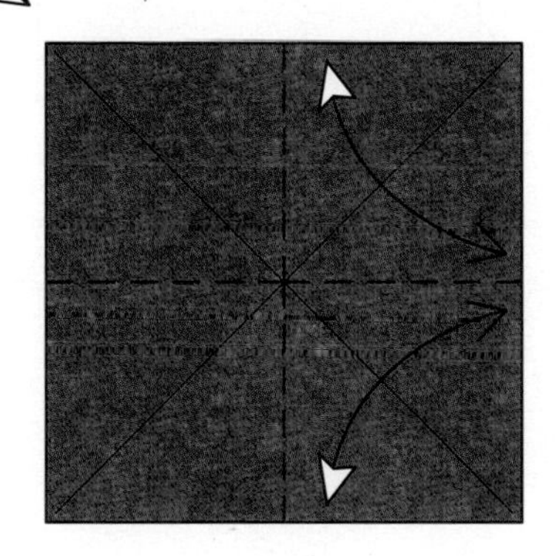

2. Fold and unfold vertically and horizontally.

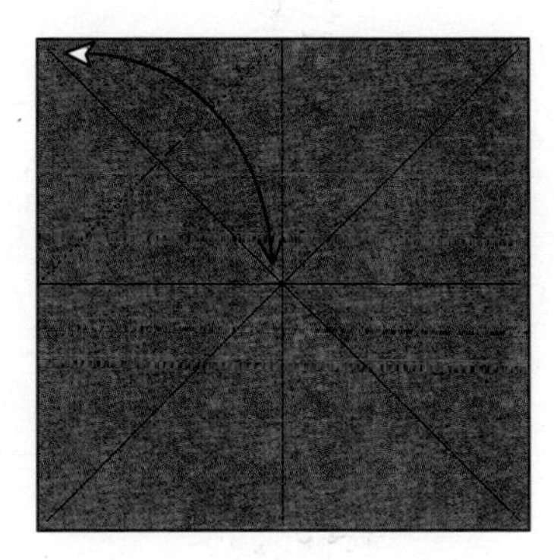

3. Fold the top left corner to the center; make a pinch that crosses the diagonal crease and unfold.

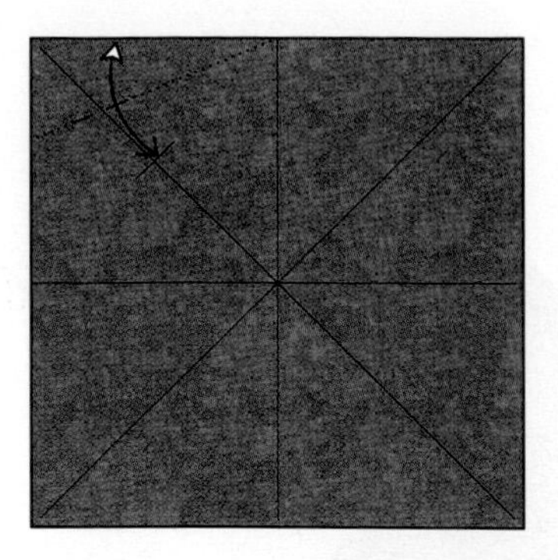

4. Fold the top edge to the pinch you just made; make another pinch on the diagonal crease and unfold.

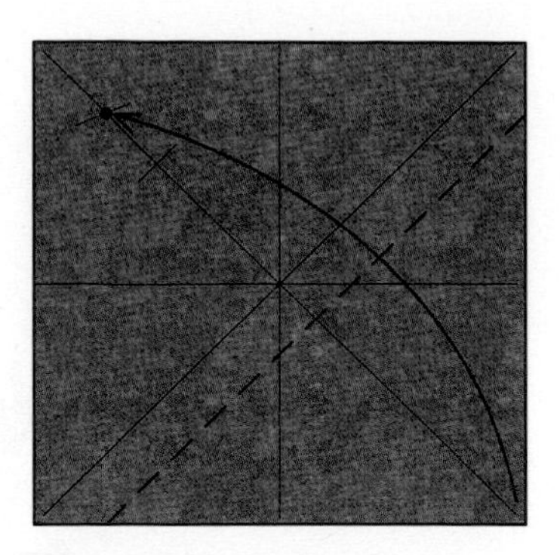

5. Fold the bottom right corner up to the crease intersection.

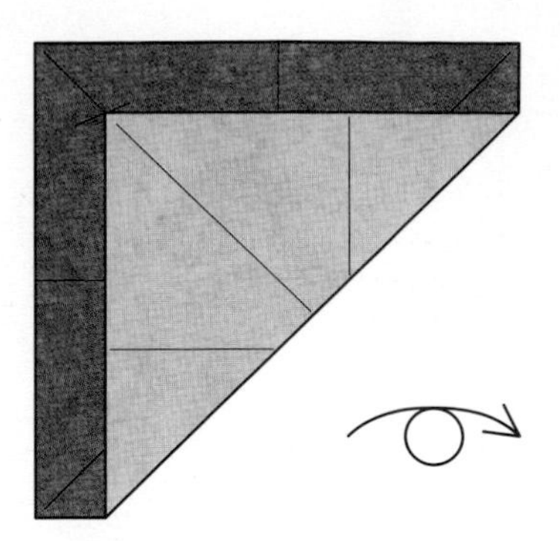

6. Like this. Turn the paper over.

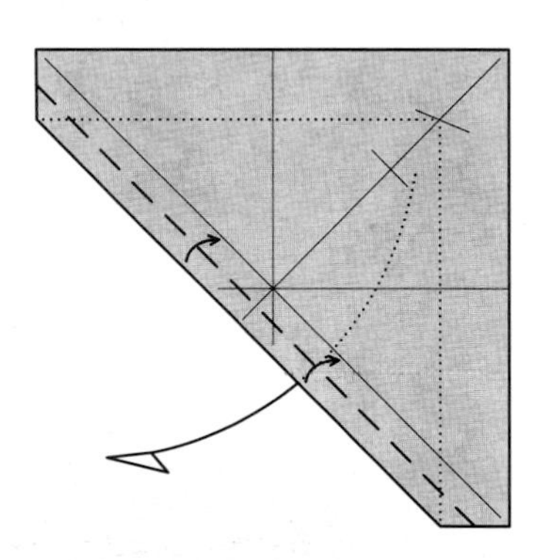

7. Fold the edge to the diagonal crease, allowing the corner behind to flip out.

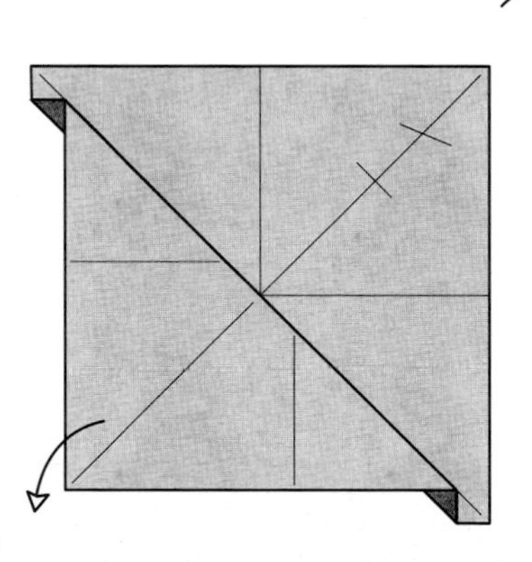

8. Unfold the pleat and turn the paper back over.

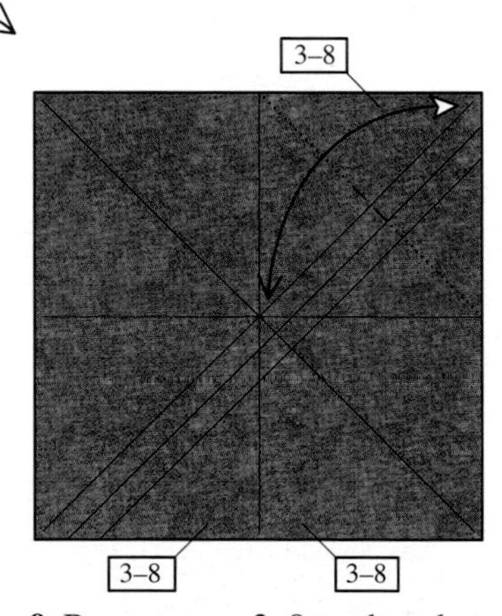

9. Repeat steps 3–8 on the other three corners. Leave the paper white side up when you're done.

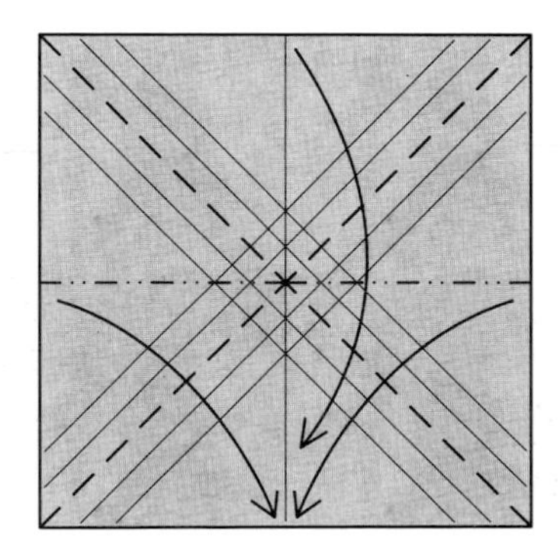

10. Fold a Waterbomb Base using the existing creases.

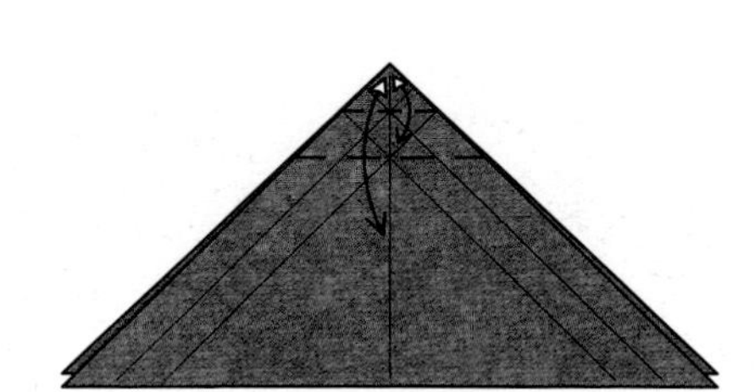

11. Fold and unfold through the crease intersections.

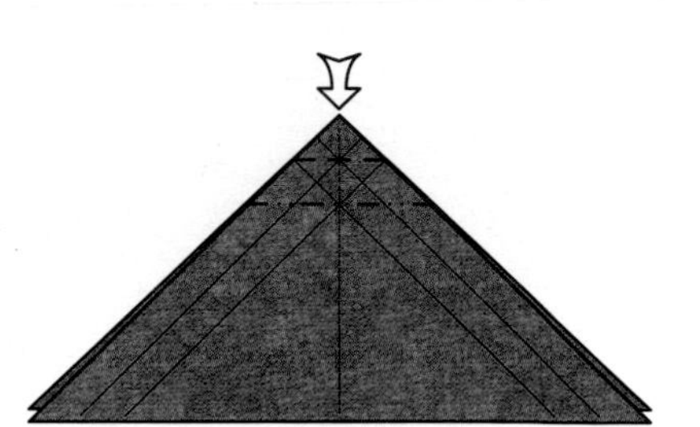

12. Open-sink the point in and out.

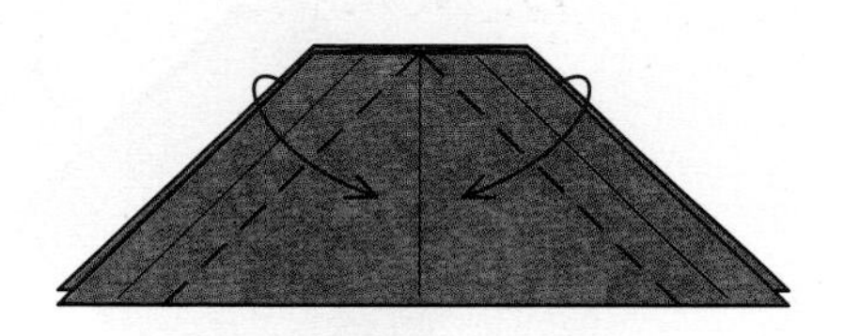

13. Fold two edges down.

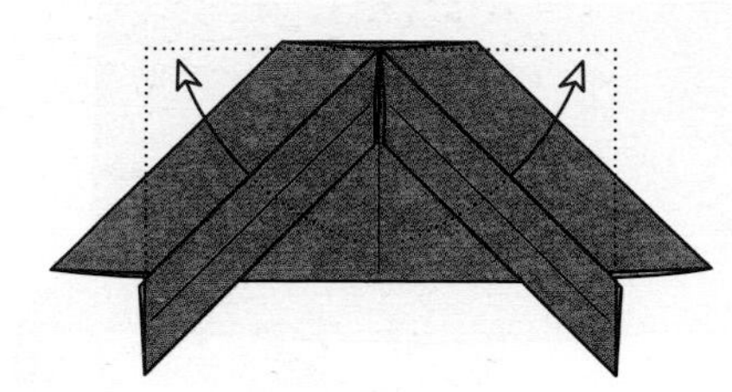

14. Reverse-fold two corners out.

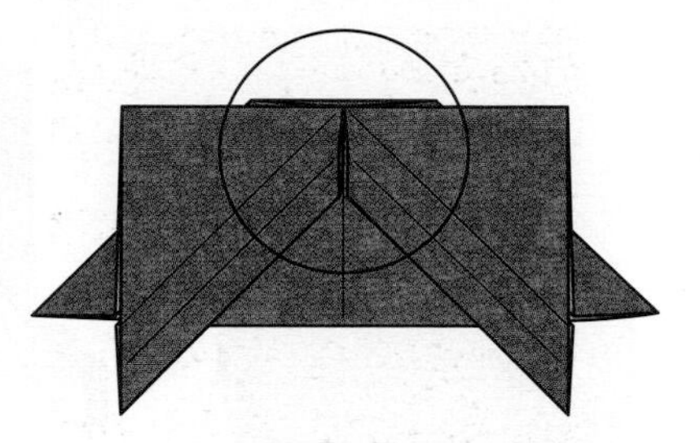

15. Steps 16–28 will focus on the region in the circle.

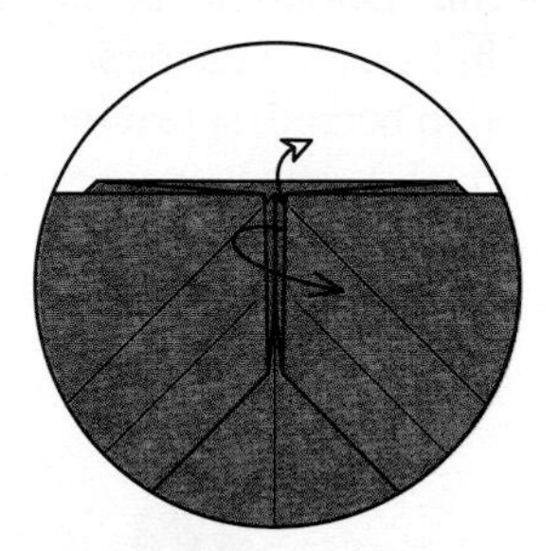

16. Pull the near layer to the right and release some trapped paper. You may find it easier to do steps 16–19 all at once.

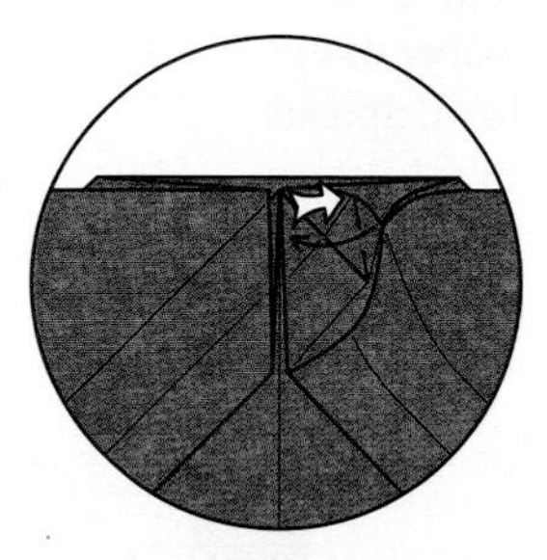

17. Close up the model.

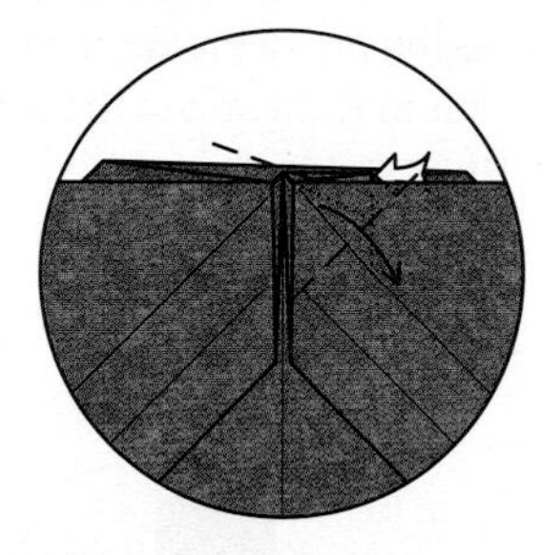

18. Fold the corner down and squash-fold the interior corner.

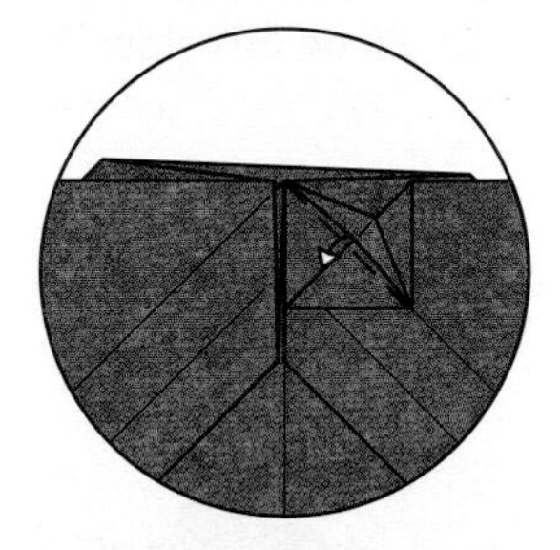

19. Unsink a layer of paper from inside the petal fold. The result should be symmetric.

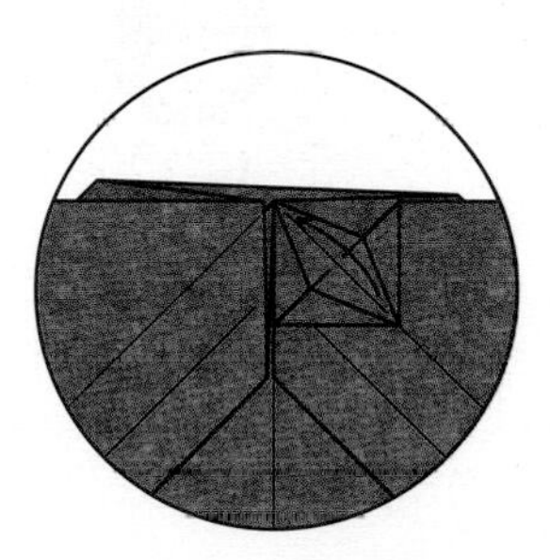

20. Close up the flap.

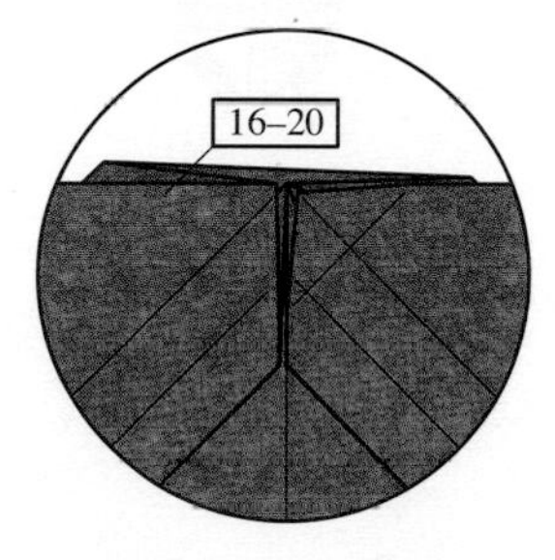

21. Repeat steps 16–20 on the left.

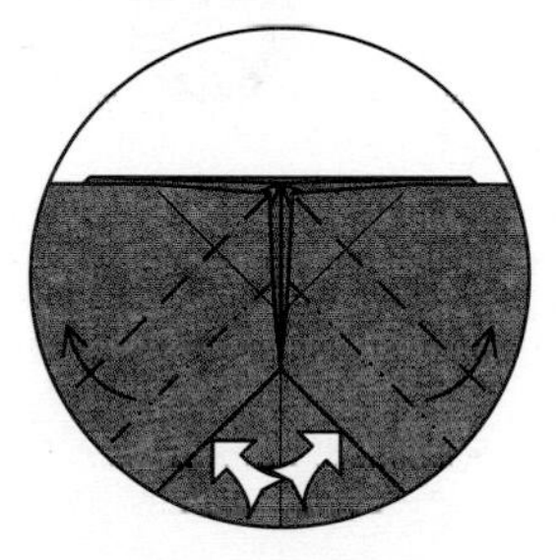

22. Spread-sink the edges.

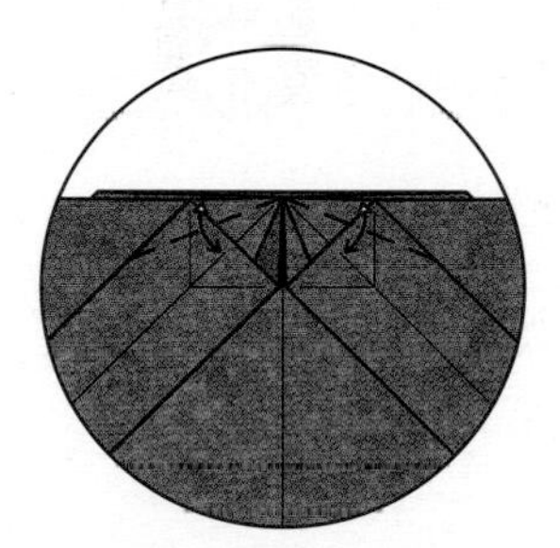

23. Fold and unfold along angle bisectors.

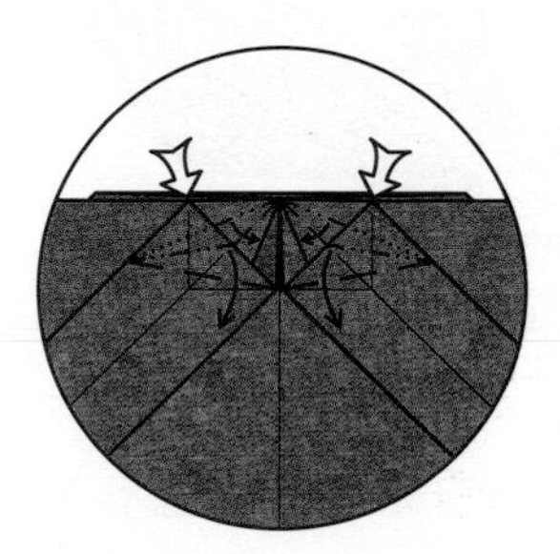

24. Spread-sink two corners.

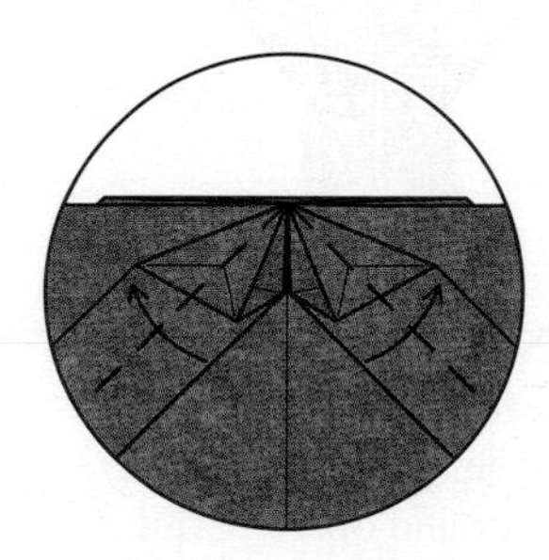

25. Close up the flaps.

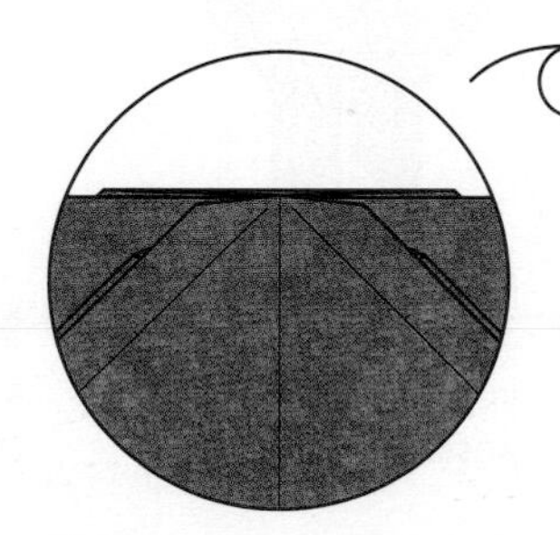

26. Turn the paper over.

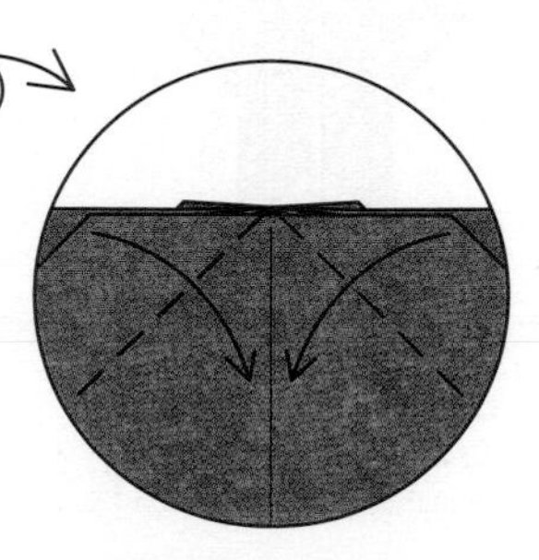

27. Fold the two flaps down.

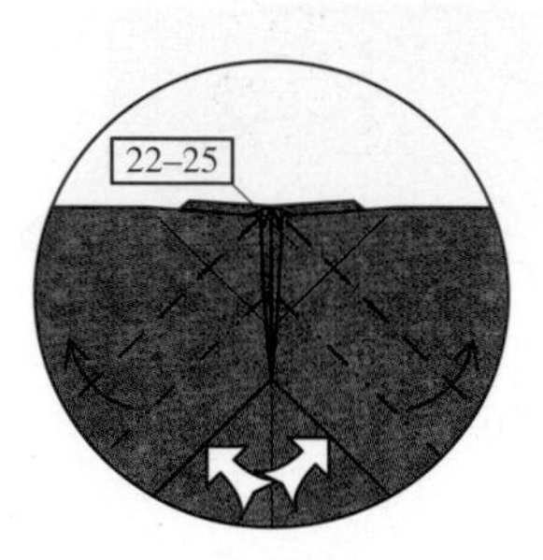

28. Repeat steps 22–25 on this side.

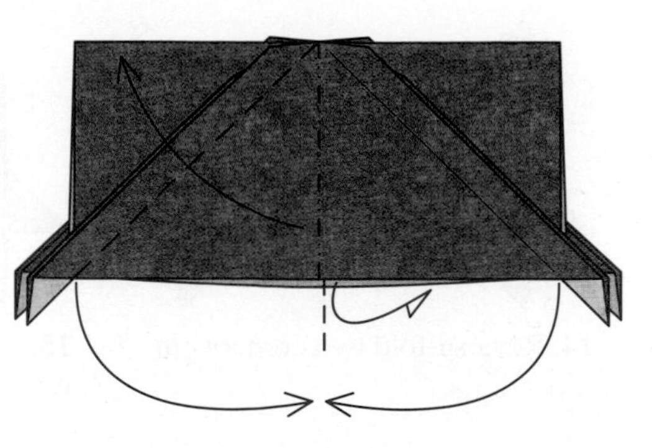

29. Bring the bottom left and right corners together, swinging the excess paper to the left in front and to the right behind. Spread the extra layers up near the tip evenly, front to back.

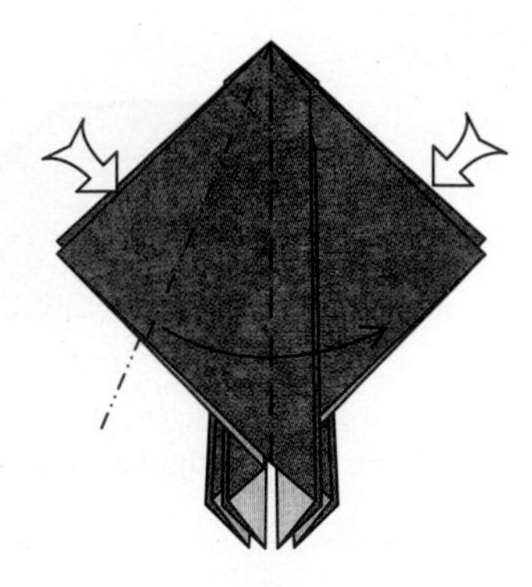

30. Squash-fold the left flap in front and the opposite flap behind, spreading the thick layers up near the tip symmetrically.

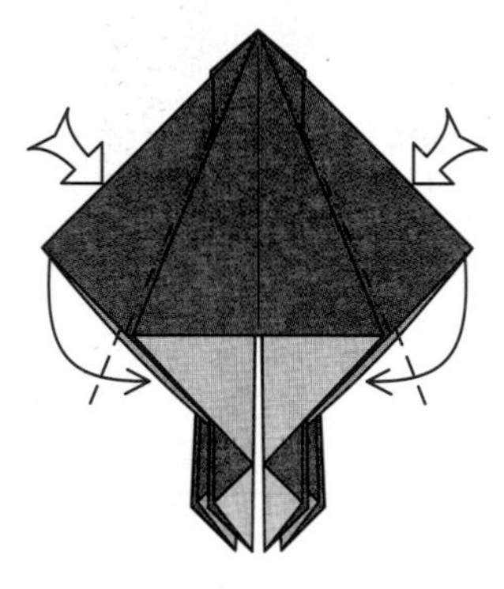

31. Reverse-fold in the two remaining flaps.

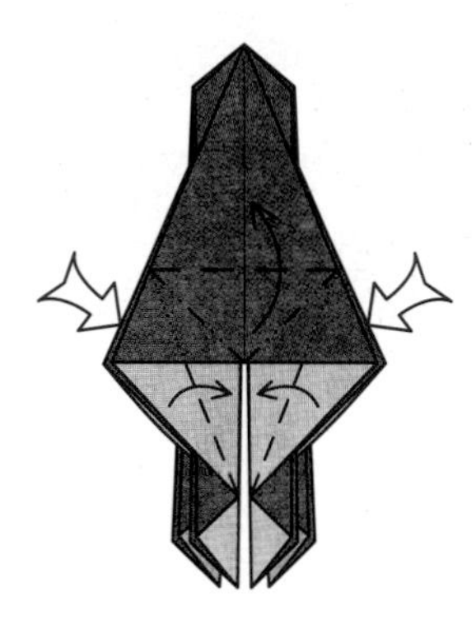

32. Petal-fold the edge in front.

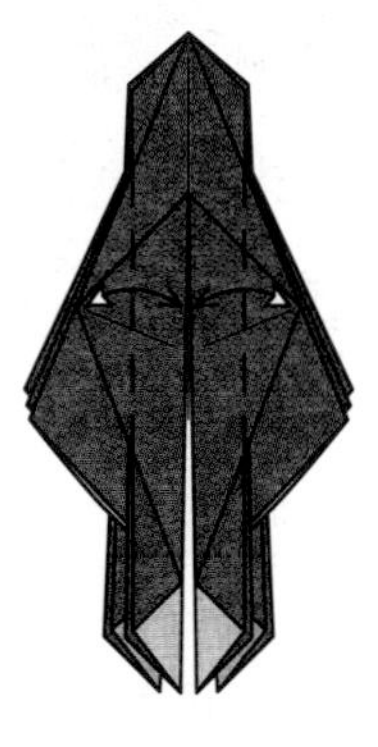

33. Fold the side corners to the center and unfold.

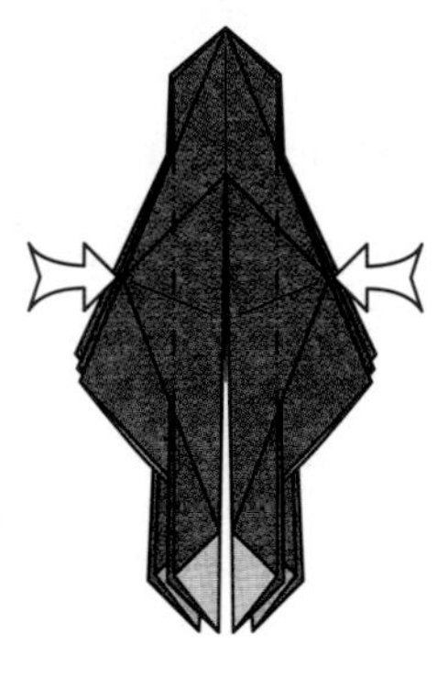

34. Open-sink the corners on the creases you just made.

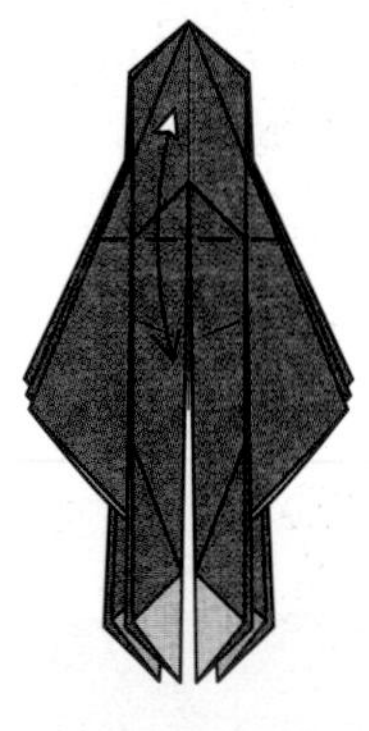

35. Fold and unfold through all layers.

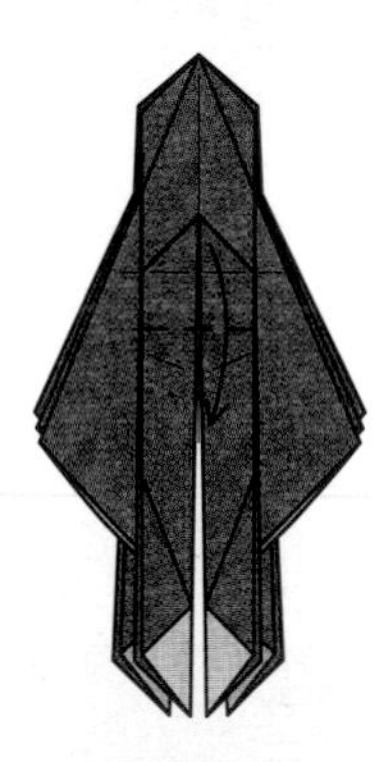

36. Fold the small point down as far as possible.

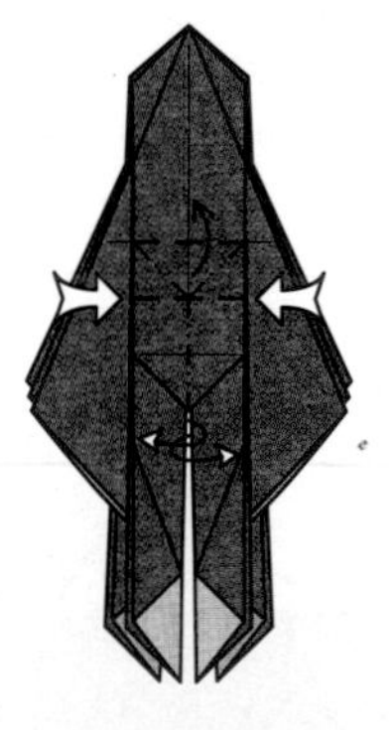

37. Lift up the point and squeeze it in half, forming a rabbit ear from a single layer of paper and opening out a layer on the lower flaps. The paper will not lie flat.

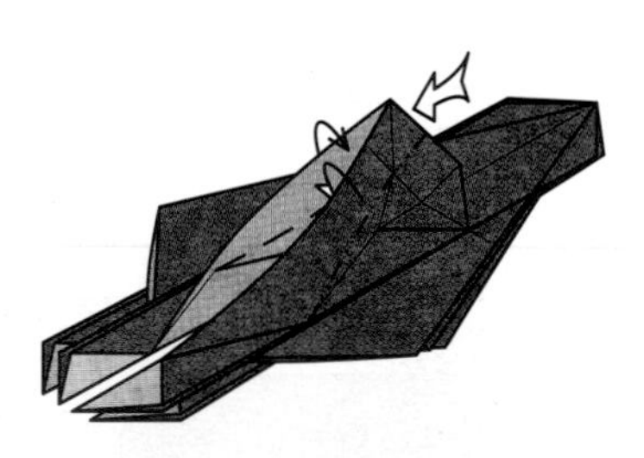

38. Fold the raw edge inside on existing creases.

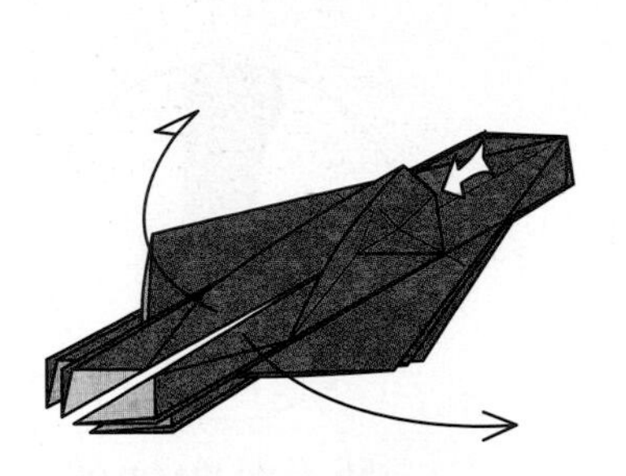

39. Stretch the two near points out to the sides.

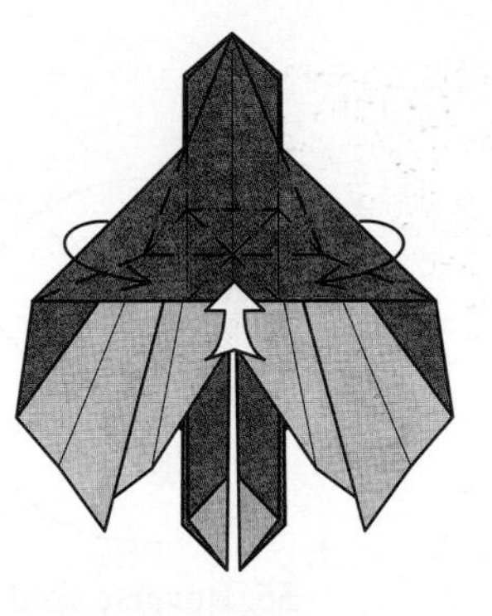

40. Tuck the paper underneath and close the model up.

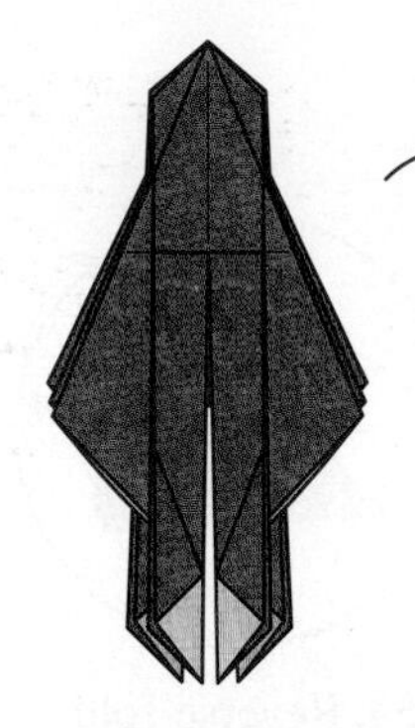

41. Turn the model over.

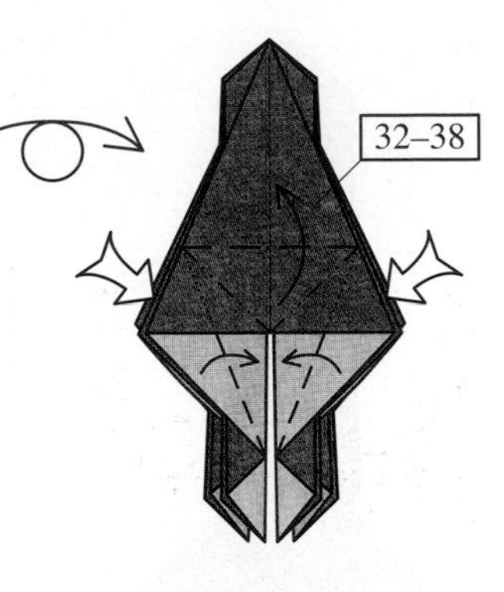

42. Repeat steps 32–38 on this side.

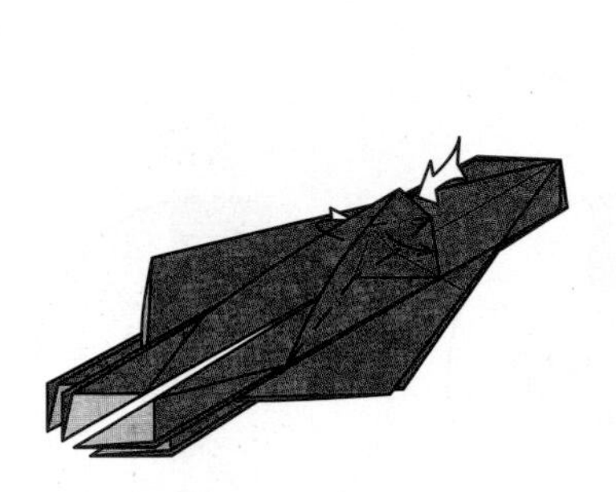

43. Squash-fold the top and flatten.

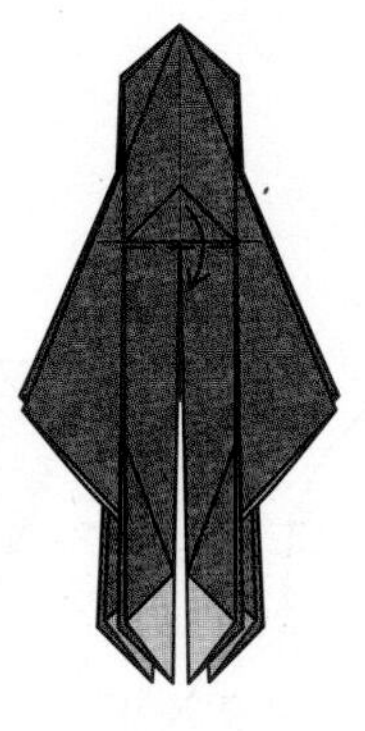

44. Fold a small point down.

45. Fold one wide layer to the right in front and one to the left behind.

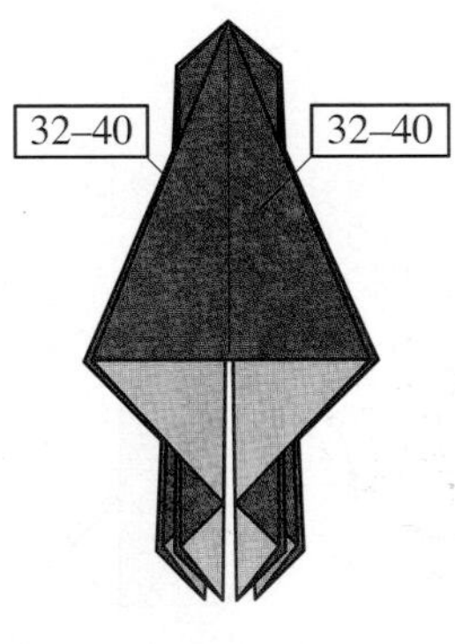

46. Repeat steps 32–40 in front and behind.

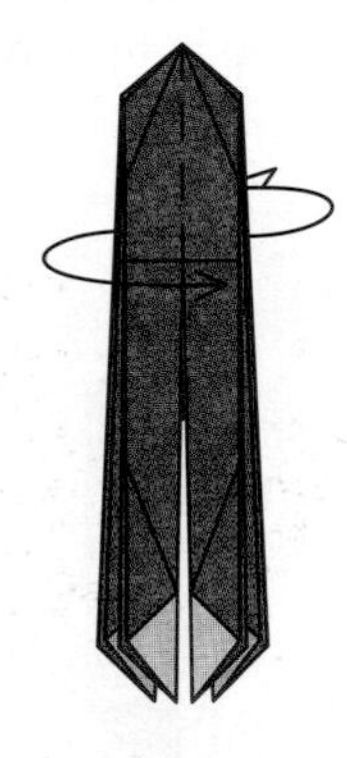

47. Fold half the layers to the right in front and half to the left behind.

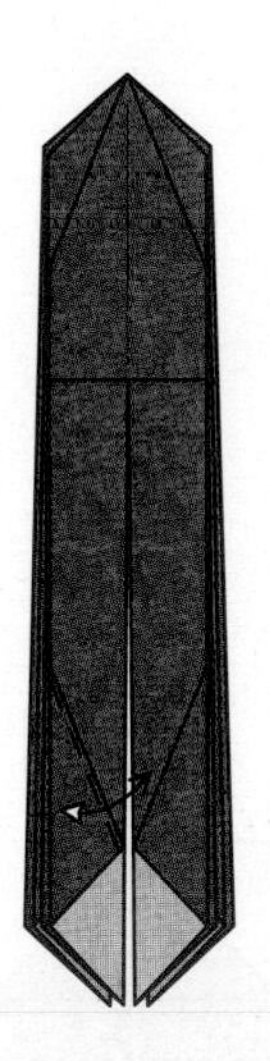

48. Fold a single flap along the folded edge and unfold.

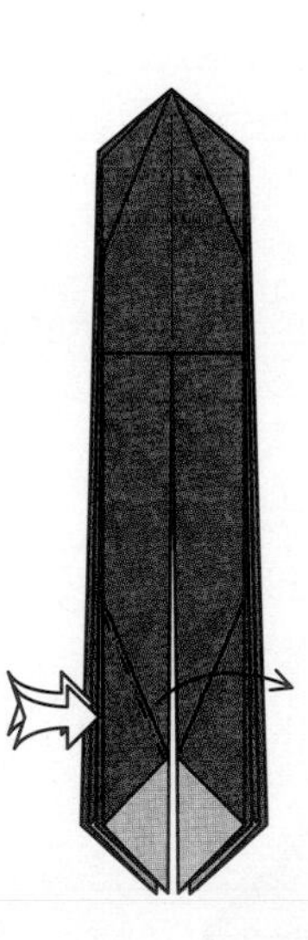

49. Reverse-fold two edges and stretch the flap over to the right. The result will not lie flat.

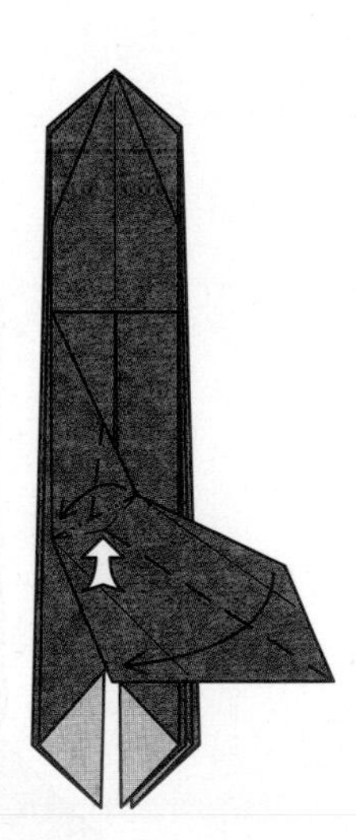

50. Make a small crimp across the flap and close it up.

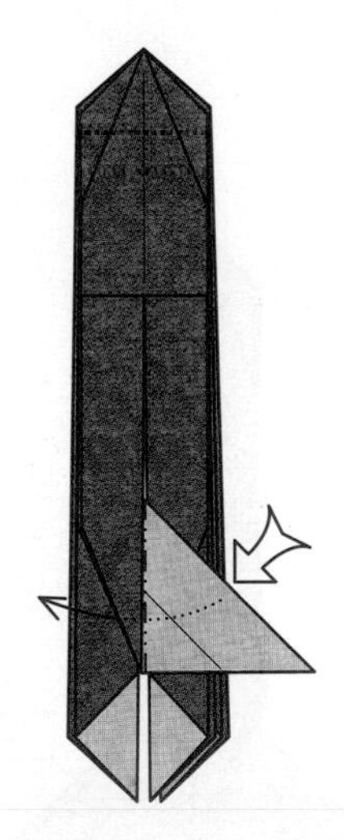

51. Reverse-fold the corner back to the left.

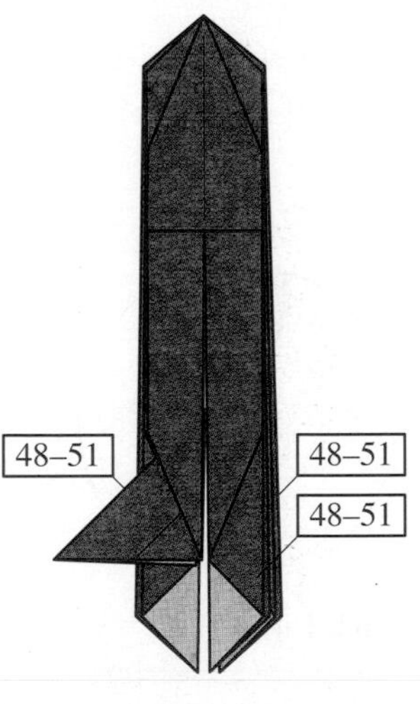

52. Repeat steps 48–51 on the other three flaps.

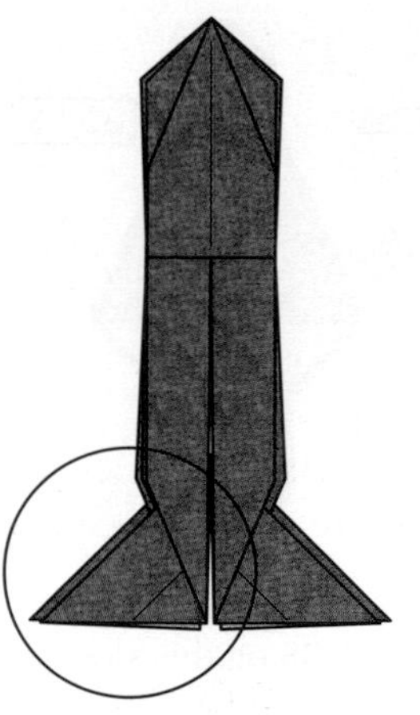

53. Steps 54–62 will focus on the flap in the circle.

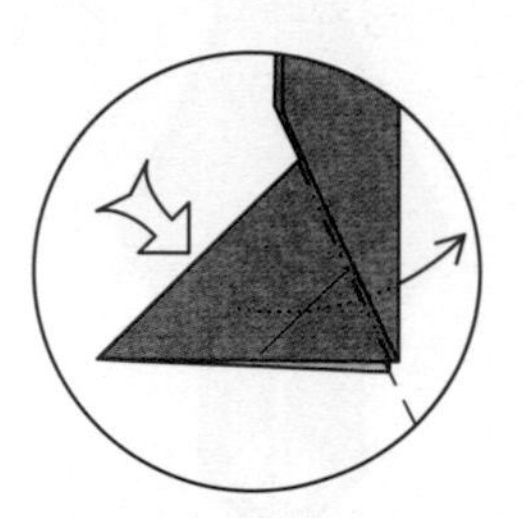

54. Reverse-fold the flap to the right along creases aligned with the folded edge.

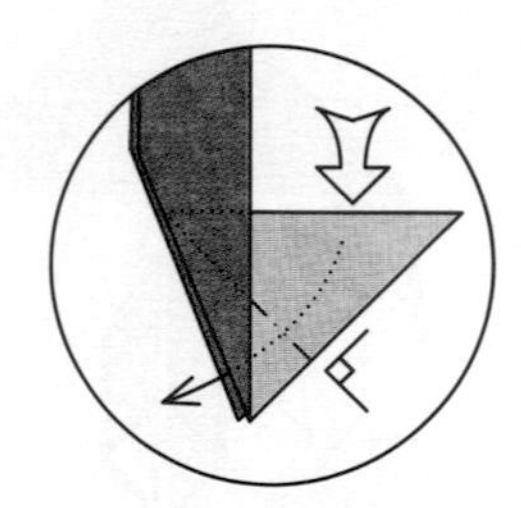

55. Reverse-fold the flap downward so that the raw edges are aligned.

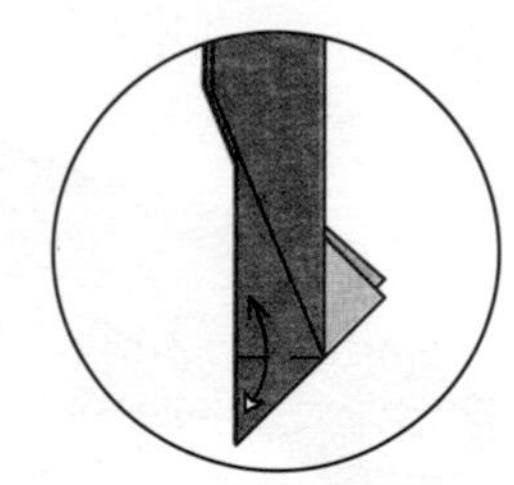

56. Fold the tip of the flap up and unfold.

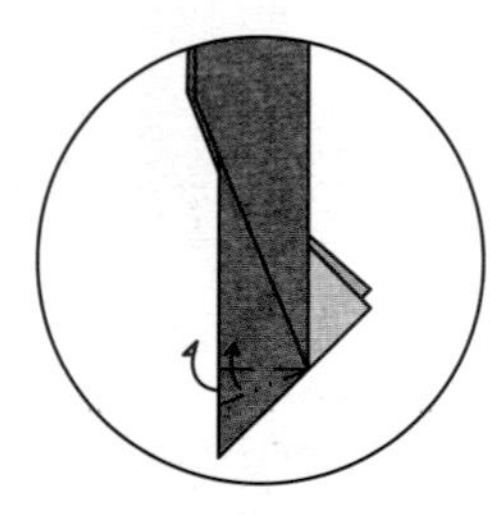

57. Crimp the tip upward so that the raw edge ends up horizontal.

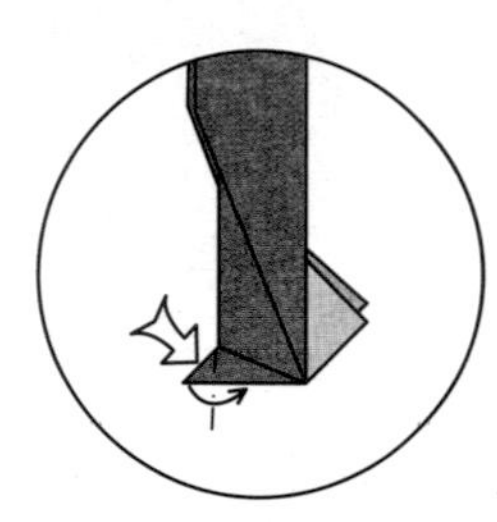

58. Reverse-fold the tip inside.

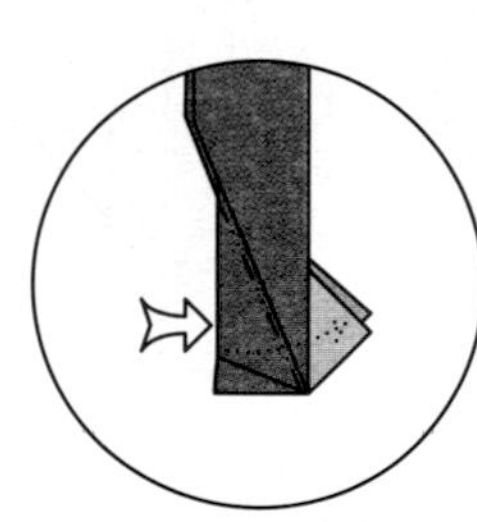

59. Reverse-fold the edge.

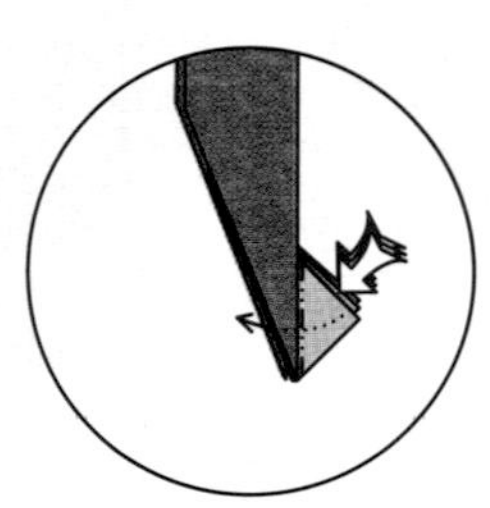

60. Reverse-fold three edges back to the left.

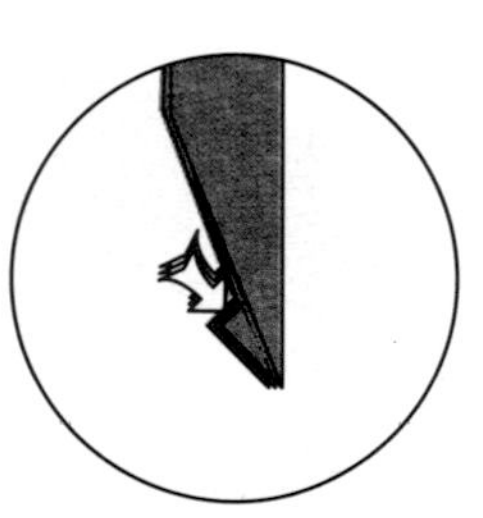

61. Reverse-fold three edges back to the right.

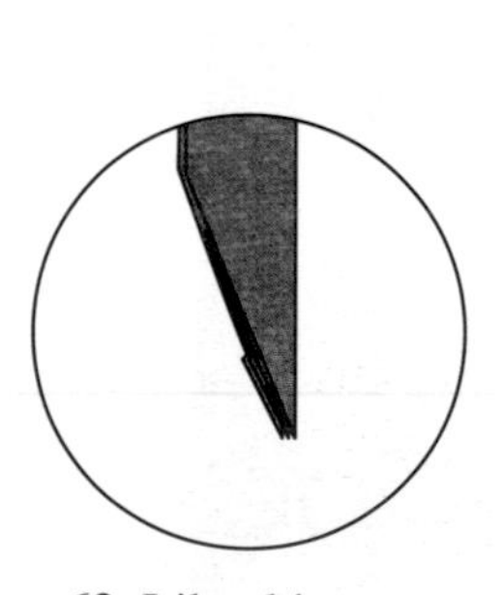

62. Like this.

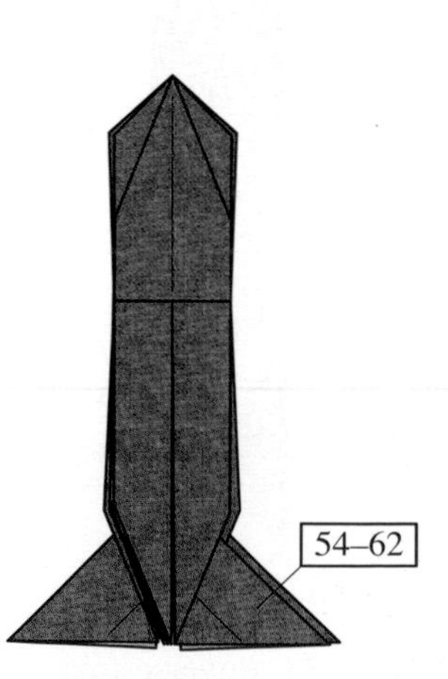

63. Repeat steps 54–62 on the right near flap.

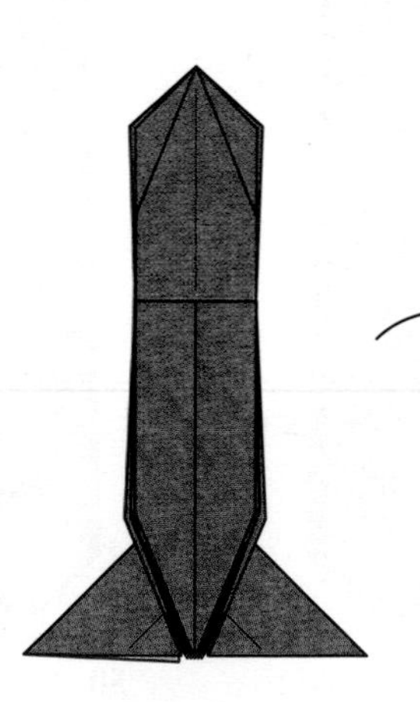

64. Turn the model over.

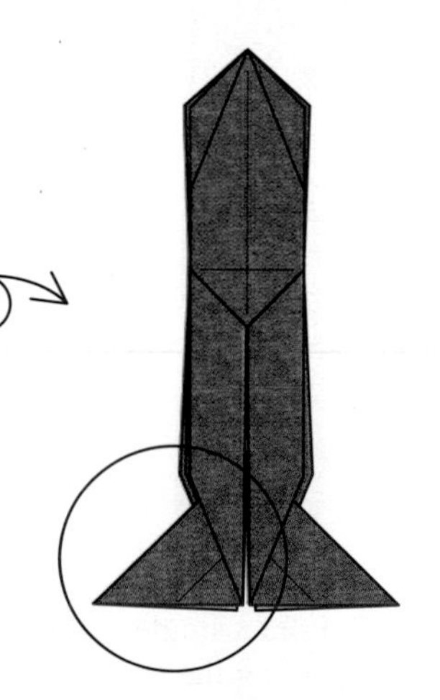

65. Steps 66–76 will focus on the flap in the circle.

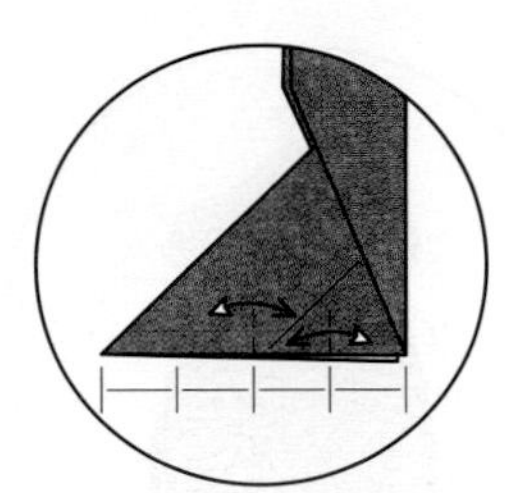

66. Divide the flap into quarters along the bottom edge.

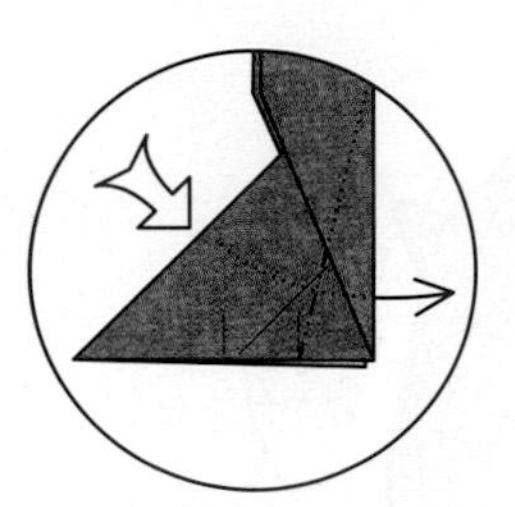

67. Reverse-fold along a line that connects to the rightmost mark.

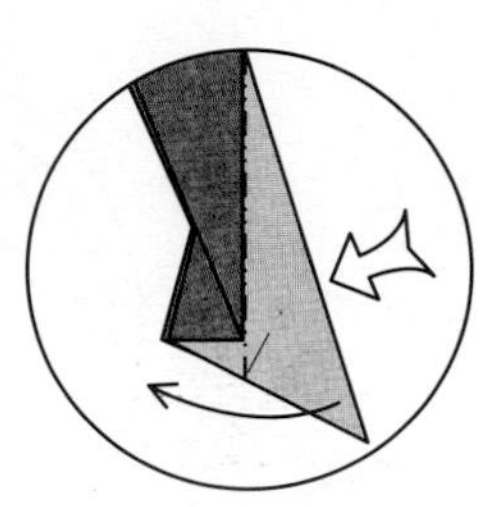

68. Reverse-fold the flap back to the left along vertical creases.

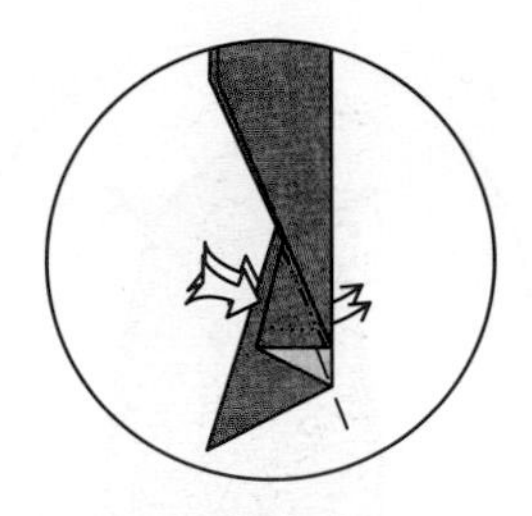

69. Reverse-fold the two small corners to the right (one in front, one behind).

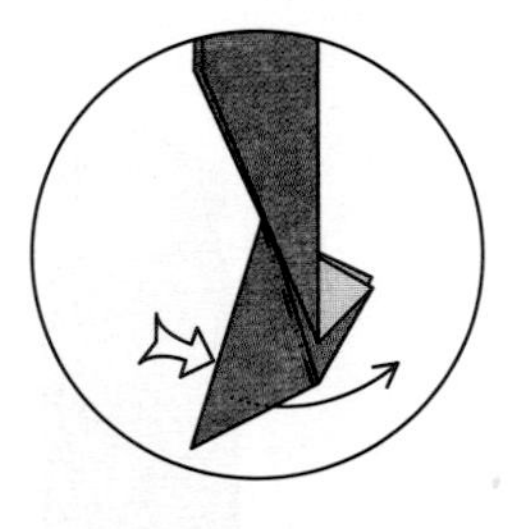

70. Reverse-fold the larger flap along creases aligned with the folded edges.

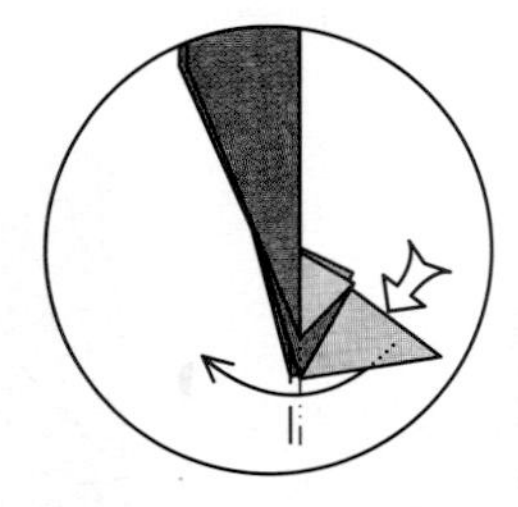

71. Reverse-fold the larger flap to the left along vertical creases.

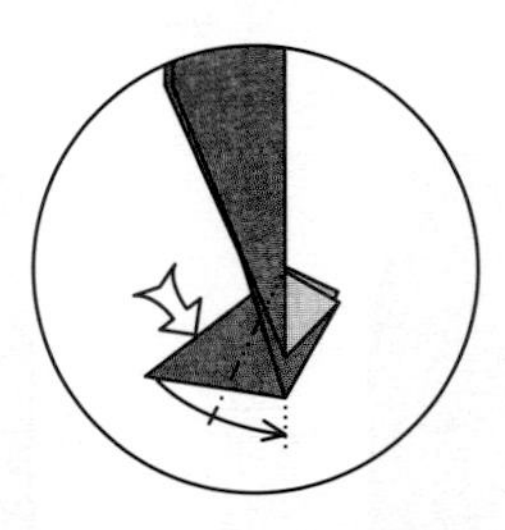

72. Reverse-fold the point so that its right edge is vertical.

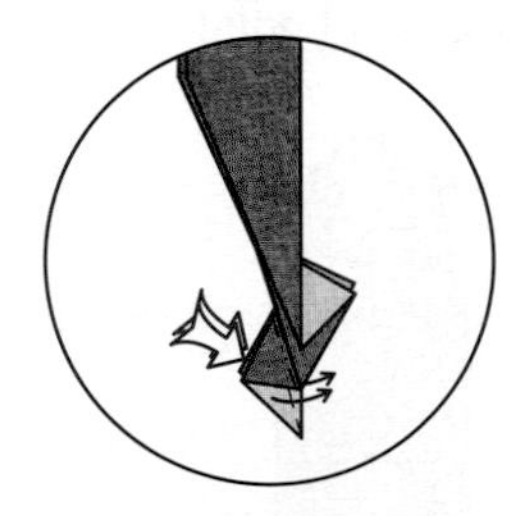

73. Reverse-fold two flaps to the right so that a crease lines up with a folded edge in each reverse fold.

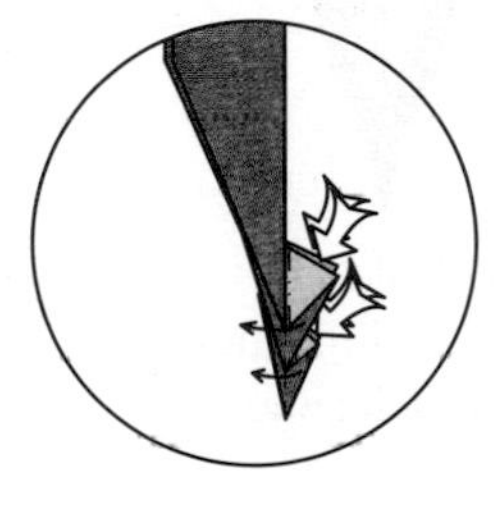

74. Reverse-fold four flaps along vertical creases.

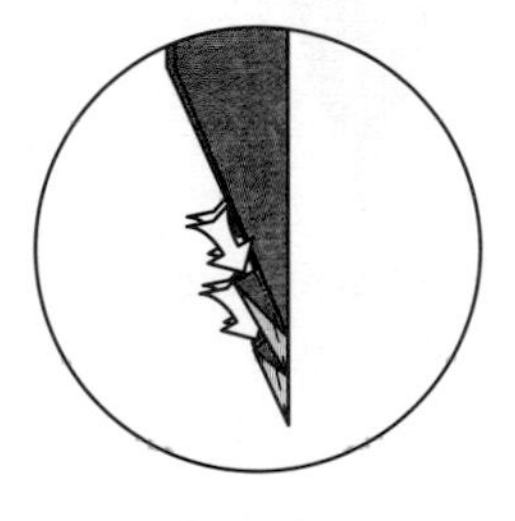

75. Reverse-fold the four flaps so that the raw edges are all vertical.

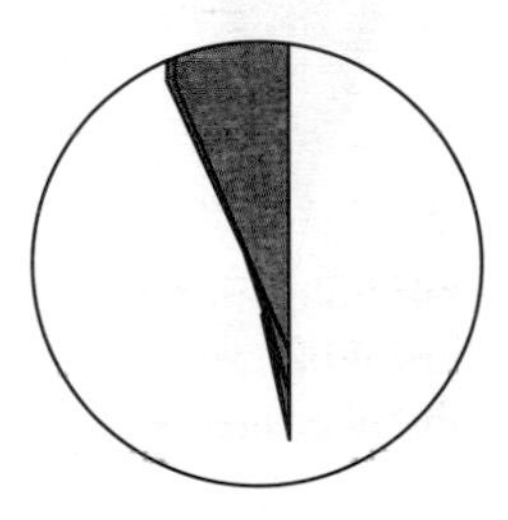

76. Like this.

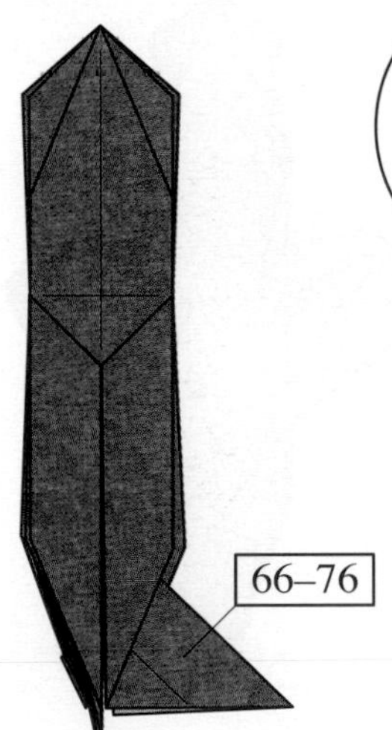

77. Repeat steps 66–76 on the right flap.

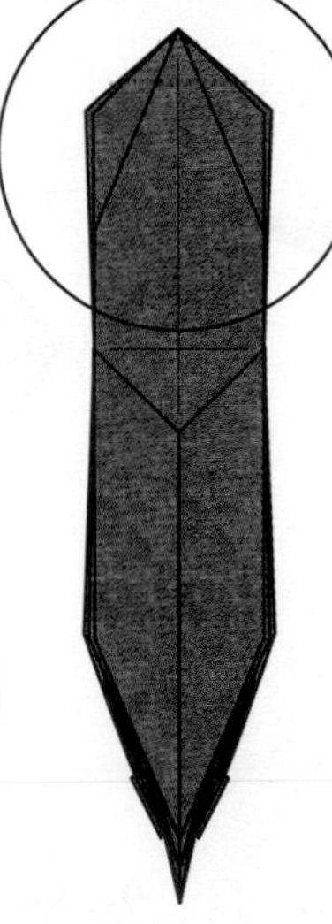

78. Steps 79–83 will focus on the top of the model.

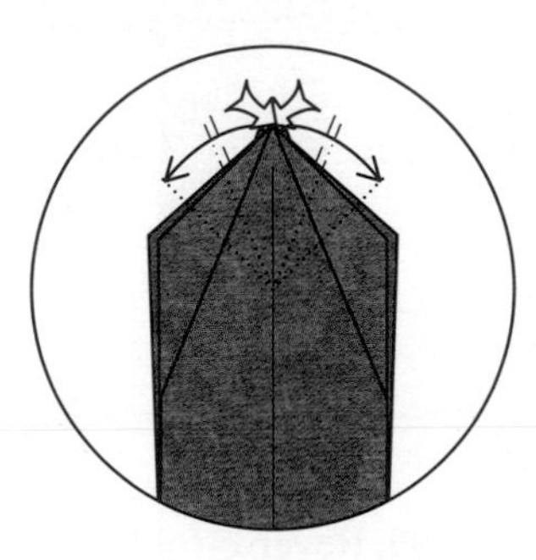

79. Reverse-fold two points out to the sides.

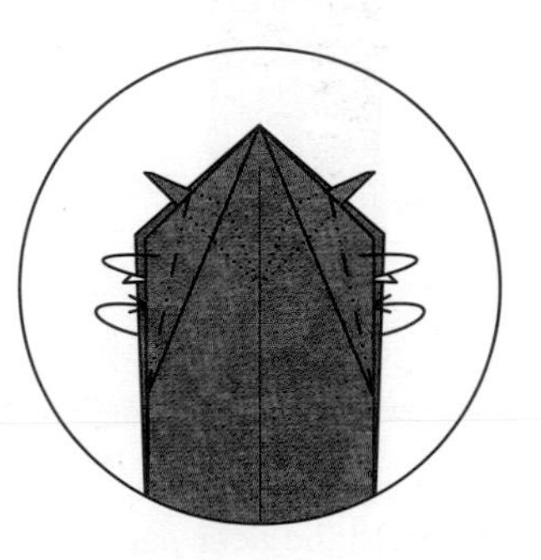

80. Mountain-fold the near layers. Valley-fold the far layers.

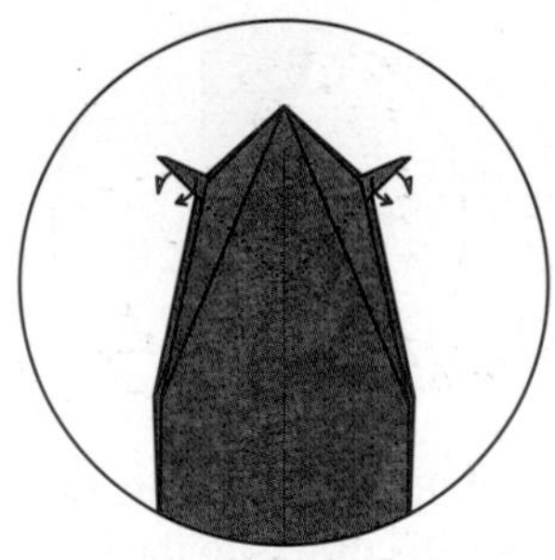

81. Fold down two layers from the front and two layers from the rear of each of the two small points.

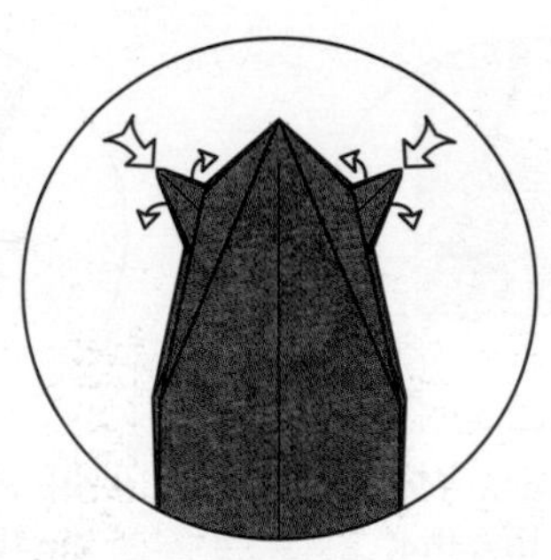

82. Pull out the layers and spread each point into a smooth bulging eye.

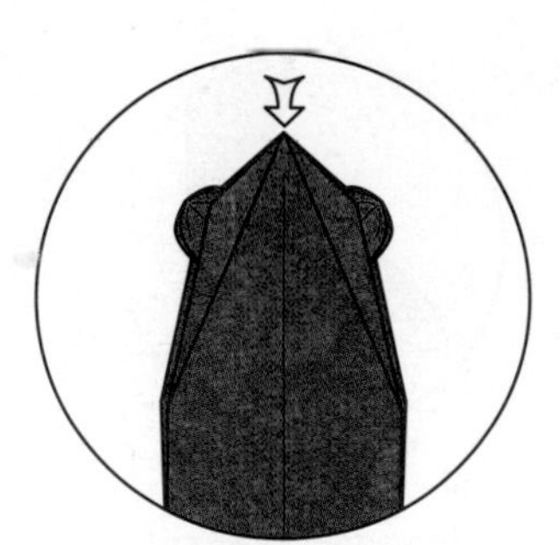

83. Blunt the nose.

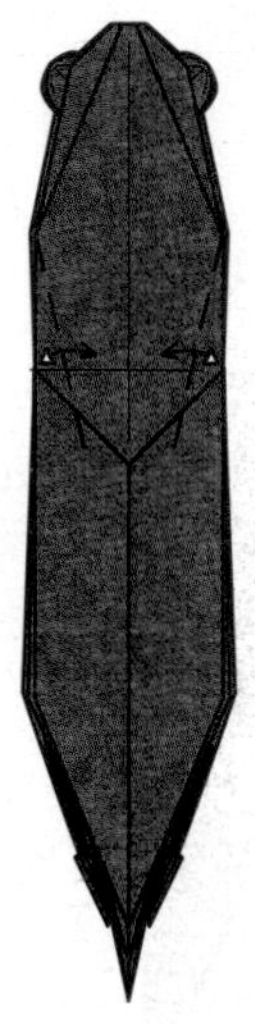

84. Fold and unfold through the near layers.

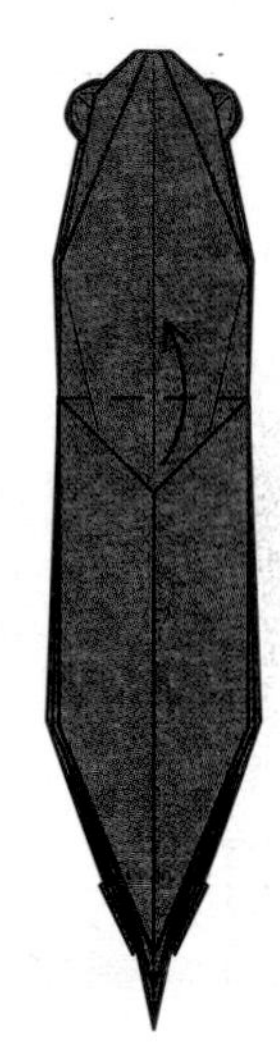

85. Fold the small point upward.

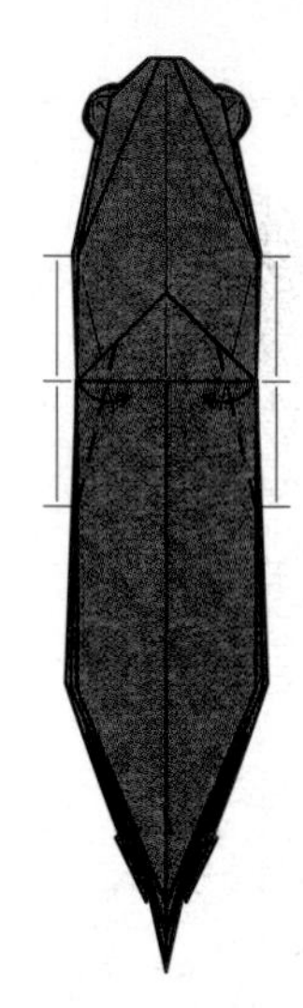

86. Fold a single layer in on each side.

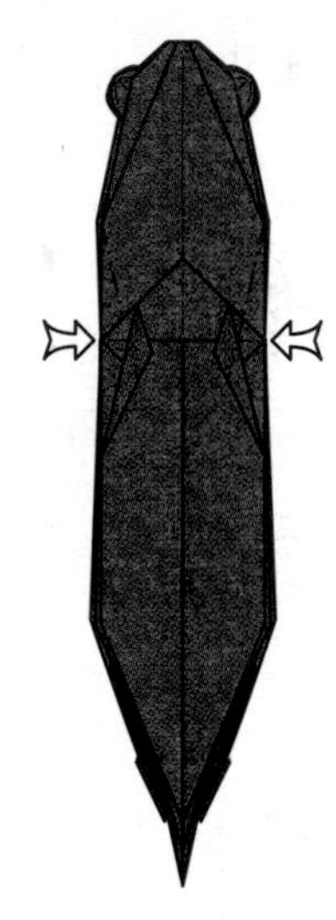

87. Sink the corners on the existing creases.

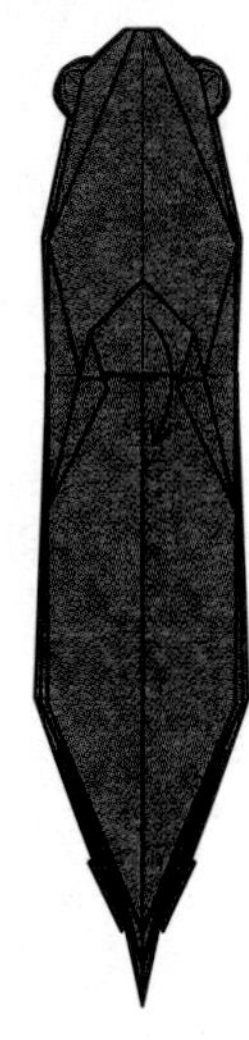

88. Fold the small point back down.

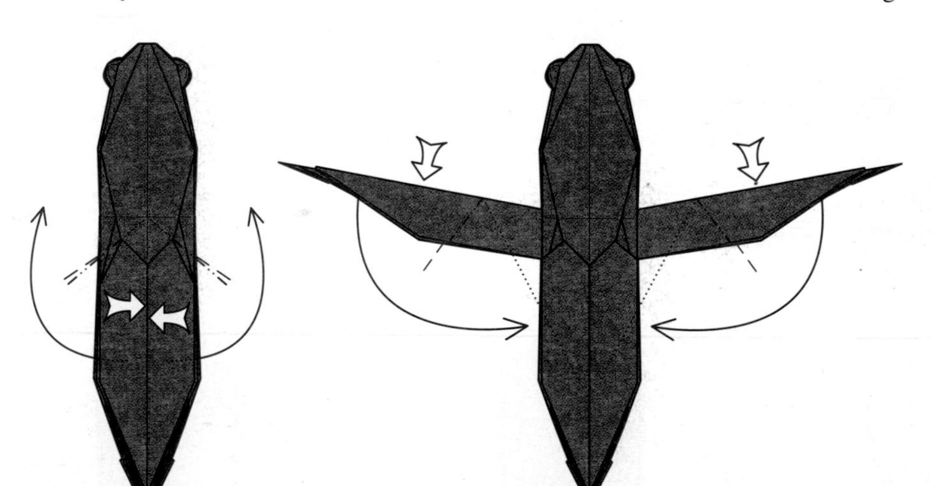

89. Fold the near pair of flaps out and slightly upward.

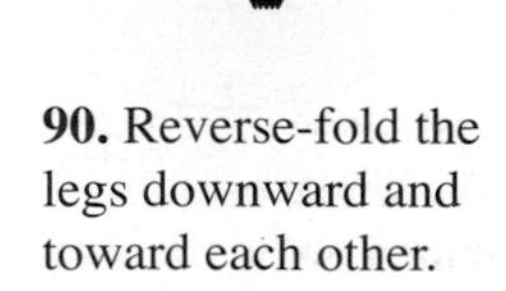

90. Reverse-fold the legs downward and toward each other.

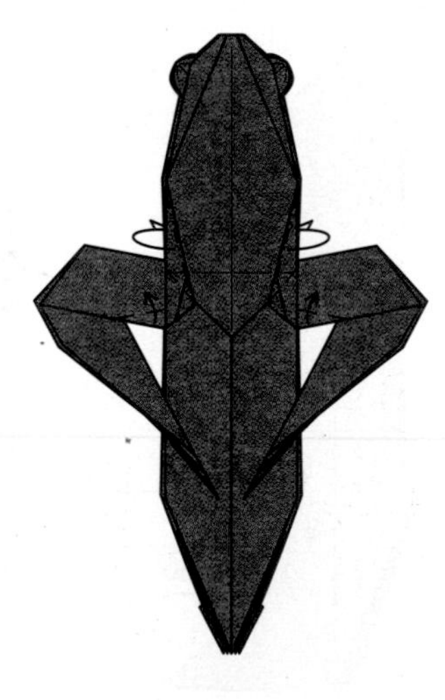

91. Mountain-fold the sides of the body and swivel the lower edges of the legs upward. Repeat on the far side of the legs.

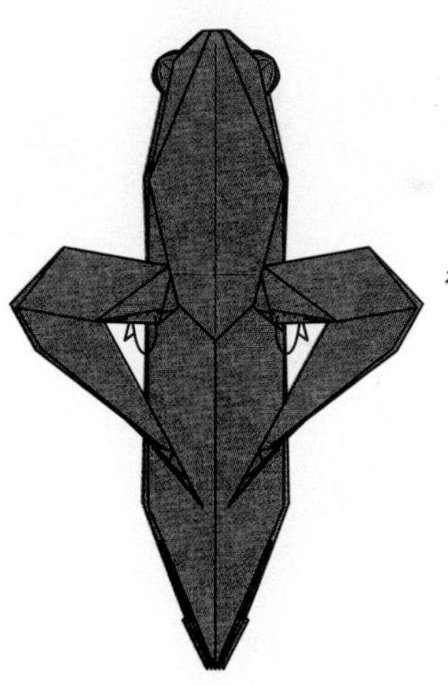

92. Mountain-fold the corners at the bottoms of the legs.

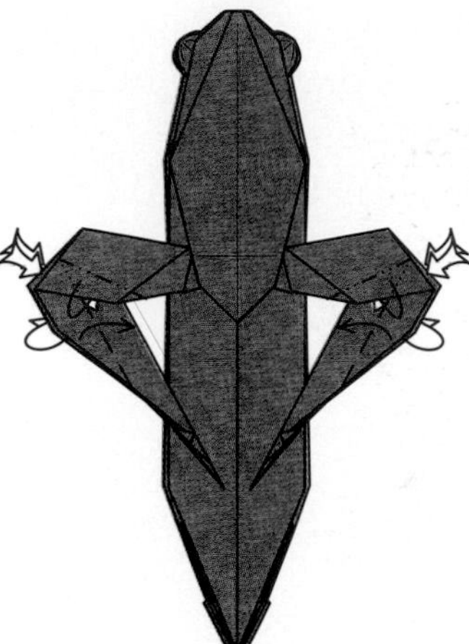

93. Reverse-fold the knees and narrow the legs on both near and far sides.

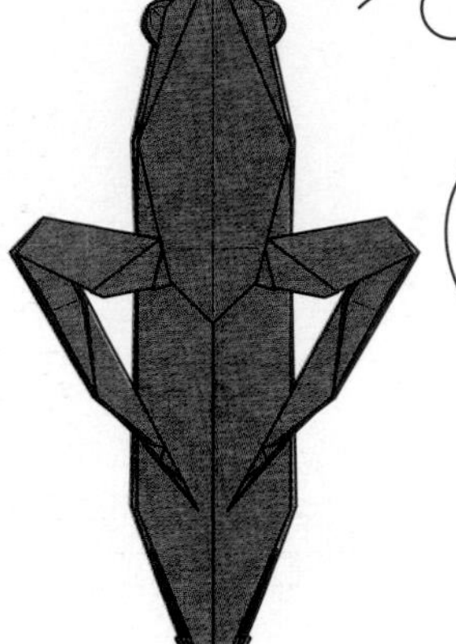

94. Turn the model over.

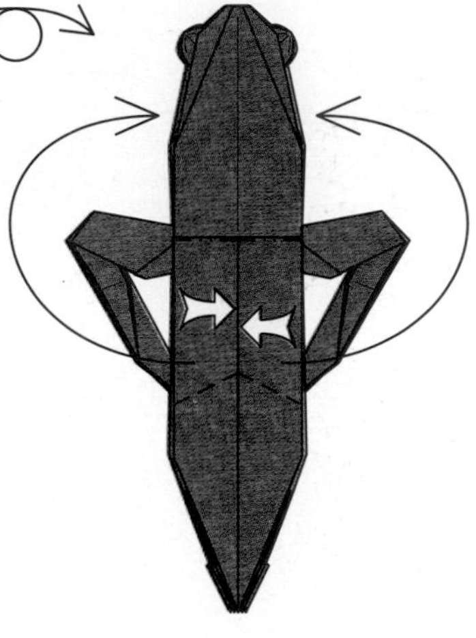

95. Reverse-fold the two bottom flaps up and then out to the sides (it's easier to do both reverse folds on each leg before flattening). Divide the layers asymmetrically, with one layer on the near side and three on the far side.

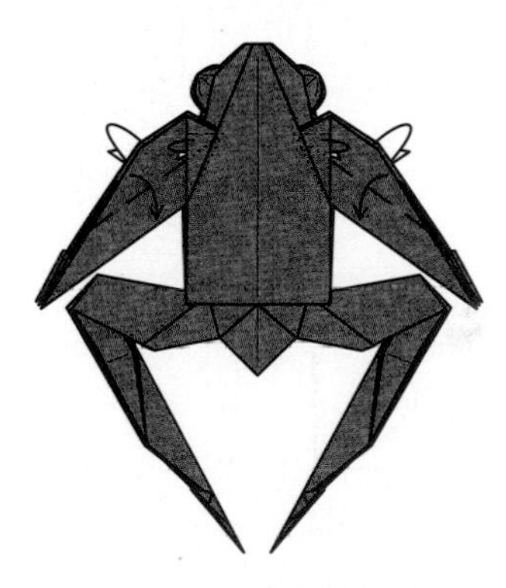

96. Fold the one near edge of each leg flap down and the three far edges behind.

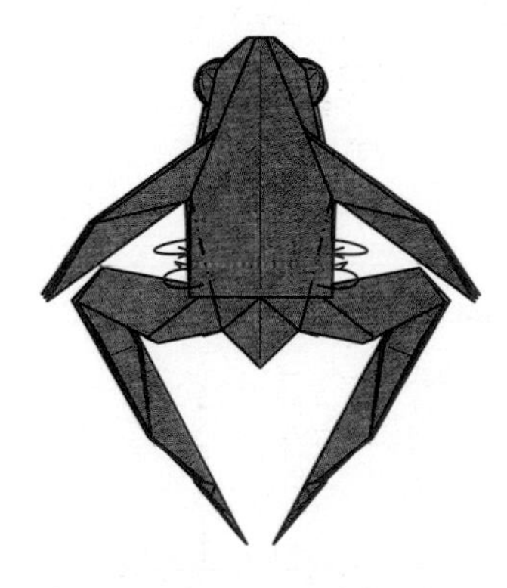

97. Narrow the body with mountain and valley folds that line up with the edges behind.

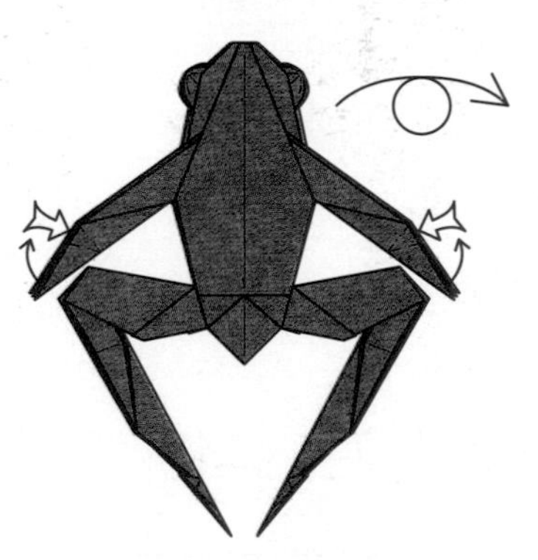

98. Crimp the forefeet, spreading the layers symmetrically. Turn the model over.

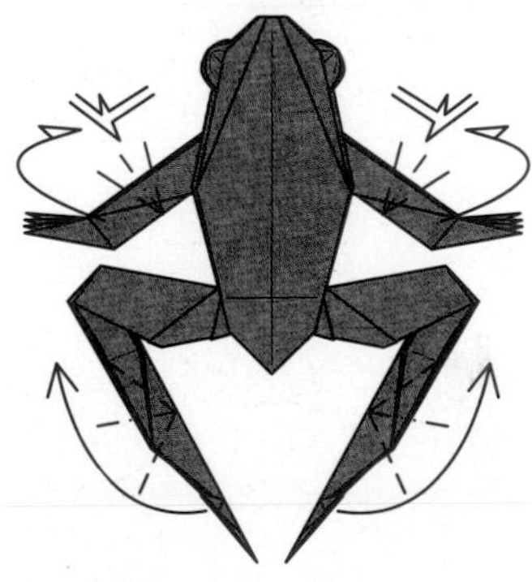

99. Crimp the forelegs and bend them downward. Crimp the feet so that they point slightly forward.

100. Spread the toes. Mountain-fold the back. Shape the body.

101. Finished Tree Frog.

7 Pattern Grafting

Simple grafting can take the form of borders (whole or partial) around the paper or strips that propagate inward from the edges of the paper. The strip grafts you've seen thus far use pleats to add paper along the edges of the square in order to expand appendages. But it's also possible to use the pleats themselves as the additional feature of the model, for example, to create a pattern in an expanse of paper. In the best of all possible worlds, one can add pleats that both create extra paper in appendages and create a useful pattern in the rest of the model. In this way, all of the added paper makes a contribution to the overall model.

7.1. Pleated Patterns

An ideal candidate for this sort of two-for-the-price-of-one design is a turtle. There are many origami turtles—not as many as there are elephants, but still quite a few—nearly all of which have smooth shells. But the pattern of plates on a turtle's shell is a distinctive feature of the animal (beyond the presence of the shell itself, of course), and in recent years, several designers have taken it as a challenge to fold the plate pattern as part of the shell, with varying results. Using strip grafting, it is a relatively straightforward process to add pleats to the shell of an otherwise smooth-shell turtle design in order to create the natural pattern of plates. As a bonus, we can use the pleats to add detail to other parts of the model.

Here's how we do it: Figure 7.1 shows the structure of a simple turtle. It's easy to fold and has a very simple structure. As we have done before, it is useful to examine the base and the crease pattern as well as the folded model in order to establish

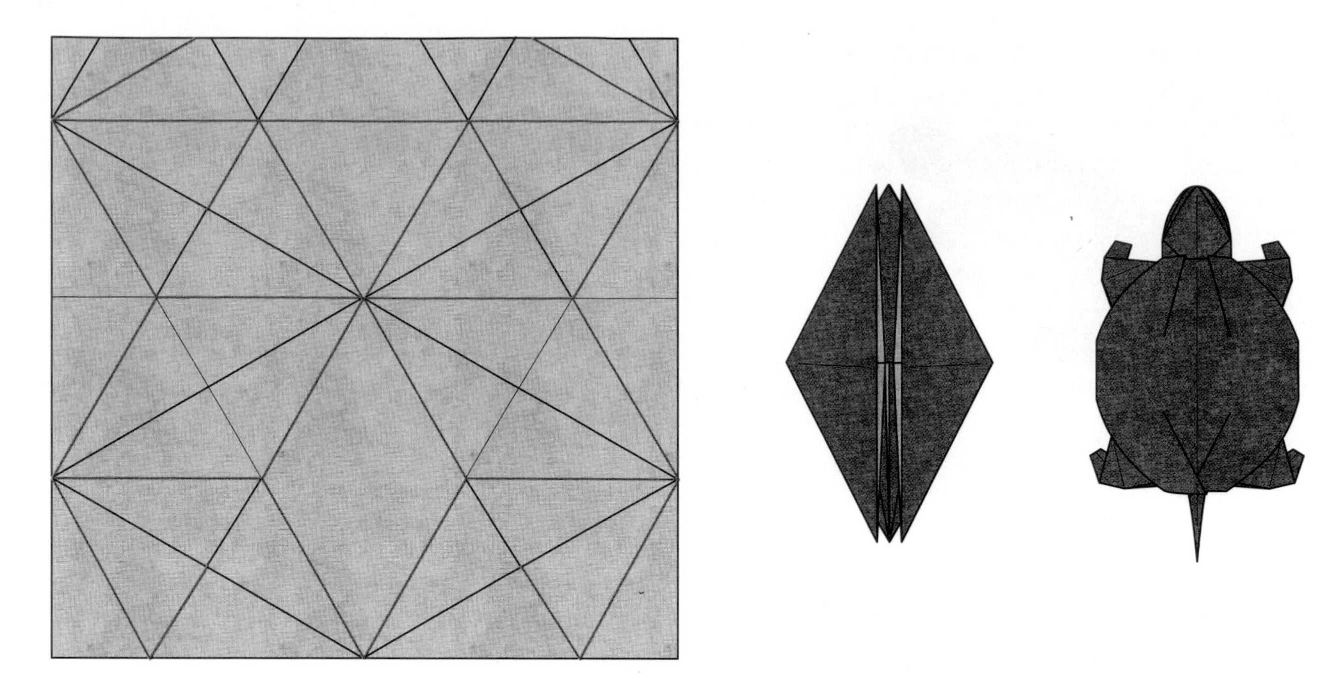

Figure 7.1.
Crease pattern, base, and folded model of the Turtle.

correspondences between features of the crease pattern and features of the model, using the base as an intermediate position.

In this case, the base isn't one of the Classic Bases, but one can apply grafting to any preexisting base. An important observation here is that the underlying structure of this base is actually a rectangle rather than a square. The strip of paper along the top of the crease pattern really isn't necessary to the base, something you can easily verify by cutting off the strip and re-folding the base, which still has all of its flaps.

If we examine the base of the Turtle and its crease pattern, we see that the flat diamond shape in the middle of the crease pattern gives rise to the shell (as well as the head and tail). It would be a fairly simple task to decorate it with lines to outline the plates of a real turtle's shell, as in Figure 7.2.

Figure 7.2.
The Turtle shell, with a shell pattern overlaid.

We can use grafts to replicate the pattern of lines, by running pleats composed of strips of paper along each of the lines; the folded edges of the pleats will then produce the shell pattern. But where should the pleats go in the crease pattern? A reasonable way to proceed with the design is to fold the simple turtle, draw the plate pattern on the back, and then unfold the shell to see where the pattern winds up on the unfolded square. The result is shown in Figure 7.3.

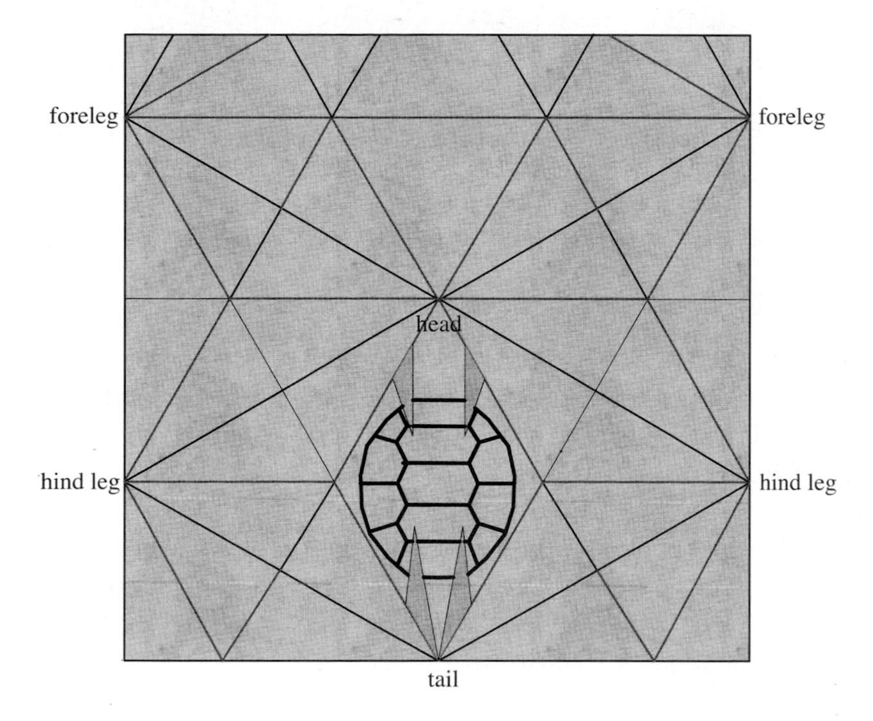

Figure 7.3.
The unfolded square with the shell pattern placed on the region that becomes the shell. Note that the colored wedges are concealed by pleats in the folded model.

Now, we could replicate this directly with pleats, but in striking a balance between exact reproduction and elegance of line, it's usually desirable to simplify the pattern, focusing attention on a smaller number of distinctive lines rather than overwhelming the viewer with a clutter of lines. It is visually pleasing and gives cleaner folding patterns to make the pattern fairly symmetric. Since the crease pattern itself has a strong 60° angle symmetry throughout, it is not unreasonable to adopt that symmetry for the pattern of plates as well. I therefore chose a simplified pleat pattern as shown in Figure 7.4.

The simplifications are twofold: First, I force all lines to lie at multiples of 60°, which makes them match up with the lines in the rest of the model; I also eliminate the oval of lines going all the way around, figuring that I can create this line by folding up the edge of the shell in the finished shape. That leaves just three wide hexagons, plus ten pleats radiating away from them. The decision to force lines to run at multiples of 60° is aesthetic; it moves the lines away from the more evenly

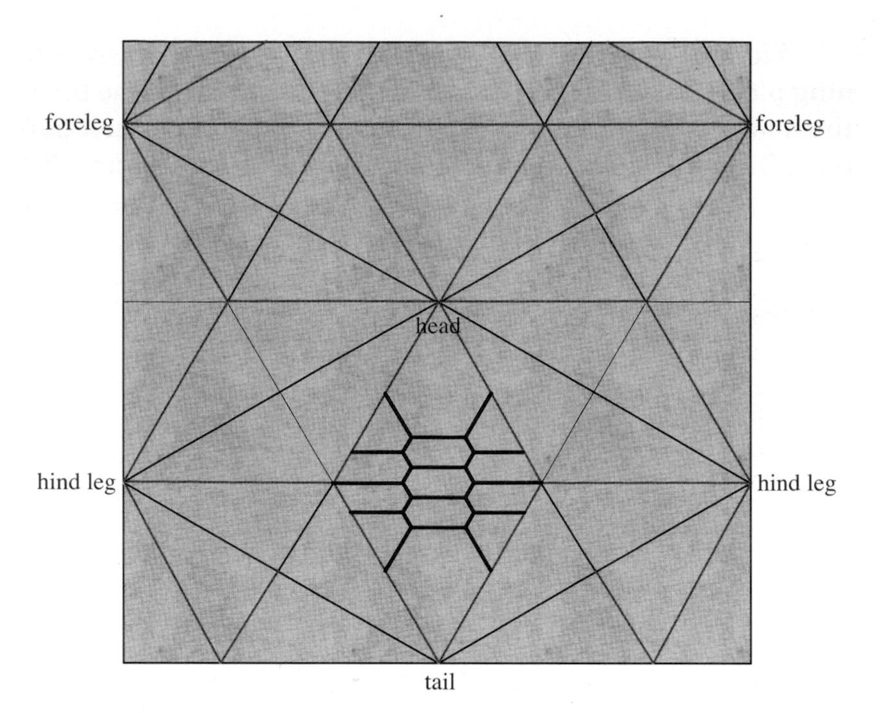

Figure 7.4.
The desired pattern of pleat lines on the shell.

distributed lines of nature, but by keeping to the natural symmetry of the underlying crease pattern, we create the possibility of fortuitous alignments of the creases, leading (we hope) to a relatively elegant folding method.

The pleats are only needed on the shell, but pleats have to propagate all the way to an edge (or terminate at a junction of other pleats), so I extend the pleat lines all the way to the edge of the paper as shown in Figure 7.5.

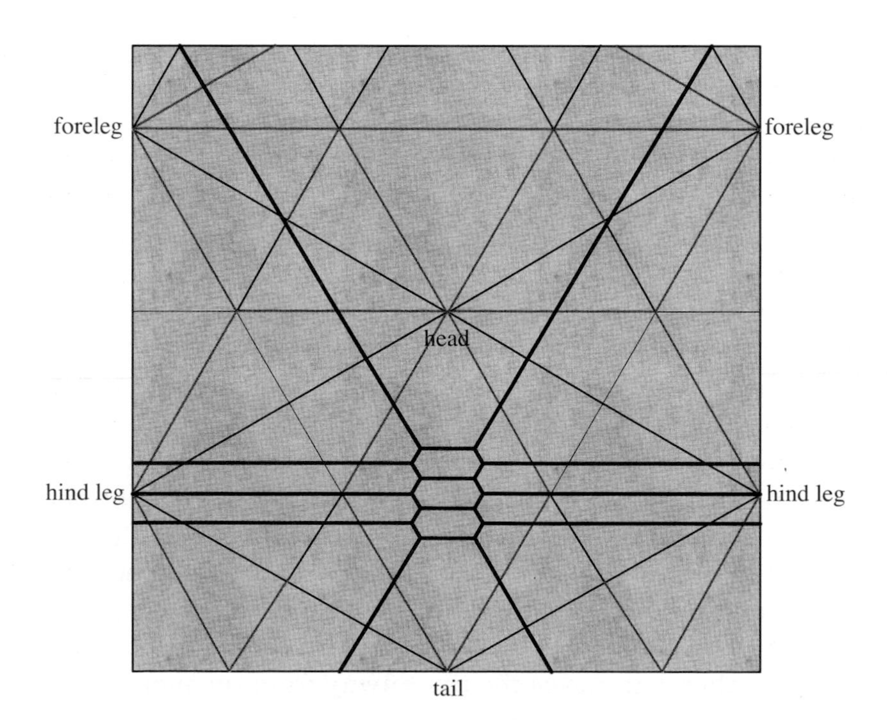

Figure 7.5.
The shell pattern, with pleats extended to the edges of the paper.

To form the pleat lines, we need to give the pleats finite width, which we do by (effectively) cutting the crease pattern apart on the crease lines and inserting finite-width strips of paper as shown in Figure 7.6.

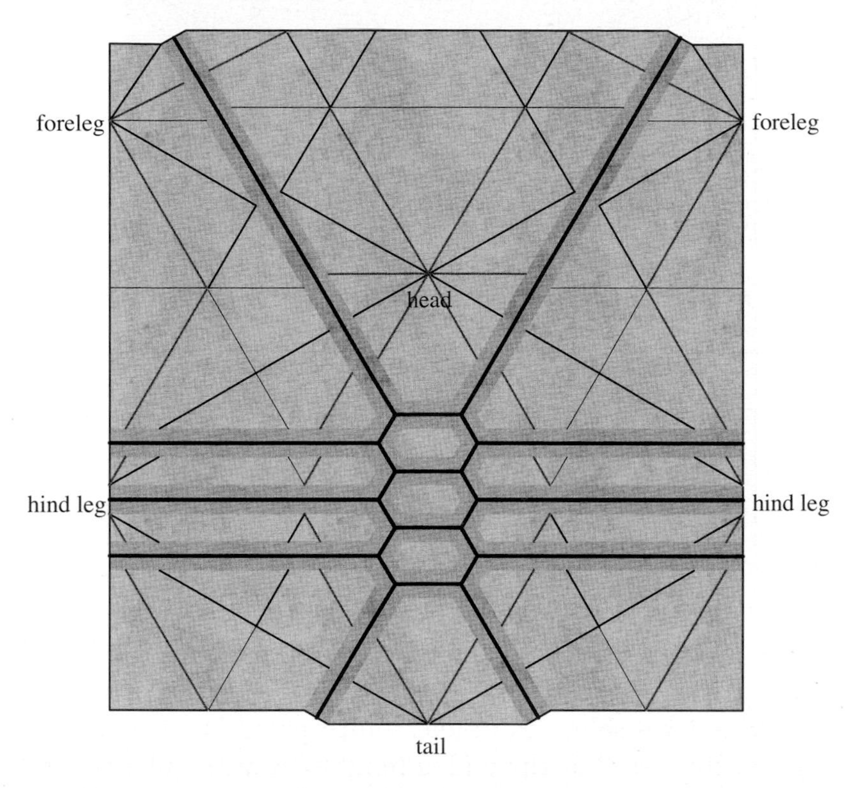

Figure 7.6.
The crease pattern with paper inserted for pleats.

Because some of the pleats hit the edge of the paper at an angle, when we insert finite-width pleats, the paper becomes no longer square. We'll fix that up in a minute. But first, let's see if we can do anything more with these pleats. Observe that one pleat already hits the edge of the paper at one of the appendages (the hind legs). This will allow us to use the paper in the pleat to make a fancier hind foot (with toes, for example); this paper comes for free. If we're going to add paper to the hind feet, we might as well do the same for the front feet, and so I add another pleat near the top of the square that comes out at the front feet, as shown in Figure 7.7.

Having added pleats to decorate the shell and produce more complex feet, the paper's overall dimensions have become roughly rectangular. To get it back to a square, we could add more paper along the sides, or we could cut some off the top or bottom. Looking back at the original crease pattern, recall that the small strip running along the top of the square wasn't used for anything in the original base. So we could cut it off without losing anything from the base; we could have folded the original turtle from a rectangle that is shorter in height than width.

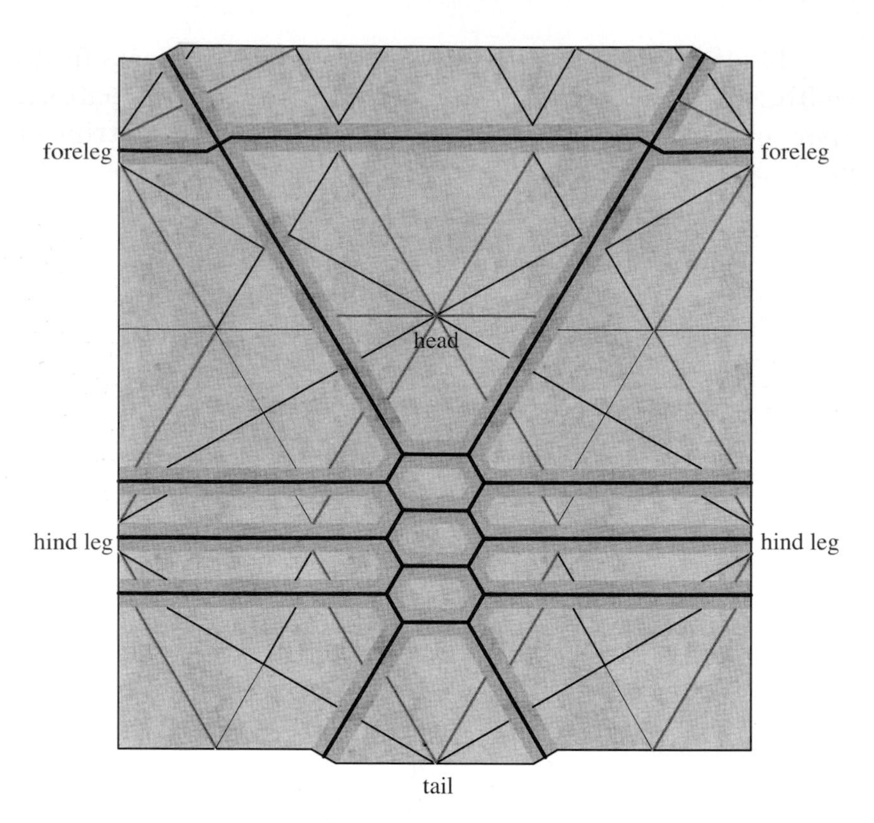

Figure 7.7.
The crease pattern with pleats for both front and hind feet.

On the other hand, the pleats we've added have increased the height of the square much more than they have increased its width. If we select the pleat width carefully, we can arrange matters so that the added height (from the pleats) and the lost height (from taking off the top strip) precisely cancel each other out, resulting in a perfect square once again, as shown in Figure 7.8.

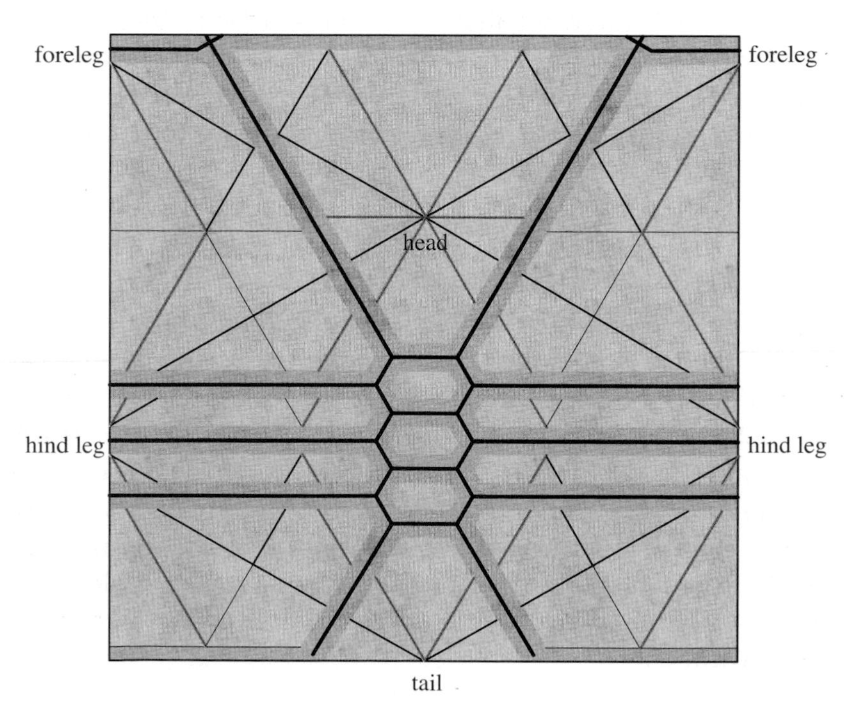

Figure 7.8.
The final crease pattern, back to square and with strip grafts for shell and feet.

Now, we can form the pleats to create the shell and use the excess paper where the pleats hit the edges to make more detailed feet; the result takes a simple model to a new level.

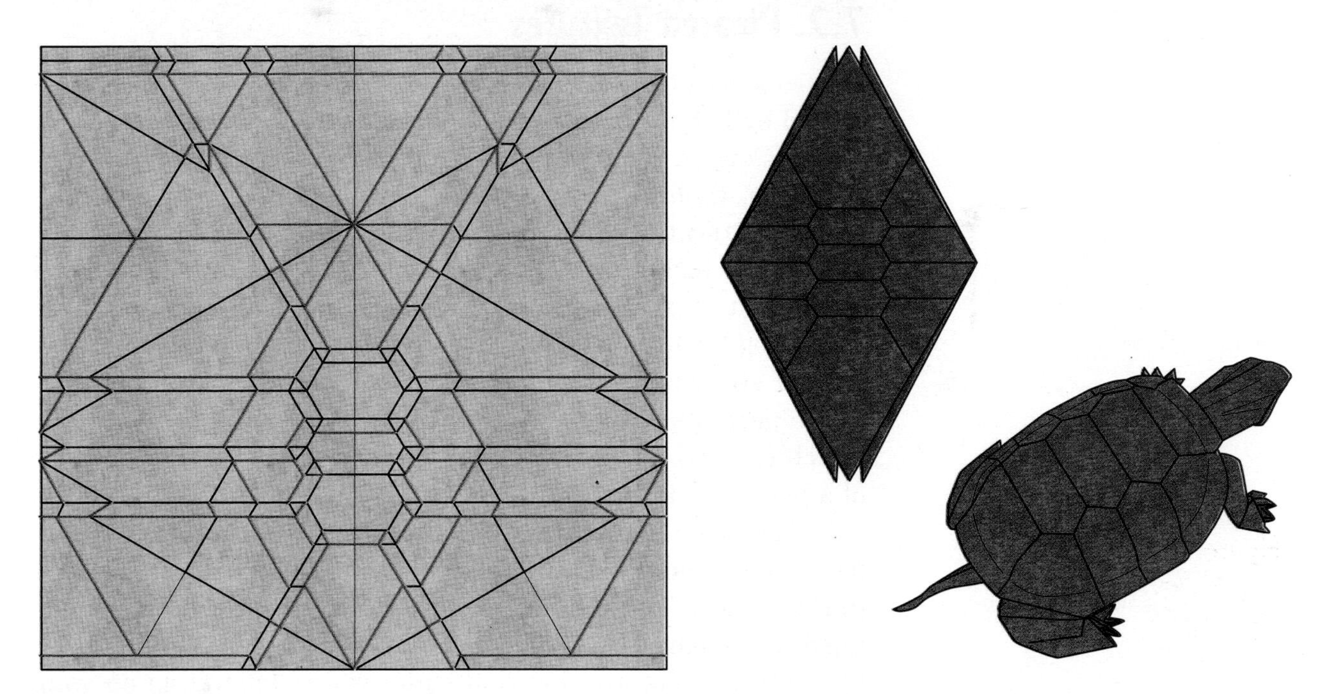

Figure 7.9.
Crease pattern, base, and folded model of the Western Pond Turtle.

I call this use of multiple intersecting pleated strips *pleat grafting*. While you can use pleat grafts on any model to add more detail here and there, there is always an aesthetic balancing act to such surgery: Are the added complexity and extra layers of paper justified by the appearance of the finished result? This balance is, ultimately, a matter of personal taste. However, as you become more accustomed to folding complex structures and/or fine detail, the perceived burden of added complexity diminishes over time with the folder's experience. And if you can use the added pleats for multiple purposes (as we did to create both shell pattern and more detailed feet) or to eliminate an inefficiency in the original base (which we also did in this example), then the balance will, more and more often, tip in the direction of detail.

Most applications of grafting serve to create distinct flaps or appendages, but the turtle shell is a bit different. Here we are not creating flaps; we are creating a patterned surface. This opens up a large range of possibilities: incorporation of patterns into origami models to represent subjects that have a strong textural visual impact. As we did with the Turtle, we can create a texture and overlay it onto the paper before folding the figure, or (better yet) incorporate the folds that create the texture into the folds that create the rest of the model.

7.2. Pleated Textures

The concept of origami texture as art in itself was widely explored by French artist and folder Jean-Claude Correia in the 1980s. Correia adopted the technique of creating crossing grids of pleats, then manipulated the excess paper created at the pleat intersections. While Correia's work was primarily abstract, the technique has been adopted by several artists to combine textures with representational origami; an early hedgehog by John Richardson used crossed pleats to make a grid of short spines on a three-dimensional body. The technique perhaps reached its zenith in Eric Joisel's Pangolin, in which crossed pleats of varying sizes created the scaled body of a primitive anteater.

The basic concept of a pattern graft is to create a regular pattern of creases that emulates some regular pattern present in the subject. The simplest possible pattern is formed by making a row of parallel pleats in one direction, then again at 90° to the first; this creates a grid of squares (or, depending on your orientation, diamonds). The resulting pattern resembles scales, which is perhaps why most patterned subjects have tended to be scaly: snakes, dragons, scaled anteaters, and the like.

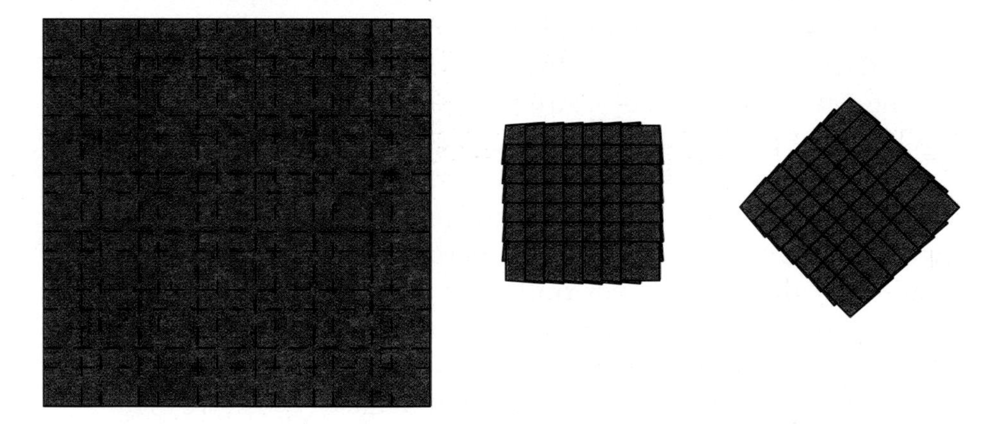

Figure 7.10.
Left: crossed pleats.
Middle: the folded structure.
Right: crossed pleats at 45° create a grid of diamonds.

There is nothing that says one can't use other patterns, however; it is possible to take many regular tiling patterns and create pleated origami representations of them. Origami artist Chris K. Palmer has single-handedly created an entire genre of origami by doing exactly that. For representational origami, however, the patterns one can create are restricted to those that resemble some subject, which tends to favor fairly simple patterns. Grids of squares or diamonds are straightforward: Make crossing sets of pleats. It's also possible to make grids of triangles and/or hexagons (you saw a small piece of the latter in the turtle shell), but these are somewhat harder to fold as they require three different directions of pleats to interact.

The pattern, or texture, grafted into a model is generally going to be dictated by the pattern in the subject. One subject that seems natural for texturizing is a fish: Like the previously done snakes and dragons, a fish has prominent scales. We'll use a particular fish model—a Koi, or Japanese carp—to illustrate the process of adding texture to a model and some of the design considerations that ensue.

The simplest way to create texture in a model is to select a simple version of the model foldable from a square, then add texture to the square in such a way that it remains square and the pattern ends up in the appropriate part of the square exposed in the folded model. We did this in the Turtle; we can apply the same approach to a Koi. The process begins with a model, of course: We'll use the Koi illustrated in Figure 7.11, which is folded from a square. (This Koi was created by putting a border graft onto a modified Kite Base to create longer fins and tail; can you identify the original base and graft?).

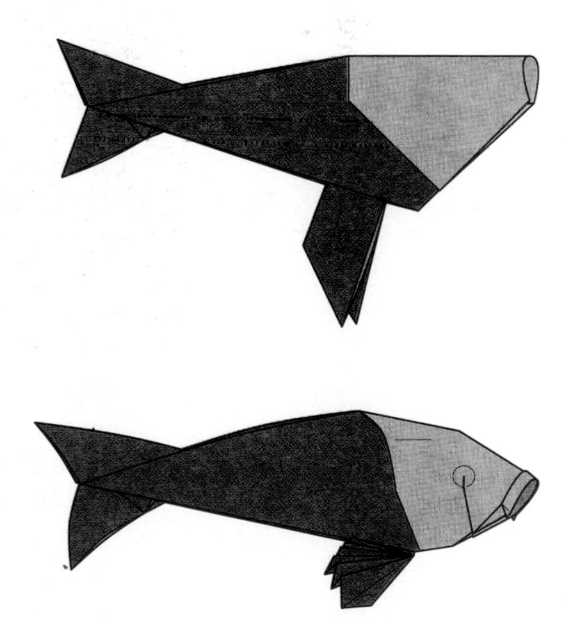

Figure 7.11.
Crease pattern, base, and folded model of the Koi.

So now, let's look at what type of pattern we'd put on this design. Fish have a distinctive pattern of overlapping scales that is very close to a pattern of overlapping half-circles, similar to the pattern shown in Figure 7.12.

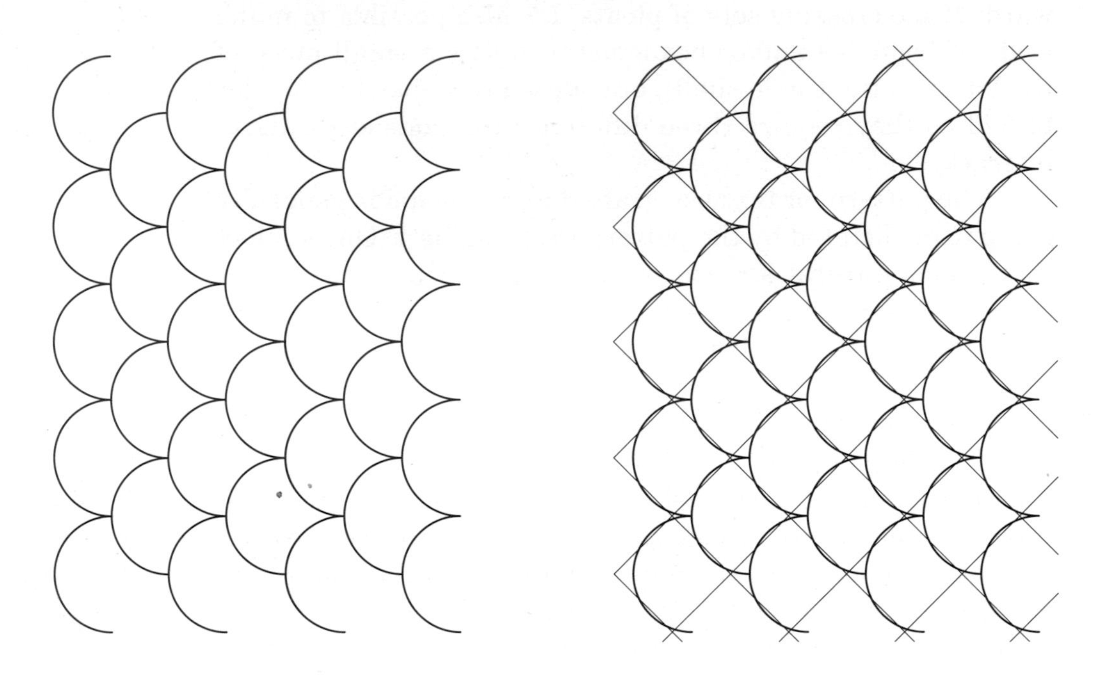

Figure 7.12.
Left: a single fish scale, represented as a half-circle.
Middle: an array of half-circles approximating fish scales.
Right: an array of crossed pleats at 45° approximating the lines of the array of scales.

If we overlay lines on top of the circles as shown in Figure 7.12, we can elucidate the underlying grid of the pattern; it is the same as the grid of crossed pleats rotated by 45°, which suggests that a grid of crossed pleats is a good place to start.

However, crossed pleats alone gives scales that are diamonds, not semicircles. A better approximation of circles can be had by blunting the tips of the squares, for example, with sink folds. But if you fold up an array of crossed pleats to work on, you will find that the tips of all squares are entangled with other layers of the pleats and need to be freed before they can be sunk. So a bit more folding is going to be necessary. In order to be efficient about it, let's make a single crossed pleat to work on, as shown in Figure 7.13.

The tip of the scale is marked A in the folding sequence in Figure 7.13. If we wish to blunt the tip, we must first free it from the entangled layers, which we can do by stretching the two pleats apart on either side of the tip as shown in Figure 7.14.

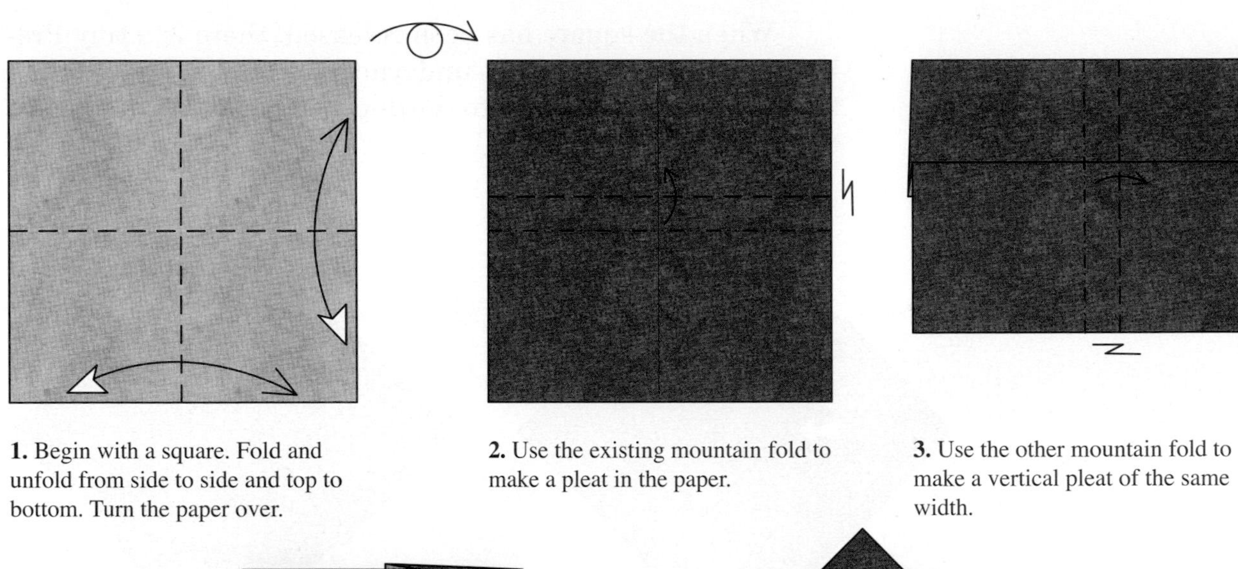

1. Begin with a square. Fold and unfold from side to side and top to bottom. Turn the paper over.

2. Use the existing mountain fold to make a pleat in the paper.

3. Use the other mountain fold to make a vertical pleat of the same width.

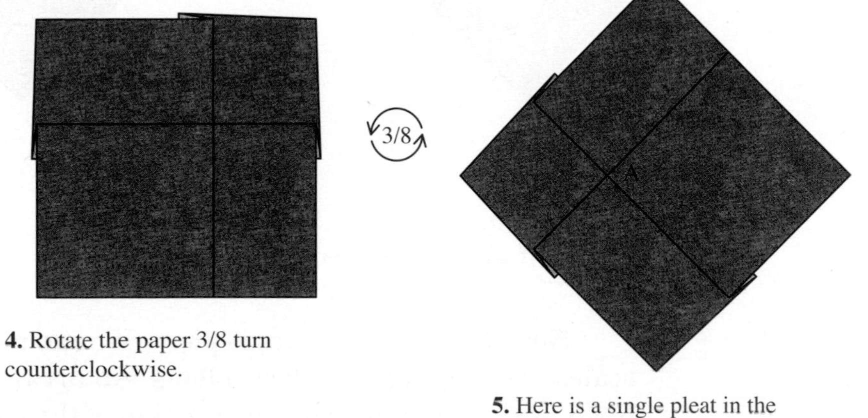

4. Rotate the paper 3/8 turn counterclockwise.

5. Here is a single pleat in the orientation of the fish scale.

Figure 7.13.
Folding sequence for a single pleat.

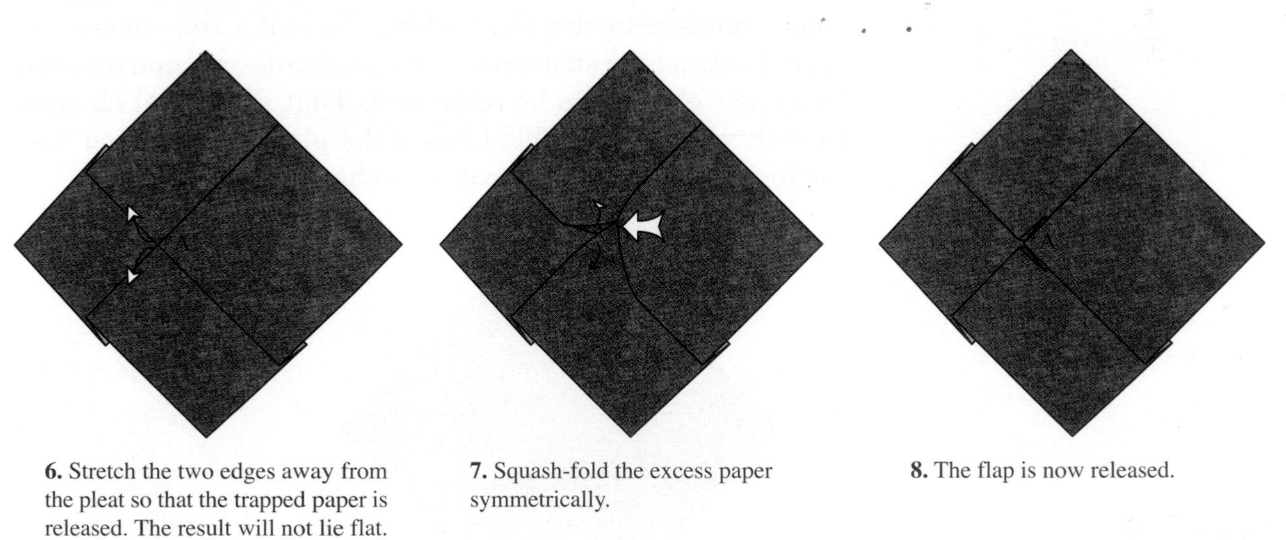

6. Stretch the two edges away from the pleat so that the trapped paper is released. The result will not lie flat.

7. Squash-fold the excess paper symmetrically.

8. The flap is now released.

Figure 7.14.
Stretching and releasing the trapped corner.

When the square has been released, there is a tiny Preliminary Fold in the layers underneath. We then can sink the tip of the square, but only to the depth allowed by the edges of the Preliminary Fold.

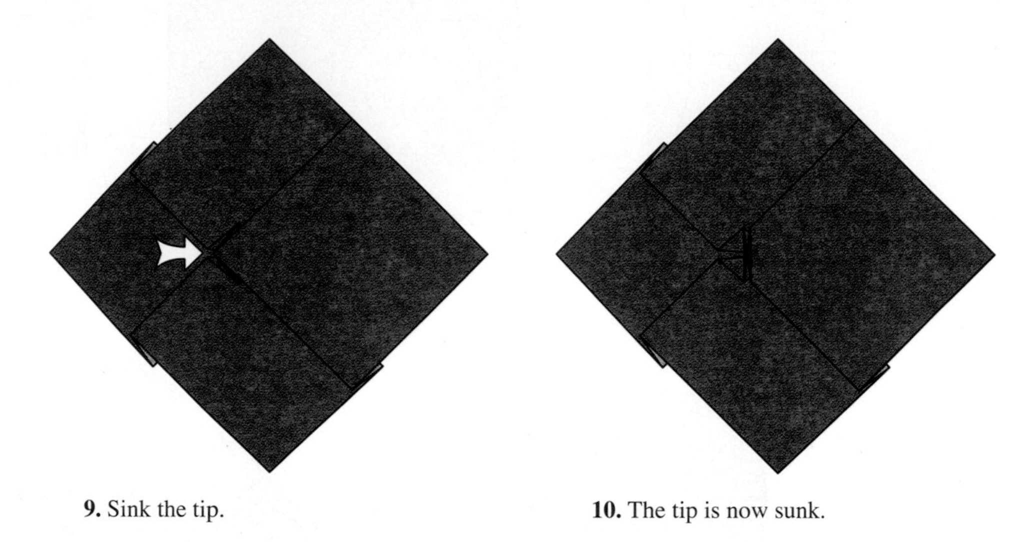

9. Sink the tip.

10. The tip is now sunk.

Figure 7.15.
Now that the tip has been freed, it can be sunk.

Now this was just a single pleat. We can make an array of scales from an array of these pleats. An array of pleats is defined by three quantities: the direction of the pleats, the width of each pleat, and the spacing from one pleat to the next. We have chosen the direction to be 45°. For a given pleat width, there is one degree of freedom left to choose: the spacing of the pleats relative to the pleat width. To make this choice, we should extract the structure of the pleat crossing, and use that as a basic element to be replicated. That structural element consists of the visible fold lines of the pleats (and, if you like, the hidden edges of the pleats), as shown in Figure 7.16.

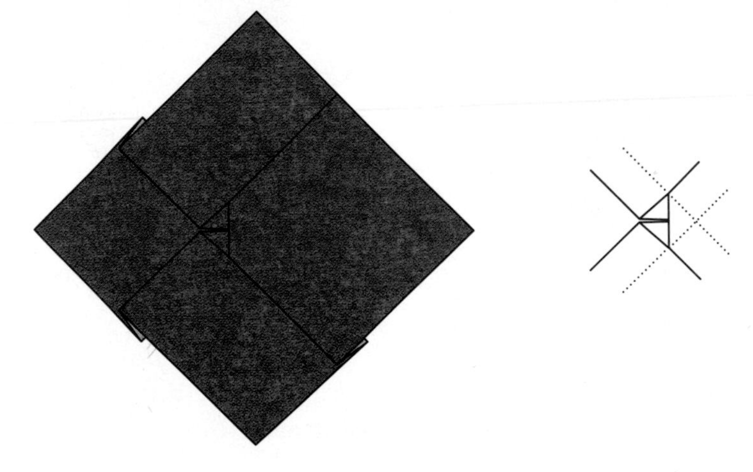

Figure 7.16.
Left: a single pleat.
Right: the structural elements of the pleat crossing.

A single pleat crossing can be thought of as an individual tile. To develop an approximation of the pattern of semicircles, we should array tiles containing the lines of the pleat crossings in such a way that they create a similar pattern. Figure 7.17 shows such an array, filling in over the semicircular array.

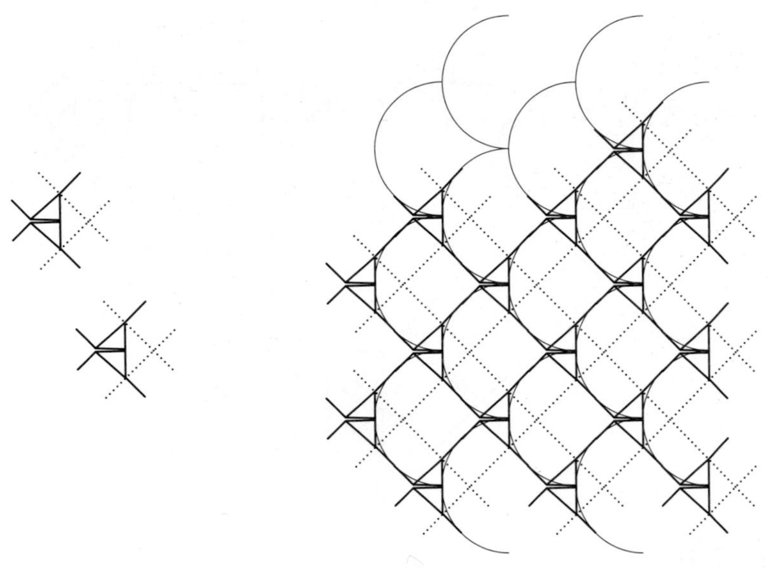

Figure 7.17.
Left: two tiles of crease pattern.
Right: the tiles arrayed over the pattern of semicircles.

One can think of this operation as cutting out small tiles of pleats, then taping them together edge-to-edge to realize the larger array. We can do this to both the folded and unfolded form of the paper. The folded form gives the folded array; the unfolded form gives the crease pattern necessary to realize the array.

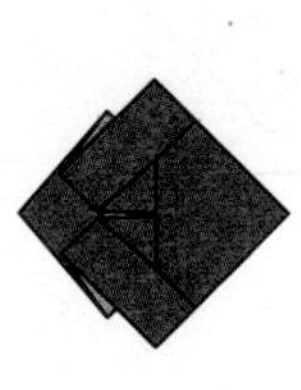

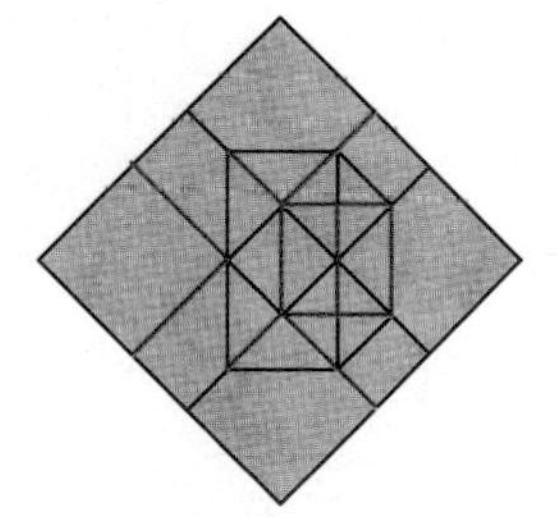

Figure 7.18.
Left: the folded tile.
Right: the crease pattern.

A concern with all pattern grafts is efficiency: How much paper is consumed by the pattern? This concern can be quantified as the ratio between the area of the pleated structure and the original paper. This ratio can be calculated from the entire area or, equivalently, from a single tile.

This comparison shows that the unfolded tile is about 60% larger in linear size (hence about 2.5 times the area) of the folded tile. That means that on average, there are two to three layers of paper everywhere in the pattern—quite a bit of thick-

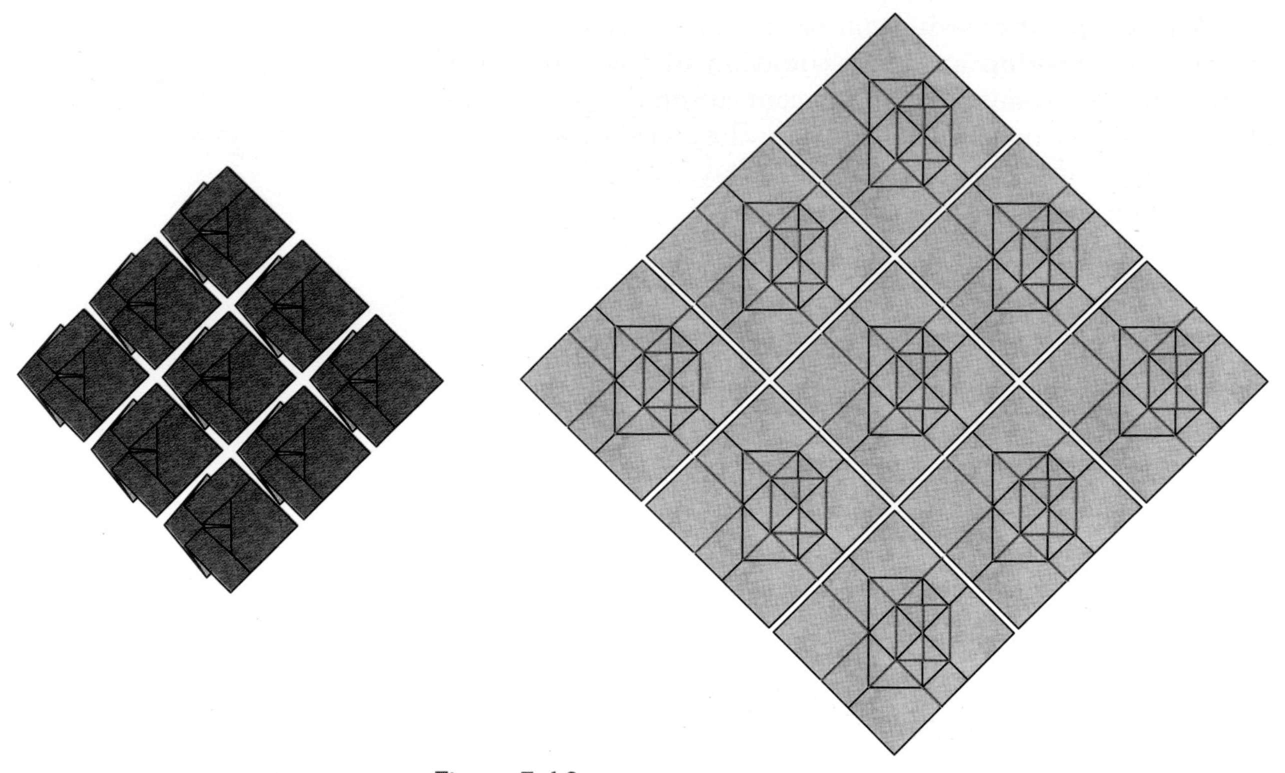

Figure 7.19.
Left: an array of folded tiles forming the scale pattern.
Right: the same array of crease patterns.

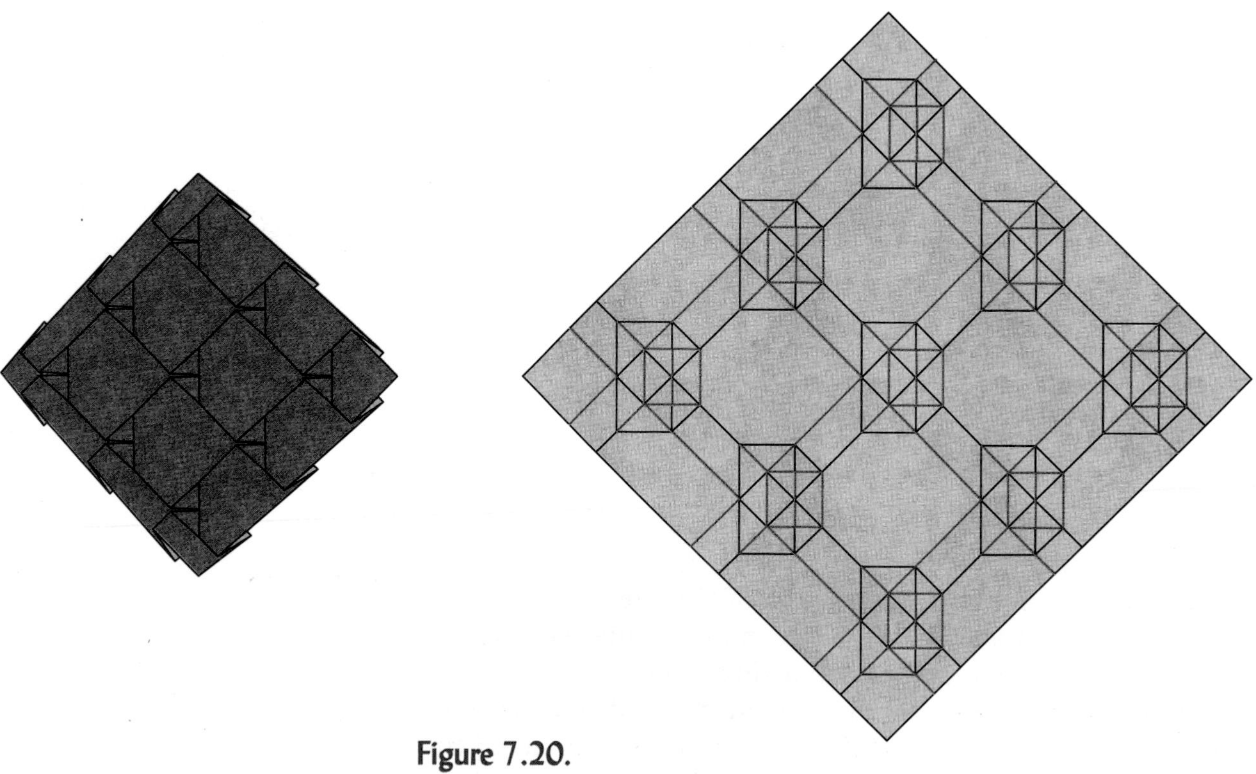

Figure 7.20.
Left: the folded array of scales.
Right: the array of creases.

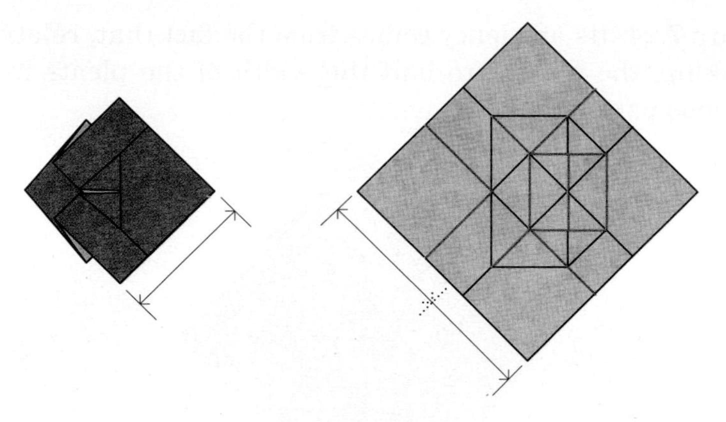

Figure 7.21.
Comparison of the tile sizes for the folded and unfolded tiles.

ness for folding. But that's the average; individual regions of the pattern can be considerably thicker, as shown in Figure 7.22, which lists the number of layers in each region of the basic tile.

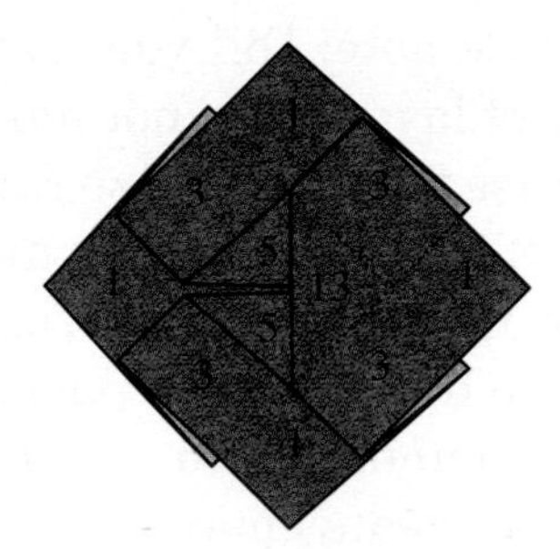

Figure 7.22.
The number of layers of paper in each region of the tile.

This shows that there are as many as 13 layers in the pattern, which means that any subsequent folding that goes on will require folding through quite thick layers.

But there's nothing particularly special about this pattern tile. There is much variety possible in creating sinks and rearrangements of layers around two crossed pleats. A bit of experimentation reveals a somewhat more efficient tile, shown in Figure 7.23 in both folded form and crease pattern.

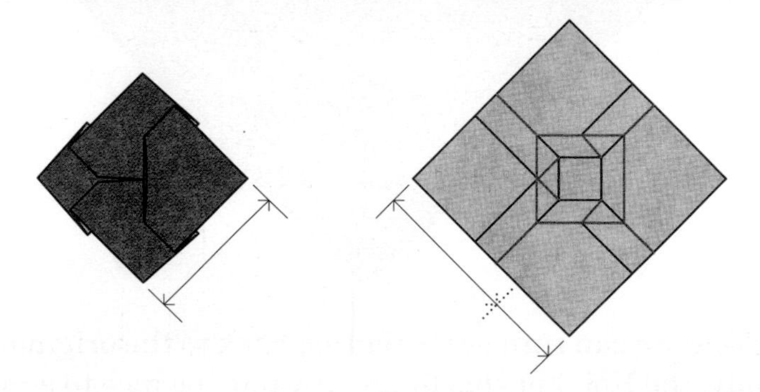

Figure 7.23.
Another tile unit for fish scales that is more efficient.

In this second tile, the crease pattern is 38% larger (1.9 times in area), so is more efficient in its use of paper. In fact, the maximum number of layers in any region (seven) is roughly half of the maximum for the sunken-tip pattern, as shown in

Figure 7.24. Its efficiency comes from the fact that, relatively speaking, the pleats are half the width of the pleats in the previous pattern.

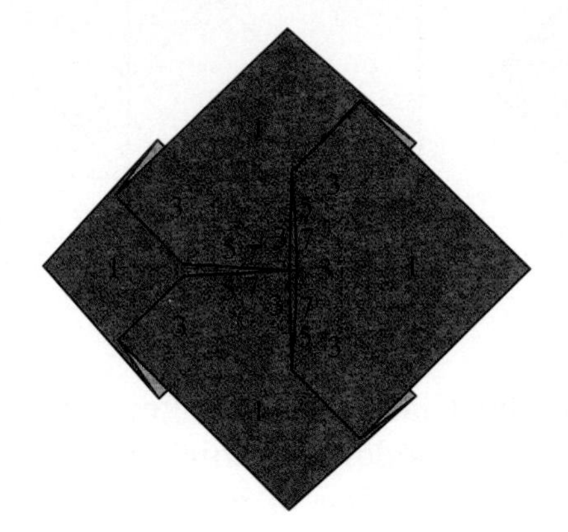

Figure 7.24.
Number of layers in each region of the efficient tile.

An interesting side note: Did you notice that all regions have an odd number of layers? It's not hard to show that this must always be the case for a tile whose raw edges are aligned along the boundary of the tile. There is an enormous body of work concerning the mathematics of origami pleat tilings—far too much to go into here. For our purposes, it is sufficient that the tiles can be combined into arbitrarily large areas of patterned regions with pleats emanating from their edges.

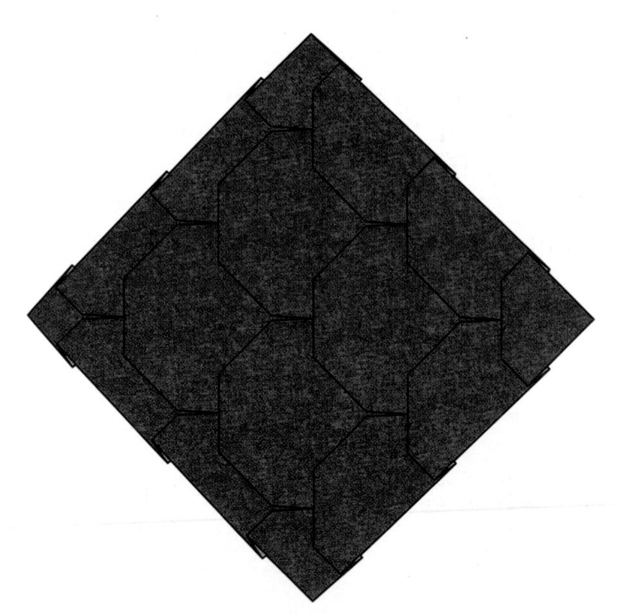

Figure 7.25.
A patch of the new scale tiling.

Now, we can turn our attention back to the original object of study: the Koi. For this figure, it would be nice to add scales to the body, but not the head, tail, or pectoral fins. So the first thing to do is to identify which parts of the paper will be exposed in the folded model. We should divide the square up into three categories: (a) those parts of the paper that become the body (these should have the pattern exposed); (b) those parts of the paper that become the head and fins (these should not have the pattern exposed); and (c) those parts of the paper that are hidden by other layers (these may or may not get the pattern, depending on how we are constrained by the pattern we choose). Obviously, it's fairly wasteful to put a lot of effort (and folding) into creating a pattern that will never see the light of exterior view, but since patterns may not be created in isolation but are part of a connected whole, it may be necessary to extend the pattern into subsequently hidden regions in order to form the entire structure. Figure 7.26 shows these regions. The body is colored. We would not like the pattern to extend onto the fins or the head, so those are colored gray; then the lightest regions are those we don't care about. Note that any region covered by another (the way the front of the body is covered by the head) falls into the "don't care" category.

Figure 7.26.
Left: color-coded crease pattern.
Right: corresponding color-coded regions of the Koi.

The task now is to fill in the colored regions with a scale pattern while avoiding the gray regions. This is not as easy as it sounds, because pleated scales don't exist in isolation: They are terminated by pleats on four sides. If we represent a pleat schematically by a single line, then an array of crossing pleats can be represented by two arrays of crossing parallel lines, as in Figure 7.27.

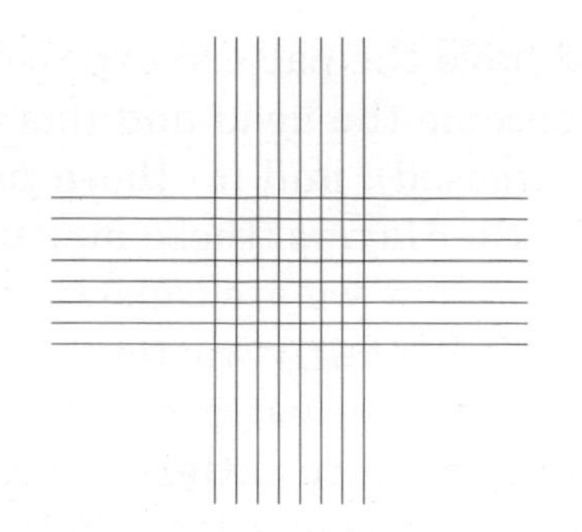

Figure 7.27.
Two sets of crossing pleats. Each pleat is represented by a single line.

We can form scales only where the pleats actually cross; conversely, anywhere the pleats cross, we will have scale patterns (or at least the busyness of crossing pleats), whether intentional or inadvertent.

So, we can overlay arrays of pleats represented by lines on the crease pattern and see what's possible.

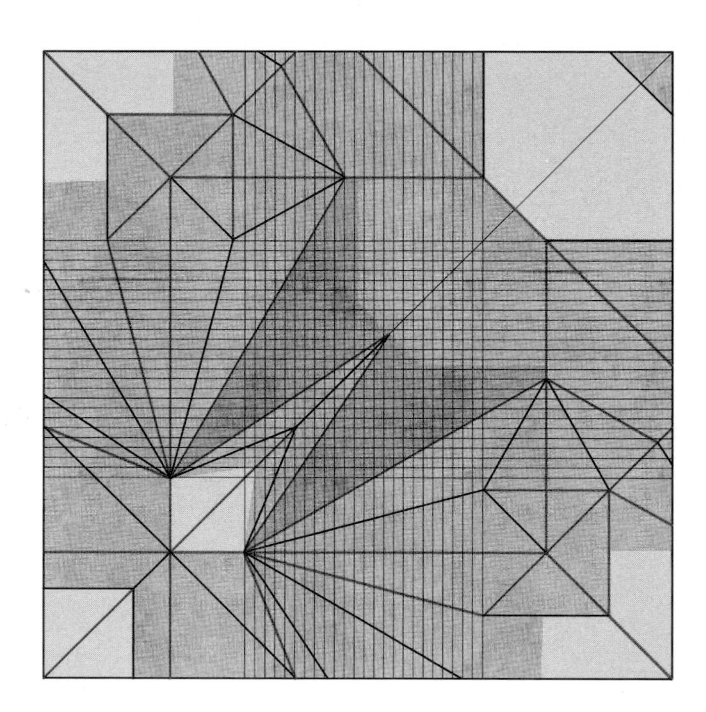

Figure 7.28.
The crease pattern with two sets of pleats arrayed across the middle.

This is close, but not ideal. There are a few regions of the body that don't get both sets of pleats: near the tail fin and near the pectoral fins. So those regions will not have enough scales. However, if we added more pleats to fill in those areas with scales, the pleats will start encroaching on the head and tail, respectively. We can reduce the uncovered region just a bit if we allow some pleats on the tail and alter the proportions of the head, to something like Figure 7.29.

In Figure 7.29, the head has been slightly reduced in size and the pleats have been allowed to creep into the edges of the

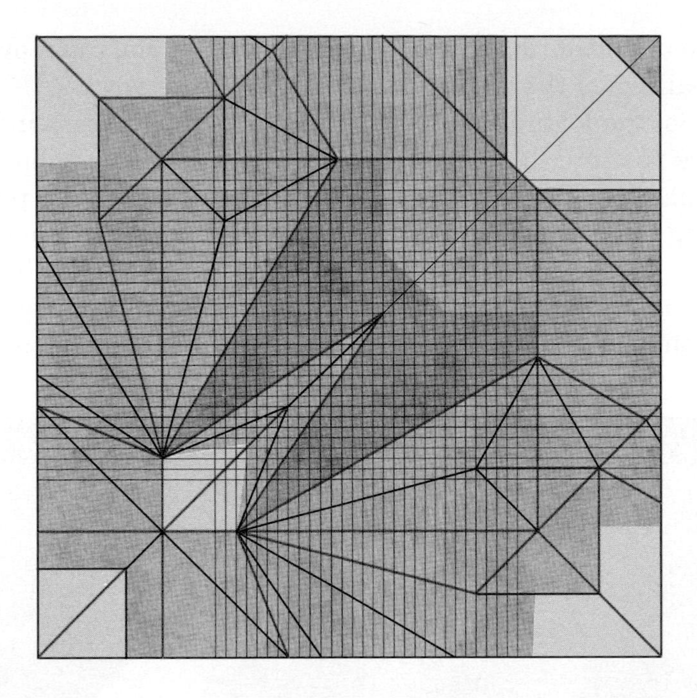

Figure 7.29.
Modified crease and pleat pattern.

head and the tail. This seems like an acceptable tradeoff to get the body nearly fully covered.

I'll leave the construction of the full folding pattern as an exercise to close this section. Here are a few guidelines. First, as we did with the Turtle, replace each of the pleat lines with a strip of paper for the pleat. How wide?

Look back at the individual scale tile. The width of the tile must be equal to the distance between pleats; the width of the inserted strip is equal to the width of the pale colored strip

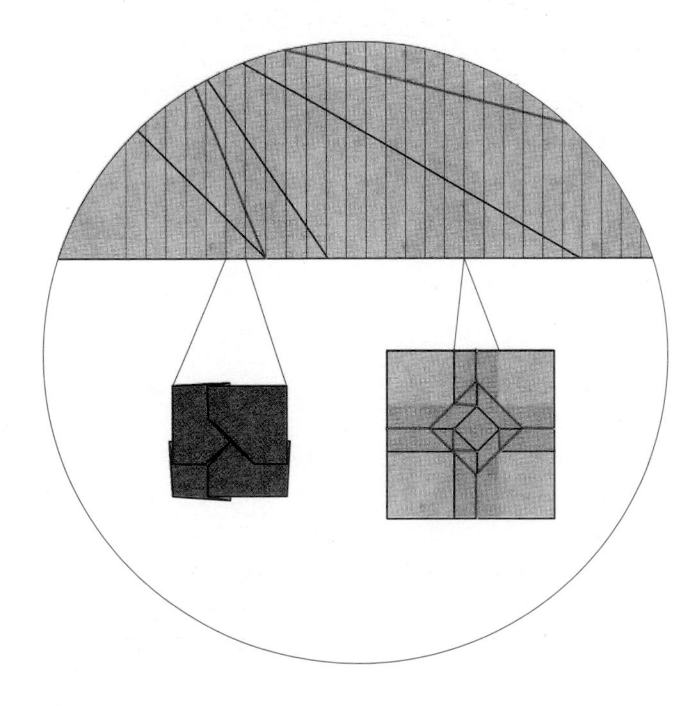

Figure 7.30.
Determination of the width of the inserted pleats and creases.

in the unfolded tile. At every pleat crossing, you must overlay the pattern of the creases that lie within the strip.

So, the folding sequence is: (a) insert the pleats into the crease pattern; (b) form all of the pleats (and the scales from the small structure in between); (c) continue with the regular folding sequence of the Koi. If you work your way through folding the entire model, you can congratulate yourself both on your understanding of the design process and, because there are some 900 individual scales to be shaped, your fortitude.

Figure 7.31.
The completed Koi with scales.

Folding Instructions

Turtle

Western Pond Turtle

Koi

Turtle

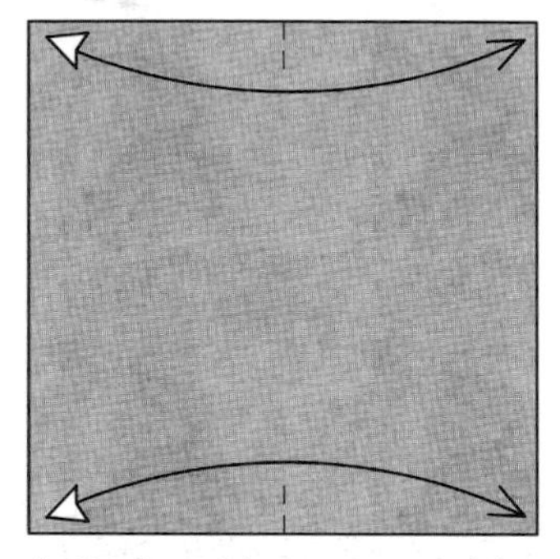

1. Begin with a square, white side up. Fold the paper in half, making a pinch along the top and bottom edges. Unfold.

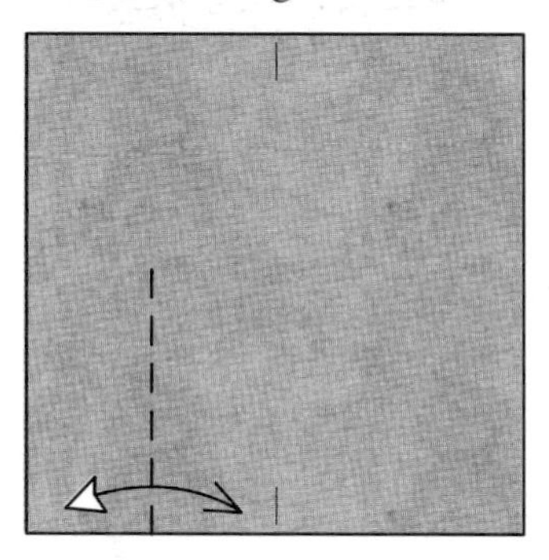

2. Fold the left side in to the center line, making a crease that extends about halfway up.

3. Fold the lower right corner over to touch the crease you just made.

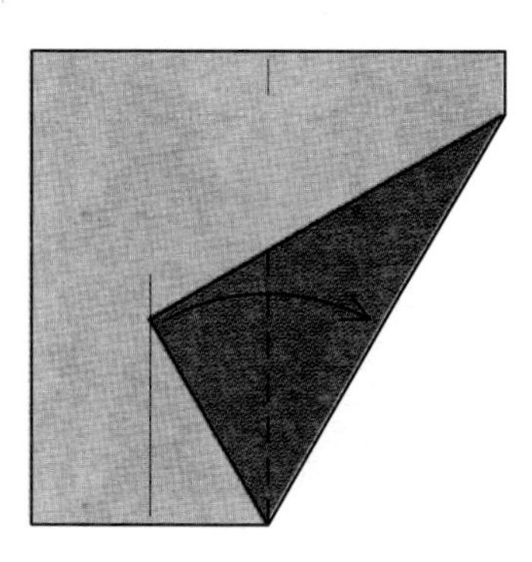

4. Fold the edge of the flap over to lie along the right edge.

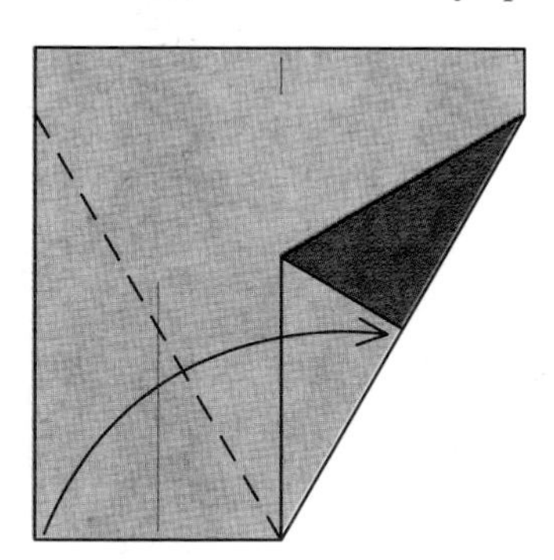

5. Fold the bottom left corner over to line up with the right corner.

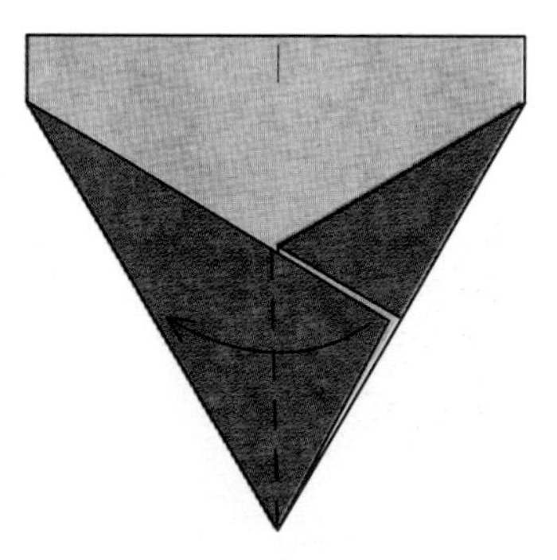

6. Fold the corner back to the left.

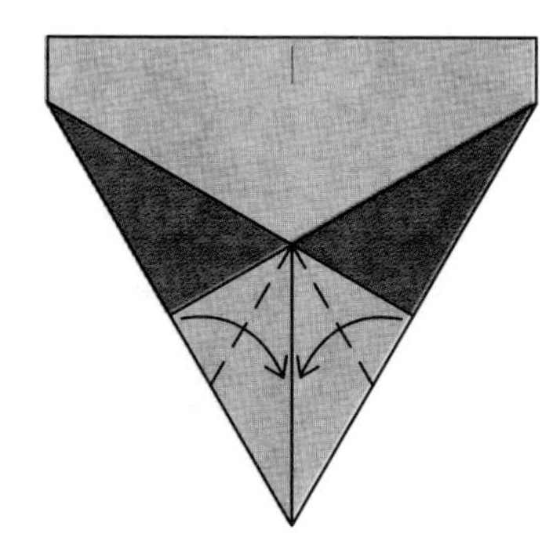

7. Fold the two corners down to lie along the center line.

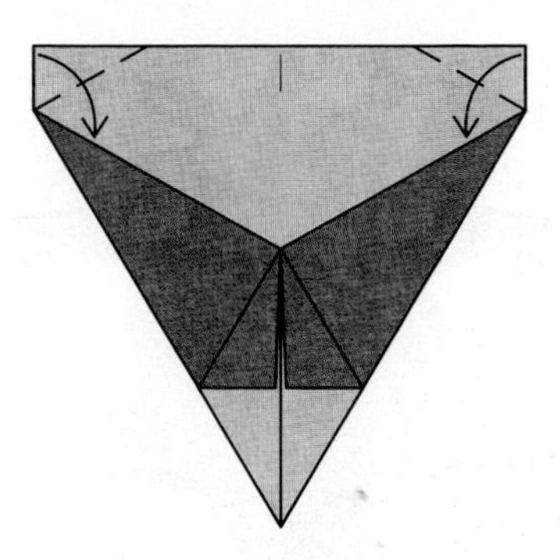

8. Fold the two top corners down to lie along the raw edges.

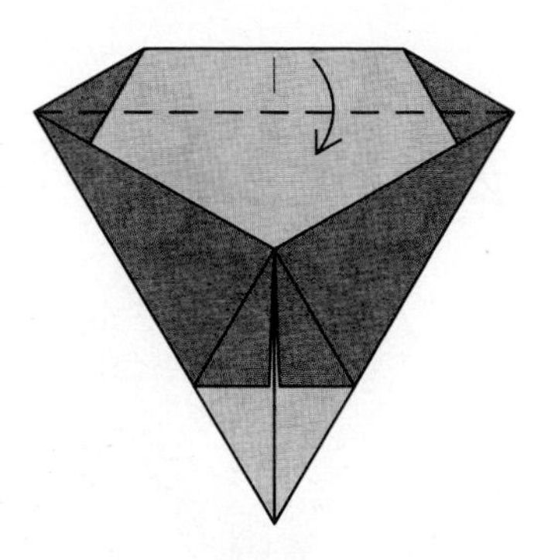

9. Fold the top edge down along a crease that runs from corner to corner.

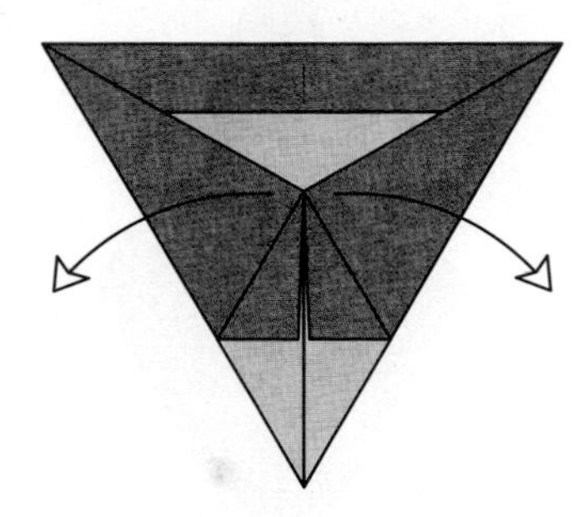

10. Unfold the two flaps.

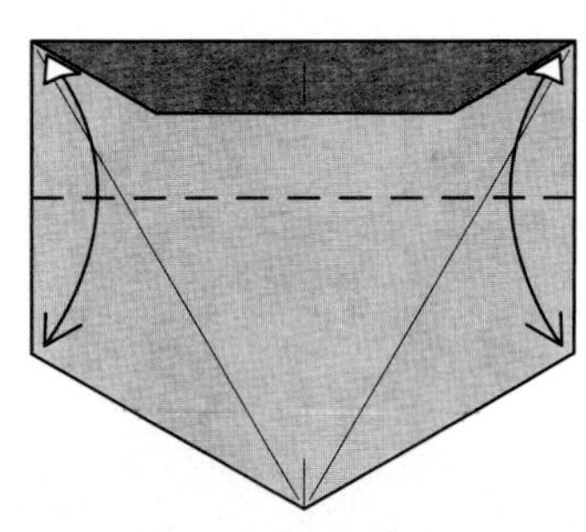

11. Fold the top corners down to the side corners; crease all the way across and unfold.

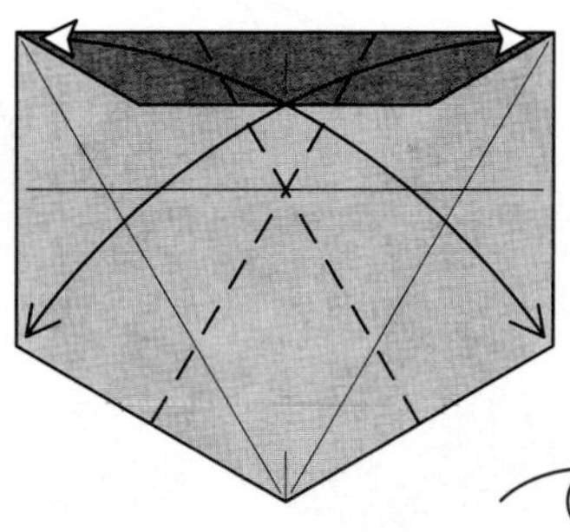

12. Fold the top left corner down to the right side corner, crease, and unfold. Repeat with the top right corner. Turn the paper over.

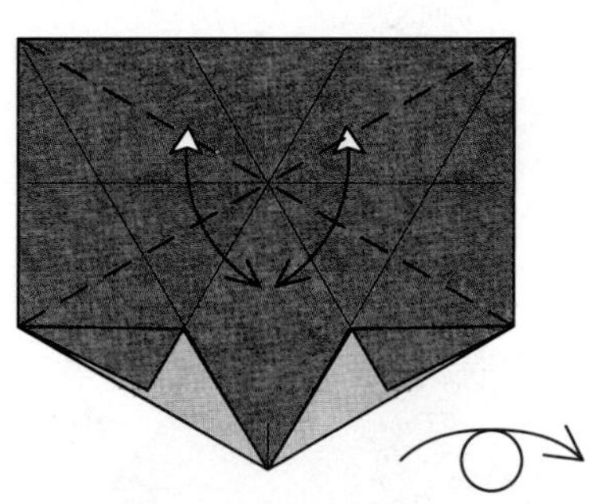

13. Crease from corner to corner in both directions. Turn the paper back over.

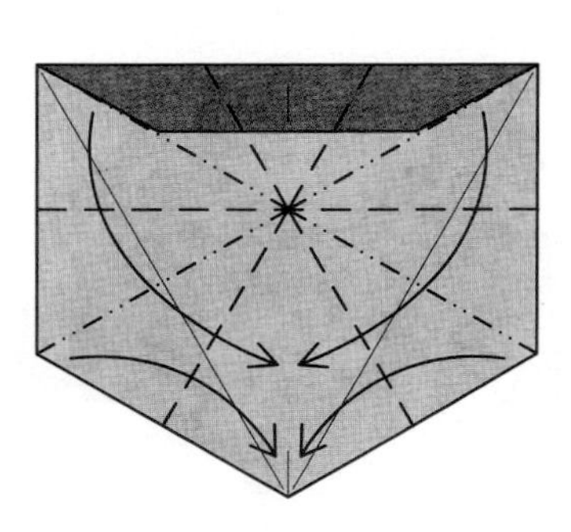

14. Collapse the model on the existing creases.

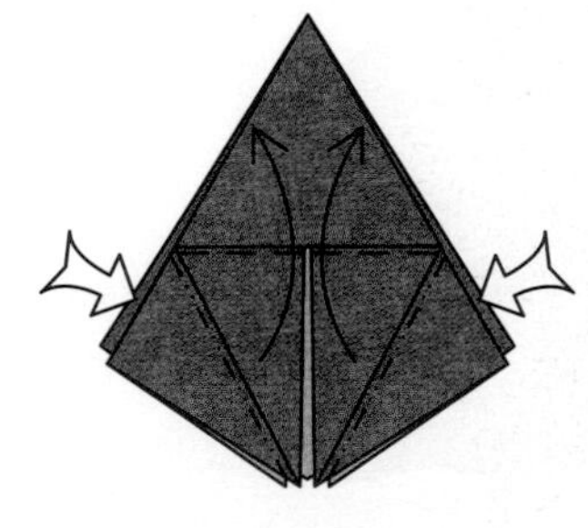

15. Squash-fold the two edges (like half of a petal fold).

16. Squash-fold the two remaining edges, including the two loose flaps.

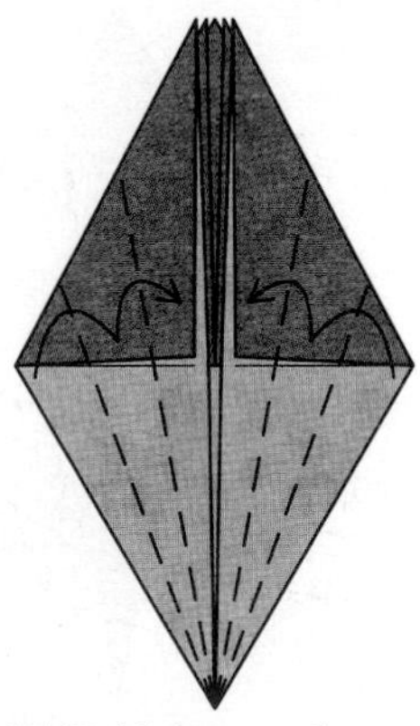

17. Fold the near layers over and over in thirds.

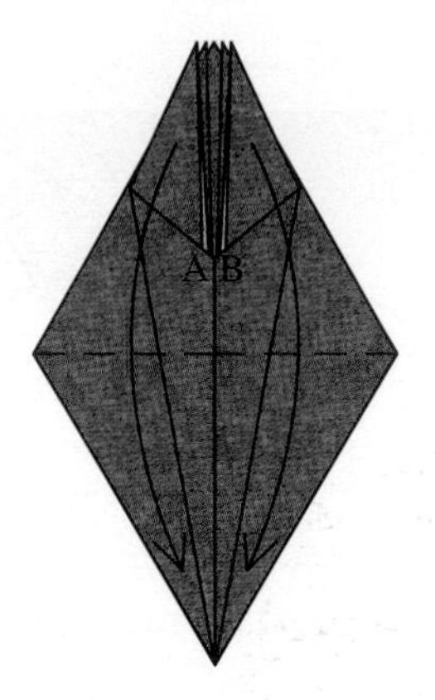

18. Fold two flaps downward. Note the corners marked A and B.

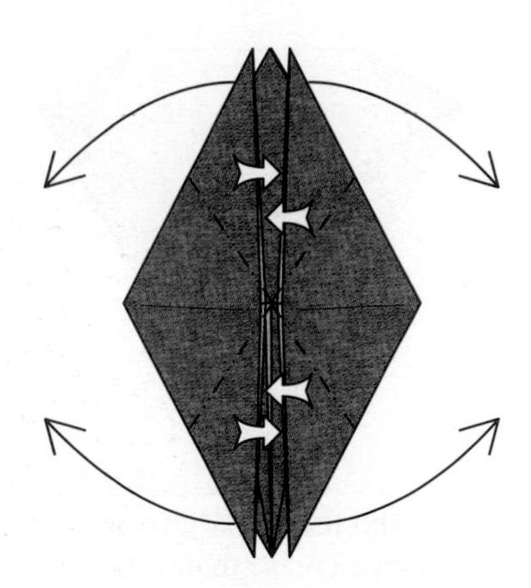

19. Reverse-fold four flaps out to the sides. Do not include corners A and B in the reverse folds.

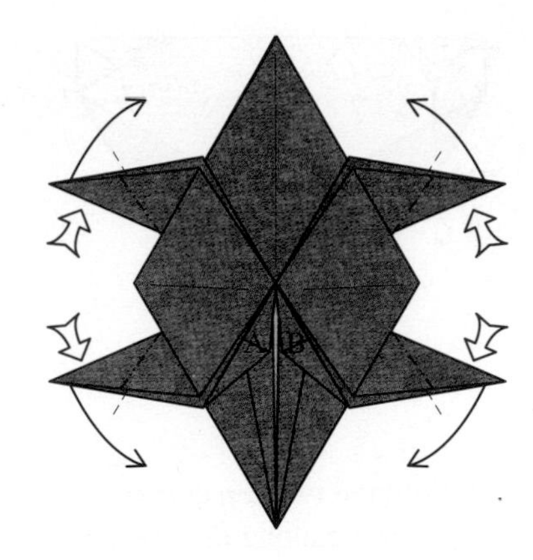

20. Observe that corners A and B remain flat. Reverse-fold the tips of the legs

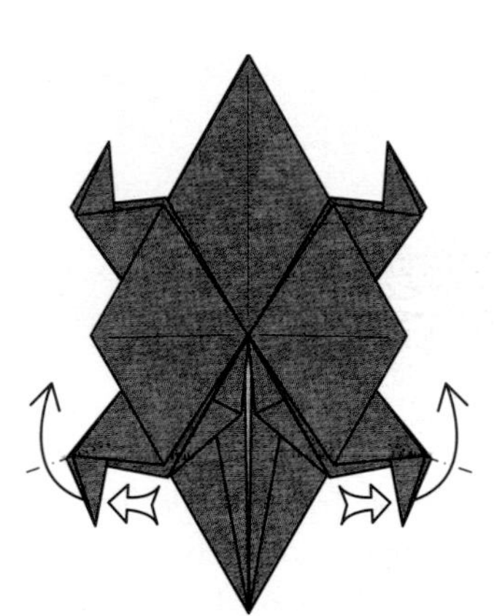

21. Reverse-fold the tips of the hind legs.

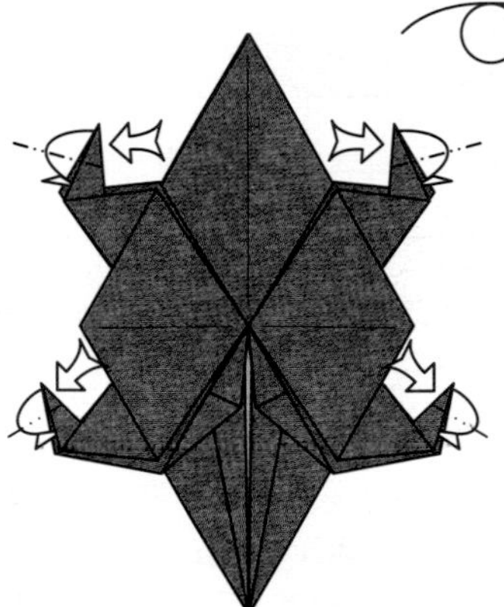

22. Reverse-fold the tips of all four legs. Turn the model over from side to side.

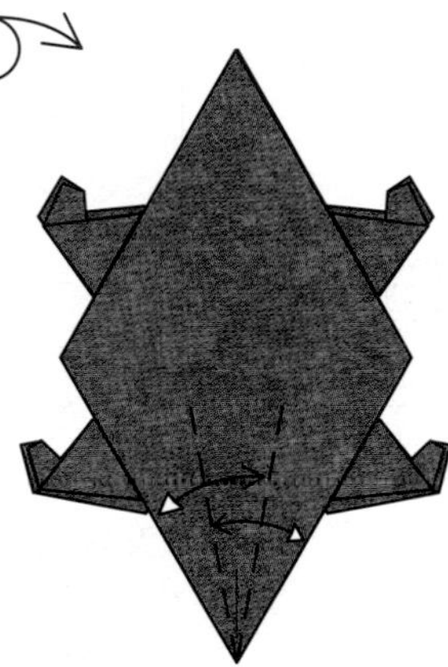

23. Divide the bottom point into thirds with creases that line up with folded edges behind.

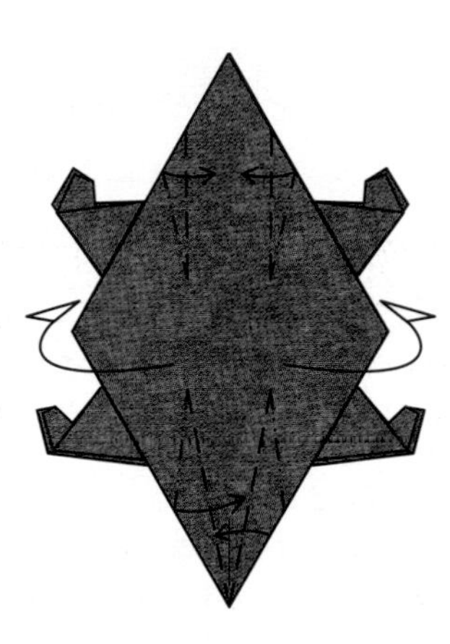

24. Pleat the top and bottom and curve the shell to make it rounded. The tail pleats are on existing creases; the head pleats have vertical valley folds.

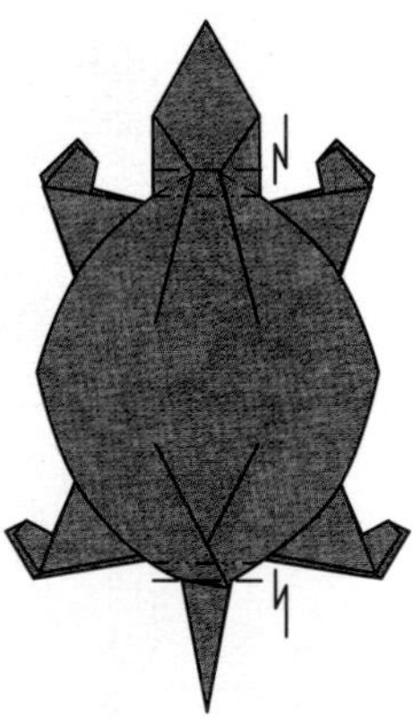

25. Pleat the head and tail; these pleats lock the pleats made in the previous step.

26. Puff out the head. Pinch the tail to make it three-dimensional. Round the shell and shape the legs.

27. Finished Turtle.

Western Pond Turtle

There are three ways to get the reference points and guidelines for this model, depending on your desire for folding purity and tolerance of leftover creases on the paper:

1. Folding only—follow all steps as instructed.
2. Marking and folding—follow steps 1–13; draw lines instead of making creases in steps 14–20; continue folding from step 21.
3. Measuring, marking, and folding—divide the top edge into 39ths and the bottom edge into 78ths; jump to step 14; draw lines instead of creases in steps 14–20; continue folding from step 21.

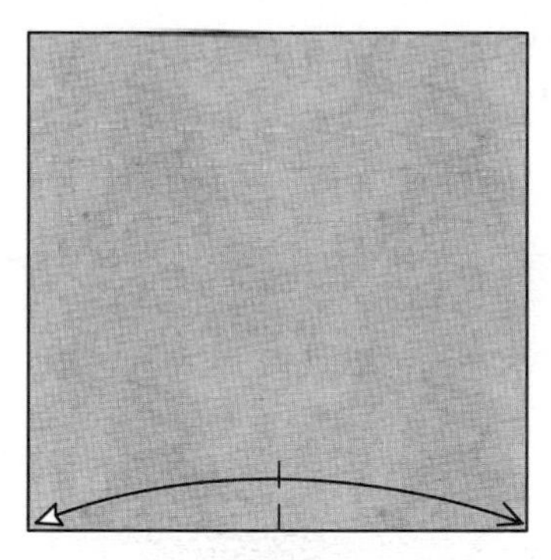

1. Begin with a square, white side up. Make a pinch along the bottom edge extending about 1/10 of the way up.

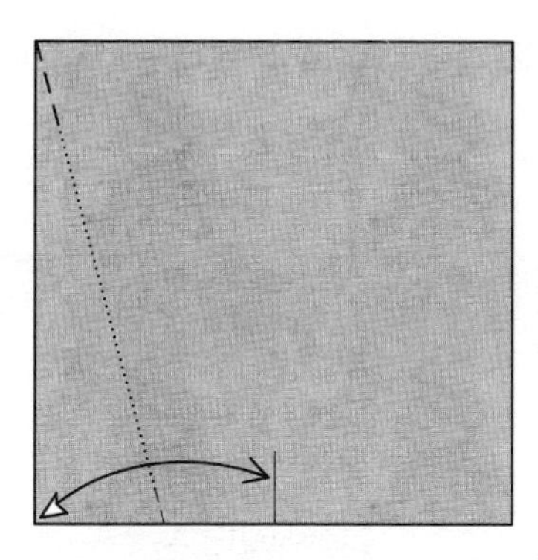

2. Fold the bottom corner over to the mark, pinch at top and bottom, and unfold.

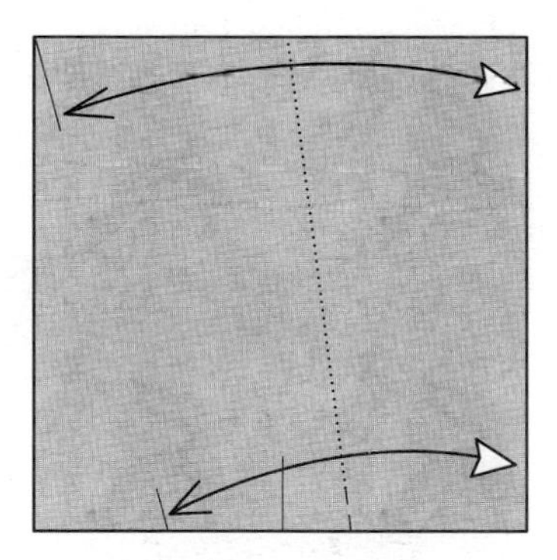

3. Fold the right edge over to the two pinches you just made; make a pinch along the bottom and unfold.

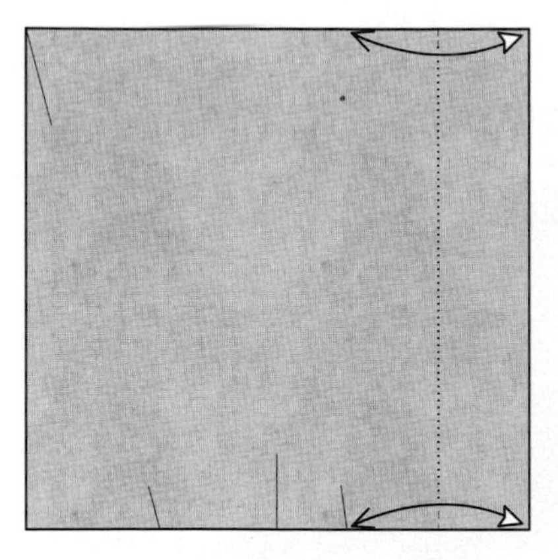

4. Fold the right edge over so that the corner hits the mark you just made; pinch at top and bottom and unfold.

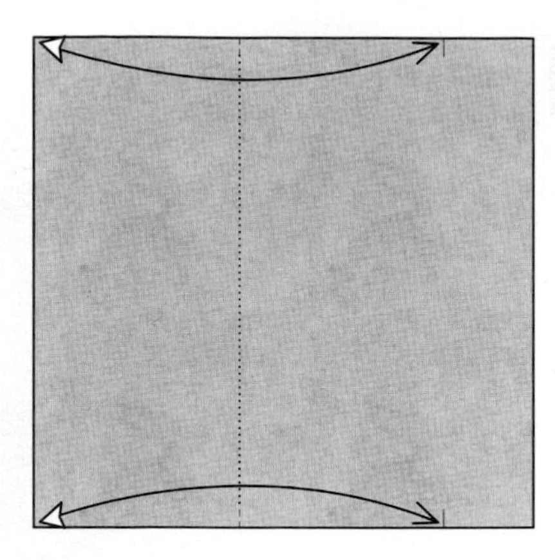

5. Fold the left two corners over to the mark you just made; pinch at top and bottom and unfold. This divides in half the distance between the corner and mark.

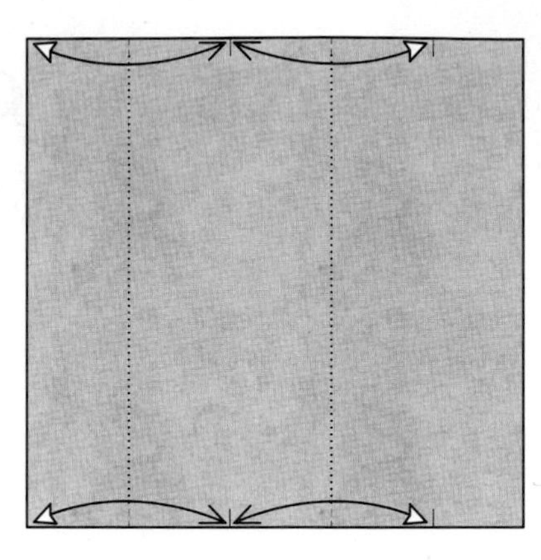

6. Make pinches between the pairs of marks at top and bottom, dividing into quarters.

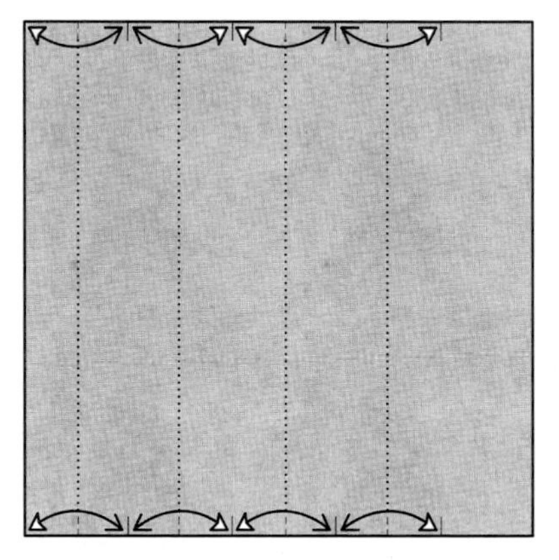

7. Make pinches between the pairs of marks at top and bottom, dividing into eighths.

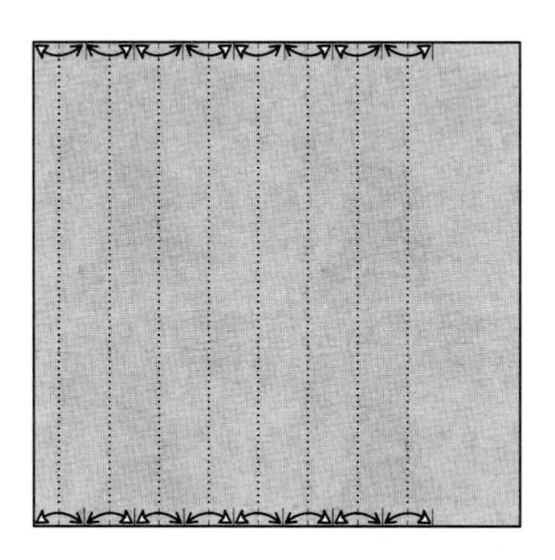

8. Divide again, getting sixteenths.

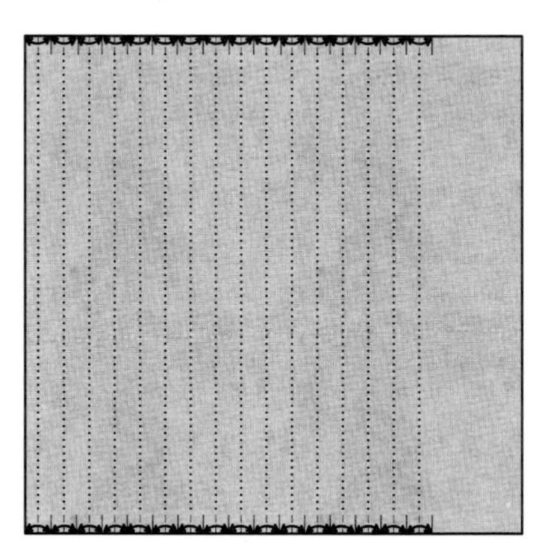

9. One more time, getting thirty-seconds.

10. Fold the right corners to the last-but-one mark, pinch, and unfold.

11. Divide the distance into quarters.

12. Divide each of the three gaps in half. When you are done, you will have divided the top and bottom edges into 39 equal divisions.

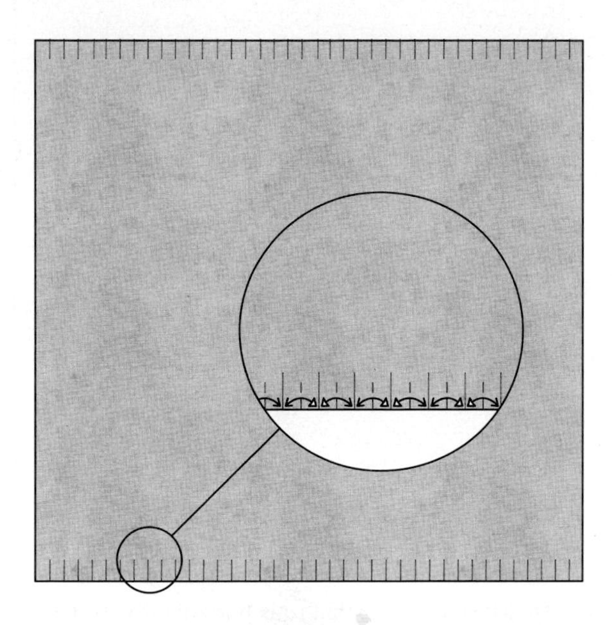

13. Divide each of the gaps along the bottom edge in half.

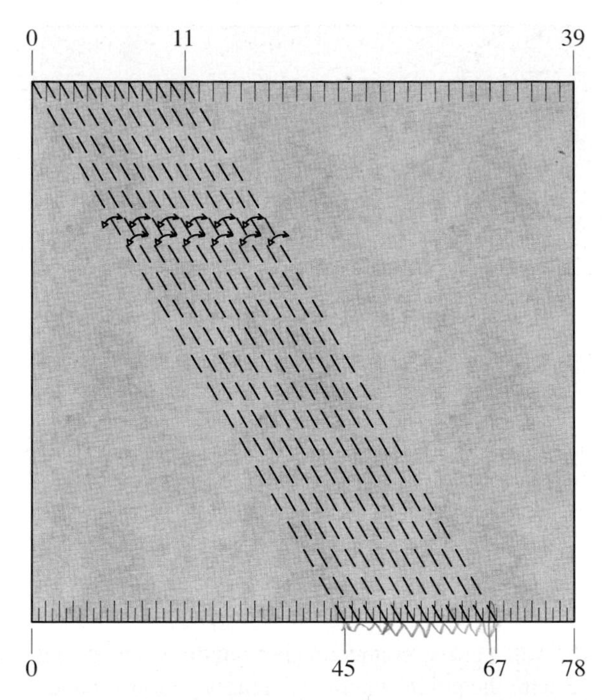

14. The top is divided into 39ths; the bottom is divided into 78ths. Number the divisions from the left. Make 12 creases that connect top-0 with bottom-45, top-1 with bottom-47, and so forth, up to top-11 with bottom-67.

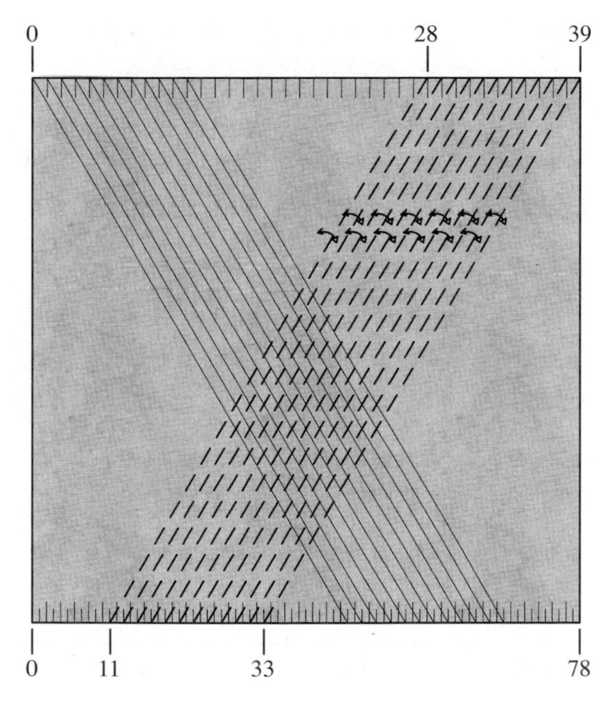

15. Do the same thing going the other direction.

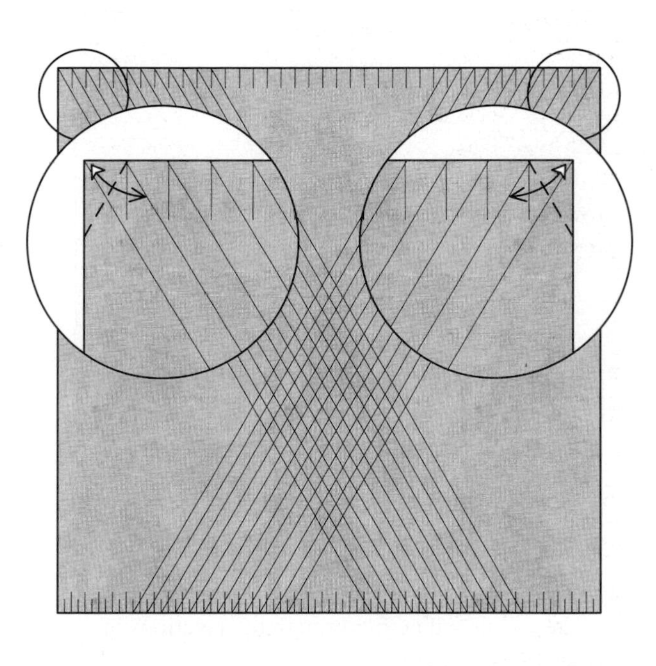

16. Fold each top corner down so that its edge lies along one of the creases and unfold.

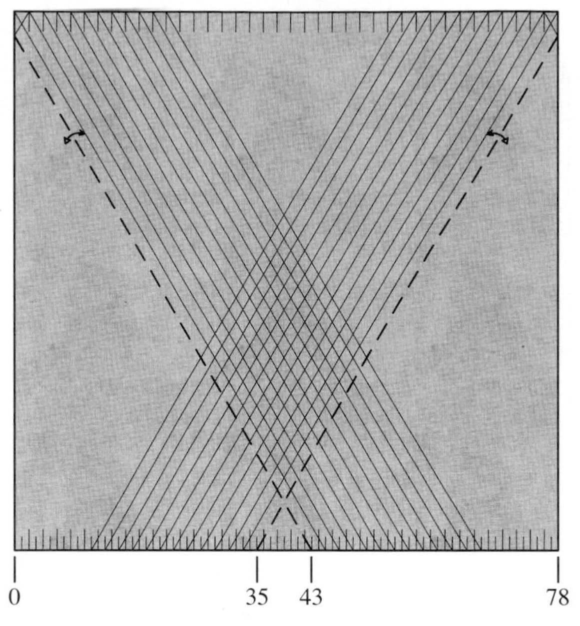

17. Add a crease connecting bottom divisions 35 and 43 with the points where the creases you just made hit the side edges.

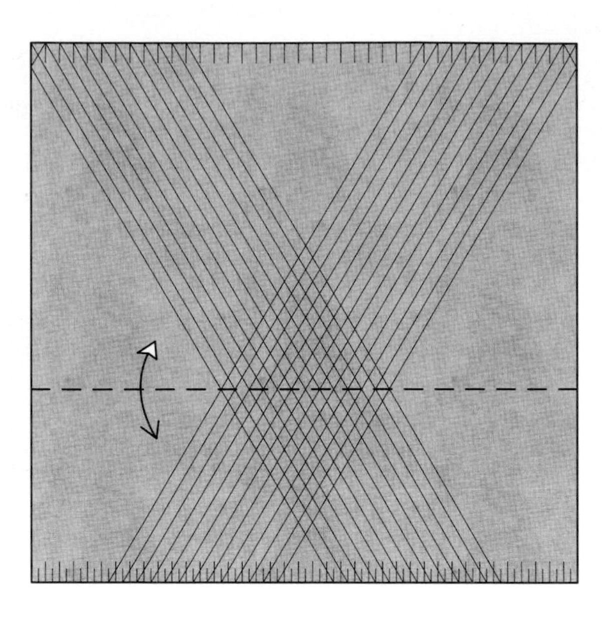

18. Make a crease that runs horizontally through the middle of the "X" formed by the creases.

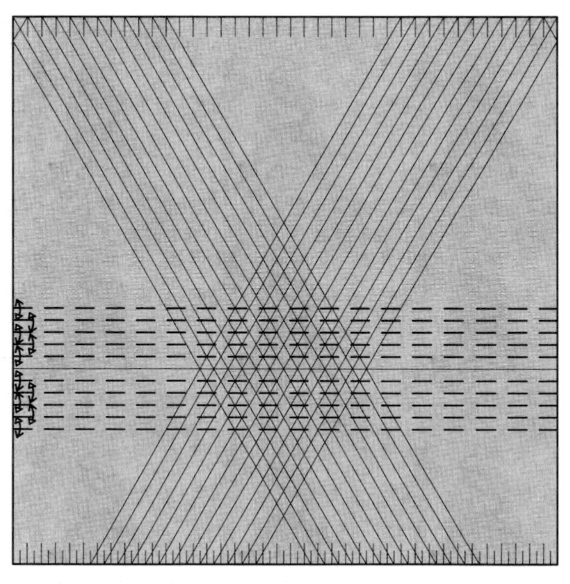

19. Add five creases above and five below the crease you just made, all going through intersections of the grid.

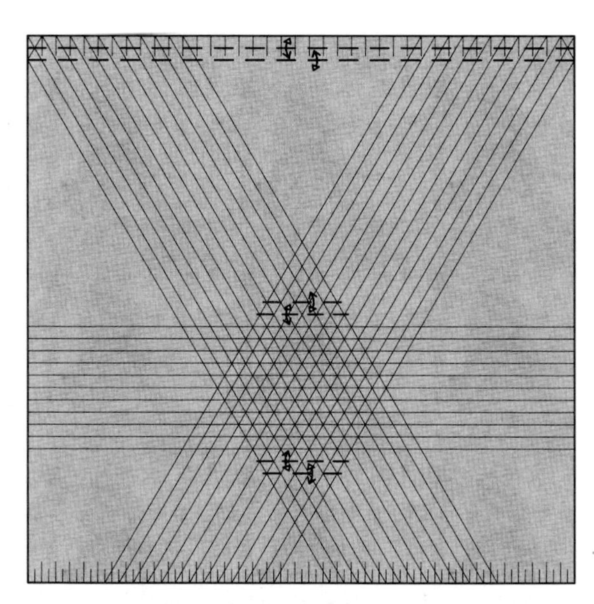

20. Add two more short creases above and below the hexagonal grid; also make two horizontal creases at the top of the model.

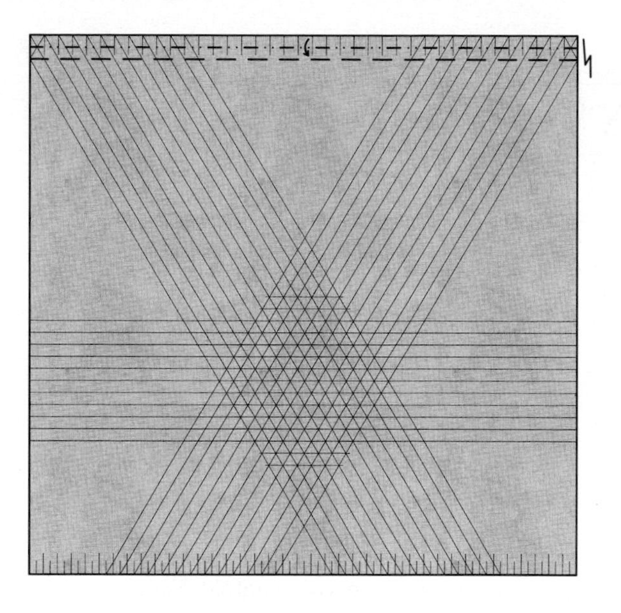

21. If you have been drawing the reference lines rather than folding them, now is the time to start folding. Pleat the top of the paper on the existing creases.

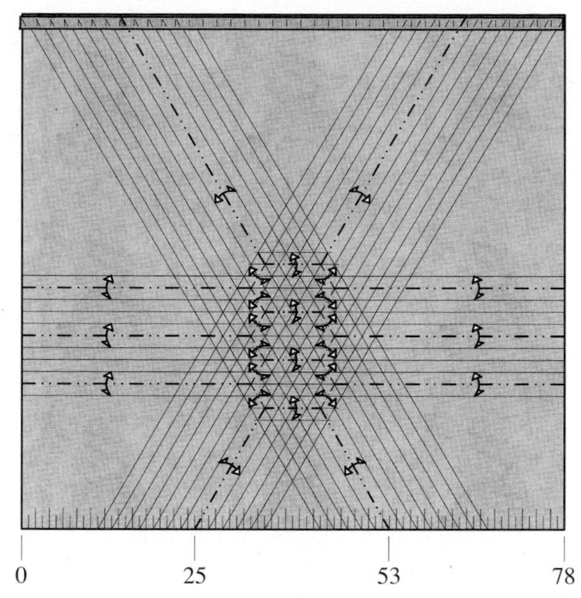

22. Form mountain folds where indicated.

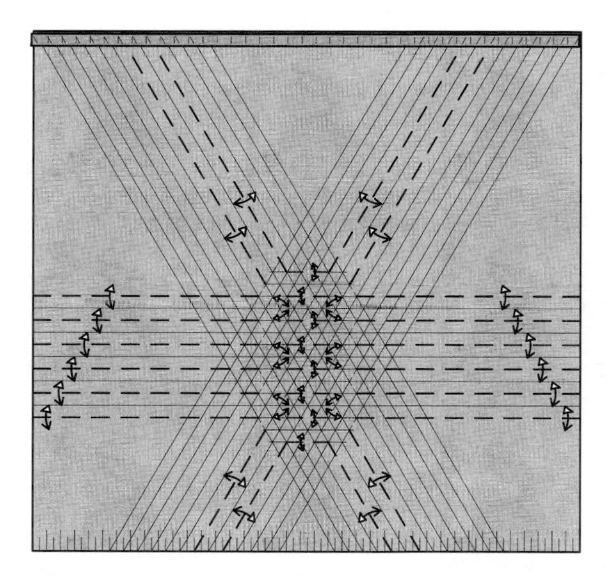

23. Form valley folds where indicated.

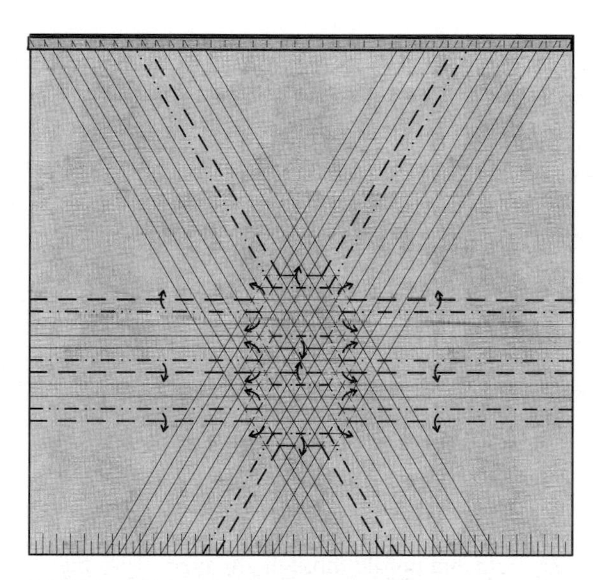

24. Make all the pleats together on the creases shown and flatten the paper completely.

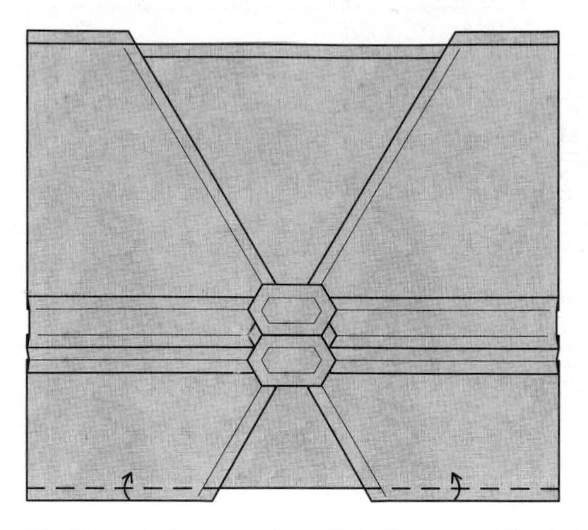

25. Fold the bottom edge of the left and right sides upward so that it runs straight across.

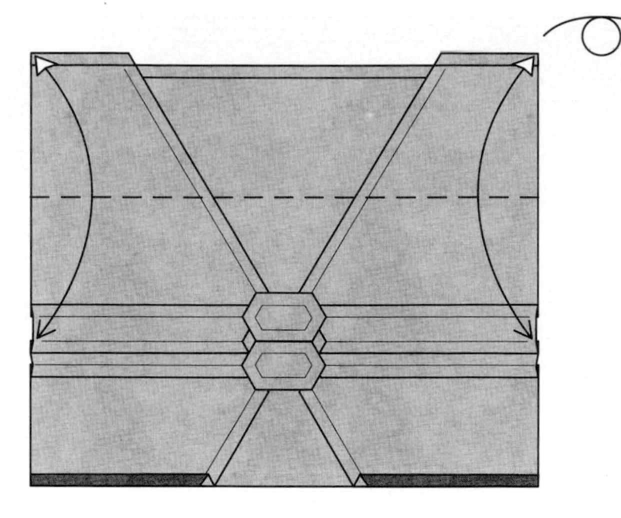

26. Fold the top edge down to the indicated points; crease all the way across and unfold. Turn the paper over.

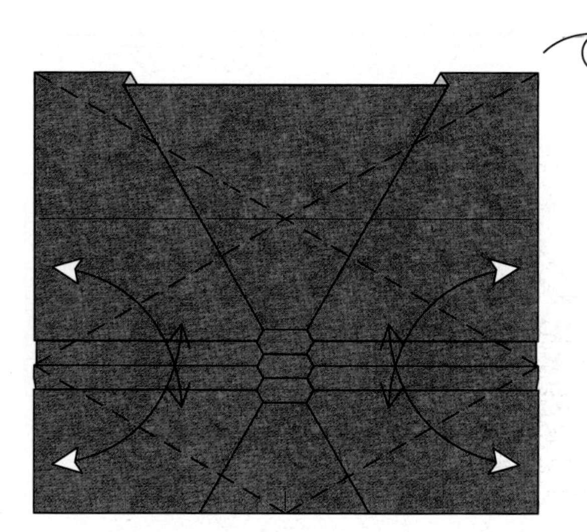

27. Fold and unfold through all layers. Turn the paper back over.

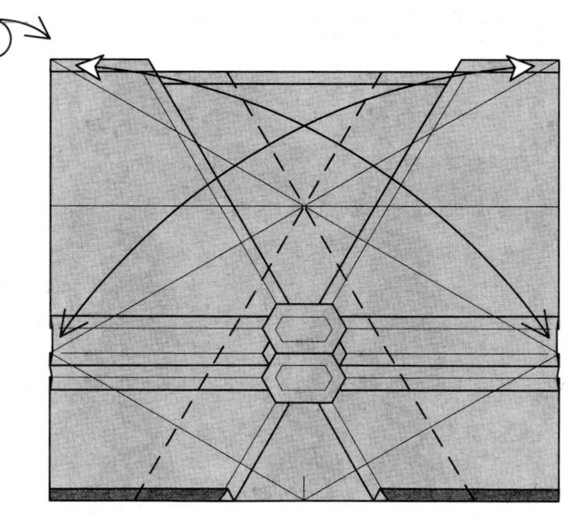

28. Fold and unfold.

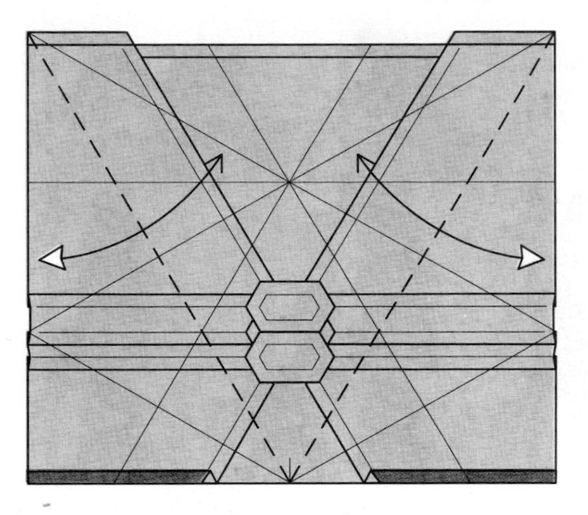

29. Fold and unfold.

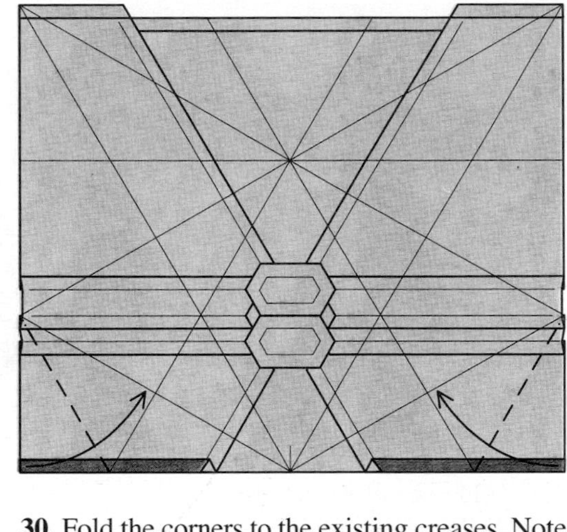

30. Fold the corners to the existing creases. Note that the creases go underneath the pleats at their upper ends.

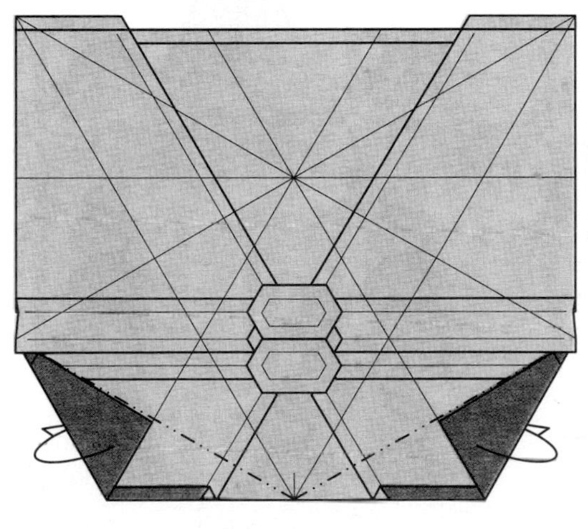

31. Mountain-fold the corners behind on the existing creases.

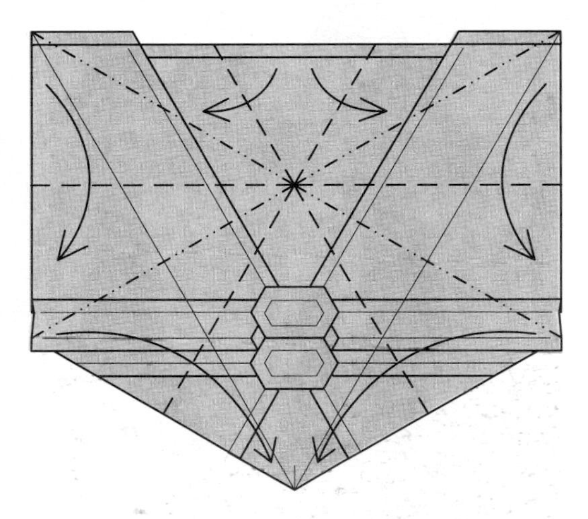

32. Bring all corners together at the bottom of the model.

33. Reverse-fold two edges in to the center line.

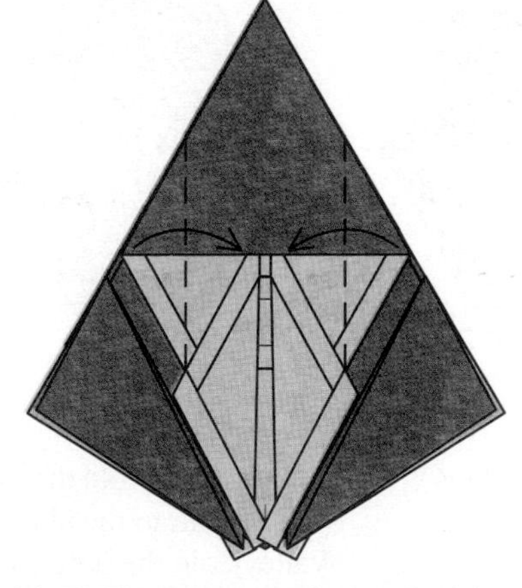

34. Fold two corners inward so that their edges align with existing folded edges.

35. Fold one layer to the right.

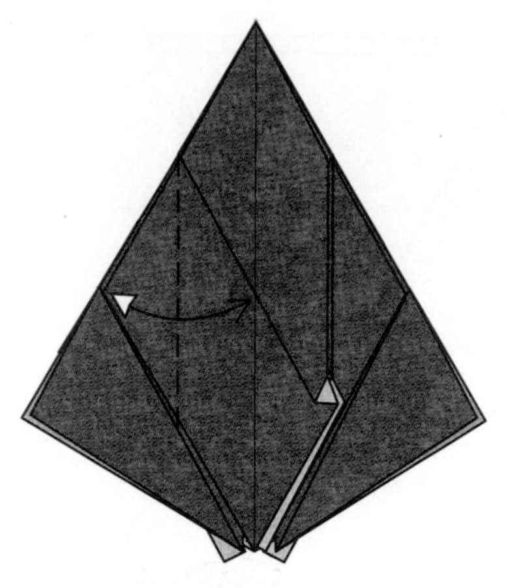

36. Fold one corner in to the center line and unfold.

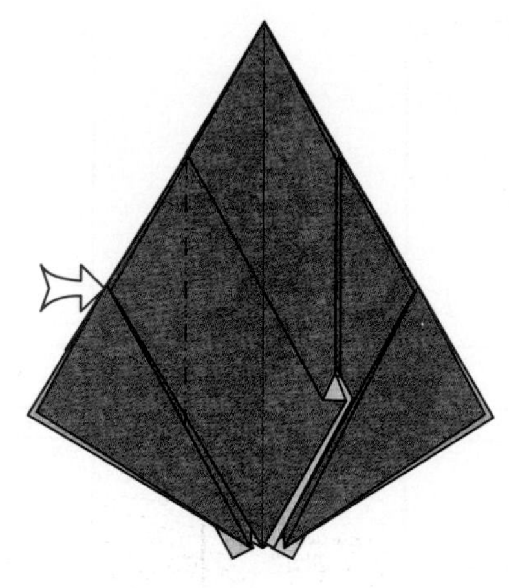

37. Sink the corner.

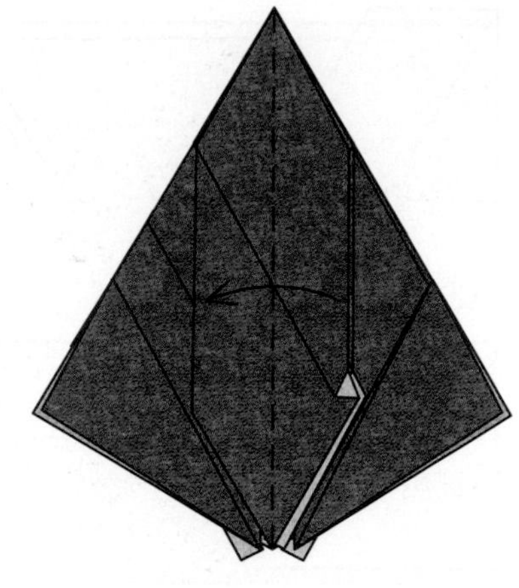

38. Fold the layer back to the left.

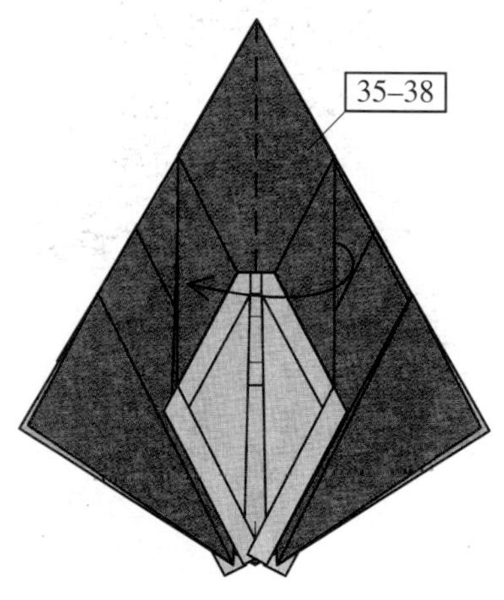

39. Repeat steps 35–38 on the right.

40. Fold two flaps up.

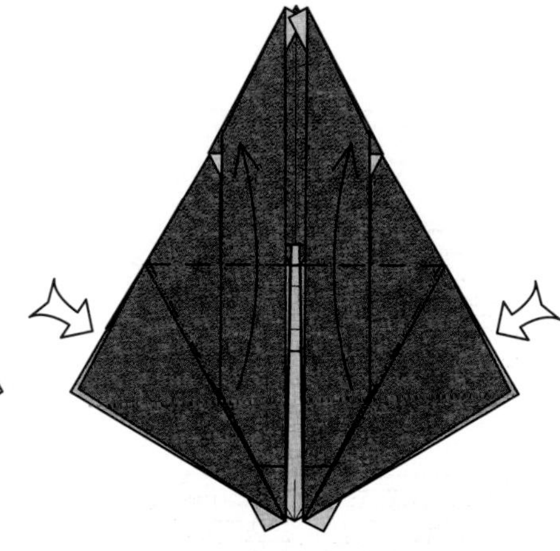

41. Squash-fold two edges, swinging the two flaps upward like a petal fold.

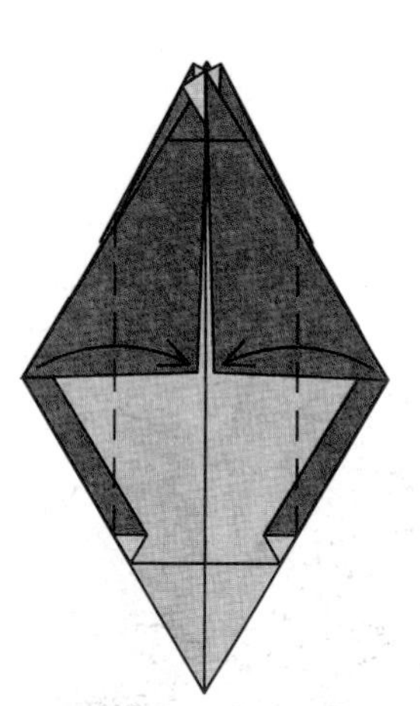

42. Fold two corners in to the center.

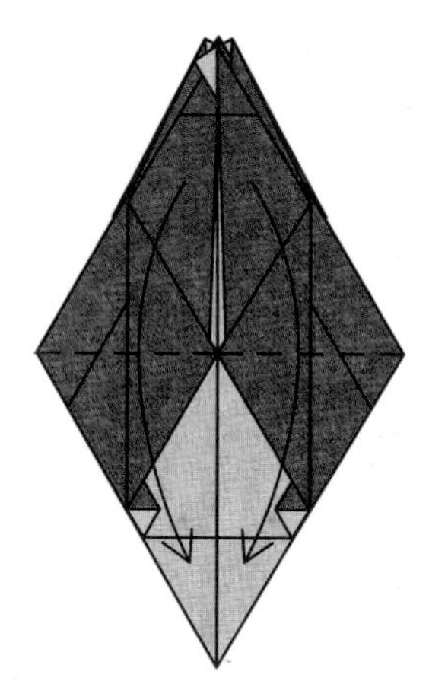

43. Fold two flaps back down.

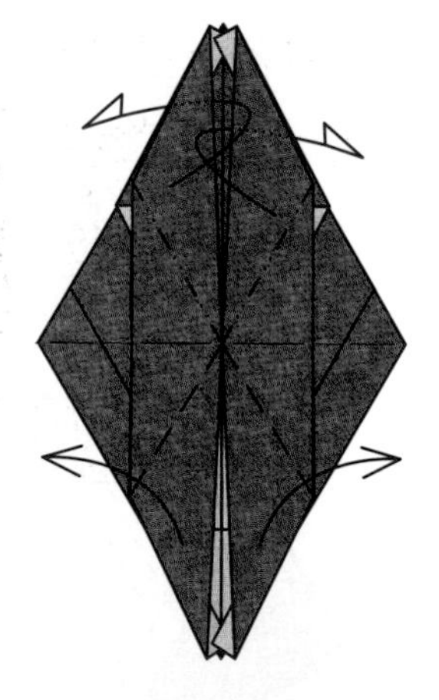

44. Mountain-fold the top pair of flaps out to the sides. Valley-fold the bottom pair. You shouldn't make either fold sharp because you will adjust these folds later.

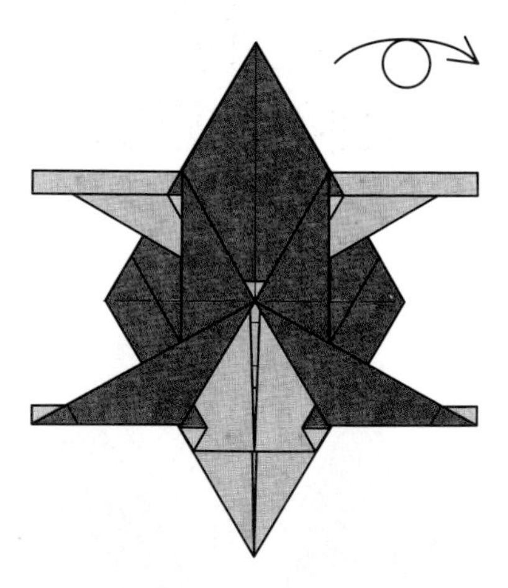

45. Turn the model over from side to side.

46. Pleat the shell and curve it around. The mountain folds are vertical; the valley folds are made so that the mountain folds line up with the existing edges.

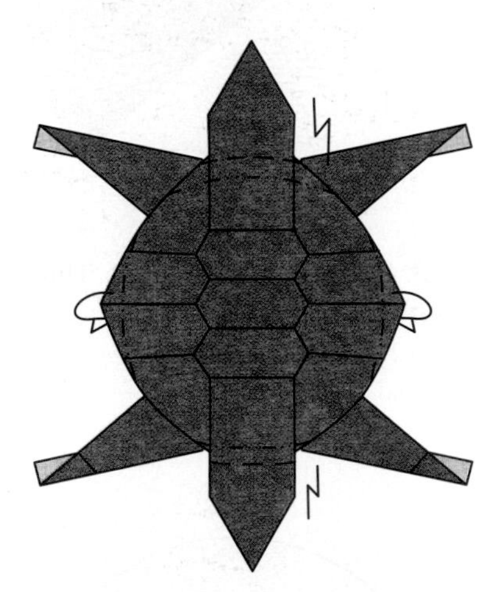

47. Pleat the head (top) and tail (bottom); note that the upper pleat extends farther into the shell. Mountain-fold the sides underneath.

48. Valley-fold all the way around the bottom of the shell to create a rim (and also further lock the pleats made in step 46).

49. Mountain-fold the edges of the head and tail through all layers.

50. Pinch and narrow the tail.

51. Steps 52–74 will focus on the legs and feet, starting with the forelegs.

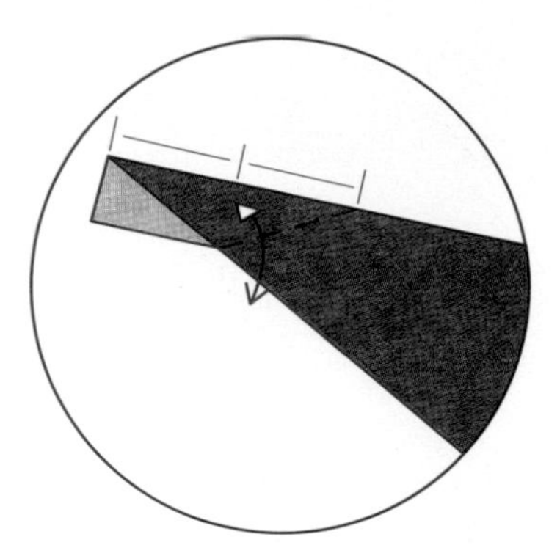

52. Detail of foreleg. Fold and unfold.

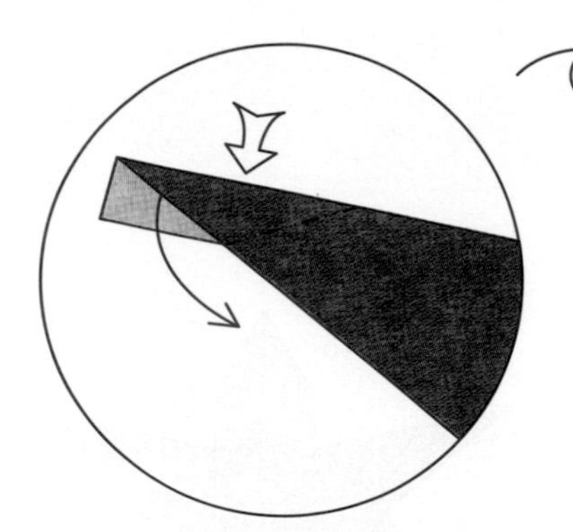

53. Reverse-fold the point. Then turn the model over.

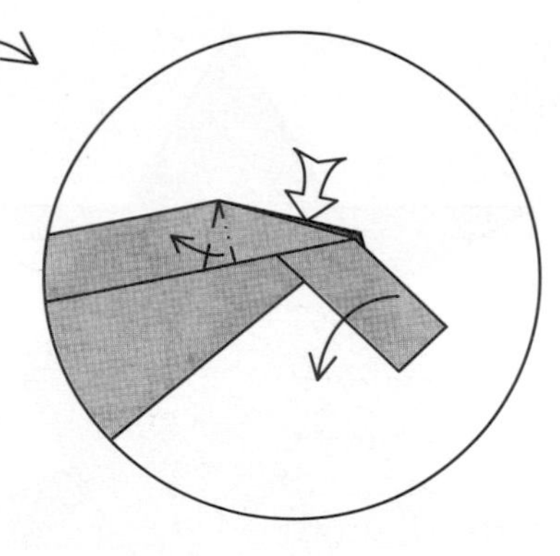

54. Open out the point, making a pleat along the folded edge.

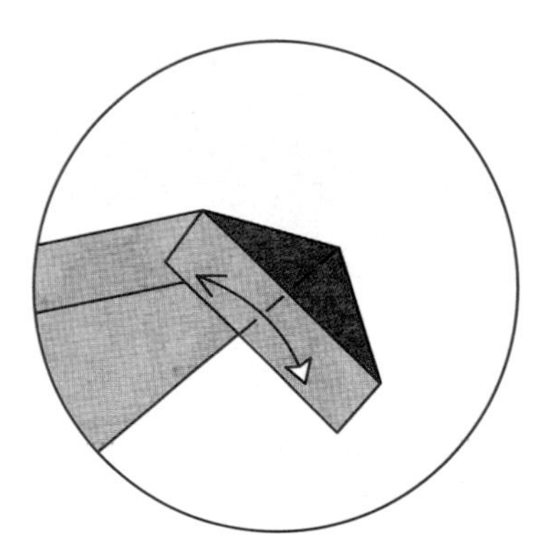

55. Fold and unfold along the edge behind the flap.

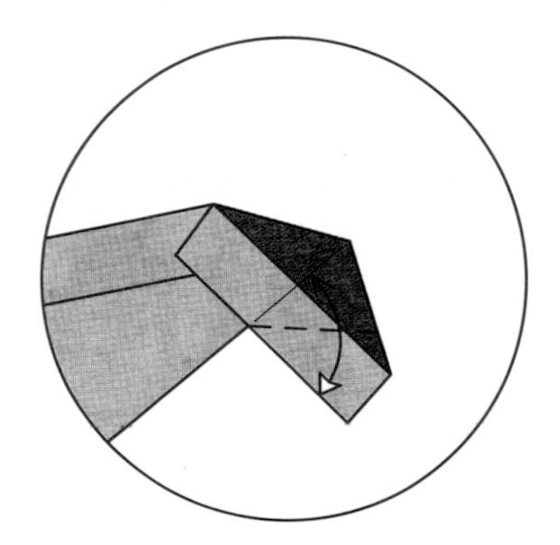

56. Fold and unfold along the angle bisector.

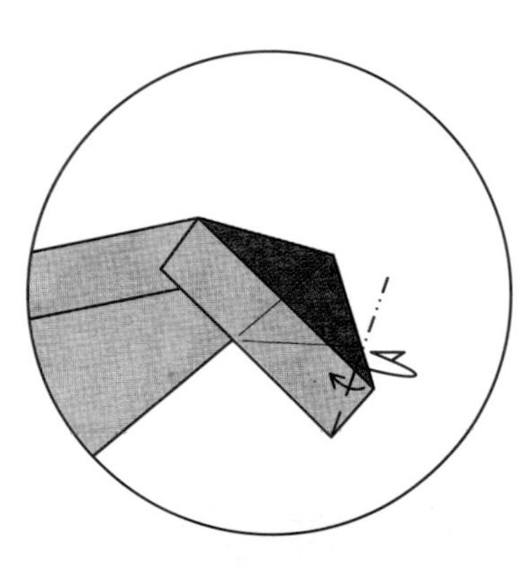

57. Valley fold the near layer and mountain-fold the far layer.

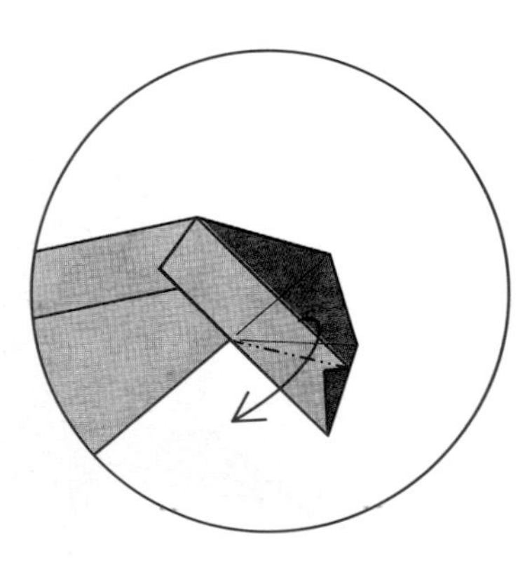

58. Form a mountain fold that runs to the corner and open out the white layer. The paper will not lie flat.

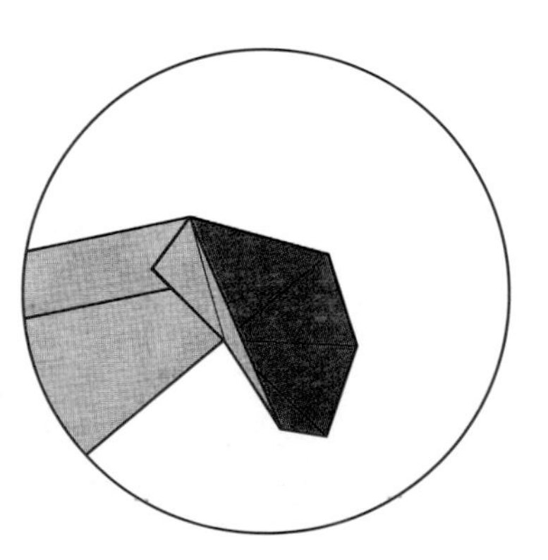

59. Turn the paper over.

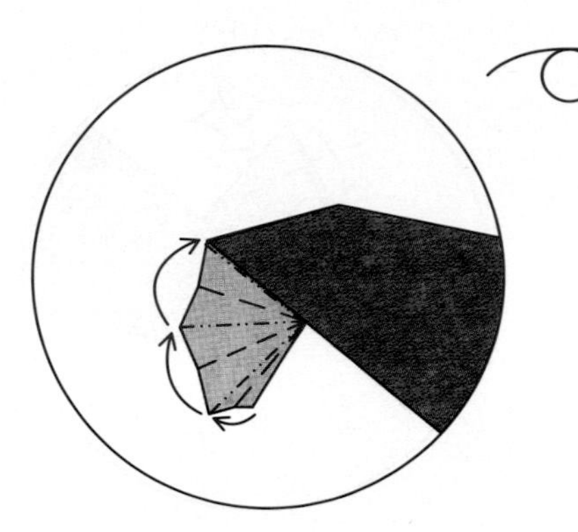

60. Form valley folds betweeen the existing mountain creases and pleat the corners all together. Then turn the paper back over.

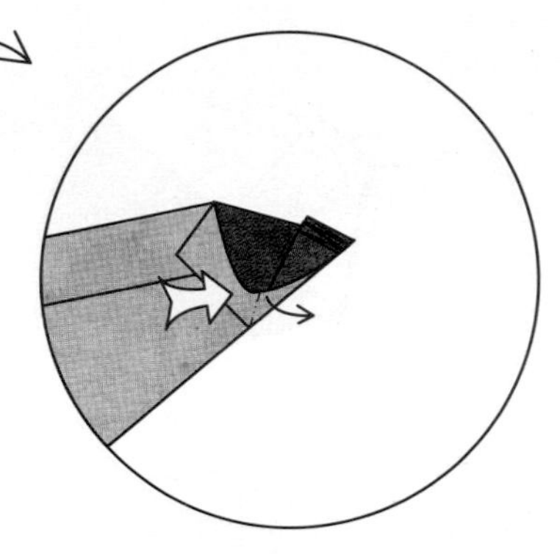

61. Squash-fold the extra paper over to the right and flatten.

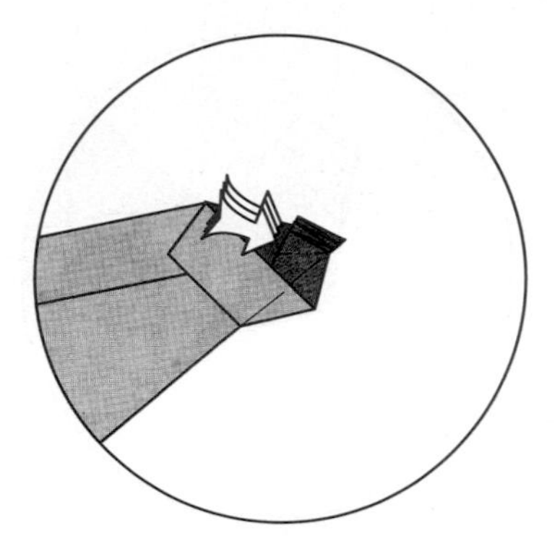

62. Reverse-fold three corners.

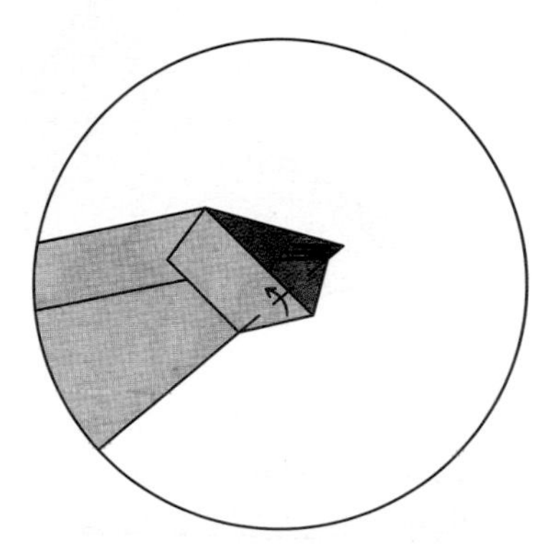

63. Valley-fold the corner.

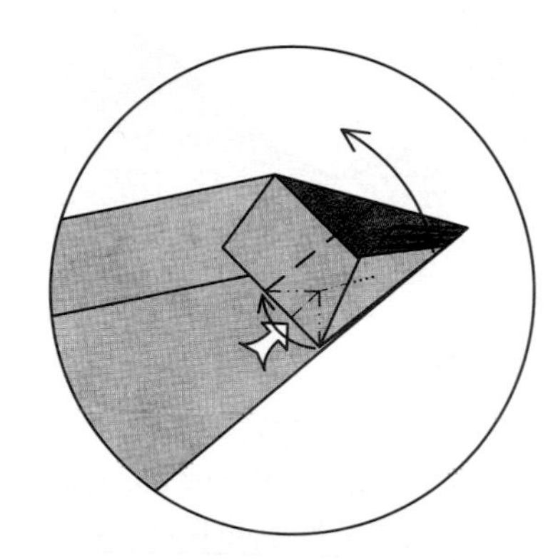

64. Spread the toes; simultaneously sink the edge of the "heel." The paper will lie almost flat, except for the pleats forming the toes, which will be vertical.

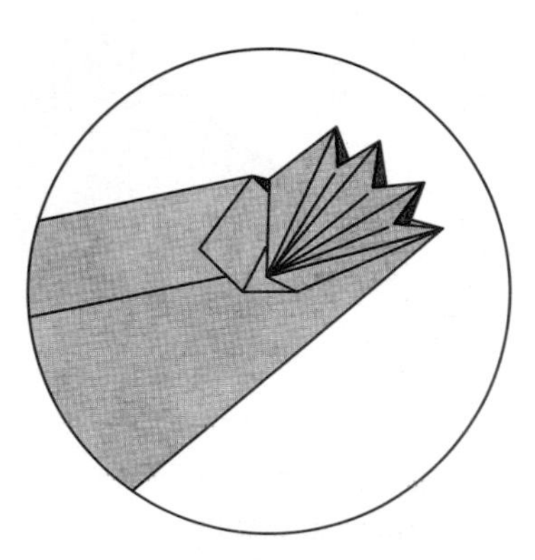

65. Finished front foot.

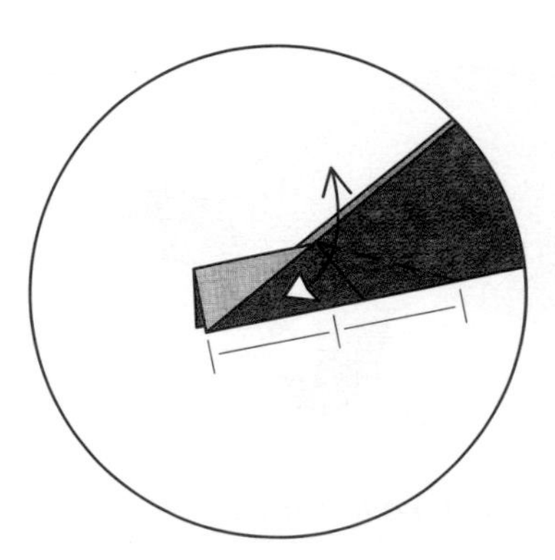

66. Detail of hind foot. Fold and unfold.

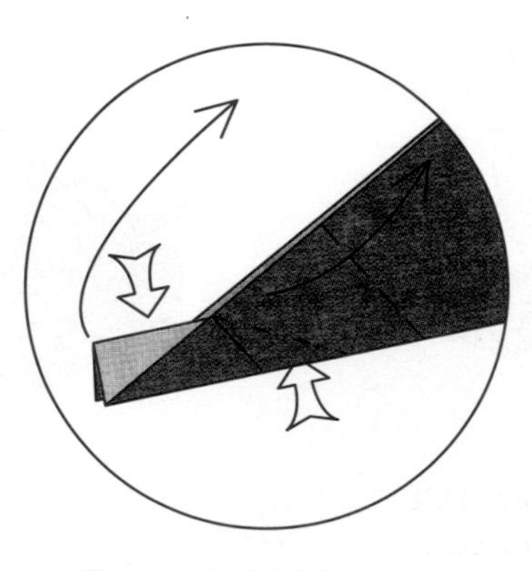

67. Squash-fold the flap, opening out the pleat.

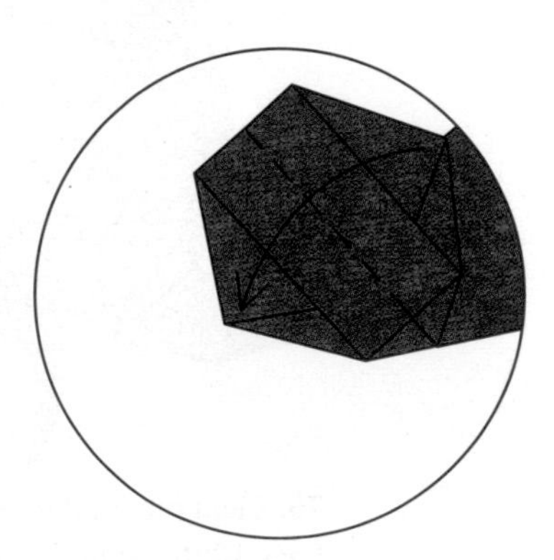

68. Close up the flap.

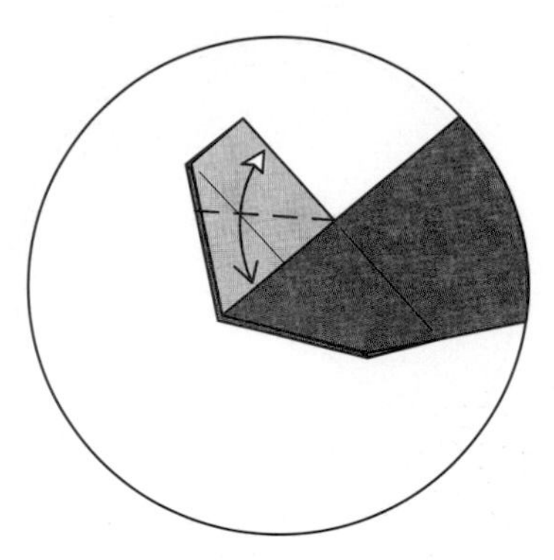

69. Fold and unfold along the angle bisector.

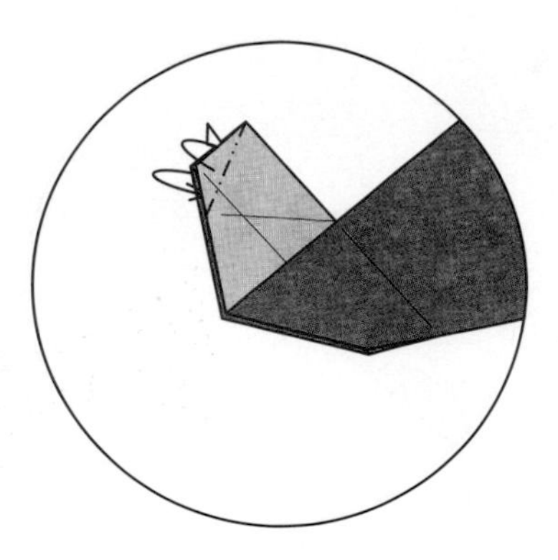

70. Mountain-fold the near layer behind; valley-fold the far layer.

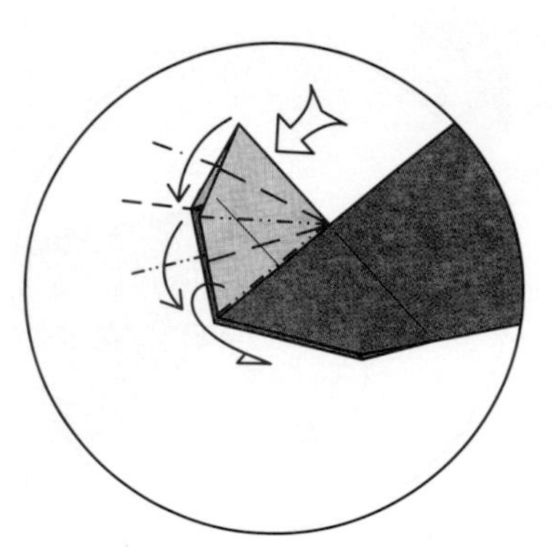

71. Crimp the white layers downward.

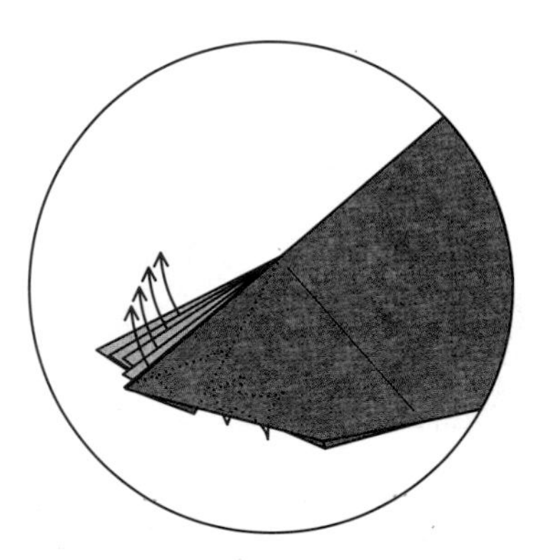

72. Reverse-fold four edges upward so that they stand above the white edges.

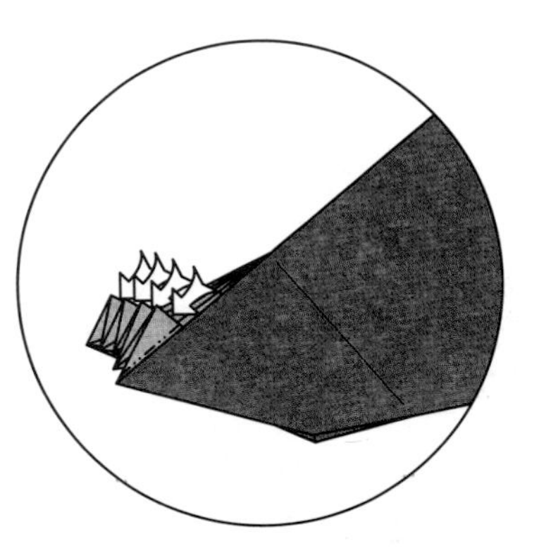

73. Reverse-fold the edges to lie flush with all the other edges.

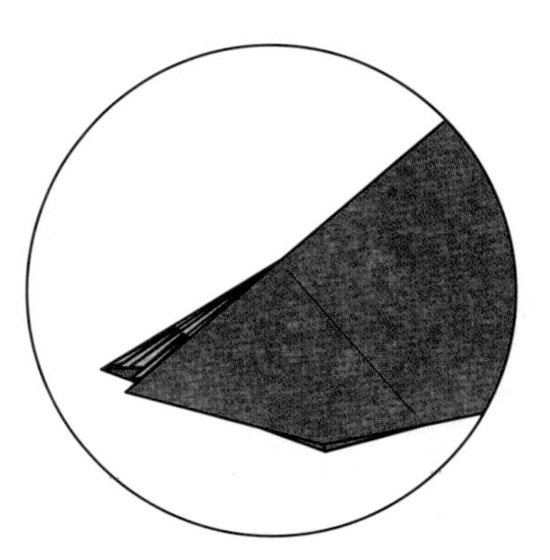

74. Finished hind foot. Repeat steps 52–73 on the other two feet.

75. Pleat the hind legs and spread the toes. Crimp and curve the forelegs. Pull out the middle edges of the head and make it three-dimensional.

76. Finished Western Pond Turtle.

Koi

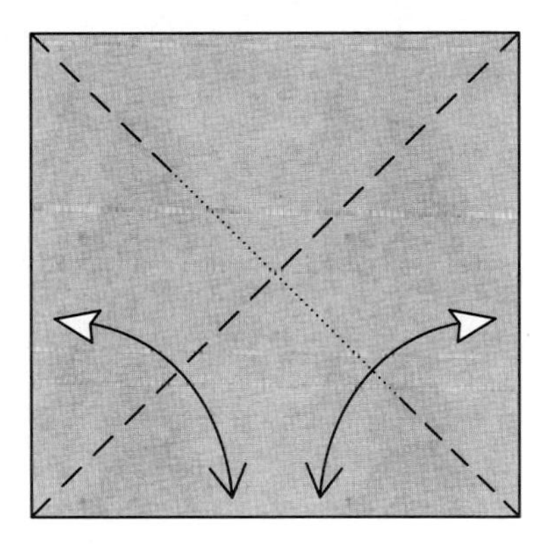

1. Begin with a square, white side up. Fold and unfold along the diagonals, but on the downward diagonal, only make the crease sharp about 1/3 of the way in from each corner.

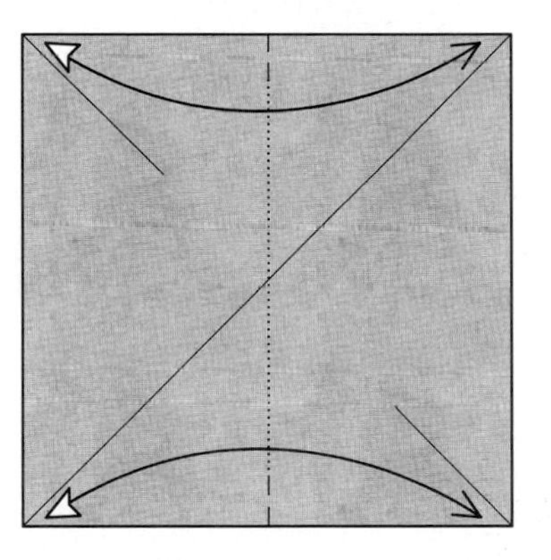

2. Fold the left side over to the right, making pinches along the top and bottom edges.

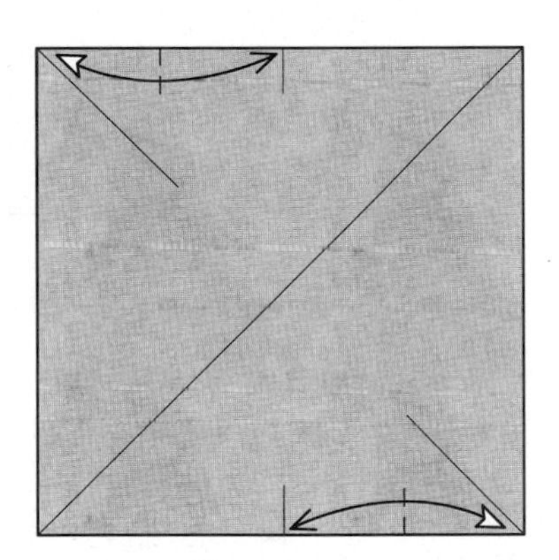

3. Fold the top left corner over to the pinch you just made; pinch along the top edge and unfold. Repeat with the bottom right corner.

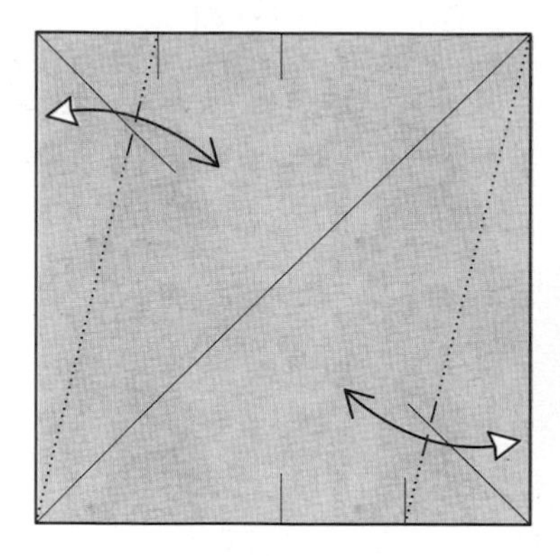

4. Fold the top left corner over along a crease connecting the pinch you just made with the bottom left corner; make a pinch where it crosses the diagonal and unfold. Repeat with the bottom right corner.

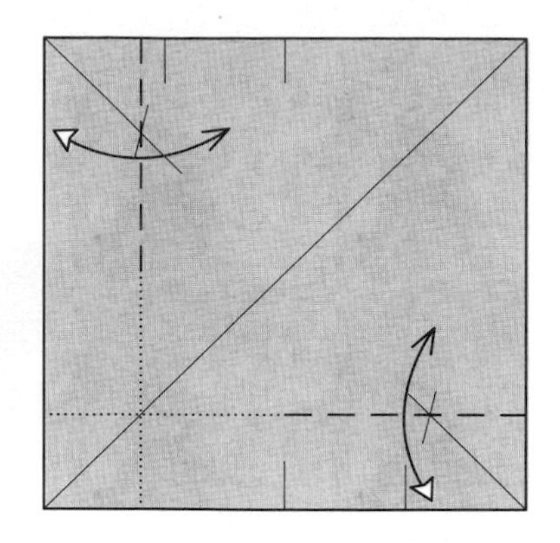

5. Fold the bottom edge up along a horizontal fold that passes through a crease intersection; make the crease sharp only from the right edge about half of the way across, then unfold. Repeat with the left edge.

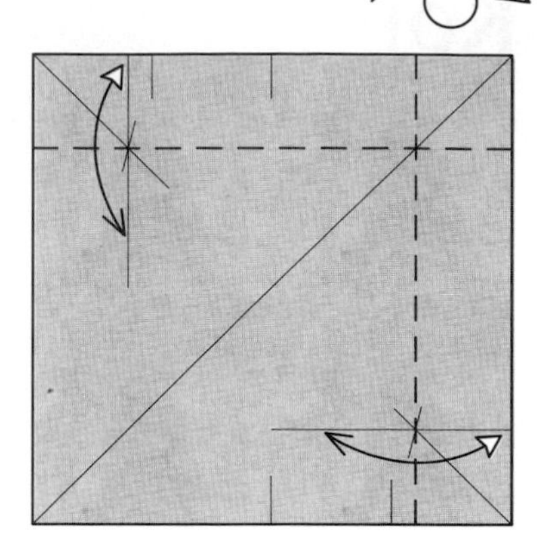

6. Fold the right edge over along a crease that passes through the crease intersection and unfold. Repeat with the top edge. Turn the paper over.

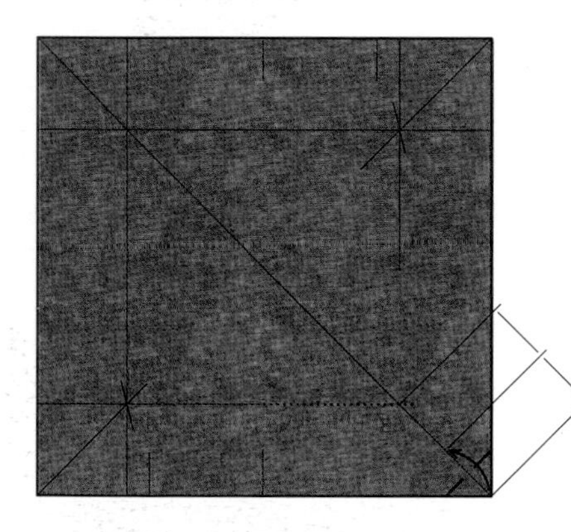

7. Fold up a bit of the lower right corner. The corner goes halfway to the imaginary crease intersection shown, but the exact amount isn't critical.

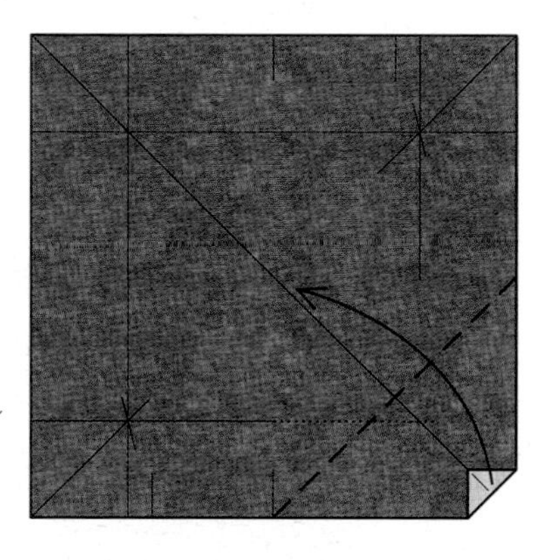

8. Fold the corner up along the diagonal. The fold hits the bottom edge at an existing crease.

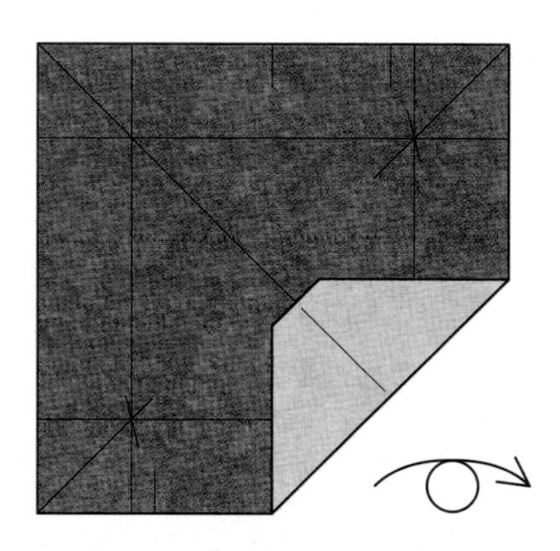

9. Turn the paper over and rotate so that it is symmetric with a corner pointing down.

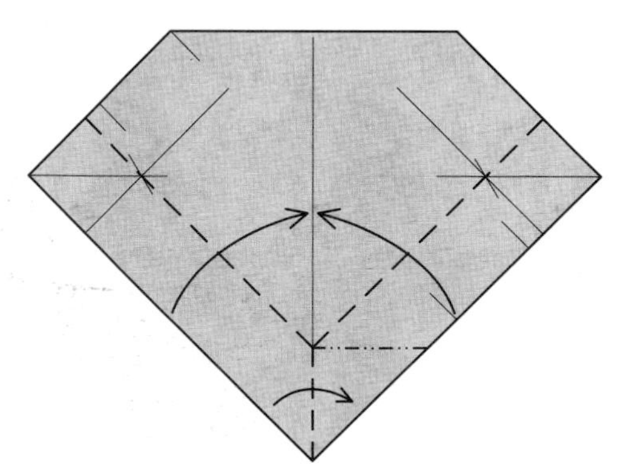

10. Fold a rabbit ear with the lower two edges using existing creases.

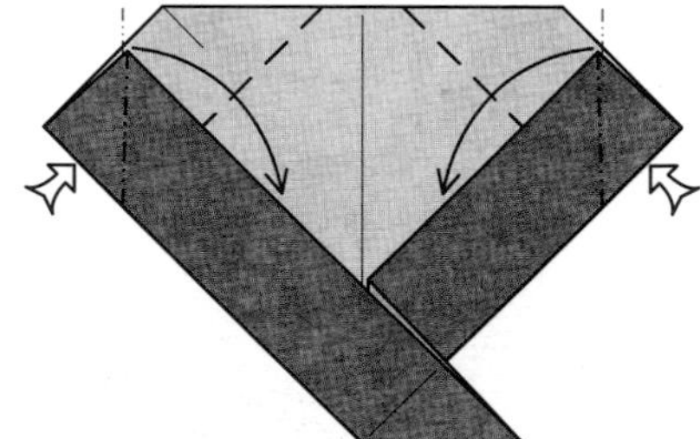

11. Reverse-fold the side corners and fold the edges down inside.

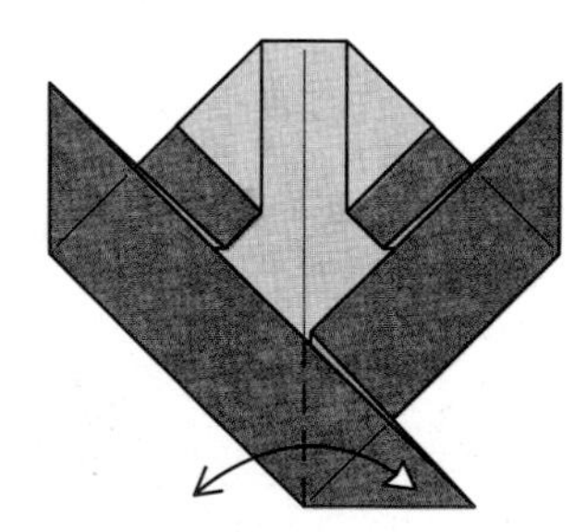

12. Fold the corner at the bottom from side to side.

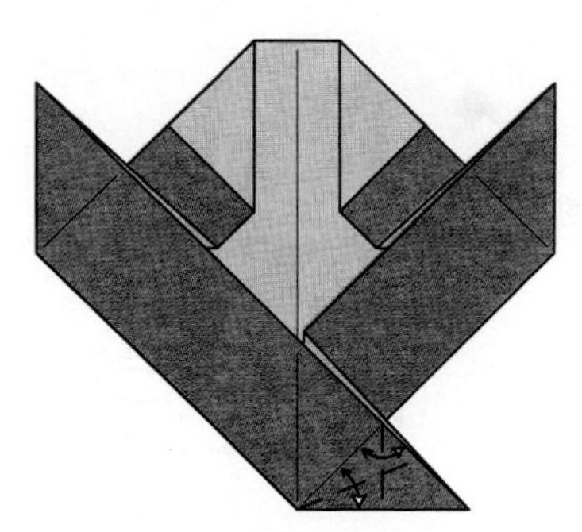

13. Fold and unfold along two angle bisectors.

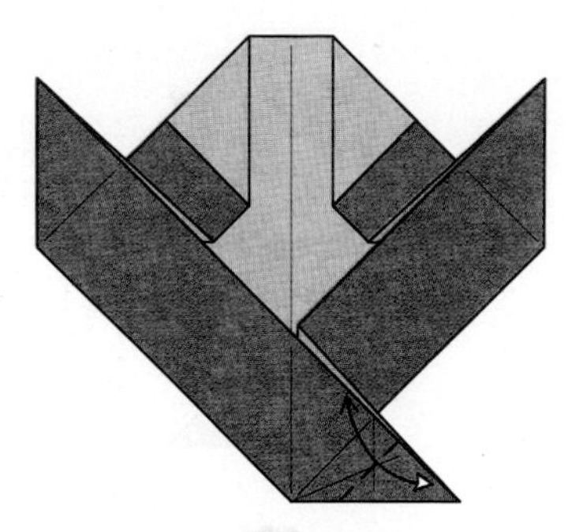

14. Fold and unfold along a crease that passes through the crease intersection.

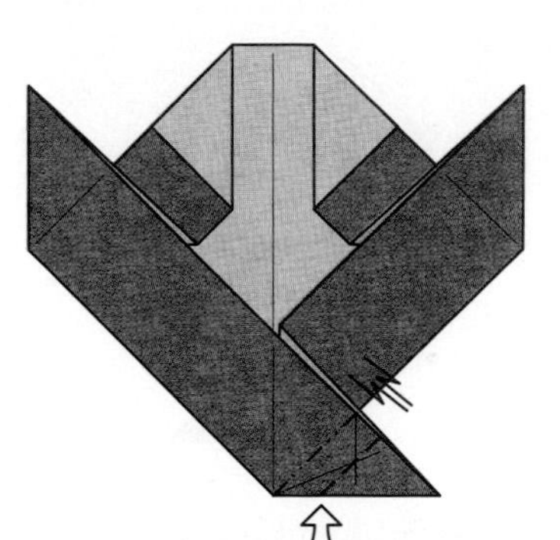

15. Crimp the corner on the two existing creases.

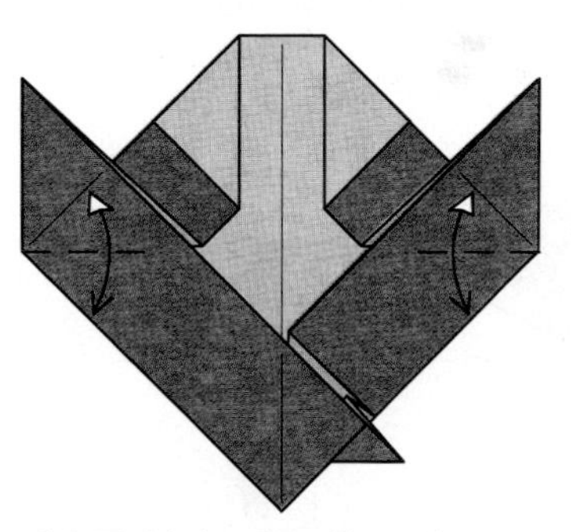

16. Fold the side flaps down and up.

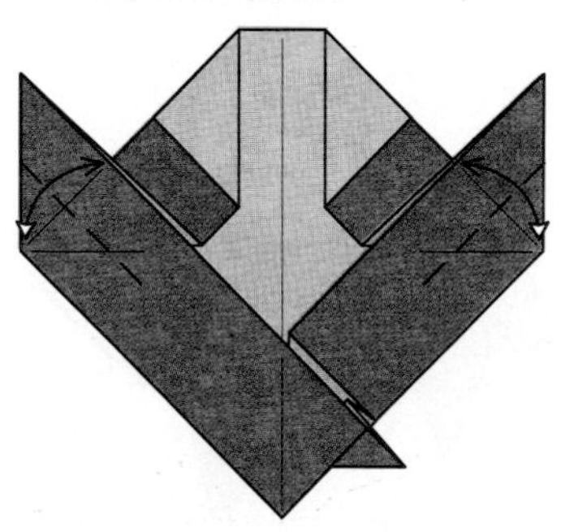

17. Fold the side corners up, crease, and unfold.

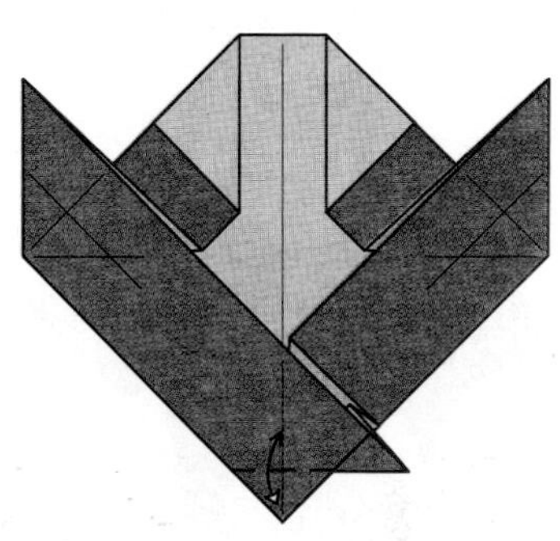

18. Lightly crease the bottom point up and down. Try not to make any crease mark on the far side.

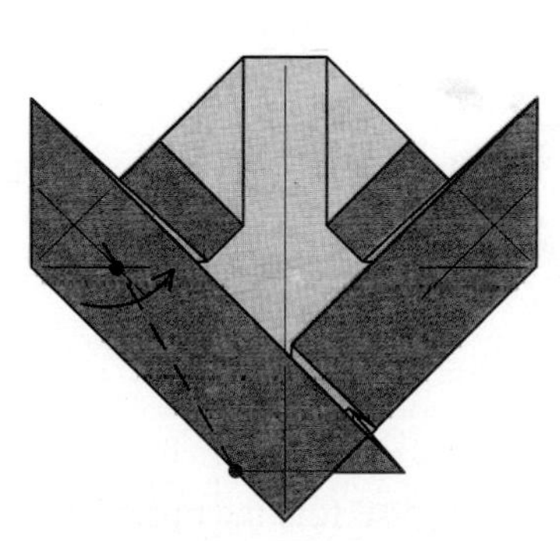

19. Fold through all layers; only make the crease sharp between the two dots.

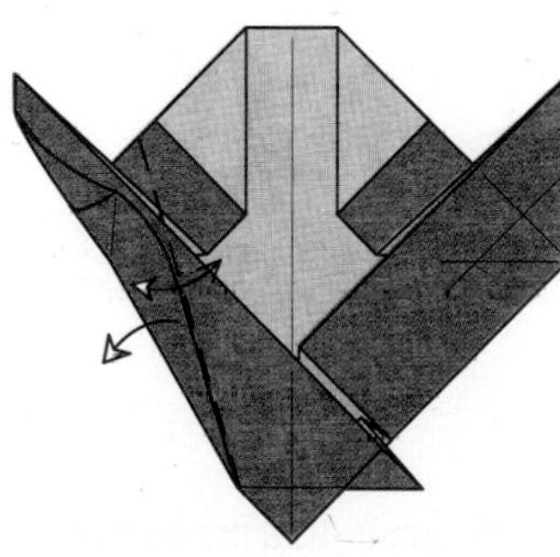

20. Fold and unfold all the way up along a crease aligned with the folded edge; unfold to step 19.

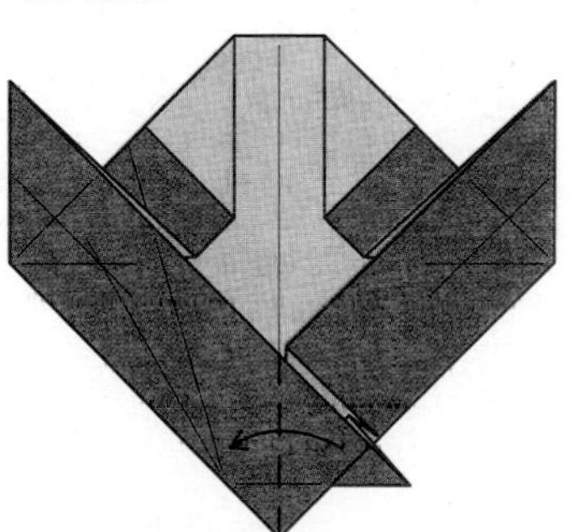

21. Swing the crimped flap to the left.

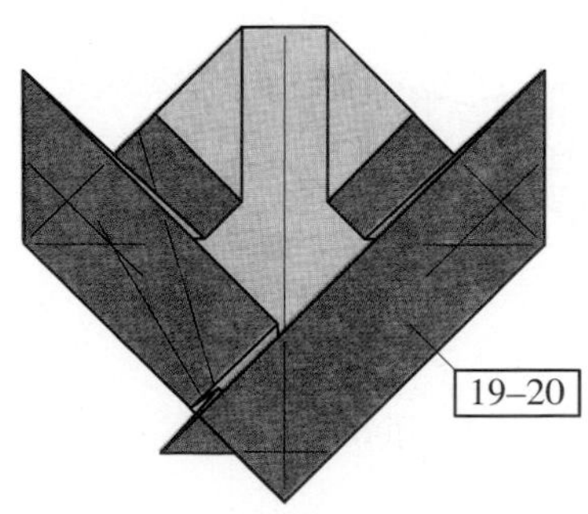

22. Repeat steps 19–20 on the right.

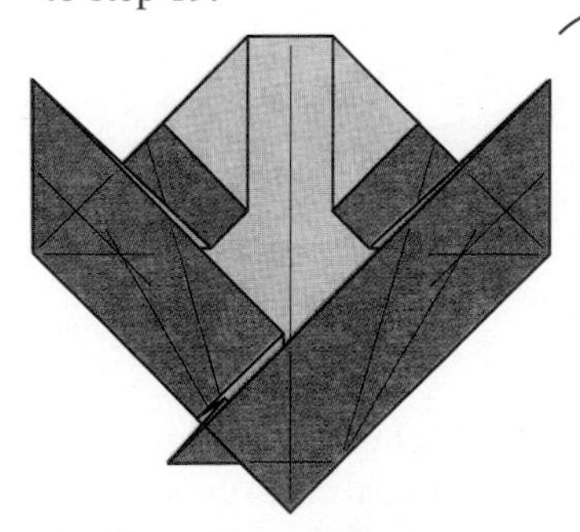

23. Turn the model over.

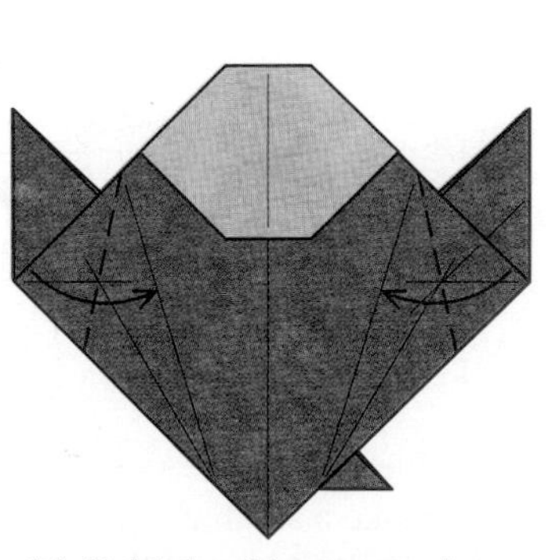

24. Fold the side corners in to lie along the crease line.

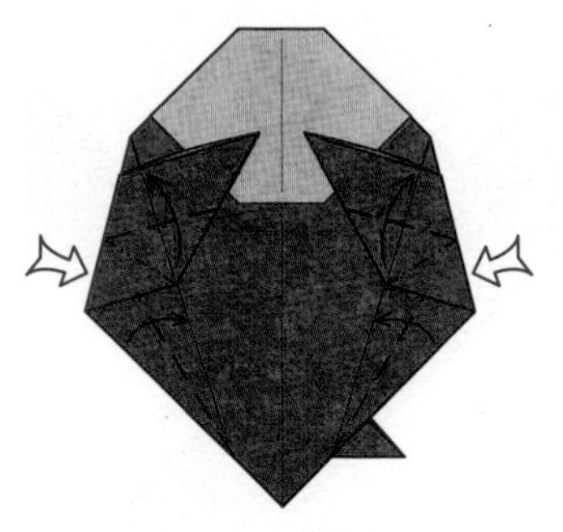

25. Swivel-fold the corners upward while folding the sides in.

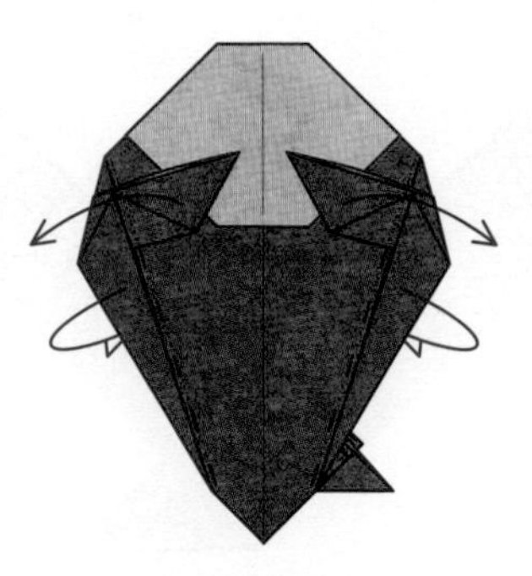

26. Fold the sides underneath on the existing creases, allowing the near flaps to swing outward.

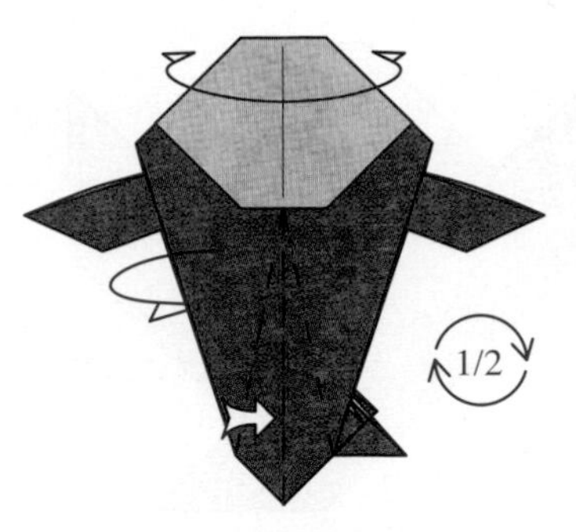

27. Fold the model in half, incorporating a reverse fold at the bottom and keeping the top gently rounded. Rotate the model 1/2 turn.

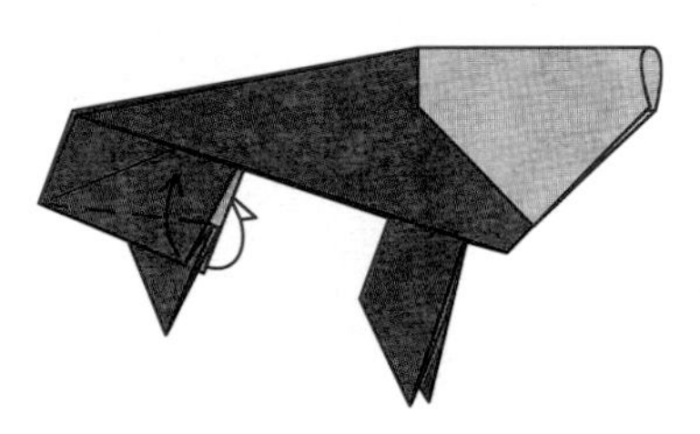

28. Fold two corners up as far as possible.

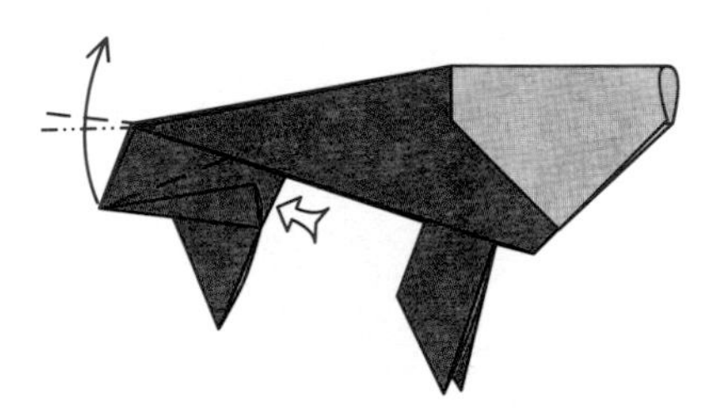

29. Crimp the tail upward.

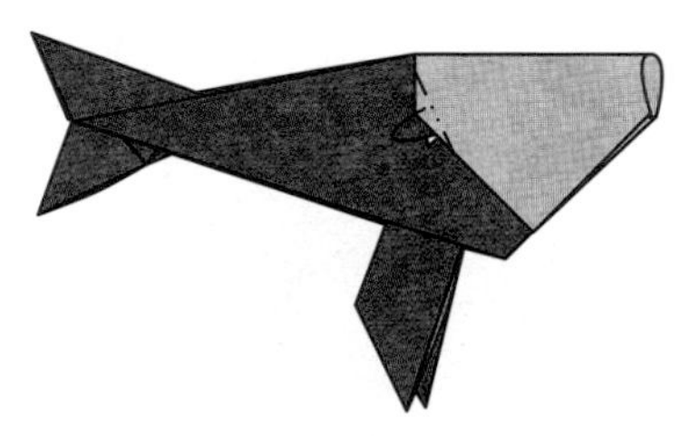

30. Mountain-fold the white corner. Repeat behind.

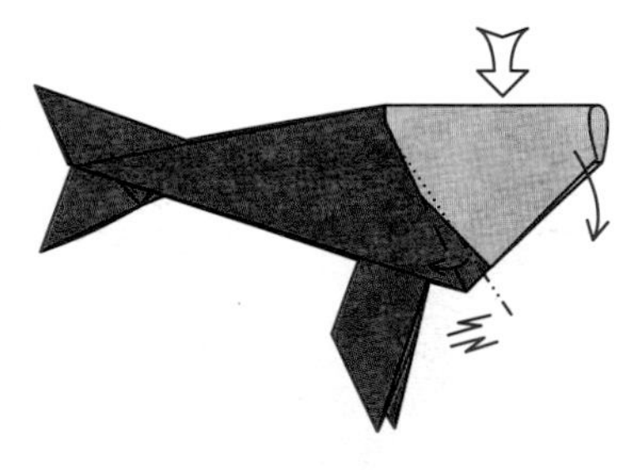

31. Crimp the head down, keeping it and the body rounded and three-dimensional.

32. Mountain-fold the corners just to the right of the fins. Pleat the edge of the nose, pinching at the corners.

33. Reverse-fold the tips of the fins.

34. Pleat the face. As you do so, make a small circular dimple at the top of the pleat to form an eye. Repeat behind.

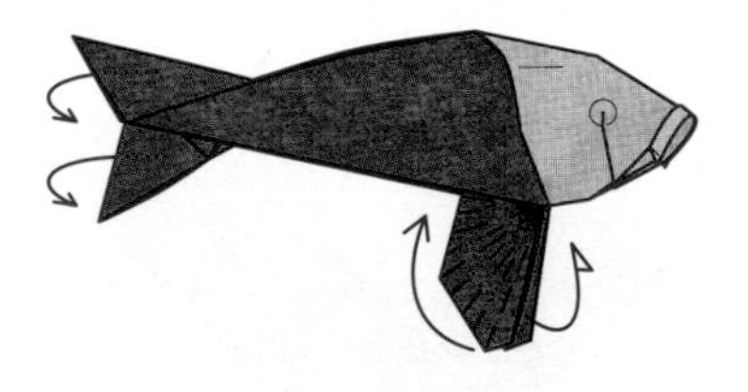

35. Pleat the fins and fold them up and out to the sides. Curve the tail slightly.

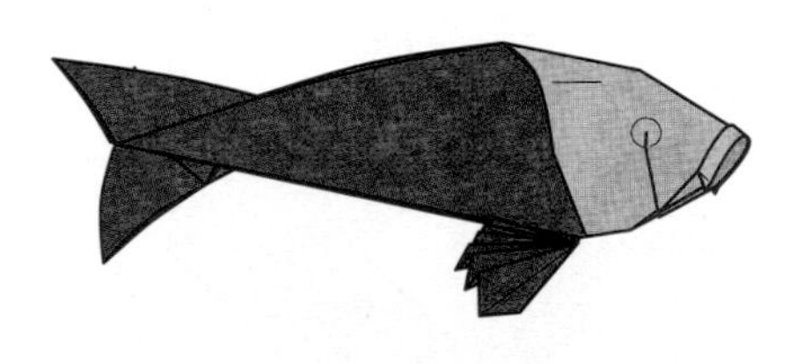

36. Finished Koi.

8 Tiling

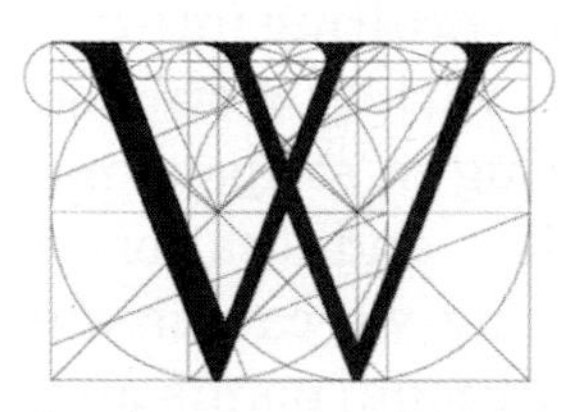

While there are many different approaches to origami design, the ones that I've shown thus far can be arranged in a rough hierarchy of complexity. We started with some simple structures—the traditional bases. Then, we modified these bases by various distortions—offsetting the crease pattern from the center of the square or distorting the entire pattern. Both types of modification leave the number of flaps unchanged; they only alter the lengths and widths of the flaps.

Then we increased the number of flaps by subdividing some flaps into smaller flaps using the various point-splitting techniques. While in principle any number of flaps can be attained, point-splitting is inherently a process of reduction; the flaps you end up with are always smaller than what you started with. Thus, there are definite limits on what you can accomplish by point-splitting.

We can escape those limitations by using grafting, by effectively adding paper to an existing crease pattern in such a way that the paper remains square after the graft. Grafting allows you to add features to an origami base without taking anything away from features that are already present. The simplest grafts are border grafts, which consist of adding paper around one or more edges of the square, but this method, too, has its limits. You can only add paper—and thus features—to flaps that are made from a raw edge, i.e., corner and edge flaps. Another limitation is that when you are border grafting, edge flaps don't offer quite the same freedom of point creation that corner flaps did; a border graft that can create four points at a corner flap only creates two points of the same size at an edge flap.

Yet more variety in added features comes when we realize that the existing crease patterns are not indivisible; we can cut them up and insert strip grafts throughout their structure. Strip grafts create points and flaps along edges just as border grafts do, but they also create extra points in the interior of the paper without diminishing the size of adjacent flaps. As an expansion of strip grafts, we can graft in pleats to create extra edges running across a face, and weave crossing groups of pleats to create scales, bristles, and other textural elements. Although they all start with an existing crease pattern, strip and pleat grafts are much more versatile than point-splitting and border grafts and come in many more variations. Strip and pleat grafting possess this great versatility because they are based on dissected crease patterns, and there are usually many different ways to dissect a given pattern.

Once we've taken the step to incorporate grafting into dissected crease patterns, an enormously richer variety of origami structures becomes accessible. When grafting in strips of paper, we can vary the width, length, direction, and location of the strips; we can insert multiple strips; and we can create branching networks of strips, all to place additional points and/or textural elements into the basic design.

In the models to which we've applied grafting—the Songbird, the Lizard, the Turtle—our grafts have taken the form of fairly narrow strips. These are still relatively small perturbations to a preexisting model. The precursor to the songbird was still a bird; the lizard with toes began life as a lizard without toes; and the turtle with a patterned shell was still recognizably a turtle when its shell was smooth. But grafts can be made much larger and more complex and can be used to create new bases so different from their predecessors that they hardly seem related at all. We will expand our palette of design techniques by exploring further the concept of dissection and reassembly. Thus far, we have treated bases and grafts as two distinctly different types of objects; we start with a base, then we add a graft. In this chapter we will learn to decompose both bases and grafts into the same underlying structures, which can be reassembled in an infinite variety of ways. We will also learn to distill origami bases down to simple stick figures; we will then use these stick figures as tools for the design of new bases.

8.1. Uniaxial Bases

Let's look at several of the bases that I've shown so far. First, we have the Classic Bases: Kite, Fish, Bird and Frog Bases; to these, we add two new bases, those used for the Lizard and the Turtle. All six are shown in Figure 8.1.

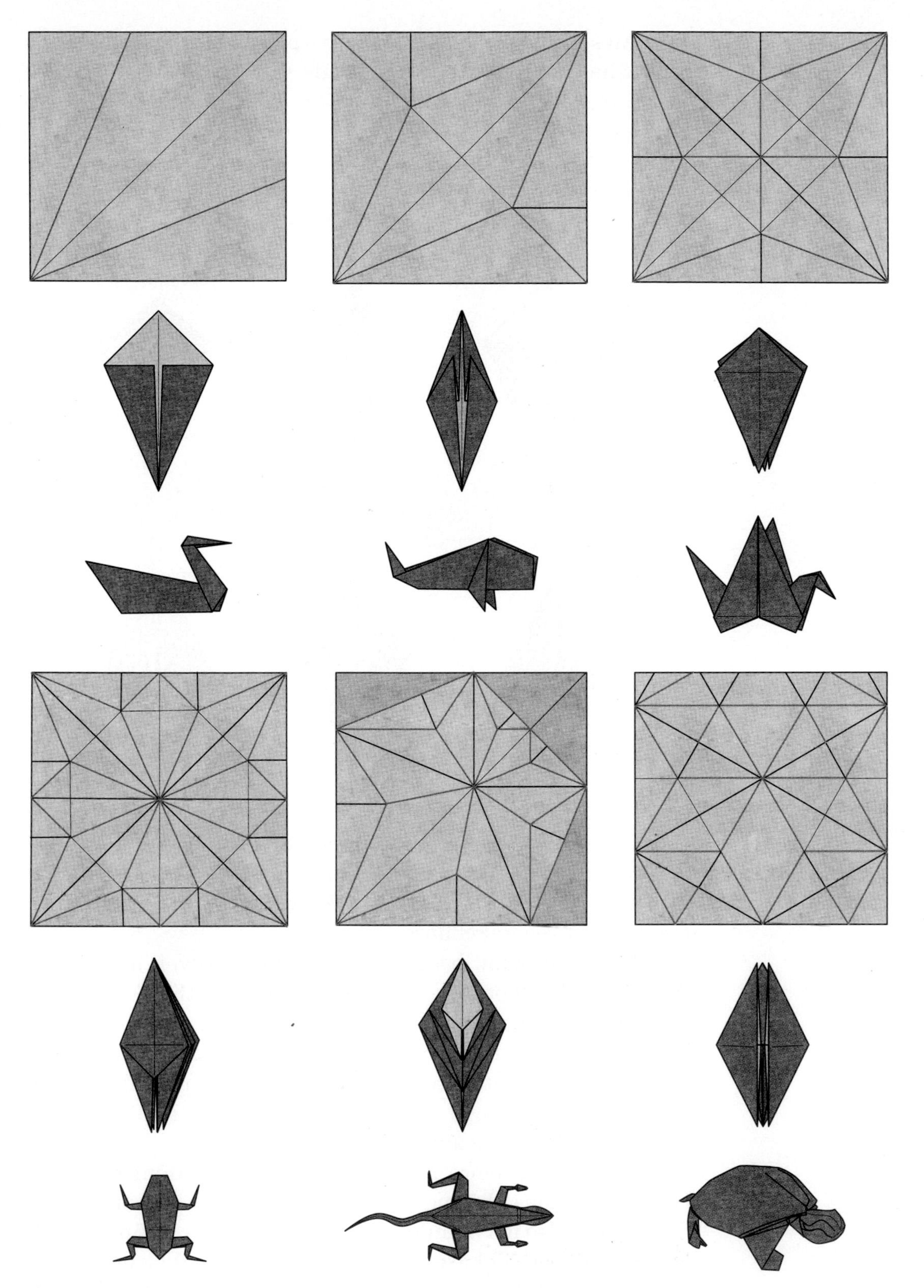

Figure 8.1.
Six bases. Top: crease patterns.
Middle: bases.
Bottom: representative designs.

All six of these bases share two properties: First, all flaps either lie along or straddle a single vertical line; second: the *hinge* at the base of any flap (i.e., the line between two adjacent flaps) is perpendicular to this line. When all flaps lie along a line, that line is called an *axis* of the base. And any base that possesses a single axis is called a *uniaxial* base. The axes of the six bases are shown by dashed lines in Figure 8.2.

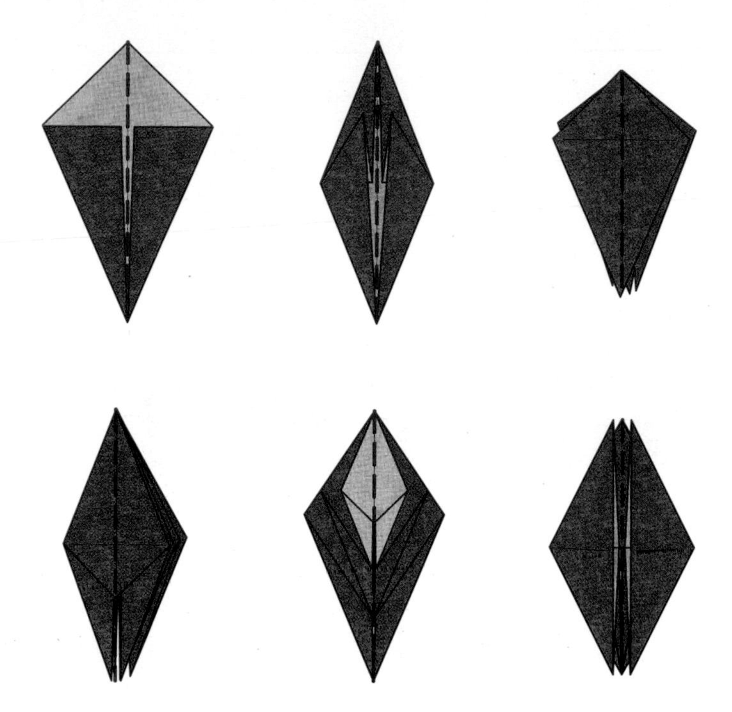

Figure 8.2.
The axes of six uniaxial bases.

Uniaxial bases are very common in origami, and they have several properties that make them relatively easy to construct, dissect, graft, and manipulate. We will study them intently for the next several chapters.

Not all origami bases are uniaxial, however, and before casting aside all other origami bases, it's worth taking a few moments to look at some exceptions.

Among the traditional bases, the Windmill Base is not uniaxial because its four flaps do not lie along a single line; instead, it has two crossed axes and the hinge creases are not perpendicular to the axis.

A base of a more recent vintage—John Montroll's Dog Base, variations of which he has used for a score of diverse figures—is also not a uniaxial base, having two distinct parallel axes. Montroll's base is remarkable for its efficiency in use of paper (and for my money, stands as the most elegant base in all of origami). So while uniaxial bases will prove to be remarkably versatile, they are not the magic solution for all origami problems.

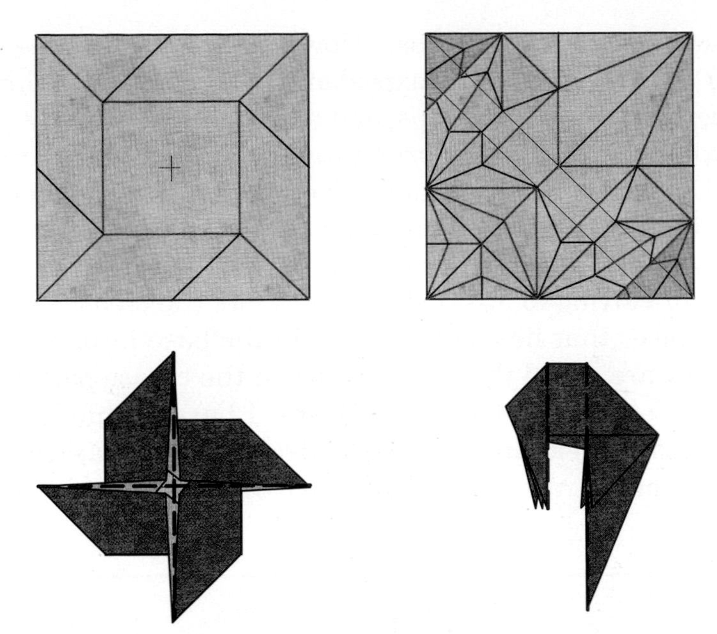

Figure 8.3.
Left: The Windmill Base has two crossed axes.
Right: Montroll's Dog Base has two parallel axes.

Montroll's Dog Base, in particular, highlights a limitation of uniaxial bases; for a given model, they may not provide the most efficient structure. However, uniaxial bases are readily constructed and quite versatile, and we will explore them thoroughly.

It should also be noted that whether or not a base is uniaxial may depend on the orientation of the base. In the six example bases I've shown, the axis lies along a line of mirror symmetry. This is usually, but not always, the case. For example, in the Waterbomb Base, if we attempt to draw the axis along the line of symmetry, we find that the raw edges of the flaps don't lie along the axis and the hinges aren't perpendicu-

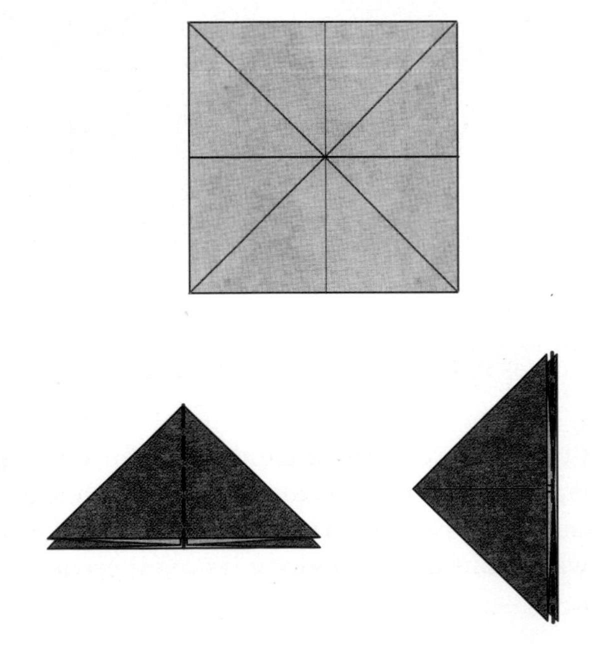

Figure 8.4.
Top: crease pattern for the Waterbomb Base.
Lower left: the Waterbomb Base is not uniaxial with respect to an axis along the symmetry line.
Lower right: it is, however, uniaxial if we draw the axis along the raw edges of the base.

lar, so it's not a uniaxial base. However, if we rotate the base by 90°, we can re-draw the axis along the raw edges, the hinges are perpendicular to the axis, and it is thereby revealed to be a uniaxial base in this new orientation, as shown in Figure 8.4.

Uniaxial bases lend themselves to strip grafting because the alignment of many folded edges along the axis of an existing base makes the creases along those edges natural candidates for cutting to insert strip grafts into the crease pattern. The creases that lie along the axis in the base form a special set; they are called the *axial creases* in the crease pattern. In the crease patterns for the six bases, I have highlighted the axial creases in Figure 8.5. I have also similarly marked those portions of the raw edge of the paper that lie along the axis.

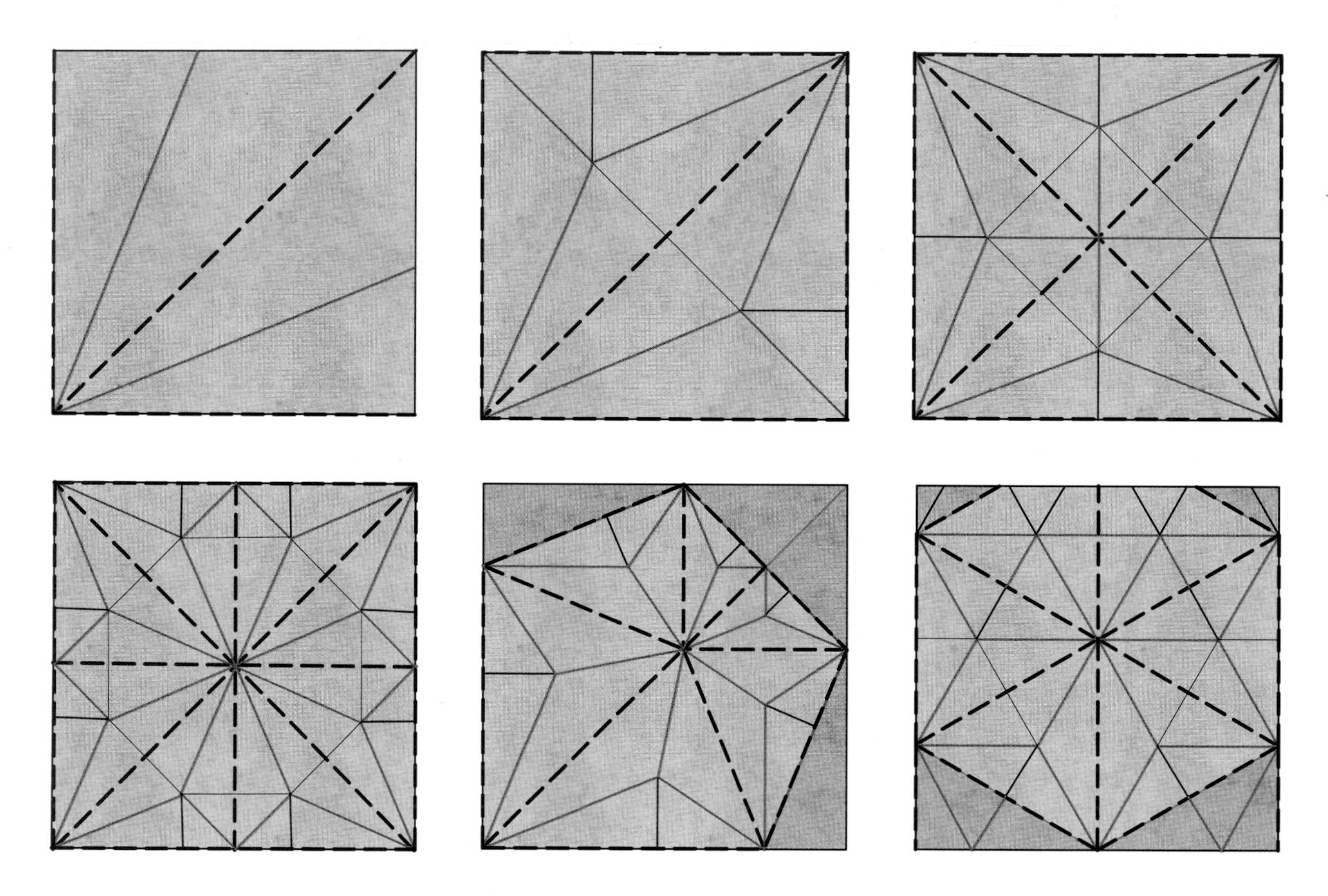

Figure 8.5.
The axial creases in the six uniaxial base crease patterns.

Axial creases are natural candidates for cutting and inserting strip grafts, in part because every flap has at least one axial crease (or a raw edge) running to its tip. Consequently, we can always split any flap along its axial crease to insert a strip graft.

Observe that the network of axial creases divides the crease pattern into a collection of distinct polygons whose boundaries are entirely composed of either axial creases or the raw edge of the paper. We will call these polygons *axial polygons*.

8.2. Splitting Along Axes

The axial polygons of the crease pattern have an interesting property in their own right: In the folded base, the entire perimeter of each polygon comes together to lie along a common line—the axis of the model. You can observe this property by taking a base and cutting it along its axis. If you give the cut a slight kerf, so that it severs folded edges that lie along the axis, both the base and the crease pattern will fall apart into distinct pieces, as shown in Figure 8.6 for the Fish, Bird, and Frog Bases.

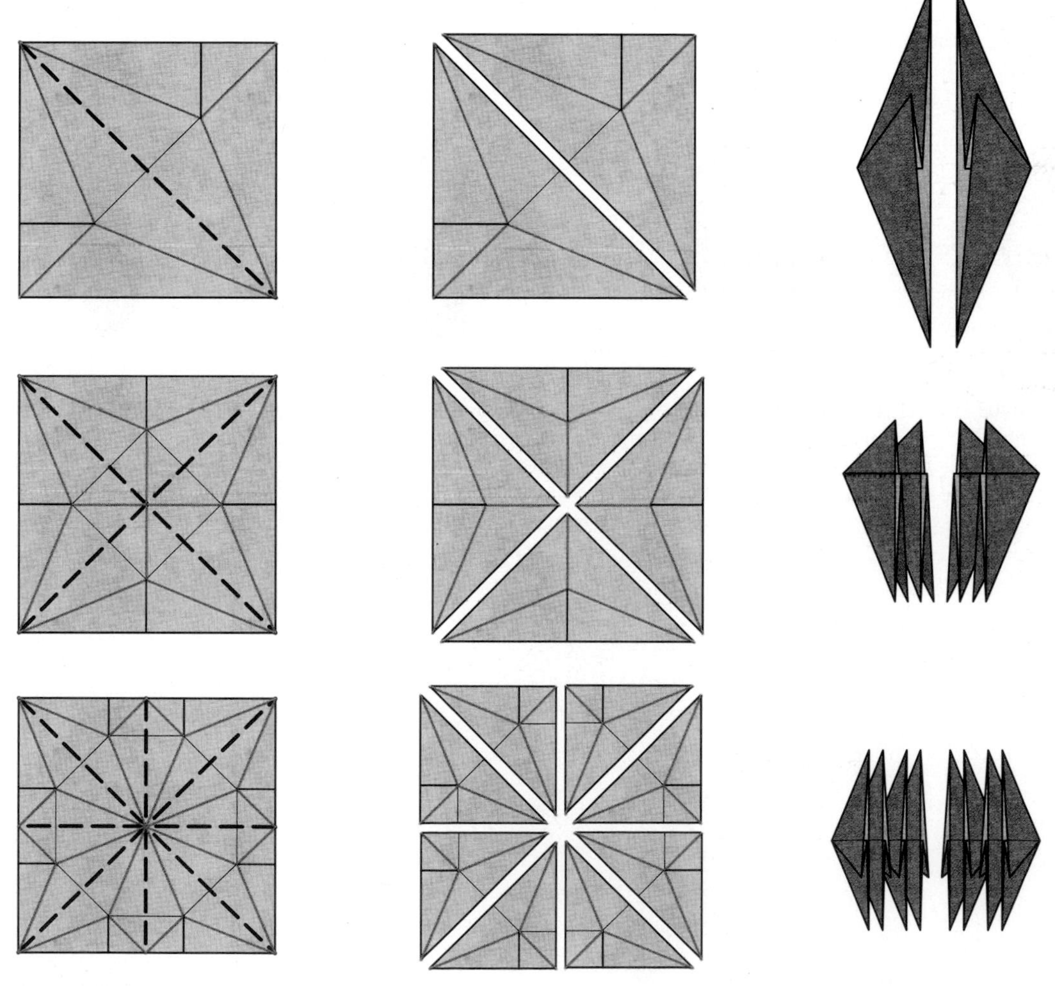

Figure 8.6.
Dissected crease patterns for the Fish, Bird, and Frog Bases.

One or more strips can be inserted along any of the gaps to split or multiply flaps. Let's look at an example.

Figure 8.7 illustrates the process of inserting a strip into the middle of a Bird Base. We cut the base down the middle, then insert a strip into the gap. The resulting shape has paired

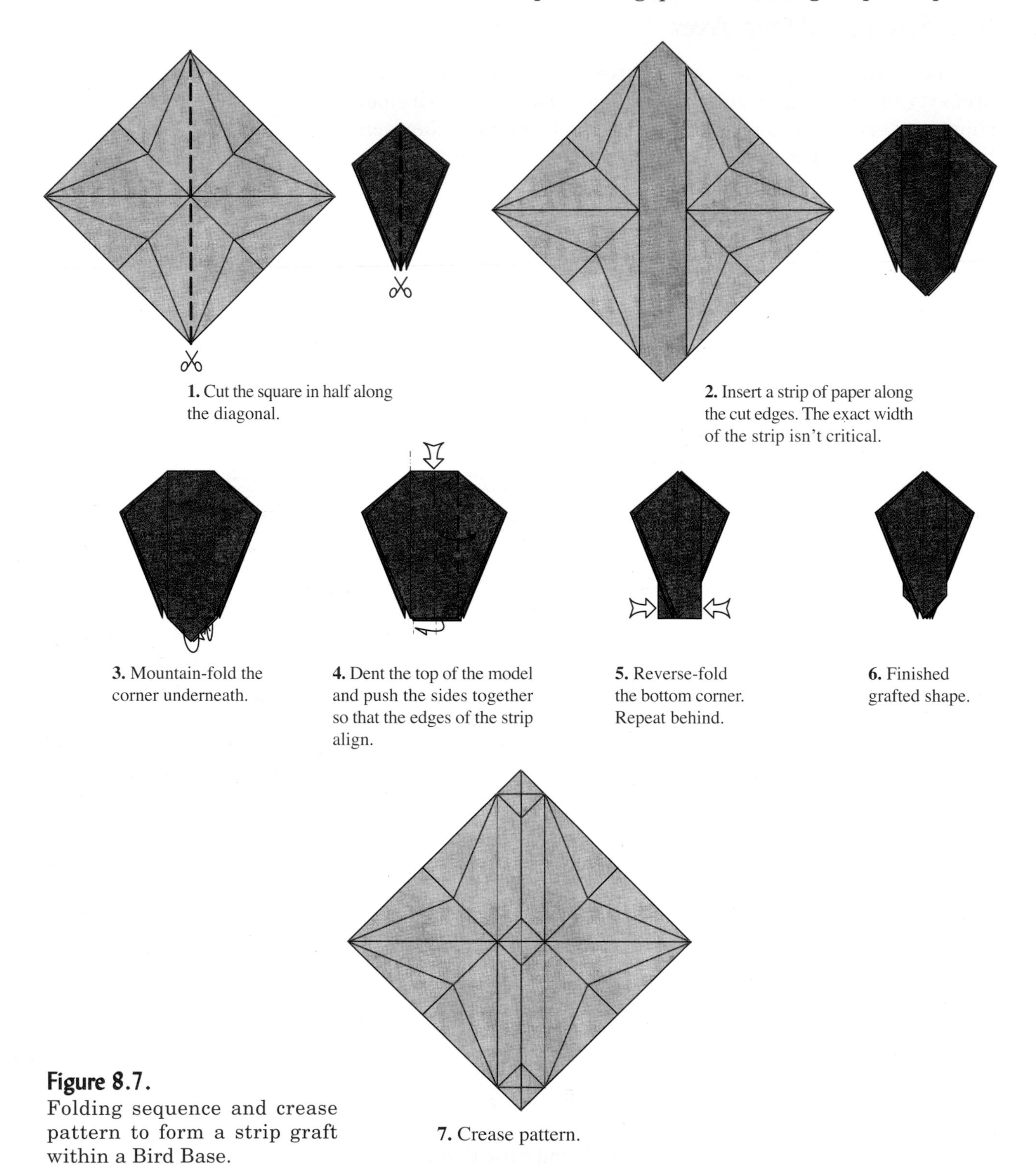

1. Cut the square in half along the diagonal.
2. Insert a strip of paper along the cut edges. The exact width of the strip isn't critical.
3. Mountain-fold the corner underneath.
4. Dent the top of the model and push the sides together so that the edges of the strip align.
5. Reverse-fold the bottom corner. Repeat behind.
6. Finished grafted shape.
7. Crease pattern.

Figure 8.7.
Folding sequence and crease pattern to form a strip graft within a Bird Base.

points at the top and bottom middle where the original base had only single points.

Now, let's look at what we've accomplished. The Bird Base that we started from had five flaps: four long ones pointing down and one short one pointing upward. Two of the long flaps at the bottom and the shorter flap at the top have now been split into a pair partway along their length. This is not entirely obvious from the final step in Figure 8.7, but if we rotate the layers so that the inserted strip stands out from the rest of the base, the gap becomes visible as shown in Figure 8.8.

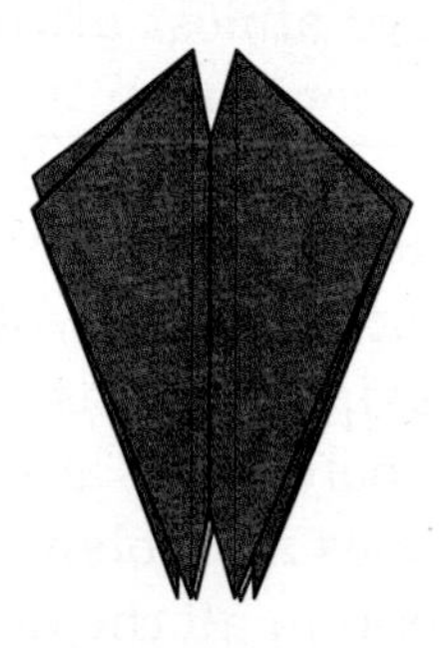

Figure 8.8.
Strip-grafted Bird Base with flaps oriented so that the gap is visible.

The interesting thing here is that after the inserted strip, we still have a uniaxial base. And it is instructive to highlight the axial creases of the new base.

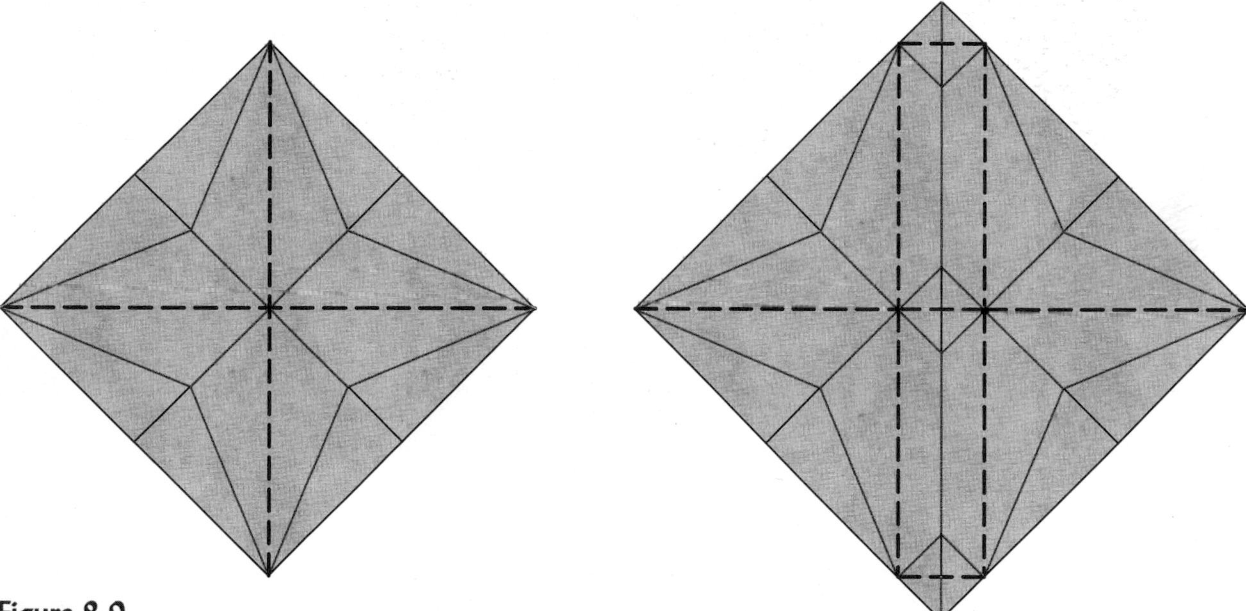

Figure 8.9.
Left: the crease pattern of the original Bird Base.
Right: the crease pattern of the strip-grafted version. The axial creases are highlighted in both.

Note that in the process of adding a vertical strip, we also created new horizontal axial creases. The Bird Base was composed of four axial polygons, which are four identical triangles.

But our inserted strip graft is similarly composed of polygons whose boundaries are axial creases (or the raw edge of the paper): In addition to the four triangles of the Bird Base, we have added two rectangles and two triangles.

We can now view grafts in a new light. While we have previously distinguished between the original base and the strip or border graft that we've added to the pattern, they are really not so different. Both the base and the graft are composed of the same fundamental elements, which are the axial polygons.

This unification allows us to approach design in a new way. In the past, we have almost always started with a base and then wrought variations upon it. But since bases are all composed of axial polygons, we can dispense with the idea of starting from a base and adding grafts; instead, we can actually build a base from scratch—maybe grafted, maybe not—simply by assembling axial polygons into a crease pattern. If we think of each axial polygon as a *tile* of creases, then the problem of design becomes a problem in fitting tiles together in such a way that we obtain all the desired flaps in our base, and the tiles fit together to make a square.

8.3. Tiles of Crease Pattern

We have already encountered several possible tiles in the Classic Bases and the grafted variants seen so far. Let's enumerate them.

First of all, there is the triangular tile that makes up the four Classic Bases. It comes in three distinct forms, depending on the orientation of the flaps (see Figure 8.10).

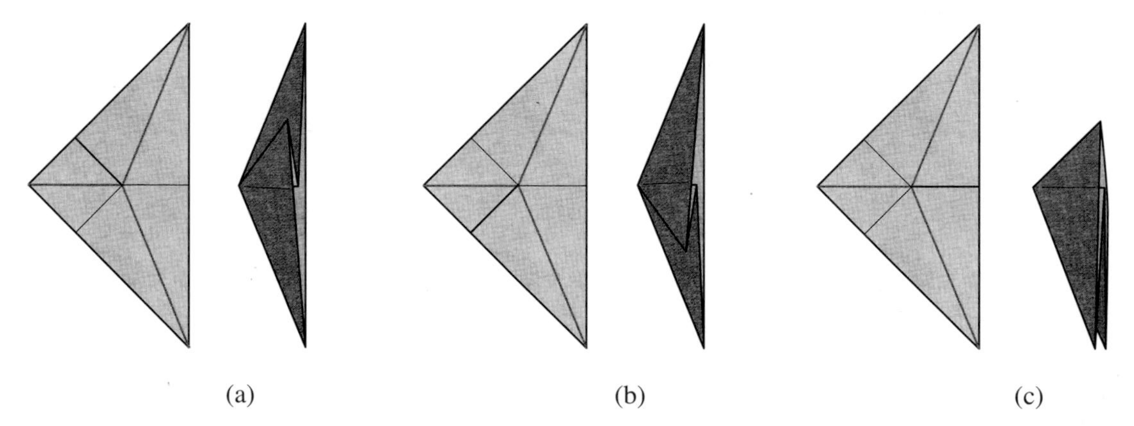

Figure 8.10.
The three orientations of the triangular tile that makes up the classic bases.

These three forms are only distinguished from one another by the location of the mountain fold in the crease pattern and the positions of the flaps in the folded form. In the crease pattern within each triangle there are four folds—one mountain fold and three valley folds—extending from the crease intersection to the corners and edges.

The Lizard and Turtle bases are also composed of triangles, but different ones: an isosceles triangle from the Lizard, and an equilateral triangle from the Turtle.

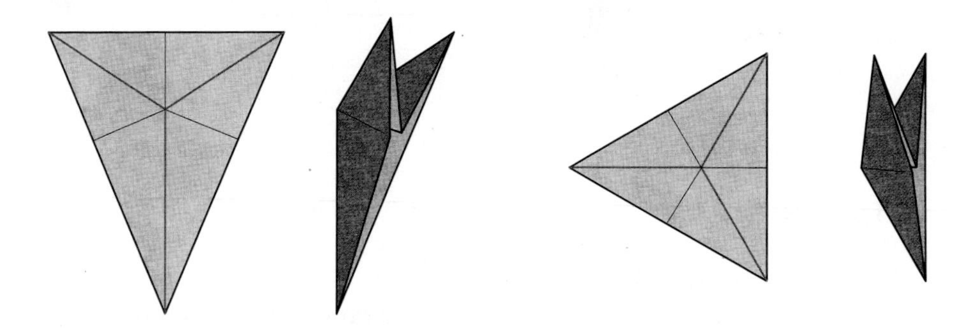

Figure 8.11.
Left: the triangle tile from the Lizard base and its folded form. Right: the equilateral triangle tile from the Turtle base.

Every such triangular tile has three possible folded forms, just like the isosceles right triangle tile shown in Figure 8.10. The creases within each tile are the three angle bisectors from each corner (which always meet at a common point) as valley folds, and a mountain fold that extends from the intersection point perpendicularly to one of the three edges. Since there are three edges, there are three possible choices for the mountain fold. When we enumerate tiles, it's not necessary to show all three forms for every triangle; you should keep in mind that for any triangle, all three are possible. The three tiles shown here are not the only possible triangular tiles, either. In fact, it can be shown that every triangle can be turned into such a tile by constructing the three angle bisectors as valley folds and dropping a perpendicular mountain fold from their intersection to an adjacent edge.

Are the only such tiles triangles? Clearly not; look again at the grafted crease pattern in Figure 8.9. The strip graft is composed of rectangles and triangles. The triangles are familiar; the rectangles are new. Rectangles, too, can be used as tiles from which crease patterns may be assembled. Figure 8.12 shows the rectangular tile from the strip graft; it, too, can be folded so that its perimeter lies along a common line. Thus, a rectangle can also serve as an axial polygon.

Just as we saw that creases can be constructed inside of any triangle to make an axial polygon, so too can creases be constructed within any rectangle, no matter what its aspect

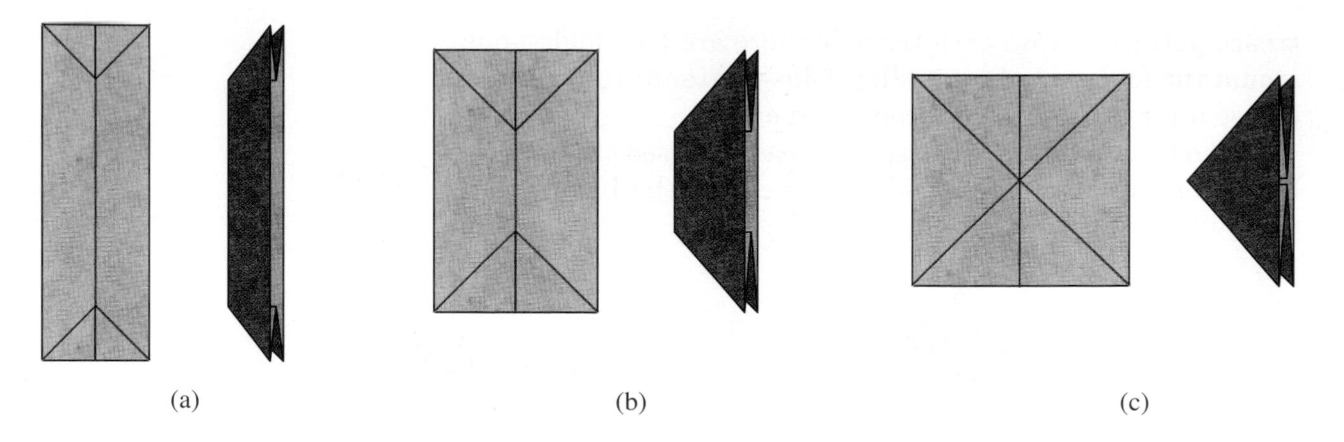

Figure 8.12.
(a) The rectangle tile from the strip graft.
(b) A wider rectangle.
(c) The limiting case of an equilateral rectangle, i.e., a square.

ratio. Figure 8.12 shows creases for three different aspect ratios, including the limiting case of a square—which gives rise to the uniaxial orientation of the Waterbomb Base as its folded form.

As we saw for the triangle, it is possible to orient the flaps of a tile in several different ways. Figure 8.13 shows several possible orientations for the flaps for one of the rectangular tiles. The variation arises in the creases that run perpendicular to an edge. We can recognize and treat the essential similarity among all such variations by simply drawing the tiles in a *generic form*, with undifferentiated creases perpendicular to all edges as in Figure 8.14. When the tiles are assembled into full crease patterns, some of those creases will get turned into mountain and/or valley folds, but we can—and will—defer that assignment until a later time.

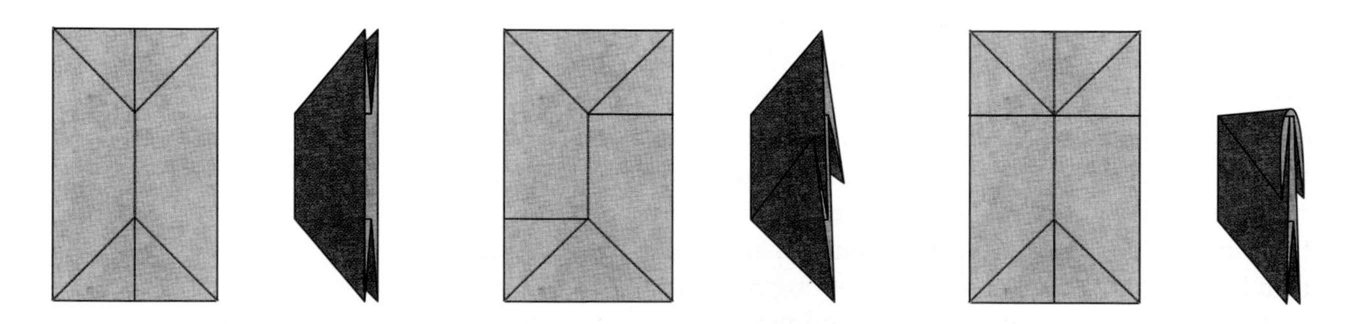

Figure 8.13.
Three different crease patterns and arrangements of flaps for a rectangular tile.

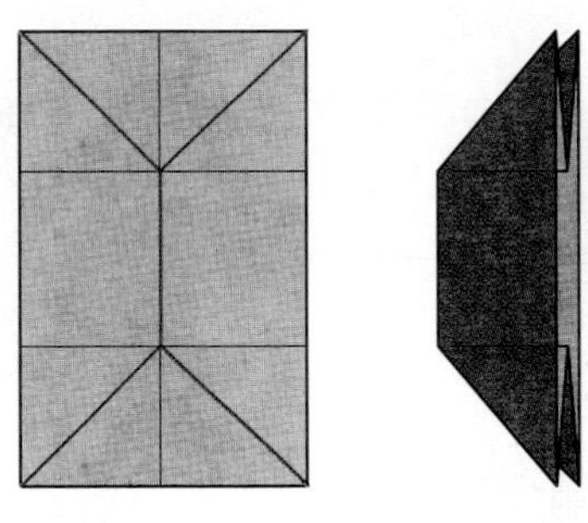

Figure 8.14.
Generic form of the rectangular tile and one possible arrangement of flaps.

Are there more possible tiles than these? Uncountably more, as it turns out. In addition to triangles and rectangles there are tiles from pentagons, hexagons, and octagons, both regular and irregular. In later chapters, we will learn how to construct new special-purpose tiles from arbitrary shapes; but even these few shapes—triangles and rectangles—allow one to construct new, custom-tailored bases.

8.4. Tile Assembly

Now, if a base can be constructed from tiles, we need some rules for their assembly. Tile assembly is not as easy as it might seem, because each tile contributes to several different flaps. When one is designing a figure, one naturally thinks in terms of flaps, and it would be very simple if any given tile corresponded to a single flap; but instead, each tile contains pieces of several flaps. So, when we assemble tiles, we need to make sure that they go together in such a way as to create entire flaps—in the right sizes, and with the right connections.

Keeping track of the correspondence between tiles and flaps is aided by decorating the tiles with circular arcs as shown in Figure 8.15 for triangular, rectangular, and square tiles.

Compare the four tile crease patterns with the folded form of each tile. Each circular arc defines a region of paper that belongs to a single flap.

The value of the circles is that when two tiles mate so that the circles line up, then the folded forms of each tile also mate so that the boundaries of adjacent flaps line up with each other. An example is shown in Figure 8.16.

In Figure 8.16, each triangular tile contains two long flaps and one short flap. If we mate the two triangles along their long edges, the long flaps merge, top and bottom, into two long flaps; but the two side flaps remain separate, so that the resulting crease pattern contains two long flaps and two short flaps: a Fish Base.

Alternatively, as shown in Figure 8.17, by mating the two tiles along their short edges, instead of long flaps merging,

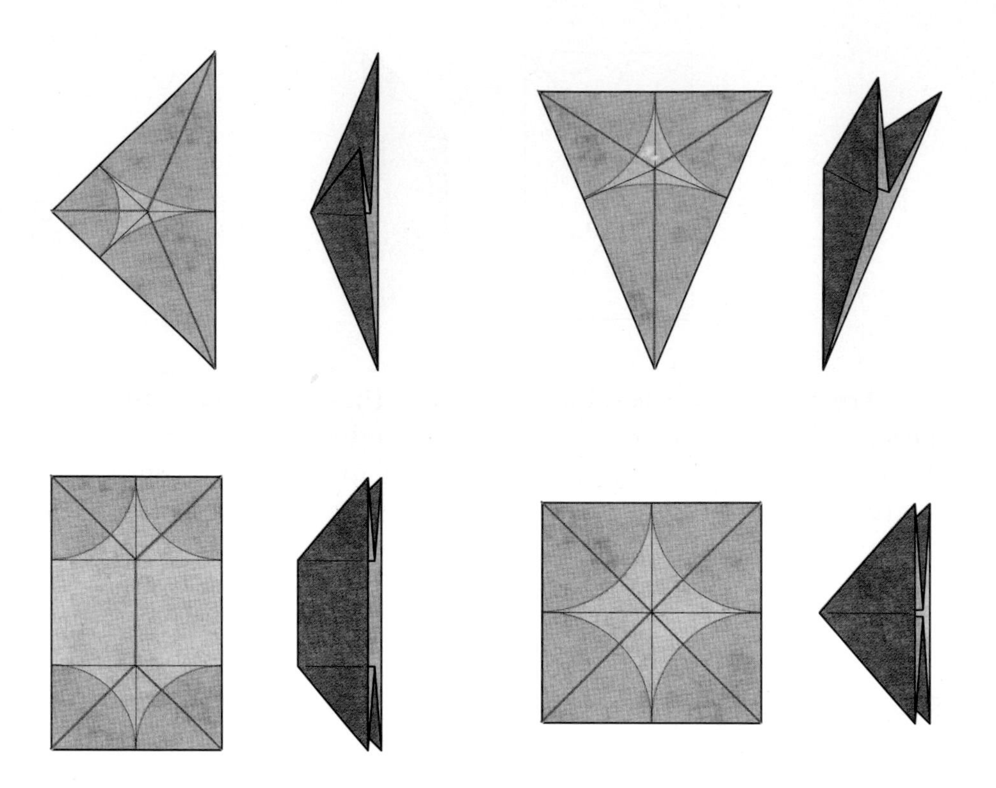

Figure 8.15.
Four tiles decorated with circular arcs and their folded forms.

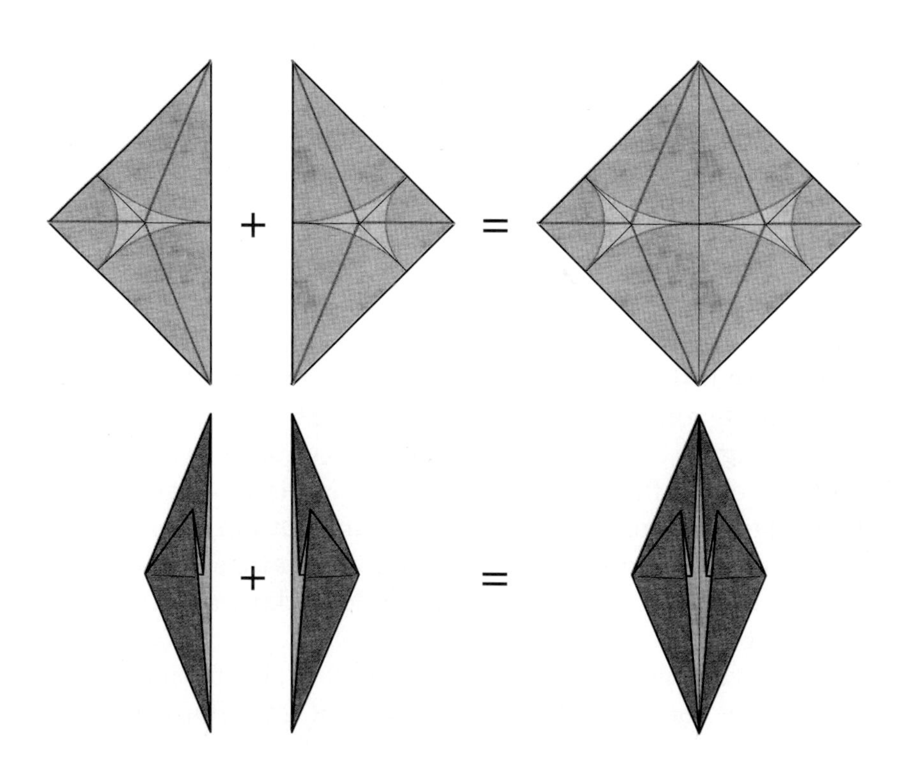

Figure 8.16.
Mating two tiles so that the circles align insures that the folded forms align as well.

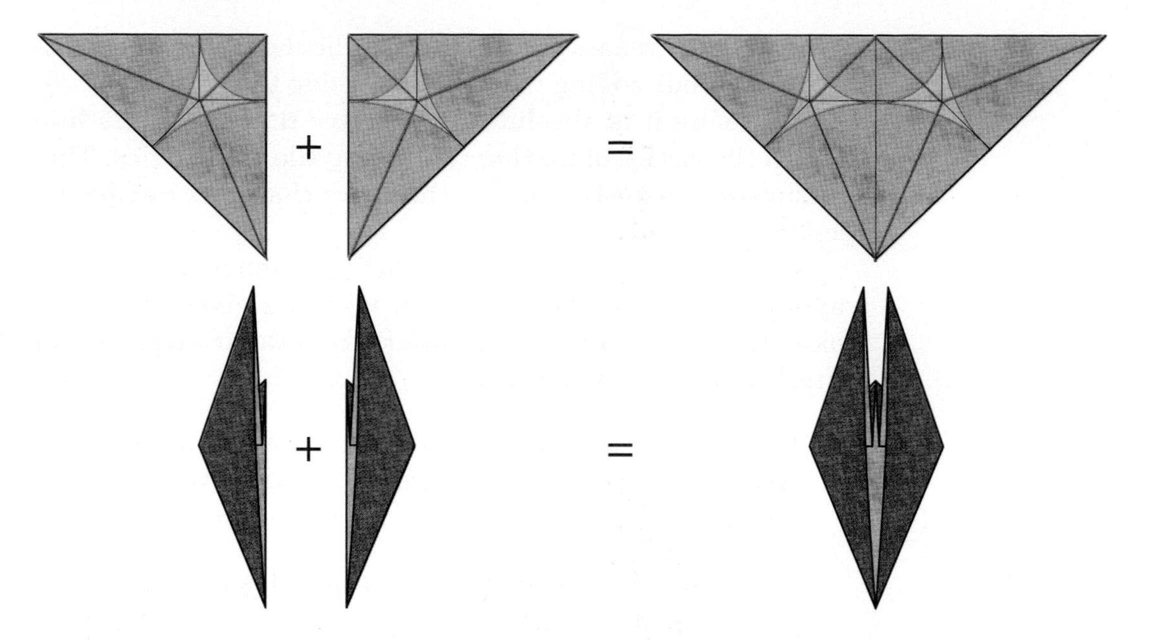

Figure 8.17.
The same two tiles can be mated along their short edges to create a different configuration of flaps.

one long and one short flap merge, and the resulting base has three long flaps and one short flap (and its crease pattern is a right triangle, rather than a square).

Observe that in each mating, distinct segments of circles correspond to distinct flaps. Thus, a simple way of determining the number of flaps created by the crease pattern is to count the number of distinct portions of circles in the crease pattern.

At this point, you might wish to explore fitting together tiles in different ways and examining the resulting crease patterns (and for a challenge, try folding the corresponding bases). The circles serve two purposes. First, they create matching rules that enforce foldability of the resulting crease patterns. If you match up two tiles with misaligned circles, you will not, in general, be able to collapse the crease pattern without adding new creases. For example, the right triangle tile and the Lizard tile cannot be mated because the circles don't line up. If you try to fold the shape in Figure 8.18, you cannot form ei-

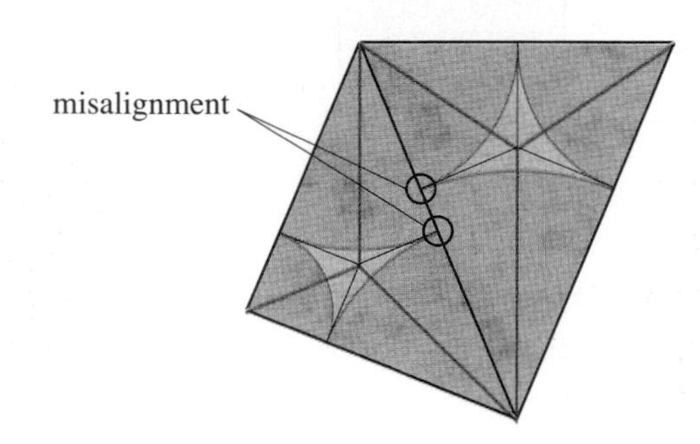

Figure 8.18.
Two tiles cannot be mated if their circles do not line up.

ther of the two creases incident perpendicularly upon the mating line without adding new creases inside the other tile.

Therefore it is absolutely necessary that all circles line up with the circles of mating tiles along tile boundaries. This is a substantial restriction on the ways that tiles can be assembled into crease patterns.

On the other hand, however, there is often more than one way that the circles can be drawn within a given tile. Let's look at the rectangular tile. It differs from the triangular and Waterbomb tiles in two ways:

- A gap in the middle of the crease pattern separates the upper pair of circles from the lower pair of circles.

- A segment in the folded form separates the upper pair of flaps from the lower pair of flaps.

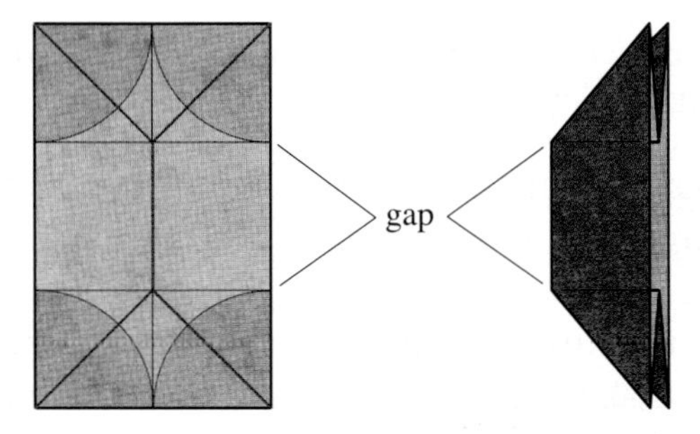

Figure 8.19.
Left: crease pattern for the rectangular tile.
Right: folded form of the tile.

It is clear from examination of the crease pattern and the folded form that the paper in the gap in the crease pattern gives rise to the paper separating the two pairs of points in the folded form. This paper is, in its own way, a kind of flap as well; but it's not a loose, isolated flap; it's a flap that connects other flaps. We can allow for such a feature in a crease pattern (and model) by inserting a stripe into the crease pattern that cuts across the rectangle, as shown in Figure 8.20.

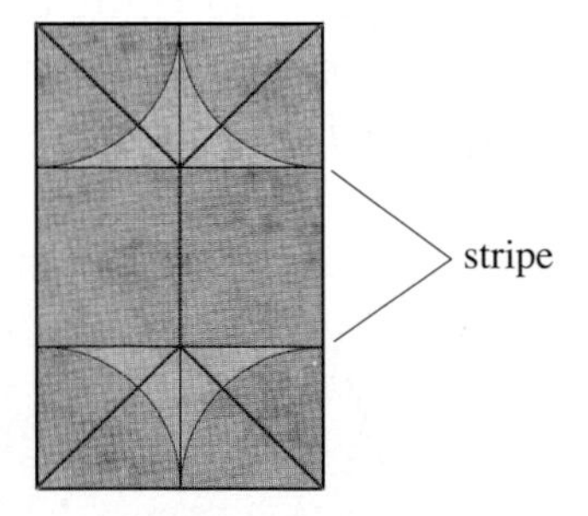

Figure 8.20.
The completed rectangular tile contains a river running across its middle.

We will give this stripe that separates groups of circles a special name: we will call it a *river*, for a reason that will shortly become apparent.

What about triangular tiles? Are there analogous structures?

The rectangular tiles give rise to two pairs of flaps separated by segments—like the body between the front and hind legs of an animal. We can similarly think of a triangular tile as giving rise to a pair of flaps separated from a third flap by a body, as shown in Figure 8.21.

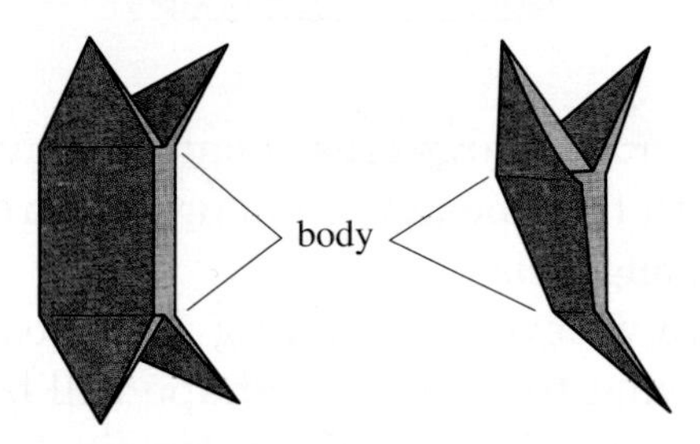

Figure 8.21.
Both a rectangle and a triangle can be folded into a shape with a body separating one or more flaps.

We can decorate the triangular tile with its own river corresponding to the body, so that the river is distinct from the circle representing the flap. It's not hard to see that while the river in the rectangular tile is a rectangle, for the triangular flap, the appropriate decoration is a segment of an annulus, i.e., a rectangle bent along a circle, as in Figure 8.22.

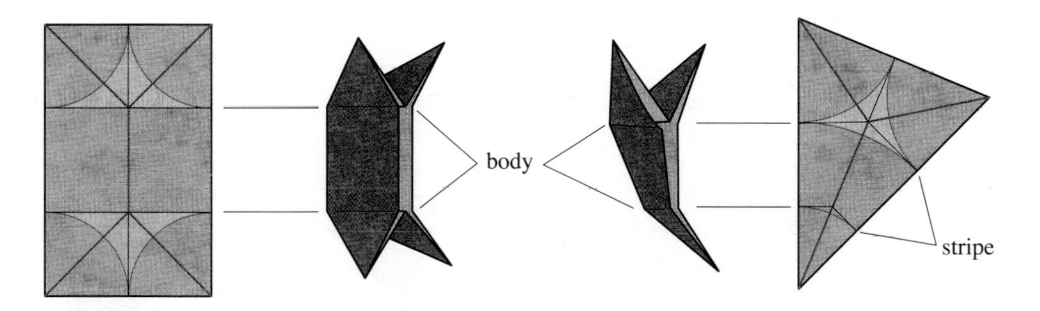

Figure 8.22.
A body can be inserted into a triangular tile by representing it as a partial annulus.

This division is not unique to the isosceles triangle tile; for any corner of any tile, the circle at that flap can be subdivided into a circle plus an annular river, thereby allowing it to be mated to a rectangular tile or to any other tile similarly divided.

So, for example, the rectangular tile and two divided isosceles triangle tiles can now be mated, one on either side, as shown in Figure 8.23. We enforce the mating of the circles on both sides, which constrains the aspect ratio of the rectangle relative to that of the two triangles.

We must enforce mating of both the circles and the rivers, as shown in Figure 8.23. And now, perhaps, you see the reason

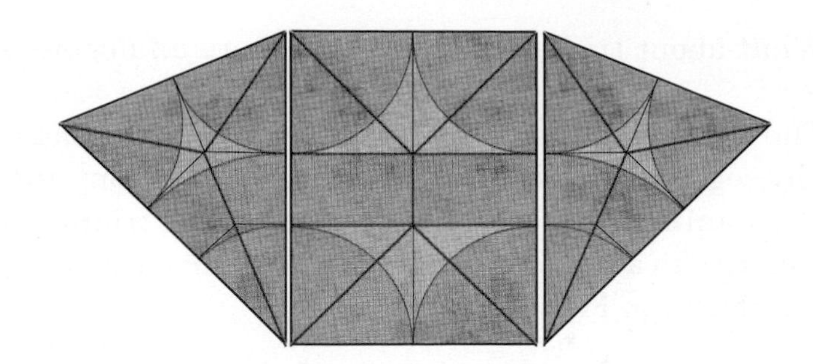

Figure 8.23.
Mating of two isosceles triangle tiles with a rectangular tile.

for the name *river*: in a large crease pattern, rivers are regions of constant width that meander among the circles like a river meandering among hills.

Now, before we even try folding this crease pattern, we can determine what the resulting shape will be simply by examining the circles and rivers. There are six distinct segments of circles; each circle will create a distinct flap. The four circles at the top are separated from the two at the bottom by a river running across the pattern; consequently, the folded shape should have six flaps with four at one end separated from two at the other by a body.

And indeed, if we fold this crease pattern, assigning crease directions as shown in Figure 8.24, that is exactly the shape we obtain.

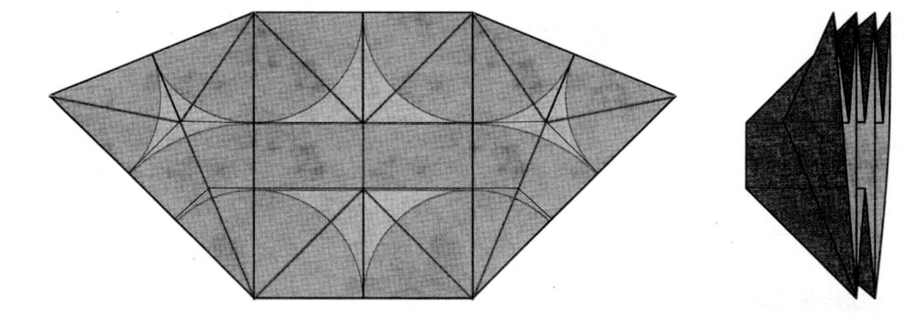

Figure 8.24.
Left: crease pattern.
Right: folded form of the resulting shape.

This crease pattern isn't a square, of course. But we can make a square pattern by packing these tiles into a square, as shown in Figure 8.25. There are two possible orientations; the axis of symmetry can be oriented along the edge of the square or along the diagonal. Packing the tiles in along the diagonal is a bit more efficient, but still leaves some unused paper at the top and bottom of the square.

No problem: we can simply add more tiles (suitably decorated by circles and/or rivers) to create more flaps and consume the rest of the paper in the square, as shown in Figure 8.25.

By enforcing circle matching, we ensure that the crease pattern can be folded flat (we will have to add some creases to the two sliver triangles along the upper edges and assign crease

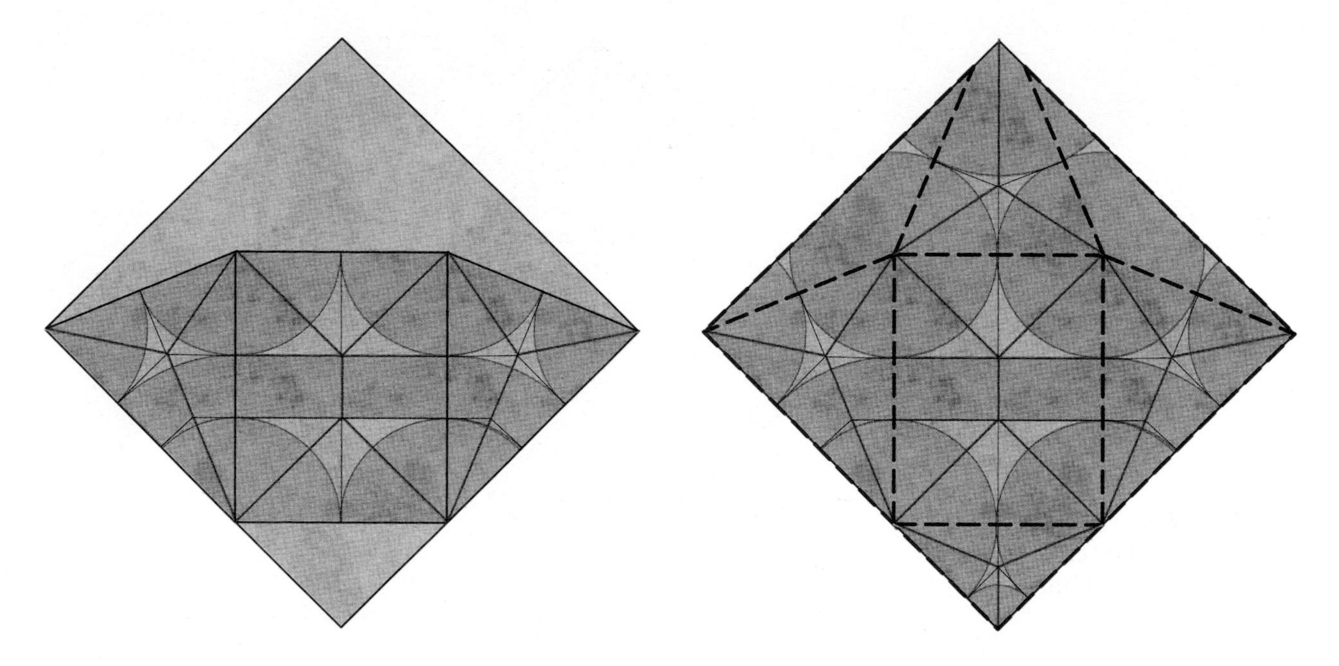

Figure 8.25.
Left: the shape fit into a square.
Right: add a few more tiles to add more flaps and consume the remaining paper in the square.

directions to the generic creases). Furthermore, by counting circles and rivers, we can elucidate the structure of the resulting base before we have even begun to fold. There are five circles at the top of the pattern, all touching; these will give rise to five flaps. The topmost circle is larger than the other four; that flap will be longer than the others. There are two circles at the bottom, separated by a river that runs across the pattern. Those two circles will give rise to two more flaps, the same length as the upper four, but separated from them by a body; and finally, the tiny circle at the very bottom will turn into a small flap, joined to the other two that it touches.

And indeed, with suitable crease assignment, this pattern can be folded into the shape shown in Figure 8.26, which matches every element of the structural description.

This structure is not just a contrived example; I have used it to realize a Pegasus. The folded model and its crease pattern is shown in Figure 8.27. Folding instructions are given at the end of the chapter.

If you compare the crease patterns in Figures 8.26 and 8.27, you will see that although the overall structure is the same, the second crease pattern has many more creases within the individual tiles. It is useful, in fact, to examine the various tiles because they are illustrative of some of the variations you can find within tiles.

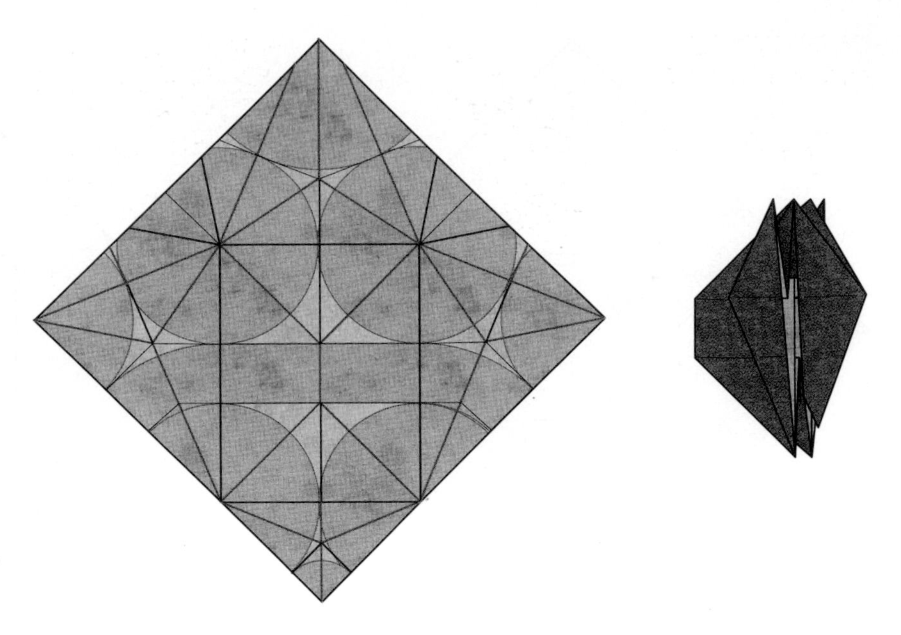

Figure 8.26.
Left: the finished crease pattern.
Right: the folded base.

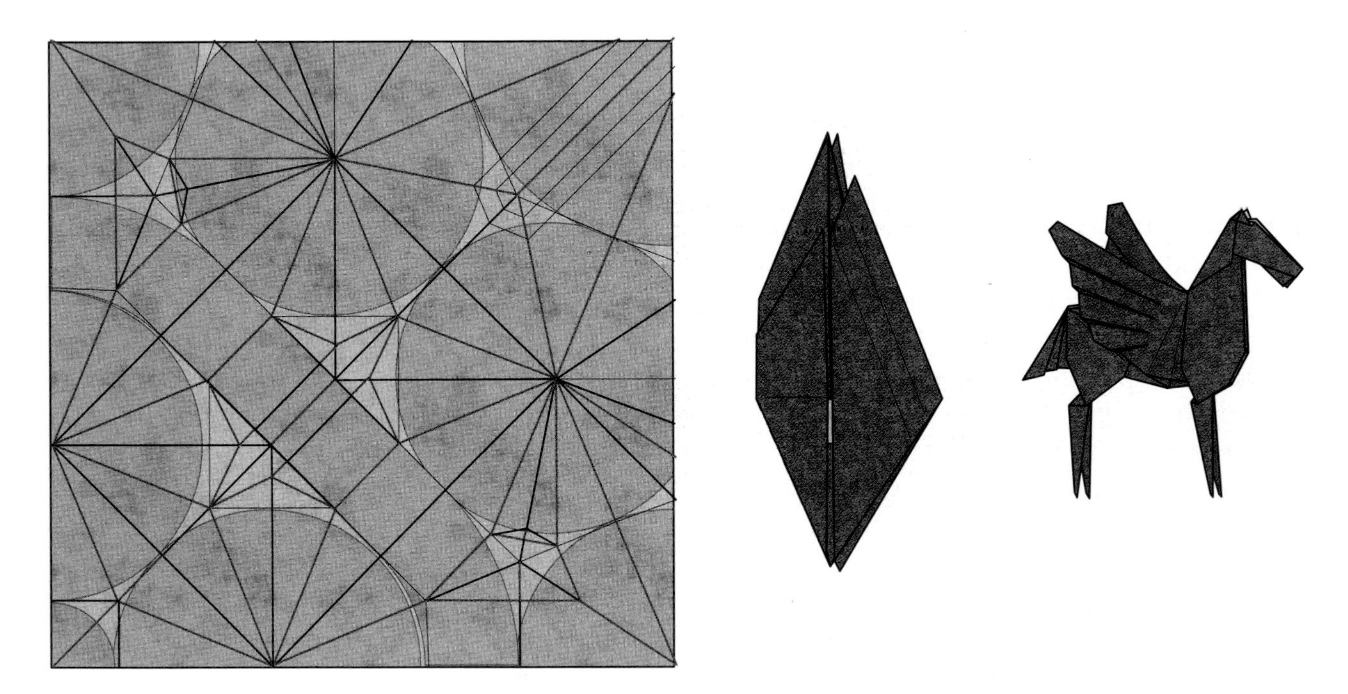

Figure 8.27.
Left: Pegasus crease pattern.
Middle: folded base.
Right: finished Pegasus.

First, let's look at the rectangular tile that forms the body and four legs of the animal. The two forms—the basic crease pattern, and the form in the folded model—are shown in Figure 8.28.

In the two tile crease patterns, the circles and rivers have the same radii and width, respectively, and the flaps have the

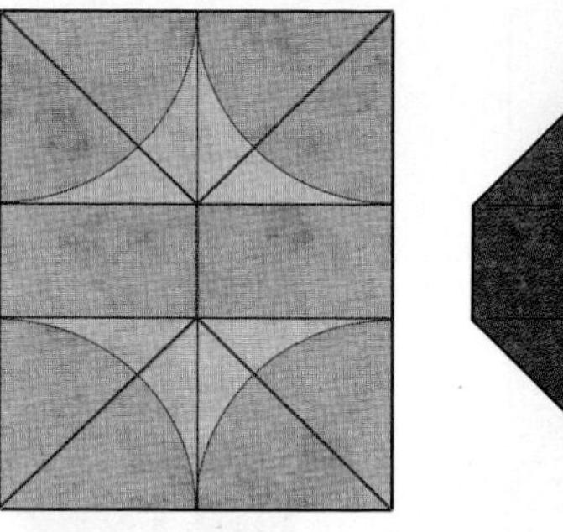
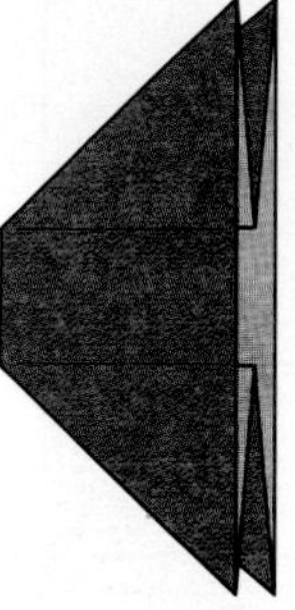
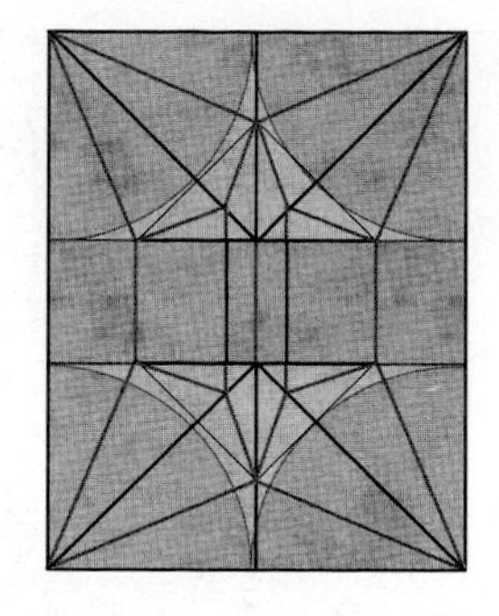
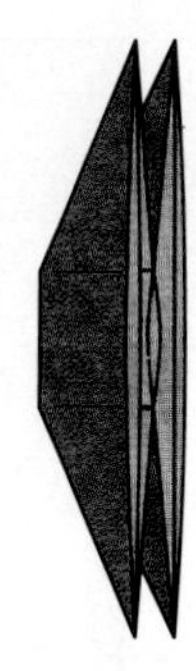

Figure 8.28.
Left: the basic rectangle tile.
Right: the tile with additional creases.

same lengths in the folded forms. The only difference lies in the widths of the folded flaps and the number of layers in each flap. The narrower flaps necessarily have more layers.

Similarly, the triangular tiles also have somewhat more complex crease patterns than we saw previously.

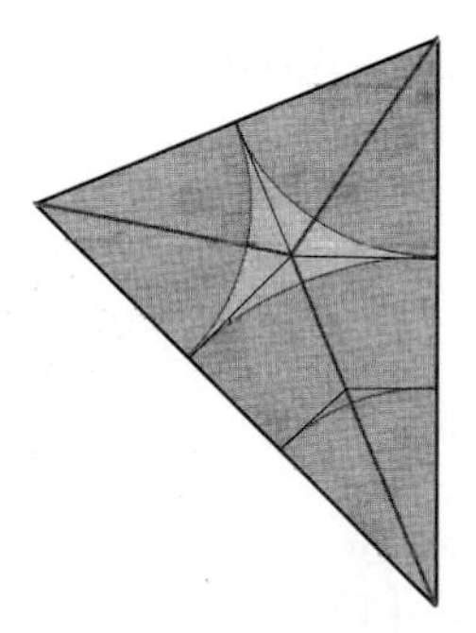
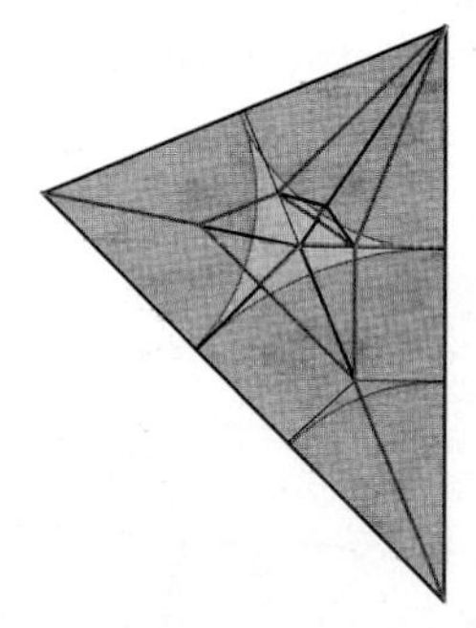

Figure 8.29.
Left: the basic isosceles triangle tile crease pattern.
Right: the same tile with additional creases.

If you fold these two patterns, you will see that the difference again lies in the width of the flap and the number of layers, rather than the length or connectivity of the flaps.

As a third example, recall that the Lizard base came in two forms: one with wide flaps, one with narrow flaps. First, let's look at the wide-flap (simpler) version of the crease pattern and base. I have highlighted the axial polygons and drawn in the circles and river on the tiles.

Observe that the river is meandering through the pattern in a way that clearly illustrates its name.

Now, look at the actual crease pattern for the Lizard and its base. I have used the same outlines for the axial polygons.

The narrow form of the Lizard base uses the same tile outlines, circles, and rivers, but there are many more creases within each tile.

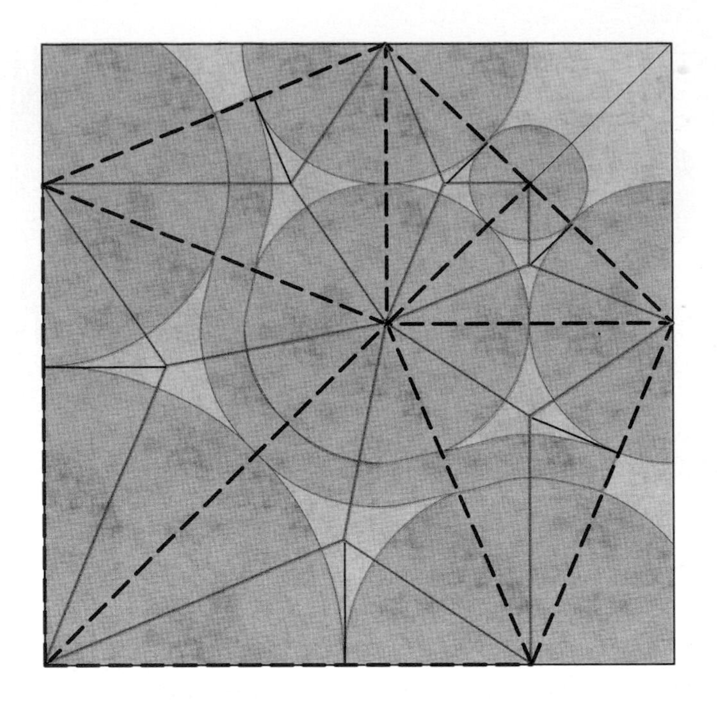

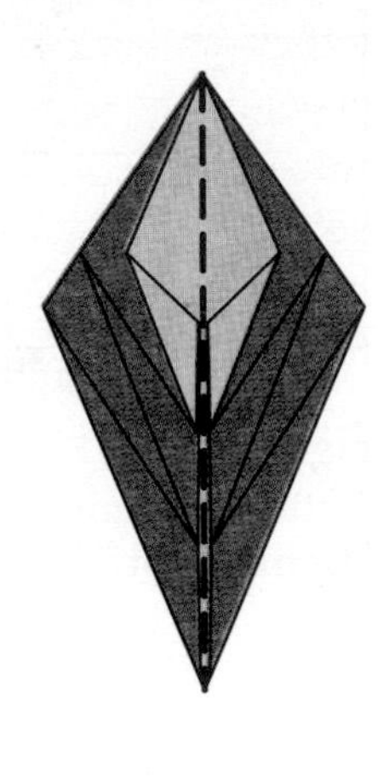

Figure 8.30.
Left: crease pattern with axial polygons highlighted and circles and rivers drawn.
Right: base for the Lizard with axis highlighted.

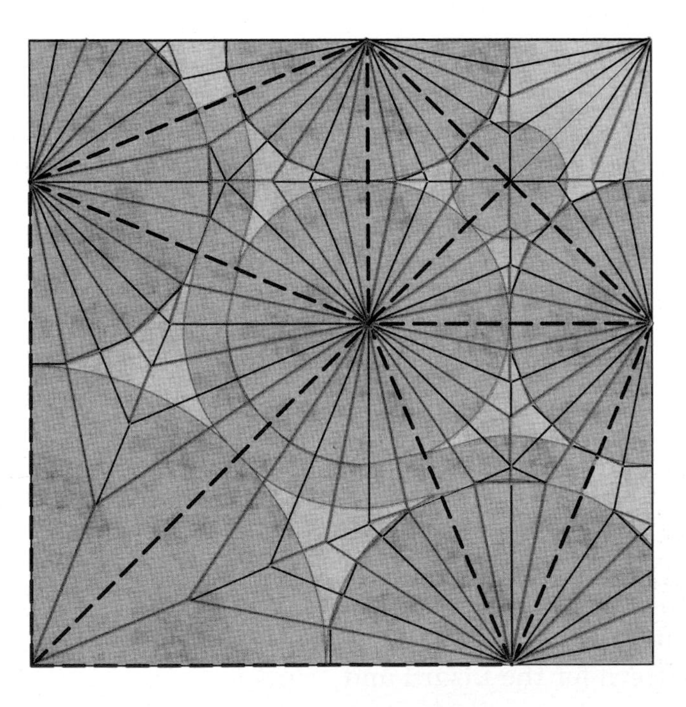

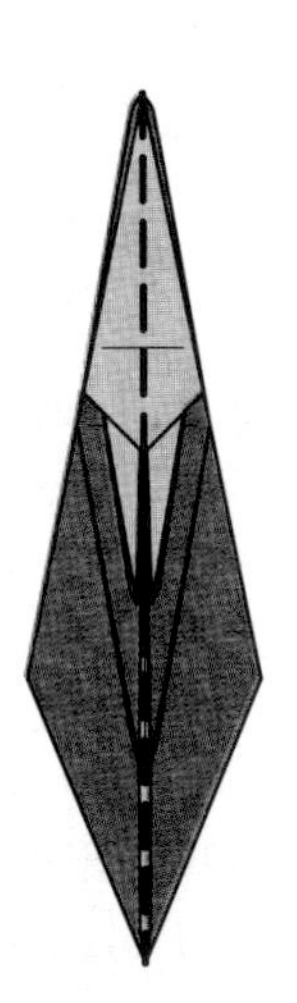

Figure 8.31.
Left: Lizard crease pattern with tile outlines, circles, and rivers.
Middle: base.
Right: finished Lizard.

8.5. A Multiplicity of Tiles

Do we need to keep track of all possible crease patterns for every possible tile? Fortunately not. The more complicated crease patterns can often be derived from simpler patterns by narrowing the flaps of the folded form of the tile in one of several different ways. The most common techniques for narrowing take the form of sink folds (which accounts, in part, for the prevalence of sink folds in complex origami designs). A simple tile with wide flaps can have its flaps narrowed by sinking one or more times. This sinking can give a much more complex crease pattern, but it should not distract you from understanding the essential simplicity of the underlying tile.

For example, let's take the isosceles triangle tile. There are several ways of narrowing the flaps. Two of the more common are shown in Figures 8.32 and 8.33. One keeps the flaps triangular while making them more acute; the other turns them into quadrilaterals.

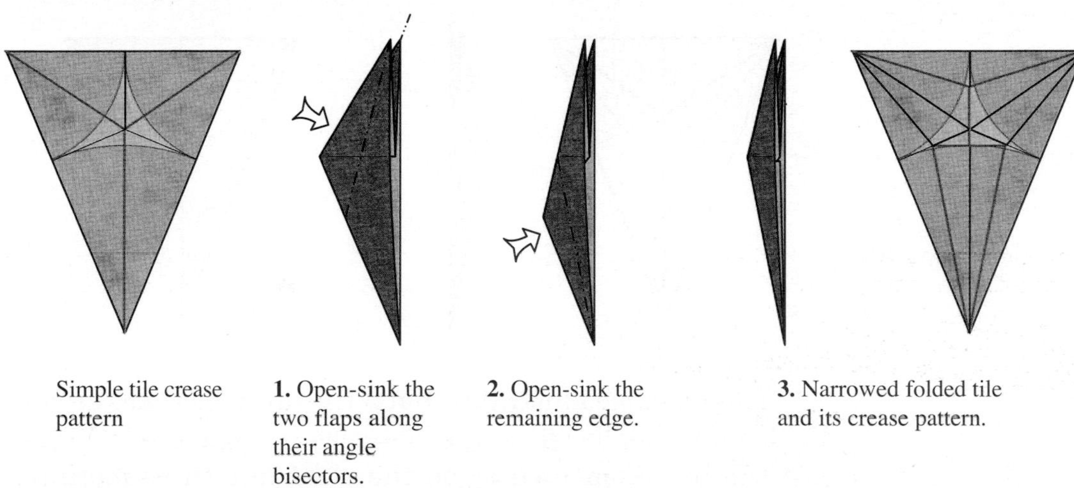

Simple tile crease pattern

1. Open-sink the two flaps along their angle bisectors.

2. Open-sink the remaining edge.

3. Narrowed folded tile and its crease pattern.

Figure 8.32.
Procedure to narrow a tile using angled sink folds.

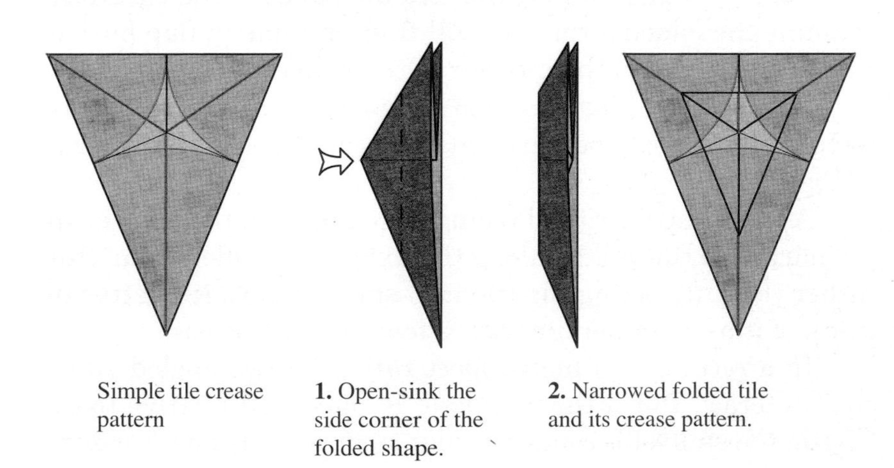

Simple tile crease pattern

1. Open-sink the side corner of the folded shape.

2. Narrowed folded tile and its crease pattern.

Figure 8.33.
Procedure to narrow a tile using sink folds parallel to the axis of the folded tile.

Which of the two you use is primarily a matter of taste. The sequence in Figure 8.32 gives flaps that taper evenly, but it is difficult to form the closely spaced creases as they converge on the tip of the flap. The sequence in Figure 8.33 distributes the layers more evenly but doesn't taper smoothly to its tip. And there are many more possibilities than these: You can sink some flaps and not others, and perform double, triple, and more complex sinks. The important thing is, this narrowing can be performed after the base is folded, so you can do all your design using the simplest possible tiles, then go back and narrow them if desired.

You might have noticed that when we narrow a tile as in Figures 8.32 and 8.33, some of the creases created within the tiles end up lying along the axis in the folded form; that is, they are also axial creases. For example, the isosceles triangle tile narrowed with angular sinks has several new creases that lie along the axis in the folded form, as shown in Figure 8.34.

Figure 8.34.
Left: the narrowed isosceles tile with all axial creases highlighted.
Right: the same tile composed of three triangular tiles and their circles.

This example illustrates that a tile can sometimes be subdivided into smaller tiles; the isosceles triangle tile in Figure 8.34 can be decomposed as on the right into three more triangle tiles, each with three circular arcs; and as the figure shows, the point where the three tiles come together can give rise to a tiny fourth flap. If you fold an example and carefully examine the folded form, you will find that fourth flap buried within the layers in the interior of the shape.

Rectangular tiles can also be narrowed. A long, skinny rectangular tile can be narrowed with angled sinks as shown in Figure 8.35.

As with the narrowed triangle tile, some of the creases in the narrowed tile will lie along the axis of the folded form. But rather than dissecting the tile into smaller tiles, it's better to think of it as a simple tile with a few extra creases.

In a rectangle of high aspect ratio, the two angled sinks don't interact. But if the rectangle is shortened relative to its length, the sink folds connect and introduce some new horizon-

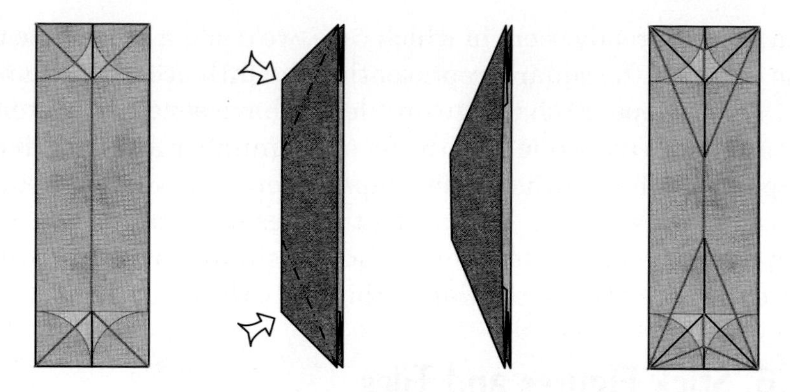

Figure 8.35.
A simple rectangular tile, narrowed with sinks, becomes a more complex tile.

tal creases. For the so-called silver rectangle, whose width-to-length ratio is $1 \times \sqrt{2}$ (this is the same proportion as European A4 letter paper, 210 × 297 mm), the narrowed form of the tile has a particularly elegant crease pattern, shown in Figure 8.36.

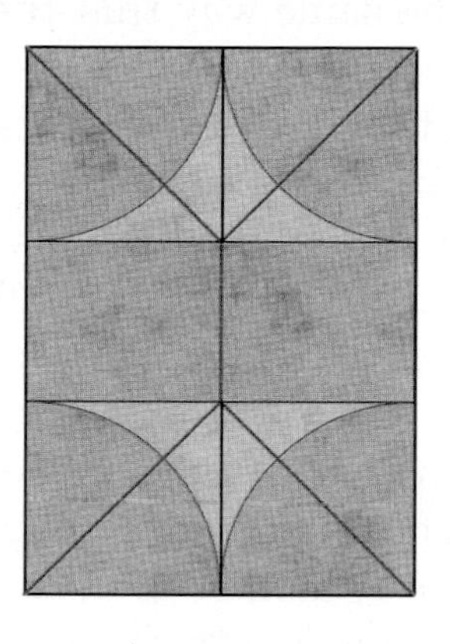

Figure 8.36.
Left: simple tile for a silver rectangle.
Right: narrowed form of the same tile.

The square, too, has a narrowed form. The simple tile for a square is, as we saw, the Waterbomb Base. The narrowed form is—surprise!—the same crease pattern as a Bird Base (see Figure 8.37).

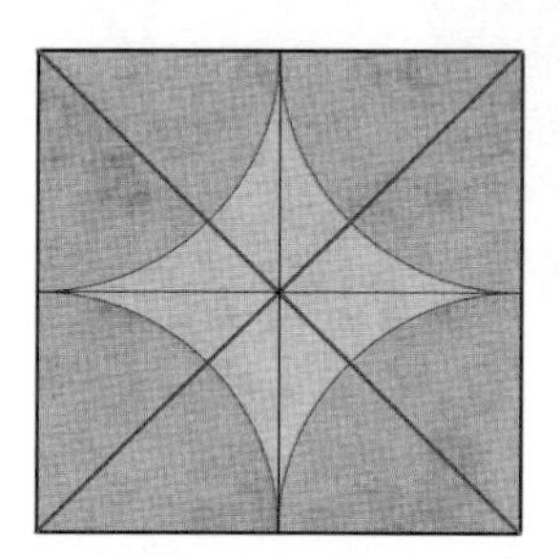

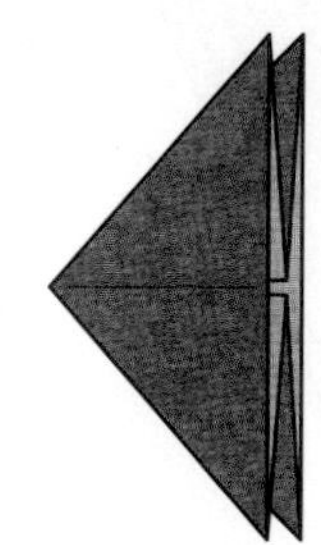

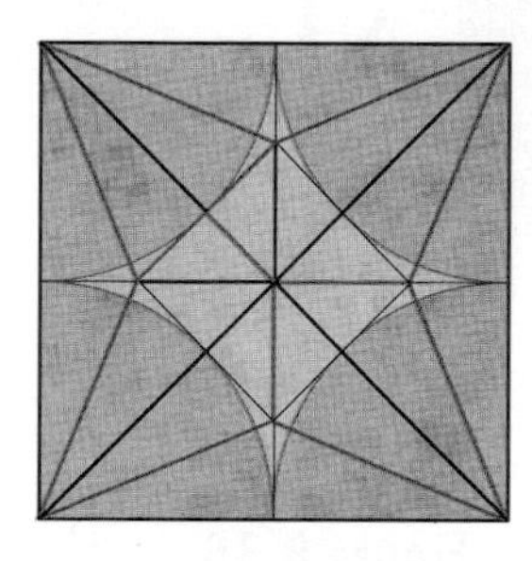

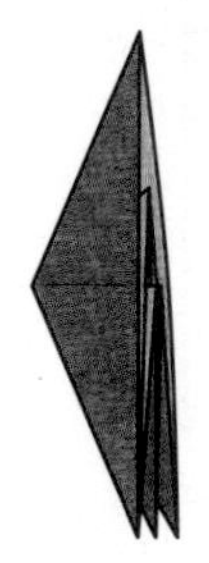

Figure 8.37.
Left: simple tile for a square.
Right: narrowed form of the same tile.

So, we could treat this tile as the narrow form of a square tile, or we could decompose it into four of the triangular tiles that we've already seen (in which case, we'd add a fifth circle to the center of the square, representing the fifth accessible flap).

So, it appears that a given tile can have several different crease patterns inside it with the same number and length of flaps; only the widths of the flaps differ. The simplest tiles have the widest flaps. By sinking the tiles in various ways, we can make the flaps narrower; in fact, by sinking over and over, we can make each of the flaps arbitrarily thin.

8.6. Stick Figures and Tiles

At this point, it is helpful to introduce a pictorial notation for the arrangement of flaps in the folded form of a tile: the *stick figure*. We represent each flap in the folded form by a line segment whose length is equal to the length of the flap, with line segments joined to each other in the same way that the flaps are joined to each other.

If two circles touch within the tile, then their corresponding flaps touch, and we will represent that connection by drawing the sticks as touching at their corresponding end. Thus, for example, the folded form of a triangular tile—three flaps—can be represented schematically by three lines coming together at a point.

Figure 8.38 illustrates this schematic form for two of the triangle tiles. A triangular tile can be folded into a shape with three flaps; we will represent the tile by a three-branched stick figure, in which the sticks are the same length as the flaps in the tile.

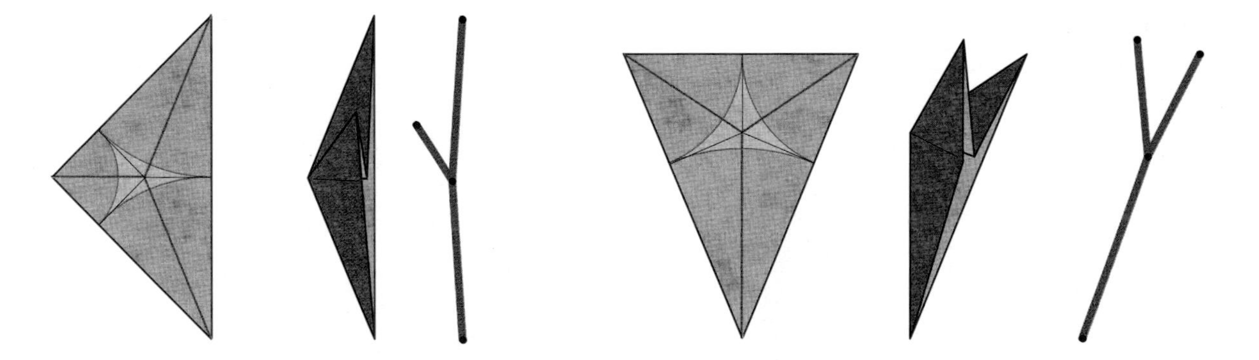

Figure 8.38.
Schematic representation of two triangular tiles. The length of each stick is equal to the length of the flap.

The stick figure can also be viewed as the limiting case as the flaps of the folded form are made narrower and narrower. However, there is an important difference between the stick figure and the base. Although we often draw the branches of the stick figure in the same orientation as the flaps of a base (as in Figure 8.38), there is no significance to the order of sticks around their common endpoint. The flaps of the base may be superimposed, one atop the other, but in order to distinguish adjacent flaps, I will always draw the stick figure schematic with the segments separated by some angle. So in the stick figure, it is not the angles between segments that are significant; only the lengths of the segments and their connections to each other matter, because the length of each segment indicates the length of the corresponding flap. This length is also equal to the radius of the inscribed circular arc for flaps represented by circles.

For a tile with a river running through it, we will represent both the flaps and the connection between them by lines as well. Thus, a rectangular tile with a river is represented schematically by four lines joined in pairs with a connection between the pairs, while a square tile composed of four circles would be represented by four lines all coming together at a point.

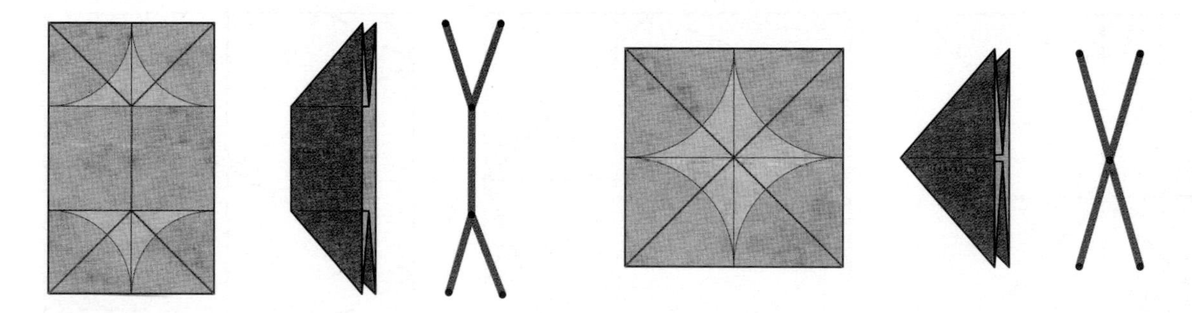

Figure 8.39.
Schematic representation of a rectangular tile with a river and a square tile.

The stick figure schematic is a useful tool because it doesn't depend on the specifics of the creases within a tile, only upon the circles and rivers within the tile. But its utility extends beyond individual tiles; we can also use the stick figure to represent the structure of an entire base.

We can use the stick figure to sketch the structure of entire crease patterns by treating the entire pattern as one large collection of circles and rivers, using a few simple rules:

- Each circle is represented by a line segment whose length is the radius of the circle. One endpoint of the segment corresponds to the center of the circle; the other corresponds to the boundary of the circle.

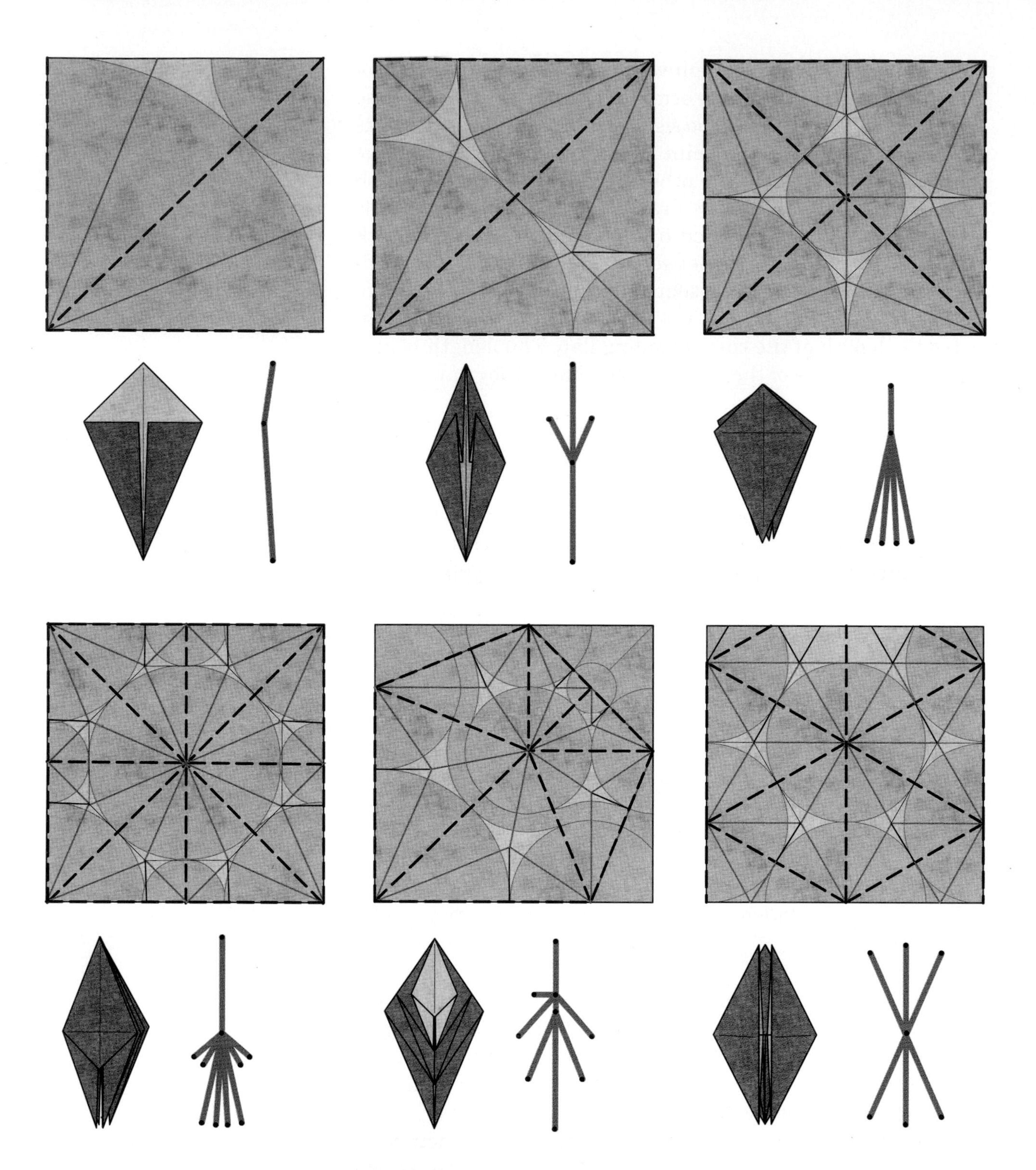

Figure 8.40.
Six bases, with inscribed circles and rivers; the bases, and corresponding stick figures.

- Each river is represented by a line segment whose length is the width of the river. One endpoint of the segment corresponds to one bank of the river; the other corresponds to the other bank of the river.

- If two features (circles or rivers) touch anywhere, their corresponding lines are connected at corresponding endpoints.

To see how this works, let's construct the stick figures for the six bases we've been working with in this section.

The circle/river patterns within the crease patterns of the four Classic Bases consist only of circles, and so their stick figures consist only of lines emanating from a common point. Thus, in the folded bases, all of the flaps emanate from a common location. The Lizard base (fifth in the row) is a bit different, however; its circle pattern contains a river. The river gives rise to a segment that separates the two groups of points in the base.

Thus, the stick figure serves as a quick, shorthand description of both the lengths of the flaps and the way they are connected to each other. You can design a crease pattern using just tiles with circles and rivers, and by drawing the stick figure, quickly ascertain whether the pattern gives rise to the necessary combination of flaps. Only after you've found a tile pattern that gives the right number of flaps with suitable lengths do you need to fill in the tiles with crease patterns.

Let's look at an example. A square can be dissected into two rectangles plus two dissimilar squares, as shown in Figure 8.41. What would be the properties of a base constructed from these four tiles?

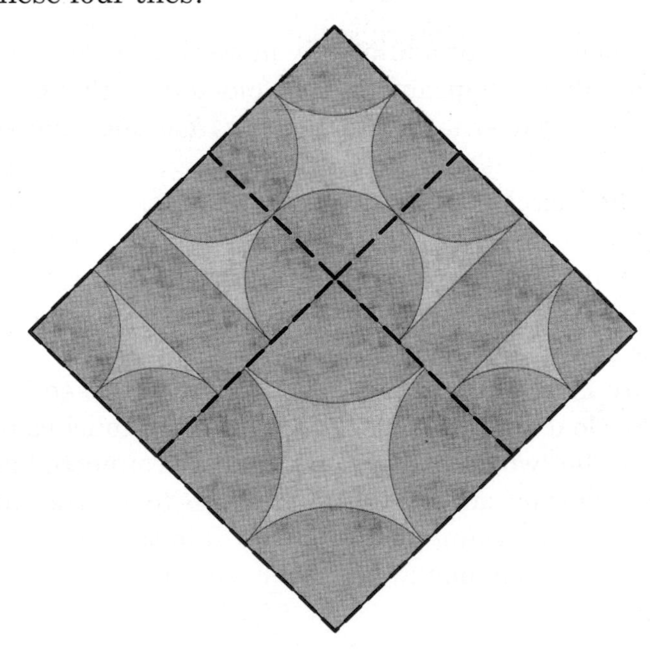

Figure 8.41.
Left: two rectangular tiles plus two square tiles fit together to make a square.

If we plug in four tiles—two squares containing four circles, plus two rectangles containing four circles and two rivers, we see that the circles in the upper square mate properly with the short sides of the rectangles, but the ones in the larger square don't mate properly with the circles and river in the rectangle. We can fix that. Recall that any circle can be subdivided into a smaller circle and an annular river; similarly, the river in each rectangle can be bisected into two rivers to mate with the newly created rivers. The result is a pattern of circles and rivers in which all matching conditions are satisfied along the edges of the tiles, as shown in Figure 8.42.

Figure 8.42.
Tiles with circles and rivers that satisfy matching conditions across tile boundaries.

And now, without adding any more creases, we can identify the number of flaps in the base folded from this structure. In the crease pattern, we label each circle and river with a letter from a to l, as in Figure 8.43.

The four circles a–d at the top are four equal-length flaps. Since a touches b and c, its corresponding line must be joined to lines b and c at the same point. Since b and c also touch circle d, that means line segment d must also be connected at the same point as well.

There is a subtlety here I don't want to speed by; even though circle d doesn't touch circle a, since d touches b and b touches a, the two corresponding flaps are connected at their base. The way you can keep this straight is to use the rule that two segments are connected at a point if in the circle pattern, you can travel from one to the other without cutting across a circle or river.

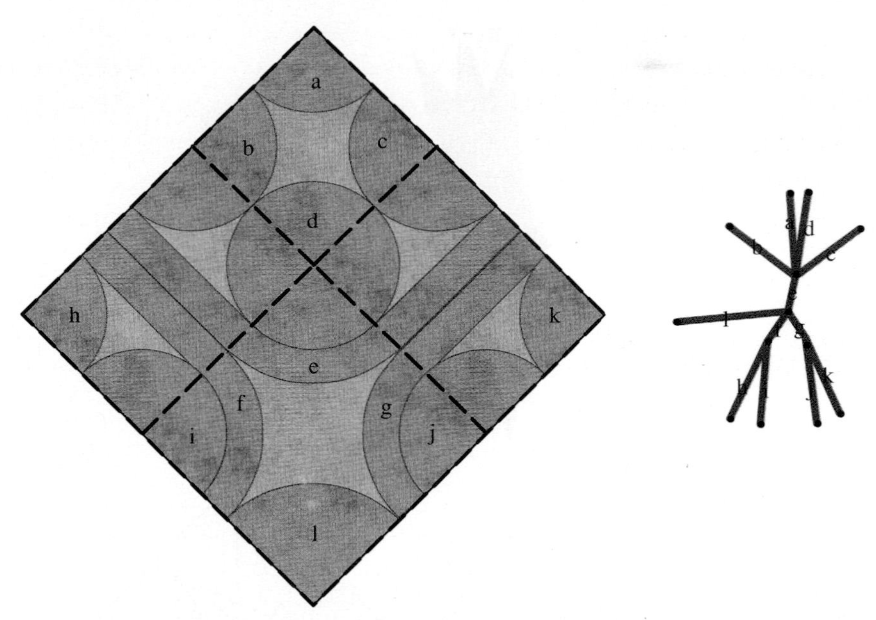

Figure 8.43.
Left: the circle/river pattern with all features labeled. Right: the schematic stick figure, illustrating the lengths and connections among flaps.

Continuing downward, flaps a–d are connected to a short segment (e), which, in turn, is connected to two more short segments (f and g) and a longer point, l. Both f and g are terminated in pairs of flaps—h and i, and j and k, respectively.

So this base will have eight longish flaps, two paired along a segment, and a single flap longer than any of them. Now, if this base meets the needs of the desired subject, we can fill it in with tile creases, as shown in Figure 8.44 with simple and narrowed tiles.

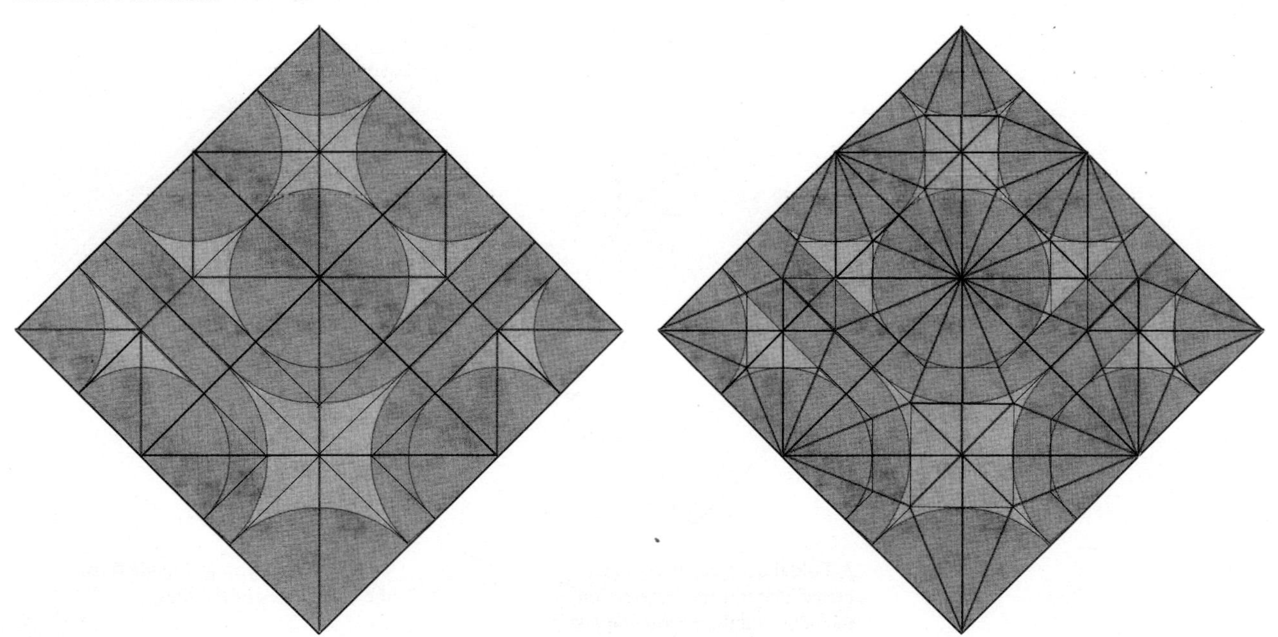

Figure 8.44.
Left: the crease pattern filled with simple tiles.
Right: the same pattern filled with narrow tiles.

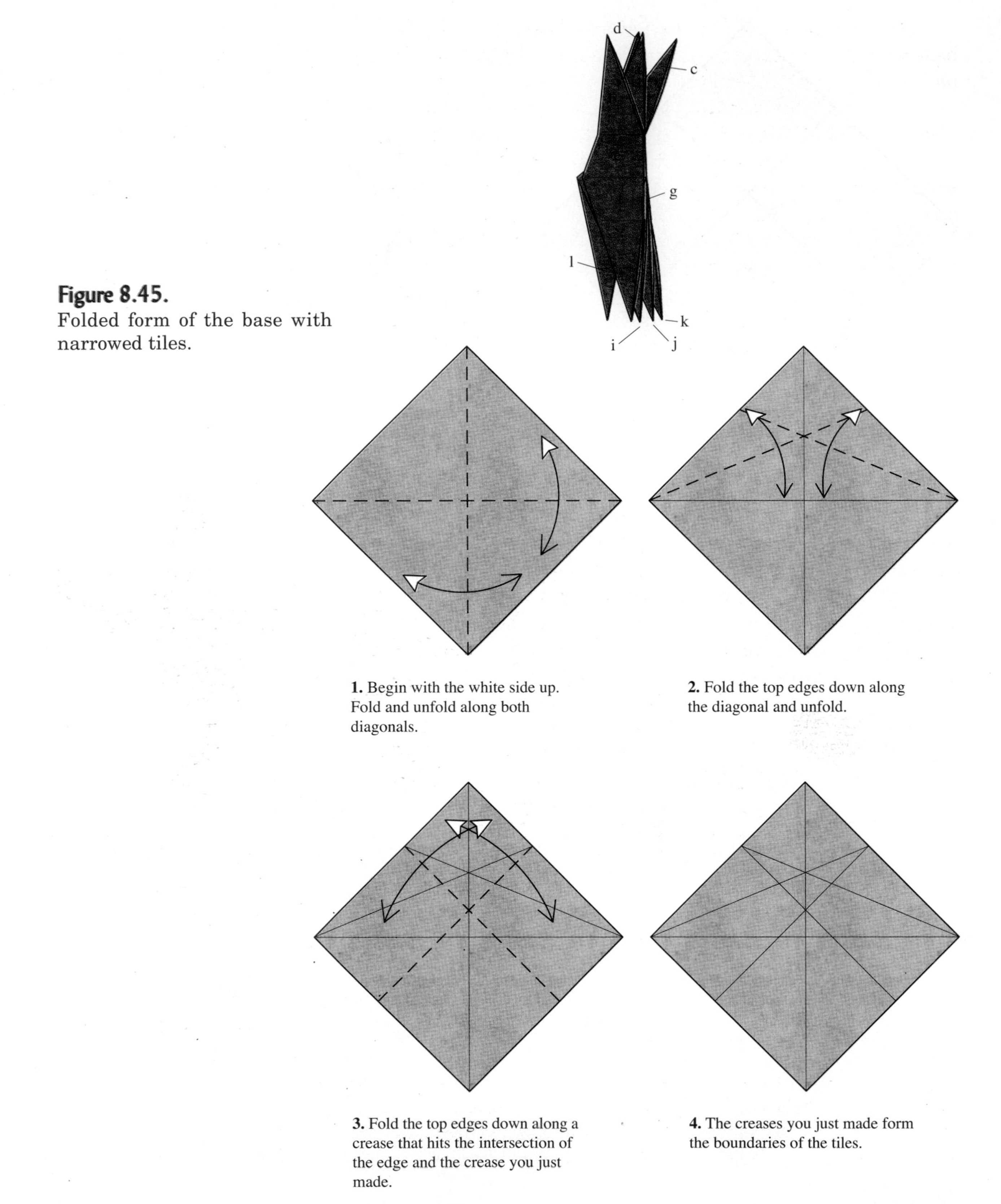

Figure 8.45.
Folded form of the base with narrowed tiles.

1. Begin with the white side up. Fold and unfold along both diagonals.

2. Fold the top edges down along the diagonal and unfold.

3. Fold the top edges down along a crease that hits the intersection of the edge and the crease you just made.

4. The creases you just made form the boundaries of the tiles.

Figure 8.46.
Folding sequence to divide the square into square and rectangular tiles.

And if we fold either pattern into a base, we will obtain a base with the same number, length, and configuration of flaps as is predicted by the circle pattern.

You might enjoy folding the base for yourself and seeing if you can identify the flaps. The folding sequence in Figure 8.46 gives the appropriate proportions for the division into squares and rectangles; from there, the other folds can be constructed by bisecting various angles.

I have used a dissection very similar to this for a model of Shiva as Nataraja, but using rectangles of proportion $2 \times (1+\sqrt{2})$, rather than the silver rectangle ($1 \times \sqrt{2}$). The crease pattern, base, and folded model are shown in Figure 8.47. Can you identify the individual tiles?

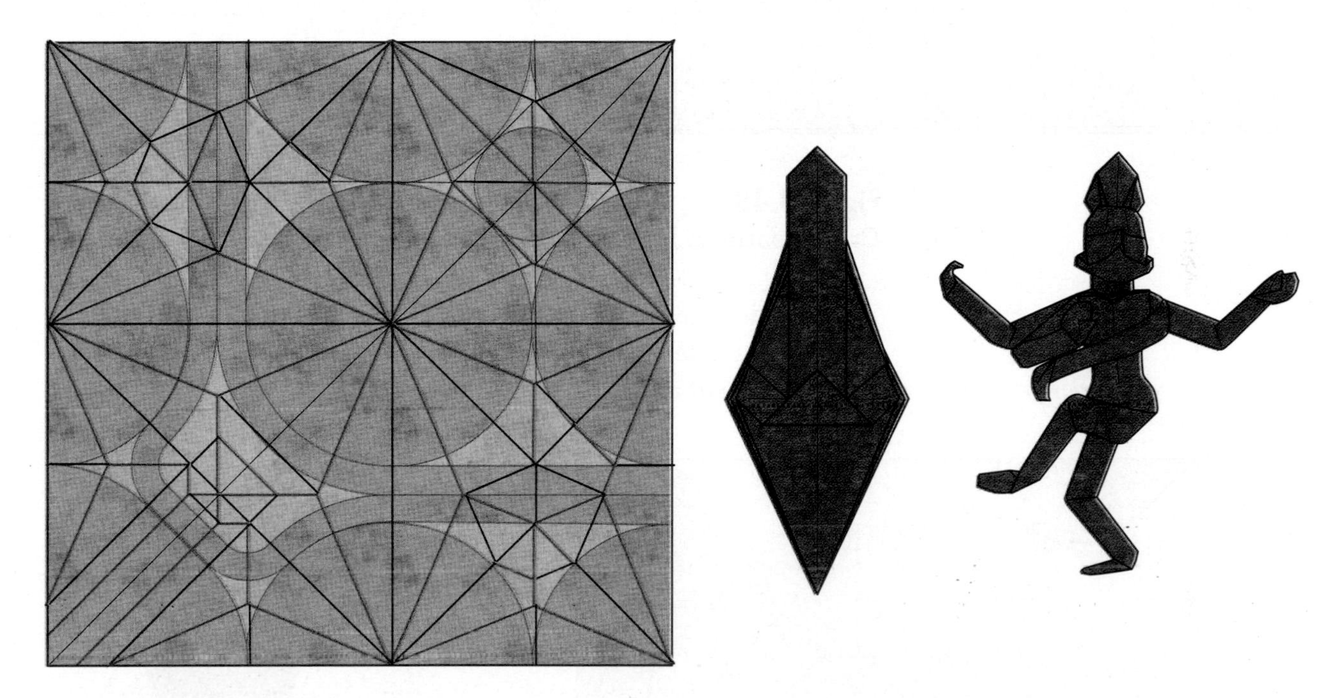

Figure 8.47.
Crease pattern, base, and folded form of Shiva.

The same base may be used in several different orientations to create distinctly different models; it is often not at all obvious from the folded form that the underlying base is the same. But if you examine the pattern of flaps—where are the long flaps, where are the short, how are they joined—you can perceive the essential similarity. The same structural base as was used in Shiva can also be used to realize a Hercules Beetle, as shown in Figure 8.48.

One can also combine techniques: construct a base by tiling, then split one or more flaps using point-splitting. The Pray-

Figure 8.48.
Crease pattern, base, and folded model of the Hercules Beetle.

ing Mantis shown in Figure 8.49 employs nearly the same base as the Hercules Beetle, but splits the middle flap into four points to form antennae.

Figure 8.49.
Crease pattern, base, and folded model of the Praying Mantis.

8.7. Dimensional Relationships Within Tiles

In any tile, every circle or river encounters two sides of the tile; this establishes a relationship between the two sides. The union of all such relationships can constrain the possible sizes of circles and rivers within the tile. In a triangle tile composed of three circles, it is clear from Figure 8.50 that each side of the triangle has a length equal to the sum of the radii of the two adjacent circles (which, you recall, are equal to the lengths of their associated flaps).

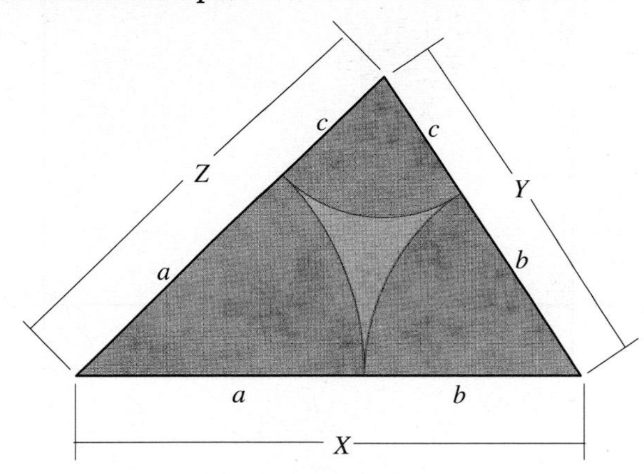

Figure 8.50.
A triangle tile composed of three different circles.

It is clear from the figure that if the sides of the triangle are X, Y, and Z and we are starting from flaps of length a, b, and c, then

$$X = a + b, \tag{8–1}$$

$$Y = b + c, \tag{8–2}$$

$$Z = a + c. \tag{8–3}$$

We can also invert the relationship to find the lengths of flaps that can be obtained from a given triangle:

$$a = \frac{1}{2}(X + Z - Y), \tag{8–4}$$

$$b = \frac{1}{2}(X + Y - Z), \tag{8–5}$$

$$c = \frac{1}{2}(Y + Z - X). \tag{8–6}$$

In the rectangle tile, because of symmetry, there are fewer variables: The circles all have the same radius, as shown in

Figure 8.51. If the circles have radius a and the river has width b, then the sides of the rectangle are given by

$$X = 2a + b, \quad (8\text{–}7)$$

$$Y = 2a. \quad (8\text{–}8)$$

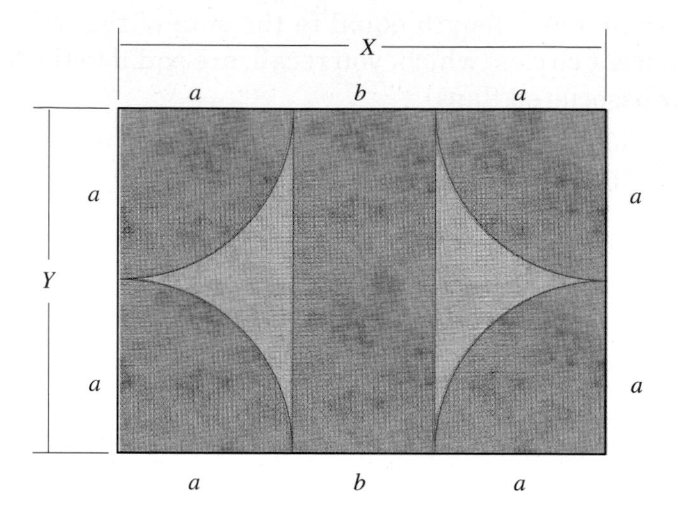

Figure 8.51.
A rectangular tile with a river.

Consequently, for a given rectangle, the dimensions of the circles and river are simply

$$a = \frac{1}{2}Y, \quad (8\text{–}9)$$

$$b = X - Y. \quad (8\text{–}10)$$

For rectangles or triangles to which we have added rivers, the radius of the circle is quite obviously reduced by the width of the added river.

These relationships can be used to construct combinations of tiles that give rise to new bases with new combinations of flaps beyond those in the standard repertoire. While this approach can be used for many origami subjects, it is particularly effective with insects, whose many appendages, often of varying lengths, have historically provided great challenge to the origami designer. By building up bases from tiles, it is possible to achieve quite complex combinations of long and short flaps. In the Periodical Cicada shown in Figure 8.52, six isosceles right triangle tiles, four isosceles triangles and four scalene triangles come together to produce six legs, two long wings, a head, thorax, and abdomen.

Figure 8.52.
Crease pattern, base, and folded model of the Periodical Cicada with a tiled crease pattern.

8.8. From New Tile to New Base

There are many possible tiles. You can search through the origami literature and catalog them, then combine existing tiles in new ways to realize new bases. Or, you can seek to construct new tiles directly. A new type of tile can inspire a new design. Squares, rectangles, and triangles are not the only possible tiles. It's possible to construct circles and rivers inside a parallelogram as well, as shown in Figure 8.53.

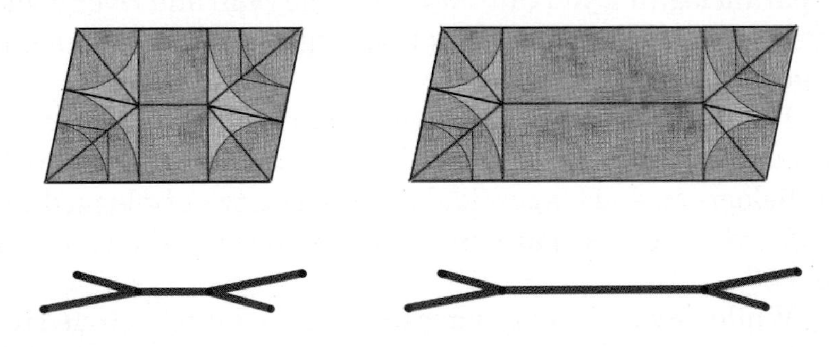

Figure 8.53.
Top left: a parallelogram tile.
Bottom left: schematic of its folded form.
Right: same for a longer parallelogram.

Like a rectangle, a parallelogram can be stretched arbitrarily; in this tile, stretching along the long direction can be taken up by increasing the width of the river running vertically.

Also like a rectangle, a parallelogram can be tiled to fill the plane. Look at what happens when we stack two of these tiles vertically or horizontally. The circles and rivers line up, so the combination can fold flat. But the tilt of the parallelograms creates an offset between adjacent points, so that the net result is a series of points evenly strung out along a common line.

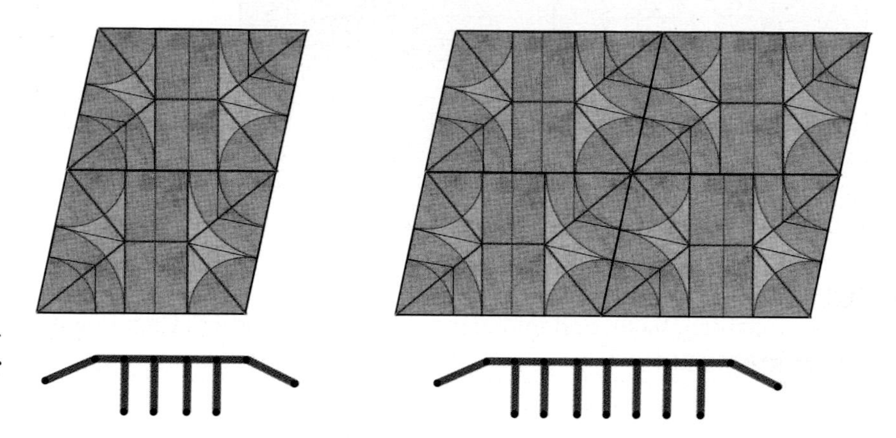

Figure 8.54.
Tiling of two and four parallelogram tiles and their schematics.

This is quite a nice trick; the circles in the crease pattern are arranged in rows and columns, while in the stick figure corresponding to the base, the flaps are distributed along a single line. Thus, the tile allows us to build an essentially one-dimensional chain of flaps while efficiently using a two-dimensional region of paper. A combination of rectangles and parallelograms gives the 14 legs and body segments of a pill bug, as shown in Figure 8.55. By varying the length and tilt angle of the parallelogram, you can vary the circle radii and river width, corresponding to the lengths of the legs and of the segments between them.

It's also possible to add circles and rivers to a trapezoid in the same way as a parallelogram. A combination of rectangle, parallelogram, and trapezoidal tiles gives a twenty-legged centipede, whose crease pattern and folded form is shown in Figure 8.56.

While any parallelogram can be turned into a circle/river tile that covers the plane, only particular proportions and tilt angles give evenly spaced legs. You might find it an interesting challenge to work out the relationship between parallelogram dimensions, leg length, body segment length, and the number of rows and columns of parallelograms and trapezoids.

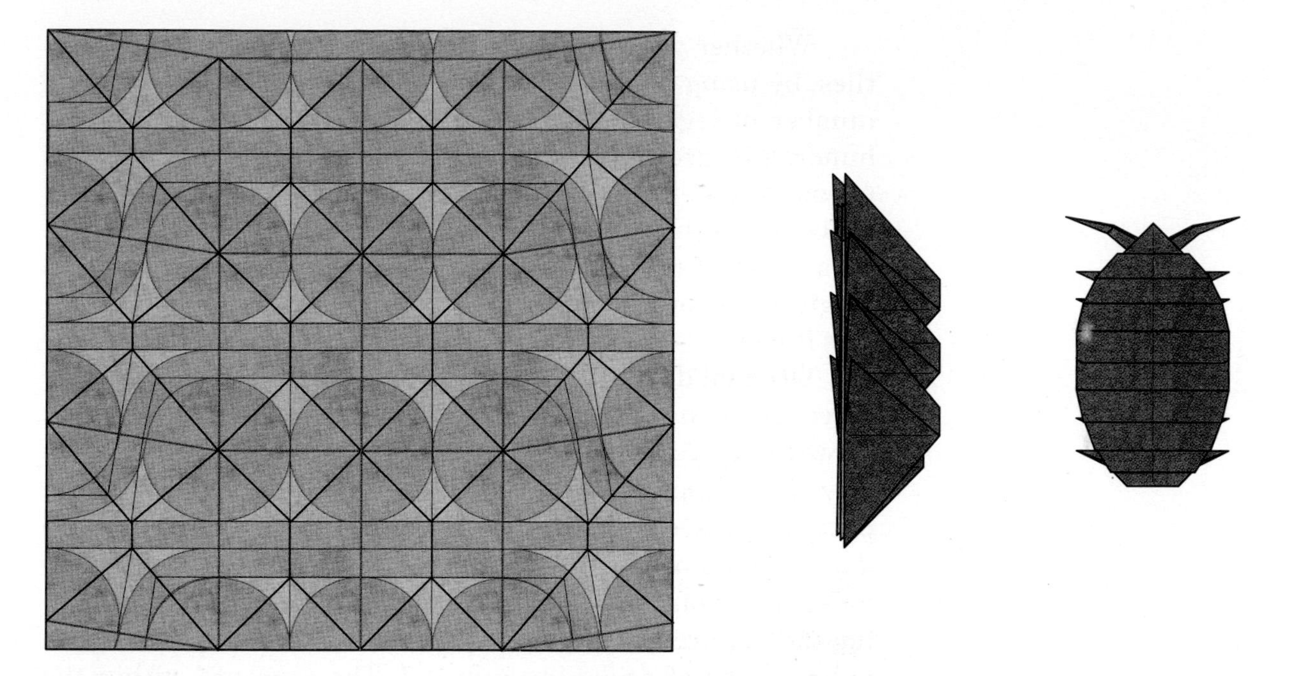

Figure 8.55.
Crease pattern with circles and rivers, base, and folded model of the Pill Bug.

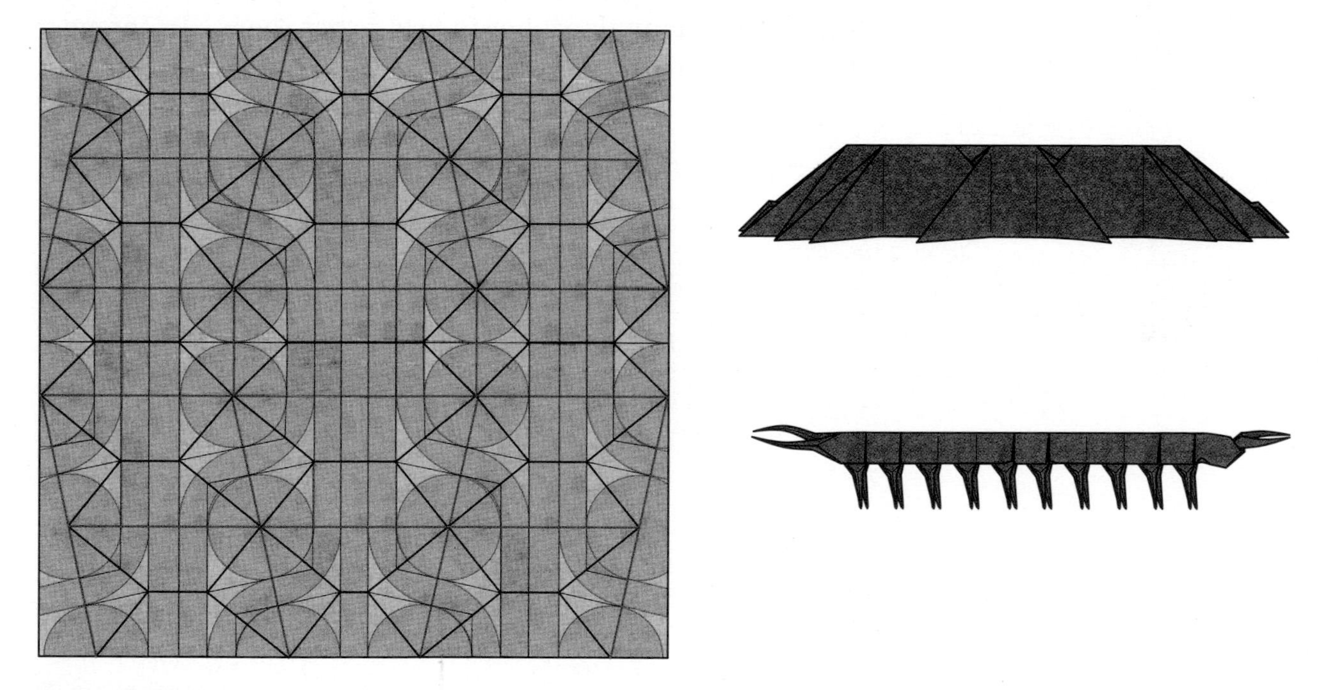

Figure 8.56.
Crease pattern with circles and rivers, base, and folded model of the Centipede.

After you've done that, you might try your hand at working out how to make multilegged centipedes using only rectangle and triangle tiles.

Whether you use triangles, parallelograms, or trapezoid tiles, by using more rows and columns, you can increase the number of legs arbitrarily; in fact, it's possible to make a hundred-legged centipede from a square. The use of tiles gives a remarkably efficient centipede. The length of the folded model is about two-thirds of the side of the square, and surprisingly, for a constant ratio between leg length and body segment, the length turns out to be about the same no matter how many legs it has.

Origami design by tiling can be a powerful technique for discovering new bases from which to fold new designs. However, there is still a bit of trial-and-error to it, in that the way we've approached it has been to assemble tiles into a pattern and see what kind of base arises. If you have built up a collection of many different types of tile, then for a particular subject, you can try fitting together different tiles comprising the required number of circles and rivers. But it's still an indirect way of designing a model. The concepts within the tiling method, however—circles, rivers, and most importantly, the uniaxial base—are fundamental. We can build off these concepts to construct several algorithms for a directed design: "I need this many flaps; here's how to get them." In the next chapter, we'll encounter the first of these algorithms.

Folding Instructions

Pegasus

Pegasus

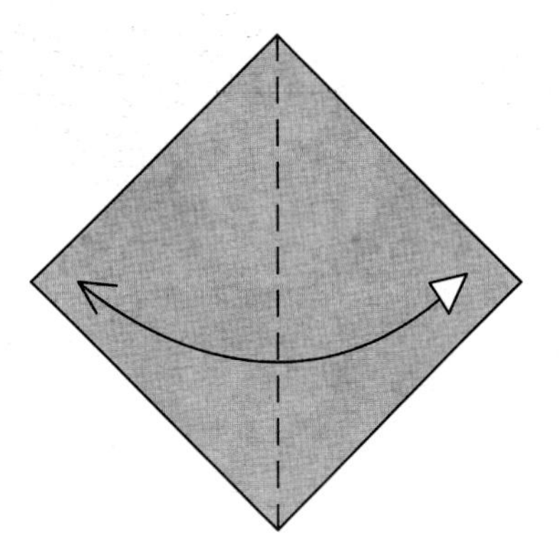

1. Begin with the white side up. Fold the paper in half along the diagonal and unfold.

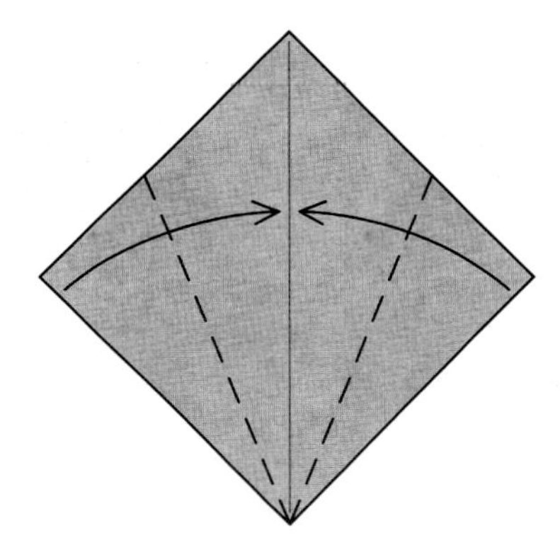

2. Fold the edges in to lie along the diagonal, but don't make the creases sharp.

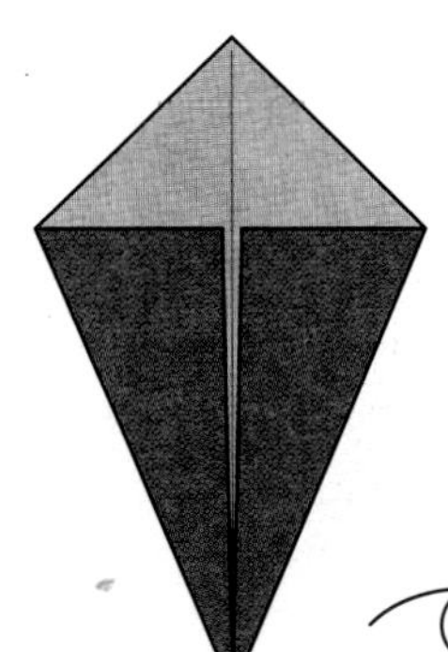

3. Turn the model over.

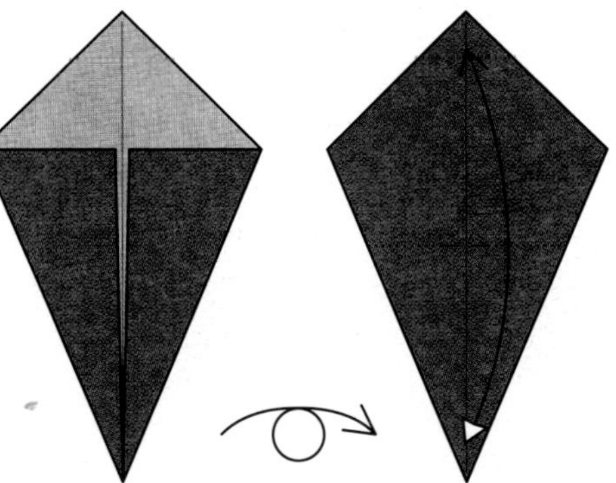

4. Fold the bottom point up to the top; make a pinch in the middle and unfold.

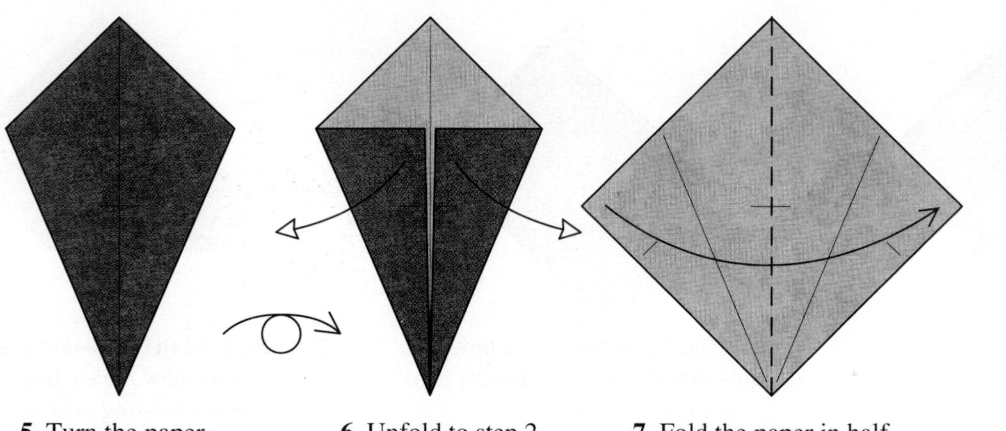

5. Turn the paper back over.

6. Unfold to step 2.

7. Fold the paper in half.

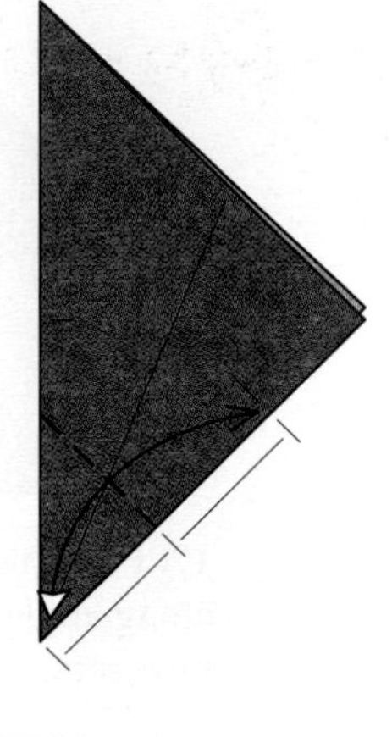

8. Fold and unfold.

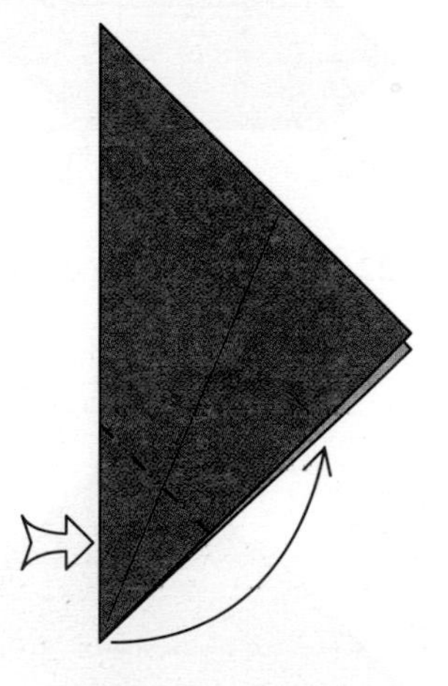

9. Reverse-fold the point on the crease you just made.

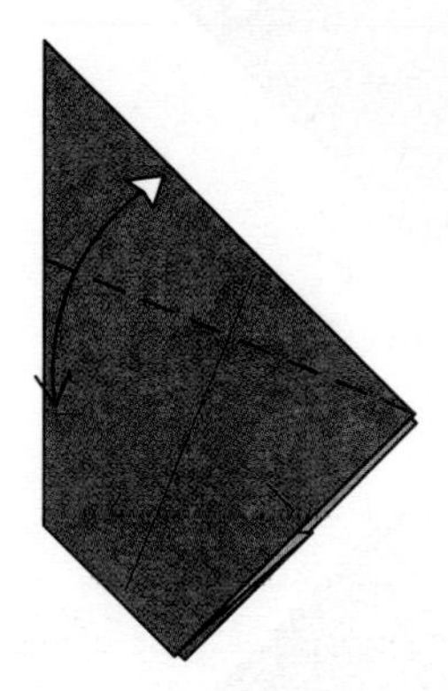

10. Fold and unfold.

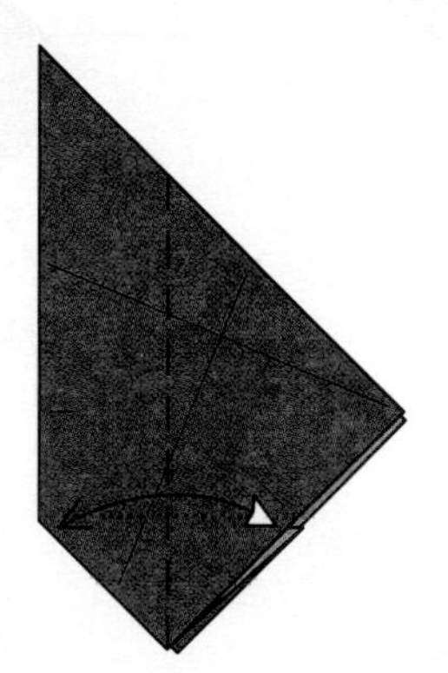

11. Fold and unfold through a single layer; repeat behind.

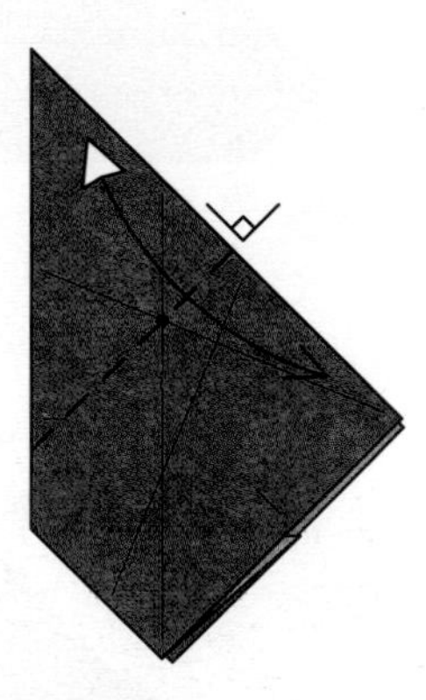

12. Fold and unfold through the intersection of two creases.

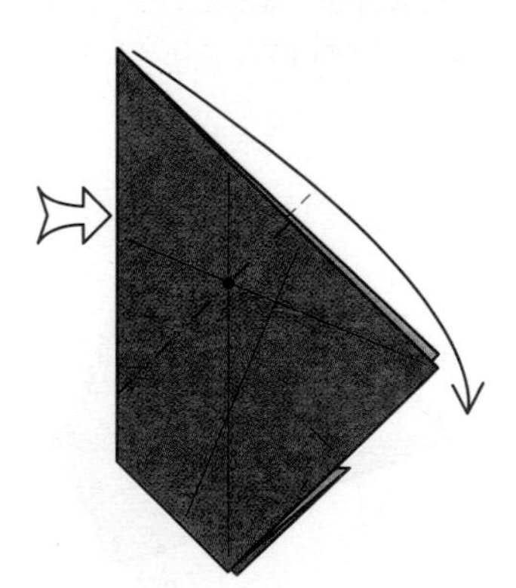

13. Reverse-fold the top point on the crease you just made.

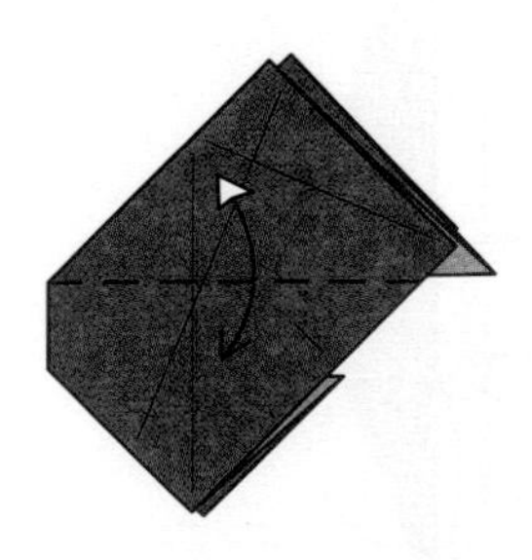

14. Fold and unfold. Repeat behind.

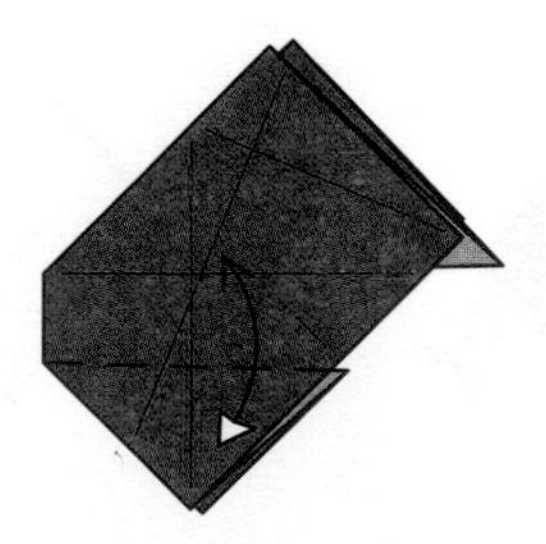

15. Fold and unfold. Repeat behind.

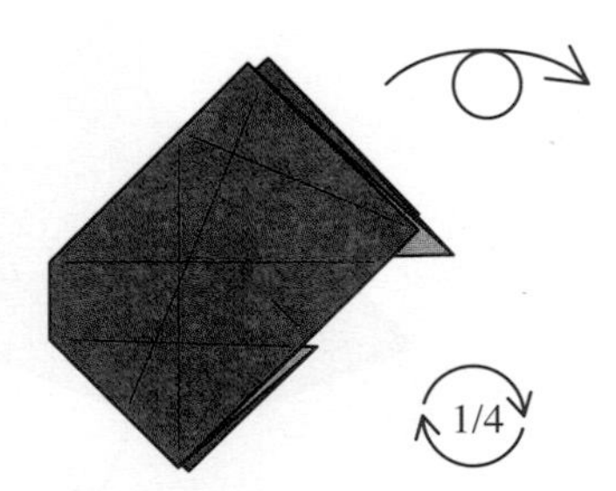

16. Turn the paper over and rotate it 1/4 turn clockwise.

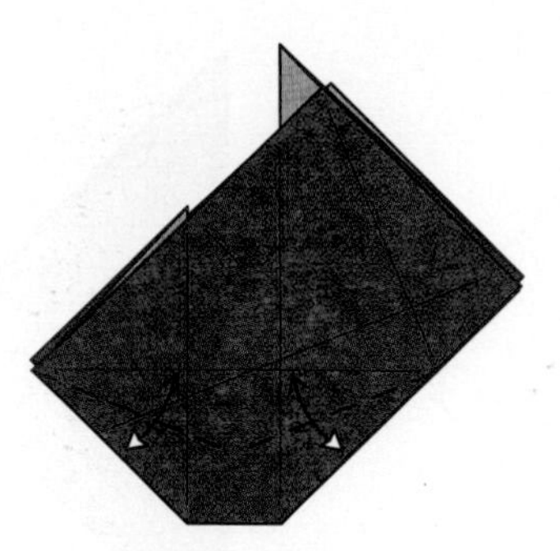

17. Fold and unfold along angle bisectors.

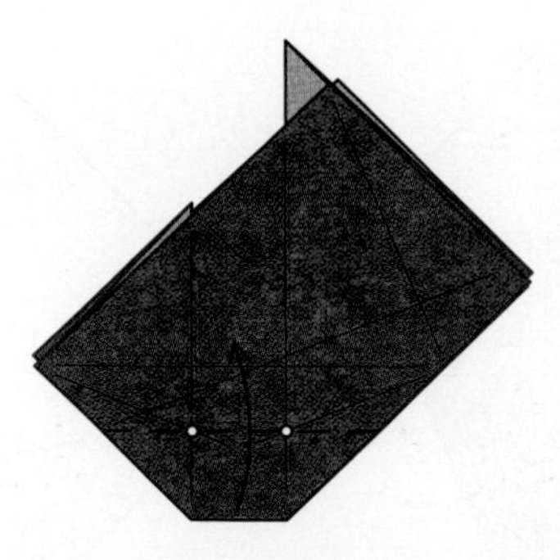

18. Fold the bottom edge upward. Note that the crease connects two crease intersections.

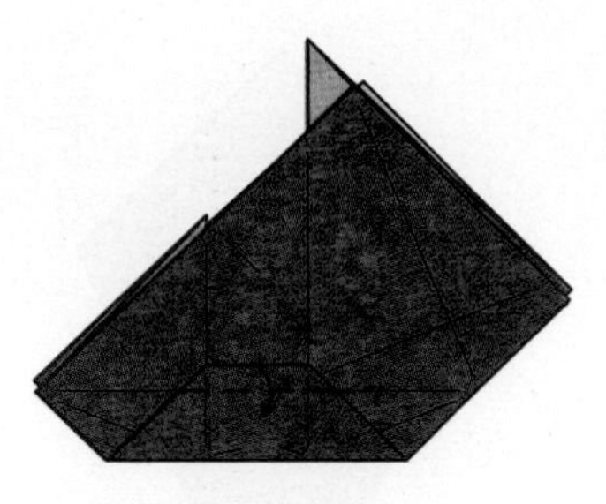

19. Fold the edge down so that the new crease lines up with an existing crease.

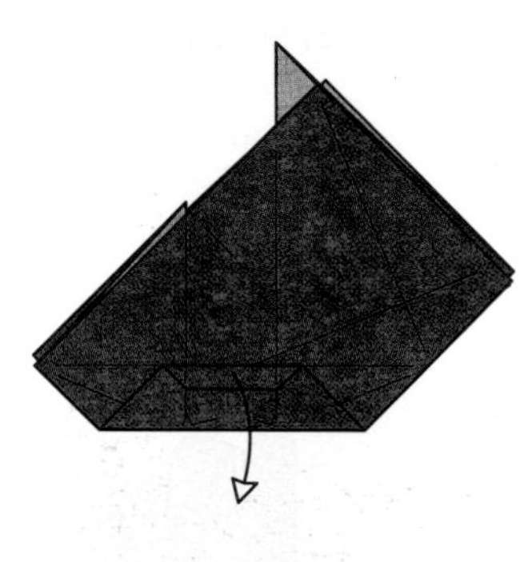

20. Unfold to step 18.

21. Sink the bottom edge in and out on the existing creases.

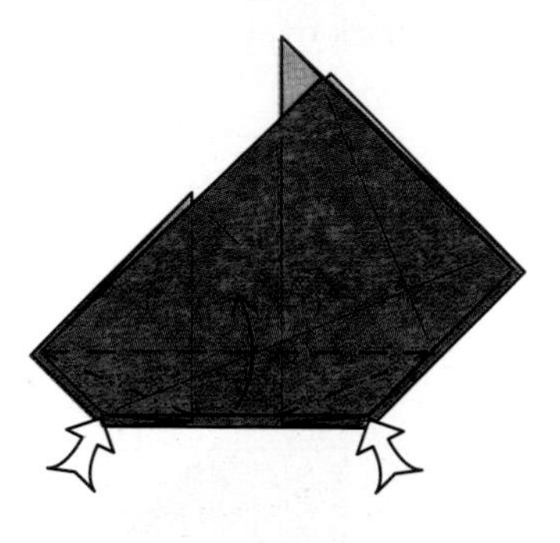

22. Spread-sink the corners.

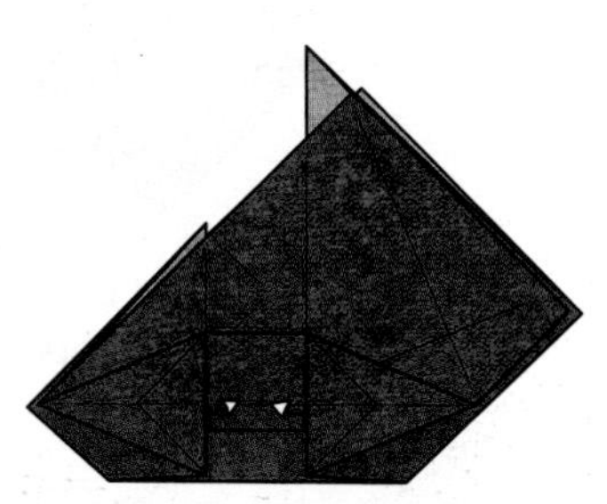

23. Pull some paper out of the pockets (unclosed sink).

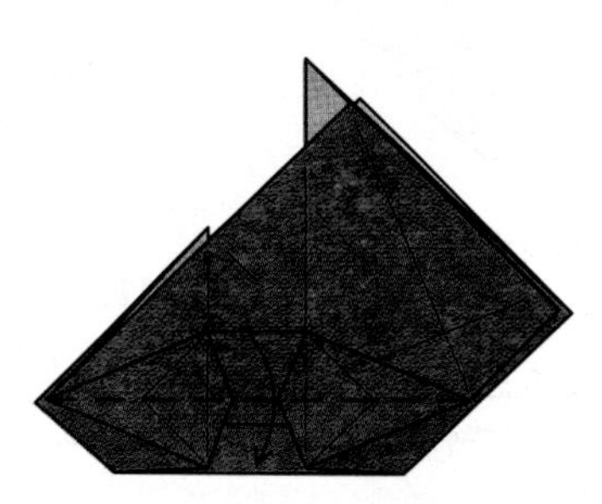

24. Fold the flap back down.

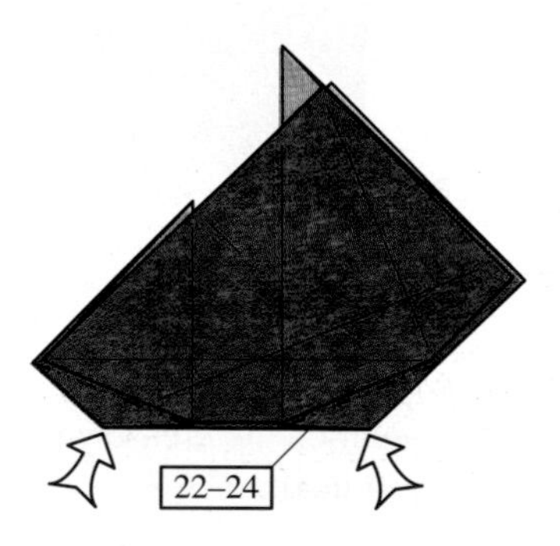

25. Repeat steps 22–24 behind.

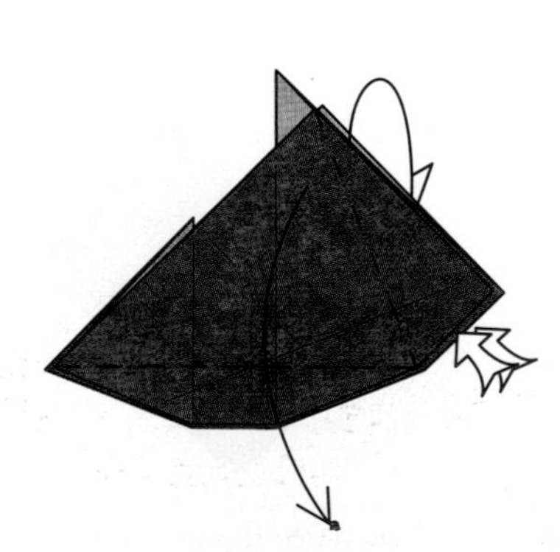

26. Squash-fold the edge. Repeat behind.

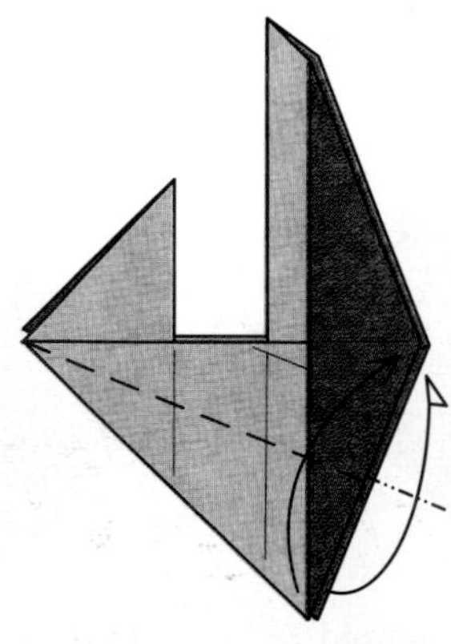

27. Fold one flap up in front and one up behind.

28. Squash-fold the corner. Repeat behind.

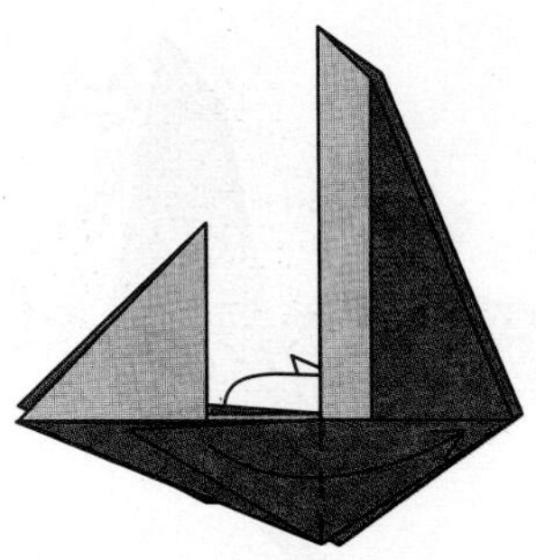

29. Fold the near flap back to the right. Repeat behind.

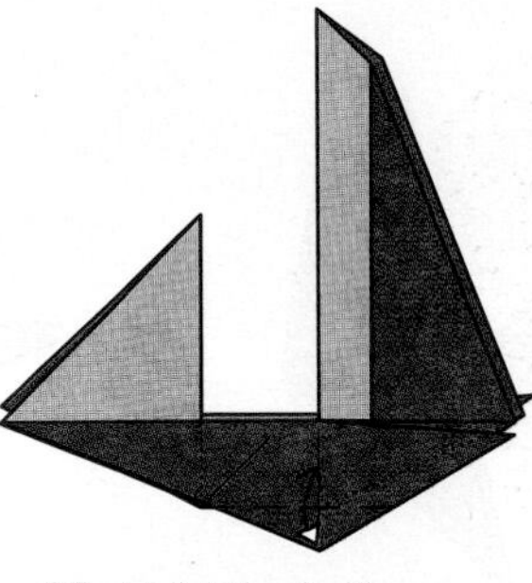

30. Fold the bottom corner up so that the crease lines up with an existing edge and unfold. Repeat behind.

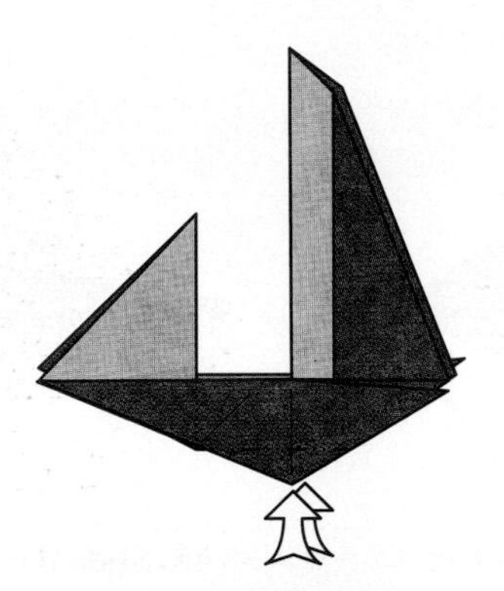

31. Sink the corner on the crease you just made. Repeat behind.

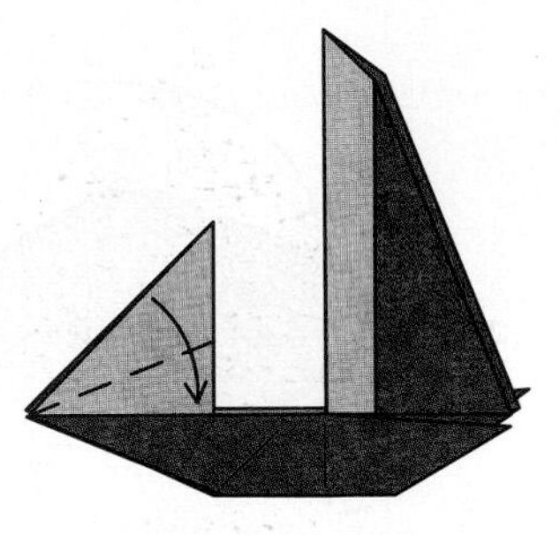

32. Fold the white flap down along the angle bisector.

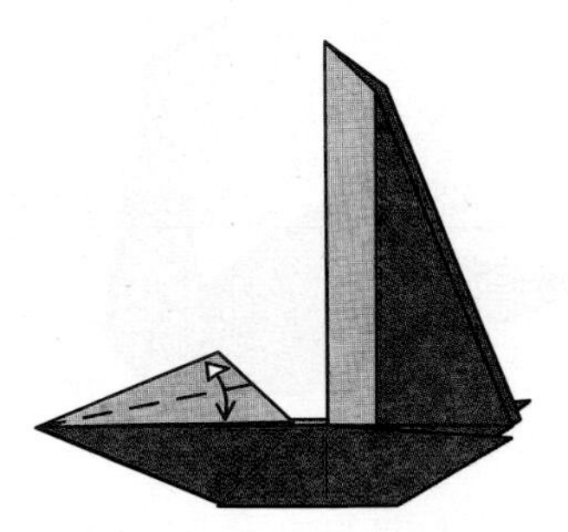

33. Fold and unfold.

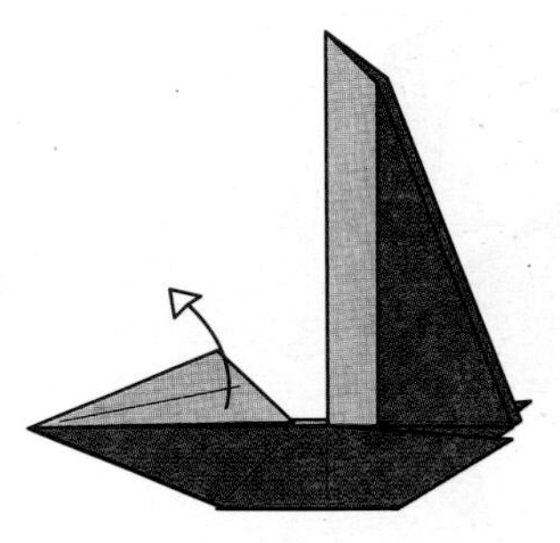

34. Unfold to step 32.

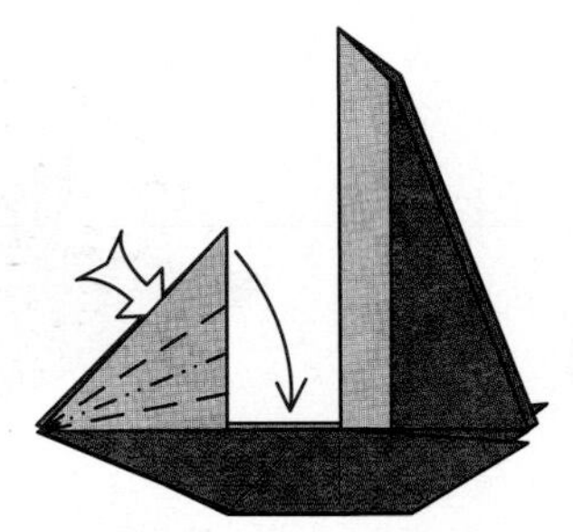

35. Reverse-fold in and out on the creases you just made.

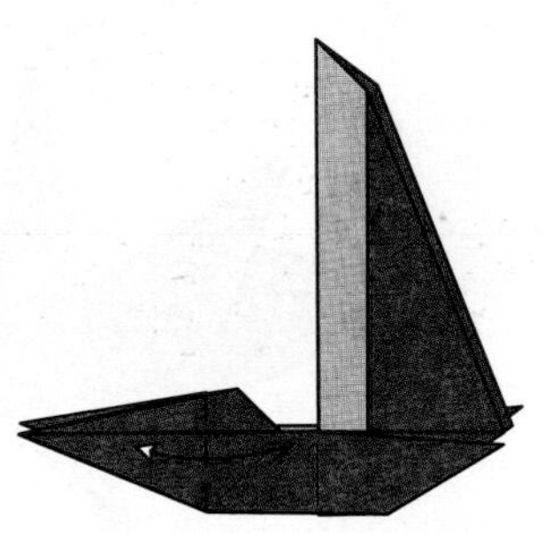

36. Fold and unfold. Repeat behind.

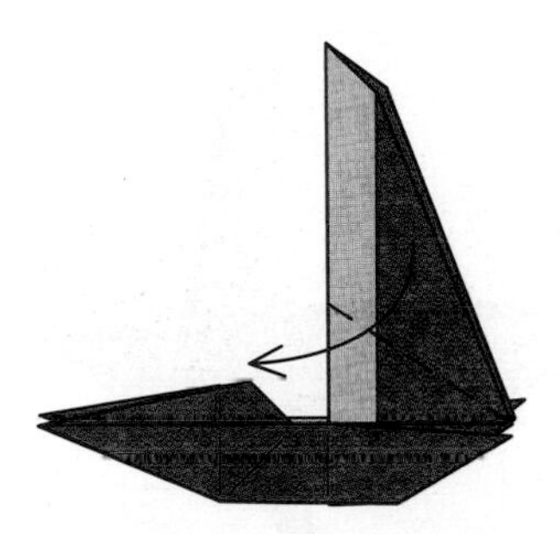

37. Fold down along an angle bisector.

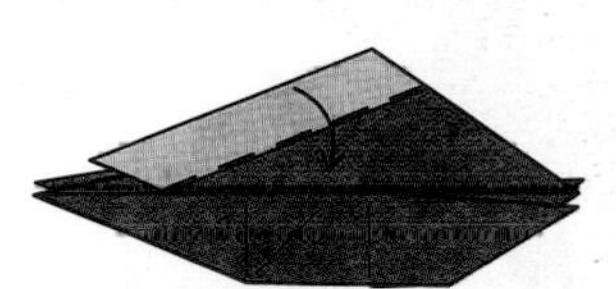

38. Fold the white edge down along the raw edge.

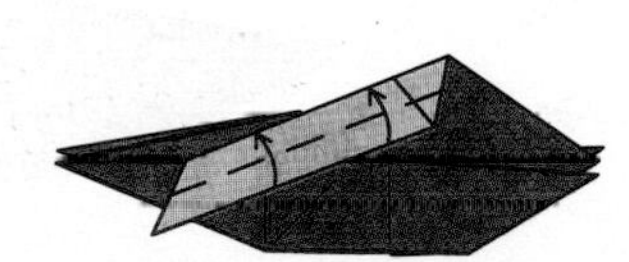

39. Fold the white edge back up to the top.

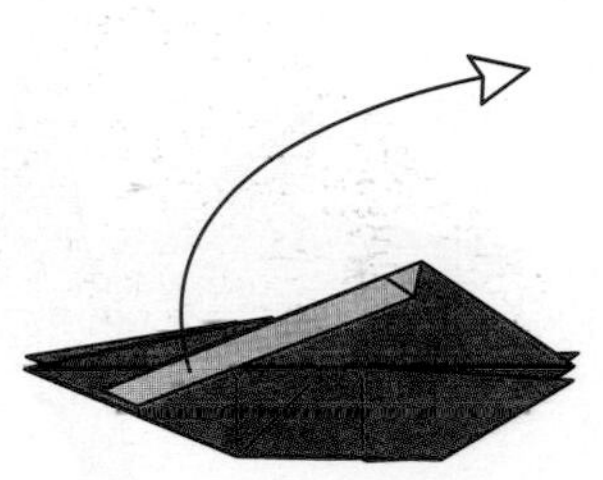

40. Unfold to step 37.

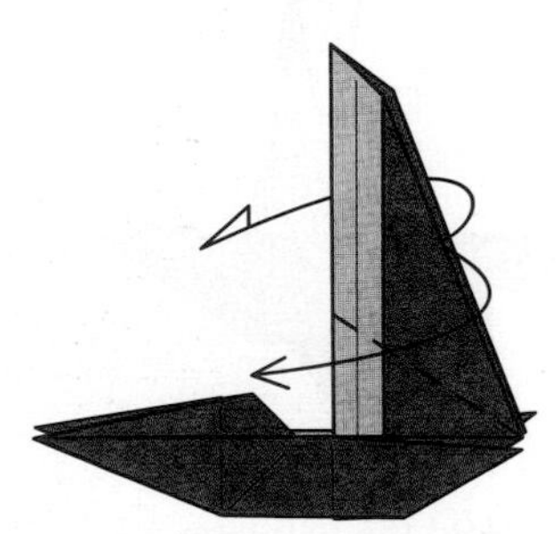

41. Reverse-fold the top flap on the existing crease.

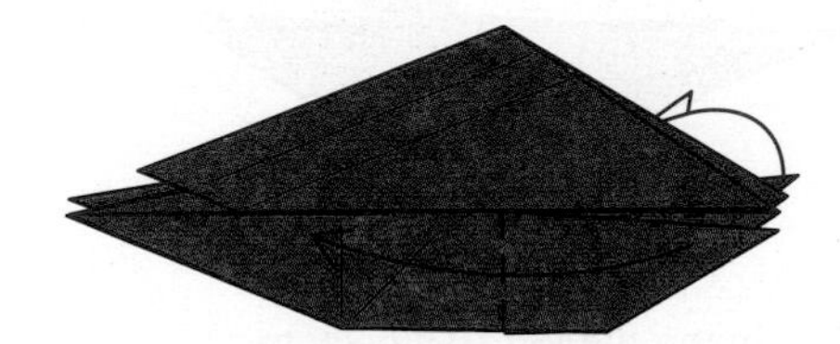

42. Fold one flap to the left. Repeat behind.

43. Sink the corner. Repeat behind.

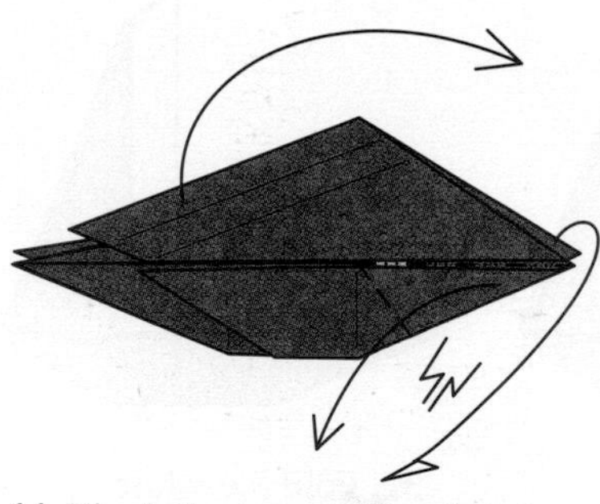

44. Pinch the two right points in half and swing them downward; at the same time, the top flap pivots over to the right, slightly beyond vertical. Flatten the model firmly.

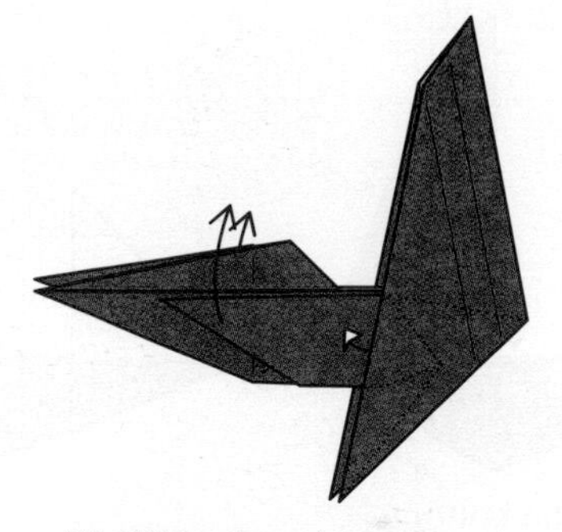

45. Slide the wings upward slightly and flatten.

46. Sink the long edge on the existing crease.

47. Reverse-fold the hidden edge upward.

48. Fold the right edge of the flap over to the left. Repeat behind.

49. Bring one layer of paper to the front. Repeat behind.

50. Swing both wings over to the right.

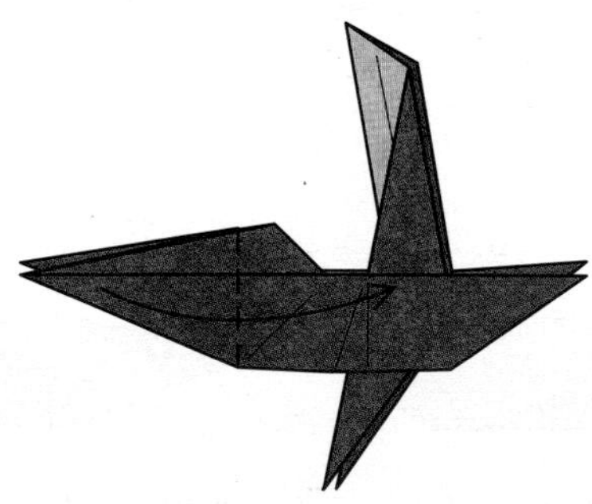

51. Fold one flap over to the right.

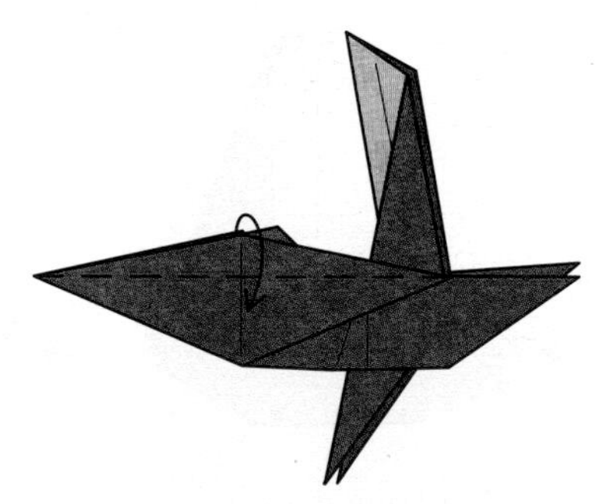

52. Fold the layers down.

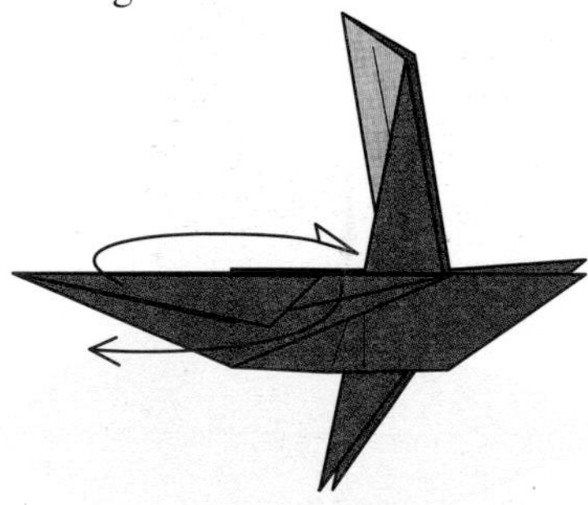

53. Swing the left flap behind to the right and bring the two right flaps (one long, one short) around to the left. Flatten firmly.

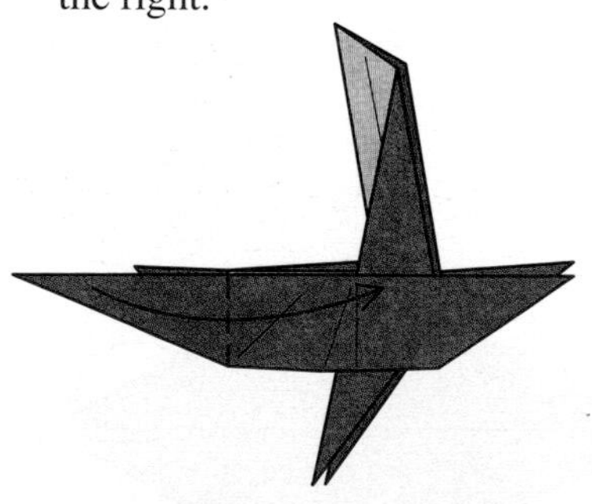

54. Swing the left flap back to the right. Flatten firmly.

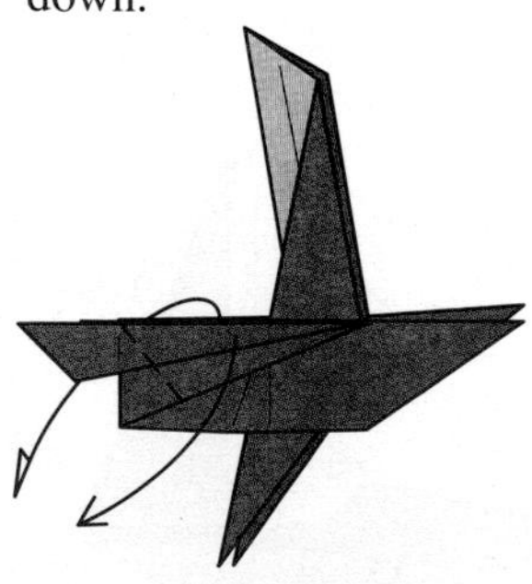

55. Fold the two legs downward. Flatten firmly.

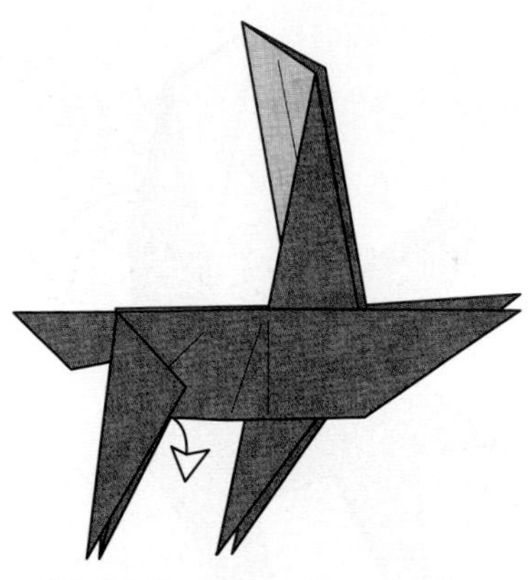

56. Reshuffle the layers of the body and legs so that they alternate.

57. Reverse-fold the centermost layer (which was freed in the previous step).

58. Swivel the tail downward.

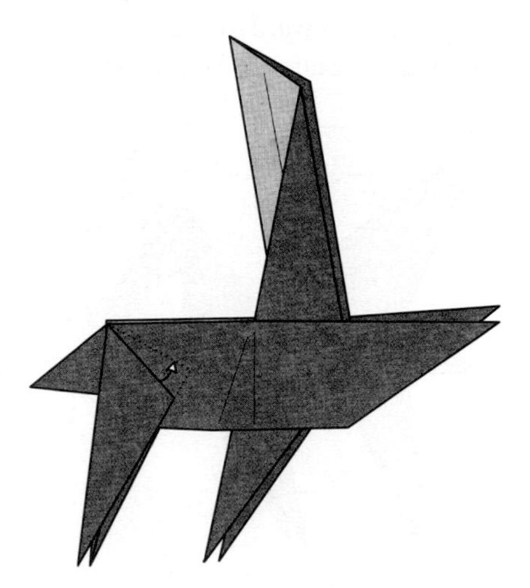

59. Pull out some loose paper. Repeat behind.

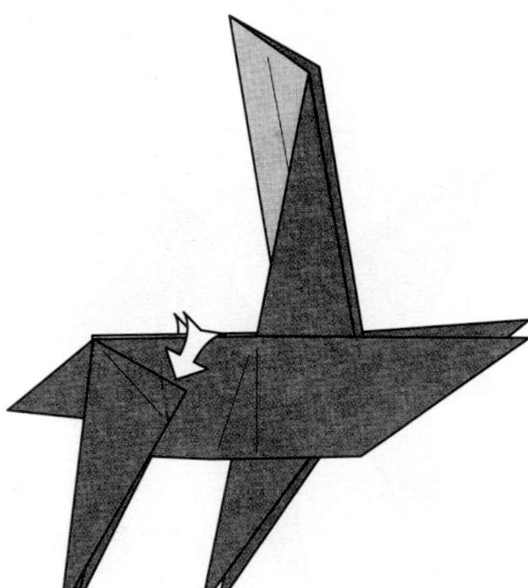

60. Reverse-fold the corner. Repeat behind.

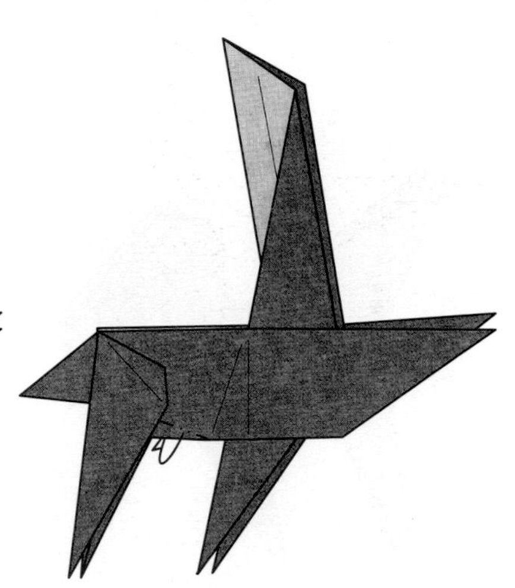

61. Mountain-fold a double layer of paper underneath. Repeat behind.

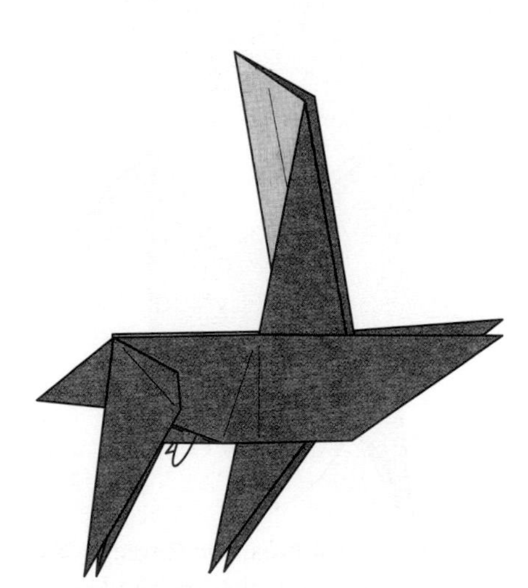

62. Mountain-fold another layer of paper underneath. Repeat behind.

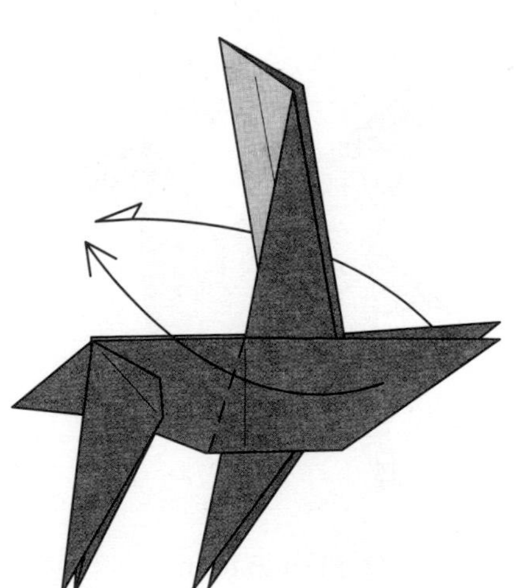

63. Swing both wings back to the left.

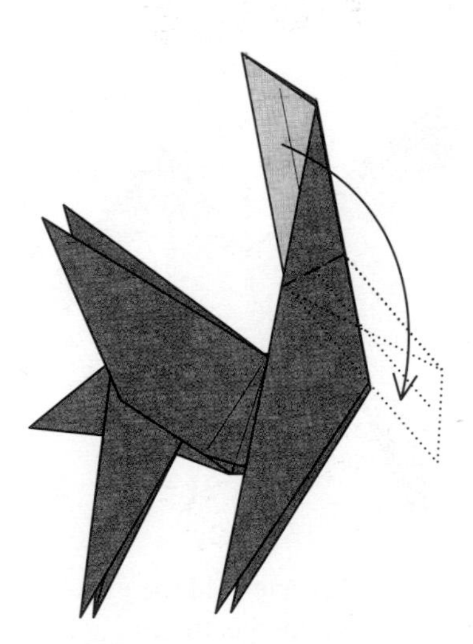

64. Fold the point down. There's no exact reference point.

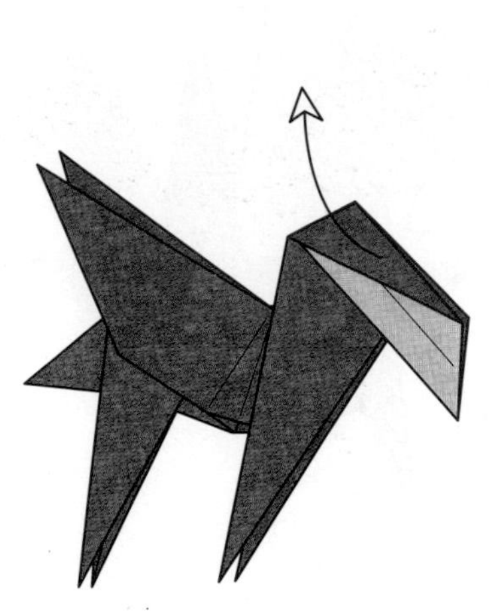

65. Unfold.

66. Mountain-fold the edges inside.

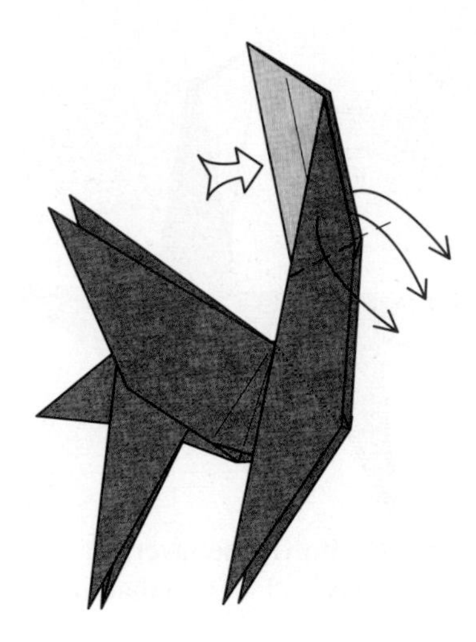

67. Double-reverse-fold the point downward on the creases you made in step 63.

68. Valley-fold one corner down in front and behind.

69. Sink the white corner; at the same time, sink the long edge upward to line up with the top of the head.

70. Outside-reverse-fold the tip of the head.

71. Reverse-fold the point underneath.

72. Pleat to form ears.

73. Double-rabbit-ear the foreleg. Note that you are spreading the layers on the right side of the point. For the cleanest results, you shouldn't flatten the paper before doing step 74.

74. Reverse-fold the thinned point downward and flatten firmly.

75. Repeat steps 73–74 behind. Note that the model is rotated slightly from step 74.

76. Double-rabbit-ear both hind legs. As with the front legs, spread the layers on the right side of the point.

77. Reverse-fold the legs downward.

78. Reverse-fold the tips of the wings.

79. Pleat the wings and curve them out from the body.

80. Pleat the tail.

81. Finished Pegasus.

9 Circle Packing

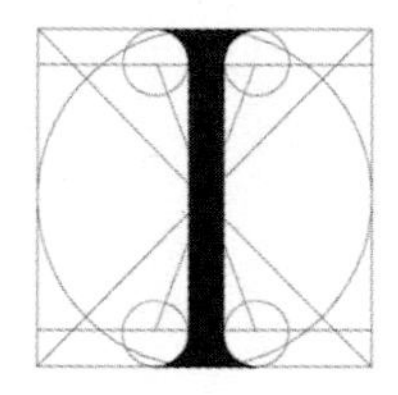

In the last chapter, we saw how new bases can be constructed by assembling tiles composed of crease patterns in ways that allow the individual tiles to fold flat. By decorating the individual tiles with circles and rivers, we created matching rules for the tiles; if two tiles mate so that their circles and rivers line up, then the union of the two tiles can fold flat without creating any new creases.

Furthermore, we are able to use the pattern of circles and rivers to divine the structure of the resulting base: how many flaps there are, how long they are, how they are connected to one another. While a given polygon may give rise to tiles with different crease patterns that have different widths of flaps, any two tiles with the same pattern of circles and rivers necessarily gives rise to the same flap configuration. We can represent this common configuration by a stick figure, in which each segment corresponds to a distinct flap.

When we have built a valid tiled crease pattern, the circular arcs of mating tiles align, creating partial or full circles. A contiguous segment of a circle in the tiled pattern corresponds to a distinct flap in the folded base. Why, you might wonder, is this so? Why use circles? The choice of circles to create matching rules is not arbitrary; there is a deep geometric connection between flaps and circles. The use of circles to represent flaps is a powerful tool within origami, and so we shall investigate it a bit further.

9.1. Three Types of Flap

As we have already seen, in origami there are three different types of flaps: corner flaps, edge flaps, and middle flaps. These types of flaps are named for the point where the tip of the flap falls on the square. A corner flap has its tip come from a corner of the square, an edge flap has its tip lie somewhere along an edge, and a middle flap, as you would expect, comes from the middle of the paper. For example, the four large flaps on a Frog Base are corner flaps; the four stubby flaps are edge flaps; and the thick flap at the top is a middle flap. All three are illustrated in Figure 9.1.

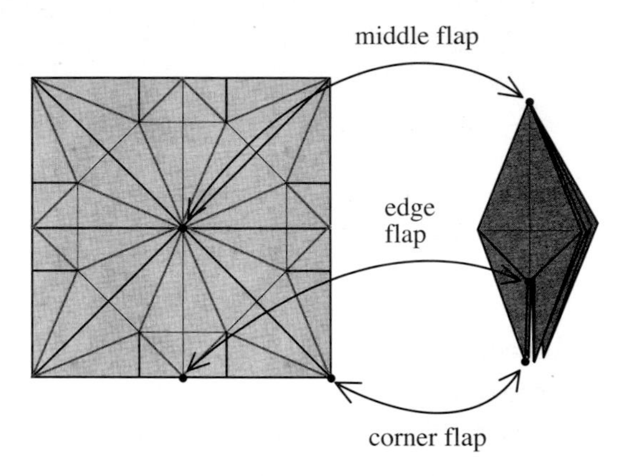

Figure 9.1.
Flaps in the base have their tips at unique points in the crease pattern. Middle, edge, and corner flaps have their tips on the middle, edge, and corner, respectively, of the square.

Paper, like people, can only be in one place at a time. Paper that goes into one flap can't be used for another. Therefore, every time you create a flap from the square, a portion of the paper gets consumed by that flap. The reason to make a distinction between the three different types of flaps—corner, edge, middle—is that for the same length flap, each of the three types of flaps consumes a different amount of paper.

One way to see this difference is to fold corner, edge, and middle flaps of exactly the same size from three different squares. Figure 9.2 shows the folding of a corner flap. If you imagine (or fold) a boundary across the base of the flap, then that boundary divides the paper into two regions: The paper above the boundary is part of the flap, and the paper below the boundary is everything else. The paper that goes into the flap is for all intents and purposes consumed by the flap; any other flaps must come from the rest of the square.

So, as Figure 9.2 shows, if you fold a flap of length L from a square so that the tip of the flap comes from the corner of the square, when you unfold the paper to the original square, you see that the region of the square that went into the flap is roughly a quarter of a circle; precisely, it's a quarter of an octagon. Suppose we made the flap half the width, as shown in Figure 9.3, before we unfolded it; then the flap becomes a quarter of a hexadecagon. If we kept making the flap thinner and thinner (using infinitely thin paper!), the boundary of the flap would approach a quarter-circle. In all cases, the polygonal segment is inscribed by a circular segment, which represents the limiting case of an infinitely narrow flap. Thus, a circle is the minimum possible boundary of the region of the paper consumed by the flap. A corner flap of length L, therefore, requires a quarter-circle of paper, and the radius of the circle is L, the length of the flap.

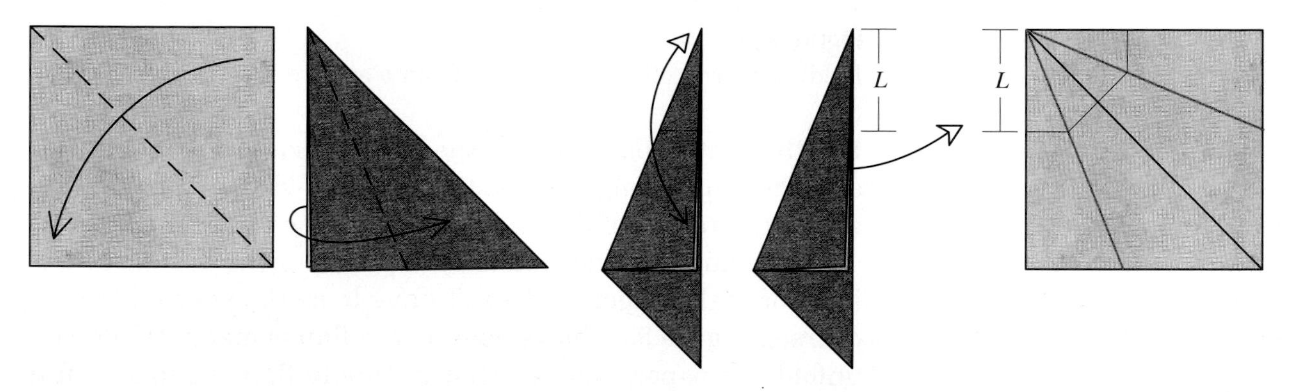

Figure 9.2.
Folding a corner flap of length L from a square.

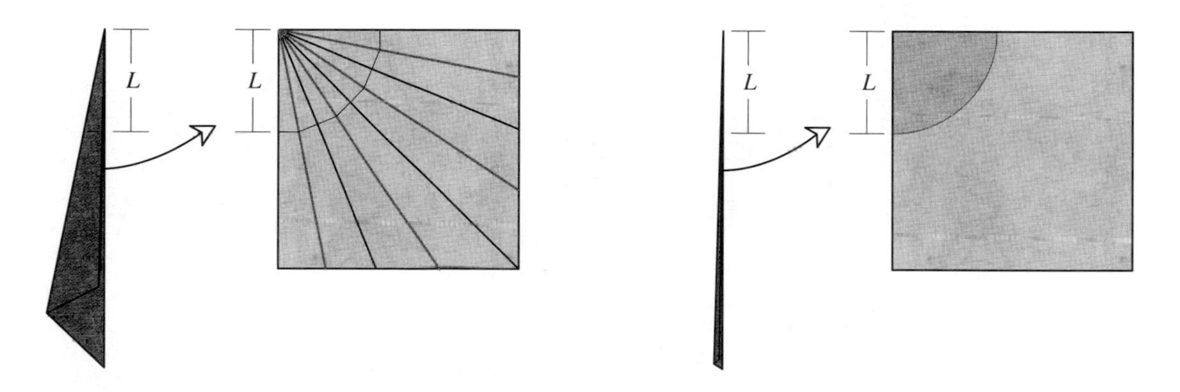

Figure 9.3.
Left: making a narrower flap makes the boundary a quarter of a hexadecagon.
Right: the limit of the boundary as the flap becomes infinitely thin approaches a quarter-circle.

Therefore, all of the paper that lies within the quarter-circle is consumed by the flap, and the paper remaining is ours to use to fold the rest of the model.

Now, suppose we are making a flap from an edge. For example, if we fold the square in half, then the points where the crease hits the edge become corners, and we can fold a corner flap out of one of these new corners, as shown in Figure 9.4. If we fold and unfold across the flap to define a flap of

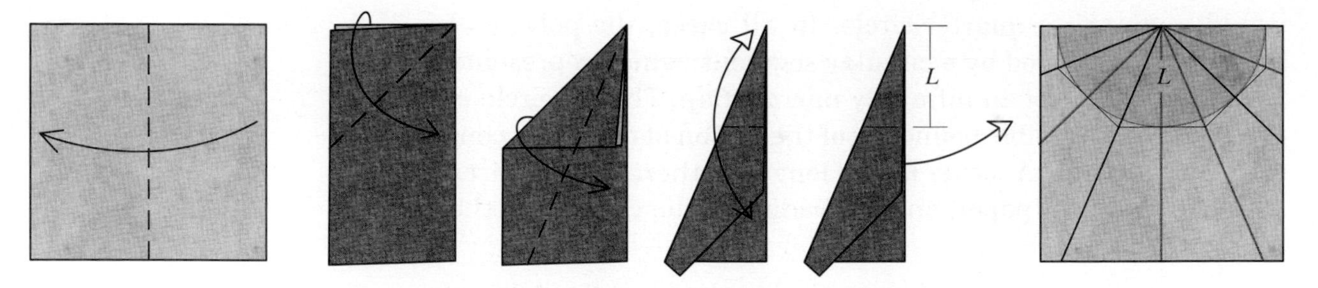

Figure 9.4.
Folding an edge flap of length L from a square.

length L and then unfold to the original square, you see that an edge flap of length L consumes a half-circle of paper, and again, the radius of the circle is L, the length of the flap.

Similarly, we can make a flap from some region in the interior of the paper (it doesn't have to be the very middle, of course). Figure 9.5 shows how such a flap is made. When you unfold the paper, you see that a middle flap requires a full circle of paper, and once again, the radius of the circle is the length of the flap.

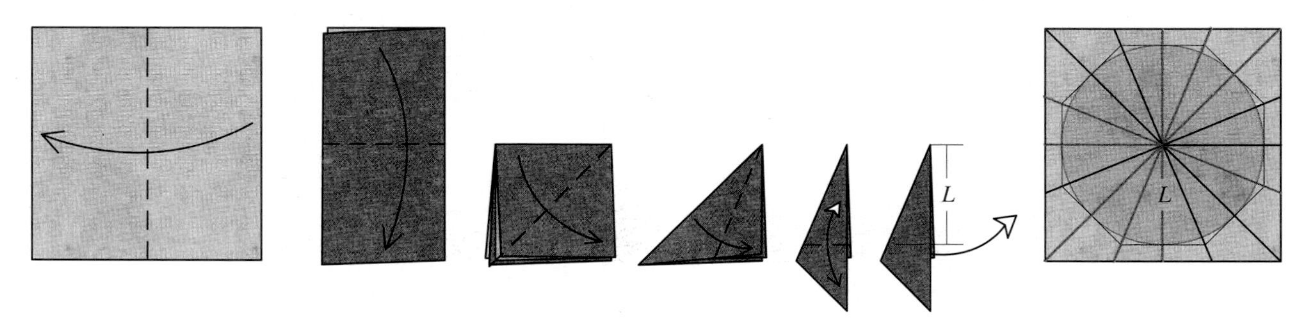

Figure 9.5.
Folding a middle flap of length L from a square.

The amount of paper consumed doesn't depend on the angle of the tip of the flap, only its length and location. So any flap in any model consumes a quarter, half, or full circle of paper, depending upon whether it is a corner flap, edge flap, or middle flap.

This relationship doesn't depend on whether the base was constructed from tiles; it doesn't depend on whether the base is a uniaxial base. It is a property of any flap that comes to a point in any origami model whatsoever. This relationship gives us a new set of tools for designing origami bases that permits a more direct approach than the assembly of preexisting tiles; we can represent each flap by a circle and work from the pattern of circles directly.

One of the goals of all scientific endeavors is the concept of unification: describing several disparate phenomena as different aspects of a single concept. Rather than thinking in terms of quarter-circles, half-circles, and full-circles for different kinds of flaps, we can unify our description of these different types of flaps by realizing that the quarter-circles, half-circles, and full circles are all formed by the overlap of a full circle with the square, as shown in Figure 9.6. The common property of all three types of flaps is that the paper for each can be represented by a circle with the center of the circle lying somewhere within the square. With middle flaps, the circle lies wholly within the square. However, with corner and edge flaps, part of the circle laps over the edge of the square. The center of the circle still has to lie within the square, though. Thus, *any* type of flap can be represented by a *full* circle whose center, which corresponds to the tip of the flap, lies somewhere within the square.

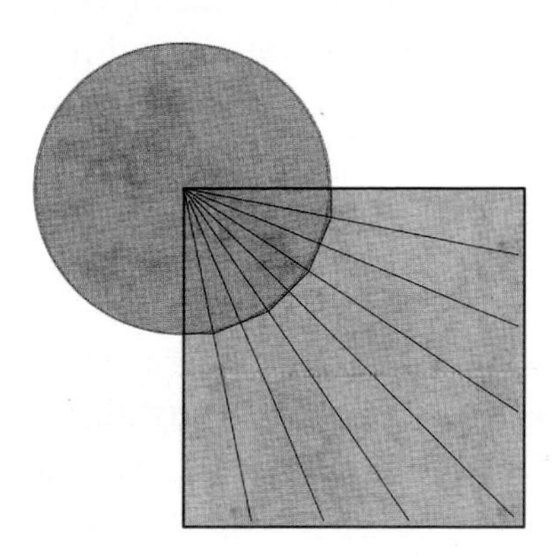
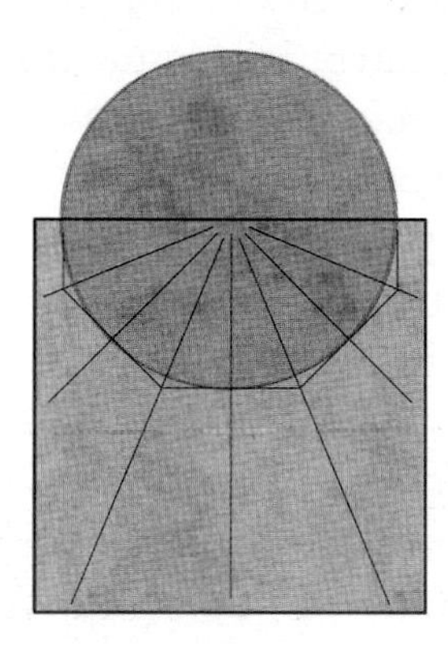
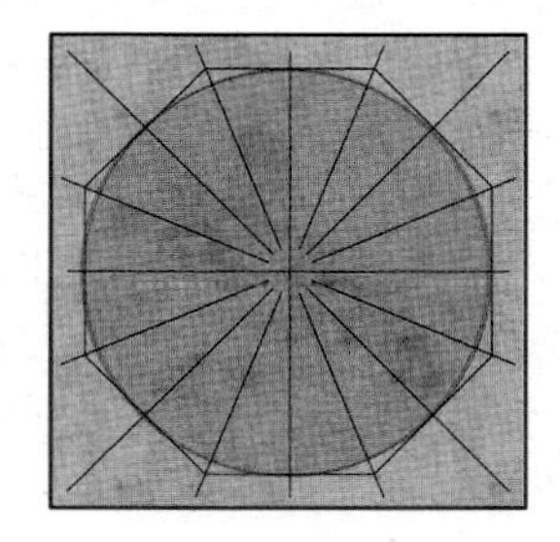

Figure 9.6.
All three types of flaps can be represented by a circle if we allow the circle to overlap the edges of the square.

9.2. Overlaps

You might have noticed an interesting feature of the circle pattern for any base; the circles corresponding to individual flaps touch, but do not overlap. In tiled crease patterns, this is, of course, by design; no circles overlapped within the tiles we started with, so no circles will overlap in the assembled crease pattern.

However, in any base, if we represent distinct flaps by circles, they can never overlap, whether the crease pattern was constructed from a tiling or not. A moment's reflection will reveal why this must be so. Each circle encloses the paper used in a single flap. If two circles overlapped, the paper shared between the circles would be included in each of the two different flaps, which is obviously impossible. Thus, we can generalize:

> *In any valid circle pattern corresponding to an origami base, the circles corresponding to two distinct flaps may not overlap.*

This condition must hold for any origami base—not just for the Classic Bases and not just for tiled bases. No matter how many flaps the base may have, the circles in the circle pattern for the base cannot overlap. Although this property seems pretty innocuous, it is in fact both restrictive and useful. Put one way, it is an interesting property of existing bases: Take any origami base, unfold it, and draw a circle for each flap of the base, each circle centered on the point that maps to the tip of the flap. No two circles will overlap. This relationship can become a useful tool for origami design, for the converse is also true: If you draw N nonoverlapping circles on a square, it is guaranteed that the square can be folded into a base with N flaps whose tips come from the centers of the circles.

If you have a pattern of N circles on a square, it's quite evident that it's *necessary* that they not overlap to fold into a base with N flaps. It's not at all obvious that this condition is *sufficient*; but it is, and we will see why in later chapters. This is an incredibly powerful result. No matter how many flaps you want to achieve in your base, all you need to do is draw a set of nonoverlapping circles, and the centers of the circles map out the tips of each and every flap in the base.

Using this fact, we can replace a somewhat abstract problem (design an origami base with N flaps of a given length) by a simpler, geometric, more easily visualized and more easily *solved* problem (draw N nonoverlapping circles whose centers lie inside a square). By representing each flap in the model by a circle and placing all of the circles on the square, we ensure that we have allocated sufficient paper to construct each flap. The problem of placing circles so that they don't overlap resembles the packing of cylindrical cans into a box; we call such a pattern a *circle packing*.

The circles need not be the same size. If we use different-sized circles, we'll get different-length flaps (recall that the length of the resulting flap is the radius of the corresponding circle). So merely by shuffling circles around on a square, you can construct an arrangement of points that can be folded into a base with the same number of flaps, no matter how complex. For that matter, by choosing different arrangements of circles, you can devise many different folding sequences giving many different bases, even though they have the same number of flaps.

While such a circle pattern is guaranteed to be foldable into a base, a guarantee is not the same as a blueprint. But here is where we can apply the tiles from the previous chapter. If the circle pattern can be cut into tiles so that the circles within each tile match up with known tiles, then we can assemble the complete crease pattern from these same tiles.

9.3. Connections to Tiles

Consider, for the moment, those tiles we have seen that contain only circles (no rivers). They are of two types: triangles and square (the Waterbomb Base tile). In both types of tile, the inscribed circles touched each other along the tile edges, which, you'll recall, were axial creases. In fact, the only places that circles touched each other were along axial creases. This is more than coincidence; it can be shown (and we will do so later) that there exist axial creases in any circle packing wherever any two circles touch. The newly created axial creases divide the square into axial polygons; if we are fortunate enough that the axial polygons are recognizable as known tiles, we can fill them in with the creases associated with each tile and use the resulting creases to fold the shape flat.

Thus, the six simple bases we used to illustrate tiling could have been derived directly from circle packings based on their desired number of flaps. We represent each desired flap by a circle; pack the circles into the square, and then construct axial creases that outline the tiles.

The circle diagrams also allow us to address the problem of folding efficiency. By representing each flap of a model by the appropriately sized circle and drawing the circles on a square, you can easily find the arrangement of points on the square that gives the most efficient base containing those flaps and, as often as not, it will be an elegant base as well.

And since the length of a flap is equal to the radius of the corresponding circle, if we design the base by laying out points on a square and represent each flap by a circle, the most efficient base will come from that layout in which the circles representing the flaps are as large as possible. In accordance with what we've shown above, the circles are to be placed according to the following rules:

- Each flap in the model is represented by a circle.
- The radius of each circle is equal to the length of the corresponding flap.

- The center of each circle lies within the square.
- The circles may overlap the edges of the square, but not each other.

We have now assembled the necessary building blocks to carry out origami design from the ground up. Throughout the history of origami, most designs were modifications, and the techniques we've learned so far—offsetting, distortion, point-splitting, grafting, tiling—have implicitly assumed that we started with something reasonably close to our final objective. But proximity is no longer needed; we can proceed directly from the desired subject to a base that contains all the structure necessary to realize our subject. Here, therefore, is an algorithm for origami design, called the *circle method*:

- Count up the number of appendages in the subject and note their lengths.
- Represent each flap of the desired base by a circle whose radius is the length of the flap.
- Position the circles on a square such that no two overlap and the center of each circle lies within the square.
- Connect the centers of touching circles to one another with axial creases, dividing the square into axial polygons.
- Identify tiles whose circles match up with the circles in the axial polygons.
- Fill in the axial polygons with tile creases.

The resulting pattern can be folded into a base with the number and dimension of flaps with which you started.

9.4. Scale of a Circle Pattern

One aspect of the circle method of design that we have already seen is that corner flaps consume less paper than edge flaps, which consume less paper than middle flaps. Turn this property around, and you find that for a given size square, you can fold a larger model (with fewer layers of paper) if you use corner flaps rather than edge flaps, and edge flaps rather than middle flaps. Seen in the light of the circle method, the traditional Crane—and the Bird Base from which it comes—is an extremely efficient design, since all four flaps are corner flaps, and almost all of the paper goes into one of the four flaps. However, add one or two more flaps, and you are forced to use edge flaps. Once you start mixing edge flaps and middle flaps, you begin to run into tradeoffs in efficiency. Sometimes, it is even better *not* to use the corners for flaps if there are additional flaps to be placed on the square.

So, for example, suppose we want to fold a base with five equal-length flaps. A little doodling with a pencil and paper (alternatively, you can cut out some cardboard circles and shuffle them around) will reveal two particularly efficient arrangements of circles, as shown in Figure 9.7.

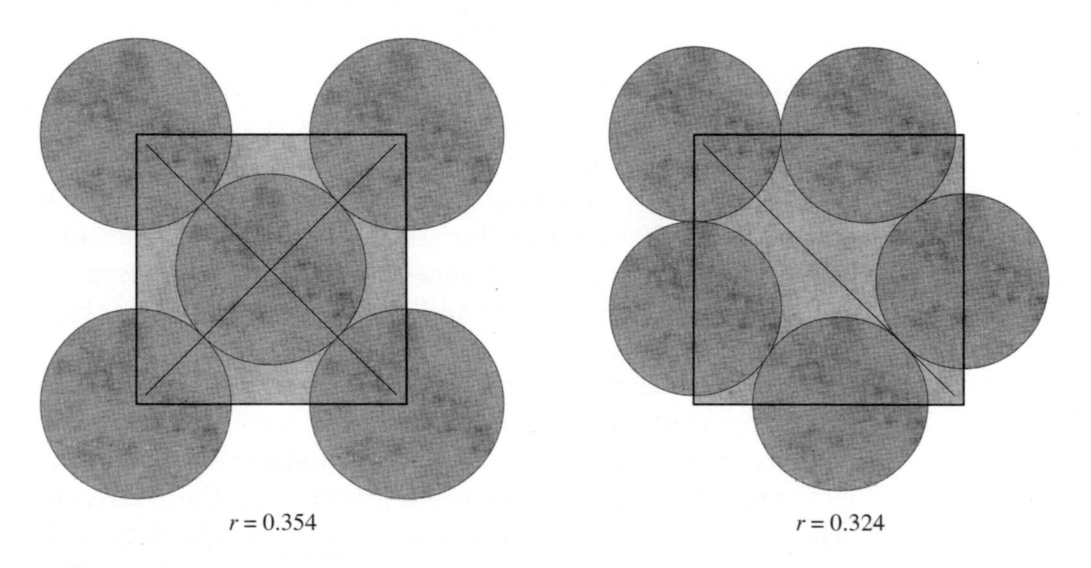

Figure 9.7.
Two circle patterns corresponding to bases with five equal-length flaps.

Now we have two possible circle patterns. Which one is better? Is there any way to quantify the quality of a crease pattern?

As mentioned earlier, one measure of the quality of a model is its efficiency, that is, the relationship between the size of the folded model and the square from which it is folded. A quantitative measure of efficiency is to compare the size of some standard feature of the base—such as the length of a flap—to the size of the original square. The most efficient base is the largest possible base for a given size square. Since the flaps of the base are represented by circles whose radius is equal to the length of the flaps, the most efficient base corresponds to the most efficient circle packing, i.e., the packing with the largest circles.

To facilitate this comparison, let's assume our square is one unit on a side. (If you're using standard origami paper, a unit is equivalent to 10 inches or 25 cm.) For the crease pattern on the left in Figure 9.7, if all of the circles are the same size, it is fairly easy to work out that the radius of each circle, and thus the length of each of the five flaps in the base, is

$$r = \frac{1}{2\sqrt{2}} \approx 0.354 \tag{9–1}$$

For the pattern on the right, it requires some algebra to calculate, but you will find that the circle radius is

$$r = 1 + \frac{1}{\sqrt{2}} - \sqrt{\frac{1}{2} + \sqrt{2}} \approx 0.324 \tag{9–2}$$

Since 0.324 is smaller than 0.354, the radius of each circle in the second pattern is about 10% smaller than in the one on the left. Since the radius of the circle is equal to the length of the flap, the flaps in a base folded from the pattern on the right are about 10% shorter than flaps folded from the pattern on the left. Thus, a five-flap base made from the pattern on the left will be slightly larger and slightly more efficient than the pattern made from the one on the right.

These two circle patterns are relatively simple. By connecting the centers of the circles with creases and adding a few more creases, you can collapse the model into a base that has the desired number of flaps. As it turns out, there already exist in the origami literature bases that correspond to both circle patterns, shown in Figure 9.8. The pattern on the left is the circle pattern for the Frog Base, while the one on the right is the circle pattern for John Montroll's Five-Sided Square.

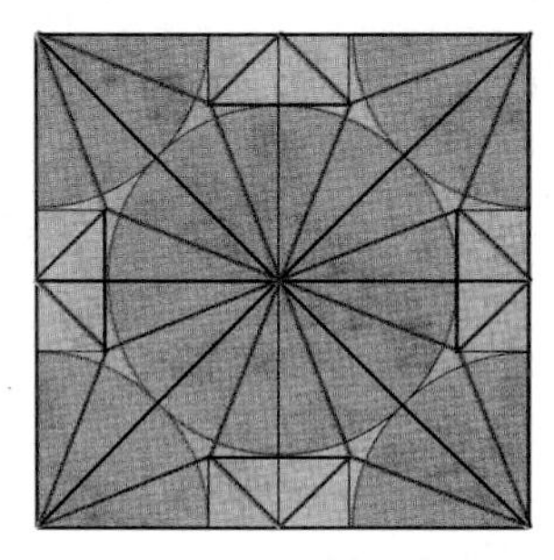
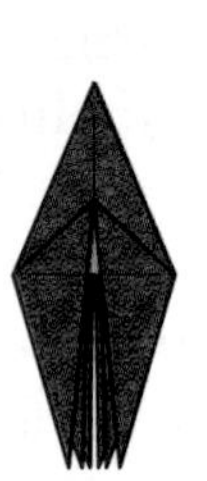
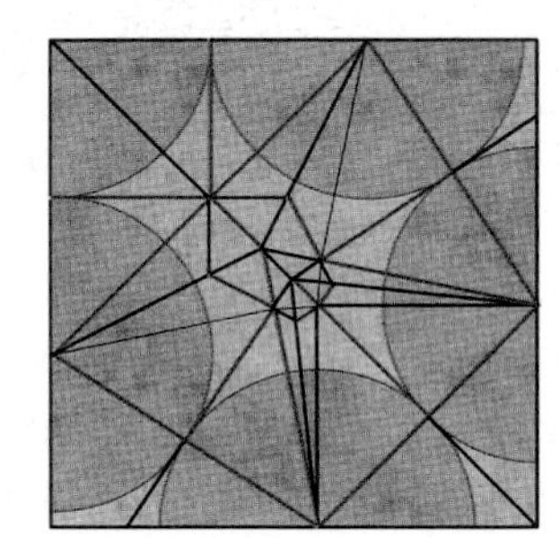

Figure 9.8.
Full crease patterns corresponding to the two circle patterns.
Left: Frog Base.
Right: Montroll's Five-Sided Square.

There is more than just a size difference between these two bases; there is also a qualitative difference between them. In the Frog Base, the fifth flap is a thick middle flap and points in the opposite direction from the four corner flaps; whereas in the Five-Sided Square, the four edge flaps and the corner flap go in the same direction and can easily be made to appear identical. (This was the original rationale for Montroll's design; the shape on the far right resembles a Preliminary Fold with one extra flap). This is an aesthetic tradeoff between efficiency and effect; it's worth a slight reduction in size to obtain the similarity in appearance for all five flaps.

Efficient circle packings also tend to require simpler tiles. At the moment, the only tiles we have at our disposal are triangles, rectangles, and parallelograms. Thus, it is most desirable to use the circles themselves to create a packing in which the induced tiles are triangles or quadrilaterals. This object is accomplished by maximizing the number of points of contact among circles; on average, the more circles touched by each circle—a number called the *valency* of the circle—the lower the number of sides in the surrounding polygons.

You can see this relationship at work in the three regular circle packings in Figure 9.9. In the triangular packing, each circle touches six others and the polygons are all triangles; in the square packing, each circle touches four others and the polygons are all quadrilaterals; and in the hexagonal packing, each circle touches only three others and the polygons are hexagons.

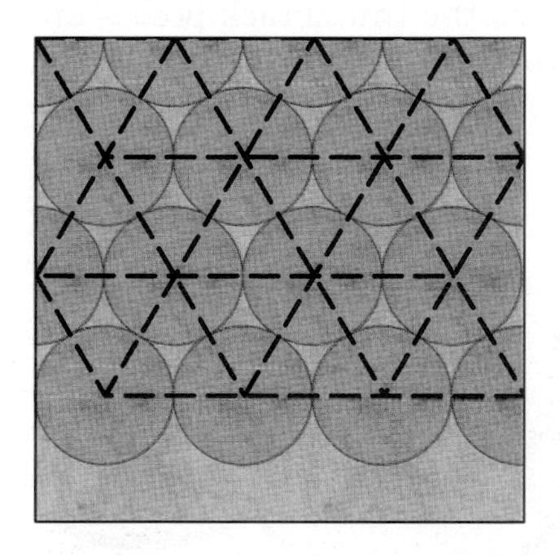
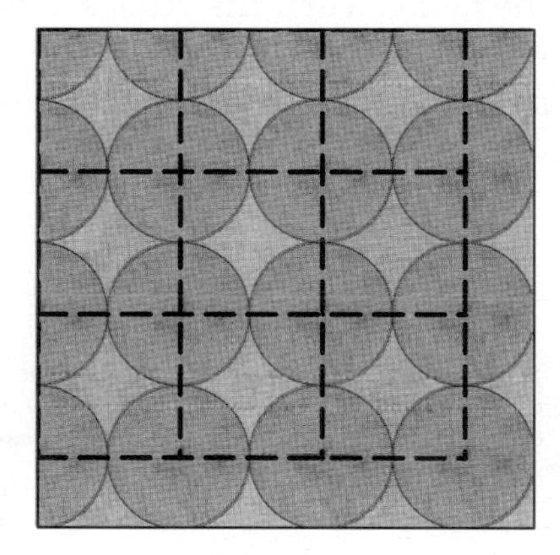
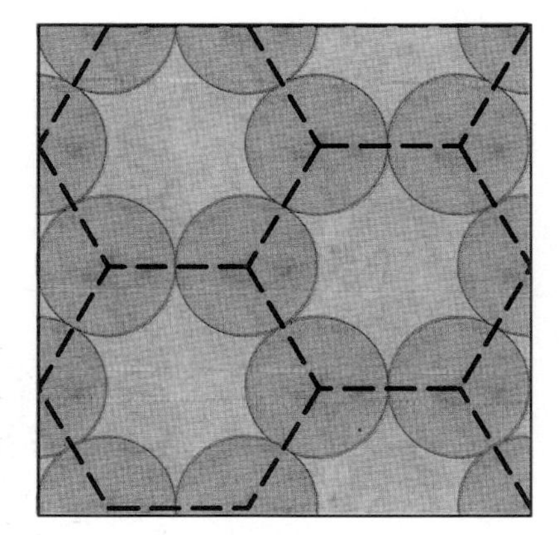

Figure 9.9.
Circle packings of varying density and valency.

As the number of neighbors declines, the amount of empty space around the circles rises. The most space-efficient packings are characterized by circles with many neighbors, which also gives polygons with relatively few sides.

9.5. The Circle Jig

A useful conceptual model for finding an efficient base using circle packing is: Represent each point by a circle whose center is confined to the square and whose radius is proportional to the desired length of the point. Initially, the circles are much smaller than the square, so that there is a lot of extra room around each; they can rattle around in the square. But then you start inflating each circle (or equivalently, shrinking the square) so that the extra room gets slowly squeezed out and the circles start bumping up against one another. Eventually, all of the room is squeezed out and each circle is pinned into place, at which point the basic structure of the crease pattern is frozen. You then fit tiles to the resulting pattern of circles, fitting tile boundaries along lines between touching circles.

You can find the optimum circle pattern several ways. The easiest is to simply draw a square, then start drawing in circles; if you can fit them all with room to spare, do the same thing again, but this time use slightly bigger circles. Repeat until all the circles you draw fit snugly.

This approach, while the simplest and quickest, does require that you have a pretty good eye for size (and that you're able to draw an accurate circle). An easier technique is to use a jig. To make the jig, cut out thick cardboard circles corresponding to the sizes of each of the flaps in your model. Then press a thumbtack through each circle in the very center of the circle. Turn the circles over (so the thumbtack points upward); you can now slide the circles around until they touch and quickly try out different arrangements of circles.

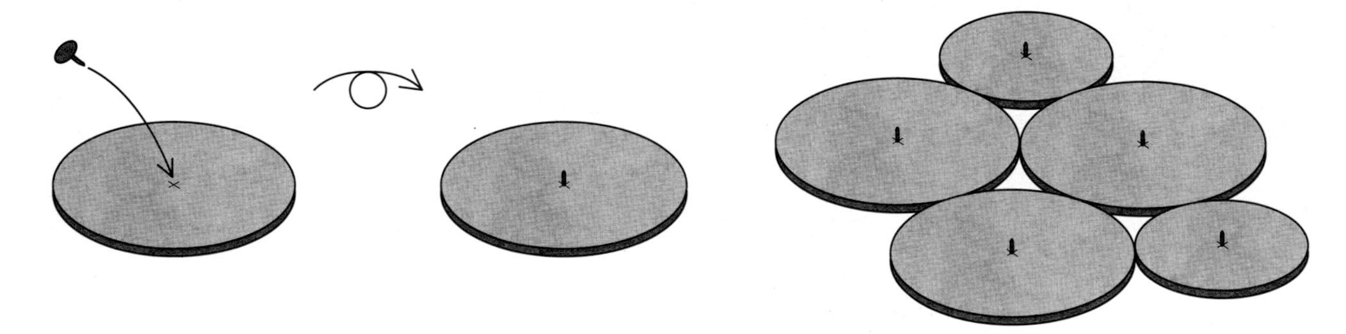

Figure 9.10.
To make a circle jig, push a thumbtack through a circle of the same size as the desired flap. Then you can slide the circles around.

But how do you insure that the centers all lie in a square? The second part of the jig is the frame. Cut out two L-shaped pieces of cardboard as shown in Figure 9.11 and mark a scale along the *inside* of each arm of the L, starting where the two arms touch.

Now you can overlap the edges of the two Ls so that their inside edges form a square; you can insure that they form a

Figure 9.11.
Make two L-shaped arms with scales along their inner edges.

square and not a rectangle by making sure that the inside edges always meet at the same points on the scale. Set the frame down over your array of circles and shrink the square by bringing the two pieces of the frame toward each other as shown in Figure 9.12. The inside edges will catch on the protruding thumbtacks, thus insuring that all of the circles keep their centers inside the square. As you shrink the frame and move the circles around, you will reach a condition in which most or all of the circles are touching each other and the two halves of the frame are held apart by a rigid network of touching circles. This pattern corresponds to the optimum circle packing for your particular model. By pressing a sheet of paper down over the thumbtacks, you can transfer the centers of all the circles to another sheet and can then, using compass and straightedge, reconstruct the optimum circle pattern for folding.

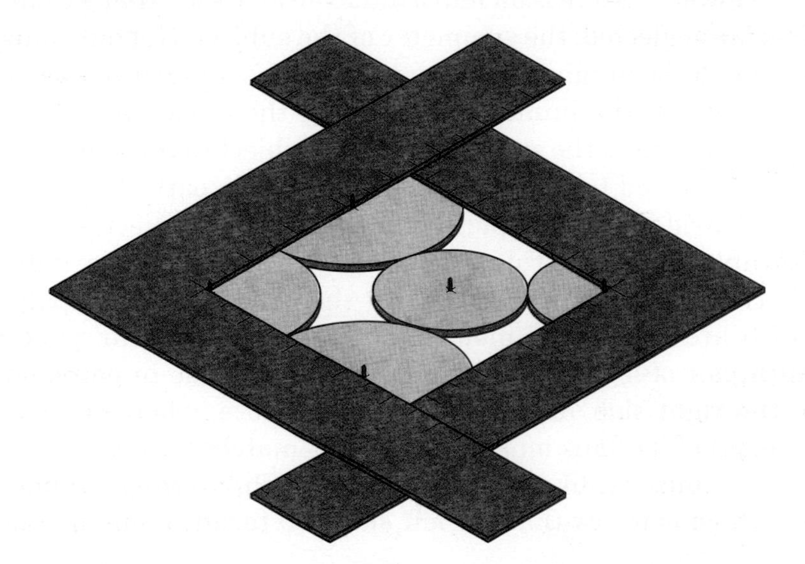

Figure 9.12.
Close the two L brackets toward each other, trapping the thumbtacks in the middle.

A third approach is to set the problem up mathematically and use a computer to solve numerically for the optimum arrangement. I will describe a numerical solution to the circle packing problem in the section on tree theory later in this book.

For small numbers of flaps, you can usually find the solution by doodling; for large numbers of flaps, the frame-and-circle jig works well. While the number of models with more than eight equal-sized flaps dwindles rapidly with increasing numbers of flaps, it is interesting to ponder the question: What are the limits on the sizes and types of flaps? One can show that as the number of flaps N becomes very large, the length r of a flap in the most efficient crease pattern approaches a value of

$$r = \frac{1}{\sqrt{2N\sqrt{3}}} \tag{9–3}$$

For 1000 flaps, this limit takes the value r=0.017; for a 25-cm square, this would imply you could make a base with 1000 flaps, each of which is 0.42 cm long. Within this base, there will be at most four corner flaps; there will be $(4/r)$, or 235 edge flaps, give or take a few, and the remaining 761 flaps will be middle flaps. The folding method for the base—as well as the choice of subject—is left as an exercise for the reader.

9.6. Symmetry

Circle packing allows one to go directly from a description of the flap configuration to the base. Each flap is represented by a circle; by packing the circles into a square and overlaying tiles, we can construct a crease pattern that folds into the desired base.

However, there is an important consideration that we have thus far neglected: the symmetry of the subject. Not only must we match the number of appendages to the number of flaps in the base and the number of circles in the circle pattern; we must also match the symmetry of the subject to the symmetry of the base and to the symmetry of the circle pattern.

Consider, for example, a ten-appendaged subject, such as a tarantula. (A tarantula, being a spider, has eight legs; it also has two prominent appendages at the head, called pedipalps, which are technically mouth parts, but that appear to be a tenth pair of legs.) The legs of a tarantula come in pairs, one on the right side, one on the left. Therefore, when we fold a base, all of its flaps must also come in matched pairs.

Tarantulas, like most animals, are bilaterally symmetric, which is to say that the left side of a tarantula is the mirror image of the right side. If we draw a line down the middle of a tarantula (or any bilaterally symmetric animal), legs and other appendages that come in pairs will lie wholly on one side or the other of this line, which we call the line of mirror symmetry of the model. On the other hand, most appendages that come singly—such as the head or abdomen—will be made from flaps that lie directly upon the line of symmetry.

The flaps of the base must show the same symmetry as the tarantula. The base should have a line of mirror symmetry; flaps that become legs should lie wholly on one side of the line of symmetry or the other. Flaps that become a head or tail should lie directly *on* the line of symmetry. These relationships are illustrated in Figure 9.13.

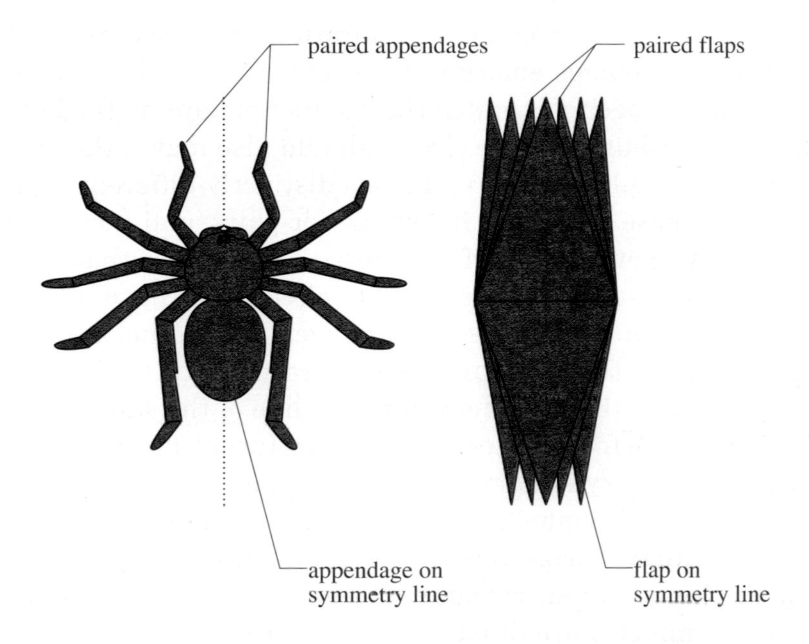

Figure 9.13.
Left: a tarantula and its line of symmetry.
Right: a hypothetical tarantula base and its line of symmetry.

If the subject has bilateral symmetry, then the base should have bilateral symmetry. And if the base has bilateral symmetry, then the crease pattern must also have the same type of symmetry. Since each flap can be identified with a particular circle in the crease pattern, we can't use just any crease pattern with the right number of flaps; we have to use a crease pattern in which each circle has the same relationship to the line of symmetry as does its corresponding flap in the base.

If the crease pattern has a line of symmetry, then (usually) that line of symmetry must be one of the lines of symmetry of the unmarked square. A square has a total of four mirror lines of symmetry, which are illustrated in Figure 9.14. But in fact, there are only two different types of symmetry possible in a crease pattern. It can be symmetric about a line between the middle of the two sides, which we call *book symmetry*; or it

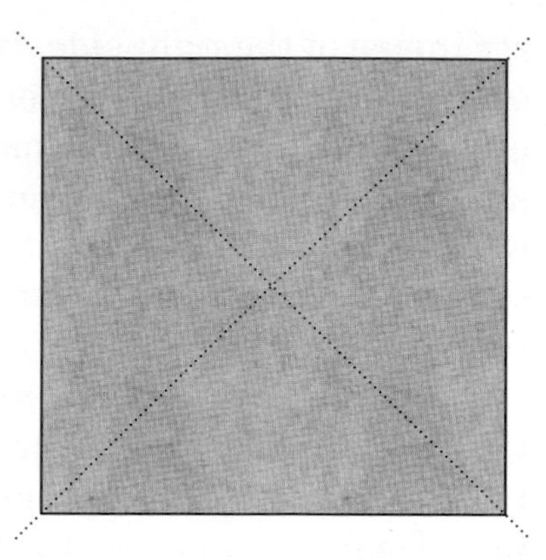

Figure 9.14.
Mirror symmetries of the square.
Left: book symmetry.
Right: diagonal symmetry.

can be symmetric about a diagonal of the square, which we call *diagonal symmetry*.

From these symmetry considerations, we see a new rule for circle placement emerging: Not only should the number and diameter of circles match the number of flaps on the base, but the distribution of the circles should also match the symmetry of the subject. There are two distinctly different types of circles: those that come in symmetric pairs, and those that lie directly upon the line of symmetry. Appendages that come in mirror-image pairs correspond to circles that have mirror-image pairs on the square. Appendages that do not come in pairs, such as the head and tail, correspond to circles that should lie directly upon the symmetry line of the square.

If we wish to fold a ten-appendaged tarantula, we should choose a line of symmetry, divide the square into two regions along the line of symmetry, and then pack five circles into each region in mirror image of each other. This task is easily done, and my (conjectured) optimal solutions for the two possible lines of symmetry are illustrated in Figure 9.15.

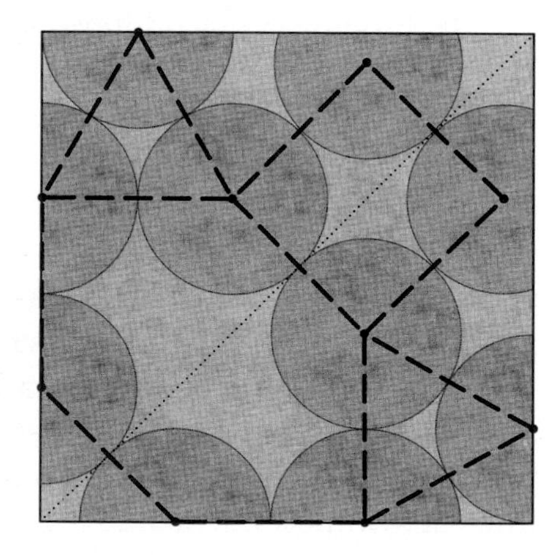

Figure 9.15.
Left: optimum ten-circle packing with book symmetry (r = 0.197).
Right: optimum ten-circle packing with diagonal symmetry (r = 0.194).

Here, the pattern with the line of symmetry parallel to a side is slightly more efficient than the one in which the line of symmetry is along the diagonal. Consequently, the corresponding base would have longer flaps, and the resulting model would be more efficient. Not by much, however. The difference is only about 1%.

The diagonal pattern is not a stable pattern; that is, we could allow two pairs of flaps to become somewhat larger without shrinking anything else. The two flaps that can grow are the two in the upper right corner of Figure 9.15. Can you see why?

A pattern that is almost as efficient is the twelve-circle packing shown in Figure 9.16, which gives twelve paired flaps with diagonal symmetry or ten paired flaps and two on the symmetry line if you use book symmetry. Even if you only need ten flaps, since the major crease lines of this pattern all run at right angles or 45° to one another, it might be simpler to make a base from this pattern than from the preceding two. With the twelve-circle pattern, the slight loss in efficiency would be offset by the ease of folding and the cleanliness of the lines of the model.

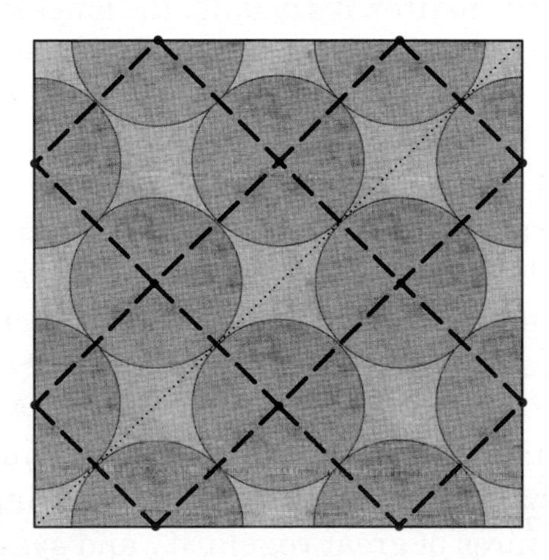

Figure 9.16.
A diagonal-symmetry twelve-circle packing ($r = 0.177$).

The mathematical study of circle packings has tended to concentrate upon packings of identical circles, corresponding to bases with circles all the same size. However, in origami, we often are seeking to construct bases in which the flaps do not have the same length. In a grasshopper, for example, the two back legs are much longer than the other four legs. When we try to find a circle packing for a grasshopper, we should use two large circles for the back legs and four smaller ones for the front legs (and perhaps a medium-sized circle for the body and another short one for the head). Ordinarily, one would choose circle dimensions that correspond precisely with desired flap lengths. However, by judicious selection of circle size, we can produce particularly elegant and symmetric crease patterns, as we will see.

9.7. Selective Inflation

We can achieve a high packing density by making use of some variability in the design process. In most origami designs, the relative lengths of the flaps are not all absolutely fixed. After all, if a point is too long, it can be shortened by reversing or sinking its tip. Conversely, if it turns out to be too short, that deficiency can sometimes be hidden by how the shaping folds are performed, for example, in a leg, by reducing or eliminating angular bends. So if we allow some variability in the lengths of flaps, we can adjust the sizes of the circles to obtain more points of contact between adjacent circles.

It's rarely desirable to allow all of the circles in a pattern to vary independently. At the very least, flaps that come in symmetric pairs—paired legs, paired wings, perhaps front-and-rear legs—should maintain the same relative sizes. The optimum is to be found not in fixing all flaps to have the same relative length, nor in allowing each flap to vary independently, but somewhere in between. To achieve this middle ground, we divide the flaps up into groups that are subject to similar scaling.

As a practical matter, it's usually the longest flaps—which have the largest circles—that dominate the structure of a crease pattern. Thus, one would typically start designing a crease pattern by putting the largest circles in the square and inflating them until they can no longer grow. One then adds the next-smallest set of circles in the spaces of the larger circles until they, too, become fixed; and then add the next set, and so on. I call this process of fixation *crystallization* of the circle packing, because the process resembles the crystallization of atoms when a liquid is cooled below freezing. And just as the atoms of a crystallized liquid form a highly symmetric arrangement, quite often, the crystallized circle packings of origami, too, form structures of great regularity and symmetry.

An example will make the process clear. Let's continue with the tarantula we introduced earlier. As I mentioned, tarantulas have in addition to their eight legs an additional pair of appendages on the head called pedipalps which resemble a tenth pair of legs (although they are typically only about half the length of the legs). Thus, we require a base with eight long flaps for legs and two somewhat shorter flaps for the pedipalps. Tarantulas also have a fairly bulbous abdomen, which we will represent by yet another flap, and a cephalothorax (which contains the head and body), which will also require a flap. Thus, our desired tarantula base would have a total of twelve flaps. We could use the twelve-equal-circle packing shown in Figure 9.16, but we can construct a more elegant and efficient base by exploiting both the variation in length and importance of the different flaps in the base.

The flaps fall into several logical groupings. The longest flaps are the eight legs. Although the legs of an actual tarantula vary in length, they are close to the same length, and the symmetry and foldability of the base will likely be higher if we choose them all to be the same length. Thus, the eight legs form the first group.

The next group would consist of the pedipalps, which are generally about half as long as the legs. We could start by choosing their length to be exactly half of the leg length, but in practice, anything from about 40% to 60% would probably give a workable model.

Next, the abdomen is also about half the length of a leg. Since there's only one abdomen flap but two pedipalp flaps that must come as a matched pair, we'll give the paired flaps a higher priority; fit them in first, then try to find a space for the abdomen.

Last comes the head flap. Since the cephalothorax of a tarantula is quite short, we almost don't need this flap at all, so we can just count on tucking it in somewhere in the finished packing.

Let's now work through the circle packing step by step. The first step is to pick the symmetry line of the base: Will it be oriented parallel to a side (book symmetry) or along the diagonal (diagonal symmetry)? Let's choose book symmetry for starters. The first step will be to pack the eight leg circles into the square.

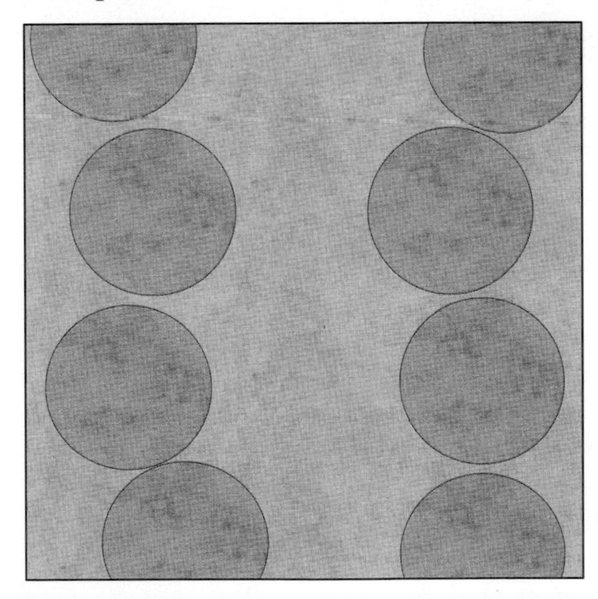

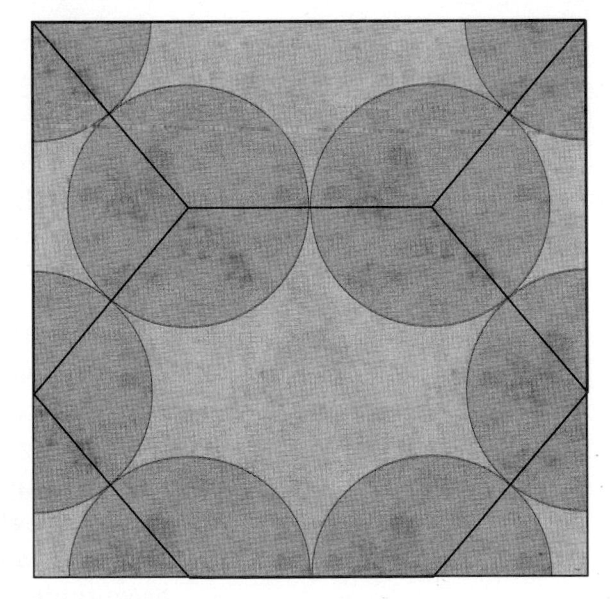

Figure 9.17.
Left: starting configuration.
Right: crystallized circle packing.

We then inflate the circles, moving them around to keep them from overlapping, until they are locked into position. Figure 9.17 shows the result of the tarantula circle crystallization, which is the largest possible book-symmetric packing of eight circles into a square. (An equivalent solution is the same pattern flipped vertically.) The radius of a circle relative to the length of the side of the square is 0.2182, so the length of the leg flaps will be about 22% of the length of the side of the square.

Now, we add the pedipalps, which would be represented by two paired circles about half the size of the leg circles. The obvious place to put them is in the center of the large hexagon in the lower half of the square. Continuing the process of inflation in Figure 9.18, we drop two small circles into an opening of the pattern, then expand them until they, too, are firmly wedged against their neighbors.

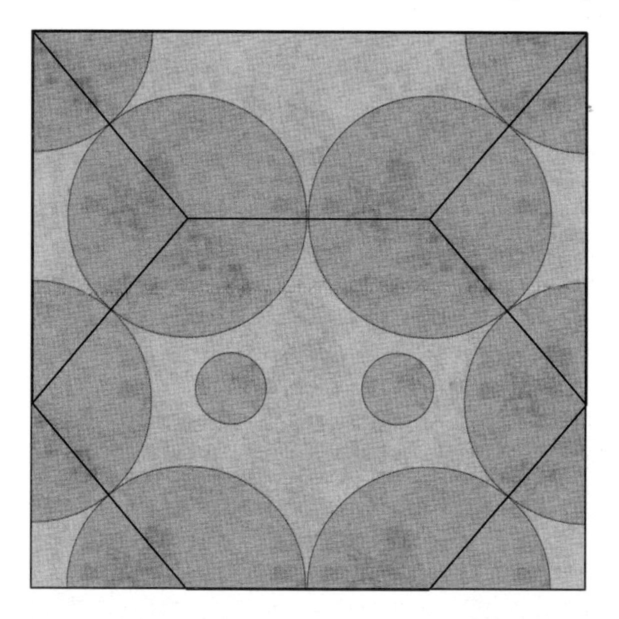

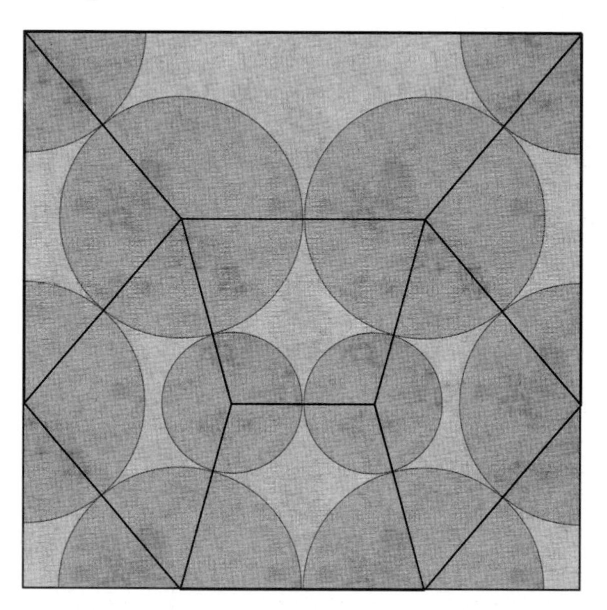

Figure 9.18.
Left: add two circles for pedipalps.
Right: crystallized circle packing.

With this configuration, the pedipalp flaps turn out to be 0.583 times as long as the leg flaps—just about right.

Next comes the abdomen, whose circle should also be about half the size of the leg circle, and so we can fit it into the top middle of the square, as shown in Figure 9.19.

When this circle is inflated, it turns out to be 0.826 times as long as a leg flap—longer than we might like, but perfectly acceptable since it is very easy to shorten a flap.

Last comes the head. There are two same-size holes remaining in the circle pattern either just above or just below

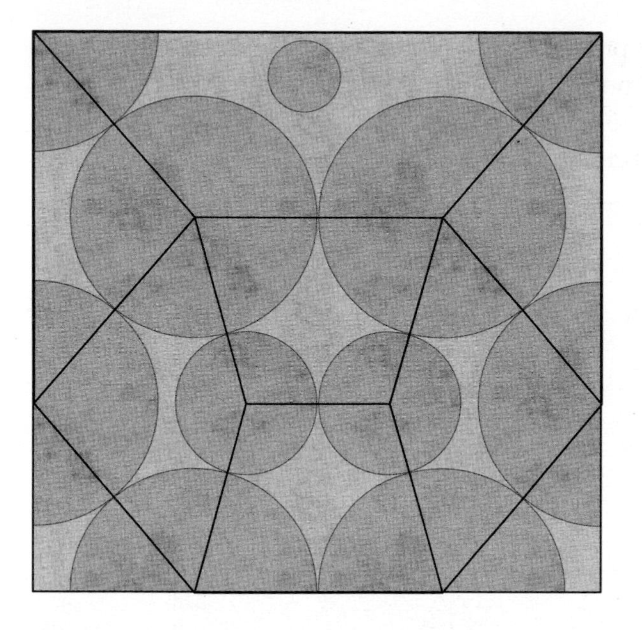

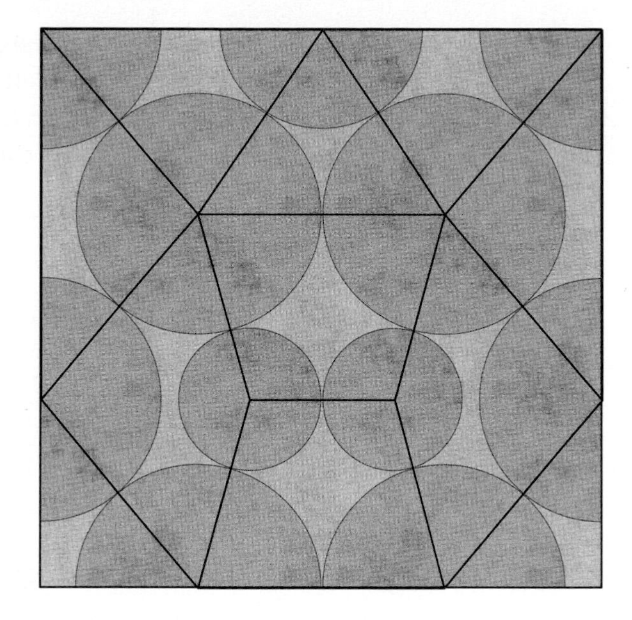

Figure 9.19.
Left: Add a circle for the abdomen.
Right: crystallized circle packing.

the pedipalps; we could put the head circle in either one. Both options are shown in Figure 9.20.

Since the head flap was the last flap that needed to be placed, you could just as well place circles in both gaps and then choose which flap becomes the head once the base is folded.

Figure 9.20.
Left: one possible choice for the head circle.
Right: the other choice.

In either case, all polygons are either triangles or quadrilaterals, and so one can now fill in the polygons with the appropriate tiles, giving the resultant crease pattern, shown superimposed over the circles in Figure 9.21.

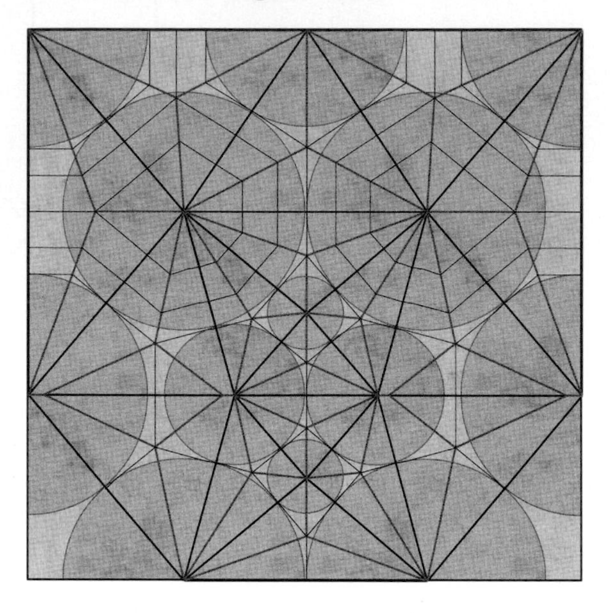

Figure 9.21.
The finished crease pattern.

Although the method of circle packing seems to be very straightforward, there are many choices to be made along the way, each giving a different result. For example, when placing the pedipalps, we could have placed them at the top of the square and put the abdomen down in the central hexagon, giving the circle packing and crease pattern shown in Figure 9.22.

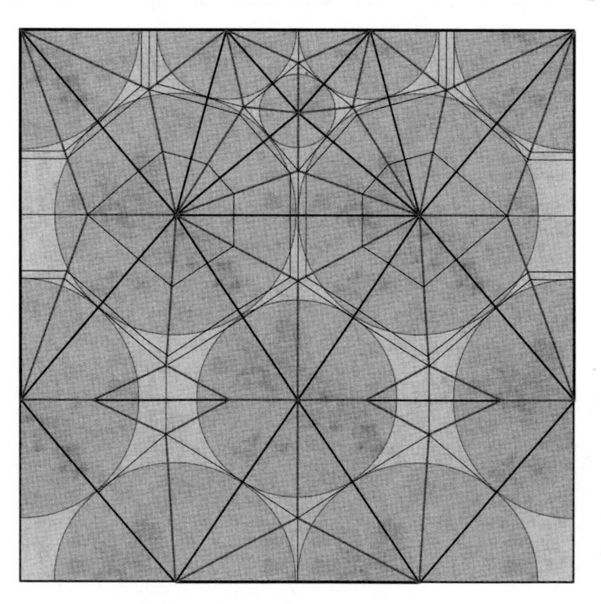

Figure 9.22.
An alternative circle packing and crease pattern.

The choice of which packing to use will be affected by various factors. For example, in the pattern of Figure 9.21, the two pedipalp flaps are middle flaps, which means that they will be wrapped inside other flaps in the folded base; this may make it difficult to achieve an elegant distribution of the legs in the folded model. Also, being middle flaps, they will have many layers of paper when they are thinned. In Figure 9.22, however, the pedipalps are edge flaps and the abdomen is a middle flap, which would probably be easier to work with. Since the abdomen is not thinned as much as a leg or pedipalp would be (it might actually be ballooned outward), it would be more tolerant of the extra layers inherent in a middle flap.

Note that in both cases, two of the legs are middle flaps; these will unavoidably be thicker than the other leg flaps, especially compared to the two corner leg flaps immediately above. This variation in leg thickness could conceivably be a weakness of any model folded from either base.

A remarkable thing about circle-packing bases is that despite the deterministic nature of their construction, there are usually many possible circle-packed bases for a given number and distribution of flaps. Consider the following sources of variety:

- There are two possible symmetric orientations for the base.
- There are typically several crystallizations of the major circles for each symmetric orientation.
- There are typically several placements of the minor circles for each crystallization.

For the tarantula configuration—eight legs, two pedipalps, an abdomen and a head—a small amount of experimentation reveals a host of possible crease patterns, some of which are shown in Figure 9.23.

Each crease pattern gives a unique base. Some are elegant, some are awkward; some can be folded in *plan view* (i.e., opened out flat, like an open book) while others only work in *side view*, i.e., in profile. All can be turned into a tarantula of one sort or another. When you add to that the infinite variations possible in thinning and shaping folds, you can see that the possibilities for exploiting circle packing are nearly limitless.

Of the nine patterns, (b) (book symmetry) and (e) (diagonal symmetry) are actually the same pattern. Both are based on an underlying octagon, which can be fit into the square with either symmetry. The presence of a valley fold running along the symmetry line for most of the model allows the base to be folded in plan view, which allows a smooth and rounded top

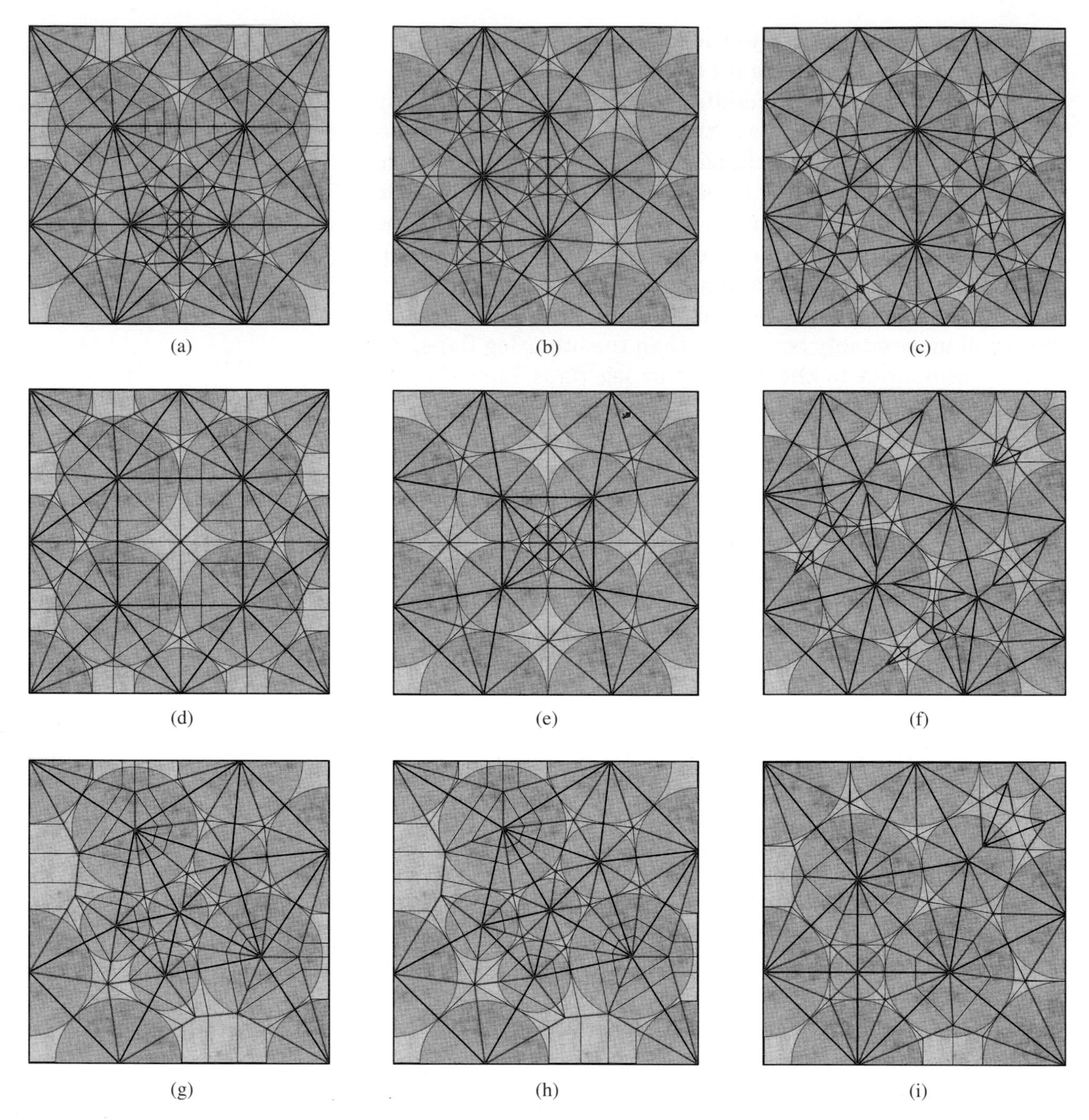

Figure 9.23.
An assortment of possible tarantula crease patterns.

surface of the tarantula. In addition, all of the legs are edge points and have exactly the same number of layers. Of all the possible bases, I think it lends itself best to the subject; and so I have carried this pattern all the way to a folded model, which is shown in Figure 9.24.

I would encourage you to try folding the other bases from their crease patterns and turn them into your own tarantula design.

Figure 9.24.
Crease pattern, base, and folded model of the Tarantula.

As we see in the tarantula design, circle packing patterns and their corresponding crease patterns are most symmetric if the circle radii (and thus flap lengths) are chosen from among only a small number of values, rather than making every flap a different length. It is common to adjust the values in order to realize particularly symmetric patterns.

The crease pattern in Figure 9.25 is based on a circle packing of six large circles and six medium ones oriented in opposite directions, corresponding to six long and six medium flaps; three small circles are added to fill in the gaps. This gives a pattern that, in the interior of the model, can be filled in entirely with triangle tiles, and works very well for flying insects, in which the six long flaps can be used for four wings, head, and abdomen, while the six medium flaps get used for legs.

We could also rotate the pattern by 90° and assign the short flaps to head, abdomen, and wings, and the long flaps to legs to make a flying insect with proportionately longer legs. Can you devise a model based on this hint?

While this circle packing is quite efficient in its use of the paper, it still leaves the corners unused, making it a great temptation to figure out something else to do with them. The Flying Ladybird Beetle in Figure 9.26 uses the same circle packing (although adding two more smaller circles), but further uses the corners of the paper to realize the spots on the wings.

Figure 9.25.
Crease pattern, base, and folded model for the Flying Cicada.

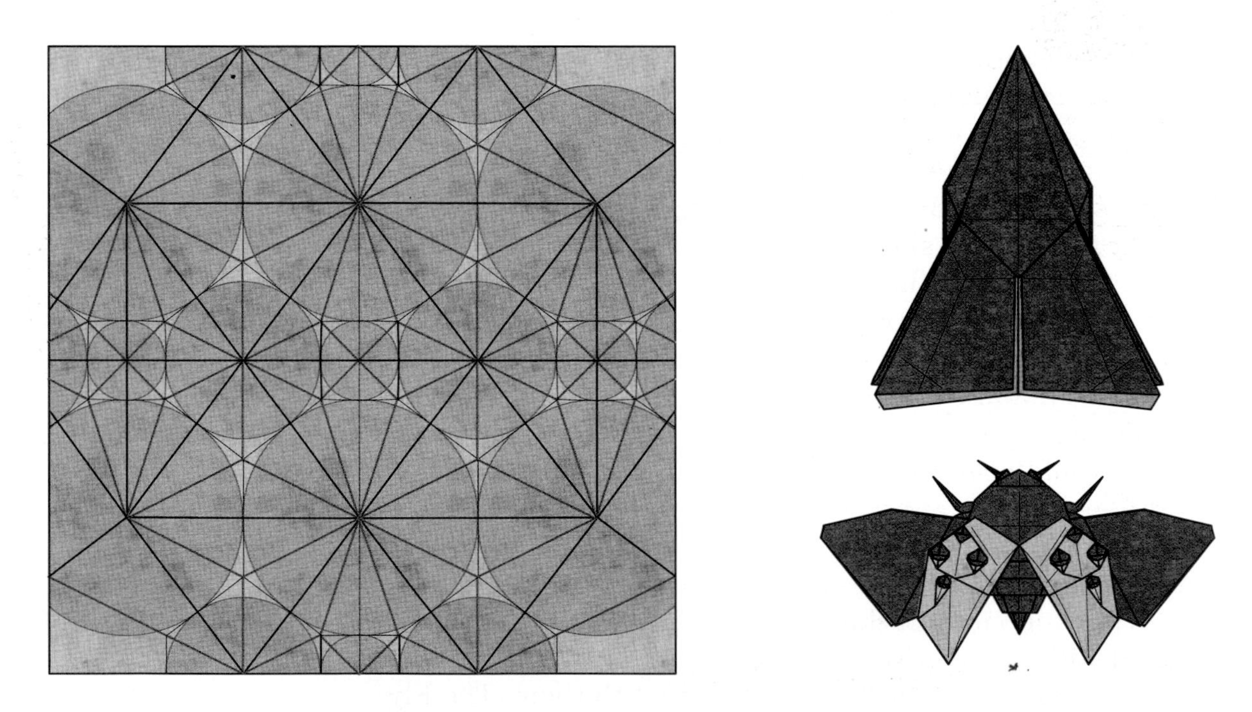

Figure 9.26.
Crease pattern, base, and folded model for the Flying Ladybird Beetle.

9.8. Circles and Rivers

Thus far, we have considered bases derived from packing of circles only. In a circle-packed base, all of the flaps emanate from a common point. But, you'll recall, we were able to construct bases from tiles that contained rivers, regions that translated into segments separating groups of flaps. It is also possible to introduce rivers into a circle-packing in an analogous fashion. That is, we introduce a river between groups of circles that corresponds to a segment separating groups of flaps. We can still construct a full crease pattern by superimposing tiles containing rivers on the pattern of axial polygons.

An example of this is illustrated in Figure 9.27, an insect with the Latin name *Acrocinus longimanus*. It contains four long flaps (corresponding to the two long forelegs and antennae) and eight shorter flaps (legs, head, abdomen). We introduce a river running around four of the legs, which introduces a segment between the two groups of flaps. We can, however, still fill in the axial polygons with tile crease patterns that incorporate the rivers; the result is a base with the desired flap configuration.

Figure 9.27.
Crease pattern, base, and folded model of the *Acrocinus longimanus*.

9.9. Mathematical Circle Packings

The circle method gives us a technique for designing a base that has any combination of flaps of different sizes by relating it to a simple geometrical construction. Specifically, the problem of designing a base with N equal-length flaps can be solved by finding a distribution of N non-overlapping circles whose centers all lie within a square.

In origami, packings of unequal circles arise more often than equal circles because origami bases are more often composed of flaps of differing lengths. When all of the flaps are the same length, however, then all of the circles are the same diameter. This problem turns out to have some interesting mathematical connections, so we will digress briefly to explore them. The problem of packing N nonoverlapping circles with their centers inside of a square is equivalent to the problem of packing N nonoverlapping circles entirely inside of a somewhat larger square; Figure 9.28 shows how the same pattern solves both problems.

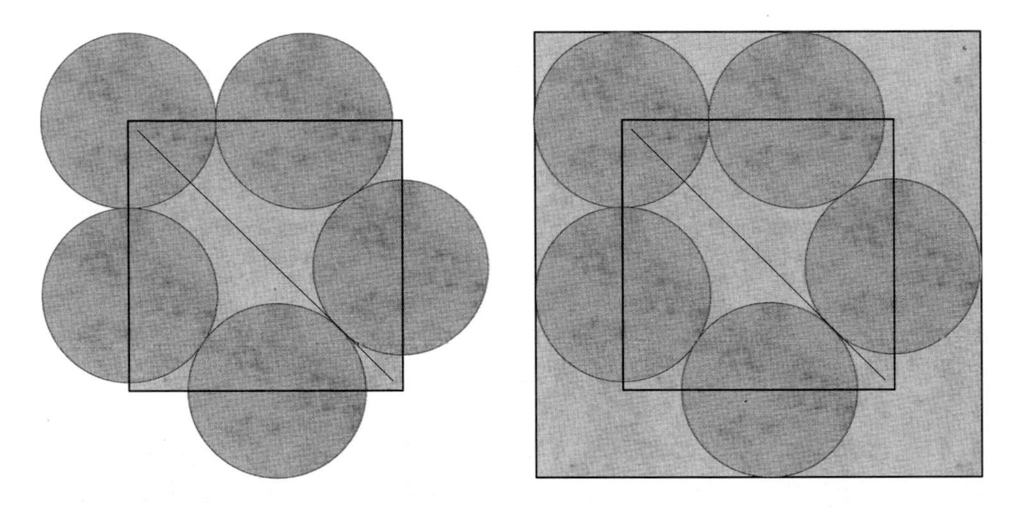

Figure 9.28.
A solution for an origami circle-packing is equivalent to a circle-packing in which the circles must lie completely inside the square.

The similarity between the two problems would be only a curiosity but for one thing; the problem of packing equal nonoverlapping circles entirely inside a geometric figure is a well-researched field in mathematics. This correspondence between the origami design problem and the mathematics of circle-packing is fortunate, because many of the solutions to circle-packing problems have already been enumerated in the open mathematical literature. Instead of rederiving a solution (not an easy task, depending upon the number of flaps), one can merely look up the optimum circle pattern for a given number of circles.

For the mathematical problem of packing N equal nonoverlapping circles into a square, the optimum solutions for $N = 1$ through 10 are known and have been mathematically proven to be optimal. Thus, for the origami problem of folding a base with N equal-length flaps, the optimum crease patterns are also known. The optimum circle patterns and lengths of each flap (as a fraction of the side of a unit square) are given in Figure 9.29 for $N = 1$ through 9. I have only drawn that portion of each circle that appears within the square.

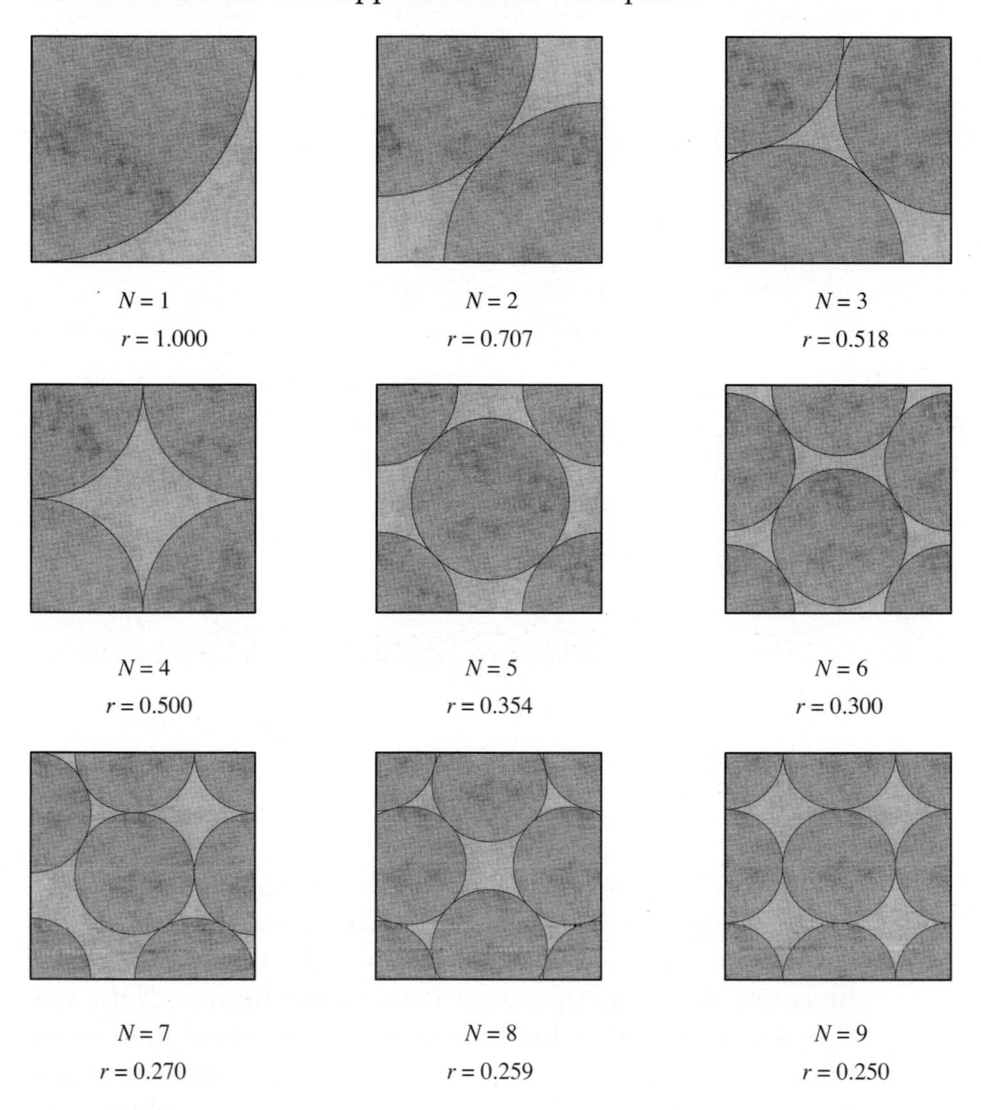

Figure 9.29.
Optimal packings for one through nine circles.

Because circle packing is a well-explored mathematical field, it is possible to look to the mathematical literature for patterns that give rise to origami bases (as I have done here). In fact, as new circle packings are discovered, new origami bases will come right along with them. The nine circle packings shown in Figure 9.29 each have corresponding origami crease patterns, which are shown in Figure 9.30 superimposed on the circle packings.

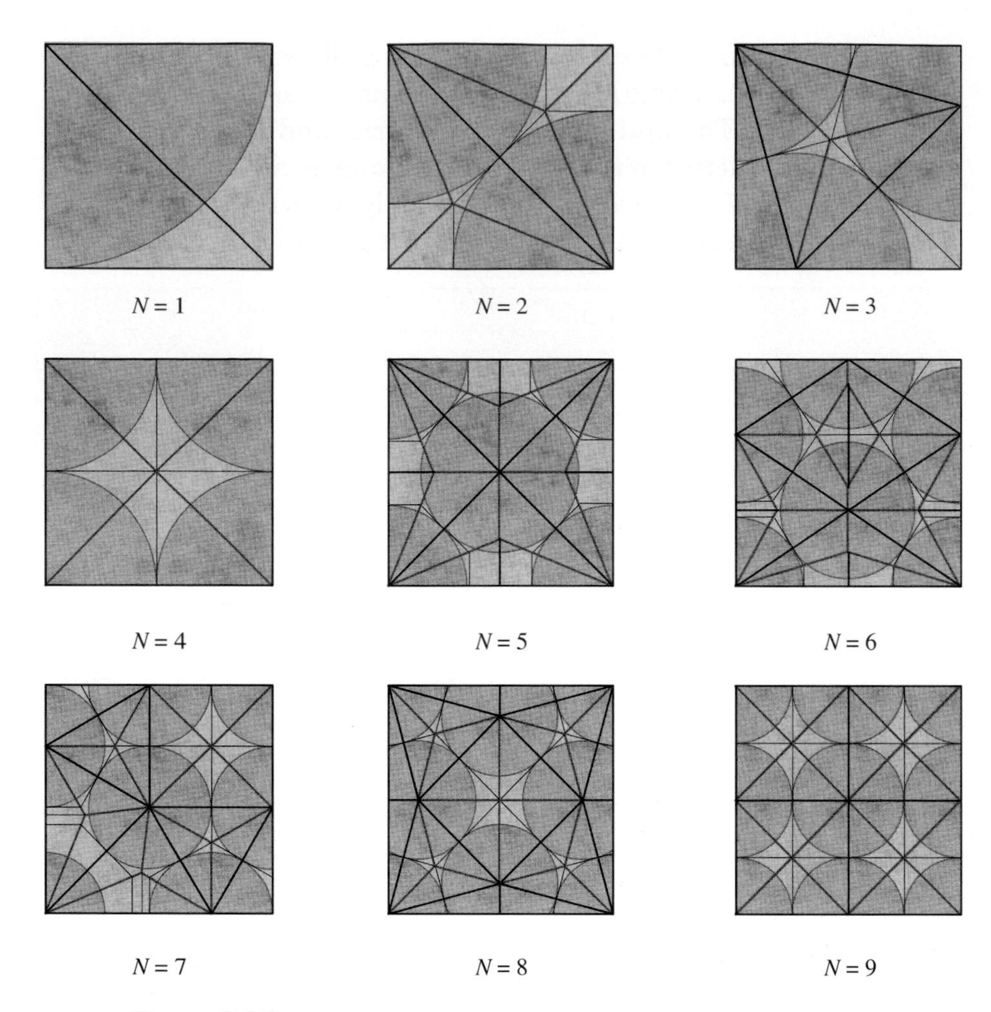

Figure 9.30.
Crease patterns for optimal bases with one through nine flaps.

In each figure, I have drawn in the major creases of the base. Notice something: Wherever two circles touch, there is a major crease connecting the centers of the two circles. These lines are shown as mountain folds in the figures. Note, too, that any crease that hits one of these mountain lines other than at a circle center hits it at a right angle. These properties hint at some deeper relationships between circle packings and their corresponding crease patterns. As we will see in later sections, there are several different types of creases that all share consistent properties.

You might find it an illuminating exercise to transfer the crease patterns shown here to origami paper squares and to fold the corresponding bases to verify that they do, indeed, have the proper number of flaps and that the flaps are all the same size within each base. You will also discover something about the coloring pattern of the creases. Most of the black creases will be mountain folds; all of the colored creases will be valley folds; and the light creases can go in either direction, depending on how you choose to orient the flaps.

It is also instructive to note what kind of flap—corner, edge, or middle—you get with each base. The distribution of flaps is enumerated in Table 9.1

Circles	Corner flaps	Edge flaps	Middle flaps
1	1		
2	2		
3	1	2	
4	4		
5	5		1
6	2	3	1
7	2	4	1
8	4		4
9	4	4	1

Table 9.1.
Numbers of corner flaps, edge flaps, and middle flaps in optimal circle-packed bases.

Given the inefficiency of middle flaps versus edge flaps—remember, a middle flap takes four times as much paper as a corner flap of the same length—it is somewhat surprising to find that for as few as five flaps, it is more efficient to use a middle flap than to create another edge flap. For only three flaps, two edge flaps are preferable to two corner flaps.

Since these patterns give the most efficient bases for the desired number of flaps, it is not surprising that several of them have been in existence for hundreds of years. The patterns for one, two, four, and five flaps correspond to four of the traditional bases—the Kite, Fish, Waterbomb, and Bird Bases, respectively (where the top of the Bird Base forms a fifth flap). The first three bases have all of their flaps on the corner; the last has one middle flap.

What is surprising, though, is that several of these circle patterns correspond to bases that have not yet been published in the origami literature! Specifically, I am aware of no published design based on the symmetry of the patterns for $N = 3$, 6, 7, and 8. I find this remarkable because there are quite a few models that require three, six, seven, or eight flaps, and these patterns correspond to the most efficient structures for getting each number of flaps. It is illuminating to examine each of these crease patterns in more detail.

9.10. Bases from Equal Circle Packings

The first unusual structure in this group is the $N = 3$ case. I suspect it is undiscovered because this pattern is not similar to any well-known base found by angle bisection. Its symmetry is based on the uncommon 30°-60°-90° right triangle rather than the more familiar 22.5° angle that you find in the four Classic Bases. Also, two of the three flaps lie on an edge, rather than at the corner; since edge flaps tend to be thicker, most early origami designers avoided using edge flaps unless absolutely necessary. I am not aware of any origami model that uses this pattern in a square to realize a subject with three major flaps, despite there being a number of such subjects around, e.g., long-legged birds. (There are, not surprisingly, models made from an equilateral triangle that utilize the creases within the triangle.) An example of a model of my own that uses this symmetry is shown in Figure 9.31 with folding instructions at the end of the chapter.

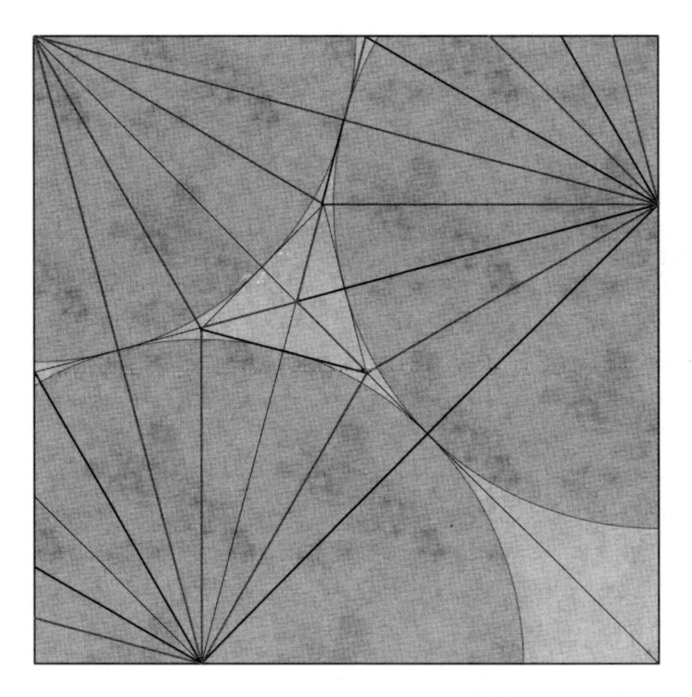

Figure 9.31.
Crease pattern, base, and folded model of the Emu based on the N = 3 circle packing.

Notice that three of the four corners of the square go unused in this base; I suspect that the profligate waste of such a large fraction of the paper has deterred folders from making use of this base, and indeed, it is difficult to resist the temptation to use some of the extra paper for wings, feathers, color-changes, or *something* in the model derived from this pattern. I used the largest unused corner to extend the tail in the emu.

The N = 4 case is quite obviously the crease pattern for the Waterbomb Base, which has been widely used for origami models. Similarly, the N = 5 case has a single flap in the middle of the square; it is the Bird Base. But that's unexpected; the Bird Base is normally considered to have four equal flaps, not five. How can this be?

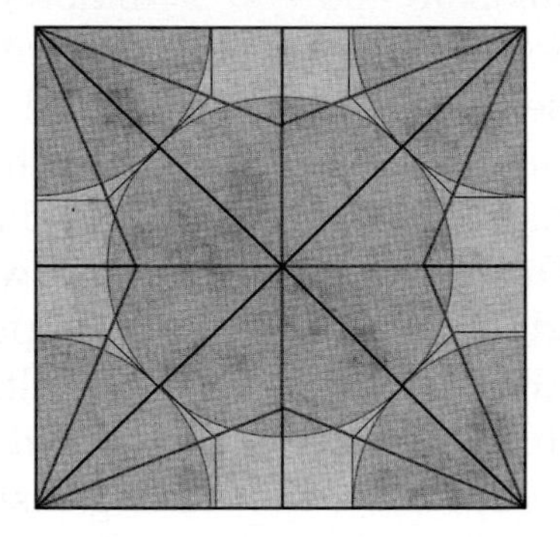

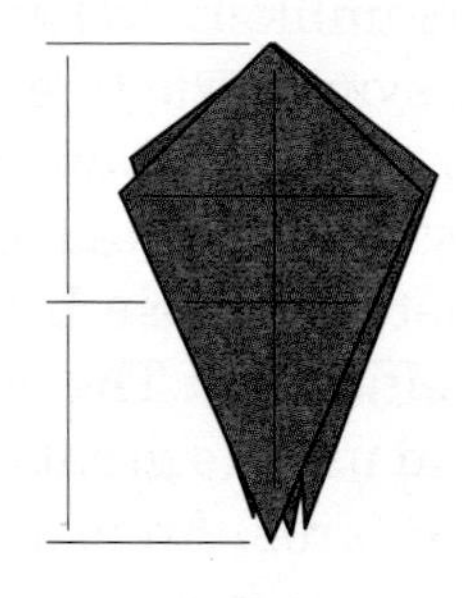

Figure 9.32.
The Bird Base as a five-flap base.

As shown in Figure 9.32, if we treat each flap as starting from a point exactly halfway between the top and bottom of the model, the four long flaps of the Bird Base become shorter and the blunt top flap becomes longer. We can get all five flaps to have the same aspect ratio by narrowing the top flap, as shown in Figure 9.33.

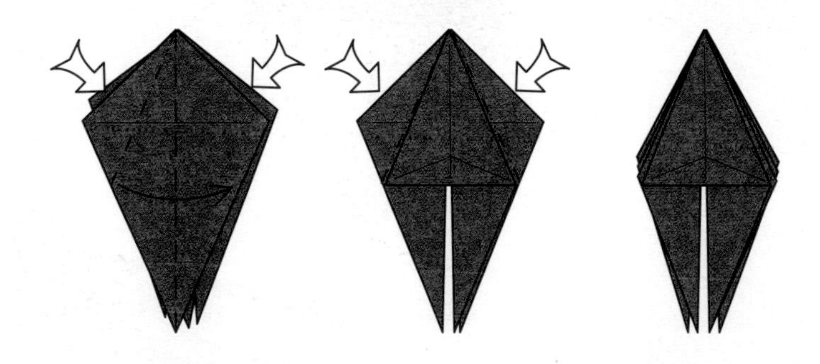

Figure 9.33.
Narrowing the flaps of a Bird Base by spread-sinking turns it into a five-equal-flap base.

And if we add four more smaller circles to the five-circle pattern corresponding to four more smaller flaps, we get the crease pattern for yet another Classic Base, the Frog Base.

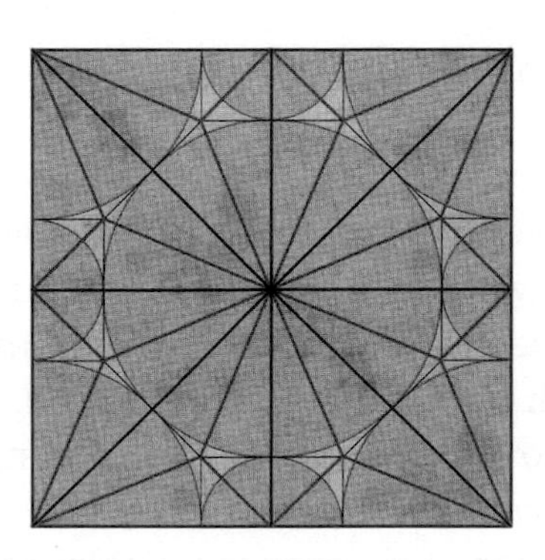

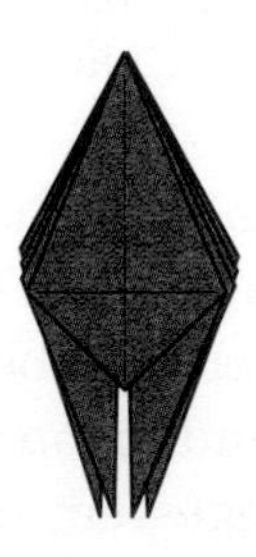

Figure 9.34.
Circle and crease pattern for the Frog Base.

The $N = 1, 2, 4$, and 5 cases correspond to Classic Bases that have been known for hundreds of years. However, the $N = 6$

solution, like the $N = 3$ pattern, has not been explored, or to my knowledge, even recognized. I suspect that it is because the $N = 6$ pattern does not incorporate either the standard 22.5° symmetry or the less-common 30°–60°–90° symmetry and so was unlikely to have been found by trial-and-error folding along symmetric lines. Because it has a line of bilateral symmetry with two flaps lying on one side and two on the other, it seems ideally suited for mammals and birds. I have used it for a general-purpose bird base that gives both legs and wings quite efficiently. The two unused corner flaps may also be pulled out and used to great effect in color changes to make multiple-colored birds. An example of this base and a two-colored bird folded from it are shown in Figure 9.35; the folding instructions are given at the end of the chapter.

Figure 9.35.
Crease pattern, base, and folded model for the Songbird, based on the $N = 6$ circle packing.

This model is the second Songbird in this book (the first was in chapter 6). The two models illustrate the fact that a single subject can be realized as an origami model in different ways, depending on which features are emphasized, which are merely suggested, and how the detail folds and shaping are rendered.

You should also compare this crease pattern and base to that of the Turtle (Chapter 7, Figure 7.1), which was also made from a six-flap base. The turtle base had all its creases running at multiples of 30° and thus had a more symmetric crease pattern and more neatly aligned edges. The songbird base here is less symmetric, but is slightly larger relative to the size of the square. Can you make a turtle from this base?

The $N = 7$ solution combines both squares and equilateral triangles into the underlying symmetry. An unusual feature of this solution is the fact that one of the circles (the one in the lower left corner) doesn't touch any other circle. The paper between the lower left circle and the rest of the model is wasted. One could put this extra paper to use, however, by enlarging the lower left circle (corresponding to lengthening the associated flap) and using this base to fold a model with one particularly long appendage, for example, an extra-long tail. We can carry out this enlargement by expanding the lower left circle until it touches one of the others (the middle circle, as it turns out). This expansion gives the circle pattern shown in Figure 9.36, corresponding to a shape with six equal-sized flaps and a seventh slightly larger one. I have also superimposed a crease pattern that gives a seven-flap base. I encourage you to enlarge the figure, draw it on a square, and fold the base; it is remarkable how all of the circles line up with each other once you have collapsed the square into the base.

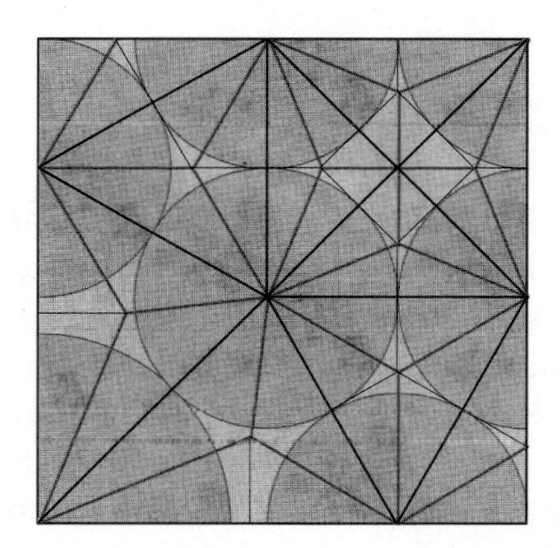

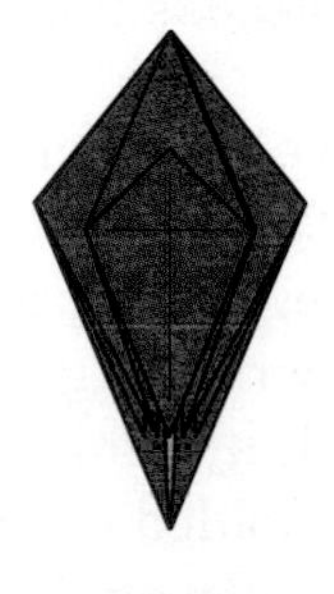

Figure 9.36.
Left: crease pattern for the $N = 7$ circle packing with one flap enlarged.
Right: seven-pointed base.

There is a mathematical term for the condition in which a circle is free to move without changing the overall scale of the model: If any circle can move without altering the scale, the pattern is said to be *unstable*; any pattern in which no circle can move is said to be *stable*. It is easy to see that a pattern is stable only if every circle touches either another circle or an edge of the square at three well-separated points. It will turn out as we explore more sophisticated design algorithms that the issue of stability plays a crucial role in the construction of crease patterns for bases.

The problem of getting eight flaps in a base is encountered when one attempts to fold the simplest insects. Beetles, for example, must have a head, abdomen, and six legs, at a minimum. Of course, it is always challenging to add more body parts: Thorax, antennae, mandibles, horns, wings, and forewings would be nice, each requiring another flap (and we will see examples that have all of them). But even the simplest insect must have six legs, which by the standards of classical folding, is no mean feat. Historically, the first published instructions for a one-piece six-legged insect of which I am aware is George Rhoads's Bug. It is made from a blintzed Bird Base, which corresponds to the $N = 9$ circle diagram.

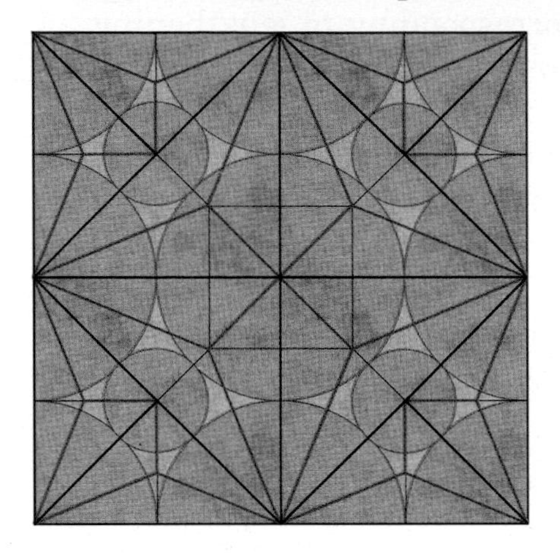

Figure 9.37.
Left: $N = 9$ circle pattern.
Right: Rhoads's Bug, made from a blintzed Bird Base.

We encountered the blintzed Bird Base back in Chapter 4. It is constructed by folding the four corners to the center of a square, folding a Bird Base from the reduced square, and then unwrapping the extra paper to form new flaps. There is also such a thing as *double-blintzing*, in which the four corners are folded to the center, and those four corners are folded to the center again, before folding a base and unwrapping all the layers. The double-blintzed Frog Base, with its thirteen equal-length flaps, was used by the Japanese master Yoshizawa as early as the 1950s for his famous Crab, and surely holds the record for the pointiest base of the Classical period.

The circle pattern provides a simple way to see the effect of blintzing a base. Although each stage of blintzing doubles the area of the square, it doesn't double the number of flaps since some of the paper is consumed turning some quarter- or half-circles into full circles. In the progression of the blintzed Frog Base shown in Figure 9.39, the original Frog Base has five long flaps; the blintzed Frog Base has nine, and the double-blintzed Frog Base has thirteen, numbers which are easily verified by examining the circle pattern.

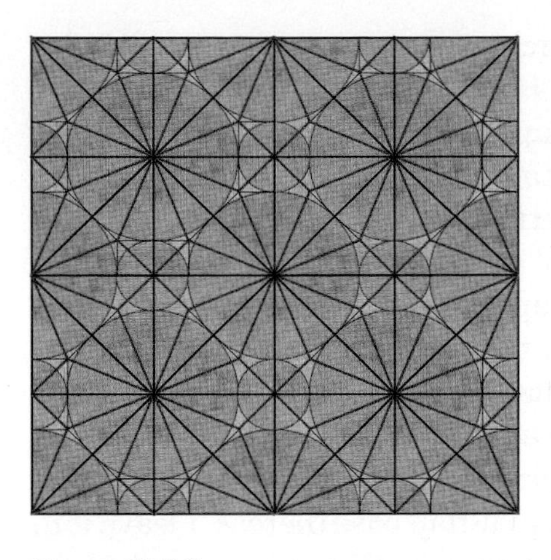

Figure 9.38.
Double-blintzed Frog Base crease pattern and Yoshizawa's Crab.

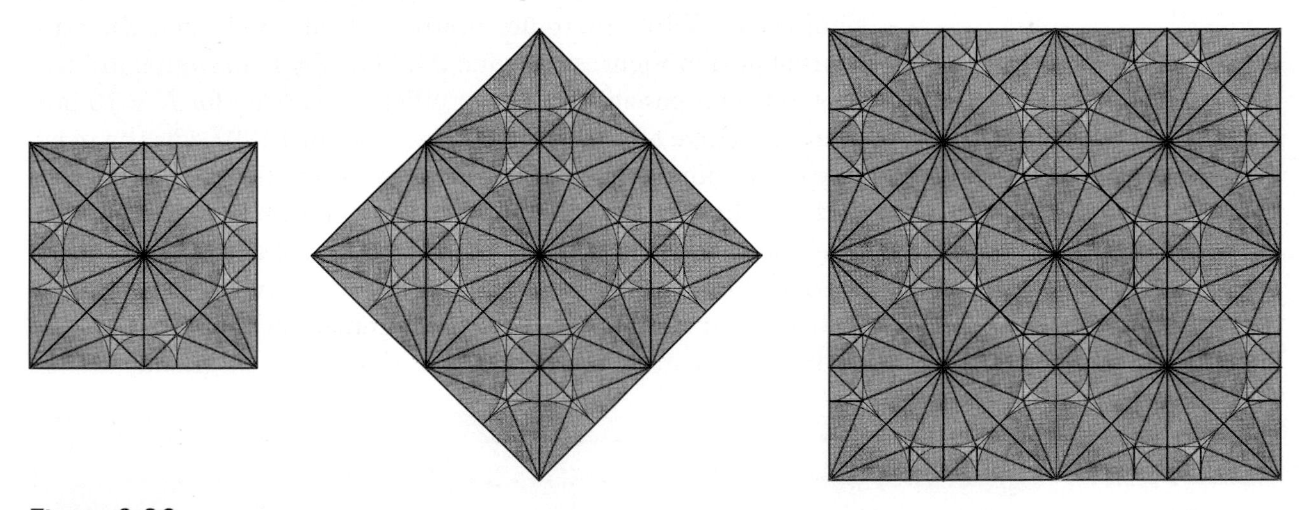

Figure 9.39.
Three stages in the progression of the blintzed Frog Base. Left: the Frog Base. Center: a blintzed Frog Base. Right: a double-blintzed Frog Base.

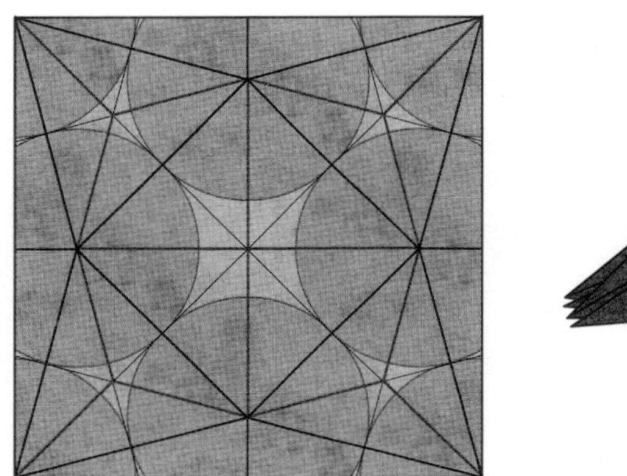

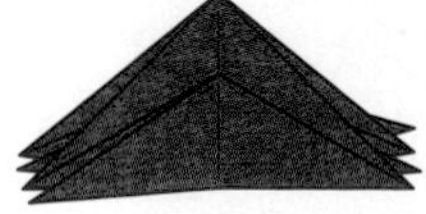

Figure 9.40.
Left: $N = 8$ circle pattern.
Right: a base made from the $N = 8$ circle packing.

For nine flaps and nine circles, the closest packing of circles gives the singly blintzed Bird Base that was used for Rhoads's Bug. But you only need eight flaps for a simple insect—head, tail, and six legs—and the $N = 8$ optimum circle packing solution, like $N = 6$, gives a crease pattern perfectly suited to the simple insect. For the same size square, the $N = 8$ pattern gives eight flaps that are about 4% longer than the flaps derived from the blintzed Bird Base. As with $N = 3$ and $N=6$, the most efficient base is rather unexpected. Four of the eight flaps are middle flaps—ordinarily, the least efficient way to make a flap—and I am unaware of any prior design based on this pattern. Nevertheless, the overall solution is the most efficient eight-equal-flap base there is. I leave it to the reader to devise a model that exploits this base.

The $N = 9$ pattern, as mentioned earlier, corresponds to the blintzed Bird Base. The crease pattern for the next case, $N = 10$ flaps, is generated by a circle packing of particular mathematical significance. While there are mathematical proofs that the patterns shown in Figures 9.29 and 9.30 for $N = 1$ through 9 are the most efficient possible, the most efficient packing for $N = 10$ has been the source of some controversy. Not until 1997 was the most efficient packing known. Over a 25-year span, five different circle packings for $N = 10$ were found, each more efficient than the previous (although the lengths of the flaps in each solution are all within 1% of one another). In each case, the discoverer conjectured that he had found the most efficient arrangement possible; in the case of all but the last, a more efficient solution was subsequently found.

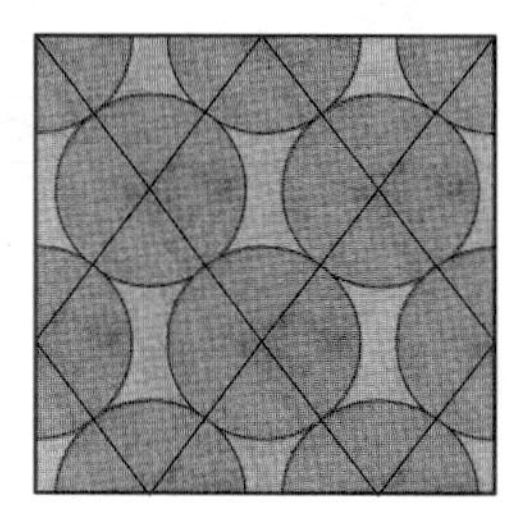

$N = 10$ (ca. 1970)
$r = 0.2083+$

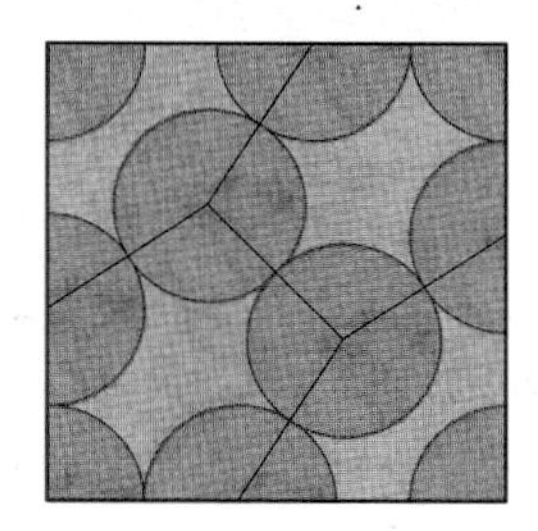

$N = 10$ (ca. 1971)
$r = 0.2096+$

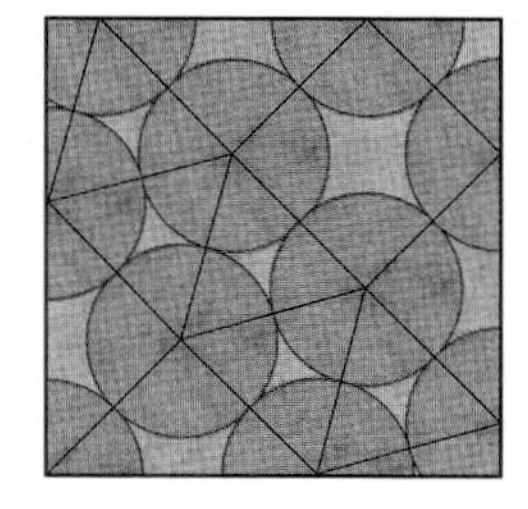

$N = 10$ (ca. 1987)
$r = 0.2100+$

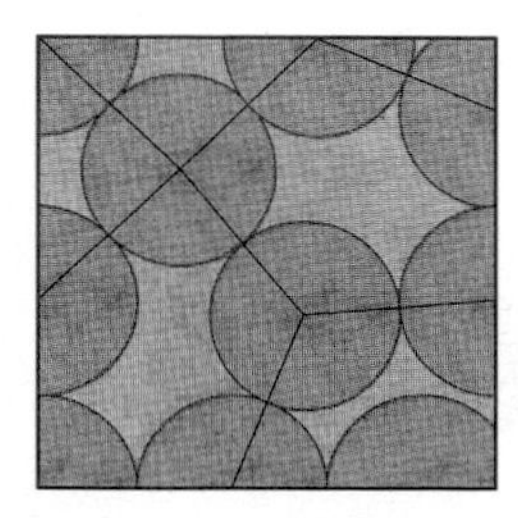

$N = 10$ (ca. 1989)
$r = 0.21059+$

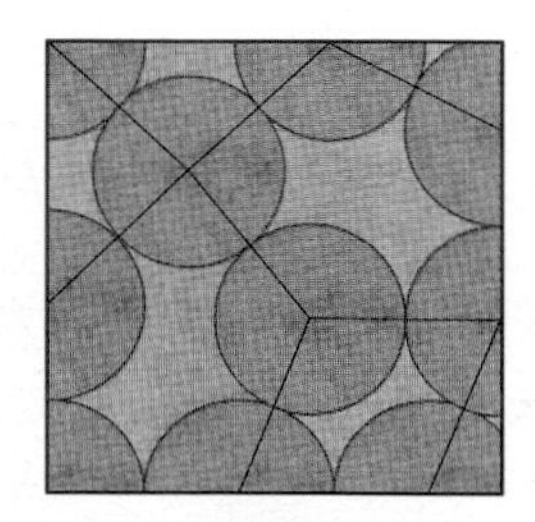

$N = 10$ (ca. 1990)
$r = 0.21063+$

Figure 9.41.
Optimal ten-circle packings giving ten-pointed bases. Upper left: Goldberg's solution. Upper center: Schaer's solution. Upper right: Milano's solution. Lower left: Valette's solution. Lower right: Mollard and Payan's solution, the proven champion.

The most recent solution, discovered in 1990 by the mathematicians Mollard and Payan, gives a flap length of 0.2106+; it was proven optimal by the Dutch mathematician Hans Melissen in 1997. The five solutions and their lines of symmetry are given in Figure 9.41.

I find it remarkable that such a simple answer to the most efficient packing of ten circles into a square should be so elusive. And if there is so much room for variation in the circle packing for this one particular origami base, think of the possibilities for arbitrary origami structures.

9.11. The Margulis Napkin Problem

We now have the machinery to design bases with any number of flaps. We also have the tools to solve an interesting mathematics problem that circulated among mathematicians in the mid-1990s, called the Margulis Napkin Problem. The problem was posed as a request for a proof:

> *Prove that no matter how one folds a square napkin, the flattened shape can never have a perimeter that exceeds the perimeter of the original square.*

That is, if you start with a square 1 unit on each side, prove that you can't fold a shape whose perimeter is greater than 4 units.

The somewhat surprising fact is that the assertion isn't true—it is indeed possible to fold a shape with a perimeter greater than 4. Figure 9.42 shows the folding sequence for a shape whose perimeter is slightly greater than 4 units—4.120 units, to be exact. Remarkably, a counterexample to this recent mathematical conjecture can be made from a 200-year-old shape: the venerable Bird Base.

A closer examination of this shape, coupled with our understanding of circle-method origami design, reveals how this can be accomplished.

The crease and circle pattern for the Bird Base is shown in Figure 9.43. The Bird Base consists of four corner flaps (modeled by the four corner circles) and one middle flap, which is provided by the single middle circle. Each of the four corner flaps is one quarter of the perimeter of the square, so if the flaps were splayed out into a star shape, the perimeter of the star would be, at most, equal to the perimeter of the square. But the middle flap is an extra flap. Thinning the entire base so that the middle flap can stick out creates extra perimeter. Thinning the base further, as shown in Figure 9.44, removes some of the overlap from the center, allowing the perimeter to get slightly larger as the thinned base approaches its limit, where each flap has zero thickness, and the perimeter approaches a value of 4.414.

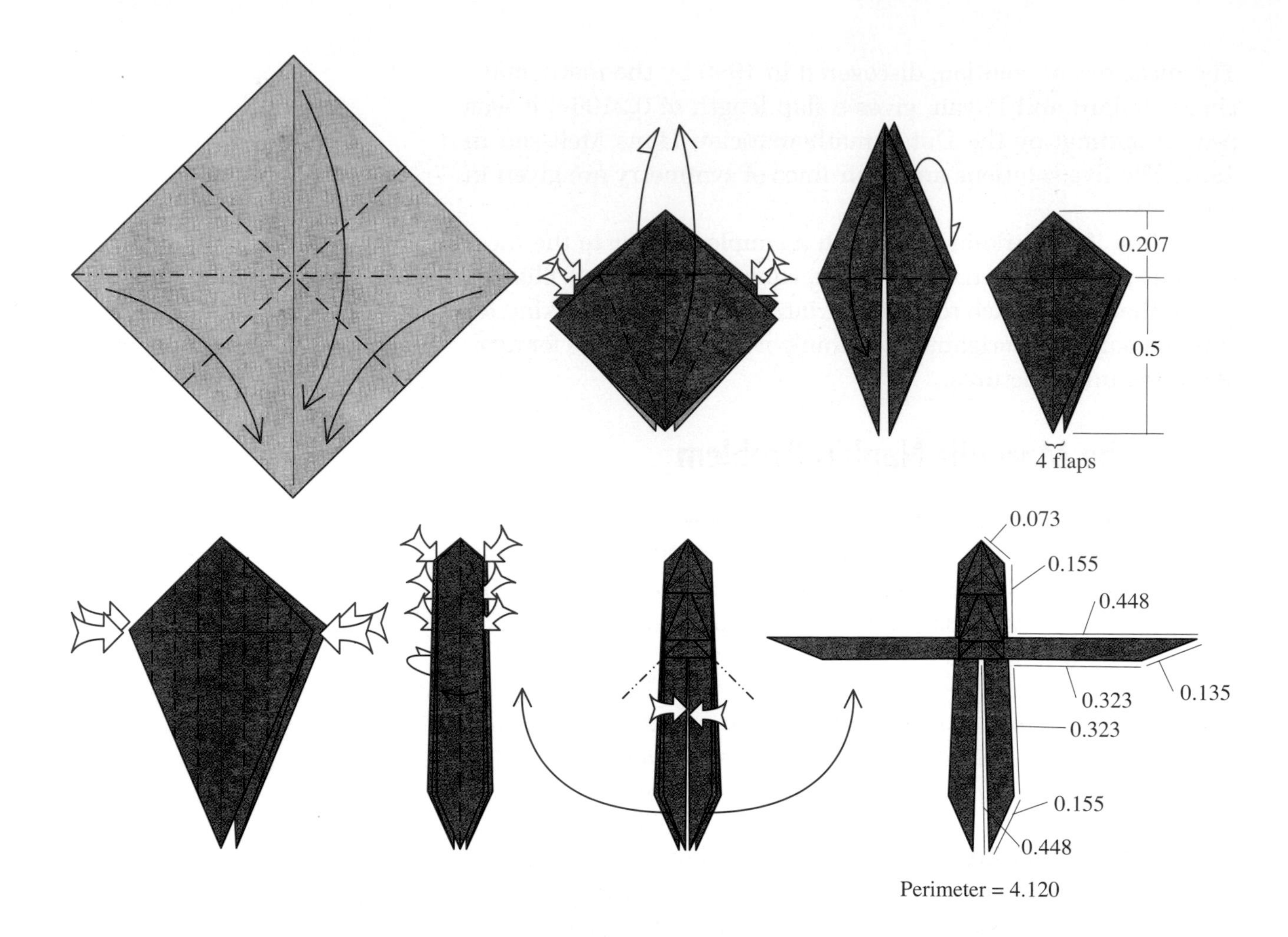

Figure 9.42.
Folding sequence for a shape that disproves the conjecture known as the Margulis Napkin Problem. The dimensions of the various segments of the perimeter are given in the last step; the total perimeter adds up to 4.120.

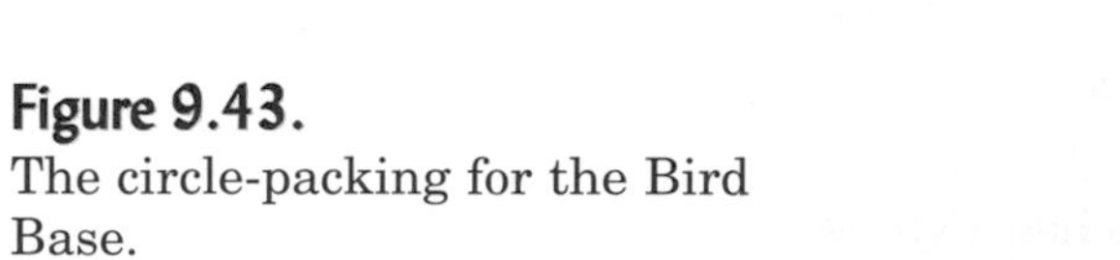

Figure 9.43.
The circle-packing for the Bird Base.

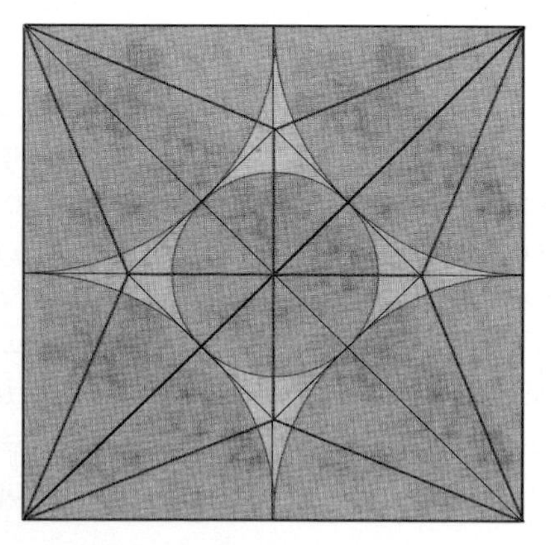

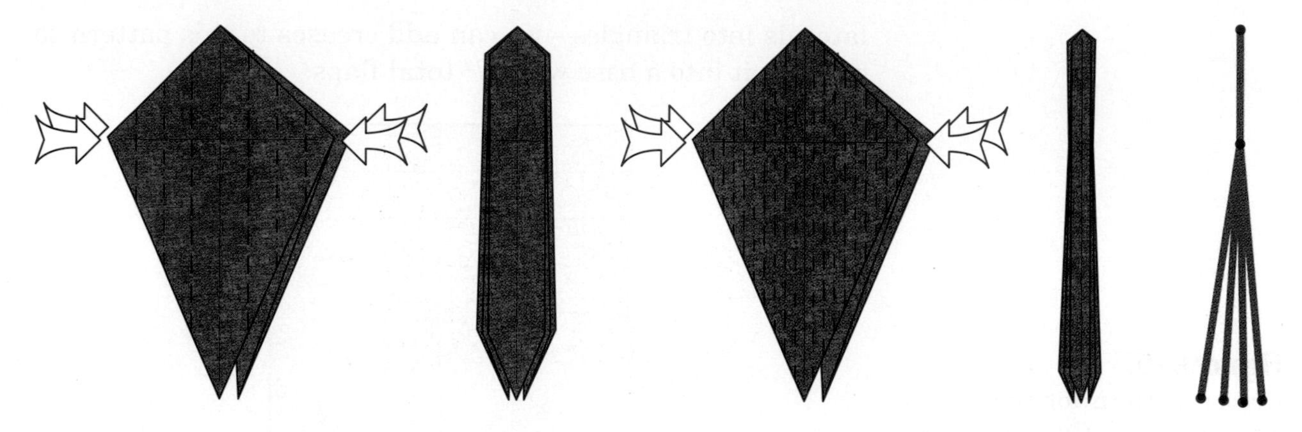

Figure 9.44.
Left: Sinking in fourths narrows the flaps.
Center: Sinking a larger number of times thins the layers further.
Right: the limiting case with zero-thickness flaps.

It was the creation of a middle flap that allowed the perimeter to exceed the conjectured limit of 4 units. But by creating more middle flaps, we can increase the perimeter even further. In fact, rather astonishingly, there is *no upper limit* to the perimeter of a flat fold—at least, one made with mathematically ideal (zero thickness) paper. You can start with as small a sheet as you like. From a postage stamp, you can theoretically fold a shape whose perimeter is the perimeter of the galaxy.

How can we do this? Circle-packing gives the key. Suppose we pack an $N \times N$ array of circles into a unit square, as shown in Figure 9.45.

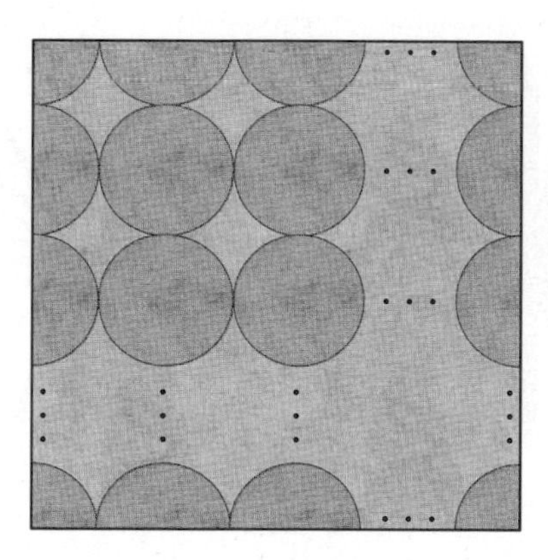

Figure 9.45.
An $N \times N$ circle packing.

Each of the circles has a radius

$$r = \frac{1}{2(N-1)}. \qquad (9\text{–}4)$$

Using our circle-packing and tiling techniques—filling in the axial creases, adding smaller circles to break up quadrilaterals into triangles—we can add creases to this pattern to collapse it into a base with N^2 total flaps.

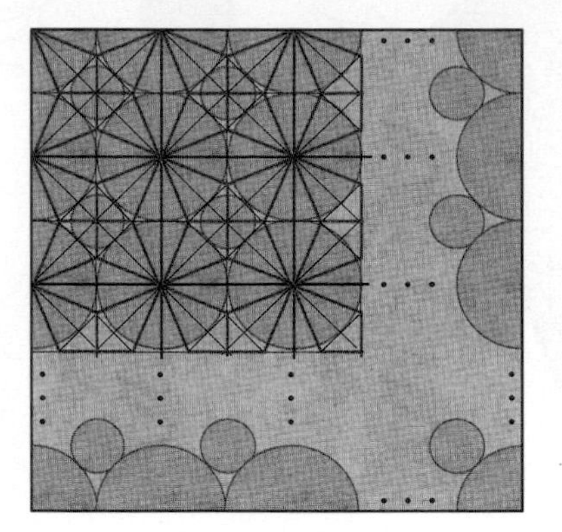

Figure 9.46.
Crease pattern for the $N \times N$ circle packing. Only the upper left portion is filled in.

The result after folding this crease pattern will be a base with N^2 points, each of length $1/(2(N-1))$. Using standard origami techniques of sinking, the points can be made arbitrarily thin. Once the points are thinned, they can be reverse-folded out in all directions, making a star with N^2 points. This sequence is shown in Figure 9.47.

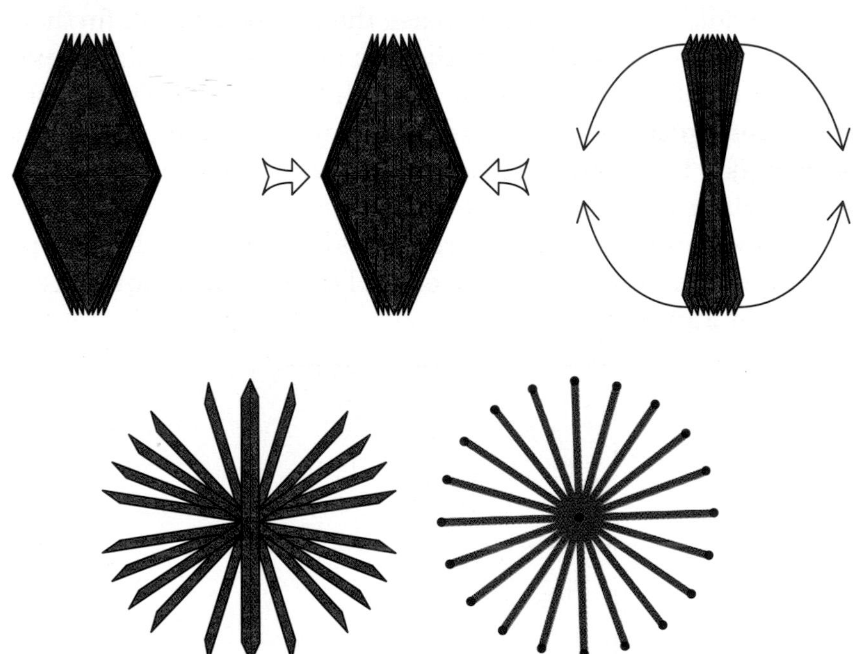

Figure 9.47.
Sequence for turning the base into a star.

Now, although the points overlap each other somewhat at their base, they can be made arbitrarily thin by making the sink folds arbitrarily close together. So the total perimeter of the star shape approaches the value

$$2\times(\text{number of points}) \times (\text{length of each point}), \qquad (9\text{–}5)$$

where the extra factor of 2 comes from the fact that each point contributes two sides to the perimeter. Thus, the total perimeter is

$$\frac{N^2}{N-1}. \qquad (9\text{–}6)$$

Let's look at the perimeter for several values of N:

N	2	3	4	5	6	7	8
Perimeter	4	4.5	5.33	6.25	7.20	8.17	9.14

In fact, as N, the number of points along one side, becomes large, the perimeter approaches N as its limiting value. Thus, the perimeter can be made arbitrarily large. We can also use this result to work backwards from the desired perimeter. For example, to fold a square postage stamp one inch on a side so that it has the same perimeter as—let's take something small—the circumference of the world (24,000 miles), we would need to make N equal to about 1.5 billion; the resulting shape would have about 2 billion billion points, and each point would be about 17 microns long—about 1/5 the diameter of a human hair. Clearly, we'd need that special zero-thickness paper that exists in mathematicians' imagination to fold such a thing! Not to mention a *lot* of patience.

Interestingly, several origami artists had created models on these principles that belied the Margulis Napkin Problem years before it had even arisen in mathematical circles (attributed, possibly apocryphally, to Margulis, a famous Russian mathematician). My own Sea Urchin, which we saw back in chapter 4 (Figure 4.8) utilizes such a square array of 25 points, and the points, properly thinned and flattened, give a star whose perimeter approaches a limit of 2 × (25) × (1/8) = 6.25. Similar urchins by others, including Toshiyuki Meguro, who pioneered circle-packing design methods in Japan, abound.

9.12. Comments

The circle method of origami design described in this chapter can be an extremely powerful tool for designing complex origami models, particularly beetles and insects. For any pattern of circles, there exists a folding method that transforms that pattern into a base with the proper number and size of points. However, although the technique of packing circles guarantees that a folding sequence exists to convert the circle pattern into a base, it doesn't provide much guidance as to how to execute a step-by-step folding sequence for that base—a shortcoming of most algorithmic origami design. So even if you work out a circle pattern, you still have some work ahead of you to figure out how to fold the crease pattern into a base.

By packing circles densely so that each circle touches several others, we can connect the centers of touching circles with creases, which turn out to be axial creases in the base. If the polygons and circle fragments outlined by the axial creases turn out to resemble the circle patterns of known tiles, we can fill in the polygons with tile crease patterns and, *voilà*, construct the full crease pattern for an origami base.

Furthermore, we can, with some further effort, add rivers of constant width to the circle packing to create bases that contain segments separating groups of flaps. These circle/river patterns, too, may be filled in with crease patterns if they happen to correspond to known tiles.

But that's a very big *if*. While we have progressed a long way in designing origami bases, so that we can start with any number, length, and connectivity of flaps we desire, we are still dependent upon the existence of tile patterns for filling in the creases of the axial polygons. There is no guarantee that such tiles exist.

At least, there is no guarantee just yet. But as we will see in the next chapter, there is a small number of general-purpose crease patterns that will allow us to fill in any circle pattern whatsoever. These patterns—some new, some old—provide the final step in the construction of a generalized uniaxial base.

Folding Instructions

Emu

Songbird 2

Emu

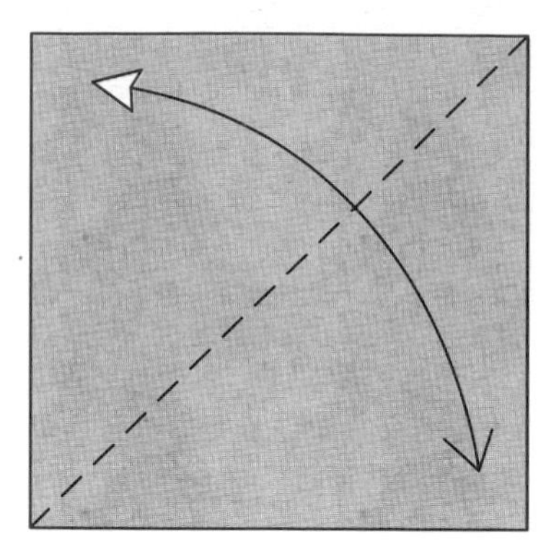

1. Begin with the white side up. Fold and unfold along a diagonal.

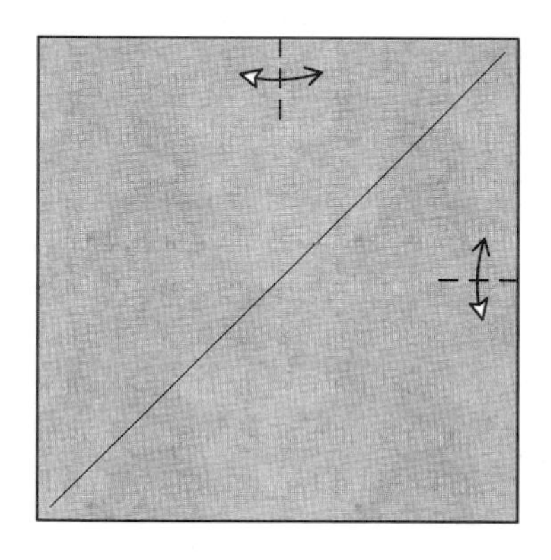

2. Fold pinch marks at the midpoints of two adjacent sides.

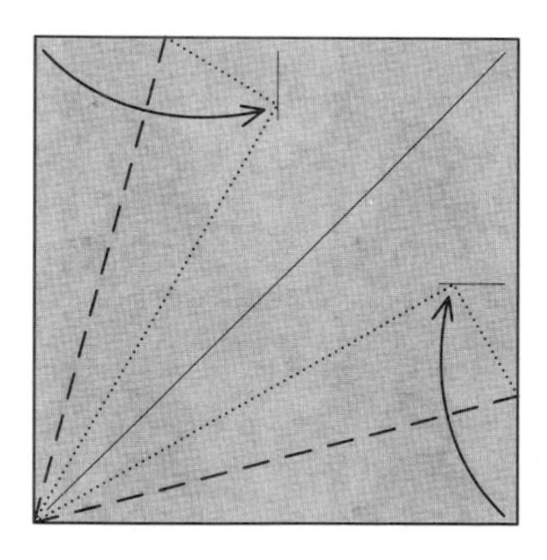

3. Fold two edges so that their corners touch the pinch marks.

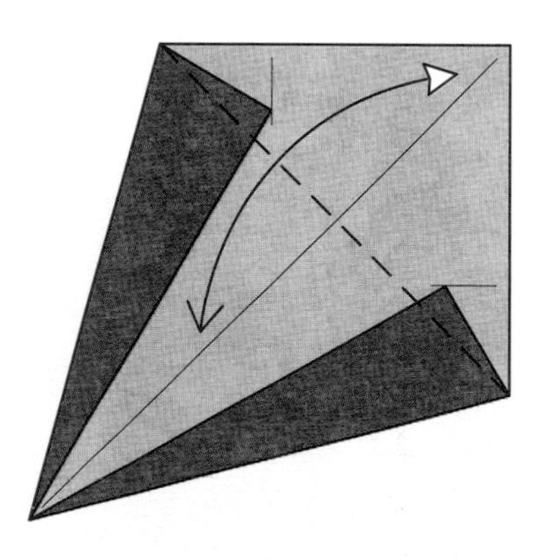

4. Fold and unfold through all layers.

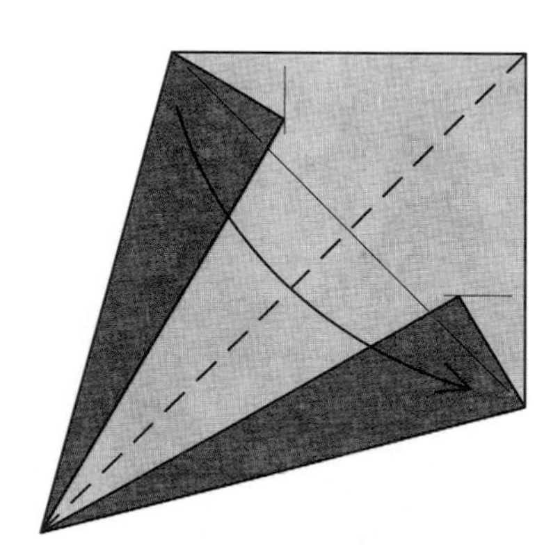

5. Fold the model in half.

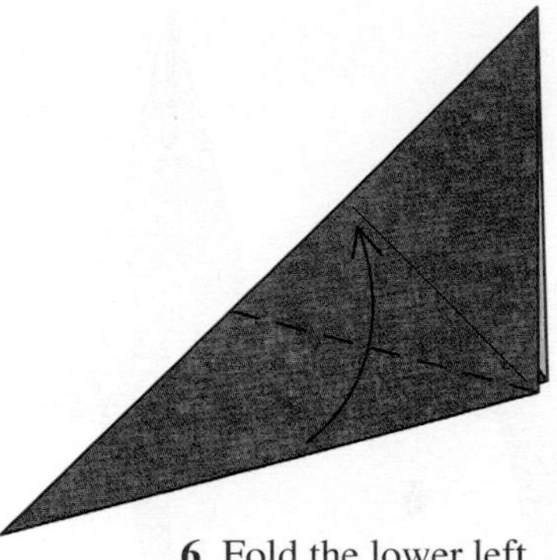

6. Fold the lower left point up to lie along an existing crease.

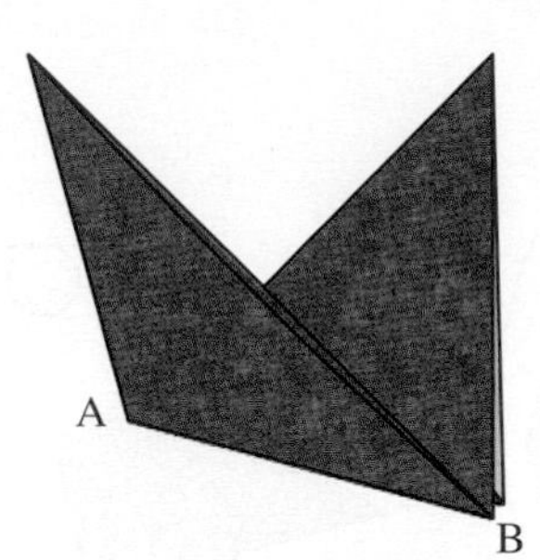

7. Rotate the model 1/4 turn clockwise, so that edge AB runs vertically.

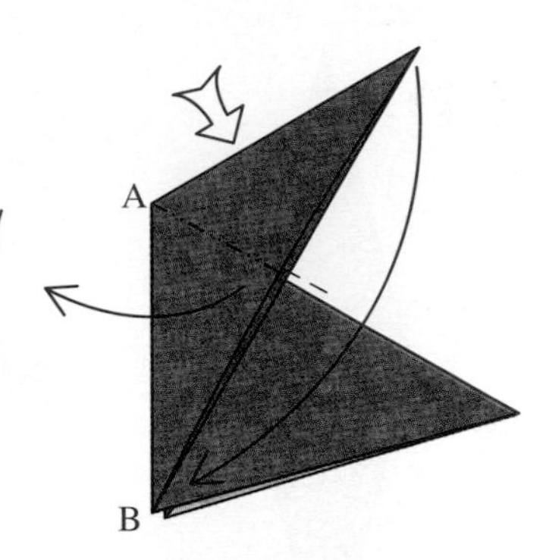

8. Squash-fold the top point down to corner B.

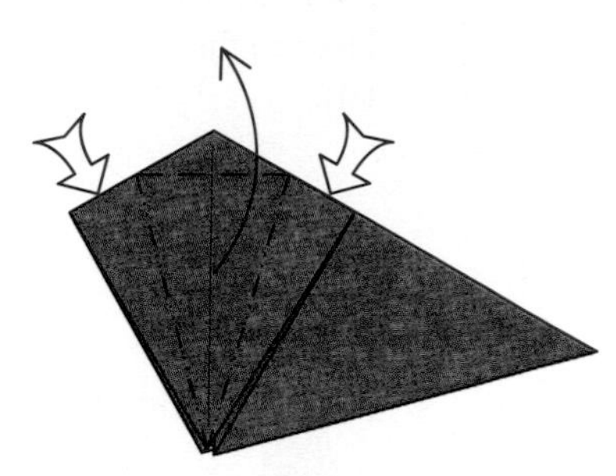

9. Petal-fold.

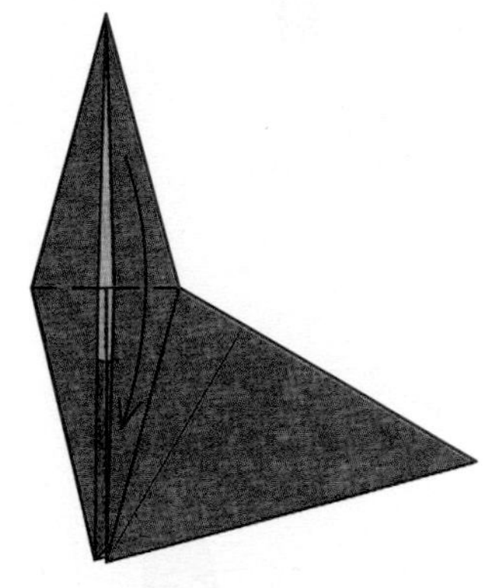

10. Fold the point down.

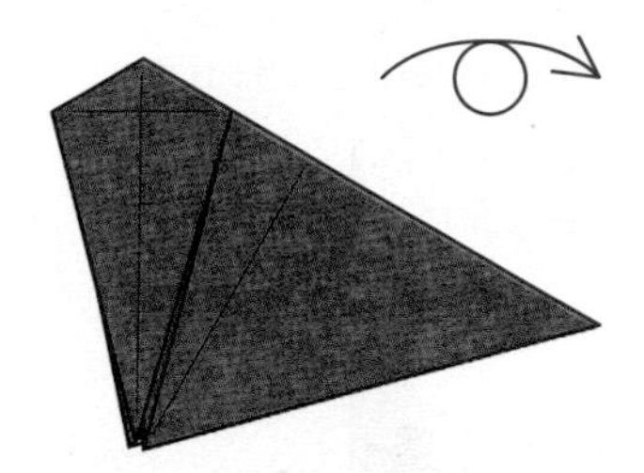

11. Turn the model over.

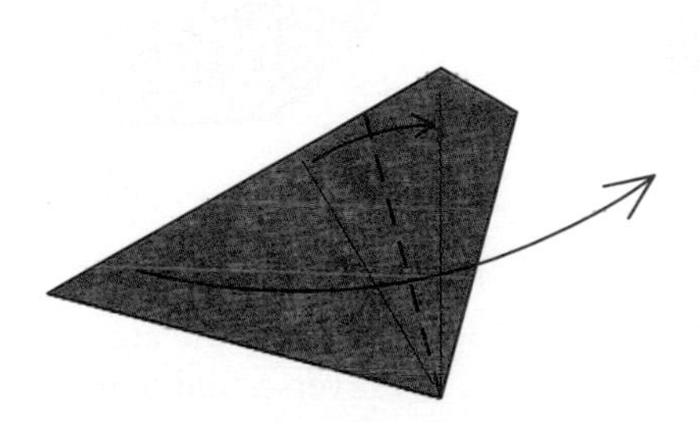

12. Fold the left flap up to the right so that the two creases in the middle line up.

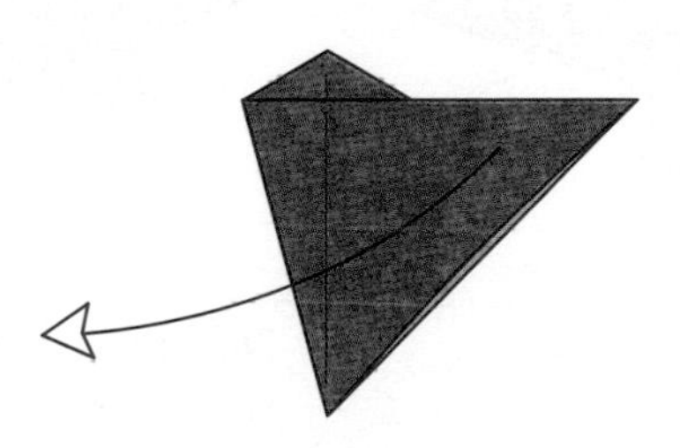

13. Unfold.

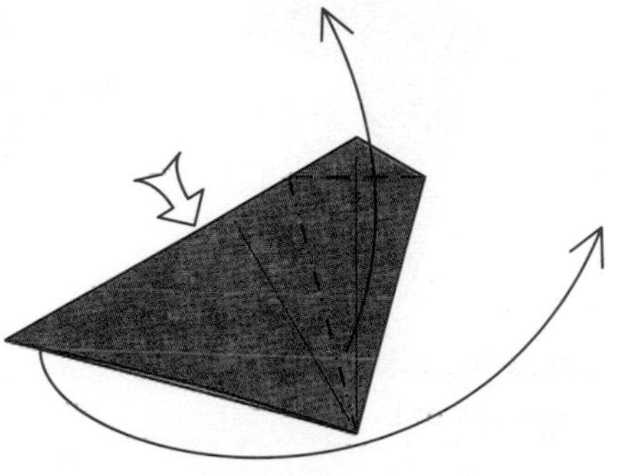

14. Lift up the near corner at the bottom and squash-fold the left point over to the right, using the creases you just made.

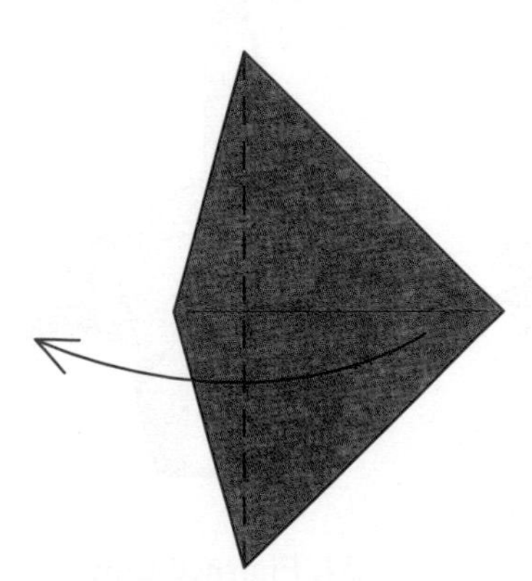

15. Fold the corner back to the left.

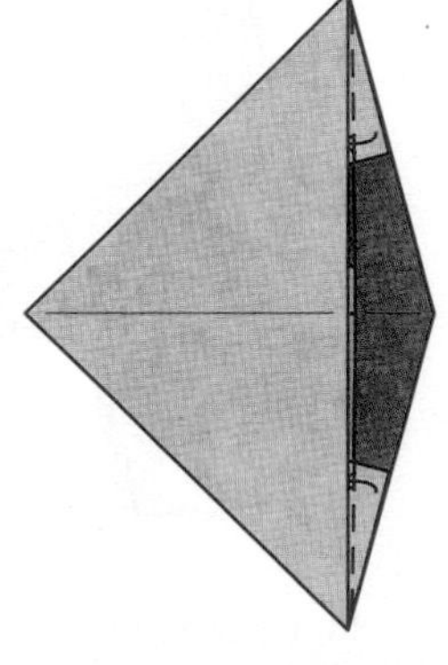

16. Tuck the white flaps inside the model.

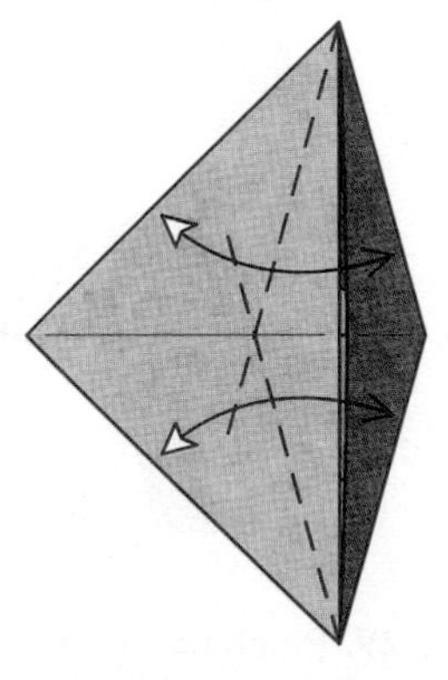

17. Fold and unfold.

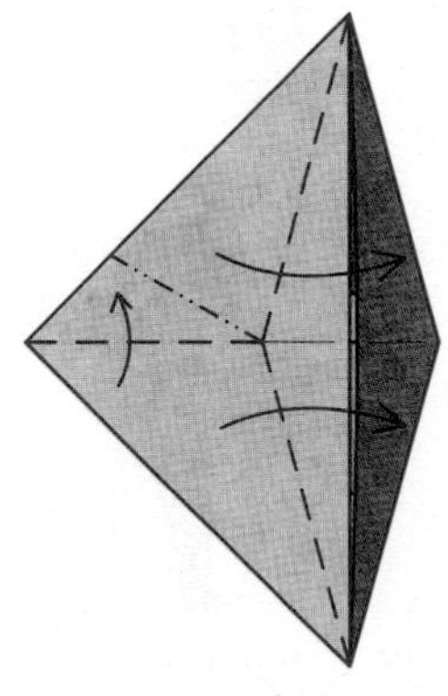

18. Fold a rabbit ear using the creases you just made.

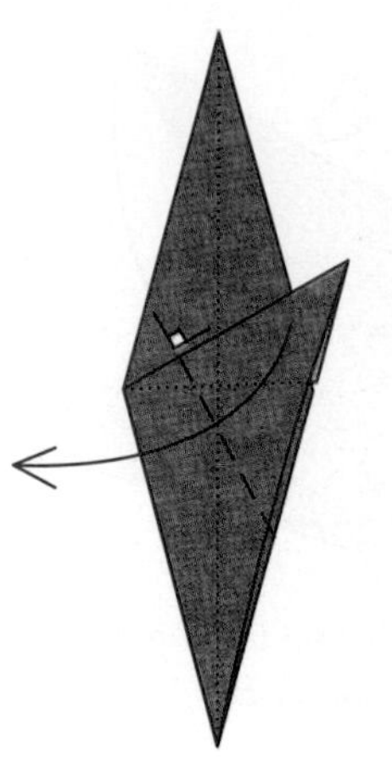

19. Fold the tip of the rabbit ear down toward the right; the model will not lie flat.

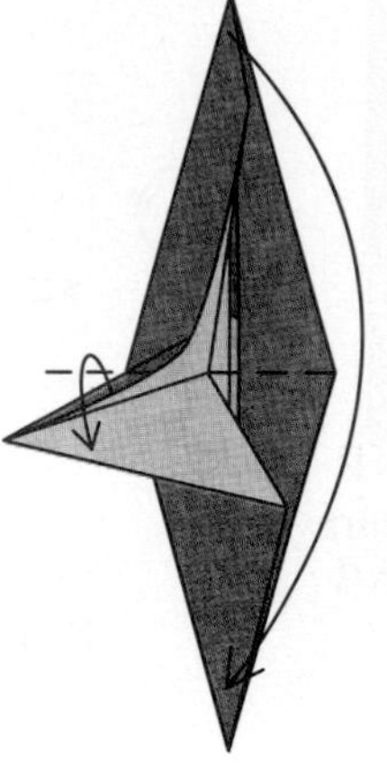

20. Fold the top flap down and flatten the model.

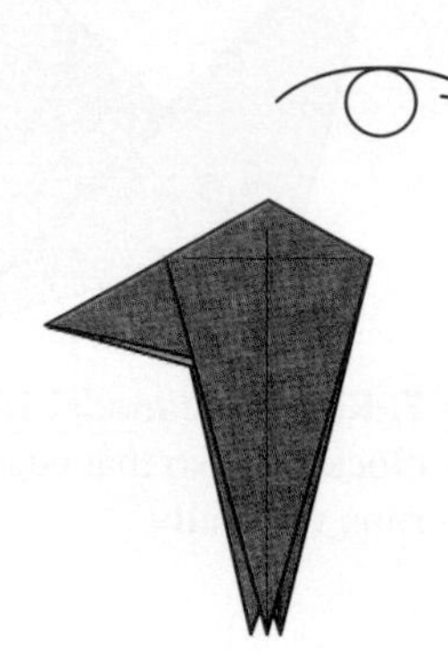

21. Turn the model over.

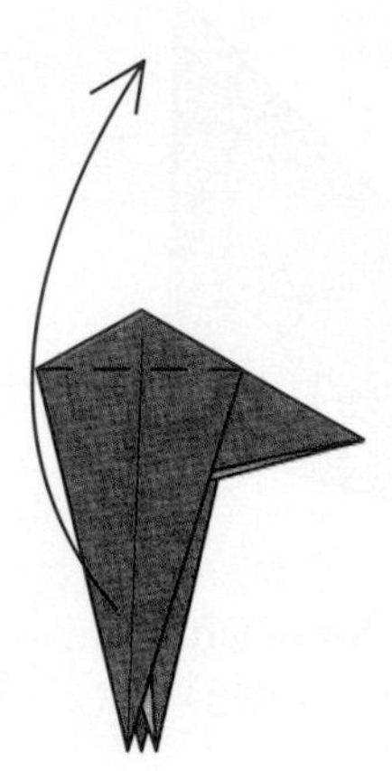

22. Lift up one point.

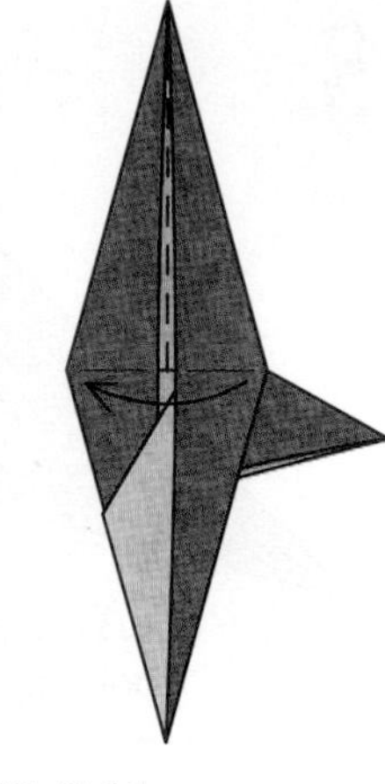

23. Fold one layer to the left.

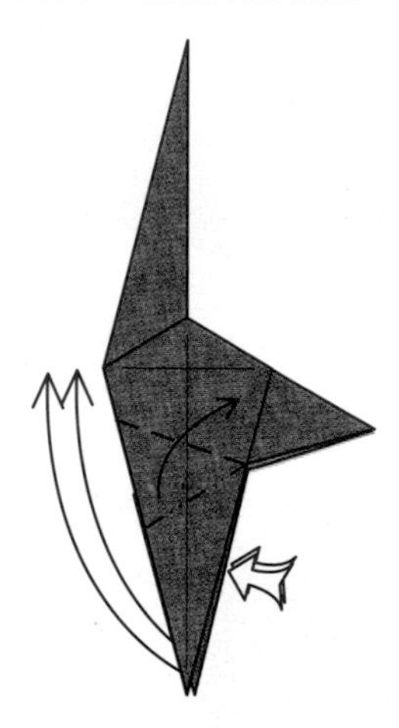

24. Squash-fold the legs upward.

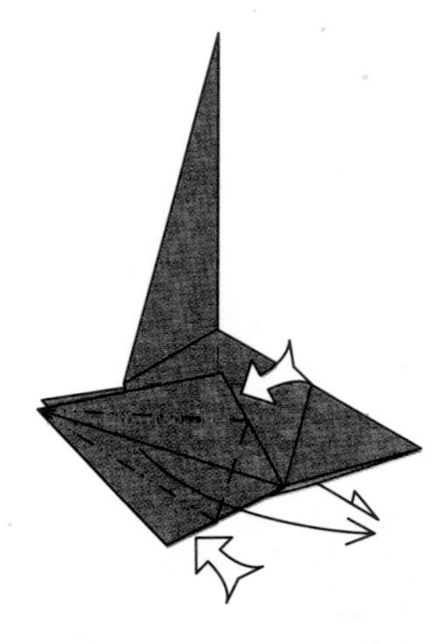

25. Petal-fold the legs.

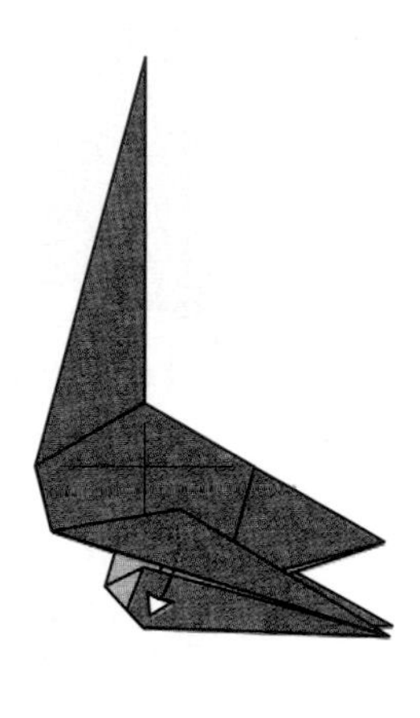

26. Pull out some loose paper.

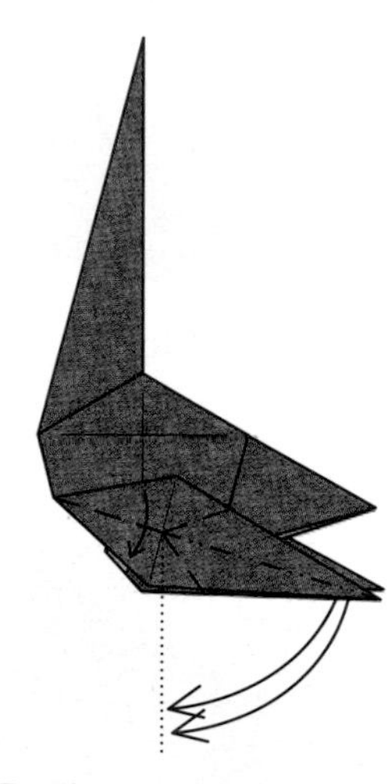

27. Close up the legs and reverse-fold them downward.

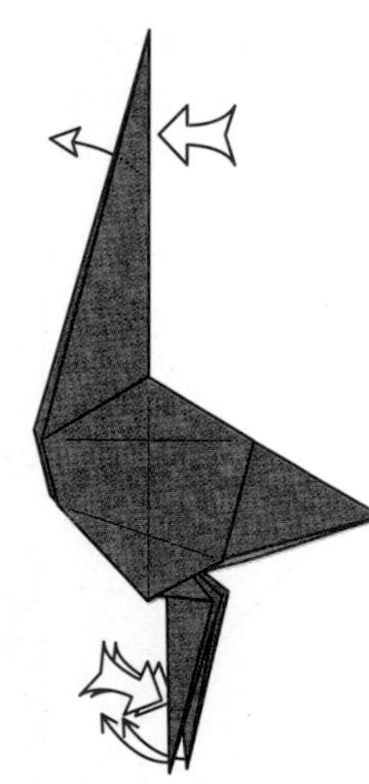

28. Crimp the feet. Pull some loose paper out of the head.

29. Pinch the head and swing it down.

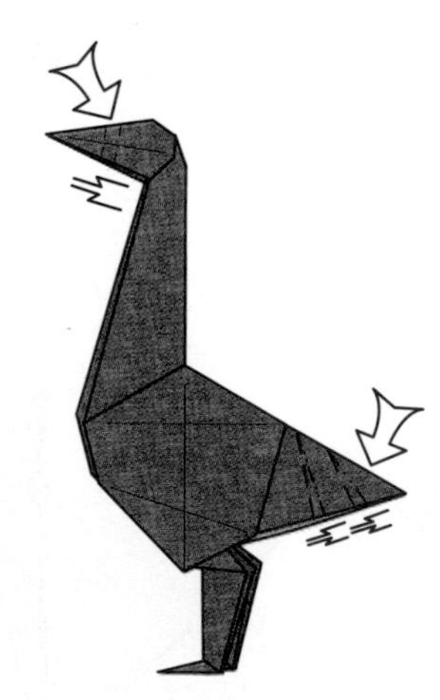

30. Crimp the beak and tail.

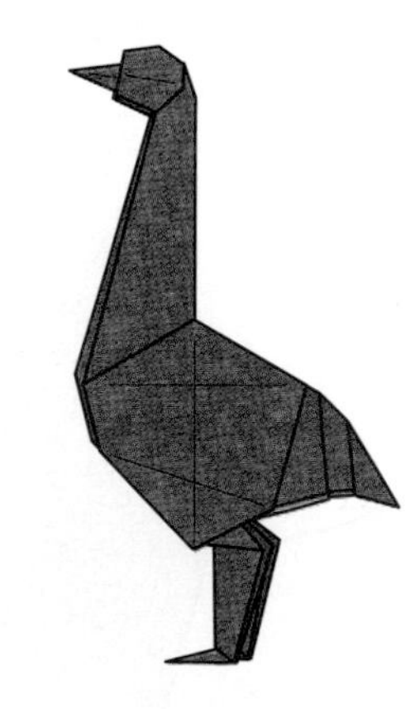

31. Finished Emu.

Songbird 2

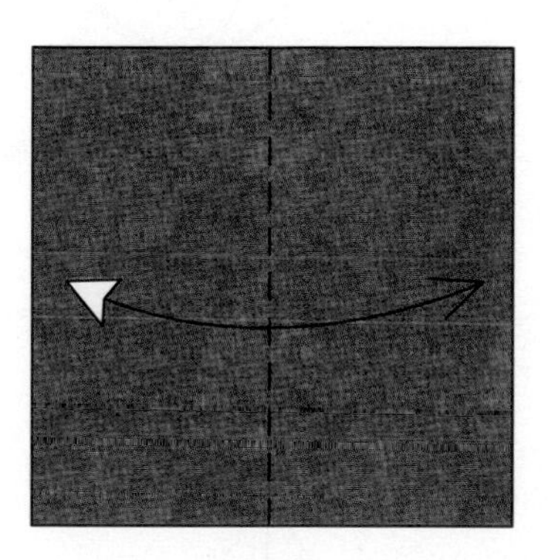

1. Begin with the colored side up. Fold and unfold from side to side.

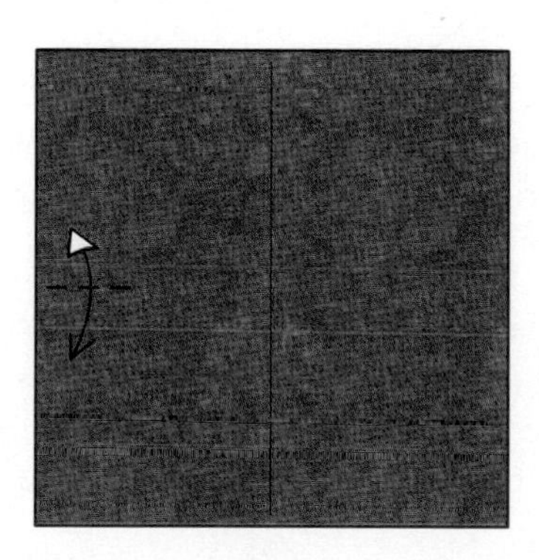

2. Fold the top down to the bottom and make a small pinch extending inward from the edge.

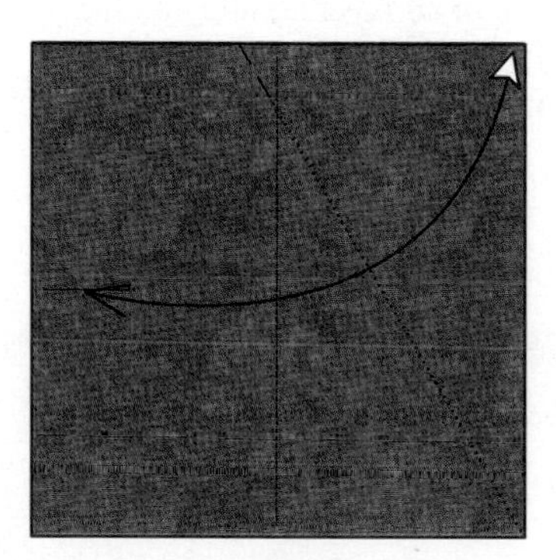

3. Fold the top right corner down to lie on the crease you just made and make a pinch along the top edge.

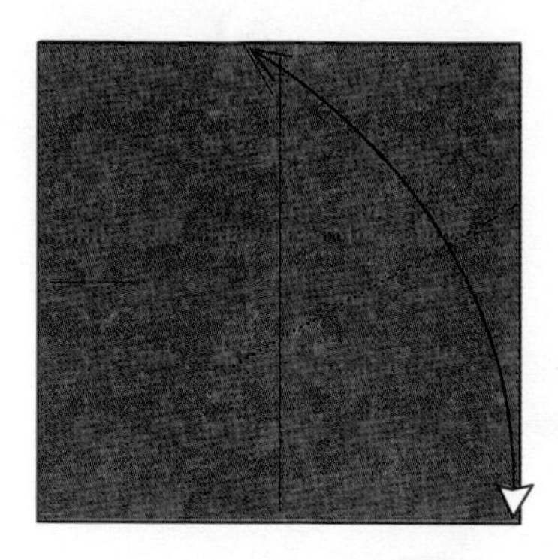

4. Fold the bottom right corner up to the pinch you just made and make another pinch along the right edge.

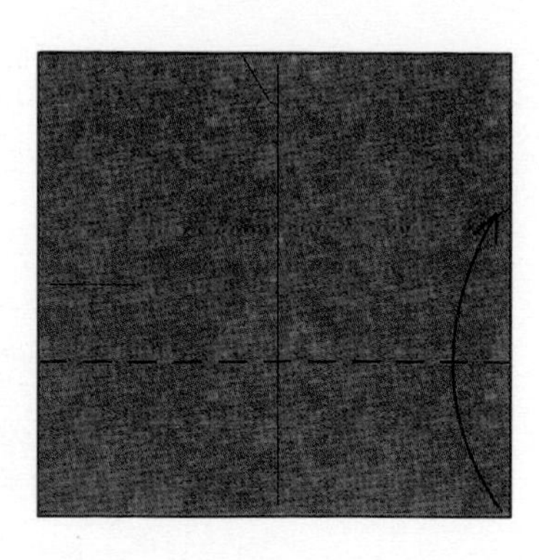

5. Fold the bottom edge up so that the corner aligns with the mark you just made.

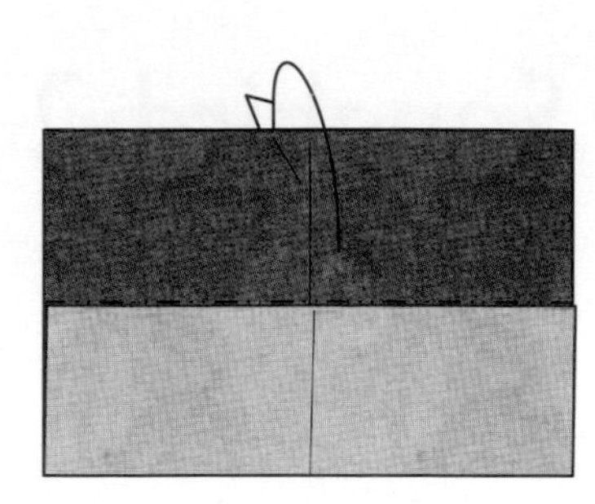

6. Fold the top edge behind along a crease that lines up with the folded edge.

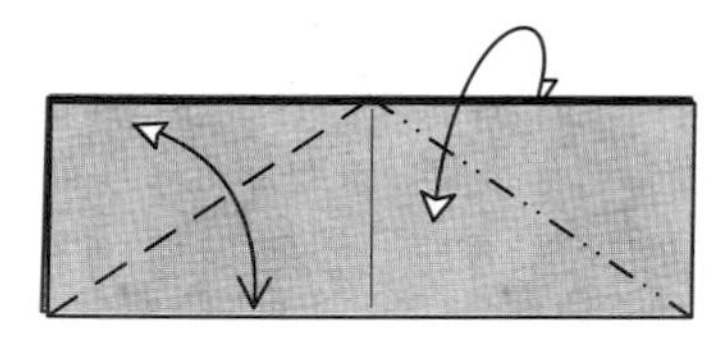

7. Fold along the diagonals through all layers (valley on the left, mountain on the right) and unfold.

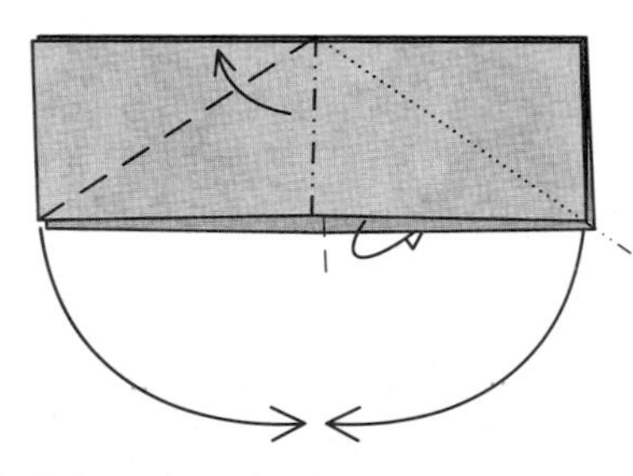

8. Crimp the middle of the model and bring the corners together, opening out a pocket on the underside.

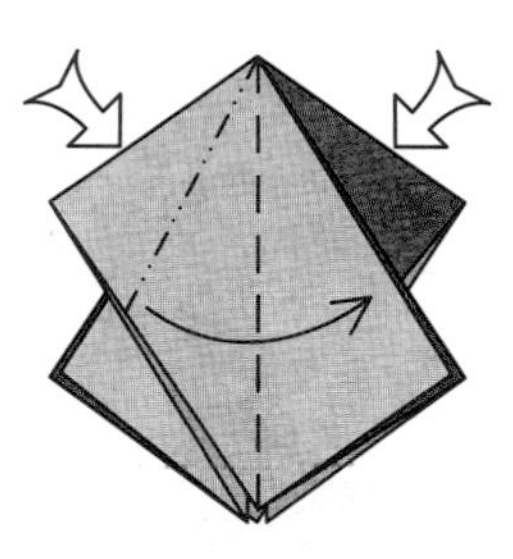

9. Squash-fold the white flap in front and the colored one behind.

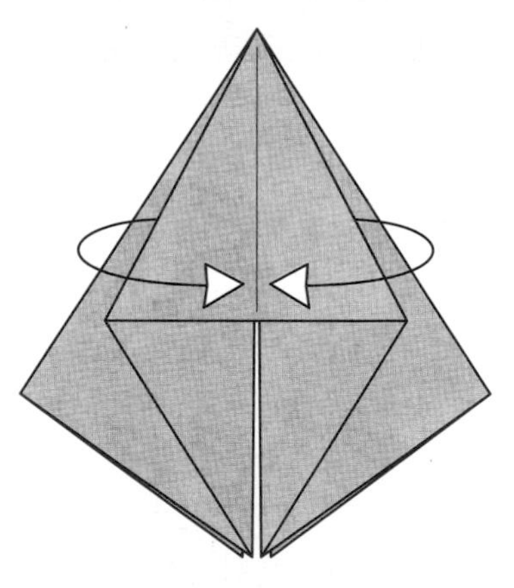

10. Bring two layers of paper to the front on each side. This is most easily done by unfolding the near layers, changing crease directions and then refolding.

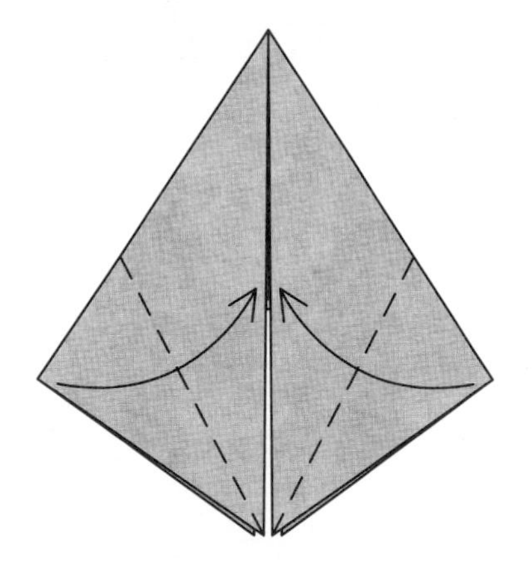

11. Fold a single layer in so that the edges meet in the middle.

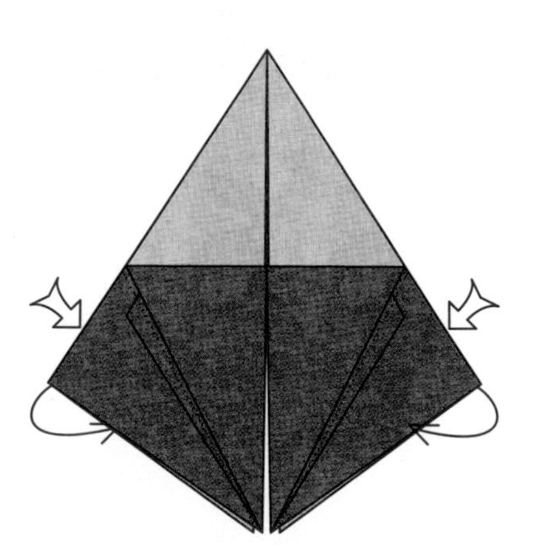

12. Reverse-fold the edges.

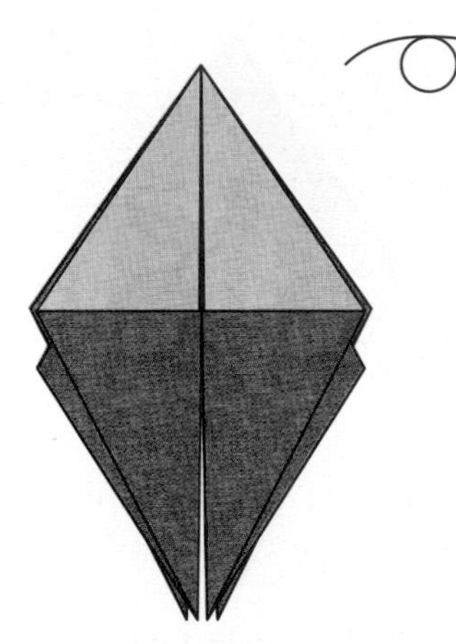

13. Turn the model over.

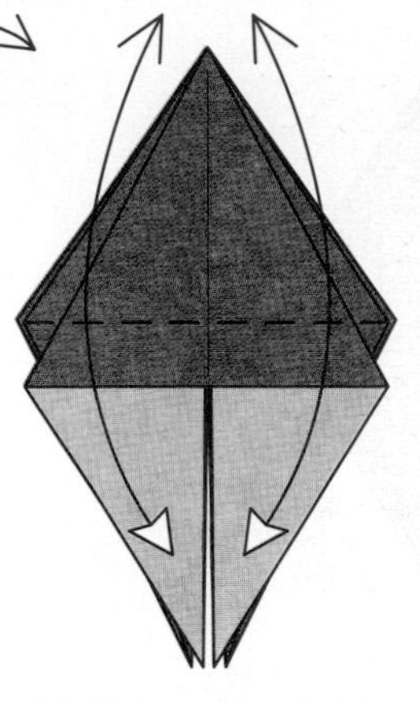

14. Fold and unfold the near flaps.

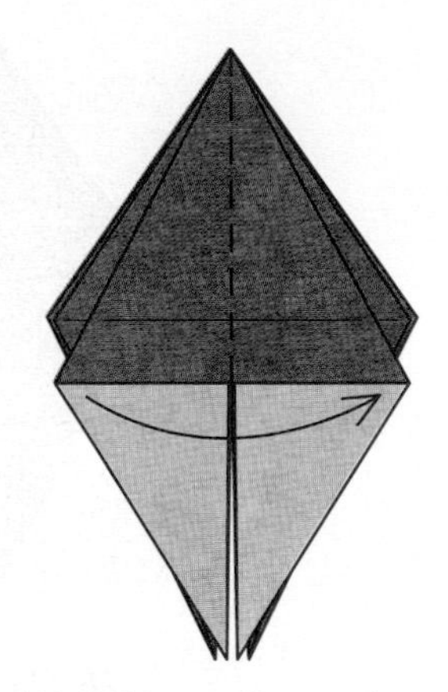

15. Fold one flap to the right.

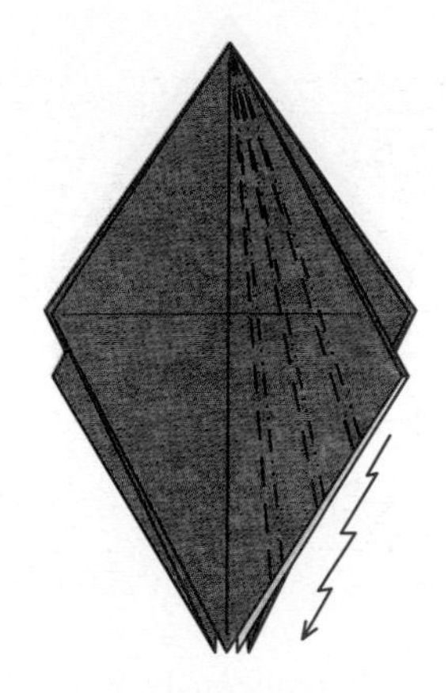

16. Pleat the flap.

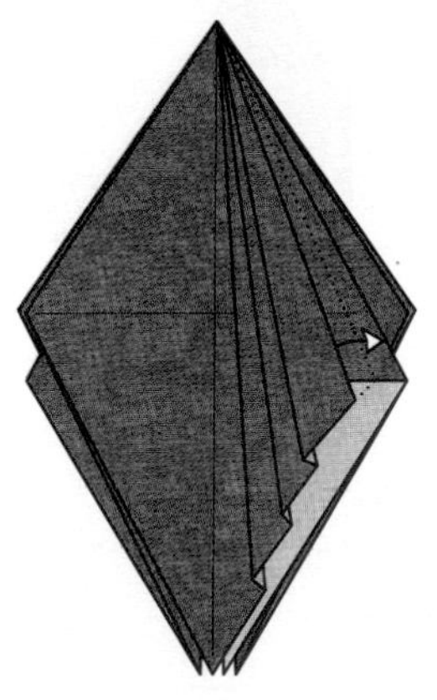

17. Release the layers of paper that are trapped in the pleats.

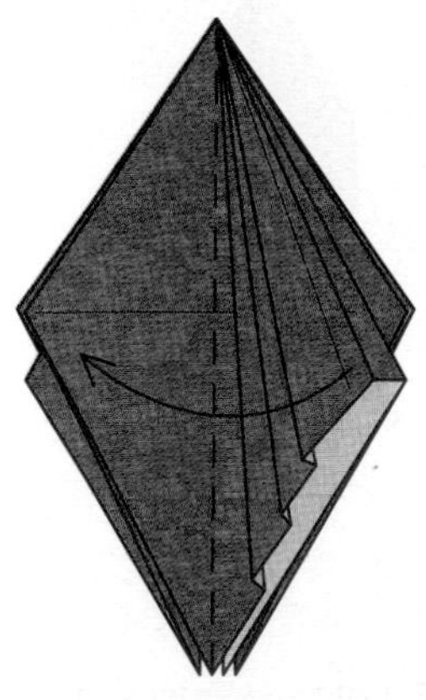

18. Fold the flap back to the left.

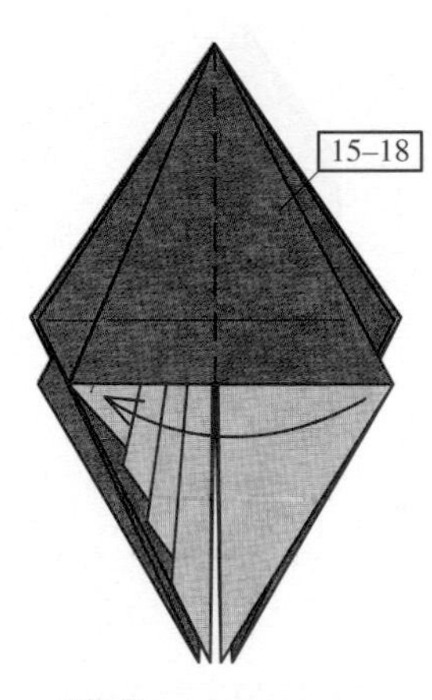

19. Repeat steps 15–18 on the right.

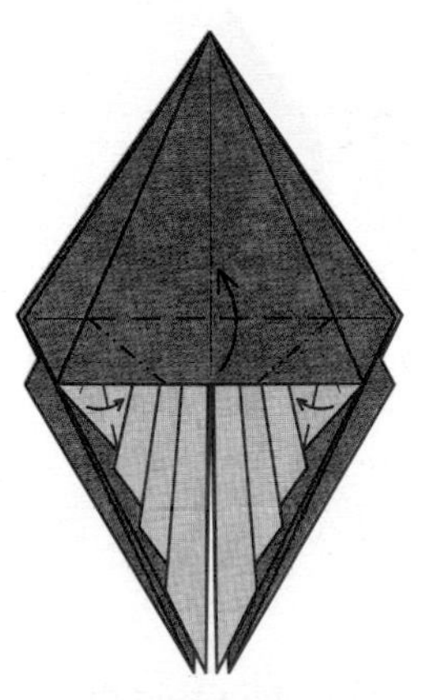

20. Petal-fold. The valley fold is on an existing crease.

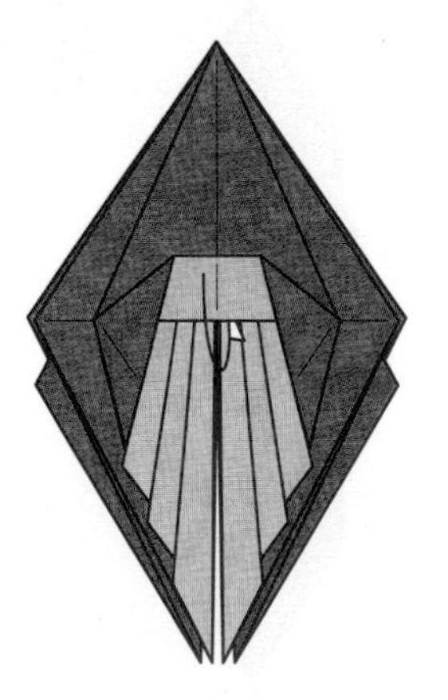

21. Tuck the flap up inside the model.

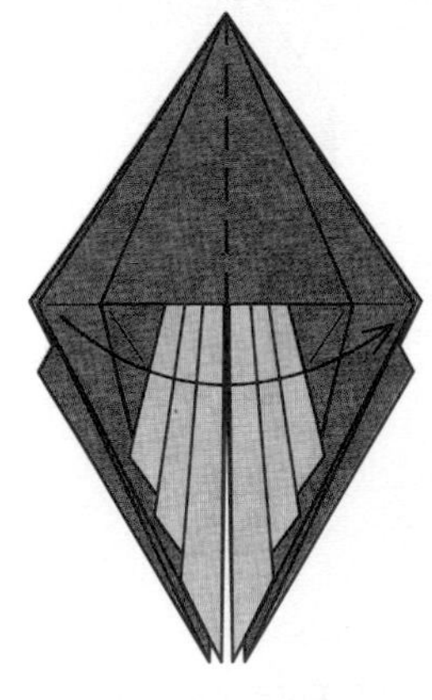

22. Fold two flaps to the right.

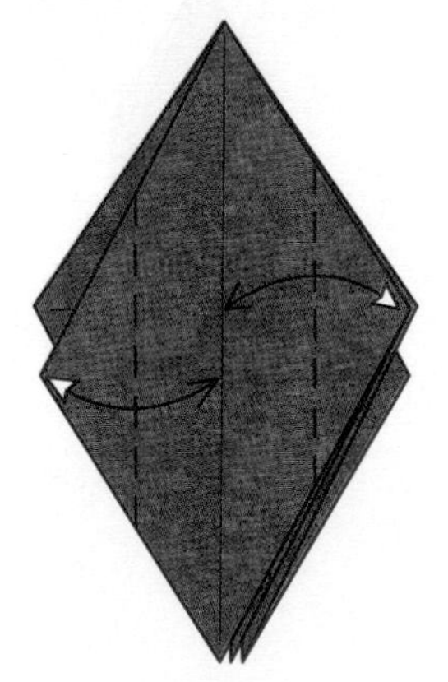

23. Fold two corners in to the center line and unfold.

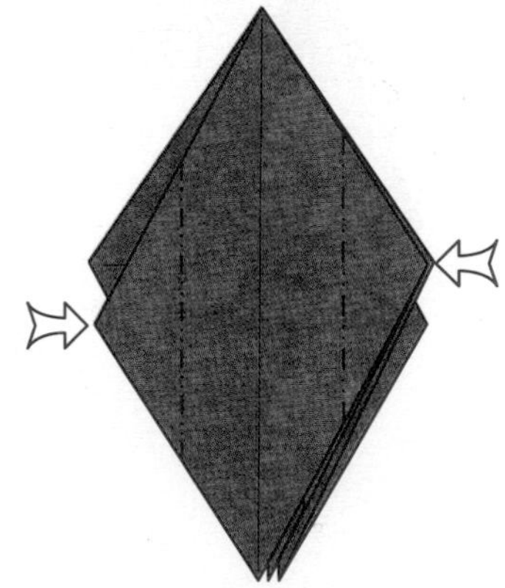

24. Sink the two corners on the creases you just made.

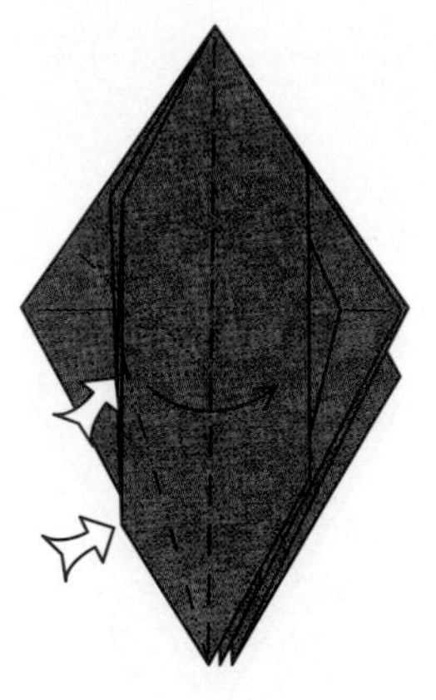

25. Simultaneously squash-fold the upper, hidden flap, and spread-sink the lower corner.

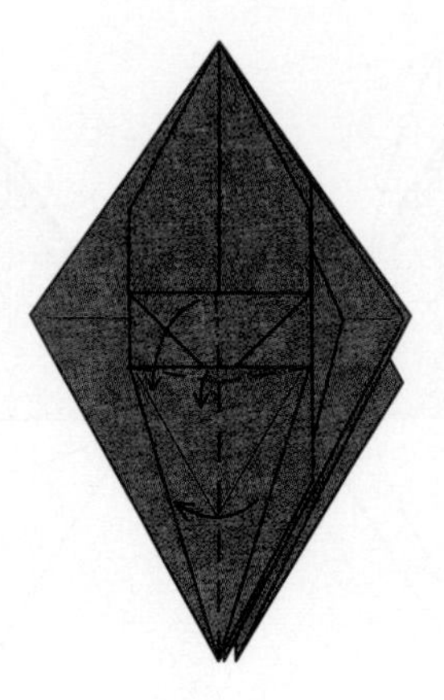

26. Close up the spread sink.

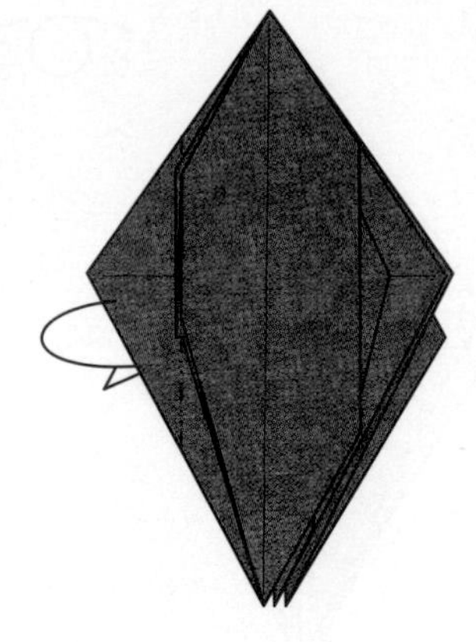

27. Mountain-fold the flap behind.

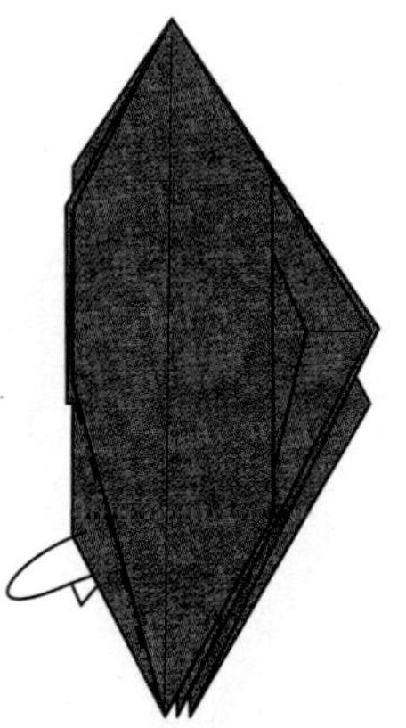

28. Mountain-fold the flap behind so the edges line up.

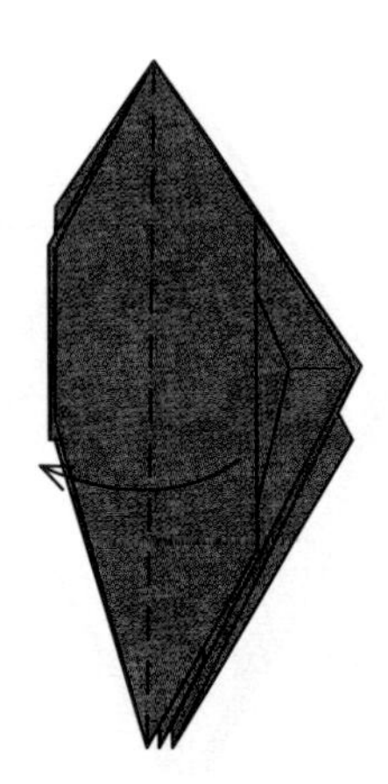

29. Fold a group of layers to the left.

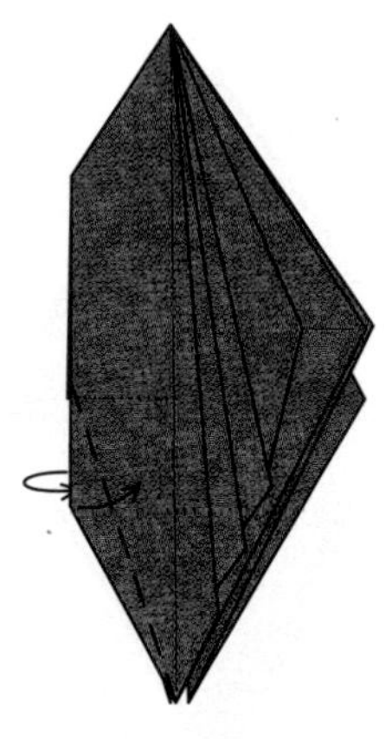

30. Fold two edges in so that the creases line up with the hidden edges behind.

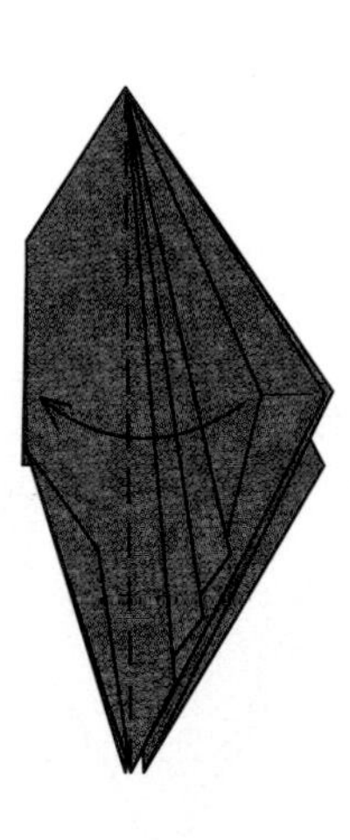

31. Fold one pleated flap back to the left.

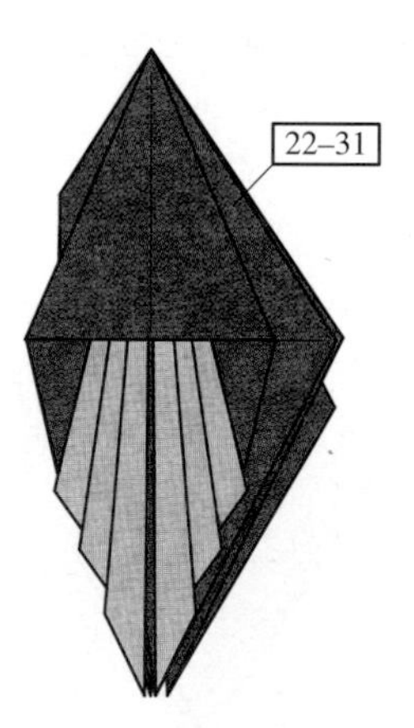

32. Repeat steps 22–31 on the right.

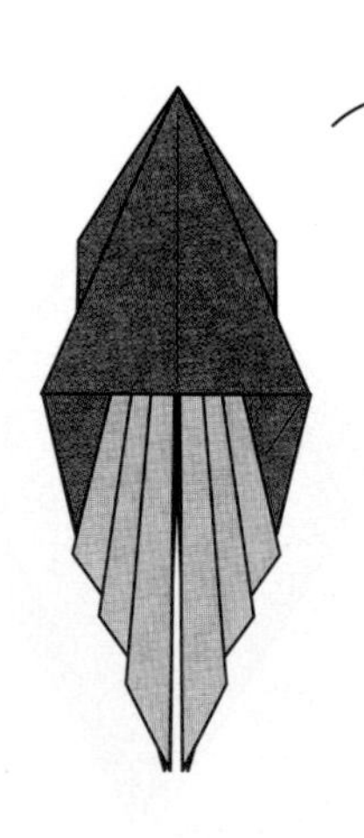

33. Turn the model over.

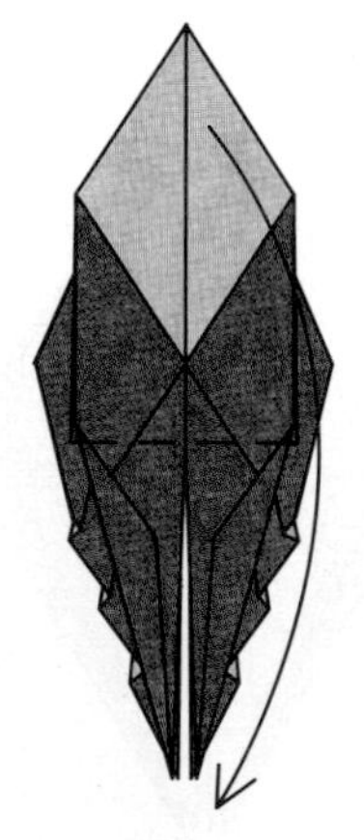

34. Fold a single flap down as far as possible.

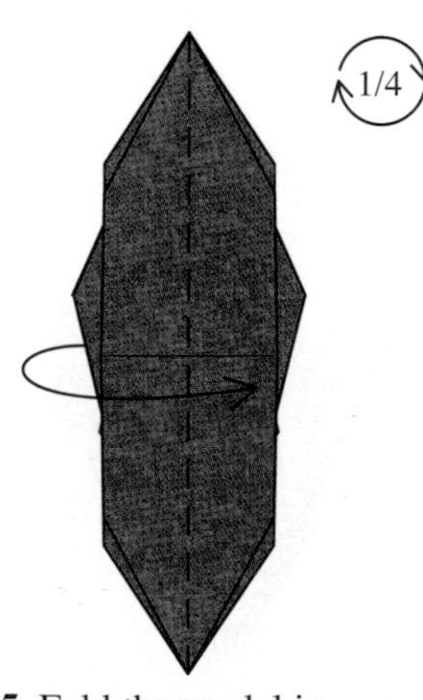

35. Fold the model in half and rotate 1/4 turn clockwise.

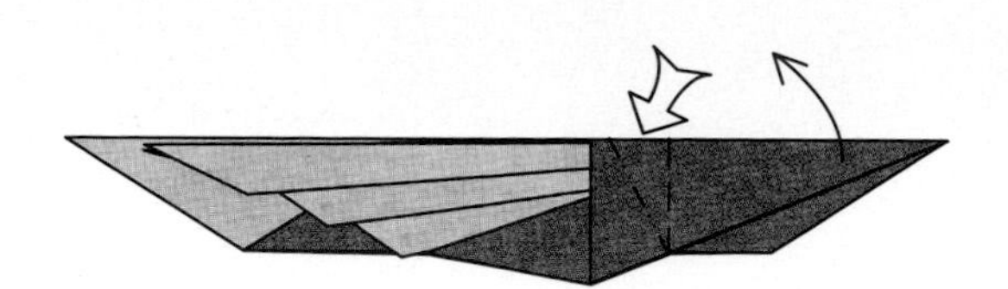

36. Crimp the flap on the right upward.

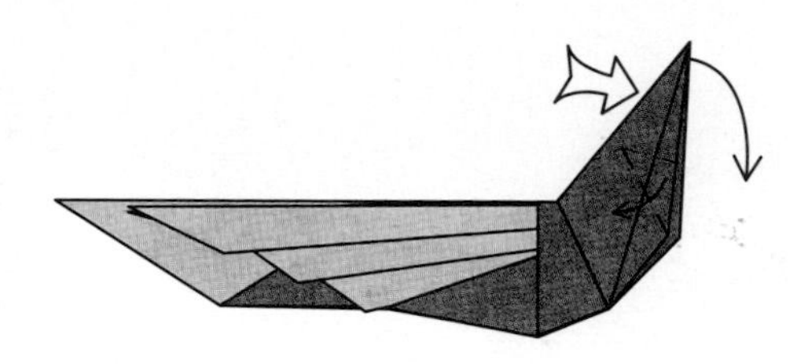

37. Crimp the head downward.

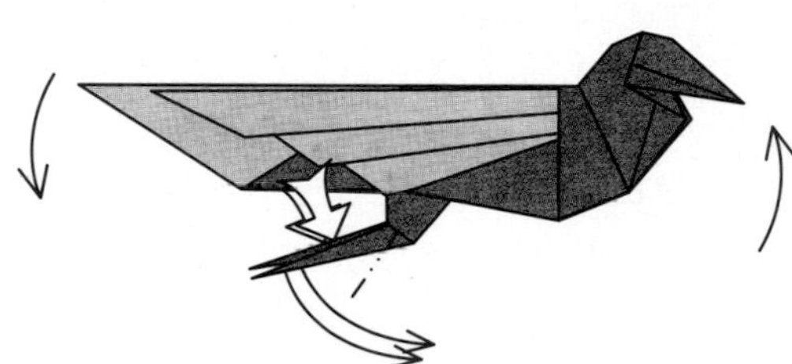

38. Reverse-fold the legs downward.

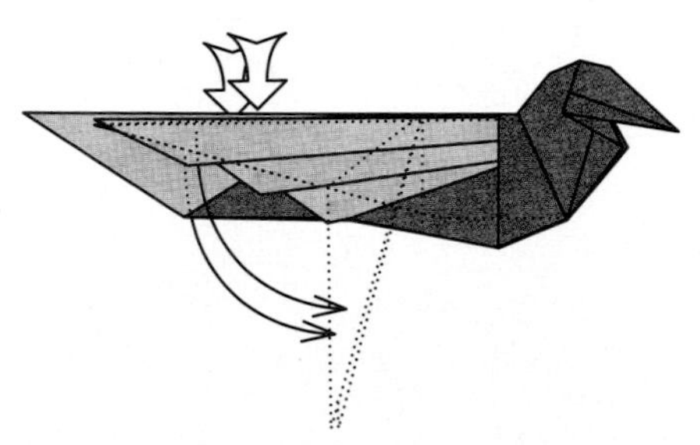

39. Double-rabbit-ear the two legs.

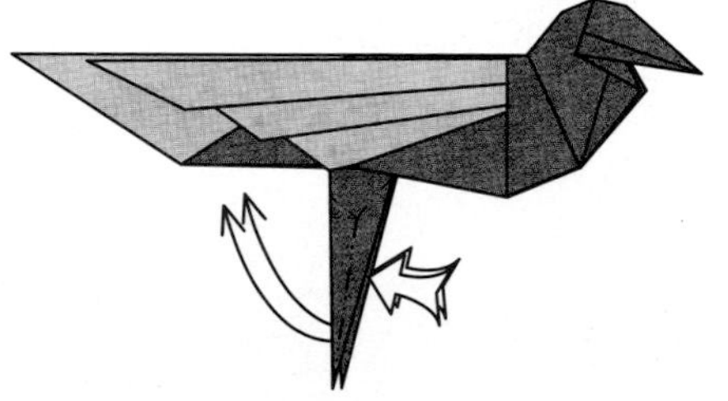

40. Reverse-fold the legs. Rotate the model slightly counterclockwise.

41. Squash-fold the tail from side to side.

42. Crimp the feet so that the model stands flat. Fold the small flap up on the side of the head and repeat behind.

43. Fold the layers underneath the head inside.

44. Open out and round the eye. Pinch the beak.

45. Finished Songbird.

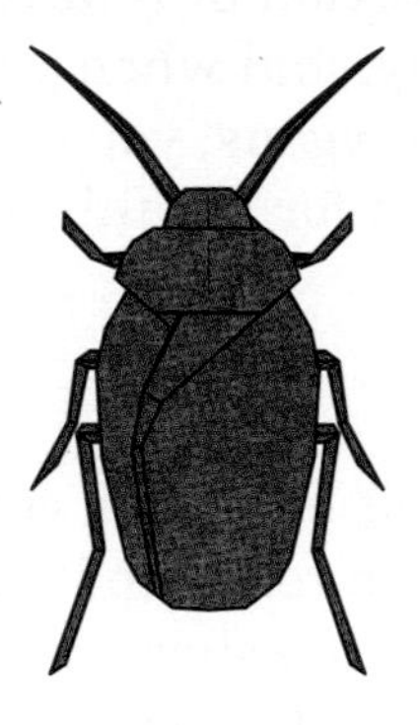

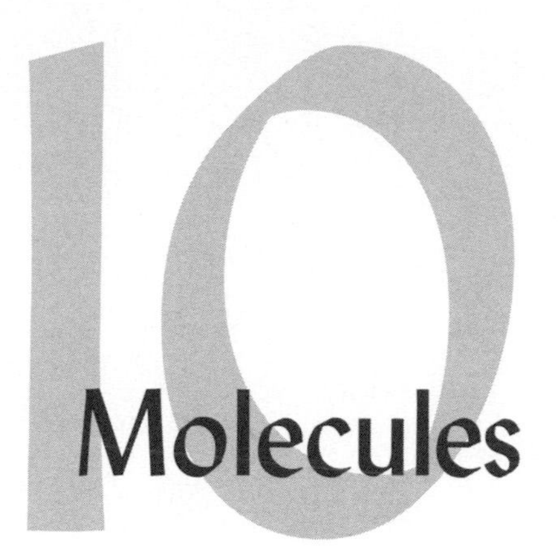

10 Molecules

We have seen that new bases can be constructed from tiles, pieces of old bases. By inscribing circular arcs within tiles and mating them according to a few simple matching rules, we can build new bases with new arrangements of flaps. However, assembly of tiles to build new bases can be a hit-or-miss proposition: You are limited to working with those tiles you have previously cataloged, and there is no guarantee than a given assembly of tiles will fit efficiently into a square of paper.

In the previous chapter, we saw how one can use circle packings to find a pattern of points within a square (or any other shape of paper) that is guaranteed to be foldable into an origami base that has a specified length and distribution of flaps. When the configuration of circles happens to match the circles (and, if needed, rivers) of known tiles, then we can fill in the crease pattern with the tile creases, and the paper can be collapsed into a base. However, a problem arises if the circle pattern matches none of the tiles we know so far. With the addition of just a few more patterns, however, we can find flat-foldable crease patterns for any circle/river packing—and for a great deal more besides.

This process is not as difficult as you might think, because there aren't that many different types of patterns that are needed. Most of the time, the polygons created by circle/river packings are triangles (as they have been in most of the examples we've seen thus far). More complicated bases may have quadrilaterals, pentagons, or higher-order polygons. All can be collapsed so that their edges lie on a line and they align with one another properly. What makes the problem of designing a base tractable is that, to a large degree, each polygon can be treated on its own. A highly complex base with numerous points can be broken up into a collection of relatively simple polygons, each analyzed individually; and when you have the crease patterns for the individual polygons, you can put them together to realize the crease pattern for the full square.

10.1. Tangent Points

Let us examine again some existing bases for common features of the crease pattern that we can relate to its underlying circle pattern. Figures 10.1 through 10.5 show the crease and circle pattern for bases with one through five equal-sized points. These crease patterns cover five different bases, but share several interesting features. In each crease pattern, I've labeled with a P the point where adjacent circles touch.

Figure 10.1 shows the crease and circle pattern for the Kite Base. The Kite Base appears to be a single flap, but we can also think of it as a base with one long flap joined to one short flap; both flaps are represented by circles in Figure 10.1. If we recognize that the horizontal crease on the right in Figure 10.1 is the boundary between the two flaps, then we see that each flap is defined by a circle; that the two circles touch at a point, that is, they are tangent circles; and that there are two creases that run through the tangent point, marked with a P in the figure. The vertical crease that connects the centers of the two circles is of a type that we have already met; it is an axial crease. The other is perpendicular to the axial crease and is tangent to both circles.

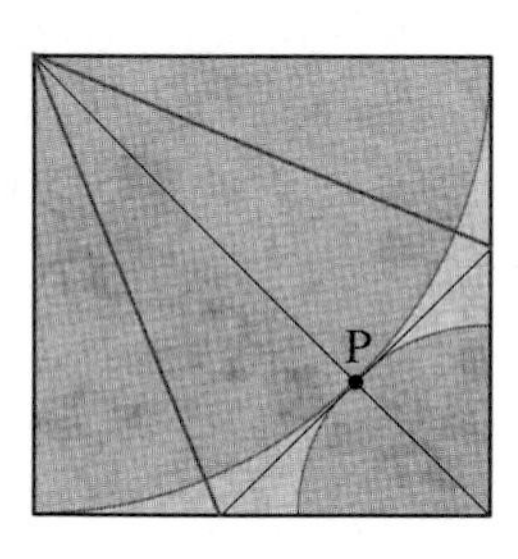

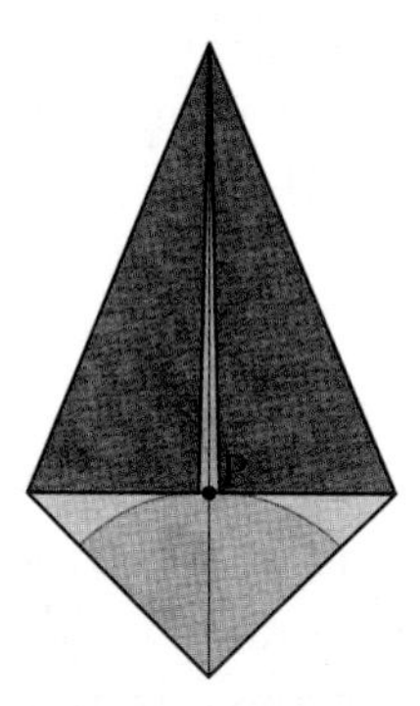

Figure 10.1.
Left: crease and circle pattern for the one-flap Kite Base. Right: the folded Kite Base.

Now look at Figure 10.2, which shows the crease and circle pattern for the Fish Base. The Fish Base has two long flaps and two short flaps. All four flaps are represented by circles, and adjacent circles touch. As with the Kite Base, there is an axial crease (or raw edge) between the centers of touching circles, and at each point of tangency, there is a crease perpendicular to the crease between centers.

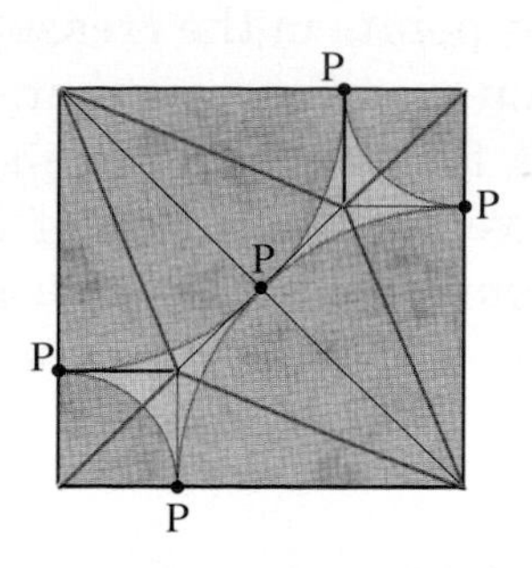

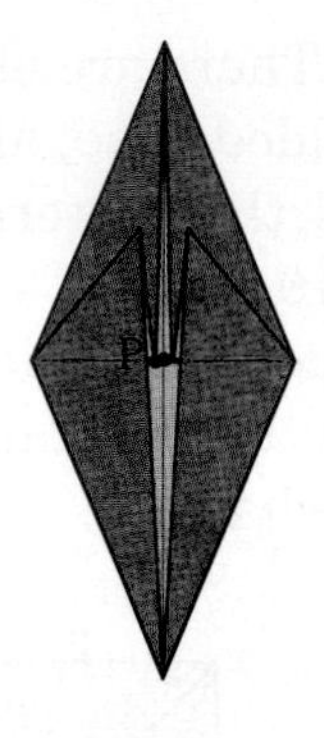

Figure 10.2.
Left: crease and circle pattern for the two-flap Fish Base.
Right: the folded Fish Base.

There's a second interesting phenomenon as well. Observe that there are 5 points where adjacent circles touch each other, called *tangent points*; I've labeled them all with a P. In the folded base, which is shown on the right, all of the tangent points lie either side-by-side or one atop the other; if you poked a pin through one of them, the pin would hit every tangent point in the model.

Now let's look at another base. Figure 10.3 shows creases and circles for a base with three equal flaps. Again, we have axial creases between touching circles and a second set of creases perpendicular to the first set that cross at the point of tangency. There are three tangent points, and in the base, all three tangent points lie on top of one another.

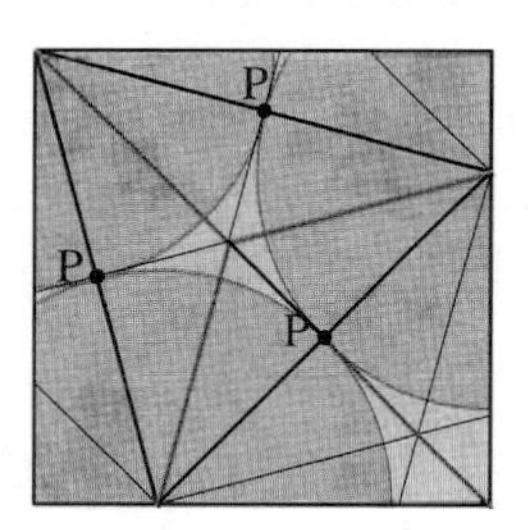

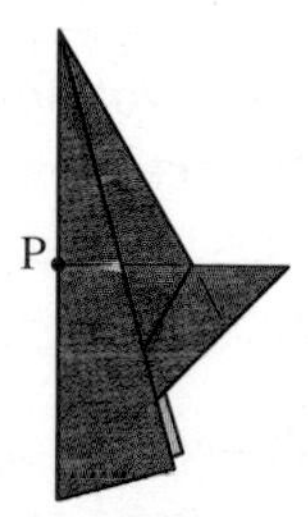

Figure 10.3.
Left: crease and circle pattern for the three-flap base.
Right: the folded base.

The first three circle bases had only edge and corner flaps. Do the patterns we observed hold for middle flaps? Indeed they do. Figure 10.4 shows the Bird Base, which has four long flaps and one short one, which is a middle flap. Again, circle centers are connected by axial creases, and creases emanate from the points where circles touch that are perpendicular to the axial

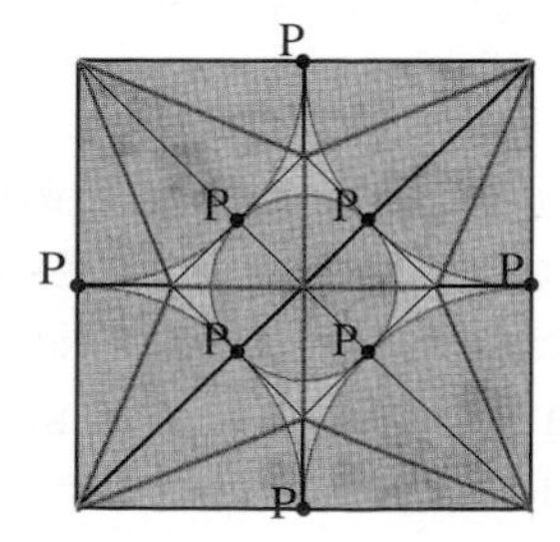

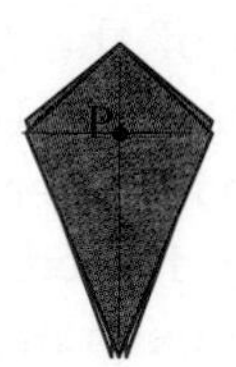

Figure 10.4.
Left: crease and circle pattern for the five-flap Bird Base.
Right: the folded base.

creases. There are eight tangent points in the crease pattern; in the folded base, all eight lie on top of one another.

And the pattern continues for the Frog Base shown in Figure 10.5: axial creases between the centers of touching circles, perpendicular creases emanating from the points of tangency, and all tangent points (this time, 16 of them) lie on top of each other.

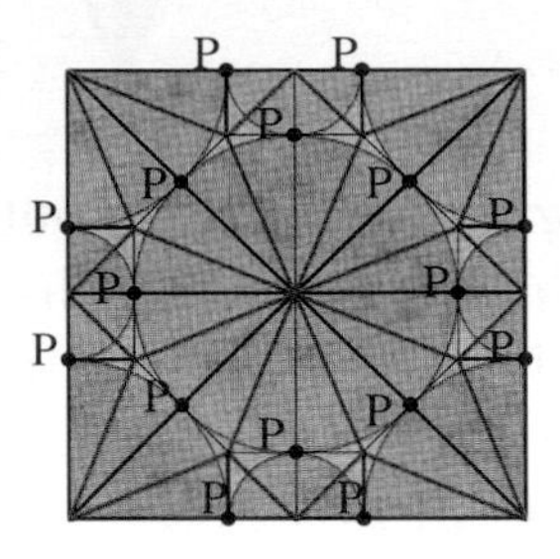

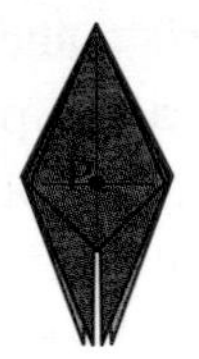

Figure 10.5.
Left: crease and circle pattern for the five-long-flap Frog Base. Right: the folded base.

Five examples don't prove universality, but they do suggest that there are features common to all circle pattern bases. In fact, there are several common attributes of circle method crease patterns:

- Where two circles touch, there is always a crease that connects the centers of the two circles. We've already encountered these; they are the *axial crease*s.

- When you fold the crease pattern into the base, all of the creases between touching circles—the axial creases—wind up lying on top of each other, i.e., along a single line, which is the axis of the base.

- Where two circles touch, there are also creases that are tangent to the two circles and perpendicular to the crease between their centers. These creases appear as horizontal lines in the bases in Figures 10.1 through 10.5. We'll call them *hinge creases*. The hinge creases form the hinges between flaps.

- In the crease pattern, the hinge creases connect to each other to make a continuous path that either starts and stops on an edge or runs all the way around each circle.

- All of the *tangent points*—the points where two circles touch (which are labeled P in the figures)—lie at the intersections of axial creases and hinge creases. In the folded form, they wind up lying precisely on top of each other in the folded base.

In the search for underlying principles, one looks for unusual coincidences. Here, we have five different bases in which the crease patterns display the same set of behaviors. They are not just coincidences; they are general principles of the circle method of design.

We can use these concepts to fill in the creases that go with an arbitrary circle-packing. There are three distinct sets of creases.

First, for any two circles that touch, there is an axial crease that runs between their centers. When the crease pattern is folded into a base, the axial creases are all parallel and lie on top of each other. Additionally, the tangent points—the points where circles touch—all lie on top of each other along the axis in the folded base.

Second, there are hinge creases perpendicular to the axial creases, which emanate from the points of tangency.

Then there is a third set, which are creases that propagate inward from the corners of the axial polygons. In the folded form, these creases form the ridges of the folded shape. We'll call them the *ridgeline creases*. The ridgeline creases bisect each of the angles at the corners of an axial polygon.

These three families are illustrated in Figure 10.6 for the Frog Base.

The three families of creases are closely related to the circles themselves. The hinge creases are conceptually the easiest to understand: They outline polygons that approximate the circles. So each polygon outlined by hinge creases delineates the boundary of a single flap of the base. We can see this by coloring one of these polygons in the crease pattern and seeing where it winds up in the folded base. Several examples are shown in Figures 10.7 through 10.9. In each case, the colored polygon provides *all* of the layers of exactly *one* flap.

While the hinge creases are most easily related to the original base and circle pattern, the axial creases—the lines between circle centers—are just as important, but in an entirely different way.

The polygons outlined by the hinge creases circumscribe the circles we have used to denote individual flaps. When we collapse an axial polygon so that its edges lie along the axis of the base, all of the tangent points come together at a single point. Thus, there are two properties that must be satisfied by the crease pattern within an axial polygon:

- Its edges must be brought to lie along a single line.
- The tangent points along its boundary for each circle must be brought together at a single point.

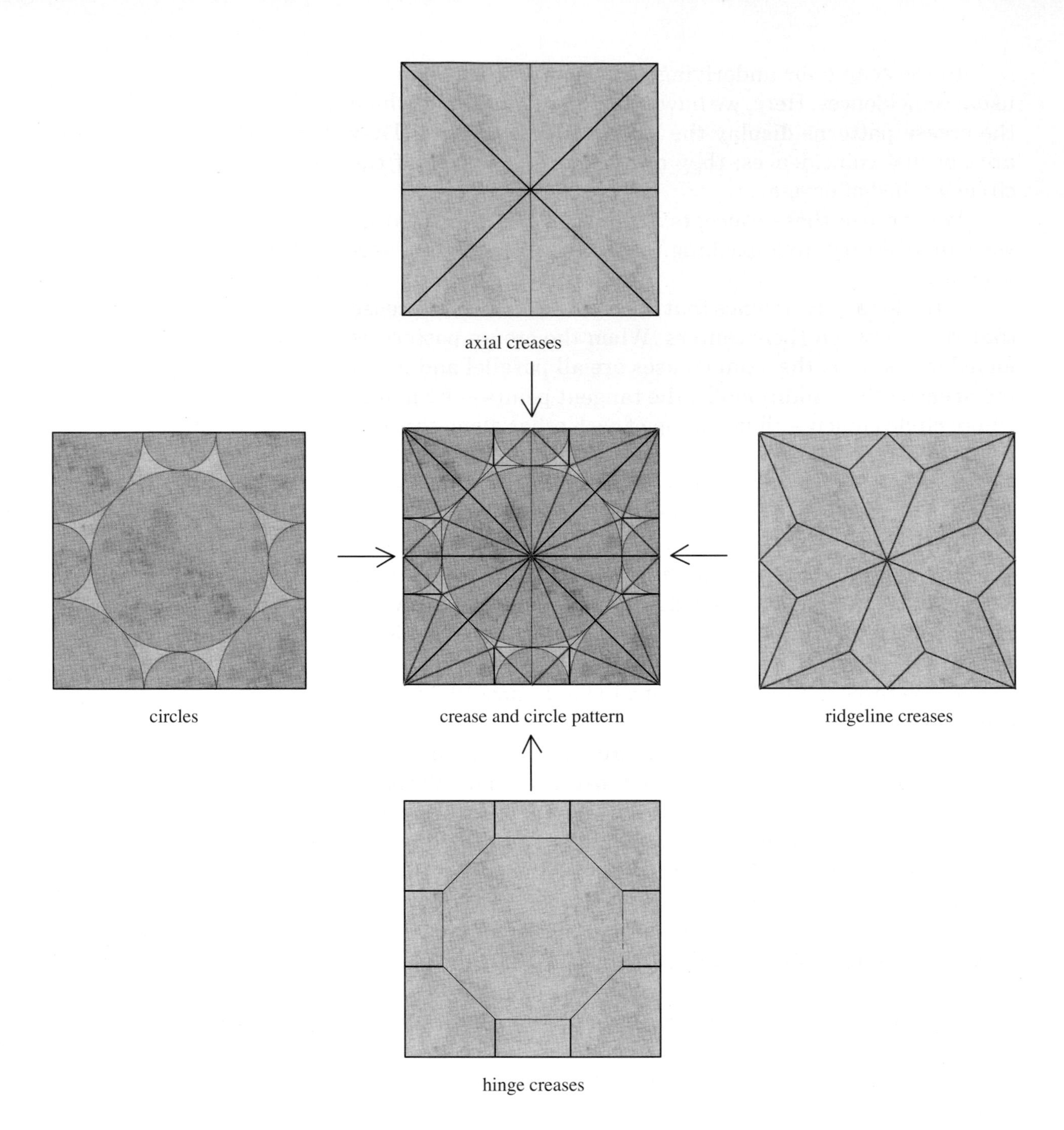

Figure 10.6.
The three families of creases that make up a crease pattern in the circle method.

The problem of collapsing a polygon so that its edges lie on a single line is well known in both origami and mathematics. In mathematics, it is related to a famous problem known as the one-cut problem: How do you fold a sheet of paper so that with a single cut, you cut out an arbitrary polygon or collection of polygons? The one-cut problem has been solved by several authors

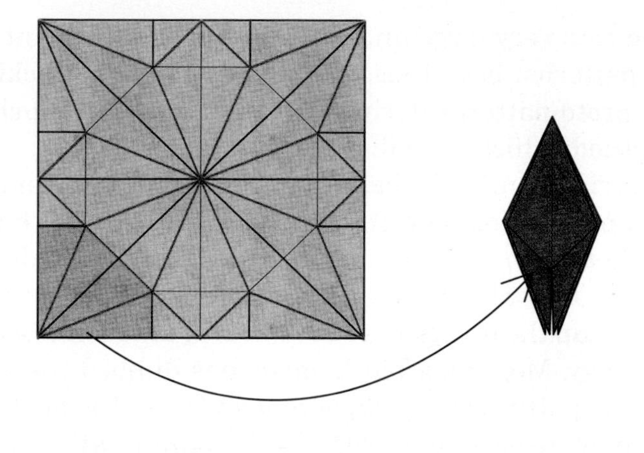

Figure 10.7.
A polygon outlined by hinge creases becomes a single flap of the base. This shows a polygon that becomes a corner flap.

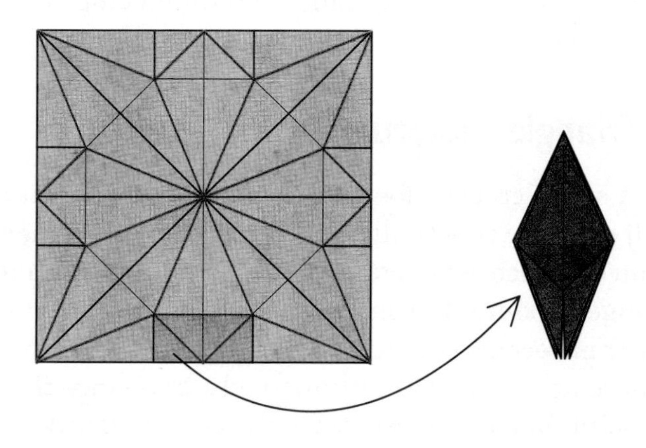

Figure 10.8.
A polygon that becomes an edge flap.

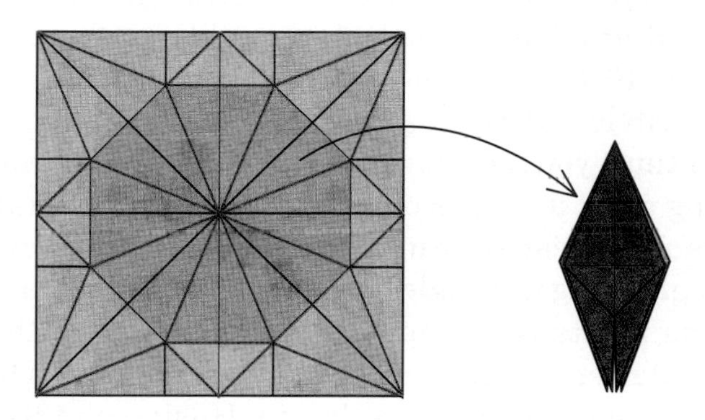

Figure 10.9.
A polygon that becomes a middle flap.

(see Appendix B for references). However, the second requirement—alignment of the tangent points—is unique to origami and leads to new and specialized crease patterns.

Within the world of origami, I and several other artists and scientists—notably Koji Husimi, Jun Maekawa, Toshiyuki Meguro, Fumiaki Kawahata, and Toshikazu Kawasaki—have studied the crease patterns that allow various polygons to be collapsed with their edges falling onto a single line. It turns out that a relatively small set of crease patterns can be assembled into very large and complex tiles, indeed, into entire crease patterns; both those derived from circle packings, as well as proto-patterns derived by other methods (such as the tree method, which we will shortly encounter).

All origami uniaxial bases can be constructed from a small set of minimal tiles. The situation is analogous to that of life itself, wherein a small number of amino acid molecules can be assembled into all the proteins that make life possible and that make up the diversity of the natural kingdom. Because of this analogy, Meguro, a biochemist, has dubbed these fundamental tile patterns *bun-shi*, or *molecules*. In the next section, we will explore origami molecules. By enumerating and identifying the molecules of origami, we will develop the building blocks of origami life.

10.2. Triangle Molecules

Finding a set of creases for folding a polygon so that all of its edges fall on a line is actually quite easy. However, there can be more than one such set of creases. Choosing the set that gets all of the tangent points to come together can be rather difficult. Meeting this second condition gets harder the more tangent points there are to align simultaneously, and since there is one tangent point for each edge of the polygon, smaller polygons are easier to find creases for than larger ones. Thus, let us start with the smallest nontrivial polygon—a triangle—and work out a crease pattern that meets the two conditions above.

Figure 10.10 shows an arbitrary triangle formed by three touching circles. If you have been folding origami for any length of time, you have already encountered a technique for collapsing all of its edges onto a line: the humble rabbit-ear fold. The rabbit ear is formed by folding all three corners of the triangle along the angle bisectors (which meet at a point); one of the points is swung over to one side and the entire structure flattened. That *any* arbitrary triangle can be folded into a rabbit ear was noted by Justin, Husimi and Kawasaki; however, the geometric relations underlying the rabbit ear (that the angle bisectors meet at a point and that adjacent triangles formed by dropping lines from the bisector intersection to all three sides are congruent) were originally proven by Euclid over 2000 years ago. Thus, the seeds of origami design were sown in antiquity.

However, for origami purposes, we need to satisfy both alignment conditions. It is not enough simply that the edges of the triangle all fall on a line. It is also essential that the tangent points all come together. Fortunately, it is not diffi-

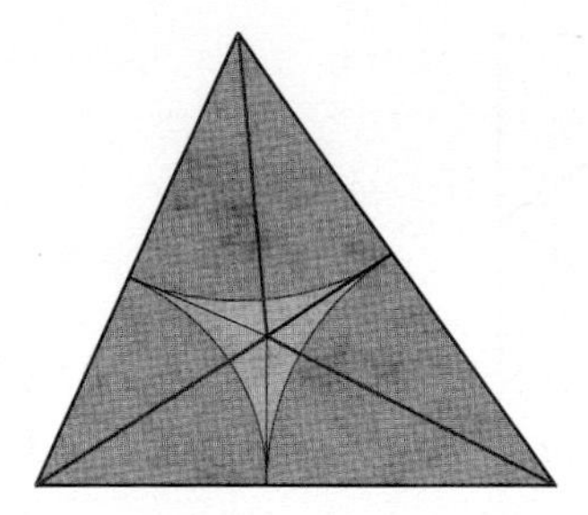
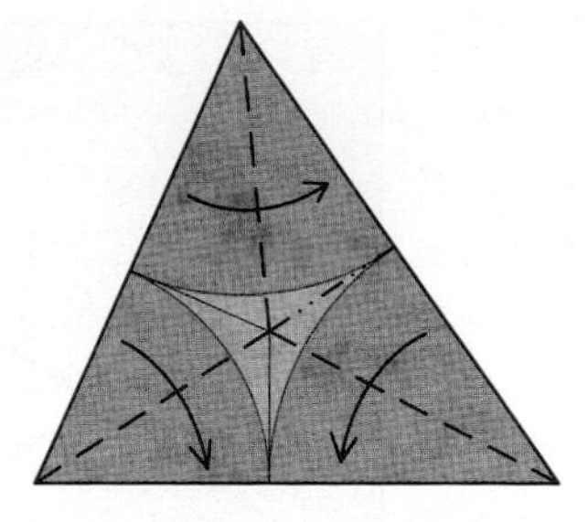
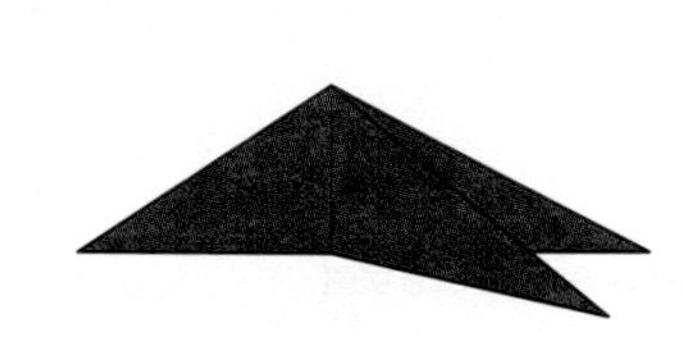

Figure 10.10.
The rabbit-ear fold brings together all edges of a triangle so that they lie on a line. Furthermore, the tangent points are all brought together to meet at a point.
Left: crease pattern.
Middle: folding sequence.
Right: folded molecule.

cult to prove mathematically that for any triangle formed by connecting the centers of three touching circles, the rabbit-ear crease pattern brings together the tangent points as well.

Therefore, we now have a construction for finding the crease patterns for any triangular polygon; just fold a rabbit ear. Or, to construct the creases without folding, construct the bisectors of each angle of the triangle, which meet at a point. Then draw a line from the tangent point on each side to the intersection of the bisectors. We will call this crease pattern the *rabbit-ear molecule*.

10.3. Quadrilateral Molecules

It is heartening that the triangle was so easy. It is further heartening that the most common polygon one encounters in circle-method bases is a triangle, and in fact, for the two- through five-flap bases seen in the previous chapter, all of the polygons were triangles. Thus, using the rabbit-ear molecule, we could find the full crease pattern for each of these bases. Wouldn't it be nice if when we diced up any circle pattern along its boundary creases, the polygons always turned out to be triangles? Alas, such is not the case. For the very next circle pattern, the pattern for six equal points shown in Figure 10.11, we find that a four-sided polygon crops up.

The crease pattern for Figure 10.11 does contain several triangles. Note that the two triangles in the upper corners of the square have only two circles inside each triangle. Any polygon with fewer than three circles in it is essentially unused paper and can be ignored. At the bottom of the model are three triangles, which can be filled with rabbit-ear molecules.

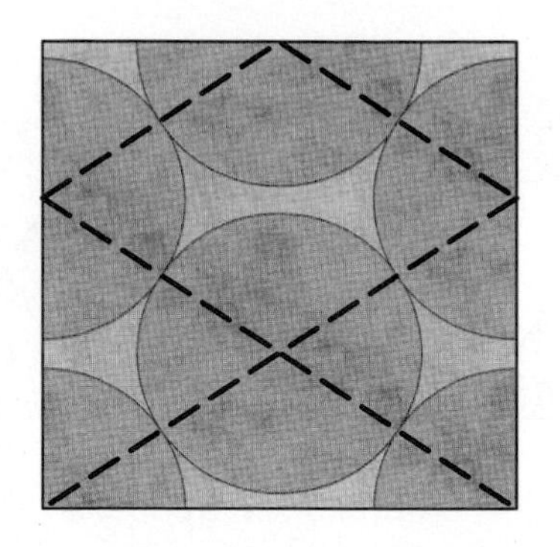

Figure 10.11.
Circle pattern for a six-pointed base.

But look at the polygon in the upper middle of the paper: the polygon is not a triangle—it is a four-sided diamond. So here we have a concrete demonstration that we will have to deal with polygons with more than three sides. Sometimes there will be four sides. So let us look at the problem of collapsing quadrilaterals so that all of their edges lie on a single line.

10.4. Waterbomb Molecule

With a triangle, there was exactly one crease pattern that put all of its edges onto a single line. Fortunately, this one crease pattern satisfied the tangent point condition—the tangent points all come together automatically. With a quadrilateral, the situation is a bit more complicated. For any quadrilateral that is formed by connecting the centers of four touching circles, the bisectors of the four angles all meet at a point as shown in Figure 10.12, which suggests one way of collapsing a quadrilateral.

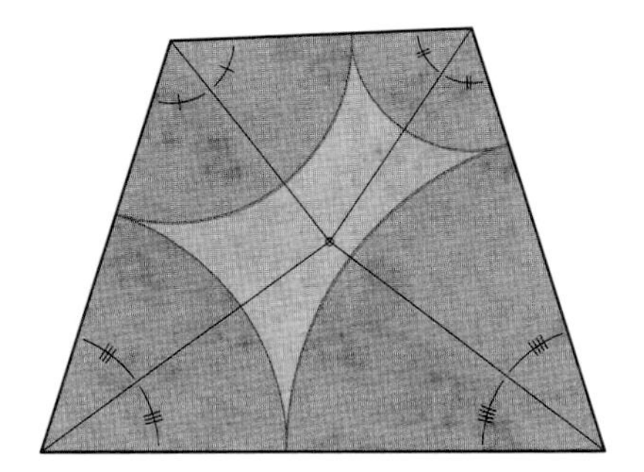

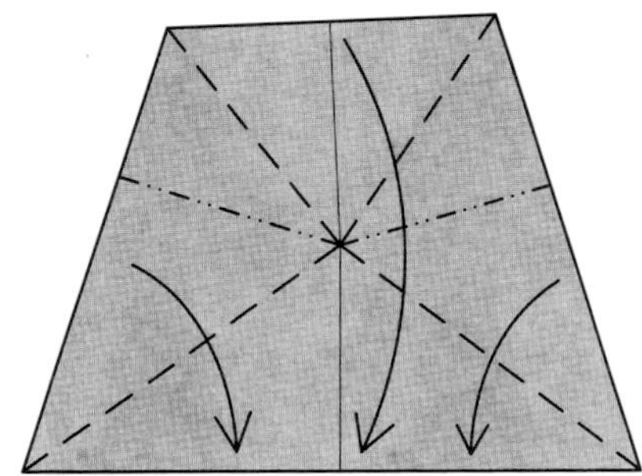

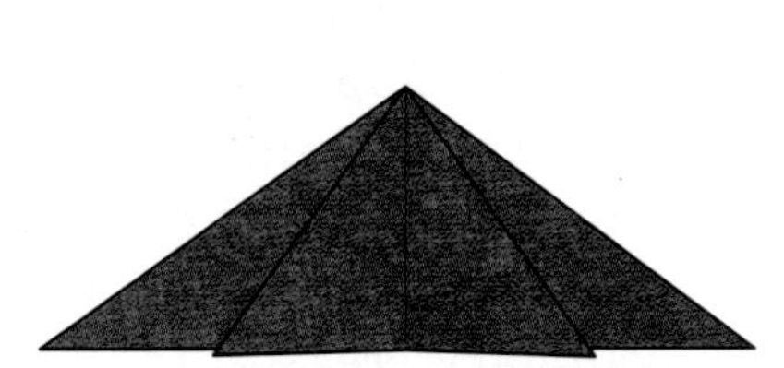

Figure 10.12.
The bisectors of a quadrilateral defined by four touching circles meet at a point, which permits the quadrilateral to be folded so that all of its edges lie on a single line.

We call this pattern the *Waterbomb molecule*, because the folded shape and the topology of the creases are those of the traditional Waterbomb.

Note, however, that not all quadrilaterals can be folded into a Waterbomb molecule; in fact, only those formed by four touching circles—called a *four-circle quadrilateral*—can be so folded. This property is fairly easy to demonstrate. As shown in Figure 10.13, if the four circles have radii a, b, c, and d, then the sides of the quadrilateral are, respectively, $(a + b)$, $(b + c)$, $(c + d)$, and $(d + a)$. The sum of the lengths of opposite sides are $(a + b + c + d)$ for both pairs of sides. We call this relationship the *Waterbomb condition*: In a four-circle quadrilateral, the sums of opposite sides are equal.

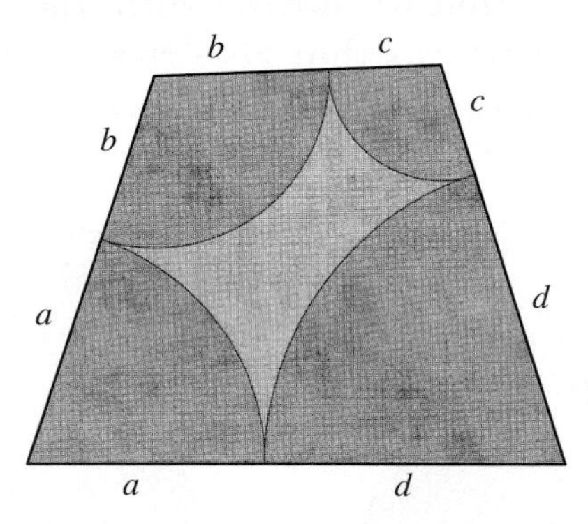

Figure 10.13.
For a four-circle quadrilateral, the sums of the lengths of opposite sides are equal.

Now, let's see if the converse is true. In a four-circle quadrilateral, one whose opposite sides sum to equal values, construct the angle bisectors from all four corners. The two bisectors on the left must meet at a point; similarly, the two on the right must also meet at a point (which may or may not be the same point). Suppose they are two different points. Drop perpendiculars from the two bisector intersections to the adjacent sides, as shown in Figure 10.14.

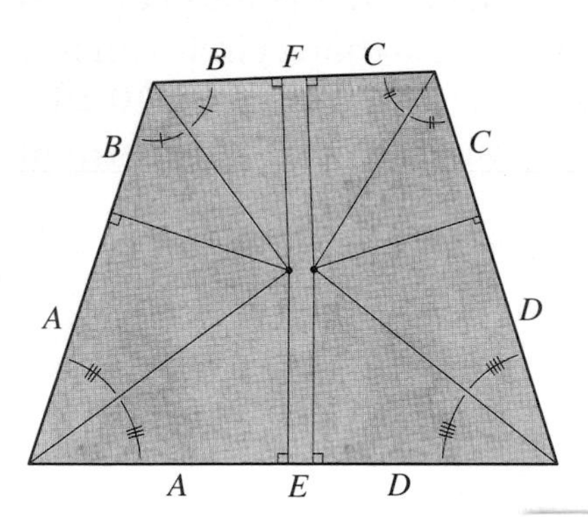

Figure 10.14.
Drop perpendiculars from the bisector intersections to all four sides.

We label the lengths of distinct segments along the edges A–F as shown; since each is a distance, all six quantities are greater than or equal to zero. If this is a four-circle quadrilateral, then the sums of opposite sides must be equal; that is,

$$(A+B)+(C+D)=(A+E+D)+(B+F+C). \quad (10\text{–}1)$$

This means that

$$E = F = 0. \tag{10–2}$$

This implies that the distance between the two bisector intersections is zero, i.e., that they are the same point. Thus, any quadrilateral that satisfies the Waterbomb condition has its angle bisectors meet at a point and can be folded into an analog of the Waterbomb Base.

Note that the distances *A–D* are not necessarily equal to the circle radii *a–d* that we started with; there are many different four-circle patterns that give rise to exactly the same quadrilateral. Three examples are shown in Figure 10.15.

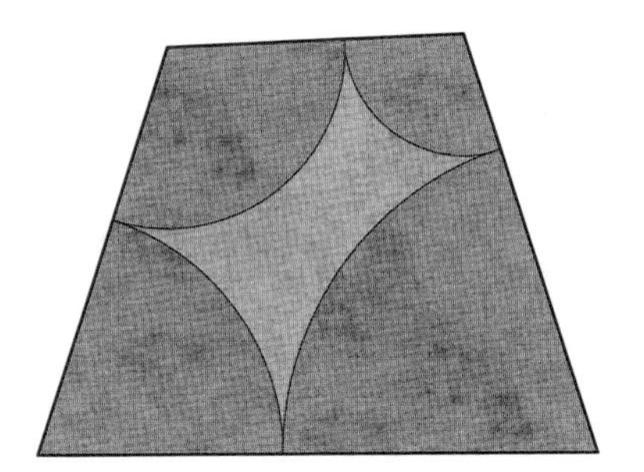

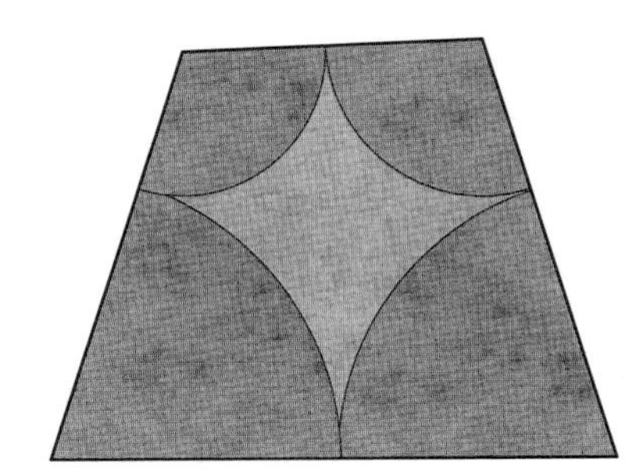

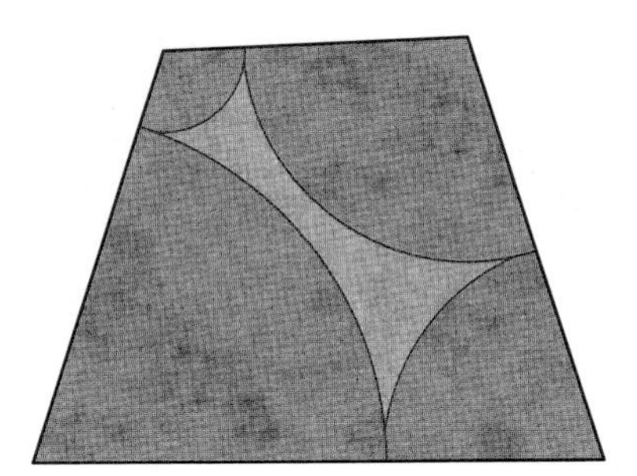

Figure 10.15.
Three identical polygons formed by four different circles.

Waterbomb molecule quadrilaterals have a couple of other interesting properties. If we draw four lines from the bisector intersection perpendicular to the four edges, they all have the same length, which means that a circle can be inscribed within the quadrilateral as shown in Figure 10.16, a property first shown by Koji Husimi, also noted by Justin and Maekawa.

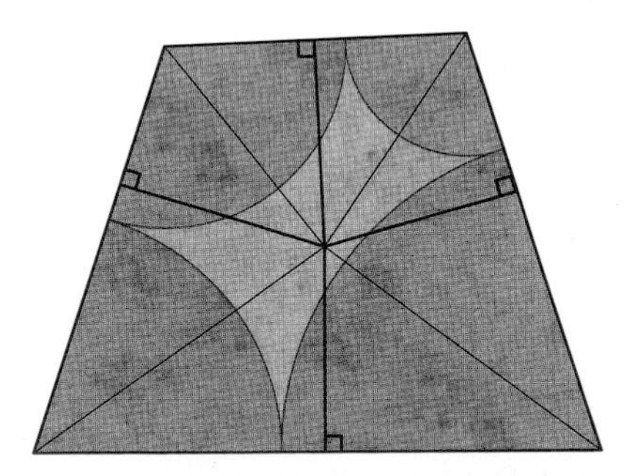

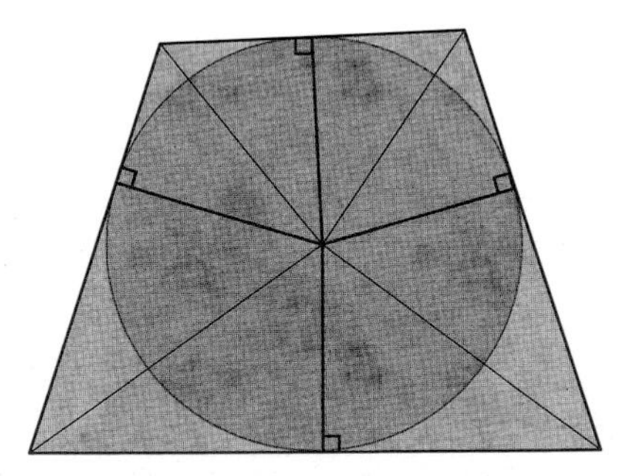

Figure 10.16.
A quadrilateral defined by four touching circles can have a circle inscribed within it that touches all four sides.

It is also quite easy to show the converse of this relationship, that the vertices of any quadrilateral with an inscribed circle tangent to all four sides are the centers of four pairwise tangent circles.

If a quadrilateral satisfies the Waterbomb condition, then the folds of the Waterbomb Base—the four bisectors, plus the four perpendiculars—are uniquely specified. So there is only one way to collapse the quad into a Waterbomb molecule. But as we saw, there are many possible circle patterns that can give rise to the same quadrilateral. Only one particular set of circles has the property that the tangent points line up with the hinge creases, as shown in Figure 10.17.

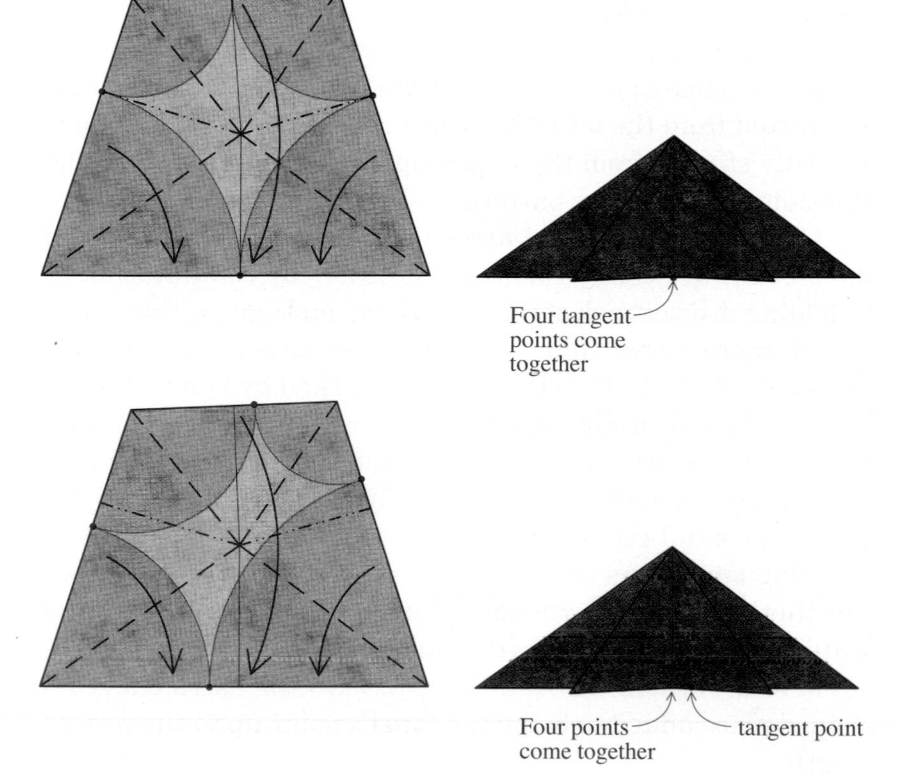

Figure 10.17.
Top: for one set of circles, the four tangent points come together.
Bottom: for all other sets of circles, the tangent points do not align.

For the rhombus that appears in the six-circle packing (Figure 10.11), this is also the situation: The Waterbomb molecule does not bring the tangent points together, as Figure 10.18 shows.

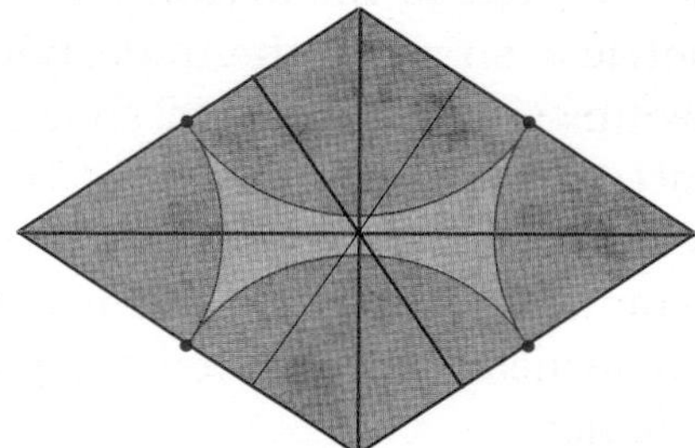

Figure 10.18.
Waterbomb molecule within a four-circle rhombus. Note that the perpendiculars do not hit the tangent points.

Thus, even if a quadrilateral satisfies the Waterbomb condition, the Waterbomb molecule may not be the appropriate crease pattern that brings together the tangent points. Fortunately, there are other quadrilateral molecules that fill this need.

10.5. Arrowhead Molecule

Although the Waterbomb molecule doesn't always bring the tangent points together, there are other crease patterns that do. One that is quite simple to construct and fold is shown in Figure 10.19. This pattern, described by Meguro and Maekawa, will always bring the four tangent points together. We call it the *arrowhead molecule*.

There is usually more than one arrowhead molecule that can be constructed from a given quadrilateral. In Figure 10.19, we started from the lower left corner; however, we could have as easily started from the upper right corner and derived the molecule whose crease pattern is shown in Figure 10.20.

A nice feature of the arrowhead molecule is that all of the creases are easily constructed either by computation or by folding. A drawback of the arrowhead molecule is that when folded, more edges than just the outer edges lie along the axis of the base. In fact, the creases marked by heavy dashed lines on the left in Figure 10.21 also lie along the axis as well as the edges when the molecule is folded up. We saw that in the full crease pattern, lines that lie along the axis of the model are axial creases, creases that connect the centers of touching circles. As shown on the right in Figure 10.21, we can think of the arrowhead molecule as a combination of a Waterbomb molecule formed from four touching circles, three out of four of them the right length, with the extra chevron-shaped piece added to bring the fourth point up to the proper length.

Any molecule that has interior creases that line up with the raw edges when the molecule is folded is called a *composite molecule*. A molecule with no interior creases is a *simple molecule*. The arrowhead molecule is a composite molecule.

Another disadvantage of the arrowhead molecule is that it can be asymmetric even when the underlying polygon and circle pattern is symmetric. Figure 10.22 shows the arrowhead molecule constructed within the diamond from the 6-circle pattern of Figure 10.11. Although the diamond and its circles have left-right symmetry—the right side is the mirror image of the left—the arrowhead molecule crease pattern (and the folded molecule) do not.

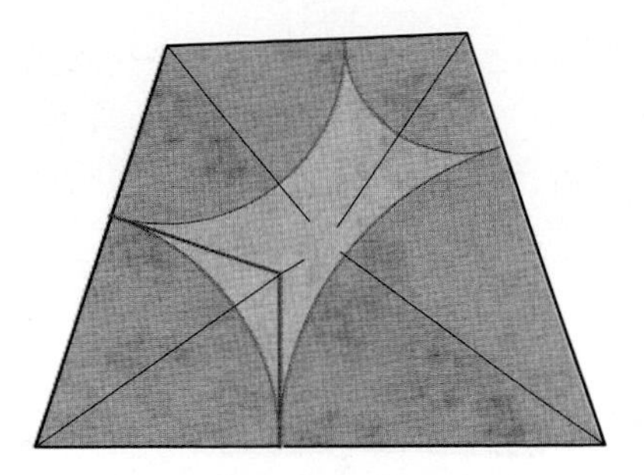

1. Begin with the four angle bisectors. Draw lines from two adjacent tangent points perpendicular to the edges until they meet at the bisector.

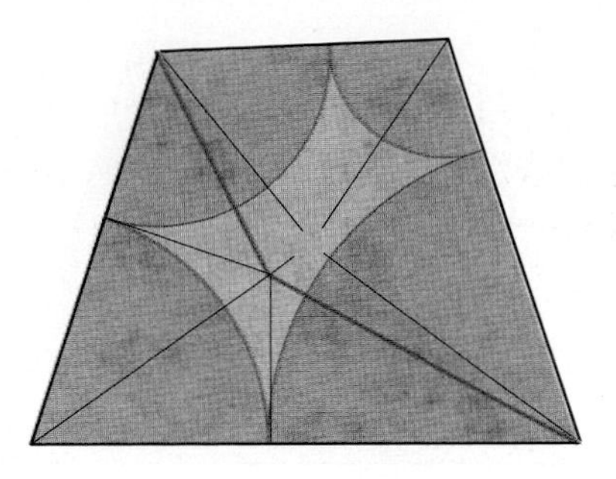

2. Draw lines from the intersection back out to two diagonally opposite corners.

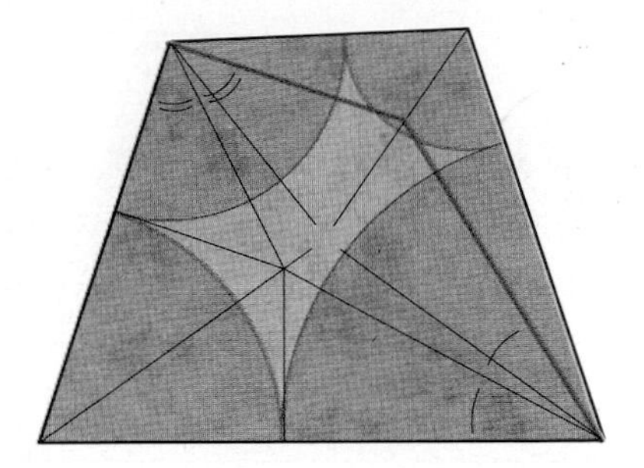

3. Draw two more lines each from the corners making equal pairs of angles at the two corners.

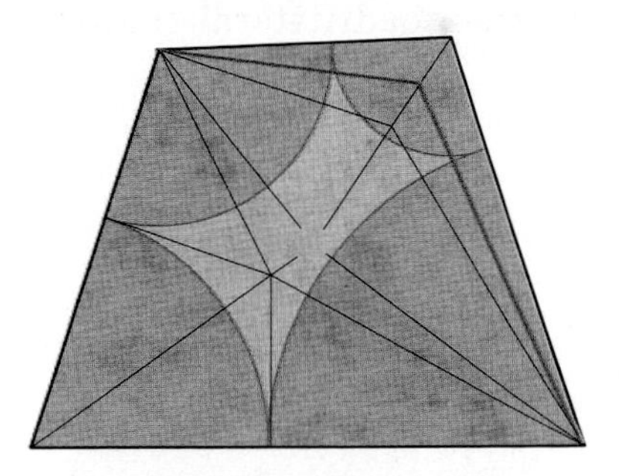

4. Bisect the remaining paper at each of the diagonal corners.

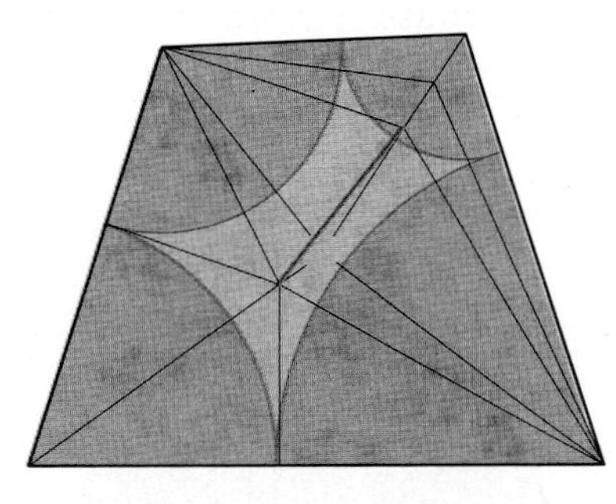

5. Add a crease connecting two crease intersections.

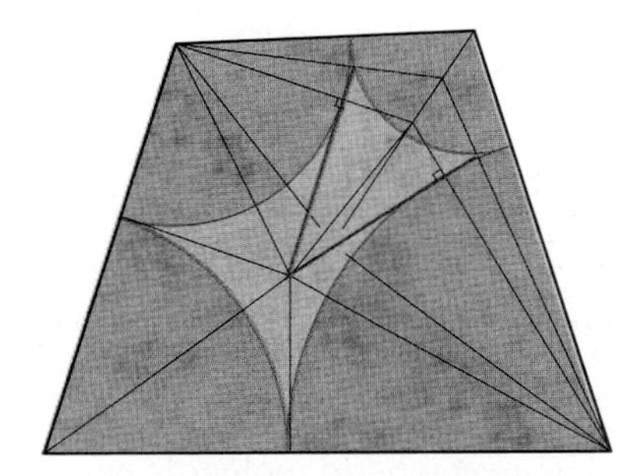

6. Add two creases emanating from the intersection and perpendicular to the two creases shown.

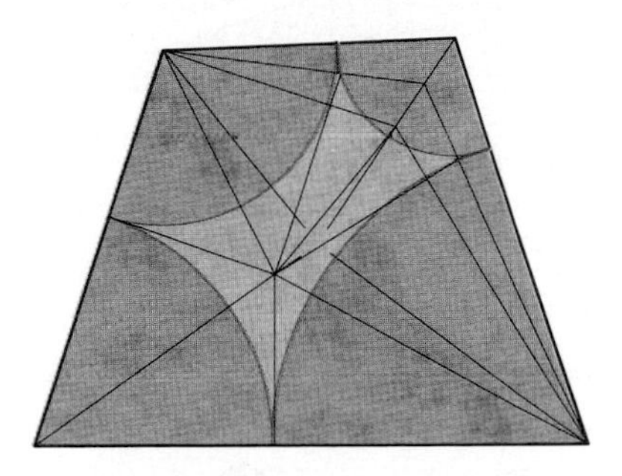

7. Connect the crease intersections with the tangent points.

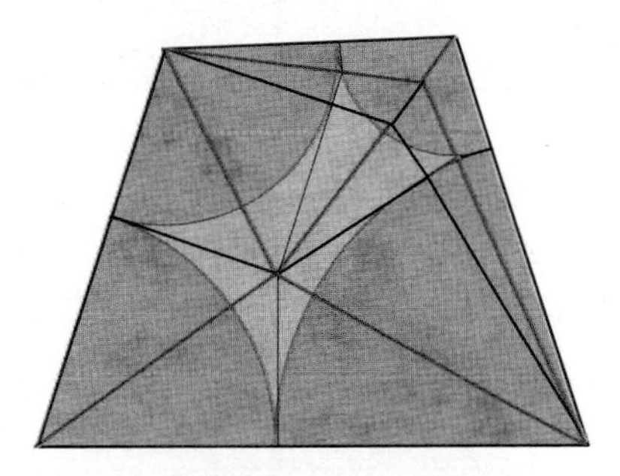

8. The completed crease pattern.

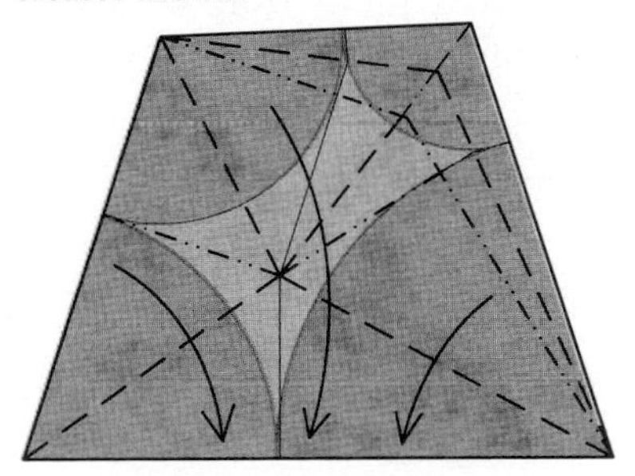

9. Using these creases, collapse the shape.

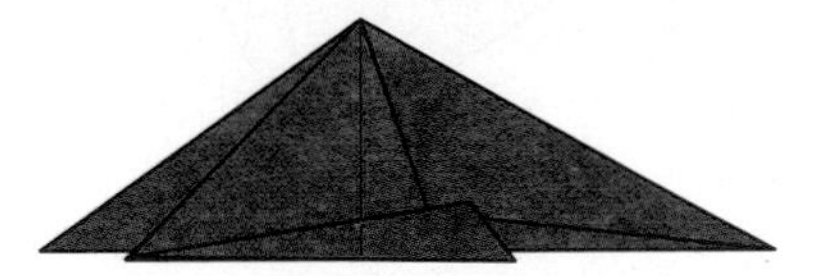

10. Finished arrowhead molecule. Note that now all four tangent points come together.

Figure 10.19.
Construction of the arrowhead molecule.

For symmetric circle patterns such as the six-circle packing, using an asymmetric molecule in a symmetric polygon will result in an asymmetric base. This may be undesirable for a symmetric subject.

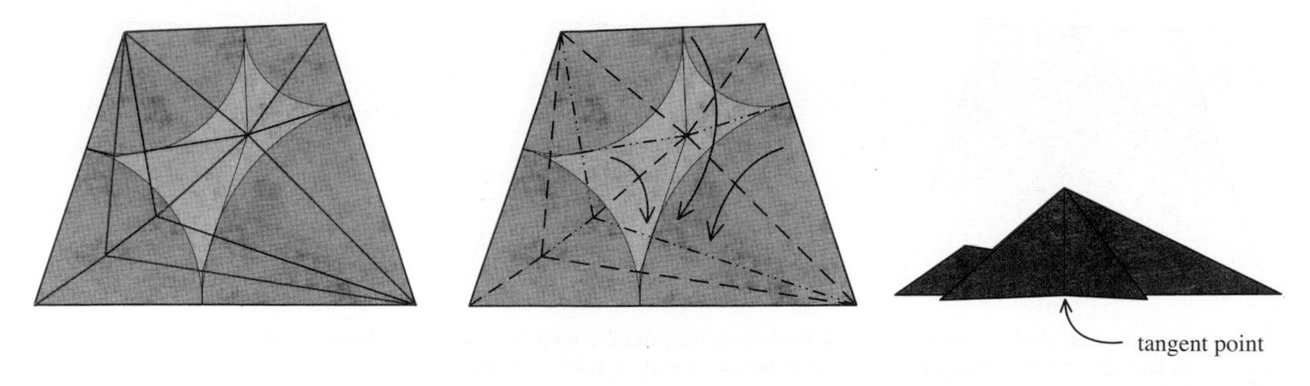

Figure 10.20.
An alternate arrowhead molecule for the quadrilateral shown in Figure 10.19.

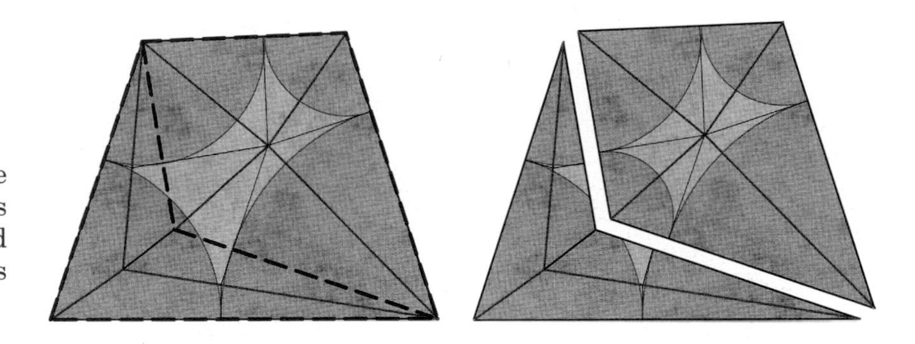

Figure 10.21.
The arrowhead molecule can be separated along axial creases into a Waterbomb molecule and an extra piece that lengthens one of the points.

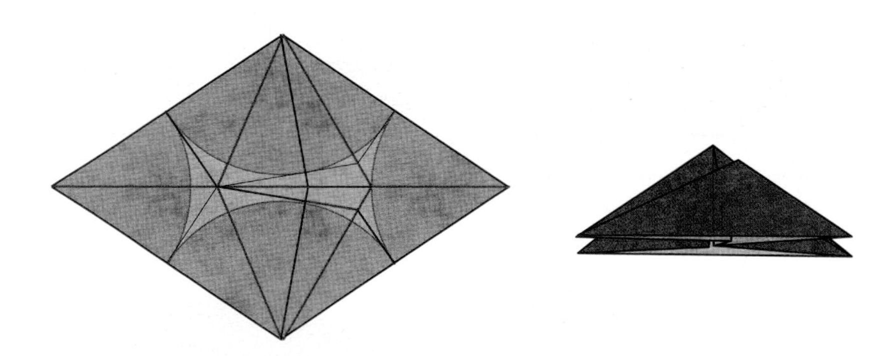

Figure 10.22.
Crease pattern and folded form of the arrowhead molecule in a four-circle rhombus.

10.6. Gusset Molecule

The arrowhead molecule is not the last molecule for quadrilaterals, however. The crease pattern shown in Figure 10.23 is a valid crease pattern for a molecule I call the *gusset molecule* that can be oriented to preserve the underlying symmetry.

Like the arrowhead molecule, the gusset molecule can be constructed for any four-circle quadrilateral. But the gusset molecule has a couple of advantages over the arrowhead

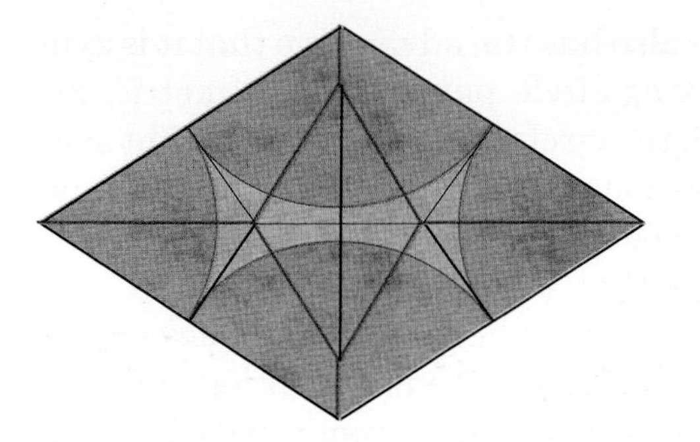

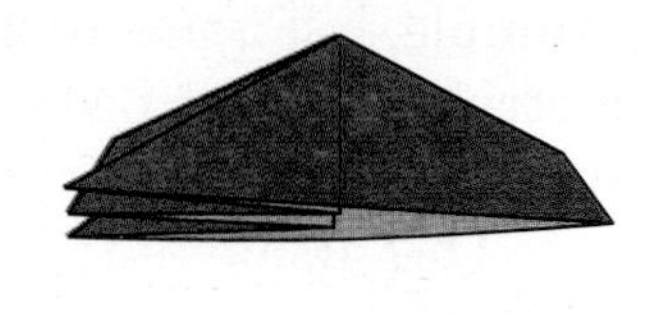

Figure 10.23.
Crease patterns and folded example of the gusset molecule. The folded molecule can have its flaps oriented to be mirror-symmetric about either symmetry of the underlying polygon.

molecule. There are no interior creases that lie along the axis when it is folded, so it is a simple molecule. Simple molecules lead to bases that have fewer layers along the axis of the model.

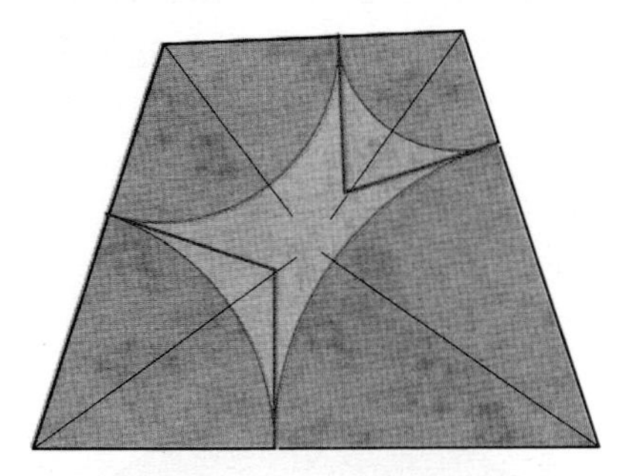

1. Begin with the four angle bisectors. Draw lines from all four tangent points perpendicular to the edges until the pairs of perpendiculars adjacent to diagonally opposite corners meet.

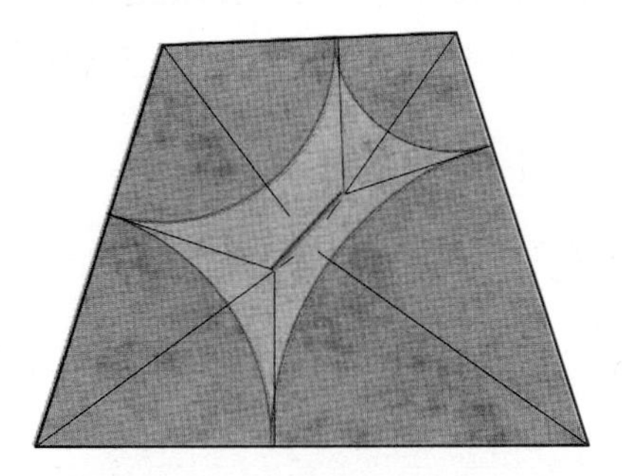

2. Connect the two points of intersection with a crease.

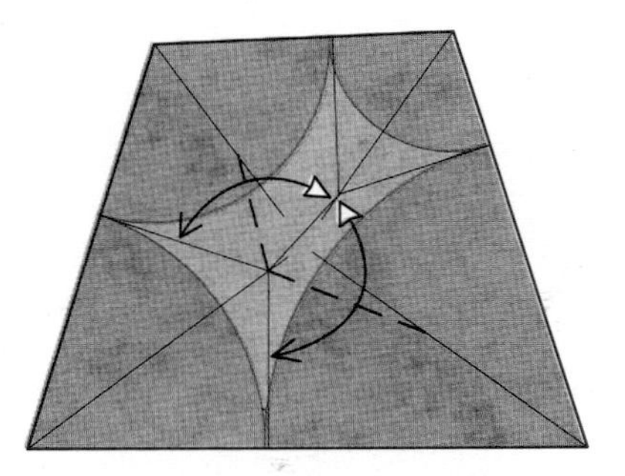

3. Fold the crease you just made to lie along two of the perpendiculars; crease and unfold.

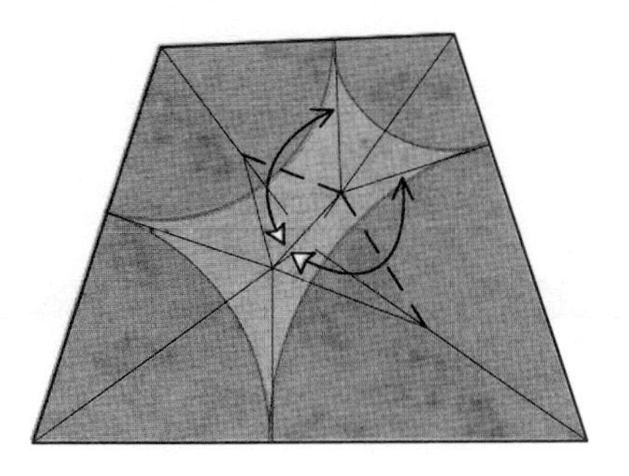

4. Repeat for the other two perpendiculars.

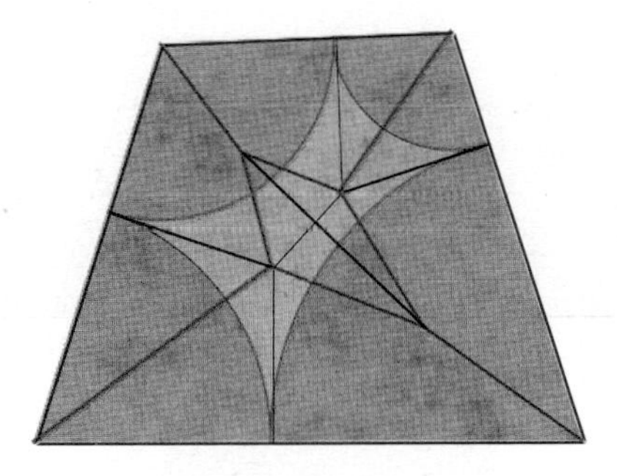

5. The finished crease pattern.

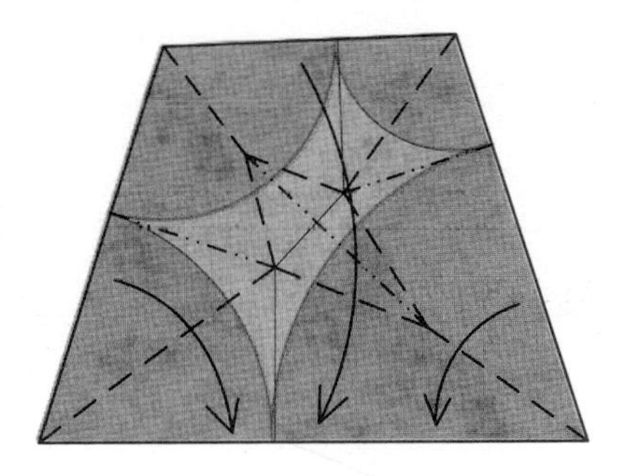

6. Collapse on the creases shown.

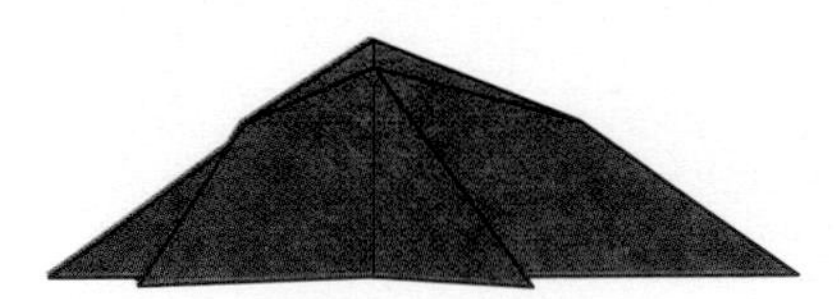

7. The finished gusset molecule.

Figure 10.24.
Construction method for the gusset molecule.

The gusset molecule also has the advantage that it is symmetric when the underlying circle pattern is symmetric. For example, in Figure 10.22, the circle pattern has left-right symmetry, but the arrowhead molecule does not have this symmetry. The gusset molecule does.

The disadvantage is that the gusset molecule is a bit harder to construct than the arrowhead molecule. However, it can be constructed by folding using the prescription shown in Figure 10.24. It can also be constructed numerically, by using analytic geometry to compute the creases shown in Figure 10.24, or as we will see in the next chapter, using the algorithms of tree theory.

In the basic gusset molecule, the baseline of the gusset (indicated by the hidden line in the final step of Figure 10.24) is parallel to the axis. However, you can vary this angle by

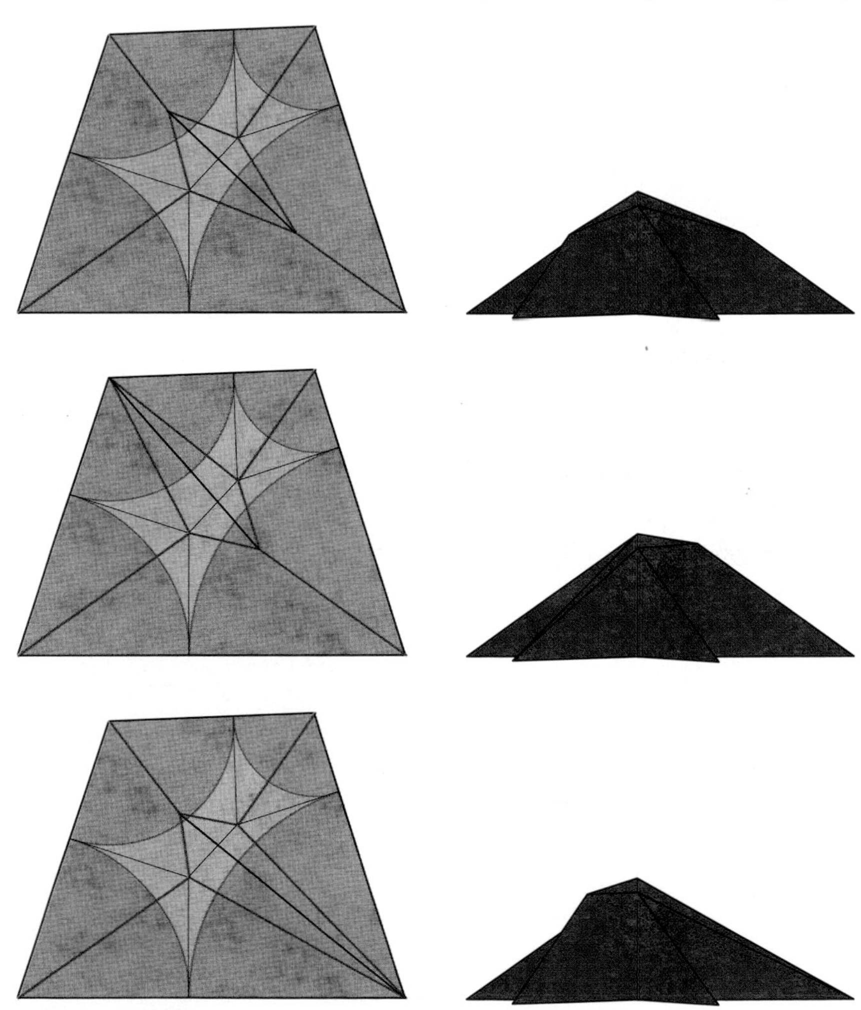

Figure 10.25.
Top: crease pattern and folded form of a basic gusset molecule (gusset parallel to axis).
Middle, bottom: two variations with tilted gussets.

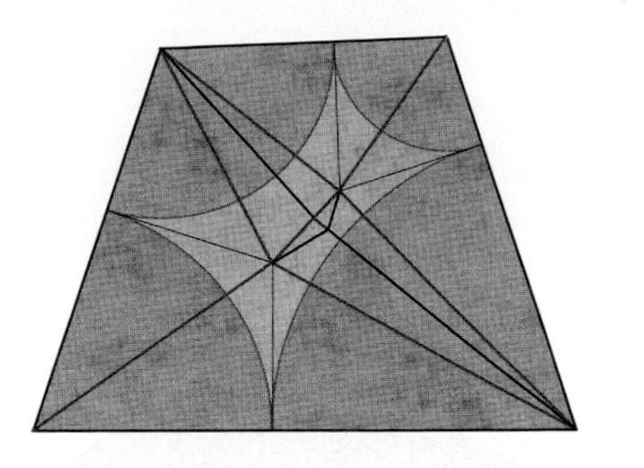
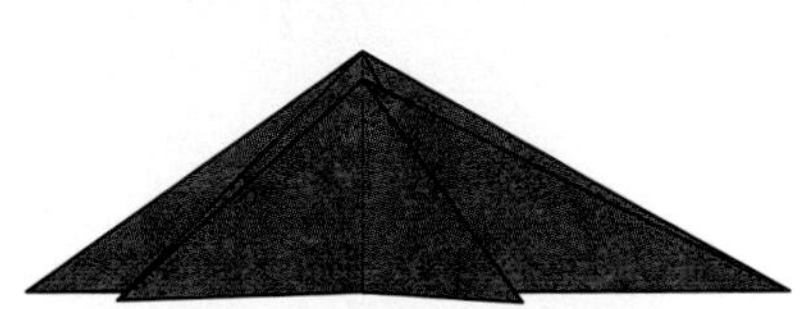

Figure 10.26.
Crease pattern and folded form of another variant of the quadrilateral gusset molecule.

tipping the gusset one way or the other. Several variations are also shown in Figure 10.25.

There is, as well, a version in which the gusset extends to both corners by addition of a crimp in its middle, as shown in Figure 10.26.

Numerous other variations are also possible. My conjecture is that for a given quadrilateral with a specified set of tangent points, the basic gusset molecule is the molecule with the minimum total crease length, but this has not yet been proven.

The gusset molecule is quite versatile. If you reexamine some of the crease patterns from the previous chapter, you will see a few gusset molecule patterns along with rabbit-ear and Waterbomb molecules.

10.7. Molecules with Rivers

When we built crease patterns from preexisting tiles, we kept track of flaps and connections between flaps by decorating the tiles with circles and rivers. Similarly, when building up a crease pattern via circle packing, we can insert segments into the base by inserting rivers into the circle packing. Breaking such a pattern down into molecules means that some of the molecules must contain rivers. The molecules we have seen thus far—rabbit-ear, Waterbomb, arrowhead, and gusset—have not contained rivers; thus, there must be additional molecules that apply to circle/river packings.

And there are, but most can be derived from the pure circle-packed molecules. Let's start with the three-circle rabbit-ear molecule and add a single river. The river must enter along one edge and exit along an adjacent edge. With no loss of generality, we can represent the situation as in Figure 10.27.

In Figure 10.27, the triangle is defined by three circles of radius a, b, and c, plus a river of width d. But viewed in isolation,

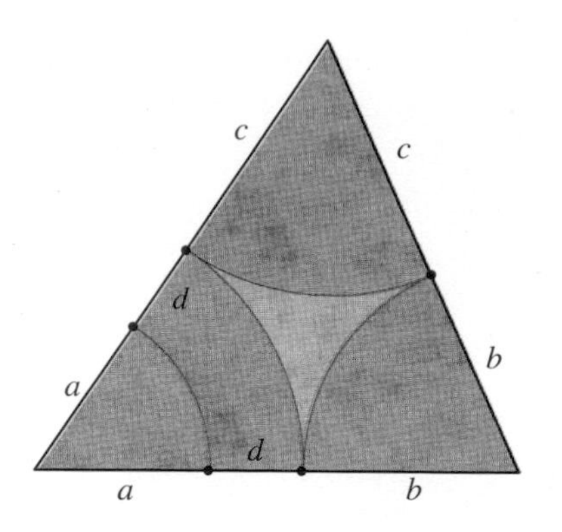

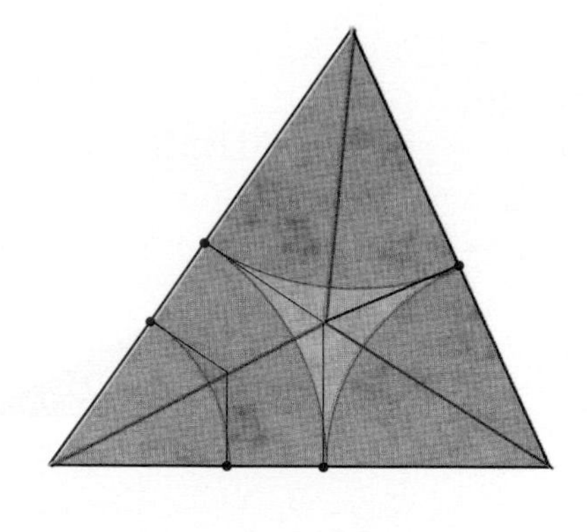
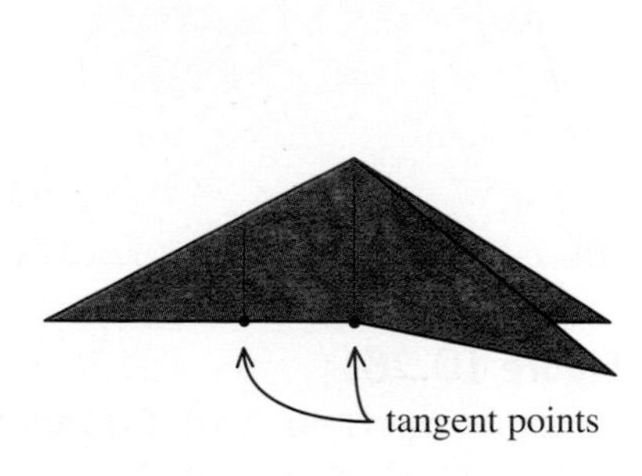

Figure 10.27.
Left: a triangle defined by three circles plus a river.
Middle: crease pattern.
Right: folded molecule with two sets of tangent points.

this is simply equivalent to a three-circle triangle, as the river can (temporarily) be absorbed into one of the circles. The crease pattern is the same as that for the three-circle triangle: the creases of a rabbit ear. The only difference is that because of the boundary between the river and circle, we have an extra set of tangent points along the flap and a set of hinge creases that denote the boundary in the folded molecule.

The situation is much the same in a quadrilateral when the river connects two adjacent edges. Just as in the triangle, the river can be absorbed into the circle it cuts off, and the crease pattern that collapses the quad is exactly the same as the pattern for the pure circle-packed version of the quadrilateral, with the addition of hinge creases to denote the boundary of the river.

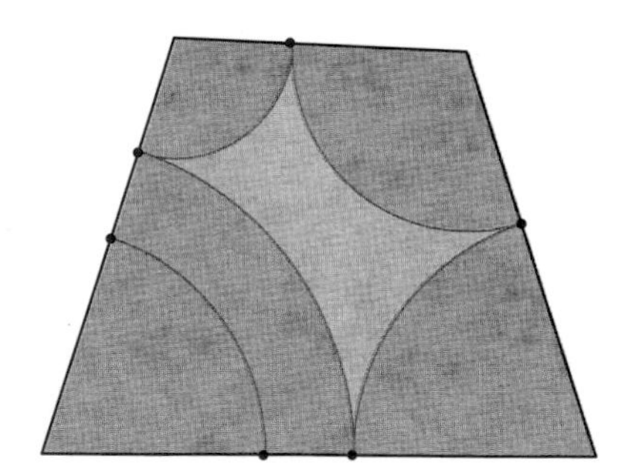
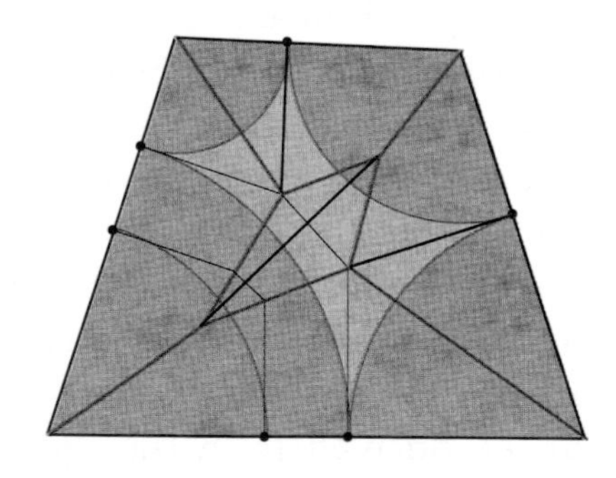
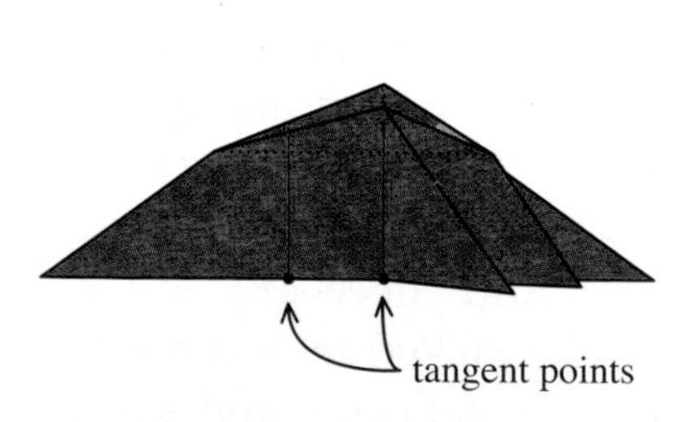

Figure 10.28.
Left: a quadrilateral defined by four circles plus a river between adjacent edges.
Middle: crease pattern for a gusset molecule.
Right: folded form with two sets of tangent points.

I leave it as an exercise for the reader to construct the arrowhead molecule for this quadrilateral.

The situation is entirely new, however, if the river cuts across the quadrilateral, connecting two opposite sides, because now the river cannot be absorbed into a single circle. In fact, a new crease pattern arises.

The simplest pattern, shown in Figure 10.29, occurs when the quadrilateral and its circles satisfy some special conditions.

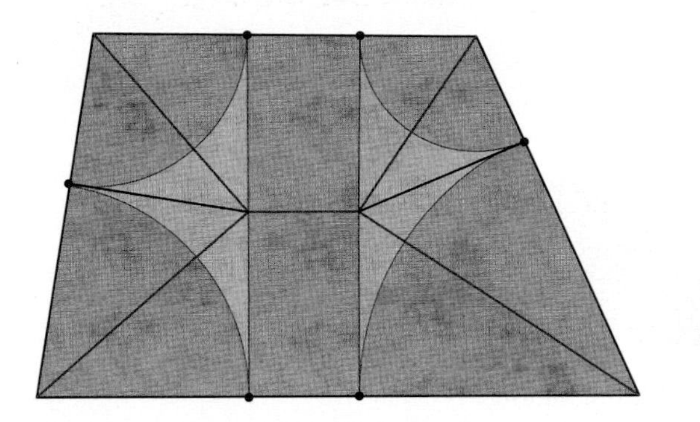

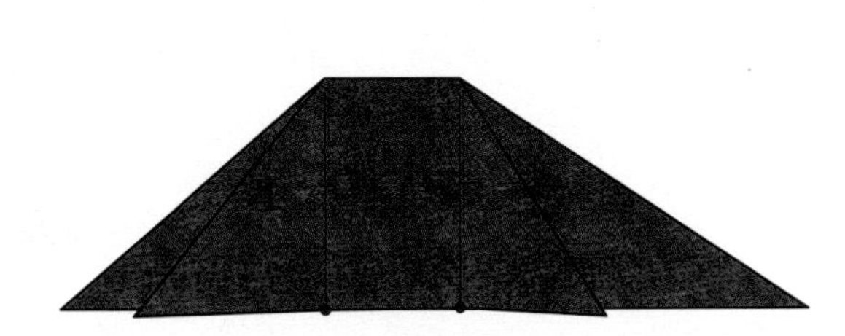

Figure 10.29.
Crease pattern and folded form for the sawhorse molecule.

This pattern, which we will call the *sawhorse molecule*, was described by Meguro and Maekawa; it can be folded from any quadrilateral quite simply, as shown in the sequence in Figure 10.30.

The Waterbomb molecule can be considered a special case of the sawhorse molecule—the limit when the central river goes to zero width.

Recall that even if a quadrilateral satisfied the Waterbomb condition (sums of opposite sides were equal), the Waterbomb molecule wasn't necessarily the molecule that aligns the tangent points. A similar situation occurs with the sawhorse molecule; even though you can fold any quadrilateral into a sawhorse molecule, the particular sawhorse molecule won't necessarily make the tangent points line up. Figure 10.31 shows the sawhorse creases superimposed on a valid circle/river pattern, and it is clear that the hinge creases do not hit the edges at the tangent points of the circles and rivers.

Once again, however, the gusset molecule comes to the rescue; it is possible to construct a version of the gusset molecule that brings all tangent points together. I have not found a simple geometric construction for the creases (a numerical prescription will be given later), but if you precrease the angle bisectors and hinge creases, you can start to collapse the pa-

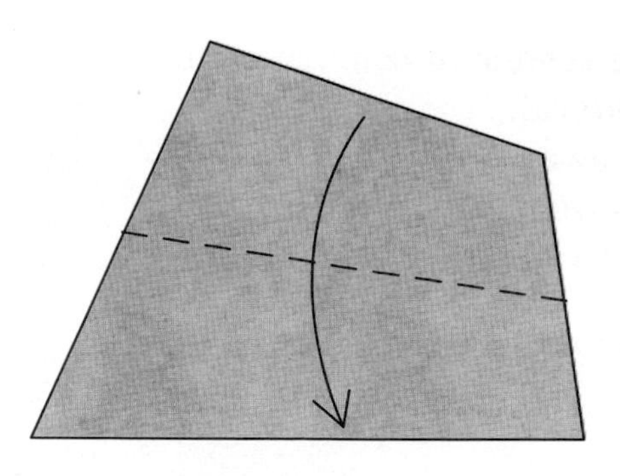

1. Fold one edge down to lie along the opposite edge.

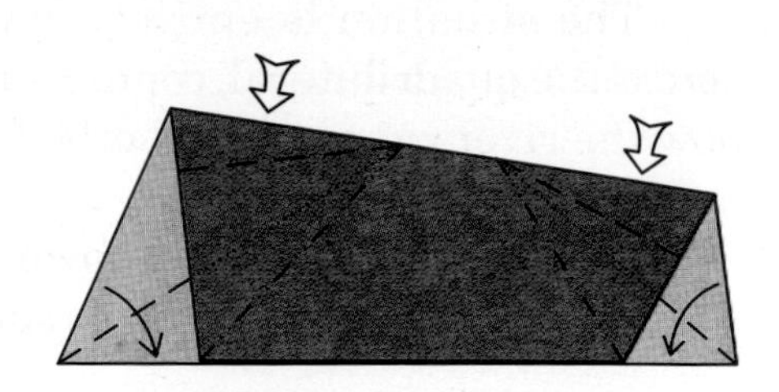

2. Reverse-fold the top corners so that all edges lie along the bottom edge.

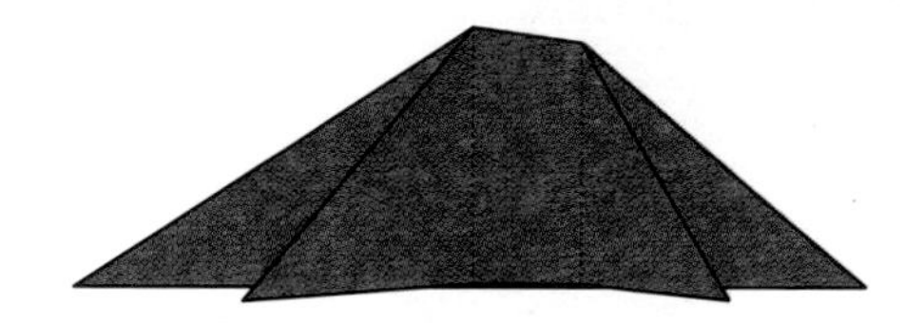

3. Finished sawhorse molecule.

Figure 10.30.
Folding sequence for the sawhorse molecule.

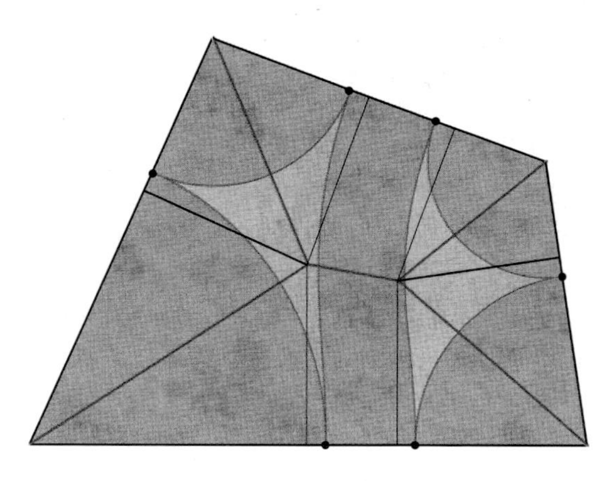

Figure 10.31.
The sawhorse molecule does not work for most circle/river quadrilaterals because the hinge creases miss the tangent points.

per and by forcing the tangent points to line up, the necessary creases will form when you flatten the paper.

These five molecules—rabbit-ear, Waterbomb, arrowhead, sawhorse, and gusset—are sufficient to fill in a flat-foldable crease pattern for any pattern of circles and rivers that define a uniaxial origami base. The combination of these molecules with circle/river packings is called the *circle/river method* of origami design.

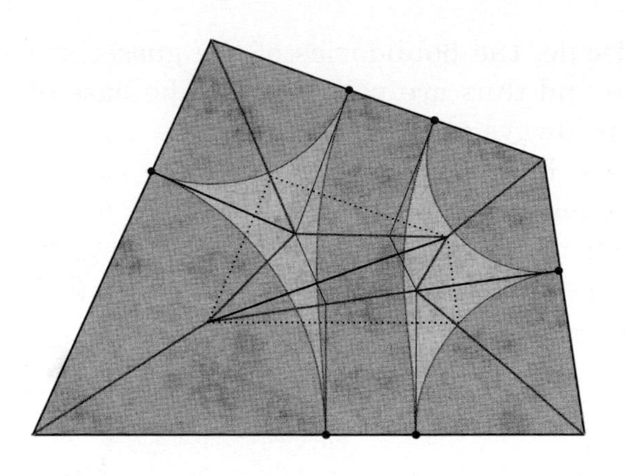

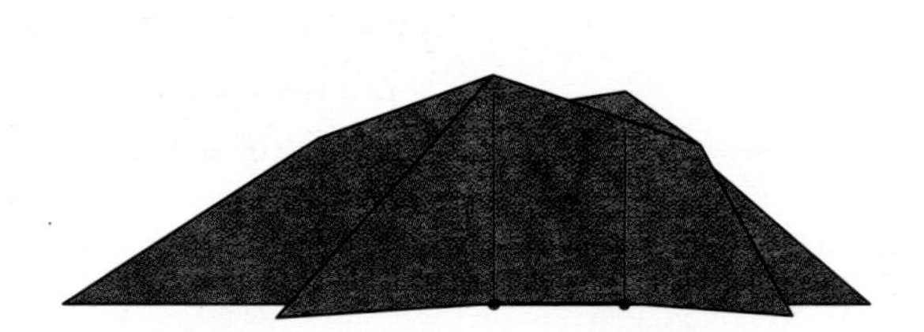

Figure 10.32.
Left: gusset molecule for a circle/river quadrilateral.
Right: the folded form. Note that the hinge creases now hit the tangent points, which are brought together along the bottom edge of the folded form.

10.8. Crease Assignment in Molecules

I have intentionally glossed over the topic of crease assignment within molecules; it is now time to straighten out the issue. When we create a base from circles and rivers, we divide up the paper into distinct polygons using the axial creases; we can then treat each axial polygon individually, filling it in with the appropriate molecular pattern. The choice of molecule is a local choice, depending only upon the pattern of circles and rivers within each axial polygon. However, the assignment of crease parity—whether each crease is a mountain, valley, or unfolded crease—is global; it depends upon the overall structure of the pattern.

Nevertheless, we can specify the parity of many—though not all—of the creases in a pattern at the local level, and it is often useful to make this identification as an incomplete approximation of the final crease pattern.

Examination of the molecular patterns we've seen thus far reveals some rules of thumb for the parity (mountain or valley) of the creases within them. Crease parity depends on one's point of view, of course; the convention I have been using (and will continue to use) is that the paper is two-colored; crease patterns are viewed from the white side of the paper, and the model is folded so that the color ends up on the outside (visible surface) of the model.

Under this convention, within any molecule, the ridgeline creases—those creases that extend inward from the corners—are always valley folds, as an examination of the molecules in the previous section will show.

In a gusset molecule, the boundaries of the gusset are also ridgeline creases and thus are valley folds. The base of the gusset, however, is always a mountain fold.

The hinge creases, however, are variable; they can be mountain, valley, or unfolded creases, depending on the orientation of the flaps of the molecule. Figure 10.33 shows several perfectly valid crease assignments for the hinge creases within a single gusset molecule.

Figure 10.33.
Six possible valid crease assignments for the hinge creases in a gusset molecule.

You might try folding up these four patterns and observing the differences in the folded forms. The choice of crease parity for the hinge creases affects the orientation of the flaps. Since, in a complete crease pattern, most flaps encompass portions of several different molecules, the choice of crease parity within a given molecule cannot be made in isolation, but only after deciding the flap orientation in the full pattern.

When constructing a crease pattern from molecules, it is helpful to assign the creases in two steps. We will first define a generic form for each molecule, in which (a) the ridgeline creases are shown as valley folds; (b) gusset baselines are mountain folds; and (c) all hinge creases are shown as unfolded creases.

The generic forms of all of the molecules we have seen are shown in Figure 10.34.

Figure 10.34.
Generic form of molecules.
Top row: rabbit-ear, Waterbomb, and sawhorse molecules.
Bottom: arrowhead and gusset molecules.

It is a curious fact that the generic form of a molecule isn't actually flat-foldable as is; but I find the generic form as illuminating, if not more so, than the true crease pattern, because the crease assignments of the generic form carry more structural information than the true crease assignment would for a specific flap orientation. It is often more useful to know whether a crease is an axial crease, ridgeline crease, or hinge crease, than to know whether it is specifically mountain or valley.

Molecules do not occur in isolation, of course; they are agglomerated into tiles and entire crease patterns. In such macro-structures, molecules are joined edge-to-edge by axial creases that must also have their parity assigned. Usually—but again, not always—the axial creases are mountain folds.

The exceptions arise when molecules completely surround a vertex in the interior of the paper; then it will be necessary to change one of the axial creases to a valley fold or, equivalently, assign two of the axial creases to be unfolded creases. This follows from a relatively famous formula within origami derived independently by Maekawa and Justin, which states:

$$V - M = \pm 2. \qquad (10\text{–}3)$$

That is, around any flat-folded interior vertex, the difference between the number of mountain folds (M) and valley folds (V) is ±2, with the choice of sign being made based on whether the folded vertex is concave (+2) or convex (–2) from the viewpoint of the observer.

If N molecules come together at an interior vertex, each contributes one ridgeline and one axial crease. The N ridgelines must all be valley folds, which means that of the N axial creases, $(N - 2)$ must be mountain folds and 2 must be unfolded or $N - 1$ are mountain folds and 1 is a valley fold.

It is helpful to define a generic form for entire crease patterns, just as we do for individual molecules, in which all axial creases are shown as mountain folds, whether they connect to interior vertices or not. A generic form crease pattern for a uniaxial base has all axial creases assigned as mountain folds; all ridgeline creases are valley folds, and all hinge creases are unfolded creases. While such a crease pattern is not, as a rule, flat-foldable, all that is needed to make it flat-foldable is to change the assignment of a handful of creases; no new creases are added. I find that in working out a design, the generic form of the crease pattern actually conveys the structural information of the base more clearly than the literal crease pattern would, for in the process of folding, one tends to flip the flaps back and forth and rearrange layers to suit aesthetic purposes. While every such change alters the literal crease pattern, the generic crease pattern for such minor variants of a base remains unchanged. And so, I will often give only the generic form of the crease pattern for molecules and models that follow.

10.9. Putting It All Together

We now have all the building blocks necessary to build a custom-made base from scratch, starting with the desired number, lengths, and connections among flaps. Let's work through such a model in detail. We'll choose an orchid blossom, which offers some interesting challenges but isn't too complicated.

Orchids come in an enormous variety. I'll pick a fairly common form. Figure 10.35 shows a sketch of an orchid blossom. Orchids typically have six petals plus a stem, but in many species, one of the petals is heavily modified. In the variety I've chosen, the bottom petal grows two distinct protrusions partway out the sides of the petal, and we'll include these as distinct flaps in the desired base. We can represent the desired configuration of flaps by a stick figure, as shown on the right.

Figure 10.35.
Left: an Orchid.
Right: its representation as a stick figure.

Now, let's count the flaps. There are five similarly-sized flaps for the ordinary petals, a sixth flap for the stem, then three smaller flaps that make up the composite petal. Those flaps are separated from the others by a short segment. Thus, our crease pattern will be made up of six large circles, three smaller circles, and a relatively narrow river, as shown in Figure 10.36.

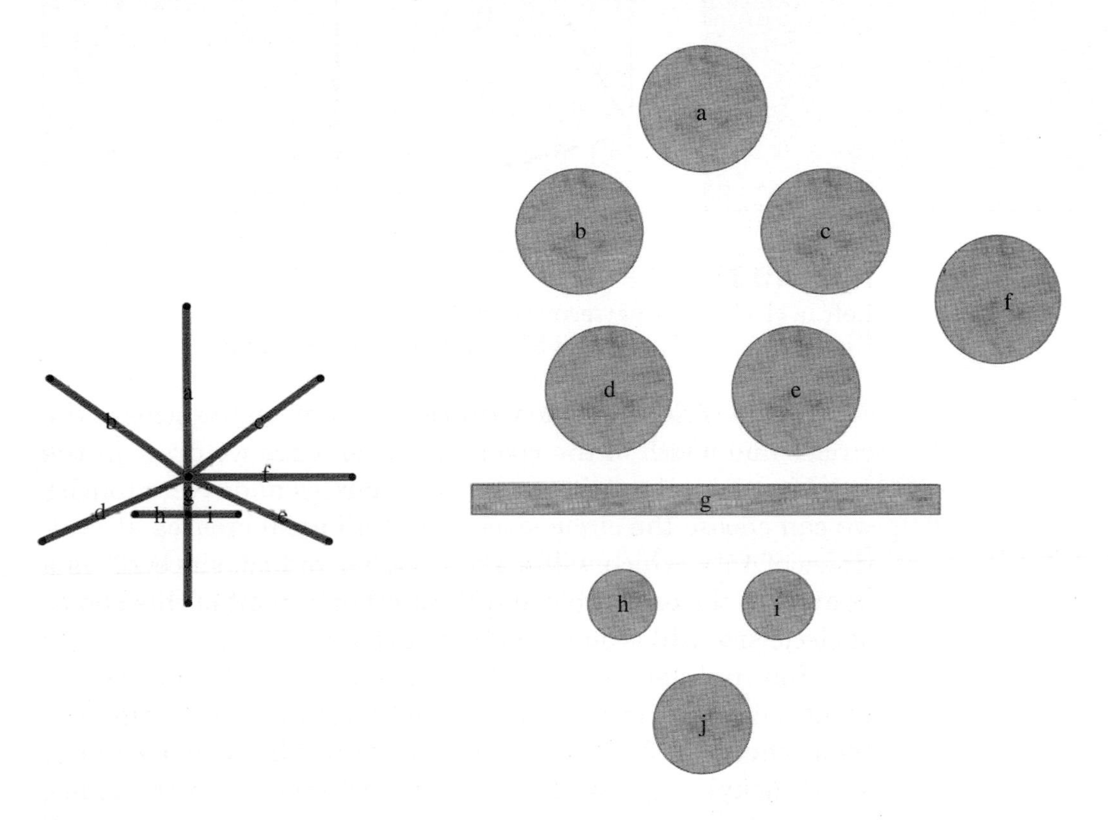

Figure 10.36.
The circles and river that correspond to the elements of the stick figure of the Orchid.

Now comes the fun part: How can we pack these items into a square in an efficient way? Recall that (a) the centers of all circles must be confined to the square and (b) circles and rivers must be connected in the same way that they are connected in the stick figure. If you like something concrete, you can cut out circles and slide them around within the circle jig shown previously; I usually just draw sketches. A bit of manipulation reveals an elegantly symmetric arrangement of circles and rivers, shown in Figure 10.37.

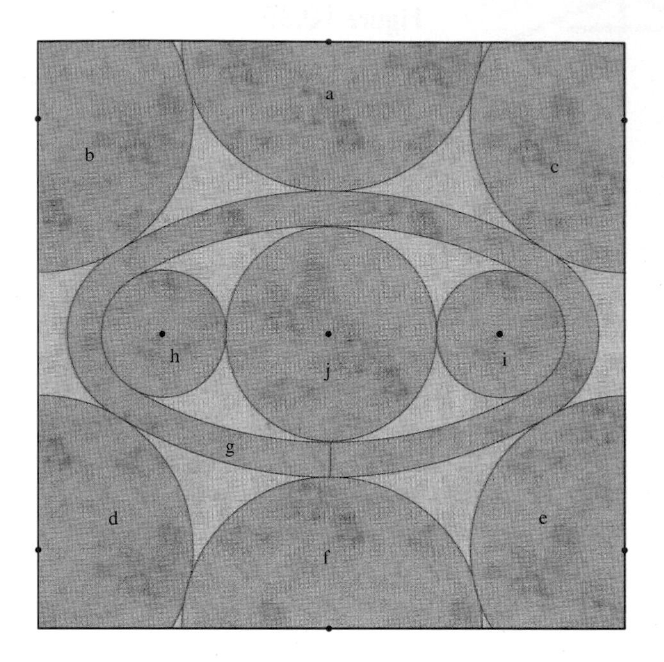

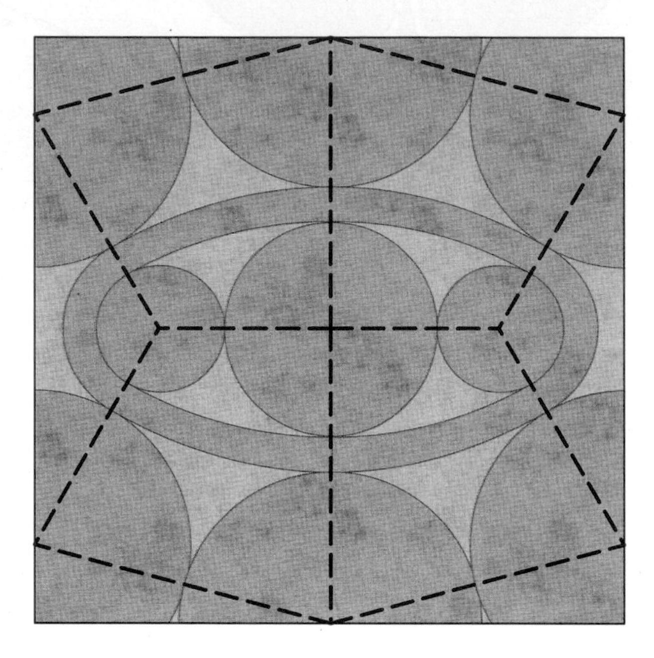

Figure 10.37.
Left: a circle/river pattern for the Orchid.
Right: the circle pattern with axial creases highlighted.

There is clearly some variation possible in the sizes of the circles and width of the river. However, once we draw in the axial creases (along lines where the circles and rivers touch), we can choose the circle sizes to put all axial creases at multiples of 15°—which will make it easier to fold, since 15° is a quarter of the easily folded 60°. Another benefit of this choice of circle size will soon become apparent.

But first, let's take stock of what we have. There are four identical quadrilaterals that are circle-plus-crossing-river type. These can be filled by either the sawhorse molecule (if we're lucky) or gusset molecule (if we're not). On the sides, we have two triangles of the circle-plus-river type; we can fill these in with rabbit-ear molecules. The rest of the paper is taken up by four triangles at the four corners of the square; since these triangles only contribute to two flaps each, they are essentially unused, and we can fold them underneath and ignore them (or pull them out later in the model if a new use arises).

With regard to the quadrilaterals, the choice of a 15° geometry was lucky (or inspired) because it allows us to use the much-simpler sawhorse molecule in the crease pattern. Filling in all six molecules with the generic form of their creases gives the pattern shown in Figure 10.38.

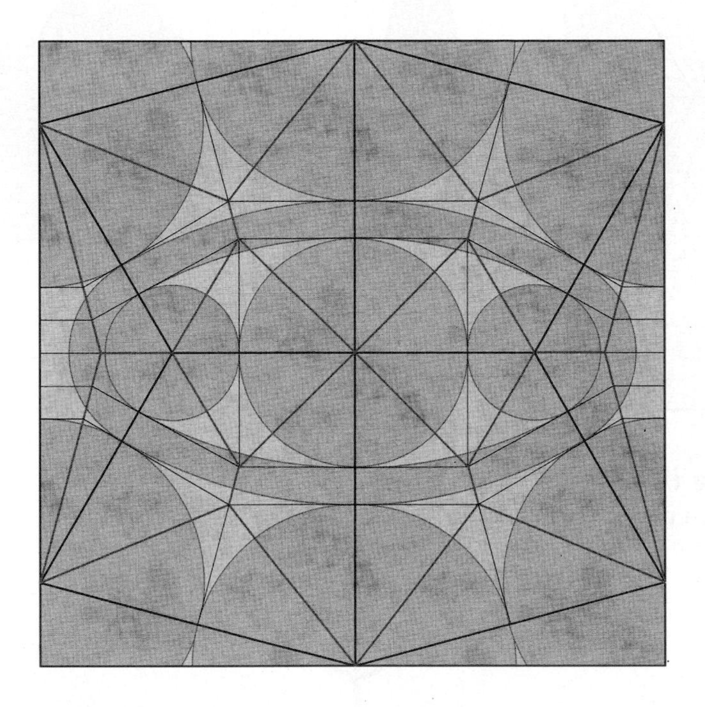

Figure 10.38.
Generic form of the filled-in crease pattern.

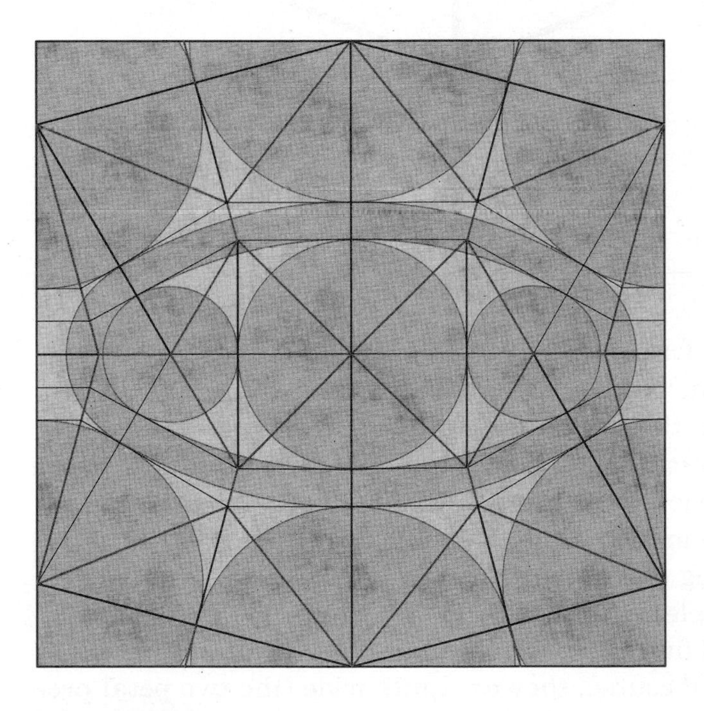

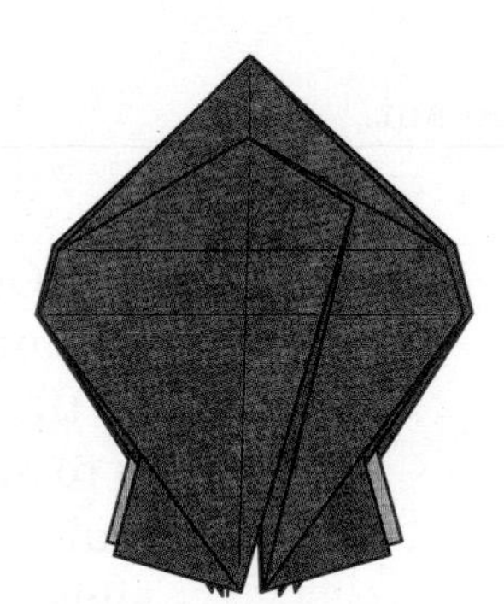

Figure 10.39.
Crease pattern and folded base.

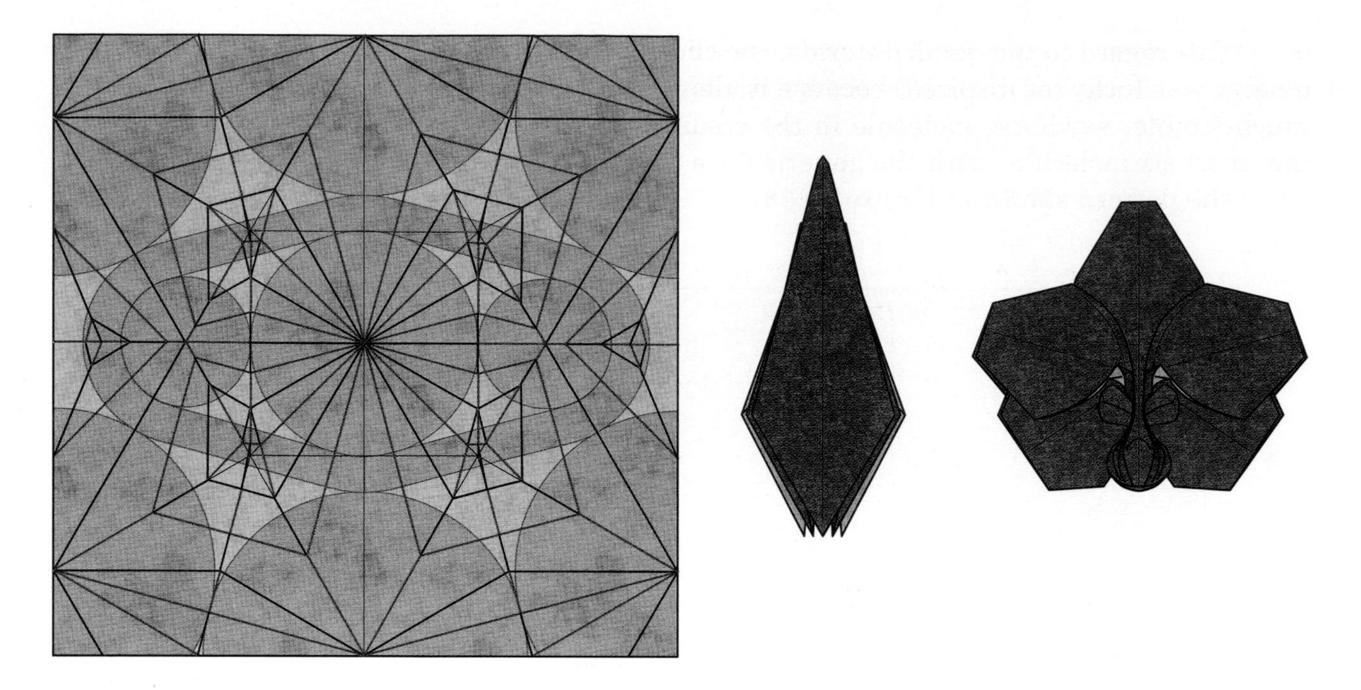

Figure 10.40.
Crease pattern, base, and folded model of the Orchid.

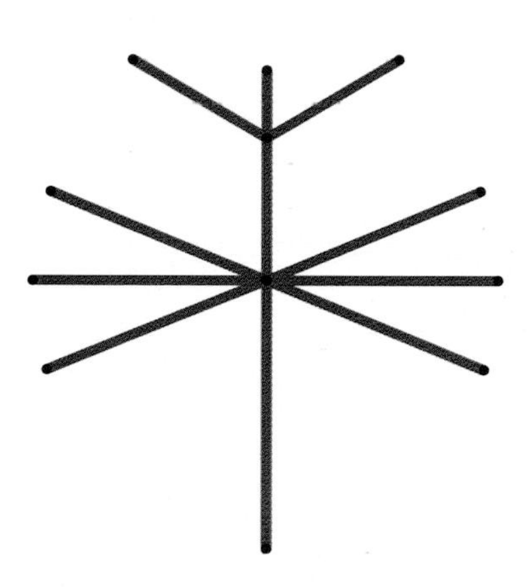

Figure 10.41.
A stick figure of an ant.

Generic-form crease patterns are not flat-foldable, but they contain all the creases necessary to make a flat-foldable base. At this point, the easiest way to finish the crease assignment is to actually cut out, precrease, and fold the generic pattern, making decisions about flap orientation as you fold. One possible arrangement of flaps is shown in Figure 10.39 with the corresponding crease pattern with proper crease assignment and the completed base.

You will find that this base contains all of the flaps we set out to fold. Of course, they are quite wide (the two petal protrusions are easy to overlook) but conventional narrowing techniques (e.g., multiple sinks) can turn them all into distinct flaps. Once the flaps are in place, the base can be turned into the desired orchid subject in many ways; my own version is shown in Figure 10.40. Folding instructions are given at the end of the chapter.

Let's do another. This time, we'll do another insect. A fairly simple ant has six legs, head and abdomen, with antennae attached to the head. A simple stick figure of an ant with all these features is shown in Figure 10.41.

From the stick figure, we can see that we'll need six circles for legs, another circle for the abdomen, a river for the connection between the legs and head, two smaller circles for the antennae, and an even smaller circle for the rest of the head.

Again, there are many possible configurations, depending on the specific sizes of the circles and rivers and their relative arrangements, but a fairly straightforward configuration is shown in Figure 10.42. Connecting the centers of touching circles with axial creases defines the polygons that we will fill in with molecules.

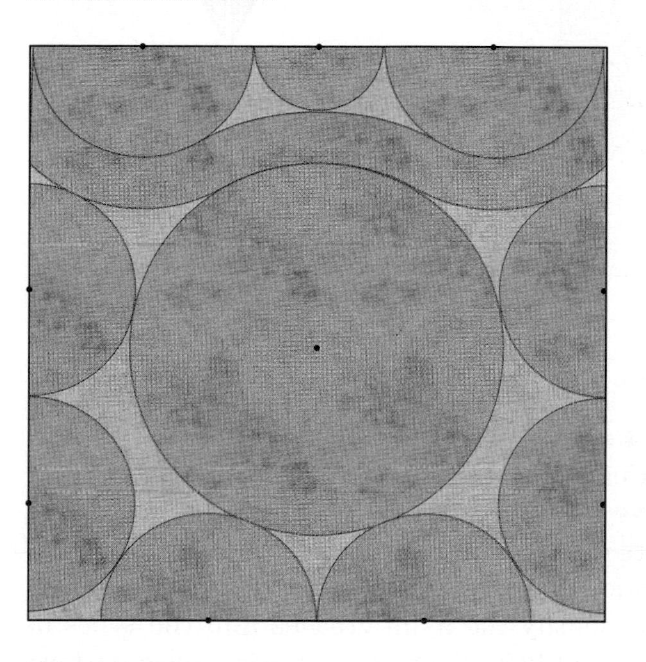

Figure 10.42.
Left: circle-river pattern for an ant.
Right: pattern with axial creases highlighted.

This pattern gives four triangles and two quadrilaterals. This time all four of the corners of the square go unused (a not uncommon occurrence with circle-packed designs). In the triangles, we have no choice: They receive rabbit-ear molecules. In the quadrilaterals, this time they don't satisfy the conditions for the Waterbomb molecule, so we can use either the arrowhead or gusset molecule.

There is no particular symmetry that would favor the gusset molecule, and the arrowhead molecule allows us to shift some extra paper toward the flap that eventually becomes the abdomen, so I chose the arrowhead molecule in my own design. (You might wish to try both yourself and see which you prefer). The generic-form crease pattern, resulting base, and a model folded from this base, are shown in Figure 10.43.

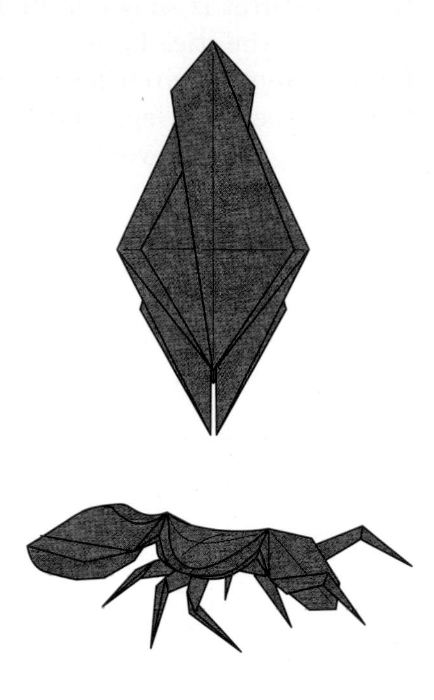

Figure 10.43.
Generic-form crease pattern, base, and folded model of the Ant.

Figure 10.44 shows one more insect design and a small challenge. This Cockroach, like the Ant, contains six legs and antennae, but I've added two more rivers (which create gaps between the pairs of legs) and varied the leg length. Can you reconstruct the stick figure from which this design is derived? Second, can you identify the axial creases and the types of molecules I used? And last, given the generic crease pattern here, can you figure out the crease assignment and fold the base? (If not, references with folding instructions for both this model and the Ant are given in the References.)

10.10. Higher-Order Polygons

We now have molecules for triangles, which are common, and quadrilaterals, which are occasional. What about higher-order polygons? Might we ever see a pentagon, hexagon, heptagon, or larger?

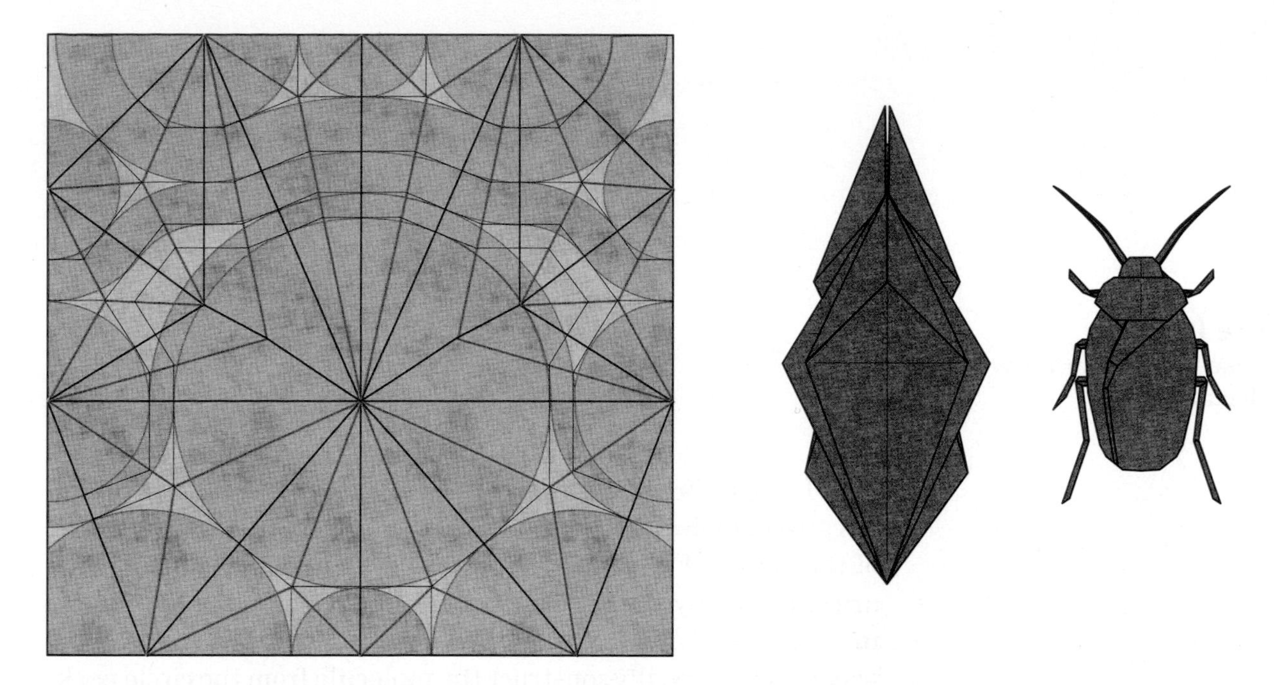

Figure 10.44.
Generic-form crease pattern, base, and folded model of the Cockroach.

Yes indeed; in fact, we have already seen one such example. If we choose to design a base with five equal flaps and require that all flaps come from the edge of the paper, we will arrive at the circle packing shown in Figure 10.45. Connecting the centers of touching circles with axial creases yields a single primary polygon with five sides. And had we desired six edge flaps, we would have ended up with a six-sided polygon. So we do indeed need to worry about molecules for higher-order polygons.

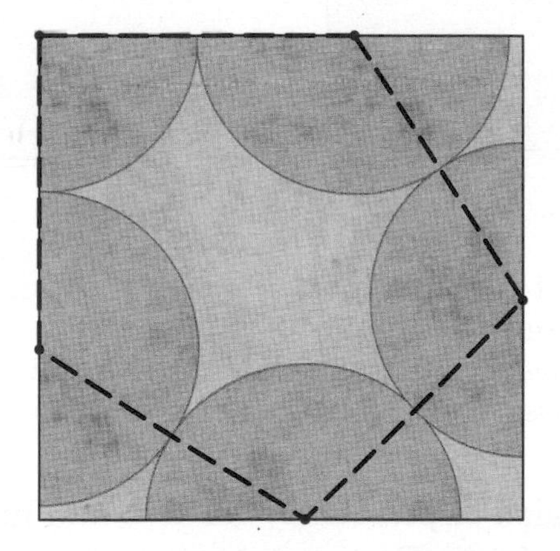

Figure 10.45.
A five-circle base with all five circles on the edge of the square yields an axial polygon with five sides.

Figure 10.46 shows a generic-form crease pattern for this five-circle polygon, which yields a five-flap molecule. It is very similar to the quadrilateral gusset molecule, which suggests that, perhaps, there is a pentagonal gusset molecule as well.

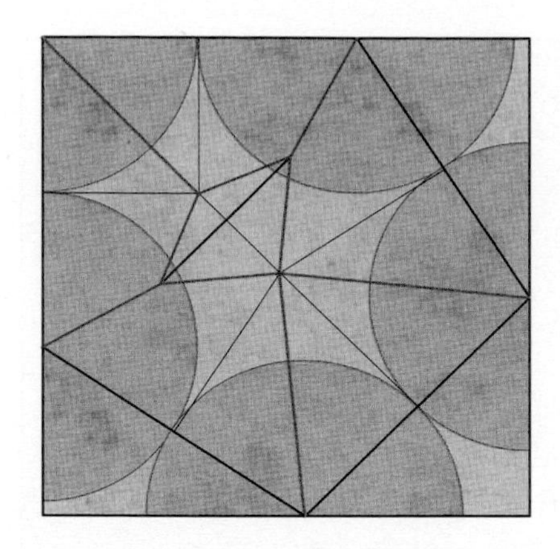

Figure 10.46.
Crease pattern for a five-circle molecule.

In fact, for five identical circles, there are many pentagonal gusset molecules, which depend on the specific arrangement of the circles. Figure 10.47 shows three such molecules, which are obtained by slight perturbations to a packing of five circles into a square. Unfortunately, there does not appear to be a simple way to geometrically construct the molecule from the circle packing; in fact, it isn't even clear where the gussets go to allow the pentagon to collapse with its edges on a line and the tangent points aligned. (The patterns in the figure were computed numerically—we will see how to do this later on.)

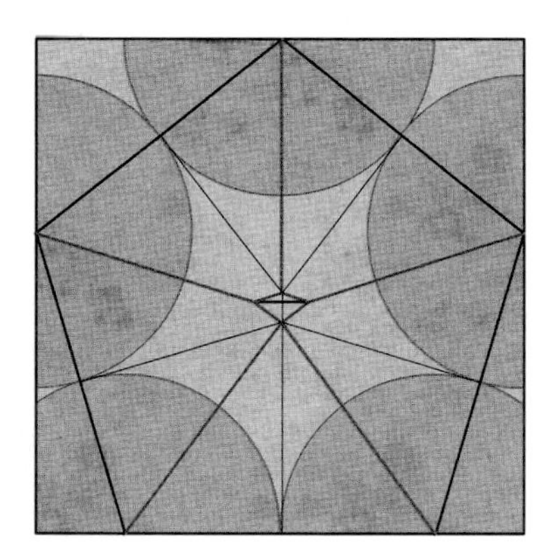

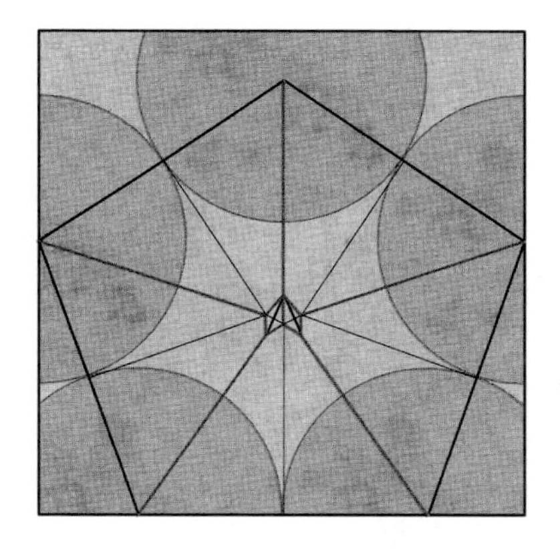

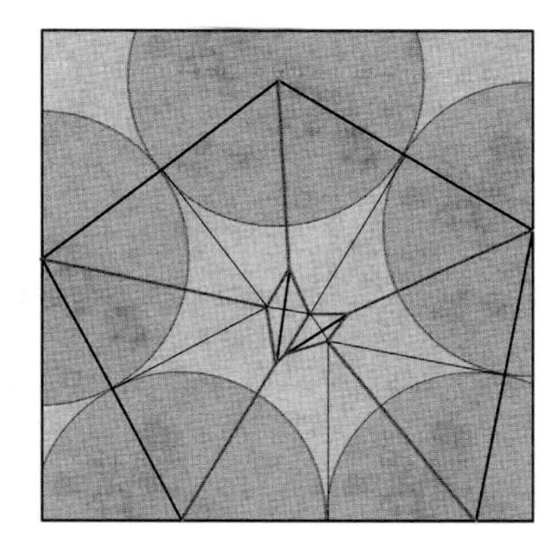

Figure 10.47.
Three pentagonal circle packings and crease patterns that collapse them. Slight changes in the arrangement of the circles can make significant changes in the arrangement of the creases and gussets. Note how the gussets vary among the three patterns.

The explosion of possibility for five and higher numbers of sides is worrisome. Fortunately, it isn't necessary to enumerate all unique molecules for higher-order polygons; there is a way to transform any higher-order polygon into a combination of triangle and quadrilateral molecules—what we called composite molecules. The basic idea is very simple. The paper that lies between the circles is, in a sense, unused. We can make use of it by adding a new circle of our own, as shown in Figure 10.48. Think of the existing circles as rigid disks; we add a small circle, then inflate it until it hits its neighbors. Once the circle contacts three others, it creates three new axial creases, which break down the higher-order polygon into several lower-order polygons.

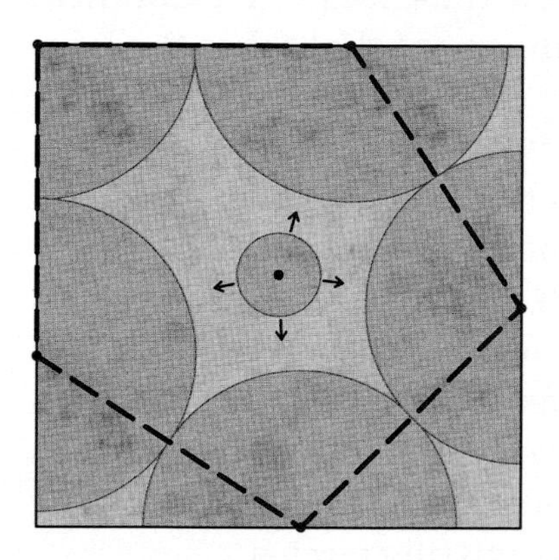

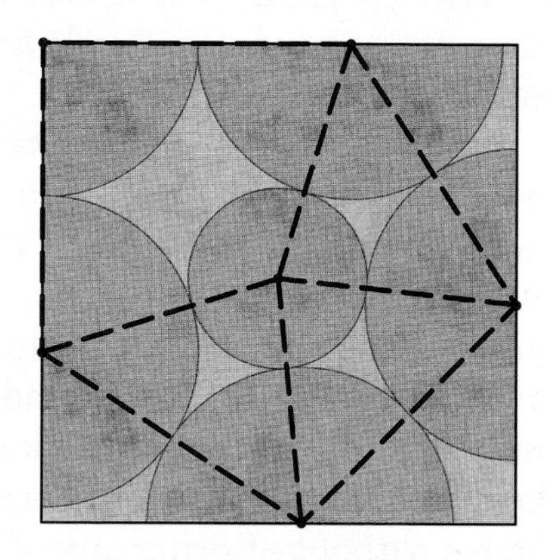

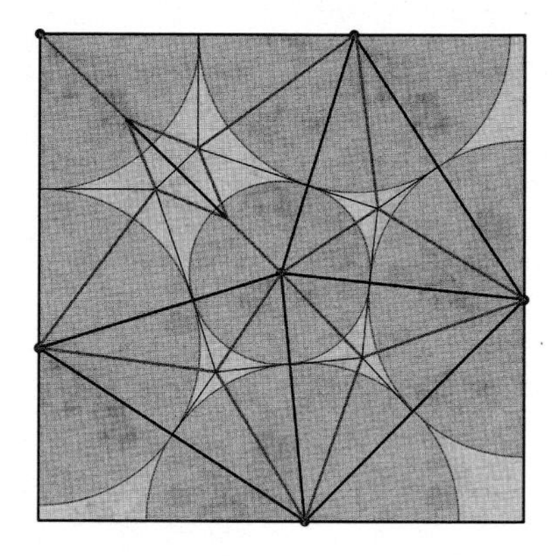

Figure 10.48.
Left: add a circle and expand it until it hits its neighbors. Center: when the circle touches its neighbors, add axial creases between touching circles.
Right: fill in the resulting triangles and quadrilaterals with rabbit-ear and gusset molecules.

Because a new circle has three degrees of freedom—the two coordinates of its center and its radius—you can always expand a circle until it hits at least three of its neighbors. (In Figure 10.48, because of the symmetry, we can actually get the new circle to touch four neighboring circles). When two circles touch, we add new axial creases. In the example shown, this has the result of dividing the pentagon into three triangles and a quadrilateral, all of which we know how to fill with molecules.

This technique always works and can be repeated over and over. Suppose we have a polygon with N sides. A circle added in the middle can always be expanded until it touches at least three others. If the three touched circles are consecutive, you will create two triangles and another N-gon, which is no help. But there is always more than one way to add another circle, and if the three touching circles are not consecutive, then the largest polygon remaining will have at most $N-1$ sides, thereby simplifying the problem. Repeatedly applying this process to every polygon of order five or larger will result in a pattern of axial creases consisting entirely of triangles and quadrilaterals, which can be filled in with rabbit-ear, arrowhead, gusset, and (where appropriate) Waterbomb molecules.

An interesting unsolved problem in circle-packed origami design is to prove that for any N-gon of touching circles with $N > 4$, it is always possible to add a circle touching at least three others so that the largest resulting polygon has, at most, $N-1$ sides. It is possible to find arrangements where the addition of a circle leaves an N-gon, but in all the cases I've examined, there has been another circle arrangement that takes the largest polygon down a notch.

It is tempting to think that we could keep applying the process to quadrilaterals and thereby reduce every uniaxial base to a collection of rabbit-ear molecules, but quadrilaterals turn out to be special. If you add a circle to the center of a quadrilateral that touches three of the four circles, you will end up with two triangles and another quadrilateral. So, it's not possible, in general, to take a circle packing crease pattern down to consist entirely of rabbit-ear molecules by adding circles without altering any of the existing circles.

Thus, in the circle packing in Figure 10.49—which corresponds to a diagonally symmetric base with thirteen equal flaps and two slightly longer flaps at the sides—the axial creases outline triangles and four pentagons. By adding another circle (meaning another flap) to each pentagon, each can be broken down into two quadrilaterals and a triangle.

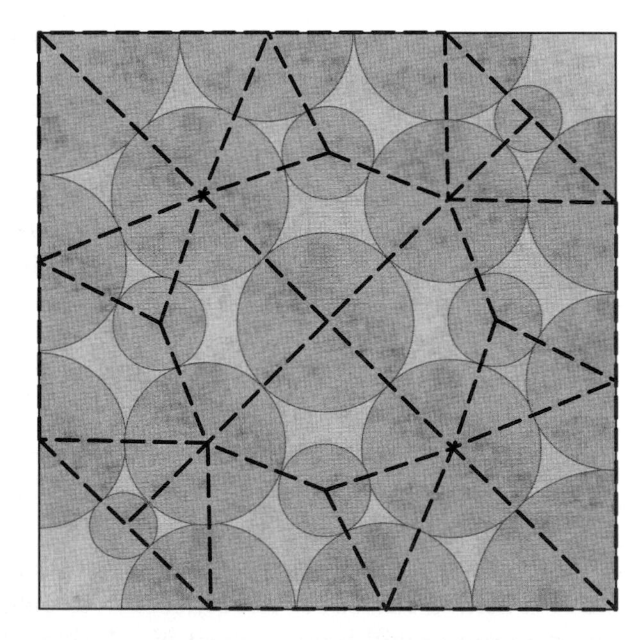

Figure 10.49.
Left: circle packing for a fifteen-flap base and axial creases.
Right: adding more circles breaks the pentagons into quads and triangles.

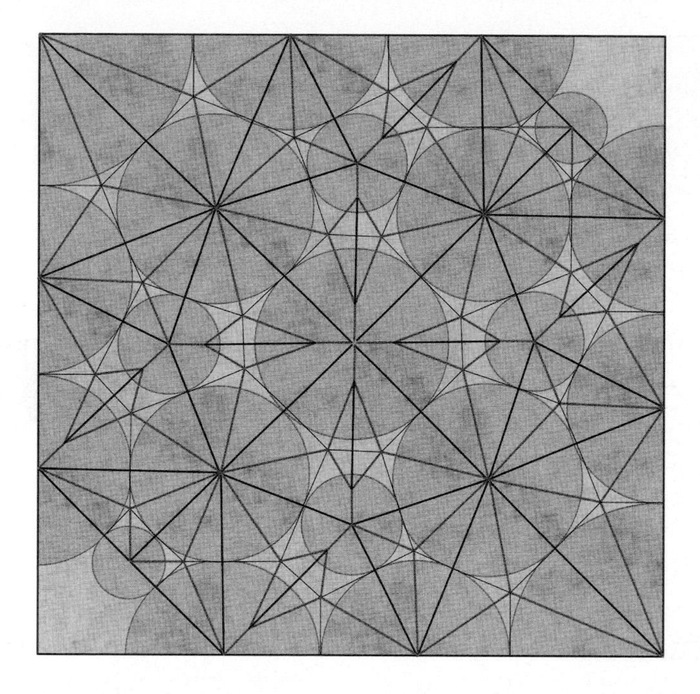

Figure 10.50.
Generic-form crease pattern, filled in with molecules.

Now all polygons can be filled in with molecular creases, giving the generic form crease pattern shown in Figure 10.50.

I shall leave it as another challenge to you to fold this pattern into a base. It is not easy, given all of the middle flaps, but you can easily derive the proportions by folding alone; many of the key lines propagate at multiples of 22.5°.

You might wonder, what would one ever make from a fifteen-flap base? I have used this circle packing for a flying *Eupatorus gracilicornus* (a horned beetle), although instead of breaking up the axial polygons in this way, I used a pentagonal analog of the arrowhead molecule. You might enjoy comparing the crease pattern in Figure 10.51 with the one in Figure 10.50 and attempting to fold a model from both.

The circle/river method of designing origami is extremely powerful. By packing circles and rivers into a square, you are guaranteed all the flaps you need; by using molecular crease patterns to fill in the axial polygons created by your packing, you are guaranteed a flat-foldable base. Using these and similar techniques, origami artists have created designs of unbelievable complexity. These techniques are at their best when the subject has many long, skinny appendages; insects, spiders, and other arthropods are prime candidates. The 1990s saw the flowering of these techniques in both the West and Japan, and launched an informal trans-Pacific competition known as the Bug Wars, in which at every origami exhibition, the chief architects of these techniques showed off their latest and greatest winged, horned, antennaed, and sometimes spot-

Figure 10.51.
Crease pattern, base, and folded model of the *Eupatorus gracilicornus*.

ted and striped creations. It was an entomologist's delight (and an arachnophobe's nightmare), and the contest is still going on with new revelations every year.

In circle/river-method designs, the packing of the circles and rivers into the square is still a bit *ad hoc*; the designer must shuffle circles on paper (or actually manipulate cardboard circles) to find an efficient arrangement; but there is no particular prescription for finding an efficient arrangement, let alone the most efficient arrangement. Circles and rivers are a wonderful tool for visualizing paper usage, but they can also be a distraction from some of the underlying principles. By reintroducing a concept we have already seen—the stick figure or *tree*—and building connections between properties of the tree and the crease pattern directly, in the next chapter we will be able to construct rigorous mathematical tools that allow the numerical solution of both locally and globally efficient crease patterns.

Folding Instructions

Orchid Blossom

Orchid Blossom

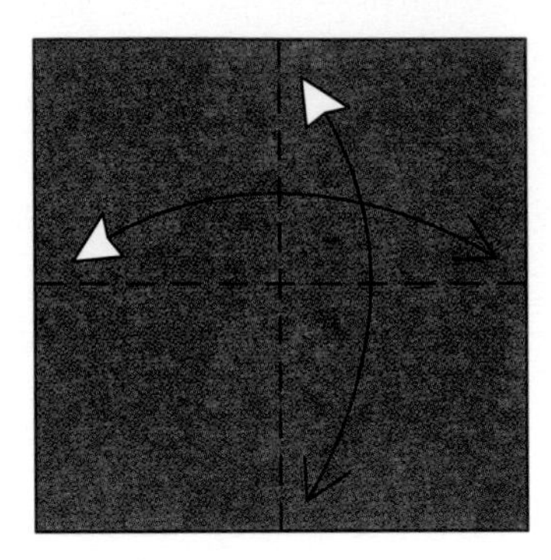

1. Begin with a square, colored side up. Fold and unfold in half vertically and horizontally.

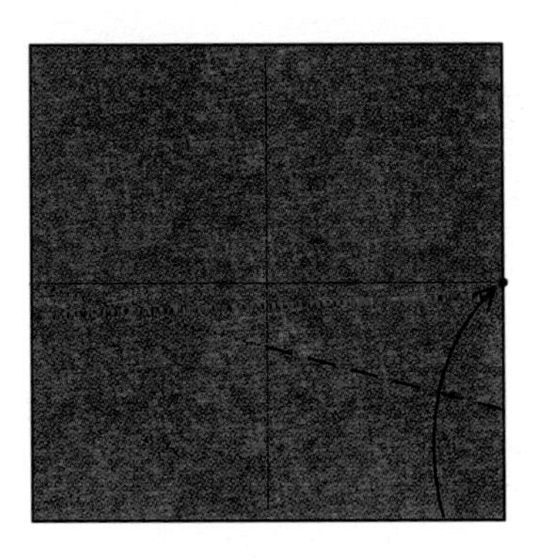

2. Fold the bottom edge up to touch the midpoint of the right edge; the crease hits the midpoint of the left edge, but don't make it sharp in the left half of the model.

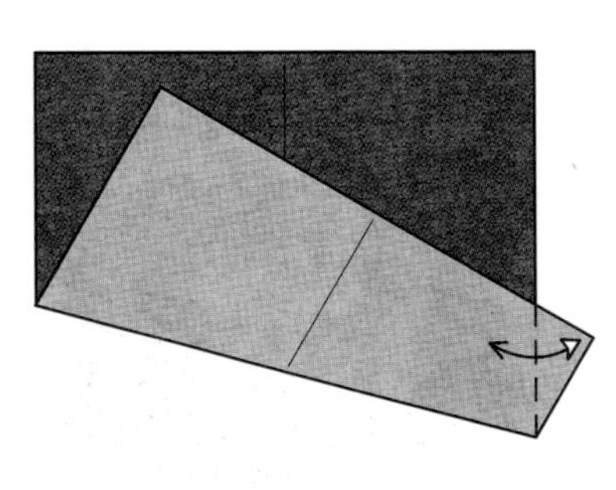

3. Fold and unfold.

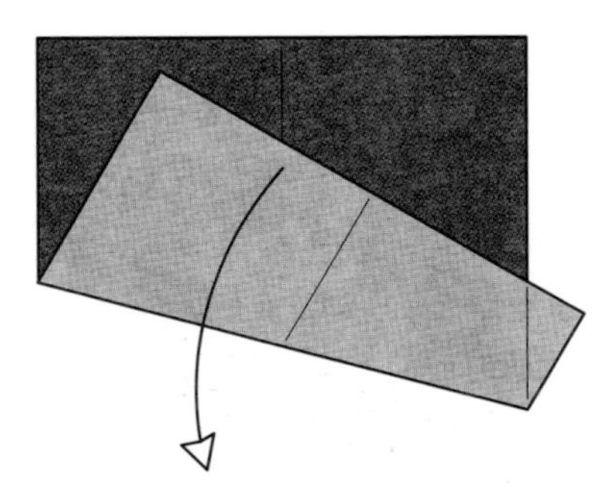

4. Unfold.

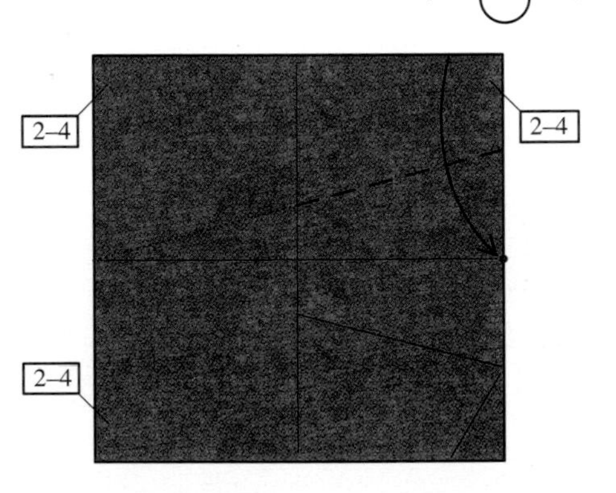

5. Repeat steps 2–4 on the other three corners. Turn the paper over.

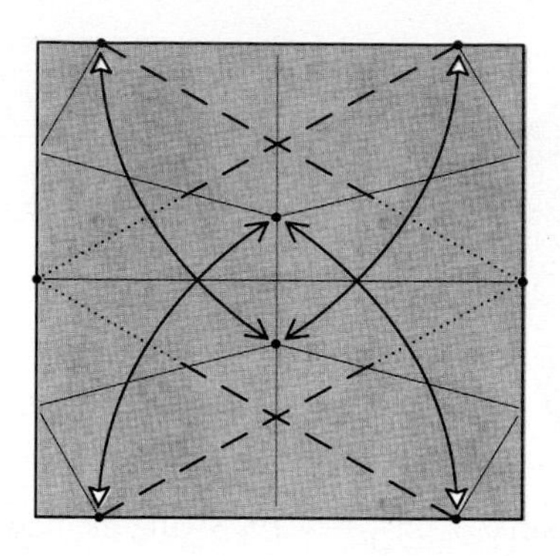

6. Fold each corner inward along a diagonal crease that connects two crease/edge intersections; make each crease sharp only where shown.

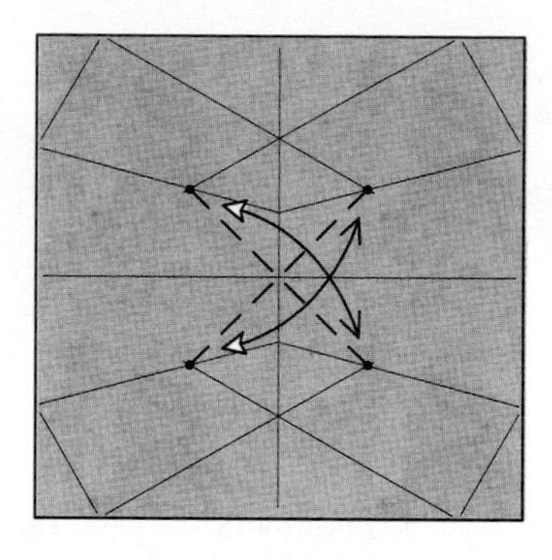

7. Make two more creases that connect pairs of crease intersections.

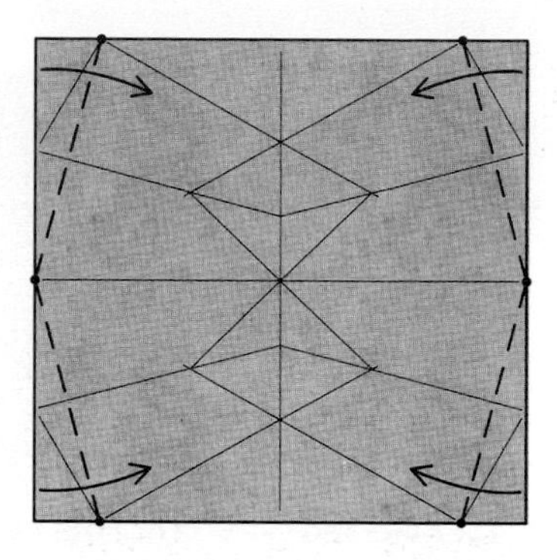

8. Fold four corners inward.

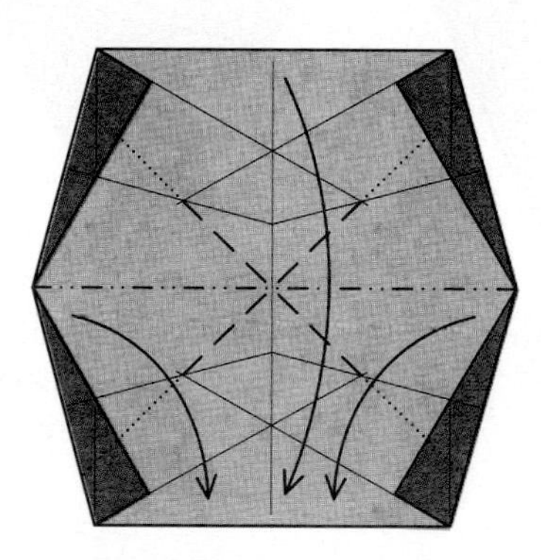

9. Fold a Waterbomb Base, but only make the creases sharp in the middle of the paper.

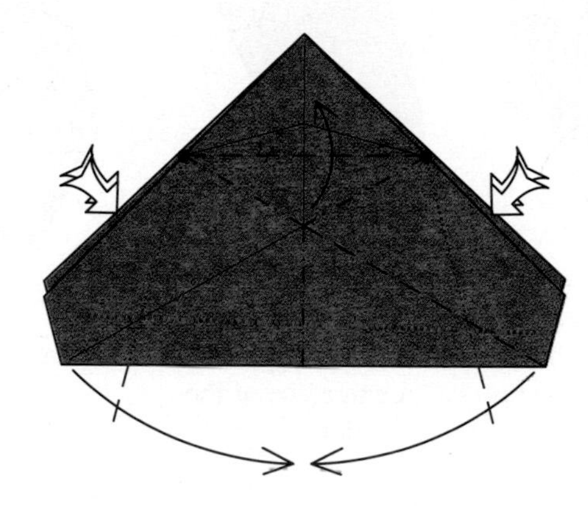

10. Squeeze the sides inward, gather the excess paper in the middle and swing it over to the right. Repeat behind.

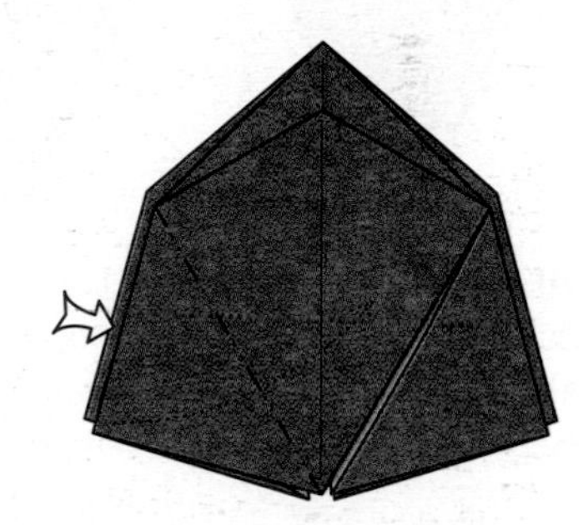

11. Reverse-fold the corner inside.

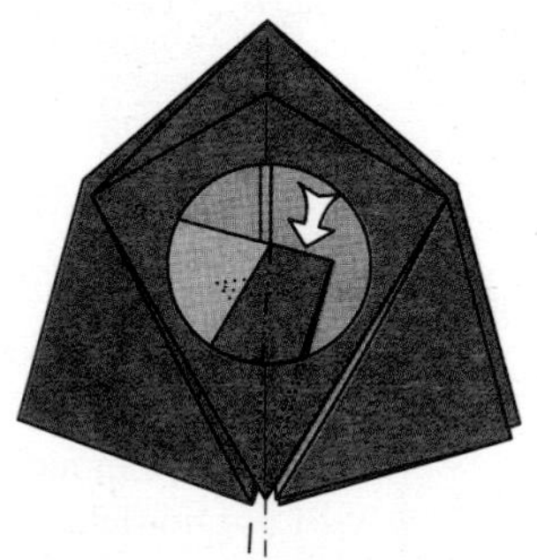

12. Reverse-fold the inside edge along the center line.

13. Swing one flap over to the left.

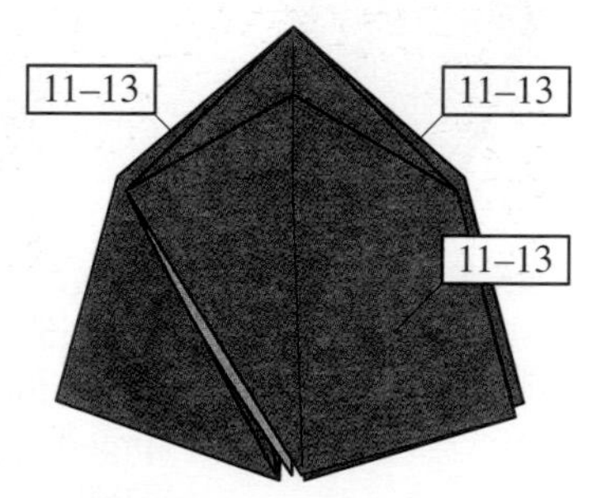

14. Repeat steps 11–13 on the right and on both sides behind.

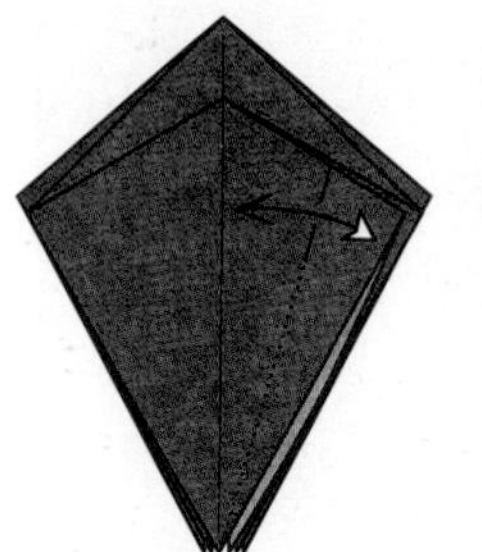

15. Fold and unfold.

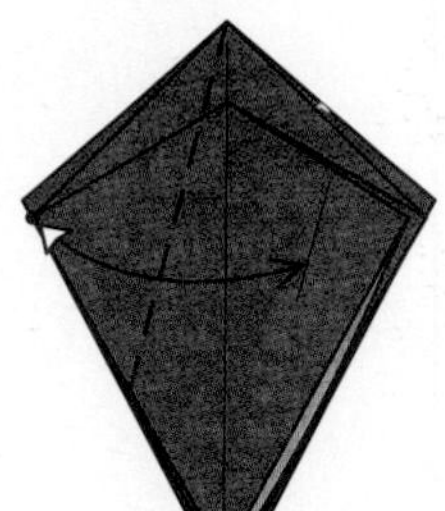

16. Fold the corner to the crease you just made and unfold.

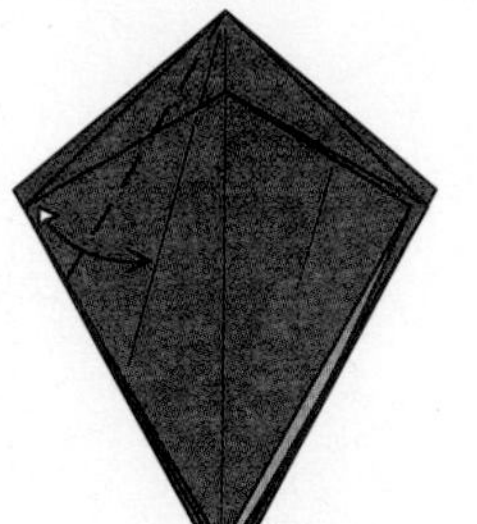

17. Fold and unfold.

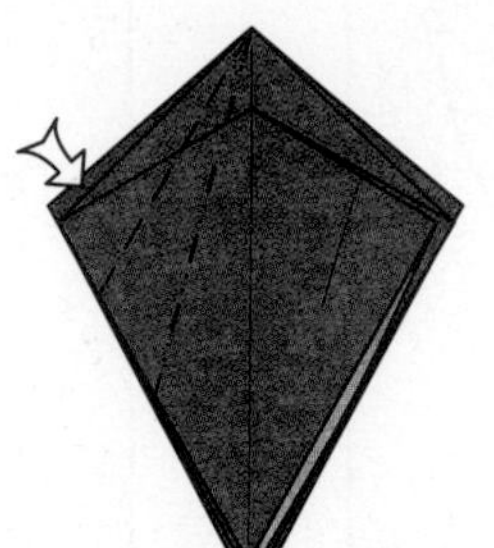

18. Open-sink in and out.

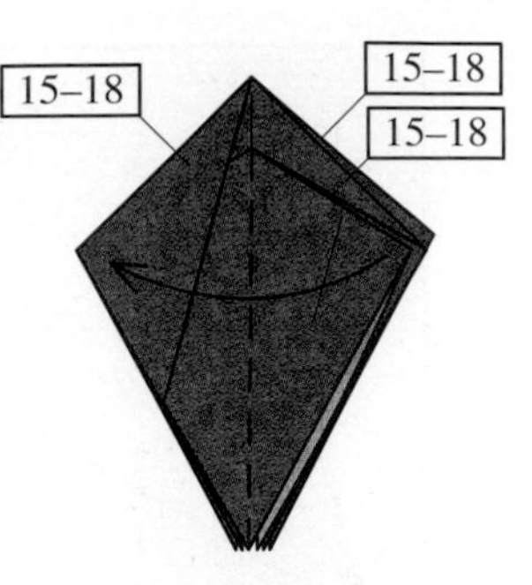

19. Swing the flap to the side and repeat steps 15–18 on the right. Repeat on both sides behind.

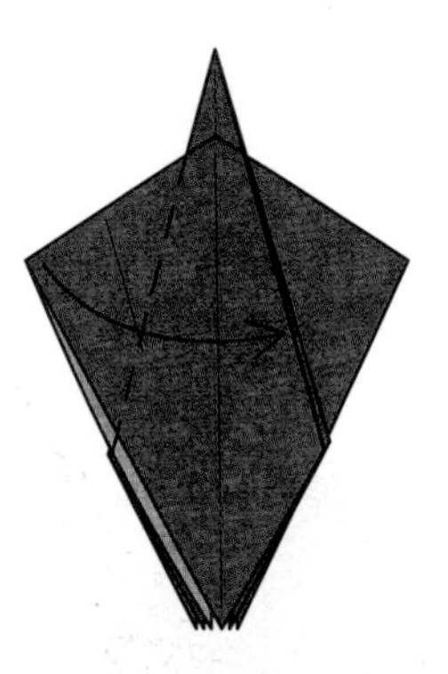

20. Fold the corner down along a crease aligned with the edges behind.

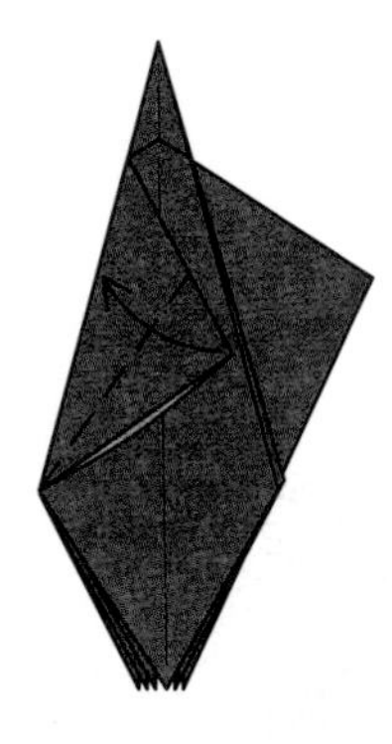

21. Fold the corner back up so that the raw edges line up with the crease you just made.

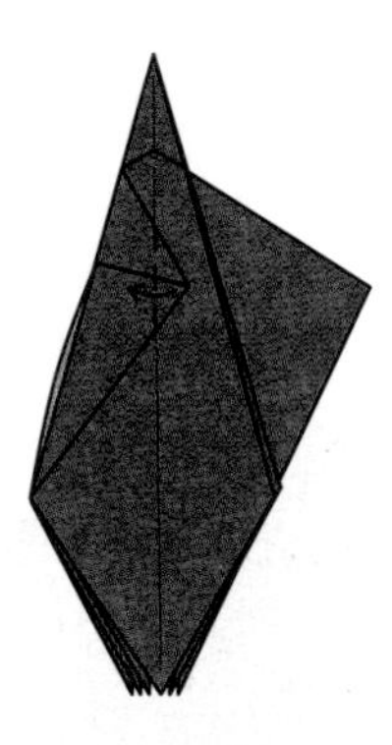

22. Fold the corner over along the center line of the model.

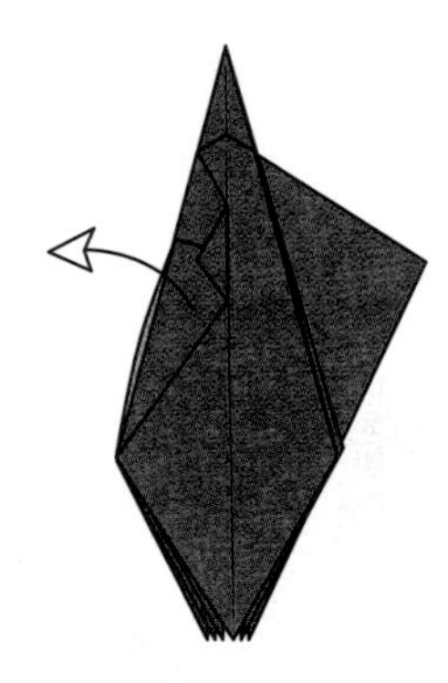

23. Unfold to step 20.

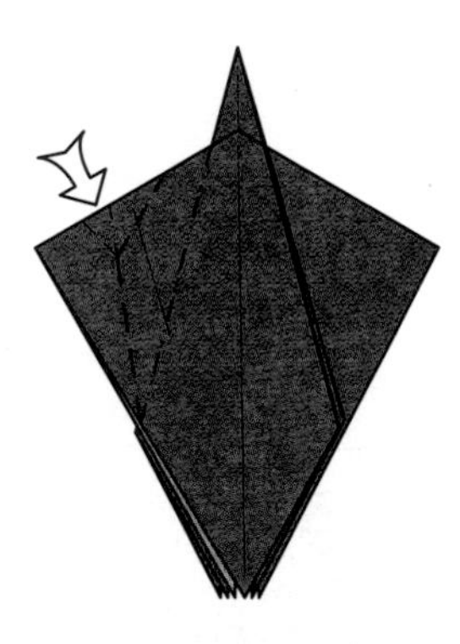

24. Reverse-fold the corner in and out on the existing creases.

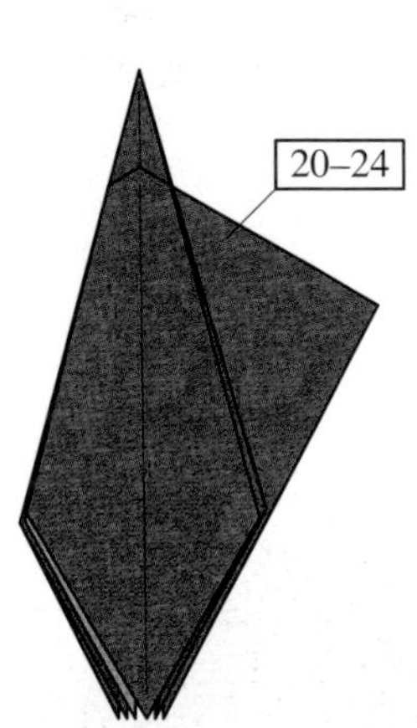

25. Repeat steps 20–24 behind.

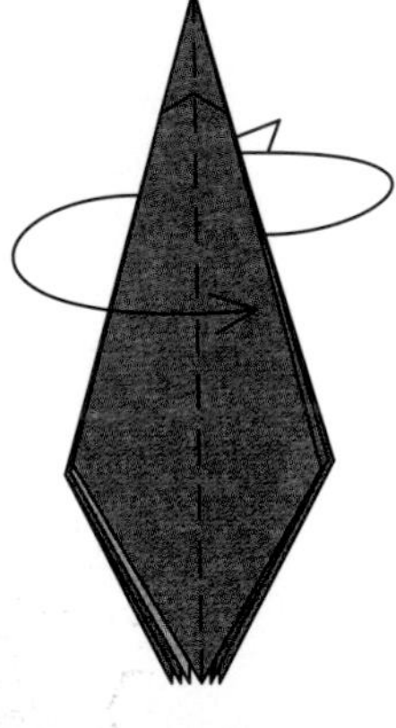

26. Fold two layers to the right in front and two to the left behind, spreading the layers symmetrically.

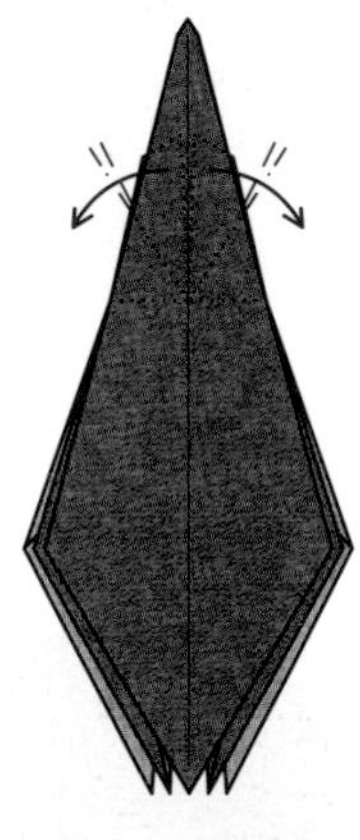

27. Reverse-fold the two hidden corners out to the sides. (There are three layers in each; it doesn't matter how you divide the layers, but divide them both the same way.)

28. Open out the two flaps to form small cups.

29. Stretch the middle pair of edges on each side apart slightly; the model will not lie flat.

30. Pleat the edges to take up the excess paper, press the layers of the tip together, and round it into a bowl.

31. Fold the model in half (except for the top, which stays rounded). Rotate 1/2 turn.

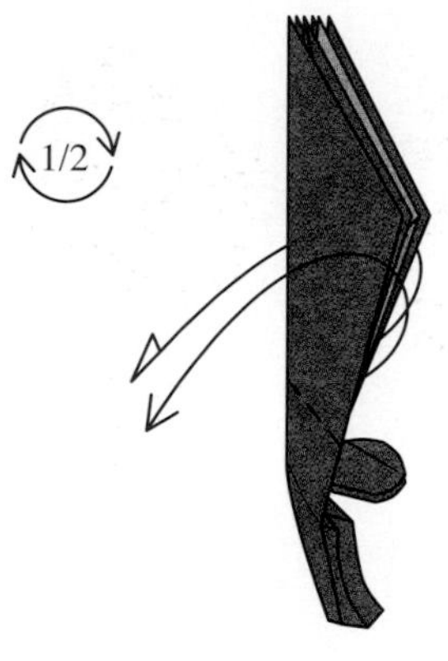

32. Reverse-fold three flaps together as one.

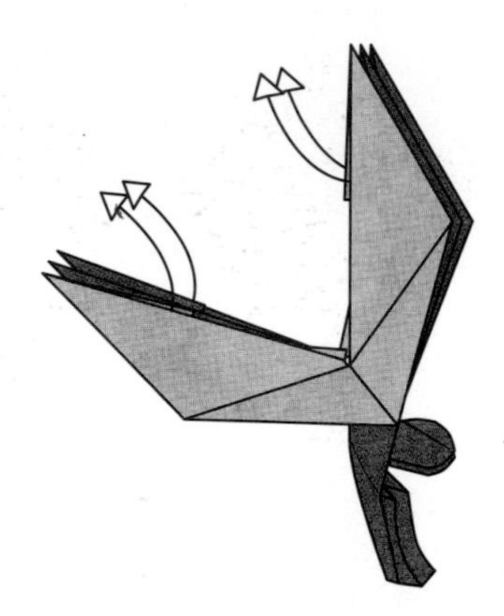

33. Pull out four loose corners completely.

34. Fold the corners over and over on existing creases.

35. Squash-fold five flaps (all but the middle flap) to stand out perpendicularly to the other layers.

36. Narrow the stem with mountain folds.

37. Pinch the stem to narrow it further. Spread the small flaps in front to the sides.

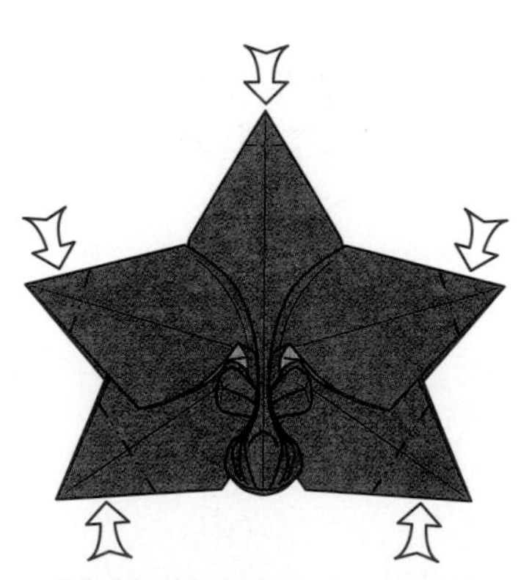

38. Reverse-fold the tips of the five flat petals. Shape the flower.

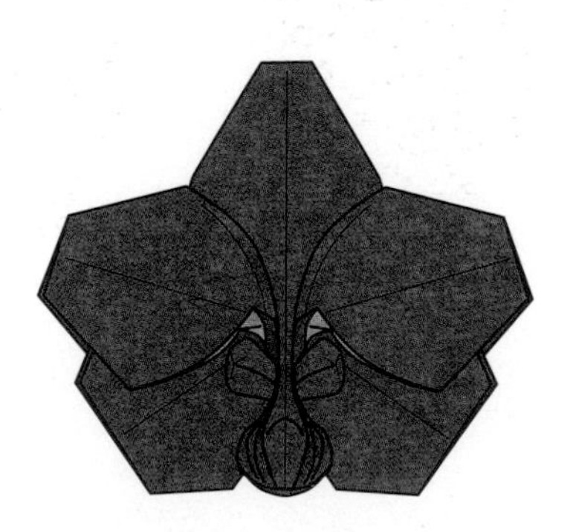

39. Finished Orchid Blossom.

11 Tree Theory

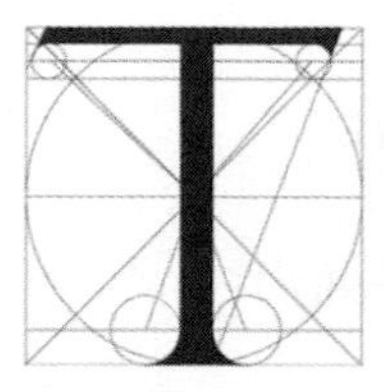

This section describes the mathematical ideas that underlie the tree method of origami design, which is a mathematical formulation of the geometric concepts that I have introduced somewhat ad hoc over the last few chapters, culminating in the circle/river/molecule method for designing uniaxial bases. The tree method does exactly the same thing—and indeed, utilizes molecules for the generation of the final crease pattern—but casts the problem in a form that is a bit less intuitive, perhaps, but is both more rigorous and is more amenable to numerical solution.

In the circle/river method, we represent flaps and connections between flaps by circles and rivers on a square of paper; we then connect the centers of touching circles to create axial polygons, which, in turn, are filled in with molecules or are subdivided by adding new circles and then filled in. The process gives a generic-form crease pattern for a base with the appropriate number, size, and configuration of flaps.

The weak point in this process was the original packing of circles and rivers; circle packings are relatively straightforward, but when we start adding rivers, the problem can get very complicated due to the many ways that rivers could meander among the circles. In tree theory, we avoid this problem by dispensing with circles and rivers entirely. Instead, we build a connection directly from a stick figure representation of the desired base to the crease pattern itself.

11.1. The Tree

We have already introduced the idea of using a small stick figure as a shorthand way of describing a base. The stick figure captures the number of flaps, their lengths, and how they are connected to each other. Using a term from graph theory, we will call such a stick figure the *tree graph* for a given (or postulated) uniaxial base, or just *tree* for short. A tree graph consists of *edges* (line segments) and *nodes* (ends of line segments).

We will also divide the nodes into two types: *leaf nodes* are nodes that come at the end of a single edge. Leaf nodes correspond to the tips of legs, wings, and other appendages. Nodes formed where two or more edges come together are called *branch nodes*. Similarly, a *leaf edge* is an edge that ends in at least one leaf node; a *branch edge* is an edge that ends in two branch nodes. These are illustrated in Figure 11.1.

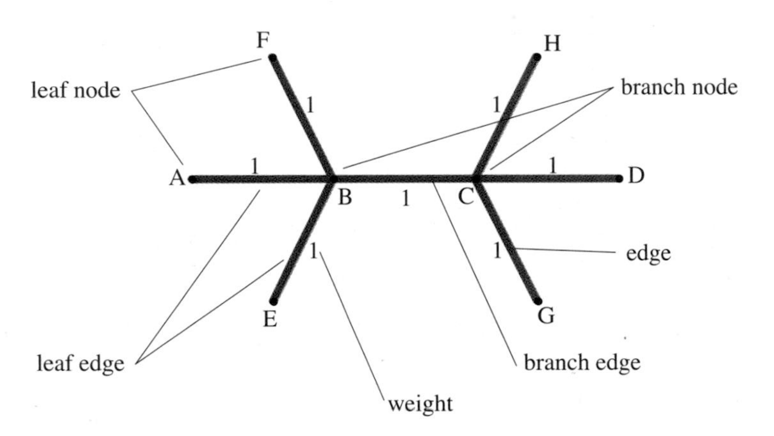

Figure 11.1.
Parts of a tree graph.

In the stick figure we drew in previous sections, the length of each segment was equal to the length of the flap or connector that it corresponded to. In a tree, we will label each edge by a *weight*, which is the numerical length of the corresponding flap. It is helpful to draw the tree with each edge length proportional to its weight, and so I will continue to do so. Thus, in the tree in Figure 11.1, each of the edges has weight 1, meaning that each corresponds to a flap or connector of unit length.

Now, the definition of a uniaxial base was a base that could be oriented so that (a) all flaps lie along a common line (the axis), and (b) the hinges between flaps were perpendicular to the axis. The perpendicularity of the hinges is an important property; it allows the flaps to be manipulated in three dimensions so that the edges of all flaps lie in a common plane, as shown for a hypothetical base in Figure 11.2. We refer to this plane as the *plane of projection*. Put formally, the plane of projection of a base is a plane that contains the axis of the base and the axial edges of all flaps, and that is perpendicular to the layers of the base.

This property allows another interpretation of the tree graph: It is the shadow cast by the base in a plane perpendicu-

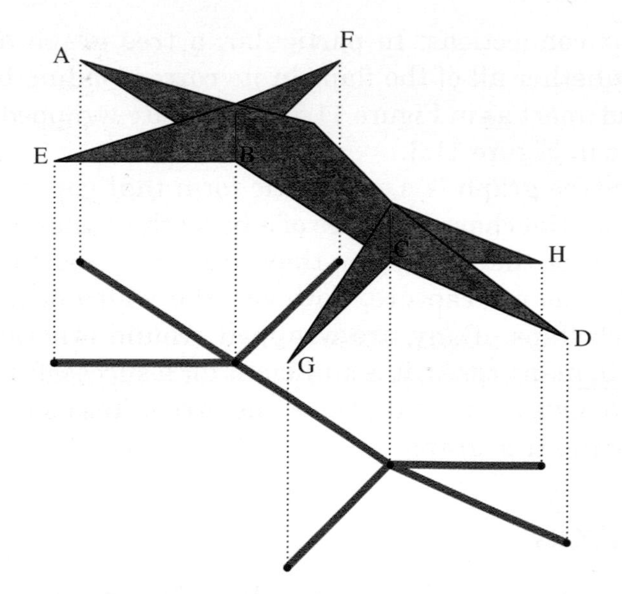

Figure 11.2.
Schematic of a hypothetical uniaxial base for an animal with four legs, a head, body, and tail. It's a uniaxial base if it can be manipulated so that all of the flaps lie in a plane and all of the layers are perpendicular to the plane. The shadow of the base consists entirely of lines.

lar to the layers of the base, as shown in Figure 11.2. This analogy can only be pushed so far, however. In many uniaxial bases—even one as simple as the Bird Base—some flaps are wrapped around others in such a way that the shadows of individual flaps are unavoidably overlapping. The true shadow would show fewer segments than the number of edges possessed by the actual tree. To avoid such ambiguities, I will always show a tree with edges (and nodes) distinctly separated, as shown in Figure 11.3.

This point emphasizes another ambiguity about trees: There is no particular significance to the orientation of the edges of the tree graph. All that matters are the edge weights

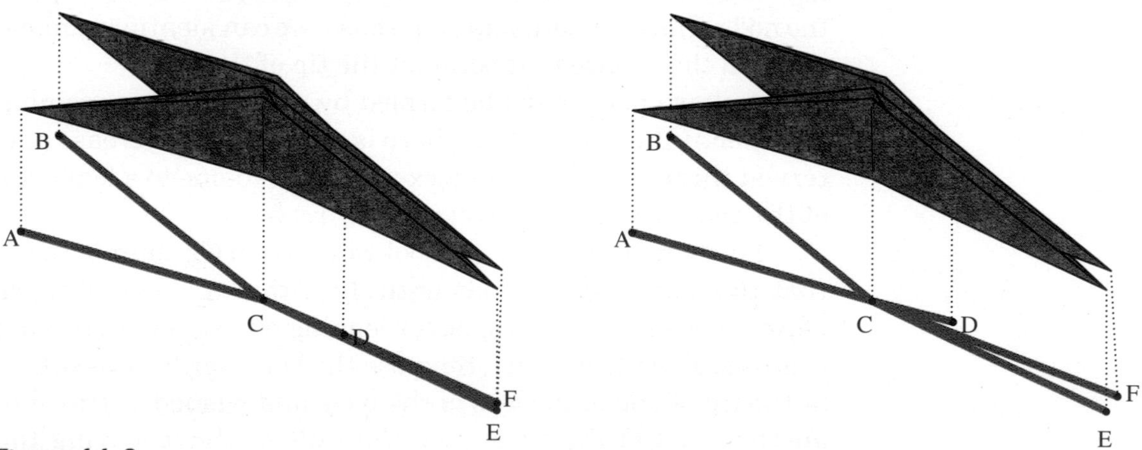

Figure 11.3.
Left: shadow cast by a Bird Base. The shadows of wrapped flaps are superimposed.
Right: base and schematic tree; the shadow is perturbed to distinguish flaps wrapped around one another.

and their connections. In particular, a tree graph does not specify whether all of the flaps in its corresponding base can be spread apart as in Figure 11.2 or some are wrapped around others as in Figure 11.3.

The tree graph is a schematic form that captures some of the essential characteristics of a base: the number of flaps, the length of the flaps, how they are connected to one another. It does not capture, however, the width of the flaps, nor which flaps, if any, are wrapped around others. Nevertheless, in many cases, it is sufficient for a successful origami design to have a base with the same attributes as those conveyed by the tree graph.

11.2. Paths

Suppose that we have a uniaxial base folded from a square and that we construct its tree graph. If we unfold the base, we get a square with a crease pattern that uniquely defines the base. The act of projecting the base into a plane—casting a shadow—can be thought of as defining a mapping between points on the square and points on the tree. In the language of mathematics, it is a *surjective* or *onto* mapping—that is, for every point on the square there is a corresponding point on the tree, but more than one point on the square can map to the same point on the tree.

That the mapping is not one-to-one is clear from Figure 11.3; wherever you have vertical layers of paper, there are many points on the base that map to the same point on the tree. However, if the flaps come to sharp points, then at the leaf nodes of the tree, there is exactly one point on the square that maps to the node. Thus, for each flap of the base, we can identify a unique point on the square that becomes the tip of the flap.

A sharp point must be formed by several creases coming together at the point. Thus, there is a vertex in the crease pattern at this point. Such a vertex maps one-to-one to a leaf node of the tree; we therefore call it a *leaf vertex*.

Let us resurrect the shy bookworm from Chapter 5; recall that this bookworm travels entirely within a sheet of paper between the two surfaces, never leaving one sheet or crossing from one sheet to another. Suppose the bookworm were sitting at the tip of one of the legs of the base and wished to travel to another part of the base—say, the tail—without leaving the paper. It would have to crawl down the foreleg to the body, down the body, and back out the tail. The distance it traveled would be (length of the foreleg) + (length of the body) + (length of the tail).

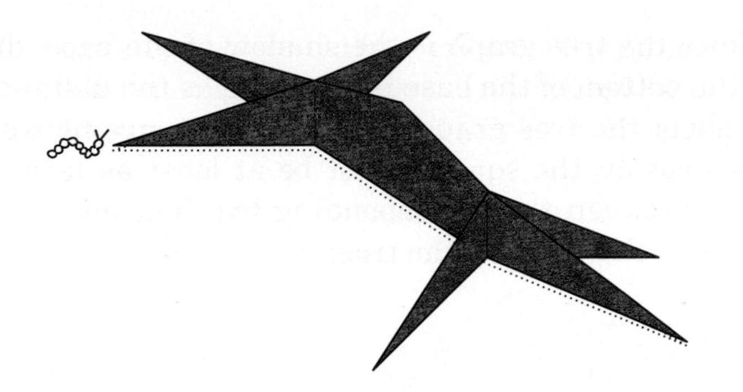

Figure 11.4.
A bookworm wishes to go from a foreleg to the tail along the base. It can take several different paths, but the most direct path is the path that lies in the plane of projection.

Now, let's think about what the path of the bookworm would look like on the unfolded square (you can imagine dipping the bookworm into ink so that it leaves a trail soaking through the paper as it crawled). Clearly, it starts and ends at a leaf vertex. On the square, the path might go directly from one leaf vertex to the other, or it might meander around a bit, or it might even backtrack. If it travels via the shortest route, then the path length on the square is equal to the length as measured along the bottom of the base. Any meandering or backtracking will make the path longer. Thus, the distance traveled on the unfolded square must be at least as long as the minimum distance traveled along the base.

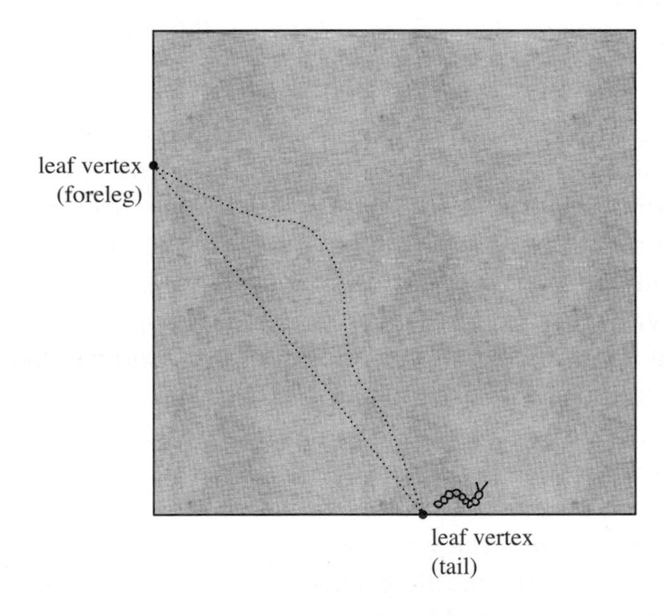

Figure 11.5.
The trail of the bookworm.

This illustrates an extremely important property of any mapping from a square to a base: Although our example went from one leaf vertex to another, the property is general: The distance between *any* two points on the square must be greater than or equal to the distance between the two corresponding points on the base.

Since the tree graph is the shadow of the base, distance along the bottom of the base is the same as the distance measured along the tree graph. Thus, the distance between two leaf vertices on the square must be at least as large as the distance between the corresponding two leaf nodes as measured along the edges of the tree.

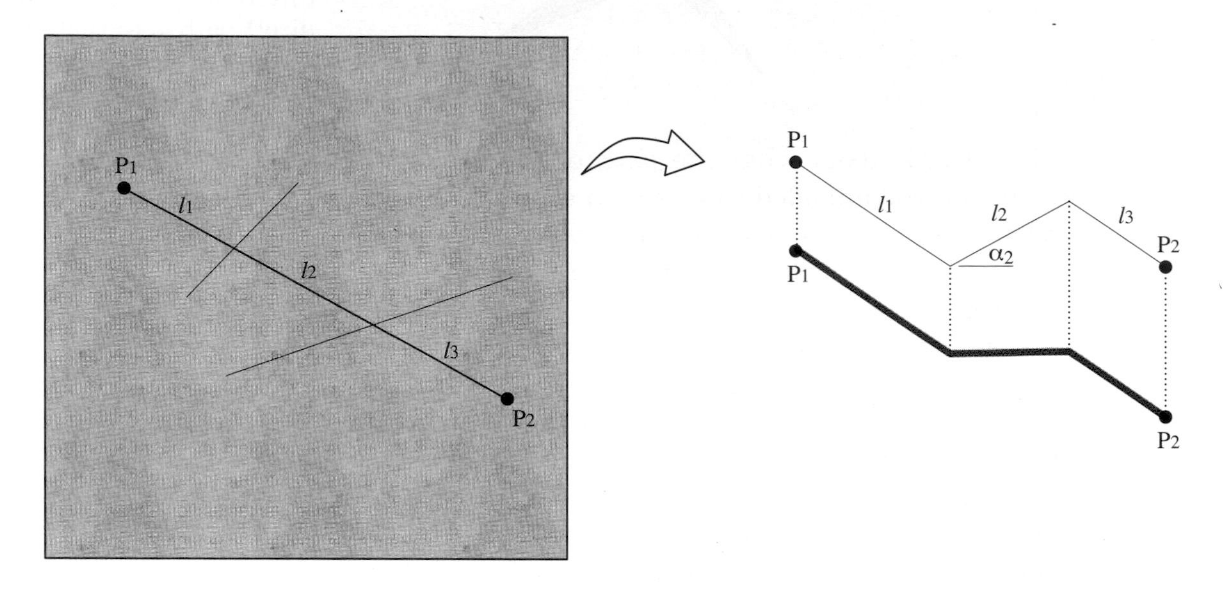

Figure 11.6.
A straight path on the square maps to a path in the base that may have uphill and downhill components.

And in particular, this relationship must hold for any two points on the base that correspond to nodes on the tree. Now while this condition must hold for *any* pair of points on the base, it turns out that if it holds for every pair of leaf nodes, it will hold for *every* pair of points on the base. That is, if you can identify a set of points on the square corresponding to all of the leaf nodes of a tree—the leaf vertices—and the leaf vertices satisfy the condition that the distance between any pair of them is greater than or equal to the distance between the corresponding nodes as measured on the tree, then *it is guaranteed that a crease pattern exists to transform the square into a base whose projection is the given tree*.

This is a remarkable property. It tells us that no matter how complex a desired base is, no matter how many points it may have and how they are connected to one another, we can *always* find a crease pattern that transforms the square (or any other shape paper, for that matter) into the base. Putting this into mathematical language, we arrive at the fundamental theorem of the tree method of design (which I call the *tree theorem* for short):

Define a simply connected tree T with leaf nodes P_i, $i = 1, 2, \ldots N$. Define by l_{ij} the distance between nodes P_i and P_j as measured along the edges of the tree; that is, l_{ij} is the sum of the lengths of all the edges between nodes P_i and P_j. For each leaf node P_i, define a leaf vertex $\mathbf{u}_i$ in the unit square $u_{i,x} \in [0,1]$, $u_{i,y} \in [0,1]$. Then a crease pattern exists that transforms the unit square into a uniaxial base whose projection is T if and only if $|\mathbf{u}_i - \mathbf{u}_j| = l_{ij}$ for every i,j. Furthermore, in such a base, P_i is the projection of $\mathbf{u}_i$ for all i.

Although the proof of the tree theorem is beyond the scope of this book (you can find a proof in the references), we will proceed to use it. The tree theorem tells us that if we can find a set of leaf vertices within a square for which the distance between any two is greater than or equal to the distance between their corresponding leaf nodes on the tree, then a crease pattern exists that can transform that pattern of vertices into a base.

Thus, for example, the tree in Figure 11.1 has six leaf nodes; there are fifteen possible pairs of leaf nodes to worry about. The distance from node A to node E is 2 units; thus, the leaf vertices on the square that correspond to nodes A and E must be at least 2 units apart. Similarly, to get from node A to node D on the tree, you must travel 3 units; and so the distance between leaf vertices A and D on the square must be at least 3 units as well. And so on, for the other thirteen possible pairs.

For a given tree, there are often several possible arrangements of leaf vertices that satisfy the tree theorem, each of which yields a different base. For our six-pointed base, a little doodling with pen and paper will reveal that the pattern of nodes shown in Figure 11.7 satisfies all such conditions if the square has side length $2\sqrt{((121 + 8\sqrt{179})/65)} \approx 3.7460$, in which

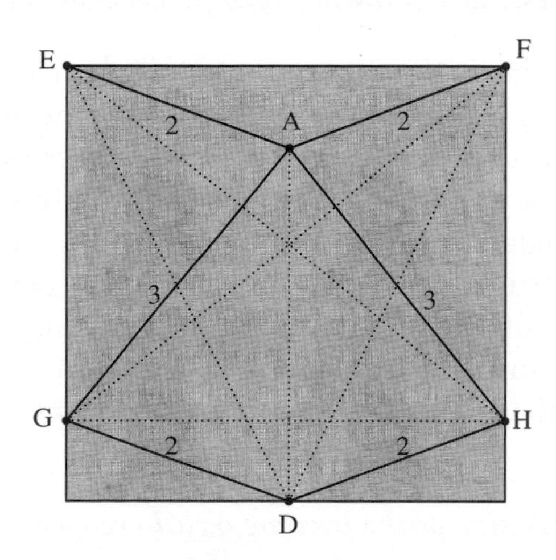

Figure 11.7. Node pattern that satisfies the tree theorem for the six-legged tree.

case the distances drawn in solid lines are equal to their minimum values, and all other paths (indicated by dashed lines) are greater than the minimum length.

The tree theorem is an existence theorem; it says that a crease pattern exists, but it doesn't tell us what this supposed crease pattern actually *is*. It does provide a strong clue, however. The tree theorem says that the leaf vertices become the tips of the flaps on the folded base. Are there any other features on the square that we can identify on the base?

Consider the inequality in the tree theorem. Two leaf vertices must be separated on the square by a distance *greater than or equal to* the distance between their corresponding nodes on the tree. In the special case where equality holds, we can uniquely identify the line between the two vertices. We will call a line on the square that runs between any two leaf vertices a *path*. Every path has a *minimum* length, which is the sum of the lengths of edges of the tree between the two leaf nodes that define the path. (In the symbolism of the tree theorem, l_{ij} is the minimum length of path *ij*.) The *actual* length of a path is given by the distance between the vertices of the crease pattern that correspond to the leaf nodes as measured upon the square ($|\mathbf{u}_i - \mathbf{u}_j|$ in the tree theorem). Any path for which its actual length in the crease pattern is equal to its minimum length as defined by the tree graph is called an *active path*.

In the base, the only route between two flap tips that is equal to the distance between the leaf nodes lies in the plane of the projection. Thus, any active path between two leaf vertices on the square becomes an edge of the base that lies in the plane of projection. Consequently, we have another important result:

> *Any active path between leaf vertices forms an edge of the base that lies in the plane of projection of the base.*

Active paths on the square lie in the plane of projection of the square, but the plane of projection is where the vertical layers of paper in the base are connected to each other. In other words, since the paper on both sides of the path lies above the path in the folded base, there must be a fold along the path. This must be true for every active path. Thus, active paths are not only edges of the base: they are major creases of the base. And not just any creases; since the plane of projection contains the axial edges of the flaps, these creases must be axial creases.

Active paths become axial creases.

So now we have the rudiments of the crease pattern for the base. We know that the points on the square that correspond to leaf nodes of the tree become the tips of the flaps of the base, and we know that active paths on the square become axial creases of the base.

We can construct further correspondence between elements of the tree and the crease pattern, namely, the branch nodes. The axial creases in the crease pattern map onto paths on the tree graph, so any point on the tree corresponds to one or more points along axial creases. Specifically, we can locate the points along each axial crease that correspond to each branch node, points we will call *branch vertices*.

If our hypothetical bookworm travels from one leaf vertex to another, encountering branch vertices at distances d_1, d_2, d_3, and so forth along the way, then when we draw the crease pattern, we can identify each branch vertex at the same distances along the active path connecting the two leaf vertices. Thus, we can add all of the branch vertices to our budding crease pattern. In Figure 11.8, I've identified all of the vertices, both leaf and branch, by a letter on the tree graph, and have added their corresponding vertices to the active paths in the crease pattern on the square. Observe that in general, a branch node may show up on more than one active path.

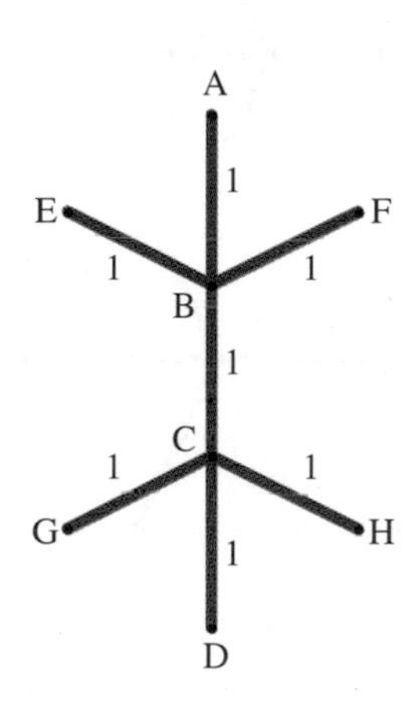

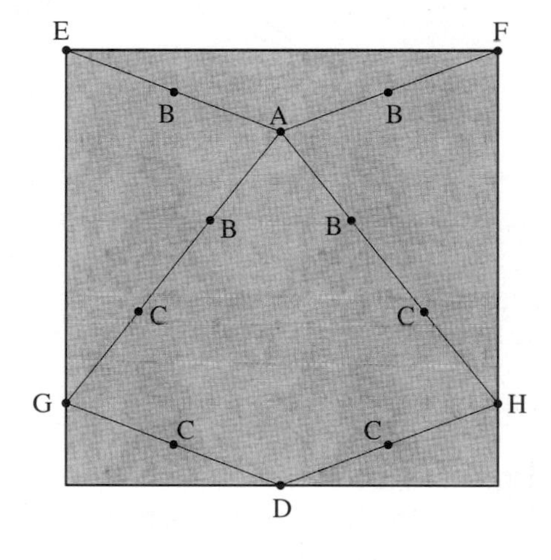

Figure 11.8.
Left: tree with all nodes lettered.
Right: crease pattern with leaf vertices, branch vertices, and active paths.

It's also worth pointing out that we don't show any leaf vertices along the edges of the square because the paths between node pairs G and E, E and F, and F and H are not active paths.

11.3. Scale

There is one more factor to consider: the relationship between the size of the tree graph and the crease pattern on the square.

In the pattern shown in Figure 11.8, we have given each stick unit length; but for this to fit within a square, the square must be larger than a unit square. In order to fit the crease pattern into a unit square, we introduce a quantity we call the *scale*, which is simply the distance on the square that corresponds to one unit in the tree graph. We can fit the crease pattern in Figure 11.8 into a unit square if we choose a scale factor $m = 0.267$; that is, one unit of length on the tree is equivalent to a distance of 0.267 in the crease pattern. Then we must modify the tree theorem to incorporate a scale factor. Our path condition becomes: For every path between leaf vertices $\mathbf{u}_i$ and $\mathbf{u}_j$, the leaf vertices must satisfy the inequality

$$\left|\mathbf{u}_i - \mathbf{u}_j\right| \leq m l_{ij} \tag{11–1}$$

for a scale factor m. We call the set of all such equations the *path conditions* for the given tree graph.

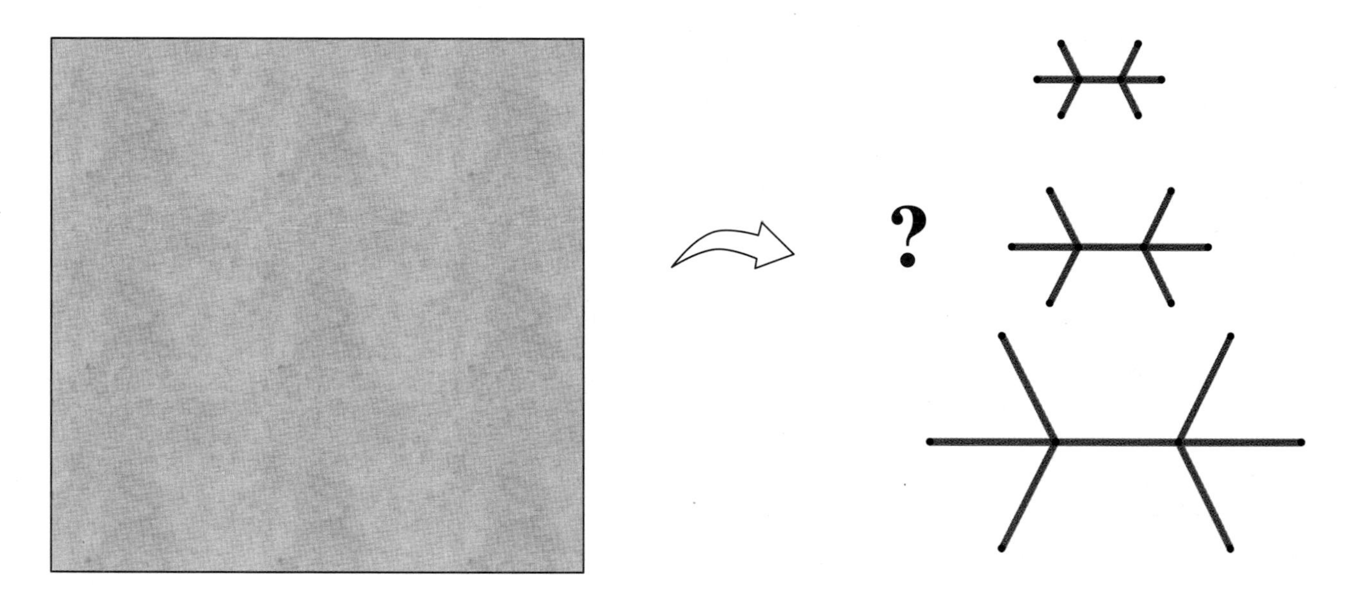

Figure 11.9.
What is the relationship between the size of the square and the scale of the tree graph?

In this way, the scale factor becomes a quantitative measure of the efficiency of the crease pattern, and the search for the most efficient crease pattern can be expressed formally as an optimization problem: Given a tree that defines a base, optimize the scale factor while varying the coordinates of the leaf vertices, subject to the constraints that (a) the path conditions are satisfied, and (b) all leaf vertices lie within a unit square.

11.4. Subtrees and Subbases

It can be shown that active paths cross each other only at leaf vertices. Since active paths become axial creases, the pattern of axial creases breaks up the square into axial polygons. In some of the polygons, all of their sides are active paths (like the inverted-kite-shaped quadrilateral in the center of Figure 11.8). If one of the sides of a polygon lies on the edge of a square, it may or may not be an active path (in Figure 11.8, each triangle has one side on the edge of the square that is not an active path.) Each axial polygon has the property that all of its sides map to the plane of projection of the base when the square is folded into a base. Consequently, to find a crease pattern that collapses the square into the base, it is necessary to find a crease pattern that maps the network of axial polygons onto the plane of projection of the base.

That problem should sound familiar; this sounds like a job for molecules. Recall that the tree is the projection of the base, which is folded from the complete square. Each polygon on the square corresponds to a portion of the overall base, and if you collapse any polygon into a section of the base—which I call a *subbase*—the projection of the subbase is itself a portion of the projection of the complete base, i.e., a portion of the original tree graph. The tree graph of a subbase is called a *subtree*. For example, Figure 11.10 shows the polygons for our six-legged base and the corresponding subtrees for each subbase. Note that since all of the corners of an axial polygon must be leaf vertices, the triangles at the bottom corners of the square are *not* axial polygons and, in fact, do not contribute to the base in a significant way.

One requirement of axial polygons that we saw in previous sections was that if two axial polygons shared a common side and that side was an axial path, any crease pattern that collapses the first polygon into a subbase must be compatible with a crease pattern that collapses the adjacent polygon into its subbase. In tiles, we enforced this matching by drawing circles and rivers within axial polygons and forcing the circles and rivers to line up. Then, when we introduced molecules, we found that circle/river alignments could be enforced by requiring alignment of the tangent points of the circles.

Let's look at the circle/river treatment of this problem. When the path conditions are written as equations, it is difficult to form an intuitive picture of them, but the value of such a treatment is that this optimization can be formulated as a set of equations capable of being solved by existing computer algorithms. (A complete set of the equations is given in Chap-

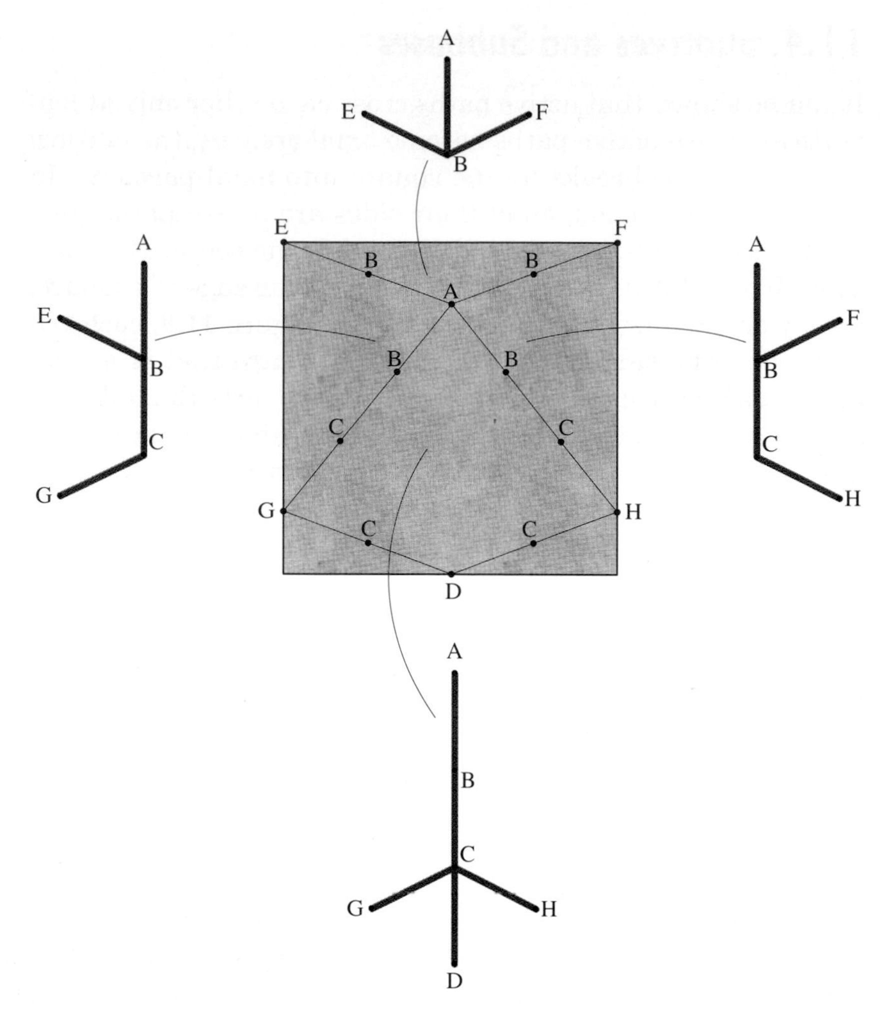

Figure 11.10.
The four axial polygons for the six-legged base and the subtrees corresponding to each subbase.

ter 14.) We could have also solved for a base corresponding to this tree by the circle/river method; if we did this, we would have arrived at a configuration of circles and rivers that we can superimpose on the rudimentary crease pattern from Figure 11.8, as shown in Figure 11.11.

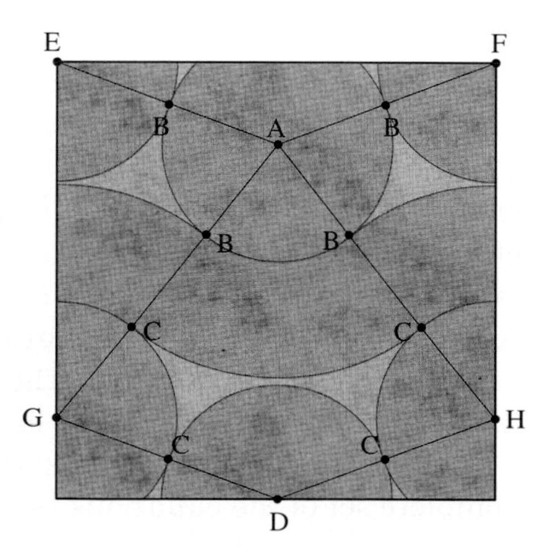

Figure 11.11.
The pattern of leaf vertices and axial paths with circles and rivers from the corresponding base.

Figure 11.11 makes it clear: The tangent points, which we introduced in an ad hoc way in the previous chapter, are simply the branch vertices, points along the axial paths that correspond to the branch nodes of the tree. The creases that fill in this structure will be those creases that collapse the individual polygons so that the branch vertices around the perimeter of each polygon are aligned. And so, the molecular crease patterns we have seen—rabbit-ear (for triangles), Waterbomb, arrowhead, gusset, and sawhorse (for quadrilaterals)—will be the patterns that fill in these axial polygons as well.

You can also see from Figure 11.12 that the use of nonoverlapping circles and rivers is simply a geometric way of enforcing the path conditions that apply to pairs of leaf vertices. For example, take the case of two leaf nodes with a single branch node between them as shown in Figure 11.12. If the two leaf nodes are separated by edges with lengths a and b, then the path condition between their corresponding leaf vertices in the crease pattern would be

$$\left|\mathbf{u}_A - \mathbf{u}_B\right| \geq m(a+b) \tag{11–2}$$

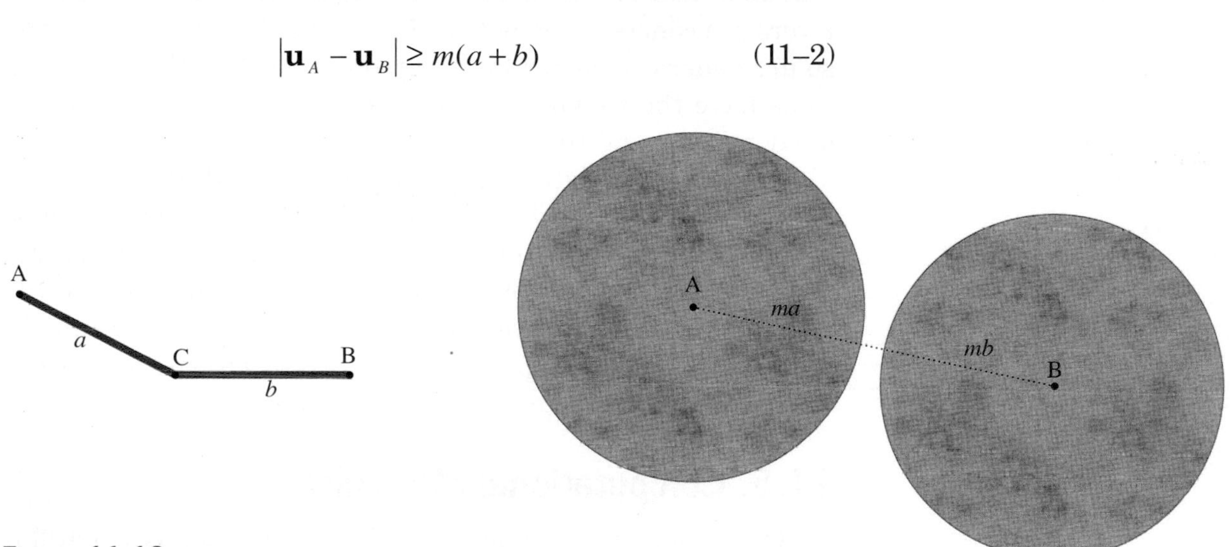

Figure 11.12.
Left: a tree with two leaf nodes.
Right: use of nonoverlapping circles to represent the path condition.

If we draw a circle around node A of radius ma (the scaled length of flap A) and one around node B of radius mb, then the path condition is satisfied if and only if the two circles do not overlap; and at equality, the two circles touch.

Similarly, if the two leaf nodes are separated in the tree by multiple edges as in Figure 11.13, we can still represent this geometrically by inserting rivers whose width is proportional (by the same scale factor m) to the corresponding segments of the tree.

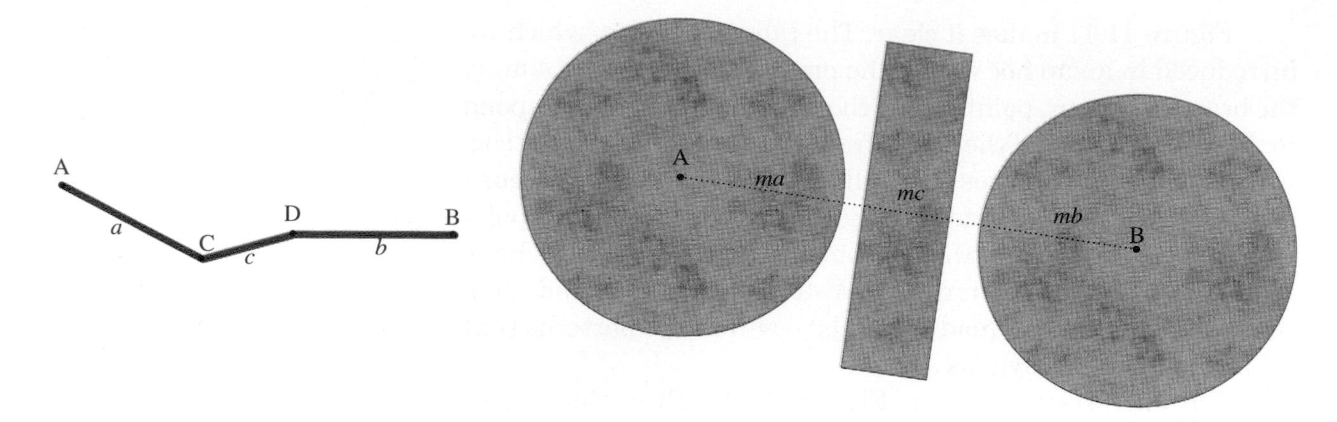

Figure 11.13.
Left: a tree with two leaf nodes and two branch nodes.
Right: use of circles and rivers to represent the path condition.

The use of circles and rivers to design a crease pattern and the solution of the path equations are completely equivalent approaches. Why use one instead of the other? Circles and rivers are concrete geometric objects, easily visualizable, and so are generally easier for a person to work with. But equations have their own value; they can be manipulated, rigorously proven, and turned into algorithms. The first computer algorithm for sophisticated origami design and the proof of its sufficiency were based on the path equations. However, most origami designers who use these techniques work with circles and rivers to do their own designs. It is still a useful aid to one's intuition when working with crease patterns found by path methods to draw in the corresponding circles (and/or rivers) to illustrate the underlying structure.

11.5. Computational Molecules

In the previous chapter on molecules, we distinguished different molecules by their number of flaps and whether or not they had connectors between groups of flaps. This distinction is concisely captured by associating with each molecule the particular tree graph (a subtree of the base's tree graph) to which it corresponds.

As we have seen, there is a single triangle molecule, the rabbit-ear molecule. It has three flaps that come to a common point; thus its tree has three leaf nodes and three edges, which are joined at a common branch node.

If you are folding the axial polygon, you can find the intersection of the angle bisectors—point E in the figure —by pinching each corner in half along the bisector and finding the point where all three creases come together. If you are calcu-

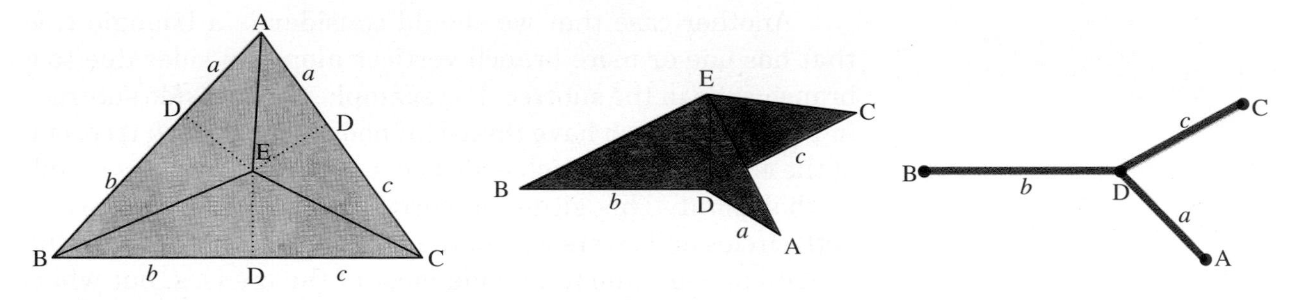

Figure 11.14.
Generic-form crease pattern, folded form, and tree graph for a rabbit-ear molecule.

lating the crease locations numerically, there is an elegant formula for the location of the intersection of the angle bisectors of an arbitrary triangle. If $\mathbf{p}_A$, $\mathbf{p}_B$, and $\mathbf{p}_C$ are the vector coordinates of corners A, B, and C and $\mathbf{p}_E$ is the coordinate of the bisector intersection, then $\mathbf{p}_E$ is given by the simple formula

$$\mathbf{p}_E = \frac{\mathbf{p}_A(b+c)+\mathbf{p}_B(c+a)+\mathbf{p}_C(a+b)}{2(a+b+c)}$$

$$= \frac{\mathbf{p}_A(s-a)+\mathbf{p}_B(s-b)+\mathbf{p}_C(s-c)}{2s} \qquad (11\text{–}3)$$

where s is the perimeter of the triangle. That is, the location of the bisector intersection is simply the weighted average of the coordinates of the three corners, with each corner weighted by the sums of the lengths of the two opposite sides.

What happens when one of the sides of the triangle is not an active path? This can happen, for example, when one of the sides of the triangle lies along an edge of the square; all of the triangles in Figure 11.8 are of this type. Since the distance between any two leaf vertices must be greater than or equal to the minimum path length, the side that isn't an active path must be slightly too long to be an active path rather than too short. Fortunately, only a slight modification of the rabbit ear is necessary to address this situation. Figure 11.15 shows the crease pattern and subbase when side BC is slightly too long.

If the triangle has two sides that aren't active paths, a similar modification will still collapse it appropriately.

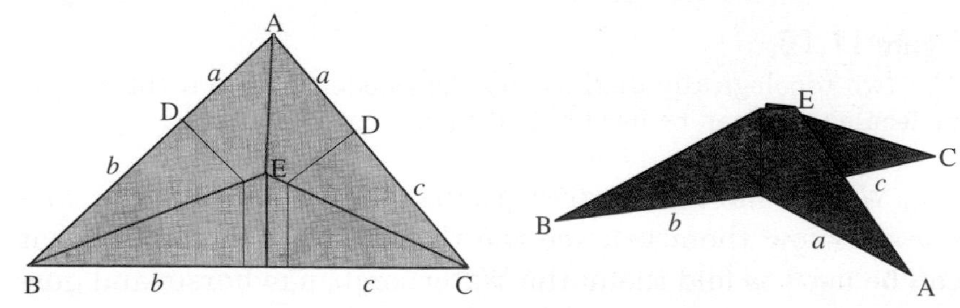

Figure 11.15.
Left: crease pattern for a triangle when side BC is not an active path.
Right: resulting subbase.

Another case that we should consider is a triangle tree that has one or more branch vertices along its sides due to a branch node in the subtree. For example, the two side subtrees in Figure 11.8 each have three leaf nodes, but in each tree, one of the edges has a branch node because the subtree has a kink at that point. This situation corresponds to the presence of both circles and rivers within the triangle. We can still use the rabbit-ear molecule to provide most of the creases, but wherever we have a branch vertex along an axial path, we need a hinge crease propagating inward from the branch node to the ridgeline crease and back down to the adjacent side.

11.6. Quadrilaterals

As we saw in the last chapter, there were two classes of quadrilateral molecules: those with no rivers or rivers connecting adjacent edges, and those with rivers running across the quadrilateral. These two classes correspond to the two topologically distinct tree graphs with four leaf nodes, which are shown in Figure 11.16.

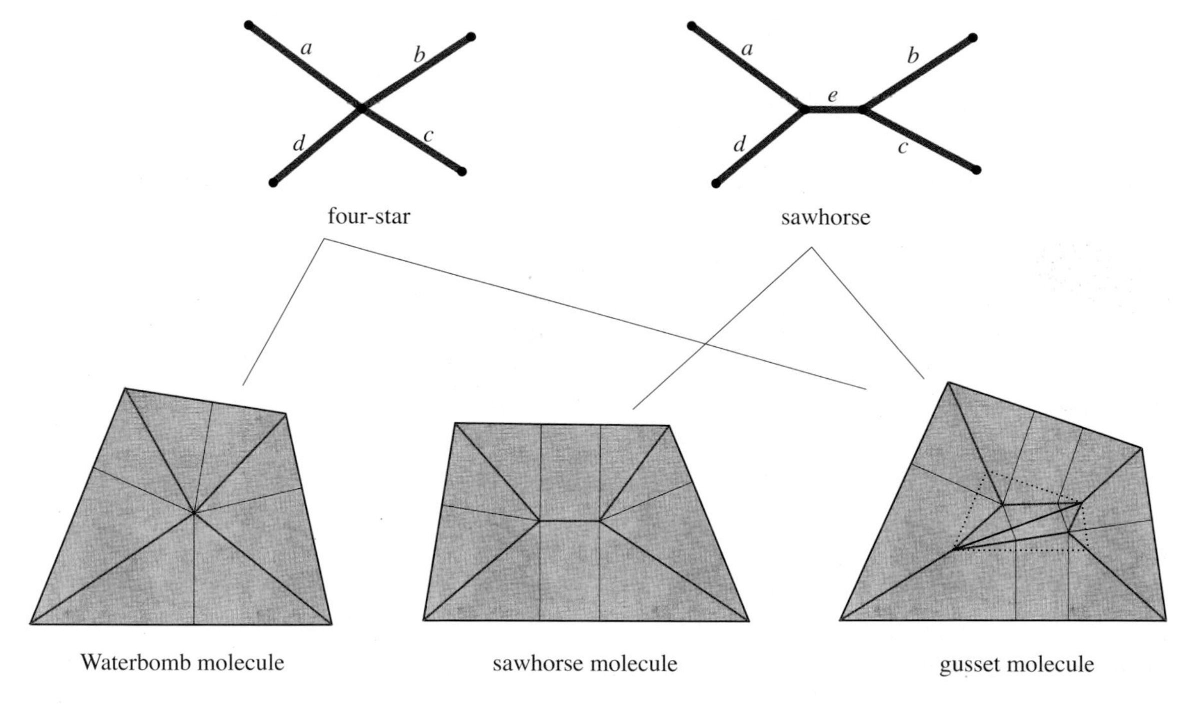

Figure 11.16.
The two topologically distinct four-leaf-node trees and the simple molecules that can be used to fold them.

We will call the two tree graphs the *four-star* and the *sawhorse*. Below them you see the three simple molecules that can be used to fold them: the Waterbomb, sawhorse, and gusset molecules. The four-star graph can be thought of as a degenerate form of the sawhorse graph, the limiting case as the central segment (e) goes to zero length. Both the Waterbomb molecule and the sawhorse molecule can be considered special cases of the gusset molecule. Since the gusset molecule serves for any quadrilateral whether the underlying tree is a four-star or sawhorse, let's go through its numerical construction.

In the previous chapter, I showed how to construct the gusset molecule by folding; here, I will show its construction by computation. Given a quadrilateral ABCD as shown in Figure 11.17, construct a smaller quadrilateral inside whose sides are parallel to the sides of the original quadrilateral but are shifted inward a distance h (the value of h is not yet determined). Denote the corners of the new quadrilateral by A′, B′, C′, and D′. Drop perpendiculars from these four corners to the sides of the original quadrilateral. Label their points of intersection A_{AB} where the line from A′ hits side AB, B_{AB} where the line from B′ hits AB, and so forth.

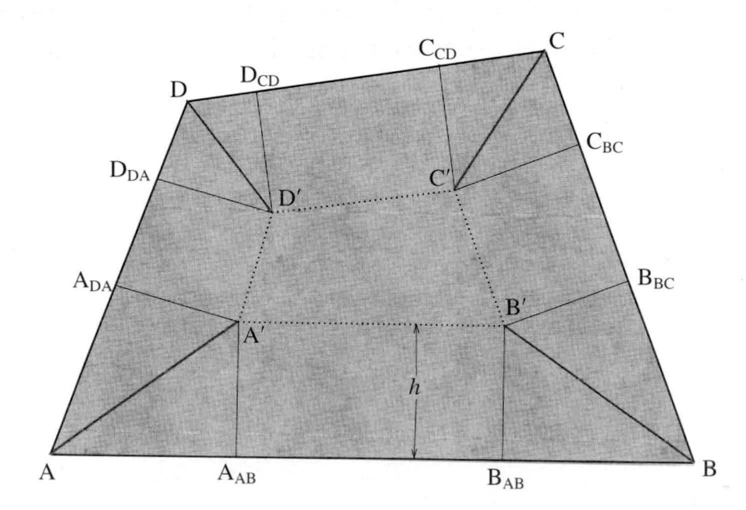

Figure 11.17. Construction of the gusset quad for a quadrilateral ABCD. Inset the quadrilateral a distance h; then drop perpendiculars from the new corners to the original sides.

Now we need some distances from the tree graph. Let l_{AC} be the distance from node A to node C on the tree and l_{BD} be the distance from node B to node D. In *most* cases (see below for the exceptions), there is a unique solution for the distance h for which one of these two equations holds:

$$AA_{AB} + A'C' + CC_{BC} = l_{AC}, \text{ or} \qquad (11\text{–}4)$$

$$BB_{BC} + B'D' + DD_{AD} = l_{BD}. \qquad (11\text{–}5)$$

Let us suppose we found a solution for equation (11–4). The diagonal A′C′ divides the inner quadrilateral into two triangles as shown in Figure 11.18. Find the intersections of the

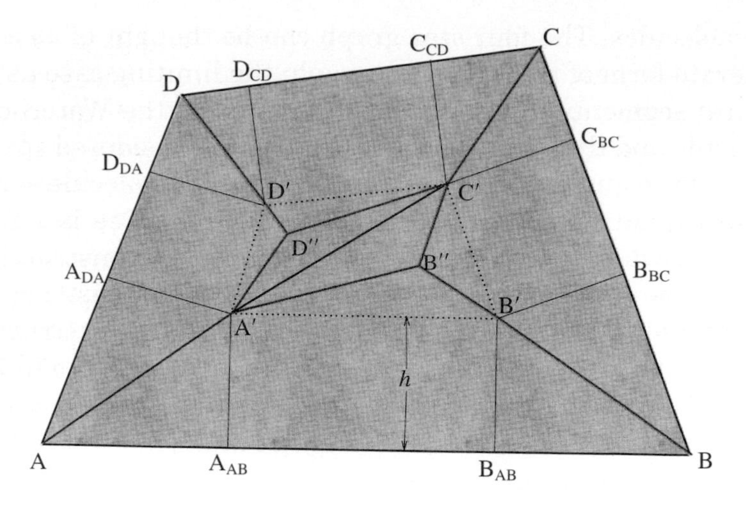

Figure 11.18.
On the inner quadrilateral, construct the bisectors of each triangle to find points B″ and D″.

bisectors of each triangle and call them B″ and D″. (If the second equation gave the solution, you'd use the opposite diagonal of the inner quadrilateral and find bisector intersections A″ and C″.)

The points A′, B″, C′, and D″ are used to construct the complete crease pattern by dropping perpendiculars to the four sides.

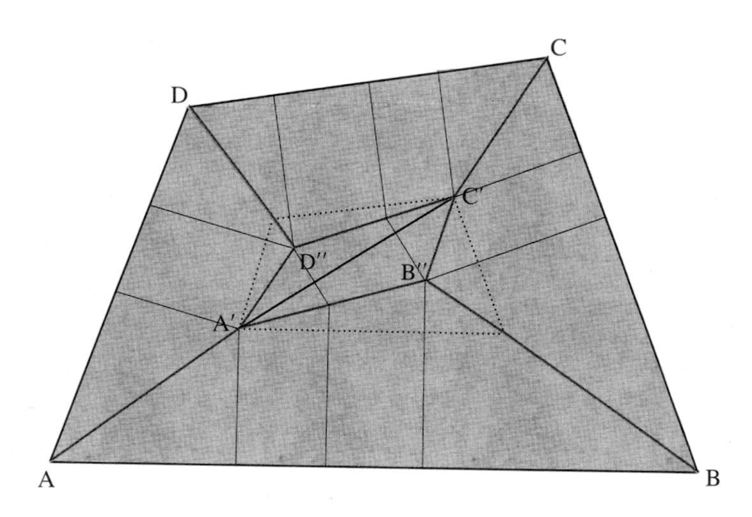

Figure 11.19.
Drop perpendiculars from the new vertices B″ and D″.

You can construct an equation for the distance h in terms of the coordinates of the four corners and the distances; it's a rather involved quadratic equation. The appropriate sequence of algebraic formulas used to solve for h is given in Chapter 14.

If you solve for the gusset quad numerically, you will see that that there are some quadrilaterals for which the points A′, B′, C′, and D′ all fall on a line or point. In these special cases, you don't get an inner quadrilateral for the gusset; instead, you get a sawhorse molecule (if a line) or a Waterbomb molecule (if a point). So the gusset molecule is, in fact, the general molecule for any quadrilateral.

Using the rabbit-ear molecule for triangles and the gusset molecule for quadrilaterals, you can fill in any tree-theorem-derived collection of axial polygons that consists of triangles and quadrilaterals to get the complete crease pattern for the base. Figure 11.20 shows the full crease pattern for the six-legged tree and the resulting base. You can easily verify the crease pattern by cutting it out and folding it on the lines. As you can see, the projection of the base into the plane is indeed the tree, and all of the flaps have the proper length.

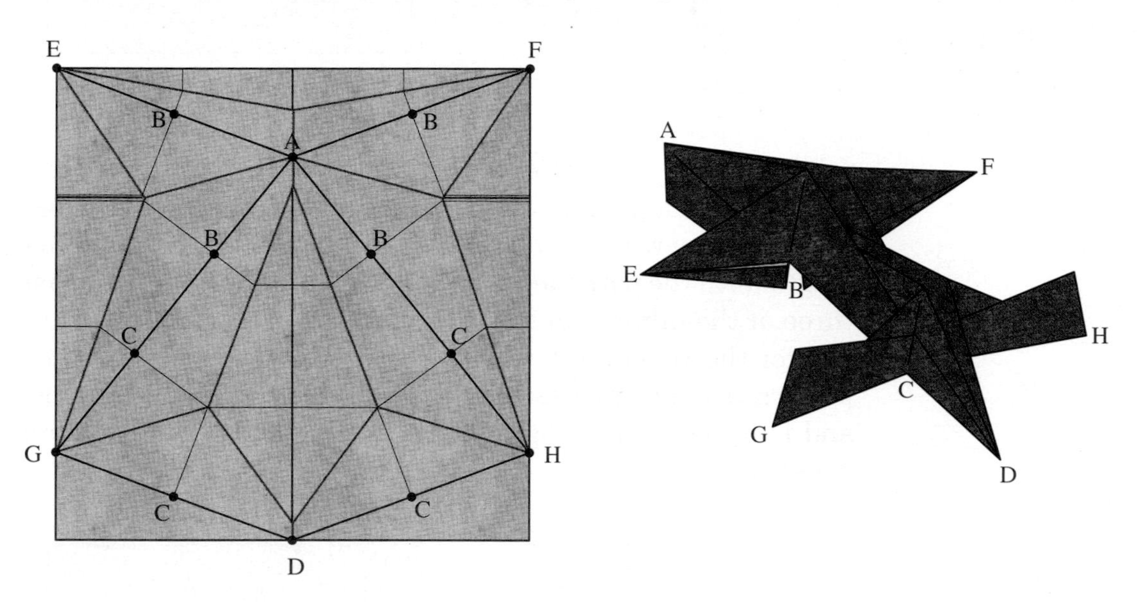

Figure 11.20.
Full crease pattern (generic form) and six-legged base.

11.7. Higher-Order Polygons

What about axial polygons with more than four sides? As we saw in the last chapter, we can reduce higher-order axial polygons formed in circle/river packings by adding a circle inside the polygon and inflating it until it contacts three other circles (or rivers). There is a corresponding procedure within tree theory.

Let's take the same example we used before: a pentagon, which would represent a five-flap base. This would have a tree containing five leaf nodes. Although there are several possible five-leaf-node trees, let's take the simplest for illustration, the one with a single branch node, i.e., a five-star. This graph and a sample axial polygon, are shown in Figure 11.21.

With circle/river patterns, we broke up higher-order polygons by adding circles within the polygon and inflating them until they contacted three (or more) of the other circles.

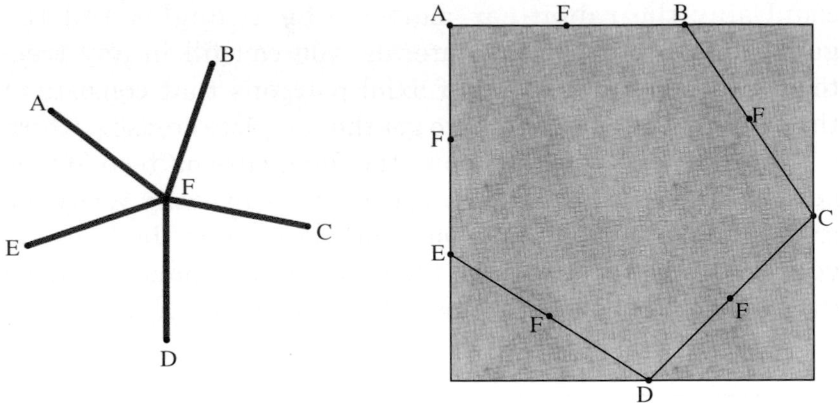

Figure 11.21.
Left: tree for a base with five equal flaps.
Right: pattern of leaf vertices and active paths corresponding to this tree.

Adding a circle to a circle/river pattern was tantamount to creating a new flap. The equivalent action in tree theory would be to add a new leaf node and edge to the tree and extend its length until the path inequalities become equalities for at least three of the other nodes (while remaining satisfied inequalities for the remaining nodes). The result would be the same pattern whether we used circles and rivers or path equations, and is illustrated in Figure 11.22, filled with rabbit-ear and gusset molecules.

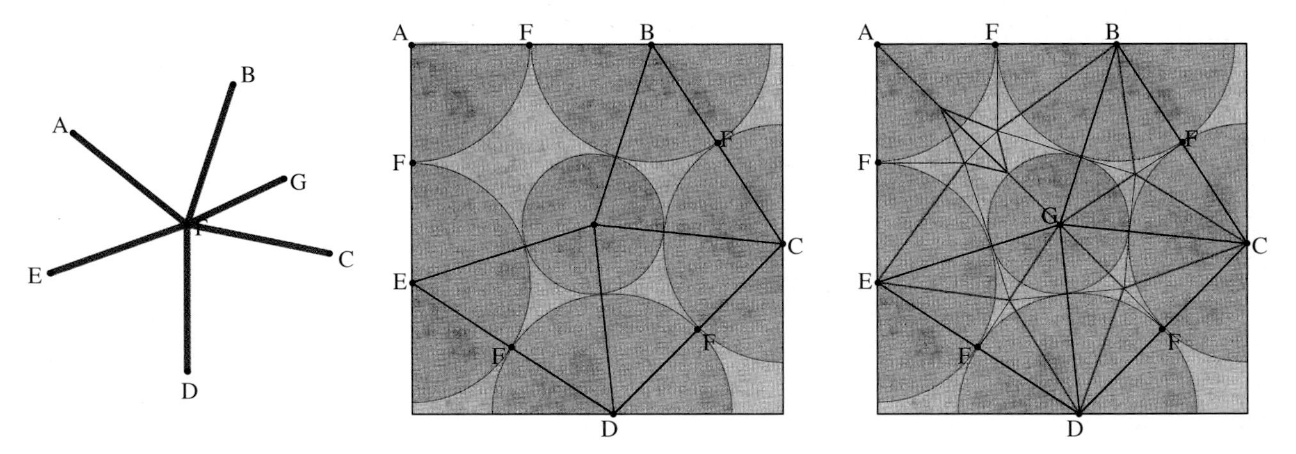

Figure 11.22.
Adding another flap to the base is equivalent to adding a new edge (and leaf node) to the tree.
Left: the modified tree.
Middle: the new circle pattern.
Right: the crease pattern with molecules in place.

In this polygon, because of the bilateral symmetry, we were able to make the new circle contact four other circles (or equivalently, turn four path inequalities into equalities). But in the general case, this is usually not possible. This can be seen by counting degrees of freedom; when we add a new circle, we have three degrees of freedom: the (two) coordinates of its center plus the radius of the circle. So we can, in general, use those three degrees of freedom to satisfy only three equalities.

Because of this limitation, we cannot usually subdivide quadrilaterals into triangles. For example, looking at quadrilateral ABGE in Figure 11.23, if we add another circle to the opening within the quadrilateral (which corresponds to adding another edge to the tree graph at node F), we will find that we divide the quadrilateral into two triangles—and another quad. Adding a circle to this new quad still leaves a quad behind. This process can continue forever, always leaving a residual quadrilateral, which is why we needed the gusset quad and other quadrilateral molecules.

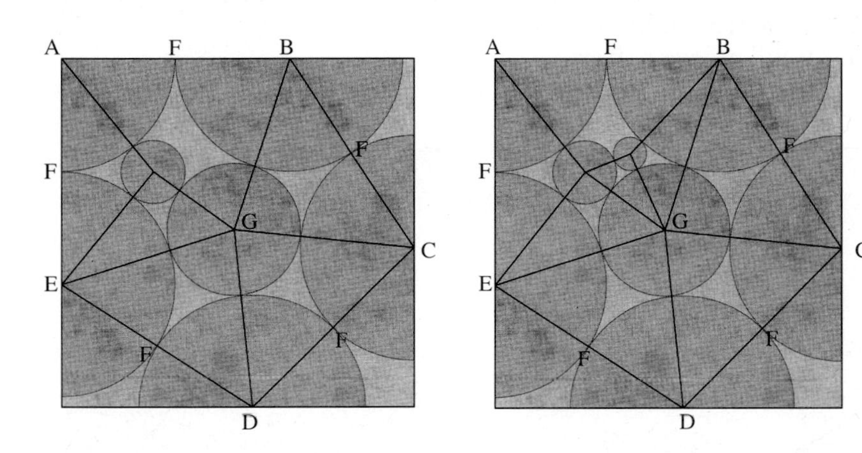

Figure 11.23.
Left: adding a circle to the quadrilateral subdivides it, leaving a new quadrilateral. Right: subdividing the new quadrilateral still leaves a smaller quad.

There is a way, however, of adding a fourth degree of freedom. We can add a new branch node along one of the existing edges of the tree and add a new edge and new leaf node to the new branch node. There are now four degrees of freedom: the two coordinates of the new leaf vertex, the length of the new edge, and the distance along the existing edge where the new branch node is placed.

With four degrees of freedom, it is, in principle, possible to satisfy four path equalities simultaneously. In the tree graph we have been working on, it turns out that we can add our new branch node to either of two edges, those connected to leaf nodes A and G. Both give solutions that satisfy the path conditions, as shown in Figures 11.24 and 11.25.

Both solutions divide the quadrilateral into four triangles, and in general, any quadrilateral can be similarly divided. I call this process adding a *stub* to the tree. By repeatedly adding stubs to a uniaxial base crease pattern, any such crease pattern can eventually be divided into axial polygons that are all triangles, whereupon they all can be filled in with rabbit-ear mol-

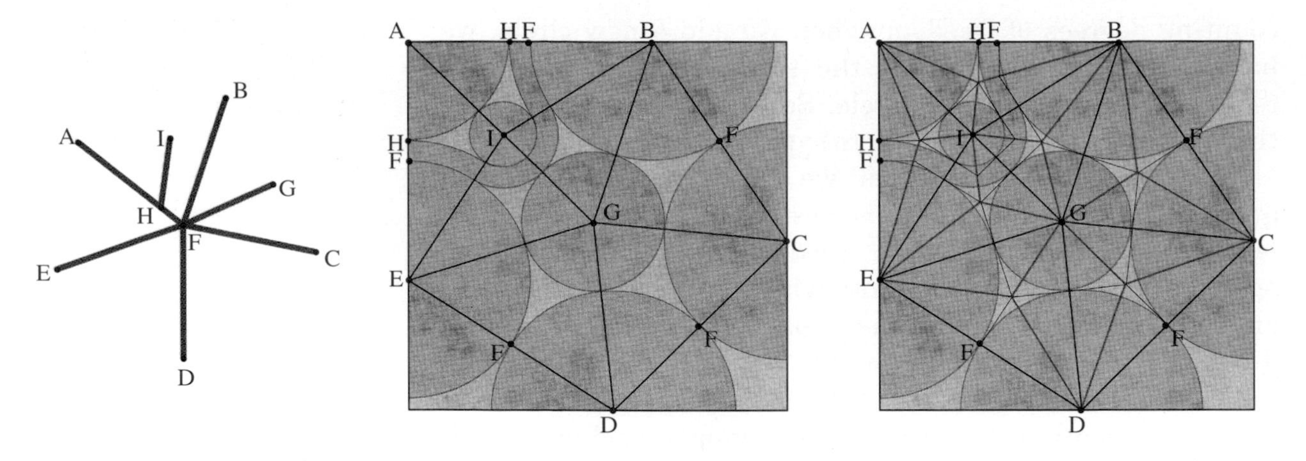

Figure 11.24.
Left: a stub added to node A's edge that satisfies four path equalities.
Middle: active (axial) paths.
Right: full crease pattern.

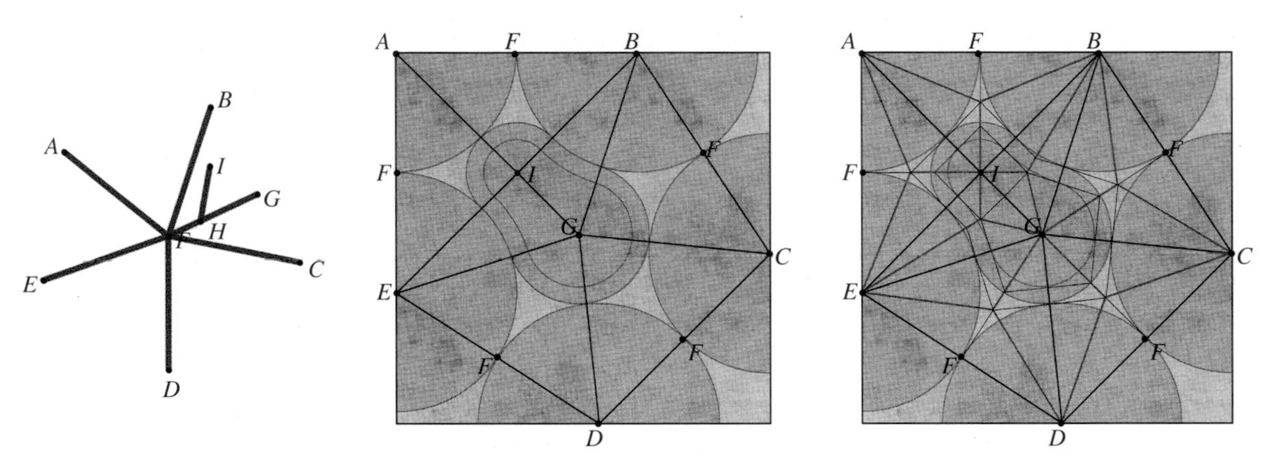

Figure 11.25.
Left: a second solution, adding the stub to node G's edge.
Middle: active (axial) paths.
Right: full crease pattern.

ecules. A crease pattern that has received this treatment, i.e., consists entirely of rabbit-ear molecules, has been *triangulated.*

There is an interesting relationship between a quadrilateral that has been quartered using a stub and the arrowhead molecule. Look at the quadrilateral crease pattern in Figure 11.26. By removing a few creases, it's possible to transform this pattern into either version of the arrowhead molecule for this quadrilateral.

Another interesting observation about stub-divided quads: The crease pattern within a stub-divided quad is topologically equivalent to a Bird Base, and by changing the

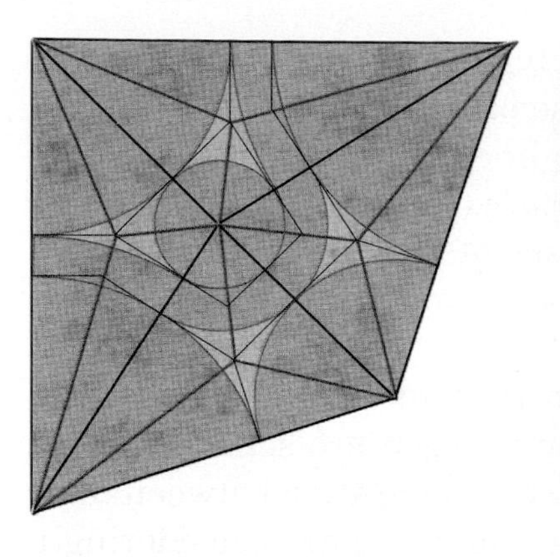
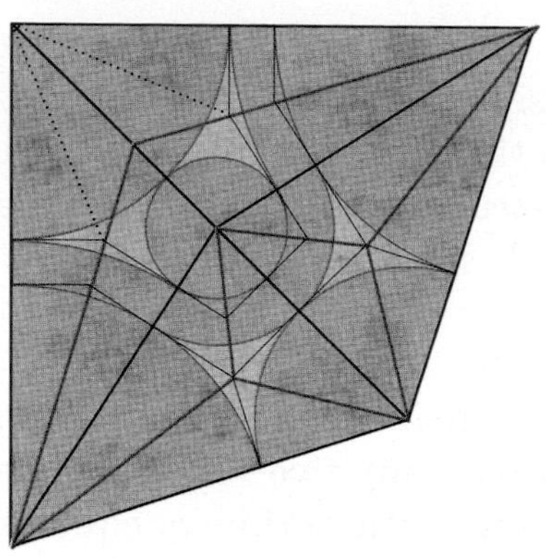
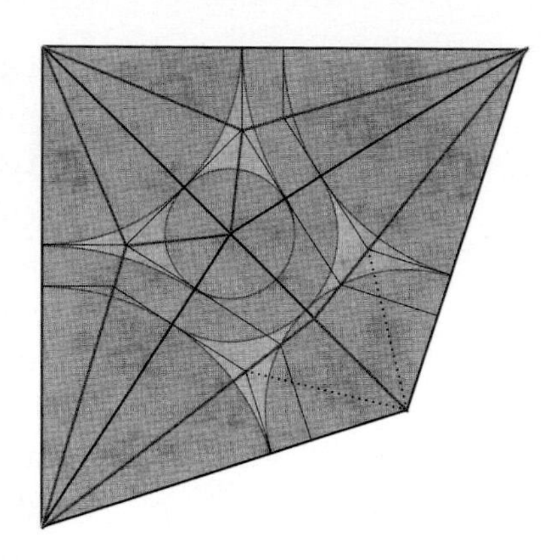

Figure 11.26.
Left: a stub-divided quadrilateral.
Middle: one version of the arrowhead molecule.
Right: another arrowhead molecule.

directions of some of the creases, it is possible to use the crease pattern of a stubbed quad to fold any such quad into an analog of the Bird Base.

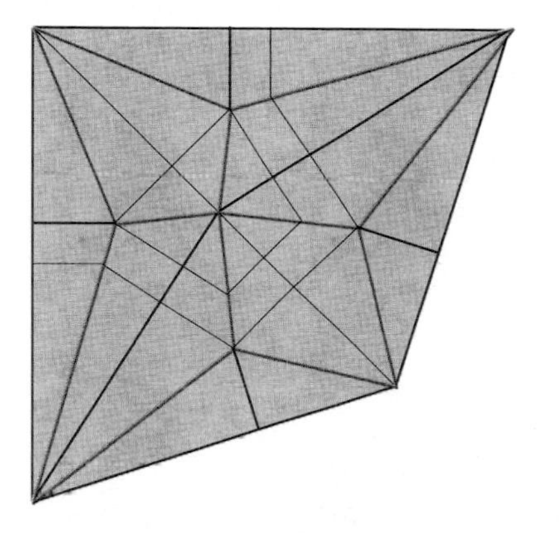
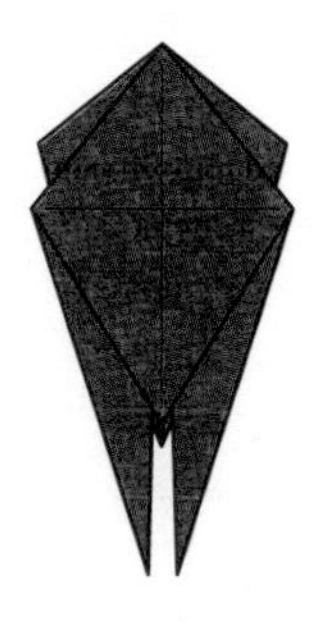

Figure 11.27.
The crease pattern from a stubbed quadrilateral can be used to fold the quadrilateral into an analog of the Bird Base.

The properties of quadrilaterals with distorted Bird Base crease patterns have been the subject of considerable investigation on their own; Justin, Husimi, and Kawasaki have all enumerated various special cases.

11.8. The Universal Molecule

Since every polygon network can be broken up into triangles and quads by the addition of extra circles, the triangle and quad molecules are by themselves sufficient for filling in the crease pattern for any tree. And if we subdivide quadrilaterals with stubs, we can get everything down to triangles, so that the rabbit-ear molecule is the only one needed. However, there are many other possible molecules, including molecules that can be used for higher-order polygons. It turns out that the gusset quad is just a special case of a more general construction that is applicable to any higher-order polygon. I call this construction the *universal molecule*. In fact, all of the known simple molecules are special cases of the universal molecule. The rest of this section describes the construction of this molecule for an arbitrary polygon.

Consider a general polygon that satisfies the tree theorem, i.e., any two vertices of the polygon are separated by a distance greater than or equal to the separation between their corresponding nodes on the tree. Since we are considering a single axial polygon, we know that of the paths between nonadjacent vertices, none are at their minimum length (otherwise it would be an active path and the polygon would have been split).

Suppose we inset the boundary of the polygon by a distance h, as shown in Figure 11.28. If the original vertices of the polygon were $A_1, A_2,\ldots$, then we will label the inset vertices $A_1', A_2',\ldots$ as we did for the gusset quad construction. I will call the inset polygon a *reduced polygon* of the original polygon.

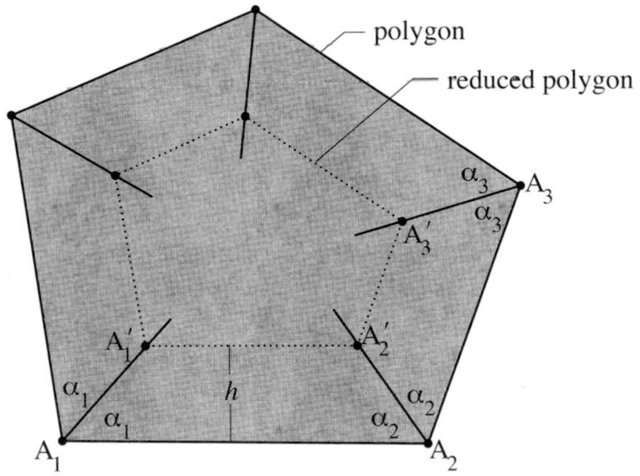

Figure 11.28.
A reduced polygon is inset a distance h inside an axial polygon. The inset corners lie on the angle bisectors (dotted lines) emanating from each corner.

Note that the points A_i' lie on the bisectors emanating from the points A_i for any h. Consider first a reduced polygon that is inset by an infinitesimally small amount. In the folded base, the sides of the reduced polygon all lie in a common plane, just as the sides of the original axial polygon all lie in a common plane. However, the plane of the sides of the reduced polygon is offset vertically from the plane of the sides of the axial polygon by a distance h. This is illustrated schematically in Figure 11.29.

As we increase h, we shrink the size of the reduced polygon. Is there a limit to the shrinkage? Yes, there is, and this limit is the key to the universal molecule. Recall that for any

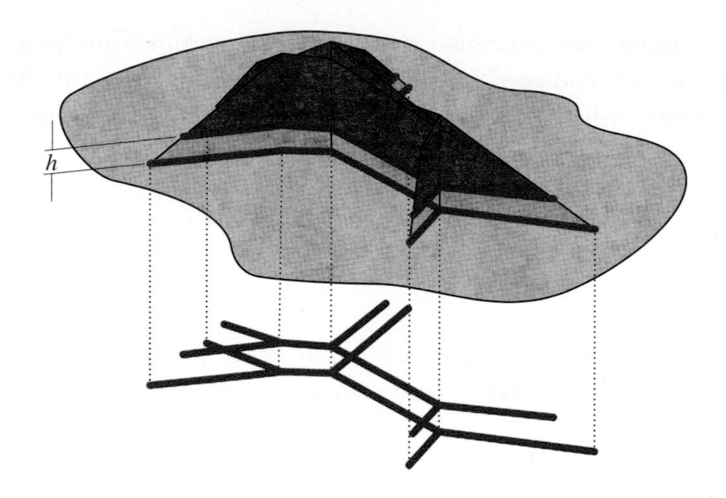

Figure 11.29.
The reduced polygon in the folded form corresponds to the original polygon cut by a plane a distance h above the original plane of projection.

polygon that satisfies the tree theorem, the path between any two vertices satisfies a path length constraint

$$\left|A_i - A_j\right| \geq ml_{ij}. \qquad (11\text{–}6)$$

where l_{ij} is the path length between nodes i and j measured along the tree. There is an analogous condition for reduced polygons; any two vertices of a reduced polygon must satisfy the condition

$$\left|A'_i - A'_j\right| \geq ml'_{ij} \qquad (11\text{–}7)$$

where l'_{ij} is a *reduced path length* given by

$$l'_{ij} = l_{ij} - h(\cot\alpha_i + \cot\alpha_j). \qquad (11\text{–}8)$$

and α_i is the angle between the bisector of corner i and the adjacent side. Equation (11–7) is called the *reduced path inequality* for a reduced polygon of inset distance h. Any path for which the reduced path inequality becomes an equality is, in analogy with active paths between nodes, called an *active reduced path*.

So for any distance h, we have a unique reduced polygon and a set of reduced path inequalities, each of which corresponds to one of the original path inequalities. We have already assumed that all of the original path inequalities are satisfied; thus, we know that all of the reduced path inequalities are satisfied for the $h = 0$ case (no inset distance). It can also be shown that there is always some infinitesimally small but positive value of h for which the reduced path inequalities are also satisfied. On the other hand, as we increase the inset distance, there comes a point beyond which one or more of the reduced path constraints is violated.

Suppose we increase h to the largest possible value for which every reduced path inequality remains true. At the maximum value of h, one or both of the following conditions will hold:

- For two adjacent corners, the reduced path length has fallen to zero and the two inset corners are degenerate; or

- For two nonadjacent corners, a path between inset corners has become an active reduced path.

These two situations are illustrated in Figure 11.30.

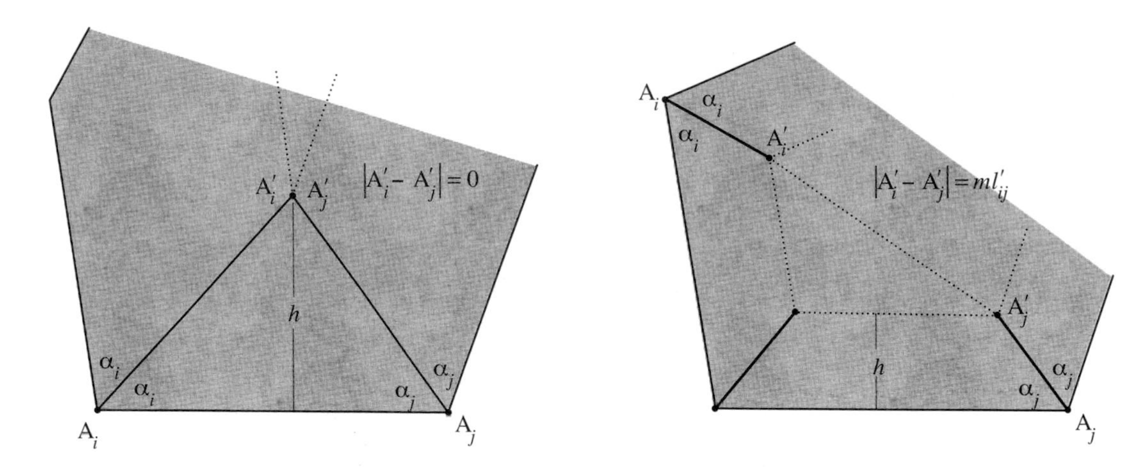

Figure 11.30.
Left: two corners are inset to the same point, which is the intersection of the angle bisectors.
Right: two nonadjacent corners inset to the point where the reduced path between the inset corners becomes active.

Again, one or the other (or both) of these situations must apply; it *is* possible that paths corresponding to both adjacent and nonadjacent corners have become active simultaneously or that multiple reduced paths have become active for the same value of h (this happens surprisingly often). In either case, the reduced polygon can be simplified, thus reducing the complexity of the problem.

In a reduced polygon, if two or more adjacent corners have coalesced into a single point, then the reduced polygon has fewer sides (and paths) than the original axial polygon. And if a path between nonadjacent corners has become active, then the reduced polygon can be split into separate polygons along the active reduced paths, each with fewer sides than the original polygon (just as in the polygon network, an active path across an axial polygon splits it into two smaller polygons).

The gusset molecule is an example of a reduced path becoming active. In the gusset molecule the reduced quadrilateral is inset until one of its diagonals becomes an active path; the reduced quad is then split along the diagonal into two triangles.

In either situation, you are left with one or more polygons that have fewer sides than the original. The process of insetting and subdivision is then applied to each of the interior polygons anew, and the process repeated as necessary.

If a polygon (active or reduced) has three sides, then there are no nonadjacent reduced paths. The three bisectors intersect at a point, and the polygon's reduced polygon evaporates to a point, leaving a rabbit-ear molecule behind composed of the bisectors.

Four-sided polygons can have the four corners inset to a single point or to a line, in which case no further insetting is required, or to one or two triangles, which are then inset to a point. Higher-order polygons are subdivided into lower-order ones by direct analogy.

Since each stage of the process absolutely reduces the number of sides of the reduced polygons created (although possibly at the expense of creating more of them), the process must necessarily terminate. Since each polygon (a) can fold flat, and (b) satisfies the tree theorem, then the entire collection of nested polygons must also satisfy the tree theorem. Consequently, *any* axial polygon that satisfies the tree theorem—no matter how many sides—can be filled with a crease pattern using the procedure outlined above and collapsed into a base on the resulting creases.

Thus, for example, the five-flap pentagon that I used to illustrate adding circles and stubs could also be turned into a molecule directly using the universal molecule construction. The pentagon ABCDE is inset, forming pentagon A′B′C′D′E′; the inset distance is chosen so that reduced path E′B′ becomes

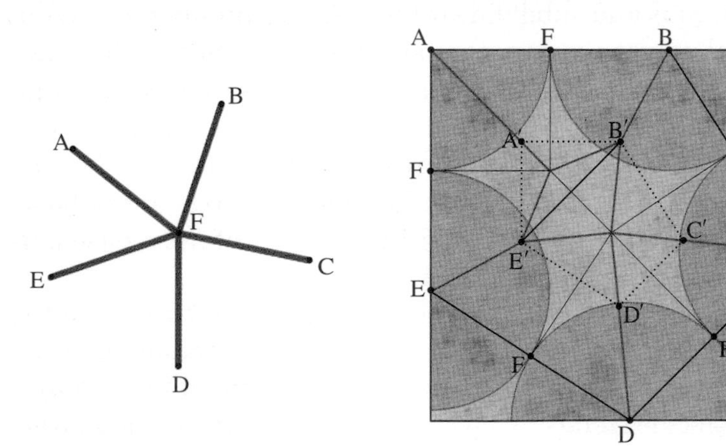

Figure 11.31.
Left: five-flap tree.
Right: universal molecule construction crease pattern (generic form).

active. This becomes a mountain fold, and splits the reduced polygon into two distinct polygons, triangle A′B′E′ and quadrilateral B′C′D′E′. Repeating the insetting process on each of these reveals that in each case, the new polygon can be inset to a common point, yielding the rabbit-ear molecule in the former and the Waterbomb molecule in the latter.

A remarkable feature of the universal molecule is that all of the simple molecular crease patterns that have been previously enumerated are just special cases of it, including the rabbit-ear molecule, the gusset quad, and both sawhorse and Waterbomb quads. So the universal molecule well deserves its name; it is the only molecule needed to turn any tree method uniaxial base into a folded base.

Unfortunately, for polygons of higher order than quadrilaterals, there is generally no easy way to construct the universal molecule by folding alone; in most cases, it must be computed. A numerical prescription for the computation is given in Chapter 14.

Faced with an axial polygon with five or more sides, you can do one of three things:

- Add a circle (equivalent to adding an edge to an existing node of the tree), which creates three or more new active paths.
- Add a stub to the tree (equivalent to adding an edge and a new node to an existing edge of the tree), which creates four or more new active paths.
- Construct a universal molecule.

Since polygon subdivision is commonly called for in several places, you can mix and match approaches; say, add a stub to fracture a polygon, then fill in the results with universal molecules. Or you could apply the universal molecule to some polygons and subdivide others. As the number of sides of the initial polygon grows, the possibilities explode. All crease patterns will be foldable into bases with the same number and length of flaps as was specified by the tree; the differences lie in the width of the flaps, the presence of extra flaps, and the number of layers of paper that lie along the axis of the base. Figure 11.32 shows the folded form for three of the crease patterns for the five-flap pentagon.

These images also illustrate some general features of the different approaches. A nice feature of the universal molecule is that it is very frugal with creases. A tree filled in with universal molecules tends to have relatively few creases and large,

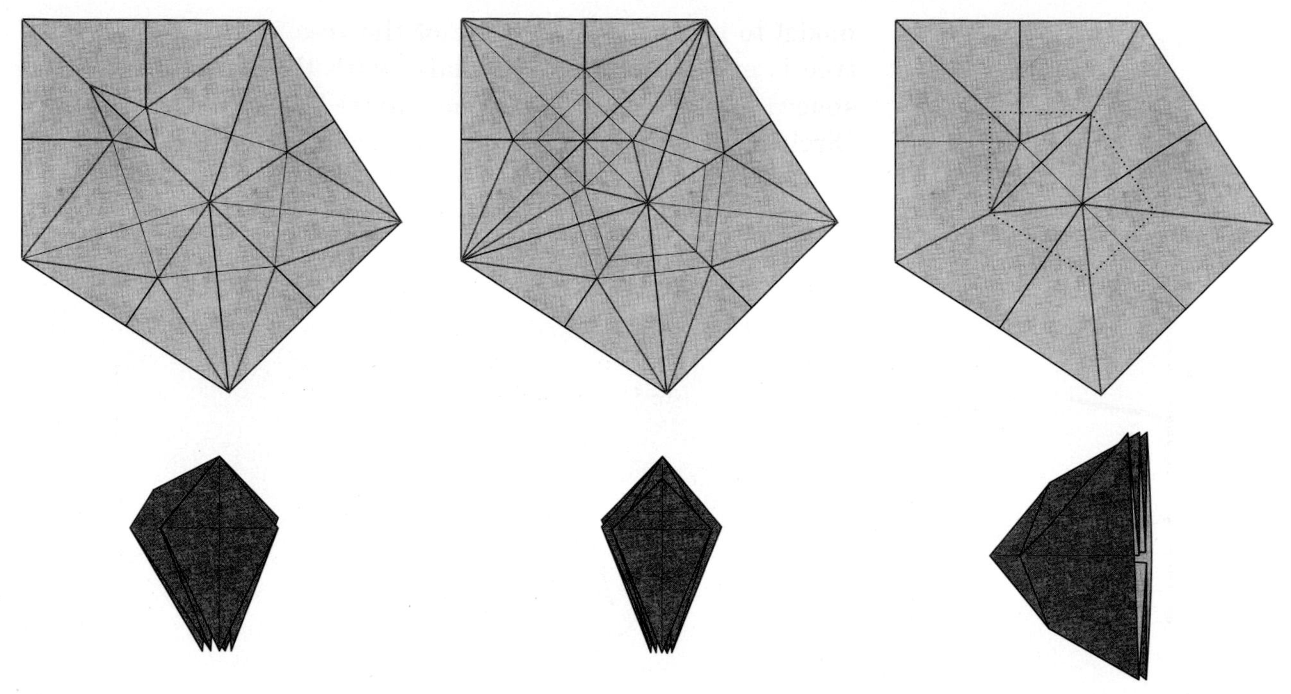

Figure 11.32.
Crease patterns and folded forms for three different molecular solutions to the five-flap pentagon.
Left: stub plus gusset quad.
Middle: two stubs.
Right: universal molecule.

wide flaps (which can, of course, be subsequently narrowed arbitrarily as desired). In fact, I conjecture that for any axial polygon, the universal molecule is the crease pattern with the shortest total length of creases that collapses that polygon to a uniaxial base. A small number of creases translates into relatively few layers in the base, at least until you start sinking edges to narrow them. A base with narrow flaps will require many folds, no matter how you design it. But with the universal molecule, because you don't have to arbitrarily add circles (and hence points) to a crease pattern to knock polygons down to quads and triangles, bases made with the universal molecule tend to have less bunching of paper and fewer layers near joints of the base, even with multiply-sunk flaps, resulting in cleaner and—sometimes—easier-to-fold models.

11.9. Other Techniques

An alternative design approach that blends aspects of the circle/river method and tree method has been described by Kawahata and Maekawa. It has been called the *string-of-beads* method of design. As in the tree method, you begin with a tree of the model to be folded. Each line of the tree is doubled and the tree is expanded to fill a square, with the nodes of the tree spaced around the edges of the square like beads on a string. Circles and circular arcs are then constructed in the square that surround each leaf vertex. The process is illustrated for a six-flap base shown in Figure 11.33.

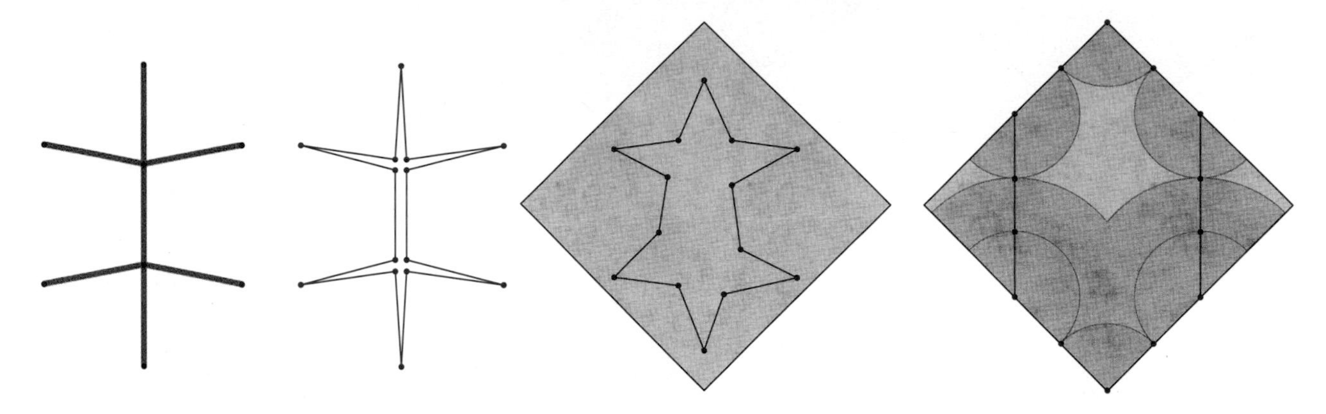

Figure 11.33.
The string-of-beads method. The tree is turned into a closed polygon, which is then inflated inside a square with straight lines between the leaf nodes. The result is a large polygon inside the square that is subsequently collapsed into the base.

In the string-of-beads method, the tree is converted into a large polygon in which each corner is one of the leaf nodes of the tree and each side is as long as the path between adjacent leaf nodes. It is clear that this distribution of leaf nodes is just a special case of the tree method in which we have constrained all of the nodes to lie on the edge of the square; it avoids creating middle flaps, but at the possible expense of efficiency.

The string-of-beads method produces a single large polygon that must be collapsed into the base. The techniques described by Maekawa involve placing tangent circles in the contours shown in the last step of Figure 11.33, which is analogous to our use of additional circles to break down axial polygons into smaller polygons in the tree method. Kawahata's algorithm projects hyperbolas in from the edges to locate reference points for molecular patterns, and produces yet another type of molecule.

One can also apply the universal molecule directly to the string-of-beads polygon, achieving another efficient crease pattern that collapses into a base.

Figure 11.34 shows the universal molecule. The initial hexagon is inset to the point that the two horizontal reduced paths become active, and the hexagon is split into two triangles

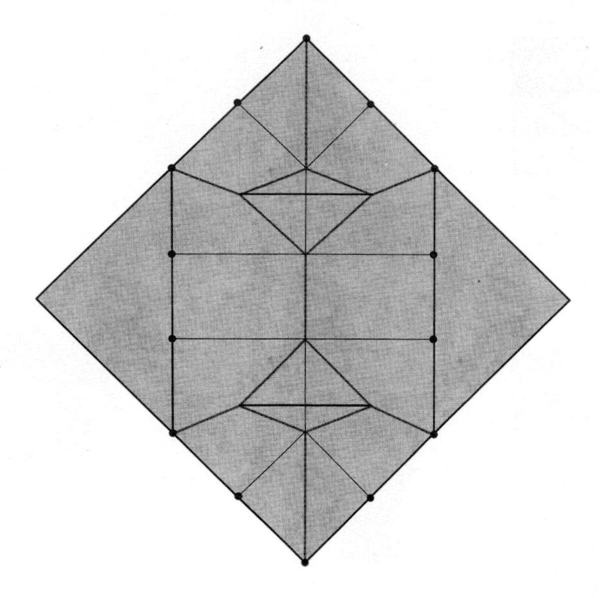

Figure 11.34.
Universal molecule for the polygon shown in Figure 11.33.

and a rectangle. The triangles are filled with rabbit ear creases; the rectangle is further inset, forming a sawhorse molecule.

The tree method of design is based on equations and has been rigorously proven to work. Rigorous proof may ease one's mind, but solving the equations can be quite difficult to do by hand. Such computationally intensive problems are best handled by computer and, indeed, the procedures described above can be cast in the mathematical and logical terms that lend themselves to computer modeling. I have written a computer program, *TreeMaker,* which implements these algorithms. Using *TreeMaker*, I've created bases for a number of subjects whose solutions have eluded me over the years—deer with varying sizes and types of antlers, 32-legged centipedes, flying insects, and more. Using a computer program accelerates the development of a model by orders of magnitude; from the tree to the full crease pattern takes less than five minutes, although folding the crease pattern into a base may take two to three hours after that!

Computerized solution offers an addition benefit: precision. It is possible to specify a different value for the length of every flap individually. This is particularly desirable when there are many flaps of unequal length in which the lengths must fall in some type of regular progression. An example requiring this is a scorpion. There are many scorpions in the origami literature; without exception, they all have legs the same length. But in the actual creature, the legs get longer from front to back; they are also spaced out along the body. By plugging in a tree with the appropriate leg lengths, it is possible to compute a base with the graduated distribution of legs, permitting a more realistic representation of the subject.

Figure 11.35.
Crease pattern, base, and folded model of the Scorpion.

However, a drawback of computed crease patterns is that it can be quite difficult to construct a linear folding sequence.

Computational techniques are also helpful in creating bases for extremely complicated subjects, such as those with many flaps in varying sizes. A flying grasshopper, for example, has six legs—two are much longer than the other four—along with antennae (of intermediate length), head and thorax (short) and abdomen (long). The legs, wings, and antennae account for six pairs of flaps. There are many possible arrangements of circles representing those flaps. One of the more symmetric and pleasing arrangements is the crease pattern shown in Figure 11.36, along with its base and the folded model. Can you identify where a pair of stubs was added in the middle of the pattern?

Computation also allows one to introduce symmetries into the crease pattern, either to make the folding sequence simpler or for aesthetic reasons. A host of symmetric requirements can be imposed as additional equations to solve: forcing flaps to be mirror-image, or requiring active paths to fall along the symmetry line. This last condition is required to fold a plan view model—one that can be oriented with half the layers to the left of the axis and half to the right—or equivalently, to fold a model with a closed back. You can also force creases to run at particular angles. In the Alamo Stallion shown in Figure 11.37, several such symmetries are imposed:

Figure 11.36.
Crease pattern, base, and folded model of the Flying Grasshopper.

- An active path runs from the head to the tail so that the back is seamless.
- The base is symmetric about a line of bilateral symmetry.
- By forcing particular fold angles, its folding sequence becomes relatively tractable and requires few arbitrary reference points.

This last symmetry is a bit subtler. Observe that the equilateral triangle in the lower right is aligned with the ridgeline creases of the adjacent triangles; among other things, this choice forces equality between the length of the tail and the length of the hind legs. You can see the effect of this choice on the ease of folding; the full folding sequence for this model is given at the end of the chapter.

11.10. Comments

Tree theory is in some ways the culmination of all of the different techniques for constructing uniaxial bases. Uniaxial bases are wonderful things, but they are by no means all of origami. While insects, arthropods, and other many-legged creatures can often be successfully addressed with a uniaxial base, there

Figure 11.37.
Crease pattern, base, and folded model of the Alamo Stallion.

are many origami subjects for which the many narrow flaps of a uniaxial base are not particularly suitable. Furthermore, the great majority of origami figures designed over the years were not constructed from uniaxial bases, and many designers—most notably John Montroll—have developed other approaches to design that are clearly not uniaxial.

However, uniaxial bases are amazingly versatile, and because they can be constructed systematically, they can be used for quite a few origami problems. Furthermore, the underlying techniques are more broadly applicable, and concepts from tree theory, circle/river packings, point-splitting, and more, can be mixed and combined with other techniques to yield efficient, novel, and sometimes beautiful structures. The next two chapters demonstrate two of the many possibilities that lie within these hybrid approaches.

Folding Instructions

Alamo Stallion

Alamo Stallion

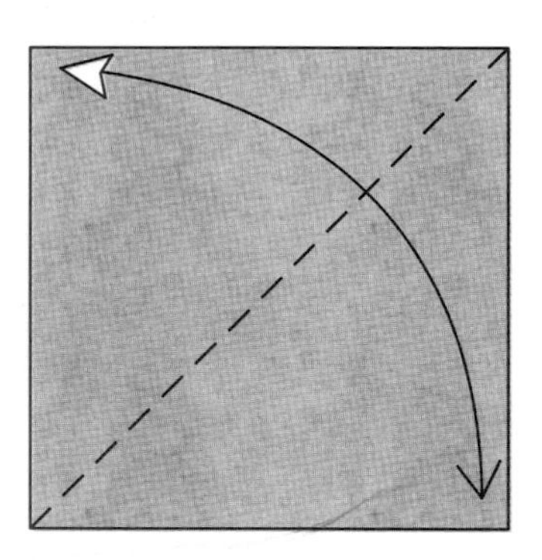

1. Begin with a square, white side up. Fold and unfold along the diagonal.

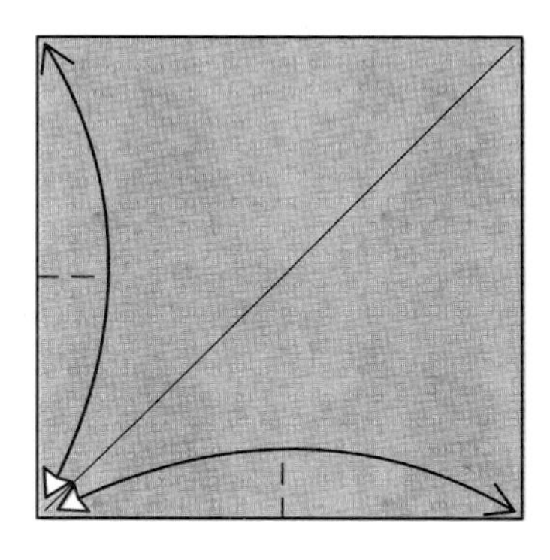

2. Make a pinch halfway along the left side and bottom.

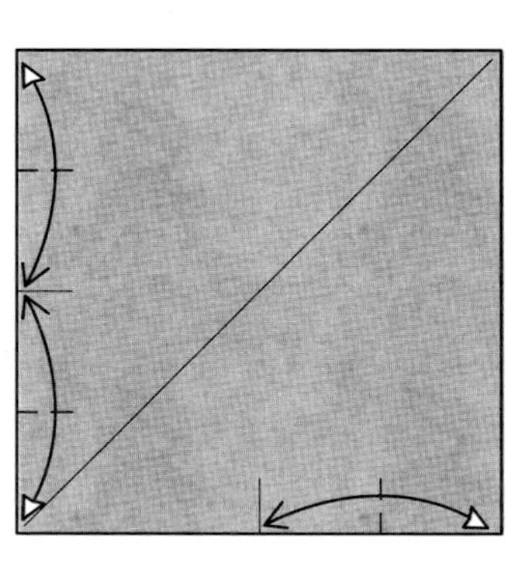

3. Make three more pinches along the edges.

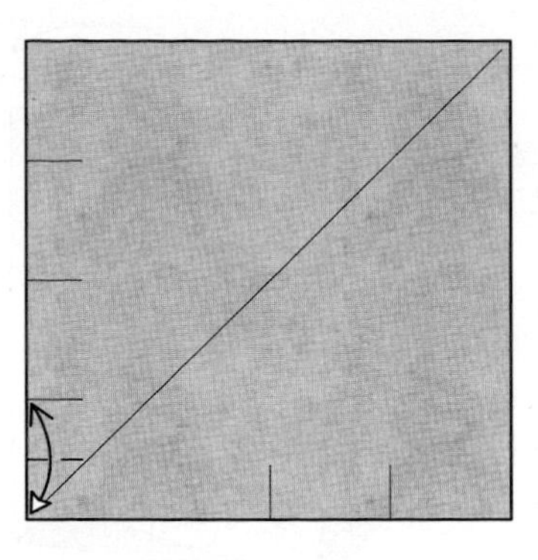

4. Make another pinch along the left edge.

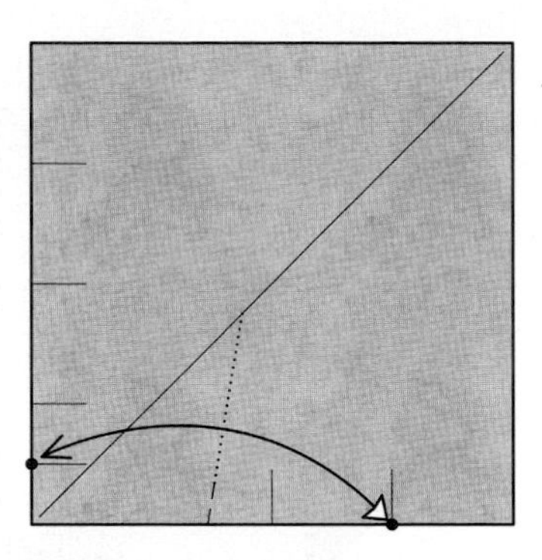

5. Fold the two indicated points together, make a pinch along the edge, and unfold.

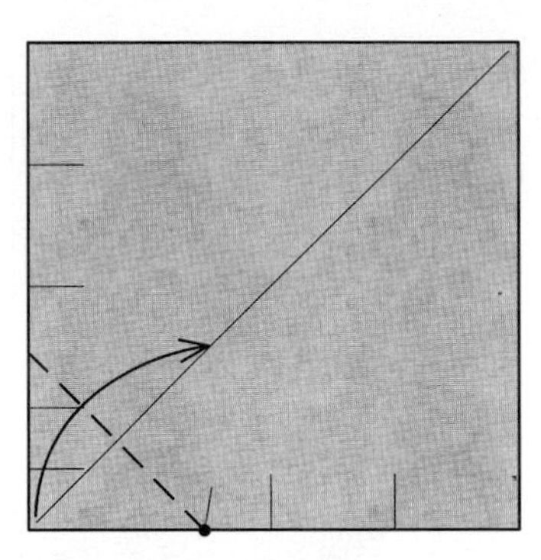

6. Fold the corner up along the diagonal so that the crease hits the pinch you just made.

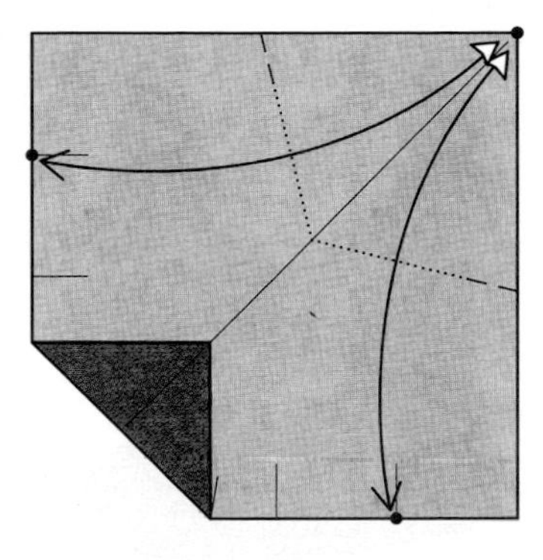

7. Fold the top right corner to the two indicated points, make pinches along the top and right edges, and unfold.

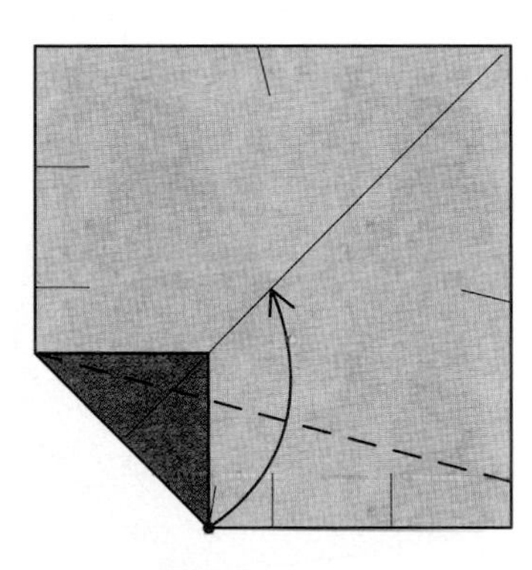

8. Fold the bottom corner up to lie on the diagonal so that the crease hits the left corner.

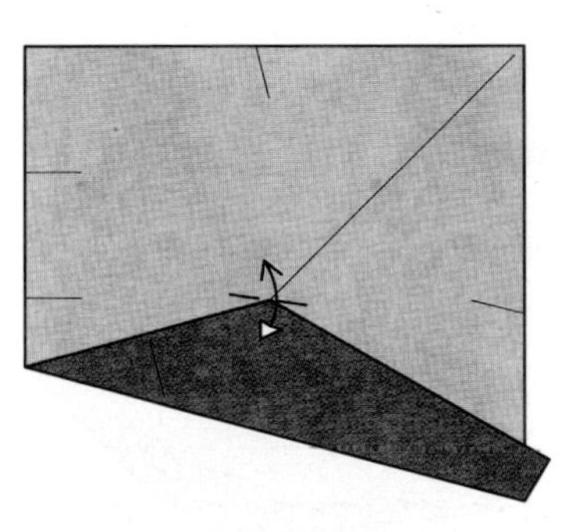

9. Make a pinch through the point where the corner touches the diagonal.

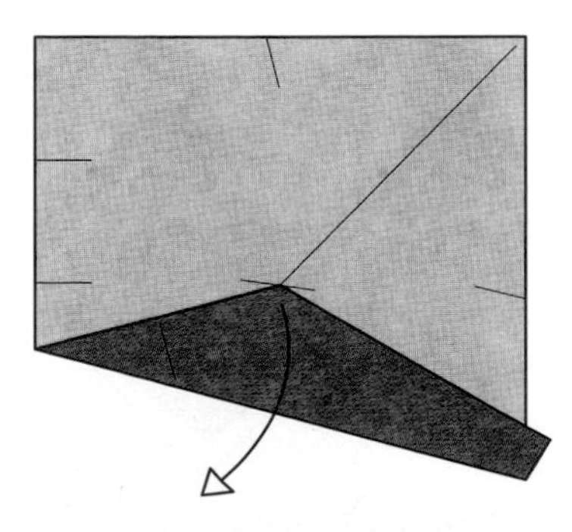

10. Unfold the paper.

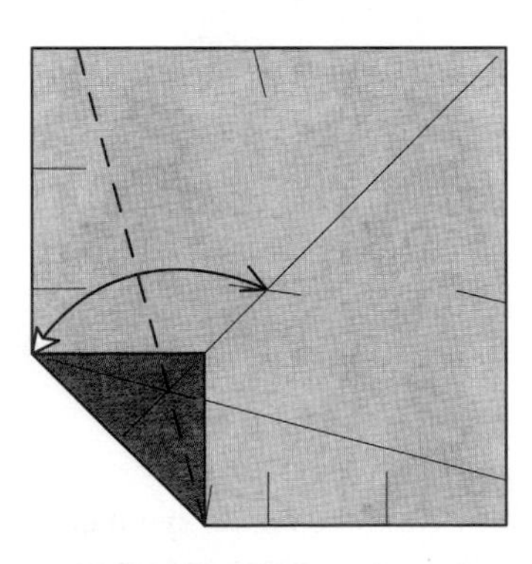

11. Fold the left corner to the pinch and unfold.

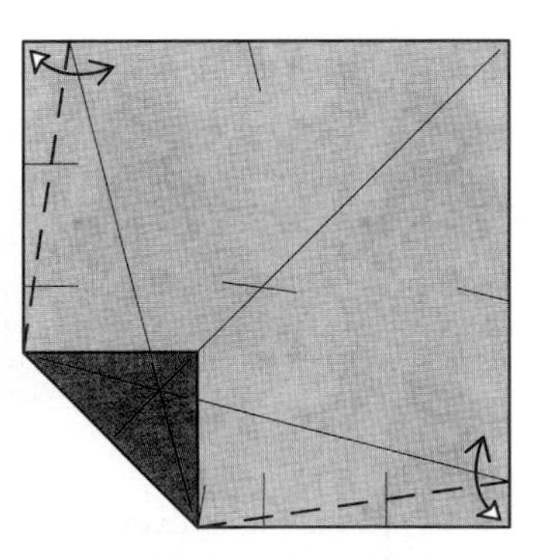

12. Fold and unfold.

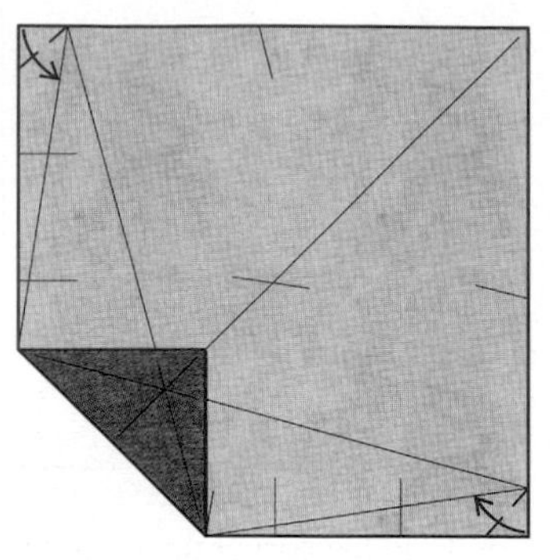

13. Fold the corner to the crease you just made.

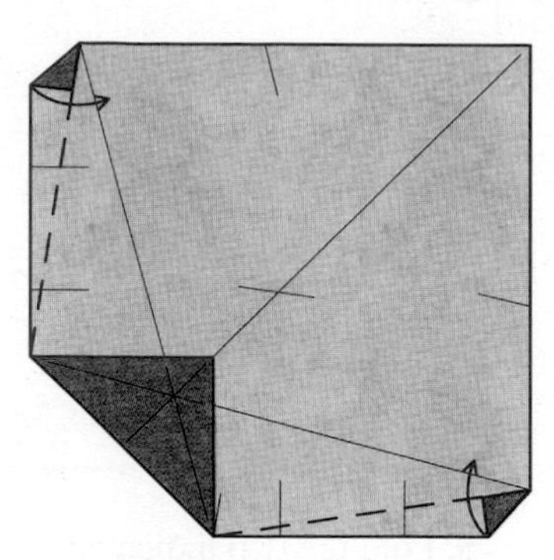

14. Refold on the creases you made in step 12.

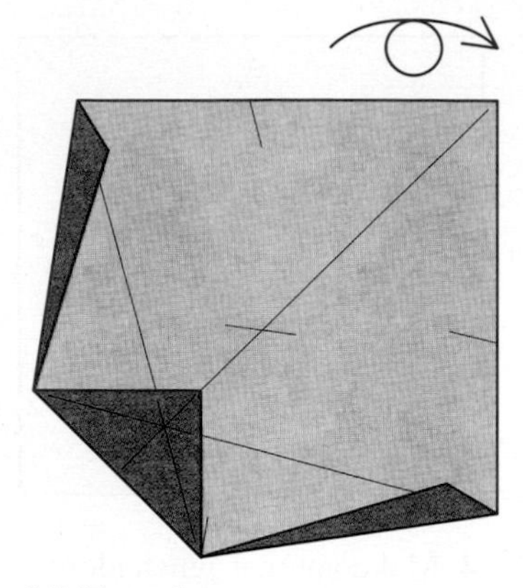

15. Turn the paper over.

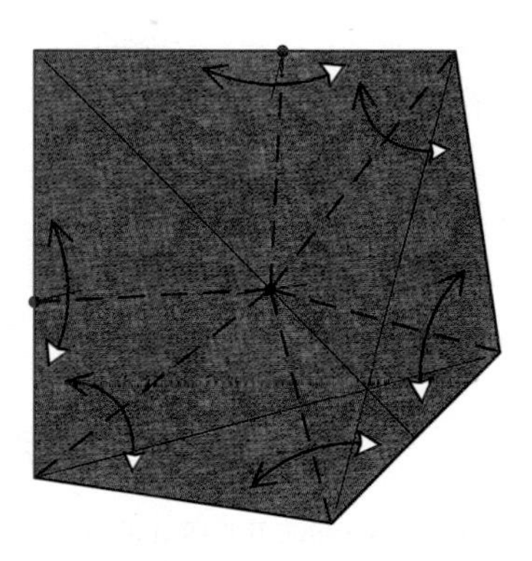

16. Fold and unfold. All six creases hit the diagonal at the mark you made in step 9.

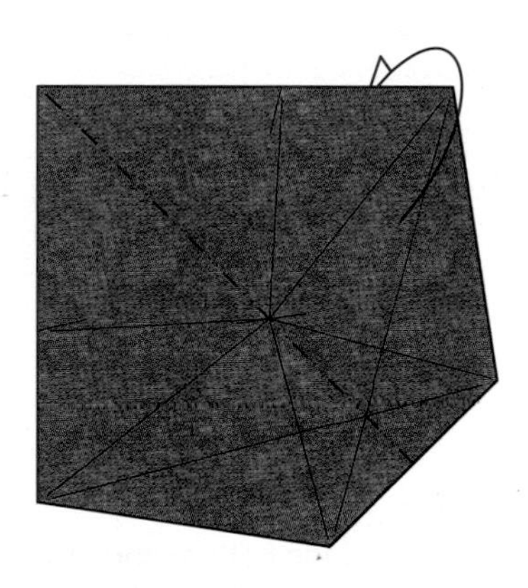

17. Mountain-fold the paper in half along the diagonal.

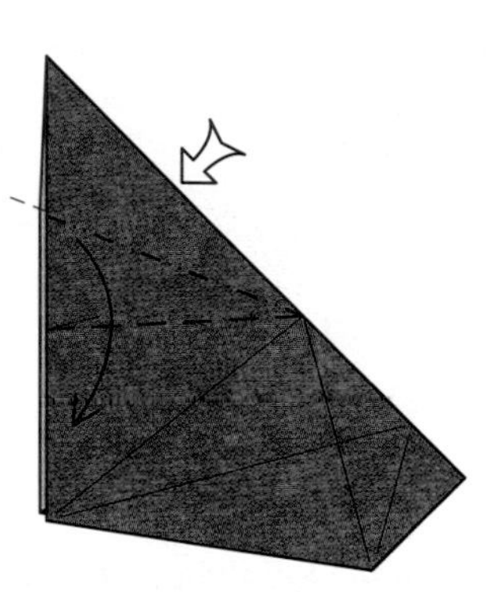

18. Squash-fold the flap symmetrically. The valley fold lies on an existing crease.

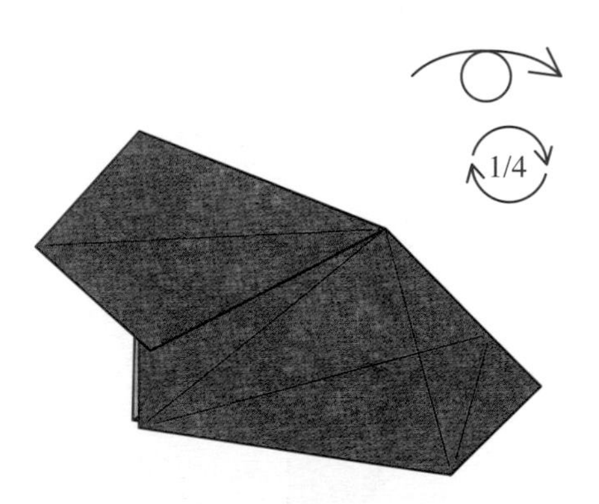

19. Turn the paper over and rotate 1/4 turn so that the white triangle is at the bottom.

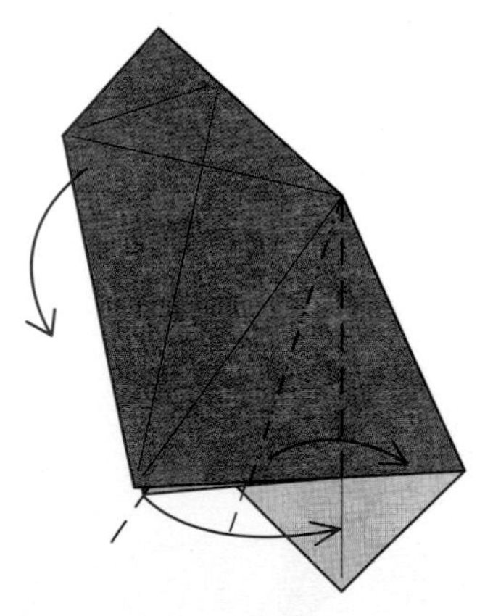

20. Crimp the model symmetrically so that two corners end up on the vertical crease.

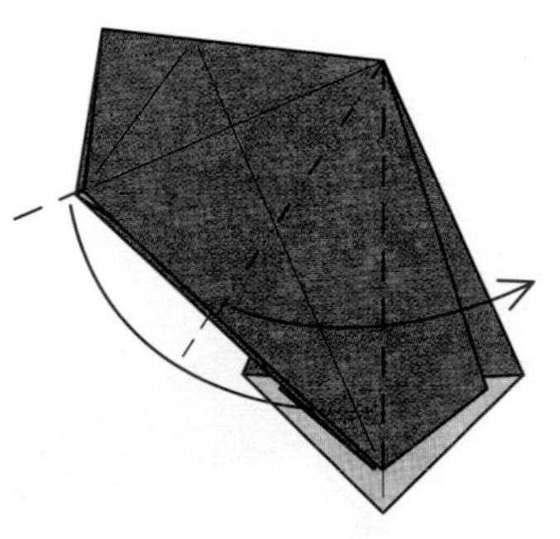

21. Crimp the model symmetrically so that the next two corners end up on the vertical crease.

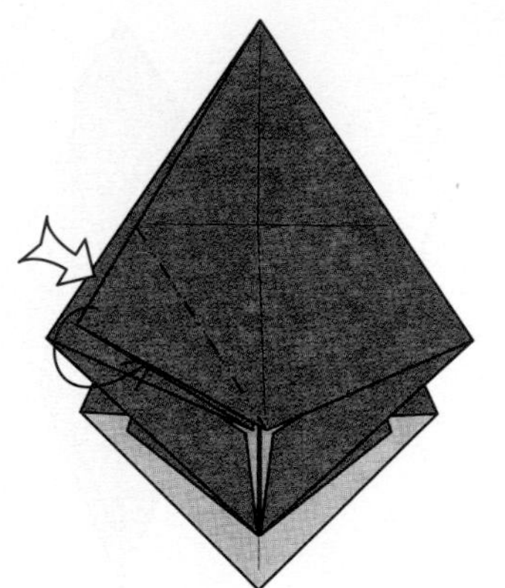

22. Reverse-fold the corner on the existing crease.

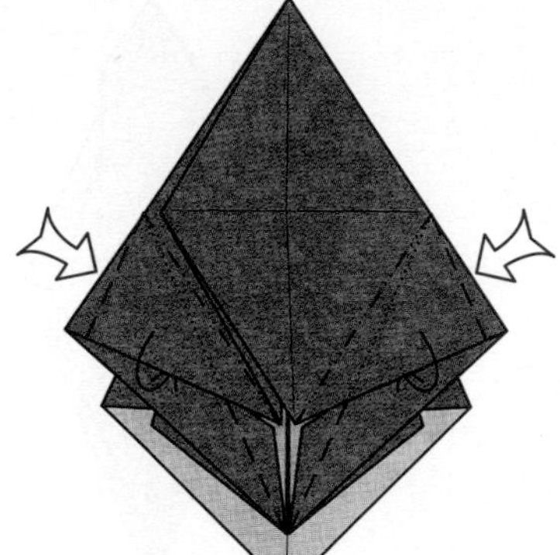

23. Reverse-fold the corners on both sides.

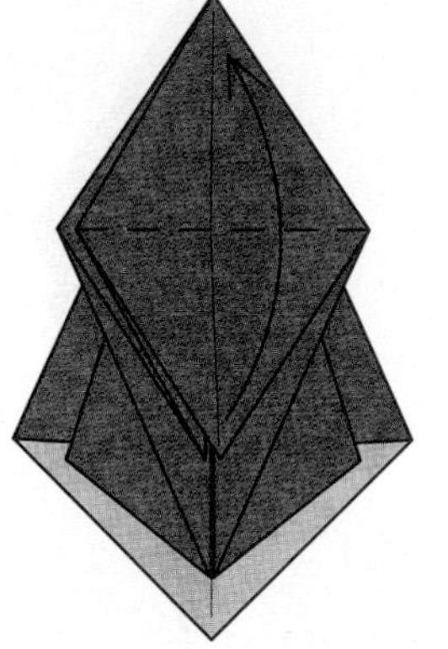

24. Fold one flap up.

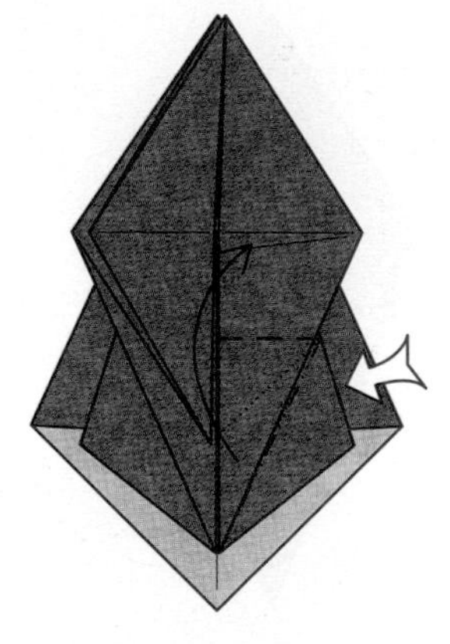

25. Squash-fold the edge.

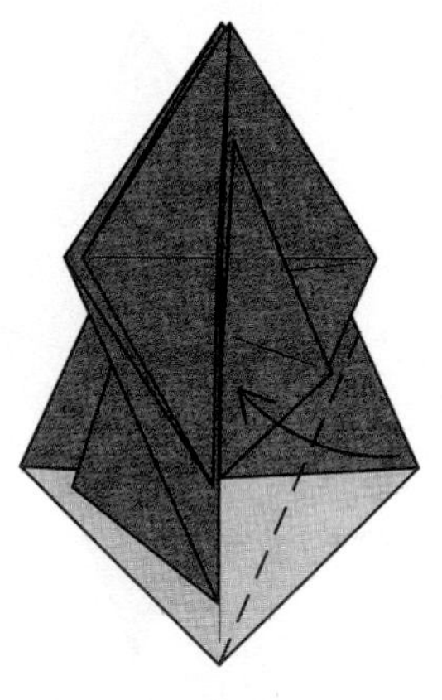

26. Fold the raw edge to the center line.

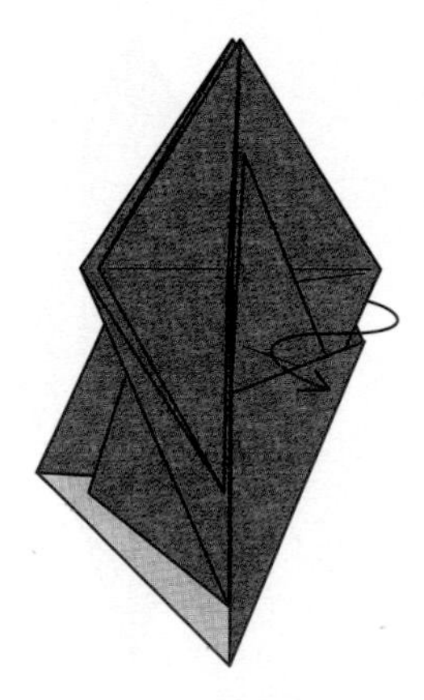

27. Bring a flap to the front.

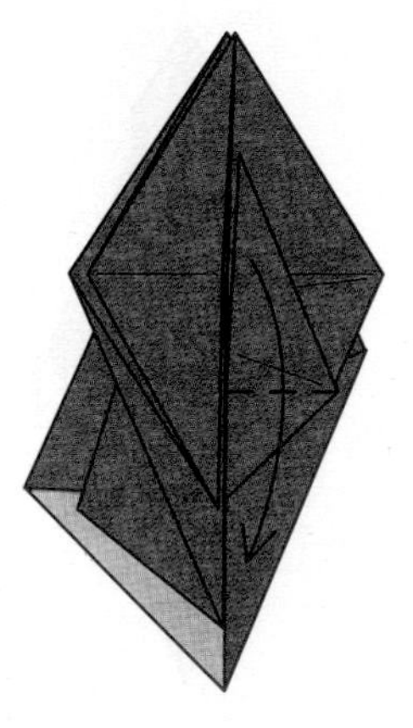

28. Fold one flap down.

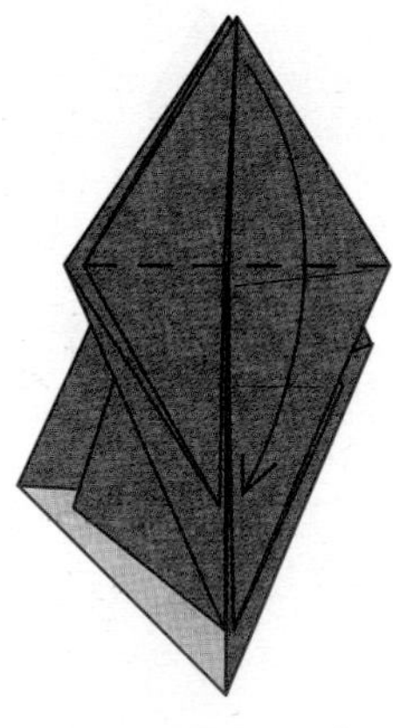

29. Fold the next flap down.

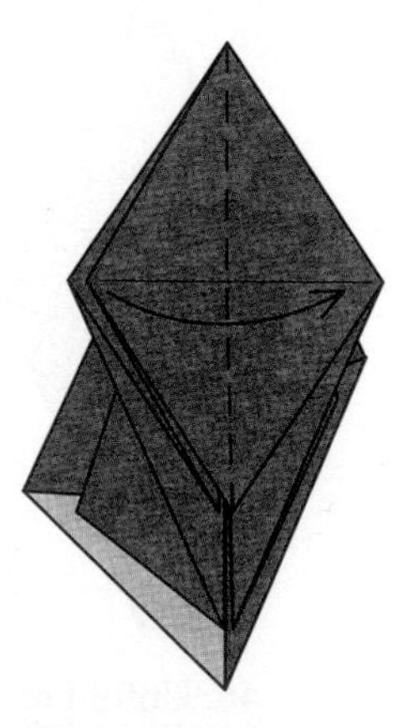

30. Swing one flap to the right.

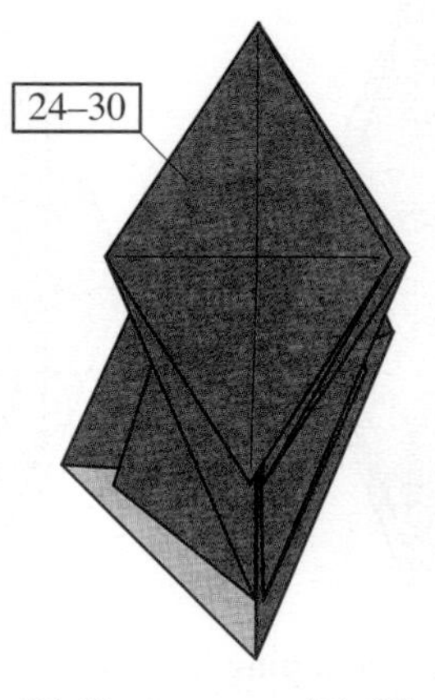

31. Repeat steps 24–30 on the left.

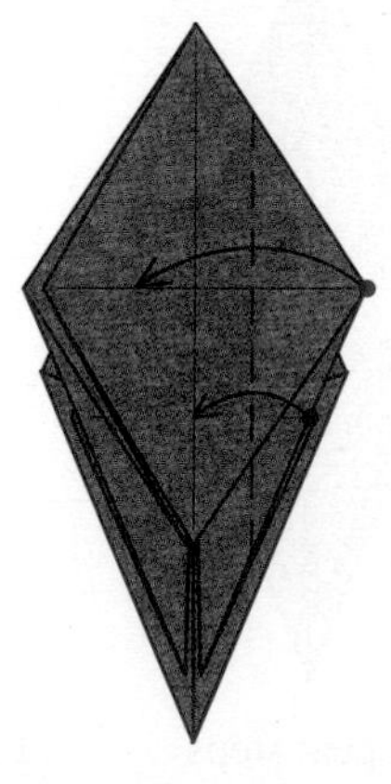

32. Fold two points to two lines.

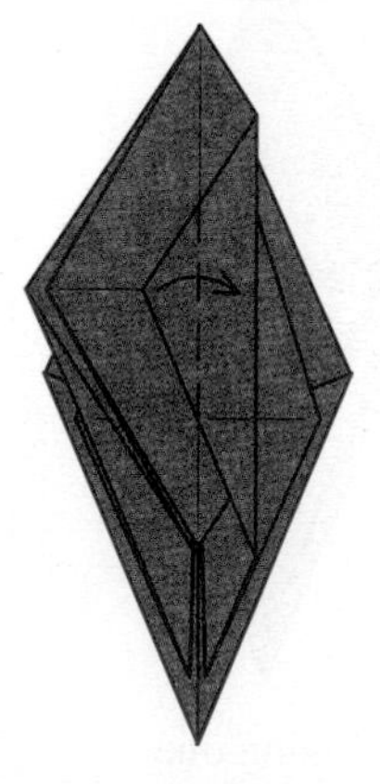

33. Fold the tip back to the right along a crease aligned with the center line.

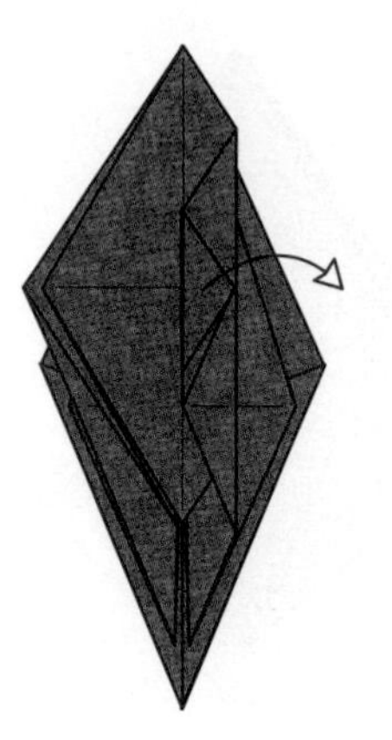

34. Unfold to step 32.

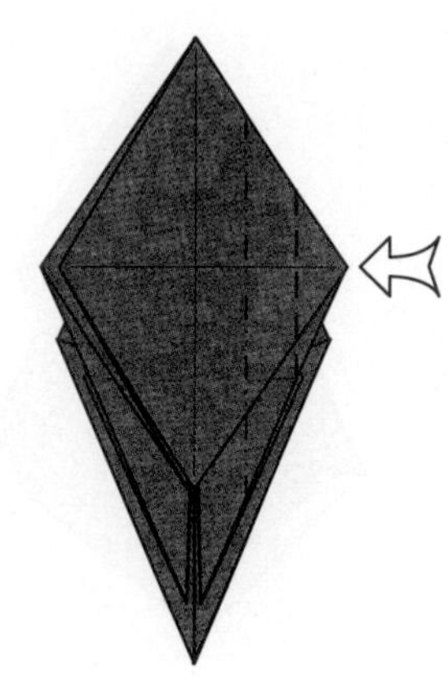

35. Open-sink in and out on the creases you just made.

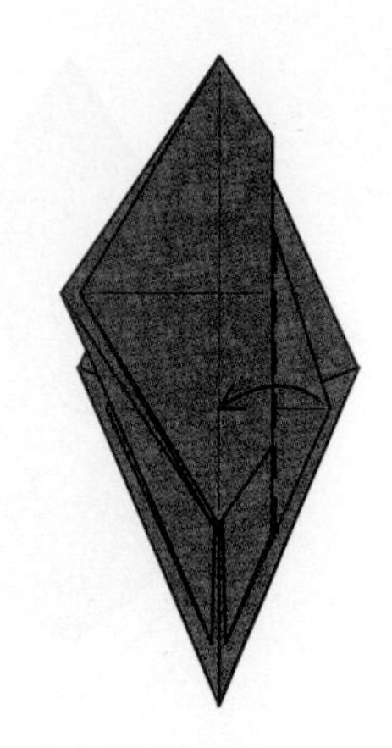

36. Fold the next flap tightly over the edges of the sink.

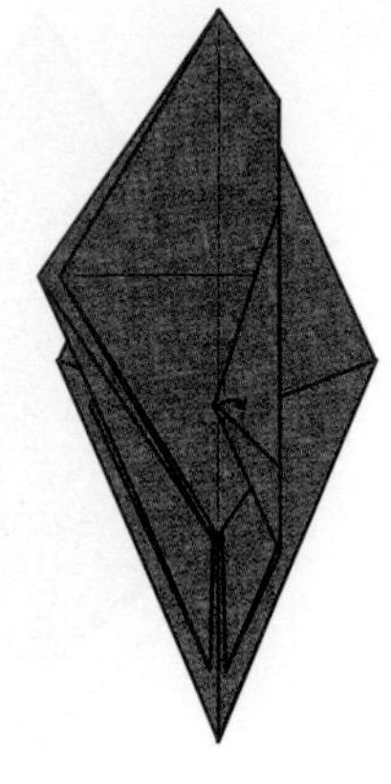

37. Fold the tip of the flap back to the right along a crease aligned with the center line.

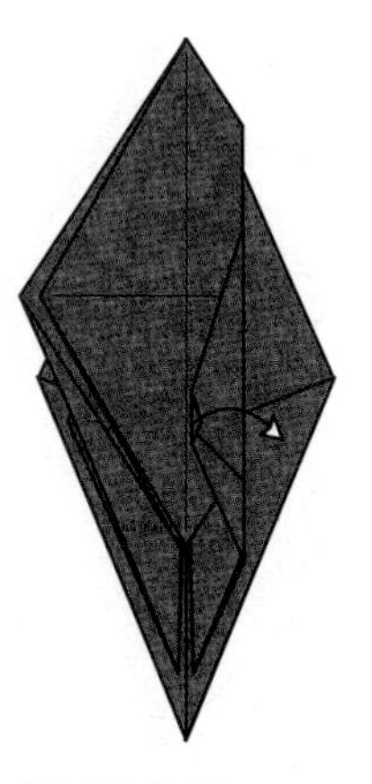

38. Unfold to step 36.

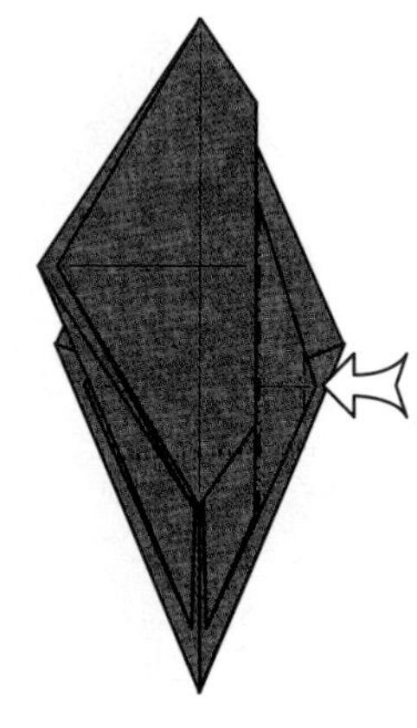

39. Open-sink in and out on the existing folds.

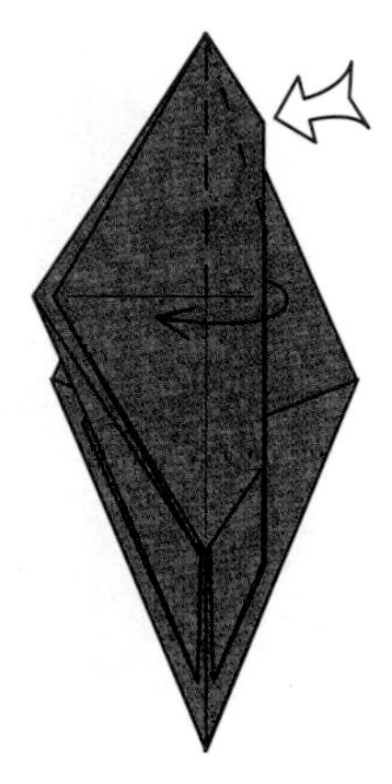

40. Open the top of the edge and spread-sink the corner.

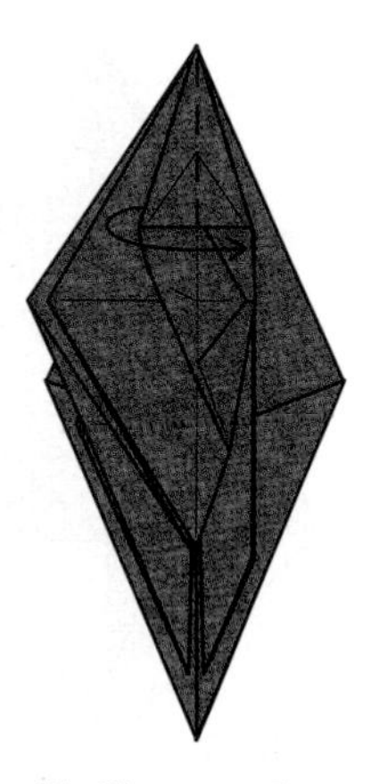

41. Close up the model.

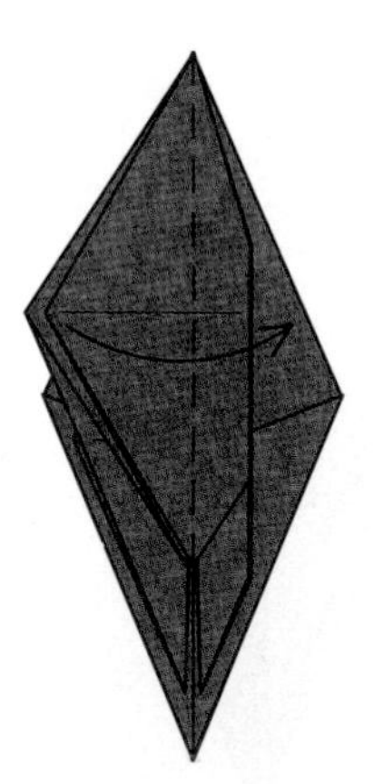

42. Swing one flap to the right.

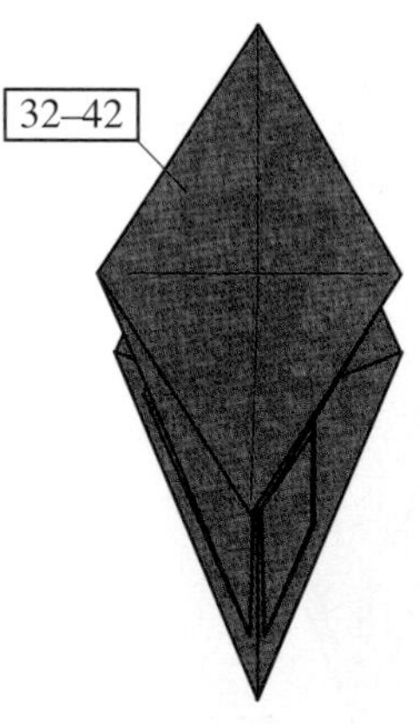

43. Repeat steps 32–42 on the left.

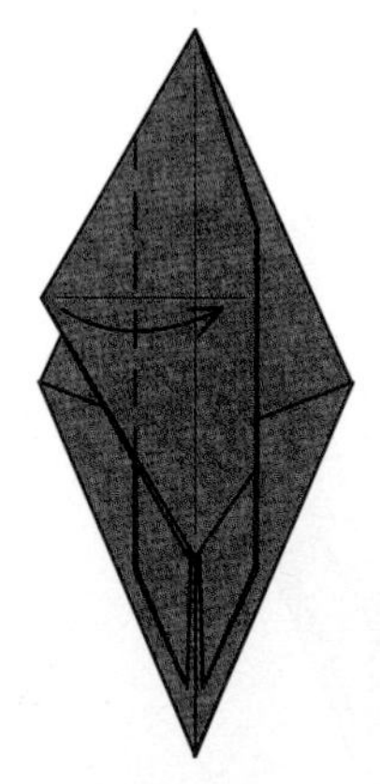

44. Fold the remaining flap to the right along a crease that lines up with the folded edges.

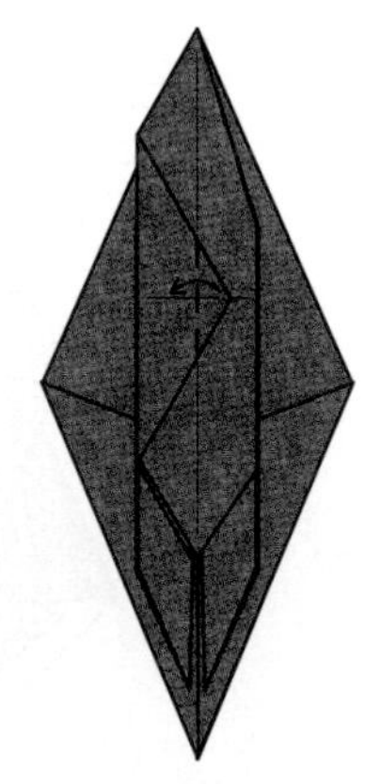

45. Fold the tip back to the left.

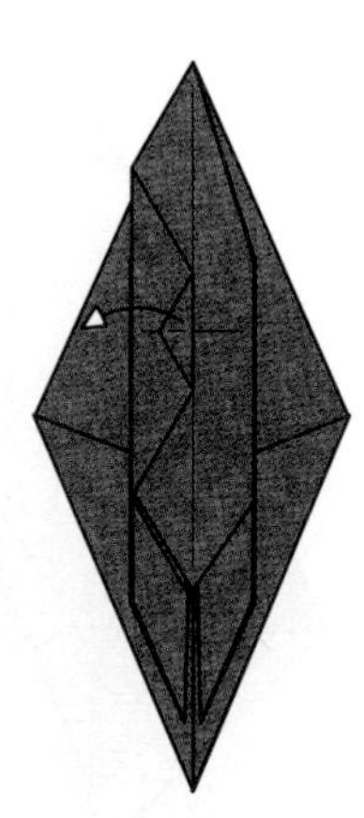

46. Unfold to step 44.

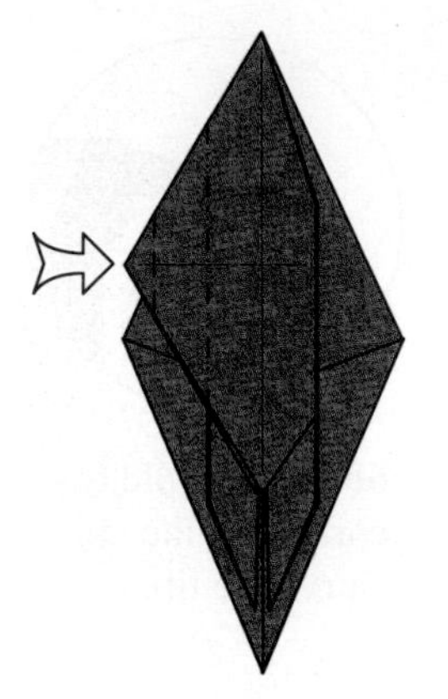

47. Open-sink the corner in and out on the existing creases.

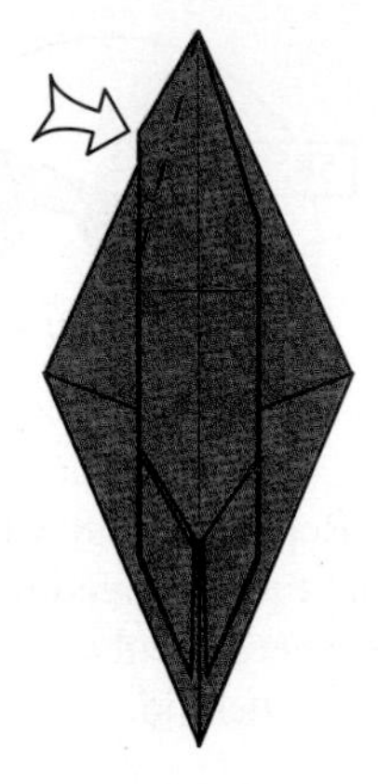

48. Spread-sink the corner as you did in steps 40–41.

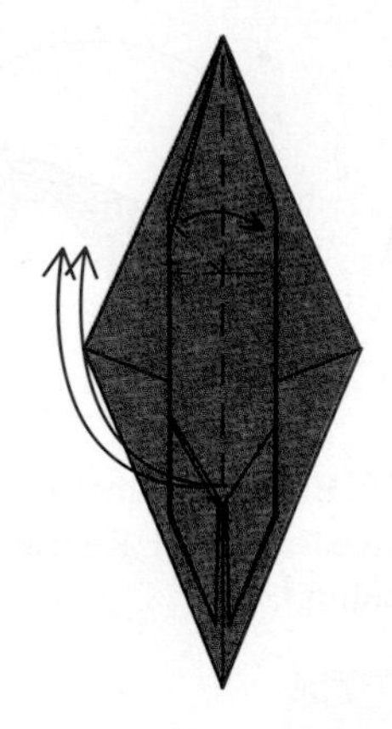

49. Spread the layers of the top point symmetrically and bring two points up to stand out away from the model.

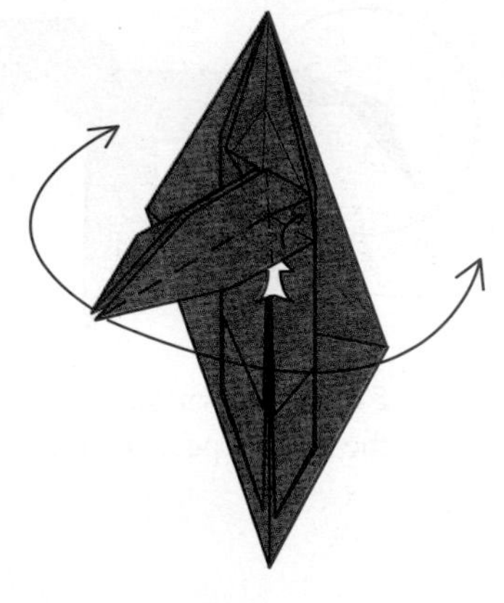

50. Fold the two points in half and swing them out to the sides.

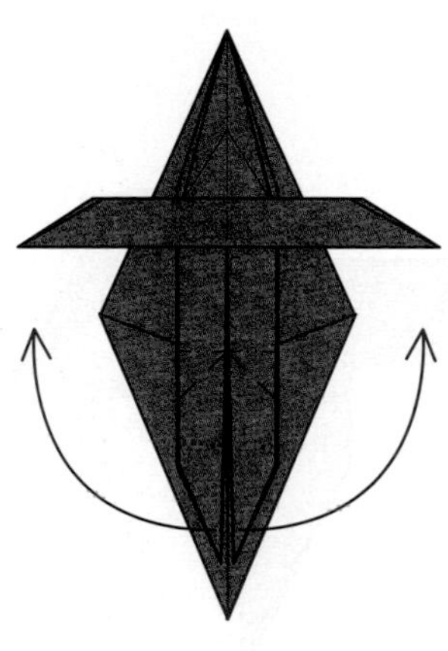

51. Crimp the two remaining flaps out to the sides.

52. Fold the sides in and tuck under the legs. There's no exact reference point for this.

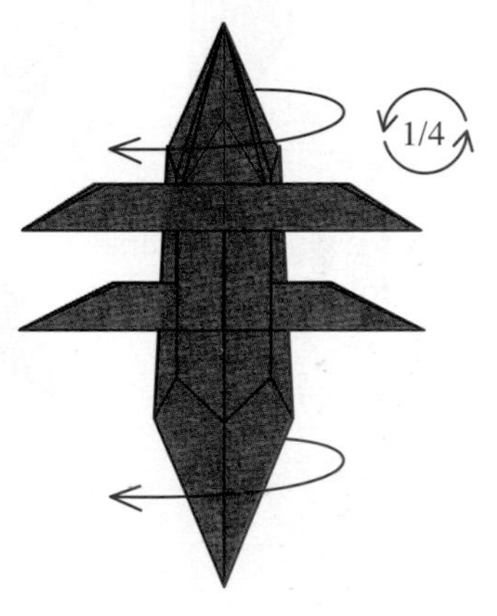

53. Curve the body into a half-cylinder and rotate 1/4 turn counterclockwise.

54. Crimp the tail, narrowing it at its base. Crimp the neck upward.

55. Reverse-fold the tail. Reverse-fold the head.

56. Narrow the tail with mountain folds. Narrow the neck with mountain folds.

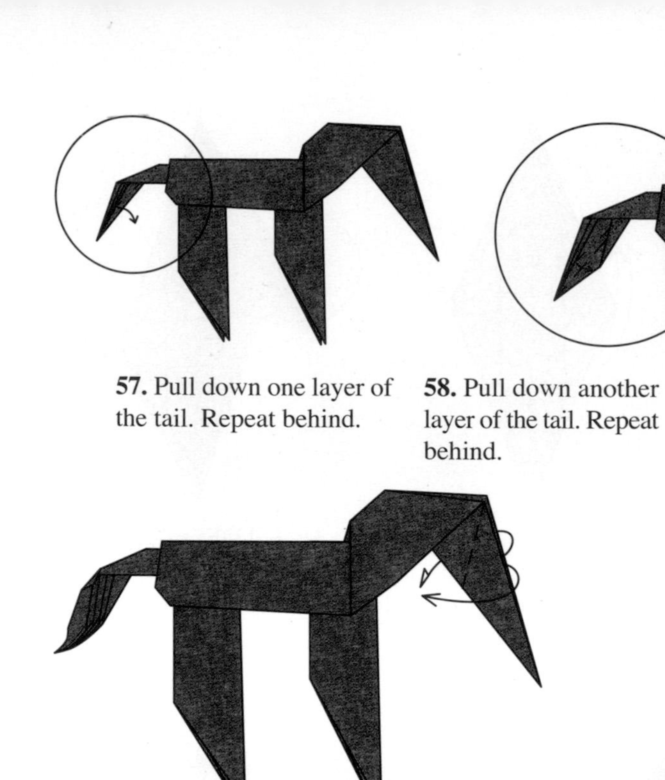

57. Pull down one layer of the tail. Repeat behind.

58. Pull down another layer of the tail. Repeat behind.

59. Repeat step 58 two more times, spreading the layers evenly. Repeat behind.

60. Valley-fold two corners of the tail. Curve the tip.

61. Outside reverse-fold the head.

62. Crimp the head upward.

63. Pleat the mane.

64. Double-rabbit-ear the hind legs to the left.

65. Reverse-fold the legs. Rotate the horse 1/8 turn counterclockwise.

66. Mountain-fold the edge of each hind leg, front and rear. Reverse-fold the tip. Repeat behind.

67. Narrow the leg. Fold the corners of the hoof underneath.

68. Crimp and open out the hooves. Shape the tail so that the tail and hooves form a stable tripod.

69. Double-rabbit-ear the forelegs.

70. Reverse-fold the tips of the forelegs. Steps 71–73 will focus on the forelegs.

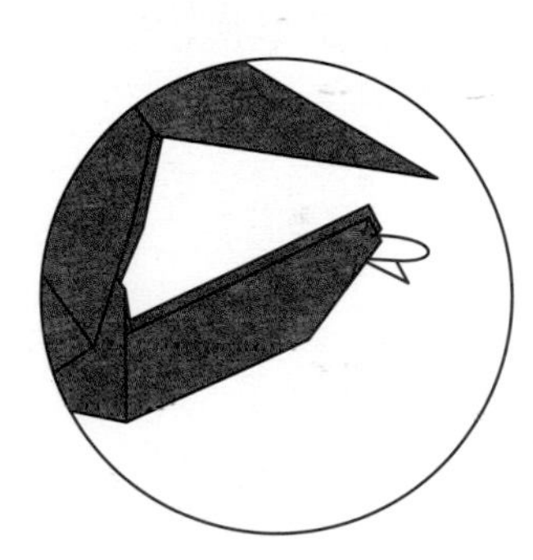

71. Mountain-fold the corners of the forelegs.

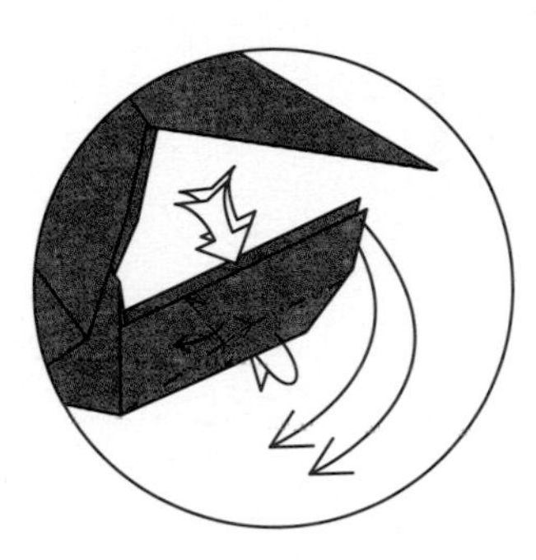

72. Simultaneously narrow and crimp the forelegs downward at slightly different angles.

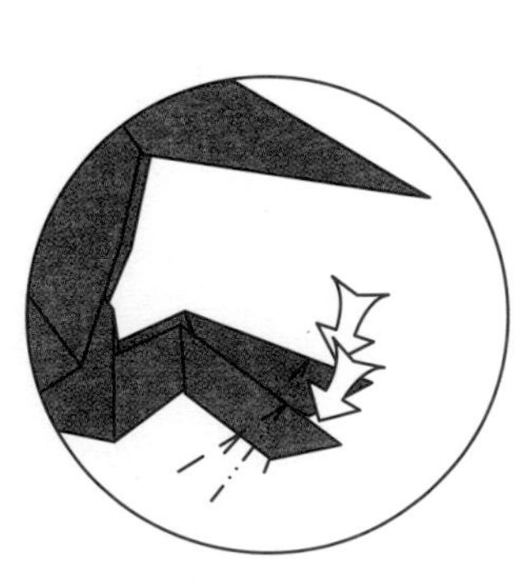

73. Crimp and open out the hooves.

74. Pleat the mane. Crimp the body. Reverse-fold the nose and mouth. Shape to taste.

75. Finished Alamo Stallion.

12 Box Pleating

One of the characteristics of many artistic endeavors—as well as science and engineering, which also possess a significant artistic component—is the presence of creative bursts. Origami is no exception. The progress of origami design through the 20th century was one of steady, incremental advance punctuated by occasional episodes of remarkable creativity. This is a universal phenomenon: It is as if some threshold is reached, that a truly new approach to design is discovered, then the technique or techniques are so rapidly explored and exploited that a jaw-dropping new field appears as if from thin air. Usually after the fact, historians can tease out the antecedents of a particular revolution, but in the days and years leading up to the critical event, no one saw it coming. This phenomenon happens in many fields of endeavor: Quantum theory revolutionized physics in the early 20th century; Impressionism changed the world of painting forever. In origami, the most outstanding example of a creative burst was the mid-1960s appearance of Dr. Emmanuel Mooser's Train, which ushered in an era of multiple subjects from a single sheet, of origami representing man-made articles, and the collection of techniques that has come to be known as *box pleating*.

12.1. Mooser's Train

In the small, loosely knit world of Western origami, Mooser's Train, shown in Figure 12.1, was something of a bombshell. While many folders had grown comfortable with the notion of using multiple sheets of paper to realize a single subject—head and forelegs from this square, hind legs and tail from that—here was the far opposite extreme: use of a single sheet of pa-

Figure 12.1.
Mooser's Train, folded by the author.

per to realize many different objects, the engine and cars of a complete train! The result was so unbelievable that folders scrambled to see how it was done.

Such a novel result was accomplished by an equally novel approach. What set Mooser's Train apart from the vast majority of origami designs was the folding style and technique, as well as the complexity of the resulting model. The difference was immediately apparent to even a superficial examination of the crease pattern. In nearly all ancient and early modern origami, the major creases were predominantly radial. They emanated, star-like, from various points in the square: the center, the corners, the midpoints of the edges.

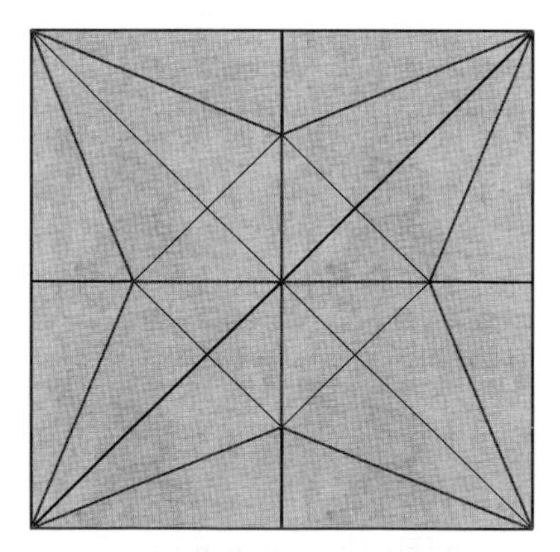

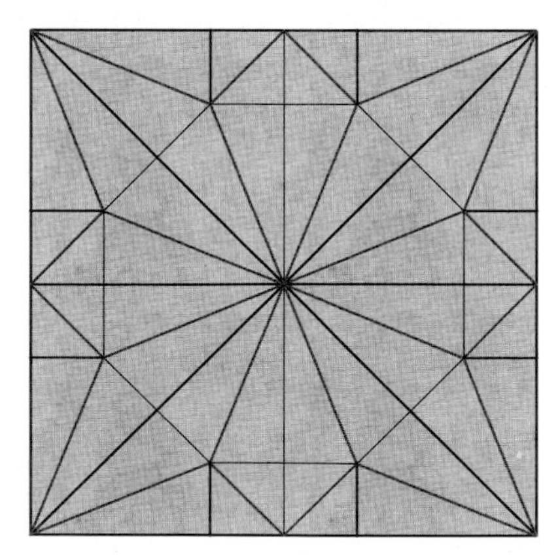

Figure 12.2.
Crease patterns of the Bird and Frog Bases, illustrating the radial pattern of creases.

But in Mooser's design, things were different. First, he started from a long rectangle; that alone was not a novelty, as several traditional models begin with a rectangle. But in contrast to most origami, the creases in Mooser's Train formed a grid of mostly evenly spaced parallel lines, occasionally broken by diagonals running at 45° to the edges of the paper. The overall appearance of the crease pattern was wholly unlike the patterns of conventional origami.

Fortunately for the curious, origami has by and large fomented a culture of sharing of both results and how-to, and it wasn't long before a hardy folder, Raymond K. McLain, had constructed and circulated an instruction sheet for the design. In lieu of formal publication—origami books were few and far between in the 1960s and 1970s—it was passed from person to person, photocopied, and recopied (this at a time when copiers were far from ubiquitous). Dauntingly, the instructions consisted of a single page containing the crease pattern, no step-by-step diagrams, and a smattering of tiny, handwritten verbal instructions wrapped around the edges of the pattern. I've redrawn McLain's instructions in Figure 12.3 if you'd like to give it a try yourself; for the adventurous soul who'd like to experience folding from the original, they are reproduced in Figure 12.4.

The challenging diagrams and their lack of widespread availability only added to the aura of mystery surrounding this model, and soon after its appearance it became one of the test pieces against which the origami-hopeful must test his or her folding skills. And any folded Mooser's Train instantly became a focal point for the origami gathering at which it appeared.

Mooser's Train fulfilled a valuable role: Its folding provided evidence that the folder had attained the pinnacle of the art. That was itself a worthy role. But Mooser's Train was not the culmination of a new style; on the contrary, it was the roadmap, leading the way to an entirely new approach to origami design and a new class of origami subject matter—the man-made object. It would inspire a small group of origami designers through a decade of creative growth, of exploration, and of pushing the boundaries of what was possible within the one sheet/no cuts origami paradigm. Their innovations, in turn, by showing that truly anything was possible with folding alone, would lead to the near abandonment of multi-sheet, or composite, origami design. And their work would go on to inspire an entire generation of origami designers, including the author of this book.

The revolution that was initiated by Mooser's model began in earnest when its techniques were adopted and expanded

Mooser's Train Crease Pattern & Order of Attack

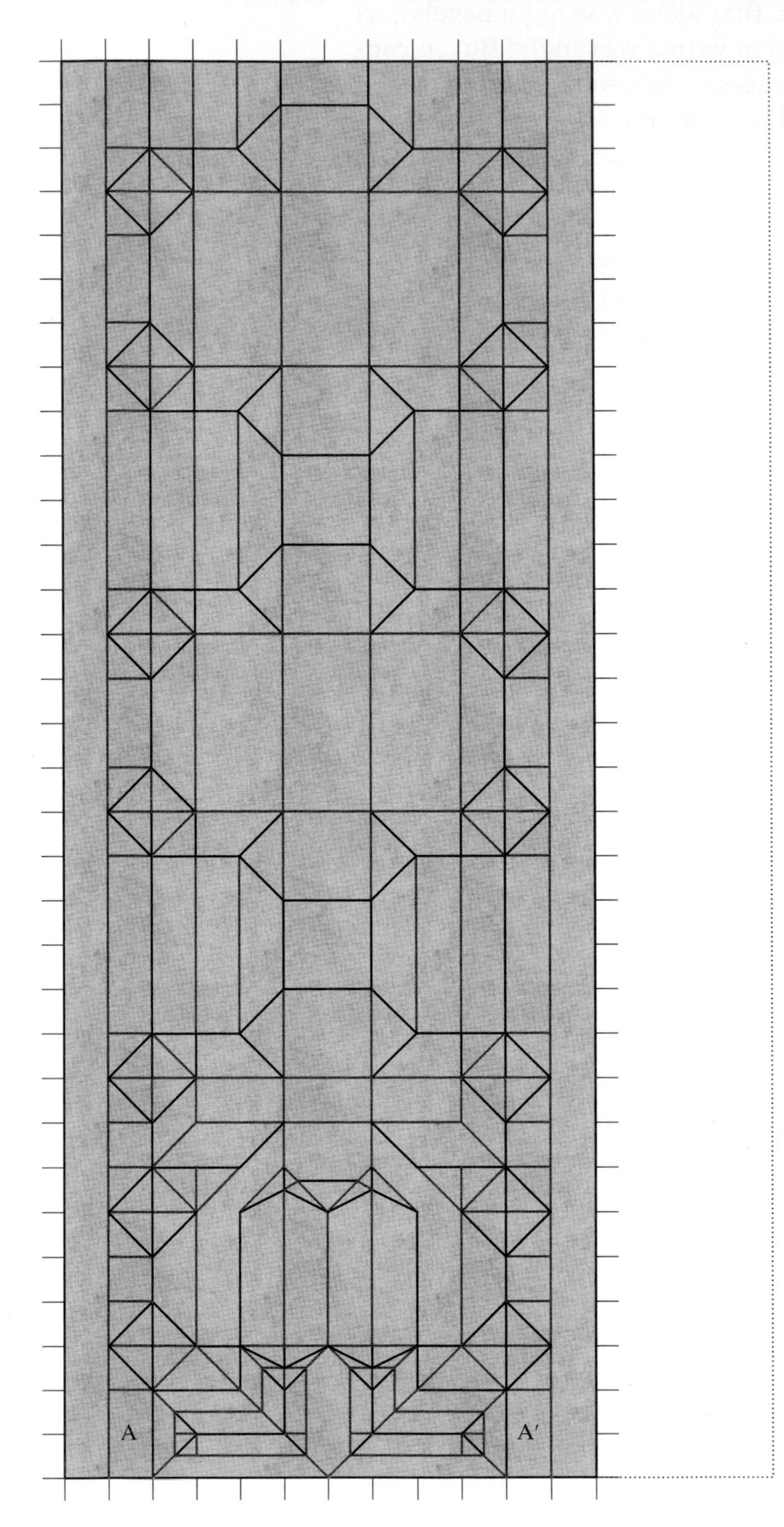

Worked out by R. K. McLain,
March 20, 1967
Hindman, KY 41822

Begin with (2) x (1) square.

Divide (2) into 32 squares.

Divide (1) into 16 squares.

Remove 4 squares the long way.

You now have 32 x 12 squares.

Mtn. fold under 1 square the long way on each side.

Now make the crease pattern as indicated. Each box car requires 10 squares long and 12 squares wide. The locomotive requires 12 x 12.

Now mould the model much as you would clay.

Several things must give at once so that a firm crease pattern without extraneous creases is helpful. Be patient & gentle.

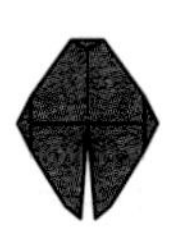

When moulding is completed, squash & partially petal fold the wheels & turn under the end points a little.

(Make catcher with A & A′.)

Dent inwards the platform between cars, lock the end of the last car by valley folding inwards the platform part, lock the underside by folding inward the extra material between & behind the wheels.

Bend the locomotive's snout upwards, penetrate (with a cut) it inwards into the boiler & bring it back outwards (with another cut) (and a valley fold) as a smoke stack. If you succeed, you get the prize for diligence! I'll take one too! This surely is a clever model & points the way to future 3D origami.

Perhaps the crease pattern could be scratched onto paper (making valley folds only on both sides of the paper) with a knife denting but not cutting through.

Figure 12.3.
Folding instructions for Mooser's Train.

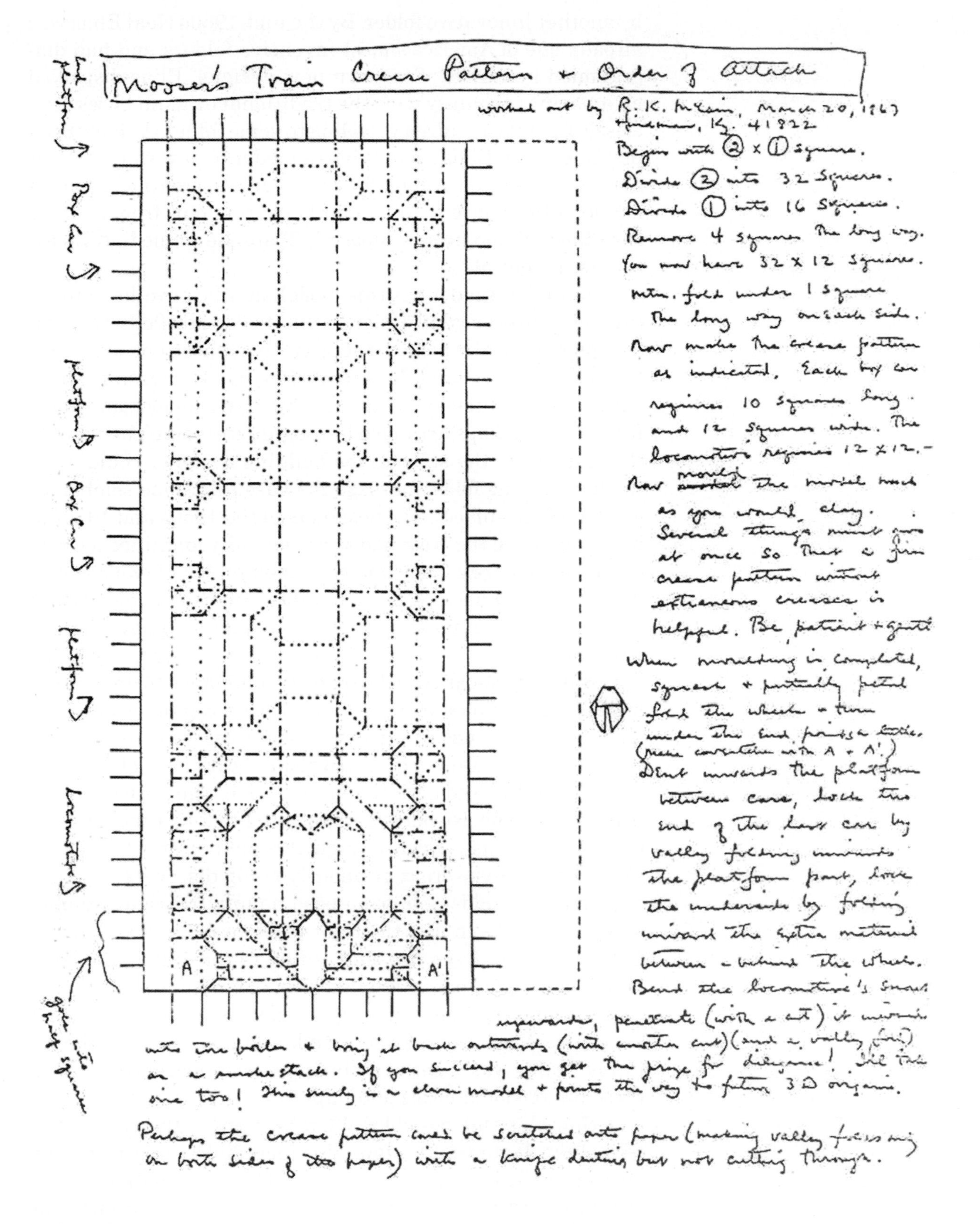
Mooser's Train Crease Pattern + Order of attack

worked out by R. K. McLain, March 20, 1967
Hindman, Ky. 41822

Begin with ② x ① squares.
Divide ② into 32 squares.
Divide ① into 16 squares.
Remove 4 squares the long way.
You now have 32 x 12 squares.
Mtn. fold under 1 square the long way on each side.
Now make the crease pattern as indicated. Each box car requires 10 squares long and 12 squares wide. The locomotive requires 12 x 12.—
Now ~~model~~ mould the model much as you would clay. Several things must go at once so that a firm crease pattern without extraneous creases is helpful. Be patient + gentle.
When moulding is completed, squash + partially petal fold the wheels + turn under the end points a little. (Make correlation with A + A'.) Dent inwards the platform between cars, lock the end of the last car by valley folding inwards the platform part, lock the underside by folding inward the extra material between + behind the wheels. Bend the locomotive's snout upwards, penetrate (with a cut) it inwardly into the boiler + bring it back outwards (with another cut) (and a valley fold) as a smokestack. If you succeed, you get the prize for elegance! I'll take one too! This surely is a clever model + points the way to future 3D origami.

Perhaps the crease pattern could be scratched onto paper (making valley folds only on both sides of the paper) with a knife denting but not cutting through.

Figure 12.4.
Raymond K. McLain's original instructions for Mooser's Train.

by another innovative folder. By the mid-1960s Neal Elias was already one of America's most inventive folders and had diagrammed hundreds of his own new designs. Elias displayed an amazing ingenuity with the traditional origami bases. The classic Bird Base—already feeling to some folders to have been played out—in Elias's hands blossomed into new shapes. Most notably, Elias had a flair for multisubject creations, for example, a birdhouse with two birds peering out, from a single Bird Base. When he saw Mooser's Train, he immediately saw its vast potential.

To understand what this model signifies, we have to recall the state of origami design in the 1950s and 1960s. Origami designers typically picked a subject, then chose one of several bases that had several flaps to work with. By choosing a base with the same number of flaps as the desired subject, and hopefully with the flaps arranged in roughly the same positions as the features of the subject, the budding designer could, with further shaping folds, massage the base into some semblance of the desired subject. The designers of the 1950s and 1960s in both Japan and the West had systematically identified a dozen or so known bases. They had combined pieces of two bases to make hybrid bases. A few—notably American folder (and friendly rival of Elias) Fred Rohm—had devised new bases of their own.

But a three-car train bears no resemblance to any known origami base, uniaxial or not. Such a train combines big, boxy shapes with the need for fourteen identical flaps to form the wheels, appropriately distributed along the bottoms of the three cars (six on the locomotive, four on each of the boxcars). This is pretty specific. No one was ever going to fold a train from a conventional base. Even though throughout the 1950s and 1960s new bases were continuously being discovered by trial and error, the odds of a given base having the right number and size of flaps in just the right place to make a train were millions to one. Even fast-forwarding to the 1990s, the techniques of uniaxial bases—circles, rivers, molecules, and trees—could handle the flaps but were not going to produce the solid elements. What Mooser had found, and displayed brilliantly in his Train, was a set of techniques for apparently making three-dimensional boxes and flaps *at will*.

How was this possible? What is it about the crease pattern of the Train that bestows this incredible versatility? The answer is not immediately obvious. The most distinctive aspect of the crease pattern of Mooser's Train is the fact that most of the creases run up-and-down or left-to-right. A smaller number run at 45°. This is to be contrasted with other origami bases in which the creases appear at first perusal to run every which way at many different angles and directions. Which pattern shows greater flexibility: the constrained, uptown/downtown/crosstown pattern of the Train, or the many-different-direction pattern of conventional origami? Clearly, the rules by which the Train was constructed were more restrictive than the rules of conventional origami. How could it be that a *more* restrictive set of rules leads to a *less* restrictive, more flexible result?

Paradoxically, it is the very tightness of the constraints of box pleating that makes it possible to fold such complex designs. The reason it has always been difficult to develop new origami bases is that a base is a *gestalt*, an inseparable whole; all parts of the pattern interact with other parts, so that it is very difficult to make a substantial change in one part of the pattern without having to change all other parts. The resemblance of a crease pattern to a spider's web is an apt analogy; pluck a single strand and it reverberates throughout the web. Perhaps a better analogy is a stack of apples: Move the wrong apple and the heap collapses. Move one circle in a circle-packing and the entire packing might need to rearrange. Change a single vertex in a crease pattern, and its effects propagate throughout the entire pattern.

And those effects may very well precipitate a descent into unfoldability. Let's take a simple example: the Frog Base, shown in Figure 12.5. Suppose that for some reason we wished to move the vertex that corresponds to the central point. Move that vertex the tiniest amount away from the center, changing nothing else, and the crease pattern becomes unfoldable (or rather, un-flat-foldable; it can no longer be pressed flat without creating wrinkles). It is possible, however, to move other vertices to return the base to flat foldability, as shown on the right in Figure 12.5; but to do so requires that we shift the location of all the other interior vertices, resulting in moving nearly every crease in the pattern.

One seemingly innocuous change in the pattern forces changes throughout the design. And this was the result of an attempt to shift the location of a single point. We have not even added any points. In the early days of origami, design was incremental, a change at a time. But if such a tiny change forces a complete redesign of the crease pattern, what hope has the designer of incrementally creating a fourteen-wheeled, three-vehicled conveyance such as a train? How would a designer of a real steel-and-wood train fare if the most minor change—say, moving a door handle—forced an unpredictable change in every dimension of every part of the structure?

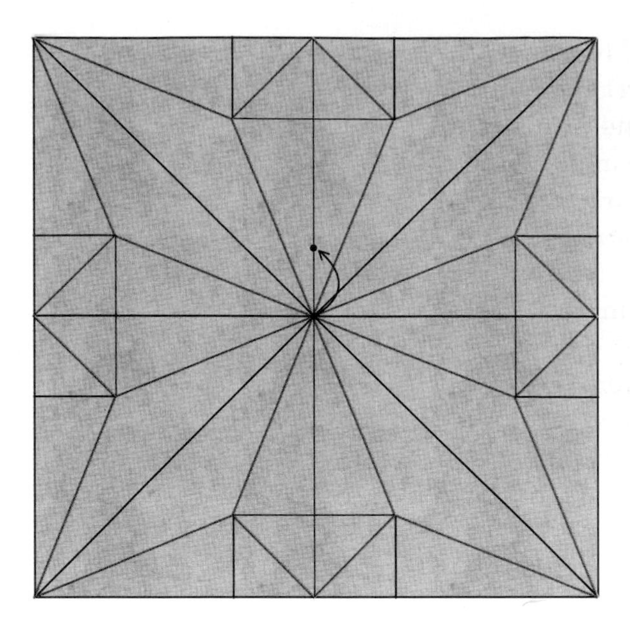

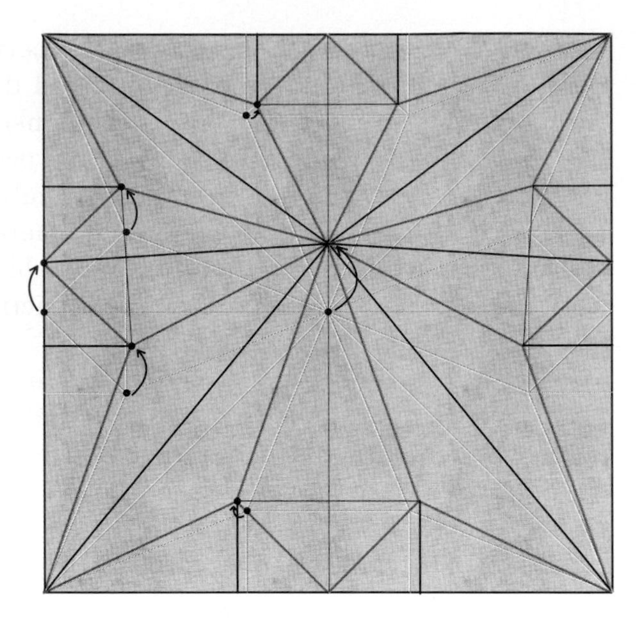

Figure 12.5.
Left: the crease pattern for a Frog Base. Suppose we move the central vertex upward.
Right: the new flat-foldable crease pattern. Note that every other interior vertex has also moved.

But in Mooser's Train, some changes don't cause so much trouble. In the Train, the creases don't run every which way. In fact, they only extend in one of four different orientations: up/down, left/right, diagonally upward, diagonally downward. And the creases don't fall just anywhere: There is an underlying grid, so that up/down and left/right creases run solely along grid lines, while diagonal creases always connect diagonal grid points. So the crease pattern is quite tightly constrained.

The constraint of the grid brings order to the crease pattern: It winnows the unimaginably vast space of possible patterns down to a manageable set. And most importantly, it limits the ways that different parts of the pattern can interact with each other. The problem with an old-style base like the Frog Base is not just that the central point interacts with the surrounding points: It's that it interacts with each surrounding point *in a different way*. So one type of change creates several types of changes in its surroundings, which then create more changes in theirs, and so forth. This means that the complexity induced by a change quickly cascades as the change propagates away from the original perturbation. But in a box-pleated pattern, by contrast, where different parts of the crease pattern correspond to different parts of the model, all interact in the same basic way. And so, a fairly small tool kit of basic techniques can be combined and built up into quite complex structures.

The basic elements of this tool kit are visible in Mooser's Train, the archetype for all the box-pleated models that followed. Those two elements are a technique for building and linking boxes (used for the bodies of the engine and the two cars), and techniques for creating flaps (used in the wheels and, especially, the smokestack). Both boxes and flaps grow out of the same rectilinear grid of creases, which allows arbitrary combinations of boxes and flaps to be created and combined at will.

12.2 Box Folding

The techniques to create box-like structures have their antecedents in well-known traditional models that include (perhaps not surprisingly) a simple box, known for decades, if not hundreds of years. The box displays the underlying mechanism that enables box pleating as a style and that makes up the overall structure of Mooser's Train. Box-pleating as a style was sitting there all along, waiting to be discovered, but the most common folding sequence for the traditional box (given in Figure 12.6) and the diagonal orientation of the model obscure the underlying structure and its relationship to the train.

This is a fairly common occurrence in origami: the published folding sequence is usually constructed for ease of foldability, or in some cases, for elegance of presentation (with a surprise move at the end). In either situation, the choice of folding sequence may well conceal, rather than illuminate, the underlying structure of the model.

Superficially, what we have here is simply a box with two handles. But let's look at it as a collection of forms. We have a linear series of forms:

- a flat form (the handle),
- a transition from a flat form to a three-dimensional form,
- the three-dimensional form (the box itself),
- another transition from the three-dimensional form to a flat form,
- and finally another flat form (the opposite handle).

How does this combination of two- and three-dimensional forms arise from the flat sheet? The best way to find out is to

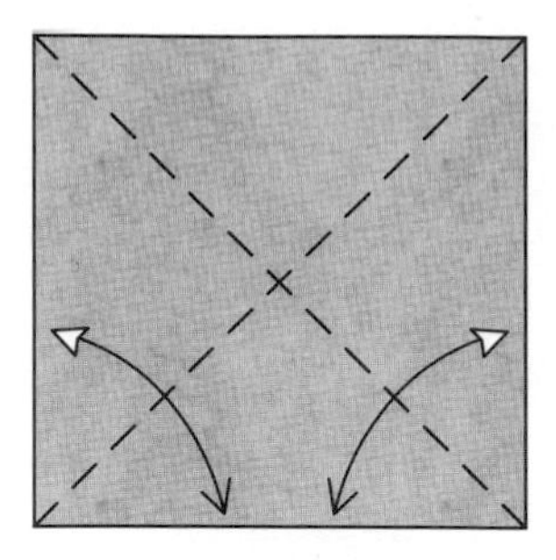

1. Begin with a square. Fold and unfold along the diagonals.

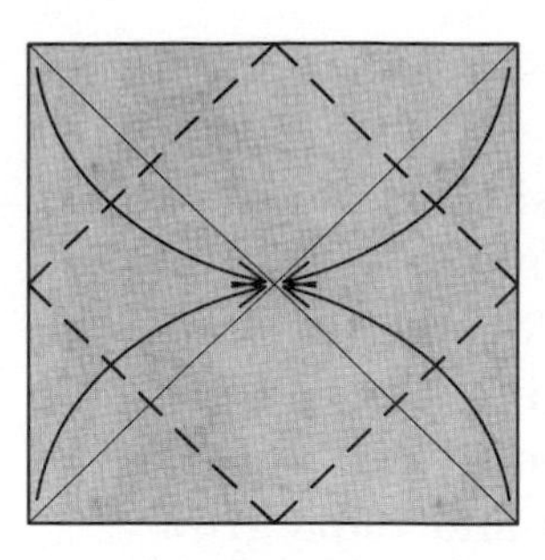

2. Fold the four corners to the center of the paper.

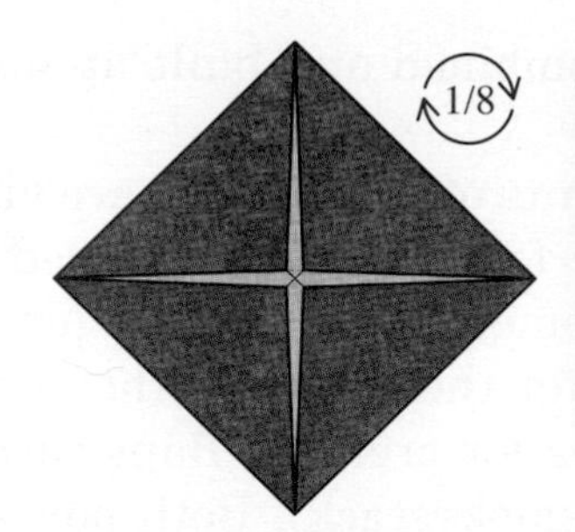

3. Rotate the paper 1/8 turn clockwise.

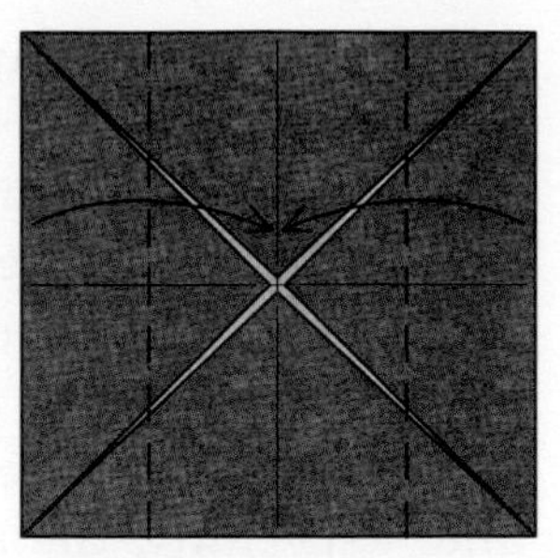

4. Fold the sides in to the center line.

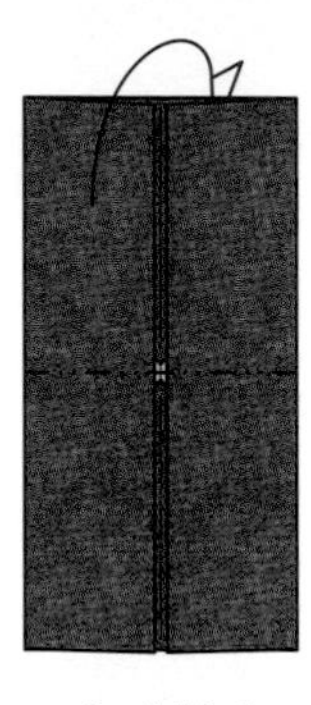

5. Mountain-fold the top half of the model behind.

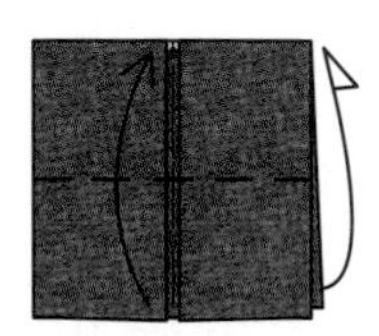

6. Fold one flap up to the top edge in front; repeat behind.

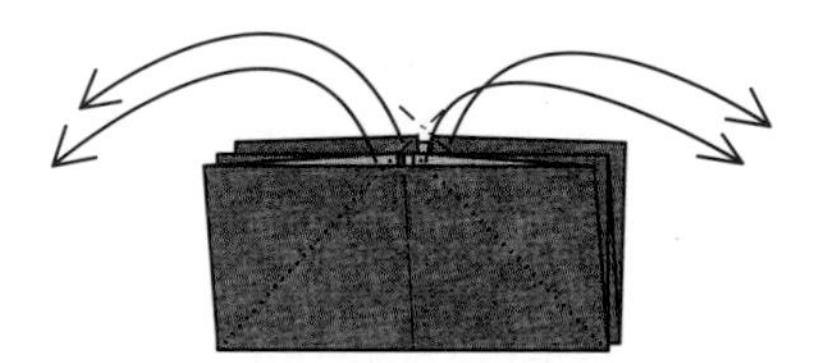

7. Pull the corners out to the sides as far as possible and flatten the model.

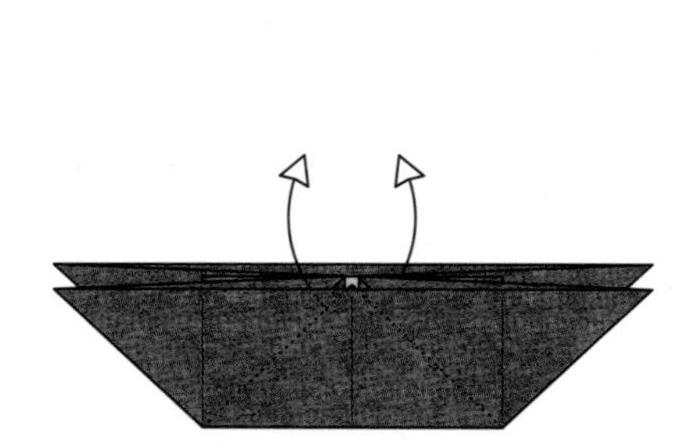

8. Pull the raw corners out completely in front and behind.

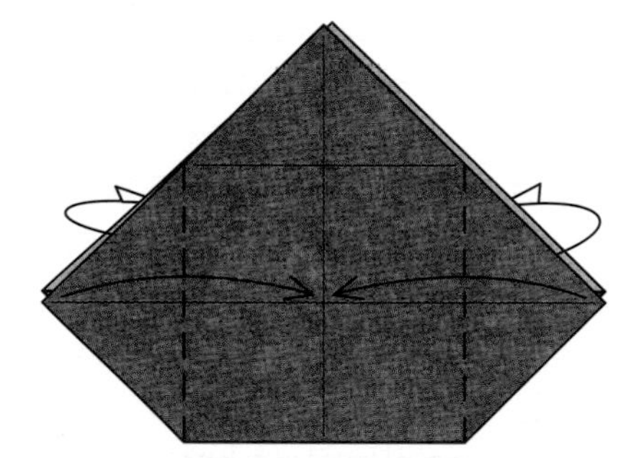

9. Fold the corners in to the center on existing creases. Repeat behind.

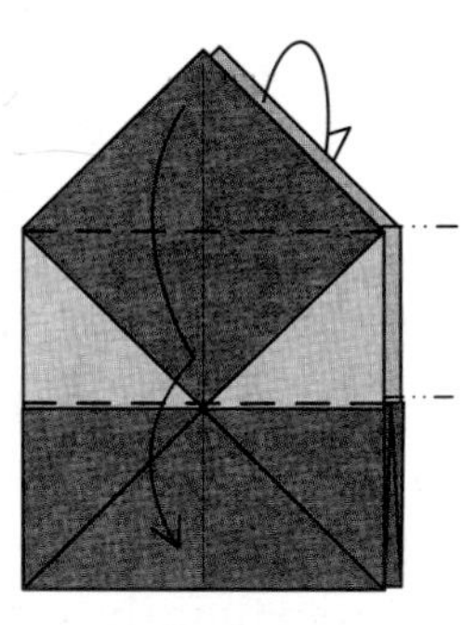

10. Fold the top corner down; fold the resulting flap down again. Repeat behind.

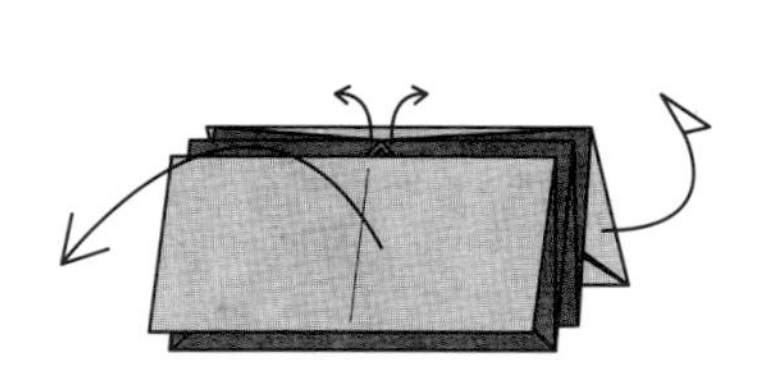

11. Grasp the two white flaps and pull them in opposite directions, opening out the model.

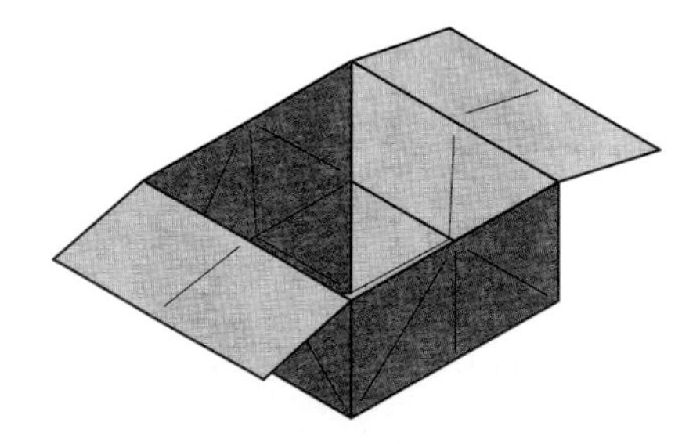

12. Finished Box.

Figure 12.6.
Folding sequence for the traditional box.

take the model back to the flat sheet, keeping track of which parts came from where. If we label the features of the box—base, side, front, rear, handle—and note where each region comes from in the unfolded sheet, we can establish a correspondence between the folded and unfolded forms of the model, as shown in Figure 12.7.

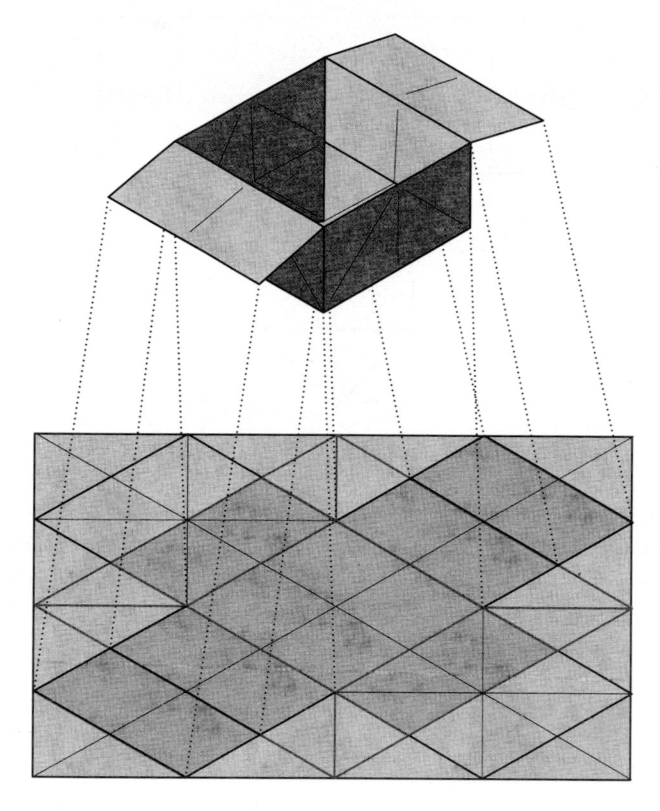

Figure 12.7.
Correspondence between the parts of the folded model and the crease pattern.

If we examine the crease pattern by itself, we see that not all of the paper is needed to make the model. In particular, the top and bottom corners (which are tucked down inside the bottom of the model) don't contribute much (other than a bit of extra stiffness, owing to the multiple layers), and the side corners are tucked underneath the handle as well.

Note that in this three-dimensional model, some of the mountain and valley folds make a *dihedral angle*—the angle between adjacent surfaces—of 90° while others are pressed flat in the folded model.

Examination of the labeled crease pattern in Figure 12.8 shows that we don't need the entire square to fold this box. In fact, we can fold what is essentially the same model from a 3 × 2 rectangle, as outlined by dotted lines in Figure 12.9.

Although a 3 × 2 rectangle is considered nonstandard in origami (or at least, less common) and is less pleasingly symmetric than a square, it is a more natural shape for folding the box, since the edges of the paper are aligned with the sides of

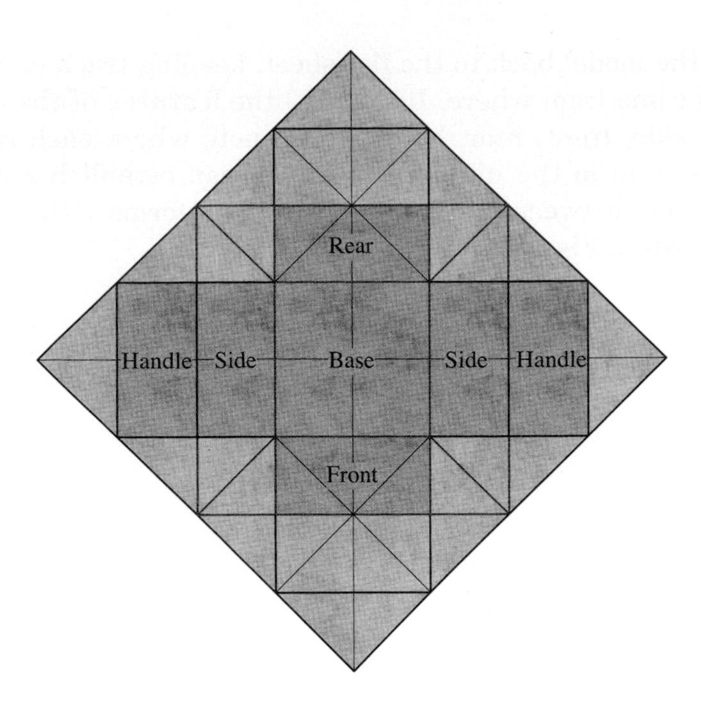

Figure 12.8.
Crease pattern for the box with features labeled.

the box and the layers are more evenly distributed. We can fold essentially the same box from a 3 × 2 rectangle, as shown in Figure 12.10. Note, however, that the folding sequence is considerably different.

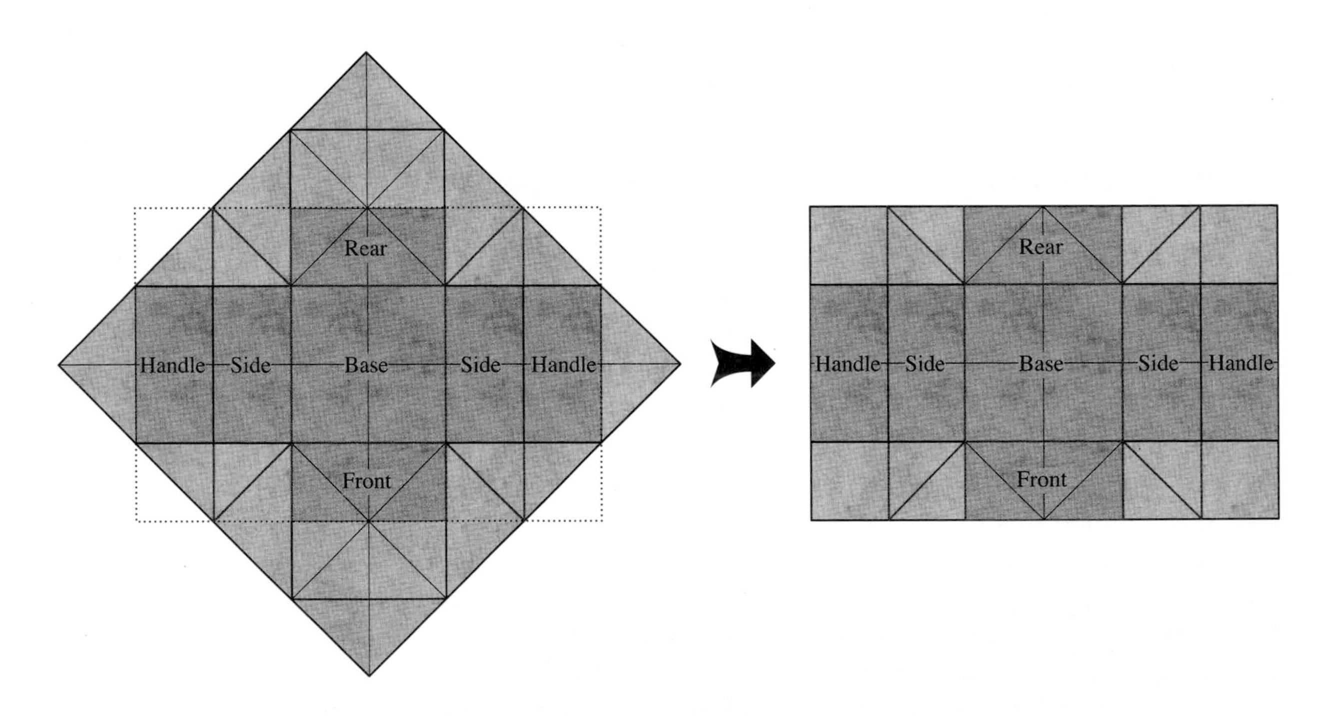

Figure 12.9.
A 3 × 2 rectangle (dotted line) encloses all the important elements of the model.

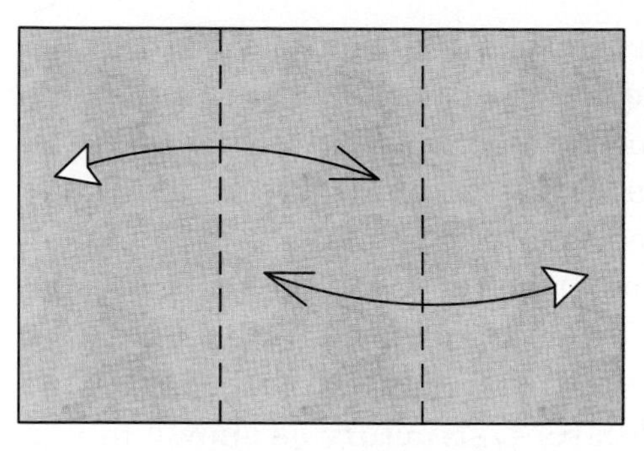

1. Begin with a 3 × 2 rectangle.
Fold and unfold in thirds.

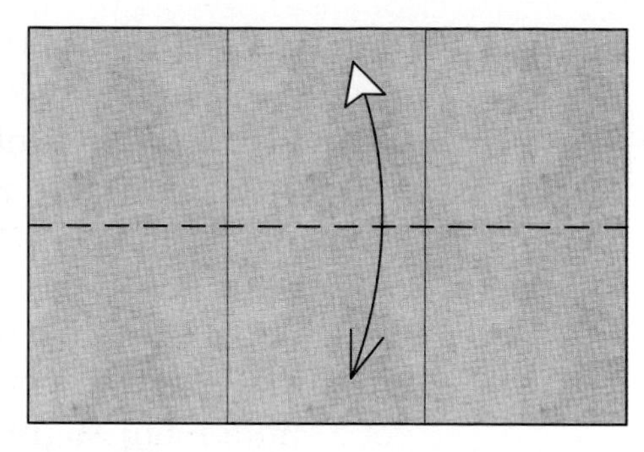

2. Fold the top down and unfold.

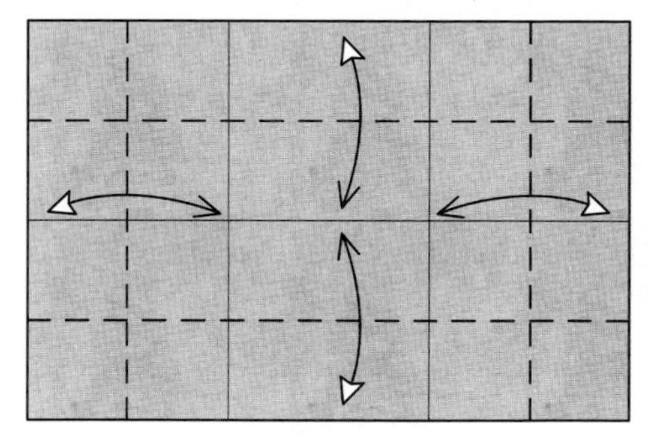

3. Fold each edge to a crease line and unfold.

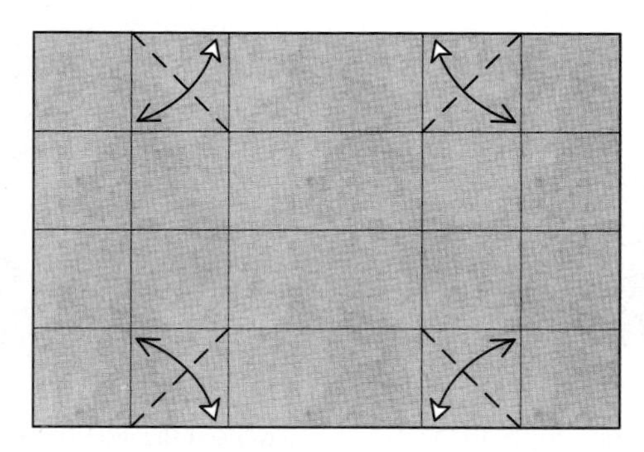

4. Bring the bottom edge to the vertical crease line, crease, and unfold. Repeat on the right, and in two places up top.

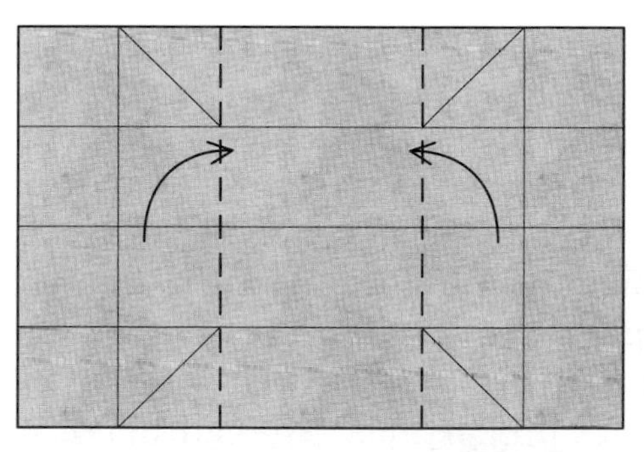

5. Fold the sides up so that they stand straight up.

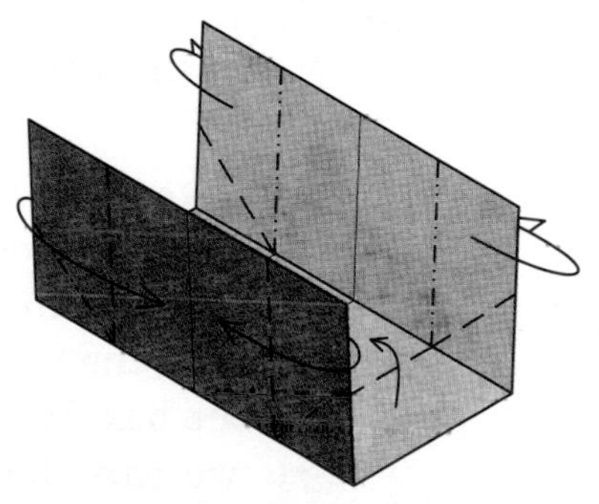

6. Fold the raw edges of each side toward each other, lifting up the front and rear edges at the same time.

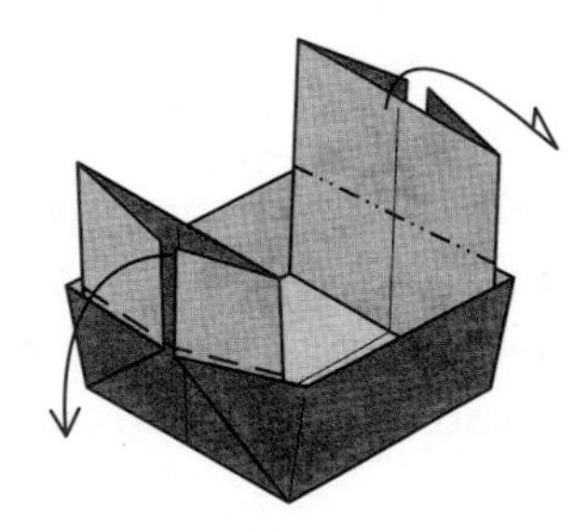

7. Fold the sides down.

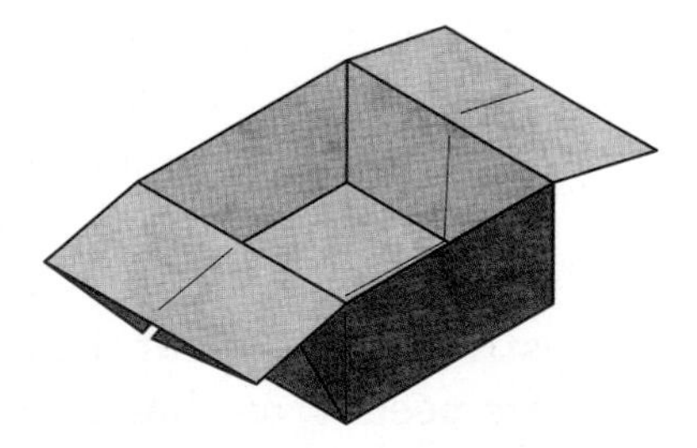

8. Finished Box.

Figure 12.10.
Folding sequence for the traditional box from a rectangle.

This simple box is one of the building blocks of box-pleated models. It is a structure that can be stretched, squeezed, modified, and most importantly, combined with duplicates and variations of itself to yield remarkably complex objects. Let's run through a few of the simplest possible variations.

The first variation stems from the fact that there are two ways to fold the same box. If you fold steps 1–6 the same, but at step 7, wrap the vertical edges around to the other side, you get a similar, but slightly different, structure as shown in Figure 12.11.

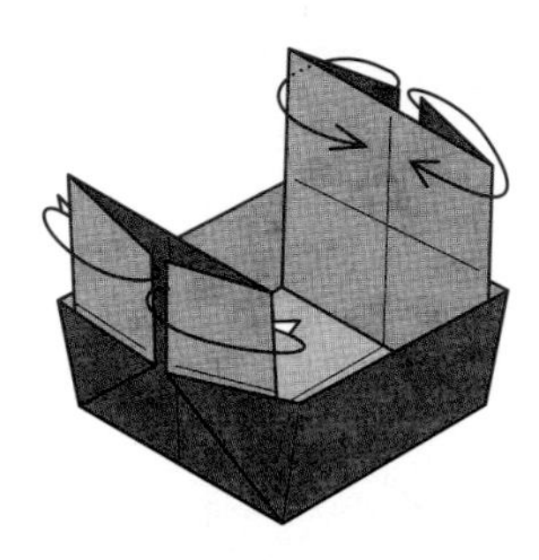

7. Wrap the raw edges to the inside, turning valley folds into mountain folds and vice versa.

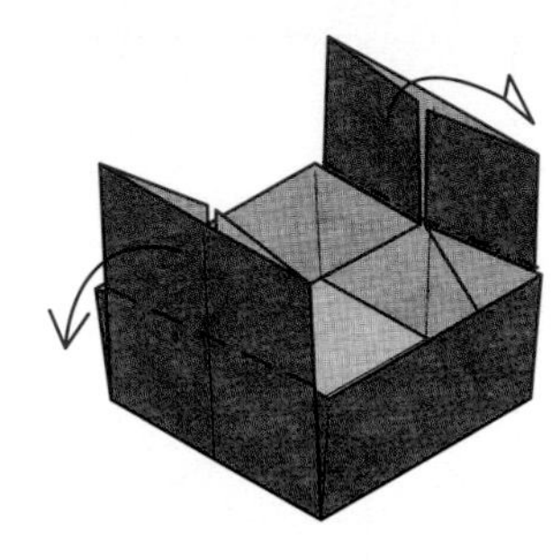

8. Fold the side flaps down.

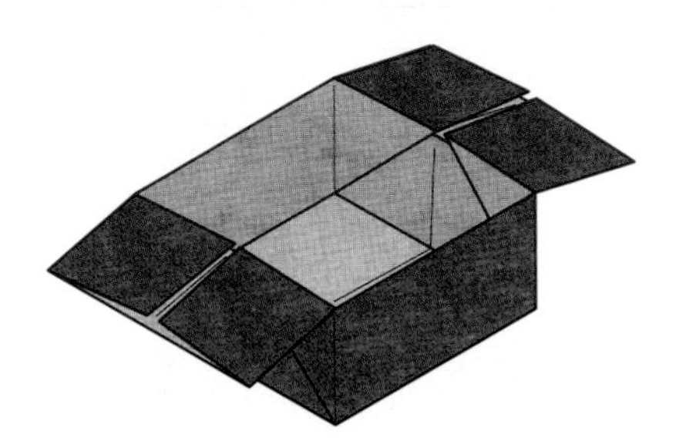

9. Finished Box.

Figure 12.11.
A different finish for the 3 × 2 box.

The two versions of the box differ slightly in the handles. In the first, the handles are white; in the second, they are colored. But there is a more important difference: In the second form of the box, the raw edges of the paper are exposed on the top side of the box. We'll make use of this a bit later.

Next, we can change its proportions. We can make it longer, wider, or taller, or any combination of the three. We can approach all three by way of a little thought experiment. Suppose we wished to make it longer (i.e., shift the handles farther from each other). If the paper were made of rubber, we could simply stretch it, as shown in Figure 12.12.

But since paper can't stretch, we need another approach. Suppose we wanted to make the box 50% longer, that is, half again as long as it is now. An approach that doesn't require stretching is to cut the model in two and add more paper where we need it, as shown in Figure 12.13.

At this point, origami purists are howling in protest: Origami is the art of folding, not cutting and taping paper! How can this be called origami? For it to be pure origami, we would have to fold this box from an uncut sheet of paper. But this is

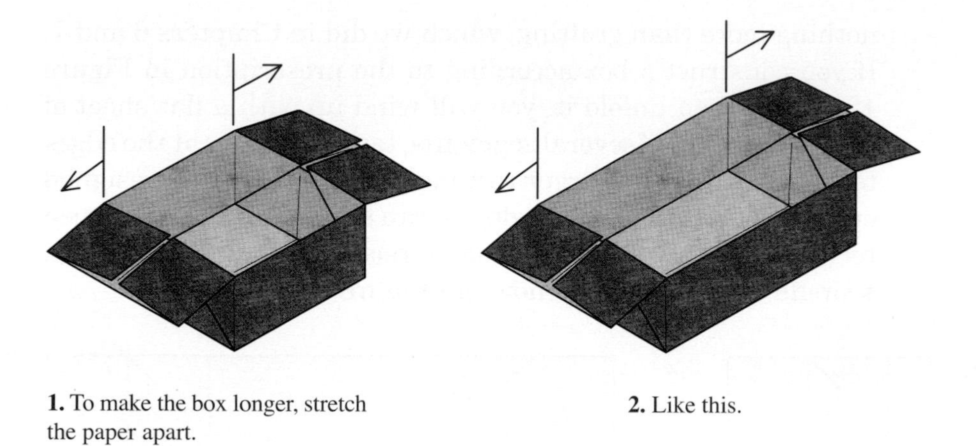

1. To make the box longer, stretch the paper apart.
2. Like this.

Figure 12.12.
Stretching the box to make it longer.

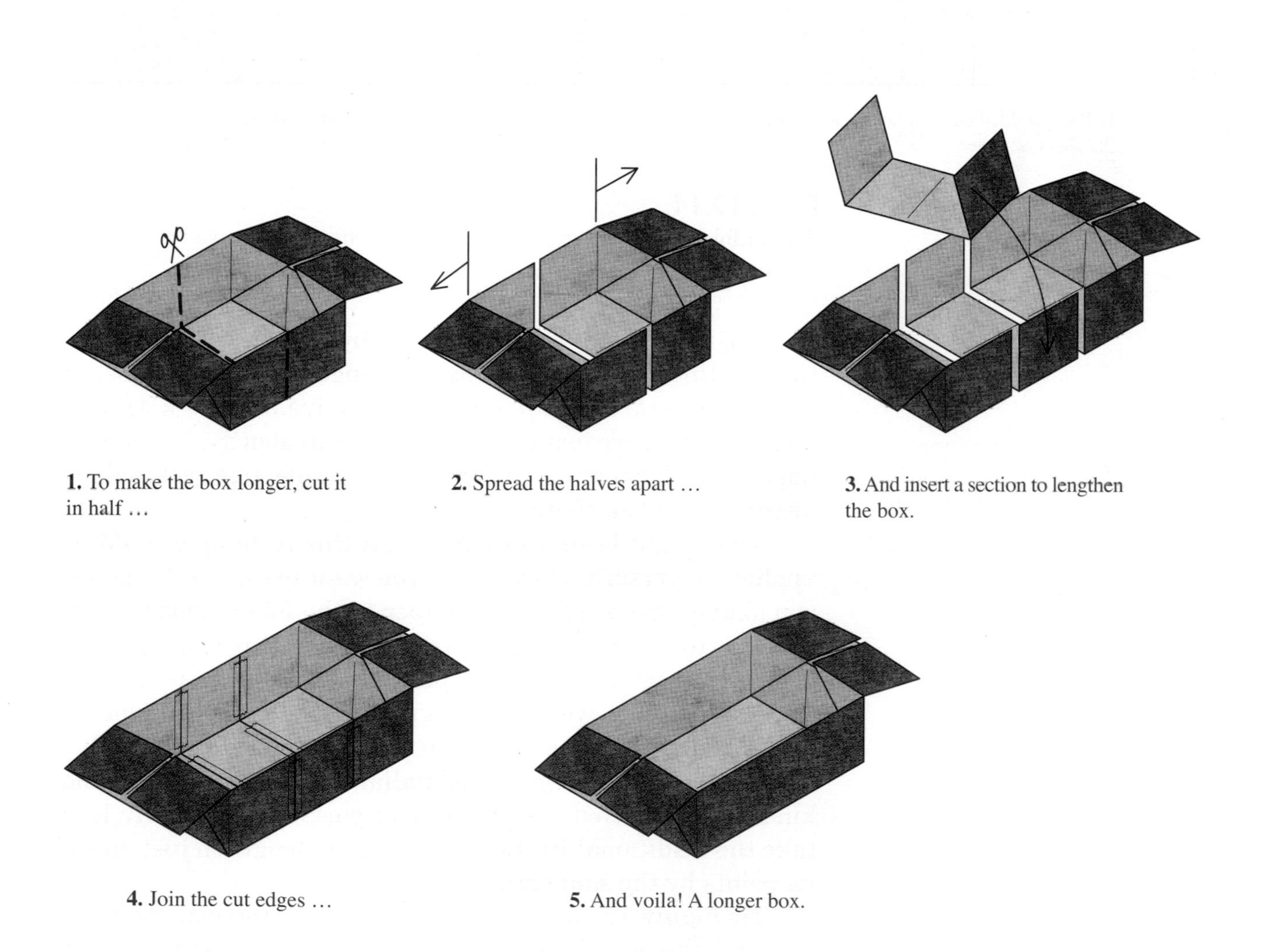

1. To make the box longer, cut it in half …
2. Spread the halves apart …
3. And insert a section to lengthen the box.
4. Join the cut edges …
5. And voila! A longer box.

Figure 12.13.
Lengthening the box by cutting and inserting more paper.

nothing more than grafting, which we did in Chapters 6 and 7. If you construct a box according to the prescription in Figure 12.13 and then unfold it, you will wind up with a flat sheet of paper composed of several segments, taped together at the edges to form a somewhat larger rectangle. Having already resigned ourselves to using a rectangle, we can simply convert the taped rectangle into a new, slightly longer rectangle that is once again, a single uncut sheet, as shown in Figure 12.14.

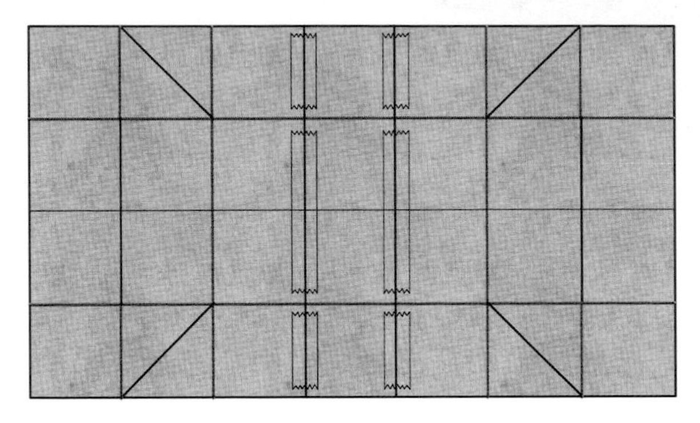

1. The unfolded cut-and-taped model is a flat sheet of paper.

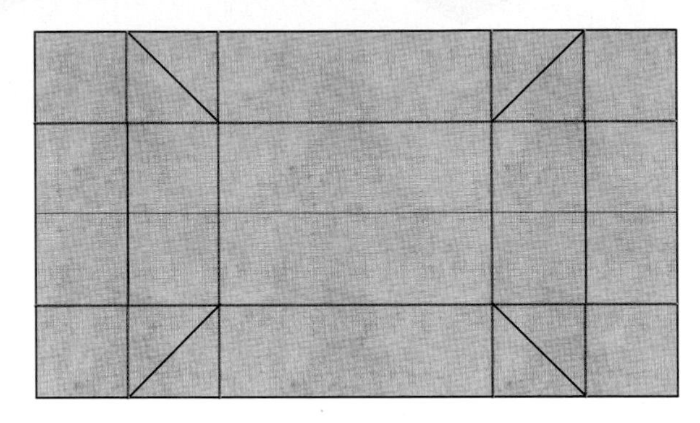

2. So the model could be folded by starting with an uncut sheet of the same size.

Figure 12.14.
The unfolded model, and an uncut sheet that can be used to fold the longer box.

So, the box can be made longer by adding more paper to the starting rectangle. We have changed its proportions, of course—we started with a 3×2 (or equivalently, 6×4) rectangle; we now are using a $3\frac{1}{2} \times 2$ (or equivalently, 7×4) rectangle. But if you're folding from a rectangle, one rectangle is nearly as good as another.

One might begin to suspect that this technique could be applied universally; everywhere you want to lengthen a point, you simply add a segment of paper to the folded model, then unfold it to get the new crease pattern. But this is not always possible; in fact, it is rarely possible with most traditional origami bases. As we saw with grafting, we were often forced to add paper that showed up in several different places. It's difficult to add a local graft in a radial-crease base, such as the kind we constructed with circle/river packing. To see why, let's take the traditional Bird Base, and try to lengthen just one of its points by the same grafting strategy.

As Figure 12.15 shows, it doesn't work. You can certainly lengthen the point by cutting and inserting a section of paper, but the resulting shape, when unfolded, cannot be flattened. Often in origami, we start with a flat sheet of paper and try to

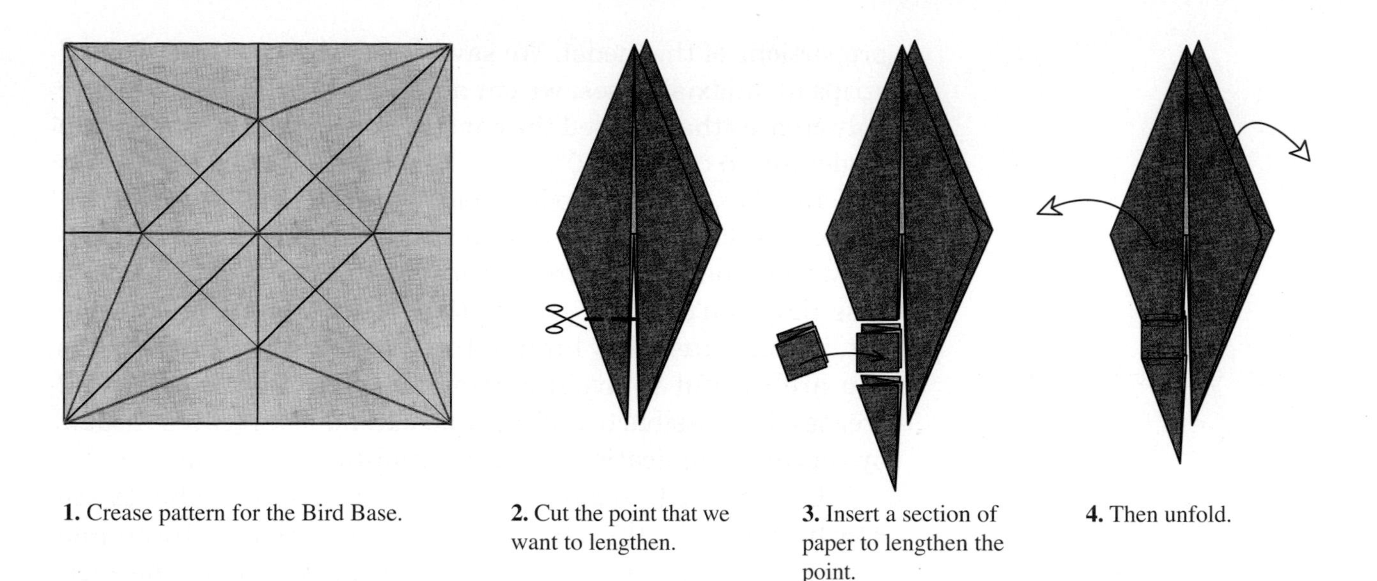

1. Crease pattern for the Bird Base.

2. Cut the point that we want to lengthen.

3. Insert a section of paper to lengthen the point.

4. Then unfold.

5. The unfolded paper cannot be flattened.

Figure 12.15.
An attempt to lengthen a single Bird Base point by an inserted graft fails.

make a model that folds flat; here, we have the opposite problem: The model is flat, but the sheet from which it springs is not! So one cannot willy-nilly use grafting as a means to change the proportions of a small portion of the model.

But with the 3 × 2 box—with box-pleated models, in general—you can often change the proportions of parts of the folded model by changing the proportions of the rectangle from which you started as if you had cut the original rectangle and inserted a strip. What makes it all possible is the angular relationship between the cuts and creases that cross the cut (and here I refer only to creases that are folded, not to crease marks left over from some prior fold-and-unfold step). If all creases that cross a cut do so at 90° to the cut, then one can, in general, add a strip of paper between the cut edges to alter the proportions of the model. We saw this when we added grafted strips to uniaxial bases; we cut along axial creases so that the only creases that crossed the cut were the hinge creases, which by definition cross at 90°.

In a box-pleated model, nearly all the creases are either vertical or horizontal. So if a cut is made vertically or horizontally, then the creases are either parallel to the cut, in which case they don't hit the cut, or they are perpendicular to the cut, in which case they hit it at the proper angle. So, as long as you are careful to avoid cutting through the few diagonal creases, it's possible to enlarge and extend box-pleated models by repeated application of the cut-and-tape technique.

Coming back to our 3 × 2 box, you should be able to see now how to make the box wider rather than longer by adding a strip running horizontally through the middle of the rectangle. This process, which changes the rectangle from 6 × 4 to 6 × 5, is shown in Figure 12.16.

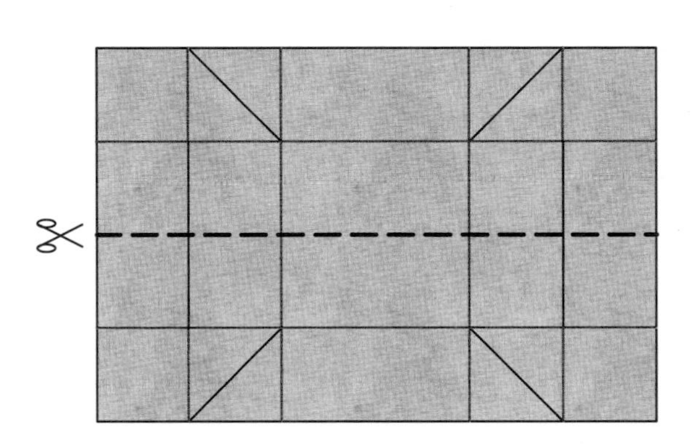

1. To make the box wider, add a strip of paper horizontally through the crease pattern.

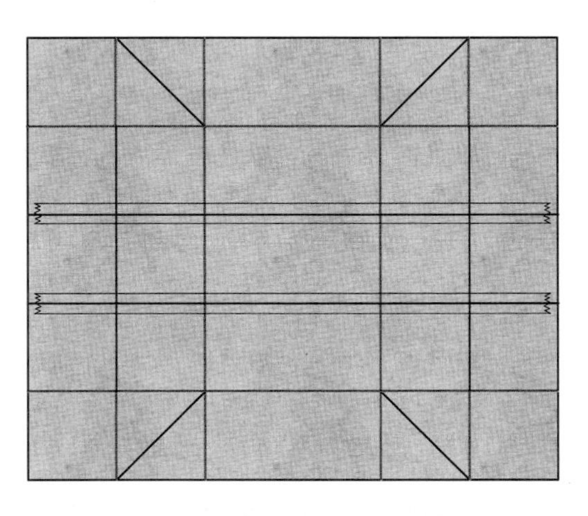

2. Like this.

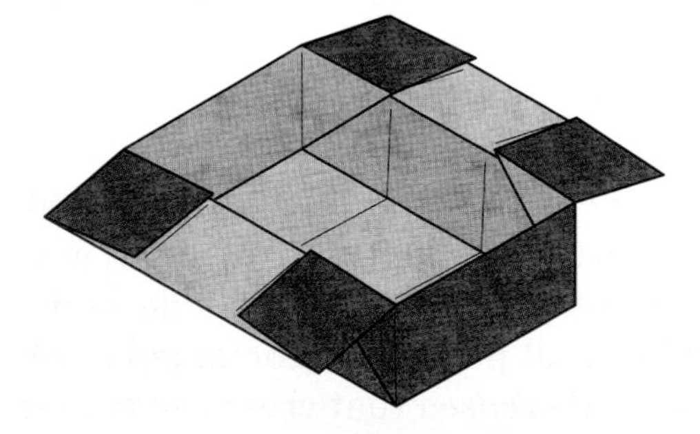

3. The wider box.

Figure 12.16.
Adding a strip to enlarge the box in the other direction.

What if we wanted to make the box smaller, not larger? Then instead of adding paper, we would take paper away. Let's reduce both the length and the width of the box by a single square in each direction. We do this by cutting out both a vertical and horizontal strip.

Because we've cut paper out, the flaps that fold toward each other in the handles now overlap. It is desirable to avoid such overlaps; we can eliminate them by adding a few extra reverse folds as shown in Figure 12.17.

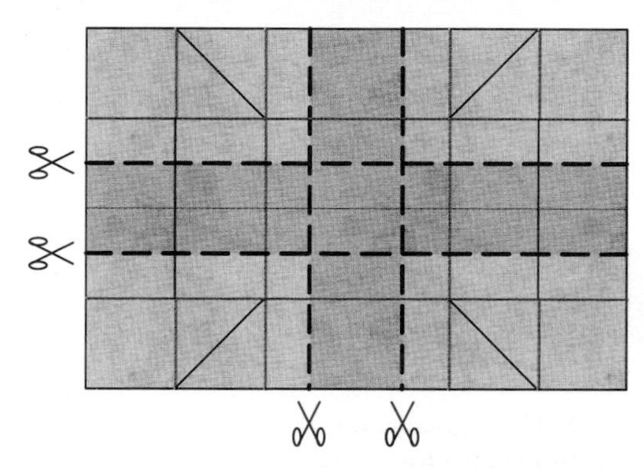

1. Cut out the shaded cross-shaped region.

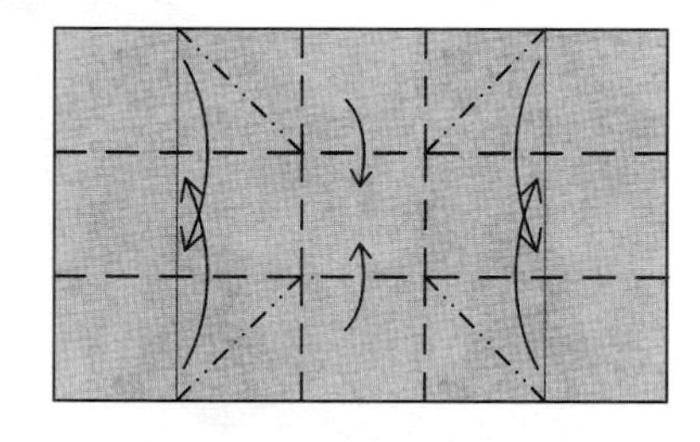

2. Now fold the box in the same way as before. Note that the two side flaps overlap one another.

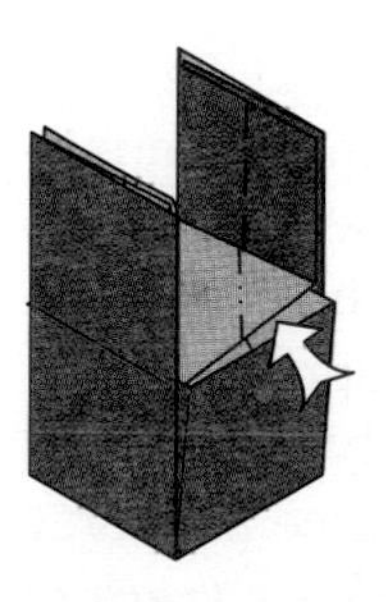

3. Reverse-fold the corner so that the raw edge lines up with the far vertical edge.

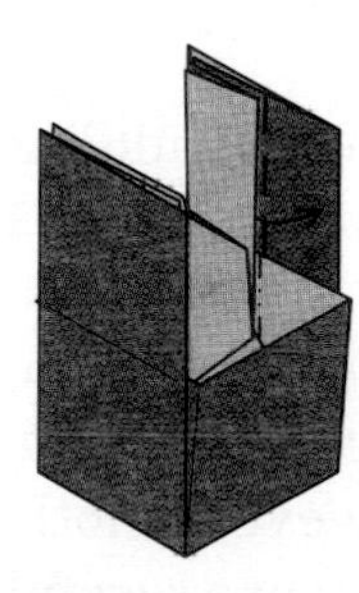

4. Reverse-fold the other corner in the same way.

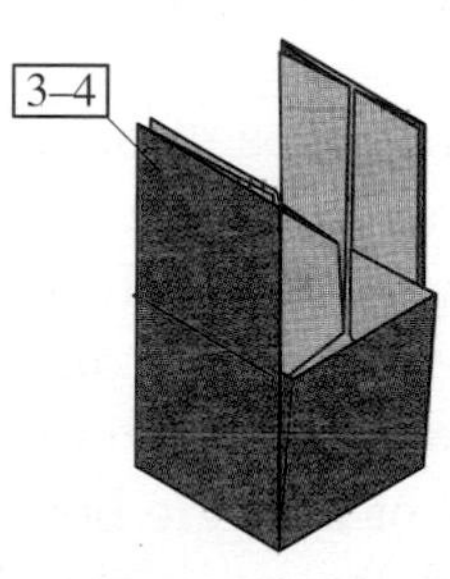

5. Repeat steps 3–4 on the near flaps.

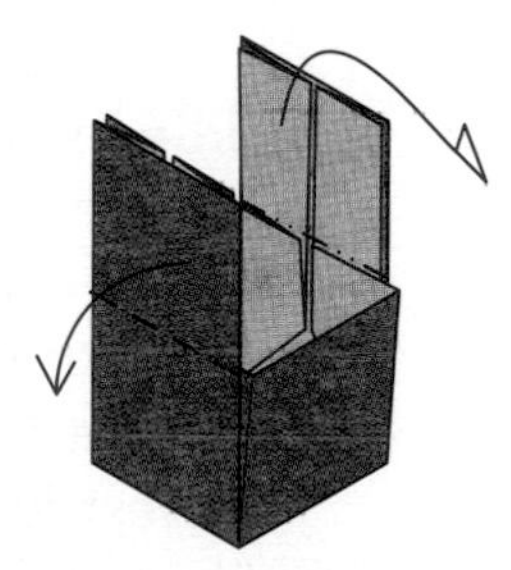

6. Fold the flaps out to the sides.

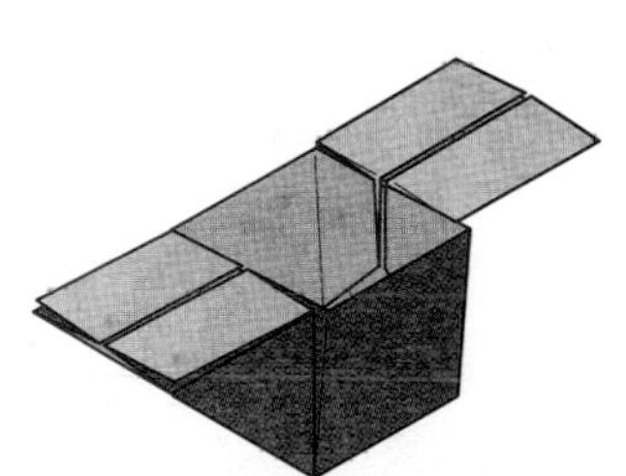

7. Finished box.

Figure 12.17.
Folding a smaller box.

The extra reverse folds add a few new folds to the crease pattern. They, too, are predominantly vertical and horizontal. If you cut out a section of this box (the shaded region in Figure 12.18), you will come back to the original 3 × 2 box exactly half the size of the original pattern (with somewhat longer handles).

Comparing steps 4 and 5 shows that the difference between a shallow box and a deeper box is precisely the shaded region in step 2. Thus, we can make a box wider or longer by

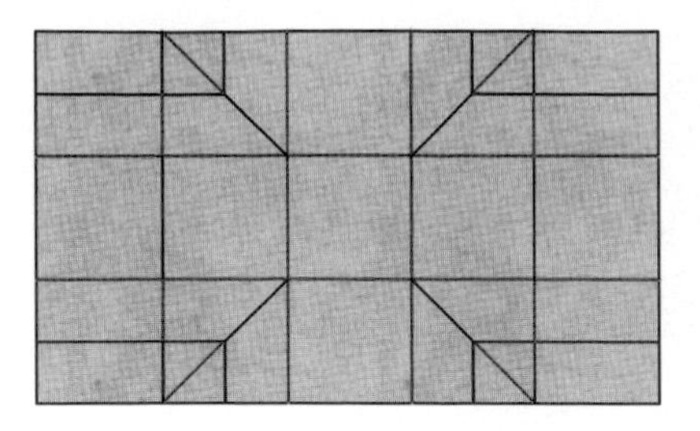

1. The crease pattern for the small box.

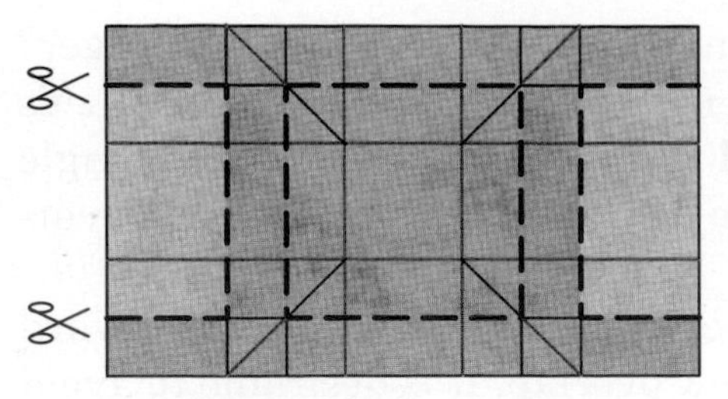

2. Cut out the shaded region ...

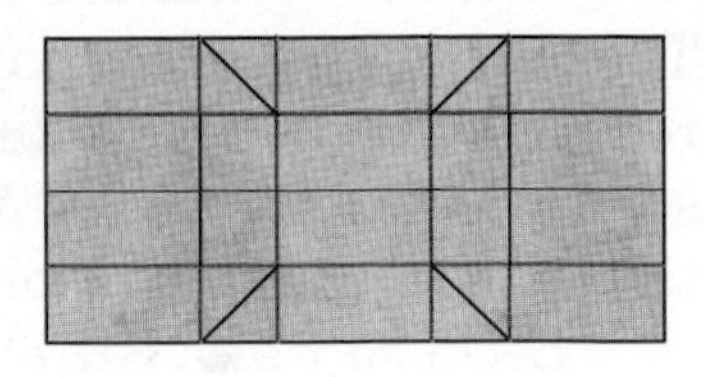

3. And we're back to the original 3 × 2 box pattern, but with longer handles.

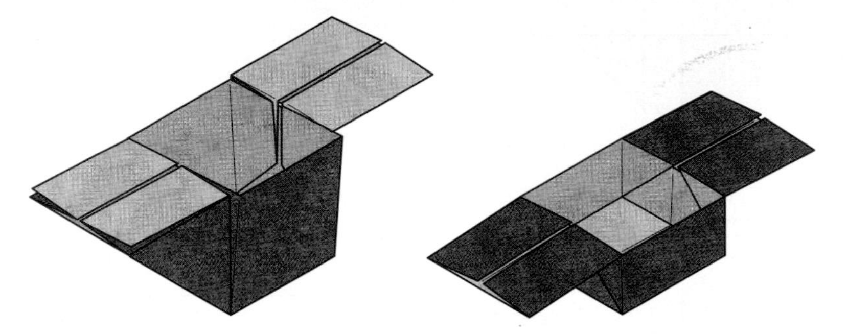

4. The crease pattern from step 1 folds this.

5. The crease pattern from step 3 folds this.

Figure 12.18.
The crease pattern for the smaller box.

adding simple strips of paper, and by adding a more complicated shape, as shown in step 2 of Figure 12.18, we can make the box deeper as well. Thus, it is possible, using basically the same structure, to make any length, width, or depth, box.

But this is still only a single box. We quickly exceed the interest level of a single box. However, another nice property of box-pleated designs is that if you are careful to keep track of the raw edges of the paper, you can easily join structures in a way very similar to the way we expanded them.

Figure 12.19 shows how two boxes can be joined at their edges to make a double-box, which can, in turn, be folded from a single 4 × 12 rectangle.

It was possible to join the two boxes because the raw edge along one side of the paper lay along a single line in the folded form of the model. That raw edge could therefore be mated to a similarly aligned edge.

It isn't necessary, however, that the raw edges lie on a single line for two shapes to be joined. The raw edge can actually take on any three-dimensional path whatsoever, as long as the mating part takes on the same path. This next structure (Figure 12.20) mates boxes and partial boxes to realize a fully enclosed box.

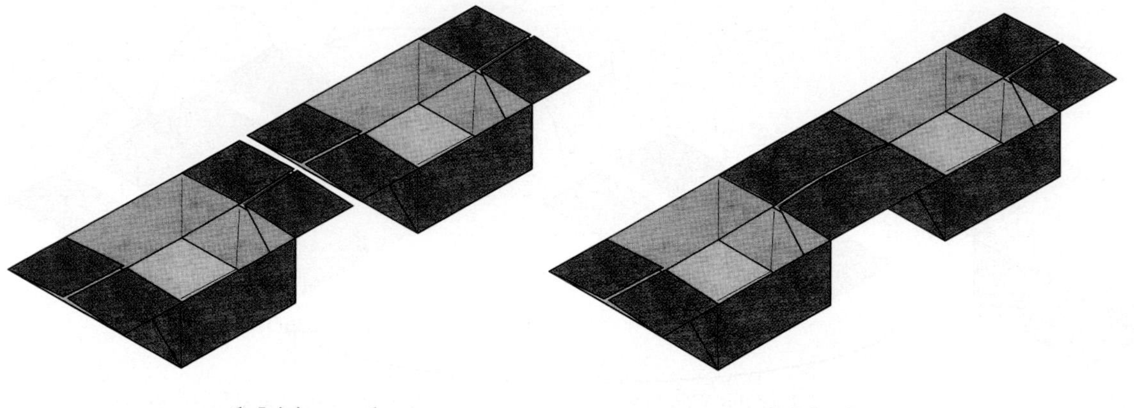

1. Joining two boxes.

2. Joined.

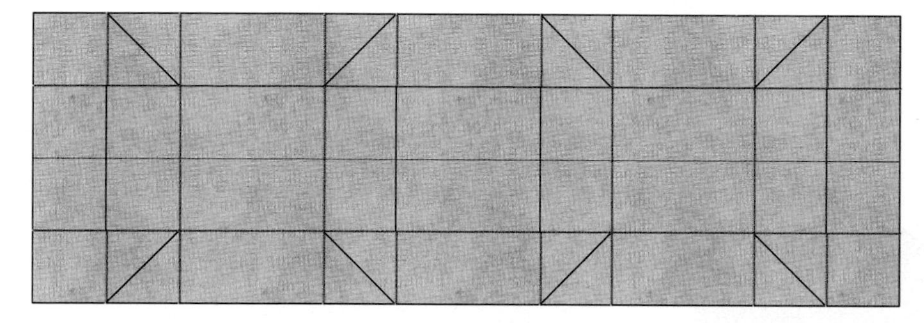

3. The crease pattern for the double-box.

Figure 12.19.
Joining two boxes, and the resulting crease pattern.

Two or three of these boxes can be joined at their ends. They can be lengthened, made taller, and butt-joined, and as the collection of boxes grows, the rectangle from which it is folded grows correspondingly.

Another way of thinking of this box is as a tube that is squeezed at the ends, as shown in Figure 12.21.

So now, we have a general-purpose way of making boxes: long boxes, wide boxes, open boxes, closed boxes, and chains of boxes. Boxes of all shapes and sizes. But as a starting point for origami, boxes are somewhat limited: you can only use them to make things that are, well, box-like. Fortunately, what could be more box-like than—a boxcar? Or, in the case of Mooser's Train, a train of boxcars! It's not hard to see how one progresses from a chain of boxes to a train of boxcars. And while Mooser's Train isn't built from precisely this type of box, the main structural element, shown in Figure 12.22, is a small modification of it.

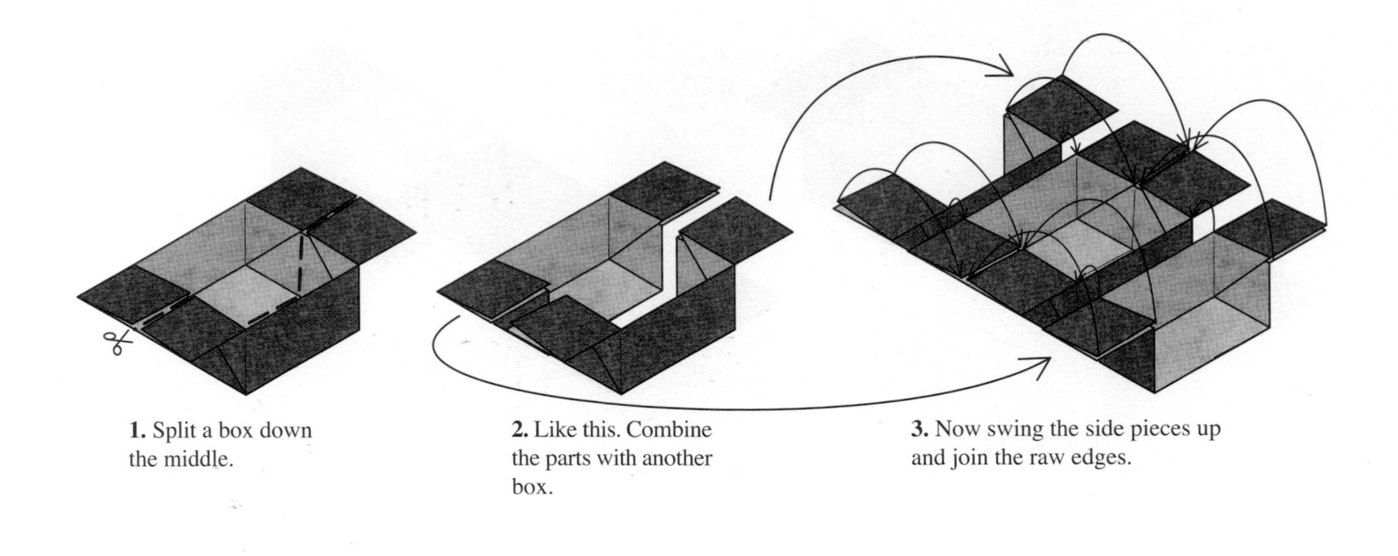

1. Split a box down the middle.

2. Like this. Combine the parts with another box.

3. Now swing the side pieces up and join the raw edges.

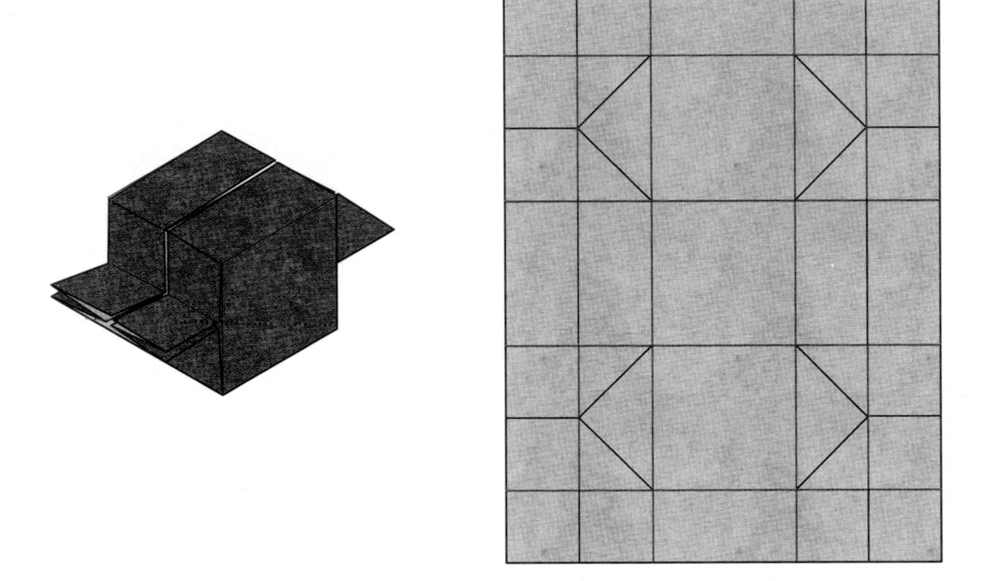

4. A three-dimensional box and its crease pattern.

Figure 12.20.
A fully three-dimensional box.

This shape doesn't look very much like a boxcar yet. But by using the techniques shown in this section, one can lengthen the car, add extra paper along the bottom, turn the excess underneath—and suddenly, the model begins to look very boxcar-like. Connecting the boxcars—by turning the single-car-square into a chain of squares, i.e., a long rectangle—yields an entire train. The use of primarily orthogonal creases allows relatively straightforward grafting of different box-like structures together. But the final element, which bloomed in the hands of Elias, Hulme, and others, was that box-pleated struc-

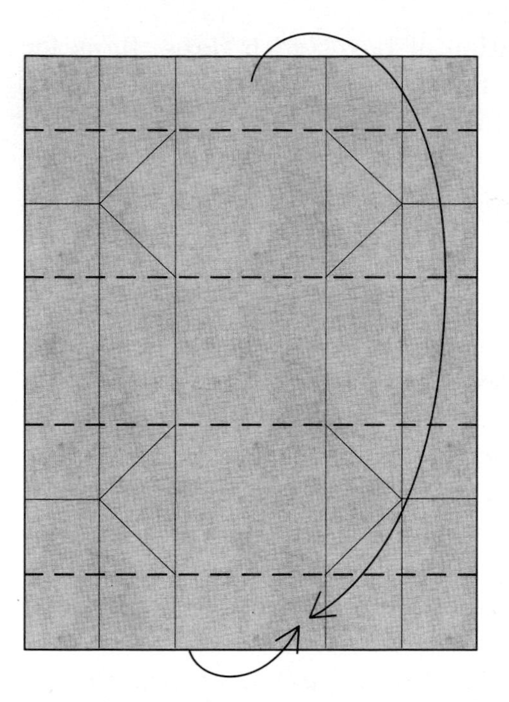

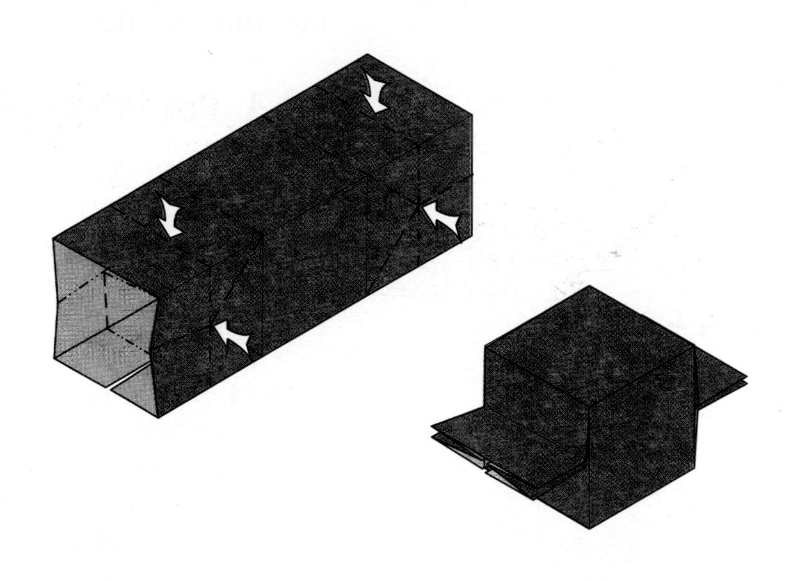

1. Fold the crease pattern into a tube.

2. Squeeze the top and bottom of the tube.

3. A three-dimensional box.

Figure 12.21.
The box can also be thought of as a pinched tube.

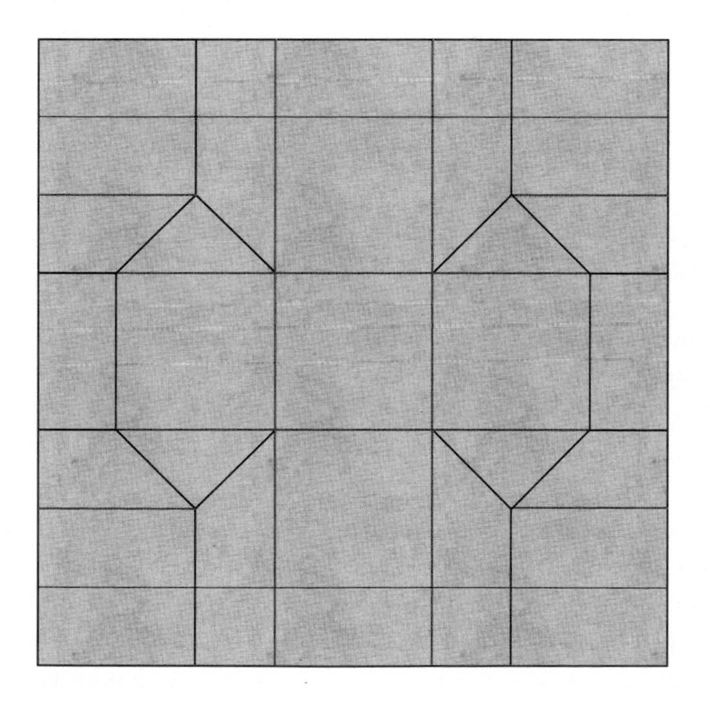

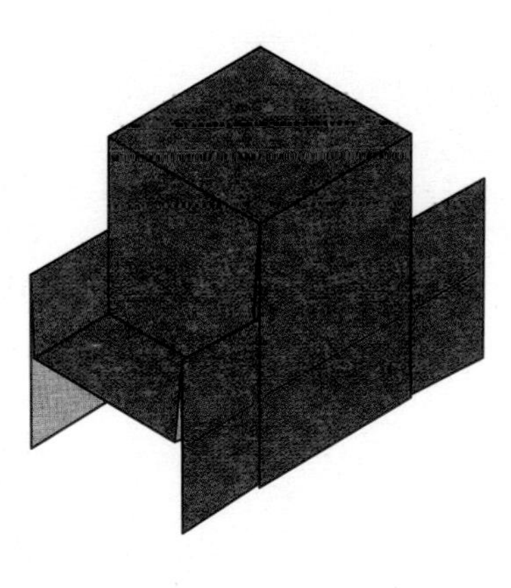

1. The basic crease pattern for Mooser's Train.

2. The basic box for Mooser's Train.

Figure 12.22.
The building-block crease pattern and box for Mooser's Train.

tures allowed the integration of boxes with flaps: flaps for wheels, for legs, for arms, for entire bodies. And so we shall now turn our attention to flaps.

12.3. Box-Pleated Flaps

Of course, we already know how to generate flaps, at least those that are part of a uniaxial base: we pack circles and rivers representing flaps and connectors, or we solve a set of path equations. In either case, we fill in the resulting grid of axial polygons with molecular crease patterns.

But when we construct molecular crease patterns, the creases often run at strange angles, particularly in gusset molecules and universal molecules; they do not, as a rule, fit the model of box pleating, in which all creases run at multiples of 45°. But could they? Is there a set of molecules that satisfy the box-pleating convention?

Every molecule is outlined by creases, so a box-pleated molecule must fill a polygon whose sides run at multiples of 45°. But the situation is actually more restrictive than this. Every simple molecule has at each corner a crease that is the angle bisector of the two sides forming the corner. If the angle between the two sides is an odd multiple of 45°, then the angle made by the bisector is an odd multiple of 22.5°—not valid in a box-pleated pattern. Thus, the angle between any two adjacent sides of a box-pleated molecule must be an even multiple of 45°, that is, a multiple of 90°. This means that all convex molecules in box-pleated patterns must be rectangles or squares. In particular, all axial creases must be either up-and-down or side-to-side.

We already know of two molecules that work in rectangles: the Waterbomb and sawhorse molecules, shown in Figure 12.23. Both have the property that the ridgeline creases lie at odd multiples of 45°, while the axial creases and hinge creases lie at even multiples of 45°. Thus, both molecules satisfy the constraints of box pleating; both could be used to construct flaps within a box-pleated model.

Molecules were defined and constructed in order to create uniaxial bases. Box-pleated bases are no longer necessarily uniaxial—they may not even lie flat—but we can still graft portions of uniaxial bases around the edges of the boxes to attach flaps to them.

Thus, it would appear that we could use circle packings to construct box-pleated collections of flaps. But we must add an additional condition: Since axial creases are formed wherever two circles (or rivers) touch, in order to insure that those

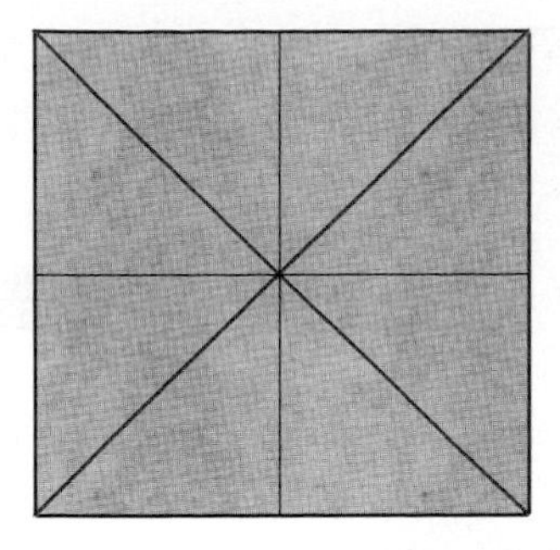
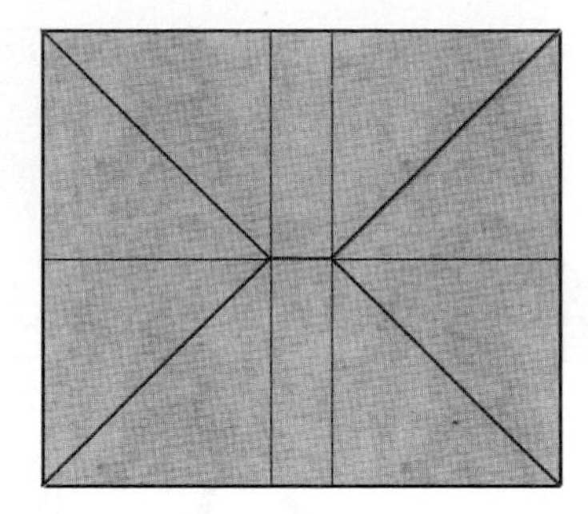

Figure 12.23.
Left: waterbomb molecule. Right: sawhorse molecule. Both satisfy the definitions of a box-pleated crease pattern.

axial creases are valid box-pleating creases, the touching circles must be either vertically or horizontally aligned so that the axial creases that connect their centers run only vertically or horizontally.

If all axial creases run vertically or horizontally, then the angle between successive axial creases must be either 90° or 180°. 90° is obviously a corner; 180° is perhaps less obvious; this situation arises when three or more circles or rivers touch along a single side, as in Figure 12.24. This puts a circle in the middle of an edge.

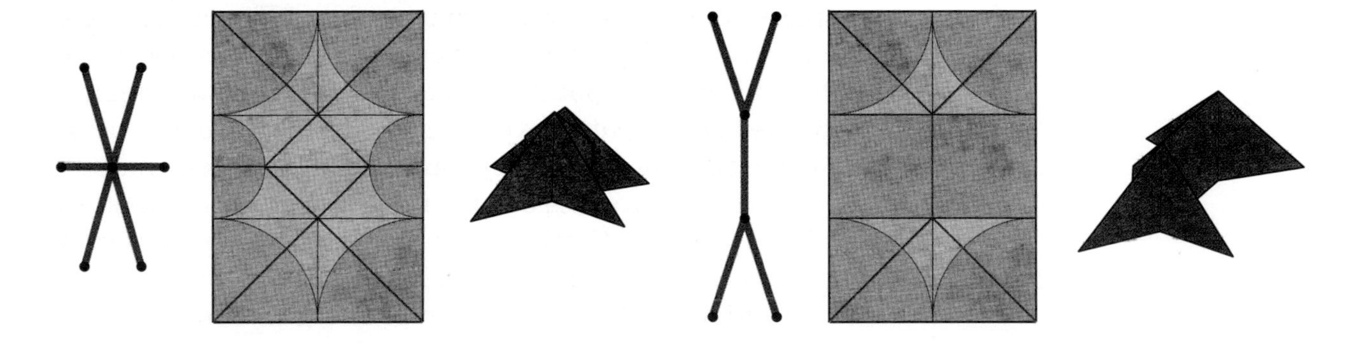

Figure 12.24.
Left: tree, crease pattern, and folded form for a six-flap molecule. Note that two circles occur in the middle of an edge.
Right: tree, crease pattern, and folded form when the edge circles are replaced by a river.

In both patterns in Figure 12.24, the crease pattern may be obtained by applying the universal molecule construction. Fortunately, all of the computed creases turn out to fulfill the box-pleating requirement.

But the universal molecule frequently does not give a box-pleated crease pattern, even if one exists. For example, introducing a bit of asymmetry into the two tree graphs in Figure 12.24 significantly distorts the universal molecule, resulting in non-box-pleated molecules shown in Figure 12.25.

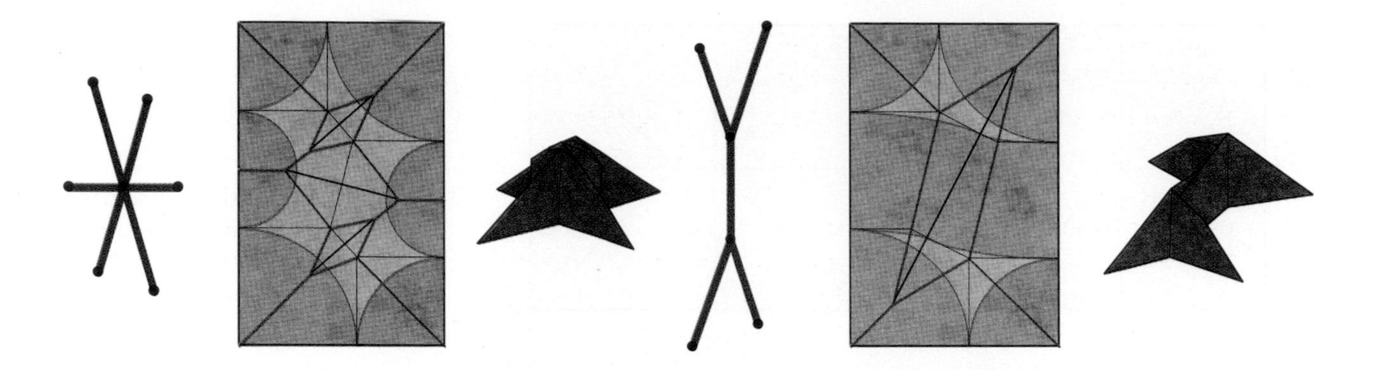

Figure 12.25.
Left: tree, crease pattern, and folded form for an asymmetric six-flap molecule.
Right: tree, crease pattern, and folded form when the circles are replaced by a river.

Both molecules, of course, are fully valid molecules and could certainly be used within origami figures; they just wouldn't be box pleated. Why the big concern? In box pleating, since all of the creases run at multiples of 45°, they are (usually) relatively easily constructed by folding alone. However, the creases in the interior of the molecules of Figure 12.24 are at odd angles, and are fairly difficult to construct without resort to numerical computation. Because of their simple angles, box-pleated crease patterns can be much easier to develop linear folding sequences for. They come with a cost, however; not all circle patterns possess box-pleatable molecules. In the two examples in Figure 12.25, the tree on the left does possess a box-pleated molecule that corresponds to its circle packing, shown in Figure 12.26. The tree on the right, however, does not permit a box-pleated molecule.

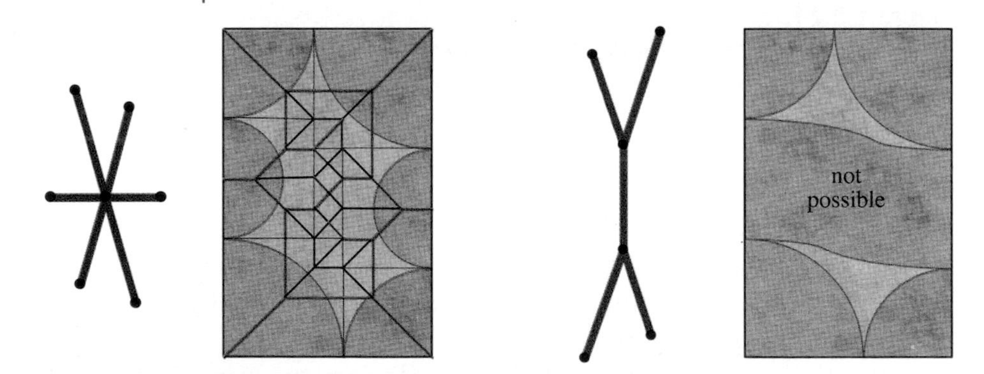

Figure 12.26.
Left: box-pleated molecule for the asymmetric six-flap tree.
Right: there is no box-pleated molecule for this tree and combination of circles.

There is much to consider in these two examples. First, the molecule on the left does not resemble any of the molecules we have seen thus far, suggesting that there is some new principle at work. Second, the pattern on the right is certainly a valid circle/river packing; its lack of a box-pleated molecule indicates that the conditions for existence of a box-pleated molecule are somehow different—and probably more restrictive—than the conditions for existence of a molecule based on circle/river packings.

Incidentally, the crease pattern in Figure 12.26 is given in generic form, i.e., without specifying the direction of all hinge creases. It is a pleasant little puzzle to cut out the pattern as shown and try to find the crease assignments that assemble it into a flat-folded base.

In fact, the conditions that produce box-pleated uniaxial bases are more restrictive than those for circle/river uniaxial bases. Consider three vertices, A, B, and C, that lie on the axis of the base as shown in Figure 12.27. In the crease pattern, A and C lie along a common vertical line (which must be an axial crease). Similarly B and C lie along a common horizontal line (which must also be an axial crease), so that A and B are diagonally offset from each other. Since vertex B lies on the axis of the base, in the folded form, vertices A, B, and C must lie on the same line (the axis).

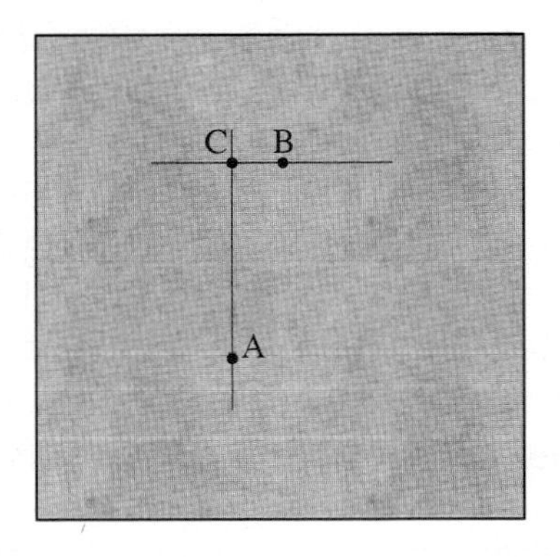

Figure 12.27.
Left: the square with axial vertices A and B.
Right: schematic of the folded base.

The question now is: If B is also an axial vertex, how far away from vertex A can it lie in the base?

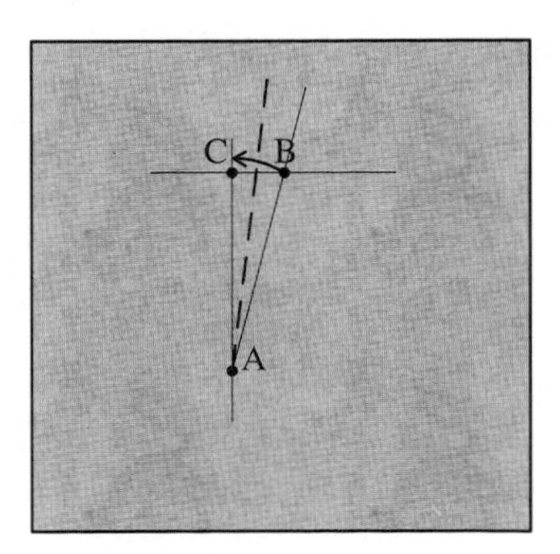

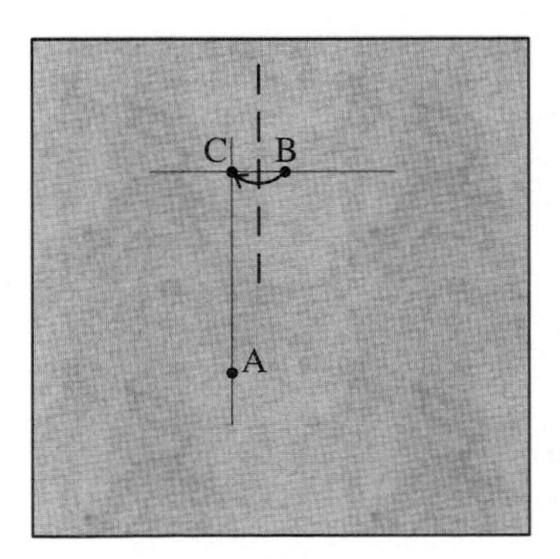

Figure 12.28.
Left: an angle bisector (not box pleated) brings point B to line AC.
Right: a vertical crease accomplishes the same.

We must consider the possible folds that could bring point B onto line AC. In Figure 12.28, we see that an angle bisector brings B to line AC, which makes the distance between points A and B in the base equal to their separation in the crease pattern; this is the limit set by tree theory. But the angle bisector is, in most cases, not a valid box-pleating crease. The only valid crease that puts B onto line AC (without shortening line AC) is the vertical crease shown on the right in Figure 12.28. But with this crease, point B is coincident with point C in the folded form, which means that even though B may be farther from A than C in the crease pattern, in the folded form it can be no farther.

This analysis applies if the vertical separation between A and B is greater than their horizontal separation; a similar analysis applies if their horizontal separation is greater. The combination of the two analyses can be stated succinctly as follows:

> *The distance between two axial vertices in the folded form is less than or equal to the greater of their vertical and horizontal separation in the crease pattern.*

This sounds very similar to the path condition that applied to circle/river bases, which was:

> *The distance between two axial vertices in the folded form is less than or equal to their absolute separation in the crease pattern.*

In fact, we can modify and then use much of the machinery of tree theory to construct valid patterns of leaf vertices for box-pleated designs. Recall that for existence of an ordinary crease pattern, we required that for every path between leaf vertices $\mathbf{u}_i$ and $\mathbf{u}_j$, the leaf vertices must satisfy the inequality

$$\left|\mathbf{u}_i - \mathbf{u}_j\right| \geq m\, l_{ij} \qquad (12\text{–}1)$$

where l_{ij} is the distance between nodes i and j along the tree and m is the scale factor. For box-pleated patterns, the analogous condition is:

$$\max\left(\left|\mathbf{u}_{i,x} - \mathbf{u}_{j,x}\right|, \left|\mathbf{u}_{i,y} - \mathbf{u}_{j,y}\right|\right) \geq m\, l_{ij} \qquad (12\text{–}2)$$

We call the set of modified equations (one for each path ij) the *box-pleating path conditions*. Valid patterns of node vertices may be solved by the same sort of optimization we used in tree theory, namely, maximize the scale factor m subject to the constraints of the box-pleating path conditions. However, just as the path conditions of tree theory possessed a visualizable geometric analog in circles and rivers, so too do the box-pleating conditions. Instead of packing circles and annular rivers, we will find that we pack squares, rectangles, and a new type of river (shortly to be described) to construct valid crease patterns.

Even better, while it was not at all uncommon for circle packings to result in distances and reference points that were defined solely as the roots of high-order sets of polynomial equations, we will see that the square-river packings that define box-pleated patterns produce dimensions that are simple linear combinations of distances that appear in the underlying tree, and thus the reference points are nearly always readily found by folding alone.

12.4. Square/River Packings

We constructed valid patterns of leaf vertices for uniaxial bases by packing circles and rivers that corresponded to the edges of a tree graph. We required that (a) the centers of the circles lie within the square of paper, and (b) the circles and rivers do not overlap. We can similarly construct valid patterns of leaf vertices for box-pleated uniaxial bases by packing squares and a new type of river subject to the same rules: the centers of the squares must lie within the paper and the squares and rivers must not overlap.

Each element of the tree graph has a box-pleating analog. In circle/river packing, for a leaf edge, we used a circle whose radius was the length of the flap. Another way of describing the circle is to say that the distance from the center of the circle to its rim is the length of the flap. In box pleating, instead of a circle, we use a square in which the perpendicular distance from the center to the rim is equal to the length of the flap. That is, the square's side is twice the length of the corresponding flap.

Branch edges, flaps that connect groups of flaps to each other, are a bit more complicated. In circle/river theory, a branch edge was represented by a river, a curve of constant width, which we commonly represented as piecewise continuous annuli. In box pleating, a branch edge is represented by an *ortholinear river*, which, roughly speaking, is a river of constant width whose sides are either vertical or horizontal. Examples of both leaf and branch objects in both circle/river packing and box pleating are shown in Figure 12.29.

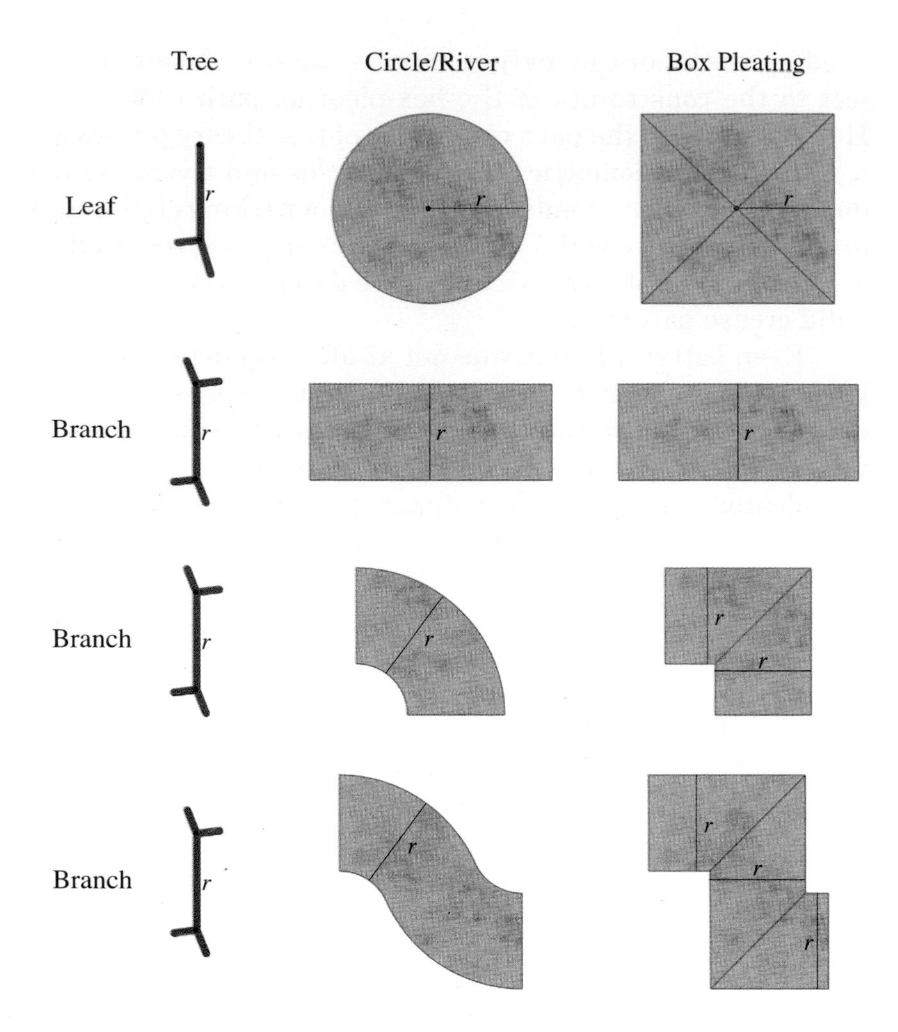

Figure 12.29.
Correspondence between circles and rivers and the squares and ortholinear rivers used in box pleating.

It was necessary to qualify the definition of ortholinear river because the width—ordinarily taken to mean the shortest distance across the river—is not always a constant in an ortholinear river. Ortholinear rivers always turn by 90°, but if two such turns are closely spaced, as in Figure 12.30, the width in the middle of the bends is actually larger than at the ends. A cleaner definition of an ortholinear river relies on the 45° lines at each bend (as shown in Figure 12.30); every straight line parallel to a bank of the river reflects at each 45° line and maintains a constant perpendicular distance from each bank between reflections.

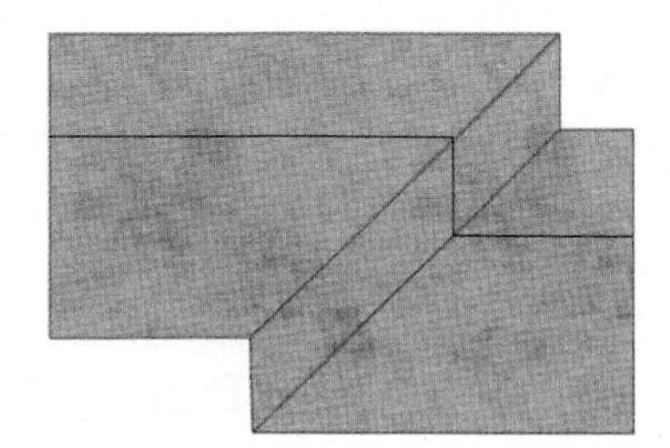

Figure 12.30.
An ortholinear river can have two closely spaced bends.

It is fairly straightforward to show that the box-pleating path conditions are satisfied for a set of node vertices if and only if a nonoverlapping packing of squares and ortholinear rivers exists. Furthermore, such a valid packing can be used as the starting point for the full crease pattern of a box-pleated uniaxial base. We call such a pattern a *square packing* or, if it contains ortholinear rivers, a *square/river packing*.

We can now use the concepts of square packing to examine the two molecular patterns shown in Figure 12.26. For the six-flap molecule, we pack six squares with their centers into the rectangle as shown in Figure 12.31. They fit perfectly without overlap, which tells us what we've already seen: There is a box-pleated molecule that corresponds to this packing.

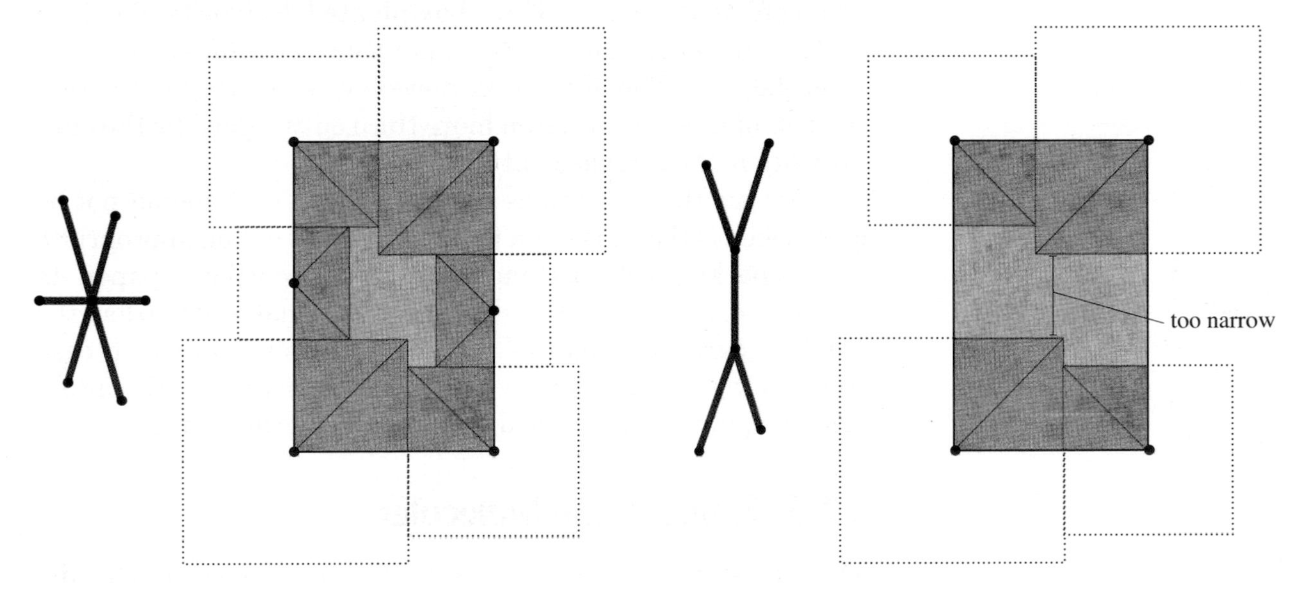

Figure 12.31.
Left: square packing corresponding to a six-flap molecule.
Right: the packing for the sawhorse molecule doesn't leave enough room for the ortholinear river.

For the second molecule, if we pack the four squares into the corners and examine the space left for the ortholinear river, you can clearly see that in the middle of the square, there is a narrow region that makes it impossible to place an ortholinear river of the necessary width. Consequently, there is no valid packing, and thus, no box-pleated molecule is possible.

It is possible, however, to expand the rectangle to allow both the packing and the box-pleated molecule; one such expansion is shown in Figure 12.32.

If you compare the areas of the shaded objects in Figure 12.29, you'll see that the squares and ortholinear rivers of box pleating take up more area than the circles and rivers of circle/

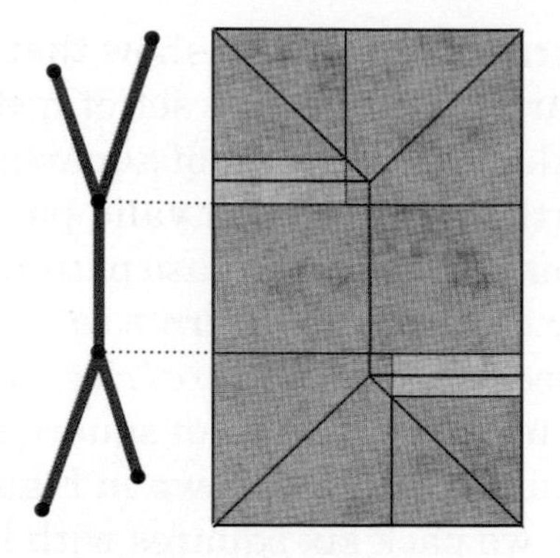

Figure 12.32.
Slightly lengthening the rectangle allows a valid square packing and box-pleated molecule.

river packing. Furthermore, the sharp corners in squares and ortholinear rivers can get in the way of what might otherwise be an efficient packing. Thus, box-pleated designs tend to be a bit less efficient in their use of paper than those based on circle/river packing. The difference, however, is not very large, and the loss of efficiency is often more than outweighed by the simplification of the crease pattern.

Yet another advantage of box pleating that should not be overlooked is that it is often possible to work out an appropriate square packing with nothing more than a pencil and paper. As we will see, it is frequently advantageous to make all flap lengths small integral multiples of a common distance, in which case the packing and the complete crease pattern fit neatly onto a grid of squares and can be drawn on ruled graph paper.

12.5. Filling in the Molecules

The first step in constructing a box-pleated uniaxial molecule is the square packing; we place all leaf vertices into a square (or other starting shape, if desired) along with the squares and ortholinear rivers that correspond to each flap of the base. Any arrangement of squares and rivers that avoids overlap gives a valid configuration of the leaf vertices, but the most efficient base is the one in which the packing elements are as large as possible relative to the square. This optimal pattern can then be turned into a complete crease pattern by filling in the creases that connect the leaf vertices.

How do we do this? Where do the creases go? We've already seen that the universal molecule is not always the appropriate algorithm for a box-pleated molecule; in fact, it is rarely appropriate. Fortunately, there is another general procedure we can use, which I will describe by way of a design example.

Consider, for example, the tree shown in Figure 12.33, which corresponds to a type of longhorn beetle of the family *Cerambycidae*. Longhorn beetles, though otherwise unremarkable, have extremely long antennae. Let's first look at what a traditional circle packing produces. If we choose the antennae flaps to be three times the length of the leg and abdomen flaps and further require that both the antennae and legs come from the corners and edges of the paper, we'll find that their circles pack neatly around the outside of the square, leaving plenty of room in the interior for the abdomen circle to rattle around in.

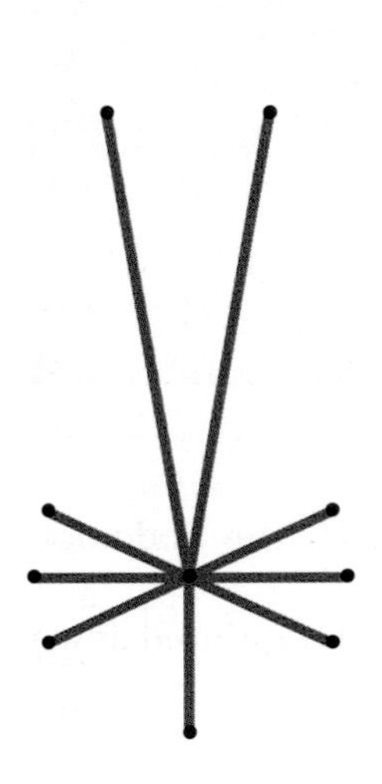

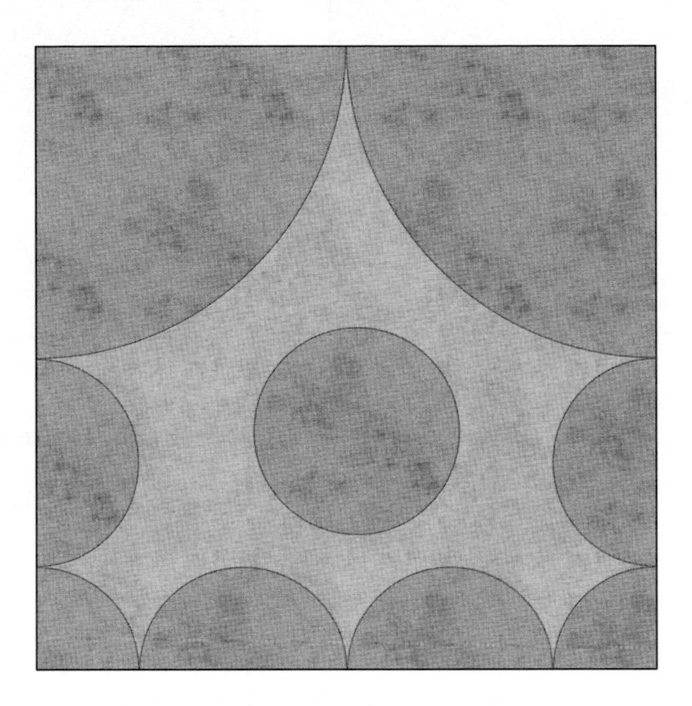

Figure 12.33.
Left: tree for a longhorn beetle.
Right: initial circle pattern.

Now, using the conventional approach, we can inflate the abdomen circle until it makes contact with four other circles, crystallizing the circle pattern into a rigid pattern. The pattern can then be filled in with molecules in several ways that we have already seen. Two are shown in Figure 12.34. Filling in each polygon with universal molecules gives the simplest crease pattern, while fully triangulating the tree by adding extra circles and stubs allows us to fill in all polygons with rabbit-ear molecules.

Neither of these crease patterns is particularly foldable. The creases run at odd angles and emanate from what appear to be fairly arbitrary reference points in the interior of the paper.

However, we can also seek to find a box-pleated pattern. There are no rivers in this design, so we will only be packing squares. Squares representing each of the flaps pack neatly

Figure 12.34.
Left: universal molecule-based crease pattern with the abdomen circle maximized.
Right: same circle packing, but with all molecules triangulated down to rabbit-ear molecules.

into the overall square as in Figure 12.35. Remember, as in circle packings, we only require that the center of each square falls within the boundaries of the paper.

The lighter regions within the square in Figure 12.35 are paper that doesn't belong to any feature of the tree graph. The first step in the algorithm is to soak up as much of this excess paper as possible by expanding squares in one of two ways. We can uniformly expand any square, which is equivalent to lengthening the associated flap. This process is analogous to inflating circles in circle/river packings.

We can also turn a square into a rectangle by stretching it either vertically or horizontally, as shown in Figure 12.36. This alteration doesn't lengthen the flap, but it does absorb more paper into the flap.

In the beetle packing, we can stretch the abdomen square horizontally, which effectively absorbs all of the unassigned paper as shown in Figure 12.37.

It is not always possible to absorb all paper in this way—I'll show an example (and what to do about it) in a minute—but in this design, stretching the abdomen square into a rectangle consumes all of the unused paper, and so we are now ready for the next step.

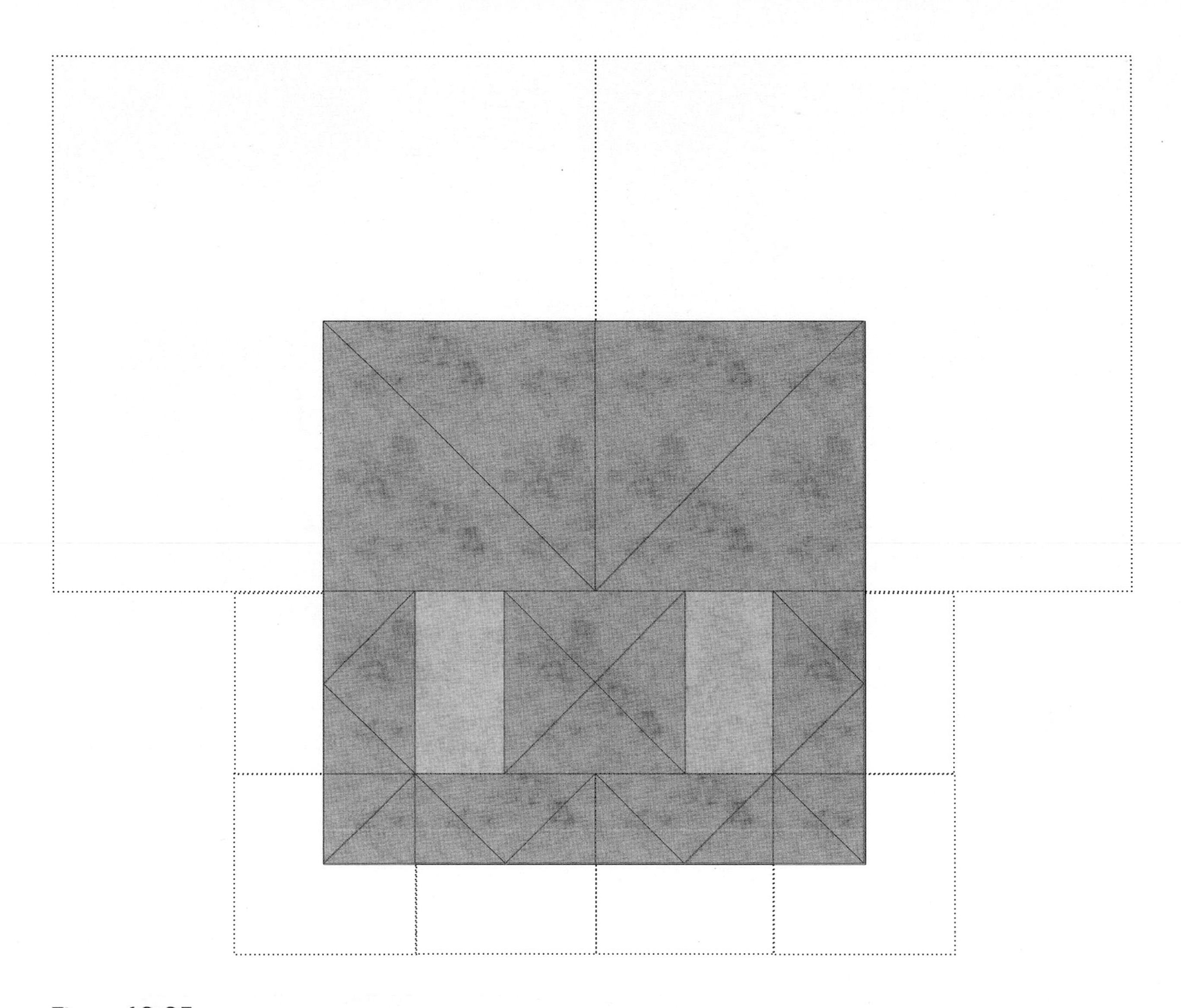

Figure 12.35.
Square packing for a longhorn beetle.

The next step is to start filling in creases. As we did with circle/river packings, we will use a generic form in which all axial creases are mountain folds and all hinge creases are unspecified creases. Each object—square, rectangle, or ortholinear river—receives a particular pattern of mountain and valley creases, as shown in Figure 12.38.

Recall that a molecule is a crease pattern that folds flat in such a way that its perimeter ends up lying along a single line. Each of the patterns in Figure 12.38 is a portion of a molecule; it is hard to resist the temptation to call them *atoms*.

Filling in the box-pleated pattern with the appropriate atoms gives a preliminary crease pattern, shown in Figure 12.39.

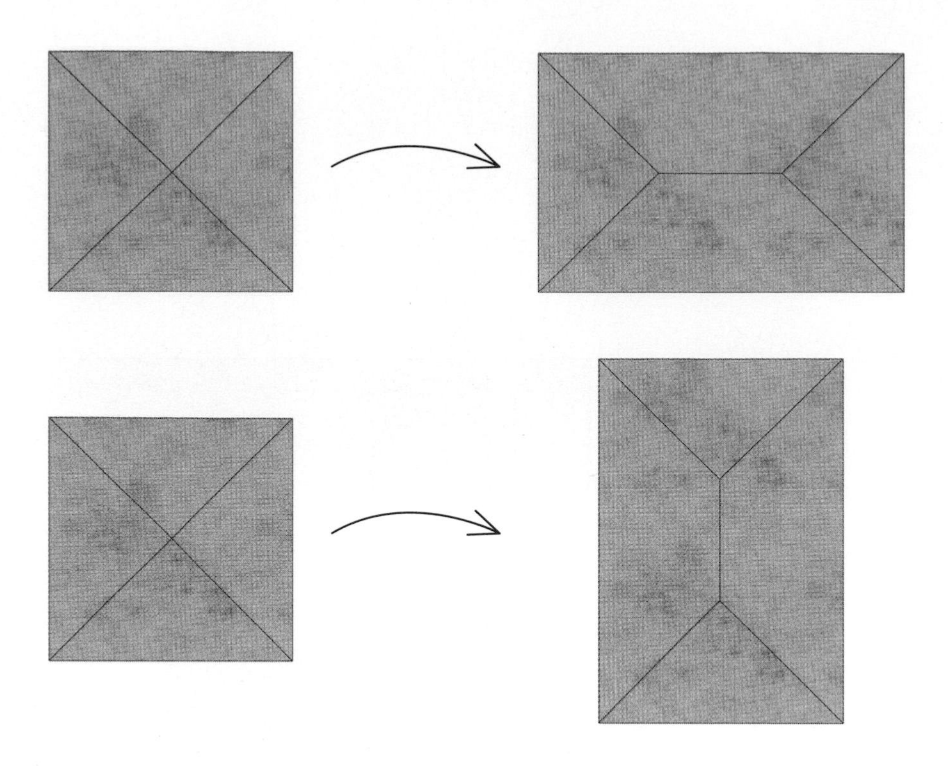

Figure 12.36.
A square can be stretched either vertically or horizontally into a rectangle to absorb unused paper in a packing.

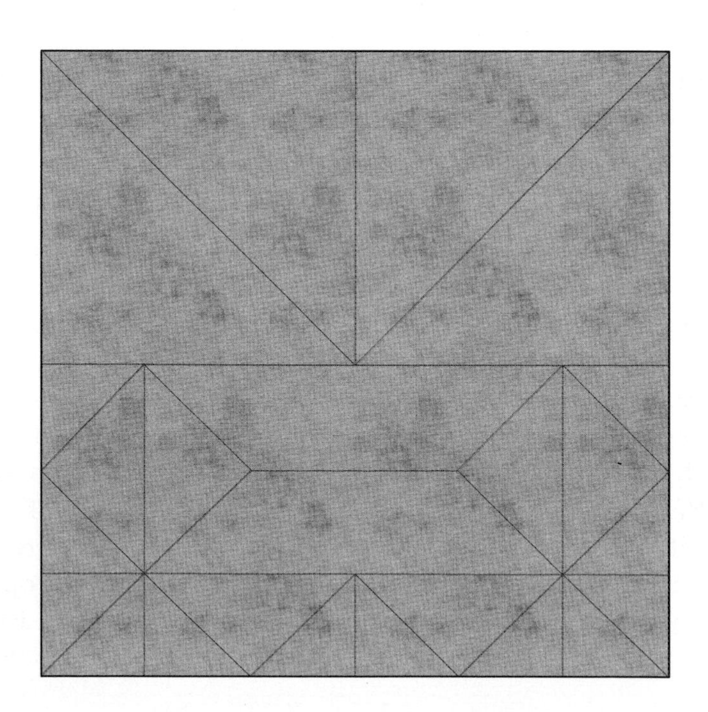

Figure 12.37.
Square (and rectangle) packing for a longhorn beetle.

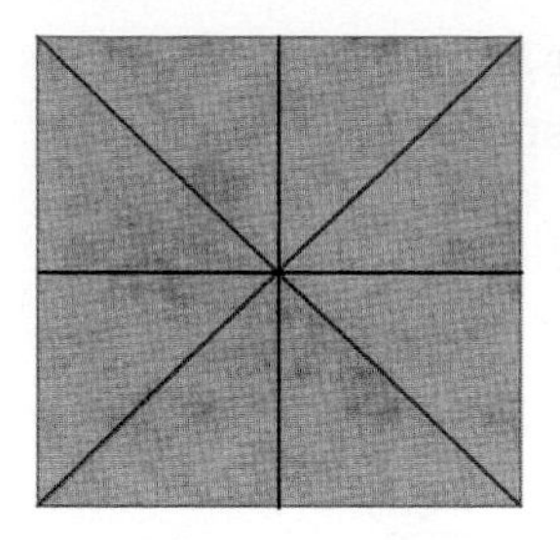
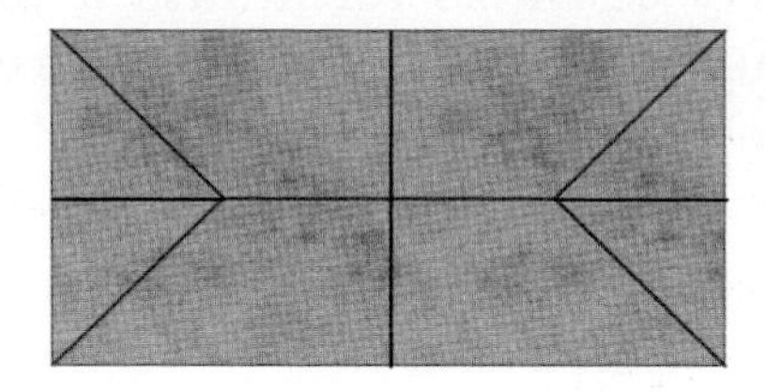
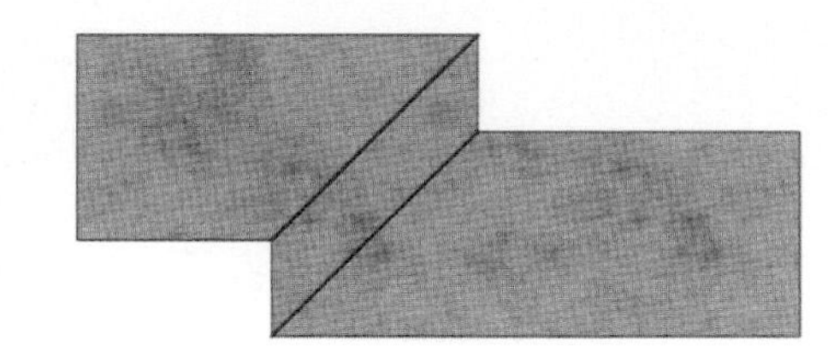

Figure 12.38.
Atoms for the elements of square/ortholinear river packings.
Left: a square atom.
Middle: a rectangle atom.
Right: an ortholinear river atom.

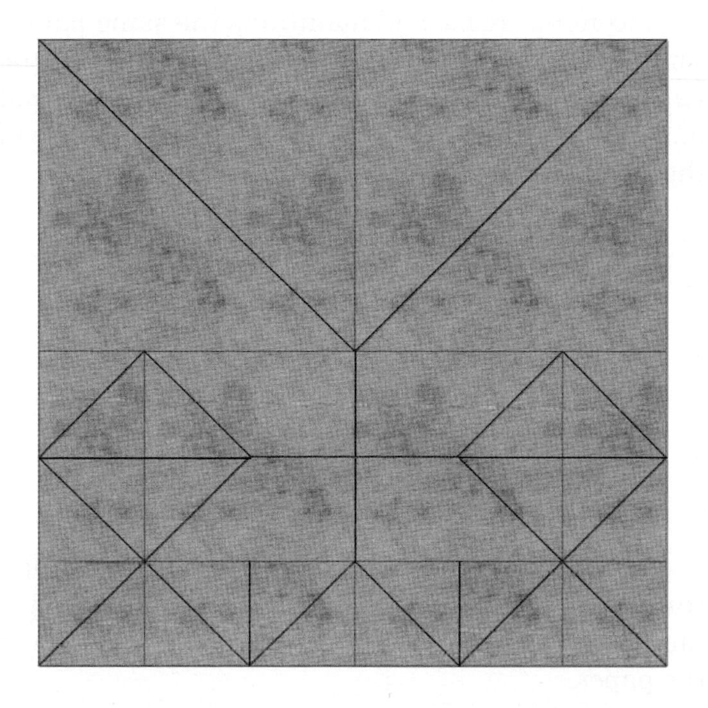

Figure 12.39.
Box-pleated pattern with atomic creases filled in.

This crease pattern is obviously incomplete: For one thing, there are creases that terminate in the middle of the paper, which we know is impossible. In other places, the pattern of creases around a vertex cannot fold flat. The next step is one unique to box pleating and its generalizations: there are features of the pattern that add, extend, and change the direction of creases according to a few simple rules.

1. Two perpendicular 45° creases that meet launch a new hinge crease of the same parity (i.e., mountain or valley) from the point of intersection. The new crease splits the obtuse angle between the two original creases. Two 45° creases that meet perpendicu-

larly can also terminate a propagating crease. Launching and termination are illustrated for valley folds in Figure 12.40.

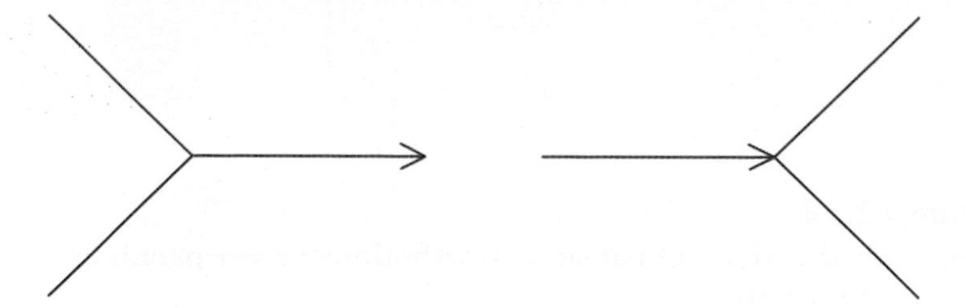

Figure 12.40.
Illustration of Rule 1.

2. If a crease hits a 45° line, it reflects across the line so that the outgoing crease makes the same angle as the incoming crease and maintains the same parity. The parity of the propagating crease matches the parity of the crease on the outside of the turn and has the opposite parity of the crease on the inside of the turn. This is illustrated in Figure 12.41.

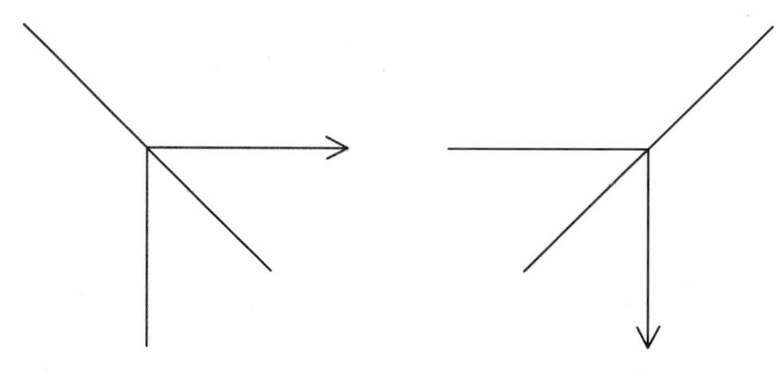

Figure 12.41.
Illustration of Rule 2.

3. A crease, once launched, propagates straight and through reflections until it either terminates (at a junction of two 45° creases) or runs off the edge of the paper.

4. A crease that crosses another 0° or 90° crease either switches its parity at the crossing or switches the parity of one side of the crossed line. The three possibilities for an incoming mountain fold are illustrated in Figure 12.42.

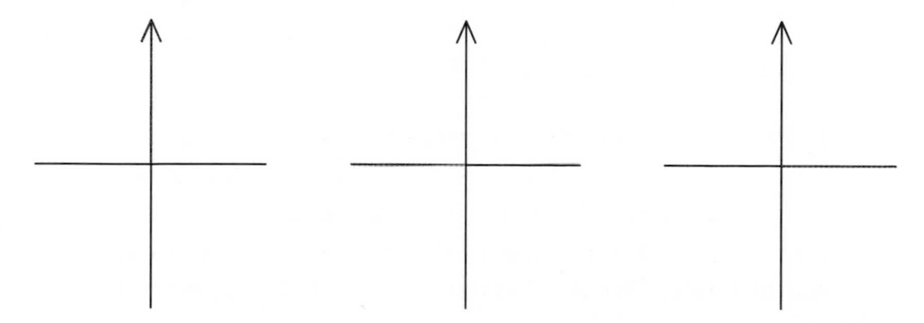

Figure 12.42.
Illustration of Rule 4.

5. A crease that crosses a Ψ intersection (branch of three creases) crosses and maintains the same parity, as shown in Figure 12.43, but flips the parity of the 3 lines on either the left or right side.

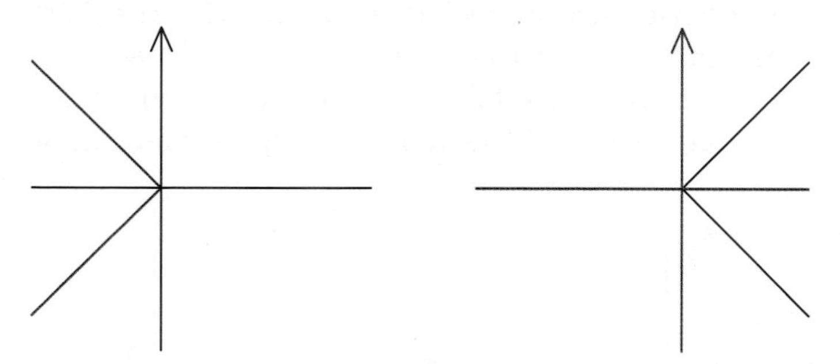

Figure 12.43.
Illustration of Rule 5.

6. Two parallel creases of the same parity launch a crease of the opposite parity that is parallel to and spaced halfway between the two original creases, as shown in Figure 12.44.

Figure 12.44.
Illustration of Rule 6.

7. The boundaries between all atoms become hinge creases.

These seven rules are probably not all that are needed (they have not been proven complete), but they work for many square/ortholinear river packings and they will work for our beetle design. Applying these rules to the preliminary crease

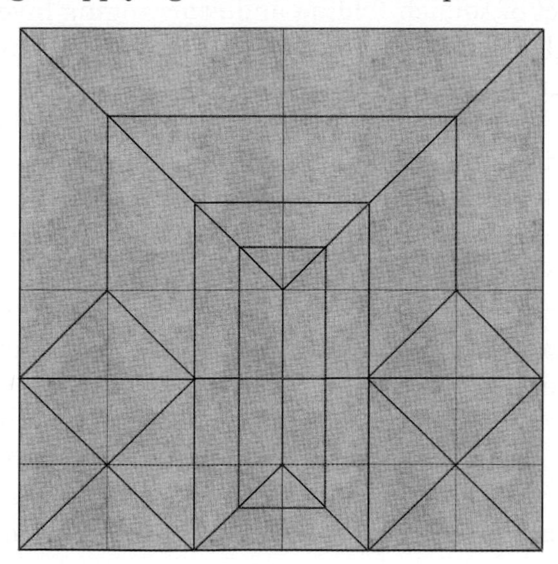

Figure 12.45.
Generic-form crease pattern for the beetle.

pattern in Figure 12.38 gives the generic-form crease pattern shown in Figure 12.45.

Recall that generic-form crease patterns do not typically fold flat; they are approximations of the base. In this pattern, the hinge creases are unspecified (as are the resulting flap directions) and some of the axial creases will need to be reversed to make the base fold flat. By further squash-folding and rearranging some of the abdomen layers, you can achieve

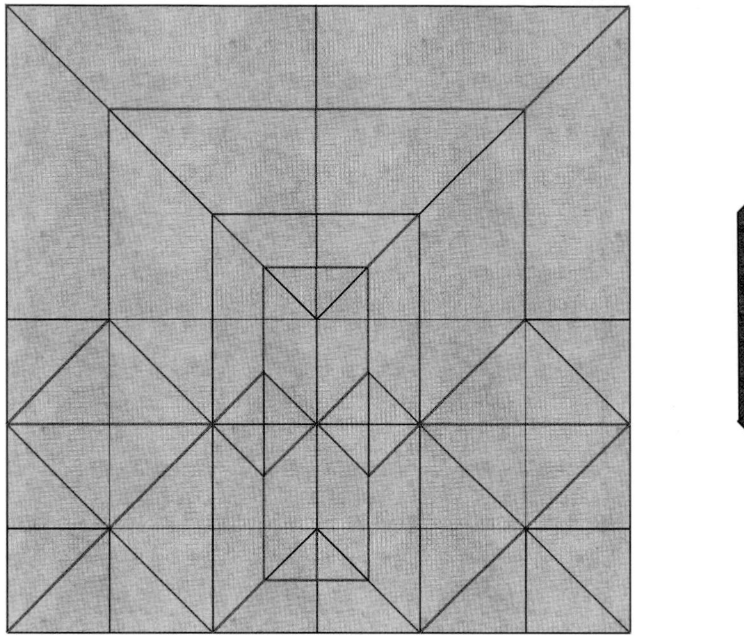

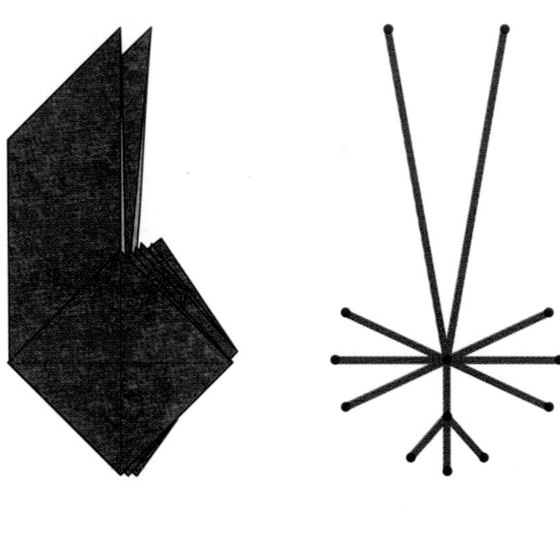

Figure 12.46.
Left: crease pattern.
Middle: folded base.
Right: the underlying tree.

the nicely flattened base whose complete crease pattern is shown in Figure 12.46.

The act of squash-folding and rearranging layers creates three smaller points at the abdomen, which you can see in the base in Figure 12.46.

This same tree and arrangement of circles and rivers also has a valid crease pattern based on the universal molecule, shown in Figure 12.47. However just like its circle/river-packed predecessor, it contains numerous creases that run at arbitrary angles.

The crease pattern gets even more complicated when it comes to narrowing flaps. In the box-pleated pattern, we can narrow the legs to a 1:4 aspect ratio (and the antennae to 1: 12) by sinking the corners of the base in and out with parallel creases. The parallel creases created by the sink fit neatly into the parallel lines of the crease pattern and permit an even distribution of layers within the flaps, as shown in Figure 12.48.

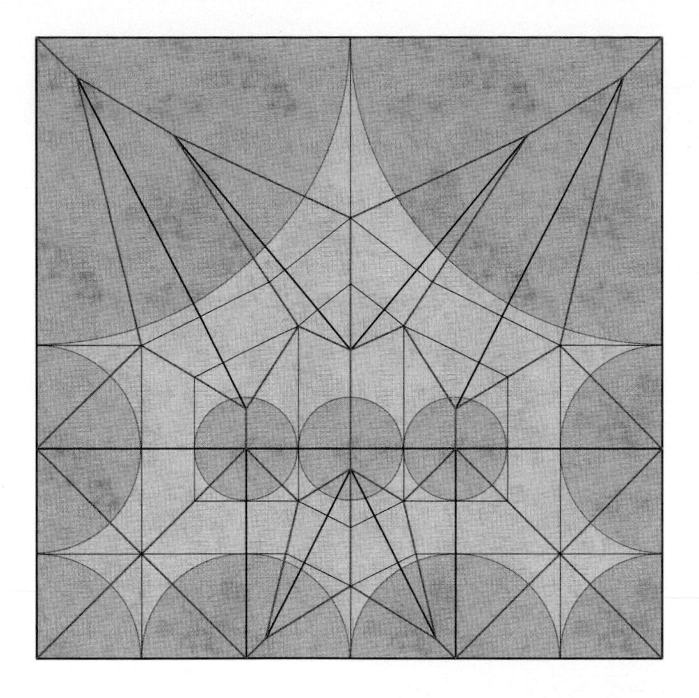

Figure 12.47.
The universal molecule crease pattern for the same number and arrangement of flaps.

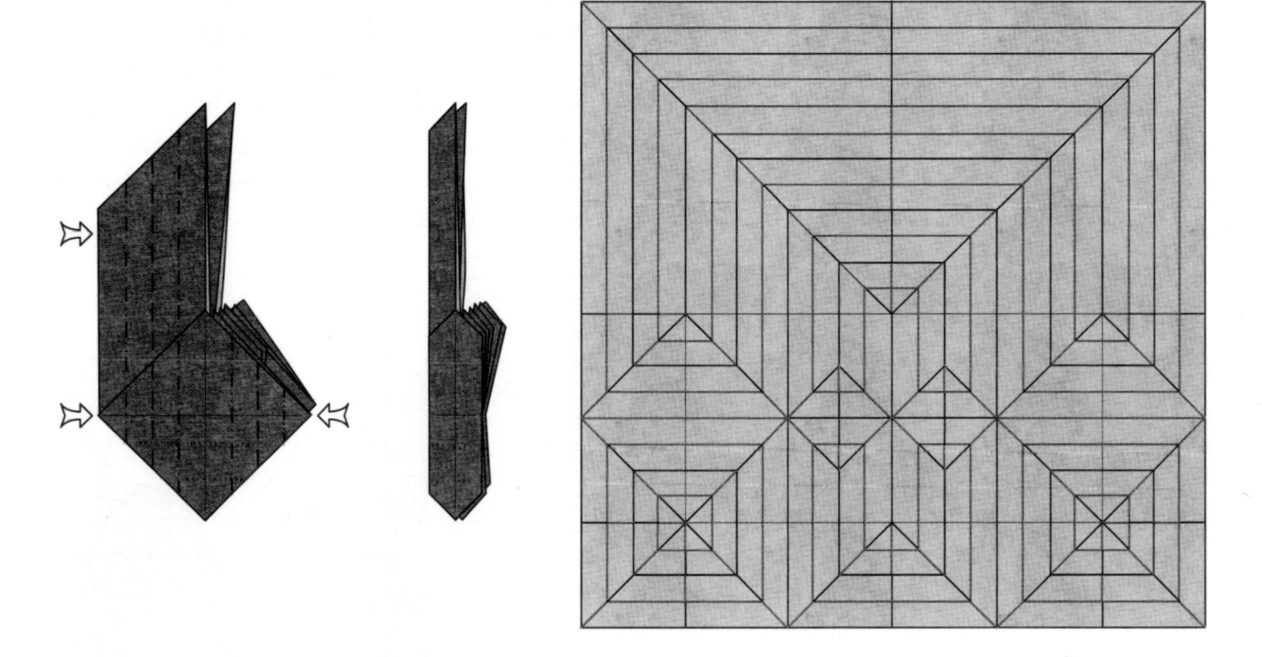

Figure 12.48.
Left: sink the flaps of the base multiple times to narrow the flaps.
Middle: the narrowed base.
Right: the resulting crease pattern.

Contrast the regularity and elegance of the crease pattern in Figure 12.48 with the universal-molecule version that has been subjected to the same set of sinks, shown in Figure 12.49.

In a box-pleated base, the edges of the flaps are usually parallel; in fact, the presence of such flaps in the folded model

Figure 12.49.
Narrowed version of the universal molecule base.

is often a sign that the underlying base was box-pleated. Parallel lines are not common in nature, and their presence in an origami model can give it an unnatural appearance. In a natural subject, however, shaping folds can remove the artificial appearance of parallel-sided flaps, and in representations of man-made objects—like a train—the parallel lines introduced by box pleating are frequently an asset.

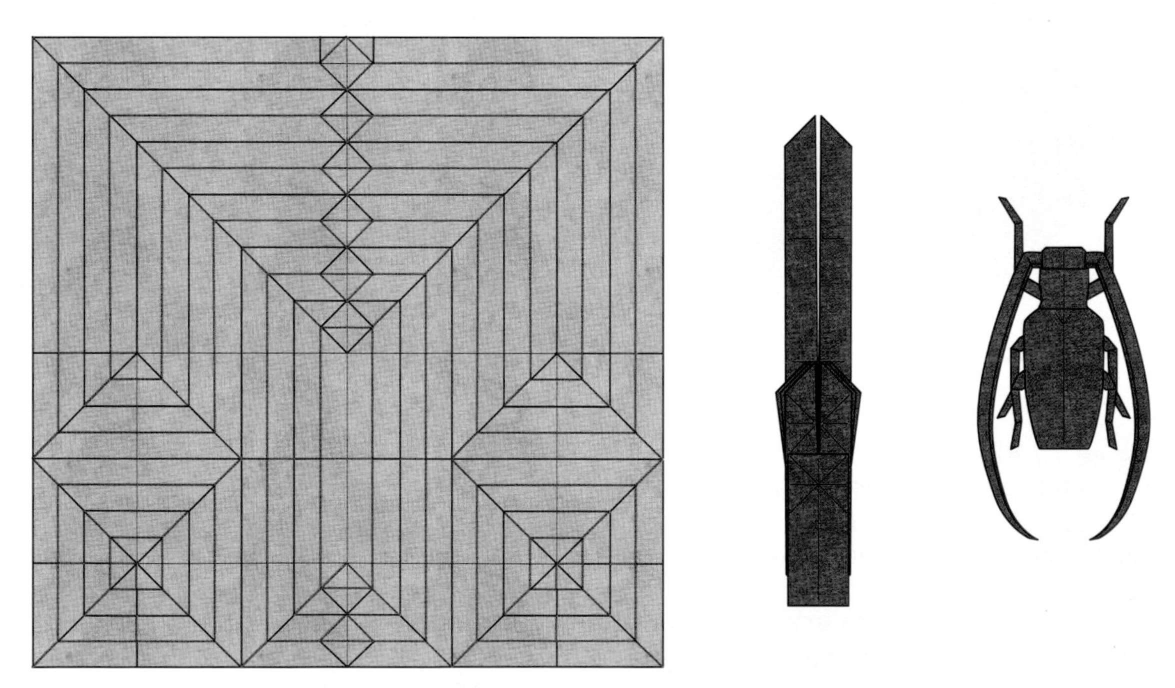

Figure 12.50.
Crease pattern, base, and folded form of the Cerambycid Beetle.

Insects and other arthropods lie somewhere between the natural and the artificial in appearance. Their long legs and antennae work well from parallel-edged flaps, and the even distribution of layers in box-pleated flaps makes them easier to thin. But the rounded head, thorax, and abdomen will need to be shaped to keep a more natural appearance. A completed Cerambycid Beetle, with a crease pattern based on the tree and square packing of Figure 12.37 is shown in Figure 12.50.

12.6. Bouncing Creases

One of the unique features of box-pleated crease patterns is an unexpected complexity that arises when we start propagating and reflecting the creases that run at 0° and 90°. Such a crease bounces from diagonal to diagonal and can describe a very circuitous path, doubling back on itself again and again before it closes.

As an example, consider the tree in Figure 12.51 and two possible packings of squares into a rectangle that give rise to it. One packing is symmetric about a horizontal line of symmetry; it is most likely the first packing that comes to mind. The packing on the right in Figure 12.51 is equally valid, however. In it, the bottom half of the packing is shifted laterally with respect to the top half. (Two of the square centers are no longer in the corners of the bounding rectangle and the rectangle must be slightly lengthened.) This packing is one of a continuous family of packings; it may be characterized by the distance d, which is the offset between the top and bottom halves of the packing. The quantity d is expressed as a fraction of the length of the six flaps, so the rectangle has proportions $2 \times (4 + d)$. For the example in Figure 12.51, $d = 0.190$.

Now, let's construct the box-pleated molecules for both packings; first, the symmetric packing on the left in Figure 12.51. Figure 12.52 shows the box-pleated molecule.

This pattern is not at all unexpected. It consists of two Waterbomb Bases joined side-to-side. Now try to construct the box-pleated crease pattern for the offset packing. Each of the four pairs of 45° creases launches new vertical creases that propagate up and down, then ricochet off the diagonals on the other side to travel from side to side, until they bounce again at another diagonal. This process continues for a considerable distance before the creases finally terminate by running off of the paper. The result, shown in Figure 12.53, is rather astonishing.

This isn't a molecule; it's a labyrinth. Each of the axis-parallel creases bounces around the rectangle at considerable

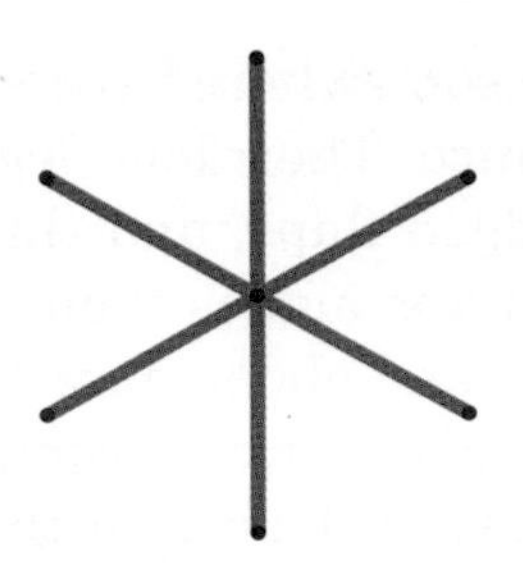

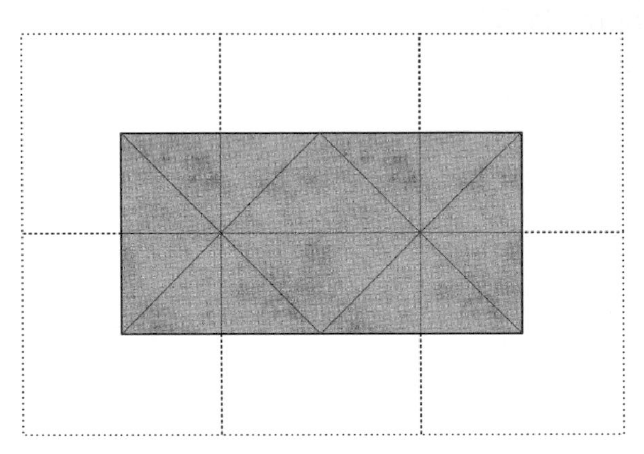

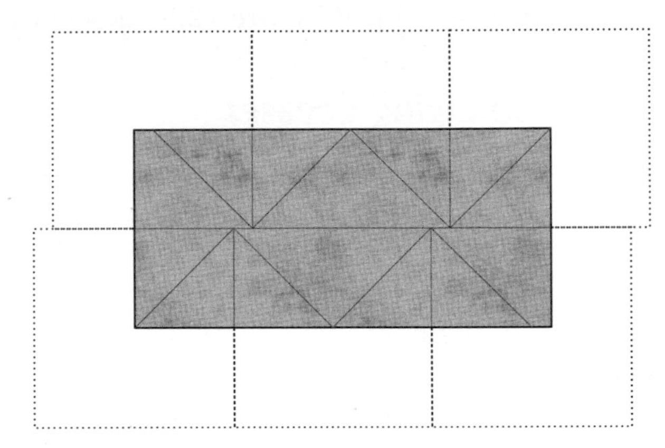

Figure 12.51.
Two packings of six squares with their centers in a rectangle.
Top: the six-star tree graph.
Left: a symmetric packing.
Right: the squares are slightly offset and fit into a slightly longer rectangle.

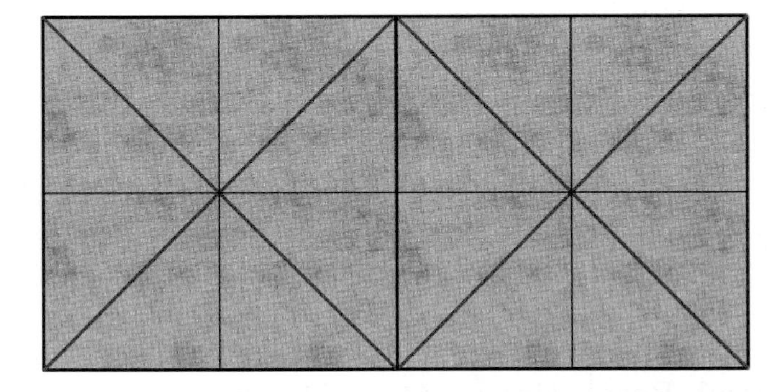

Figure 12.52.
Box-pleated molecule for the symmetric packing.

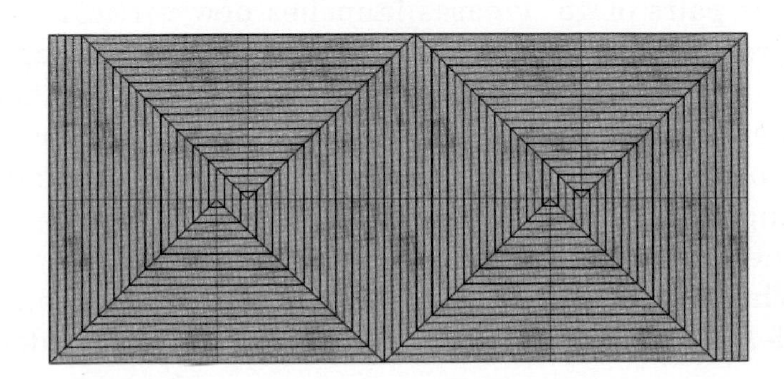

Figure 12.53.
Box-pleated molecule for the offset square packing with $d = 0.190$

length; you might try tracing any one of them to see where and how far it travels. It is also surprising that the total length of crease in the pattern does not vary smoothly with the size of the offset. In this example, changing the offset by just a bit cuts the number of creases nearly in half, as in Figure 12.54.

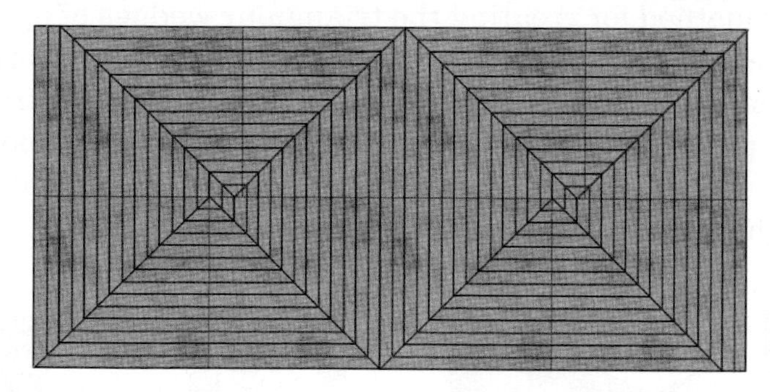

Figure 12.54.
Box-pleated molecule for offset $d = 0.143$.

The number of creases in the molecule does not vary smoothly with the size of the offset. As you vary the offset d between 0 and 1, simple and complex patterns are finely interspersed. It turns out that dense patterns arise whenever d approaches values $\{2/3, 2/5, 2/7,\ldots\}$, i.e., $d = 2/(2n + 1)$ for integer $n > 0$. Conversely, relatively simple patterns arise when d takes on the values $\{1/2, 1/3, 1/4,\ldots\}$, i.e., $d = 1/n$ for $n > 0$.

In order to keep the crease patterns as simple as possible, it is desirable to keep any offsets between adjacent squares as simple fractions of the square size. This is most easily done by making all distances—all flap lengths—small multiples of a common unit; for example, instead of making flaps 1, $\sqrt{2}$, and 2 units long as we might in a circle/river packing, we could make them 2, 3, and 4 units (or 5, 7, and 10), keeping nearly the same relative ratios. Another benefit from such a choice is that the crease pattern fits neatly onto grid paper and can easily be sketched. For this reason, it is quite natural in box-pleated designs to make all flaps integral multiples of the same length. The crease pattern falls naturally onto a grid of squares and the major creases and reference points can usually be constructed by dividing the paper into even divisions by folding. Historically, the most common divisions have been powers of 2 and/or 3, due, perhaps, to the existence of simple folding methods for their division, but in fact, any number of divisions may arise from a square-packed design.

12.7. Elias Stretch

As you have seen, box-pleated bases are typically characterized by rows of parallel creases, or pleats, composed of alternating mountain and valley folds. The pleats in the crease pattern run vertically and horizontally, and groups of vertical and horizontal creases meet each other along diagonal creases at odd-numbered multiples of 45°. Often, there is a dominant direction to the pleats; more run vertically than horizontally or vice versa. When this is the case, there is a fairly simple folding method for creating the triangular wedges of opposing creases. All pleats are first formed running in the dominant direction, say, the vertical direction. Then, by separating particular pairs of pleats and stretching them apart, it is possible to add, one by one, the pleats that run in the horizontal direction. This process is shown in Figure 12.55.

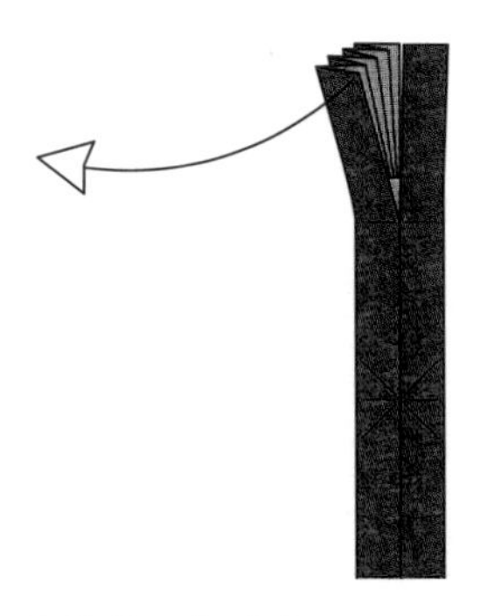

1. Pull a single layer out to the left, stretching out the pleat.

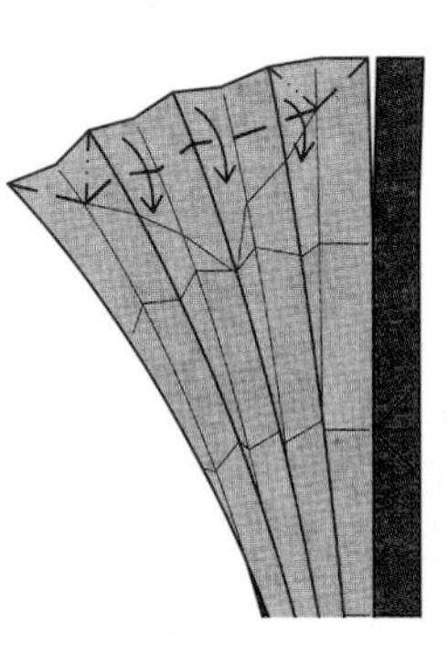

2. Fold down the top edge, making diagonal creases at 45°.

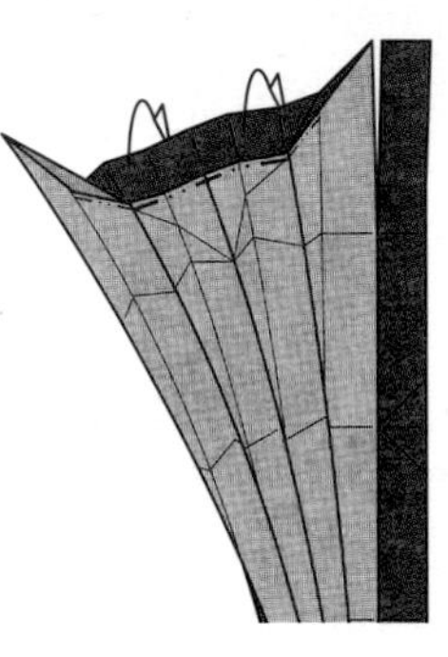

3. Mountain-fold the top edge behind.

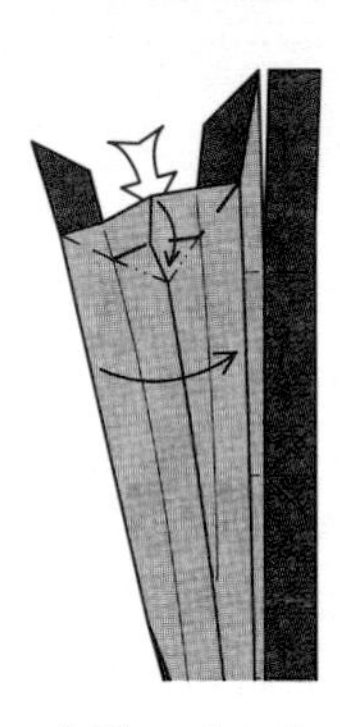

4. Pleat the edge again and close up the model.

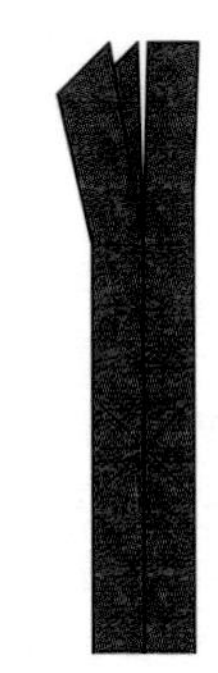

5. Finished Elias stretch. The pleat has now been turned into a gap between two flaps.

Figure 12.55.
Folding sequence for an Elias stretch.

This maneuver occurs often in the box-pleated designs of Neal Elias and Max Hulme. Elias popularized the style of box pleating in the 1960s and early 1970s; for this reason, the maneuver in Figure 12.55 has come to be known as the *Elias stretch*. You will find many examples and variations of Elias stretch in the two models whose folding sequences appear at the end of the chapter.

12.8. Ortholinear Rivers

The beetle example was relatively straightforward because its tree was a star tree with only a single branch node and no branch edges. In general, a tree will have one or more branch edges, which must be represented in the packing by an ortholinear river. As we have seen, ortholinear rivers can turn corners, go straight, or form a dogleg. With both a single corner or a straight river, the form of the river is completely specified; there are no more degrees of freedom. The dogleg river, however, can be shifted from side to side, as shown in Figure 12.56.

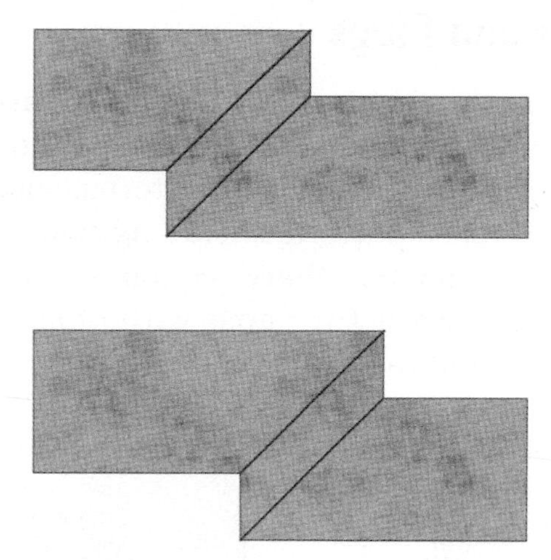

Figure 12.56.
Two equivalent dogleg rivers.

The specific dogleg configuration can often be fixed into position by stretching adjacent squares into rectangles, as shown in Figure 12.57, where we show which objects can be stretched. Note that an ortholinear river cannot be stretched arbitrarily, but rather both sides have to move together.

There is frequently more than one way to stretch rectangles to absorb all the paper. Different choices will result in

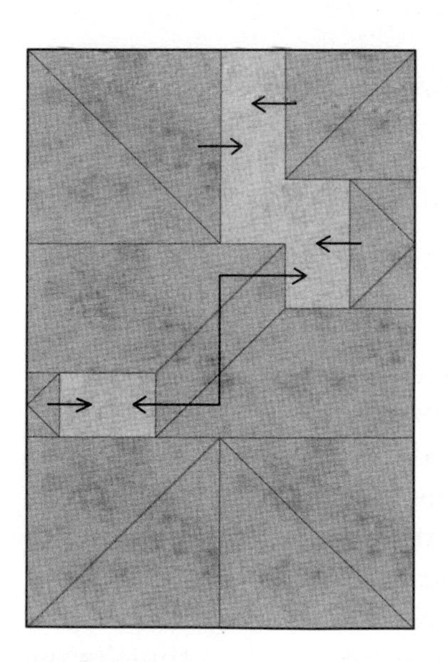

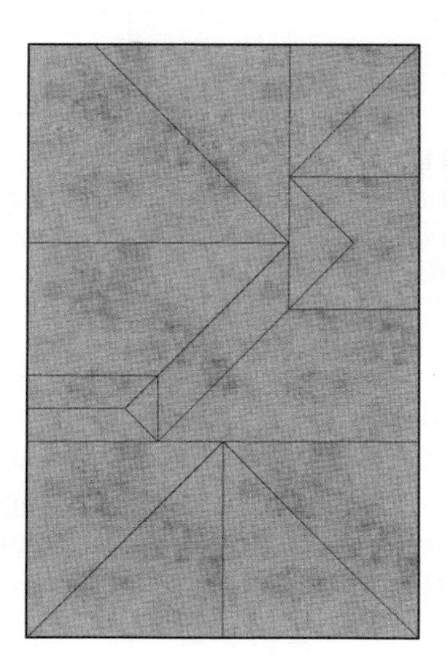

Figure 12.57.
Left: a square packing that includes an ortholinear river.
Middle: the pattern with stretched rectangles. The river is now pinned in place.
Right: the resulting crease pattern.

different width flaps, and sometimes (as in the examples in the previous section) more or less complex crease patterns.

12.9. Holes and Plugs

Much of the time, by choice of flap length (square size) and by stretching squares into rectangles, it is possible to completely cover the square with atom crease patterns; once this is done, the rest of the crease pattern, including bouncing ridges, is fully determined. However, there are times when it is not possible to completely cover the paper with atoms. A simple example is the tree and rectangle shown in Figure 12.58.

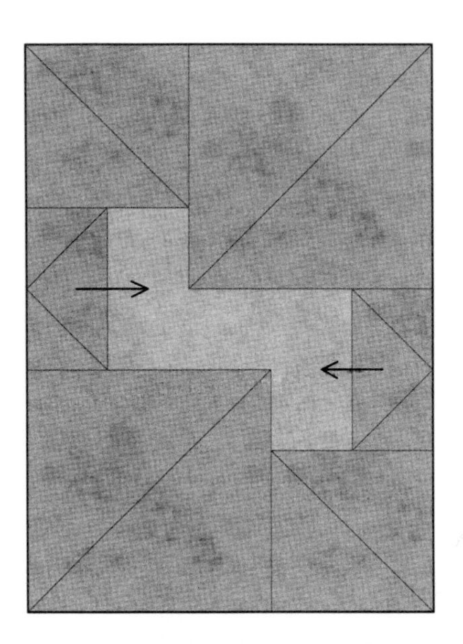

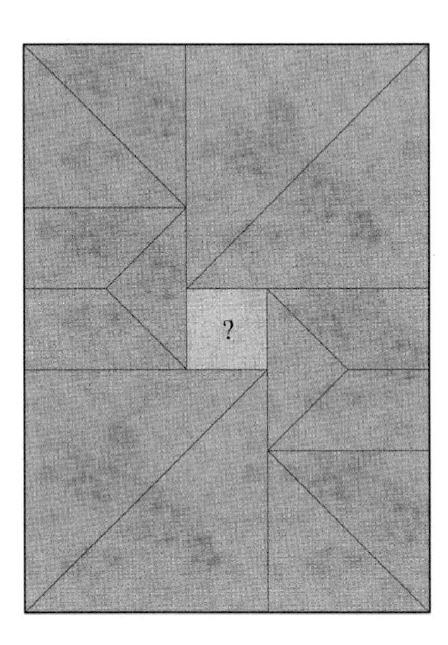

Figure 12.58.
Left: a six-flap square packing into a rectangular molecule. Only two of the squares are stretchable.
Right: there remains an unaccounted-for patch inside the molecule.

The question then arises: With what do we fill the hole? A clue is provided by examining the perimeter of the hole; it consists of the boundary of various atoms. But the boundaries between atoms turn into hinge creases in the folded base; thus, in the folded molecule, the boundary of any hole must fold so that it lies along a single hinge crease.

This is very similar to the original definition of a molecule, which is a crease pattern that folds flat so that its perimeter lies along the axis of the base. So we could use any known molecule to fill the hole, but in the folded form, it will be rotated 90° relative to the axis; its perimeter will lie along a hinge crease, perpendicular to the axis.

Fortunately, in box-pleated patterns, residual holes tend to be fairly simple shapes, typically squares and rectangles. Such holes can be plugged by combinations of the two simplest rectangular molecules, Waterbomb and sawhorse molecules, shown in Figure 12.59.

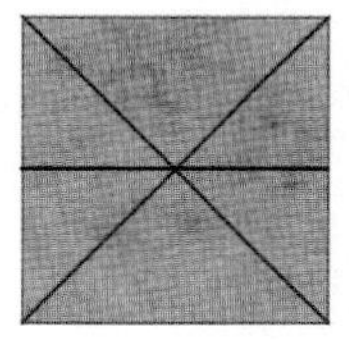
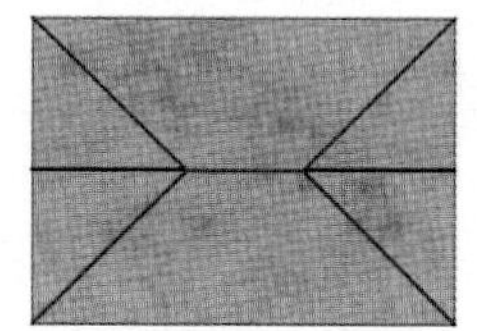

Figure 12.59.
Left: the Waterbomb molecule.
Right: the sawhorse molecule.

These shapes launch and terminate creases in the same way that atoms do; by joining, propagating, and reflecting their creases with those emanating from the other atoms, you can construct full, valid molecules and bases.

If you make your crease pattern fall on a grid (by making all flaps integral multiples of the same common distance), then any holes will be assemblies of unit squares—often, only a single unit square. In this case, only the Waterbomb molecule is needed to plug each square hole. In fact, there is an even simpler shape that launches no 0° or 90° creases; shown in Figure 12.60, it is the crease pattern of a square folded in half along both diagonals.

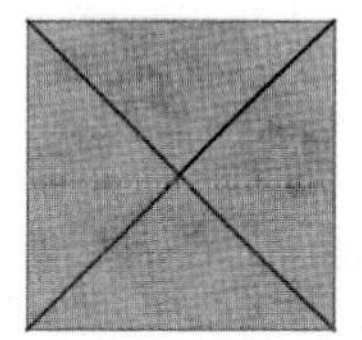
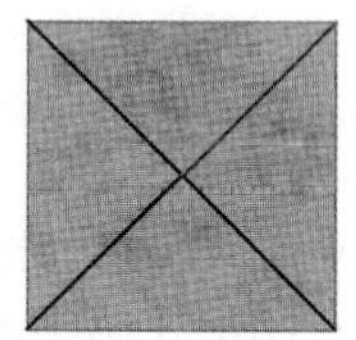

Figure 12.60.
The simplified plug for square holes in a box-pleated molecule or base.

In the simple square plug, the diagonal creases come three of one parity and one of the other. With two possible parities and four orientations of each, there are eight possible patterns. However, the only difference in the folded base lies in the direction of a tiny pleat that is usually buried within the interior of the base. Figure 12.61 shows the complete (generic-form) crease pattern for the example from this section using a square plug. If you cut the pattern out and fold it up, you can see for yourself how different configurations of the plug creases affect the configuration of the folded base.

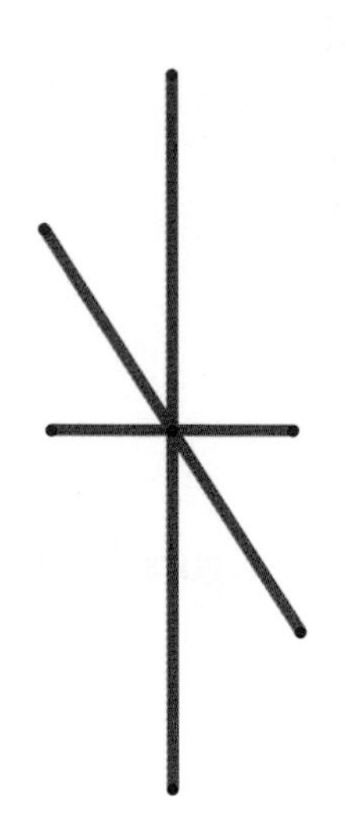
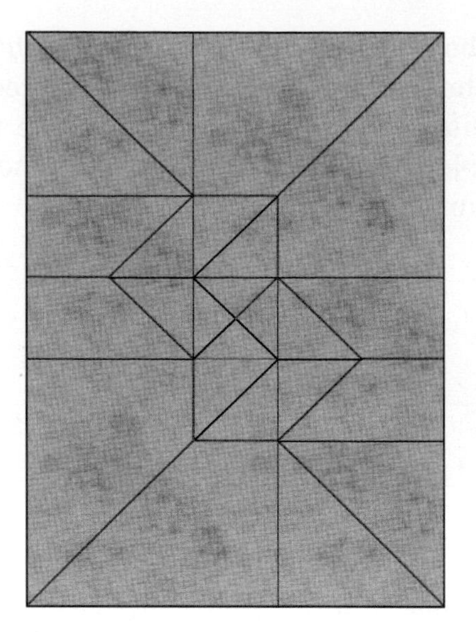
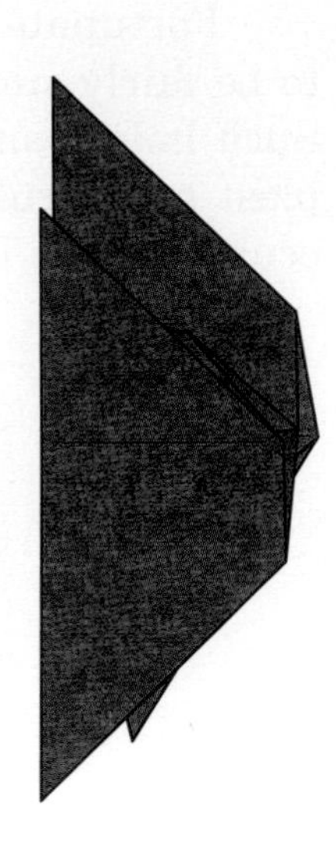

Figure 12.61.
Tree, generic-form crease pattern, and folded form for a rectangular molecule with a plug in the interior. The X-ray line shows the folded edges of the plug buried in the interior.

12.10. Box-Pleating versus Optimal Bases

Box-pleated bases can be used to realize highly complex structures with many flaps. They are not, however, usually the most efficient bases; as we have seen, a base constructed by the method of path optimization will be the most efficient uniaxial base. The difference in efficiency is often quite small, however, and box-pleated bases are much more easily folded by a linear folding sequence, as I will show in this section.

Another example of a box-pleated design is the Bull Moose, shown in Figure 12.62.

It is not obvious from the folded model, but the Bull Moose is, in fact, made from a uniaxial base. This can be more readily seen from the base, shown at the middle of Figure 12.62. All of the flaps of the base clearly lie along the central axis, and closer inspection reveals that the hinges are also perpendicular to the axis, the two hallmarks of a uniaxial base.

If the base is uniaxial, there is an underlying tree for the model. Further examination of the base and the crease pattern reveals the tree shown in Figure 12.63.

We could use this same tree to carry out a design using path optimization. If we use the circle/river method of design (or rather, use the numerical tree method; the circle/river packing problem is probably too complex to solve manually), then apply the universal molecule, we arrive at the complicated crease pattern in Figure 12.64.

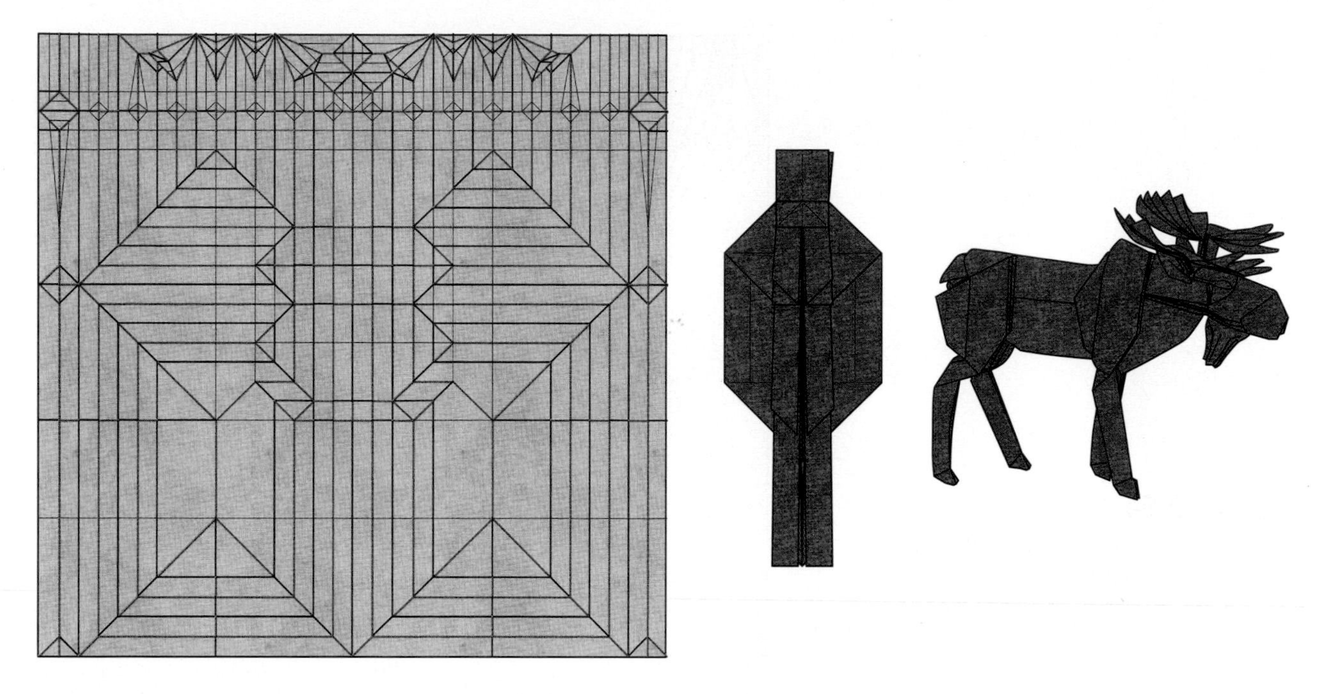

Figure 12.62.
Crease pattern, base, and folded model of the Bull Moose.

Figure 12.63.
Tree and uniaxial base for the Bull Moose.

Beauty is in the eye of the beholder, of course. From the perspective of efficiency, this is actually a pretty nice base; it is definitely more efficient than the box-pleated base. If we define the length of a leg as 1 unit, then the scale—the relationship between 1 tree unit and the side of the square—is, in the box-pleated pattern, 7/32 (since a leg is seven small squares long and each small square is 1/32 of the side of the square). That fraction comes to 0.2188. For the tree method crease pat-

Figure 12.64.
A moose base crease pattern derived by the tree method, filled in with universal molecules.

tern, the scale turns out to be 0.2423, or about 10% larger. So the tree method, as expected, gives a more efficient solution.

But at what cost? The tree method pattern is quite irregular and has several aesthetic deficiencies; note the long, skinny gussets extending from the forelegs to the small antler points. Moreover, as is typical with tree method patterns, most of the vertices of the crease pattern have no simple reference points associated with them. There are a total of 17 coordinate values that are required just to specify the vertices of the axial creases (not including those points that can be obtained by reflection of another in the symmetry line). Given that accurately locating a single reference point may take three to five creases, one is looking at several tens of folds just to identify these reference points, let alone the creases that lie in the interior of each axial polygon.

But, of course, it is rare that one works with a raw tree method pattern. As noted in the previous chapter, one will typically make small adjustments to the dimensions of the tree in order to massage the crease pattern into something that is simpler and more symmetric. If we do this with the moose tree and its circle/river packing, we can arrive at the pattern in Figure 12.65, which is considerably simpler.

However, in addition to distorting the branches of the tree (by amounts ranging up to about 6% of their original values), we have also reduced the scale, to 0.2240—now only slightly more efficient than the scale of the box-pleated design. (You can see where some of the lost scale turned into lost paper:

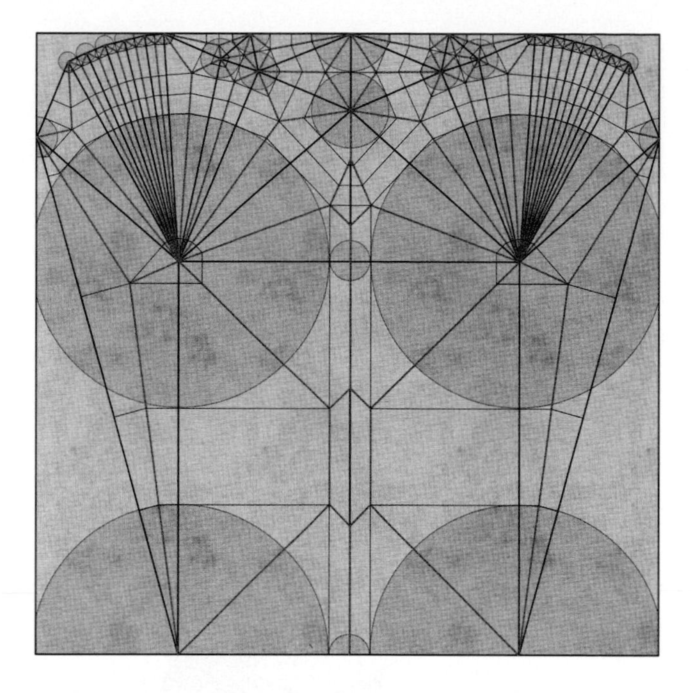

Figure 12.65.
Revised crease pattern for a moose base.

The triangles along the bottom left and right are no longer used.) Thus, we have traded efficiency for foldability. In the end, the box-pleated design is nearly as efficient as the cleaned-up tree method design, but it is much easier to fold, as you can see for yourself. You will find instructions for the box-pleated Bull Moose at the end of this chapter. Finding a folding method for the tree method design—I shall leave that as an exercise for the reader!

12.11. Comments

Box-pleating offers an alternative design approach for generating bases with specified structure in which both the design and the folding method can be simpler than those generated by the tree method; in fact, the design can often be worked out in its entirety with no more than a pencil and paper.

The payoff of using box pleating is twofold. First, the resulting crease pattern can, due to its regularity, often be constructed by a linear folding sequence with well-defined reference points. The crease pattern is simplified and the foldability is further enhanced if all flaps have lengths that are integer multiples of a common small quantity; in this case, the crease pattern lies within a regular square grid, as in the Cerambycid Beetle and Bull Moose patterns (24 × 24 and 32 × 32, respectively). In such models, one can start the folding sequence by creasing the paper into equal divisions one way and/or the other, at which point many of the creases of the model will exist.

Figure 12.66.
Black Forest Cuckoo Clock, a box-pleated design from a 1 × 10 rectangle.

If you crease the paper into, say, a 32 × 32 grid, you will have created many of the creases in the model. But you will have also created many creases that aren't part of the base, and in fact, every flap or surface will be covered with the grid of creases. These extraneous creases can be distracting to the eye in the folded model. Although it is harder to devise such a folding sequence, it's preferable to minimize the number of unnecessary creases when precreasing the model. Take a look at the folding sequence for the Bull Moose for an example of this.

The second payoff for using box pleating is that box-pleated structures for obtaining flaps are compatible with box-pleated structures for constructing boxes. Thus, one can make complex three-dimensional structures containing both two-dimensional flaps and three-dimensional solids. Some of the most fantastic and downright unbelievable origami structures are designed using box pleating: hundreds of designs by Neal Elias, including human figures and compositions of several figures (a bull, bullfighter, and cape from a single sheet); various vehicles by Max Hulme (a Stephenson Rocket train engine, a double-decker bus); and of course, the model that started it all, Mooser's Train. Box pleating also combines nicely with the use of pleats to define textures, a combination that has been exploited brilliantly by modern masters such as Eric Joisel and Satoshi Kamiya.

In recent years, the ethic of one square for complex models has grown strong, but during the 1960s, 70s, and 80s, the use of rectangles was still reasonably common. Mooser's Train was folded from a rectangle, of course, as were many of the designs of Neal Elias. In keeping with this tradition, in the early 1980s, I developed a Cuckoo Clock from a rectangle, which I subsequently enhanced with many of the techniques I've described in this section. I will close this section with this model and its instructions. It illustrates all of the techniques of box pleating: the creation of three-dimensional boxes, numerous flaps, and their combination and connection. Its folding sequence—at 216 steps, and never before published—is not for the faint of heart! But if you succeed in folding it, you may find within in it inspiration for your own box-pleated designs.

Folding Instructions

Bull Moose

Black Forest Cuckoo Clock

Bull Moose

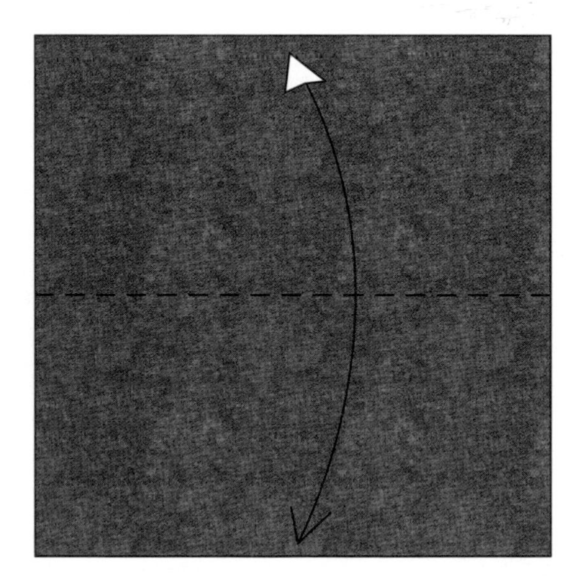

1. Begin with a square, colored side up. Fold in half vertically and unfold.

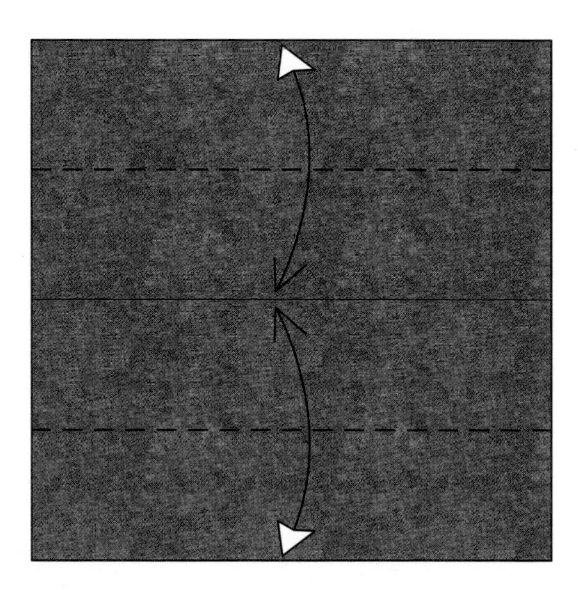

2. Fold the edges to the center line; crease and unfold.

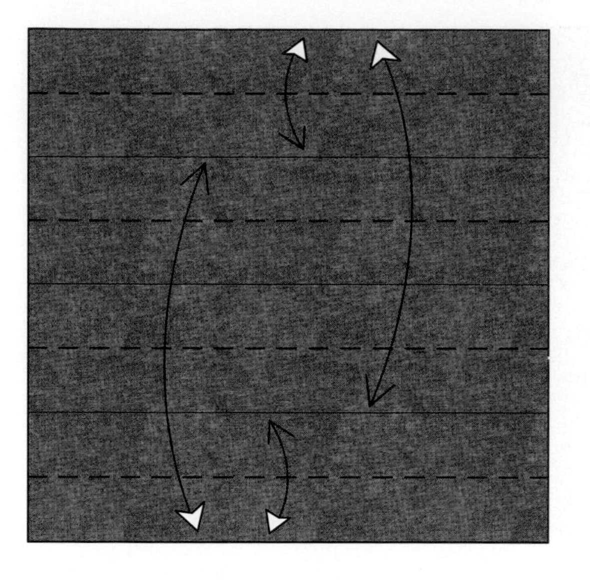

3. Fold four times, dividing the paper into eighths.

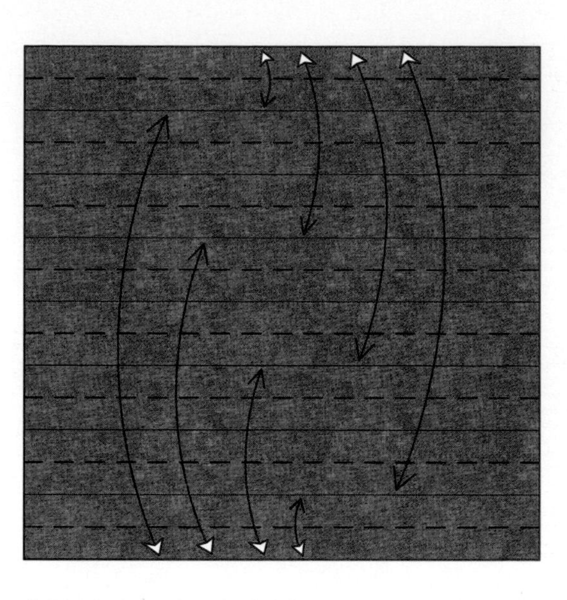

4. Fold eight times, dividing the paper into sixteenths. Turn the paper over.

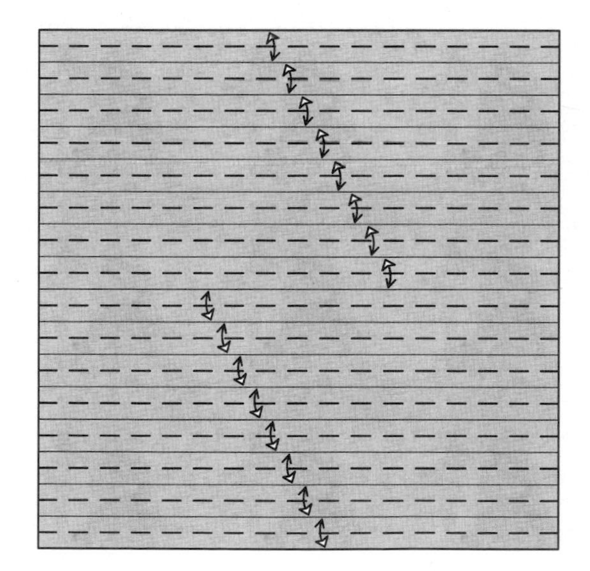

5. Fold 16 times, dividing the paper into 32nds. Make each fold by bringing the mountain fold just below it to the one just above it (or vice-versa); this will insure later that the pleats all line up.

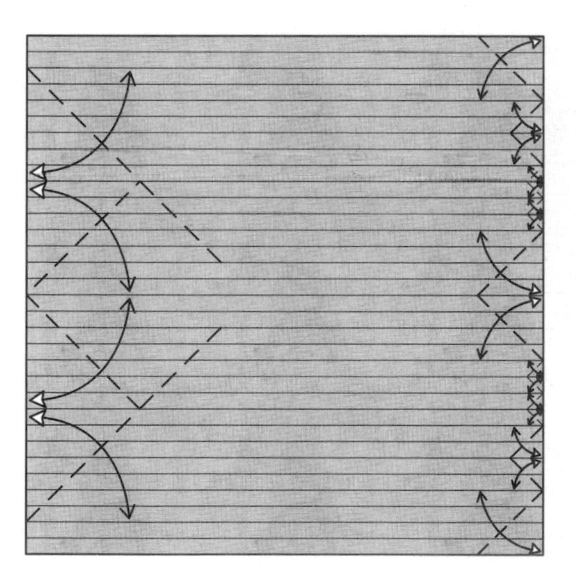

6. Make a bunch of diagonal creases.

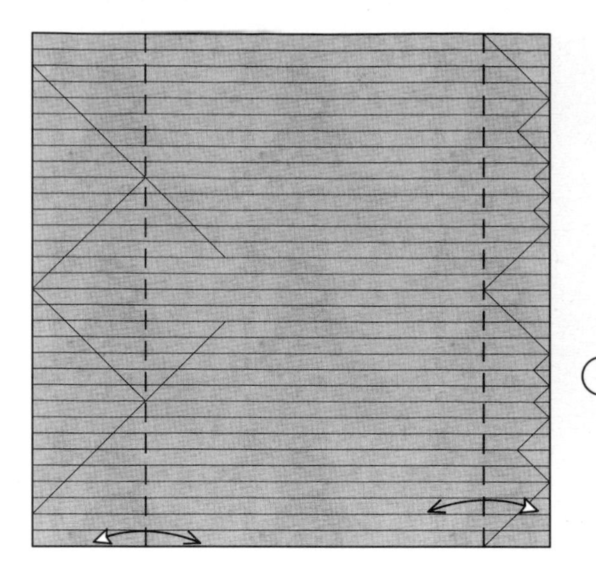

7. Make two vertical folds through existing crease intersections. Turn the paper over from top to bottom.

8. Make another vertical fold through existing crease intersections. Turn the paper back over.

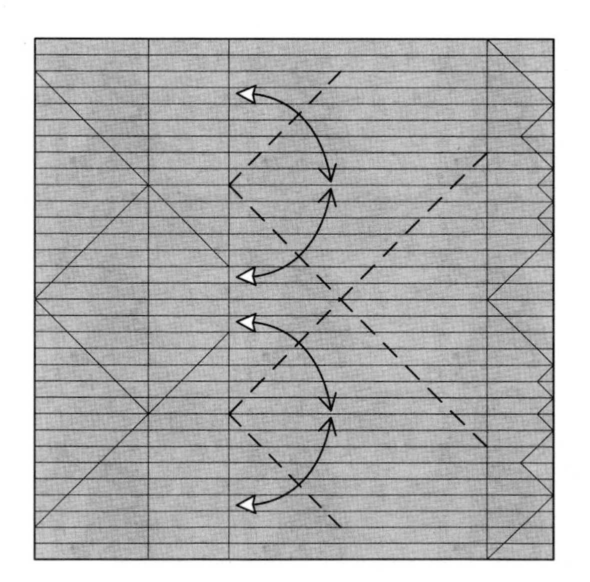

9. Make four diagonal folds.

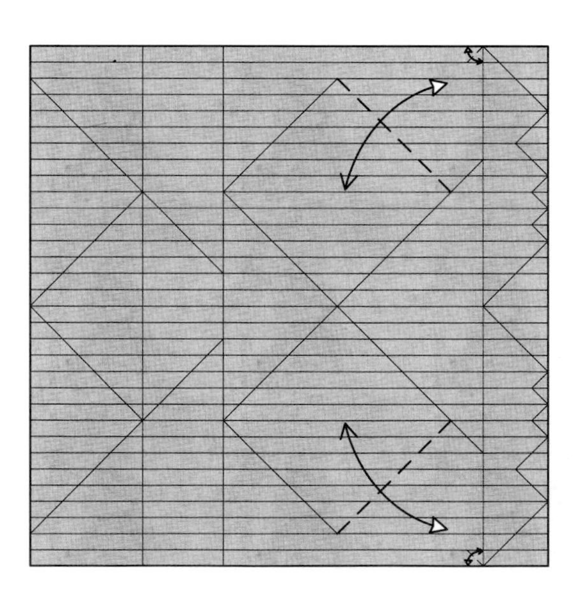

10. Make four more diagonal folds.

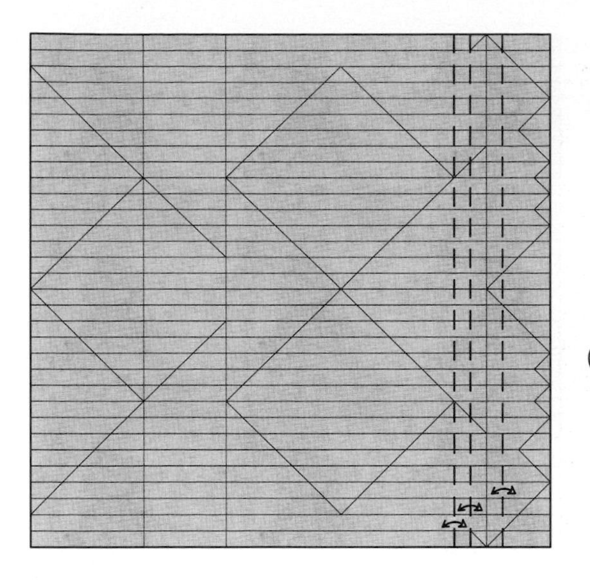

11. Make three more vertical creases, again through existing crease intersections. Turn the paper over from top to bottom.

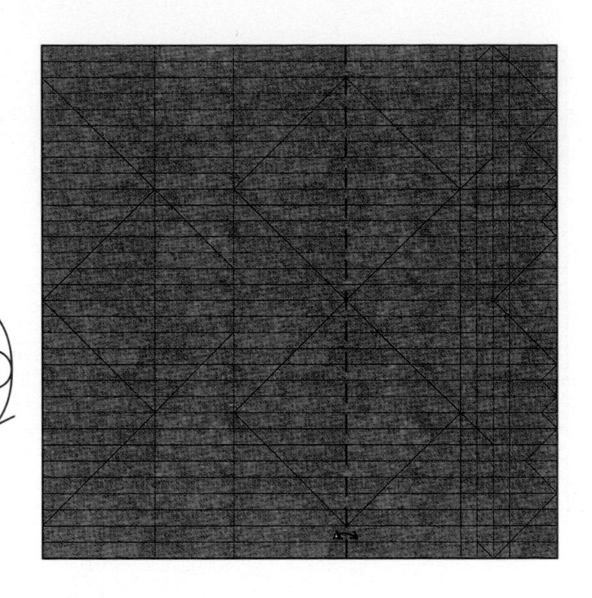

12. Make another vertical crease.

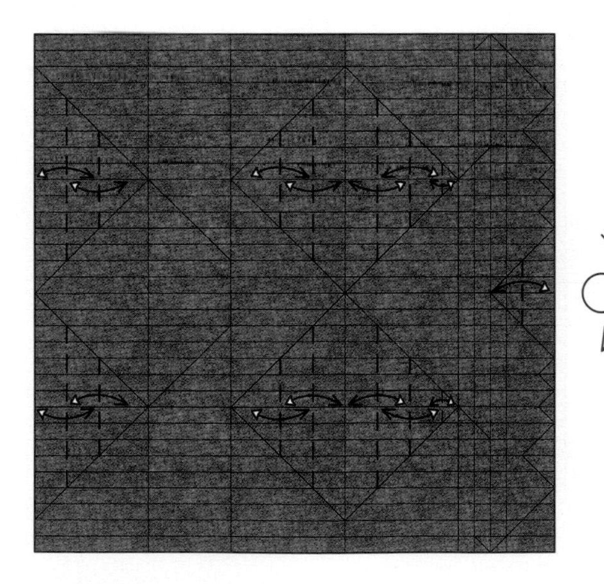

13. Make 15 vertical creases, connecting the crease intersections shown. Turn the paper back over from top to bottom.

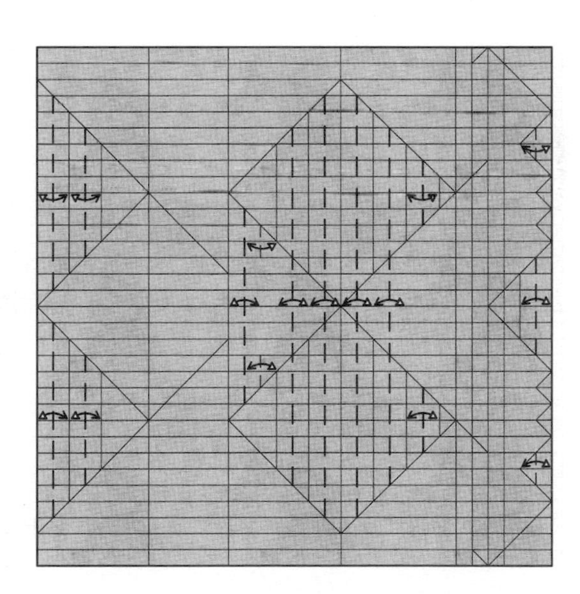

14. Make 16 more vertical creases, connecting the crease intersections shown.

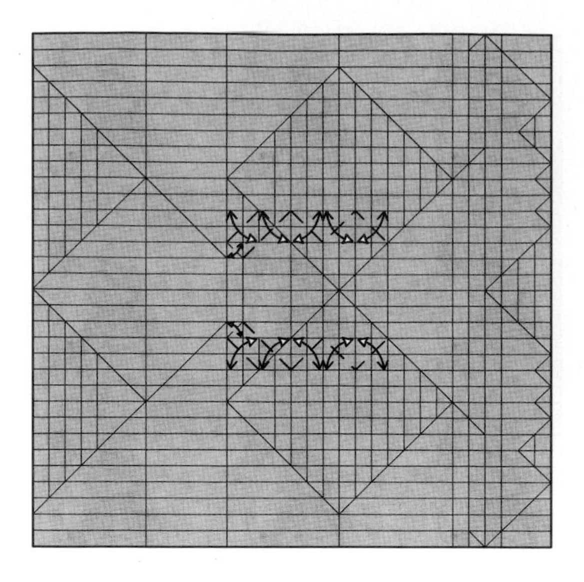

15. Add 12 diagonal creases.

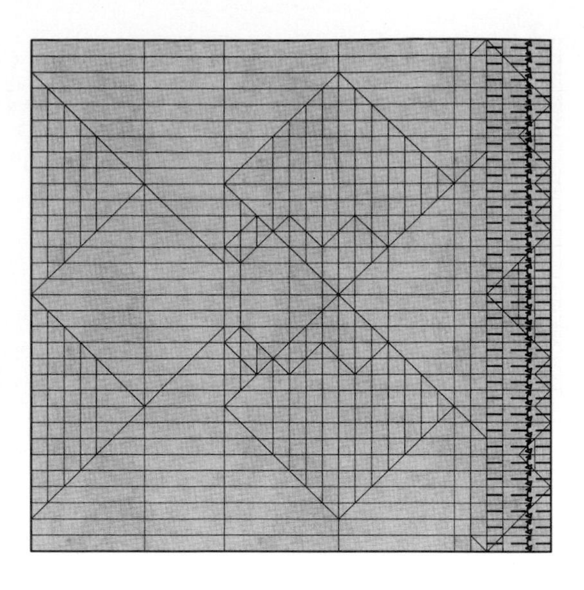

16. Fold and unfold in 32 places along the right edge.

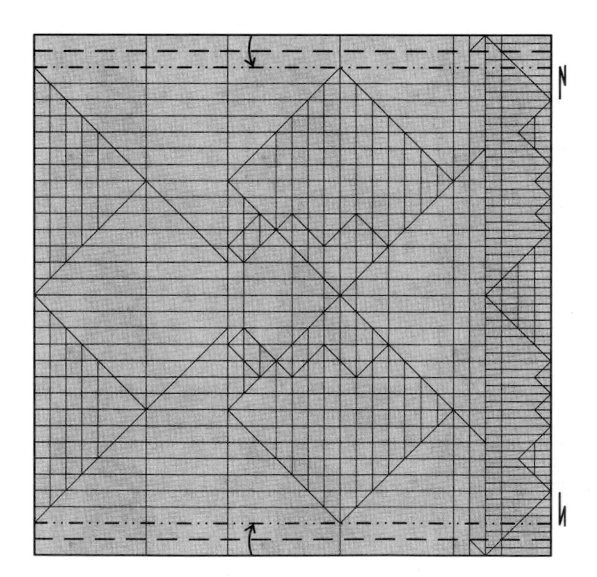

17. Precreasing is finished. Pleat the top and bottom edges on existing creases.

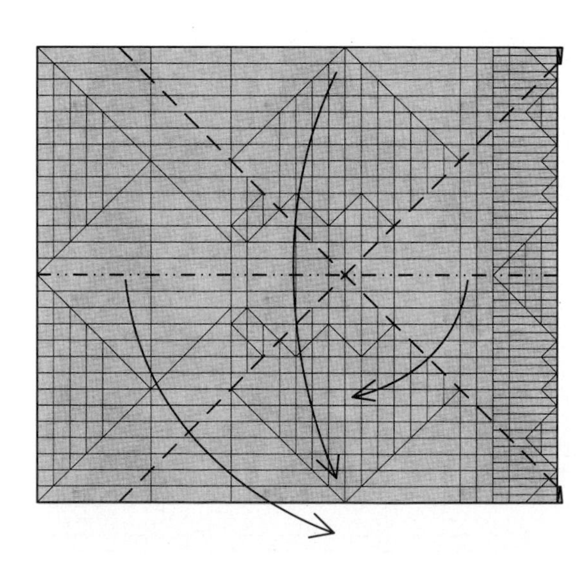

18. Form a Waterbomb Base-like shape using the existing creases. You don't need to press it fully flat yet.

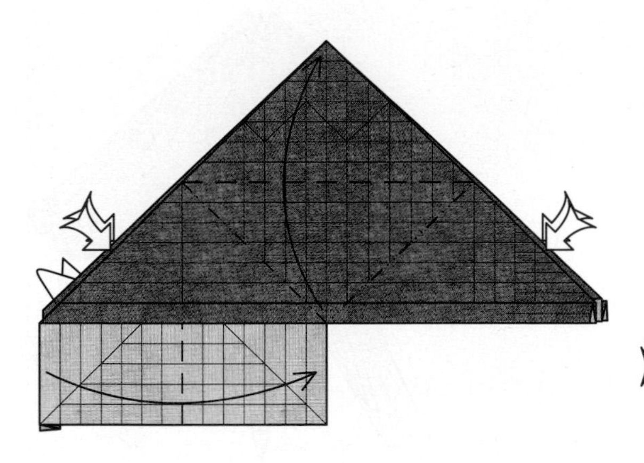

19. Petal-fold the edge in front and behind on the existing creases.

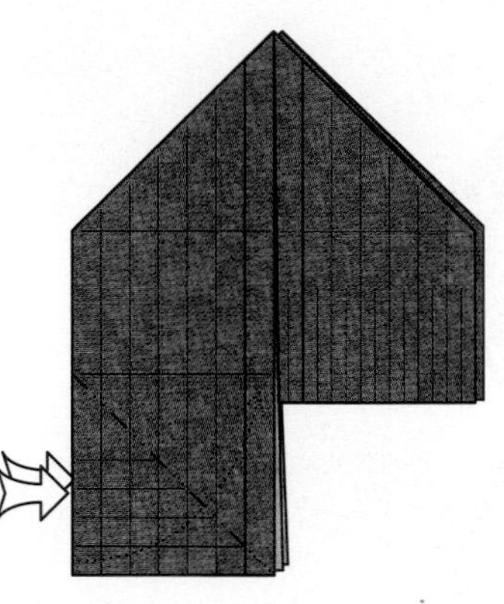

20. Reverse-fold two edges on existing creases.

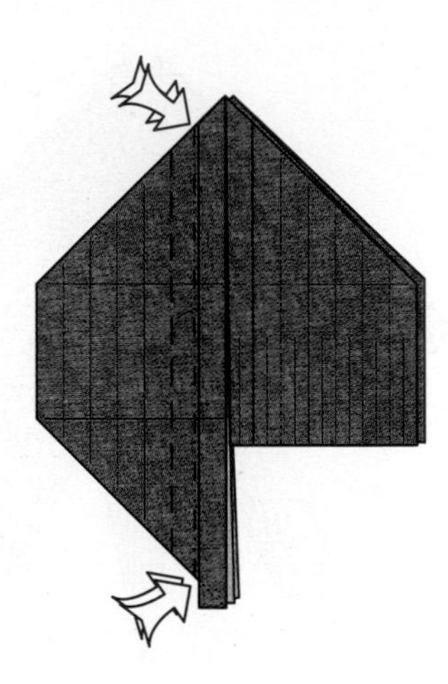

21. Open-sink in and out. Repeat behind.

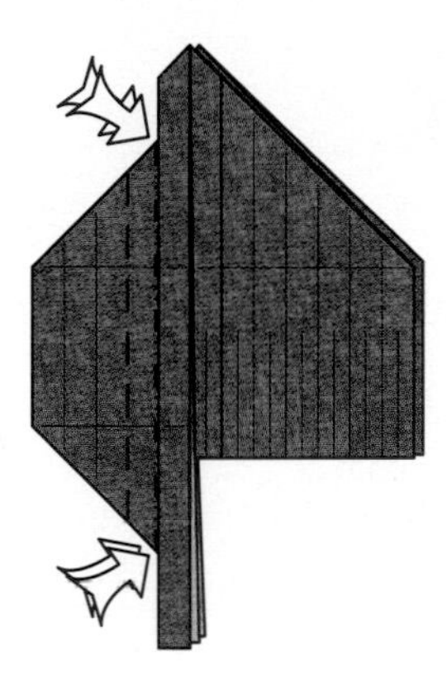

22. Open-sink in and out again. Repeat behind.

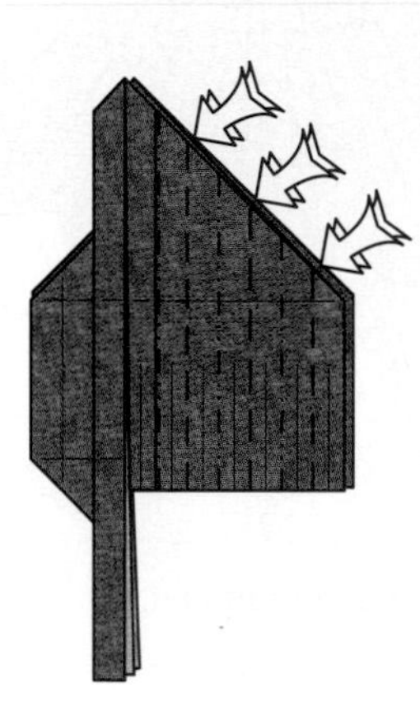

23. Open-sink the edge in and out (and in and out, and in and out!) on the existing creases. Repeat behind.

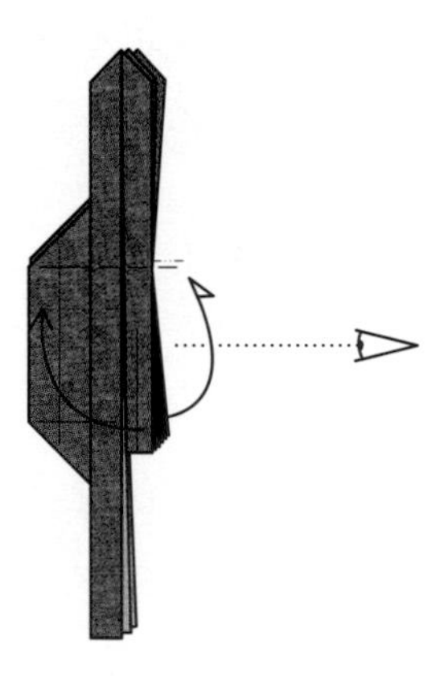

24. Fan the edges of the pleated layers out to the sides. The next step will be a side view.

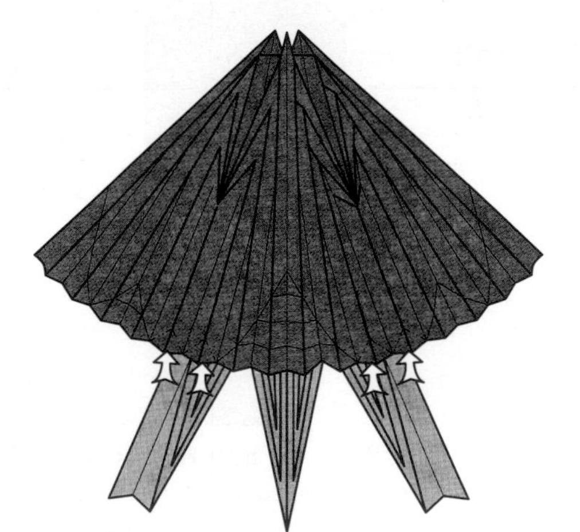

25. Reverse-fold four corners; the diagonal creases already exist.

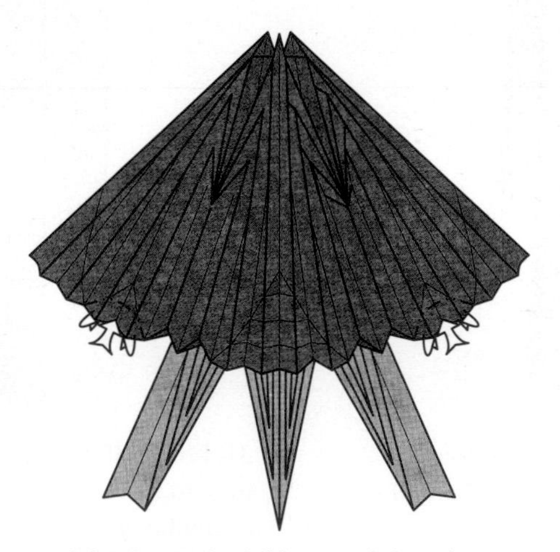

26. Mountain-fold part of the edge underneath using the existing creases.

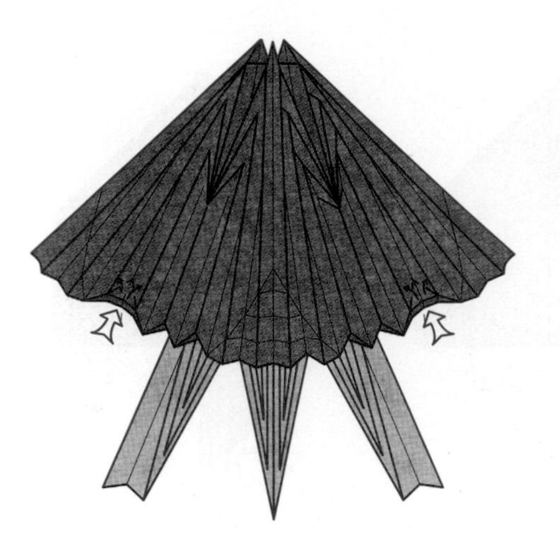

27. Reverse-fold the edge inward, again using the existing creases.

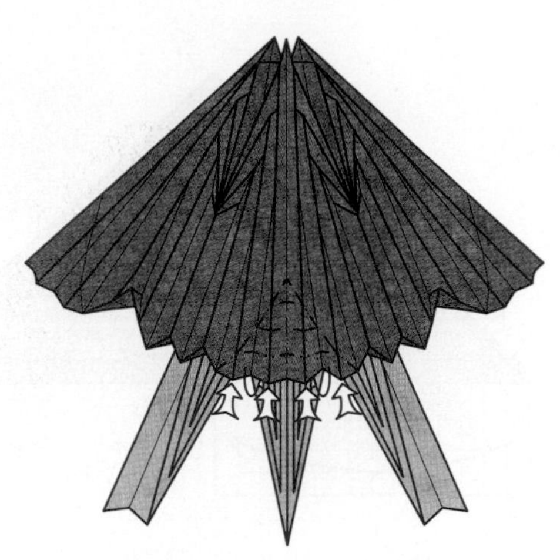

28. Pleat the edge underneath as you did in steps 26–27, but making the dented region deeper (again, follow the existing creases).

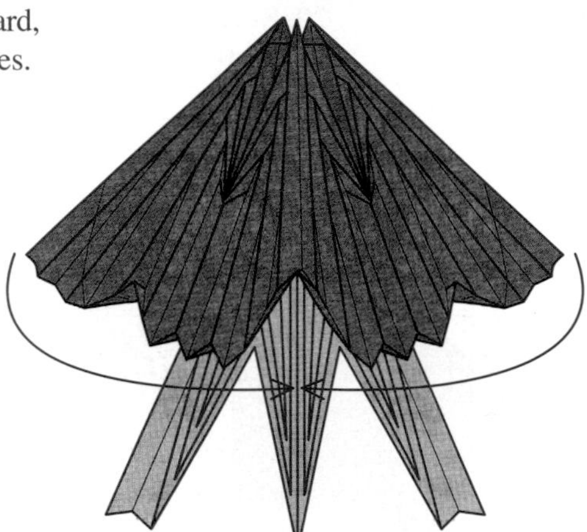

29. Close up the model.

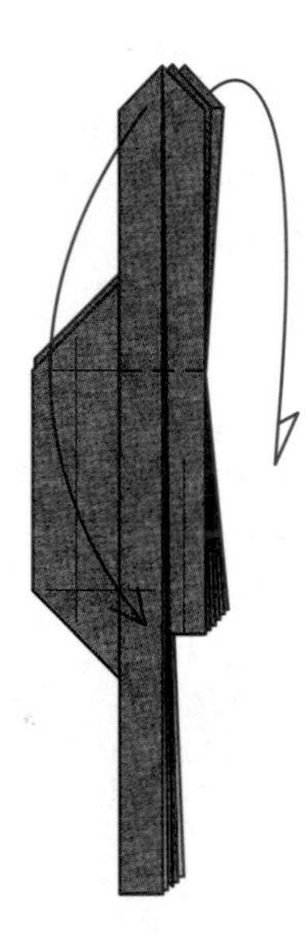

30. Fold one flap down in front and one down behind.

31. Fold half the layers on the right toward the left in front, and half of the layers on the left behind, spreading the layers symmetrically.

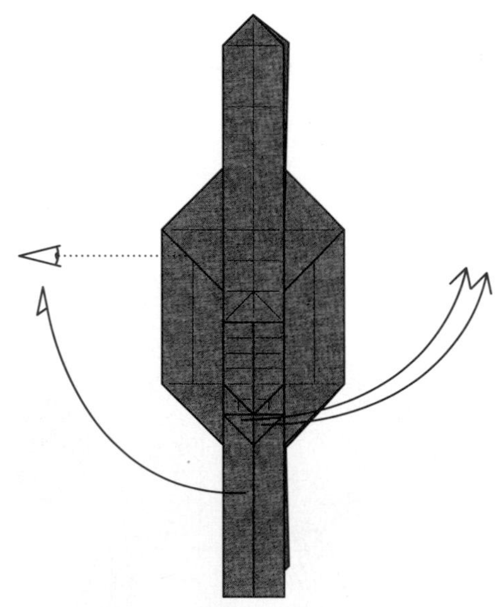

32. Bring the near flaps to the right and the far flaps to the left, stretching out the model. The next view will be a side view.

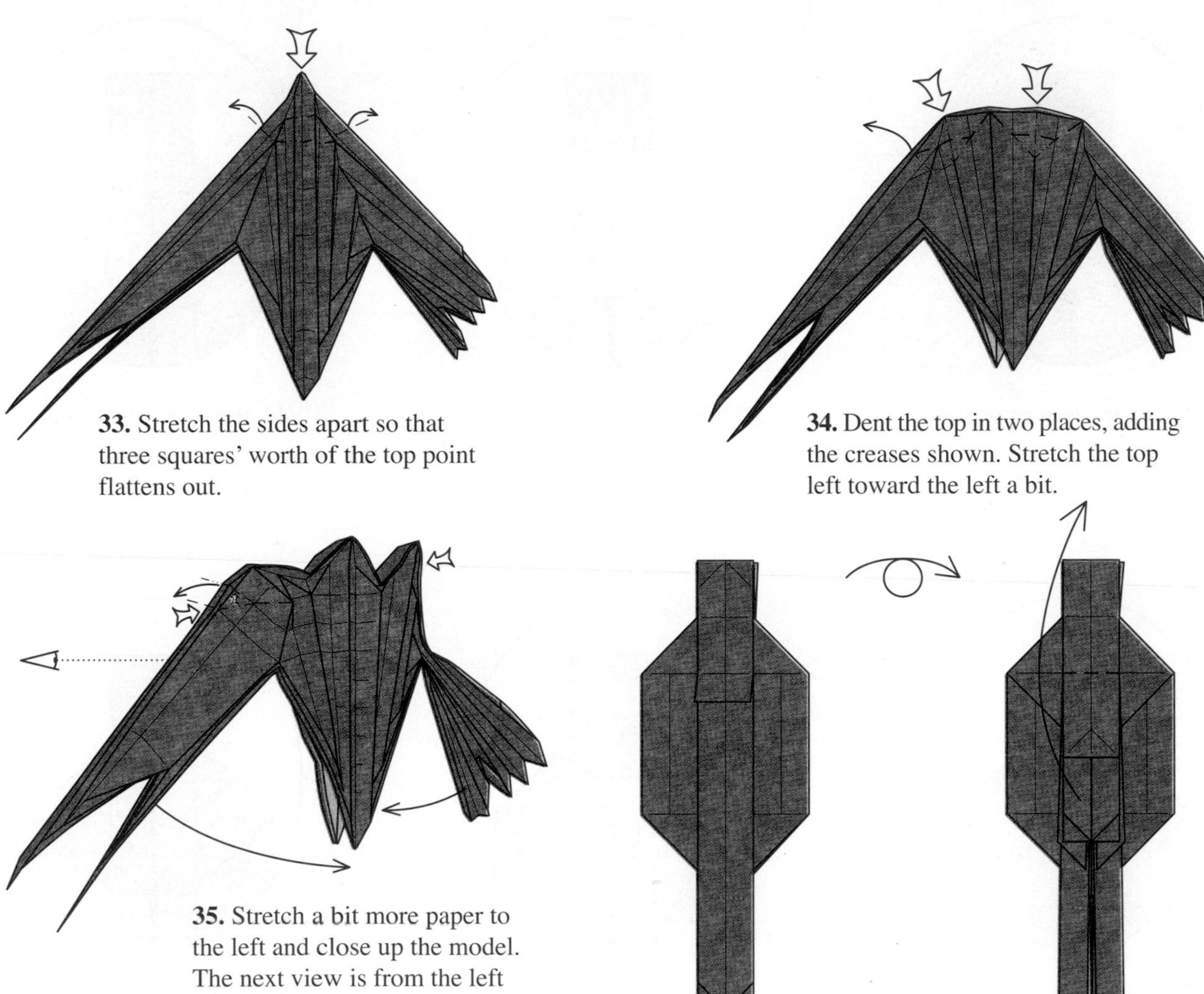

33. Stretch the sides apart so that three squares' worth of the top point flattens out.

34. Dent the top in two places, adding the creases shown. Stretch the top left toward the left a bit.

35. Stretch a bit more paper to the left and close up the model. The next view is from the left side.

36. Turn the model over.

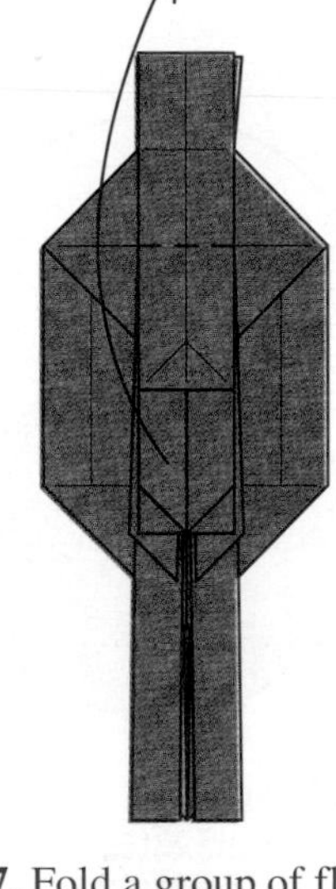

37. Fold a group of flaps upward as far as possible.

38. Pull out a single layer of paper, folding the pair of flaps up at right angles to the rest of the model.

39. Fold a rabbit ear from the vertical layers and fold the upright flaps down.

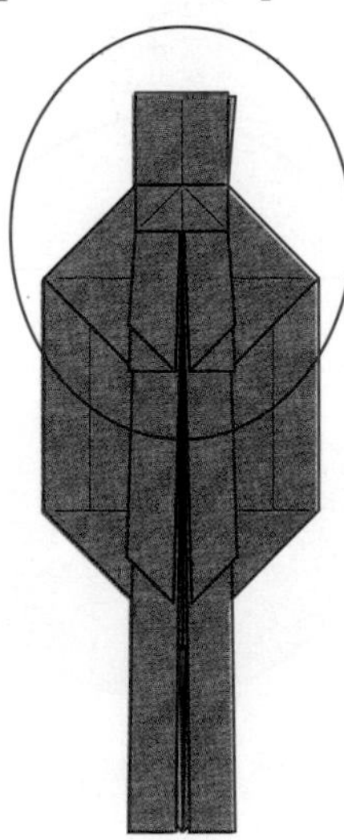

40. Steps 41–56 will focus on the top of the model.

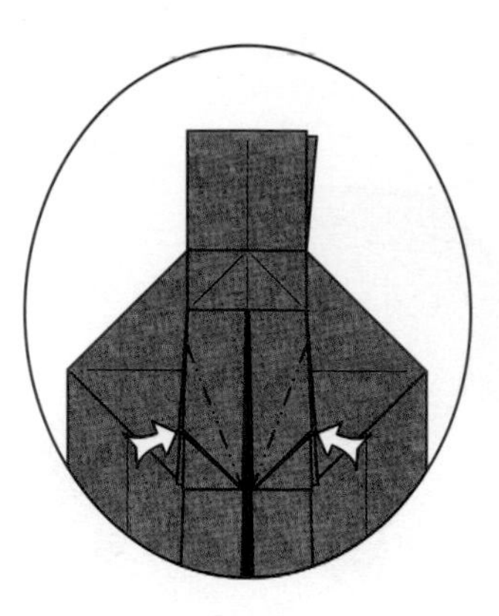

41. Sink the near corner on each side.

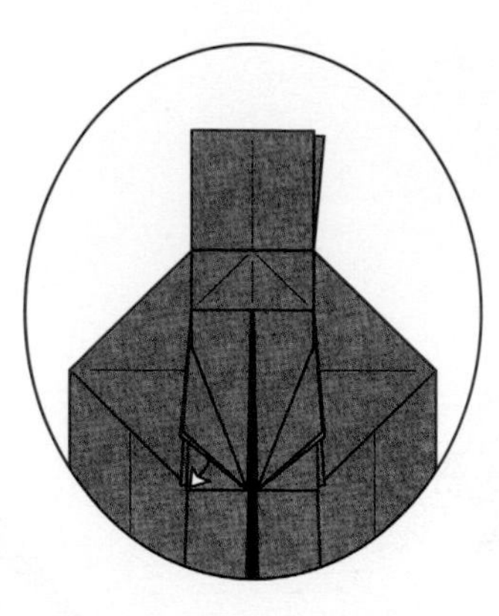

42. Pull out some loose paper from the left near flap.

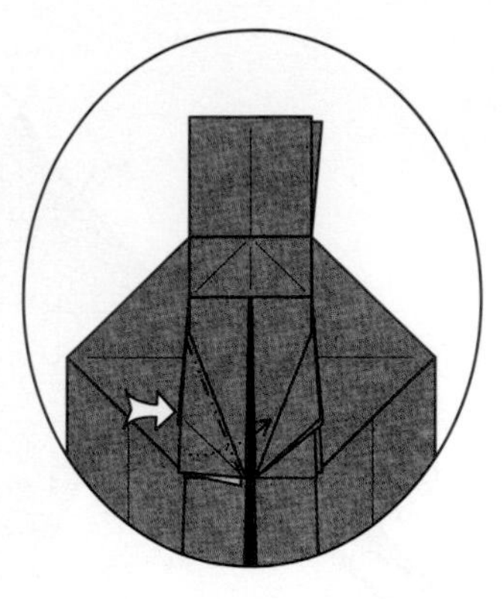

43. Reverse-fold the edge.

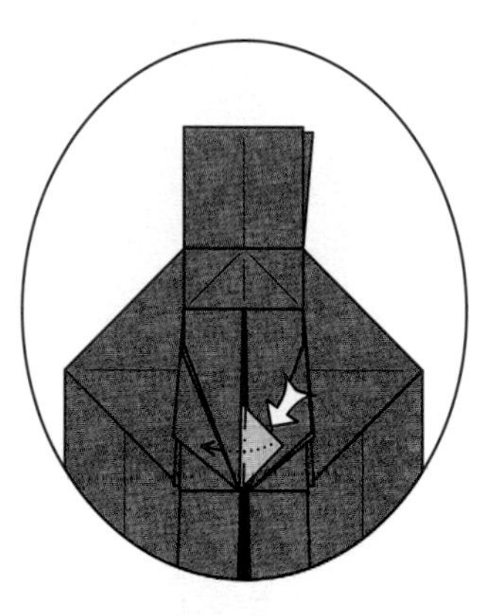

44. Reverse-fold the edge back to the left.

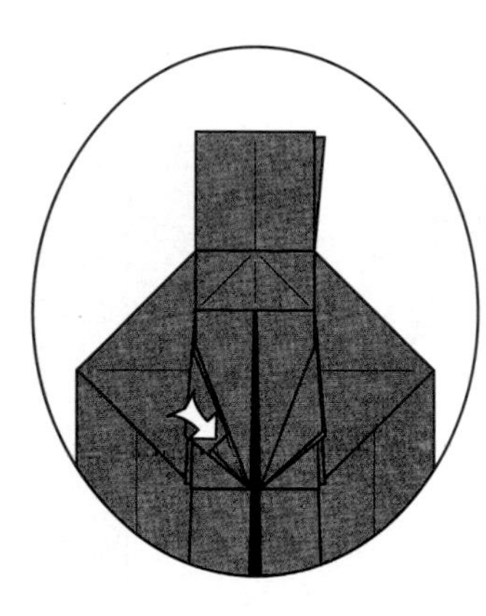

45. Reverse-fold the edge back to the right so that the edges are aligned with the other layers.

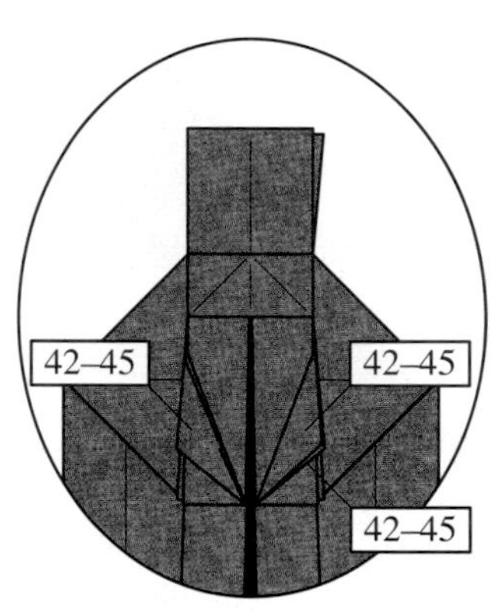

46. Repeat steps 42–45 on the next flap on the left and on both flaps on the right.

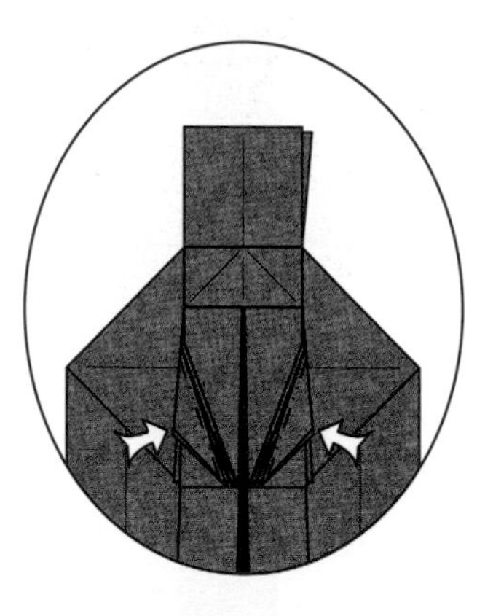

47. Sink the next pair of corners.

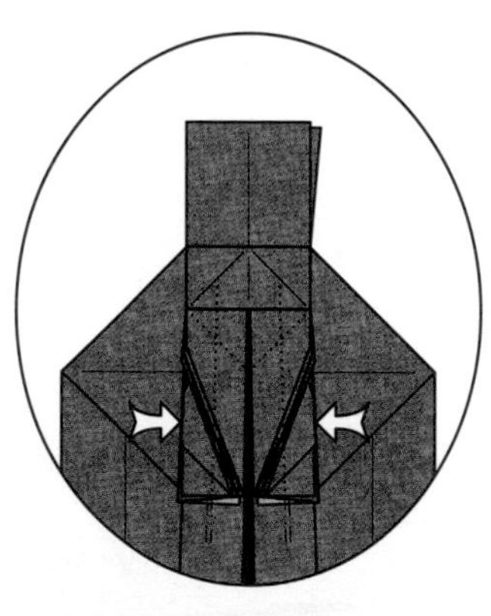

48. Open-sink the far edge on each side. Over the next 8 steps, you will start with the far edges and work your way to the near ones.

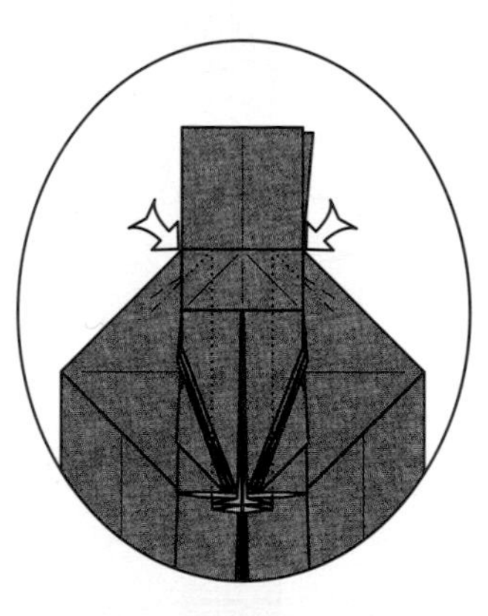

49. Reverse-fold the top hidden corner of the next nearest layer on each side.

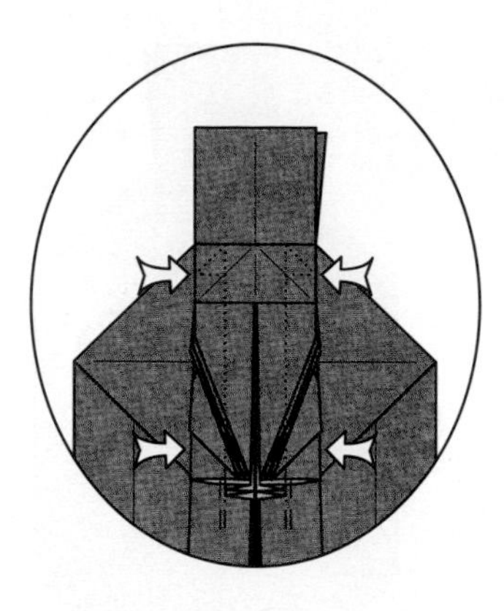

50. Reverse-fold the long edge on each side (which terminates with a closed sink at the reverse folds you just made).

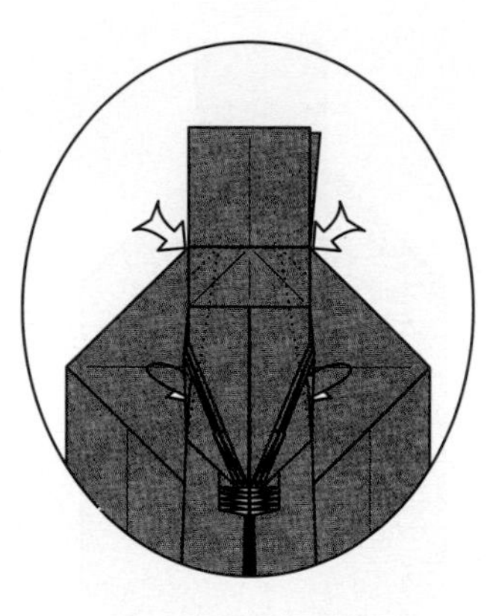

51. Mountain-fold the next edge, reverse-folding it at the top to align with the previous two reverse folds.

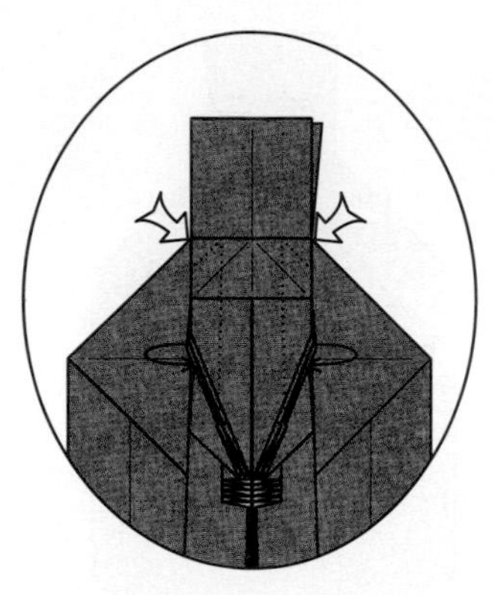

52. Valley-fold the next pair of edges, again reverse-folding at their top.

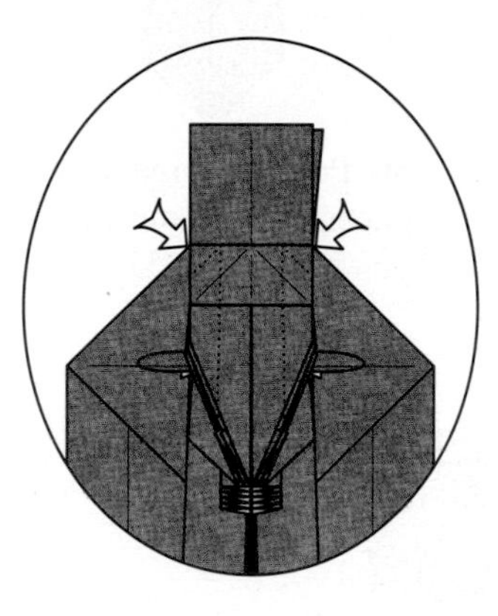

53. Mountain-fold the next pair of edges, again reverse-folding at their top.

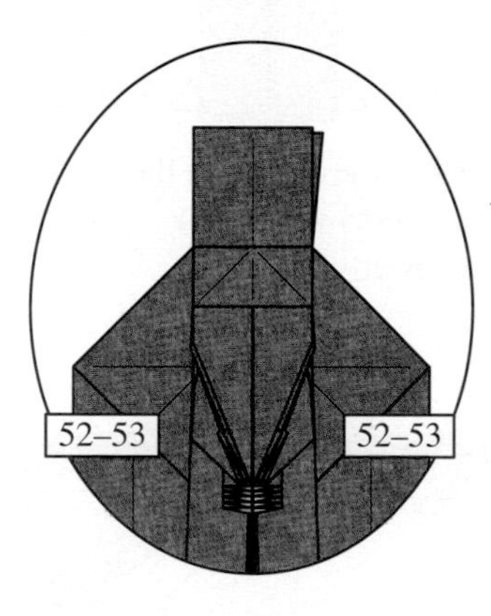

54. Repeat steps 52–53 on the next pair of layers.

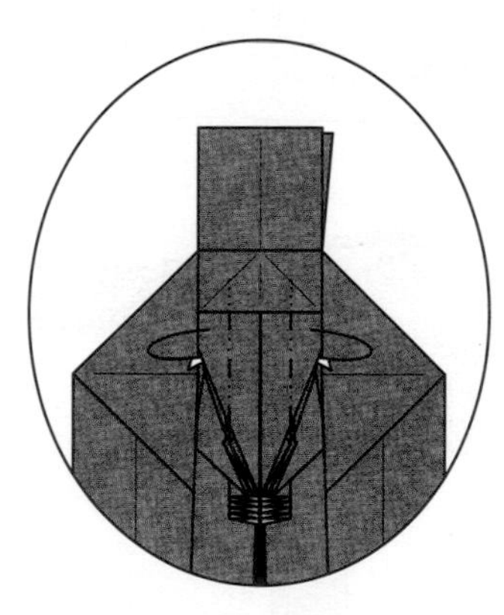

55. Mountain-fold the near edges underneath.

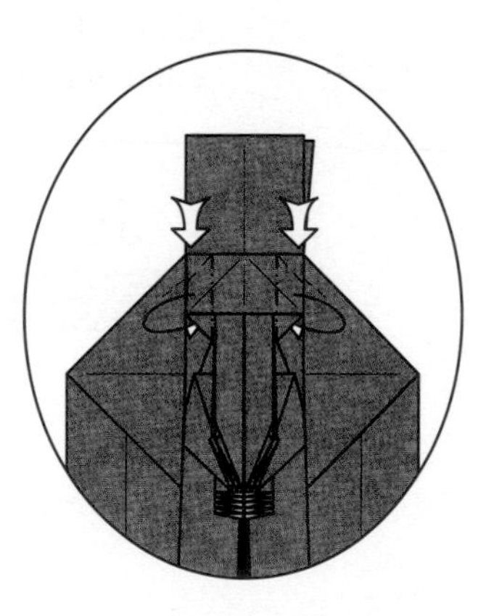

56. Mountain-fold the near edges underneath; reverse-fold the (final) pair of corners at the top.

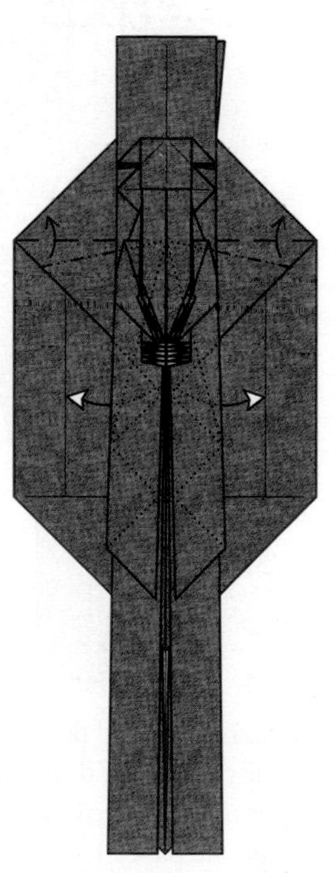

57. Pull out some excess paper from the long near flaps and crimp the excess paper upward at the top.

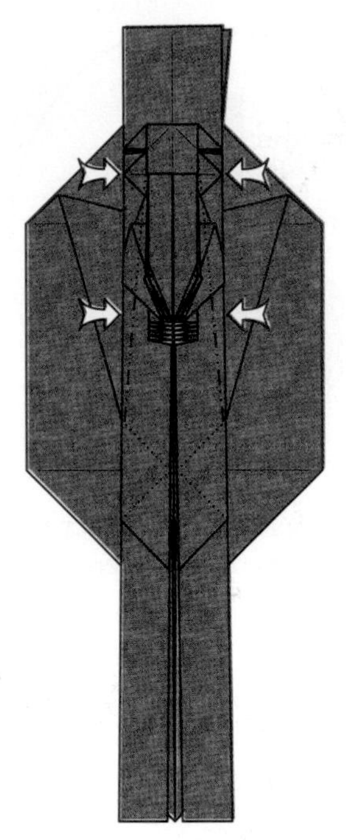

58. Open-sink the long edges up to the corners.

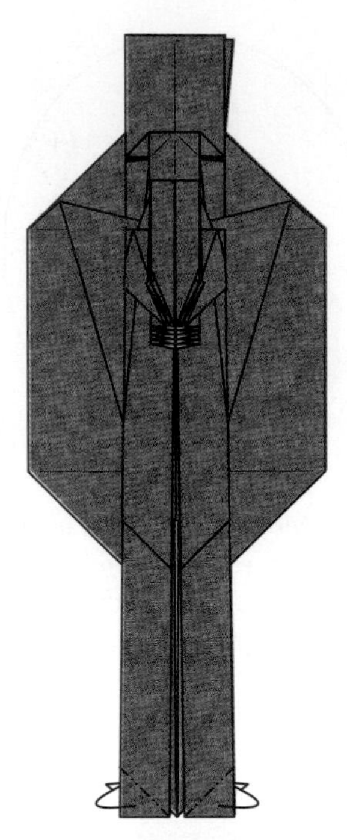

59. Mountain-fold the corners inside their respective flaps.

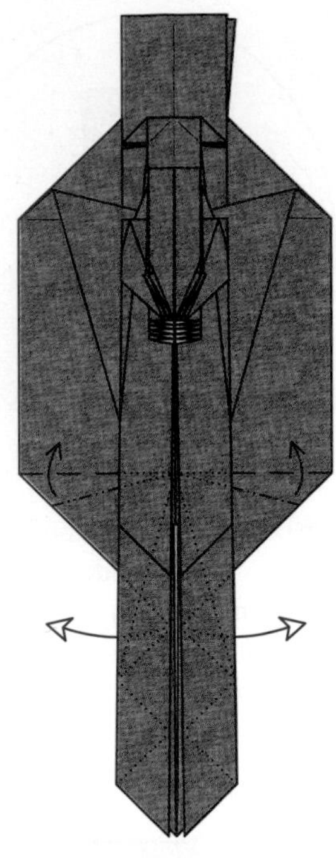

60. Pull out some loose paper in the same way you did in step 57.

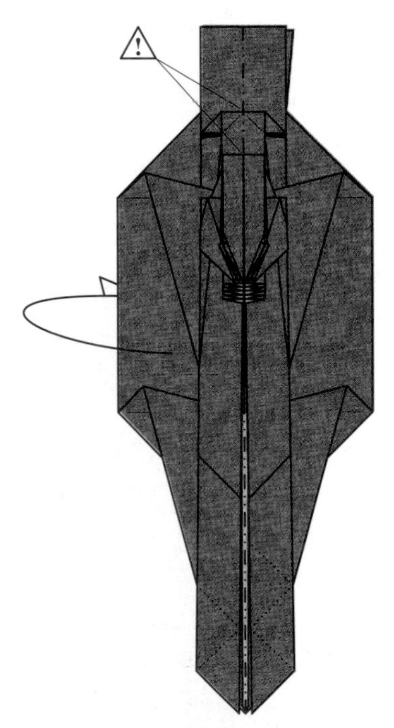

61. Carefully fold the model in half, making sure you don't split the paper at the two marked points. Rotate the paper 1/4 turn clockwise.

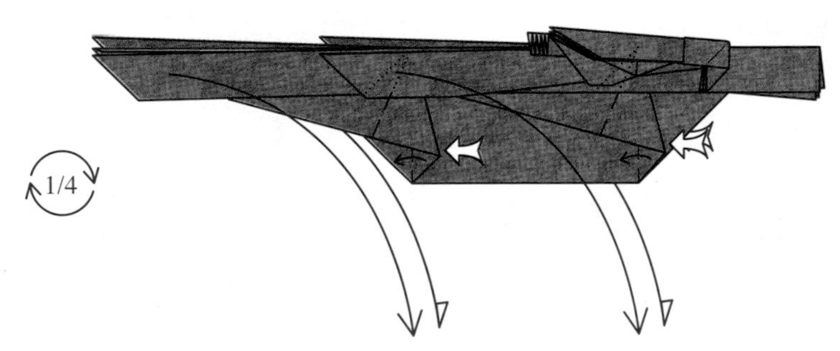

62. Fold all four legs down, removing the crimp at their base.

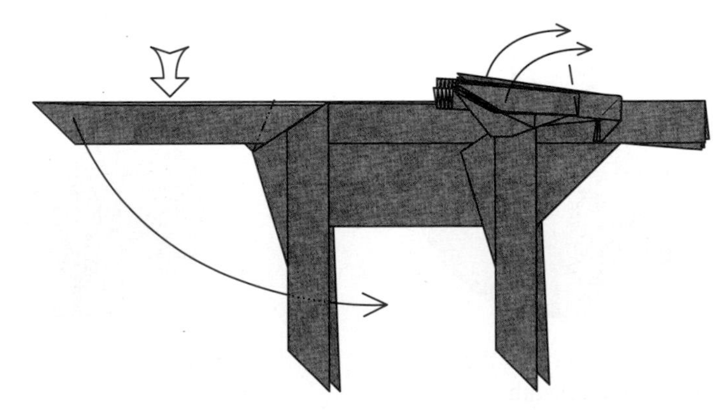

63. Reverse-fold the tail downward. Lift up the antlers so that they stand up and out from the head.

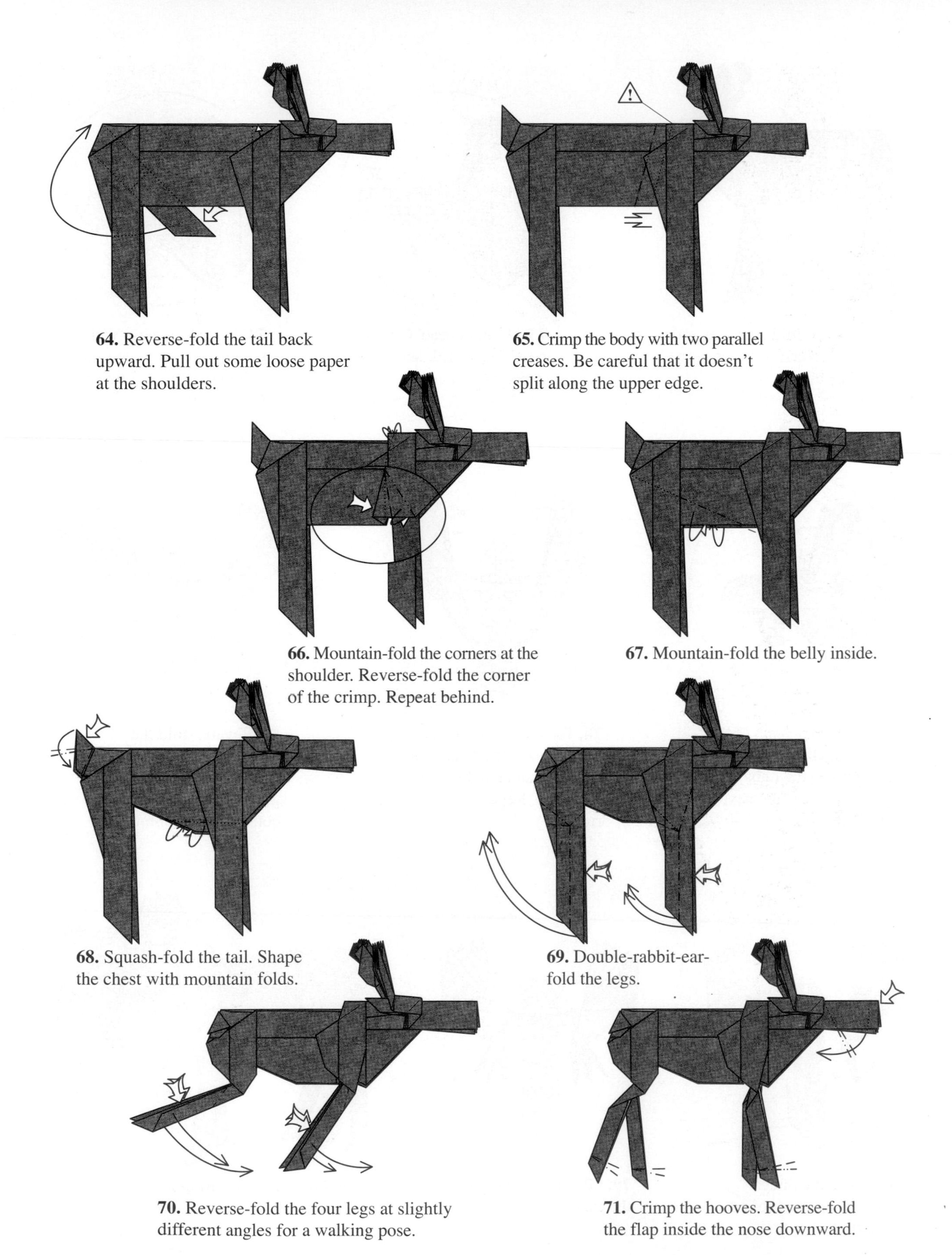

64. Reverse-fold the tail back upward. Pull out some loose paper at the shoulders.

65. Crimp the body with two parallel creases. Be careful that it doesn't split along the upper edge.

66. Mountain-fold the corners at the shoulder. Reverse-fold the corner of the crimp. Repeat behind.

67. Mountain-fold the belly inside.

68. Squash-fold the tail. Shape the chest with mountain folds.

69. Double-rabbit-ear-fold the legs.

70. Reverse-fold the four legs at slightly different angles for a walking pose.

71. Crimp the hooves. Reverse-fold the flap inside the nose downward.

72. Mountain-fold the edge of the shoulder. Reverse-fold the corner under the chin.

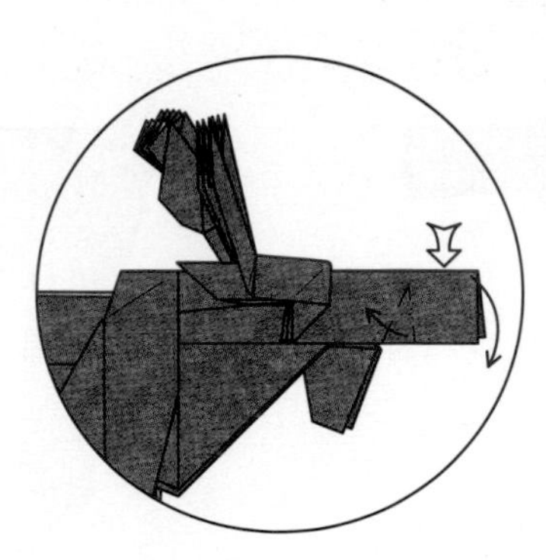

73. Detail of head. Crimp the nose downward, leaving its upper edge rounded.

74. Round the nose with reverse and mountain folds. Crimp an edge to form an eye. Crumple the dangling flaps slightly.

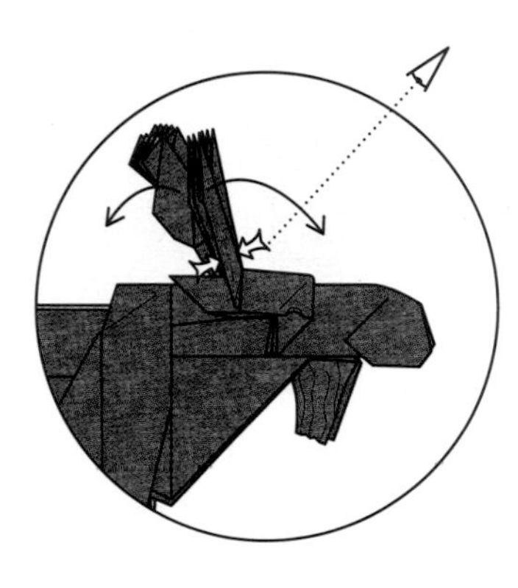

75. Narrow the stalk of the antlers; fold three points toward the right and the remaining pleats to the left. The next view is from the top.

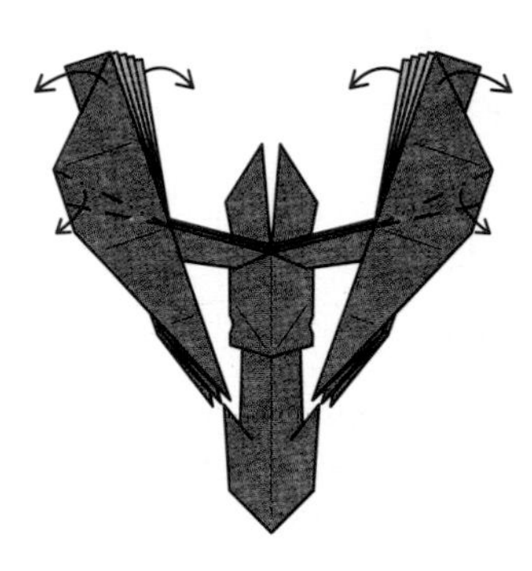

76. Top view of head and antlers. Pleat the middle of the antler and fan the pleats at the top. Repeat on the right.

77. Reverse-fold the corners along the top edge of each antler. Pinch the group of three points at the bottom into a rabbit ear.

78. Spread the three points. Shape and round the antlers.

79. Like this. Open out the ears.

80. Finished Bull Moose.

Black Forest Cuckoo Clock

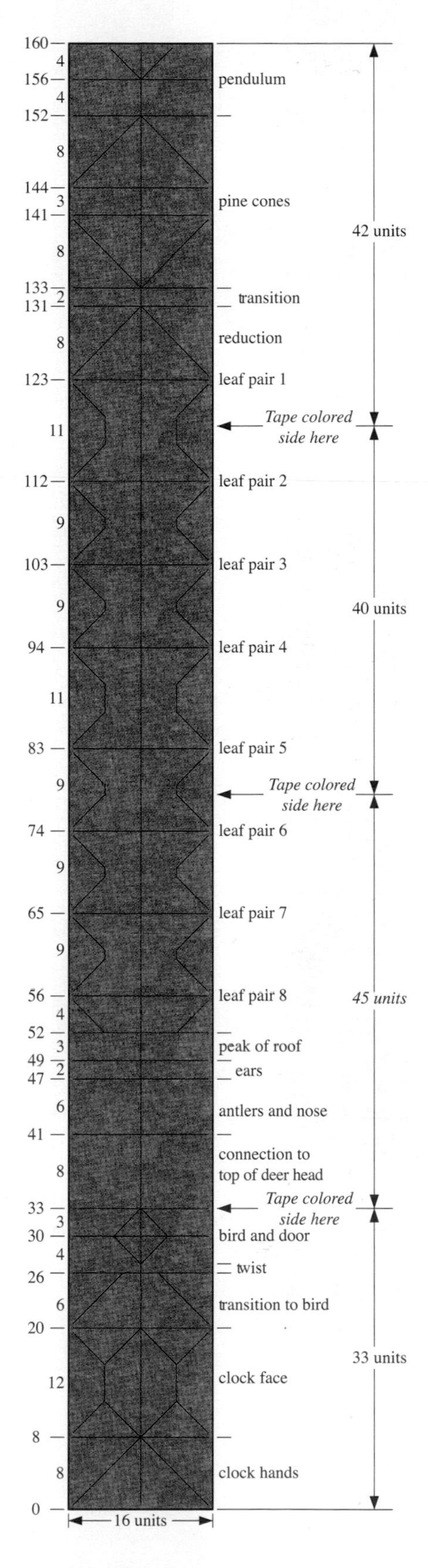

1. Begin with a strip of paper in the proportions 10:1, at least 16 by 160 cm (or 8 by 80 inches), colored side up.

For larger models, it is difficult to find paper of the necessary length. However, you can cut four strips that are, respectively, 16 by 33, 16 by 45, 16 by 40, and 16 by 42 and tape them together where noted by the italicized text. If you use these lengths to make your starting rectangle, then the tape seams will be hidden inside the model.

Steps 2–10 of the model are devoted to locating the major horizontal creases. There are two ways of getting these starting proportions. If you want to locate all the creases only by folding, begin with step 2. However, you will create many unnecessary creases during the folding process that will show up on the final model and that may be confusing later in the folding sequence. The easiest and cleanest method of locating the major horizontal creases is to measure and mark the locations of the creases; if you choose to do this, then you may skip to step 11.

The notes along the right side indicate from which part of the paper each part of the model is derived.

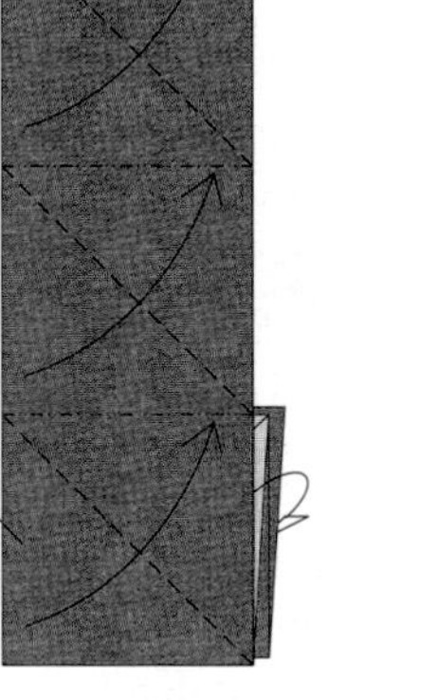

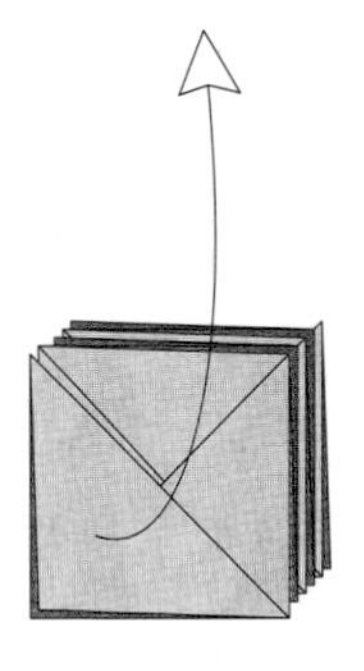

2. Fold the lower left corner up to lie along the right edge.

3. Mountain-fold the bottom corner behind.

4. Repeat steps 2–3.

5. Repeat steps 2–3 eight more times (until you run out of paper).

6. Unfold to step 2.

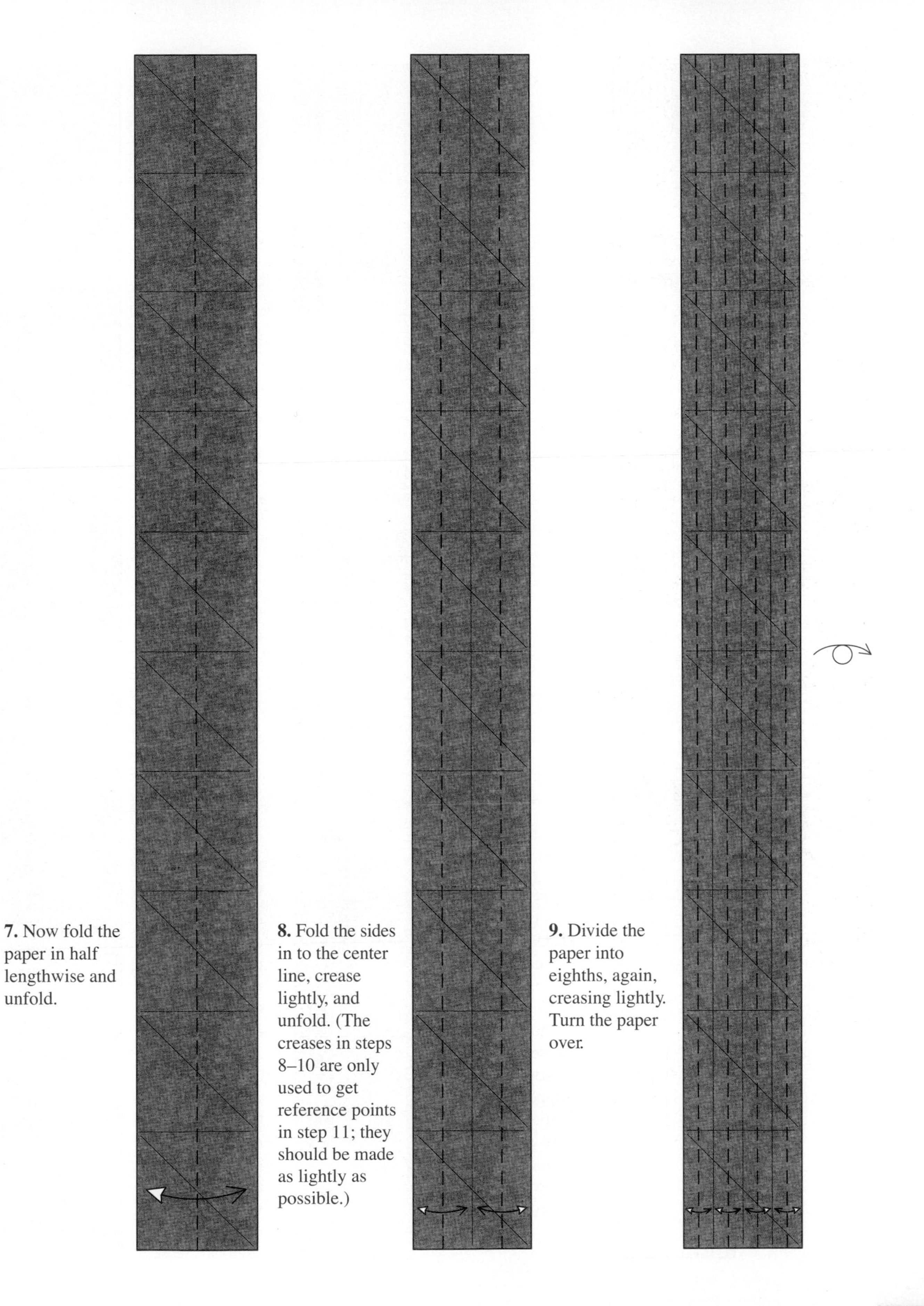

7. Now fold the paper in half lengthwise and unfold.

8. Fold the sides in to the center line, crease lightly, and unfold. (The creases in steps 8–10 are only used to get reference points in step 11; they should be made as lightly as possible.)

9. Divide the paper into eighths, again, creasing lightly. Turn the paper over.

10. Divide the paper into sixteenths, again, creasing lightly, and turn the paper back over.

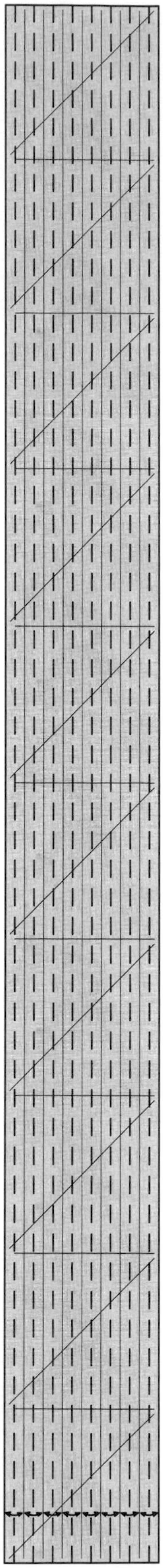

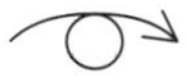

11. Using the vertical and diagonal creases as guides, add the horizontal creases shown. (If you have skipped steps 2–10, you may measure the locations of the horizontal creases.)

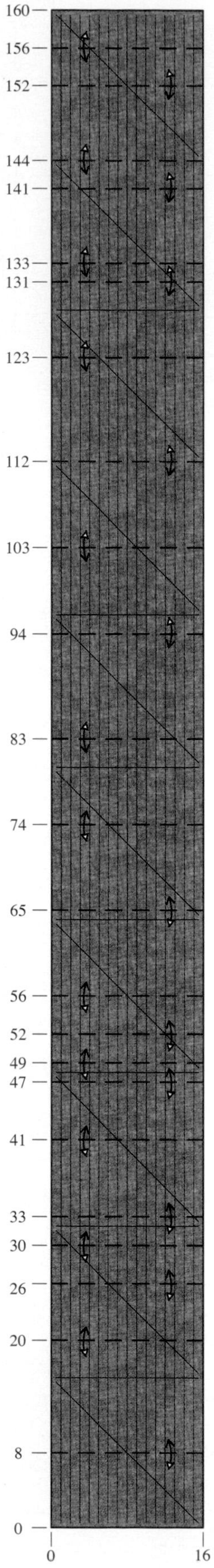

12. For the rest of the model, ignore the light creases you made in steps 2–10 and use the horizontal creases from step 11 as your major landmarks. Fold and unfold vertically, then add the indicated diagonal creases.

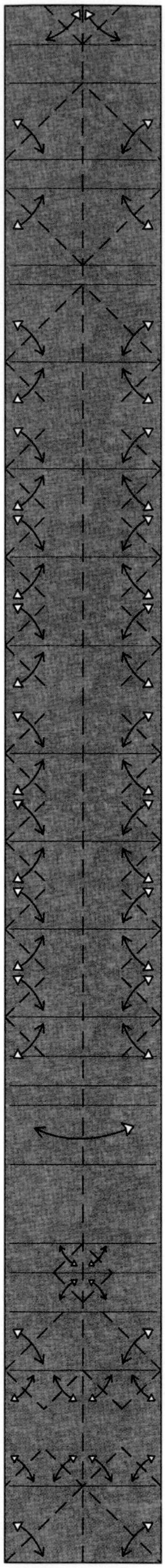

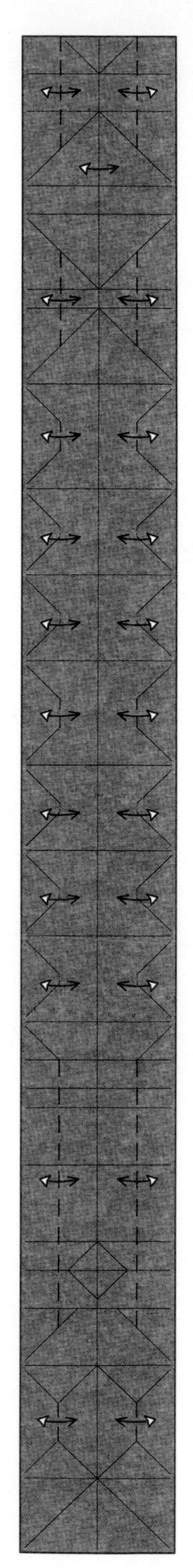

13. Now add some vertical creases.

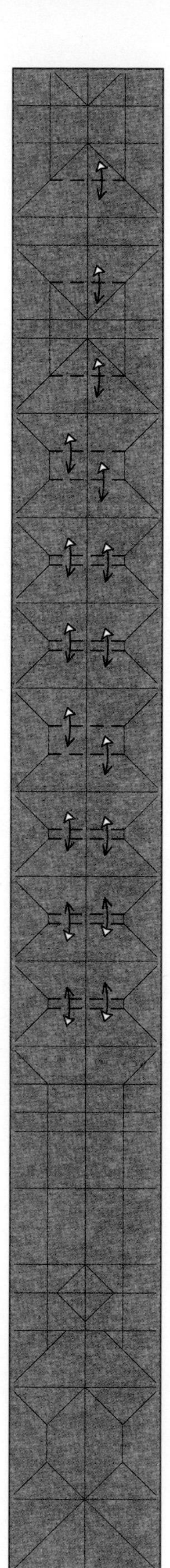

14. Now add some horizontal creases.

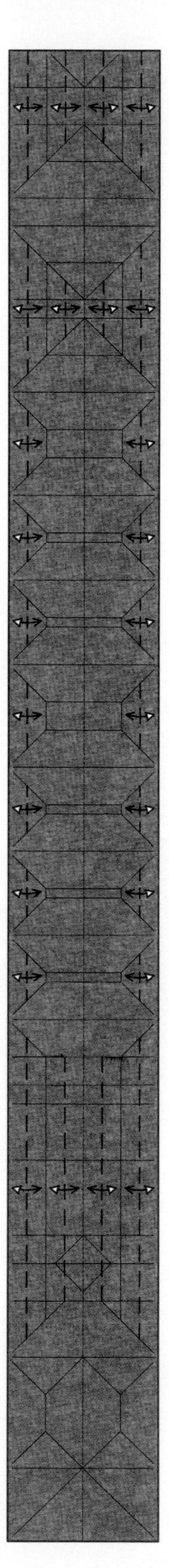

15. Add more vertical creases.

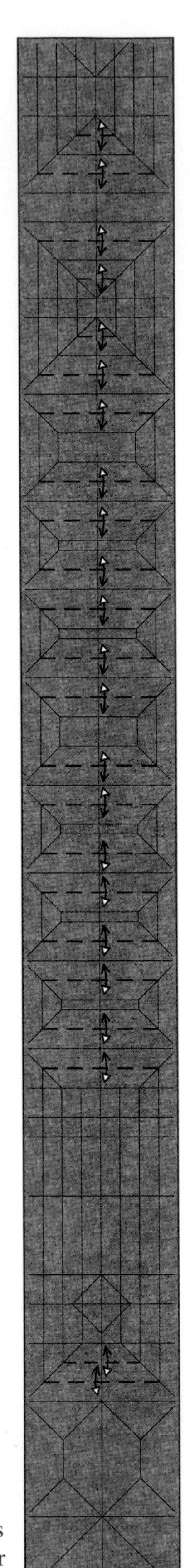

16. Add more horizontal creases and turn the paper over.

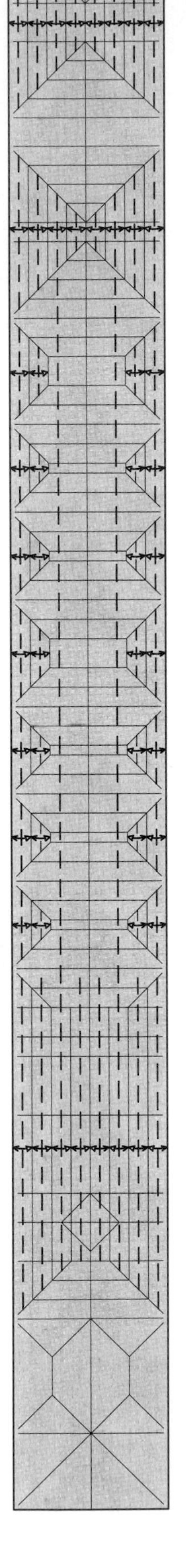

17. Add more vertical creases.

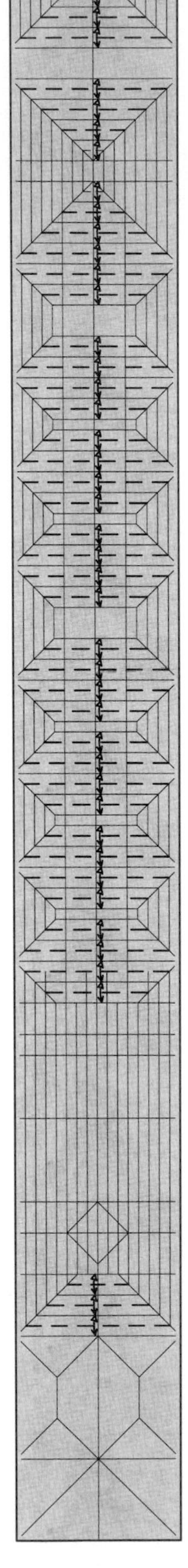

18. Add more horizontal creases.

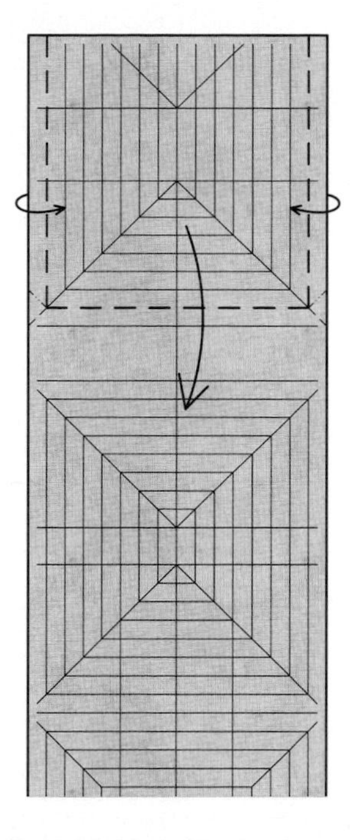

19. The precreasing is finished; now we begin folding in earnest. Beginning at the top and using the existing creases, fold the sides in as you fold the top of the rectangle down.

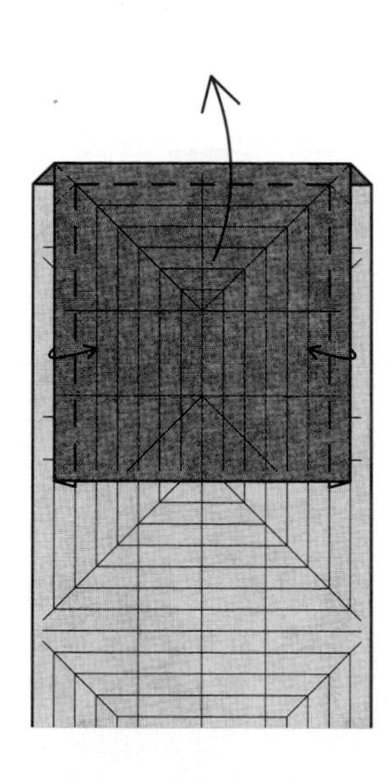

20. Fold the sides in as you fold the flap upward.

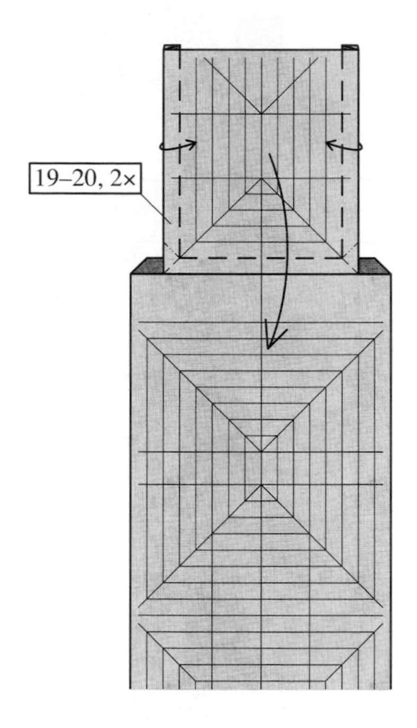

21. Continue narrowing the flap by repeating step 19–20 twice, then step 19 once more.

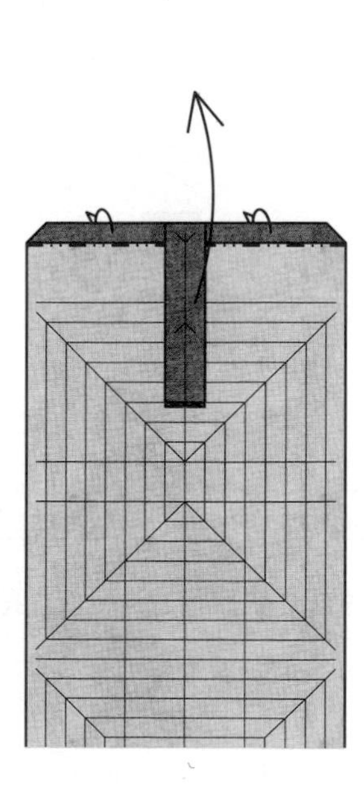

22. Swing the narrow flap (which will be the pendulum) upward.

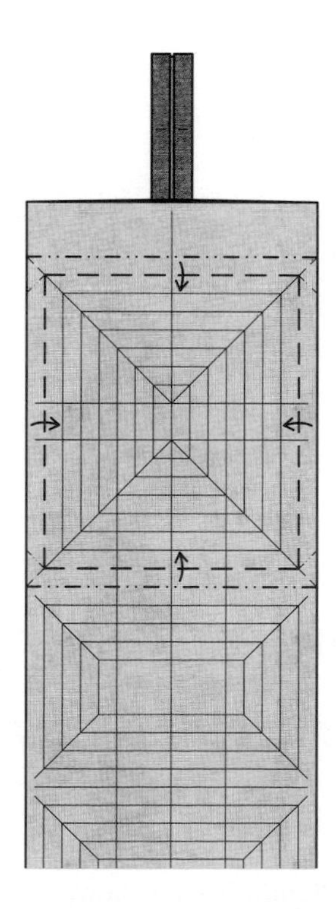

23. Simultaneously fold the sides in and pleat horizontally on existing creases.

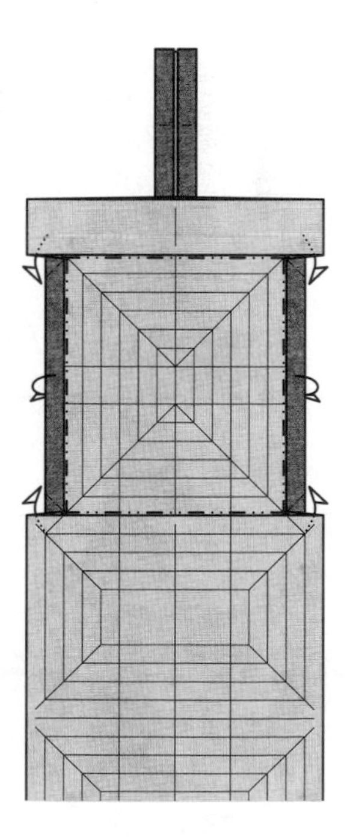

24. Simultaneously fold the sides back as you swing a layer down (above) and up (below).

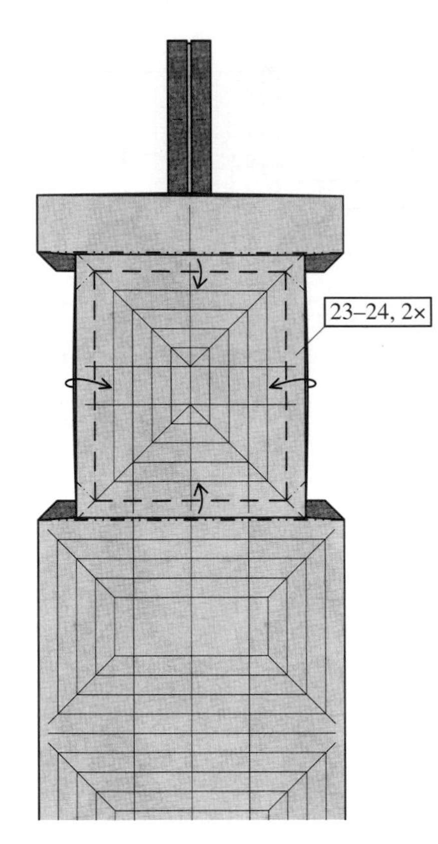

25. Repeat steps 23–24 twice, then step 23 once more.

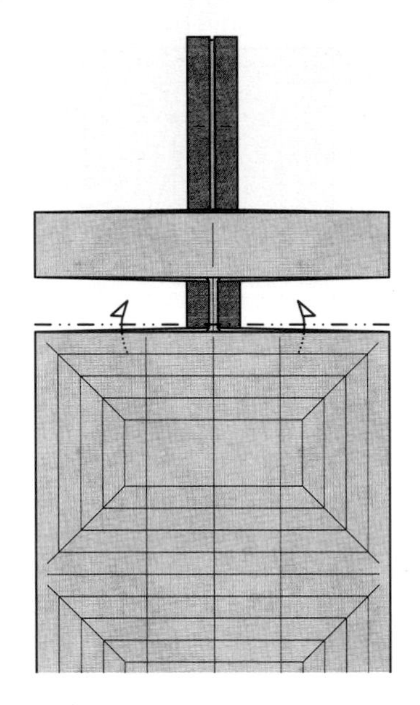

26. Fold the thick narrow flap up from behind.

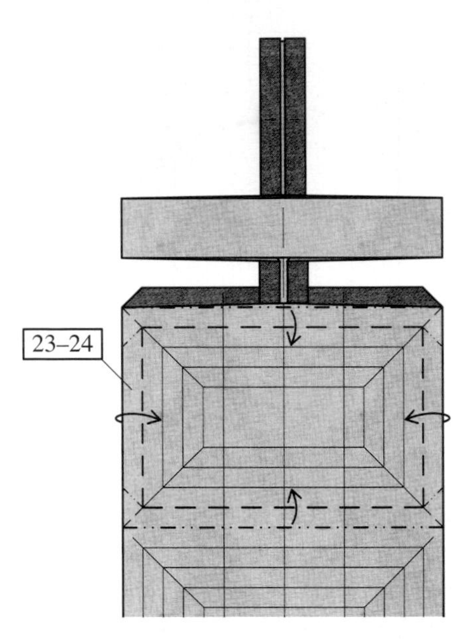

27. Repeat steps 23–24.

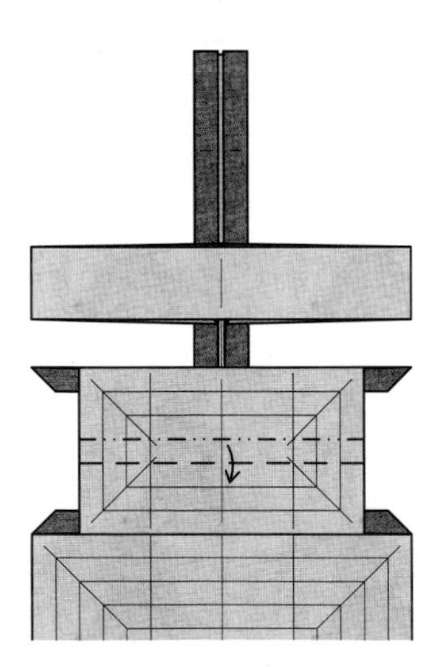

28. Pleat.

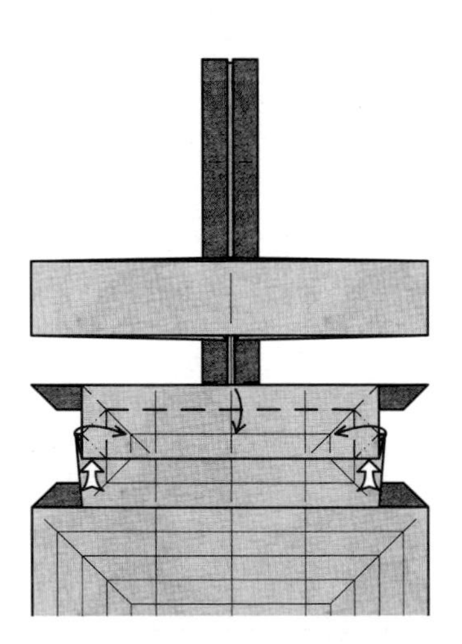

29. Squash-fold the lower corners; simultaneously fold the top down and the sides above the squash folds inward.

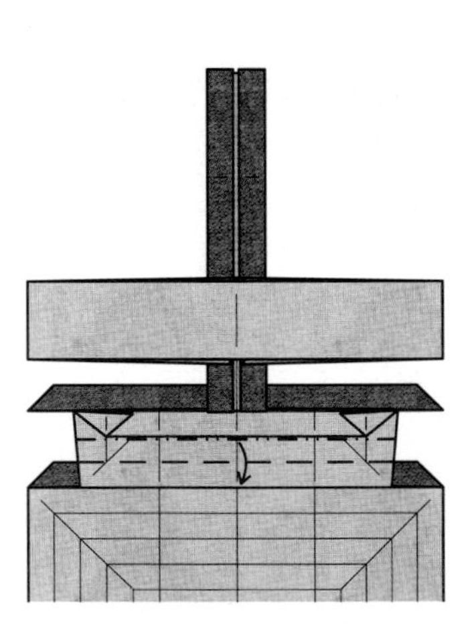

30. Pleat again.

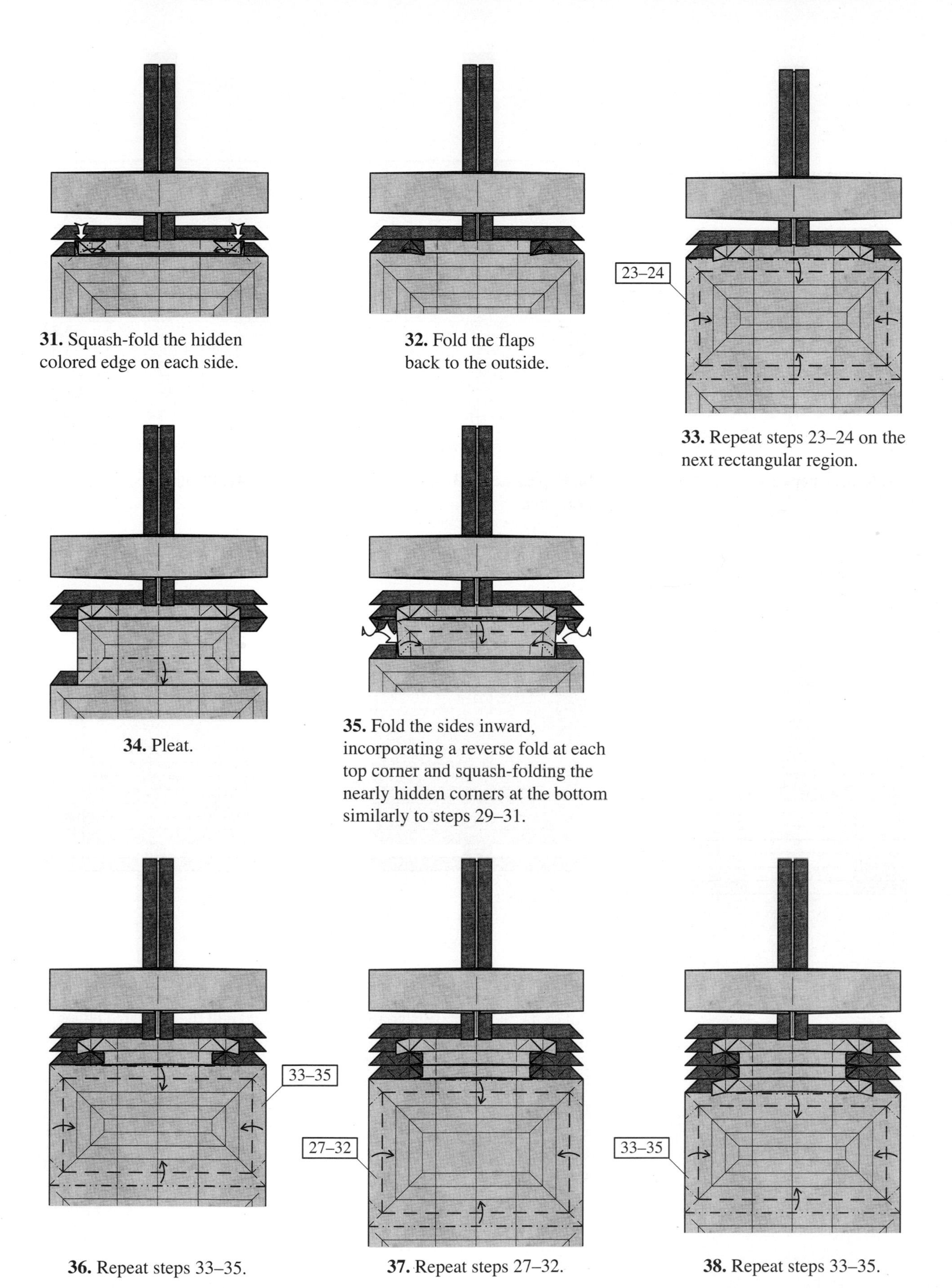

31. Squash-fold the hidden colored edge on each side.

32. Fold the flaps back to the outside.

33. Repeat steps 23–24 on the next rectangular region.

34. Pleat.

35. Fold the sides inward, incorporating a reverse fold at each top corner and squash-folding the nearly hidden corners at the bottom similarly to steps 29–31.

36. Repeat steps 33–35.

37. Repeat steps 27–32.

38. Repeat steps 33–35.

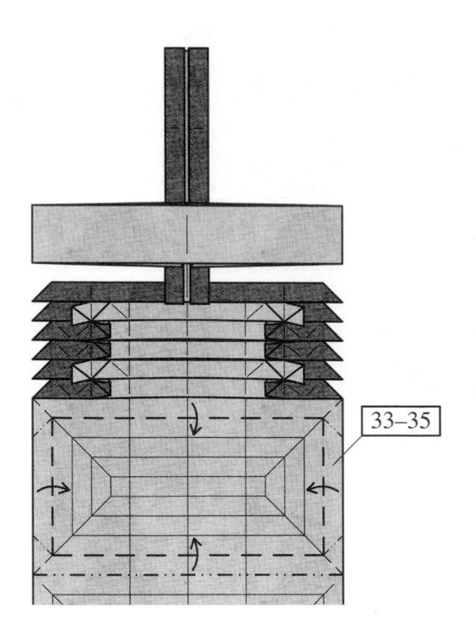

39. Repeat steps 33–35 again.

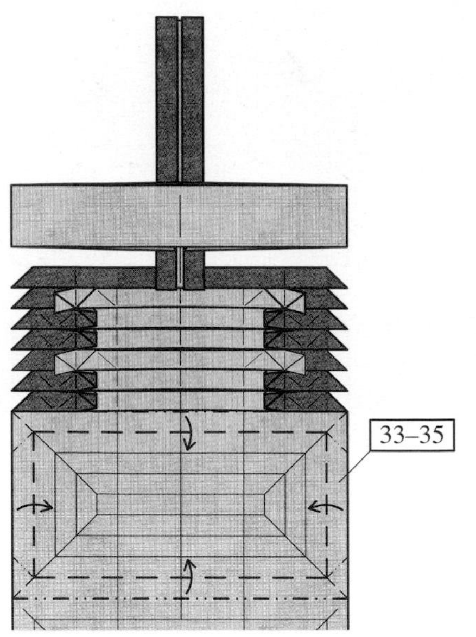

40. Repeat steps 33–35 one more time.

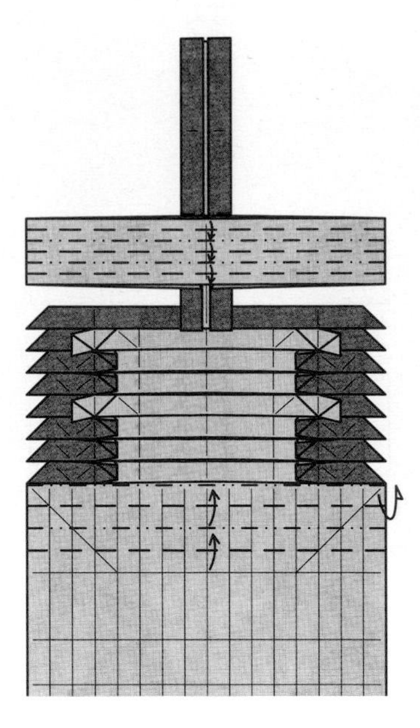

41. Pleat in two places.

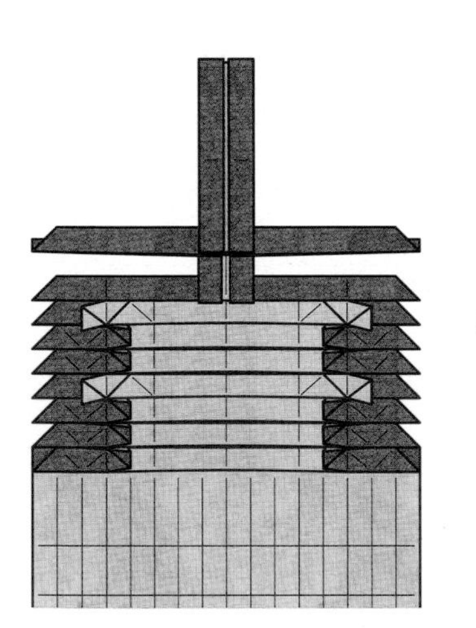

42. Turn the paper over.

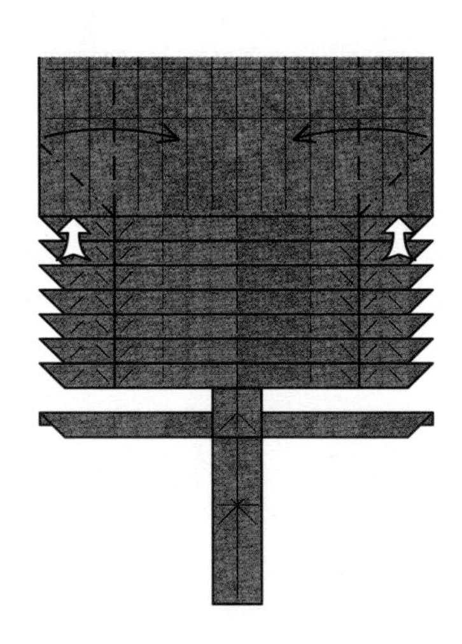

43. Squash-fold the corner. All of the pleated layers go to the left. In this and succeeding steps, don't extend the vertical creases any farther than you have to, so that the top of the paper remains unfolded (and the model does not lie flat).

44. Squash-fold the corners and pull the indicated edges downward.

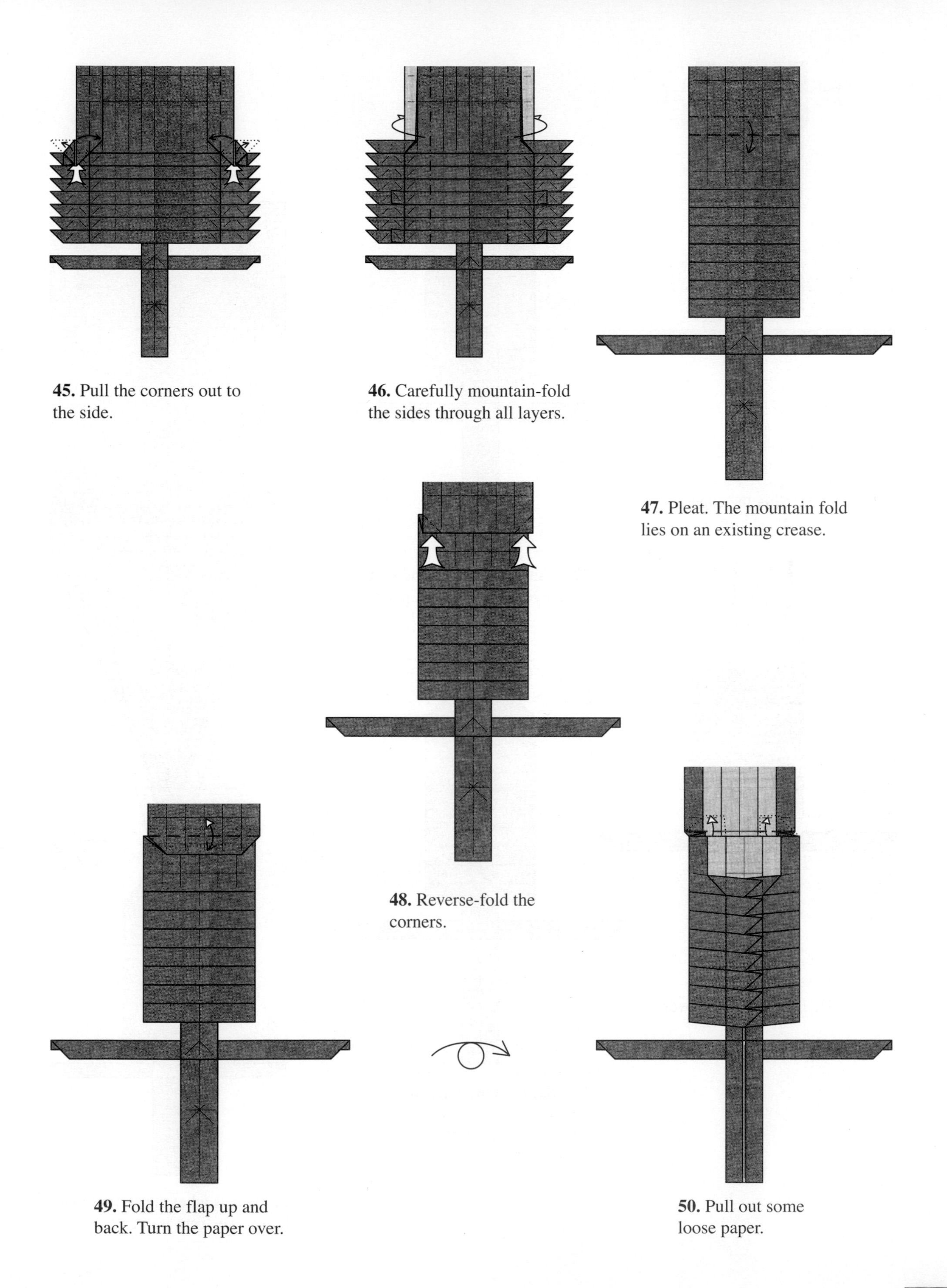

45. Pull the corners out to the side.

46. Carefully mountain-fold the sides through all layers.

47. Pleat. The mountain fold lies on an existing crease.

48. Reverse-fold the corners.

49. Fold the flap up and back. Turn the paper over.

50. Pull out some loose paper.

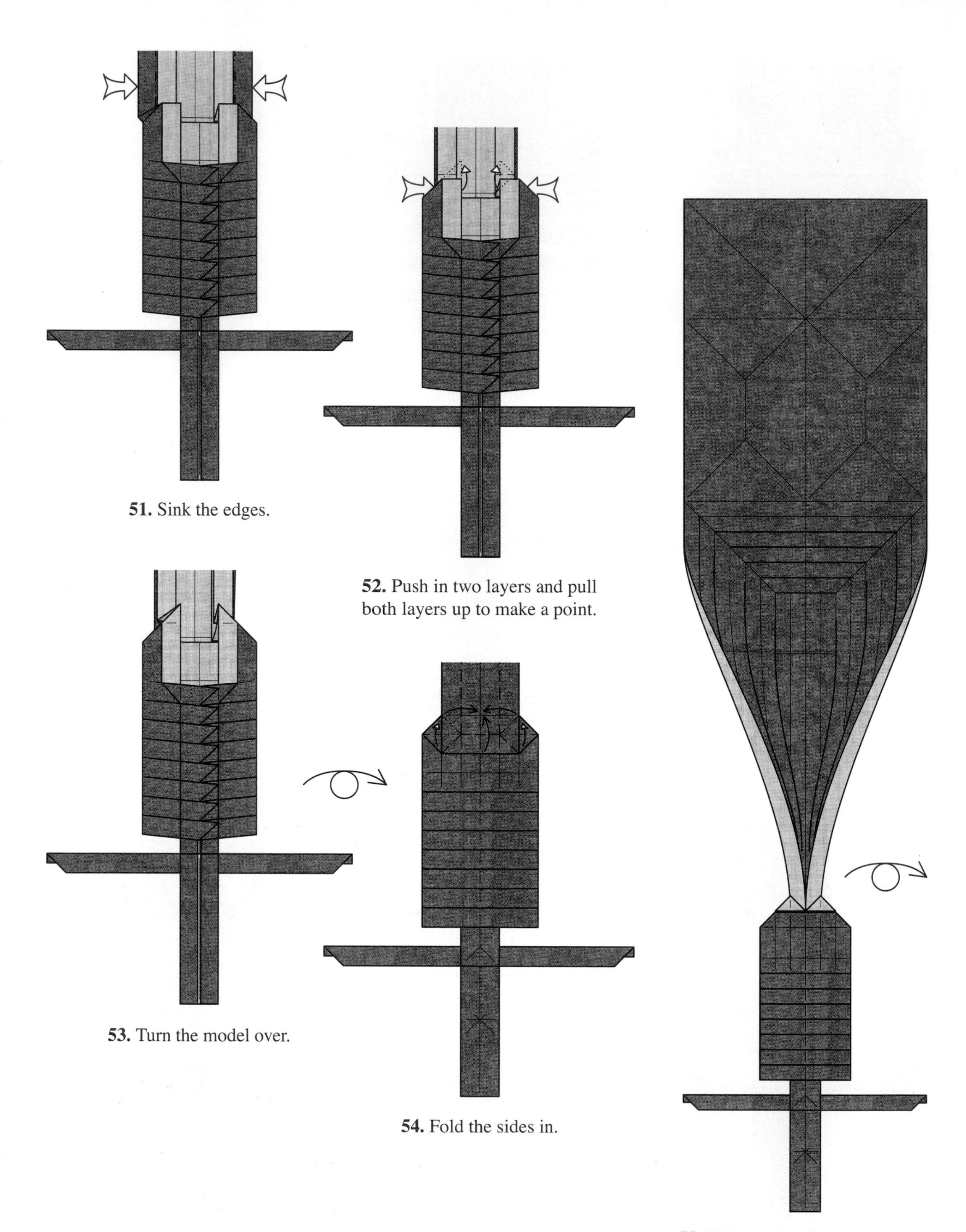

51. Sink the edges.

52. Push in two layers and pull both layers up to make a point.

53. Turn the model over.

54. Fold the sides in.

55. This is what the entire model looks like. Turn it over.

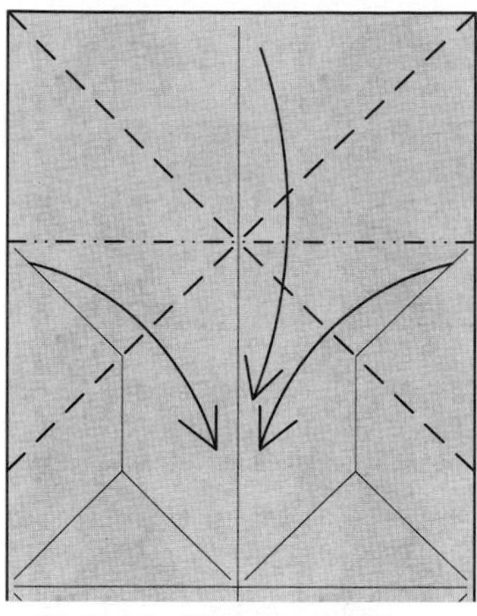

56. Fold a Waterbomb Base.

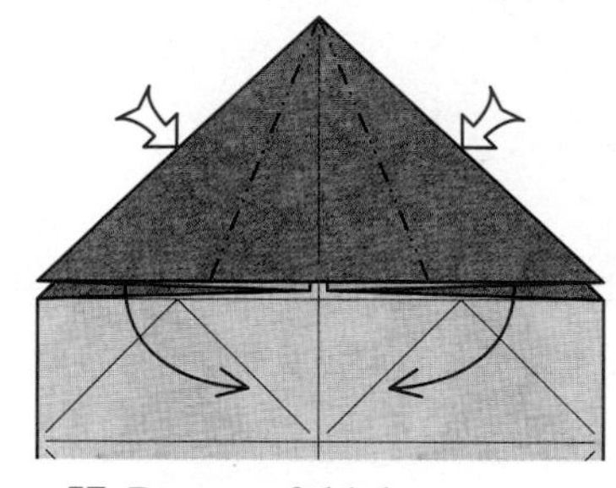

57. Reverse-fold the corners.

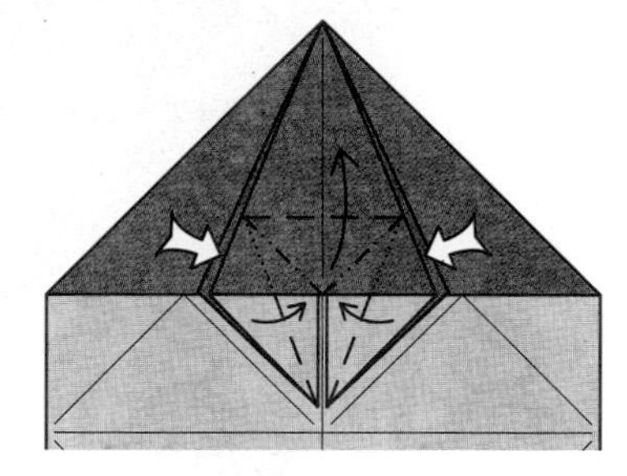

58. Petal-fold.

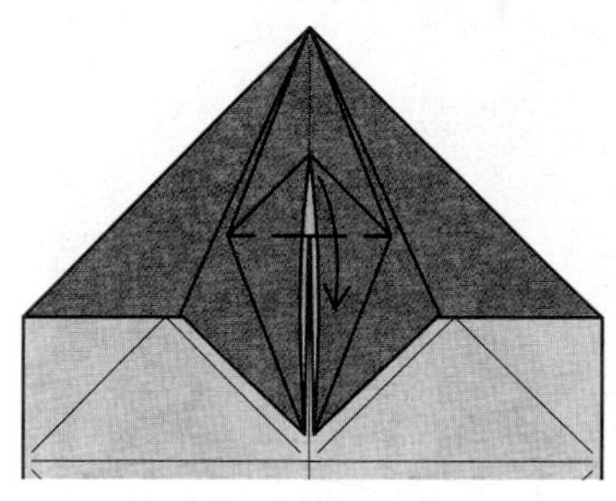

59. Fold the tip down.

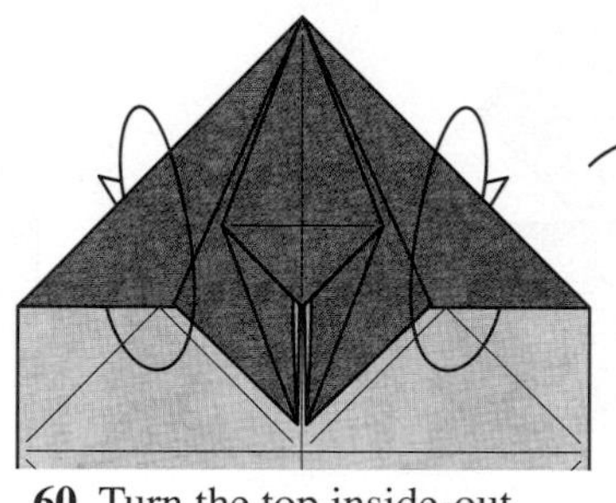

60. Turn the top inside-out and turn the paper over.

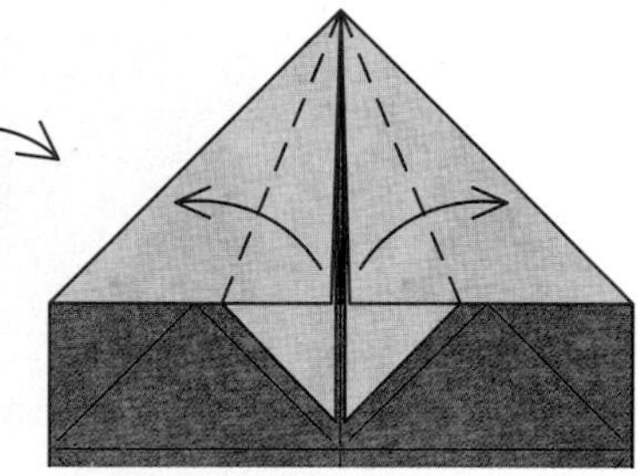

61. Fold the flaps out to the sides.

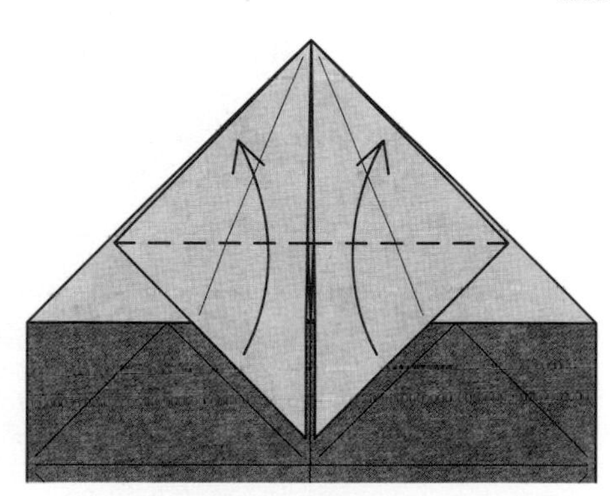

62. Fold the two points upward.

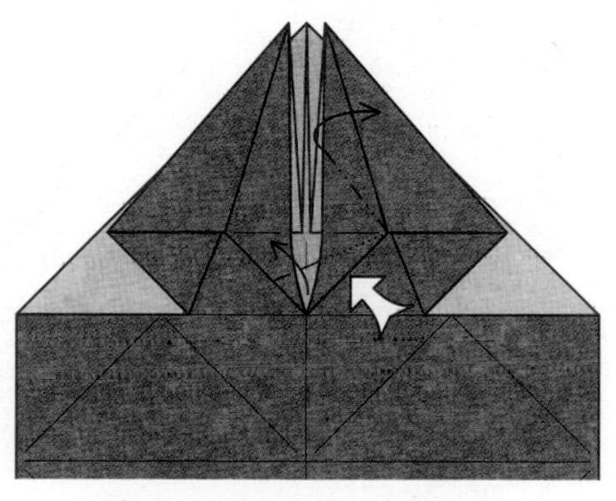

63. Swivel one layer upward.

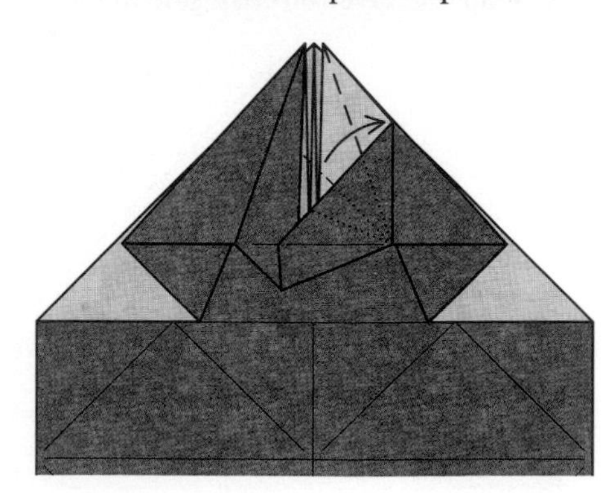

64. Pleat the next layer to match.

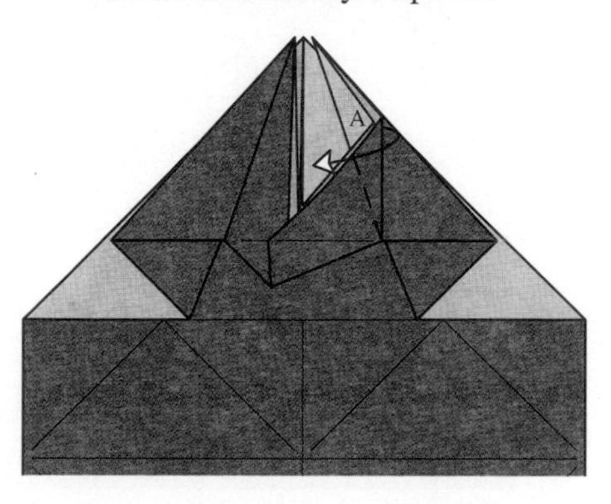

65. Wrap one layer around to the front. Layer A stays in place.

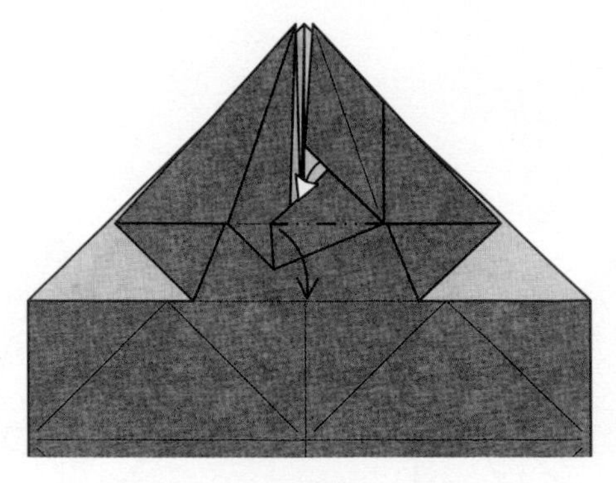

66. Pull out some loose paper.

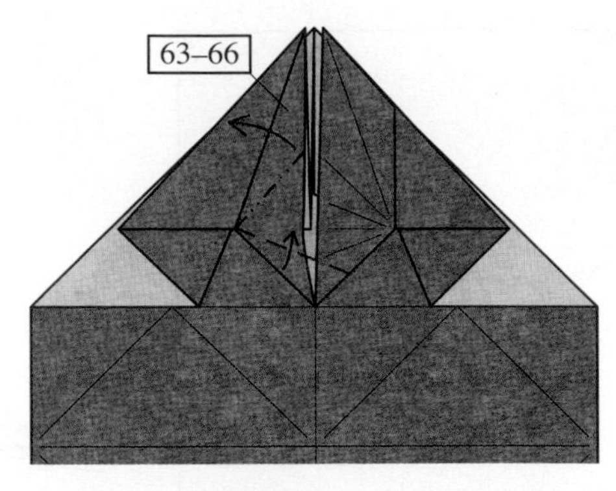

67. Repeat steps 63–66 on the left.

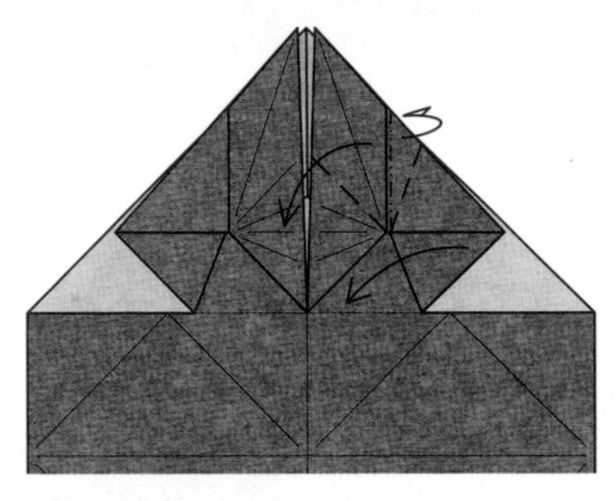

68. Fold the right point downward on an existing crease and swing it over to the left.

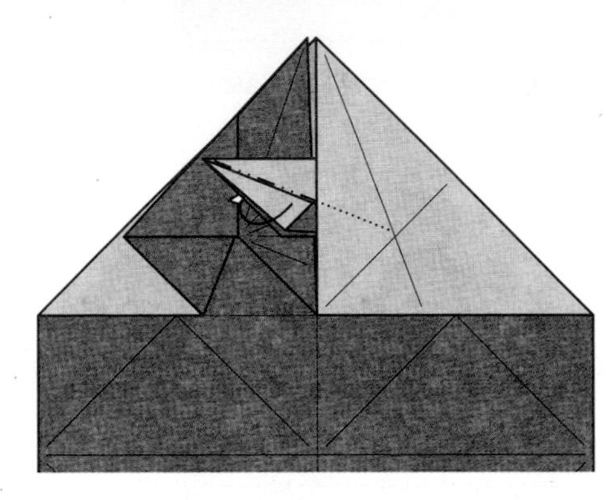

69. Mountain-fold both layers underneath.

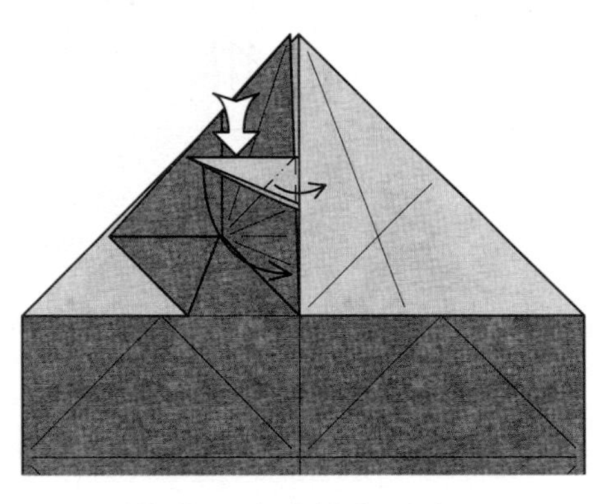

70. Squash-fold the point.

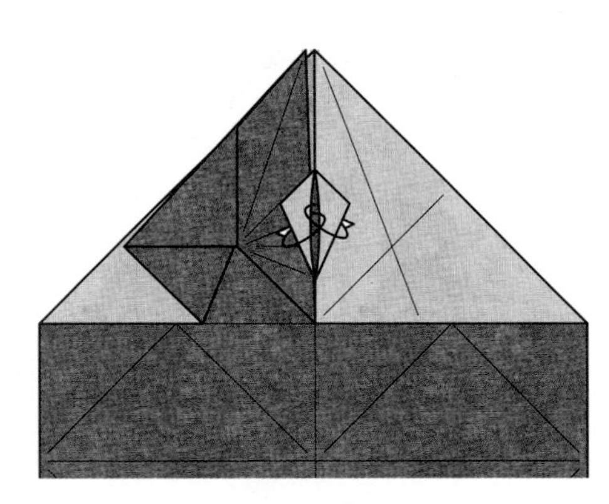

71. Wrap one layer on each side of the point.

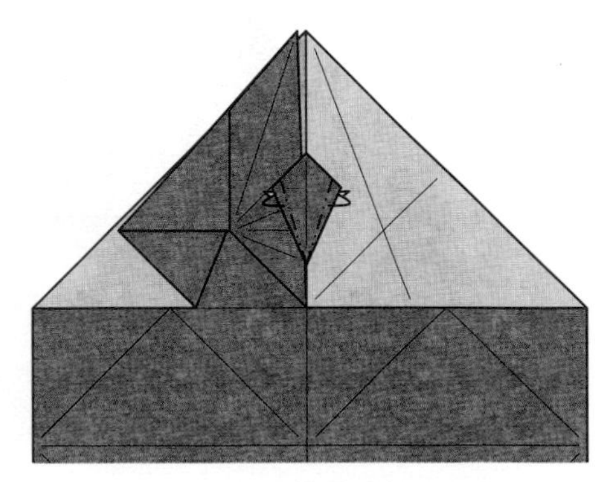

72. Fold one-third of the edge underneath on each side.

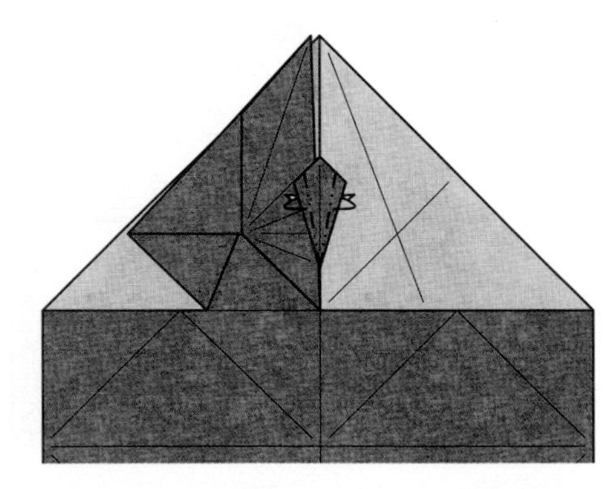

73. Narrow further with mountain folds.

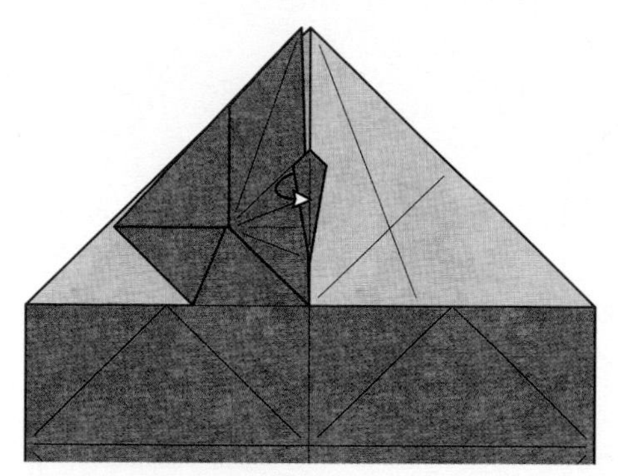

74. Bring the layers on the left to the front.

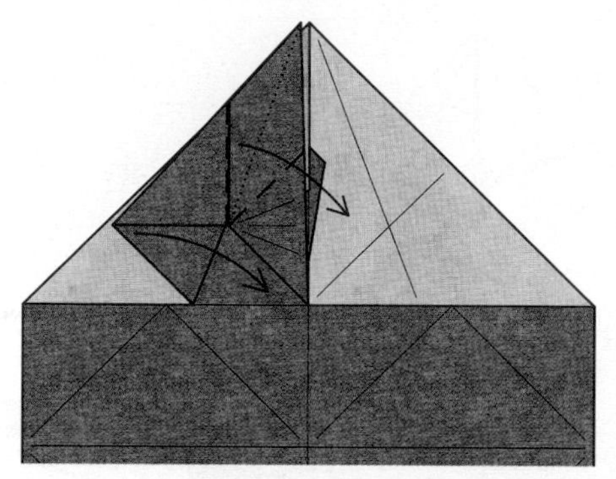

75. Fold the left point down. Note that this is slightly different from step 68.

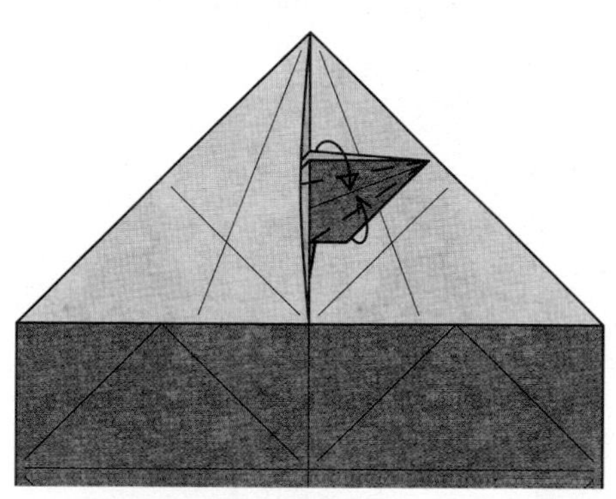

76. Narrow the point with valley folds.

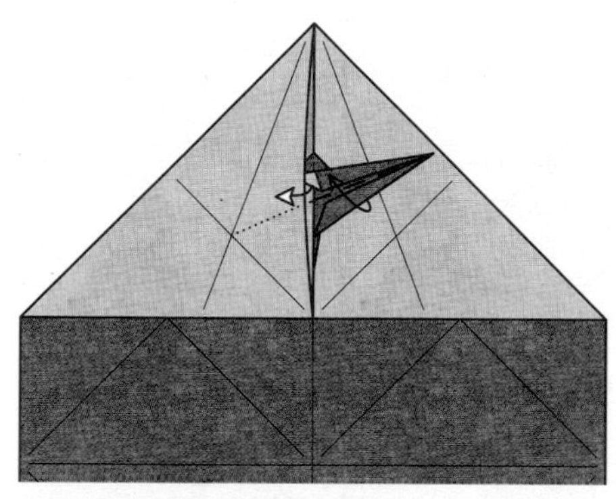

77. Fold the lower half upward and tuck its left side under the white paper.

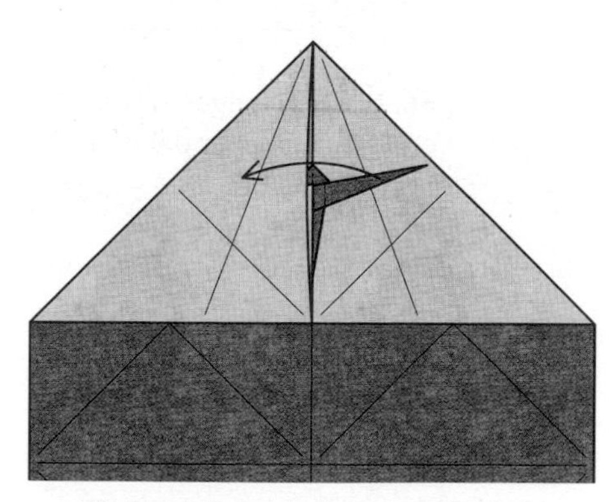

78. Fold the point over to the left.

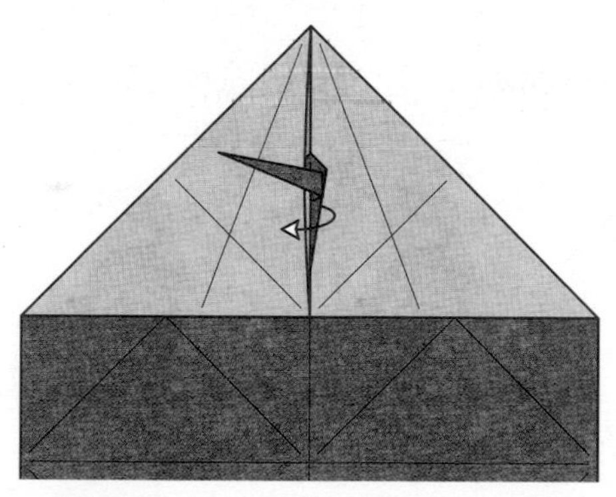

79. Bring the other point to the front.

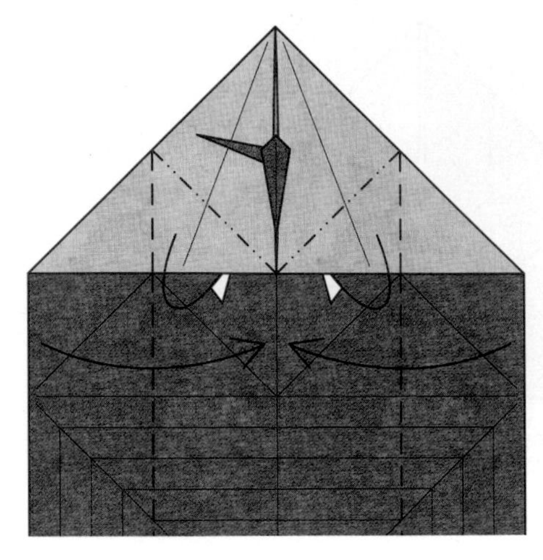

80. Fold the sides in. The creases don't go all the way to the other end.

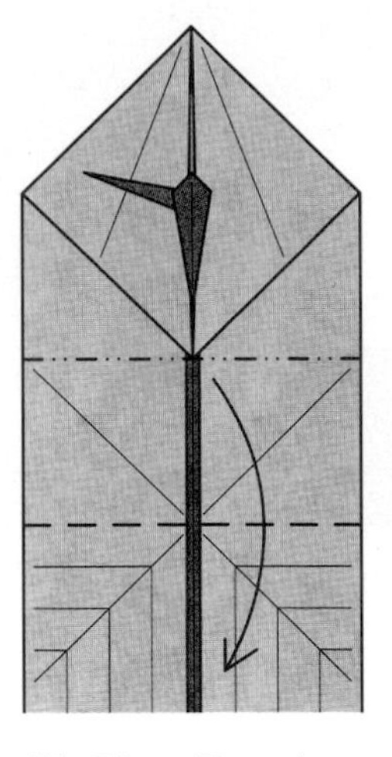

81. Pleat. Turn the paper over.

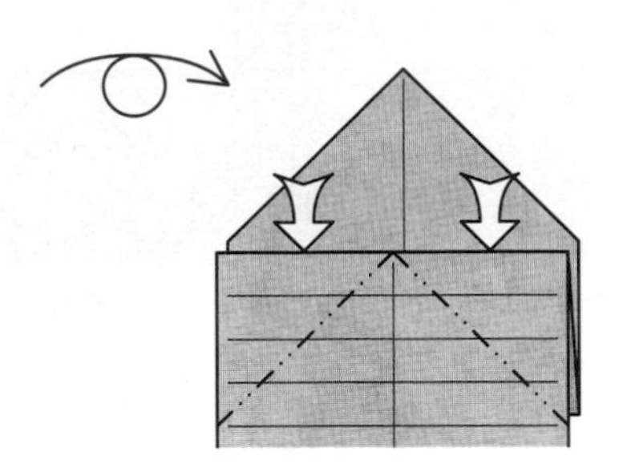

82. Reverse-fold the corners.

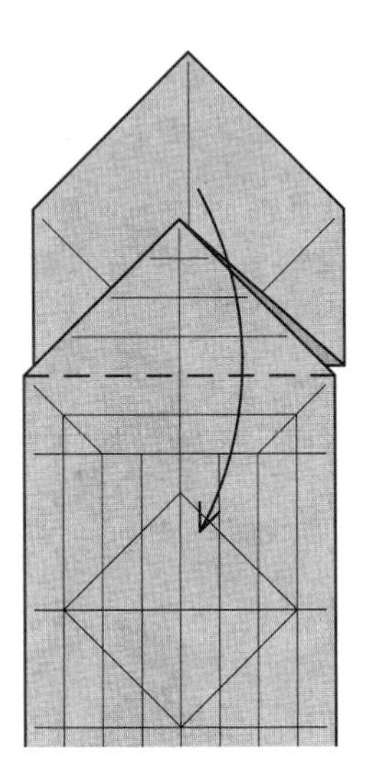

83. Fold the top down.

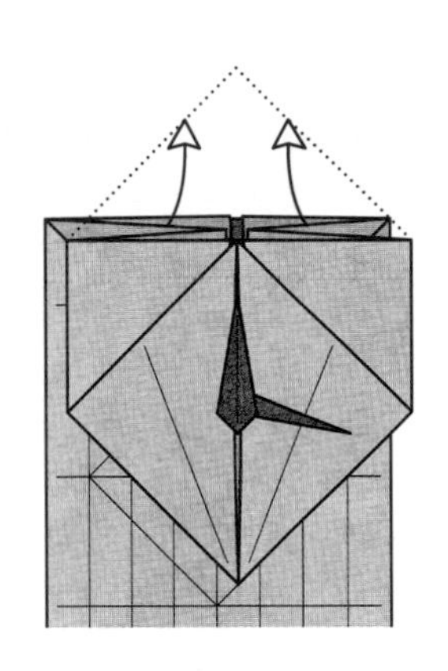

84. Pull out one layer from each side.

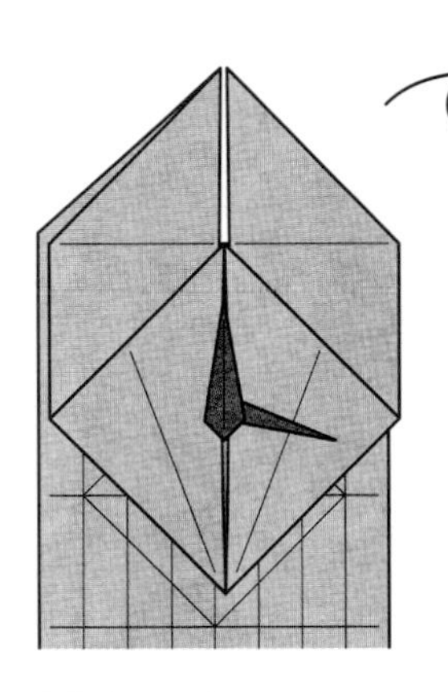

85. Turn the paper over.

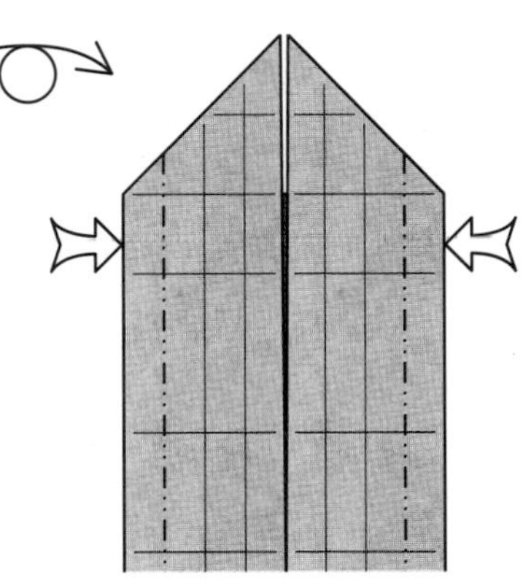

86. Sink the edges. These creases connect up with the ones on the bottom.

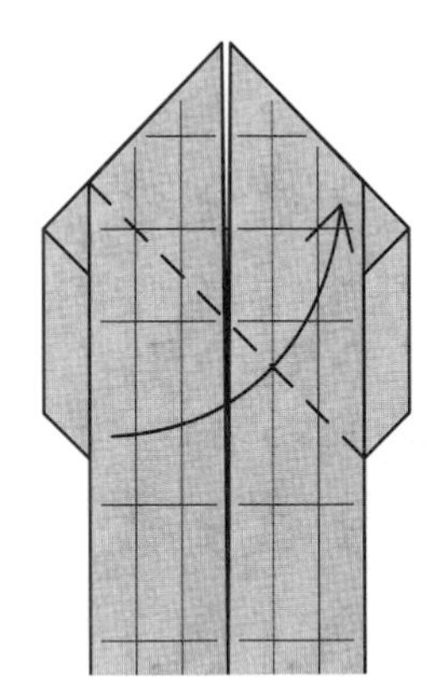

87. Swing the lower portion of the model up to the right. The model will not lie flat.

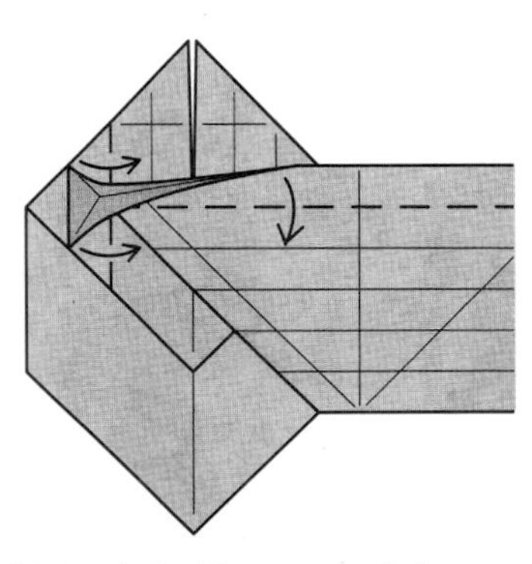

88. Fold the left side over and the top edge down and flatten. This fold connects to a crease on the rest of the model.

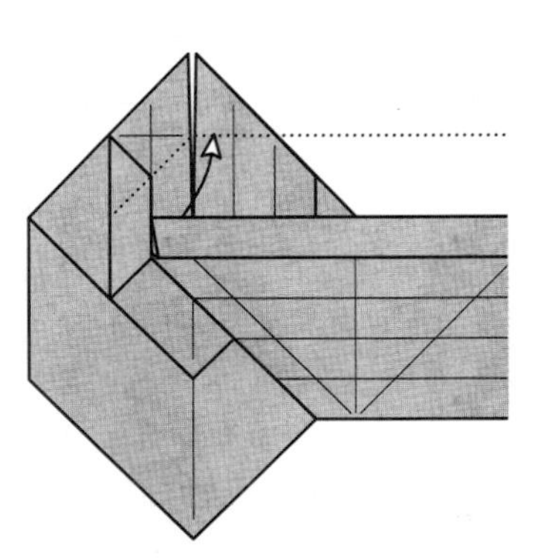

89. Pull out a single layer of paper.

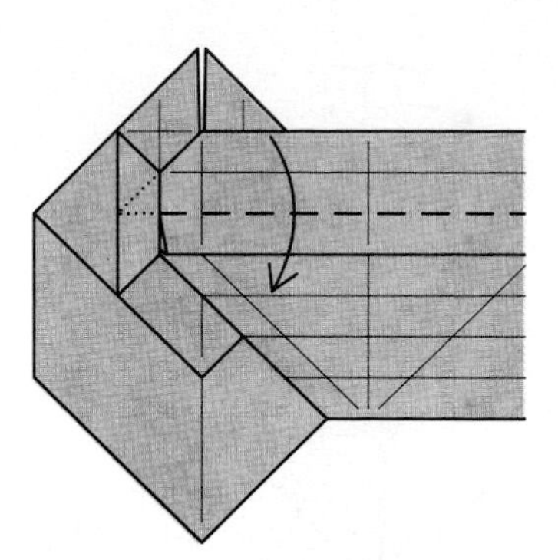

90. Fold the edge back down.

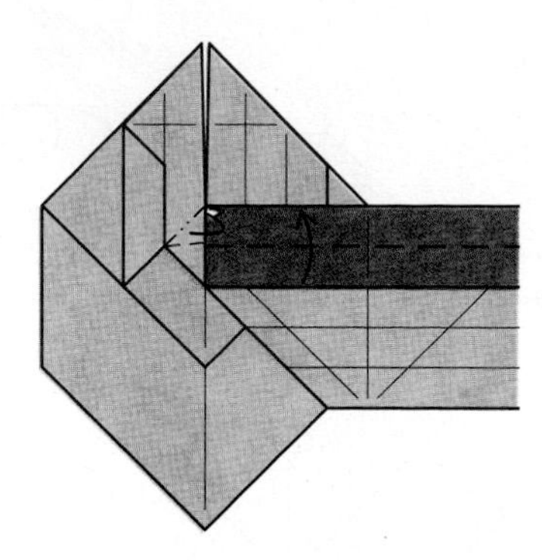

91. Fold the edge up again and tuck paper inside at the left.

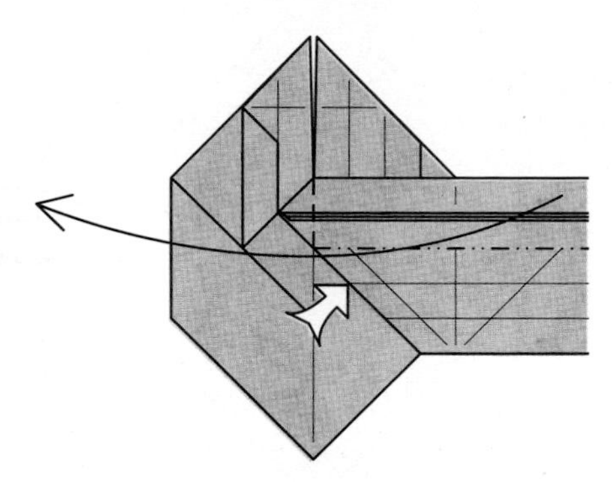

92. Swing the rest of the model over to the left.

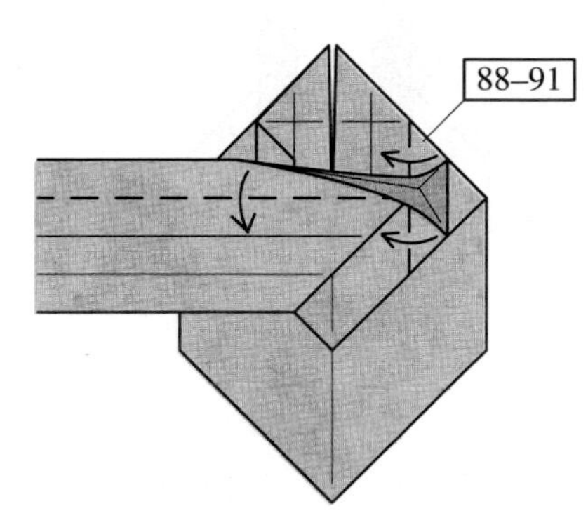

93. Repeat steps 88–91 on this side.

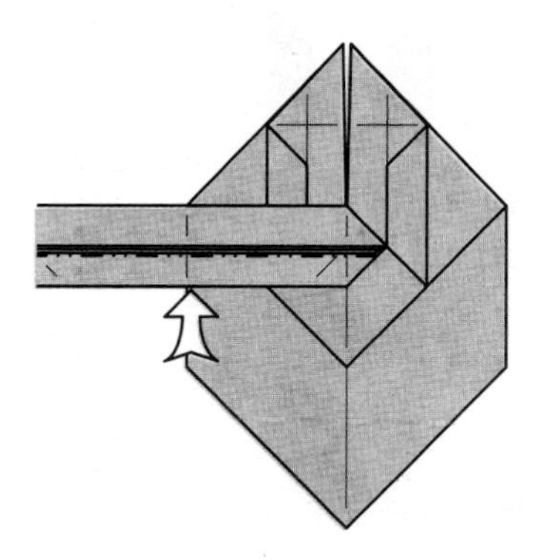

94. Sink the bottom edge (this connects with creases on the rest of the model, too).

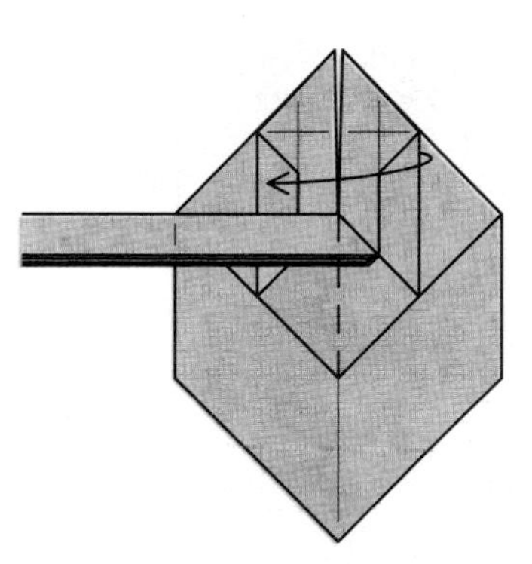

95. Fold the flap to the left.

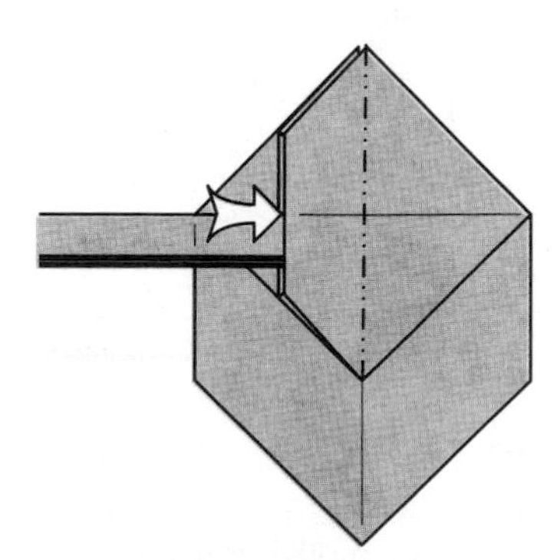

96. Closed-sink the flap.

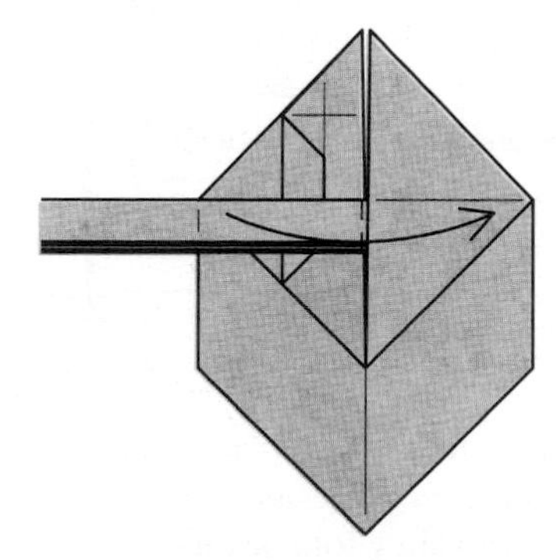

97. Swing the rest of the model over to the right.

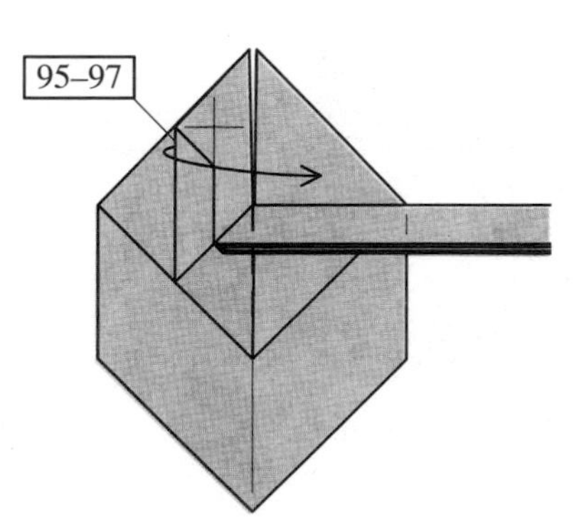

98. Repeat steps 95–97.

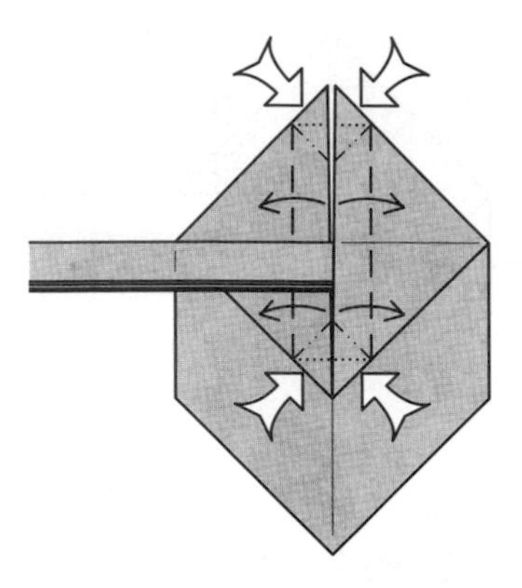

99. Fold the middle pair of edges out to the sides (two layers together at the top). Simultaneously, squash-fold the top two points and the bottom point.

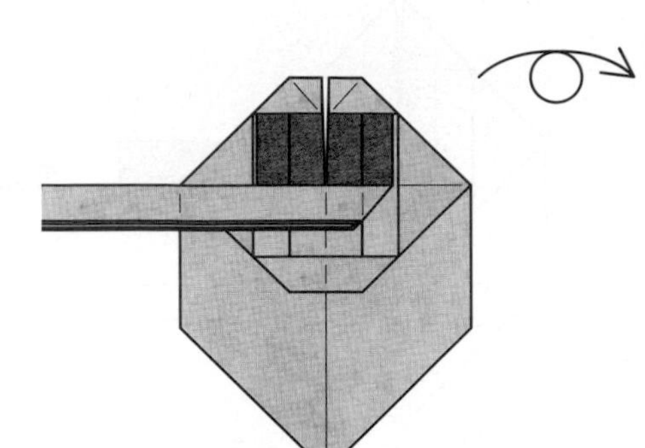

100. Turn the paper over.

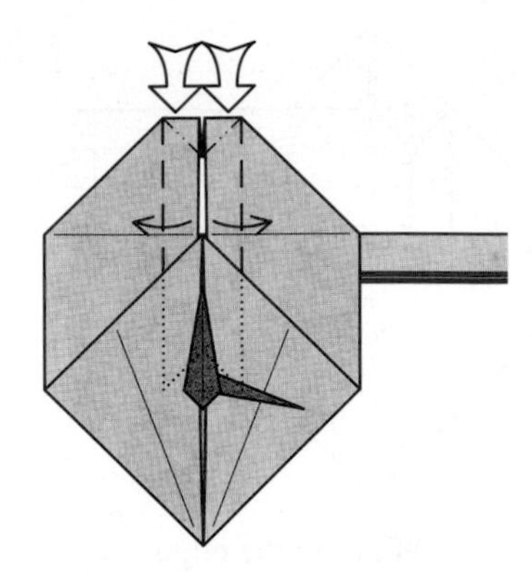

101. Squash-fold the top points and fold the edges outward (the vertical creases line up with the edges underneath).

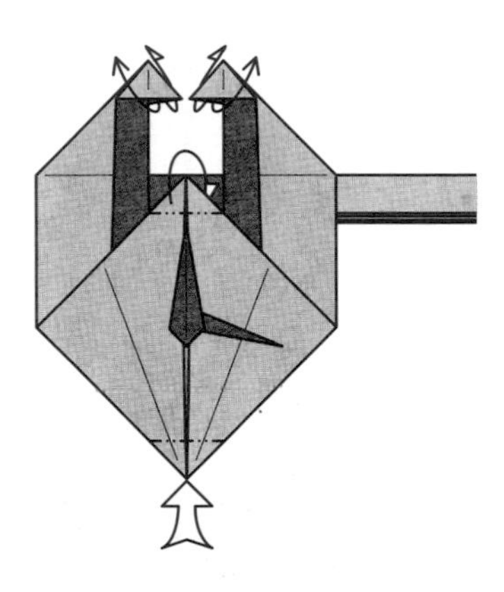

102. Unwrap the two top points. Mountain-fold the top of the clock face. Closed-sink the bottom point.

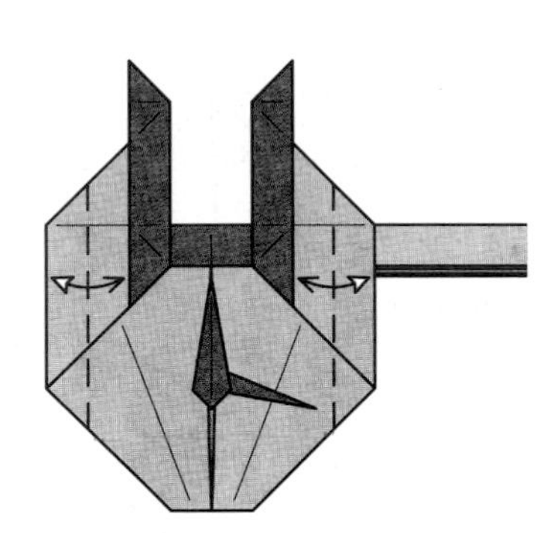

103. Fold and unfold.

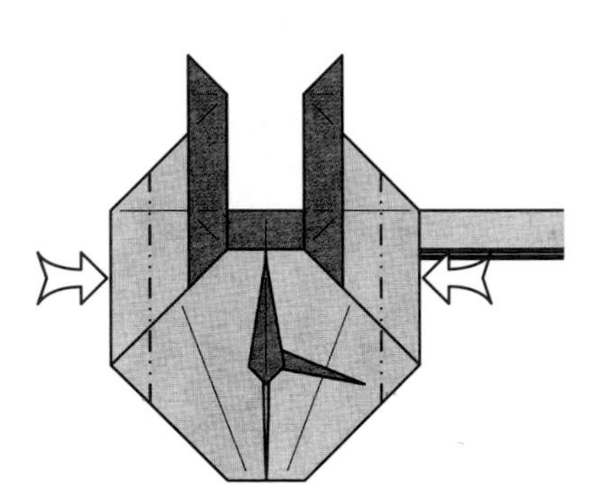

104. Closed-sink the edges (it is important that the sinks be closed, not open).

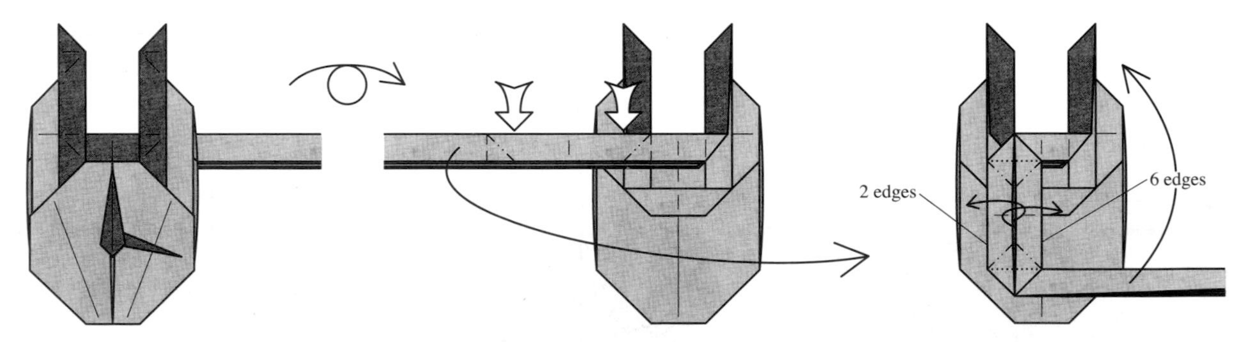

105. Turn the model over.

106. Squash-fold in two places and swing the rest of the model around to the right. The leftmost valley fold occurs on an existing crease. In the squash fold, two layers go to the left and six go to the right.

107. Fold the vertical edges out to the side and swing the rest of the model upward. Two edges go to the left; four edges go to the right.

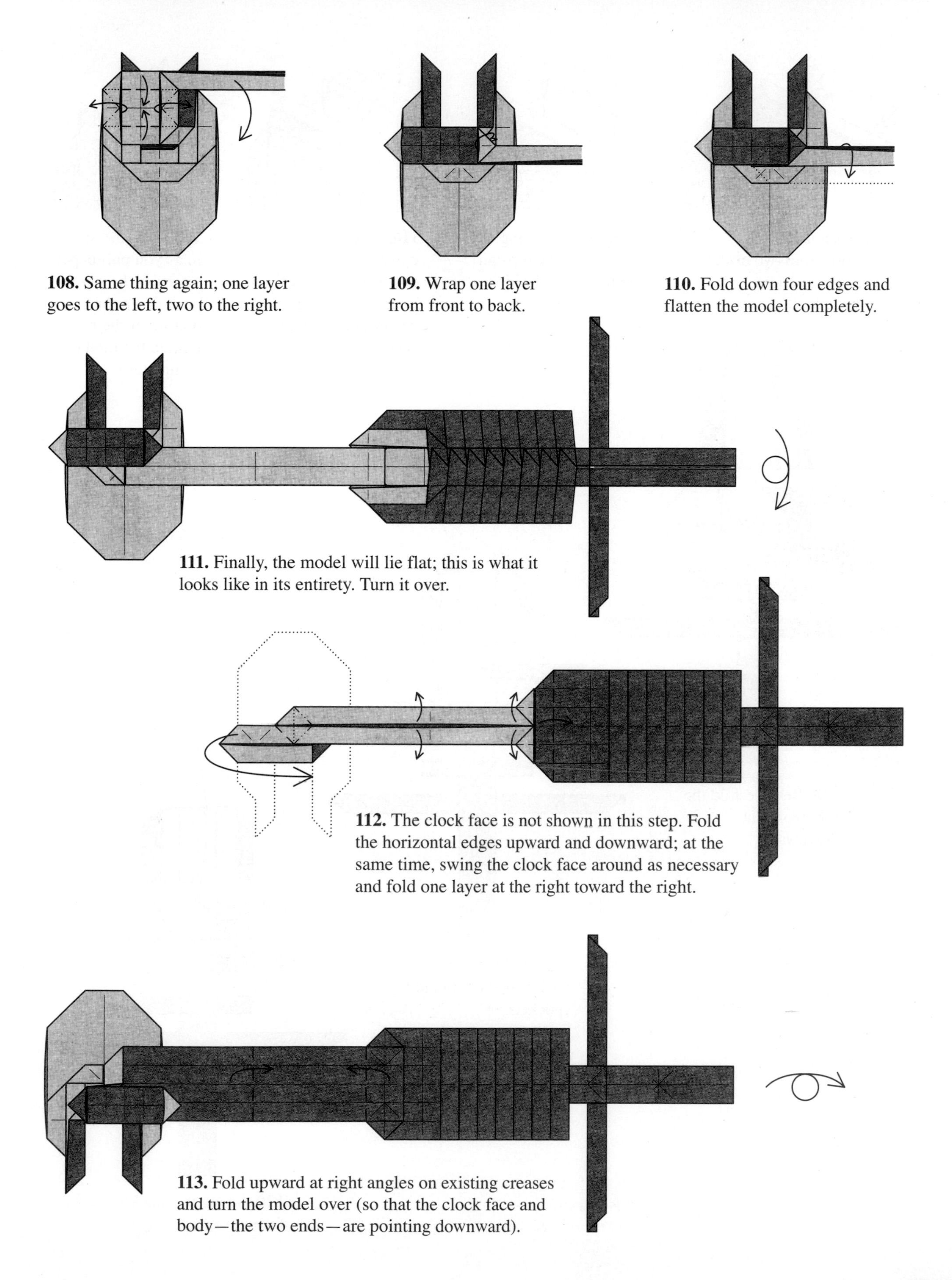

108. Same thing again; one layer goes to the left, two to the right.

109. Wrap one layer from front to back.

110. Fold down four edges and flatten the model completely.

111. Finally, the model will lie flat; this is what it looks like in its entirety. Turn it over.

112. The clock face is not shown in this step. Fold the horizontal edges upward and downward; at the same time, swing the clock face around as necessary and fold one layer at the right toward the right.

113. Fold upward at right angles on existing creases and turn the model over (so that the clock face and body—the two ends—are pointing downward).

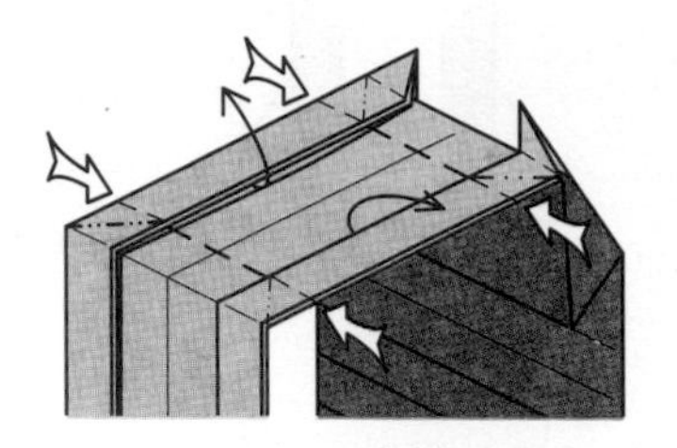

114. Push in the sides; on each side, fold two of the edges upward, one downward.

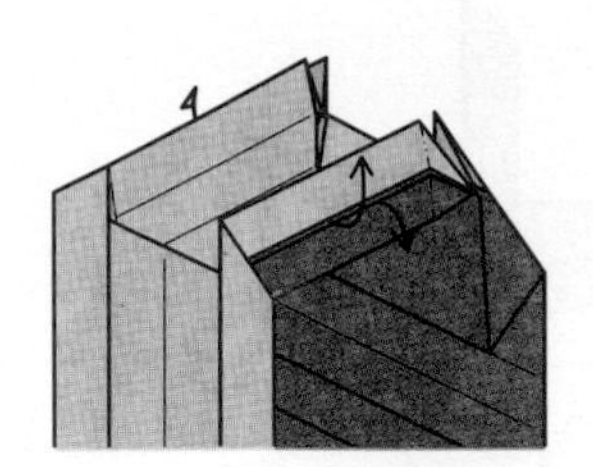

115. Pull out a single layer from the top and bottom of each side.

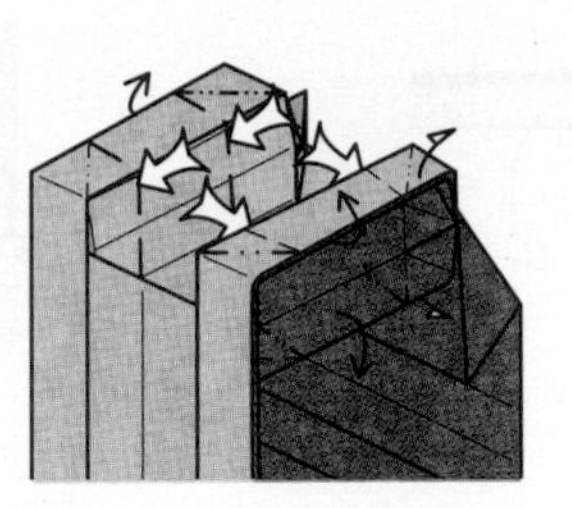

116. Fold two layers upward and one downward on each side. The box-like region becomes taller and deeper.

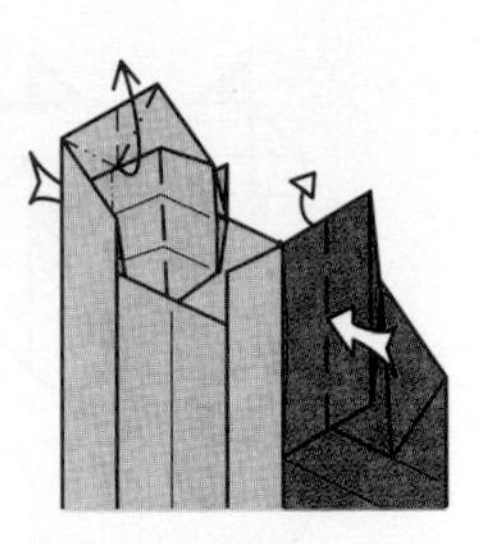

117. Push the sides in as you pull paper out from the top and push down the bottom of the box. Flatten the model completely.

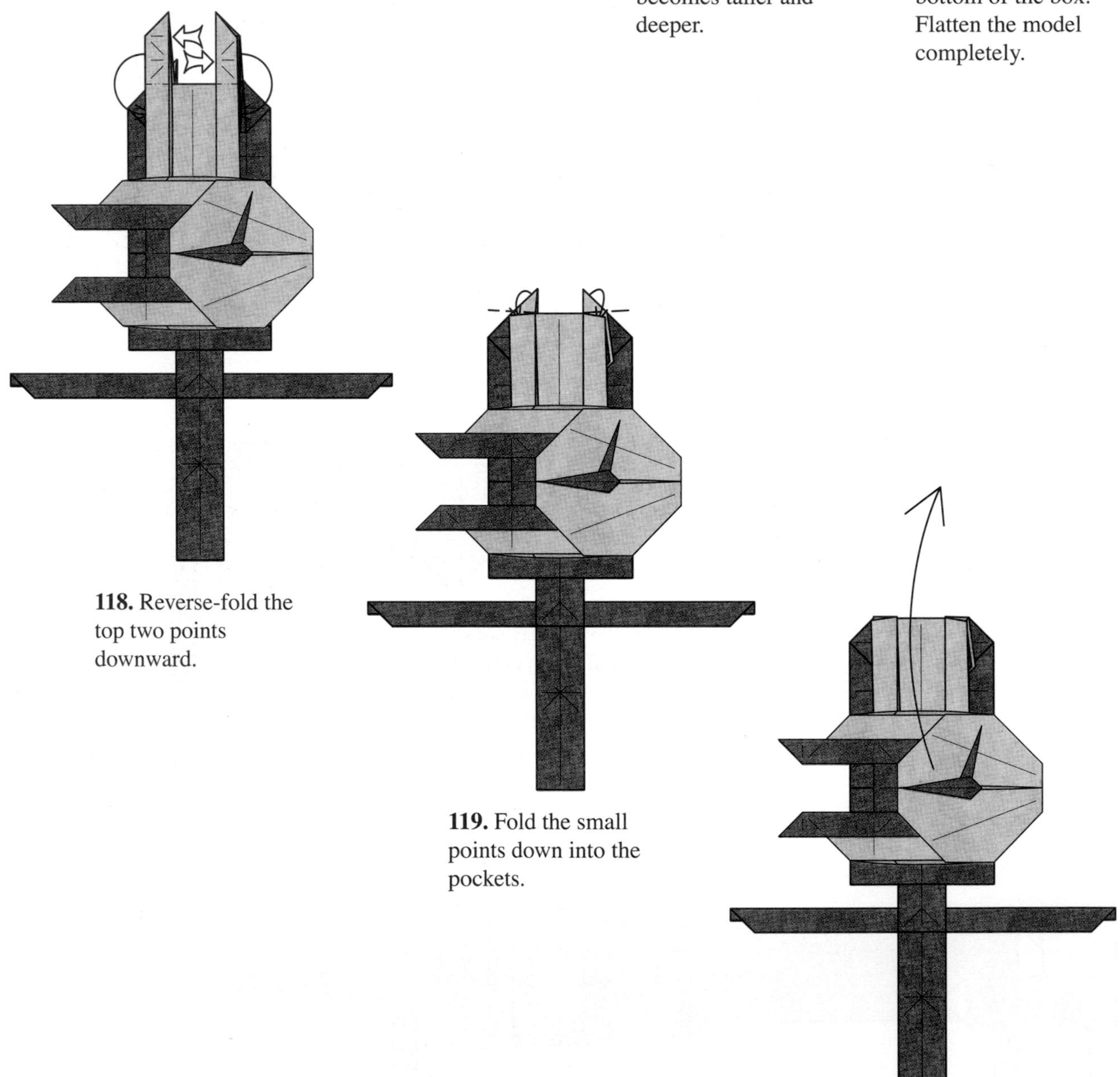

118. Reverse-fold the top two points downward.

119. Fold the small points down into the pockets.

120. Swing the clock face upward.

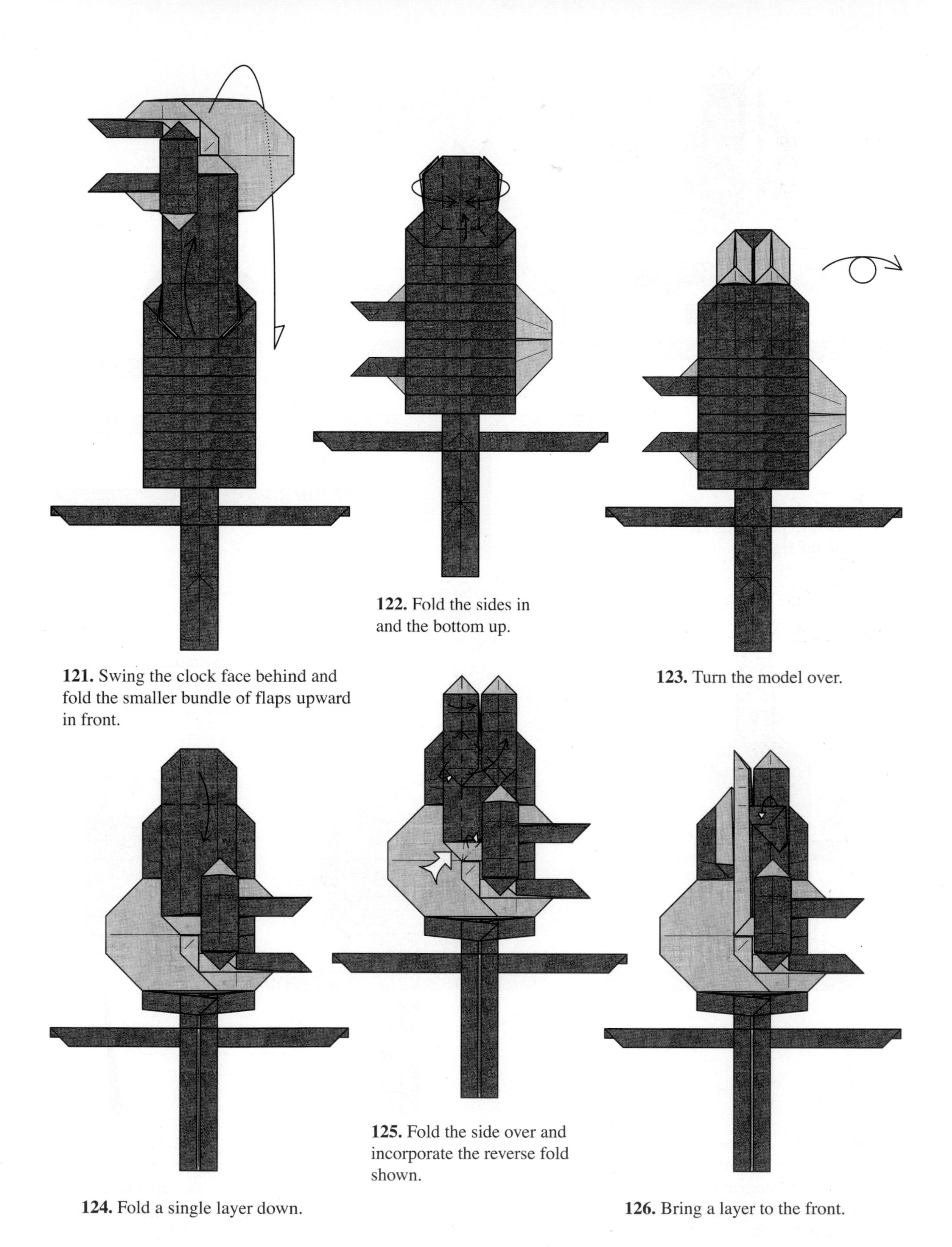

121. Swing the clock face behind and fold the smaller bundle of flaps upward in front.

122. Fold the sides in and the bottom up.

123. Turn the model over.

124. Fold a single layer down.

125. Fold the side over and incorporate the reverse fold shown.

126. Bring a layer to the front.

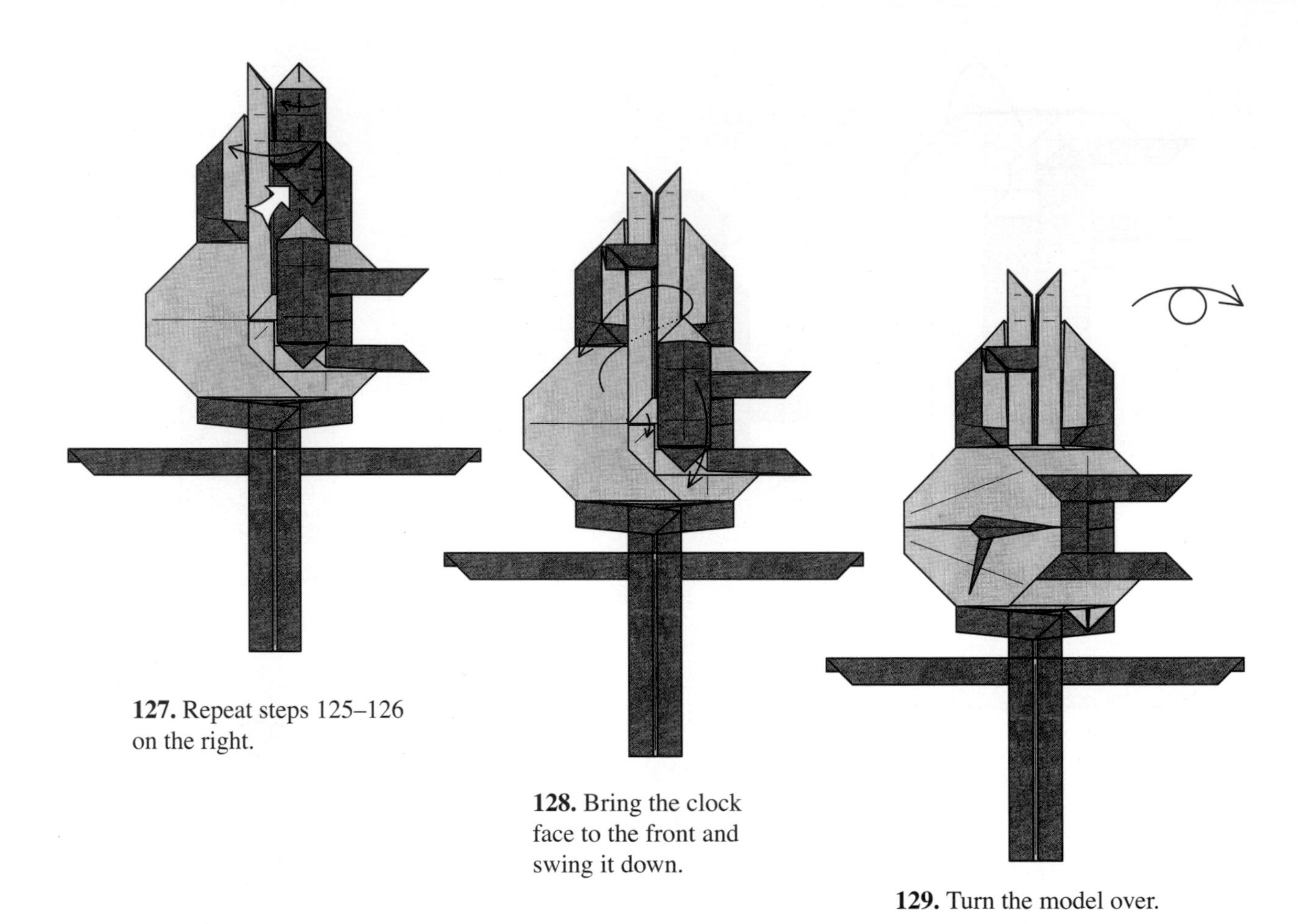

127. Repeat steps 125–126 on the right.

128. Bring the clock face to the front and swing it down.

129. Turn the model over.

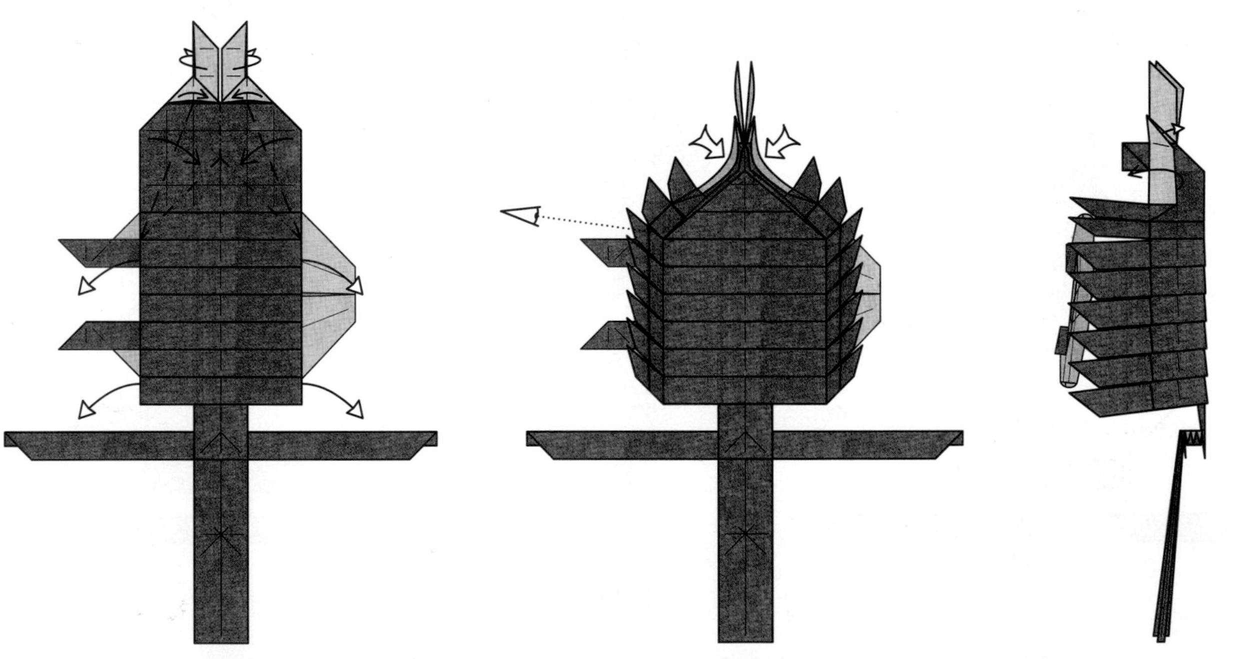

130. Pleat the sides of the top and fold the top in half. At the same time, pull the sides of the clock body out to stand at right angles to the clock back.

131. Squeeze the top of the model together and smooth out the layers along the roof.

132. Side view. Fold two layers over to the left and release the trapped paper at the top.

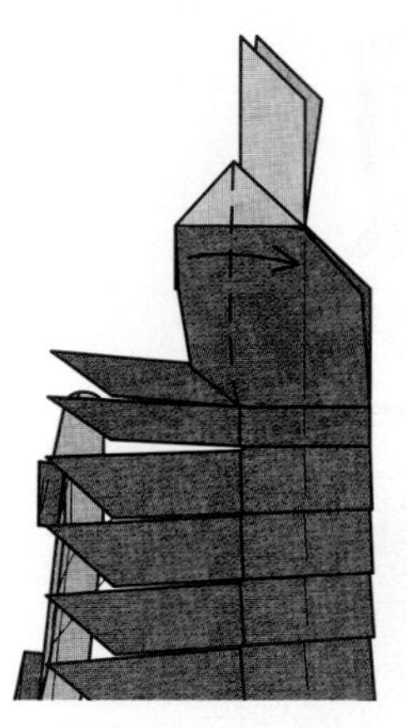

133. Close up again.

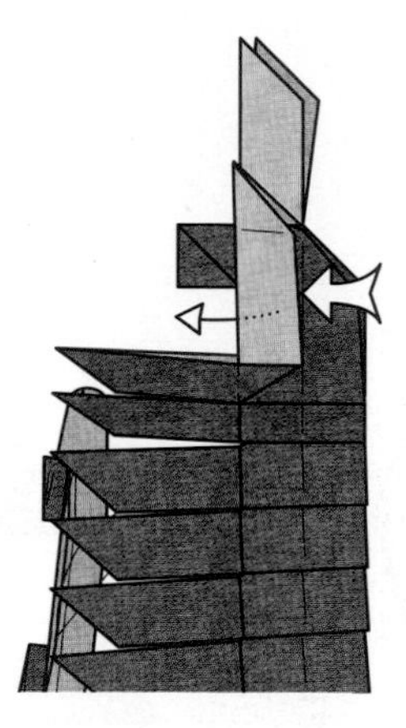

134. Closed-sink the edge.

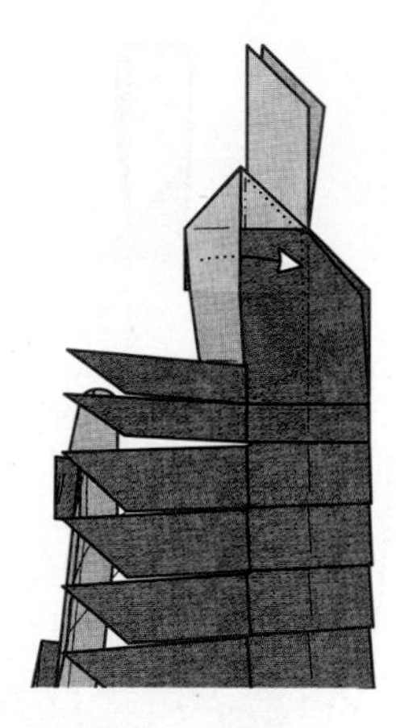

135. Pull out a single layer.

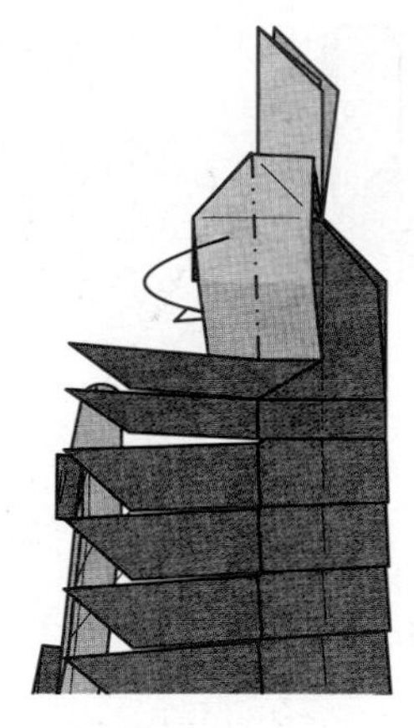

136. Mountain-fold the white layer inside.

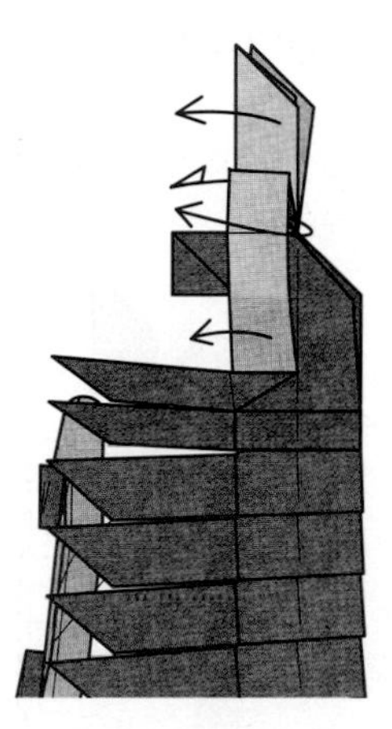

137. Unwrap the white hood and reverse-fold the bottom corner.

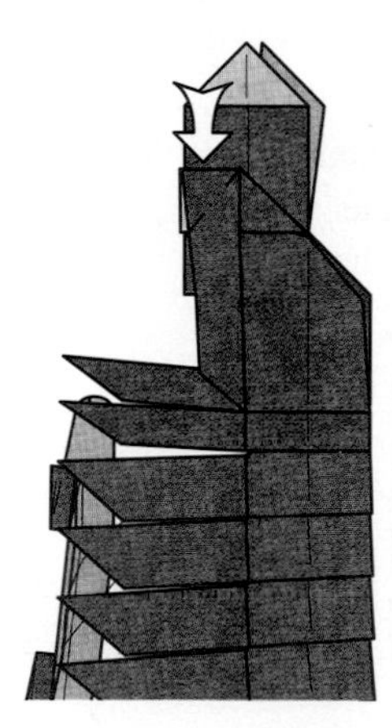

138. Reverse-fold the corner.

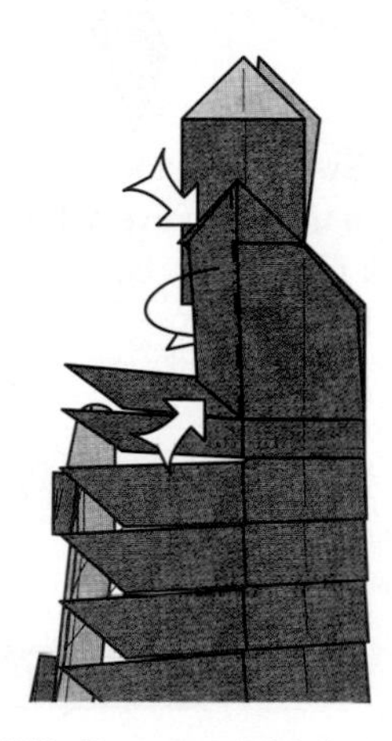

139. Reverse-fold the top corner and mountain-fold the edge inside.

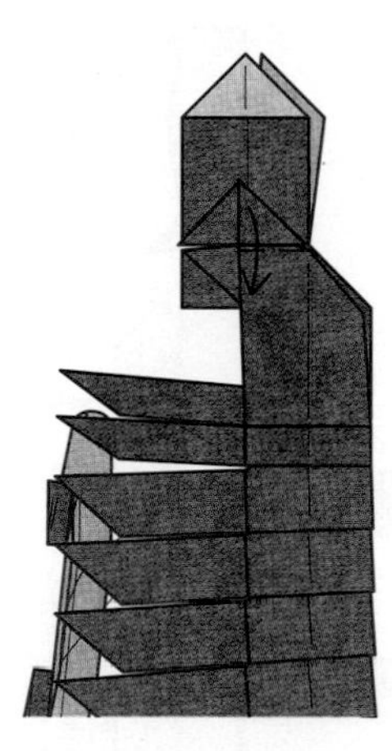

140. Fold the point down.

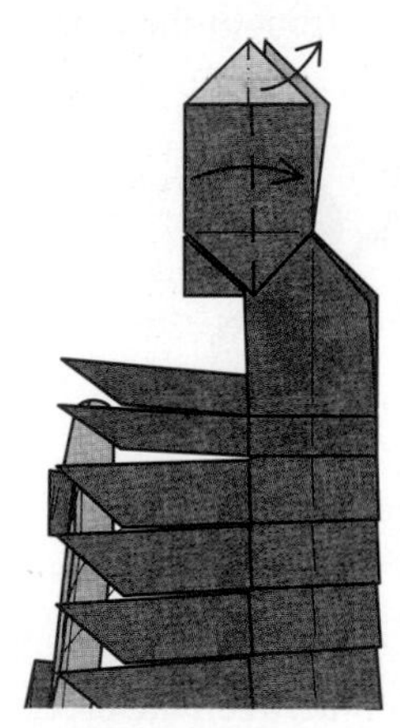

141. Fold the left edge over to the right and pull up the loose paper at the top to make a hood.

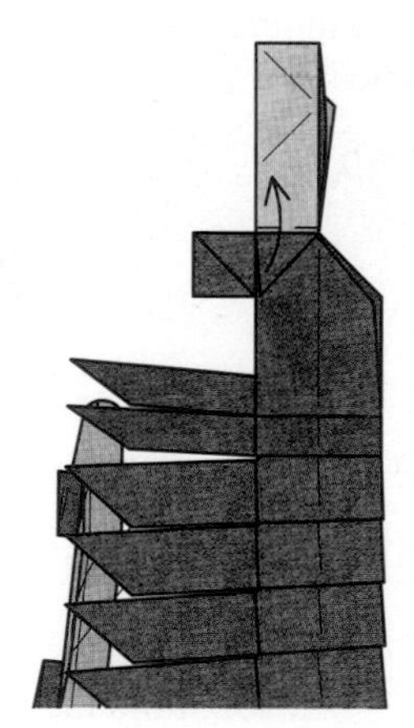

142. Fold the small colored point upward.

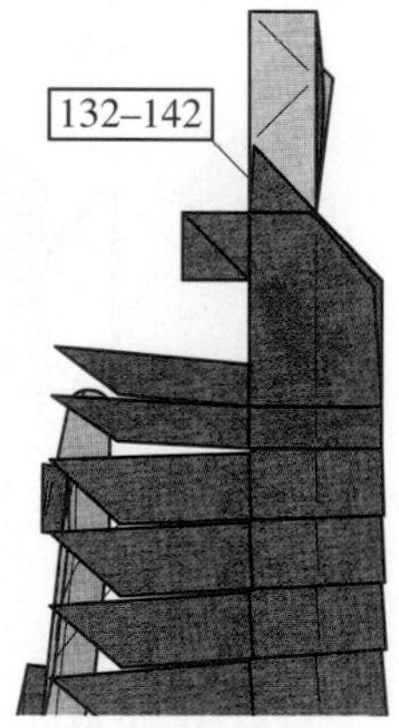

143. Repeat steps 132–142 on the far side.

144. Mountain-fold the white strip (which connects the clock face to the deer head) behind. (The back of the clock will interfere with this, but don't worry about that.)

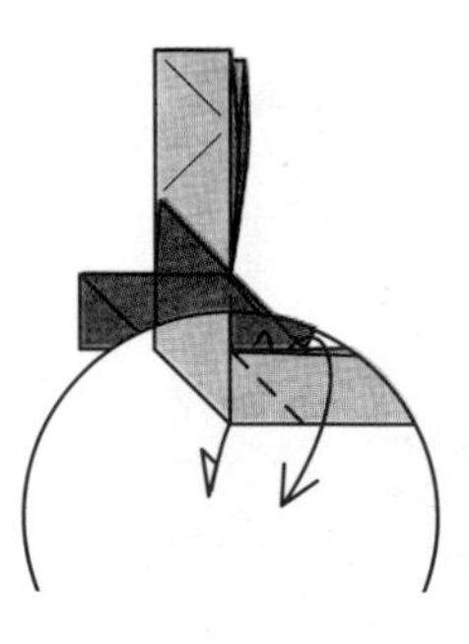

145. Outside reverse-fold the white strip.

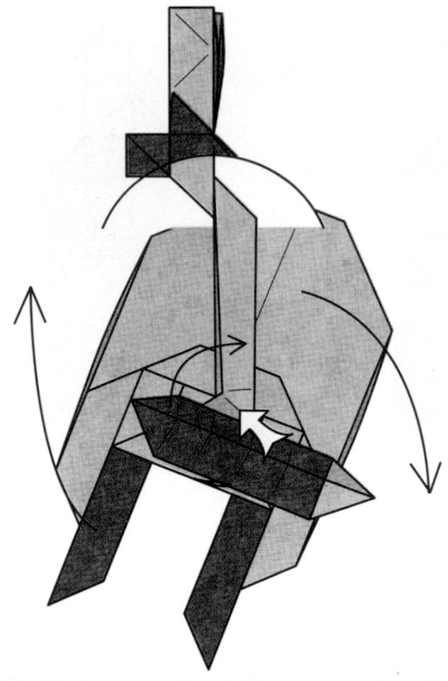

146. Swing the clock face around and flatten the strip; everything from the clock face to just below the deer's head will lie flat.

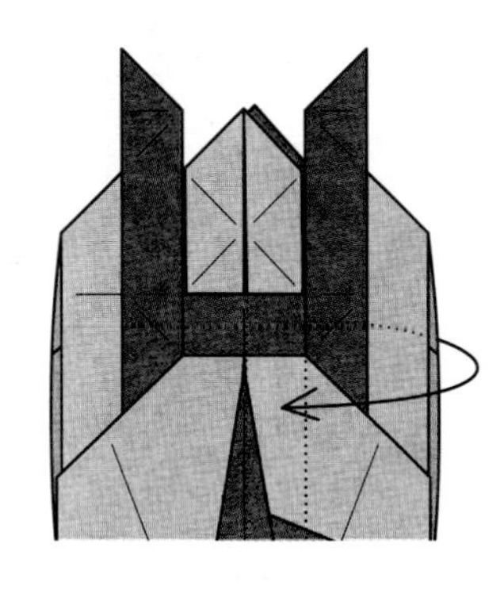

147. Fold the flap down.

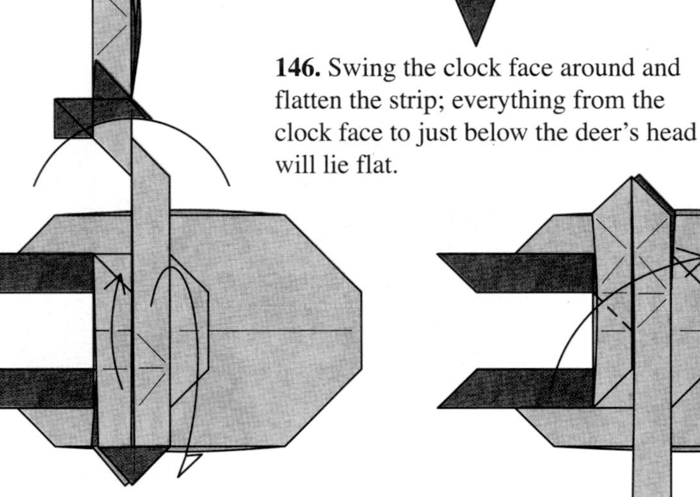

148. Swing the paired flaps upward and swing the rest of the model down between the paired flaps and the clock face.

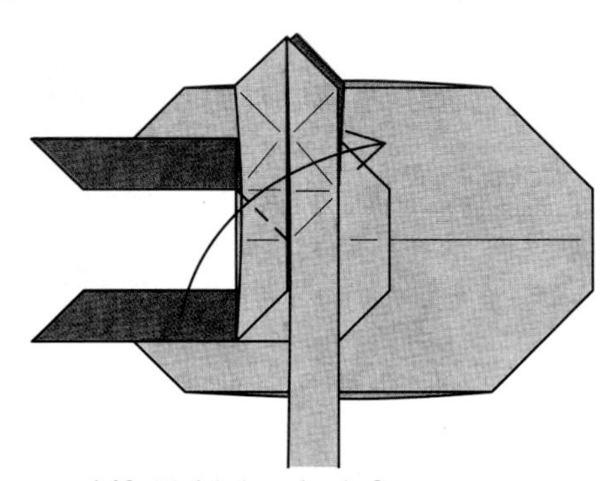

149. Fold the clock face up to the right, swinging it around from behind.

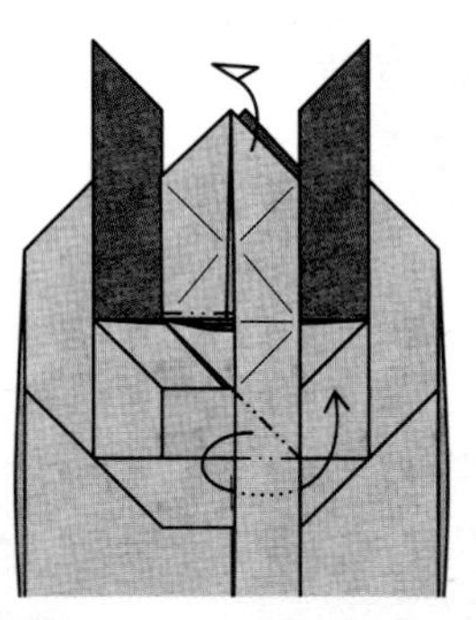

150. Bring the vertical strip in front of the clock face.

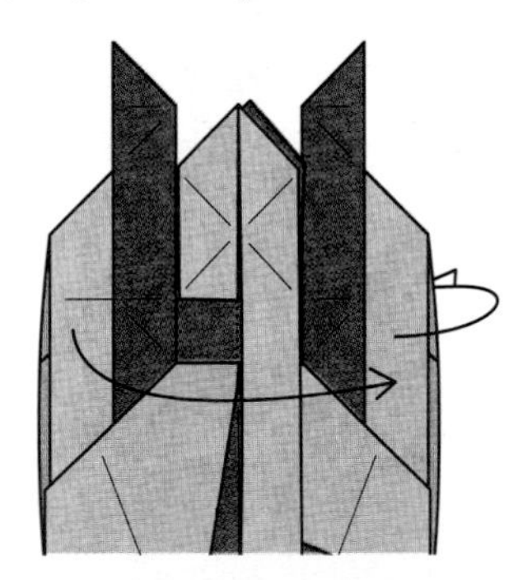

151. Flip the clock face around its center axis.

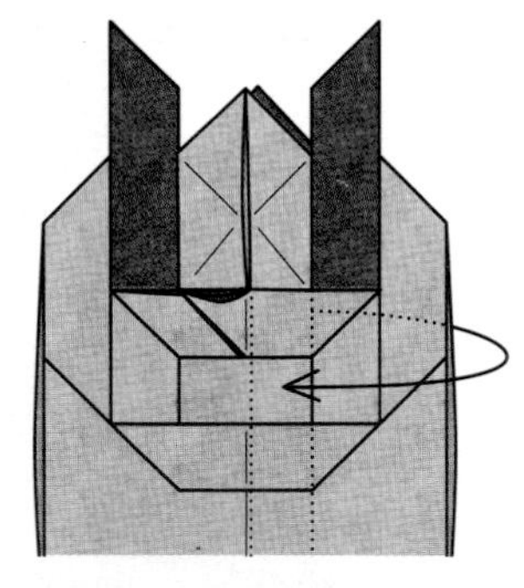

152. Again bring the vertical strip in front of the clock face.

153. Fold the connection between the clock face and deer head around and behind; at the same time, fold the paired flaps down through the opening in the clock face. At this point, the clock face no longer interferes with the clock body, but should sit more or less within the clock body.

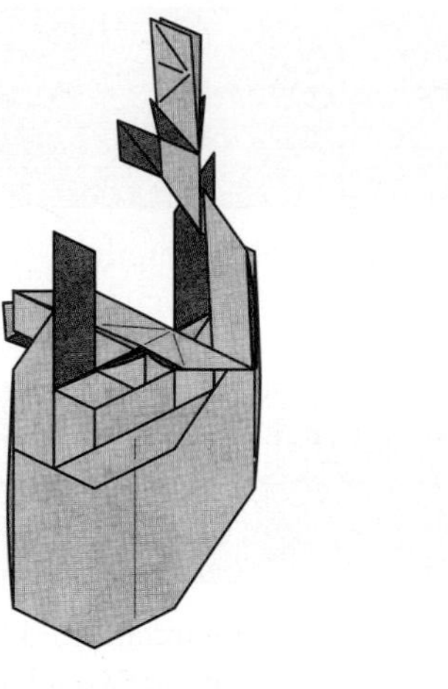

154. This view shows the relative positions of the clock face and the deer head. The clock body is not shown, and the view is of the back of the clock face. Turn over.

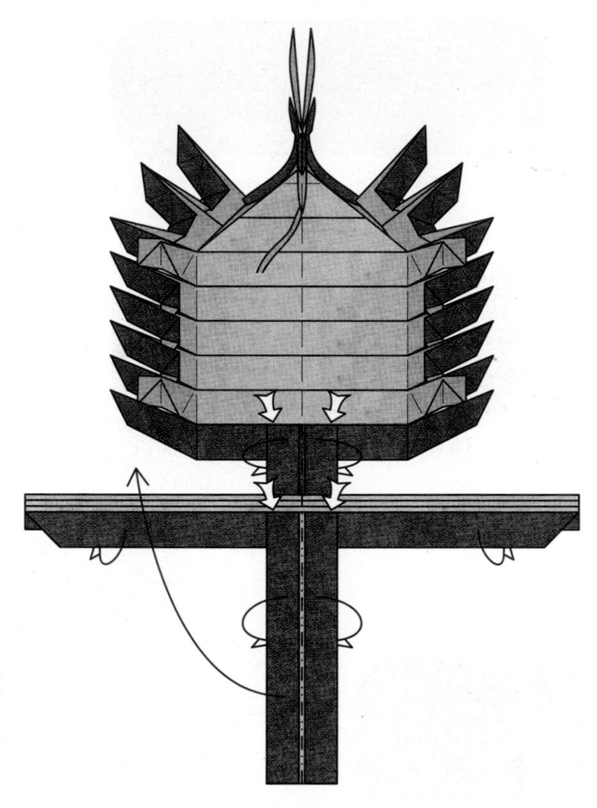

155. The clock face is not shown in this view. Mountain-fold the vertical part in half and swing the resulting flap up to stand out at right angles to the clock body.

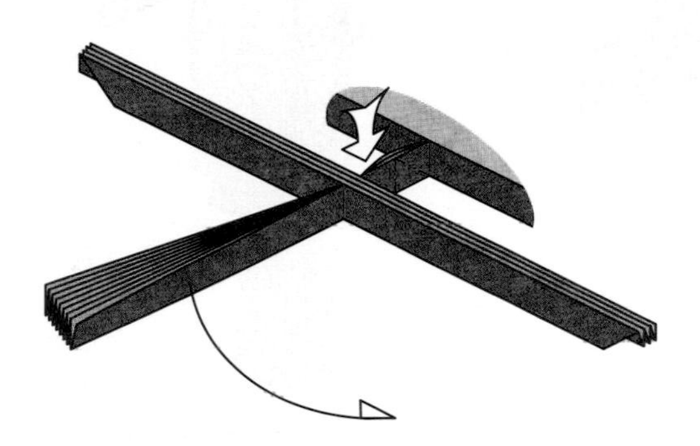

156. Crimp the T-shaped flap downward.

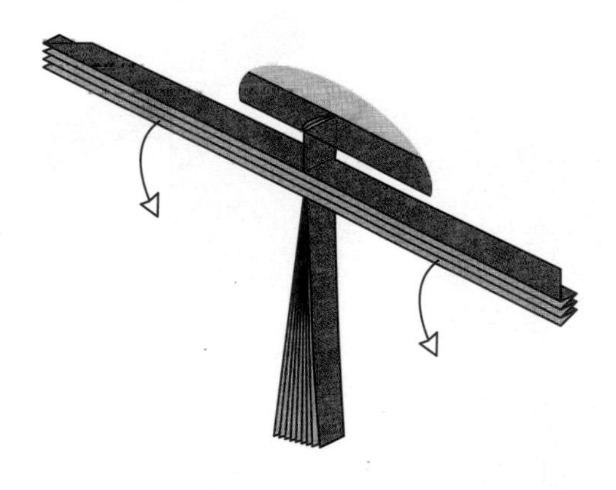

157. Open out the pleated part.

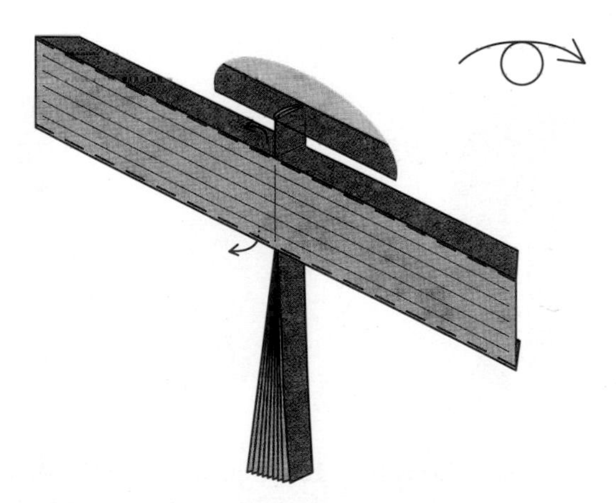

158. Fold the colored flaps along the top and bottom of the white region up and down, respectively; turn the model over. We will be working on the colored back side of the white pleated region for steps 159–165.

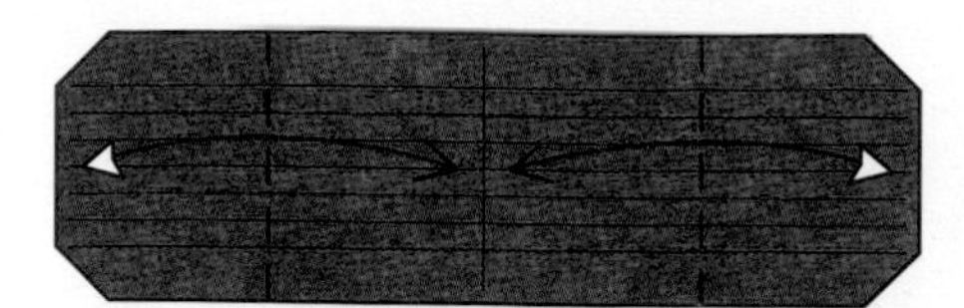

159. Fold and unfold.

160. Fold and unfold.

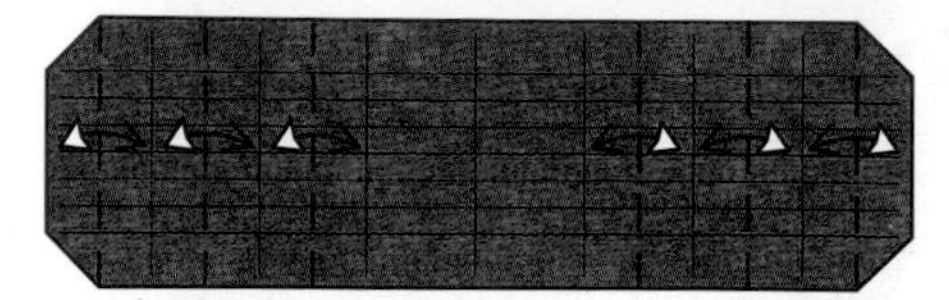

161. Fold and unfold.

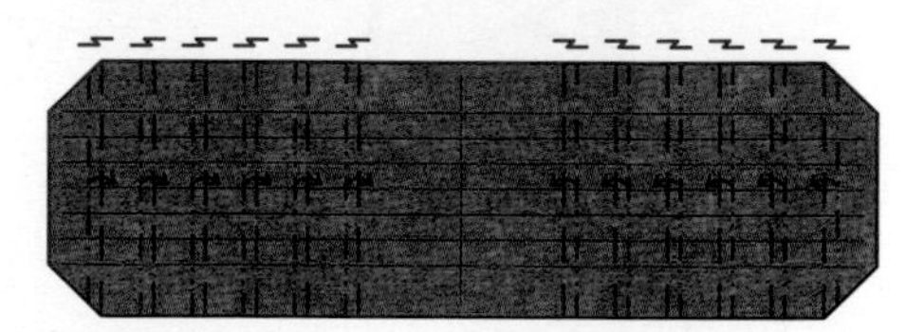

162. Pleat. Each mountain fold lies 1/4 of the way between the two adjacent valley folds.

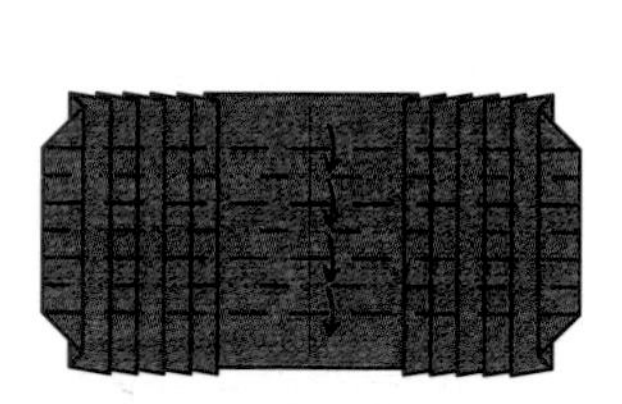

163. Refold the pleats you undid in step 157 and turn the model over again.

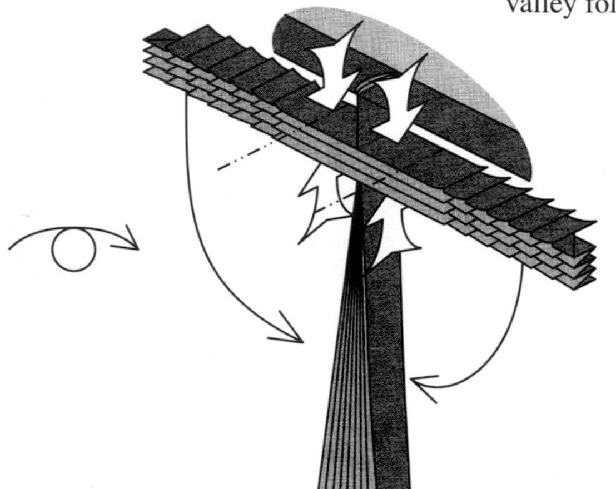

164. Pinch the pleated flaps at their base and fold them downward.

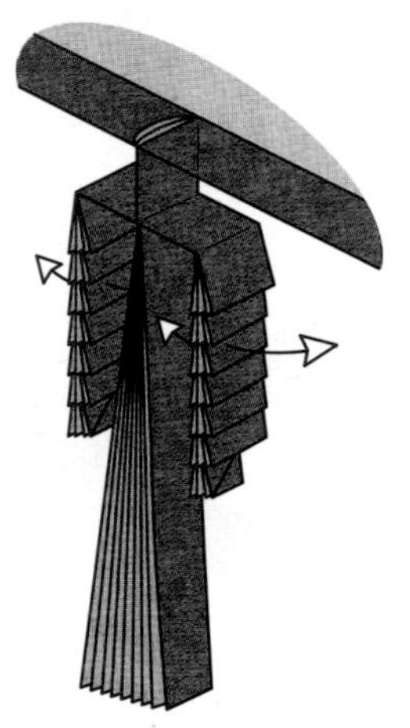

165. Spread the sides of the pleated flaps apart. They will become the pine cone weights.

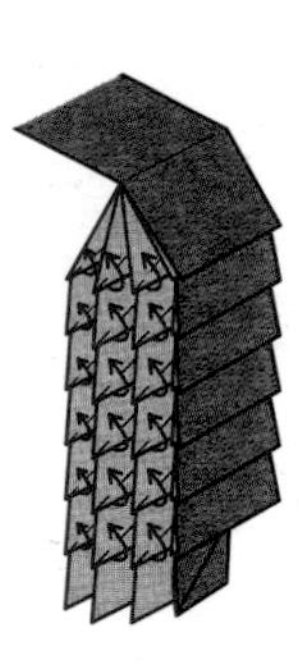

166. Reverse-fold each horizontal pleat upward between the vertical pleats. There are 18 such folds on each of the two pine cones.

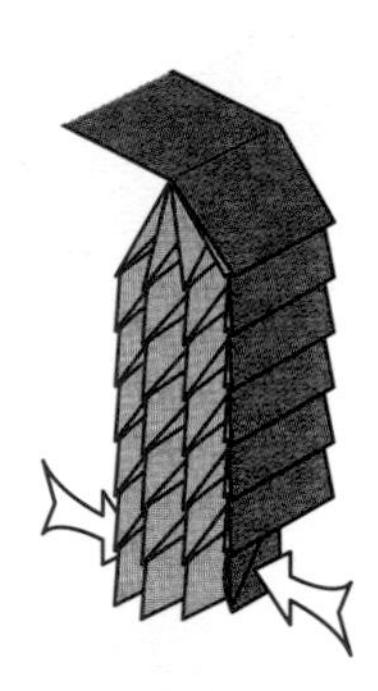

167. Squeeze the bottom of the pine cone together.

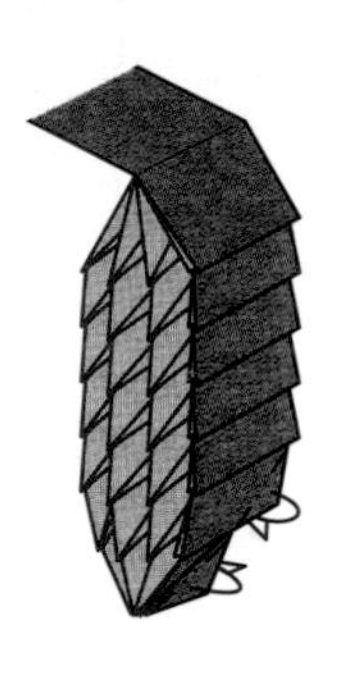

168. Mountain-fold the corners at the bottom of the pine cone to lock it. Repeat on the other pine cone.

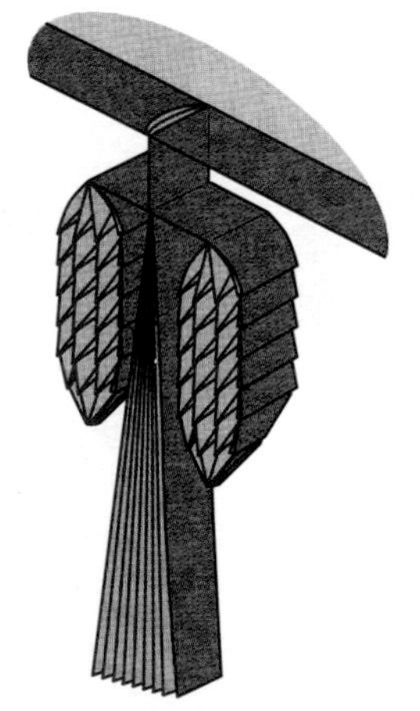

169. Adjust the position of the pine cone weights so that they hang straight.

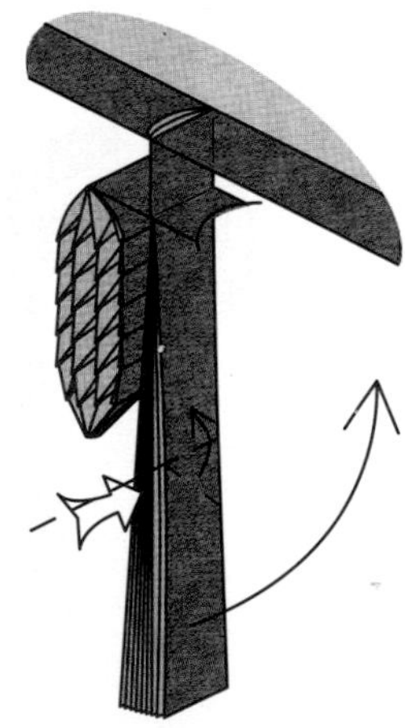

170. Squash-fold the bottom point symmetrically. The valley fold lies on an existing crease.

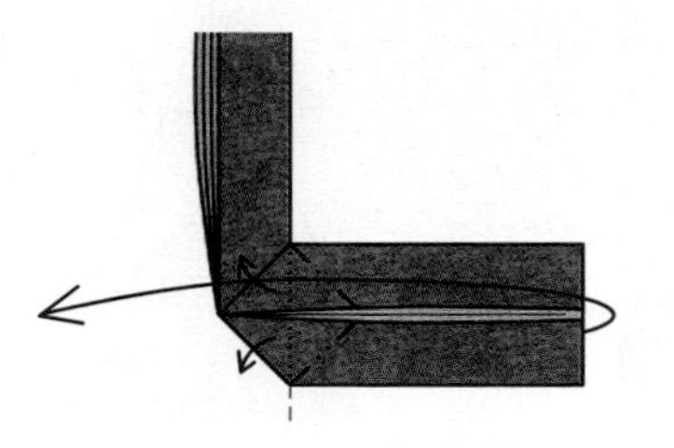

171. Squash-fold the flap over to the left and swing all of the layers outward.

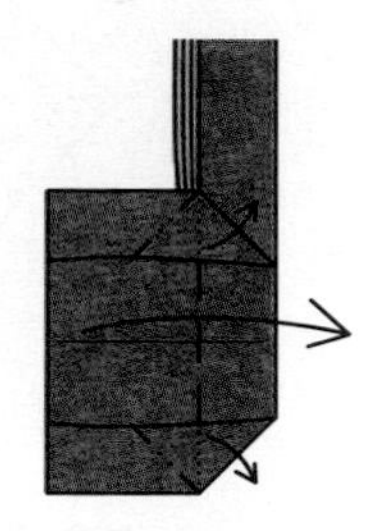

172. Squash-fold the flap over to the right.

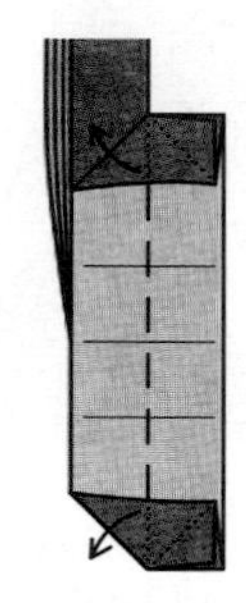

173. Again.

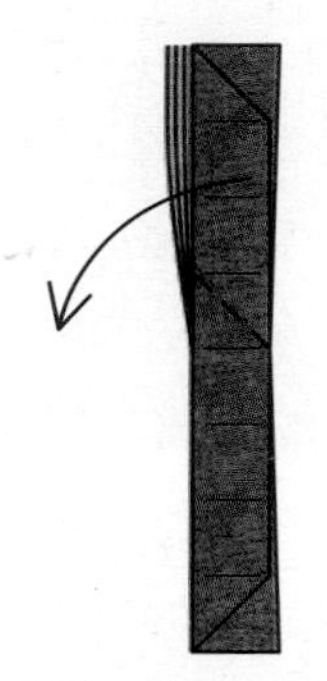

174. Valley-fold one point down to the left.

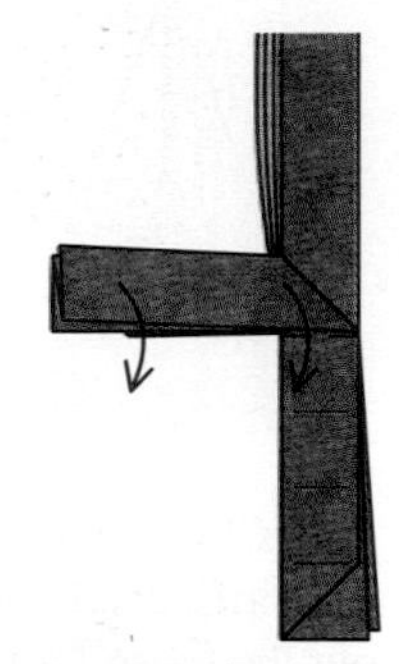

175. Valley-fold the flap downward.

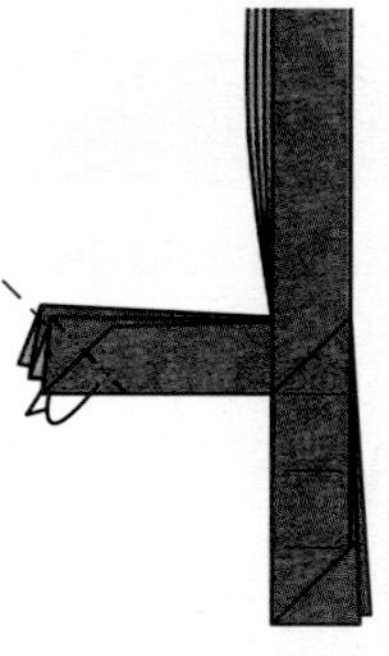

176. Fold the corners into the interior (fold the two far layers together as one).

177. Fold the bottom corners into the interior.

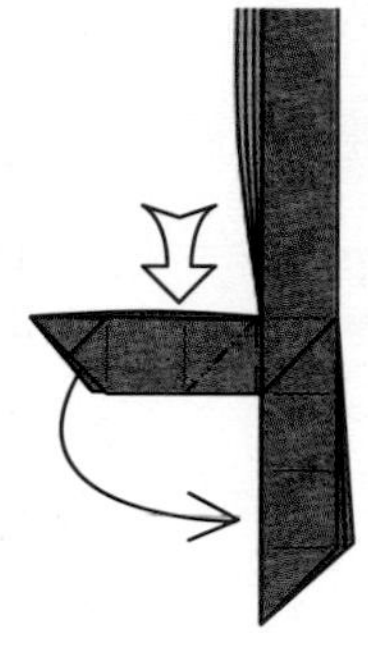

178. Squash-fold the point so that it stands perpendicular to the pendulum.

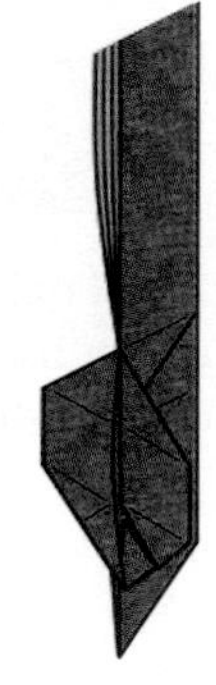

179. Like this (perspective view).

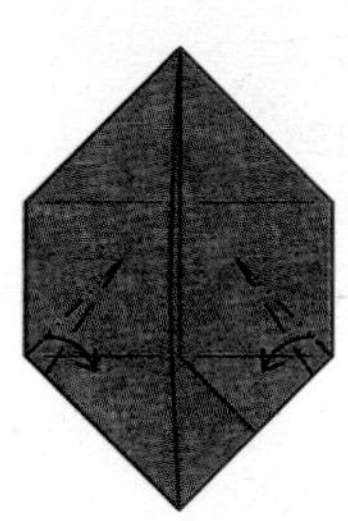

180. Crimp the sides of the leaf downward; it will not lie flat.

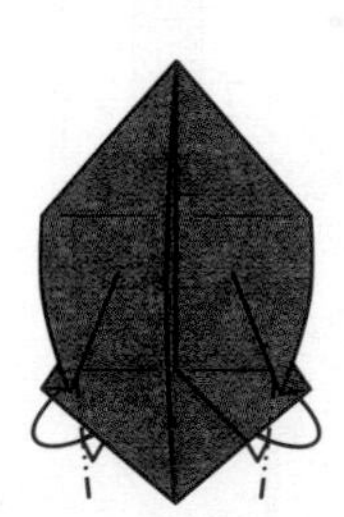

181. Mountain-fold the edges to lock the crimps into place.

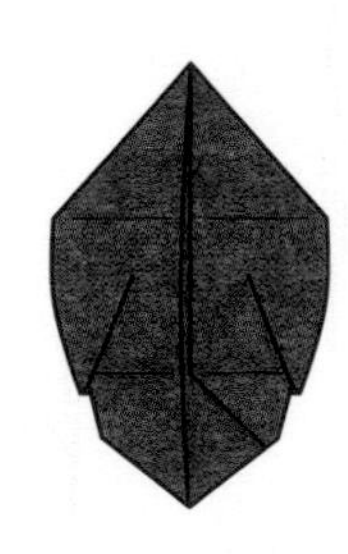

182. Like this.

183. The pendulum and pine cone weights are now complete.

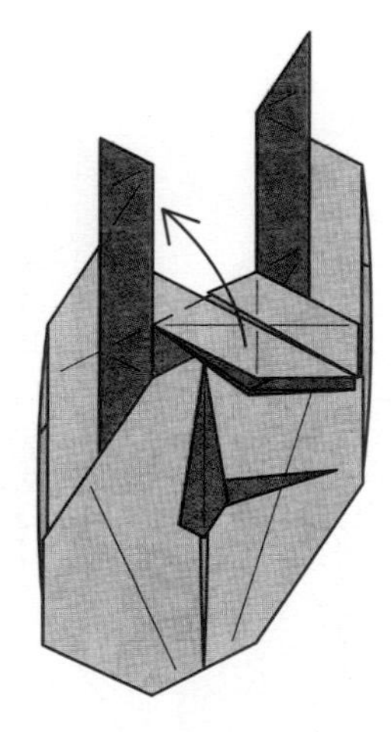

184. Now we'll work on the clock face. Fold one of the two flaps standing out from the face upward.

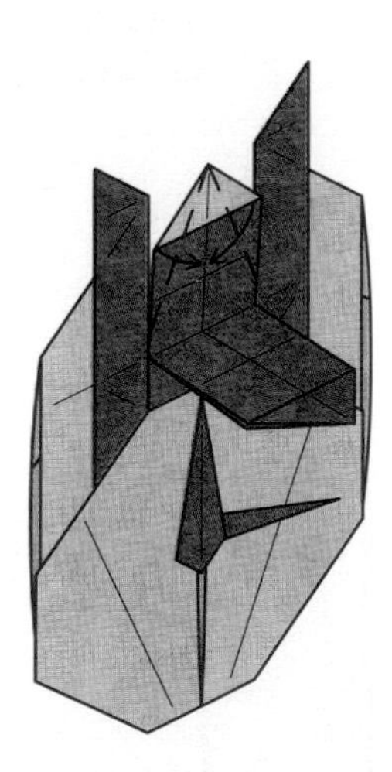

185. Fold the corners in to meet at the center line.

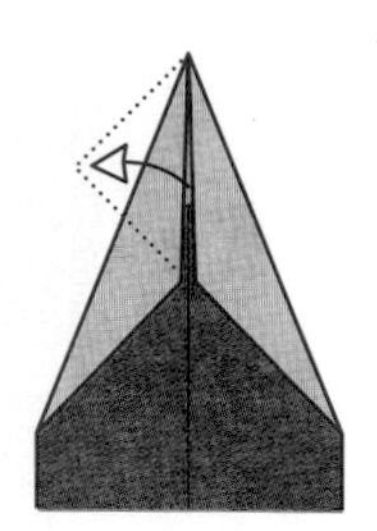

186. Close-up of the bird's head. Pull out the loose flap of paper.

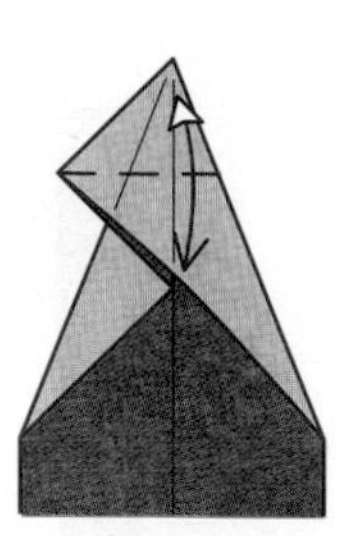

187. Fold and unfold.

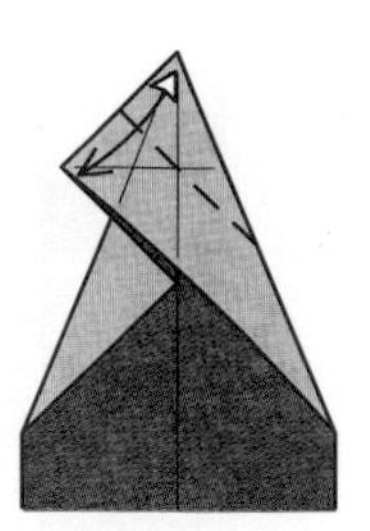

188. Fold and unfold.

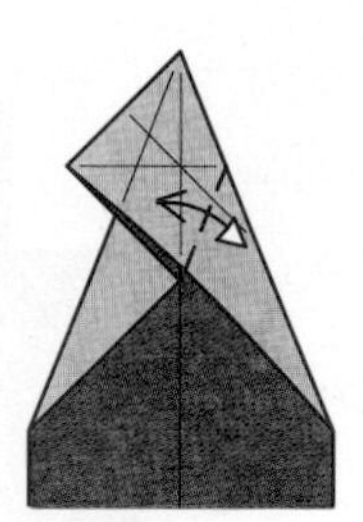

189. Fold and unfold.

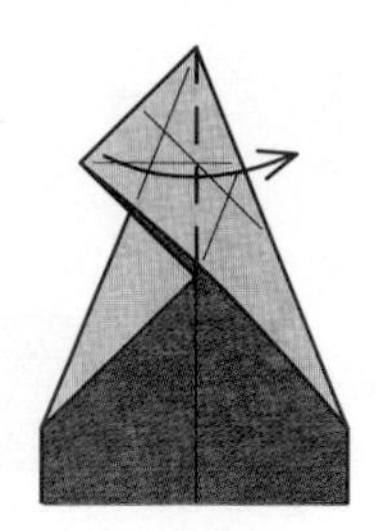

190. Fold the flap over to the right.

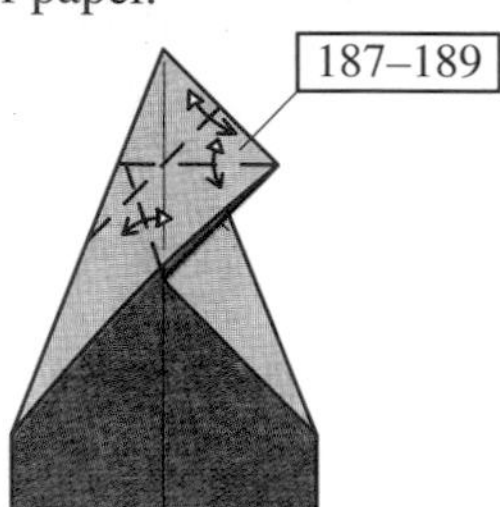

191. Repeat steps 187–189.

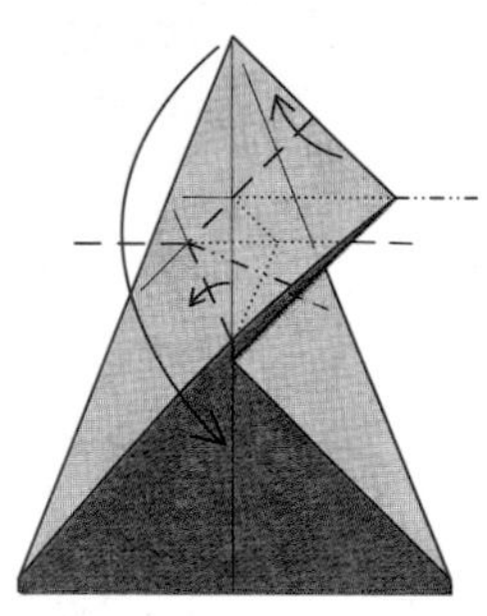

192. Fold the top point down, incorporating the creases shown.

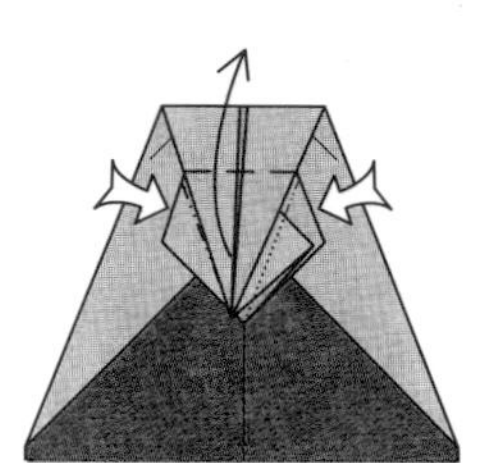

193. Petal-fold. Bring the two points upward together.

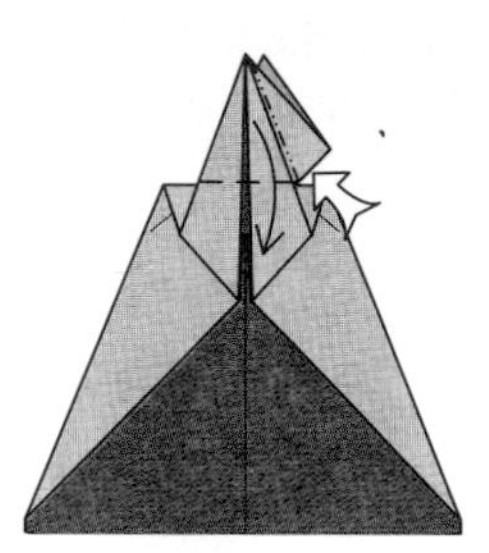

194. Squash-fold (like half of a petal fold).

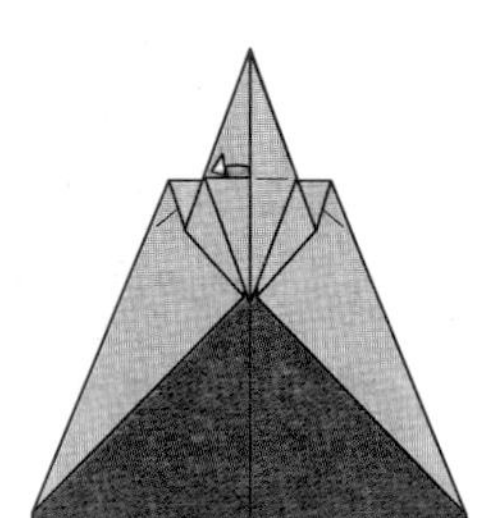

195. Pull out a single layer of paper.

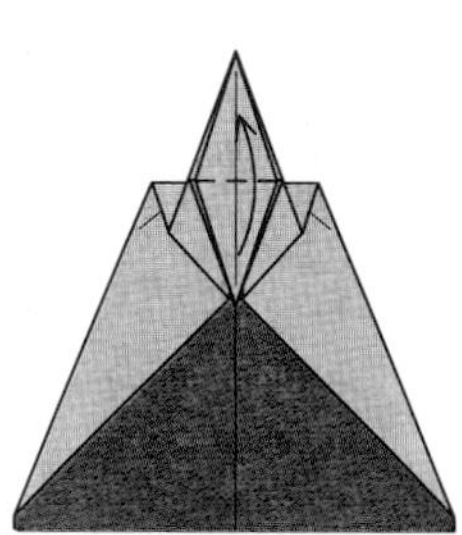

196. Fold the point back upward.

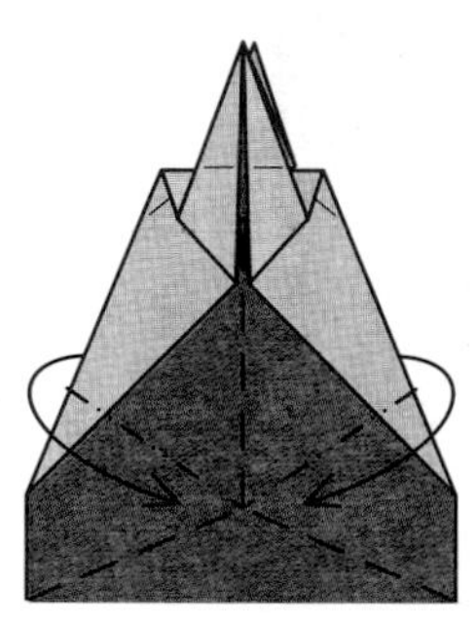

197. Fold a rabbit ear that stands straight out from the clock face.

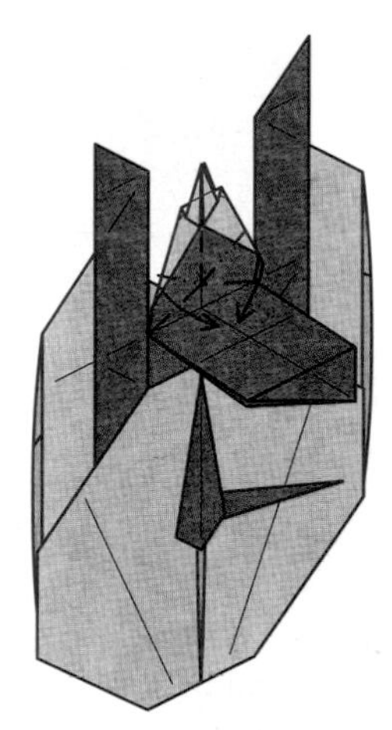

198. Perspective view of step 197.

199. Crimp the head downward.

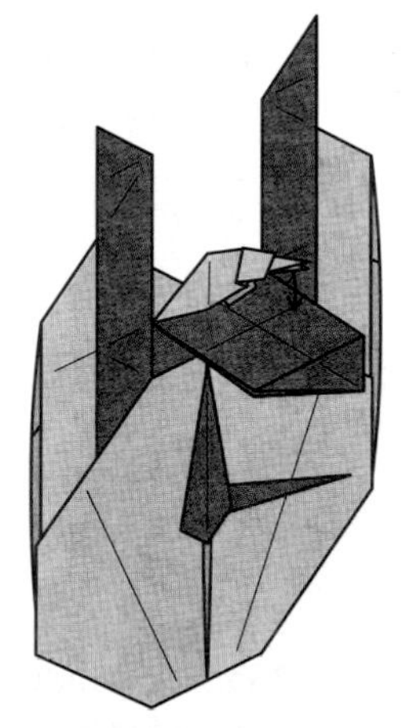

200. Slide the lower half of the beak downward.

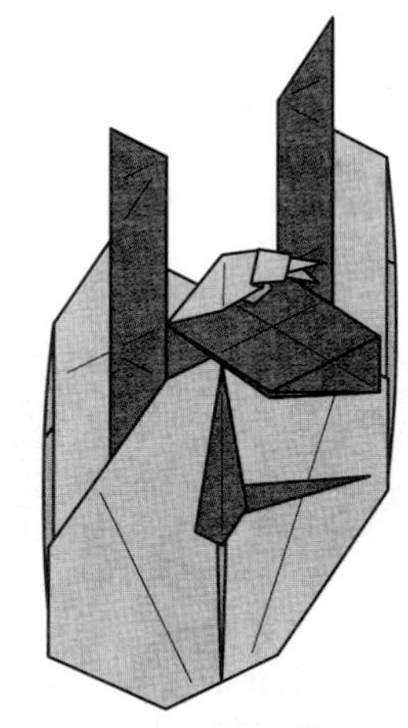

201. Finished bird. Now, we will attach the clock face to the clock body.

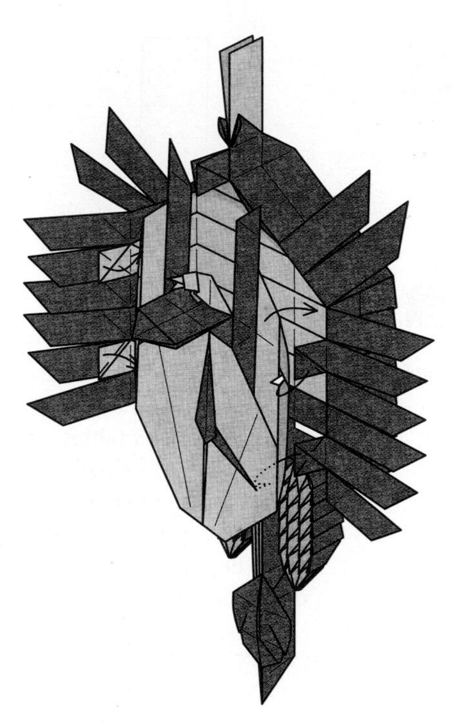

202. Place the clock face over the opening in the body. Tuck the four tabs along the sides into the slots in the clock face and hook them over the edges inside the clock face. This will lock the face securely to the body.

203. Push in the side of the top point and hook it around the uppermost leaf-flap, between the layers of the roof. The right side is shown completed.

204. This shows the entire model thus far. Now, we will work on the deer's head.

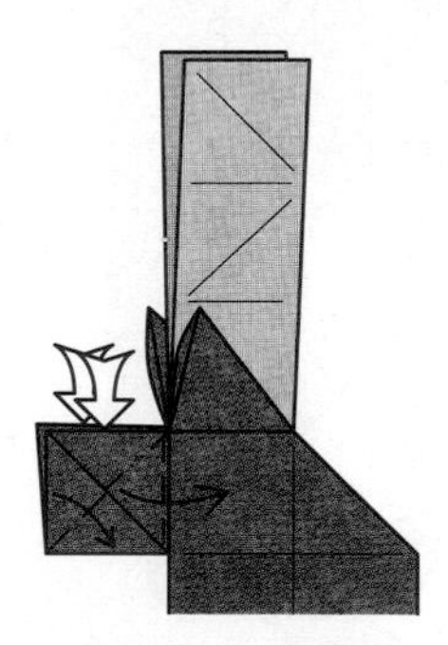

205. Squash-fold the top of the head, wrapping the excess paper around the thicknesses below the ears. Repeat behind.

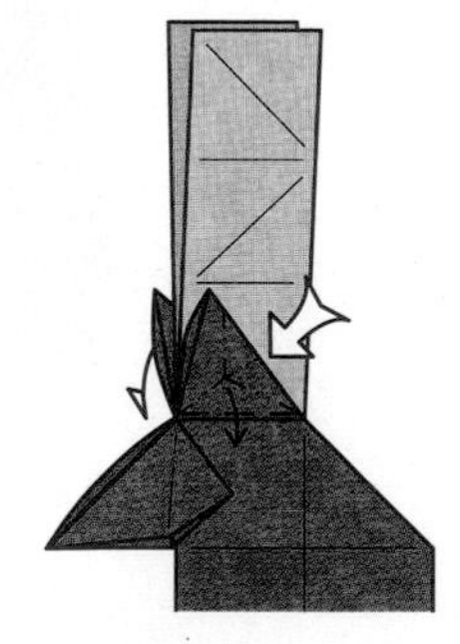

206. Push in the back of the ear (which is quite thick) and slightly squash- and petal-fold it, leaving it three-dimensional. Repeat behind.

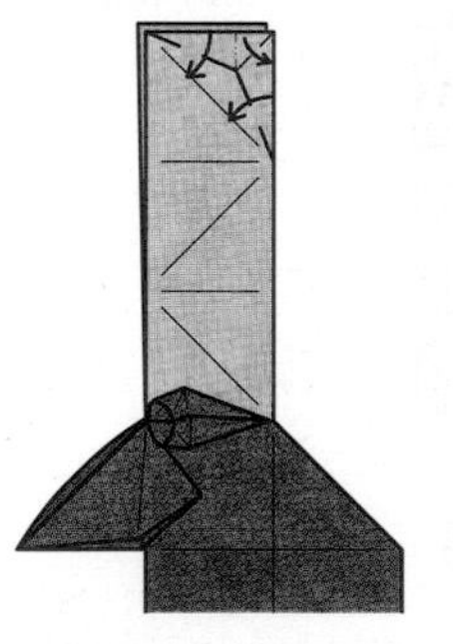

207. Fold a rabbit ear from the top of the white flap (which will become an antler).

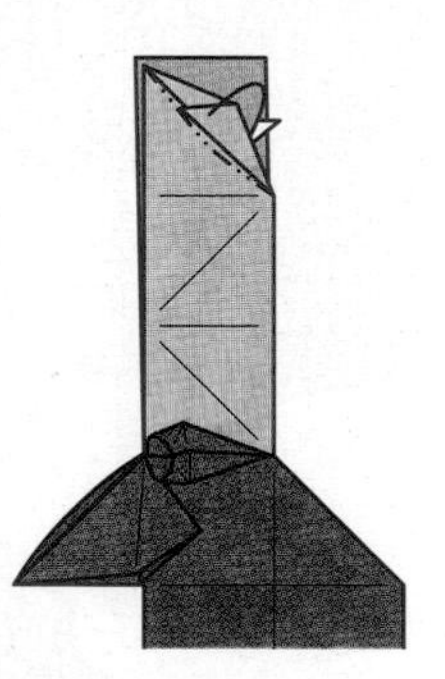

208. Mountain-fold the point behind.

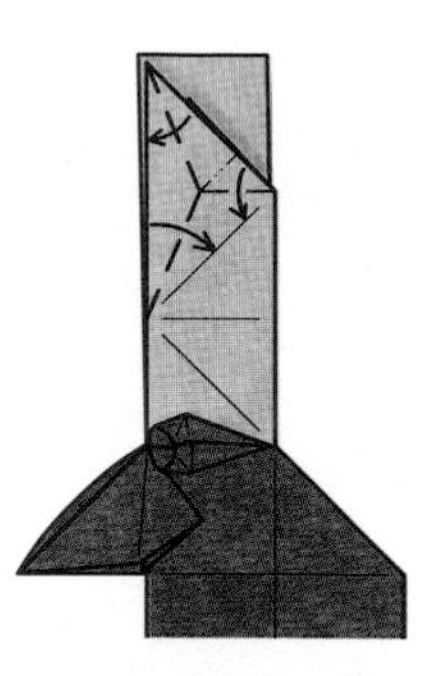

209. Fold another rabbit ear.

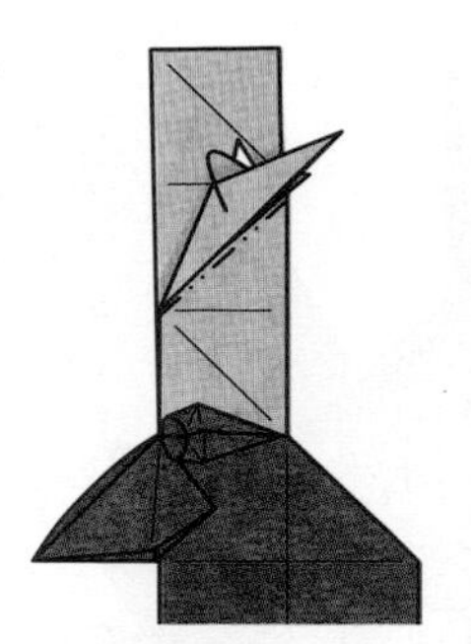

210. Mountain-fold the rabbit ear to the rear.

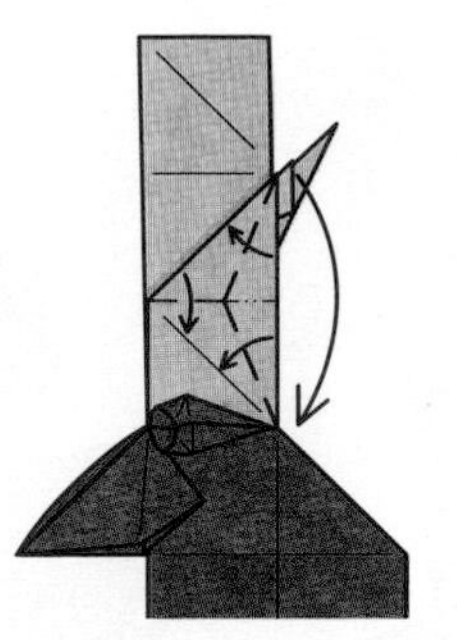

211. Fold another rabbit ear.

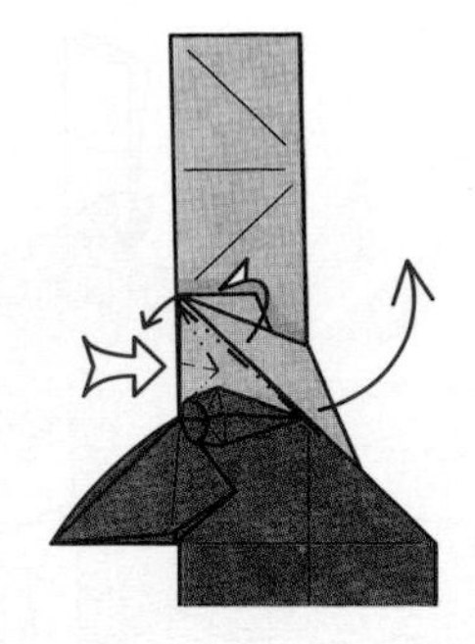

212. Push in the front of the antler (at the left) by making a partial rabbit ear. At the same time, lift up the points in back (at the right) and spread them out. They will be three-dimensional.

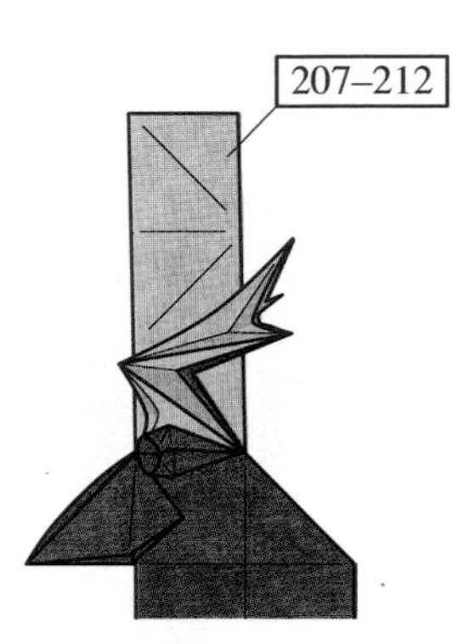

213. Repeat steps 207–212 on the other antler.

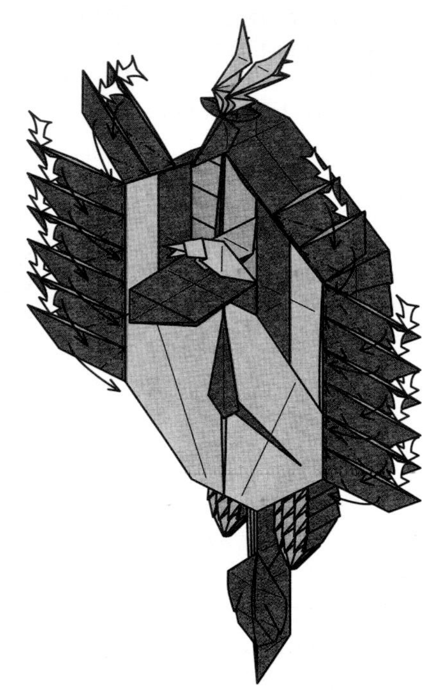

214. Squash-fold each of the flaps downward (they will be the leaves). Offset each squash fold, so the leaves alternate and overlap each other.

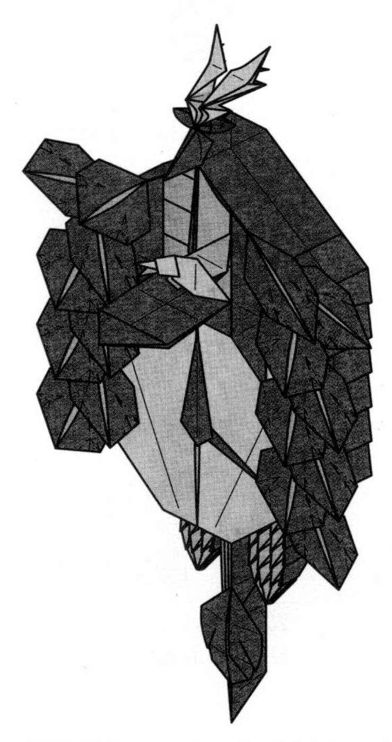

215. Crimp and swivel-fold each leaf as you did in steps 180–182.

216. Finished Black Forest Cuckoo Clock.

13 Hybrid Bases

Box pleating is a specialization of the circle/river method of flap generation and uniaxial bases, but it is also a way of extending them; it allows one to easily combine two-dimensional flaps and three-dimensional objects in the same model. It also illustrates a general principle of origami design: that one can mix and match different styles and techniques in the same model, using particular design elements where they are needed. Many—perhaps most—origami designs are of this hybrid type. Circle packing techniques are powerful, but focusing solely on flap generation can be limiting; there are only so many subjects out there with 23 pairs of appendages. Nearly all of the techniques I've shown so far are based on the concept of a uniaxial base, but there are many potential origami subjects that do not fit into the stick-figure abstraction that is part and parcel of the uniaxial base.

When faced with a subject that does not fit into the uniaxial mold, rather than starting over from scratch, one can often adapt elements of uniaxial bases and combine them with other folding techniques to form a *hybrid base*, one that provides both a better representation of the chosen subject and a more visually interesting physical structure.

The question then arises: In a hybrid structure, for what should we use circle packing and tree methods? All of these techniques are good for generating flaps, particularly long, skinny ones. A subject that is composed primarily of long, skinny appendages is a perfect candidate for a pure circle/river or box-pleated design. But then the counter-question also arises: For what should we NOT use circle packing? And the answer is, anything that isn't approximated reasonably well by a stick figure. Most notably, large, flat regions are not usually produced by a circle-packed base, because the process of maximizing the length of a flap often minimizes its width. Additionally, large, bulbous, three-dimensional shapes do not typically arise from circle packing, first, because uniaxial bases lie flat by design, and second, because (again) the process of optimizing the length of flaps tends to cut down on their width. With subjects that have large, two-dimensional expanses of surface, other techniques must be employed. Attempting to design such a subject using circle packing is akin to using a pair of pliers to pound nails: It can be done, but the results are often unsatisfactory.

However, circle packing can have a place in such a design, if you use it when it's appropriate. In a design that combines large, flat expanses with many narrow flaps, you can allocate polygons of paper for the flat regions and then tie them all together with regions of circle packing to generate the required flaps.

13.1. Flats and Flaps

Here is an example of this hybrid approach. While circle packing is ideal for the design of insects and other arthropods (as you might expect from the many arthropodic examples I've shown), it does not work particularly well for a butterfly or moth. In the members of the order *Lepidoptera*, the wings are the dominant structure in the model; indeed, for many years, the only origami butterflies consisted of wings only, plus, perhaps, a few crimps and/or blunt points to suggest a body. Legs and antennae were not even considered.

As the new geometric design techniques were discovered during the early 1980s, however, several folders cast their eye on the butterfly for its unique challenge: how to create large wings, plus small body, legs, and antennae (and, in some cases, even faceted eyes and proboscis!). Artist and architect Peter Engel devised the first (and still perhaps the best in terms of its usage of the paper); by the end of the 20th century, several other folders, including myself, had followed in his path.

The problem of combining large flat wings with small legs and features provides a nice challenge. Both butterflies and moths have four large wings, but since the fore and hind wings inevitably overlap, one always has the choice of representing the pair by one or two distinct panels of paper.

The observation that all four of a butterfly's wings are roughly triangular suggests one approach: Create each wing flap from a folded-in-half square region, as shown in Figure 13.1. We will allocate four such squares (one for each wing) at

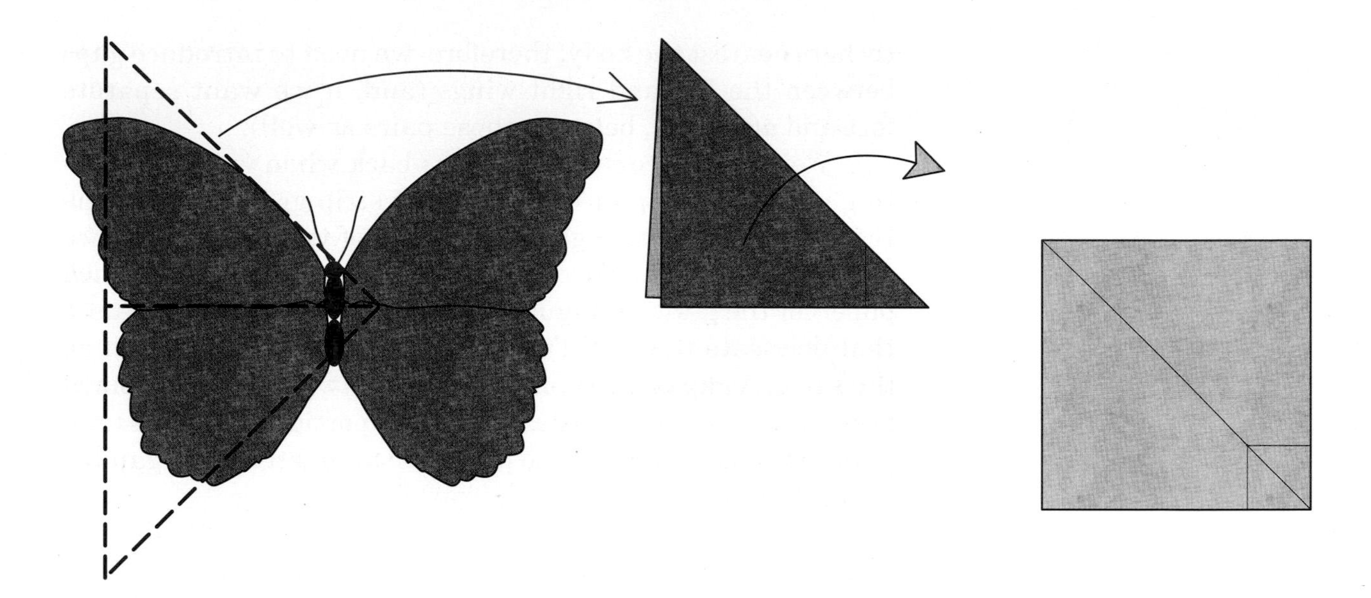

Figure 13.1.
The four wings of a butterfly are roughly triangular; each can be folded from a square region of paper.

each of the four corners of the square as in Figure 13.2. The rest of the paper between the four wing flaps is then available to create head, thorax, abdomen, antennae, and legs.

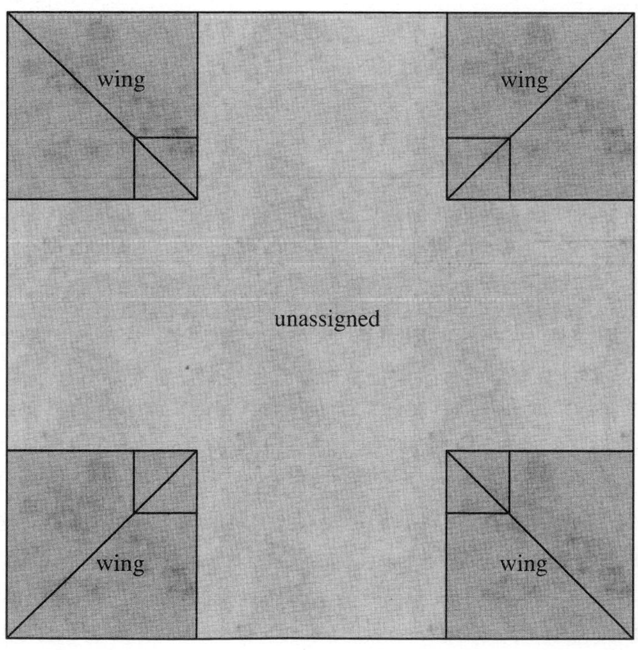

Figure 13.2.
The four wings can be obtained by placing the four wing-squares in the four corners of the paper.

Now, having assigned the four corners regions to become wings, what to do with the rest of the paper. We will need flaps, of course; but just as important: We need gaps. The four triangular wings of a butterfly are joined to each other only at the corners nearest the body; therefore, we need to introduce gaps between the left and right wings (and, if we want separate fore and aft wings, between those pairs as well).

We saw how to introduce gaps back when we were splitting points in Chapter 6; we added a strip graft between the regions that needed a gap. The width of the strip was twice the depth of the gap. We can do that here using the unassigned paper for the graft. In Figure 13.3, I've added diagonal creases that delineate the gap. I've also added half-circles, which do the same. A gap can be considered to be two half-flaps, joined at their base; consequently, we can use portions of circles (and portions of molecular crease patterns) to construct the gaps as well as the flaps.

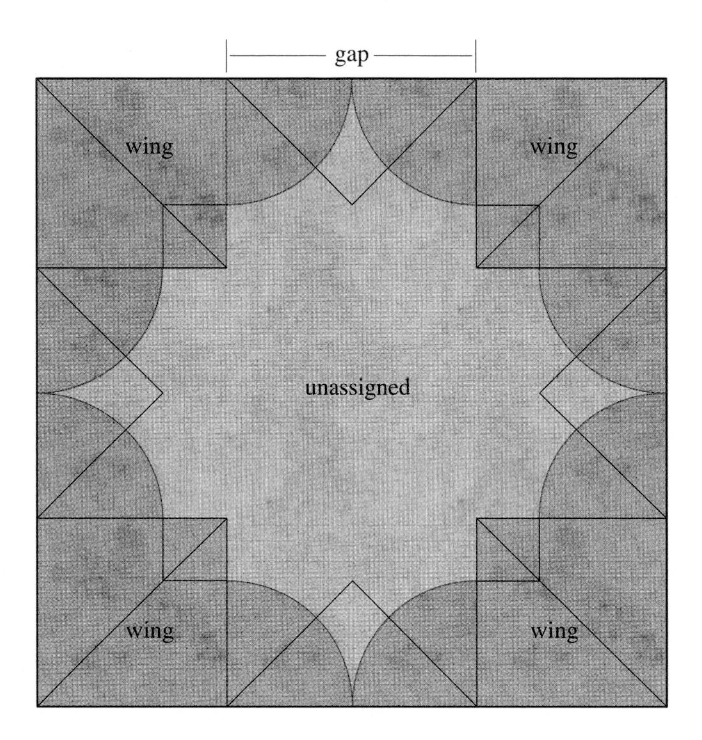

Figure 13.3.
Between adjacent pairs of wings, we introduce gaps (pairs of half-points). The paper required to form the gaps is indicated by the half-circles.

In this model, I've made the gaps two-thirds of the length of the side of the wing triangles. It's possible, of course, to extend the gaps all the way to the tips of the wing triangles, but if I extend it only partway, then I can use the corners of the wing triangles in a different way, as four points of the cluster of points forming body, legs, and antennae.

The head, legs, and abdomen all emanate from the same point. To a reasonable approximation, the antennae can also be treated as emanating from the same location, which means that all ten flaps—two antennae, six legs, abdomen and head—can be represented by a simple circle packing. But we will now require that the circles not just fit into the square, but that they not intrude into the wing regions—at least, not beyond the circles that delineate the gaps.

A bit of manipulation reveals that nine circles fit neatly into the space available, as shown in Figure 13.4. Unfortunately, that's one circle too few. The obvious next step is to reduce the circles and rearrange them to add a tenth circle. But the nine-circle packing is so elegant, it would be nice to find a way to make use of it. Rather than rearranging, we can jettison one flap; a separate flap for the head isn't really needed if we use the presence of the two antennae flaps to suggest a head. This means that a final packing of circles for the appendages and the square facets for the wings can be solidified as shown in Figure 13.4.

Figure 13.4.
Circle packing for butterfly with flat regions allocated for the wings.

In this packing, all of the axial creases are orthogonal, which suggests that a box-pleated crease pattern is possible, and indeed it is. We have a choice of how many divisions to use in the box-pleated sections. In the published version of this model, I chose to use 12, as shown in Figure 13.5 in the crease pattern, base, and folded model.

The same packing and arrangement of flaps can also be folded using more divisions, which give narrower flaps. You might enjoy the challenge of working out what the crease pattern (and folded result) would be using 20 divisions, rather than 12, in the box-pleated portion.

Note that our nine flaps would ideally come in the form of four symmetric pairs of flaps for legs and antennae, plus a single

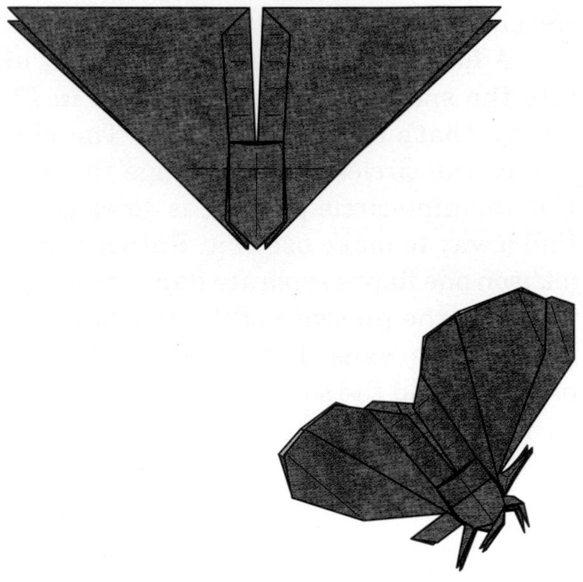

Figure 13.5.
Crease pattern, base, and folded model of the Butterfly.

flap on the line of symmetry for the thorax. In this pattern, three of the flaps fall on the line of symmetry, which means that two of them have to be manipulated to lie side-by-side in the folded model. Fortunately, the layers allow this rearrangement.

In retrospect, creating four separate flaps for the wings was probably representational overkill. The hazard of attempting to create too much in the way of appendages is that inevitably, some other aspect of the model is compromised. It does no good artistically to get the point count correct if the result is misshapen, clunky, or lifeless. In subsequent butterfly designs, I have gone back to representing both fore and aft wings by a single flap. Even with that simplification, the case can still be argued that adding legs is an aesthetic mistake. Because they are almost never as thin as a real butterfly's legs (which are almost impossible to see without an extreme close-up or still photo), explicitly created legs are frequently more of a distraction than an enhancement to the model. But perhaps this is not an inherent limitation of the subject, merely a statement that an accurately representational, yet artistically graceful, butterfly origami figure has yet to be designed. Perhaps by applying some of the techniques I've outlined here, yours could be the first.

Yet another example of allocating extra paper to widen flaps is illustrated in the Dragonfly design in Figure 13.6. The construction of the abdomen and legs is classic circle packing.

However, by adding a rectangular segment into the middle of each of the four wing flaps, we create extra paper that allows a uniform width to the wings along their length. Can you find the added paper in the crease pattern?

Figure 13.6.
Crease pattern, base, and folded model of the Dragonfly.

13.2. Multiaxial Bases

One of the biggest mismatches between technique and subject that arises in the use of uniaxial bases is that multiflapped bases tend to be skinny, while many subjects have parts that are thick and chunky. In particular, many animals have relatively stout bodies and hindquarters relative to their limbs: mice and squirrels, hippos and elephants. A purely uniaxial base, while possessing flaps for all of the major appendages, may not provide enough width in the flaps that are used for the body. Furthermore, the very efficiency that makes circle-packed bases so desirable usually means that there isn't much, if any, excess paper available to pull out to widen the desired flap.

Another problem is a bit more subtle. If we create an animal subject from a uniaxial base that is represented in side view as opposed to plan view, we will typically fold the leg flaps out to the sides, then fold the model in half, as, for example, was done with the Bull Moose in the previous chapter.

When we fold a uniaxial base in half, the fold line occurs on the axis of the base, and this naturally becomes the back of the animal. The leg flaps extend downward from the axis, as in Figure 13.7. This means that the legs need to traverse the entire height of the body before they extend beyond it, and the portion of the flap that extends beyond the body is shorter than the original flap. In effect, a portion of the hard-won flap length gets used up inside the body, where it serves no useful purpose.

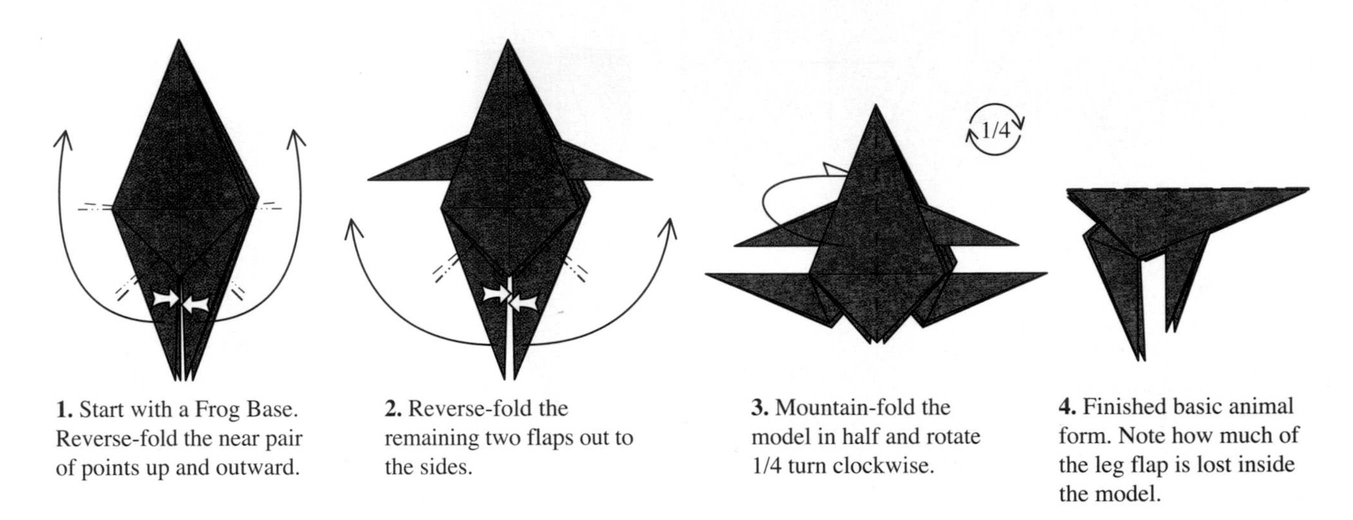

1/4
1. Start with a Frog Base. Reverse-fold the near pair of points up and outward.
2. Reverse-fold the remaining two flaps out to the sides.
3. Mountain-fold the model in half and rotate 1/4 turn clockwise.
4. Finished basic animal form. Note how much of the leg flap is lost inside the model.

Figure 13.7.
Folding sequence for a basic animal. Since the axis runs along the spine, portions of the leg length are lost inside the model running from the spine down to the point where they emerge.

The wider the body region, the greater the fraction of the leg flaps that gets consumed. Obviously, we could reduce the waste by narrowing the body, but if we need a particular body width, that option is not available. To compensate, the leg flaps must be lengthened in the original design, which ends up reducing the relative size of everything else, and making the overall model less efficient than it needs to be.

Ideally, the leg flaps wouldn't emanate from the spine of the subject. This goal can be realized in several ways, by reorganizing the model so that the axis is no longer along the spine, or by moving away from uniaxial bases entirely. Several artists, notably John Montroll, have over the past few decades devised numerous clever alternatives to uniaxial bases that sidestep this problem with remarkable efficiency. One approach used by many artists is a natural outgrowth of two of the concepts I have described in this book, grafting and uniaxial bases. As we did with the Butterfly example, we combine portions of uniaxial bases with folded structures that provide the portions of the subject that don't fit neatly into the uniaxial mold.

In the case of a vertebrate animal, we can create a large flat polygon for the wide body of the animal and pack pieces of uniaxial bases around it to create flaps for the appendages; instead of distributing those flaps along the centerline of the polygon as in a uniaxial base, we can distribute them around the periphery, thus reducing or eliminating the wide-body penalty.

The simplest way of accomplishing this would be to cut the base along some axial creases and insert a strip graft, as we did in chapter 6, but instead of pleating the strip and turning it into more points, we leave it relatively unfolded. Figure 13.8 illustrates the surgical process performed on the Frog Base of Figure 13.7.

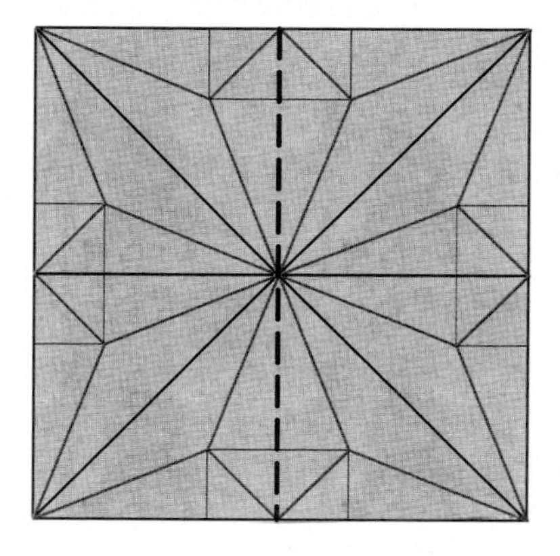

1. Here is the crease pattern for a Frog Base. We cut it down the center …

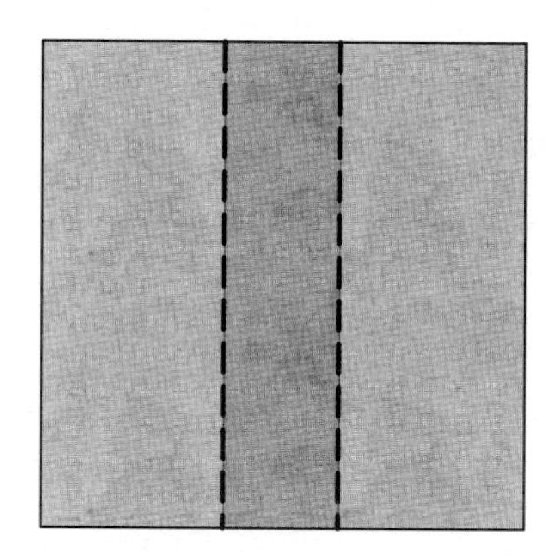

2. … and insert a strip, bounded on both sides by two axes.

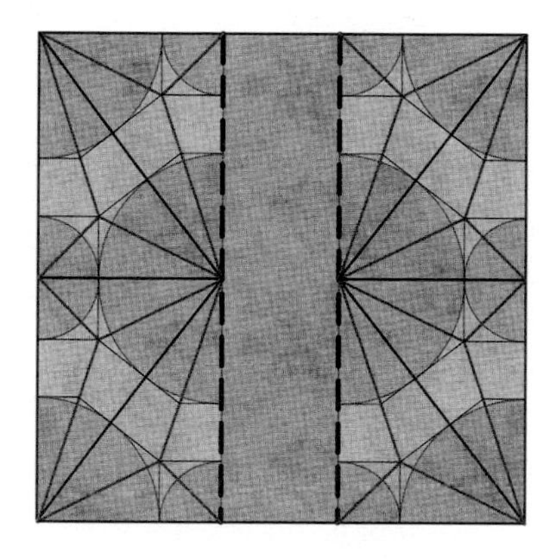

3. Construct a modified Frog Base (using circle-river packing) in the paper remaining outside the strip.

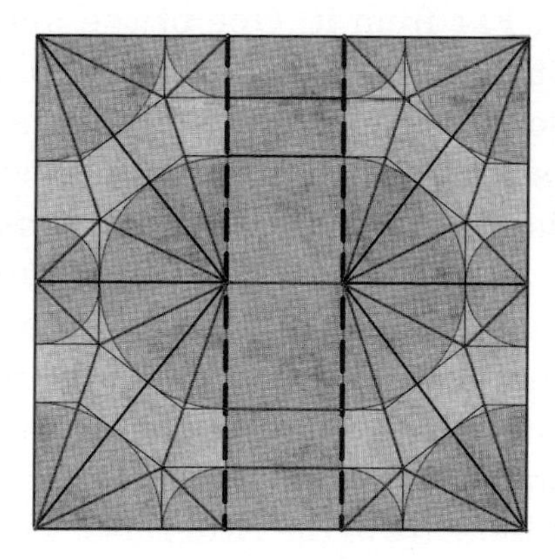

4. Extend creases across the strip and fold it into a base.

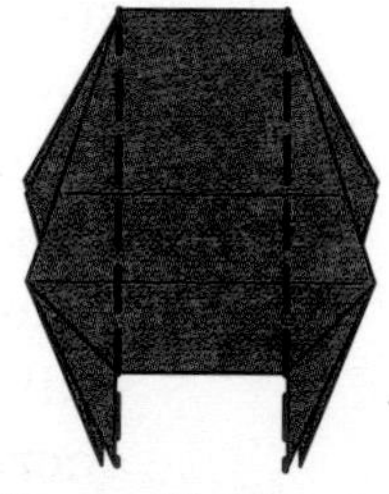

5. The resulting base is no longer uniaxial.

Figure 13.8.
Construction of a multiaxial base.

By inserting the strip, we have created two axes within the base; it is now *multiaxial*. By using the inserted strip for the body, we can utilize nearly the full length of the leg flaps by narrowing the uniaxial portions, while the central strip retains its full width, as shown in Figure 13.9.

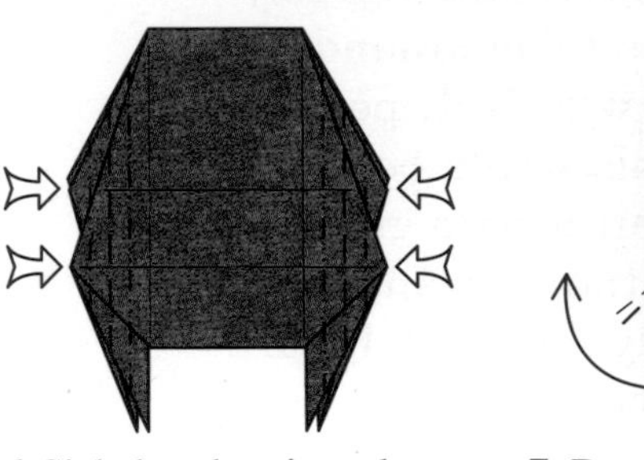

6. Sink the edges in and out to narrow them.

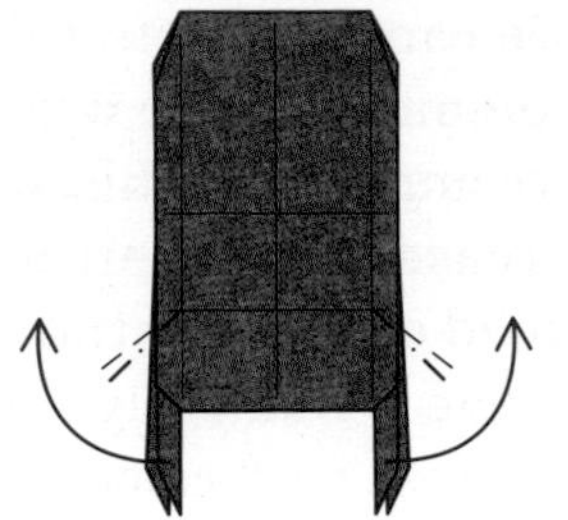

7. Reverse-fold the flaps out to the sides.

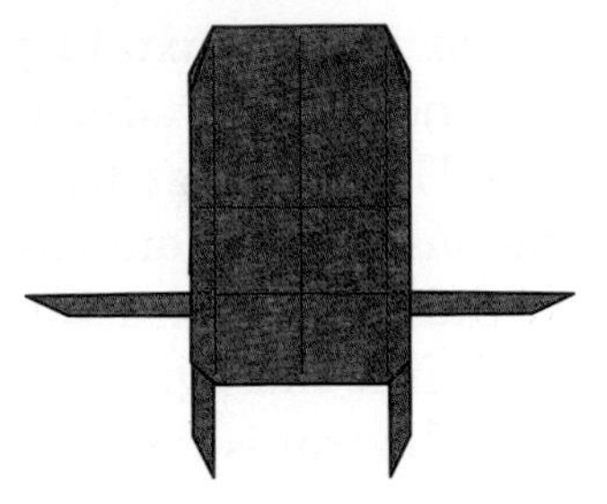

8. Observe that very little of the horizontal flap is now hidden inside the model.

Figure 13.9.
Narrowing the edges that would conceal the leg flaps still leaves paper available for a wide body.

The example in Figure 13.9 is a bit contrived to illustrate the principle. But you can use this technique in many ways, varying the width of the inserted strip relative to the paper remaining to vary the ratio of body width to flap length while still preserving efficiency.

13.3. Grafted Kite Base

The region that you insert does not have to be a rectangular strip, of course. Far from it: One of the most versatile techniques for creating animal forms, used in designs by numerous artists, inserts a Kite Base (or modification thereof) into the corner of a square. Or, viewed another way, it consists of a strip graft added to two sides of a Kite Base, similar to the strip graft that created the KNL Dragon in Chapter 6. But now, rather than simply using the strip to create small features at the corners of the model, the strip is made wide enough that, when filled with flap-creating molecules, it contributes a collection of flaps around the periphery of the triangle that makes up the silhouette of the Kite Base. This added material thereby produces much of the overall structure of the model. Better yet, it is highly variable: By varying the width of the grafted strip, you can add more or fewer flaps, make them larger or smaller, and create a remarkable variety of flat and three-dimensional fauna. I call the family of structures the *grafted Kite Base*.

The concept of the grafted Kite Base is illustrated in Figure 13.10. The basic structural form is the Kite Base, whose crease pattern is embedded in the square. The central triangle of the Kite Base will be preserved in the final model, giving a large, flat region from which to form the body. Instead of running the axis of the model down the center of the square, we can treat the perimeter of the preserved triangle as consisting of axis; we then use conventional techniques, such as circle/river packing, to create flaps from the region outside the preserved triangle.

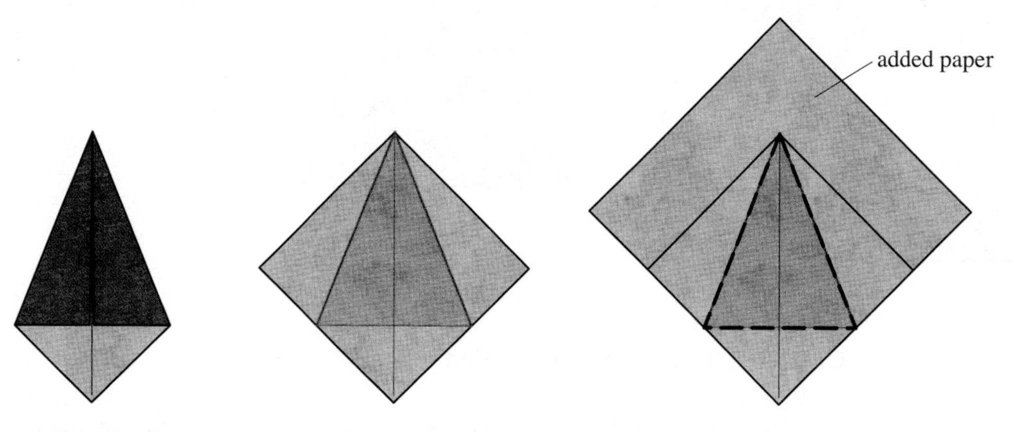

Figure 13.10.
Left: Kite Base.
Middle: crease pattern for the Kite Base.
Right: Kite Base embedded within a larger square.

Not all of the theory carries over; the molecular crease patterns we constructed were based on the assumption that all axial creases wind up collinear in the folded model. This will assuredly not be the case if we keep the colored triangle from folding flat. However, we can still use circles to allocate paper to the flaps that will lie along the creases.

We can also incorporate portions of the colored triangles into flaps, by allowing our circles to partially overlap the triangle; however, these flaps will not be axial flaps. That may not be a problem; in fact, it may be quite desirable. Thus, for example, in the Rabbit shown in Figure 13.11, the two bottom corners of the embedded triangle become the rear legs of the animal. Obviously, they are not axial flaps, but for this figure, axial flaps would not be very useful as the rear legs. On the other hand, axial flaps work very well for the head and ears, and the four-circle-packing—and the crease pattern that results—should, by now, be very familiar to you.

The ratio between the size of the embedded Kite Base and the original square is a design variable that changes continu-

Figure 13.11.
Crease pattern, base, and folded model of the Rabbit.

ously (which is why the grafted Kite Base is a family of bases, rather than a single base). The smaller the Kite Base is relative to the full square, the more paper is available for other flaps. Thus, in Figure 13.12, where I have drawn three different sizes, you can see that in the image on the left, the four circles at the top of the square (and thus their corresponding flaps in the base) are relatively small compared to the lower flaps; compare circles A, B, and C in Figure 13.12. Reducing the Kite Base relative to the larger square allows the four type A flaps to enlarge, as you can see in the progression in the figure.

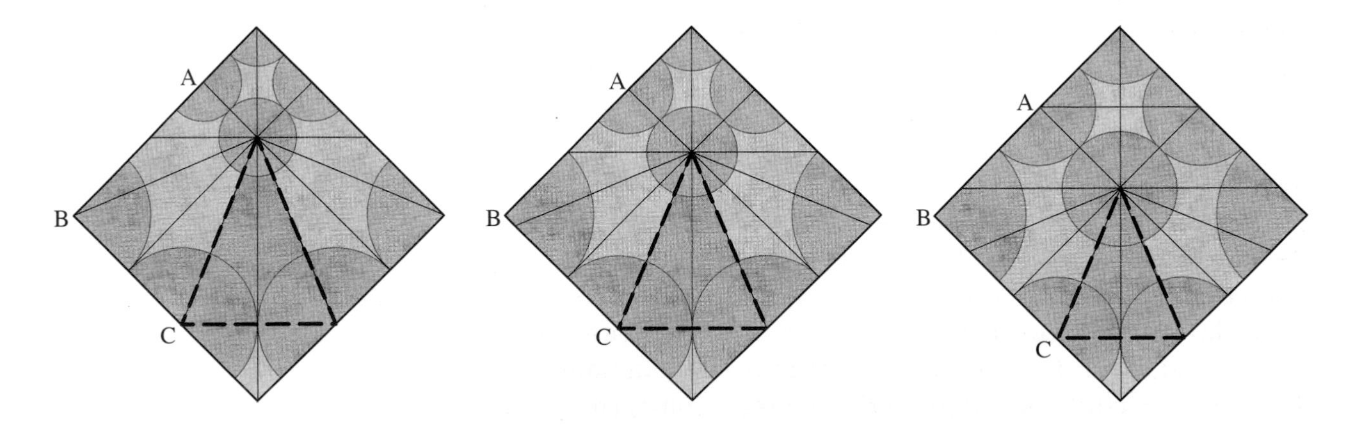

Figure 13.12.
Three different ratio embeddings of the grafted Kite Base.

What is less obvious but also a consideration is the length of the gap between circle pairs A–B and B–C. In the first two patterns, the B–C circles are touching, indicating that their corresponding flaps are joined at their base. Conversely, in the last pattern, circles A and B are touching with a gap between B and C, which means that flaps A and B are joined at their base. By adjusting the size of the Kite Base embedded within the square and manipulating the circles that allocate paper for flaps, you can adjust not only the length of the flaps, but also their topology.

You can also graft other shapes into squares in a similar way. The design shown in Figure 13.13, for example, grafts the diamond of a Fish Base into a square.

Figure 13.13.
Crease pattern, base, and folded model of the Mouse.

In all of the grafted Kite Base examples, the top point of the Kite Base becomes a relatively thick middle flap. In the previous two models, this middle flap ends up unused, sunken down into the model. But it would also be possible to use it for features, for example, by point-splitting, as we will see shortly.

One of the things you should always do when you learn a new technique is to ask: How can this be generalized? In the grafted Kite Base, an obvious generalization is to vary the size of the Kite Base relative to that of the bounding square. Another generalization, perhaps less obvious, but equally powerful, is to vary the apex angle of the Kite Base. Different angles give a different aspect ratio to the embedded triangle. Perhaps more interesting, other angles allow crease patterns with different symmetries. You might find it interesting to explore the possibilities of some of these other angles; an apex angle of 60°, in particular, offers several fruitful possibilities.

13.4. Mixing and Matching

Throughout this book, I have chosen examples that were pure illustrations of the various mathematical design techniques. The real world of design, however, is rarely so pure. More often than not, an origami design is best served by employing a mixture of techniques: box pleating here, circle packing there, grafting, molecules, point-splitting, pleated textures—and others beyond the ones shown here.

The various design techniques are, at the end, tools; and just as a painter may use an assortment of brushes and pigments to realize his design, the origami artist can employ a variety of design techniques within the same model to realize a single unified vision of the subject.

This last design brings together several of the design techniques I have shown. As in the Rabbit and Mouse in this chapter, I use the grafted Kite Base to embed a large triangle into the crease pattern, from which the massive hindquarters come; I employ point splitting to turn the large middle flap into a

Figure 13.14.
Crease pattern, base, and folded model of the African Elephant.

forehead and flapping ears; I use circle packing to specify the creases in the forelegs, trunk, and elsewhere; and even hearkening back to the first designs in the book, I make use of the various elephant head designs from Chapter 2. The result is, of course, yet another elephant. From the simplest to the complex, the African Elephant spans the spectrum of origami technique, and serves as a fitting final example for our foray into origami design.

13.5. Wrapping It Up

During the great westward migration of the mid-19th century, a saying arose among the American pioneers who were setting out on the Oregon Trail: "I am going to see the elephant." The elephant was a metaphor for all of their goals, their hopes, their dreams, their aspirations. They did not set out unequipped; they brought with them the tools with which to make a new life, break new ground, and with luck, make their fortune.

Despite its antiquity, the art of origami is still in its pioneering days. The practice of new creation began within the last century, via the works of Yoshizawa, Uchiyama, and Unamuno, then spread around the world in its own westward expansion. It was led by names that have become legendary in origami: Oppenheimer, Harbin, Randlett, Solorzano Sagredo, Montoya, Rohm, Elias, Crawford, Cerceda, and others too numerous to mention. The early pioneers of origami creation had little more than a handful of traditional designs and their own intuition to guide them. But as the art and the knowledge spread, a collection of lore and technique has arisen, akin to the blazing of the westward trails.

What I have attempted to provide in this book is a collection of tools to help you on your way down the path of origami design. These tools, like any others, are only useful with the knowledge of how to wield them. And they become more useful with practice. You can apply the concepts I've shown by deconstructing the things you see. If you fold a clever or appealing model, pull it apart, examine the crease pattern, look for signs of structure. What paper goes into the flaps? Is there an axis? Are there multiple axes? Are some creases more important than others?

Just as tools become more useful with practice, as they become more widely used, they get improved, extended, and even replaced. I have no doubts that the mathematical methods of origami design that once seemed strangely foreign—splitting, grafting, tiles, circles, rivers, square packing and trees—will eventually be augmented, if not superseded, by more powerful and more general techniques. We now look upon the origami designers of the 1950s and 1960s as the pioneers, but we may find in the future that the entire 20th century is seen as the era of origami pioneers as new and wondrous creations arise through the use of these new techniques.

While the early American pioneers blazed the trails, the next wave turned them into roads using better equipment and the knowledge of what was possible. Each wave of origami designers takes the art to new heights, creating not just more complex structures, but utilizing the inherent capabilities of the folded paper in new and unexpected ways. In this work, I have focused on a fairly narrow set of concepts, tied together by the common theme of obtaining a base with a specified configuration of flaps in a controlled way. But new designs go far beyond this narrow concept; some—such as the intricate geometric patterns of Chris Palmer, the curved and swirling masks of Eric Joisel, and the organic crumpled forms of Vincent Floderer—redefine the boundaries of origami itself.

Each journey into origami design is personal and original. It is my hope that the mathematical ideas in this book—the tools, geometry, structures, and equations—will help you on your own journey into design. At the very least, they perhaps offer a new way to look at origami, a way of looking beyond the final appearance, beyond the linear folding sequence, to understand the structure, its constituent elements, the building blocks of folding.

To the California Forty-Niners, "seeing the elephant" was their grand, glorious goal. Those who were ill-equipped or unlucky were turned back, saying that they had seen no more than the elephant's tracks or tail. On your origami journey, the tools of systematic design can equip you to overcome the challenges posed by any origami subject and bring you success in your own quest to see the elephant.

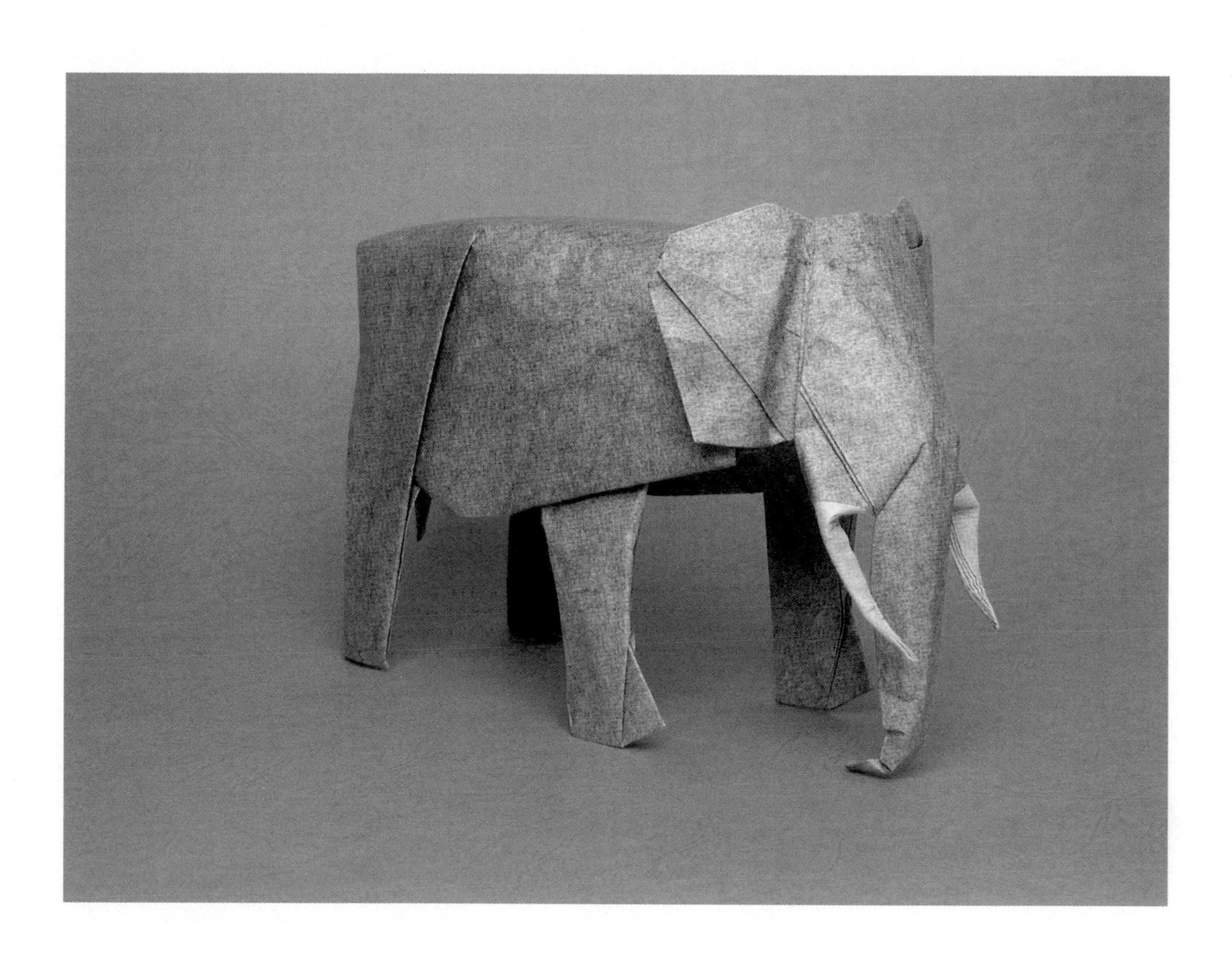

Folding Instructions

African Elephant

African Elephant

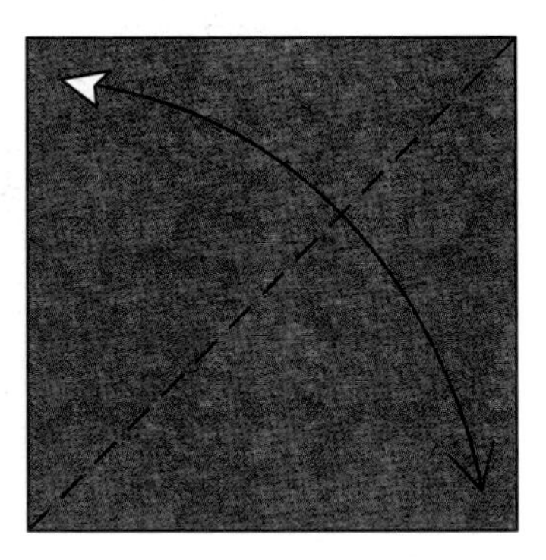

1. Begin with a square, colored side up. Fold and unfold along one diagonal.

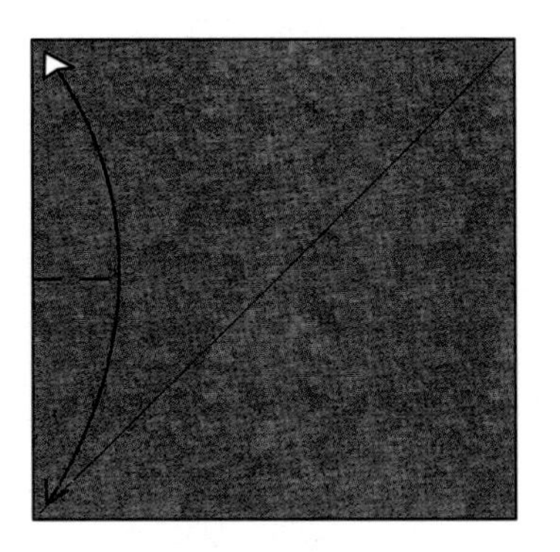

2. Fold the top corner to the bottom and unfold, making a small pinch along the left side.

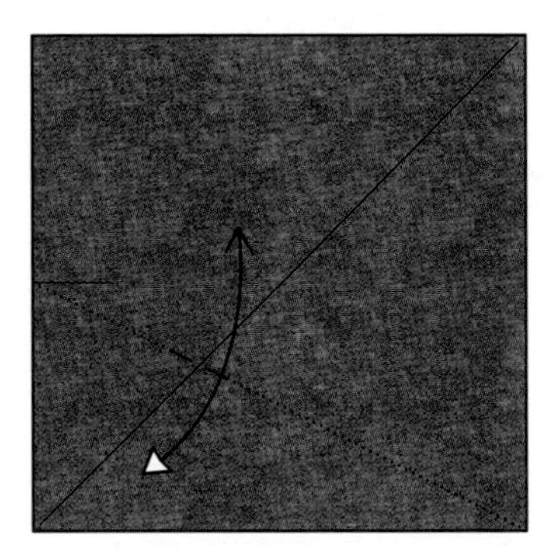

3. Make a fold that connects the lower right corner with the pinch you just made; make the crease sharp where it crosses the diagonal and unfold.

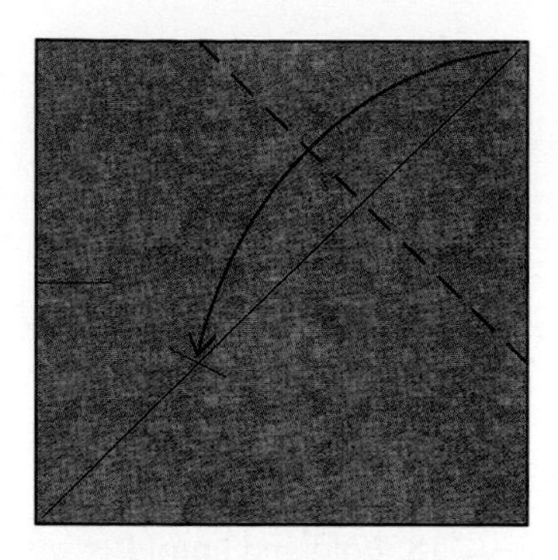

4. Fold the top right point down to touch the crease intersection.

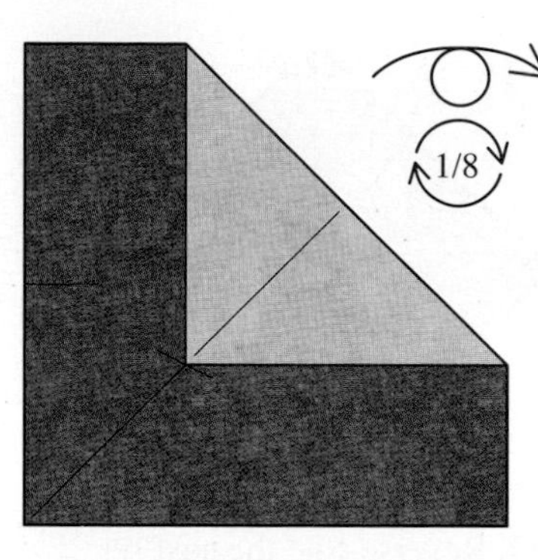

5. Turn the paper over and rotate it 1/8 turn.

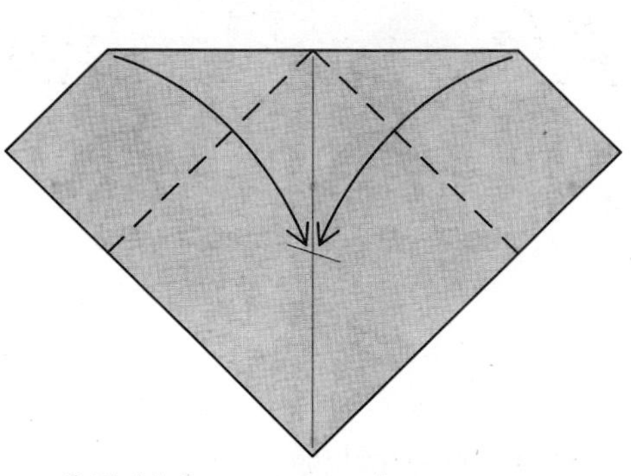

6. Fold the top edges down to meet along the centerline of the paper.

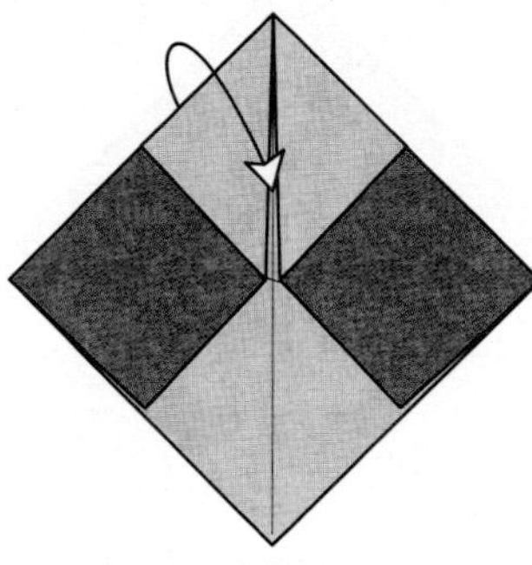

7. Wrap the corner of the paper from back to front and flatten symmetrically.

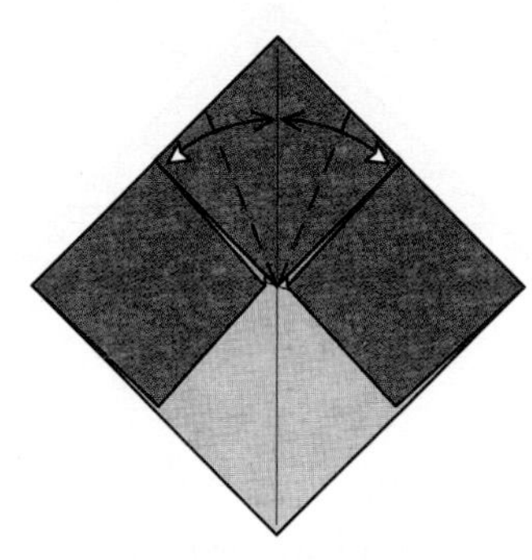

8. Fold the raw edges in to lie along the center line and unfold.

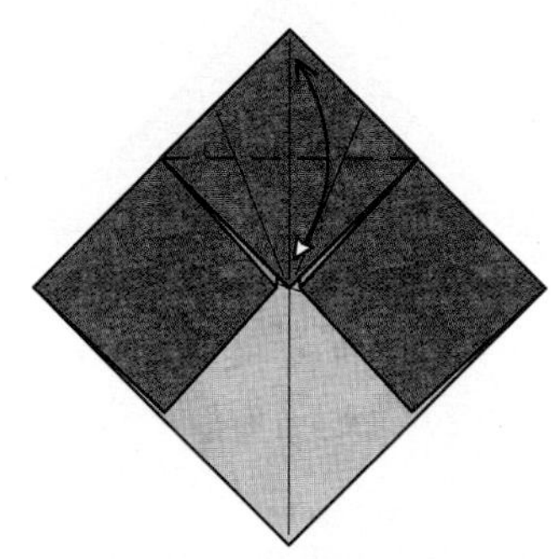

9. Fold and unfold through a single layer.

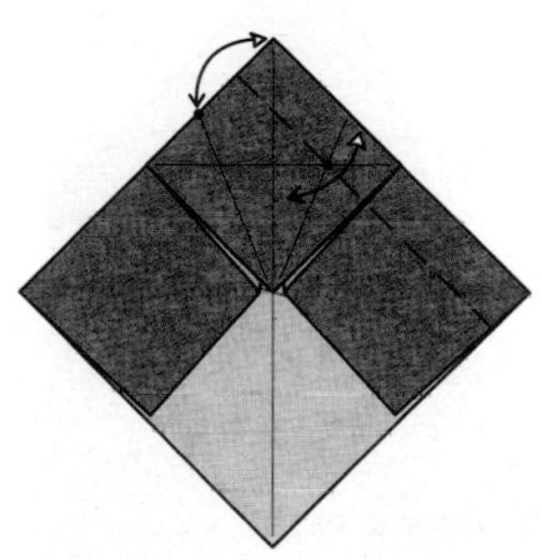

10. Fold and unfold through all layers.

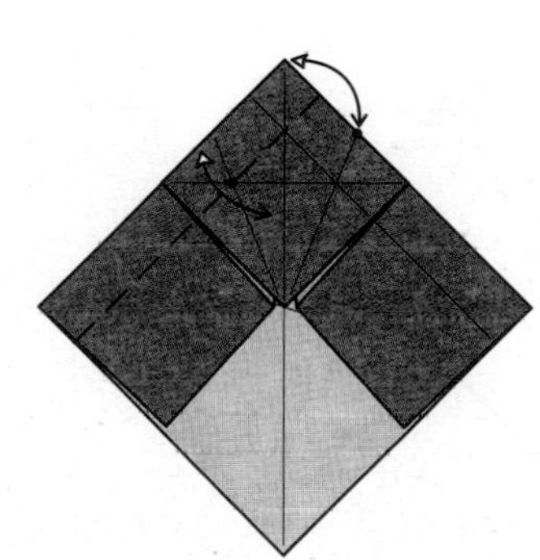

11. Repeat on the left.

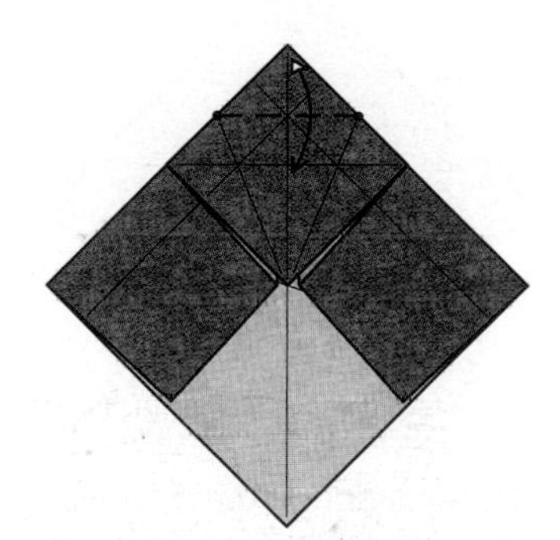

12. Fold and unfold.

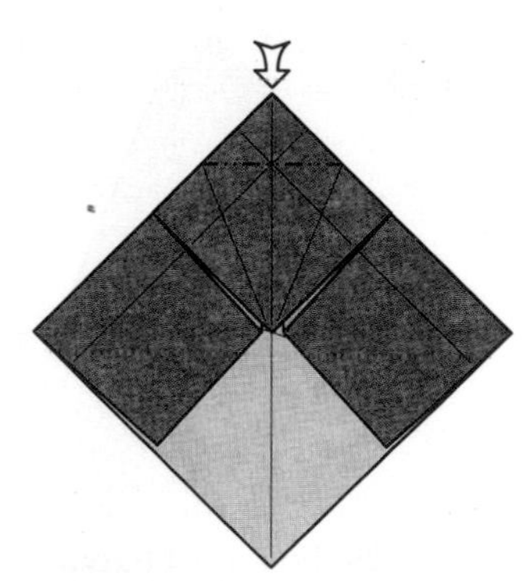

13. Sink the corner on the creases you just made.

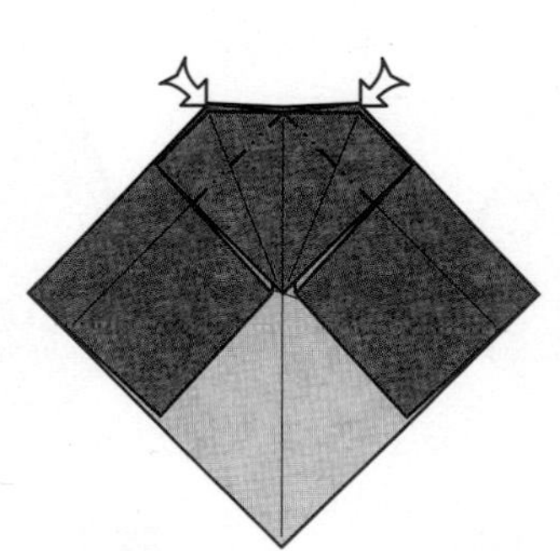

14. Open-sink the two edges on existing creases.

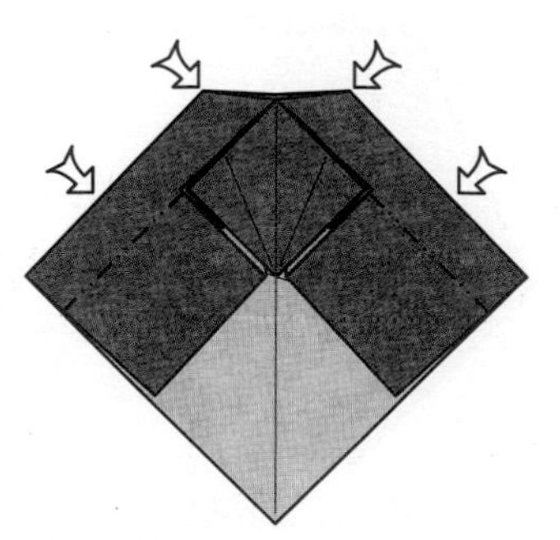

15. Open-sink the two far edges in the same way.

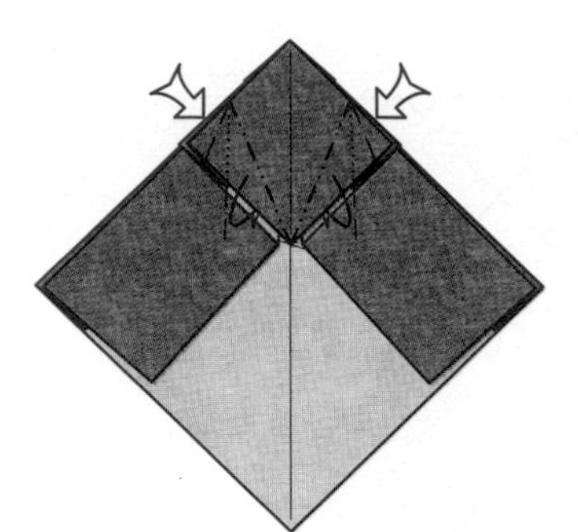

16. Reverse-fold the edges underneath.

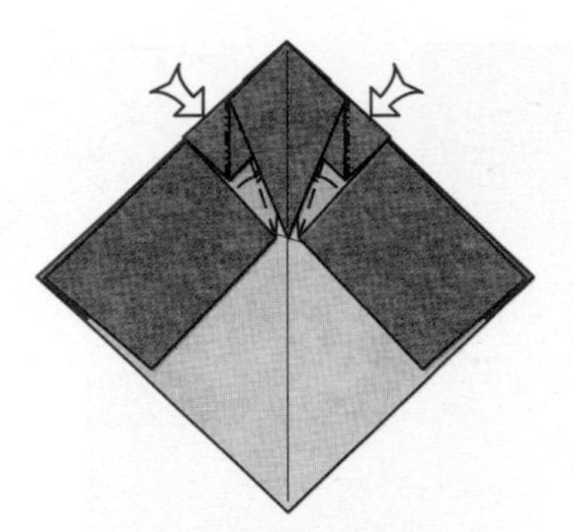

17. Reverse-fold the next set of edges in the same way.

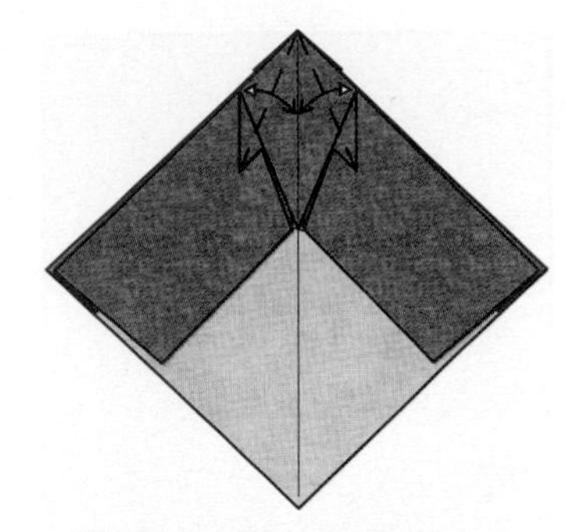

18. Fold and unfold through the near layers.

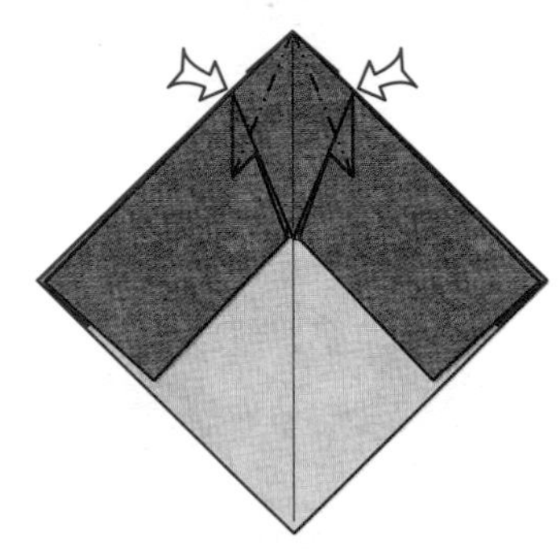

19. Open-sink the corners.

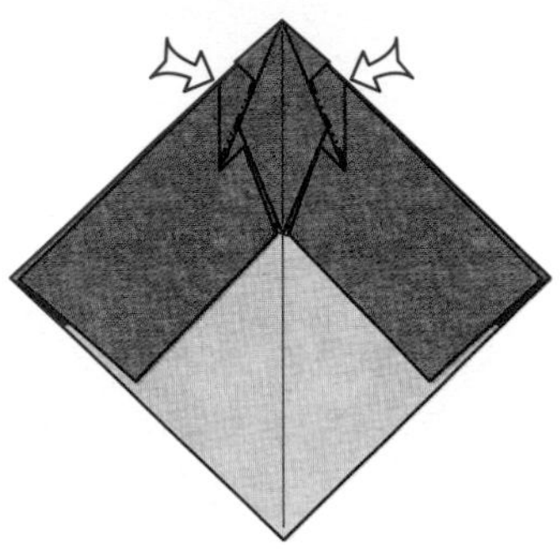

20. Open-sink the next pair of edges in the same way.

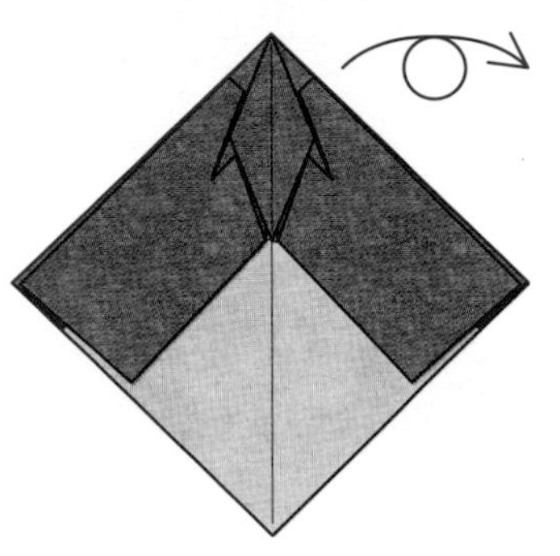

21. Turn the model over.

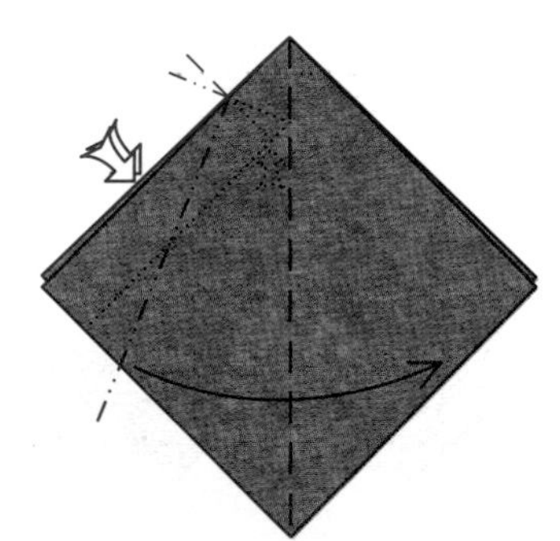

22. Squash-fold the double-edge, pushing up from inside and flattening symmetrically.

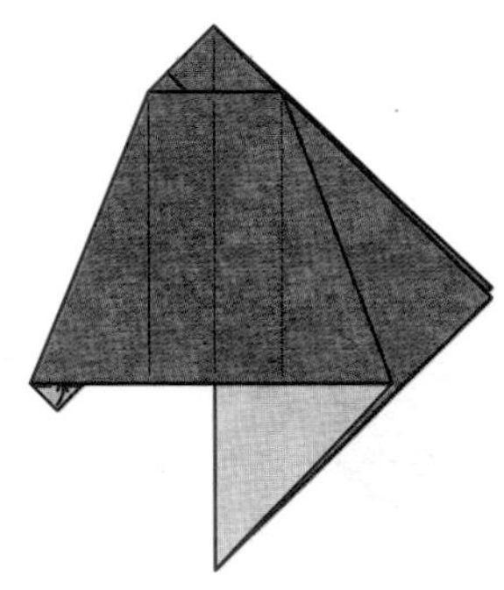

23. Tuck the small white corner up inside.

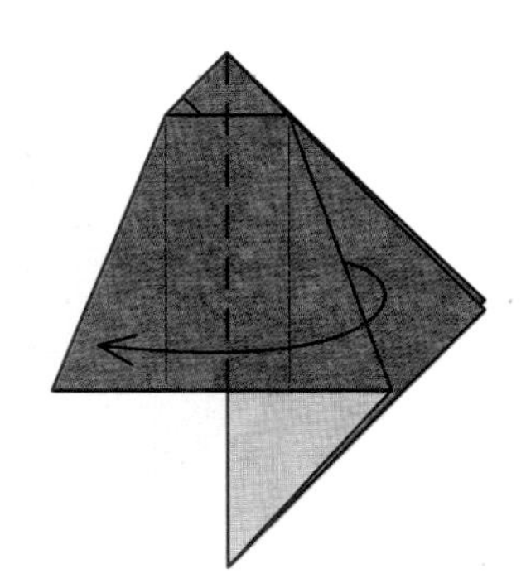

24. Fold one flap back to the left.

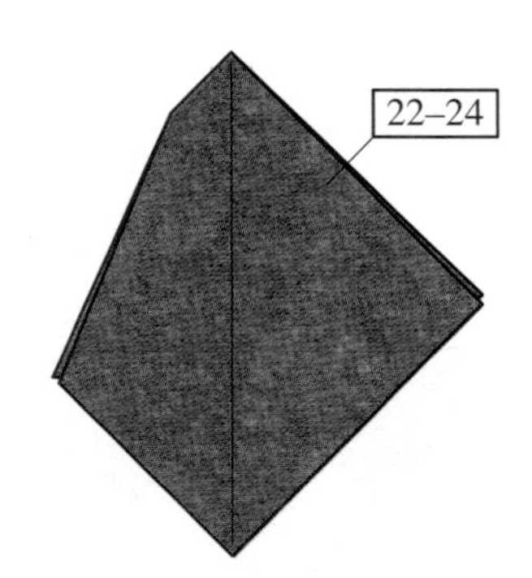

25. Repeat steps 22–24 on the right.

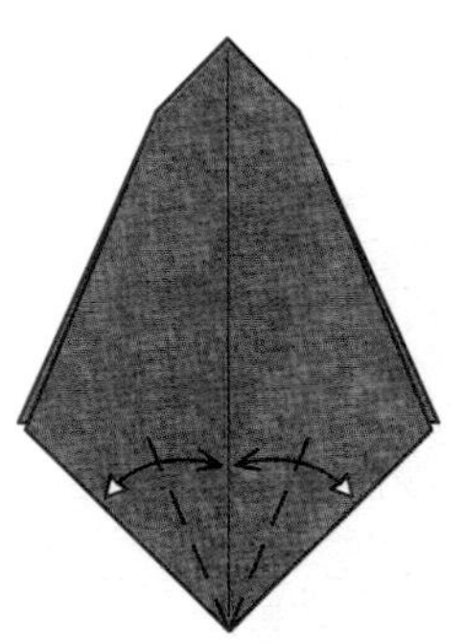

26. Fold and unfold along angle bisectors.

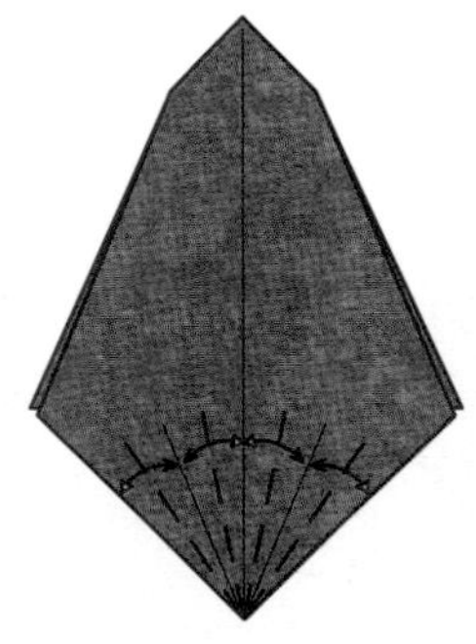

27. Fold and unfold along four more angle bisectors.

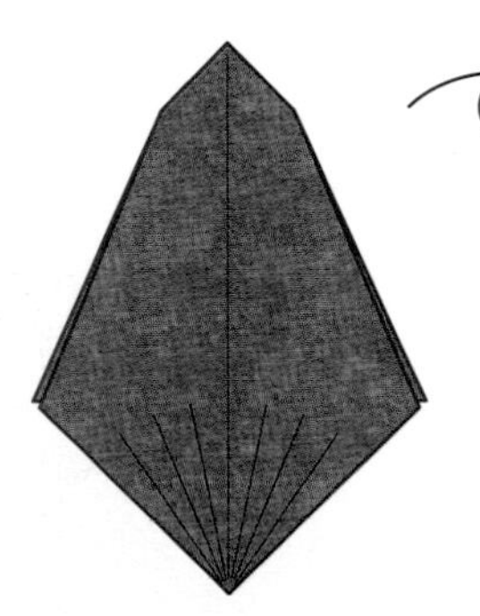

28. Turn the paper over.

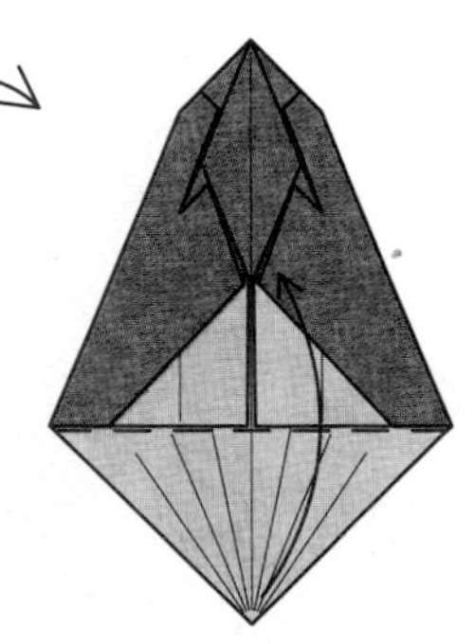

29. Fold the bottom point up over all the layers.

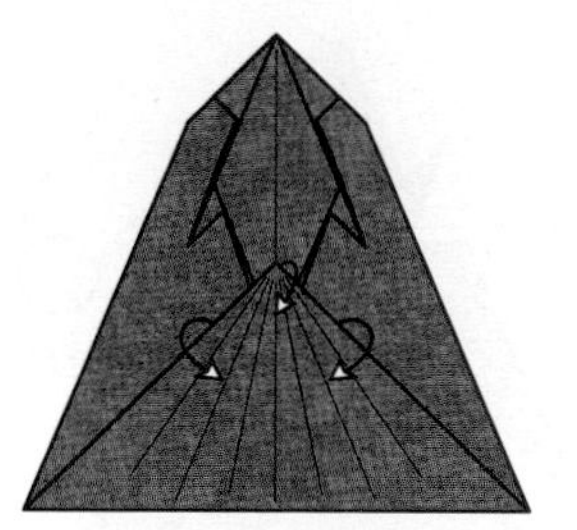

30. Bring three layers in front of the triangular flap.

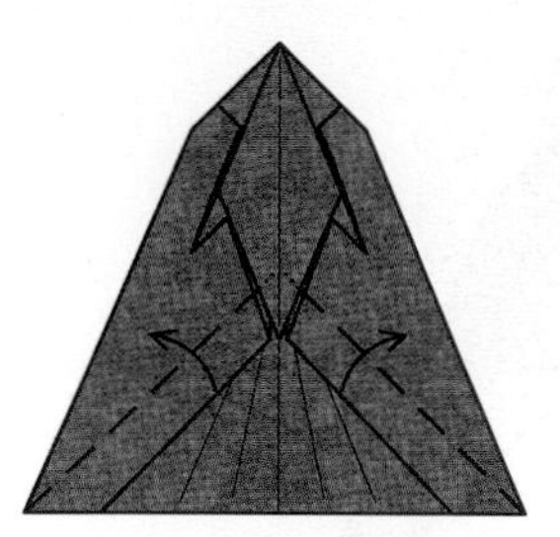

31. Fold the raw edges up along a crease aligned with the hidden raw edges.

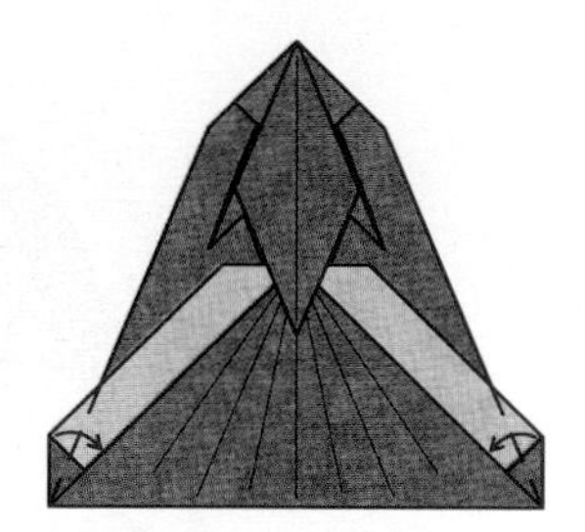

32. Fold down the side corners.

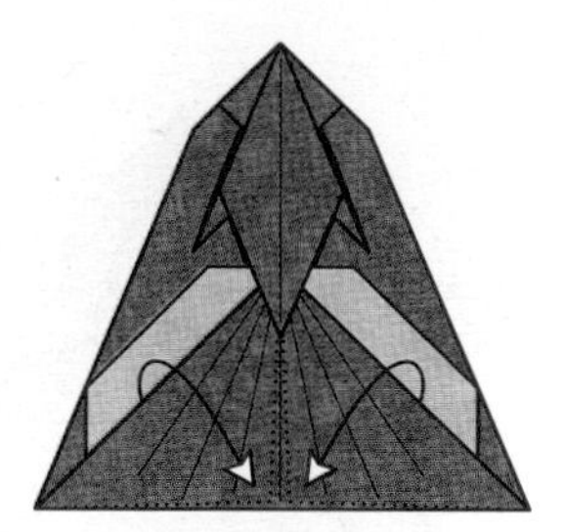

33. Bring two white flaps in front of the colored triangle.

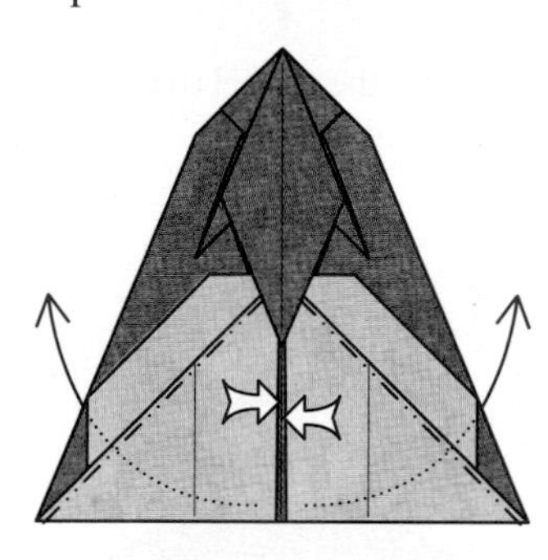

34. Reverse-fold the two white flaps up along creases aligned with raw and folded edges.

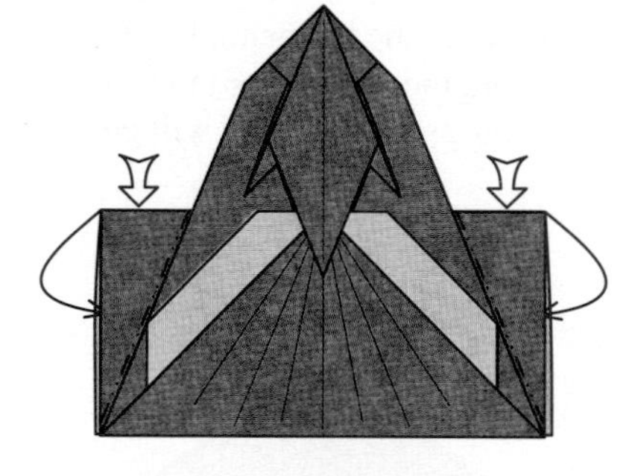

35. Reverse-fold the two corners.

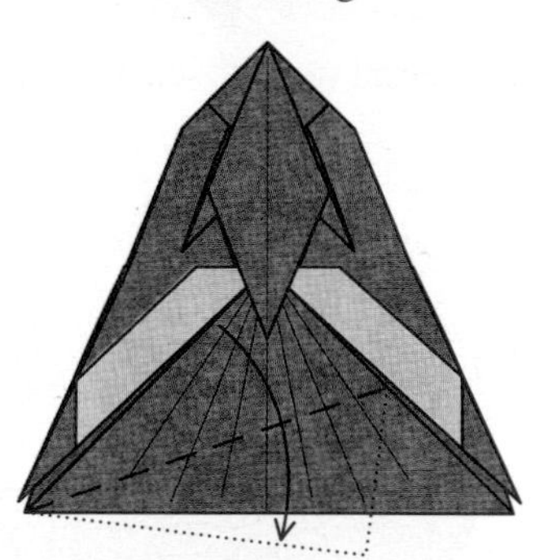

36. Fold the corner down. Look at the next step for the precise reference point.

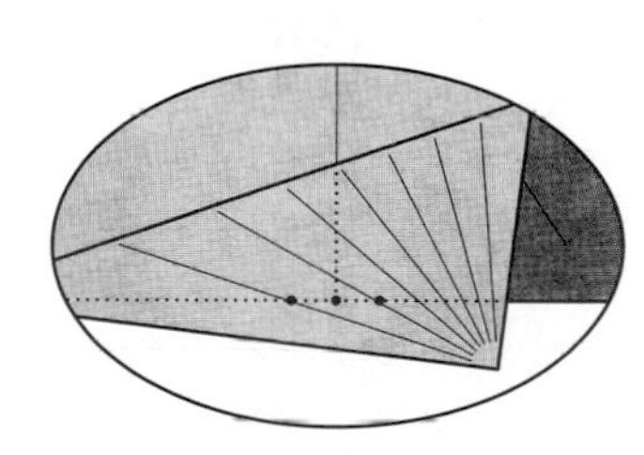

37. The hidden crease (visible on the back side) hits the edge halfway between two crease intersections.

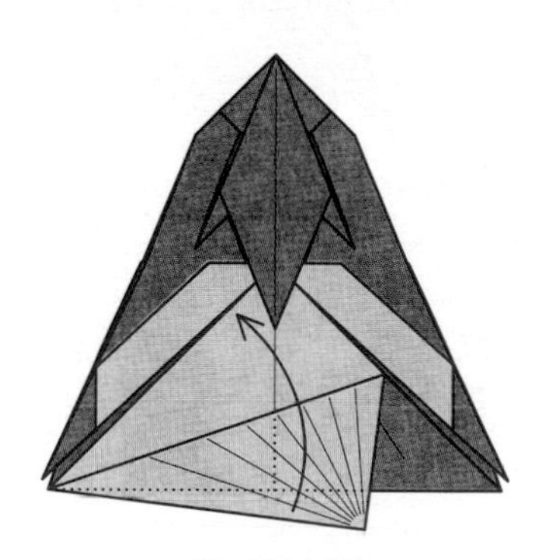

38. Unfold.

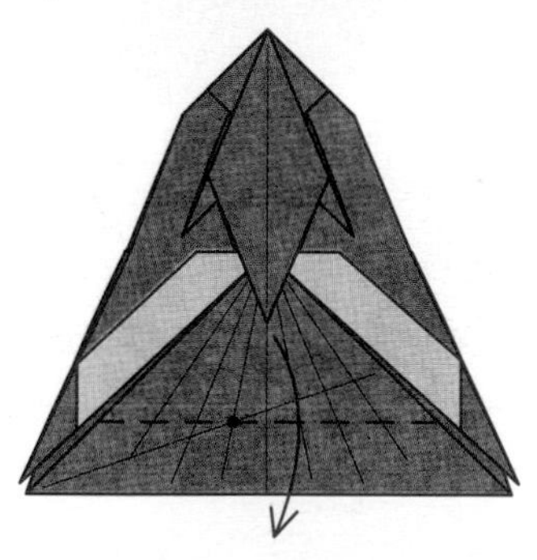

39. Fold the corner down along a crease that runs through the indicated crease intersection.

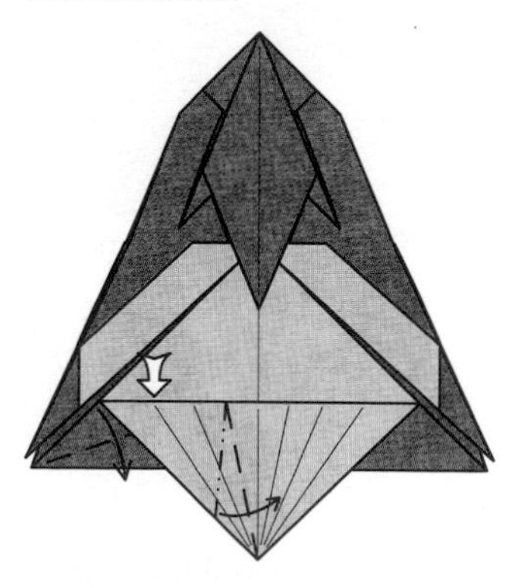

40. Swivel-fold, using the existing creases.

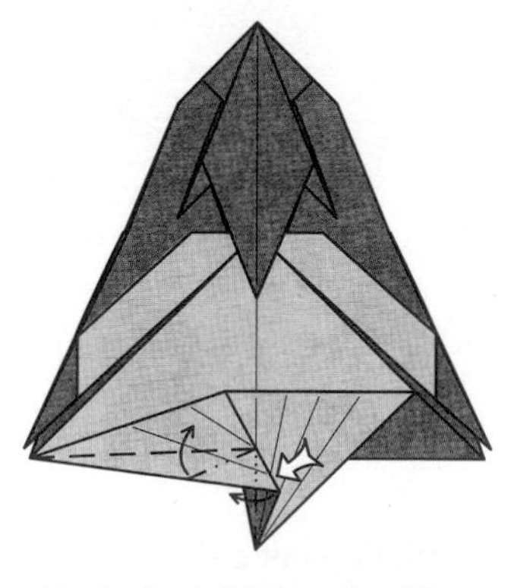

41. Swivel-fold again. The vertical crease already exists.

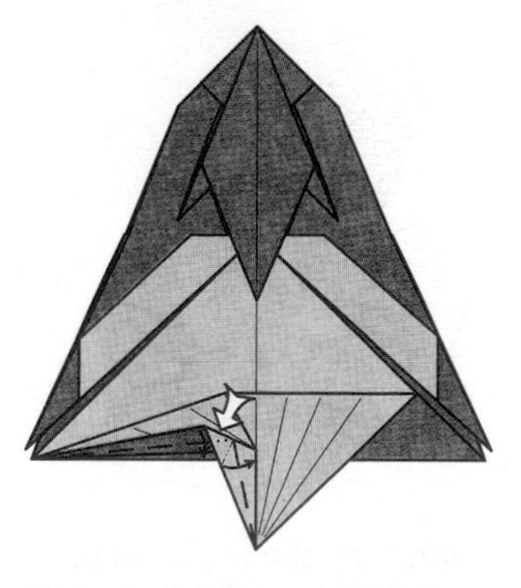

42. Swivel-fold one final time.

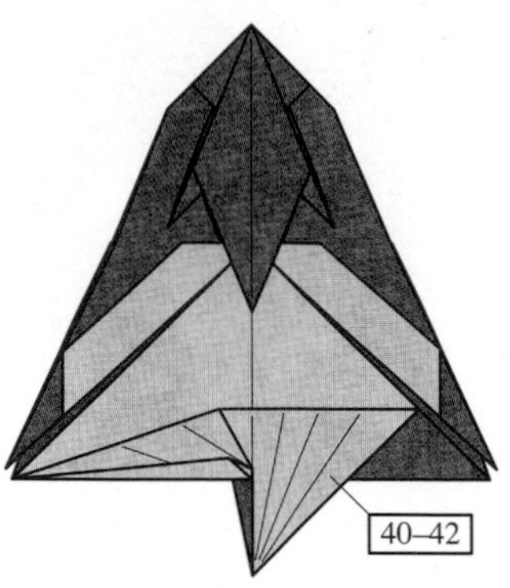

43. Repeat steps 40–42 on the right.

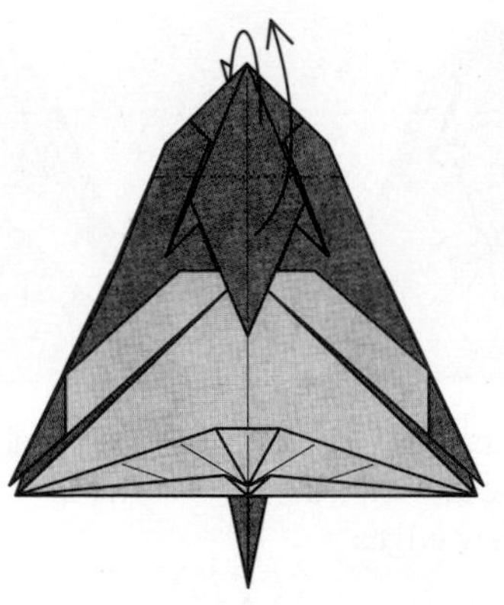

44. Fold the top behind and swing the three flaps in front as far upward as they will go.

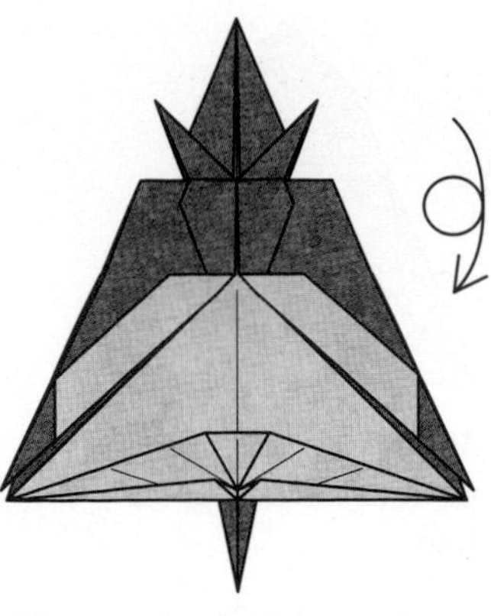

45. Turn the model over from top to bottom.

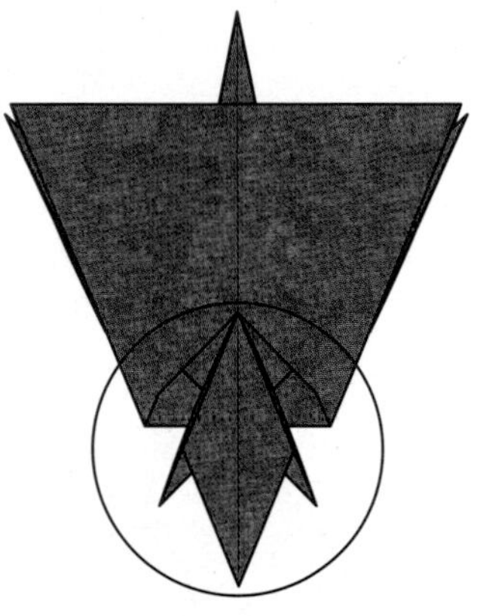

46. Steps 47–60 will focus on the head.

47. Fold the edges in toward the centerline; the edges should be vertical and parallel.

48. Unfold the two flaps.

49. Sink the edges on the creases you just made.

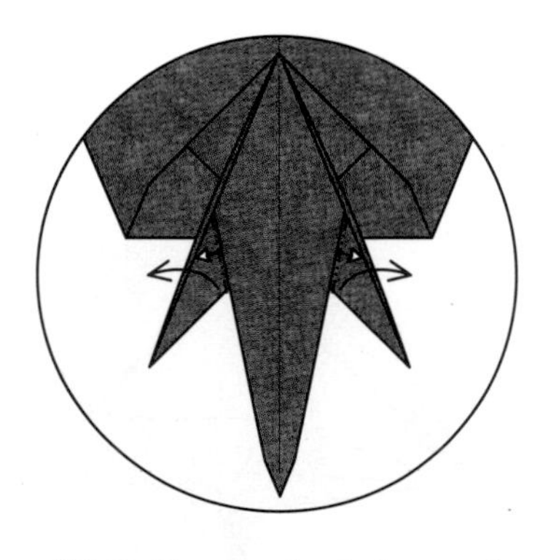

50. Pull out a single layer of paper partway on each side.

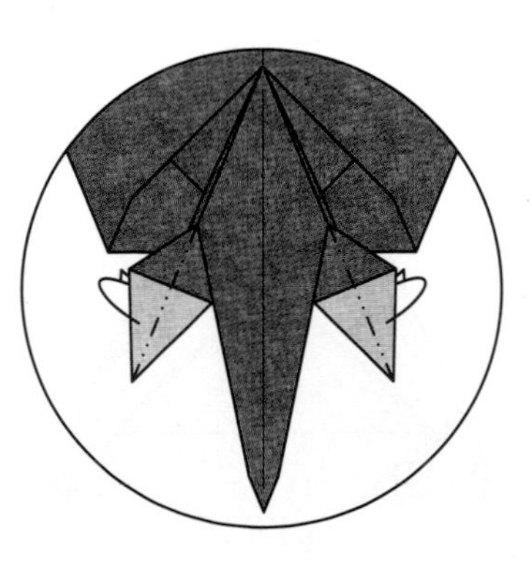

51. Mountain-fold the edges underneath on each side.

52. There is a small Preliminary Fold at the top of the head. Valley-fold the near edges and mountain-fold the far edges.

53. Pleat a single layer near the bottom and swing the two points out to the sides. Flatten firmly.

54. Reverse-fold the corners. Repeat on the far layers.

55. Mountain-fold the near point as far down as possible.

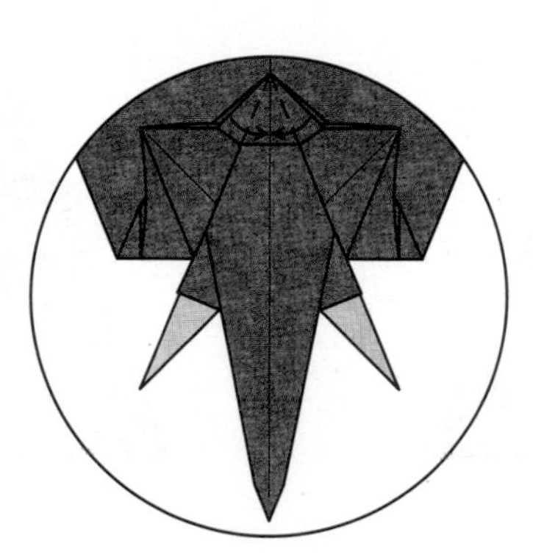

56. Mountain-fold two edges to the center line.

57. Mountain-fold the point as far behind as possible.

58. Sink the corners of the ears a bit.

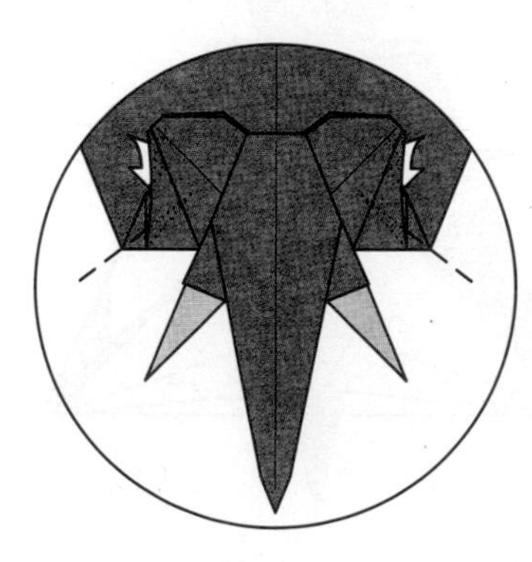

59. Reverse-fold two corners.

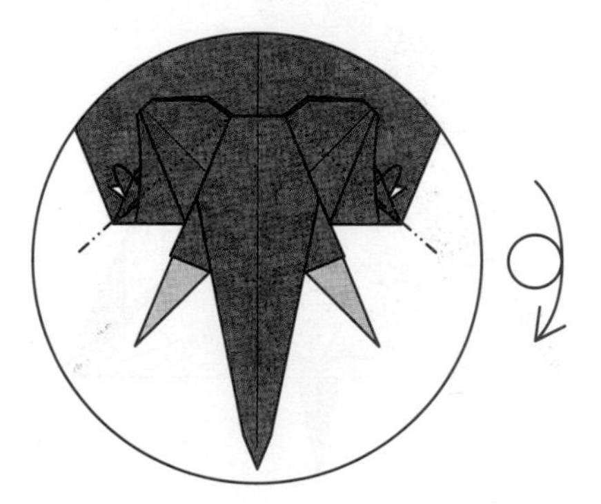

60. Mountain-fold two edges. Turn the paper over from top to bottom.

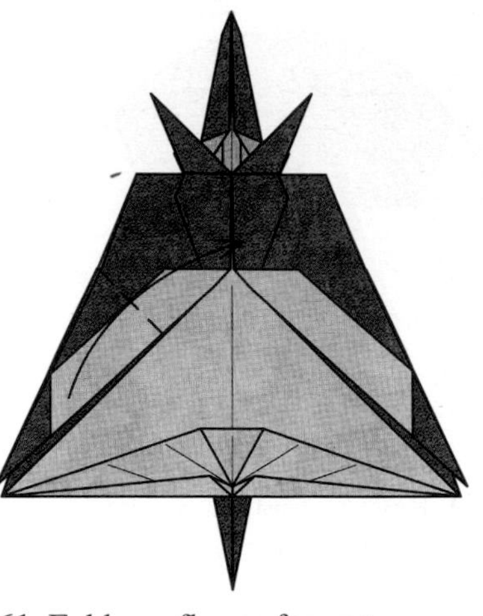

61. Fold one flap as far up to the right as possible.

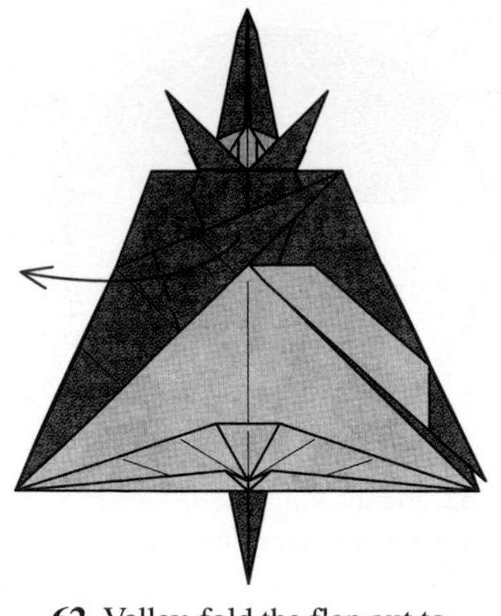

62. Valley-fold the flap out to the side.

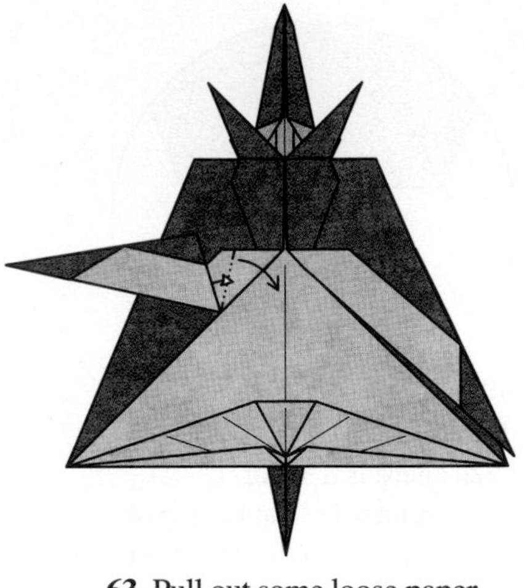

63. Pull out some loose paper.

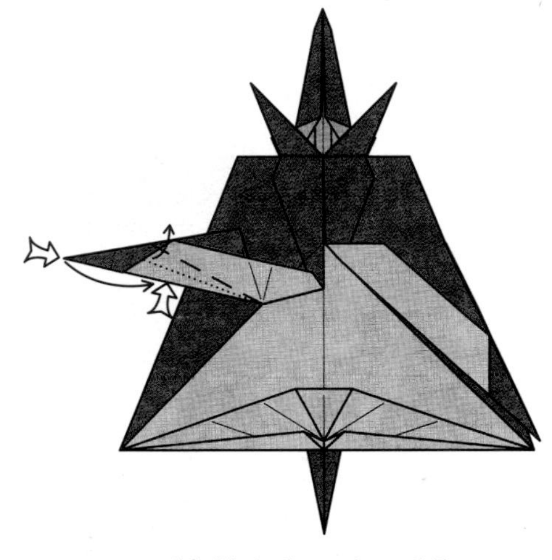

64. Sink the point while squash-folding one of the white folded edges.

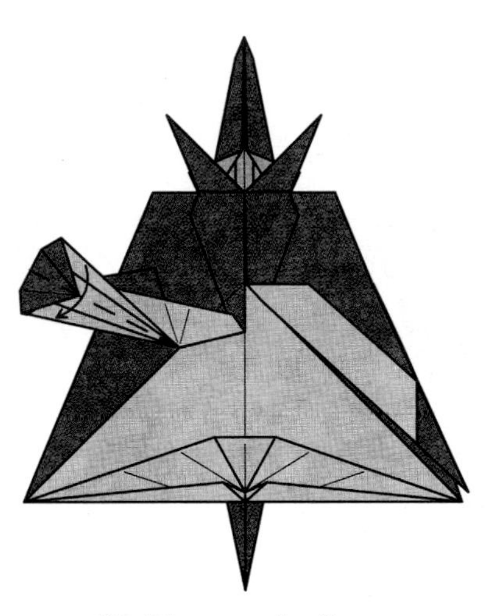

65. Close up the flap.

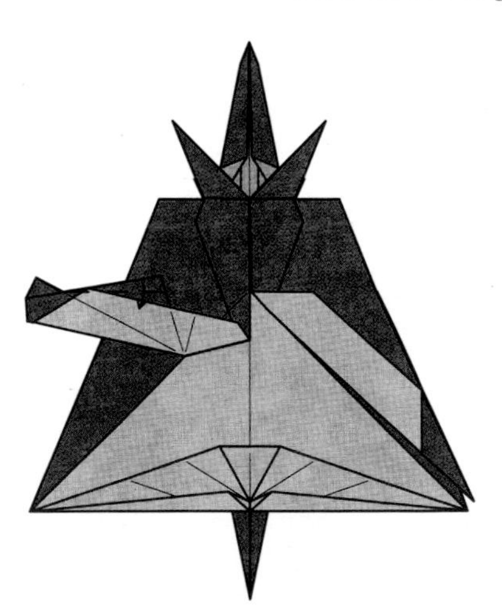

66. Narrow the leg a bit with a valley fold.

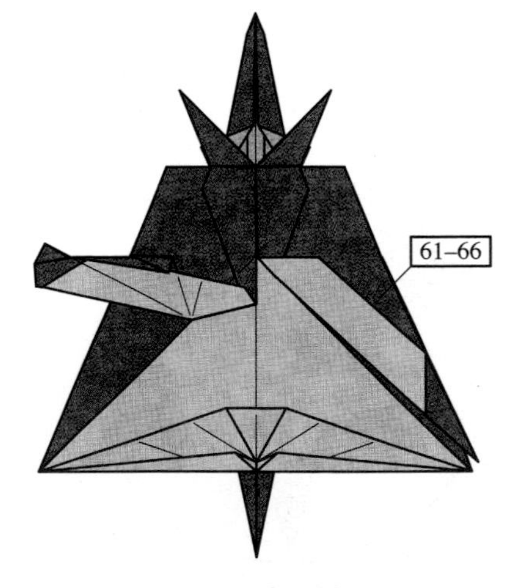

67. Repeat steps 61–66 on the right.

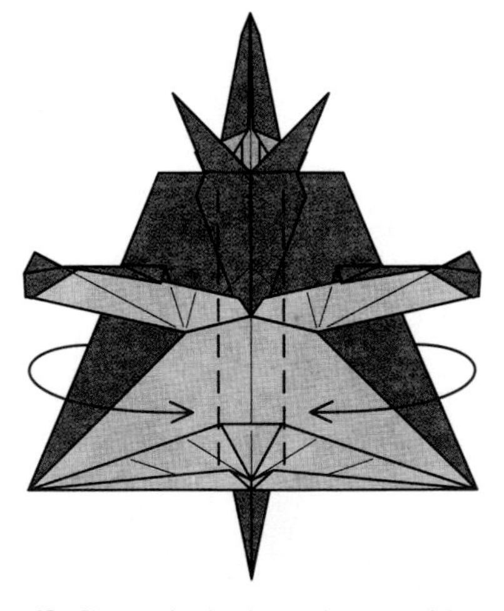

68. Curve the body so that the sides are parallel and the middle is U-shaped.

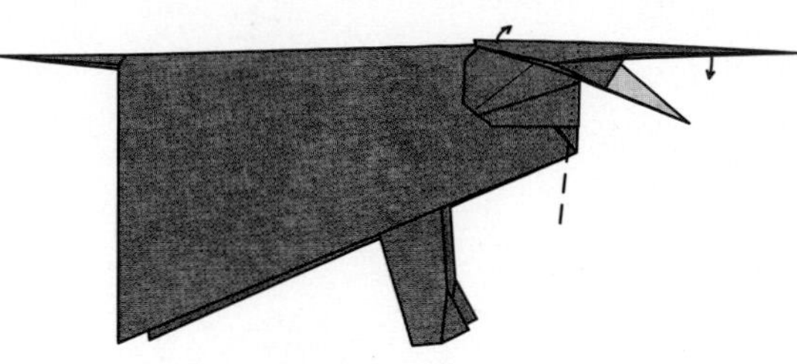

69. Rotate the head slightly by adjusting the location of the valley fold where it joins the body.

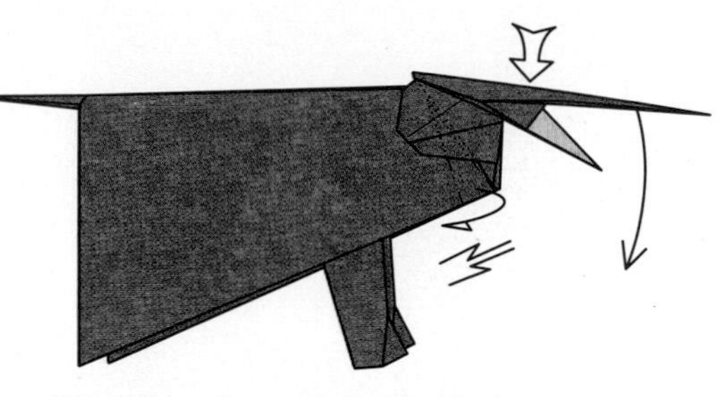

70. Crimp the neck just behind the ears (the pleats tuck under the ears) and rotate the head downward.

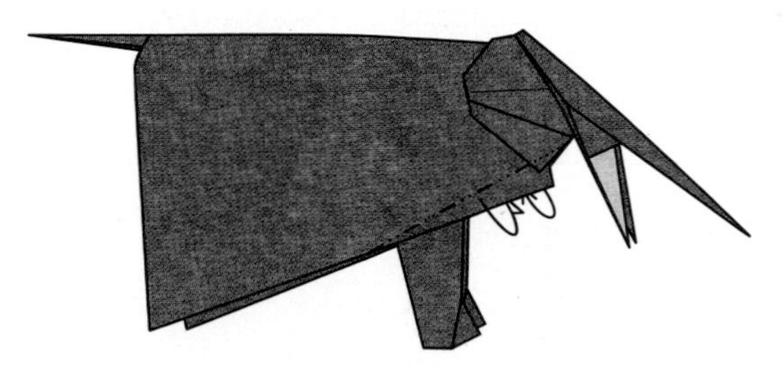

71. Mountain-fold the edges of the body underneath.

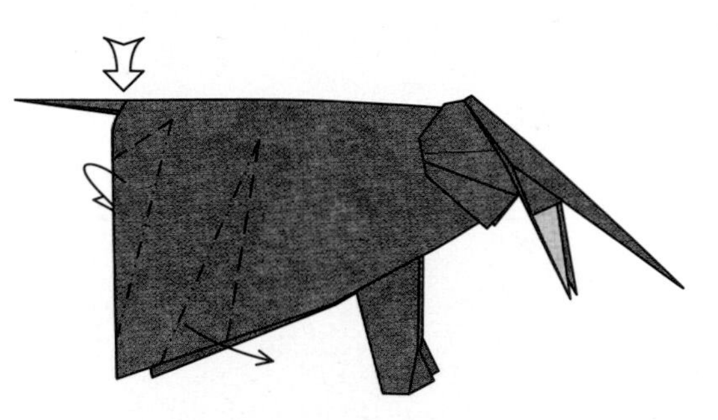

72. Crimp the rear portion of the body in two places to form legs.

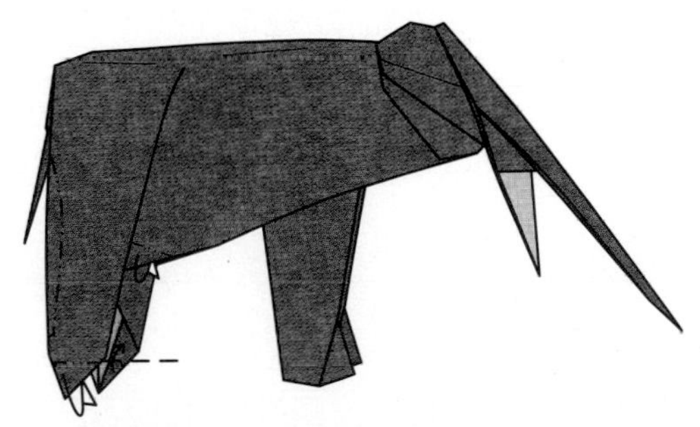

73. Fold the tips of the hind feet underneath. Round the belly and shape the backs of the legs.

74. Crimp the trunk downward and spread the layers at its tip. Shape the legs with slight mountain folds. Adjust the overall position of the limbs to a natural one.

75. Finished African Elephant.

14 Algorithms

14.1. Mathematical Model

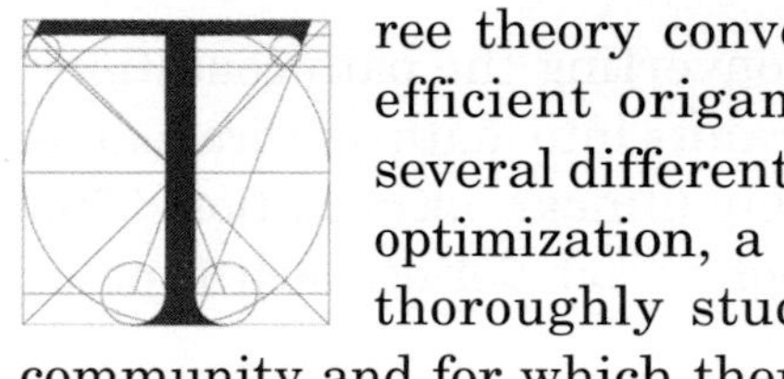

Tree theory converts the problem of finding an efficient origami crease pattern into one of several different types of nonlinear constrained optimization, a type of problem that has been thoroughly studied by the computer science community and for which there are many known algorithms for solution.

The key quantity here is efficiency; the goal is to make the most efficient model (i.e., largest in proportion to the starting paper). The quantity to maximize is therefore the scale, which is the size of a one-unit flap compared to the size of the square. There are two families of constraints that must be satisfied for any valid crease pattern:

- The coordinates of every vertex must lie within the square.
- The separation between any two vertices on the square must be at least as large as the scaled length of the path between their corresponding two nodes as measured along the tree.

If there are N leaf nodes in a figure, the first condition sets $4N$ linear inequality constraints while the second sets $N(N-1)/2$ quadratic inequality constraints on the $2N$ vertex coordinates. In addition to these constraints, the designer can set a number of other constraints to enforce various symmetries in the model:

- A vertex can have its position set to a fixed value.

- A vertex can be constrained to lie on a line of bilateral symmetry.

- Two vertices can be constrained to lie symmetrically about a line of symmetry.

- Three vertices can be constrained to be collinear.

- An edge can be constrained to a fixed length.

- Two edges can be constrained to have the same distortion (strain).

- A path can be forced to be active.

- A path can have its angle set to a fixed value.

- A path can have its angle quantized to a multiple of a given angle.

The problem is solved by converting the path conditions of the tree theorem and all constraints into mathematical equations, specifically, a constrained nonlinear optimization. This section enumerates the equations that define each type of optimization.

14.2. Definitions and Notation

Define U to be the set of all vector-valued *vertex* coordinates $\mathbf{u}_i$, $i \in I^n$, where I^n is a set of vertex indices: $I^n = \{1 \ldots n_n\}$. Each vertex $\mathbf{u}_i$ has coordinate variables $u_{i,x}$ and $u_{i,y}$.

Define E to be the set of all *edges* e_i, $i \in I^e$ where I^e is a set of edge indices $I^e = \{1 \ldots n_e\}$. Each edge contains exactly two nodes $n_i, n_j \in e_k$. Each edge has a length and a fractional distortion, called *strain*, σ_i.

Each vertex $\mathbf{u}_i$, corresponds one-to-one with the node n_i.

Define U^t to be the set of *leaf vertices*, which are those vertices that correspond to a node connected to exactly one edge. Define I^{tn} to be the set of leaf node indices. Clearly, $U^t \subseteq U$ and $I^{tn} \subseteq I^n$.

Define P to be the set of all *paths*, P_{ij}, $i, j \in I^n$, $i \neq j$. Each path is identified by the indices of the nodes at each end of the path. Each path has a length, l_{ij}, which is given by the sum of the strained lengths of the edges in the path; that is,

$$l_{ij} \equiv \sum_{e_k \in p_{ij}} (1+\sigma_k) l_k \qquad (14\text{–}1)$$

Define P^t to be the set of *leaf paths*, which are those paths that connect two leaf nodes.

Define m to be the overall *scale* of the tree.

Define w, h to be the *width* and *height* of the paper. The paper is a rectangle whose lower left corner is the origin (0,0) and whose upper right corner is the point (w, h).

14.3. Scale Optimization

The most basic optimization is the optimization of the positions of all leaf vertices and the overall scale of the design. This is equivalent to solving the problem:

$$\text{minimize } (-m) \text{ over } \{m, \mathbf{u}_i \in U^t\} \text{ s.t.:} \tag{14–2}$$

$$0 \le u_{i,x} \le w \text{ for all } \mathbf{u}_i \in U^t \tag{14–3}$$

$$0 \le u_{i,y} \le h \text{ for all } \mathbf{u}_i \in U^t \tag{14–4}$$

$$m \sum_{e_k \in p_{ij}} [(1+\sigma_k) l_k] - \sqrt{(u_{i,x} - u_{j,x})^2 + (u_{i,y} - u_{j,y})^2} \le 0$$

$$\text{for all } p_{ij} \in P^t. \tag{14–5}$$

14.4. Edge Optimization

Edge optimization is used to selectively lengthen flaps by the same relative amount to fill out a crease pattern. The scale m is fixed, and a subset of the edges E^s is subjected to the same variable strain σ. A subset of the leaf vertices U^s is allowed to move. The edge optimizer solves the problem:

$$\text{minimize } (-\sigma) \text{ over } \{\sigma, \mathbf{u}_i \in U^s\} \text{ s.t.} \tag{14–6}$$

$$0 \le u_{i,x} \le w \text{ for all } \mathbf{u}_i \in U^s \tag{14–7}$$

$$0 \le u_{i,y} \le h \text{ for all } \mathbf{u}_i \in U^s \tag{14–8}$$

$$m \left[\sum_{\substack{e_k \in p_{ij} \\ e_k \in E^s}} [(1+\sigma) l_k] + \sum_{\substack{e_k \in p_{ij} \\ e_k \in E^s}} [(1+\sigma_k) l_k] \right] - \sqrt{(u_{i,x} - u_{j,x})^2 + (u_{i,y} - u_{j,y})^2} \le 0$$

$$\text{for all } p_{ij} \in P^t. \tag{14–9}$$

Equation (14–9) can be broken into a fixed part and a variable part:

$$m \cdot \underbrace{\sum_{\substack{e_k \in p_{ij} \\ e_k \in E^s}} [l_k]}_{fixed} + \underbrace{\sigma}_{variable} \cdot \underbrace{m \sum_{\substack{e_k \in p_{ij} \\ e_k \in E^s}} [l_k]}_{fixed} + \underbrace{m \sum_{\substack{e_k \in p_{ij} \\ e_k \notin E^s}} [(1+\sigma_k) l_k]}_{fixed} - \underbrace{\sqrt{(u_{i,x} - u_{j,x})^2 + (u_{i,y} - u_{j,y})^2}}_{fixed\ or\ variable} \leq 0. \quad (14\text{–}10)$$

Note, however, that the last term may or may not be a mixture of fixed and variable parts depending upon which vertices are moving.

14.5. Strain Optimization

Strain optimization is used to distort the edges of the tree minimally in order to impose other global constraints, e.g., symmetry, particular angles, etc., on the overall crease pattern. As with edge optimization, the overall scale is fixed, but in this case, rather than *maximizing* the same strain for all affected edges, the strain optimization should *minimize* the RMS strain for a large set of edges with each edge potentially having a different strain. As with the edge optimization, there is a set of strainable edges E^s and a set of moving vertices U^s. The strain optimizer solves the problem:

$$\text{minimize } \sum_{e_i \in E^s} \sigma_i^2 \text{ over } \{\sigma_i\big|_{e_i \in E^s}, \mathbf{u}_j \in U^s\} \text{ s.t.} \quad (14\text{–}11)$$

$$0 \leq u_{i,x} \leq w \text{ for all } \mathbf{u}_i \in U^s \quad (14\text{–}12)$$

$$0 \leq u_{i,y} \leq h \text{ for all } \mathbf{u}_i \in U^s \quad (14\text{–}13)$$

$$m\left[\sum_{\substack{e_k \in p_{ij} \\ e_k \in E^s}} [(1+\sigma_k) l_k] + \sum_{\substack{e_k \in p_{ij} \\ e_k \notin E^s}} [(1+\sigma_k) l_k]\right] - \sqrt{(u_{i,x} - u_{j,x})^2 + (u_{i,y} - u_{j,y})^2} \leq 0. \text{ for all } p_{ij} \in P^t. \quad (14\text{–}14)$$

Equation (14–14) can also be broken into fixed and variable parts:

$$\sum_{\substack{e_k \in p_{ij} \\ e_k \in E^s}} \left[\underbrace{\sigma_k}_{variable} \underbrace{m l_k}_{fixed}\right] + \underbrace{m\left[\sum_{\substack{e_k \in p_{ij} \\ e_k \in E^s}} [l_k] + \sum_{\substack{e_k \in p_{ij} \\ e_k \notin E^s}} [(1+\sigma_k) l_k]\right]}_{fixed} - \underbrace{\sqrt{(u_{i,x} - u_{j,x})^2 + (u_{i,y} - u_{j,y})^2}}_{fixed\ or\ variable} \leq 0. \quad (14\text{–}15)$$

14.6. Optional Conditions

In addition to the conditions specified above, which are always required to be satisfied, one will commonly impose additional conditions on the crease pattern that are typically strict equalities. These additional conditions are placed to enforce various symmetries, either symmetries of the subject (that the model is bilaterally symmetric) or symmetries of the crease pattern (that major creases, i.e., active paths, lie at fixed angles). These conditions are easily incorporated into the nonlinear constrained optimization machinery as additional equality constraints. Several conditions and related equations commonly implemented in tree theory are listed below.

1. Vertex position fixed.

A vertex $\mathbf{u}_i$ that has one or both of its coordinates fixed to a point $\mathbf{a}$ must satisfy one or both of the equations

$$u_{i,x} - a_{i,x} = 0\,, \tag{14–16}$$

$$u_{i,y} - a_{i,y} = 0\,. \tag{14–17}$$

2. Vertex fixed to corner of paper.

A vertex $\mathbf{u}_i$ that is constrained to lie on a corner of the paper must satisfy both equations

$$(u_{i,x} - w)\cdot u_{i,x} = 0\,, \tag{14–18}$$

$$(u_{i,y} - h)\cdot u_{i,y} = 0\,. \tag{14–19}$$

3. Vertex fixed to edge of paper.

A vertex $\mathbf{u}_i$ that is fixed to lie on the edge of the paper must satisfy the equation

$$(u_{i,x} - w)\cdot u_{i,x}\cdot(u_{i,y} - h)\cdot u_{i,y} = 0\,. \tag{14–20}$$

4. Vertex fixed to line.

A vertex $\mathbf{u}_i$ that is constrained to lie on a line through point $\mathbf{a}$ running at angle α, such as a line of symmetry, must satisfy the equation

$$(u_{i,x} - a_x)\cos\alpha - (u_{i,y} - a_y)\sin\alpha = 0\,. \tag{14–21}$$

5. Two vertices paired about a line.

Two vertices $\mathbf{u}_i$ and $\mathbf{u}_j$ that are constrained to be mirror-symmetric about a line through point **a** running at angle α, such as a line of symmetry, must satisfy the two equations

$$(u_{i,x} - u_{j,x})\cos\alpha - (u_{i,y} - u_{j,y})\sin\alpha = 0 \qquad (14\text{–}22)$$

$$(u_{i,x} + u_{j,x} - 2a_{i,x})\sin\alpha - (u_{i,y} + u_{j,y} - 2a_{i,y})\cos\alpha = 0\,. \qquad (14\text{–}23)$$

6. Three vertices collinear.

Three vertices $\mathbf{u}_i$, $\mathbf{u}_j$, and $\mathbf{u}_k$ that are constrained to be collinear must satisfy the equation

$$(u_{j,y} - u_{i,y})(u_{k,x} - u_{j,x}) - (u_{k,y} - u_{j,y})(u_{j,x} - u_{i,x}) = 0\,. \qquad (14\text{–}24)$$

7. Edge length fixed

An edge e_k whose length is fixed must have its strain satisfy the equation

$$\sigma_k = 0. \qquad (14\text{–}25)$$

8. Edges same strain.

Two edges e_j and e_k that have the same strain must satisfy the equation

$$\sigma_j - \sigma_k = 0\,. \qquad (14\text{–}26)$$

9. Path active.

A path p_{ij} between two vertices $\mathbf{u}_i$ and $\mathbf{u}_j$ that is constrained to be active must satisfy the equation

$$m \sum_{e_k \in p_{ij}} [(1+\sigma_k) l_k] - \sqrt{(u_{i,x} - u_{j,x})^2 + (u_{i,y} - u_{j,y})^2} = 0. \qquad (14\text{–}27)$$

10. Path angle fixed.

A path p_{ij} between two vertices $\mathbf{u}_i$ and $\mathbf{u}_j$ that is constrained to lie at an angle α must satisfy the equation

$$(u_{i,x} - u_{j,x})\cos\alpha - (u_{i,y} - u_{j,y})\sin\alpha = 0\,. \qquad (14\text{–}28)$$

11. Path angle quantized.

A path p_{ij} between two vertices $\mathbf{u}_i$ and $\mathbf{u}_j$ that is constrained to lie at an angle of the form $\alpha_k = \alpha_0 + k \cdot \delta\alpha$ where

$$\delta\alpha \equiv \frac{180^\circ}{N}$$

must satisfy the equation

$$2^{N-1} \cdot \left[\left[(u_{i,x} - u_{j,x})^2 + (u_{i,y} - u_{j,y})^2 \right]^{-N/2} \right] \cdot \left[\prod_{k=0}^{N-1} \left[(u_{i,x} - u_{j,x}) \sin\alpha_k - (u_{i,y} - u_{j,y}) \cos\alpha_k \right] \right] = 0. \quad (14\text{–}29)$$

This function has the property that it goes to zero when the path is quantized, goes to ±1 in between quantized paths, and has no gradient component in the direction of shortening the path (which can cause problems when there are many such constraints). However, the code for the gradient of this function is rather complicated.

14.7. Stubs

A *stub* is a leaf node and edge added to the tree—usually emanating from the middle of an existing edge—such that the new leaf node creates exactly four active paths to other nodes in the tree.

When a stub is added inside of an existing N-sided polygon it breaks the polygon into four new polygons that all have fewer than N sides. Thus, by addition of stubs, high-order polygons are converted to lower-order polygons, until eventually all polygons in the crease patterns are triangles and can be filled with rabbit-ear molecules. This process is called triangulation of a crease pattern. By repeatedly adding stubs, any crease pattern can be fully triangulated.

I introduce a few more definitions:

A polygon Q is defined by a set of leaf vertices U^{tQ} that form the vertices of the polygon and a set of leaf paths P^Q, which are all paths that span U^Q.

Define the set of vertices U^Q and edges E^Q as all vertices and edges whose nodes are contained within one or more of the paths P^Q; U^Q and E^Q constitute the *subtree* of polygon Q. Define U^{tQ} as the vertices in U^Q that are also leaf vertices, i.e., $U^{tQ} = U^Q \cap U^{tm}$.

Let e_{ab} be an edge of the subtree with $\mathbf{u}_a$ the vertex at one end of e_{ab} and the $\mathbf{u}_b$ vertex at the other end.

In general, for every edge e_{ab} and set of four distinct vertices $\mathbf{u}_i$, $\mathbf{u}_j$, $\mathbf{u}_k$, $\mathbf{u}_l$, there is a stub that terminates on a new vertex $\mathbf{u}_m$; the stub is defined by four quantities:

- The vertex coordinates $u_{m,x}$ and $u_{m,y}$;
- The distance d_m from node $\mathbf{u}_a$ at which the new stub emanates from edge e_{ab};
- The length l_m of the new stub.

These four variables are found by solving the four simultaneous equalities

$$m\left[\sum_{e_k \in p_{ia}} l(e_k) + \begin{Bmatrix} d_m & \text{if } e_{ab} \in p_{ia} \\ -d_m & \text{if } e_{ab} \notin p_{ia} \end{Bmatrix} + l_m\right] - \sqrt{(u_{i,x} - u_{m,x})^2 + (u_{i,y} - u_{m,y})^2} = 0 \qquad (14\text{–}30)$$

for each of the four vertices $\mathbf{u}_i$, $\mathbf{u}_j$, $\mathbf{u}_k$, $\mathbf{u}_l$. Although a solution can potentially be found for any four vertices, not all are valid; the only valid combinations of vertices $\mathbf{u}_i$, $\mathbf{u}_j$, $\mathbf{u}_k$, $\mathbf{u}_l$ and edges e_{ab} are those for which (a) the four vertices are distinct, and (b) *both* signs of d_m are represented among the four equations. In addition, solutions for d_m that are negative or greater than the length of e_{ab} are nonphysical and must be discarded.

Note that for these equations to be used as written above, there can be no unrelieved strain in the system. They can clearly be modified to include strain.

14.8. Universal Molecule

The universal molecule is a crease pattern that is constructed by a series of repeated reductions of polygons. The construction is carried out by insetting the polygons and constructing reduced paths and fracturing the resulting network into still smaller polygons of lower order. As with triangulation, the process is guaranteed to terminate.

We use the same polygon definitions as were used in the description of stubs. In addition:

Assume the vertices $\mathbf{u}_i \in U^Q$ are ordered by their index, i.e., the N-sided polygon contains the indices $\mathbf{u}_1, \mathbf{u}_2, \ldots \mathbf{u}_N$, ordered as you travel clockwise around the polygon. Construct the following:

α_i is half of the interior angle at vertex $\mathbf{u}_i$. We define the quantity $\varsigma_i \equiv \cot\alpha_i$.

Define R_{90} as the operator that rotates a vector clockwise by 90°, i.e., $R_{90} \circ (u_x, u_y) \equiv (u_y, -u_x)$.

Define N as the operator that normalizes a vector, i.e.,

$$N \circ \mathbf{u} \equiv \frac{\mathbf{u}}{|\mathbf{u}|}. \tag{14-31}$$

$\mathbf{r}_i$ is the scaled bisector of the angle formed at $\mathbf{u}_i$, pointing toward the interior of the polygon with magnitude $\csc \alpha_i$. The vector $\mathbf{r}_i$ as well as ς_i can be constructed according to the following prescription:

$$\mathbf{r}' \equiv N \circ (\mathbf{r}_{i-1} - \mathbf{r}_i), \tag{14–32}$$

$$\mathbf{r}'' \equiv N \circ (\mathbf{r}_{i+1} - \mathbf{r}_i), \tag{14–33}$$

$$\mathbf{r}''' \equiv N \circ R_{90} \circ (\mathbf{r}'' - \mathbf{r}'), \tag{14–34}$$

$$\mathbf{r}_i = \frac{\mathbf{r}'''}{\mathbf{r}' \cdot [R_{90} \circ \mathbf{r}''']} \tag{14–35}$$

$$\varsigma_i = \mathbf{r}_i \cdot \mathbf{r}'. \tag{14–36}$$

The inset distance h is the largest value such that

1. For every path p_{ij} of length l_{ij} between nonadjacent vertices $\mathbf{u}_i$ and $\mathbf{u}_j$,

$$\sqrt{(u_x + hr_x)^2 + (u_y + hr_y)^2} \le m[l_{ij} - h(\varsigma_i + \varsigma_j)]. \tag{14–37}$$

2. For every path p_{ij} between adjacent vertices $\mathbf{u}_i$ and $\mathbf{u}_j$,

$$h \le \frac{\mathbf{u} \cdot \mathbf{u}}{\mathbf{u} \cdot \mathbf{r}}, \tag{14–38}$$

where

$$\mathbf{u} \equiv \mathbf{u}_i - \mathbf{u}_j, \tag{14–39}$$

$$\mathbf{r} \equiv \mathbf{r}_i - \mathbf{r}_j. \tag{14–40}$$

Although this problem is also a nonlinear constrained optimization, since there is only one variable and the number of paths is typically small, it can be solved directly by simply solving for each equality for all possible paths p_{ij} and taking the smallest positive real value of h found. The second relation, Equation (14–38), gives the value for h by replacing the inequality by equality. The solution for equality for (1), Equation (14–37), which is quadratic in h, is given by the following sequential substitutions:

$$w = \varsigma_i + \varsigma_j, \tag{14–41}$$

$$a = \mathbf{r} \cdot \mathbf{r} - w^2, \tag{14–42}$$

$$b = \mathbf{u} \cdot \mathbf{r} + l_{ij} w, \tag{14–43}$$

$$c = \mathbf{u} \cdot \mathbf{u} - l_{ij}^2, \tag{14–44}$$

$$h = \frac{-b + \sqrt{b^2 - ac}}{a}. \tag{14–45}$$

Obviously, negative or complex values of h should be ignored.

Once a solution is found, we create a set of new reduced vertices $\mathbf{u}'_{ij}$ and reduced paths p'_{ij} of length l'_{ij}:

$$\mathbf{u}'_i = \mathbf{u}_i + h\mathbf{r}_i, \tag{14–46}$$

$$l'_{ij} = l_{ij} - h(\varsigma_i + \varsigma_j). \tag{14–47}$$

The reduced vertices and reduced paths may then be checked for active status and subdivided into polygons, and the process repeated.

References

1. Introduction

While the number of origami elephants runs in the hundreds, there are published folding instructions for only a fraction of them. Folding instructions for the models in Figure 1.1 may be found in the following publications:

Lionel Albertino, *Safari Origami*, Gieres, l'Atelier du Grésivaudan, 1999, p. 5 (Albertino's Elephant).

Steve and Megumi Biddle, *The New Origami*, New York, St. Martin's Press, 1993, p.156 (Biddle's Elephant).

Dave Brill, *Brilliant Origami*, Tokyo, Japan Publications, 1996, p.148 (Brill's Elephant).

Vicente Palacios, *Fascinating Origami*, New York, Dover Publications, 1996, pp. 53, 57, 144, 147 (Cerceda's Elephants 1–4).

Paulo Mulatinho, *Origami: 30 Fold-by-Fold Projects*, Grange Books, 1995, p. 32 (Corrie's Elephant).

Robert Harbin, *Origami 4*, London, Coronet Books, 1977, p. 81 (Elias's Elephant).

Peter Engel, *Folding the Universe: Origami from Angelfish to Zen*, New York, Vintage Books, 1989, p. 277 (Engel's Elephant).

Rick Beech, *Origami: The Complete Guide to the Art of Paperfolding*, London, Lorenz Books, 2001, p. 96 (Enomoto Elephant).

Thomas Hull, *Russian Origami*, New York, St. Martin's Press, 1998, p. 81 (Fridryh Elephant).

Isao Honda, *The World of Origami*, Tokyo, Japan Publications, 1965, p. 168 (Honda Elephant).

Kunihiko Kasahara, *Origami Omnibus*, Tokyo, Japan Publications, 1988, p. 162 (Kasahara Elephant).

Toyoaki Kawai, *Origami*, Tokyo, Hoikusha Publishing Co., 1970, p. 10 (Kawai Elephant).

Mari Kanegae, ed., *A Arte Dos Mestres De Origami*, Rio de Janeiro, Aliança Cultural Brasil-Japão, 1997, p. 119 (Kobayashi Elephant).

Robert J. Lang, *The Complete Book of Origami*, New York, Dover Publications, 1988, p. 68 (Lang Elephant).

John Montroll, *Origami for the Enthusiast*, New York, Dover Publications, 1979, p. 67 (Montroll 1 Elephant).

John Montroll, *Animal Origami for the Enthusiast*, New York, Dover Publications, 1985, p. 70 (Montroll 2 Elephant).

John Montroll, *Origami Sculptures*, New York, Dover Publications, 1990, p. 130 (Montroll 3 Elephant).

John Montroll, *African Animals in Origami*, New York, Dover Publications, 1991, p. 79 (Montroll 4 Elephant).

John Montroll, *Bringing Origami to Life*, New York, Dover Publications, 1999, p. 90 (Montroll 5 Elephant).

John Montroll, *Teach Yourself Origami*, New York, Dover Publications, 1998, p. 109 (Montroll 6 Elephant).

John Montroll, *Origami Inside Out*, New York, Dover Publications, 1993, p. 75 (Montroll 7 Elephant).

Robert Harbin, *Secrets of Origami*, London, Octopus Books, 1971, p. 224 (Neale Elephant).

Thomas Hull and Robert Neale, *Origami Plain and Simple*, New York, St. Martin's Press, 1994, p. 72 (Neale Elephant Major).

Thomas Hull and Robert Neale, *Origami Plain and Simple*, New York, St. Martin's Press, 1994, p. 89 (Neale Elephant Minor).

Robert Harbin, *Origami 3*, London, Coronet Books, 1972, p. 121 (Noble Elephant).

Samuel L. Randlett, *Best of Origami*, New York, E. P. Dutton, 1963, p. 134 (Rhoads Elephant).

Hector Rojas, *Origami Animals*, New York, Sterling Publishing Co., 1993, p. 37 (Rojas Elephant).

Robert Harbin, *Origami: the Art of Paperfolding*, New York, Funk & Wagnalls, 1969, p. 182 (Ward & Hatchett Elephant).

Robert J. Lang and Stephen Weiss, *Origami Zoo*, New York, St. Martin's Press, 1990, p. 95 (Weiss Mammoth).

An extensive and continually updated list of published instructions for elephants (and many other subjects) may be found at an online origami model database, currently at:

www.origamidatabase.com

2. Building Blocks

The now-standard system of origami lines and arrows is called the Yoshizawa-Harbin-Randlett system. It was devised by Yoshizawa, modified and popularized in the West by Robert Harbin and Samuel L. Randlett, and is described in

Akira Yoshizawa, *Origami Dokuhon*, Tokyo, Kamakura Shobo, 1957.

Robert Harbin, *Secrets of Origami*, ibid.

Samuel L. Randlett, *The Art of Origami*, New York, E. P. Dutton, 1961.

3. Elephant Design

Dave Mitchell's One-Fold Elephant, along with several other minimalist elephant designs, may be found in:

Paul Jackson, "An Elephantine Challenge: Part 3," *British Origami #161*, August, 1993, pp. 4–7.

4. Traditional Bases

The Sea Urchin is contained in

John Montroll and Robert J. Lang, *Origami Sea Life*, New York, Dover Publications, 1990, p. 147.

Eric Kenneway's column, "The ABCs of Origami," which originally appeared in *British Origami* in 1979–1980, has been expanded and reprinted as

Eric Kenneway, *Complete Origami*, New York, St. Martin's Press, 1987.

More on the system of triangle dissections and their relationship to origami design can be found in

Peter Engel, *Folding the Universe*, ibid.

Robert J. Lang, "Albert Joins the Fold," *New Scientist*, vol. 124, no. 1696/1697, December 23/30, pp. 38–57, 1989.

Robert J. Lang, "Origami: Complexity Increasing," *Engineering & Science*, vol. 52, no. 2, pp. 16–23, 1989.

Jun Maekawa, "Evolution of Origami Organisms," *Symmetry: Culture and Science*, vol. 5, no. 2, pp. 167–177, 1994.

Several novel treatments of the traditional bases may be found among the work of Neal Elias and Fred Rohm. See, for example:

Robert Harbin, *Secrets of Origami*, ibid., p. 212 (Rohm's Hippopotamus).

Robert Harbin, *Origami 4*, London, Coronet Books, pp. 132–133 (Elias's Chick Hatching).

Robert Harbin, *Origami 4*, ibid., pp. 134–135 (Elias's Siesta).

Pete Ford (ed.), *The World of Fred Rohm (BOS Booklet #49)*, London, British Origami Society, 1998.

Pete Ford (ed.), *The World of Fred Rohm (BOS Booklet #50)*, London, British Origami Society, 1998.

Pete Ford (ed.), *The World of Fred Rohm(BOS Booklet #51)*, London, British Origami Society, 1998.

The offset Bird Base has been thoroughly explored by Dr. James Sakoda in

James Minoru Sakoda, *Modern Origami*, New York, Simon and Schuster, 1969.

Yoshizawa's Crab, an example of a double-blintzed Frog Base, may be found in

Akira Yoshizawa, *Sosaku Origami*, Tokyo, Nippon Hoso Shuppan Kyokai, 1984, pp. 72–73.

Rhoads's Elephant, an example of a blintzed Bird Base, may be found in

Samuel L. Randlett, *Best of Origami*, ibid., p. 134.

5. Splitting Points

The Yoshizawa split is shown in his Horse in

Akira Yoshizawa, *Origami Dokuhon*, ibid., p. 61.

For an example of the middle-point split shown in Figure 5.15, see the Praying Mantis in

Robert J. Lang, *Origami Insects and their Kin*, New York, Dover Publications, 1995, p. 106.

A full folding sequence for the Walrus of Figure 5.28 may be found in

John Montroll and Robert J. Lang, *Origami Sea Life*, ibid., pp. 31–33.

A full folding sequence for the Grasshopper of Figure 5.29 may be found in

Robert J. Lang, *Origami Insects & Their Kin*, ibid., pp. 59–65.

6. Grafting

The technique of folding from squares joined only at their corners is described in a two-volume set that includes a reproduction of the 1797 original text of *Sembazuru Orikata*:

Masao Okamura, *Hiden Sembazuru Orikata: Fukkoku to Kaisetsu*, Tokyo, NOABooks, 1992.

See also

Masaki Sakai and Michi Sahara, *Origami Roko-an Style*, Tokyo, Heian International Publishing, 1998.

Kasahara's Dragon may be found in

Kunihiko Kasahara, *Creative Origami*, Tokyo, New York, Japan Publications, 1967, p. 86.

Robert Neale's Dragon may be found in

Robert Neale, "Dragon," *The Flapping Bird/An Origami Monthly*, Chicago, Jay Marshall, vol. 1, no. 5, p. 27, 1969.

A full folding sequence for the Crawfish of Figure 6.30 may be found in

Robert J. Lang, *Origami Animals*, New York, Crescent Books, 1992, pp. 52–55.

A full folding sequence for the Treehopper of Figure 6.31 may be found in

Robert J. Lang, *Origami Insects & Their Kin*, ibid., pp. 10–13.

A full folding sequence for the Japanese Horned Beetle of Figure 6.31 may be found in

Robert J. Lang, *Origami Insects & Their Kin*, ibid., pp. 132–142.

A bird with individual toes that appears to have been constructed using point splitting techniques appears in:

Akira Yoshizawa, *Origami Dokuhon II*, Tokyo, New Science Sha, 1998, p. 3.

7. Pattern Grafting

John Richardson's Hedgehog may be found in

Eric Kenneway, *Origami: Paperfolding for Fun*, London, Octopus, 1980, pp. 86–87.

Eric Joisel's Pangolin may be found in

Michael G. LaFosse, *Origamido: Masterworks of Paper Folding*, Gloucester, Rockport, 2000, pp. 15–16.

Examples of Chris K. Palmer's tessellation patterns may be found in

Chris K. Palmer, "Extruding and tessellating polygons from a plane," *Origami Science & Art: Proceedings of the Second International Meeting of Origami Science and Scientific Origami*, Koryo Miura, ed., Otsu, Japan, Nov. 29–Dec. 4, 1994, pp. 323-331.

Michael G. LaFosse, *Paper Art: The Art of Sculpting with Paper*, Gloucester, Rockport, 1998, pp. 26–33.

Further examples may be found at

http: //www.shadowfolds.com

Other patterns of intersecting pleats may be found in

Paulo Taborda Barreto, "Lines meeting on a surface: the 'Mars' paperfolding," *Origami Science & Art: Proceedings of the Second International Meeting of Origami Science and Scientific Origami*, ibid., pp. 343-359.

Alex Bateman, "Computer tools and algorithms for origami tessellation design," in *Origami*^{3}, Thomas Hull, ed., Natick, A K Peters, 2002, pp. 121–127.

8. Tiling

Two origami masters who have extensively utilized tiling as a design methodology in their work are Peter Engel and Jun Maekawa. You can find both theory and many elegant examples of different tiles and models incorporating them in the following:

Peter Engel, *Folding the Universe: Origami from Angelfish to Zen*, ibid.

Kunihiko Kasahara, *Viva! Origami*, Tokyo, Sanrio, 1983.

See also examples of grafting in

Peter Engel, "Breaking Symmetry: origami, architecture, and the forms of nature," *Origami Science & Art: Proceedings of the Second International Meeting of Origami Science and Scientific Origami*, ibid., pp. 119–145.

Both tiling and grafting are described in

Jun Maekawa, "Evolution of Origami Organisms," *Symmetry: Culture and Science*, vol. 5, no. 2, 1994, pp. 167–177.

A design example using circles for allocation of points may be found in

Fumiaki Kawahata, "Seiyaku-eno chosen: kado-no oridashikata [Challenge to restrictions: how to make points]", *Oru*, no. 2, Autumn 1993, pp. 100–104.

A full folding sequence for the Shiva of Figure 8.47 may be found in

Jay Ansill, *Mythical Beings*, New York, HarperPerennial, 1992, pp. 70–75.

A full folding sequence for the Hercules Beetle of Figure 8.48 may be found in

Robert J. Lang, *Origami Insects and their Kin*, ibid., pp. 82–89.

A full folding sequence for the Praying Mantis of Figure 8.49 may be found in

Robert J. Lang, *Origami Insects and their Kin*, ibid., pp. 106–113.

A full folding sequence for the Periodical Cicada of Figure 8.52 may be found in

Robert J. Lang, *Origami Insects II*, Tokyo, Gallery Origami House (to be published).

A full folding sequence for the Pill Bug of Figure 8.55 may be found in

Robert J. Lang, *Origami Insects II*, ibid.

A full folding sequence for the Centipede of Figure 8.56 may be found in

Robert J. Lang, *Origami Insects II*, ibid.

9. Circle Packing

Montroll's Five-Sided Square may be found in

John Montroll, *Animal Origami for the Enthusiast*, ibid., pp. 21–22.

Toshiyuki Meguro describes a circle/river design of a flying insect in

Toshiyuki Meguro, "'Tobu Kuwagatamushi'-to Ryoikienbunshiho ['Flying Stag Beetle' and the circular area molecule method]", *Oru* no. 5, Summer 1994, pp. 92–95.

See also

Seiji Nishikawa, "'Tora' Saiko ['Tiger' Reconsidered]", *Oru* no. 7, Winter 1994, pp. 89–93.

A full folding sequence for the Tarantula of Figure 9.24 may be found in

Robert J. Lang, *Origami Insects II*, ibid.

A full folding sequence for the Flying Cicada of Figure 9.25 may be found in

Robert J. Lang, *Origami Insects II*, ibid.

A full folding sequence for the Flying Ladybird Beetle of Figure 9.26 may be found in

Robert J. Lang, *Origami Insects II*, ibid.

A full folding sequence for the *Acrocinus longimanus* of Figure 9.27 may be found in

Robert J. Lang, *Origami Insects II*, ibid.

Various mathematical circle packings may be found in

Jonathan Schaer and A. Meir, “On a geometric extremum problem,” *Canadian Mathematical Bulletin*, 8, 1965, pp. 21–27.

Jonathan Schaer, “The densest packing of nice circles in a square,” *Canadian Mathematical Bulletin*, 8, 1965, pages 273–277.

Michael Goldberg, “The packing of equal circles in a square,” *Mathematics Magazine*, 43, 1970, pages 24–30.

Benjamin L. Schwartz, “Separating points in a square,” *Journal of Recreational Mathematics*, 3, 1970, pp. 195–204.

Jonathan Schaer, “On the packing of ten equal circles in a square,” *Mathematics Magazine*, 44, 1971, pages 139–140.

Benjamin L. Schwartz, “Separating points in a rectangle,” *Mathematics Magazine*, 46, 1973, pages 62–70.

R. Milano, “Configurations optimales de disques dans un polygone régulier,” *Mémoire de Licence*, Unversité Libre do Bruxelles, 1987.

Guy Valette, “A better packing of ten equal circles in a square,” *Discrete Mathematics*, 76, 1989, pages 57–59.

Michael Molland and Charles Payan, “Some progress in the packing of equal circles in a square,” *Discrete Mathematics*, 84, 1990, pages 303–305.

Martin Gardner, “Tangent Circles,” *Fractal Music and Hypercards*, W. H. Freeman, 1992, pp. 149–166.

Hans Melissen, *On the Packing of Circles*, Ph.D. Thesis, University of Utrecht, 1997.

George Rhoads’s Bug, made from a nine-circle-packing base, may be found in:

Samuel L. Randlett, *The Best of Origami*, ibid., pp. 130–131.

Most of the discussion, counterexample, and solution strategy for the Margulis Napkin Problem is captured at David Eppstein's Geometry Junkyard,

http: //www.ics.uci.edu/~eppstein/junkyard/napkin.html

10. Molecules

A full folding sequence for the Ant of Figure 10.43 may be found in

Robert J. Lang, *Origami Insects II*, ibid.

A full folding sequence for the Cockroach of Figure 10.44 may be found in

Robert J. Lang, *Origami Insects II*, ibid.

A full folding sequence for the *Eupatorus gracilicornus* of Figure 10.51 may be found in

Robert J. Lang, *Origami Insects II*, ibid.

The Maekawa-Justin theorem is described in

Thomas Hull, "The combinatorics of flat folds: a survey," in *Origami*^{3}, ibid., pp. 29–37.

Toshiyuki Meguro describes circle packing and several types of molecules in:

Toshiyuki Meguro, "Jitsuyou origami sekkeihou [Practical methods of origami designs]," *Origami Tanteidan Shinbun*, nos. 7–14, 1991–1992.

11. Tree Theory

A partial description of tree theory is given in

Robert J. Lang, "Mathematical algorithms for origami design," *Symmetry: Culture and Science*, vol. 5, no. 2, 1994, pp. 115–152.

Robert J. Lang, "The tree method of origami design," *Origami Science & Art: Proceedings of the Second International Meeting of Origami Science and Scientific Origami*, ibid., pp. 73–82.

A more complete and more formal treatment may be found in

Robert J. Lang, "A computational algorithm for origami design," *Computational Geometry: 12th Annual ACM Symposium*, Philadelphia, Pennsylvania, May 24–26, 1996, pp. 98–105.

A full folding sequence for the Scorpion of Figure 11.35 may be found in

Robert J. Lang, *Origami Insects II*, ibid.

A full folding sequence for the Flying Grasshopper of Figure 11.36 may be found in

Robert J. Lang, *Origami Insects II*, ibid.

The properties of distorted Bird Base crease patterns and associated quadrilaterals are summarized in

Toshikazu Kawasaki, "The geometry of orizuru," in *Origami³*, ibid., pp. 61–73.

Fumiaki Kawahata's string-of beads method and the associated molecules are described in

Fumiaki Kawahata, "The technique to fold free flaps of formative art 'origami,'" in Koryo Miura (ed.), *Origami Science & Art: Proceedings of the Second International Meeting of Origami Science and Scientific Origami*, ibid., pp. 63–71.

Fumiaki Kawahata, *Fantasy Origami*, Tokyo, Gallery Origami House, 1995, pp. 174–179.

Additional papers on the underlying mathematics of origami include the following:

Esther M. Arkin, Michael A. Bender, Erik D. Demaine, Martin L. Demaine, Joseph S. B. Mitchell, Saurabh Sethia, and Steven S. Skiena, "When can you fold a map?," *Proceedings of the 7th Workshop on Algorithms and Data Structures,* edited by F. Dehne, J.–R. Sack, and R. Tamassia, *Lecture Notes in Computer Science,* volume 2125, Providence, Rhode Island, August 2001, pp. 401–413.

Hideki Azuma, "Some mathematical observations on flat foldings (abstract)," *Abstracts for the Second International Meeting of Origami Science and Scientific Origami,* Otsu, Japan, 1994, pp. 45–46.

Marshall Bern and Barry Hayes, "On the complexity of flat origami," *Proceedings of the 7th ACM-SIAM Symposium on Discrete Algorithms*, Atlanta, GA., 1996, pp. 175–183.

Marshall Bern, Erik Demaine, David Eppstein, and Barry Hayes, "A disk-packing algorithm for an origami magic trick," in *Origami³*, ibid., pp. 17–28.

Marshall Bern, Erik Demaine, David Eppstein, and Barry Hayes, "A disk–packing algorithm for an origami magic trick," *Proceedings of the International Conference on Fun with Algorithms,* Isola d'Elba, Italy, June 1998, pp. 32–42.

Therese C. Biedl, Erik D. Demaine, Martin L. Demaine, Anna Lubiw, and Godfried T. Toussaint, "Hiding disks in folded polygons," *Proceedings of the 10th Canadian Conference on Computational Geometry,* Montreal, Quebec, Canada, August 1998.

Chandler Davis, "The set of non-linearity of a convex piecewise-linear function," *Scripta Mathematica,* vol. 24, 1959, pp. 219–228.

Erik D. Demaine and Martin L. Demaine, "Folding and unfolding linkages, paper, and polyhedra," *Proceedings of the Japan Conference on Discrete and Computational Geometry: Lecture Notes in Computer Science*, Tokyo, Japan, November 2000.

Erik D. Demaine, Martin L. Demaine, "Planar drawings of origami polyhedra," *Proceedings of the 6th Symposium on Graph Drawing, Lecture Notes in Computer Science,* volume 1547, Montreal, Quebec, Canada, August 1998, pp. 438–440.

Erik D. Demaine, Martin L. Demaine, and Anna Lubiw, "Folding and cutting paper," *Revised Papers from the Japan Conference on Discrete and Computational Geometry,* edited by Jin Akiyama, Mikio Kano, and Masatsugu Urabe, *Lecture Notes in Computer Science,* volume 1763, Tokyo, Japan, December 1998, pp. 104–117.

Erik D. Demaine, Martin L. Demaine, and Anna Lubiw, "Folding and one straight cut suffice," *Proceedings of the 10th Annual ACM–SIAM Symposium on Discrete Algorithms,* 1999, pp. 891–892.

Erik D. Demaine, Martin L. Demaine, and Anna Lubiw, "The CCCG 2001 Logo," *Proceedings of the 13th Canadian Conference on Computational Geometry,* Waterloo, Ontario, Canada, August 2001, pp. iv–v.

Erik D. Demaine and Joseph S. B. Mitchell, "Reaching folded states of a rectangular piece of paper," *Proceedings of the 13th Canadian Conference on Computational Geometry,* Waterloo, Ontario, Canada, August 2001, pp. 73–75.

Erik D. Demaine, Martin L. Demaine, and Joseph S. B. Mitchell, "Folding flat silhouettes and wrapping polyhedral packages: new results in computational origami," *Computational Geometry: Theory and Applications,* 16 1, : 3–21, 2000. Preliminary versions in *Proceedings of the 15th Annual ACM Symposium on Computational Geometry* 1999, 105–114 and *Proceedings of the 3rd CGC Workshop on Computational Geometry* 1998.

Erik D. Demaine, "Folding and unfolding linkages, paper, and polyhedra," *Revised Papers from the Japan Conference on Discrete and Computational Geometry JCDCG 2000,,* edited by Jin Akiyama, Mikio Kano, and Masatsugu Urabe, *Lecture Notes in Computer Science,* volume 2098, Tokyo, Japan, November 2000, pp. 113–124.

J. P. Duncan and J. L. Duncan, "Folded developables," *Proceedings of the Royal Society of London, Series A,* vol. 383, 1982, pp. 191–205.

P. Di Francesco, "Folding and coloring problems in mathematics and physics," *Bulletin of the American Mathematical Society,* vol. 37, no. 3, July 2000, pp. 251–307.

D. Fuchs and S. Tabachnikov, "More on paperfolding," *The American Mathematical Monthly,* vol. 106, no. 1, Jan. 1999, pp. 27–35.

David A. Huffman, "Curvatures and creases: a primer on paper," *IEEE Trans. on Computers*, Volume C-25, 1976, pp. 1010–1019.

Thomas Hull, "On the mathematics of flat origamis," *Congressus Numerantium* 100, 1994, pp. 215–224.

Thomas Hull, "Origami math, parts 1, 2, 3 and 4," *Newsletter for Origami USA,* nos. 49–52, Fall 1994–Fall 1995.

Koji Husimi and M. Husimi, *The Geometry of Origami*, Tokyo, Nihon Nyoron-sha, 1979.

Jacques Justin, "Mathematics of origami, part 9," *British Origami,* June 1986, pp. 28–30.

Jacques Justin, "Aspects mathematiques du pliage de papier," *Proceedings of the First International Meeting of Origami Science and Technology,* H. Huzita, ed., 1989, pp. 263–277.

Jacques Justin, "Mathematical remarks about origami bases," *Symmetry: Culture and Science*, vol. 5, no. 2, 1994, pp. 153–165.

Jacques Justin, "Towards a mathematical theory of origami," *Origami Science and Art: Proceedings of the Second International Meeting of Origami Science and Scientific Origami*, K. Miura (ed.), Otsu, Japan 1997, pp. 15–30.

Toshikazu Kawasaki, "On the relation between mountain-creases and valley-creases of a flat origami," in *Proceedings of the First International Meeting of Origami Science and Technology*, ibid., pp. 229–237.

Toshikazu Kawasaki, "On high dimensional flat origamis," *Proceedings of the First International Meeting of Origami Science and Technology,* ibid., pp. 131–141.

Toshikazu Kawasaki, "On solid crystallographic origamis [in Japanese]," *Sasebo College of Technology Report,* vol. 24 1987, pp. 101–109.

Toshikazu Kawasaki, "On the relation between mountain–creases and valley–creases of a flat origami [abridged, English translation]," *Proceedings of the First International Meeting of Origami Science and Technology,* ibid., pp. 229–237.

Toshikazu Kawasaki, "On the relation between mountain–creases and valley–creases of a flat origami [unabridged, in Japanese]," *Sasebo College of Technology Report,* Vol. 27 1990, pp. 55–80.

Toshikazu Kawasaki, "R(gamma) =1," *Origami Science and Art: Proceedings of the Second International Meeting of Origami Science and Scientific Origami*, K. Miura ed., Otsu, Japan 1997, pp. 31–40.

Toshikazu Kawasaki and Masaaki Yoshida, "Crystallographic flat origamis," *Memoirs of the Faculty of Science, Kyushu University, Series A,* vol. 42, no. 2, 1988, pp. 153–157.

J. Koehler, "Folding a strip of stamps," *Journal of Combinatorial Theory,* vol. 5, 1968, pp. 135–152.

W. F. Lunnon, "A map–folding problem," *Mathematics of Computation,* vol. 22, no. 101, 1968, pp. 193–199.

W. F. Lunnon, "Multi–dimensional map folding," *The Computer Journal,* vol. 14, no. 1, 1971, pp. 75–80.

Jun Maekawa, "Evolution of origami organisms," *Symmetry: Culture and Science,* vol. 5, no. 2, 1994, pp. 167–177.

Jun Maekawa, "Similarity in origami (abstract)," *Abstracts for the Second International Meeting of Origami Science and Scientific Origami,* Otsu, Japan 1994, pp. 65–66.

Koryo Miura, "A note on intrinsic geometry of origami," *Proceedings of the First International Meeting of Origami Science and Technology,* ibid., pp. 239–249.

Koryo Miura, "Folds—the basis of origami," *Symmetry: Culture and Science,* vol. 5, no. 1, 1994, pp. 13–22.

Koryo Miura, "Fold—its physical and mathematical principles," *Origami Science and Art: Proceedings of the Second International Meeting of Origami Science and Scientific Origami,* K. Miura (ed.), Otsu, Japan 1997, pp. 41–50.

Ileana Streinu and Walter Whiteley, "The spherical carpenter's rule problem and conical origami folds," *Proceedings of the 11th Annual Fall Workshop on Computational Geometry,* Brooklyn, New York, November 2001.

Kunio Suzuki, "Creative origami 'snowflakes': some new approaches to geometric origami (abstract)," *Abstracts for the Second International Meeting of Origami Science and Scientific Origami,* Otsu, Japan 1994, pp. 37–38.

The program *TreeMaker* runs on Macintosh computers and is available with documentation at

http: //origami.kvi.nl/programs/treemaker/index.htm

12. Box Pleating

A full folding sequence for the Cerambycid beetle of Figure 12.48 may be found in

Robert J. Lang, *Origami Insects II*, ibid.

Many box-pleated figures from Neal Elias and Max Hulme may be found in the following:

Dave Venables, *Max Hulme: Selected Works 1973–1979 (BOS Booklet #15)*, London, British Origami Society, 1979.

Dave Venables, *Focus on Neal Elias (BOS Booklet #10)*, London, British Origami Society, 1978.

Dave Venables (ed.), *Neal Elias: Miscellaneous Folds I (BOS Booklet #34)*, London, British Origami Society, 1990.

Dave Venables (ed.), *Neal Elias: Miscellaneous Folds II (BOS Booklet #35)*, London, British Origami Society, 1990.

Dave Venables (ed.), *Neal Elias: Faces and Busts (BOS Booklet #36)*, London, British Origami Society, 1990.

Eric Kenneway, *Origami: Paperfolding for Fun*, London, Octopus, 1980, pp. 90–91 (Hulme's Fly).

13. Hybrid Bases

Engel's Butterfly may be found in

Peter Engel, *Folding the Universe: Origami from Angelfish to Zen*, ibid., pp. 292–311.

A full folding sequence for the Butterfly of Figure 13.5 may be found in

Robert J. Lang, *Origami Insects and their Kin*, ibid., pp. 40–45.

A full folding sequence for the Dragonfly of Figure 13.6 may be found in

Robert J. Lang, *Origami Insects II*, ibid.

A full folding sequence for the Rabbit of Figure 13.11 may be found in

Robert J. Lang and Stephen Weiss, *Origami Zoo*, ibid., pp. 115–119.

A full folding sequence for the Mouse of Figure 13.13 may be found in

Robert J. Lang and Stephen Weiss, *Origami Zoo*, ibid., pp. 89–92.

Origami Societies

Many countries have origami societies that hold conventions and exhibitions, sell origami supplies, and publish new and original designs. Four of the larger societies are:

Origami USA
15 W. 77th St.
New York, NY 10024
http: //www.origami-usa.org

British Origami Society
c/o Penny Groom
2a The Chestnuts
Countesthorpe
Leicester LE8 5TL
http: //www.britishorigami.org.uk/

Japan Origami Academic Society
c/o Gallery Origami House
1-33-8-216, Hakusan
Bunkyo-ku, Tokyo
113-0001, JAPAN
http: //www.origami.gr.jp/

Nippon Origami Association
2-064, Domir-Gobancho
12 Gobancho
Chiyoda-ku, Tokyo
102-0076 JAPAN
http: //www.origami-noa.com/

There are many other national origami societies and other origami-related resources on the Internet. I will not give links here (Internet links tend to have a short half-life), but any good search engine will turn up numerous sites for origami supplies, pictures, commentary, and diagrams.

Glossary of Terms

A

Active path (page 382): a path whose length on the crease pattern is equal to its minimum length as specified by the tree graph.

Active reduced path (page 399): a reduced path within a universal molecule whose length on the crease pattern is equal to its minimum length as specified by the tree graph.

Arrowhead molecule (page 344): a crease pattern within a quadrilateral that consists of a Waterbomb molecule combined with an angled dart; it allows an arbitrary four-circle quadrilateral to be collapsed while aligning the four tangent points.

Atom (page 453): a portion of a crease pattern that corresponds to a segment of a single flap within a molecule.

Axial crease (page 232): a crease in a crease pattern that lies along the axis in the folded form of a uniaxial base.

Axial polygon (page 233): a polygonal region of paper in a crease pattern outlined by axial creases. In the folded form, the entire perimeter of an axial polygon lies along the axis of the base.

Axis (page 230): a line on a base along which the edges of flaps lie and to which the hinges of flaps are perpendicular.

B

Base (page 51): a regular geometric shape that has a structure similar to that of the desired subject.

Bird Base (page 52): one of the Classic Bases, formed by petal-folding the front and back of a Preliminary Fold.

Blintzing (page 56): folding the four corners of a square to the center.

Blintzed base (page 56): any base in which the four corners of the square are folded to the center prior to folding the base.

Branch edge (page 376): in a tree graph, an edge that is connected to two branch nodes.

Branch node (page 376): in a tree graph, a node connected to two or more edges.

Branch vertex (page 383): a point in the crease pattern that corresponds to a branch node on the tree graph.

Book symmetry (page 291): the symmetry of a crease pattern that is mirror-symmetric about a line parallel to an edge and passing through the center of the paper.

Border graft (page 133): modifying a crease pattern as if you added a strip of paper along one or more sides of the square in order to add features to the base.

Box pleating (page 419): a style of folding characterized by all folds running at multiples of 45°, with the majority running at multiples of 0° and 90° on a regular grid.

C

Circle/river method (page 352): a design technique for uniaxial bases that constructs the crease pattern by packing nonoverlapping circles and rivers into a square.

Circle packing (page 282): placing circles on a square (or other shape) so that they do not overlap and their centers are inside the square.

Classic Bases (page 52): the four bases of antiquity (Kite, Fish, Bird, and Frog) that are related by a common structure.

Closed sink fold (page 35): a sink fold in which the point to be sunk must be popped from convex to concave; it cannot be entirely flattened.

Composite molecule (page 344): a molecule that contains axial creases in its interior.

Corner flap (page 103): a flap whose tip comes from one of the corners of the square.

Crease (page 11): a mark left in the paper after a fold has been unfolded.

Crease assignment (page 22): determination of whether each crease is a mountain fold, valley fold, or flat (unfolded) crease. Also called crease parity.

Crease pattern (page 21): the pattern of creases left behind on the square after a model has been unfolded.

Crimp fold (page 31): a fold formed by two parallel or nearly parallel mountain and valley folds on the near layers of a flap with their mirror image folds formed on the far layers.

Crystallization (page 294): the process of fixing the locations of circles in a circle packing by enlarging some of the circles until they can no longer move.

Cupboard Base (page 55): a traditional base consisting of a square with two opposite edges folded toward each other to meet in the middle.

D

Decreeping (page 143): rearranging several trapped layers of paper so that no layer is wrapped around another.

Detail folds (page 51): folds that transform the flaps of a base into details of the finished subject.

Diagonal symmetry (page 292): the symmetry of a crease pattern that is mirror-symmetric about one of the diagonals of the square.

Dihedral angle (page 429): the angle between the two surfaces on either side of a crease, defined as the angle between the surface normals.

Distorted base (page 67): a modified base formed by shifting the vertices of the crease pattern so that the paper can fold flat; the number of creases and vertices remains the same, but the angles between them change.

Double-blintzing (page 312): folding the four corners of a square to the center twice in succession.

Double rabbit-ear fold (page 26): a fold in which the creases of a rabbit ear are made on the near layer of a flap and the mirror-image creases are made on the far layer.

Double sink fold (page 35): two sink folds formed in succession on the same flap.

E

Edge (page 376): in a tree graph, a single line segment. Each edge corresponds to a unique flap or connector between flaps in the base. See **leaf edge**, **branch edge**.

Edge flap (page 103): a flap whose tip comes from one of the edges (but not a corner) of the square. An edge flap has twice as many layers as a same-size corner flap.

Edge weight (page 376): a number assigned to each edge of a tree graph that represents the length of the associated flap.

Efficiency (page 41): a measure of how much paper is used to obtain features of the subject versus extra paper that is merely hidden away.

Elias stretch (page 464): A maneuver used in box pleating to create flaps from a pleated region of paper, by changing the direction of the pleats by 90° within wedges of paper.

F

Fish Base (page 52): one of the Classic Bases, formed by folding all four edges of a square to a common diagonal and gathering the excess paper in two flaps.

Flap (page 52): a region of paper in an origami shape that is attached only along one edge so that it can be easily manipulated by itself.

Folded edge (page 15): an edge created by folding.

Four-circle quadrilateral (page 341): a quadrilateral formed by connecting the centers of four pairwise tangent circles; such a quadrilateral can be folded so that all edges lie on a line and the tangent points between pairs of circles touch.

Frog Base (page 52): one of the Classic Bases, formed by squash- and petal-folding the four edges of a Preliminary Fold.

G

Generic form (page 238): a crease pattern within a molecule or group of molecules in which (a) all axial creases are shown as valley creases; (b) all ridgeline creases are shown as mountain creases; and (c) all hinge creases are shown as unfolded creases. The generic form is an approximation of the actual crease pattern of a folded base.

Grafting (page 133): modifying a crease pattern as if you had spliced into it a strip or strips of paper in order to add new features to an existing base.

Grafted Kite Base (page 528): a family of bases composed by adding a border graft to two sides of a Kite Base.

Gusset (page 32): one or more narrow triangles of paper, usually formed by stretching a pleat or crimp.

Gusset molecule (page 346): a crease pattern within a quadrilateral that resembles a partially stretched Waterbomb molecule with a gusset running across its top. The gusset molecule, like the arrowhead molecule, allows any four-circle quadrilateral to be collapsed while aligning the tangent points.

H

Hinge (page 230): a joint between two flaps.

Hinge creases (page 334): creases that in a uniaxial base are perpendicular to the axis. Hinge creases define the boundaries of flaps or segments of a base.

Hole (page 466): a region of paper in a box-pleated crease pattern that does not belong to any atom.

Hybrid base (page 519): a base that is constructed using multiple design techniques.

Hybrid reverse fold (page 24): a more complicated form of reverse fold that combines aspects of both inside and outside reverse folds.

I

Ideal split (page 98): a technique for splitting a Kite Base flap, which gives the longest possible pair of flaps.

Inflation (page 294): the process of adding circles to a crease pattern (corresponding to adding flaps to a base) and expanding the circle (lengthening the flap) until it touches three or more others.

Inside reverse fold (page 23): a method of changing the direction of a flap, wherein the moving layers are inverted and tucked between the stationary layers.

K

Kite Base (page 52): the simplest of the Classic Bases, formed by folding two adjacent edges of a square to the same diagonal.

L

Leaf edge (page 376): in a tree graph, an edge connected to at least one leaf node.

Leaf node (page 376): in a tree graph, a node connected to only a single edge.

Leaf vertex (page 378): a point in the crease pattern that corresponds to a leaf node on the tree graph.

M

Middle flap (page 103): a flap whose tip comes from the interior of the square. A middle flap has twice as many layers as a same-sized edge flap and four times as many as a corner flap.

Mixed sink fold (page 37): a sink fold containing aspects of both open and closed sinks.

Molecule (page 338): a crease pattern which when folded flat has its perimeter lie along a common line and for which specified points along the perimeter (the tangent points) become coincident in the folded form.

Mountain fold (page 18): a crease that is concave downward. Usually indicated by a dot-dot-dash line (black line in crease patterns).

N

Node (page 376): in a tree graph, an endpoint of a line segment. See **leaf node, branch node**.

O

Offset base (page 66): a modified base formed by shifting the entire crease pattern on the square while preserving angles between creases, so that extra paper is created in some locations while others lose paper.

Open sink fold (page 34): a sink fold in which the point to be sunk can be entirely flattened during the course of the sink.

Origami (page 1): the art of folding paper into decorative shapes, usually from uncut squares.

Origami sekkei: see **Technical folding**.

Ortholinear river (page 447): the analog of a river in box-pleated designs. The river has constant vertical and horizontal width and bends only at 90° angles.

Outside reverse fold (page 23): a method of changing the direction of a flap, wherein the moving layers are inverted and wrapped around the stationary layers.

P

Parity: see **Crease assignment**.

Path (page 382): a line between two leaf vertices in the crease pattern.

Path conditions (page 384): the set of all inequalities relating the coordinates of the leaf vertices, the distances between their corresponding nodes, and a scale factor. The distance between any two vertices must be greater than or equal to the scaled distance between their corresponding nodes as measured along the tree.

Petal fold (page 28): a combination of two squash folds in which a corner is lengthened and narrowed.

Plane of projection (page 376): a plane containing the axis of the base and the axial edges of all flaps, and that is perpendicular to the layers of the base.

Plan view (page 299): a model is folded in plan view if when it lies flat you are looking at the top of the subject.

Pleat fold (page 30): a fold formed by two parallel or nearly parallel mountain and valley folds formed through all layers of a flap.

Pleat grafting (page 193): adding one or more pleats that run across a crease pattern in order to add features or textures formed by the intersections of the pleats.

Plug (page 467): a crease pattern that is used to fill in holes in box-pleated patterns.

Precreasing (page 12): folding and unfolding to create the creases required for a (usually complex) step.

Point-splitting (page 91): any of a variety of techniques for folding a single flap so that it turns into two or more smaller flaps.

Preliminary Fold (page 54): a traditional base formed by bringing the four corners of the square together.

R

Rabbit-ear fold (page 25): a way of turning a triangular corner into a flap, consisting of folding along all three angle bisectors of the triangle and gathering the excess paper into a flap.

Rabbit-ear molecule (page 339): the pattern of creases within a triangle that collapses its edges to lie on a single line.

Raw edge (page 15): the original edge of the paper, as opposed to an edge created by folding.

Reduced path (page 399): a path between two inset vertices created during the construction of the universal molecule.

Reduced path inequality (page 399): an inequality condition analogous to the path condition that applies to inset vertices and paths in the universal molecule.

Ridgeline crease (page 335): a crease within a molecule that propagates inward from the corners of the molecule. Ridgeline creases are always valley folds when viewed from the interior of a molecule.

River (page 242): an annular segment or rectangular colored region in a tile or crease pattern that creates a segment between groups of flaps in the folded form.

S

Sawhorse molecule (page 351): a crease pattern within a quadrilateral similar to the Waterbomb molecule, but with a segment separating the two pairs of flaps. Also known as the Maekawa molecule.

Scale (page 284): a quantitative measure of efficiency. The scale of a crease pattern is the ratio between the length of a folded flap and the length of its corresponding edge in the tree graph.

Side view (page 299): a model is folded in side view if when the model lies flat you are looking at the side of the subject.

Sink fold (page 32): inversion of a point. Sink folds come in several different types.

Splitting points: see **point-splitting**.

Spread sink fold (page 33): a sink fold in which the edges of the point are spread and the point flattened. Similar to a squash fold.

Square/river packing (page 449): the analog of circle and river packing that allows box-pleated crease patterns.

Squash fold (page 27): a fold in which the edges of a flap are spread, usually symmetrically, and the edges flattened.

Standard bases (page 54): the most common origami bases, usually taken to include the Classic Bases plus the Windmill Base, Cupboard Base, Preliminary Fold, and Waterbomb Base.

Stretched Bird Base (page 55): a form of the Bird Base in which two opposite corners are pulled apart to straighten out the diagonal that connects them.

Strip graft (page 139): modifying a crease pattern as if you spliced in one or more strips of paper running across a crease pattern in order to add features to the base.

Stub (page 395): a new edge added to the tree graph attached to a new node introduced into the middle of an existing edge and associated creases added to the crease pattern. Adding a stub allows four path conditions to be simultaneously satisfied as equalities.

Subbase (page 385): a portion of a base, usually consisting of a single axial polygon.

Subtree (page 385): the tree graph that is the projection of a subbase.

Swivel fold (page 27): an asymmetric version of a squash fold in which the two valley folds are not collinear.

T

Tangent points (page 333): points along axial polygons where circles (or rivers) touch each other and are tangent to the hinge creases.

Technical folding (page 46): origami designs that are heavily based on geometric and mathematical principles.

Tile (page 236): a portion of a crease pattern, usually consisting of one or more axial polygons and decorated by circles and rivers, that can be assembled into crease patterns by matching circle and river boundaries.

Tree (page 376): short for tree graph.

Tree graph (page 376): a stick figure that represents a uniaxial base, in which each edge of the tree represents a unique flap or connection between flaps.

Tree theory (page 375): the body of knowledge that describes the quantitative construction of crease patterns for uniaxial bases based on a correspondence between features of a tree graph and features in the crease pattern.

Tree theorem (page 380): the theorem that establishes that satisfying the path conditions is both necessary and sufficient for the construction of a crease pattern for a given tree graph.

Triangulation (page 396): the process of decomposing high-order axial polygons in a crease pattern into smaller polygons that are all order-3, i.e., triangles.

U

Unfold (page 11): removing a valley or mountain fold, leaving behind a crease.

Uniaxial base (page 230): a base in which all flaps lie along a single axis and all hinges are perpendicular to the axis.

Universal molecule (page 398): a generalization of the gusset molecule that can be applied to every valid axial polygon.

Unsink (page 38): removing a sink fold, or turning a closed sink from concave to convex.

V

Valley fold (page 18): a crease that is concave upward. Usually indicated by a dashed line (solid colored line in crease patterns).

Vertex (page 22): a point in a crease pattern where multiple creases come together. See **leaf vertex**, **branch vertex**.

W

Windmill Base (page 54): a traditional base that looks like a windmill.

Waterbomb Base (page 54): a traditional base formed by bringing the midpoints of the four edges of a square together.

Waterbomb condition (page 341): a quadrilateral satisfies the Waterbomb condition if and only if the sums of opposite sides are equal. A quadrilateral that satisfies this condition can be folded into an analog of the traditional Waterbomb Base.

Waterbomb molecule (page 340): a crease pattern within a quadrilateral that resembles the traditional Waterbomb. Also called the Husimi molecule.

Y

Yoshizawa split (page 92): a technique for splitting a Kite Base flap, in which the point is first sunken, followed by two spread sinks.

Index

A

B

G

H

I

J

K

L

M

N

O

P

R

S

T

U

V

W

Y